Organic Coatings

Organic Coatings
Science and Technology

Fourth Edition

Frank N. Jones
Emeritus Professor, Eastern Michigan University
Coatings Consulting Services, LLC

Mark E. Nichols
Ford Motor Company

Socrates Peter Pappas
Consultant

WILEY

This edition first published 2017

Edition History
John Wiley & Sons Inc, (1e 1994), John Wiley & Sons Inc, (2e 1999), John Wiley & Sons Inc, (3e 2007).

Registered Office
John Wiley & Sons, Inc., 111 River Street, Hoboken, NJ 07030, USA

Editorial Office
111 River Street, Hoboken, NJ 07030, USA

For details of our global editorial offices, customer services, and more information about Wiley products visit us at www.wiley.com.

Wiley also publishes its books in a variety of electronic formats and by print-on-demand. Some content that appears in standard print versions of this book may not be available in other formats.

Library of Congress Cataloging-in-Publication Data

Names: Jones, Frank N., 1936– author. | Nichols, Mark E., 1965– author. |
Pappas, S. Peter (Socrates Peter), 1936– author.
Title: Organic coatings : science and technology / Frank N. Jones, emeritus professor, Eastern Michigan University, Coatings Consulting Services, LLC, Mark E. Nichols, Ford Motor Company ; Socrates Peter Pappas, consultant.
Description: Fourth edition. | Hoboken, NJ, USA : John Wiley & Sons, Inc, 2017. | Revised edition of: Organic coatings : science and technology / Zeno W. Wicks, Jr., Frank N. Jones, and S. Peter Pappas. 2nd ed. 1999. | Includes bibliographical references and index. | Description based on print version record and CIP data provided by publisher; resource not viewed.
Identifiers: LCCN 2017011370 (print) | LCCN 2017011578 (ebook) | ISBN 9781119337157 (pdf) | ISBN 9781119337218 (epub) | ISBN 9781119026891 (cloth)
Subjects: LCSH: Plastic coatings.
Classification: LCC TP1175.S6 (ebook) | LCC TP1175.S6 W56 2017 (print) | DDC 667/.9–dc23
LC record available at https://lccn.loc.gov/2017011370

Cover design: Wiley
Cover image: Courtesy of Mark E. Nichols

Set in 10.25/12pt Times by SPi Global, Pondicherry, India

Printed in Singapore

10 9 8 7 6 5 4 3 2 1

Zeno W. Wicks, Jr., 1920–2007

Zeno was the lead author of the first three editions of this book. Two of us (Jones and Pappas) remember him fondly as an outstanding scientist, a charismatic teacher, a mentor, a marvelous colleague, and a gentleman. Zeno influenced hundreds, more likely, thousands, of students, many of whom have made careers in coatings. His favorite advice to them was "Don't park your brains at the door."

Being in a younger generation, Mark Nichols missed out on meeting Zeno. "My loss," he says, and he is right.

Zeno got his Ph.D. in Chemistry at the University of Illinois. He joined Inmont Corporation, where he advanced to vice president of research and development during a 28-year career. (Inmont was a leading coating and ink producer, acquired by BASF in 1985.) For the next 11 years, he was professor and chair of the Department of Polymers and Coatings at North Dakota State University (NDSU). He then became a consultant. Among other activities, he traveled worldwide to teach about coatings. He received the Mattiello Memorial Award, the Roy W. Tess Award, and four Roon Awards.

Zeno was the best teacher we ever saw. He could teach all day, and when he invited a class to return after dinner for optional discussion, they came. This book originated as a set of lecture notes Zeno prepared during his last year at NDSU, where he taught a full-year course in coatings for upperclassmen and graduate students. He thought, rightfully so, that the notes might be helpful to his successors.

Contents

Preface

Coatings science and technology advance in a continuous stream of improvements with an occasional breakthrough. This year's house paint may look the same as that of 10 years ago, but it is a lot better. Thus, it is time to revise the third edition of *Organic Coatings: Science and Technology*, published in 2007. Here, the third edition has been completely updated. Our purpose remains the same—to provide a reference and textbook that interrelates coatings technology with current scientific understanding.

For the fourth edition, Mark Nichols joined the team of authors. For the first time, we have a real materials scientist involved—and a very good one. As editor-in-chief of the *Journal of Coatings Technology and Research*, Mark has a broad view of contemporary coatings technology and is a leading authority on automotive coatings. His contributions are reflected in major revisions. Entire books could be written about the subject of each chapter, and many have been. To be as comprehensive as possible in the limited space available, we have summarized each topic and have provided references for readers seeking more detailed information. We have striven to enhance the usefulness of this edition both as a classroom textbook on coatings science and as a reference book. The reader will benefit from having taken college level chemistry courses through organic chemistry, but no coursework in polymer or materials science is assumed.

Some chapters include brief descriptions of coating compositions and applications, supported by references, which could be omitted in a classroom or used for outside-of-class assignments, such as term papers. We hope that these specific examples enhance the value of the volume as a reference book and self-teaching text. We understand that the first three editions were widely used for this purpose. We have also defined the jargon of coatings to help newcomers to the field understand its specialized language. While this book is written specifically about coatings, many of the principles apply to the related fields of printing inks, adhesives, and parts of the plastics industry.

Coatings technology evolved empirically by trial and error. Directions on how to make and apply paint have been published for at least 2000 years. Since about 1900, scientific understanding of the applicable principles has evolved. In 1905 Einstein published an equation applicable to flow of pigmented paints, and before 1920, pioneers such as H. A. Gardner, E. Ladd, C. B. Hall, and M. Toch applied scientific methods to testing. However, the coatings field is extremely complex, and scientific understanding remains incomplete. Empirical formulation and experimentation is still essential in developing and using coatings. The often conflicting needs for sustainability, reduced impact on the environment and health, reasonable cost, and improved coating performance require continuing innovation. Our conviction is that understanding the underlying science can help formulators work more effectively and that an appreciation of the formulators' craft is essential for scientists and engineers working in the field. Knowledge should flow both ways.

A complete literature review for each chapter would fill much of the book. We only cite key references and those that support specific information. Many of the references in older editions were replaced with newer ones, but many old references remain because they describe significant contributions to the evolution of coatings technology. Various sources of additional information are available to investigators. These include refereed journals such as the *Journal of Coatings Technology and Research* and *Progress in Organic Coatings*, as well as books, trade journals, conference proceedings, academic dissertations, internal company reports, and information from suppliers and customers. Patents are sometimes overlooked, but they often include informative reviews of the "state of the art" and specific examples including formulas, test procedures, and results. Patents are also free and readily searchable online.

We thank Dean Webster and Carole Worth for their editorial assistance and helpful suggestions.

Preface

Chapter 1

Introduction to Coatings

Coatings have been used since prehistoric times to protect objects and convey information, and they are ubiquitous in modern society as they serve to both protect substrates and impart aesthetic qualities to improve objects' appearance. If you are reading this text in a traditional paper book, the paper is coated. Look up and the walls of your room are coated, as are the windows. If you are wearing glasses, the lenses are likely coated to improve the plastic's scratch resistance and absorb UV radiation. If you are reading this text on a computer screen, the screen is coated to prevent glare and perhaps reduce fingerprints. The CPU inside your computer exists because of coatings used during the printing of nanometer-sized circuits. If you are outside, the buildings, cars, airplanes, roads, and bridges are all coated. Objects without coatings are less common than those with coatings!

Just because coatings science is an ancient technology does not mean that innovation has ceased. Today many coatings scientists and formulators are working diligently to improve the performance of coatings, reduce the environmental impact of their manufacture and application, and create coatings that provide functionality beyond today's coatings.

1.1 DEFINITIONS AND SCOPE

Coatings are typically thought of as thin layers that are applied to an object, which is often referred to as the *substrate*. Thus, one of the defining characteristics of a coating is its thinness. While the thickness of a coating depends on the purpose it serves, typical coating thicknesses range from a few microns to a few hundred microns, but of course, exceptions to this are common. Historically, the thickness of a coating was often quoted in terms of mils, where 1 mil equals one thousandth of an inch or 25.4 μm.

While coatings can be made from any material, this book is primarily concerned with organic coatings. Thus, we leave for other books coatings such as the zinc coatings used to galvanize steel, ceramic coatings that are formed from metal oxides or when metals such as aluminum are anodized, and the many other inorganic coatings used to impart hardness, scratch resistance, or corrosion protection. While these coatings are both technically and economically important, they lie mostly beyond the scope of this book.

Organic coatings are often composite materials in that they are composed of more than one distinct phase. The matrix, called the *binder*, holds the other components of the coating composition together and typically forms the continuous phase in the dry coating. As stated previously, we are mostly concerned with organic coatings, where the binder is typically an organic polymer.

A confusing situation results from multiple meanings of the term *coating*. As a noun *coating* is used to describe both the material (usually a liquid) that is applied to a substrate and the resultant "dry" film. As a verb, *coating* means the process of application. Usually, the intended meaning of the word coating can be inferred from the context. The terms *paint* and *finish* often mean the same thing as *coating* and also are used both as nouns and verbs. What is the difference between a coating and a paint? Not much—the terms are often used interchangeably. However, it is fairly common practice to use "coatings" as the broader term and to restrict "paints" to the familiar architectural and household coatings and sometimes to maintenance coatings for bridges and tanks. Some prefer to call sophisticated materials that are used to coat automobiles and computer components "coatings," and others call them "paints." Consumers are often familiar with the terms *varnish* or *stain*. These are types of coatings that are used to protect and beautify wood

Organic Coatings: Science and Technology, Fourth Edition. Frank N. Jones, Mark E. Nichols, and Socrates Peter Pappas.

and are certainly within the scope of this book as they are typically made from polymeric binders with or without pigments.

Because we are limiting the scope of this book to organic coatings that are historically associated with paints, we are also choosing not to cover important materials such as coatings applied to paper and fabrics, decals, laminates and cosmetics, and printing inks, even though one could argue that these coatings share much in common with traditional paints. However, readers interested in those materials will find that many of the basic principles discussed in this text are applicable to such materials. Restrictions of scope are necessary if the book is to be kept to a reasonable length, but our restrictions are not entirely arbitrary. The way in which we are defining coatings is based on common usage of the term in worldwide business. For classification purposes, coatings are often divided into three categories: architectural coatings, original equipment manufacturer (OEM) coatings, and special purpose coatings.

As the coatings industry is a relatively mature industry, its growth rate typically paces that of the general economy. Like many other industries, growth has slowed in North America and Europe and has dramatically increased in Asia and South America as those economies have boomed. An estimate of the value of coatings used in each region is shown in Figure 1.1. The total value of the global coatings market was estimated to be approximately $112 billion in 2014 (American Coatings Association and Chemquest Group, 2015).

Figure 1.2 summarizes the estimated value and volume of coating shipments in the United States for a recent 10-year period. The effect of the economic downturn in 2008–2009 is evident (Data from American Coatings Association and Chemquest Group, 2015).

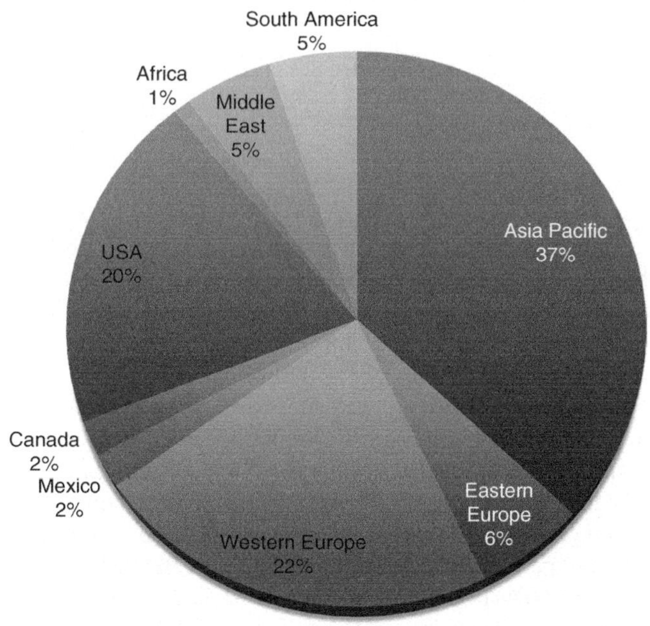

Figure 1.1 The value of coatings used in 2014. Source: Reproduced with permission of American Coatings Association.

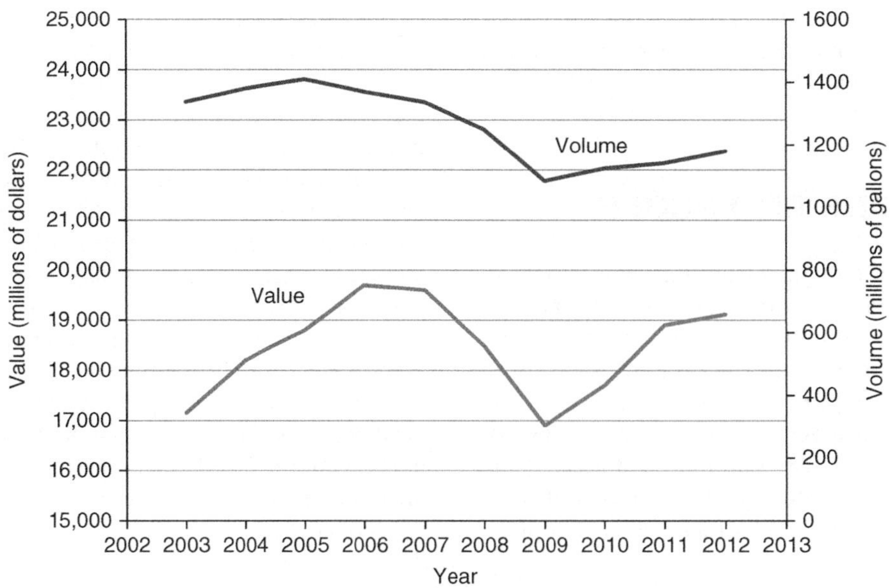

Figure 1.2 Ten-year trend in coating shipments in the United States (both gallons and dollar value). Source: Reproduced with permission of American Coatings Association.

1.2 TYPES OF COATINGS

Architectural coatings include paints and varnishes (transparent paints) used to decorate and protect buildings, outside and inside. They also include other paints and varnishes sold for use in the home and by small businesses for application to such things as cabinets and household furniture (not those sold to furniture factories). Architectural coatings are often called *trade sales paints*. They are sold directly to painting contractors and do-it-yourself users through paint stores and other retail outlets. In 2014 in the United States, architectural coatings accounted for about 60% of the total volume of coatings; however, the unit value of these coatings was lower than for the other categories, so they made up about 49% of the total value. This market is the least cyclical of the three categories. While the annual amount of new construction drops during recessions, the resulting decrease in paint requirements tends to be offset by increased repainting of older housing, furniture, and so forth during at least mild recessions. Latex-based coatings make up about 77% of architectural coatings. Interior paints are approximately 2/3 of all architectural coatings, exterior paints 23%, and stains 7%, with the remained split among varnishes, clear coats, and others.

OEM coatings are applied in factories on products such as automobiles, appliances, magnet wire, aircraft, furniture, metal cans, and chewing gum wrappers—the list is almost endless. In 2014 in the United States, product coatings were about 29% of the volume and 31% of the value of all coatings. The volume of product coatings depends directly on the level of manufacturing activity. This category of the business is cyclical, varying with OEM cycles. Often, product coatings are custom designed for a particular customer's manufacturing conditions and performance requirements. The number of different types of products in this category is much larger than in the others; research and development (R&D) requirements are also high.

Special purpose coatings are industrial coatings that are applied outside a factory, along with a few miscellaneous coatings, such as coatings packed in aerosol containers. This category includes *refinish coatings* for cars and trucks that are applied outside the OEM factory (usually in body repair shops), *marine coatings* for ships (they are too big to fit into a factory), and striping on highways and parking lots. It also includes *maintenance paints* for steel bridges, storage tanks, chemical factories, and so forth. In 2012 in the United States, special purpose coatings made up about 11% of the total volume and 20% of the total value of all coatings, making them the most valuable class. Many of today's special purpose coatings are the product of sophisticated R&D, and investment in further improvements remains substantial.

Coatings are used for one or more of three reasons: (1) for decoration, (2) for protection, and/or (3) for some functional purpose. The low gloss paint on the ceiling of a room not only fills a decorative need but also has a function. It reflects and diffuses light to help provide even illumination. The coating on the outside of an automobile adds beauty to a car and also helps protect it from rusting. The coating on the inside of a beverage can have little or no decorative value, but it protects the beverage from the can. (Contact with metal affects flavor.) In some cases, the interior coating also protects the can from the beverage. (Some soft drinks are so acidic that they can dissolve the metal.) Other coatings reduce the growth of algae and barnacles on ship bottoms, protect optical fibers for telecommunications against abrasion and guide the light within the fiber, retard corrosion of bridges, protect wind turbine blades from erosion due to the impact of raindrops, and so on. While the public most commonly thinks of house paint when talking about coatings, all kinds of coatings are important throughout the economy, and they make essential contributions to most high-tech fields. As already mentioned, computer technology depends on microlithographic coatings to pattern the circuits in CPU and memory chips.

1.3 COMPOSITION OF COATINGS

Organic coatings are complex mixtures of chemical substances that can be grouped into four broad categories: (1) *binders*, (2) *volatile components*, (3) *pigments*, and (4) *additives*.

Binders are the materials that form the continuous film that adheres to the *substrate* (the surface being coated), bind together the other substances in the coating to form a film, and present an adequately hard outer surface. The binders of coatings within the scope of this book are organic polymers—some made via synthetic organic chemistry and some derived from plant oils. In some cases, these polymers are prepared and incorporated into the coating before application; in other cases, lower molecular weight organic materials (*monomers* or *oligomers*) are mixed with the other components of the coating, and final polymerization takes place after the coating has been applied. Binder polymers and their precursors are often called *resins*. The binder governs, to a large extent, the properties of the coating film. The major resin types used in coatings as percentages of the total are given in Table 1.1. These numbers should be taken as approximations as different coating suppliers name their resins somewhat differently, and some coating contain more than one resin type.

Volatile components are included in a large majority of coatings and are often referred to as *solvents*. They play a major role in the synthesis, mixing, and application of coatings. They are liquids that make the coating fluid enough for

Table 1.1 Breakdown of Major Resin Types for the US Coatings Market

Resin type	Percent
Acrylic	31
Vinyl	20
Urethane	14
Epoxy	8
Alkyd	7
Silane	5
Polyester	4
Amino	3
PVC	2
SBR	1
Phenolic	1
Cellulosic	1
Other	3

Source: Reproduced with permission of American Coatings Association.

application, and they evaporate during and after application. Until about 1935, almost all of the volatile components were low molecular weight organic compounds that dissolved the binder components. However, the term solvent has become potentially misleading because many coatings have been developed for which the binder components are not fully soluble in the volatile components but instead act as a carrier to reduce viscosity, but not fully solvate the binder. Because of the need to reduce the environmental impact of coating manufacture and application, a major continuing drive in the coatings field is to reduce the use of *volatile organic compounds* (VOCs) by making the coatings more highly concentrated (higher solids coatings), by using water as a major part of the volatile components (waterborne coatings), and by eliminating solvents altogether.

Vehicle is a commonly encountered term. It usually means the combination of the binder and the volatile components of a coating. Today, most coatings, including waterborne coatings, contain at least some volatile organic solvents. Exceptions are powder coatings, certain solventless liquid coatings (also called 100% solids coatings), radiation-curable coatings, and a small but growing segment of architectural coatings.

Pigments are finely divided, insoluble solid particles, ranging from a few tens of nanometers to a few hundred microns in size, that are dispersed in the vehicle and remain suspended in the binder after film formation. Generally, the primary purpose of pigments is to provide color and opacity to the coating film. Additionally, pigments can provide other functions, such as corrosion-inhibiting pigments, which enhance the corrosion protecting properties of the coatings. Pigments also play a major role in the application characteristics and the mechanical behavior of coatings. While most coatings contain pigments, there are important types of coatings that contain little or no pigment, commonly called *clear coats*, or just *clears*. Clear coats for automobiles and transparent varnishes are examples. *Coating solids* typically refer to the proportion of binder and pigment and are the part of the paint that remains after the volatile components have left the coating. Pigments are distinct from dyes, which are typically soluble in their binder and/or solvent and exist as individual molecules in that vehicle. Dyes are rarely used in the types of coatings discussed in this book.

Additives are materials that are included in small quantities to modify some property of the coating. Examples are catalysts for polymerization reactions, light and heat stabilizers, rheology modifiers, defoamers, and wetting agents.

1.4 COATING HISTORY

The chemistry of most coatings used today bears little resemblance to the coatings used prior to the industrial revolution. For centuries coatings were based on naturally occurring oils and pigments. 40 000 years ago ochre was processed for use as a pigment in Africa (Rosso et al., 2016). Cave paintings in northern Spain date from over 40 000 years ago and contain depictions of animals and people. While their true purpose is impossible to ascertain, the paintings demonstrate that even in prehistoric times people were using coatings to decorate their surroundings and to convey information to others.

In Asia, a traditional coating made from urushiol, the resin from a native tree, has been used since at least 1200 B.C. to produce beautiful clear lacquers for art objects. Egg yolk was often used as the binder for paintings in the West until the fourteenth or fifteenth century, when certain plant oils, such as linseed (also known as flax) and walnut oils, were introduced to protect and beautify wood. Those oils were also used as the binder for many of the great oil paintings made by famous artists such as Michelangelo, and they continue to be favored by many artists today. During the nineteenth and early twentieth centuries, most architectural coatings employed linseed oil as the binder.

Early pigments were made from ground bones or charcoal and other minerals such as iron oxide, ochre, and calcium carbonate. Simple chemical reactions were later used to produce other pigments such as lead white (lead carbonate) and red lead (lead oxide). More chromatic pigments such as ultramarine blue were rare and expensive for centuries owing to their limited supply.

These simple binders and pigments formed the basis for almost all coatings up until the twentieth century when an explosion in our knowledge of synthetic organic chemistry multiplied the number of binders, pigments, and additives that were used in coatings. Naturally sourced binders gave way to nitrocellulose lacquers and later enamels based on synthetic polymers. Other organic and synthetic inorganic pigments displaced some, but not all, of the naturally occurring pigments. For example, white lead carbonate gave way to titanium dioxide due to TiO_2's superior *hiding* and reduced toxicity; and highly chromatic red pigments based on quinacridone chemistry were developed to provide colors that were previously difficult to achieve. *Hiding* refers to the capability of a coating to screen or hide the substrate from view, which is generally desirable from both aesthetic and protective standpoints.

Most people's interaction with coatings occurs when they paint the walls of their house or refinish an old piece of furniture; and they likely believe that coatings have changed little over the course of their lives. In many ways their thoughts are justified, as the process of painting with a brush has changed little over the past 100 years. However, as shown previously, advances in chemistry have resulted in dramatic changes in the formulation of paints. In addition, since 1965 the reduction of VOCs has been a major driving force because of the detrimental effect of VOCs on air quality. Coatings have been second only to the gasoline–automobile complex as a source of VOC pollutants responsible for excess ozone in the air of many cities on many days of the year. This situation has resulted in increasingly stringent regulatory controls on such emissions. The drive to reduce VOC emissions has also been fueled by the rising cost of organic solvents. Other important factors have also accelerated the rate of change in coatings. In particular, the increasing concern about toxic hazards has led to the need to change many raw materials that were traditionally used in coatings.

1.5 COMMERCIAL CONSIDERATIONS

The person who selects the components from which to make a coating is a *formulator*, and the overall composition he or she designs is called a *formulation*. Throughout history, formulators have been trying to understand the underlying scientific principles that control the performance of coatings. Most coating systems are so complex that our understanding of them today is still limited. Real progress has been made, but the formulator's art is still a critical element in developing high performance coatings. Demands on suppliers of coatings to develop new and better coatings are accelerating. Therefore, time is now too limited to permit traditional trial-and-error formulation. Understanding the basic scientific principles can help a formulator design better coatings more quickly. In the chapters ahead, we present, to as great an extent as present knowledge permits, the current understanding of the scientific principles involved in coatings science.

We also identify areas in which our basic understanding remains inadequate and discuss approaches to more efficient and effective formulation despite inadequate understanding. In some cases, in which no hypotheses have been published to explain certain phenomena, we offer speculations. Such speculations are based on our understanding of related phenomena and on our cumulative experience acquired over several decades in the field. We recognize the risk that speculation tends to increase in scientific stature with passing time and may even be cited as evidence or adopted as an experimentally supported hypothesis. It is our intent, rather, that such speculations promote the advancement of coatings science and technology by stimulating discussion that leads to experimentation designed to disprove or support the speculative proposal. We believe that the latter purpose outweighs the former risk, and we endeavor to identify the speculative proposals as such.

Cost is an essential consideration in formulation. Novice formulators are inclined to think that the best coating is the one that will last the longest time without any change in properties, but such a coating may be very expensive and unable to compete with a less expensive coating, which provides adequate performance for particular application. Furthermore, it is seldom possible to maximize all of the performance characteristics of a coating in one formulation. Some of the desirable properties are antagonistic with others; formulators must balance many performance variables while keeping costs as low as possible.

REFERENCES

American Coatings American Coatings Association and Chemquest Group, ACA Industry Market Analysis, American Coatings Association, Washington, DC, **2015**.

Rosso, D. E., et al., *PLoS ONE*, **2016**, *11*(*11*), e0164793.

Chapter 2

Polymerization and Film Formation

This chapter introduces basic concepts of polymer chemistry and film formation with particular emphasis on aspects related to organic coatings. Many excellent books (Sperling (2001), Odian (2004d), Billmeyer (2007), Young (2014), Young and Lovell (2001), and Fried (2014), among others) provide more comprehensive coverage.

2.1 POLYMERS

A polymer is a substance composed of large molecules. Some authors reserve the term *polymer* to describe a substance and use the term *macromolecule* for the molecules making up the substance. This usage distinguishes between the material and the molecules but is not common in the coatings field. We use the term polymer for both meanings. Depending on the context, the term refers to either the molecules or the substance. The structure of polymers is a multiple repetition of units (mers) derived from molecules of relatively low molecular weight (MW) (monomers). (The more rigorous designation of MW is *molar mass*, but we use MW because it is much more commonly used in the coatings field.)

There is disagreement about how high the MW has to be for a material to qualify as a polymer. Some people refer to materials with MWs as low as 1 000 as polymers; others insist that only materials with MWs over 10 000 (or even 50 000) qualify. The term *oligomer*, meaning "few mers," is often used for materials having MWs of a few hundred to a few thousand. This additional term does not help the definition problem much because there is no clear-cut boundary between an oligomer and a polymer, but the term can be useful because it provides a name with which most can agree for materials containing 2 to about 20 mers.

Polymers occur widely in nature; *biopolymers* are produced by living organisms. Examples are proteins, starch, cellulose, and silk. In the coatings field, we are concerned mainly with synthetic polymers, although some chemically modified biopolymers are also used.

Synthetic polymers and oligomers are prepared by *polymerization*, a sequence of chemical reactions in which small molecules are joined by covalent bonds. A polymer made from a single monomer is called a *homopolymer.* If it is made from a combination of monomers, it is often, but not always, called a *copolymer.* An example of a homopolymer is provided by the polymerization of vinyl chloride:

$$\underset{\text{Vinyl chloride monomer}}{CH_2{=}CHCl} \qquad \underset{\text{Poly(vinyl chloride)}}{X{-}(CH_2CHCl)_n{-}Y}$$

In this example, the $-(CH_2CHCl)-$ repeating unit is the mer, and n represents the number of mers joined together in the molecule. X and Y represent terminal groups on the ends of the chain of mers.

Polymers are described by the chemical compositions of their monomers. In addition, they can be synthesized in various structures (*topologies*) (Krol and Chmielarz, 2014). Three topologies are especially important in coatings:

- When the mers are linked in chains, the polymers are called *linear polymers*, a term that is potentially misleading because the large molecules seldom form a straight line, but rather twist and coil. In linear copolymers, the different monomers may be distributed more or less at random throughout the chain (*random copolymers*), they may tend to alternate (*alternating copolymers*), or they may be separated into groups of the same monomer (*block copolymers*).

Organic Coatings: Science and Technology, Fourth Edition. Frank N. Jones, Mark E. Nichols, and Socrates Peter Pappas.

- If there are forks in the chains, the polymers are called *branched* polymers. A polymer chain of one type of monomer having polymer branches of a different monomer is called a *graft copolymer*. *Comb* and *brush polymers* have a large number of branches.
- A third topology results from the bonding of chains with each other at several sites to form *cross-linked*, or *network*, polymers, also called *gels*. These are branched polymers where the branches are covalently bound to other molecules, so the mass of polymer consists mainly of a single, interconnected molecule.

Other topologies, such as *dendritic* and *hyperbranched* polymers, are gaining importance in coatings.

Reactions that join polymer or oligomer molecules together are called *cross-linking reactions*. Polymers and oligomers that can undergo such reactions are frequently called *thermosetting* polymers. Some confusion can result because the term thermosetting is applied not only to polymers that cross-link when heated but also to those that can cross-link at ambient temperature or even below. A polymer that does not undergo cross-linking reactions is called a *thermoplastic polymer*, because it becomes plastic (softens) when heated.

Polymers can also be formed from mers that contain multiple H-bonding sites, in which case the mers are held together by H-bonds rather than by covalent bonds. Such polymers are called *supramolecular* polymers, which are of particular interest when the H-bonding sites, are four-centered owing to greatly enhanced strength relative to three-center H-bonds (Brunsveld et al., 1999). Coatings compositions in which multicenter H-bonds complement covalent cross-links can enhance coatings properties owing to thermal reversibility of the H-bonds, as observed for polyurethanes (Chapter 12). Unlike covalent bonds, H-bonds are readily broken and readily reestablished.

Another term commonly, but loosely, used in the coatings field is *resin*. This term overlaps the meanings of polymer and oligomer. Historically, the term meant hard, brittle materials derived from tree exudates, such as rosin, dammar, and elemi. A variety of these naturally occurring resins were used since prehistoric times to make coatings. In the nineteenth and early twentieth centuries, such resins were dissolved in drying oils to make *varnishes* (Section 14.3.2). The first entirely synthetic polymers used in coatings were phenol–formaldehyde polymers (Section 13.6), which replaced naturally occurring resins in many applications. Accordingly, they were called *phenol–formaldehyde resins* or *phenolic* resins. As more synthetic products were developed to replace naturally occurring resins, these products were also called resins.

When words do not have precise meanings, it is important to understand the context in which they are placed. Commonly, it is assumed, without much thought, that information that has been learned about high MW polymers is also applicable to low MW polymers or oligomers, because all are often called polymers. Many characteristics, however, depend on MW. While much of the information available from studies of high MW polymers can be useful in the coatings field, it must be used with caution, because the resins used in making coatings are commonly low MW polymers or oligomers, even though they are frequently called polymers. In the next set of subsections, we describe some of the key characteristics of synthetic polymers and oligomers.

2.1.1 Molecular Weight (MW)

For most pure organic compounds, the concept of MW is straightforward—each compound has **a** MW. For synthetic polymers, however, the situation is more complex. All methods of synthesis lead to mixtures of molecules with different numbers of mers and, therefore, with different MWs. Even relatively simple thermoplastic homopolymers, such as polystyrene or poly(vinyl chloride), contain molecules with hundreds of different chain lengths. With copolymers, the number of different molecules present is much larger. There is a distribution of MWs in a synthetic polymer; accordingly, MWs can be defined only by a statistical calculation. In the simplest cases, the distribution of the number of molecules of each MW resembles a skewed Gaussian distribution, but in other cases, the distribution may be quite complex. While many types of average MW can be calculated, the two most widely used are number and weight average MWs.

Number average MW $(\bar{M}_n)$ is the MW average based on summing the products of the numbers of molecules and their MWs and dividing by the sum of the number of molecules in the sample. Mathematically, it is expressed by Eq. 2.1, where M_1, M_2, and M_i are the MWs of the first, second, and ith species, respectively, and the N values are the numbers of molecules of each species present:

$$\bar{M}_n = \frac{\sum N_1 M_1 + N_2 M_2 + \cdots}{\sum N_1 + N_2 + \cdots} = \frac{\sum_i N_i M_i}{\sum_i N_i} \tag{2.1}$$

$$\bar{P}_n = \frac{\sum_i N_i P_i}{\sum_i N_i} \tag{2.2}$$

A similar equation (Eq. 2.2) represents the *number average degree of polymerization* $\bar{P}_n$, where P is the number of mers in a molecule and P_i is the number of mers in the ith polymer. For homopolymers, $\bar{M}_n = \bar{P}_n$ times the MW of each mer; for copolymers, a weighted average MW of the mers is used. The differing weights of end groups can be neglected in calculating $\bar{M}_n/\bar{M}_w$ of high polymers, but not of oligomers, for which the effect can be appreciable.

Weight average MW $(\bar{M}_w)$ is defined by Eq. 2.3, in which w_1, w_2, and w_i are the weights of molecules of species 1, 2, and *i*th; since $w_1 = N_1M_1$, $\bar{M}_w$ can also be calculated from the numbers of molecules of the different species, as shown in the equation. *Weight average degree of polymerization* $\bar{P}_w$ is defined by analogous equations:

$$\bar{M}_w = \frac{w_1M_1 + w_2M_2 + \cdots}{w_1 + w_2 + \cdots} = \frac{\sum_i w_iM_i}{\sum_i w_i} = \frac{\sum_i N_iM_i^2}{\sum_i N_iM_i} \quad (2.3)$$

Higher-order MWs such as M_z and M_{z+1} give additional weight to the larger molecules. M_v correlates with solution viscosity of many polymers.

Figure 2.1 shows an idealized plot of weight fraction of molecules of each MW as a function of degree of polymerization for oligomers made from the same monomer by three different processes (Hill and Wicks, 1982). In relatively simple distributions of MWs, the value of $\bar{P}_n$ is at, or near, the peak of the weight fraction distribution curve. $\bar{M}_w$ and $\bar{P}_w$ are always larger than $\bar{M}_n$ and $\bar{P}_n$.

The breadth of the MW distribution can have an important effect on the properties of a polymer and is often critical to achieving satisfactory performance of a coating. The ratio $\bar{M}_w/\bar{M}_n$ is widely used as an index of the breadth of distribution. In the case of high MW polymers, $\bar{M}_w/\bar{M}_n = \bar{P}_w/\bar{P}_n$, but in the case of oligomers, differences in end groups can be significant and affect the equality of the ratios. These ratios are called *polydispersity* (PD), or sometimes, *polydispersity index* (PDI). We use the symbols $\bar{M}_w/\bar{M}_n$ and $\bar{P}_w/\bar{P}_n$. The ratios provide a convenient way to compare the MW distributions of different polymers. However, one must be cautious in the use of a single value to describe a possibly complex distribution. As shown in Figures 2.1 and 2.2, synthetic polymers commonly have broad distributions of MWs. As $\bar{M}_w/\bar{M}_n$ increases, the fractions of polymer at the extremes above and below the number average MW increase. Even the oligomer with a number average of 12 mers and with $\bar{M}_w/\bar{M}_n = 1.07$ has substantial numbers of molecules containing 7–18 mers, and a polymer with a more typical $\bar{M}_w/\bar{M}_n = 3$ has molecules spanning several orders of magnitude of MW.

$\bar{M}_n$ is the MW of most importance for relating stoichiometric ratios of reactants and for comparing certain physical properties. $\bar{M}_w$ often proves more useful than $\bar{M}_n$ when considering the relationship between MW and many physical properties of polymers, including some of the properties that are crucial to coating performance. M_z and even M_{z+1} provide useful correlations with certain film properties in some thermoset coatings.

The classical ways of measuring $\bar{M}_w$ and $\bar{M}_n$ are difficult and are beyond the scope of this book; Elias (1984) is but one of many books that describe them. In practice, most scientists in the coatings field use *gel permeation chromatography* (GPC), more properly called *size exclusion chromatography* (SEC), to measure MWs. In this convenient

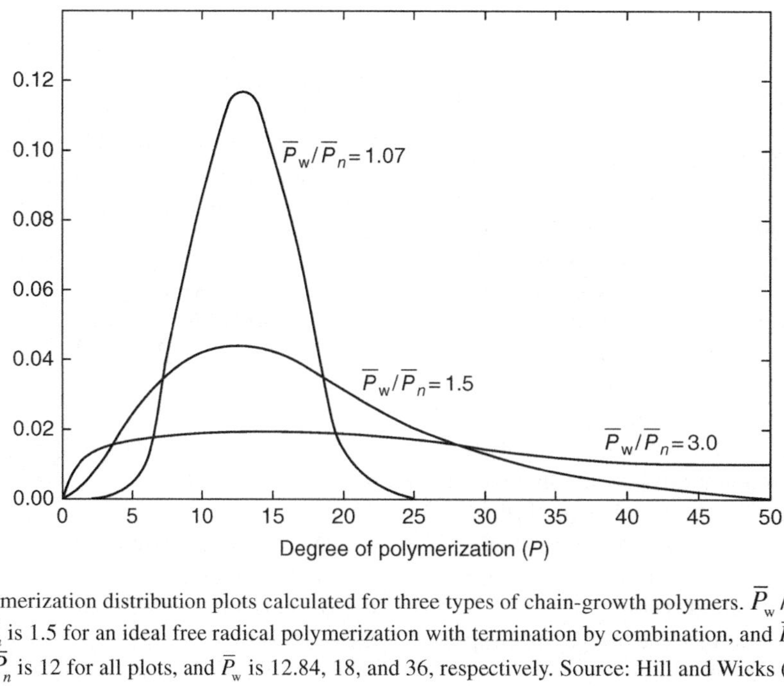

Figure 2.1 Degree of polymerization distribution plots calculated for three types of chain-growth polymers. $\bar{P}_w/\bar{P}_n = 1.07$ is for an ideal anionic polymerization, $\bar{P}_w/\bar{P}_n$ is 1.5 for an ideal free radical polymerization with termination by combination, and $\bar{P}_w/\bar{P}_n$ is 3.0 for a typical free radical polymerization. $\bar{P}_n$ is 12 for all plots, and $\bar{P}_w$ is 12.84, 18, and 36, respectively. Source: Hill and Wicks (1982). Reproduced with permission of Elsevier.

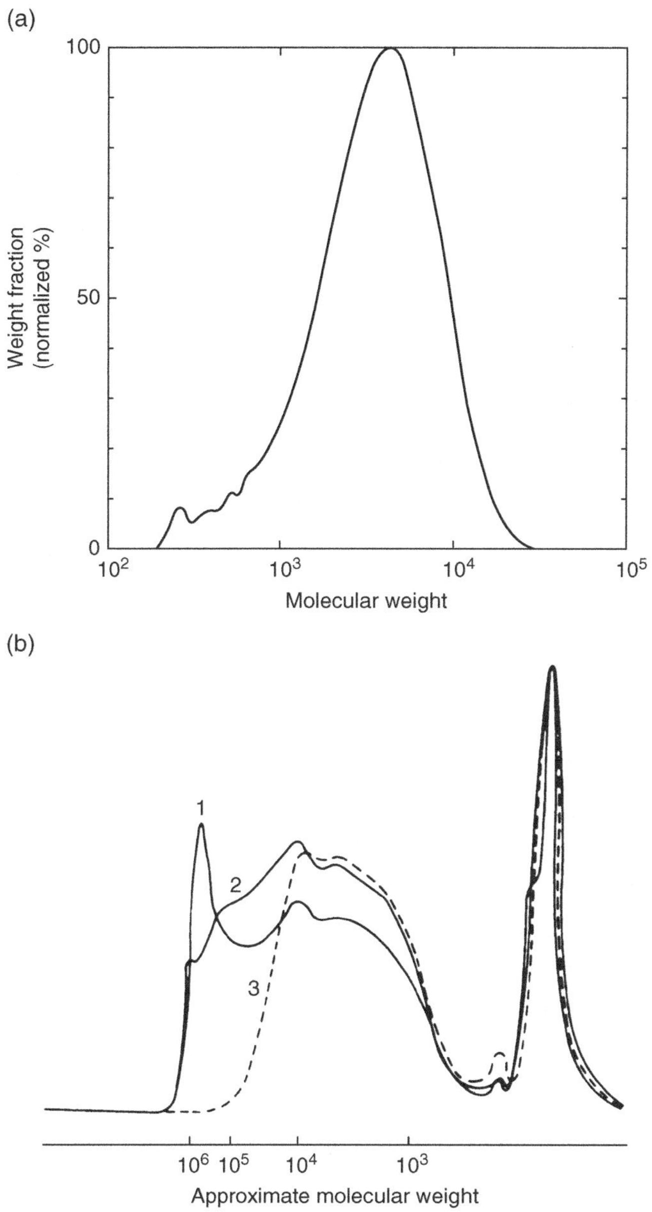

Figure 2.2 (a) Molecular weight distribution of a typical polyester resin. *Source*: Sullivan et al. (1990) Reproduced with permission of American Coatings Association. (b) Molecular weight distributions of three alkyd resins, as measured by GPC with a UV detector. Source: Kumanotani et al. (1984). Reproduced with permission of Marcel Dekker Inc.

method, a dilute solution of an oligomer or a polymer is pumped at high pressure through a series of columns containing porous gels. The molecules are "sorted" by sizes, wherein the largest ones elute first and the smaller ones, which are slowed by entering and leaving more of the gel pores, elute later. The concentration of polymer in the solvent is analyzed as it leaves the column and is plotted as a function of time. A computer program compares the plot to plots of standard polymers of known MWs and calculates $\bar{M}_n$, $\bar{M}_w$, M_z, and M_{z+1} of the entire polymer sample. The results appear precise, but they may not be accurate; errors of +10% are common, and much larger errors are possible. Errors can result because the MW is not measured directly. Rather, the size of the polymer molecules in solution is measured, and the calculations are based on differences in detector response to different compositions. Despite its inaccuracy, GPC is a standard tool, especially valuable for comparing polymers of similar structure. Instrumentation is steadily improving.

The $\bar{M}_n$ of oligomers can be accurately measured by colligative methods, such as freezing point depression and vapor pressure osmometry. However, the accuracy decreases as MW increases, and colligative methods are of little use above $\bar{M}_n = 50\,000$. Mass spectroscopic methods are available that can accurately measure the MWs of individual molecules in oligomers, and even in fairly high polymers (Section 10.2 gives examples).

Some polymers and oligomers have MW distributions approaching the idealized distributions shown in Figure 2.1, as illustrated by the GPC trace of a polyester oligomer in Figure 2.2a. However, many polymers used in coatings have complex distribution patterns as exemplified by the alkyd resins in Figure 2.2b. The $\bar{M}_w$ and $\bar{M}_n$ can be calculated for the entire trace or for portions of complex traces. But, such PD numbers must be used with caution for complex traces.

The MW of resins is an important factor affecting the viscosity of coatings made with solutions of the resins: generally, the higher the MW, the higher the viscosity. The MW of oligomers used in higher solids coatings is especially critical. It is often desirable to prepare oligomers with as narrow a range of MW as possible, in order to minimize the proportions of very low and very high MW molecules. The low MW fraction is generally undesirable from the standpoint of film properties, whereas the high MW fraction increases the viscosity of a resin solution disproportionately. However, alkyd resins having broad, complex MW distributions often perform better than alkyds with similar compositions having less broad distributions (Kumanotani et al., 1984).

MW is often a critical factor controlling the strength of films that are not cross-linked. In general, the higher the MW, the higher the tensile strength of such films, at least up to a point. The acrylic copolymer in automotive acrylic lacquers must have an $\bar{M}_w$ greater than about 75 000 for acceptable film properties, but less than 100 000 for acceptable application properties. For other lacquers the required MWs depend on polymer composition and application methods. Film property considerations favor using high MW polymers in formulating solution coatings, but viscosity considerations favor low MWs. As is often the case in coatings, compromises are needed.

An important advantage of many waterborne coatings is that MW of the polymer generally does not directly affect viscosity, since the polymers are dispersed rather than dissolved in the solvent.

2.1.2 Morphology and Glass Transition Temperature, T_g

Morphology is the study of the physical forms of materials. Like MW, morphology is more complex with polymers than with smaller molecules. Pure small molecules generally solidify to crystals if the temperature is sufficiently low. In contrast, few synthetic polymers crystallize completely, and many do not crystallize at all. Non-crystalline materials that appear to be solids are called *amorphous solids*. There are at least two reasons that synthetic polymers are at least partly amorphous. In general, synthetic polymers are not pure compounds, so it is difficult to achieve the completely regular structure characteristic of a crystalline material. In addition, the molecules are so large that the probability of complete crystallization is low. Part of a molecule can associate with a part of a different molecule or with another part of the same molecule, reducing the probability of pure crystal formation. However, small crystalline domains are common in synthetic polymers; polymers with fairly regular structures, usually homopolymers, are most likely to crystallize partially. In these crystalline domains, fairly long segments of molecules associate with each other in a regular way. The remaining parts of the same molecules are unable to fit together regularly and remain amorphous. While polymers used in fibers and films (e.g., polyethylene and nylon) are often partly crystalline, polymers used in coating applications are, with few exceptions, amorphous.

An important difference between crystalline and amorphous materials is shown in Figure 2.3a and b, which schematically compare the changes in specific volume of these materials with temperature. In the case of a pure crystalline material (Figure 2.3a), as temperature increases, initially there is a slow increase in specific volume, owing to increasing vibrations of the atoms and molecules. Then, at a specific temperature, the substance melts. The melting point T_m is the lowest temperature at which the vibrational forces pushing molecules apart exceed the attractive forces holding them together in crystals. With almost all substances, the molten compound occupies more volume at the same temperature than the crystals; because the molecules are freer to move in a molten compound, they "bounce" their neighbors out of the way, leading to an abrupt increase in specific volume at T_m. Above T_m, the specific volume of a liquid slowly increases with further increase in temperature. Water is a notable exception to this behavior—what might be the consequences if ice was denser than water?

Amorphous materials behave differently, as shown in Figure 2.3b. Starting from a low temperature, there is a slow increase in specific volume as temperature increases, but there is no temperature at which an abrupt change in volume occurs because there is no melting point. Rather, there is a temperature at which there is a change in the rate of increase of specific volume with temperature. Above that temperature, the thermal expansion coefficient is larger than below it. This change of slope is not a phase change; it is a second-order transition, that is, there is a discontinuity in a plot of the *derivative of volume change* as a function of temperature. The temperature at which it occurs is called the *glass transition temperature* (T_g). T_g is properly defined as the temperature at which there is an increase in the thermal expansion coefficient. By comparison, T_m is a

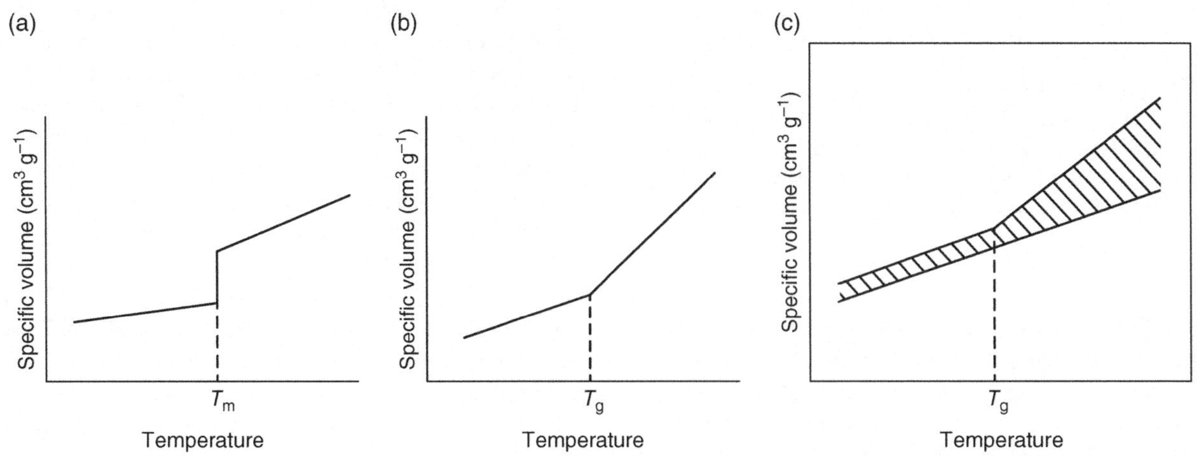

Figure 2.3 Specific volume as a function of temperature (a) for a crystalline material and (b) for an amorphous material; (c) shows free volume within an amorphous material as a function of temperature. Units of specific volume are volume per mass (usually cubic centimeter per gram). Source: Hill and Wicks (1982). Reproduced with permission of Elsevier.

first-order transition, that is, there is a discontinuity in change of volume as a function of temperature, corresponding to the solid–liquid phase change. Unfortunately, T_g is often improperly defined as the temperature below which a material is brittle and above which it is flexible. While there are many cases when this is true, there are other cases when this definition is misleading (Section 4.2). Perhaps, part of the reason for the misunderstanding is the connotation of the word *glass*, which we associate with a brittle material. Like all amorphous materials, glasses undergo a second-order transition. In fact, the phenomenon was first observed in the study of glasses—hence, the name *glass transition temperature*. The idea has proliferated that T_g is a phenomenon associated only with polymers. This is not true. Many small molecules can be supercooled without crystallization to form amorphous glasses that have a T_g. For example, the T_g of *m*-xylene is 125 K (Wicks, 1986). The T_g is always lower than T_m. Partially crystalline polymers show both a T_m and a T_g (Chapter 4).

What is physically happening at T_g? As an amorphous material is heated, atoms in the molecules vibrate with increasing energy, colliding with neighbors and shoving molecules apart for very short periods of time. At T_g, a few of the short-lived "holes" between the molecules become large enough such that an adjacent molecule or a segment of a polymer molecule can fit between two molecules. Thus, T_g can be considered the lowest temperature at which segments of polymer molecules can move with facility in cooperation with neighboring segments. The increase of the coefficient of thermal expansion above T_g results from the greater degree of freedom available to the molecule segments. The larger volume between molecules gives more degrees of freedom, so the same increase in temperature gives a greater increase in volume. As temperature rises, specific volume increases, but there is no more material—just the same material occupying more space. What is in this "extra" volume? Nothing. It is called *free volume*, represented by the hatched area in Figure 2.3c. The molecular motions involved can be detected by spectroscopic techniques, such as solid-state nuclear magnetic resonance (NMR), and change as polymers are heated through T_g (Dickinson et al., 1988; Mathias and Colletti, 1989).

While it is difficult to overemphasize the importance of the concepts of T_g and free volume in coatings science, our understanding of these parameters and our ability to measure them are limited. Salez et al. (2015) describe recent progress toward theoretical understanding, but according to Philip Anderson (a Nobel Prize winner in solid-state physics), "the deepest and most interesting unsolved problem in solid-state physics is probably the glass transition" (Salez et al., 2015). Coatings scientists who do not fully understand T_g are in good company.

T_g values of a material that are measured by different methods may not agree with each other by 20°C or even higher. Obviously, one must be careful in comparing T_g values of different materials to be sure they are based on consistent test methods. The classical method of measuring T_g is *dilatometry* (measurement of specific volume as a function of temperature). Nowadays, T_g is usually measured as described in Section 4.5 by *differential scanning calorimetry* (DSC), by *dynamic mechanical analysis* (DMA), or by *thermal mechanical analysis* (TMA). Roe (1987) and Mengqiu and Xin (2015) introduce these and other methods. The measured T_g depends on the measurement method and the conditions under which the measurement was made. Heating rate is an important variable. The faster the rate of heating during the determination, the higher the apparent T_g. When free volume is small, the rate of movement of molecules or segments is slow. If the rate of heating is slow, there is more time for movement and, hence, the expansion, and the measured T_g is lower.

Some scientists argue that T_g is not a real thermodynamic parameter. They point out that if the determination of specific volume were done at a slow enough heating rate, no transition would be observed and that rather than two straight lines, as shown in Figure 2.3c, there would be a smooth curve. Despite this controversy, T_g is a very useful concept and is well understood in qualitative terms. Relationships between polymer structure and T_g are understood well enough that it is often possible to make reasonable predictions of T_g from knowledge of composition and $\bar{M}_n$. Beyond that, knowing the T_g tells a lot about the film properties expected of a polymer. Important factors affecting the T_g of thermoplastic polymers include the following:

1. *Number average MW*. T_g increases with increasing $\bar{M}_n$, approaching a constant value at $\bar{M}_n$ in the range of 25 000–75 000, depending on the polymer structure. It is logical that T_g is related to $\bar{M}_n$ because decreasing $\bar{M}_n$ results in an increasing proportion of chain ends to chain middles, since chain ends have more freedom of movement than the middles. The relationship of T_g to $\bar{M}_n$ is approximated by Eq. 2.4, where $T_{g\infty}$ is the T_g at infinite MW and A is a constant (T_g is in Kelvin).

$$T_g = T_{g\infty} - \frac{A}{\bar{M}_n} \quad (2.4)$$

2. *Polymer backbone flexibility*. T_g is affected by the ease of rotation about bonds in the polymer backbone. For example, the siloxane bond, Si—O—Si, rotates easily; the T_g of poly(dimethylsiloxane) is 146 K (−127°C) (Andrews and Grulke, 1999). Aliphatic polyethers, such as poly(ethylene oxide), $-(CH_2-CH_2-O)_n-$, also have low T_g, generally in the range of 158–233 K, because there is considerable ease of rotation around the ether bond. The T_g of polyethylene varies because,

although we commonly think of polyethylene as being just chains of methylene groups, the backbone is actually substituted to varying degrees with alkyl side chains, such as ethyl groups. Also, most grades of polyethylene are partially crystalline; only the amorphous areas show a T_g. However, all would agree that the T_g of a long linear aliphatic chain is low, perhaps less than 200 K. The presence of rigid aromatic or cycloaliphatic rings in the polymer backbone substantially increases T_g.

3. *Side chains*. Pendant aromatic rings also lead to high T_g, for example, 373 K for polystyrene, since ease of rotation is decreased. Similarly, pendant methyl groups and carboxymethyl groups increase T_g. For example, T_g increases from 281 K for poly(methyl acrylate) to 378 K for poly(methyl methacrylate), which has both methyl and carboxymethyl groups on alternate carbons of the chain. If the side chains are several atoms long and flexible, T_g is reduced, for example, to 219 K for poly(*n*-butyl acrylate). However, if the side chain is short, bulky, and inflexible, it has less effect and, in some cases, raises T_g, for example, to 314 K for poly(*t*-butyl acrylate).

Care must be taken when comparing T_g values to be sure that the determinations have been carried out under consistent conditions and that the MWs are high enough to eliminate MW effects. Table 2.1 provides the T_g of high MW homopolymers of a group of acrylic and methacrylic esters, as well as other monomers often used as comonomers in polymers for coatings (Lesko and Sperry, 1997; Andrews and Grulke, 1999; Neumann et al., 2004).

Synthetic copolymers often have a disorderly distribution of mers within the chain, in which case, they are called *random copolymers*, although few of them are strictly random in the pure mathematical sense. Such copolymers have T_g values intermediate between those of the homopolymers. It is common to use the *Fox equation* (Eq. 2.5) to estimate the T_g of "random" copolymers, where w_1, w_2, w_3, and so on are the weight fractions of the various monomers in the copolymer and T_{g1}, T_{g2}, T_{g3}, and so on are the T_g (Kelvin) of their high MW homopolymers:

$$\frac{1}{T_{g(\text{copolymer})}} = \frac{w_1}{T_{g1}} + \frac{w_2}{T_{g2}} + \frac{w_3}{T_{g3}} + \cdots \qquad (2.5)$$

Somewhat better approximations can be calculated using a different mixing equation, 2.6, also devised by Fox, in which v_1, v_2, v_3, and so on are the volume fractions of the various monomers in the copolymer; this equation is not widely used because some of the homopolymer densities needed to calculate v_1, v_2, v_3, and so on are not readily available:

$$T_{g(\text{copolymer})} = v_1 T_{g1} + v_2 T_{g2} + v_3 T_{g3} + \cdots \qquad (2.6)$$

Gupta (1995) reports an extensive study on estimating the T_g of acrylic copolymers. He recommends the use of the van Krevelen equation, 2.7, for estimation of T_g, where M is the MW of the repeat unit and Y_g is a molar glass transition factor. Gupta's values for the T_g of *n*-butyl methacrylate (10°C) and of 2-ethylhexyl acrylate (−63°C) vary considerably from the values given in Table 2.1, illustrating that different values are often found in the literature:

$$T_g = \frac{Y_g}{M} \qquad (2.7)$$

Block copolymers sometimes have two or more distinct T_gs.

The T_g of cross-linked polymers is controlled by several factors and their interactions:

- T_g of the segments of polymer between the cross-links
- The cross-link density (XLD)

Table 2.1 Glass Transition Temperatures (°C) for Homopolymers of Various Monomers

Monomer	Methacrylate	Acrylate	Monomer	Methacrylate	Acrylate
Free acid	185	106	*n*-Tridecyl[a]		−46
Methyl	105	9	*iso*-Tridecyl		−39
Ethyl	65	−23	2-Hydroxyethyl	55	
Isopropyl	81	−8	2-Hydroxypropyl	73	
n-Butyl	20	−54	Other monomers		
Isobutyl	53	−40	Styrene	100	
t-Butyl	114	74	Vinyl acetate	29	
n-Hexyl	−5	−57	Vinyl chloride	81	
2-Ethylhexyl	−10	−50	Vinylidene chloride	−18	
iso-Decyl		−30			

[a] Mixture of C_{12-14}.

- The presence of dangling ends
- The presence of cyclic segments (Stutz et al., 1990)
- The structure of the cross-links when XLD is high

While generalized equations showing the relationships of the first four of these factors with T_g have been developed, the complex relationships are not fully understood. The T_g of the polymer segments between cross-links is governed by the chemical structures of the resin and the cross-linking agent, by the ratio of these components, and by the extent of the cross-linking reaction. The factors discussed in connection with thermoplastic polymers apply in terms of their effects on the T_g of the chain segments between cross-links. Since cross-links restrict segmental mobility, T_g increases as XLD increases. On the other hand, T_g decreases with an increasing proportion of dangling ends—that is, chain segments that are connected to the cross-linked network at only one end.

Solutions of polymer in solvent and of solvent in polymer have T_g values intermediate between the T_g of the polymer and that of the solvent. The T_g of solutions increases with increasing polymer concentration. When the weight fraction of solvent w_s is less than about 0.2, a simple mixing equation (Eq. 2.8) gives reasonable correlation between experimental and predicted results (Ferry, 1980). Over a wider range of concentrations, this simple equation gives poor correlations:

$$T_{g(\text{solution})} = T_{g(\text{polymer})} - kw_s \tag{2.8}$$

For solutions of oligomeric *n*-butyl methacrylate in *m*-xylene (Wicks et al., 1986), Eq. 2.9 gave a good fit between observed and predicted data over the whole range from pure solvent to solvent-free oligomer. Here, w_s and w_o are weight fractions and T_{gs} and T_{go} are the T_g of the solvent and the oligomer, respectively. While Eq. 2.1 accurately describes a limited number of oligomer and polymer solutions, its generality is not fully established:

$$\frac{1}{T_{g(\text{solution})}} = \frac{w_s}{T_{gs}} + \frac{w_o}{T_o} + Kw_s w_o \tag{2.9}$$

T_g is discussed further in Section 4.2 in the context of film mechanical properties.

2.2 POLYMERIZATION

In coatings there are two major classes of polymerization reactions: *chain-growth* and *step-growth*. The mechanisms and kinetics of both classes have been extensively studied. The common denominator of chain-growth polymerization is that reactions are chain reactions. Frequently, chain-growth polymerization is called *addition polymerization*, but this terminology is inadequate. While all chain-growth polymerizations involve addition reactions, not all addition polymerizations involve chain-growth reactions—some are step-growth reactions.

2.2.1 Chain-Growth Polymerization

Chain-growth polymerization, initiated by free radicals, is the most commonly used chain-growth polymerization for making vinyl copolymers (often acrylics) for coatings. Odian (2004a) provides an extensive review of the topic, particularly the kinetics of the reactions. Free radical chain-growth polymerizations of most interest to coating applications are *solution polymerization* (Chapter 8) and *emulsion polymerization* (Chapter 9). A related process of importance in coatings is the autoxidation involved in cross-linking drying oils and drying oil derivatives (Chapters 14 and 15). The discussion in this section applies to solution polymerization, although many of the principles are apply to emulsion polymerization as well.

Three types of chemical reactions—*initiation, propagation,* and *termination*—are always involved in chain-growth polymerization; and a fourth, *chain transfer*, often plays a significant role. Initiation occurs when an *initiator* (I) reacts to form an initiating free radical (I·) (Eq. 2.10), which, in turn, adds rapidly to a monomer molecule to form a second free radical (Eq. 2.11):

$$I_2 \longrightarrow I\cdot \tag{2.10}$$

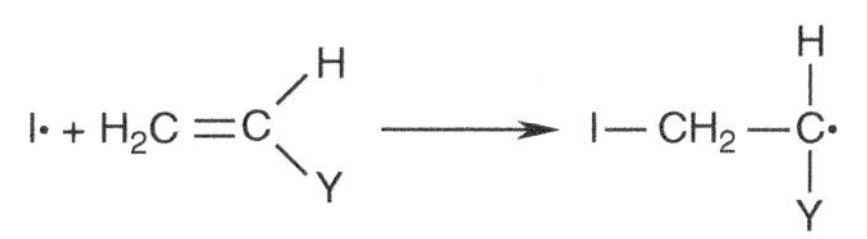

(2.11)

The polymer chain grows by the propagation reaction, in which the monomer free radical adds to a second monomer molecule to extend the chain and form a new free radical (Eq. 2.12):

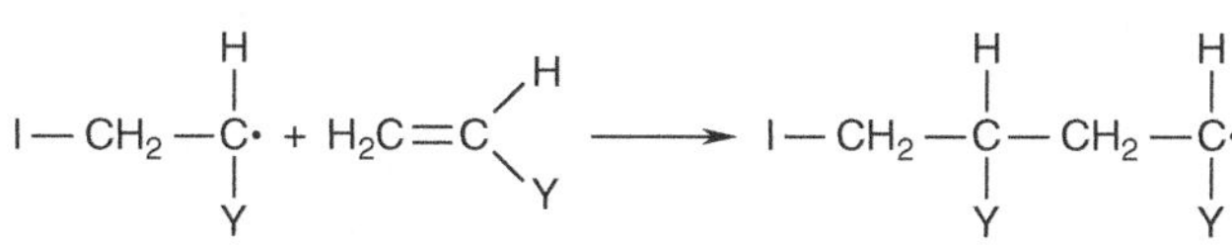

(2.12)

Propagation reactions are very fast—so fast that a chain with hundreds of mers can grow in a fraction of a second. At any moment, the concentrations of monomer and polymer greatly exceed the concentration of growing polymer molecules, which is about $10^{-6}\,\text{ML}^{-1}$. (Exceptions are controlled radical polymerizations (CRP), Section 2.2.1.1.)

The final stage is termination of the growing chain. Two common types of termination reactions are *combination* (Eq. 2.13) and *disproportionation* (Eq. 2.14). In most free radical initiated polymerizations, the rate of propagation (Eq. 2.12) is faster than the rate of initiation, which is limited by the rate of Eq. 2.10:

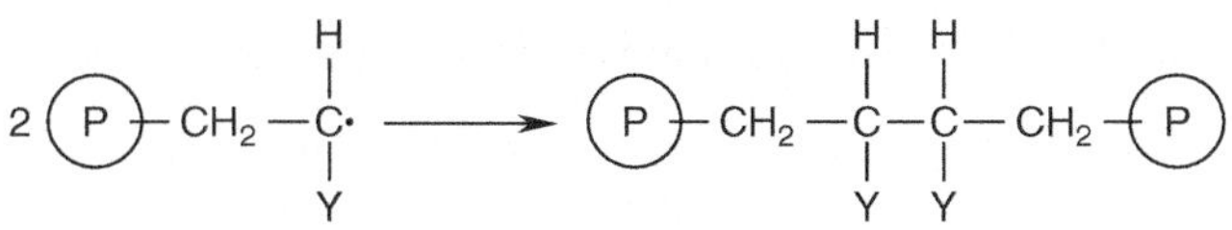

(2.13)

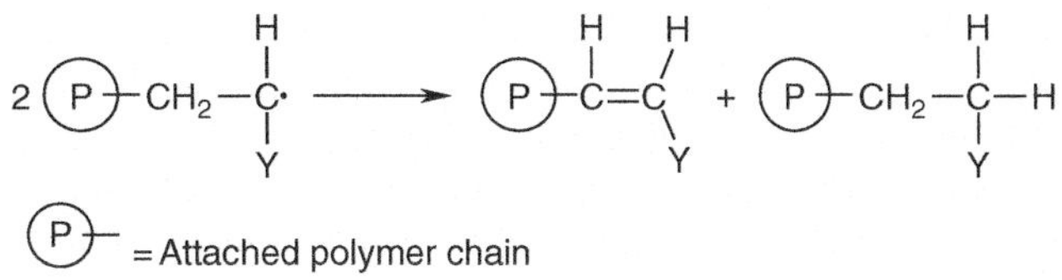

(2.14)

Side reactions also occur; among the most important are chain transfer reactions, in which the free radical on the end of the propagating polymer chain abstracts a hydrogen atom from some substance X—H present in the polymerization reaction mixture (Eq. 2.15):

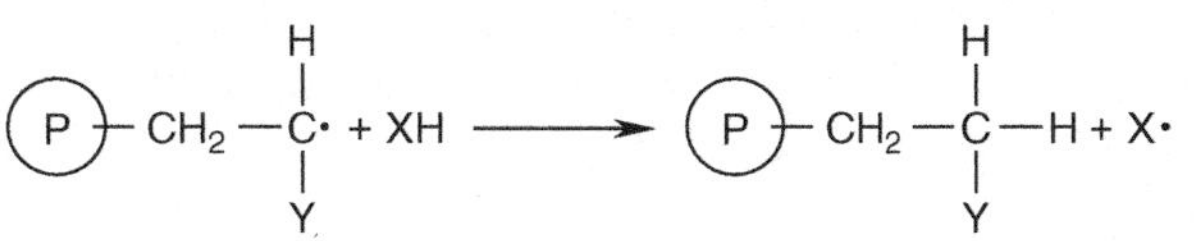

(2.15)

The net effect of chain transfer is to terminate the growing chain while generating a free radical, which may initiate a second chain. X—H may be a solvent, a monomer, a molecule of polymer, or a *chain transfer agent*, a reactant that is added to the polymerization reactants to cause chain transfer. When chain transfer is to a solvent or a chain transfer agent, MW is reduced. When chain transfer is to a polymer molecule, growth of one chain stops, but a branch grows on the polymer molecule; the result is a higher $\bar{M}_w/\bar{M}_n$.

Note that the structures of the propagating polymer chains show substitution on alternate carbon atoms. This structure results from the favored addition of free radicals to the CH_2 end of most monomer molecules, corresponding to *head-to-tail addition*. Head-to-tail addition predominates in almost all monomers, but a small fraction of *head-to-head addition* also occurs. The result is a polymer with most of the substitution on alternating carbons in the chain, but with a few chain segments having substitution on adjacent carbons. The effect of a small fraction of head-to-head structure is generally negligible, but it sometimes has significant consequences with regard to exterior durability and thermal stability.

Initiators, sometimes, incorrectly called catalysts, are used in low concentration (usually in the range of 0.5–4 wt% (weight percent), but sometimes higher when low MW is desired). A variety of free radical sources has been used. Two common classes of initiators are azo compounds, such as azobisisobutyronitrile (AIBN) and peroxides such as benzoyl peroxide (BPO) or *t*-amyl peracetate. AIBN is fairly stable at 0°C but decomposes relatively rapidly when heated at 70–100°C to generate free radicals. A substantial fraction of the resulting radicals initiate polymerization, although some combine to form a coupling product. The half-life of AIBN is about 5 h at 70°C and about 7 min at 100°C:

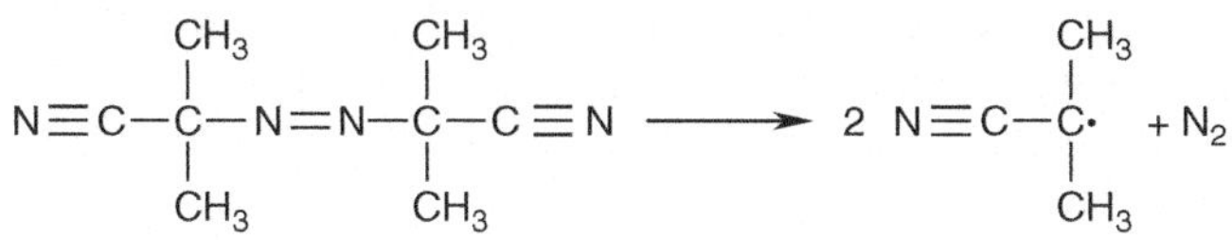

BPO decomposes at similar temperatures—its half-life is about 20 min at 100°C. The reactive benzoyloxy free radical generated can initiate polymerization; also, it can dissociate (rapidly at higher temperatures, such as 130°C) to yield a very highly reactive phenyl free radical and CO_2:

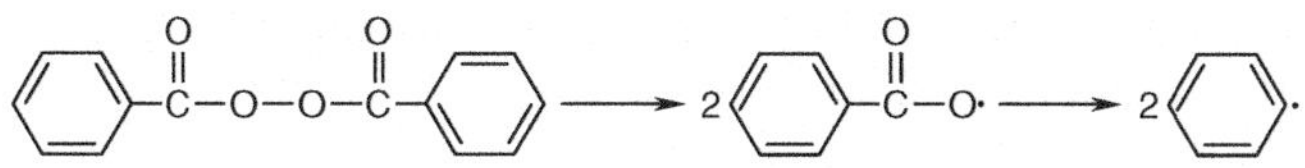

A range of monomers is capable of propagating a radical initiated chain reaction. Most are alkenes having an electron-withdrawing group; methyl acrylate (MA) and methyl methacrylate (MMA) are important examples:

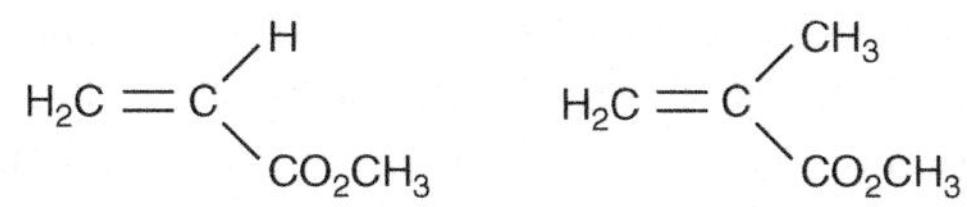

MA MMA

Copolymers containing a preponderance of acrylic and methacrylic ester monomers are called *acrylic polymers*, or often, just *acrylics*. They are extensively used in coatings. Control of MW and MW distribution is critical in preparing polymers for coatings. There are three major factors that affect MW when using the same monomer, initiator, and solvent:

1. *Initiator concentration.* Higher initiator concentration reduces the MW. When the initiator concentration is higher, more initiating free radicals are generated to react with the same total amount of monomer. More chains are initiated and terminated, thereby reducing both the $\bar{M}_n$ and $\bar{M}_w$ of the resulting polymer.
2. *Temperature.* At higher temperatures, more initiator is converted into initiating free radicals in a given time,

thereby increasing the concentration of growing chains and the probability of termination. As with increasing initiator concentration, the result is lower $\bar{M}_n$ and $\bar{M}_w$.

3. *Monomer concentration*. Higher monomer concentration increases $\bar{M}_n$ and $\bar{M}_w$. The highest MW is obtained in a solvent-free reaction mixture. With the same concentration of growing free radical ends, a higher monomer concentration increases the probability of chain-growth relative to termination.

To the extent that any of these factors change during a polymerization process, $\bar{M}_n$ and $\bar{M}_w$ of the polymer molecules also change. The usual result is a broader MW distribution. Changes in monomers also change MW distribution. Consider the difference between MA and MMA. Since the free radicals at the ends of growing chains of poly(methyl methacrylate) (PMMA) are sterically hindered, termination by combination is impeded, and termination by disproportionation predominates. On the other hand, with poly(methyl acrylate) (MA), a major fraction of the termination reactions occur by combination. Theoretical calculations show that for high MW polymers, the lowest $\bar{M}_w/\bar{M}_n$ attainable with termination by combination is 1.5, while the minimum with termination by disproportionation is 2.0, corresponding to higher PD. In actual polymerization processes, $\bar{M}_w/\bar{M}_n$ is usually higher, although with very high initiator concentrations, polydispersities tend to be lower. No basic studies have been reported to account for the low PDs with high initiator concentrations.

Chain transfer to polymer must also be considered. This reaction occurs to a degree in the polymerization of MMA but is more important in the polymerization of MA. The tertiary hydrogen on the carbon to which the carboxymethyl group is attached in PMA is more susceptible to abstraction by free radicals than any other hydrogen in PMA or PMMA. When this H atom is abstracted, growth of the original chain is terminated, and a new free radical on a PMA chain is formed. This free radical can now add to a monomer molecule, initiating growth of a branch on the polymer molecule. The result is a polymer containing branched molecules and having a larger $\bar{M}_w/\bar{M}_n$ than predicted for ideal linear polymerization. In extreme cases, chain transfer to polymer results in very broad MW distributions and, ultimately, to formation of gel particles through cross-linking. The preceding discussion supposes that chain transfer to polymer occurs mainly between different molecules. Another possibility is that the growing radical may abstract a nearby hydrogen from the same molecule, a process called *back biting*, as discussed further in Chapter 8.

Branching can also result from the abstraction of hydrogen atoms from a polymer chain by initiating free radicals. Phenyl free radicals from high temperature decomposition of BPO are so reactive that they will abstract almost any aliphatic hydrogen, leading to substantial branching. Accordingly, if branching is desired, initiation using BPO at high temperature (e.g., 130°C) is a good choice. However, in most cases, minimization of branching is more desirable. In these cases, azo initiators, such as AIBN, or aliphatic peroxy initiators are preferred over BPO.

Since initiator residues remain attached to the polymer chain ends, they may affect polymer properties. For high MW polymers, the effect on most properties is usually negligible. Exterior durability (Chapter 5) is an exception. However, for oligomers, the effect may be appreciable, particularly on exterior durability (Section 8.2.1).

MW and MW distribution also depend on solvent structure. For example, substituting xylene for toluene, with other variables constant, leads to a decrease in MW. Since each xylene molecule has six abstractable benzilic hydrogen atoms, while toluene has only three, the probability of chain transfer is higher for xylene and M_n decreases.

To prepare a low MW polymer or oligomer, one can add a compound that undergoes facile hydrogen abstraction as a chain transfer agent. If the hydrogen atoms are readily abstracted, the addition of even relatively low concentrations of a chain transfer agent can lead to a substantial reduction in MW. Mercaptans (RSH) are widely used as chain transfer agents owing to the readily abstractable SH hydrogen atom, as well as the high initiating capability of the resulting thiyl radical.

Other variables affecting MW and MW distribution are the decomposition rate of the initiator and the reactivity of the resulting free radicals. To achieve a low $\bar{M}_w/\bar{M}_n$, concentrations of reactants must be kept as constant as possible throughout the polymerization. It is undesirable to simply charge all of the monomers, solvents, and initiators into a reactor and heat the mass to start the reaction. This procedure is sometimes used in small-scale laboratory reactions, but almost never in production. At best, it yields a high $\bar{M}_w/\bar{M}_n$; at worst, the reaction may run violently out of control, because free radical polymerizations are highly exothermic. Instead, one charges some of the solvent into the reactor, heats to reaction temperature, and then adds monomer, solvent, and solutions of initiator to the reactor at rates such that the monomer and initiator concentrations are kept as constant as possible. Adding monomer at a rate that maintains a constant temperature leads to a fairly constant monomer concentration. The appropriate rate of addition of the initiator solution can be calculated from the rate of its decomposition at the temperature being used. Maintaining constant solvent concentration is more complex since as the polymerization proceeds, polymer is accumulating; in a sense, the polymer becomes a part of the "solvent" for the polymerization. Solvent is added at a decreasing rate so that the other concentrations stay as constant as possible. Perfect control is not possible, but careful attention to details makes an important difference in the $\bar{M}_w/\bar{M}_n$ of the polymer produced.

Bulk copolymerization of mixtures of unsaturated monomers further complicates the situation. The rates of reaction involved in the various addition reactions depend on the structures of the monomers. If the rate constants for all of the possible reactions were the same, the monomers would react randomly and the average composition of molecules of substantial length would all be the same. However, the rate constants are not equal. If polymerization is carried out by putting all of the reactants in a flask and heating, the first molecules formed would contain more than proportional amounts of the most reactive monomer, and the last molecules formed would have an excess of the least reactive monomer. This situation is usually undesirable. Such effects have been extensively studied, and equations have been developed to predict the results with different monomer combinations. (See Odian (2004a) for a detailed discussion of copolymerization.)

In actual practice, the problem is less complex, since reactions are not run in bulk. Rather, as mentioned earlier, monomers, solvent, and initiator solution are added gradually to the reaction mixture. If the additions are carefully controlled so that the rate of addition equals the rate of polymerization, copolymers having reasonably uniform composition corresponding to the feed ratio are obtained with most monomers. This procedure, called *monomer-starved conditions*, results in copolymerization under conditions in which the concentration of monomers is low and fairly constant. Further process refinements are possible by adding individual reactants or mixtures of reactants in two or three streams at different rates. Computer modeling of the processes can help achieve the desired results.

2.2.1.1 Living Polymerizations: Controlled Radical Polymerizations (CRP)

For several decades, considerable effort has been directed to the preparation of acrylic and other chain-growth (co) polymers with narrow MW distributions and controlled structures. Narrow MWD can only be accomplished when the rate of initiation is much faster than the rate or propagation and when the rates of termination reactions are slow, which corresponds to the opposite of the kinetics described in Section 2.2.1. Under these circumstances, almost all polymer chains start growing early in the process and grow at about the same rate under about the same conditions. In these processes the polymer chain ends often remain reactive even after all monomer has been consumed, in which case they are called *living polymers* (Darling et al., 2000).

Early living polymerizations included *anionic polymerizations* and *group transfer polymerizations* (Sogah et al., 1987; Webster, 2000). These methods require highly purified monomers and very dry conditions, as well as the absence of proton donor (active hydrogen) groups such as —OH groups on the monomers. GTP can produce polymers with PDIs as low as 1.03; however, because of process costs, commercial use in coatings has been limited to specialty applications, such as pigment dispersants.

Attention has shifted to CRP. We prefer using CRP to the frequently used acronym CFRP for *controlled free radical polymerization*, since CFRP is also a common acronym for carbon fiber reinforced *plastics*.

Boyer et al. (2016) reviewed the history of CRP starting in 1982 by pioneers such as Otsu and Georges and their coworkers. Since then researchers have produced many thousands of papers and patents on the subject because of its great scientific interest and its enormous potential for diverse applications.

In general, CRP methods mediate the rates of propagation and termination by including some substance that reversibly bonds to the radical at the growing end of the polymer chain. By this means, it is possible to slow the propagation rate by many orders of magnitude, satisfying the kinetic requirements for living polymerization. Since these polymerizations are free radical processes, they are relatively insensitive to impurities, and they can be used to copolymerize monomers with proton donor (active hydrogen) groups. These methods enable the preparation of block, alternating block, sequence-controlled, and gradient copolymers by sequential addition of monomers; they can also be adapted to synthesize an enormous variety of linear, graft, and star polymers, copolymers, and *macromonomers* (polymerizable oligomers) from many of the common acrylic and styrenic monomers. The literature about CRP processes lacks consistency about how the processes are named. One possible classification follows:

1. *Stable free radical polymerization* (SFRP), also called *nitroxide-mediated polymerization* (NEM), in which polymerization is mediated by nitroxide compounds (Auschra et al., 2002). This method works well with monomers having hydroxyl functionality.
2. *Reversible addition–fragmentation chain transfer* (RAFT) polymerization, which involves mediation by certain dithioesters or xanthates (trithiocarbonates) (Perrier et al., 2004).
3. *Transition metal-mediated living radical polymerizations* (TMMLRP); several methods can be distinguished, although some authors lump them together. Here is a simple classification:

   3a. *Atom transfer radical polymerization* (ATRP) where the mediator is a metal salt, often of copper, with a carefully chosen organic ligand and an organic halide that can undergo a redox reaction with the metal to trigger the polymerization (Matyjaszewski, 2012; Krol and Chmielarz, 2014; Boyer et al., 2016).

3b. *Catalytic chain transfer polymerization* (CCTP), which uses a conventional initiator, usually an azo type, in the presence of a chelated cobalt (or other) metal salt. CCTP is especially useful for preparing macromonomers having relatively low color (Chiefari et al., 2005; Smeets et al., 2012; Boyer et al., 2016).

Descriptions of CRP techniques are provided by Matyjaszewski (1999), Muller and Matyjaszewski (2009), and Lutz et al. (2014); Odian's (2004b) book includes a lengthy discussion about it. The subject has become so vast that not one book covers it all. Aspects of the subject have also been extensively reviewed, for example, by Boyer et al. (2016) and by Krol and Chmielarz (2014). The latter work emphasizes applications (realized and potential), including coatings. Journals and especially patents are rich with specific procedures for CRP synthesis of acrylic polymers having various structures. In some cases, the procedures are straightforward.

CRP development has largely been driven by applications in medical and biomedical technologies. In coatings, TMMLRP methods have been the focus of most researches. Block acrylic copolymers prepared by ATRP are useful aqueous pigment dispersants (White et al., 2002), as are graft copolymers made with CCTP (Viosscher and McIntryre, 2003). Krol and Chmielarz (2014) cite several more recent examples. CRP methods can be used in aqueous media, notably in emulsion polymerization. See Section 21.3.1 for further examples and discussion of the use of the copolymers in pigment dispersion. It is technically feasible to use CRP to synthesize desirable resins for large-scale uses such as automotive clear coats, where the technology is slowly making commercial inroads (Chapter 30).

Astonishing progress has been achieved in controlling polymer structures, but polymer chemists are still a long way from an answer to the chemists' prayer:

> Oh Lord, I fall upon my knees,
> And pray that all my syntheses,
> Will not always be inferior,
> To those effected by bacteria.

2.2.2 Step-Growth Polymerization

A second class of polymerization that is important in the coatings field is *step-growth polymerization*. As the name indicates, the polymer is built up a step at a time. The term *condensation polymerization* has been used for this process because early examples involved condensation reactions—reactions in which a small molecule by-product, such as water, is eliminated. While both terms are still used, *step-growth polymerization* is more appropriate because many step-growth polymerizations are not condensation reactions. Step-growth polymerization reactions are used in two ways in coatings. One is to prepare resins for use as vehicles, and the other is for cross-linking after the coating has been applied to a substrate.

The formation of polyesters, which are discussed more broadly in Chapter 10, is used here to illustrate the principles involved. Of the many reactions that form esters, three are commonly used to prepare polymers and oligomers for coatings by step-growth polymerization: direct esterification of an acid with an alcohol, transesterification of an ester with an alcohol, and reaction of an anhydride with an alcohol. A fourth, less common, method is ring-opening polymerization of a lactone. The first two of these reactions proceed rapidly only at elevated temperature; and process temperatures of 200°C and higher are common.

In order to form a polymer from two reactants, both must have two or more functional groups. When both reactants are difunctional, linear polymers form. High MW linear step-growth polymers are commonly used in fibers, films, and plastics. Most polyester resins used in coatings, however, have relatively low MWs and are branched, requiring that at least one reactant has at least three or more functional groups. After application of the coating, the terminal groups on the branch ends are reacted with a cross-linker to form the cured coating. Note that in this section, the terms *reactant* and *monomer* are used interchangeably.

When a difunctional acid (AA) reacts with a difunctional alcohol (BB) in a direct esterification reaction, the MW builds up gradually. Under ideal conditions, polymer chains averaging hundreds of mers per molecule can be made, but this can occur only if (a) the reactants AA and BB contain no monofunctional impurities, (b) the amounts of AA and BB are exactly equimolar, (c) the reaction is driven virtually to completion, and (d) side reactions are negligible. If one reactant is present in excess, terminal groups of the excess monomer predominate. The MW of the completely reacted system is progressively lower as the difference from equal equivalents is increased. For example, if 7 mol of dibasic acid are completely reacted with 8 mol of a dihydroxy compound (a diol), the *average* molecule will have terminal hydroxyl groups as shown in the following equation (here, for convenience, AA and BB represent both the reactants and the mers in the polymer):

$$7AA + 8BB \longrightarrow BB\left(AA-BB\right)_6 AA-BB + 14H_2O$$

Common monomers (reactants) are as follows:

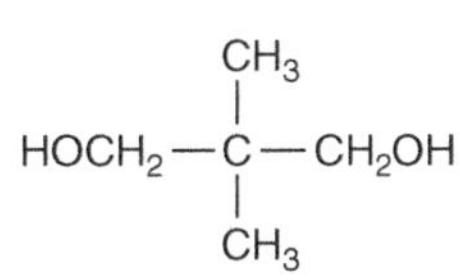
Neopentyl glycol (NPG) $F = 2$

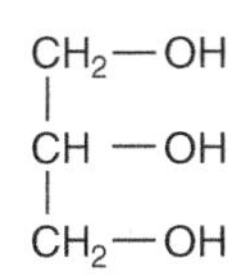
Glycerol (Gly) $F = 3$

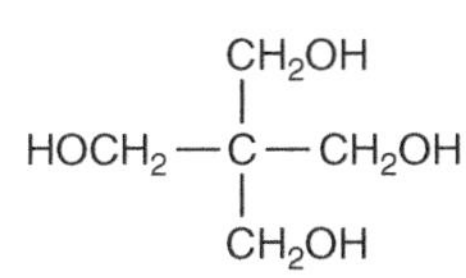
Pentaerythritol (PE) $F = 4$

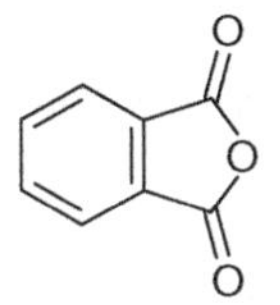

Phthalic anhydride (PA) $F=2$

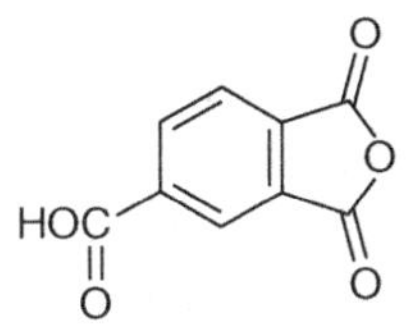

Trimellitic anhydride (TMA) $F=3$

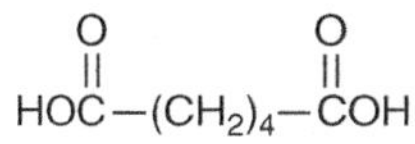

Adipic acid (AA) $F=2$

The symbol F is used for the functionality of monomers, which is the number of reactive groups per molecule. The anhydride groups in PA and TMA count as two functional groups since they can form two ester groups during polymerization.

The average functionality, represented by $\bar{F}$, of a mixture of monomers containing equal equivalents of hydroxyl and carboxyl groups is calculated as follows:

$$\bar{F} = \frac{\text{total equivalents}}{\text{total moles}}$$

Most polyester resins for coatings are hydroxyl-functional and are made using monomer mixtures having excess hydroxyl groups. Since some of the hydroxyl groups, thereby, have no carboxyl groups to react with, the equation must be modified to reflect only the total number of equivalents that can react. In a resin having excess hydroxyl groups prepared from dicarboxylic acids, the total equivalents that can react correspond to twice the number of equivalents of carboxylic acid groups:

$$\bar{F} = \frac{\text{total equivalents that can react}}{\text{total moles}} = \frac{2(\text{equivalents of COOH})}{\text{total moles}}$$

A simple formulation for a polyester oligomer is given in Table 2.2.

An additional important consideration in designing resins is the functionality of the resin. To distinguish the functionality of the resin from that of the monomers and the monomer mixture, the symbol f is used. Since almost all coating polyester resins are made using some trifunctional and/or tetrafunctional monomers, a number average functionality $\bar{f}_n$ is more appropriate:

Table 2.2 Polyester Formulation

Component	Moles	Equivalents
Adipic acid	0.9	1.8
Phthalic anhydride	0.9	1.8
Neopentyl glycol	1.0	2.0
Glycerol	1.0	3.0
	3.8	8.6

$F = 8.6/3.8 = 1.89$

$$\bar{f}_n = \frac{\text{number of functional groups in a sample}}{\text{number of molecules in the sample}}$$

The value of $\bar{f}_n$ can be calculated from $\bar{M}_n$ and the number of functional groups per sample weight obtained by analysis.

Esterification of a carboxylic acid with an alcohol is acid catalyzed. In the absence of a catalyst, the rate r is approximately of third order in reactants, as shown in Eq. 2.8, with one carboxyl group reacting with the alcohol and v second catalyzing the reaction. Since water is generally removed rapidly from the reaction mixture, it is reasonable to use Eq. 2.16, which disregards the reverse reaction:

$$r = k[\text{RCOOH}]^2[\text{R'OH}] \tag{2.16}$$

Because of the second-order dependence on acid concentration, the rate decreases precipitously as reaction proceeds. For example, polyesterification of equal moles of diethylene glycol with adipic acid at 160°C without a catalyst is 60% complete in 1 h, but requires 27 h to reach 94.5% conversion and would require years to reach 99.8% conversion (Flory, 1939). The reaction is accelerated by strong acid catalysts, but in many cases, conventional strong acids cause side reactions and discoloration. Therefore, the usual catalysts are organotin compounds, such as monobutyltin dioxide, or titanate esters. It has been shown that both the organotin compound and the carboxylic acid act as catalysts (Chang and Karalis, 1993).

The kinetics of ideal step-growth polyesterification for difunctional reactants can be analyzed in terms of p, the fractional extent of reaction; n_p, the number fraction of differing degrees of polymerization; P_n, the degree of polymerization; and w_p, the weight fraction of molecules. As p increases, the degree of polymerization builds up slowly at first—at $p=0.5$ (corresponding to 50% conversion), $\bar{P}_n$ is only 2. The $\bar{P}_n$ is only 10 at $p=0.9$, and $p=0.998$ is required to reach a $\bar{P}_n$ of 500. Thus, with difunctional monomers, high MW can only be attained when the mole ratio of COOH/OH is 1.00 and when esterification is driven beyond $p=0.99$. This is difficult because of the decreasing reaction rate at high values of p. Note that, as shown in Figure 2.4a, the number of unreacted monomer molecules remains higher than that of any other single species in the reaction mixture, no matter how high p becomes. As shown in Figure 2.4b, $\bar{P}_n$, the peak of the P distribution curve, only reaches substantial values at high p values. In the case of high MW linear polymers, under ideal conditions, the $\bar{M}_w/\bar{M}_n$ obtained in step-growth polymerizations is 2.

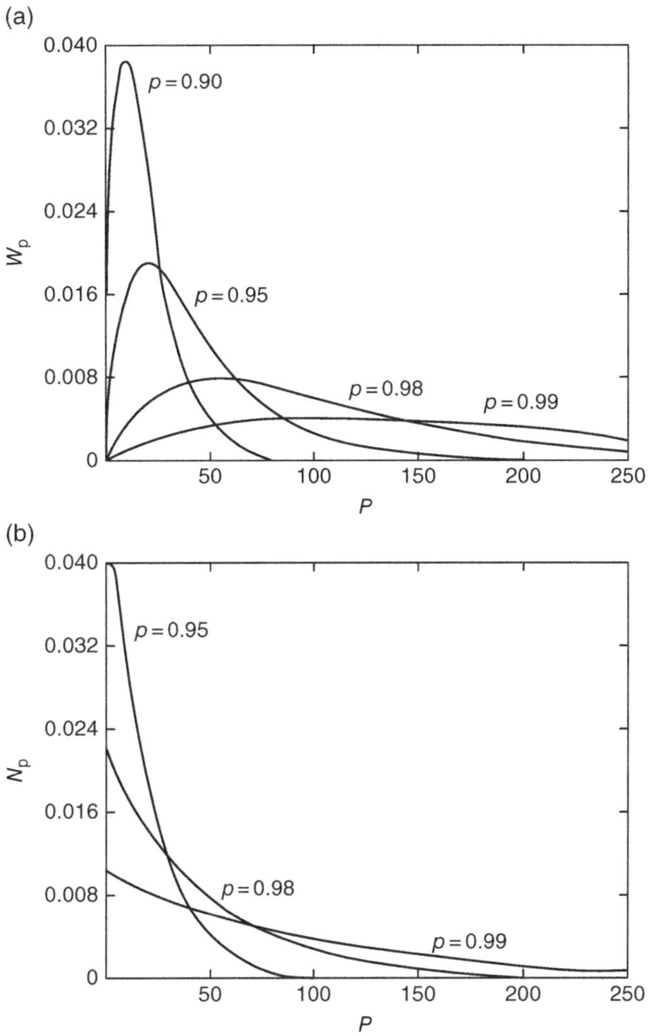

Figure 2.4 (a) Weight fraction distribution w_p of molecules in a linear step-growth polymer for several extents of reaction p. (b) Number, or mole fraction, distribution n_p. Source: Odian (2004c) with permission.

2.3 FILM FORMATION

Most coatings are liquids having a viscosity appropriate for the application method, generally in the range of 0.05–1 Pa·s at high shear rates. After application, the liquid is converted to a "dry," that is, solid film. In powder coatings, the powder is liquefied after application and then converted to a solid film. The chemical and physical changes that occur in these processes are called *film formation*, which critically determines the ultimate appearance and performance of the coating.

If the polymers of the applied coating were crystalline, there would be no difficulty in defining a solid film. The film would be solid if the temperature were below its freezing point; however, binders of coatings are almost always amorphous, thereby having no melting point or sharp demarcation between a liquid and a solid. A useful definition of a solid film is that it is not significantly damaged under the pressures to which it is subjected during use. Thus, one can define whether a coating is a solid by stating the minimum viscosity required to resist flow sufficiently to satisfy a particular test requirement under a specified set of conditions. For example, it is reported that a film is *dry-to-touch* if its viscosity is greater than about 10^3 Pa·s (Burrell, 1962). However, if the definition of a solid is that the film resists *blocking*—that is, sticking together when two coated surfaces are put against each other for 2 s under a pressure of 1.4 kg cm^{-2} (20 psi)—the viscosity has to be greater than about 10^7 Pa·s.

For thermoplastic binders, we can use this information to predict polymer structures that could meet such tests. Using a simplified form (Eq. 2.17; in the equation T is in Kelvin) of the Williams–Landel–Ferry (WLF) equation (Section 3.4), using "universal constants" and assuming that the viscosity at T_g is 10^{12} Pa·s, one can estimate the T_g of a binder required so that a film does not flow under some set of circumstances:

$$\ln\eta = 27.6 - \frac{40.2\left(T - T_g\right)}{51.6 + \left(T - T_g\right)} \tag{2.17}$$

Using Eq. 2.17, we can estimate the appropriate $(T - T_g)$ value required for a film to be dry-to-touch, that is, to have a viscosity of 10^3 Pa·s. The calculated $(T - T_g)$ value is 54°C, which corresponds to a T_g of −29°C for a film to be dry-to-touch at temperature T of 25°C. The T_g calculated for block resistance (at 1.4 kg cm^{-2} for 2 s at 25°C, i.e., for a viscosity of 10^7 Pa·s) is 4°C. This is near the optimum T_g for many architectural paints. Because there is considerable variation in the WLF "universal constants," these T_g values are not exact, but they can serve as a formulation guide. Since we have a reasonable idea of the relationships between structure and T_g (Section 2.1.2), we can approximate the requirements to make a binder with the viscosity necessary to pass a particular test. If the coating has to pass a test at a higher temperature than 25°C, the T_g of the binder must be higher, since the free volume dependence is on $(T - T_g)$. If the pressure to which the film is to be subjected is higher or the time under pressure is to be longer, the T_g must be higher.

2.3.1 Film Formation by Solvent Evaporation from Solutions of Thermoplastic Binders

Films can be formed in a variety of ways. One of the simplest methods is to dissolve a polymer in solvent(s) at a concentration needed for application requirements, apply the coating, and allow the solvent to evaporate. Let us illustrate with a copolymer of vinyl chloride, vinyl acetate, and a hydroxyl-functional vinyl monomer with $\bar{M}_n$ of 23 000 that is reported to give coatings having good mechanical properties without cross-linking (Mayer and Kaufman, 1984). The T_g of the copolymer is 79°C. A solution in methyl ethyl ketone (MEK) with a viscosity of 0.1 Pa·s required for spray application would have about 19 NVW (nonvolatile weight, i.e., weight percent solids) and about 12 NVV (nonvolatile volume, i.e., volume percent (vol%) solids). MEK has a high vapor pressure at room temperature and evaporates rapidly from a thin layer. In fact, a sizable fraction of the MEK evaporates from the atomized spray droplets between the time they leave the spray gun and reach the substrate. As solvent evaporates from a film, viscosity increases, and the film will be dry-to-touch soon after application. Also, in a short time, the coating will not block under the conditions mentioned previously. Nevertheless, if the film is formed at 25°C, the "dry" film contains several percent of retained solvent. Why?

In the first stage of solvent evaporation from a film, the rate of evaporation is essentially independent of the presence of the polymer. Evaporation rate depends on the vapor pressure of the solvent at the particular temperature, the ratio of surface area to volume, and the rate of air flow over the surface. However, as solvent evaporates, viscosity increases, T_g increases, free volume decreases, and the rate of loss of solvent is no longer dependent on its vapor pressure, but rather becomes limited by how rapidly solvent molecules can diffuse to the surface of a film. The solvent molecules must jump from one free-volume hole to another to reach the surface. As solvent loss continues, T_g increases, free volume decreases further, and solvent loss slows. If the film is formed at 25°C from a solution of a polymer that, when solvent free, has a T_g greater than 25°C (in this example, it is 79°C), the film retains considerable solvent even though it is a hard "dry" film. Solvent slowly leaves such a film, but it has been shown experimentally that 2–3% of solvent remains after several years at ambient temperature. To assure nearly complete removal of solvent in a reasonable period of time requires baking at a temperature significantly above the T_g of the solvent-free polymer. Solvent loss from films is discussed in more detail in Section 18.3.4.

2.3.2 Film Formation from Solutions of Thermosetting Resins

High MW thermoplastic polymers are required for good film properties, which, in turn, require high solvent levels (often on the order of 80–90 vol% solvent) to achieve the necessary viscosity for application. Considerably less solvent is needed for coatings based on solutions of lower MW thermosetting resins. After application, the solvent evaporates, and chemical reactions occur resulting in polymerization and cross-linking, which imparts good film properties. A goal is to attain an optimal XLD in the film. Many combinations of chemical reactions are used in thermosetting coatings, as discussed in Chapters 8–17. A critical aspect of the design of a coating is the selection of components that provide required mechanical properties (Chapter 4). In this section, we discuss the general principles of cross-linking reactions.

A dilemma with thermosetting systems is the relationship between coating stability during storage and the time and temperature required to cure a film after application. Generally, it is desirable to be able to store a coating for many months, or even several years, without a significant increase in viscosity resulting from a reaction during the storage period. On the other hand, after application, one would like to have the cross-linking reactions proceed rapidly at the lowest possible temperature.

As formulations are shifted to higher solids to reduce volatile organic compound (VOC) emissions, there are higher concentrations of functional groups, resulting in greater difficulty in formulating storage stable coatings. The problem results not only from the presence of less solvent but also from the lower MWs and lower equivalent weights needed to achieve an acceptable XLD. Both factors increase the concentration of functional groups in a stored coating. Concentration of reactants in a film increases after application and evaporation of solvent, which increases reaction rates; but, since the MW is lower, more reactions must occur to achieve the desired cross-linked film properties.

What controls the rate of a reaction? We can consider this question broadly as a reaction between two groups, represented by the symbols A and B that react to form a cross-link A–B:

$$A + B \rightarrow A - B$$

In the simplest cases, one can express the rate of reaction r of A and B by Eq. 2.10, where k is the rate constant for the reaction between A and B at a specified temperature, and [A] and [B] represent the concentration of the functional groups in terms of equivalents per liter. The rate constant is the reaction rate when $[A]\times[B]$ equals 1 equiv.^{2}L^{-2}:

$$r = k[A][B] \tag{2.18}$$

To minimize the temperature required for curing while maintaining adequate storage stability, it is desirable to select cross-linking reactions for which the rate depends strongly on temperature. This dependence is reflected in the rate equation by the dependence of k on temperature. It is commonly taught in introductory organic chemistry classes that rate constants double with each 10°C rise in temperature. That generalization is true for only a limited number of reactions within a narrow temperature range near room temperature. A better estimate, but still an estimate, of the temperature dependence of k is given by the empirical Arrhenius equation, Eq. 2.19, where A is the *pre-exponential term*, E_a is the thermal coefficient of reactivity (commonly labeled *activation energy*), R is the gas constant, and T is temperature (in Kelvin):

$$\ln k = \ln A - \left(\frac{E_a}{RT}\right) \tag{2.19}$$

Reaction rate data that fit this equation give straight lines when ln k is plotted against $1/T$, as illustrated in Figure 2.5. As seen in plot a of competing reactions (1) and (2), where $A(1)=A(2)$ and $E_a(1)>E_a(2)$, the temperature dependence of the reaction rate increases with increasing values of E_a. However, the rate of reaction (2) is slower than that of reaction (1) at all temperatures. This effect can be counteracted by selecting a reaction having a higher A value, as shown in plot b, where $A(3)>A(1)$ and E_a for the two reactions is equal. If both A and E_a are sufficiently greater for one reaction than for another, the rate constant at storage temperature would be smaller, while the rate constant at a higher temperature would be larger, as shown schematically in plot c.

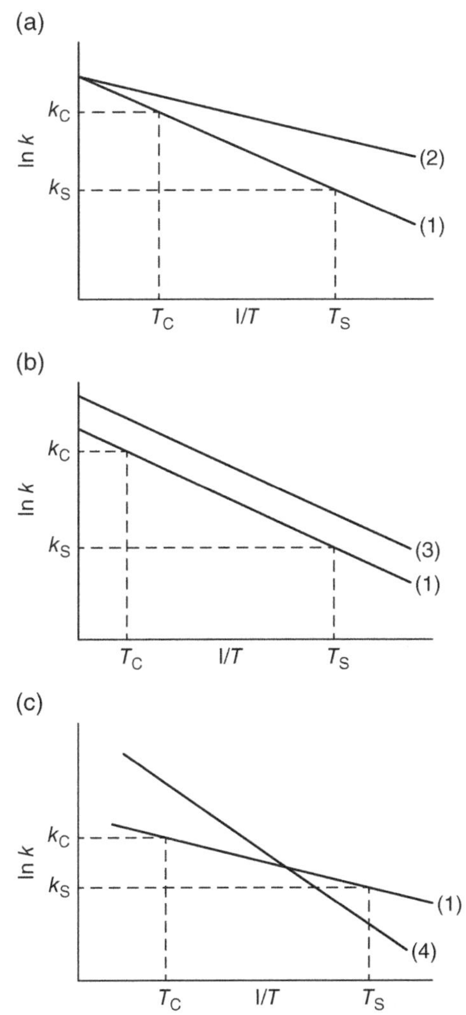

Figure 2.5 Arrhenius plots for competing reactions: (a) $A(1)=A(2)$, $E_a(1)>E_a(2)$; (b) $A(3)>A(1)$, $E_a(1)=E_a(3)$; (c) $A(4)>A(1)$, $E_a(4)>E_a(1)$. Source: Pappas and Feng (1984). Reproduced with permission of Technology Marketing Corporation.

Term A is controlled predominantly by entropic factors, or more specifically, by changes in randomness or order, as the reaction proceeds to the activated complex in the transition state. Three important factors to consider in selecting appropriate reactions are as follows:

1. Unimolecular reactions tend to exhibit larger A values than those with a higher molecular order.
2. Ring opening reactions tend to have high A values.
3. Reactions in which reactants become less polar exhibit larger A values.

The importance of these factors—in particular, factor (3)—depends on the reaction medium; accordingly, solvent selection can have a significant effect on storage stability.

While unimolecular reactions are desirable for high A values, cross-linking reactions are necessarily bimolecular. A way around this problem is to use a *blocked* reactant BX that thermally releases a reactant B by a unimolecular reaction—most desirably, with ring opening and decreasing polarity—followed by cross-linking between A and B:

$$\mathrm{BX} \rightleftarrows \mathrm{B} + \mathrm{X}$$

$$\mathrm{A} + \mathrm{B} \longrightarrow \mathrm{A-B}$$

Another approach is to use a blocked catalyst CX, where C catalyzes the cross-linking of A and B:

$$\mathrm{CX} \rightleftarrows \mathrm{C} + \mathrm{X}$$

$$\mathrm{A} + \mathrm{B} \xrightarrow{\mathrm{C}} \mathrm{A-B}$$

Here an important consideration is that the cross-linking reaction, which follows unblocking, should be faster than the reverse reaction, which regenerates the blocked reactant or catalyst. While one often encounters "threshold" or "unblocking" temperatures for reactions in the literature, such minimum reaction temperatures do not exist in the kinetics of reactions. Reactions proceed at some rate at any temperature. Threshold or unblocking temperatures are actually the temperatures at which some observable extent of reaction has occurred within a specified time interval.

These concepts are helpful for understanding the differences we see between storage stabilities, but another reason for understanding these kinetic considerations is that they can be used to predict whether any chemical reaction will ever be found to meet some combination of stability and cure schedule that might be desired for a coating. Pappas and Hill (1981) carried out calculations to permit such predictions. They made reasonable assumptions about reactive group concentrations during storage and in the applied film, permissible degrees of reaction during storage, and required extents of reaction during curing. Using these assumptions, they calculated the ratio of rate constants needed to permit any specified time of storage together with any specified cure time. In turn, this allowed calculation of E_a and A values as a function of any combination of storage and curing temperatures. Kinetic parameters calculated for unimolecular blocked reactant systems that proceed at 5% over a 6 month period at 30°C (storage temperature) and 90% in 10 min at various curing temperatures are listed in Table 2.3 (Pappas and Feng, 1984).

Rate constants and kinetic parameters are independent of concentration for unimolecular, but not for bimolecular reactions. Kinetic parameters calculated for bimolecular (i.e., second order) reactions using concentrations corresponding to high solids coatings were of similar magnitude to those for unimolecular reactions. Although the values in Table 2.3 represent order of magnitude calculation, they provide useful guidelines for avoiding development projects aimed at impossible goals and provide insights for design of thermosetting coatings. Kinetic parameters are known for many chemical reactions. As a point of reference, a reasonable upper limit of A values for unimolecular reactions is $10^{16}\,s^{-1}$, which corresponds to an upper value for the frequency of simple vibrations. For bimolecular reactions, A values tend to be less than $10^{11}\,l\,mol^{-1}\,s^{-1}$, an upper limit for the rate constant of diffusion, which must precede reaction. However, as shown in Table 2.3, a coating stable at 30°C would require an A value of $10^{24}\,s^{-1}$ if it were to cure within 10 min at 100°C. No such reaction is known or even conceivable. Users would like to have package stable coatings that cure in a short time at 80°C, a convenient temperature for low pressure steam heat, but it is pointless to pursue this goal via kinetic control. That is not to say that it is impossible to make such a coating, it means that the problem must be solved by an approach other than kinetic control.

Table 2.3 Kinetic Parameters as a Function of Cure Temperature

T (°C)	A (s^{-1})	E_a ($kJ\,mol^{-1}$)[a]
175	10^{10}	109
150	10^{12}	121
125	10^{17}	146
100	10^{24}	188

Source: Pappas and Feng (1984).

[a] 1 kJ = 0.239 kcal.

Storage life can be extended by refrigeration, but users are seldom willing to incur the expense. More reactive combinations can be used in *two package coatings*, in which one package contains a resin with one of the reactive groups and the second package contains the component with the other reactive group. Alternatively, the second package could contain a catalyst for the reaction. The packages are mixed shortly before use. Two package coatings are often called *2K coatings* and single package coatings are sometimes called *1K coatings*. The K stands for the German word for component. Two package coatings are used on a large scale commercially, but they pose substantial additional problems for the user; mixing and cleaning equipment take extra time, some of the material is usually wasted, they are generally more expensive, and error may occur in mixing. Even 2K coatings have the analogous problem of *pot life*, corresponding to the length of time after the two packages are mixed that the viscosity stays low enough for application. Pappas and Hill (1981) made similar calculations of A and E_a values for the shorter times involved.

There are several approaches to increasing package stability while permitting cure at ambient or moderately elevated temperature. One or more of these approaches are included in the discussions of many of the cross-linking reactions that are covered in later chapters. The following list gives some of the approaches to solving this "impossible" problem:

1. Use of a radiation-activated cross-linking reaction instead of a thermally activated reaction.
2. Use of a cross-linking reaction requiring an atmospheric component as a catalyst or reactant; reactions involving oxygen or water vapor in the air are examples. (In a way, these are 2K coatings, but the second component is free.) A similar principle is involved in passing a coated article through a chamber containing a catalyst vapor.
3. Use of a volatile component that inhibits the reaction when the coating is stored in a closed container, but volatilizes after application as a thin film, permitting the reaction to proceed. Examples include use of a volatile antioxidant in a coating that dries by oxidation and use of oxygen as an inhibitor in a composition that cures anaerobically.
4. Use of a cross-linking reaction that is a reversible condensation reaction involving loss of a volatile reaction product. If the volatile reaction product is included in the solvent mixture of the coating, the noncross-linked

side of the equilibrium is favored during storage but shifts to the cross-linked side after application, when the solvent evaporates. A similar approach can be used with blocked reactants and catalysts in which the blocking group is volatile.

5. Use of an encapsulated reactant or catalyst, whereby the capsules are ruptured during application. Encapsulation has been useful in adhesives, but applications in coatings are limited because residual capsule shells interfere with appearance and/or performance. An extension of this idea is to put encapsulated reactants in the coating that do not break during application but break later when the coating suffers mechanical damage; the goal is to make the coating *self-healing* (Section 34.3).
6. Use of a reactant that undergoes a phase change. While there are no threshold temperatures for kinetic reactions, phase changes can occur over narrow temperature ranges. A crystalline blocked reactant or catalyst, insoluble in the vehicle, could give an indefinitely stable coating; heating above the melting point permits the unblocking reaction to occur, releasing a soluble reactant or catalyst. Over a somewhat wider temperature range, the same idea can be used for an amorphous blocked reactant, having a T_g about 50°C above the storage temperature and, perhaps, 30°C below the desired cure temperature.

Mobility considerations. Another consideration in selecting components for thermosetting coatings is the potential effect of the availability of free volume on reaction rates and reaction completion. For reactions to occur, the reacting groups must diffuse into a reaction volume to form an activated complex that can then form a stable bond. If the diffusion rate is greater than the reaction rate, the reaction will be kinetically controlled. If the diffusion rate is slow compared with the kinetic reaction rate, the rate of the reaction will be mobility controlled. The major factor controlling the diffusion rate is the availability of free volume. If the reaction is occurring at a temperature well in excess of T_g, the free volume is large and the rate of reaction is controlled by concentrations and kinetic parameters. If, however, the temperature is well below T_g, the free volume is so limited that the polymer chain motions needed to bring unreacted groups close together are very slow, and reaction virtually ceases. At intermediate temperatures, the reaction can proceed, but the reaction rate is controlled by the rate of diffusion, that is, by the mobility of the reactants.

Since cross-linking often starts with low MW components, T_g increases as the reaction proceeds. If the reaction temperature is well above the T_g of the fully reacted polymer, there will be no mobility effect on the reaction rate. However, if, as is often the case in ambient cure coatings, the initial T_g is below the ambient temperature and the T_g of the fully reacted polymer is above the ambient temperature, the reaction rate will become mobility controlled as cross-linking proceeds. As further reaction occurs, cross-linking may essentially stop before reaching completion. As T_g approaches the temperature (T) at which the reaction is occurring, reactions become slower. When T_g equals T, reactions become very slow and *vitrification* (glass formation) is said to occur. Unless the experiment is continued for a relatively long time, the reactions have been said to cease (Aronhime and Gilham, 1984). However, with extended reaction times, it can be seen that reactions continue slowly. Blair (1985) reports that reaction rate constants drop by about three orders of magnitude when T_g equals T, but that the reaction continues at a slow rate until T_g increases to $T+50$°C. It is interesting to consider the similarity of this value to the universal B constant of 51.6 in the WLF equation (Eq. 2.9). T_g-B is the temperature at which viscosity goes to infinity and free volume theoretically approaches zero, corresponding to a T_g of $T+B$. Dusek and Havlicek (1993) studied the effect of variables on mobility control of reaction rates. In the reaction of bisphenol A diglycidyl ether and 1,3-propanediamine, they determined the effects of temperature, polymer–solvent interaction, and solvent volatility on reaction rates and extent of reaction. They also reviewed the theories involved.

It seems reasonable for a formulator to assume that cross-linking reactions begin to slow as T_g increases to about 10°C below the curing temperature and get progressively slower until T_g is about 50°C above the curing temperature, where reaction essentially ceases. The slow rates of reaction mean that properties of many ambient cure coatings can change substantially for several weeks, and change may continue indefinitely. Caution is required, since T_g values are dependent on the method of determination and the rate of heating used. For this purpose, T_g values obtained at very slow rates of heating and at low rates of application of stress are most appropriate. Dusek and Havlicek (1993) point out that, if reactions are very fast, equilibrium T_g may not be reached. Another factor that may affect the development of mobility control is the size of the diffusing reactants. Small molecules may diffuse more readily to reaction sites than functional groups on a polymer chain. Yet another consideration is that water plasticizes coatings such as polyurethanes and epoxy–amines, lowering their T_g.

If the initial reaction temperature is well below the T_g of the solvent-free coating, it is possible that little or no reaction can occur after solvent evaporation and that a "dry" film forms merely due to solvent evaporation without much cross-linking. The result is a weak, brittle film. One must be careful when defining what is meant by a dry film, especially when dealing with ambient temperature cure coatings. One consideration is whether the film is dry to handle. This stage could be reached with little cross-linking if the T_g of the solvent-free binder is high enough. Another consideration is whether the required extent of cross-linking has occurred. This must be tested by some method other than hardness, most easily by determining resistance to solvent rubbing or by the extent of solvent swelling (Section 4.2).

Mobility limitations are less likely to be encountered in baking coatings because in most cases, the final T_g of the film is below the baking temperature. Furthermore, the T_g is usually well above ambient temperatures so even if there are some unreacted groups, reactions essentially stop after cooling to ambient temperature. In moderate temperature cure powder coatings, mobility control of reaction can be a limitation, since the initial T_g of the reactants has to be above 50°C so that the powder will not sinter during storage (Section 28.3). To achieve a high extent of reaction, the baking temperature must be above the T_g of the fully reacted coating. Gilham and coworkers extensively studied factors influencing reaction rates in high T_g epoxy–amine systems; Simon and Gilham (1993) summarize Gilham's work and, in particular, point out its applicability to powder coatings.

Other complications arise from the likelihood that cross-linking does not yield uniform films. Formation of gel particles early in the process can cause inhomogeneities. Also, surface effects may alter structures of parts of the cross-linked network near the film's surfaces and near the surfaces of pigments, fillers, and nanoparticles. Such complications are difficult to study experimentally but can be addressed by computer modeling. An example has been provided by Song et al. (2015), who modeled the cure of nanoparticle-filled automotive clear coats and demonstrated the model's effectiveness in a case study.

2.3.3 Film Formation by Coalescence of Polymer Particles

In contrast to the processes of film formation from solutions of thermoplastic or thermosetting polymers, dispersions of insoluble polymer particles form films by *coalescence* (fusion) of particles. After application and loss of volatile components, the particles form a continuous film. How does this happen? It is a very complex process; here we will outline the key considerations. Readers seeking more detail are referred to a book by Keddie and Routh (2010) and books edited by Provder et al. (1996) and Provder and Urban (2001). The largest volume of coatings that form films by coalescence is latexes, which are a dispersion of high MW polymer particles in water (Chapter 9).

For a given latex, the lowest temperature at which coalescence occurs sufficiently to form a continuous film is called its *minimum film formation temperature* (MFFT); some authors call it the *minimum filming temperature* (MFT). MFFT is measured by placing samples on a heated metal bar with a temperature gradient. A major factor controlling MFFT is the T_g of the polymer in the particles. The T_g of PMMA is about 105°C, and one cannot form a useful film from a PMMA latex at room temperature; instead, one gets a layer of material that powders easily. Many latexes are designed to have layers of material with different T_gs within each particle (Sections 9.1.3 and 9.2), making it difficult to directly relate MFFT to T_g.

The mechanism of film formation from latexes has been extensively studied but still is not fully understood. A simplified picture of the steps in film formation used by many authors is as follows:

1. *Evaporation* of water and water-soluble solvents that leads to a close-packed layer of latex particles
2. *Deformation* of the particles from their spherical shape that leads to a more or less continuous, but weak, film
3. C*oalescence,* a relatively slow process in which the polymer molecules *interdiffuse* across the particle boundaries and entangle, strengthening the film

It should be emphasized that these steps, especially 2 and 3, overlap during the process. This picture is called *vertical drying. Horizontal drying* is discussed in the following text. A further complication is that different authors use different terms for the three steps: some call step 2 "coalescence" and step 3 "diffusion" or "interpenetration."

During step 1, repulsive forces between the particles favor uniform packing, but other forces, such as convection, may cause irregularities (Gromer et al., 2015). The close-packed array of particles usually still contains water. The internal phase volume at this stage depends on particle size distribution: the broader the distribution, the higher the internal phase volume.

If the particle size distribution of the latex particles is narrow, the end result of step 1 can be formation of ordered *colloidal crystals.* In such cases, small angle X-ray scattering (SAXS) can be used to study step 2 by measuring the rate of disappearance of the crystalline structure (Sulyanova et al., 2015).

As the polymer particles come together, they begin to deform. T_g of the latex particles is an important factor controlling the rates of particle deformation. Lower T_g particles have a lower modulus—that is, they are softer (Section 4.2) and, hence, are more easily deformed. Surfactant stabilizers can increase water absorption and also act as plasticizers for the polymer (Vandezande and Rudin, 1996). In general, higher surfactant content reduces MFFT. Structure of the surfactant also affects MFFT; for example, nonyl phenol ethoxylates having fewer than 9 ethoxylate units reduced MFFT further than those with 20 or 40 units. Furthermore, MFFT can be reduced by as much as 5°C by forming the film in a humid atmosphere (Eckersley and Rudin, 1990). Reduction of T_g by water is largest with hydrophilic polymers.

Deformation of the close-packed particles increases the area of contact between particles and promotes coalescence. For deformation to occur it is necessary overcome the stabilizing repulsion. What are the driving forces for deformation? Since 1950, various plausible theories have been advanced (Dobler and Holl, 1996). These theories include the following:

- Dry sintering driven by particle–air interfacial tension
- Wet sintering driven by particle–water interfacial tension
- Capillary pressure within the still-wet close-packed particles
- Compression forces caused by evaporation of water after a surface film has formed
- Reduction of surface free energy of the particles (Another name for sintering?)

Each of these theories is supported by experimental evidence in some cases. Furthermore, different deformation mechanisms have been observed for the same latex under different conditions (Gonzales et al., 2013). Thus, it seems likely that, in a given situation, several driving forces may be at work to differing extents during film formation (Routh and Russel, 2001). Furthermore the contributions of different forces probably vary as drying conditions change.

Lin and Meier (1996) strongly adhere to the position that capillary pressure is the dominant driving stress for film formation. Croll (1987) estimated that capillary forces could generate pressure as much as 3.5 MPa (5000 psi), but he pointed out that they act only briefly. His data on drying rates support a proposal of Kendall and Padget (1982) that a major driving force for coalescence is surface free energy reduction. The surface area of a coalesced film is only a small fraction of the surface area of the particles, so the driving force resulting from reduction in surface area must be significant.

Consistent with the differences in surface area, small particle size latexes often form films at somewhat lower temperatures than larger particle size ones (Eckersley and Rudin, 1990). However, Kan (1999) showed that this is not always the case—with other latexes, larger size promotes deformation. Still other work indicates that in some cases there is no effect of particle size on MFFT. Since broader distribution of particle size increases close-packed volume fraction, it leads to more rapid film formation. Possibly, these different conclusions are caused not just by particle size but also by particle size distribution and differences in the compositions of the latexes used. It is speculated that the larger proportion of surface active material in small particle size latexes may also play a role.

The third step, coalescence, begins during the process of deformation. With the large number of variables involved in coalescence, one would expect interactions among the different variables. The rate of interdiffusion is primarily driven by $T-T_g$. Adequate interdiffusion will occur only when the T_g of the particles is lower than the temperature at which the film is to be formed. Theoretical and experimental studies of coalescence suggest that molecules only need to interdiffuse a distance comparable to the radius of gyration of one molecule to develop maximum film strength (Winnik, 1997). This distance is considerably less than the diameter of a typical latex particle. The rate of interdiffusion is directly related to T_g and is, therefore, controlled by free volume availability (Winnik, 1997). The major factor affecting free volume is the difference between the temperature of film formation and the T_g of the particles. Presumably, the T_g of the material near the surface of the original particle is most important. As a general rule, coalescence will be very slow unless the temperature is at least slightly higher than T_g. Latexes for paints are generally copolymers of acrylic and vinyl esters that have a T_g well below room temperature so that they can coalesce readily within the usual range of application temperatures.

Most studies of the mechanism of film formation use carefully prepared uniform films that are dried in a laboratory under conditions of intermediate humidity and little air flow over the surface. Real-life conditions are quite different: the latex has been formulated into a paint with pigments and many additives, and dried under a wide variety of conditions. In many cases film thickness is not uniform. A further complication is that water evaporation is seldom uniform across the whole area of a film. The result is often *horizontal* (or *lateral*) *drying* in which a drying front starts, perhaps at a thin spot or edge, and spreads laterally across the panel (Salamanca et al., 2001). When this happens, both water and polymer particles move sideways near the drying front, and smaller particles appear less likely to give uniform films. The practical implications of this finding are only partly understood. It may partly explain why film formation is poor when exterior house paint is applied on a hot windy day with low humidity.

Plasticizers, which dissolve in the polymer, can be added to lower T_g and MFFT of the formulation. Since nonvolatile plasticizers permanently reduce T_g, most latex paints contain volatile plasticizers, which are called *coalescing solvents*, although they accelerate deformation as well as coalescence. A coalescing solvent must be soluble in the polymer and have a low, but appreciable, evaporation rate. The coalescing solvent acts as a plasticizer to lower the MFFT, but after the film has formed, it diffuses to the surface of the film and evaporates. Since free volume in the film is relatively small, the rate of loss of the last of the coalescing solvent is very slow. Though the films feel dry in a short time, they will still block for days, or even weeks, after application. Effects of coalescing solvents on film formation have been quantitatively studied by fluorescence decay measurements to follow the extent of polymer diffusion in films as they coalesce (Winnik et al., 1992). The efficiency of coalescing solvents varies over a wide range; an example of a relatively efficient solvent is the acetate ester of propylene glycol monobutyl ether (P*n*BA) (Geel, 1993). The rate of evaporation from films also varies. For example, dipropylene glycol dimethyl ether leaves a film more rapidly than P*n*BA, but somewhat more is required for film formation. A widely used coalescing solvent is Texanol®, the isobutyric ester of 2,2,4-trimethylpentane-1,3-diol. Some of the common coalescing solvents are counted as VOC in North America but not in Europe (Chapter 18).

While films form rapidly from latexes when the temperature is above MFFT, complete coalescence is a relatively slow process; in many cases, the film probably never equilibrates to become completely uniform. The rate of equilibration is affected by $(T - T_g)$. From the standpoint of rapid coalescence, it is desirable to have a latex having T_g well below the temperature at which the film is to be formed. Review papers discuss factors affecting the development of cohesive strength of films from latex particles (Daniels and Klein, 1991; Winnik, 1997). The extent of coalescence has been studied by small angle neutron scattering, direct energy transfer of particles labeled with fluorescent dyes, and scanning probe microscopy (Butt and Kuropka, 1995; Rynders et al., 1995). Berce et al. (2015) employed electrochemical impedance spectroscopy (EIS) and atomic force microscopy (AFM) to observe the course of deformation and coalescence in styrene/acrylic and acrylic latexes.

Usually, architectural paints are formulated so that film formation occurs at temperatures as low as 2°C. Film formation at such a low temperature requires a low T_g latex. However, as discussed in the introductory paragraphs of Section 2.3, $(T - T_g)$ also affects whether the film obtained will be solid. It was estimated that for a film to withstand the relatively mild blocking test described, $(T - T_g)$ would have to be on the order of 21°C. If the film is to be exposed to the blocking test when the temperature is 50°C (not an unreasonable expectation during direct exposure to summer sun), the T_g should be about 29°C or higher. Thus, the paint formulator is faced with a difficult challenge to design a system that can form a film when applied at 2°C and yet resist blocking at 50°C.

Coalescing solvents help solve this problem, but environmental regulations are limiting permissible emissions of VOC. A second approach is to design latex particles so that there is a gradient of T_g from a relatively high T_g in the center of the particles to a relatively low T_g at the outer periphery (Hoy, 1979; Sections 9.1.3 and 9.2). The low T_g of the outer shell permits film formation at low temperature. Over time, the T_g of the coalesced film approaches the average T_g of the total polymer. The higher average T_g reduces the probability of blocking. Relatively small amounts of coalescing solvents can be sufficient with such a latex.

A third approach is to use blends of high and low T_g latexes, which can reduce MFFT without the presence of coalescing solvents (Winnik and Feng, 1996). It has been proposed that the films contain particles of high T_g polymer dispersed in a matrix of lower T_g polymer, reinforcing the matrix, increasing its modulus, and, hence, decreasing blocking. There must be enough soft polymer to form a continuous film enclosing the hard particles. These studies were done without pigment, which can also reinforce a film. A fourth approach involves use of core–shell latexes (Juhue and Lang, 1995; Section 9.1.3).

Another complication is the potential for development of stresses within the coalesced films. Price et al. (2014) showed that stresses in films derived from core–shell latex particles can influence important film properties.

Polyurethane dispersions (PUDs; Section 12.7.1) have low MFFT values in relation to dry film T_g owing to the plasticizing effect of H-bonding with water. Accordingly, coalescing solvents are not needed with PUDs. When PUDs are blended with high T_g acrylics, MFFT is reduced without reducing hardness (Section 12.7.2).

Cross-linkable (thermosetting) latexes are another approach to the problem of film formation at low temperatures and block resistance. This topic will be introduced here and discussed fully in Section 9.4. In general, a lower T_g latex can be used in such coatings, which permits coalescence with less or even no coalescing solvent. The cross-linking may not increase T_g significantly but does improve mechanical properties and solvent resistance of the films. Taylor and Winnik (2004) reviewed the subject in detail, describing the theory as well as a variety of possible cross-linking chemistries. At that time, two package coatings were often required, and they were only useful for industrial applications.

Chemists and formulators have now learned to make package stable cross-linking latex coatings that are suitable as architectural and special purpose coatings. These are often called *self-cross-linking paints.* While this terminology suggests that the cross-linking reactants are copolymerized into the latex, this is often not the case. Typically, the cross-linking results from reaction of one reactive group within the latex and a separate bi- or polyfunctional cross-linker. An example of such a combination is a latex copolymer that includes a ketone or aldehyde functional monomer and a separate, water-soluble dihydrazide cross-linker (Section 17.11).

For such systems to be practical, rates of many different processes much be controlled. In addition to the rates of evaporation, deformation, and coalescence (polymer molecule interdiffusion) required of thermoplastic latexes, the rates of cross-linker diffusion into the polymer and the rates of the chemical cross-linking reactions must be properly timed (Winnik, 2002). Premature cross-linking is avoided by having the cross-linker in the aqueous phase and coreactant imbedded in the polymer particles. In order to develop good properties, significant interdiffusion of polymer molecules during coalescence and diffusion of cross-linker into the polymer must occur before extensive cross-linking reaction occurs. The rates of interdiffusion and diffusion are governed by $(T - T_g)$ and the chain lengths of the latex molecules. Thus, thermosetting latexes are designed with somewhat lower T_g and MW to facilitate interdiffusion. T_g and shortens chain length. Straight chains diffuse more rapidly than branched chains.

Caution: Most studies of the mechanism of film formation have been done with latexes, not with fully formulated paints. It is to be expected that other components of paints, such as pigments, pigment dispersing agents, and water-soluble polymers used as thickening agents, among others, affect MFFT, rate of film formation, and blocking resistance. See Chapter 31 for further discussion of latex paints.

Other types of coatings that involve coalescence of particles are discussed in later chapters, including aqueous PUDs, organosols, water-reducible resins, and powders.

GENERAL REFERENCES

Mark, J. E., Ed., *Physical Properties of Polymers Handbook*, American Institute of Physics, Woodbury, **1996**.
Odian, G. W., *Principles of Polymerization*, 4th ed., Wiley-Interscience, New York, **2004a**.
Provder, T.; Urban, M. W., *Film Formation in Coatings: Mechanisms, Properties, and Morphology*, American Chemical Society, Washington, DC, **2001**.
Provder, T.; Winnik, M. A.; Urban, M. W., *Film Formation in Waterborne Coatings*, American Chemical Society, Washington, DC, **1996**.

REFERENCES

Andrews, R. J.; Grulke, E. A., Glass Transition Temperatures of Polymers in Brandrup, J.; Immergut, E. H.; Grulke, E. A., Eds., *Polymer Handbook*, 4th ed., John Wiley & Sons, Inc., New York, **1999**, pp 193–198.
Aronhime, M. T.; Gilham, J. K., *J. Coat. Technol.*, **1984**, *56(718)*, 35.
Auschra, C., et al., *Prog. Org. Coat.*, **2002**, *45*, 83.
Berce, P.; Skale, S.; Slemnik, M., *Prog. Org. Coat.*, **2015**, *82*, 1–6.
Billmeyer, F. W., Jr., *Textbook of Polymer Science*, 3rd ed., Wiley-Interscience, New York, **2007**.
Blair, H. E., *Polym. Prepr.*, **1985**, *26(1)*, 10.
Boyer, C., et al., *Chem. Rev.*, **2016**, *116(4)*, 1803–1949.
Brunsveld, L., et al., *J. Polym. Sci. A Polym. Chem.*, **1999**, *37*, 3657.
Burrell, H., *Off. Digest*, **1962**, *34(445)*, 131.
Butt, H.-J.; Kuropka, R., *J. Coat. Technol.*, **1995**, *67(848)*, 101.
Chang, W. L.; Karalis, W. L., *J. Polym. Sci. A Polym. Chem.*, **1993**, *31*, 493.
Chiefari, J., et al., *Macromolecules*, **2005**, *38*, 9037.
Croll, S. G., *J. Coat. Technol.*, **1987**, *58(734)*, 41.
Daniels, E. S.; Klein, A., *Prog. Org. Coat.*, **1991**, *19*, 359.
Darling, T. R., et al., *J. Polym. Sci. A Polym. Chem.*, **2000**, *38*, 1706.
Dickinson, L., et al., *Macromolecules*, **1988**, *21*, 338.
Dobler, F.; Holl, Y., Mechanisms of Particle Deformation During Latex Film Formation in Provder, T. et al., Eds., *Film Formation in Waterborne Coatings*, American Chemical Society Symposium Series 648, American Chemical Society, Washington, DC, **1996**.
Dusek, K.; Havlicek, I., *Prog. Org. Coat.*, **1993**, *22*, 145.
Eckersley, S. T.; Rudin, A., *J. Coat. Technol.*, **1990**, *62(780)*, 89.
Elias, H. G., Structure and Properties in *Macromolecules*, Plenum Press, New York, **1984**, pp 301–371.
Ferry, J. D., *Viscoelastic Properties of Polymers*, 3rd ed., John Wiley & Sons, Inc., New York, **1980**, p 487.
Flory, P. J., *J. Am. Chem. Soc.*, **1939**, *61*, 3334.
Fried, J. R., *Polymer Science and Technology*, 3rd ed., Prentice Hall, Englewood Cliffs, **2014**.
Geel, C., *J. Oil Colour Chem. Assoc.*, **1993**, *76*, 76.
Gonzales, E., et al., *Langmuir*, **2013**, *29(6)*, 2044–2053.
Gromer, A., et al. *Langmuir*, **2015**, *31(40)*, 10983–10994.
Gupta, M. K., *J. Coat. Technol.*, **1995**, *67(846)*, 53.
Hill, L. W.; Wicks, Z. W., Jr., *Prog. Org. Coat.*, **1982**, *10*, 55.
Hoy, K. L., *J. Coat. Technol.*, **1979**, *51(651)*, 27.
Juhue, D.; Lang, J., *Macromolecules*, **1995**, *28*, 1306.
Kan, C. S., *J. Coat. Technol.*, **1999**, *71(896)*, 89.
Keddie, J. L.; Routh, A. F., *Fundamentals of Latex Film Formation*, Springer, Dordrecht, **2010**.
Kendall, K.; Padget, J. C., *Int. J. Adhes. Adhes.*, **1982**, *2(3)*, 149.
Krol, P.; Chmielarz, P., *Prog. Org. Coat.*, **2014**, *77*, 913–948.
Kumanotani, J., et al., *Org. Coat. Sci. Technol.*, **1984**, *6*, 35.
Lesko, P. M.; Sperry, P. R., Acrylic and Styrene-Acrylic Polymers in Lowell, P. A.; El-Aasser, M. S., Eds., *Emulsion Polymerization and Emulsion Polymers*, John Wiley & Sons, Inc., New York, **1997**, pp 622–623.
Lin, F.; Meier, D. J., *Prog. Org. Coat.*, **1996**, *29*, 139.
Lutz, J.-F., et al., Eds., *Sequence-Controlled Polymers: Synthesis, Self-Assembly and Properties*; SCS Symposium Series, American Chemical Society, Washington, DC, **2014**.
Mathias, L. J., Colletti, R. F., *Polym. Prepr.*, **1989**, *30(1)*, 304.
Matyjaszewski, K., Ed., *Controlled Radical Polymerization*; ACS Symposium Series, *685*, American Chemical Society, Washington, DC, **1999**.
Matyjaszewski, K., *Macromolecules*, **2012**, *45(10)*, 4015–4039.
Mayer, W. P.; Kaufman, L. G., *XVII FATIPEC Congress Book I*, Federation of Associations of Technicians for Industry of Paints in European Countries, Paris, **1984**, p 110.
Mengqiu, Z.; Xin, X., *Zou Mengqiu Int. J. Eng. Res. Appl.*, *5(7)*, **2015**, 121–123.
Muller, A. H. E.; Matyiaszewski, K., *Controlled and Living Polymerizations*, Wiley-VCH, Weinheim, **2009**.
Neumann, C.; et al., Proceedings of the International Waterborne, High-Solids, and Powder Coatings Symposium, New Orleans, LA, **2004**, Paper No. 32.
Odian, G. W., *Principles of Polymerization*, 4th ed., Wiley-Interscience, New York, **2004a**, pp 39–197.
Odian, G. W., *Principles of Polymerization*, 4th ed., Wiley-Interscience, New York, **2004b**, pp 313–330.
Odian, G. W., *Principles of Polymerization*, 4th ed., Wiley-Interscience, New York, **2004c**, pp 40–62.
Odian, G. W., *Principles of Polymerization*, 4th ed., Wiley-Interscience, New York, **2004d**.
Pappas, S. P.; Feng, H.-B., *International Conference in Organic Coatings Science and Technology*, Athens, Greece, **1984**, pp 216–228.
Pappas, S. P.; Hill, L. W., *J. Coat. Technol.*, **1981**, *53(675)*, 43.

Perrier, S., et al., *Macromolecules*, **2004**, *37*, 2709.
Price, K., et al., *J. Coat. Technol. Res.*, **2014**, *11(6)*, 827.
Provder, T.; Urban, M. W., *Film Formation in Coatings: Mechanisms, Properties, and Morphology*, American Chemical Society, Washington, DC, **2001**.
Provder, T.; Winnik, M. A.; Urban, M. W., Eds., *Film Formation in Waterborne Coatings*; Symposium Series, *648*, American Chemical Society, Washington, DC, **1996**.
Roe, R. J., *Encyclopedia of Polymer Science and Technology*, 2nd ed., John Wiley & Sons, Inc., New York, **1987**, Vol. *7*, pp 531–544.
Routh, A. F.; Russel, W. B., *Ind. Eng. Chem. Res.*, **2001**, *40*, 4302.
Rynders, R. M., et al., *J. Coat. Technol.*, **1995**, *67(845)*, 59.
Salamanca, J. M., et al., *Langmuir*, **2001**, *17*, 3202.
Salez, T., et al., Proc. Natl. Acad. Sci. U. S. A., **2015**, *112(27)*, 8227–8231.
Simon, S. L.; Gilham, J. K., *J. Coat. Technol.*, **1993**, *65(823)*, 57.
Smeets, N. M. B., et al., *Polym. Chem.*, **2012**, *3*, 514–524.
Sogah, D. Y.; Hertler, W. H.; Webster, O. W.; Cohen, G. M., **1987**, *Macromolecules*, *20*, 1473.
Song, H.; Xiao, J.; Huang, Y., *Ind. Eng. Chem. Res.*, **2015**, *55(12)*, 3351–3359.
Sperling, L. H., *Introduction to Physical Polymer Science*, 3rd ed.,Wiley-Interscience, New York, **2001**.
Stutz, H., et al., *J. Polym. Sci. B Polym. Phys.*, **1990**, *28*, 1483.
Sullivan, C. J., et al., *J. Coat. Technol.*, **1990**, *62(791)*, 37.
Sulyanova, E. A., et al., *Langmuir*, **2015**, *31*, 5274–83.
Taylor, J. W.; Winnik, M. A., *JCT Res.*, **2004**, *1*, 3.
Vandezande, G. A.; Rudin, A., *J. Coat. Technol.*, **1996**, *68(860)*, 63.
Viosscher, K. B.; McIntyre, P. F., US patent publication 6599973 B1 (**2003**).
Webster, O. W., *J. Polym. Sci., A Polym. Chem.*, **2000**, *38*, 2855.
White, D., et al., US patent 6,462,125, **2002**.
Wicks, Z. W., Jr., *J. Coat. Technol.*, **1986**, *58(743)*, 23.
Wicks, Z. W., Jr., et al., *J. Coat. Technol.*, **1986**, *57(725)*, 51.
Winnik, M. A., The Formation and Properties of Latex Films in Lovell, P. A., El-Aasser, M. S., Eds., *Emulsion Polymerization and Emulsion Polymers*, John Wiley & Sons, Inc., New York, **1997**, pp 467–518.
Winnik, M. A., *J. Coat. Technol.*, **2002**, *74(925)*, 49.
Winnik, M. A.; Feng, J., *J. Coat. Technol. Res.*, **1996**, *68(852)*, 39.
Winnik, M. A., et al., *J. Coat. Technol.*, **1992**, *64(811)*, 51.
Young, J. R., *Polymer Science and Technology*, 3rd ed., Prentice Hall, Upper Saddle River, **2014**.
Young, R. J.; Lovell, P. A., *Introduction to Polymers*, 3rd ed., CRC press, Boca Raton, **2001**.

Chapter 3

Flow

Rheology is the science of flow and deformation. This chapter deals only with the flow of relatively liquid materials; deformation of relatively solid materials, another aspect of rheology, is discussed in Chapter 4.

The flow properties of coatings are critical for proper application and appearance of films. For example, in brush application of a paint, the flow properties govern settling of pigment during storage, how much paint is picked up on the brush, film thickness applied, leveling of the applied film, and control of sagging of the film. Depending on how stress is applied to a fluid, there are several types of flow. Of major importance in coatings is the flow under a shear stress. We consider shear flow first and then, more briefly, other types of flow.

3.1 SHEAR FLOW

To understand and define *shear flow*, consider the model shown in Figure 3.1 (Patton, 1979). The lower plate is stationary, and the upper parallel plate is movable. The plates are separated by a layer of liquid of thickness x. Lateral force F is applied to the top movable plate of area A, so the plate slides sidewise with velocity v. The model assumes that there is no slip at the interfaces and that there is no fluid inertia. When the plate moves, the liquid near the top moves with a velocity approaching that of the movable plate and the velocity of the liquid near the bottom is close to zero. The velocity gradient dv/dx for any section of the liquid is constant and, therefore, equals v/x. This ratio is defined as *shear rate*, $\dot{\gamma}$. The units of shear rate are reciprocal seconds, s^{-1}:

$$\dot{\gamma} = \frac{dv}{dx} = \frac{v}{x}; \quad \frac{\text{cm s}^{-1}}{\text{cm}} = \text{s}^{-1}$$

Force F acting on the top plate of area A results in *shear stress* τ. The units of shear stress are pascals (Pa):

$$\tau = \frac{F}{A}; \quad \text{units: m kg s}^{2}\,\text{m}^{-2} = \text{Nm}^{-2} = \text{Pa}$$

A liquid exerts a resistance to flow called *viscosity*, η, defined as the ratio of shear stress to shear rate. This type of viscosity is more correctly called *simple shear viscosity*, but since it is the most widely encountered type of viscosity, it is usually just called viscosity. The separation of molecules in flow dissipates energy, primarily as heat. Thus, viscosity is the energy per unit volume dissipated to attain a unit velocity gradient. The units are pascal seconds (Pa·s). The older, and still commonly used, unit is the poise (P). One Pa·s equals 10 P, and 1 mPa·s equals 1 cP:

$$\eta = \frac{\tau}{\dot{\gamma}}; \quad \frac{\text{Pa}}{\text{s}^{-1}} = \text{Pa}\cdot\text{s}$$

When a liquid flows through a hole or a capillary, part of the energy is diverted into kinetic energy; then the resistance to shear flow is called *kinematic viscosity*, ν, with units of $m^2\,s^{-1}$, formerly called stokes, where $1\,m^2\,s^{-1} = 10^4$ stokes. When the acceleration results from gravity, kinematic viscosity equals simple shear viscosity divided by the density, ρ, of the liquid:

$$\nu = \frac{\eta}{\rho}$$

Organic Coatings: Science and Technology, Fourth Edition. Frank N. Jones, Mark E. Nichols, and Socrates Peter Pappas.

3.2 TYPES OF SHEAR FLOW

When the ratio of shear stress to shear rate is constant, liquids are *Newtonian*; and viscosity is independent of shear rate (or shear stress). A plot of shear rate as a function of shear stress is linear (Figure 3.2a), and the slope equals the inverse of viscosity. Such plots sometimes appear in the literature with axes opposite to those shown; then the slope is the viscosity. Newtonian flow is exhibited by liquids composed of miscible small molecules. Many solutions of oligomeric resins also approximate Newtonian flow.

Many liquids are non-Newtonian; that is, the ratio of shear stress to shear rate is not constant. One class of non-Newtonian liquids exhibits decreasing viscosity as shear rate (or shear stress) increases; these liquids are *shear thinning*. As stress is applied, the molecules line up parallel to the direction of the flow, reducing the energy required to displace them further. In the case of liquids having a disperse phase, the particles line up as bead chains, again reducing the energy required to displace them further. Shear thinning has been called *shear-induced order*. When the molecules or particles are all aligned in the direction of the flow, the viscosity is again independent of shear rate and the curve becomes linear (Figure 3.2b).

With some liquids, no detectable flow occurs unless a minimum shear stress is exceeded. Such materials exhibit *plastic flow* and are sometimes called *Bingham bodies*. The minimum shear stress required is called the *yield value*, or *yield stress*, and is designated by the symbol τ_0:

$$\tau - \tau_0 = \eta_p \dot{\gamma}$$

A schematic plastic flow diagram is shown in Figure 3.2c. The yield value depends strongly on the rate at which stress is increased. The faster the rate of stress is increased, the higher the measured value that is obtained. It is fairly common to extrapolate the linear part of the curve to the intercept with the shear stress axis and call the intercept a *yield value*.

Another class of liquids exhibits increasing viscosity as shear rate (or shear stress) increases. A schematic plot of such behavior is given in Figure 3.2d. Such liquids are *shear thickening*. Shear thickening has been called *shear-induced disorder*. If shear thickening liquids also increase in volume under shear, they are called *dilatant fluids*. Dilatant fluids have dispersed phases that become less ordered and, hence, occupy more volume. Examples are

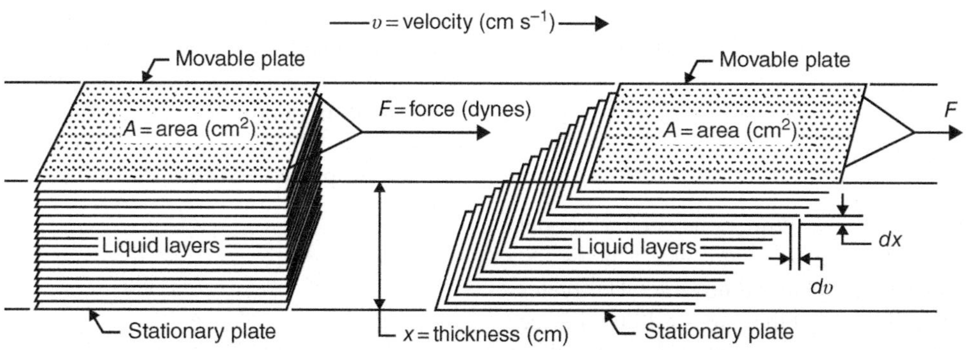

Figure 3.1 Model of shear flow of an ideal liquid. (In current usage, the symbol for shear stress is τ and its units are Pa; the symbol for shear rate is γ and the units of η are Pa·s.) Source: Patton (1979). Reproduced with permission of John Wiley & Sons.

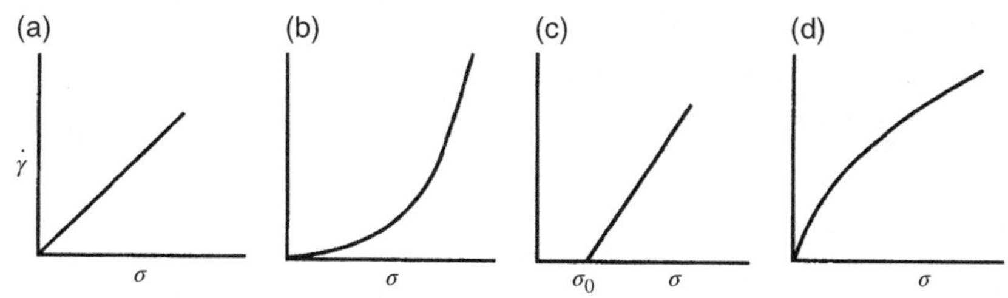

Figure 3.2 Plots of the flow of various types of liquids. (a) Newtonian; (b) shear thinning; (c) plastic; (d) shear thickening.

pigment and resin dispersions in which the dispersed phase is sufficiently concentrated so that the particles approach being randomly close packed. If sufficient stress is applied for flow, microscopic voids are created, increasing the volume and increasing the energy necessary to induce flow. Thus, the viscosity increases. Another example is quicksand. Mewis and Vermant (2000) further describe the factors affecting dilatant flow.

The Casson equation, Eq. 3.1, linearizes viscosity/shear rate data of shear thinning or thickening fluids; the slope of the line is the yield stress, and extrapolation gives the viscosity at infinite shear rate η_∞. In many cases, the value of n is 0.5, and commonly, the Casson equation is shown with just the half-power relationship. In plots of log viscosity against shear rate, the degree of curvature is related to the value of τ_0. In Figure 3.3, the values of η and η_∞ are held constant to show the effect of changes in τ_0 on flow response (Hester and Squire, 1997). For a Newtonian fluid, τ_0 equals zero and the plot is a straight line parallel to the shear rate axis:

$$\eta^n = \eta_\infty^n + \left(\frac{\tau_0}{\dot{\gamma}}\right)^n \tag{3.1}$$

Some fluids show time history dependence of viscosity, as illustrated in Figure 3.4a. The curves in the figure result from shear stress readings taken at successively higher shear rates to some upper limit (right-hand curve), followed immediately by shear stress readings taken at successively lower shear rates (left-hand curve). At any shear rate on the initial curve, the stress would decrease with time to an equilibrium value between the two curves; that is, the viscosity would decrease. On the other hand, if such a system

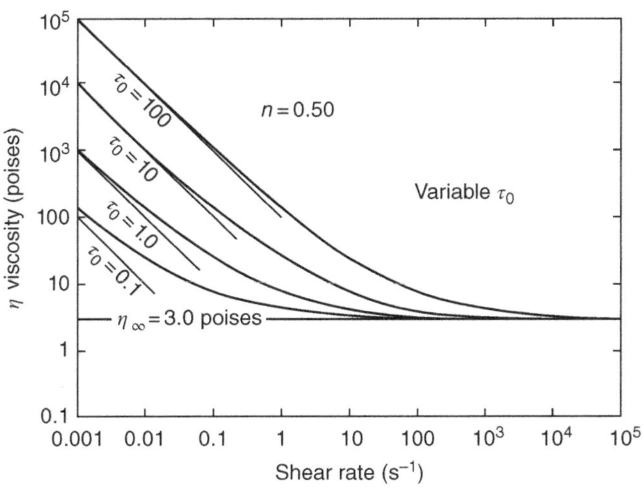

Figure 3.3 Casson plot of viscosity as a function of shear rate showing the dependence τ_0 with constant η and η_∞. Source: Patton (1979). Reproduced with permission of John Wiley & Sons.

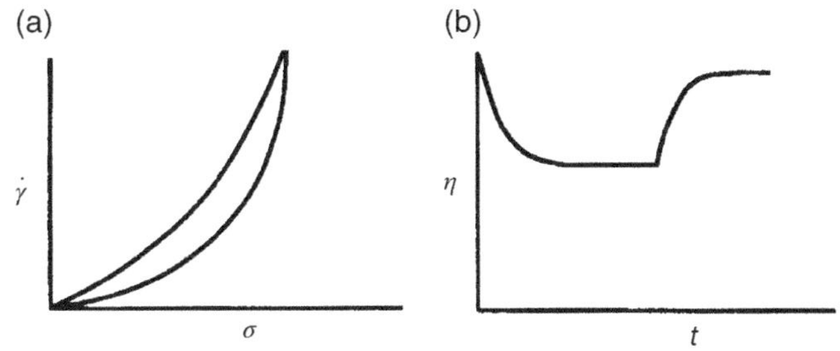

Figure 3.4 Schematic plots of systems exhibiting thixotropic flow. (a) The curve to the right is based on readings taken as shear rate was being increased, and the curve to the left is based on readings taken as shear rate was then being decreased. (b) The viscosity drops as shear continues and then increases as the shear rate is decreased.

had been exposed to a high rate of shear and then the shear rate decreased, the shear stress would increase to an equilibrium value as the measurement was continued; that is, the viscosity would increase with time. This behavior is called *thixotropic* flow.

Armstrong et al. (2016) define thixotropy as "... the continuous decrease of viscosity with time when flow is applied to a sample that has been previously at rest, and the subsequent recovery of viscosity when the flow is discontinued." Thixotropic fluids are shear thinning fluids. In addition, their viscosity depends on time and prior shear history. All thixotropic fluids are shear thinning fluids, but not all shear thinning systems are thixotropic. Unfortunately, the term thixotropy is often improperly used as a synonym for shear thinning.

Thixotropy is a desirable attribute of many paints because it enables paint formulators to retard settling, optimize the application characteristics, and minimize sagging and dripping after application. Thixotropy usually results from reversible formation of a structure within a fluid; an example is association of dispersed particles held together by weak forces. Thixotropic structure is broken down by applying shear for a sufficient time and re-forms over time when shear is stopped. Some thixotropic fluids undergo viscosity reduction to equilibrium values in short time periods and recover their viscosity rapidly when shearing is stopped; others change more slowly with time. In early work, areas within hysteresis loops, as shown in Figure 3.4a, were compared to estimate the degree of thixotropy. However, such comparisons may be misleading because the areas of such loops depend on the shear history, the peak shear, and the time intervals between successive measurements. A more appropriate way to represent the effect is to plot the viscosity at a series of shear rates as a function of time (t), as illustrated in Figure 3.4b. The time for recovery after shifting from a high shear rate to a lower shear rate is a useful method for comparing different paints. It is generally desirable to formulate such coatings so that the rate of recovery is rapid enough to avoid sagging but slow enough to permit leveling.

Additives to impart thixotropy to coatings include clays treated with quaternary ammonium compounds, which form salts with the clay platelets and make them organophilic. Such additives are widely used in solvent-borne coatings. Attapulgite clay is used in both solvent-borne and waterborne coatings. The needles of this clay associate with each other, providing an increase in viscosity that reverses with agitation. In waterborne coatings, the clay also acts by absorbing water to make swollen particles that distort in shape with stress. Fine particle size silica has also been used for many years. Castor oil derivatives and powdered polyethylene are effective thixotropic agents in baking coatings. Polyamides such as alkyds partly gelled with aminoalcohols are used as thixotropes. Basic calcium sulfonate derivatives are liquid thickeners that do not reduce gloss and retain their effectiveness when a coating is heated. They are particularly useful in moisture-curable urethane coatings since their basicity neutralizes the CO_2 produced by reaction isocyanates with water, thereby reducing the bubbling problem as well as imparting thixotropy. In latex-based paints, water-soluble polymers and associative thickeners are used as discussed in Section 3.5.1. Hare (2001) provides a review of thixotropes and additives for modifying paint flow.

Rheologists frequently discuss the properties of thixotropic fluids in terms of *viscoelasticity* and consider that they flow by a combination of viscous flow and elastic deformation (viscoelasticity is discussed in Section 4.2.2). Such interpretations are valid and useful, but historically they have been seldom applied in the coatings industry. Time dependency can best be measured in terms of viscoelasticity; see Hester and Squire (1997) and Boggs et al. (1996) for such analyses. The physics of thixotropy is quite complicated; for a recent overview see Armstrong et al. (2016), where various models are described and tested experimentally.

Another way to show the effect of shear on a thixotropic fluid is by a different type of Casson plot, as shown in Figure 3.5 (Schoff, 1988). The square root of the viscosity is plotted against the square root of the reciprocal of the shear rate; the steeper the slope, the greater the degree of shear thinning. If the sample had been sheared until all of the thixotropic structure was broken down, and if the measurements could be made before any structure buildup occurred, the sheared plot would be linear and parallel to the x axis. Although comparisons of the differences between the slopes of such lines give a qualitative expression of the extent of thixotropy, the slopes of the curves may depend on prior shear history, the rate of acceleration of shear, and the length of time that the sample was exposed to the highest shear rate.

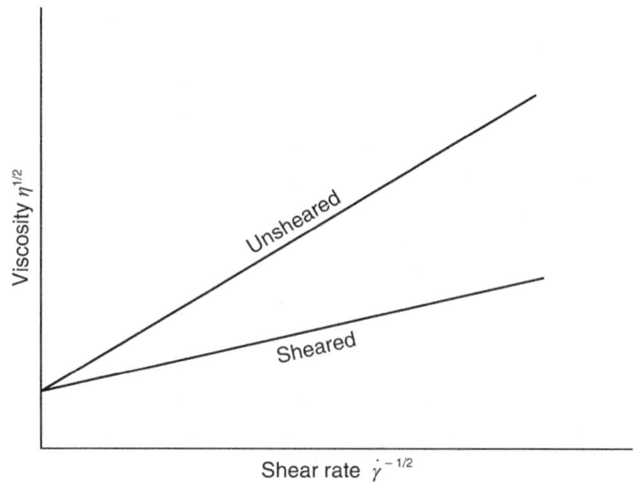

Figure 3.5 Schematic Casson plots of a sheared and unsheared thixotropic coating. The degree of divergence gives an estimate of the degree of thixotropy. Source: Schoff (1988). Reproduced with permission of John Wiley & Sons.

3.3 DETERMINATION OF SHEAR VISCOSITY

A variety of instruments is available to determine viscosity. They vary in cost, time required for measurements, operator skill required, sturdiness, precision, accuracy, and ability to measure shear rate variability or time dependency effects. Data obtained on the same sample with different instruments and by different operators with the same instruments can vary substantially, especially for shear thinning liquids at low rates of shear (Anwari et al., 1989). Some of the variation can result from lack of attention to details, especially temperature control and possible solvent evaporation; major errors can result from comparing samples that have had different shear histories. Recent improvements in instrumentation have helped the situation, but care is still needed.

Temperature must be carefully controlled. Since viscosity depends strongly on temperature (Section 3.4.1), it is critical that the measurement is made while the sample is at a constant, known temperature. When high viscosity fluids are sheared at high shear rates, heat is evolved and the temperature of the sample increases unless the heat exchange efficiency of the viscometer is adequate. If viscosity is determined while both shear rate and temperature are increasing, it may be impossible to tell whether or not the fluid is shear thinning.

Mezger (2014) describes dozens of instruments that measure rheology of liquids. Viscometers are most suitable for measuring shear viscosity of Newtonian liquids, whereas rheometers are most suitable for profiling the flow behavior of non-Newtonian fluids. Viscometers can be placed into three broad classes: (1) those that permit quite accurate viscosity determinations, (2) those that permit determination of reasonable approximations of viscosity, and (3) those that provide flow data marginally related to viscosity. We restrict our discussion to major examples of each class. The general references at the end of this chapter are good sources of further information.

3.3.1 Capillary Viscometers

Figure 3.6 shows an example of a *capillary viscometer* (Schoff, 1991). The time required for a known amount of liquid to flow through a capillary tube is measured (Poiseuille, 1840). While viscosities can be calculated based on the diameter of the capillary, usually each instrument is standardized with liquids of known viscosity; then, calculation is simply based on instrument constants and time.

Since capillary flow is driven by gravity, kinematic viscosity (Section 3.1) is measured. Simple shear viscosity can be calculated by correcting for density. Capillary viscometers with a range of diameters permit the determination of viscosities from 10^{-7} to $10^{-1}\,m^2\,s^{-1}$. For liquids with a density of 1, these values correspond to a range of 1 mPa·s–1000 Pa·s.

Historically, capillary viscometers have been the instruments of choice for research work because of their accuracy. They are applicable only to Newtonian fluids and are not appropriate for routine work, because determinations are relatively time consuming, especially when temperature dependence of viscosity data is desired. Temperature equilibration is slow because of the relatively large sample sizes and the low rate of heat transfer by glass. Careful cleaning is essential. Capillary viscometers are particularly appropriate for use in determining the viscosity of volatile liquids or solutions containing volatile solvents, since they are essentially closed systems.

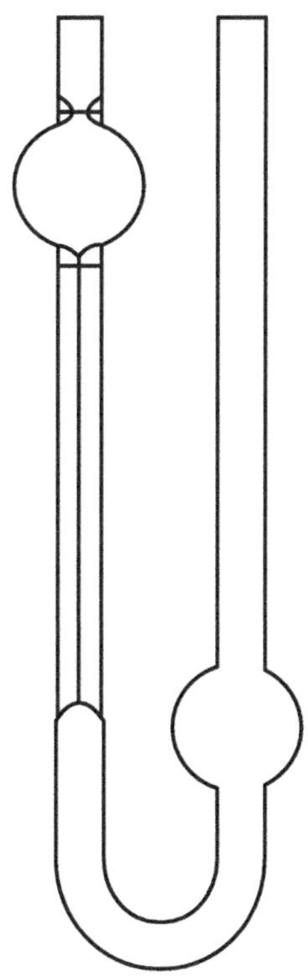

Figure 3.6 Ostwald capillary viscometer. Source: Schoff (1991). Reproduced with permission of John Wiley & Sons.

Modified capillary viscometers are available that overcome limitations of the standard types. Vacuum viscometers suck up the sample into the capillary tube, and the time to pass from the lower marker to the upper one is measured. Since the flow is not driven by gravity, density does not affect the time, and viscosity is measured in Pa·s, avoiding the need to determine density. Similar viscometers can be used for opaque liquids since the passage of liquid from the lower mark to the upper mark can easily be seen.

3.3.2 Rheometers

For non-Newtonian liquids, including pigmented liquids, the highest accuracy over a wide range of shear rates is obtained with rotational rheometers, exemplified by *cone and plate rheometers*. A schematic diagram is shown in Figure 3.7. The sample is placed on the plate that is then raised to a level with small clearance from the cone. The cone can be rotated at any desired number of revolutions per minute (rpm), and the torque is measured. The shallow angle of the cone is designed to make the shear rate constant throughout the sample in the gap. Shear rate is proportional to the rpm; and shear stress is related to the torque. Temperature is controlled by passing temperature-controlled water through the plate; temperature control problems are minimized by the small sample size.

Rheometers vary in the range of shear rates that can be used and the time required to increase or decrease shear rate. (See general references and Schoff (1988 and 1991) for further discussion.) The least expensive instruments are sufficiently rugged, simple to use, and sufficiently fast for quality control applications. The most versatile ones are sensitive scientific instruments that require skill in use and are most appropriate for research applications. For solutions containing volatile solvents, the cone and plate unit is shrouded in an atmosphere saturated with solvent vapor to prevent evaporation.

Rheometers are available that control strain or stress. The latter type offers advantages for coatings, as it is generally superior for measurements at very low shear rates. Hybrid rheometers that operate in both modes are also available.

The viscosity of highly viscous materials can be determined at high rates of shear by the use of *mixing*

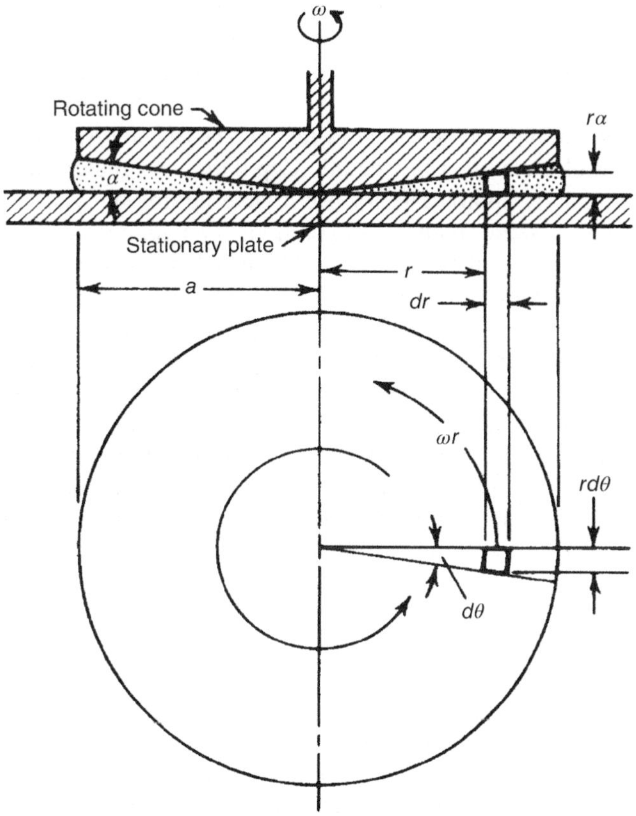

Figure 3.7 Schematic representation of cone and plate viscometer geometry. Source: Schoff (1991). Reproduced with permission of John Wiley & Sons.

rheometers that are small, heavy-duty mixers. The test sample is confined to a relatively small space and subjected to intense mixing by dual rotors in the form of sigma-shaped blades. A dynamometer measures the work input. The speed is set by a tachometer. The newer instruments are computerized. These instruments were originally designed for studying the molding of plastics but are also used in studying the effect of pigments on viscosity. Heat buildup can be substantial with high viscosity fluids. A micro-twin-screw rheometer is available in which the rpm of the screws can be varied over a wide range, thereby permitting measurement of shear stress as a function of rpm.

3.3.3 Rotating Disk Viscometers

Rotating disk viscometers have a motor that rotates a disk immersed in a liquid over a range of rpm, and the resulting torque is measured. A schematic diagram is shown in Figure 3.8. The instruments must be calibrated with standards. Measurements of the sample and standards should be made in containers having the same dimensions, since the response can be affected by the distance of the disk below the surface of the liquid, above the bottom of the container,

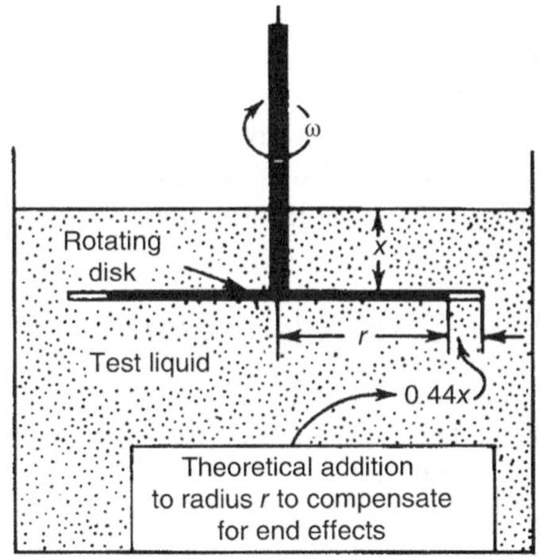

Figure 3.8 Schematic drawing of a disk viscometer. Source: Patton (1979). Reproduced with permission of John Wiley & Sons.

and from the side walls. In reporting results, the rpm setting should always be specified. The instruments are sturdy and relatively inexpensive, although they must be recalibrated periodically. Newer models have touch screen controls and computer interfaces. When properly used, rotating disk viscometers provide relatively accurate viscosity measurements for Newtonian liquids. For non-Newtonian liquids, the viscosity reading represents an average response corresponding to the viscosities resulting from a span of shear stresses.

Rotating disk viscometers can detect whether a liquid is shear thinning or shear thickening by measurements carried out at different rpm settings. They can detect thixotropy by a change in response over time at the same rpm setting. Useful comparisons of thixotropic properties can be obtained by applying a high shear rate, then suddenly reducing the rate of shear to a lower value, and measuring the time required for the viscosity to reach equilibrium at that lower rate.

Several types of viscometers are designed for production use; in some circumstances viscosity can be monitored continuously.

3.3.4 Bubble Viscometers

Bubble viscometers are widely used in quality control to approximate the viscosity of solutions of resins (Section 15.5.2). The measurement is based on the rate of rise of an air bubble in a tube of liquid; the higher the viscosity, the slower the bubble rises. A glass tube is filled with a liquid to a graduation mark and stoppered so that a definite amount of air is enclosed at the top as shown in Figure 3.9. The tube is placed in a thermostatic bath and kept there long enough for the temperature to equilibrate. Equilibration is slow but is essential if meaningful measurements are to be made. The tube is then inverted, and the

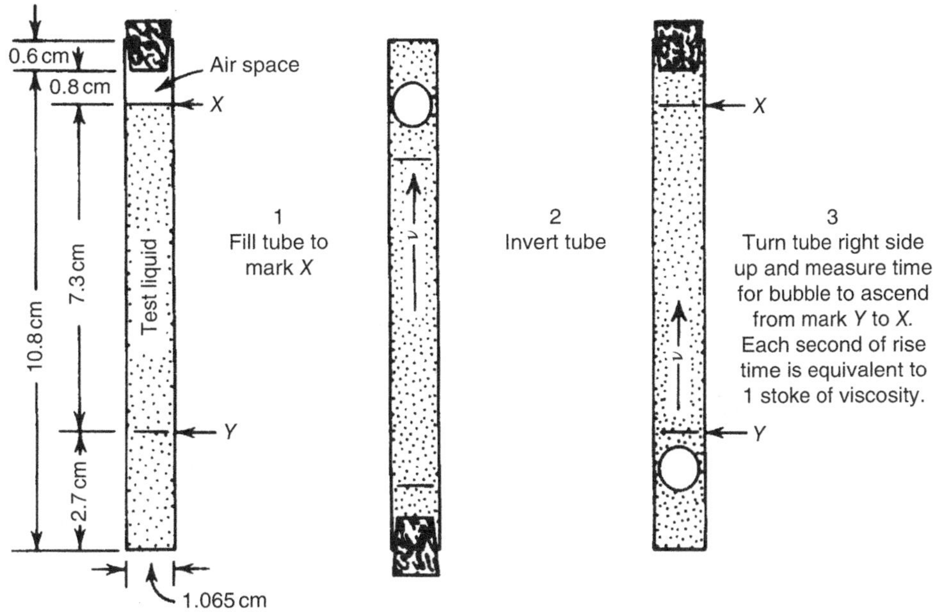

Figure 3.9 Determination of viscosity with a bubble tube. Source: Patton (1979). Reproduced with permission of John Wiley & Sons.

time required for the air bubble to travel between two calibration marks on the tube is measured. Provided the length of the bubble is greater than its diameter, the rate of rise is independent of the bubble size. Density of the liquid affects the rate of rise of the bubble, so kinematic viscosity is measured. Sets of standard tubes are designated as A, B, C, and so on; after Z, tubes are designated as Z_1, Z_2, and so on. Kinematic viscosities range from about 10^{-5} to $0.1\ m^2 s^{-1}$. Bubble viscometers are only appropriate for Newtonian transparent fluids. They are low in cost and simple to use. If the bubble has a tail, the resin is approaching gelation.

3.3.5 Efflux Cups

The most widely used control device for measuring flow of industrial coatings, especially for spray application, is the *efflux cup*. A variety of efflux cups is available. Schoff (1988, 1991) compares about two dozen of them. A schematic diagram of one of the most common efflux cups, the Ford No. 4 cup, is shown in Figure 3.10. One holds a thumb over the hole in the bottom of the cup, fills the cup with coating, removes the thumb, and then determines the time at which there is a break in the stream of coating flowing through the hole. The result is expressed in seconds. The data should not be converted into kinematic viscosity numbers, since a significant amount of the force is converted into kinetic energy, especially with low viscosity coatings. The method is not appropriate for non-Newtonian liquids, although efflux cups are frequently used for coatings that exhibit a small degree of shear thinning. Despite their limitations, efflux cups are useful quality control devices. They

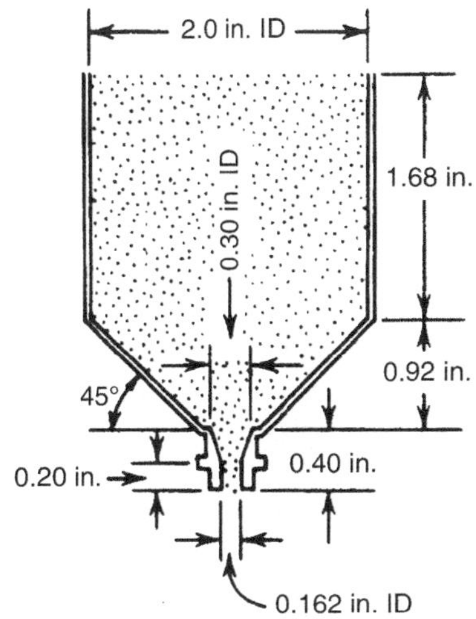

Figure 3.10 Schematic diagram of a Ford No. 4 efflux cup. Source: Patton (1979). Reproduced with permission of John Wiley & Sons.

are low in cost, rugged in construction, and easily cleaned. Results are simply and quickly obtained, but reproducibility is poor, reportedly in the range of only 18–20% (Schoff, 1991).

The proper way to use an efflux cup to control viscosity for spraying, for example, is to adjust the viscosity of the coating by solvent addition until the coating sprays properly and then measure the time it takes to flow through the efflux cup (see aforementioned paragraph). This time can then be used as the standard for spraying that particular coating through that spray gun at that distance from the object being sprayed. Proper efflux cup flow times for spraying vary with different coatings for a given application system and with different application systems for a given coating.

3.3.6 Paddle Viscometers

The *Stormer viscometer* has been widely used for architectural paint formulation. The instrument paddle is immersed in the paint and rotated at 200 rpm. The force required to maintain this rotation rate is measured by adding weights to a platform at the end of a cord over a pulley connected by a gear train to the paddle. A schematic diagram is shown in Figure 3.11. The weight loaded is converted into *Krebs units* (KU) by an arbitrary conversion scale. Originally, a KU value of 100 was considered to correspond to good brushing consistency. In current practice, paints are usually formulated with somewhat lower KU. The instrument is of little utility with Newtonian fluids; and the readings have no real meaning for non-Newtonian fluids, which include most architectural paints. Even when used only for quality control, the paddle viscometer is not satisfactory.

The proper way to establish the flow properties for a trade sales paint is to apply the paint with a brush or roller and adjust the viscosity until the best compromise of such properties as ease of brushing, leveling, sagging, settling, and so forth has been achieved. Having done this work, then a quality control test can be set up. Appropriate quality control instruments are cone and plate viscometers (some of which are reasonable in cost) or, lacking that, rotating disk viscometers. Unfortunately, Stormer viscometers are widely used. The director of research of a large manufacturer of trade sales paints in the United States said some years ago that the Stormer viscometer was responsible for setting back the formulation of one-coat hiding latex paints by 20 years.

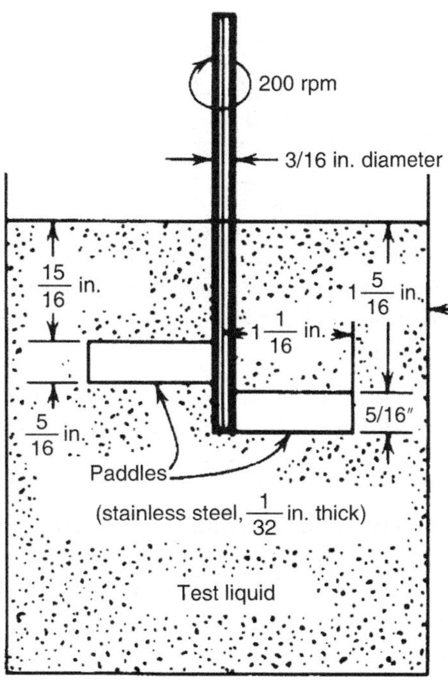

Figure 3.11 Schematic diagram of a paddle viscometer. Source: Patton (1979). Reproduced with permission of John Wiley & Sons.

3.4 SHEAR VISCOSITY OF RESIN SOLUTIONS

The viscosity of liquids depends on free volume availability. To put it somewhat simplistically, there are free volume holes rapidly opening and closing in a liquid; molecules move randomly through these free volume holes. When a stress is applied, movements in the direction that relieves the stress are favored, and the liquid flows. Therefore, factors that control viscosity of resin solutions are those that control the availability of free volume. Many coatings are based on solutions of polymers or oligomers. The variables affecting the flow behavior of these concentrated solutions are not fully understood. The variables that govern the flow of very dilute polymer solutions have been extensively studied and are better understood. Factors affecting the flow of dilute solutions are discussed in Section 3.4.2, and factors affecting the flow of more concentrated solutions are discussed in Section 3.4.3.

3.4.1 Temperature Dependence of Viscosity

Temperature dependence of viscosity for a range of low molecular weight (MW) resins and their solutions has been shown to fit a Williams–Landel–Ferry (WLF) equation (Section 2.3) (Wicks et al., 1985; Toussaint and Szigetvari, 1987; Jones, 1996; Haseebuddin et al., 1997). In Eq. 3.2, T_r, the reference temperature, is the lowest temperature for which experimental data are available, and η_r is the viscosity at the reference temperature, T_g. Except for very dilute solutions, data fit Eq. 3.2 when η_r is assumed to be 10^{12} Pa·s (Wicks et al., 1985):

$$\ln\eta = \ln\eta_r - \frac{c_1(T-T_r)}{c_2+(T-T_r)} \cong 27.6 - \frac{A(T-T_g)}{B+(T-T_g)} \quad (3.2)$$

For higher MW polymers at temperatures above $T_g + 100°$, the temperature dependence of viscosity approximately fits an Arrhenius equation, shown in Eq. 3.3, where E_v is the activation energy for viscous flow. (Note that the Arrhenius *A* value *in Eq.* 3.3 is not the same as constant A in Eq. 3.2.)

$$\ln \eta = K + B/T = \ln A + \frac{E_v}{RT} \tag{3.3}$$

Using the data available for low MW resins and their solutions, Arrhenius plots of ln η as a function of $1/T$ have been found to be curved and not linear (Wicks et al., 1985; Jones, 1996). On the other hand, the data do fit a WLF equation. From a practical viewpoint, the differences in the models are small if the temperature range is small. However, over a wide range of temperatures, the differences are relatively large.

A major factor controlling viscosity of resin solutions is $(T-T_g)$, but it is not the only factor. When the differences between T_gs are small, differences in WLF constants A and B may overshadow the small difference in $(T-T_g)$. Constant A depends on the difference in thermal expansion coefficients above and below T_g, but no studies have been reported on the structural factors that control these coefficients. Constant B is the value of (T_g-T) at which viscosity goes to infinity. The so-called universal value of this constant is 51.6°C, but the "constant" varies considerably with composition. No studies have been reported on the relationship between structure and the value of constant B.

Generally, in designing resins, it is reasonable to predict that a lower T_g will lead to a lower viscosity of the resin and its solutions. (See Section 2.1.2 for discussion of the factors controlling the T_g of polymers.) Linear poly(dimethylsiloxanes) have low T_gs and low viscosities. Linear polyethylene glycols have almost as low T_gs and viscosities. Poly(methyl methacrylate) resin solutions have higher T_g values and viscosities than comparable poly(methyl acrylate) resin solutions. BPA epoxy resins have higher T_g values and viscosities than corresponding hydrogenated derivatives. Exceptions to this generalization on the effect of T_g have been reported for some high solids acrylic resins (Section 8.2.1). Acrylic resins made with a comonomer that has a bulky group, such as 3,3,5-trimethylcyclohexyl methacrylate (Kruithof and van den Haak, 1990) or isobornyl methacrylate (Wright, 1996), are reported to have low viscosities at high solids even though their homopolymers have high T_g values; no explanation of this effect has been advanced.

3.4.2 Dilute Polymer Solution Viscosity

Measurement of viscosities of very dilute solutions of polymers played an important part in the early development of polymer science, since they were the earliest method of estimating MW. However, this method is obsolete in coatings technology, so the rather complex subject is not covered here. It is described in previous editions and in polymer chemistry textbooks, such as Allcock and Lampe (1990).

3.4.3 Concentrated Polymer Solution Viscosity

Concentrated solutions of polymers behave quite differently than dilute solutions. The binders of solventborne coatings are usually concentrated solutions of polymers, and their viscosity affects application behavior. Viscosity of such solutions is influenced by many factors—a partial list:

- Concentration
- MW of the polymer and its MW distribution
- Polymer structure—linear versus branched
- Viscoelastic effects
- Viscosity of the solvent(s)—surprisingly an important variable
- Hydrogen bonding between polymer molecules and with solvent
- Temperature
- Changes in solvent composition during and after application
- Effects of pigments
- Effects of additives such as thixotropes

It is no wonder that there have been relatively few fundamental studies of the factors controlling viscosity of more concentrated solutions of polymers and resins, such as those used in the coatings field. Several empirical relationships have been proposed, such as the concentration dependence of *relative viscosity*, shown in Eq. 3.4, in which w_r is the weight fraction of resin and k's are the constants:

$$\ln \eta_r = \frac{w_r}{k_1 - k_2 w_r + k_3 w_r^2} \tag{3.4}$$

Relative viscosity is a unitless number that expresses the ratio of solution viscosity to solvent viscosity. By studying relative viscosity, researchers can eliminate the important variable of solvent viscosity (see succeeding text) and gain some insight into the behavior of more concentrated solutions. Nonlinear regression analysis of the limited number of sets of data available in the literature in 1985 fit Eq. 3.4 over a wide range of concentrations (Wicks et al., 1985). Even with this many constants, there is some systematic deviation from the model at very low concentrations. Constant k_1 is the reciprocal of *weight intrinsic viscosity*, $[\eta]_w$, which, although formally unitless, is based on the number of grams of solution containing a gram of resin. Weight intrinsic viscosity can be converted into the more familiar volume intrinsic viscosity $[\eta]$ by dividing the density of the solution at the concentration $w_r = k_1$. No physical significance of the other two constants, k_2 and k_3, has been elucidated; they are presumably related to further solvent–resin interactions and to free volume availability.

Over narrower ranges of concentration, the experimental data reasonably fit the simpler Eq. 3.5. The even simpler Eq. 3.6 has been extensively used to calculate approximate relative viscosities over a range of concentrations in which viscosities range from around 0.01 to 10 Pa·s:

$$\ln \eta_r = \frac{w_r}{k_1 - k_2 w_r} \tag{3.5}$$

$$\ln \eta_r = \frac{w_r}{k_1} = [\eta]_w w_r \tag{3.6}$$

Relative viscosity of dilute solutions of polymeric resins increases as the solvent gets "better" in the sense of their solvency. However, in concentrated solutions, relative viscosity is often higher in poor solvents than it is in good solvents. In good solvents, there are stronger interactions between solvent molecules and resin molecules than in poor solvents. In very dilute solutions, this signifies that the chains become more extended and sweep out larger hydrodynamic volumes in good solvents than in poor solvents. However, in more concentrated solutions, flow of resin molecules is constrained by the hydrodynamic volumes swept out by neighboring resin molecules. In theory, when interaction between solvent and resin is stronger than resin–resin interactions, the molecules can flow readily through the hydrodynamic volumes of neighboring molecules (provided the free volume is adequate), and relative viscosity is lower. When, on the other hand, resin–resin interactions are stronger, transient polymer clusters form and relative viscosity increases. In solutions containing good solvents, flow is generally Newtonian. In many cases, more concentrated resin "solutions" flowing in poor solvents behave somewhat like dispersed systems; they are non-Newtonian because shear can break up or distort resin clusters.

Although the difference in the viscosity of resin solutions in good and poor solvents is reasonably well understood, there is little definitive work in the literature on comparisons between solutions in various good solvents in which some of the solvents are "better" than others. Erickson (1976) studied relative viscosities of solutions of several low MW resins in a range of solvents. He concluded that relative viscosities decrease as one changes from a very good solvent to a good solvent, pass through a minimum, and then increase rapidly in very poor solvents. Equations 3.4–3.6 relate relative viscosity to concentration, the hydrodynamic volume of the isolated resin molecule plus its associated solvent molecules. Hydrodynamic volume is a factor in determining the viscosity not just of very dilute solutions but also of more concentrated ones. In changing from a very good to a less good solvent, the equations predict that intrinsic viscosity and, hence, relative viscosity should decrease; this prediction conforms with Erickson's hypothesis. The range of error in Erickson's work is not small enough to establish his conclusions beyond doubt.

Solvent effects on hydrogen bonding between resin molecules can be substantial (Schoff, 1999). Figure 3.12 shows the viscosities of solutions of an acrylated epoxidized linseed oil in three solvents chosen because of their similar viscosities but very different hydrogen-bonding characteristics (Hill and Wicks, 1982). The resin molecules have multiple hydroxyl groups. Note that the viscosities of the xylene solutions are highest. Xylene is a poor hydrogen-bond acceptor and hence promotes intermolecular hydrogen bonding between the resin molecules. Methyl ethyl ketone (MEK) is a good hydrogen-bond acceptor and reduces the viscosity more effectively than xylene by reducing intermolecular hydrogen bonding. Although methanol is a much stronger hydrogen-bonding solvent than MEK, it is only marginally better at reducing viscosity. Since methanol is both a hydrogen-bond donor and acceptor, methanol can possibly bridge resin molecules by functioning as a hydrogen-bond donor with one resin molecule and a hydrogen-bond acceptor with the other. Such bridging would counteract the effectiveness of viscosity reduction.

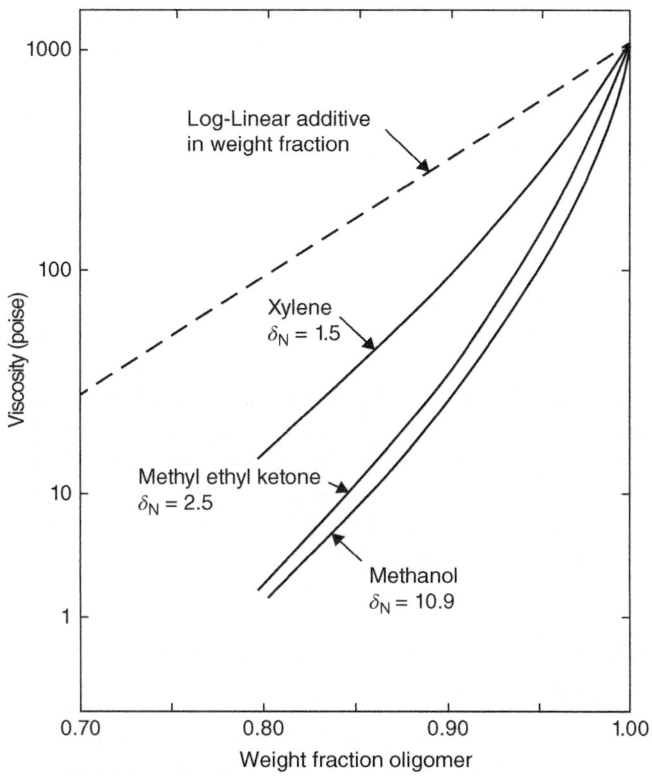

Figure 3.12 Viscosity reduction of a hydroxy-functional UV-curable oligomer with xylene, MEK, and methyl alcohol compared with predicted viscosity if the viscosity reduction were a log-linear additive relationship by weight. Source: Hill and Wicks (1982). Reproduced with permission of Elsevier.

Intermolecular hydrogen bonding involving carboxylic acid groups is particularly strong in poor hydrogen-bonding solvents (Sherwin et al., 1981). Simple carboxylic acids, such as acetic acid, exist as dimers in poor hydrogen-bond acceptor solvents, such as benzene, whereas the dimers are dissociated in good hydrogen-bond acceptor solvents, such as acetone. An analogous effect in resin solutions was demonstrated in a study of a primarily monocarboxylic acid-substituted acrylic oligomer (Wicks and Fitzgerald, 1985).

In the few cases reported in the literature, log of relative viscosity increases with the square root of MW of resins dissolved in good solvents at higher concentrations (Wicks et al., 1985; Mewis and Vermant, 2000).

A *theta solvent* (or mixture of solvents) for a specific resin is one in which the free energies of solvent–solvent, resin–resin, and solvent–resin interactions are equal. In a theta solvent solution, the relative viscosity is simply related to the weight fraction of resin and the square root of MW as shown in Eq. 3.7:

$$\ln\eta_{\mathrm{r}} = Kw_{\mathrm{r}}M^{1/2} \quad \text{or} \quad \ln\eta = \ln\eta_{\mathrm{s}} + Kw_{\mathrm{r}}M^{1/2} \tag{3.7}$$

This relationship appears to be true for solutions in good solvents of resins having relatively narrow MW distributions and viscosities between about 0.01 and 10 Pa·s; but further research is needed.

It has been postulated that intrinsic viscosities of oligomers exhibit a *theta* condition response; that is, they follow Eq. 3.7.

An important conclusion from Eq. 3.7 is that the viscosity of the solvent is a major factor affecting the viscosity of resin solutions. At first glance, it might appear that a small difference in the viscosity of the solvent would have a trivial effect on the much higher viscosity of the resin solution. However, there are examples in which solvents differ in viscosity by as little as 0.2 mPa·s, whereas 50 wt% resin solutions in those same solvents differ in viscosity by as much as 2 Pa·s, corresponding to a thousandfold difference in viscosities.

Another solvent effect on viscosity is the solvent T_g. The T_g of resin solutions depends on the concentration and the T_g of both the resin and the solvent. This effect has been recognized in the addition of plasticizers to polymers, but has not been widely studied in resin solutions having concentrations and viscosities in the range of interest for coatings. In one study, it was found that the data fit Eq. 3.8, where T_{gs} is the T_g of the solvent and T_{gr} is the T_g of the solvent-free resin (Wicks et al., 1985). In this study, Eq. 3.8 fits the data over the whole range of concentrations from pure solvent to pure resin. Equation 3.8 needs to be tested with other systems:

$$\frac{1}{T_{\mathrm{g}}} = \frac{w_{\mathrm{s}}}{T_{\mathrm{gs}}} + \frac{w_{\mathrm{r}}}{T_{\mathrm{gr}}} + kw_{\mathrm{r}}w_{\mathrm{s}} \tag{3.8}$$

The relationships are further complicated because solvent–resin interactions can be further affected by resin MW and the number of polar groups per molecule. For example, consider the data in Table 3.1 on the viscosity of solutions of a pair of styrene/allyl alcohol (SAA) copolymers in MEK and toluene (Hill and Wicks, 1982). SAA-I has a higher MW, but a lower functional group content than SAA-II. Comparison of the data illustrates that the number of polar groups is the dominating factor that determines viscosity in toluene, whereas in MEK, an effective H-bonding solvent, the viscosity of both resins is reduced dramatically with MW having a substantial effect at high concentrations. Also see Table 18.5 for examples of effects of solvents on viscosity of solutions of a high solids acrylic resin.

As expected from the data in Table 3.1, viscosity solutions of hydrogen-bonding resins in mixed solvents will depend on the hydrogen-bonding capacity of the mixture in a complex way (Rocklin and Edwards, 1976; Erickson and Garner, 1977).

In reducing viscosity to spray viscosity by adding solvent to a coating, it is desirable to have an equation that combines the effects of temperature and concentration on viscosity. For the limited ranges of temperature and concentration involved, Eqs. 3.3 and 3.6 can be combined to give such a relationship in Eq. 3.9 (Eiseman, 1995). (In Eq. 3.9, log of solvent viscosity is incorporated into constant *K*.)

$$\ln\eta = K + \frac{B}{T} + \frac{w_{\mathrm{r}}}{k_1} \tag{3.9}$$

The flow within coatings during film formation has a major influence on final film appearance and uniformity but has proven to be a very difficult area to study because of the many variables involved, including nonuniformities within the forming film. Progress is being made however. For example, Eley and Schwartz (2002) used mathematical modeling and numerical simulation to develop a theory that correlated with the behavior of two architectural paints.

Table 3.1 Effects of Molecular Weight and Functional Group Content on Viscosity

Characteristic	SAA-I	SAA-I I
$\bar{M}_n$	1600	1150
$\bar{M}_w/\bar{M}_n$	1.5	1.5
OH content (wt%)	5.7	7.7
Viscosities (mPa·s)		
80 wt% in MEK	10000	6500
70 wt% in MEK	300	230
60 wt% in MEK	80	65
50 wt% in MEK	34	30
50 wt% in toluene	760	3840

3.5 VISCOSITY OF LIQUIDS WITH DISPERSED PHASES

Since many coatings contain dispersed pigment and/or resin particles, it is important to consider the effect of dispersed phases on the viscosity of liquids. When a small amount of a dispersed phase is present, the effect is small (unless the dispersed phase is flocculated). However, as the volume of dispersed phase increases, there is a sharply increasing effect. More energy is needed to rotate particles, and the presence of the particles increasingly interferes with the ability of other particles to move. When the system becomes closely packed with particles, viscosity approaches infinity.

Numerous equations have been developed that model the effect of a dispersed phase on viscosity; the earliest was proposed by Einstein. Here we use a treatment by Mooney (1951). Equation 3.10 is a useful form of the Mooney equation for understanding the effects of variables on viscosity where η_e is the viscosity of the continuous or external phase, K_E is a shape constant, V_i is the volume fraction of internal phase, and ϕ is the packing factor. Two major assumptions are involved in the Mooney equation: (1) the particles are rigid and (2) there are no particle–particle interactions other than physical collisions:

$$\ln\eta = \ln\eta_e + \frac{K_E V_i}{1-(V_i/\phi)} \tag{3.10}$$

The packing factor (ϕ) is the maximum volume fraction of internal phase that can be accommodated by an external medium, when the particles are randomly close packed and the external phase fills all the interstices between the particles. When V_i equals ϕ, the viscosity of the system approaches infinity. Figure 3.13 shows an example of a plot of the relationship between log viscosity and V_i for a dispersed-phase system.

The shape constant K_E for spheres is 2.5. Some of the particles in coatings are reasonably close to being spheres, but many pigments are irregularly shaped. In the case of uniform diameter spheres—that is, monodisperse systems—the value of ϕ is 0.637. This value is the packing factor that has been calculated for a random mixture of cubical and hexagonal close-packed spheres and has been confirmed experimentally. To the surprise of many people first considering the question, the packing factor of monodisperse spheres is independent of particle size. The packing factor for basketballs is 0.637; the packing factor for marbles is 0.637; the packing factor for monodisperse latex particles is 0.637. However, marbles fit into the interstices between close-packed basketballs, and latex particles fit into the interstices between close-packed marbles. In other words, the packing factor depends strongly on particle size distribution—the broader the particle size distribution, the higher the packing factor.

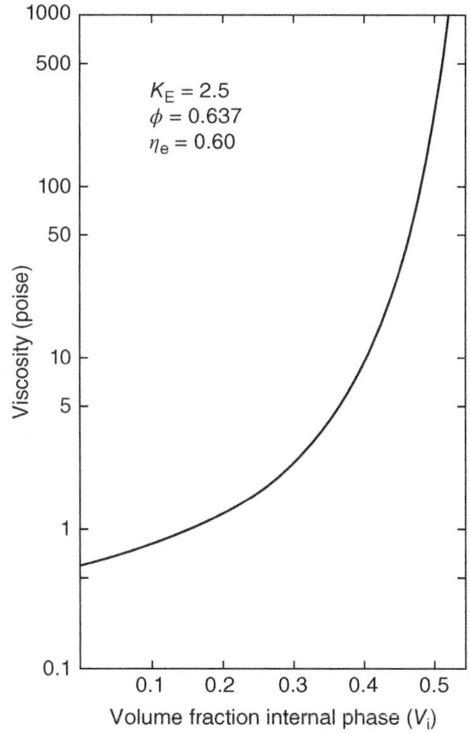

Figure 3.13 The effect of increasing volume fraction of noninteracting spherical particles on the viscosity of a dispersion. Source: Hill and Wicks (1982). Reproduced with permission of Elsevier.

The viscosity of dispersions of nonrigid particles does not follow the Mooney equation. When a shear stress is applied to such a dispersion (e.g., an emulsion), the particles can distort. When the particles are distorted, the shape constant changes to a lower value and the packing factor increases (Jones et al., 1992); both changes lead to a decrease in viscosity. Commonly, such systems are thixotropic. This is logical since, depending on the difference between the viscosities of the internal and external phases, there would be time dependency of the distortion of the particles and, hence, a decrease in viscosity as a function of time at a given shear rate. Nielsen (1977) modified the Mooney equation to account for viscosities of the two phases, but not time dependency. Time dependency can be studied using viscoelastic deformation analysis (Boggs et al., 1996; Schoff, 1988) or by large amplitude oscillatory shear using a controlled-strain rheometer (Armstrong et al., 2016).

Fluids with readily distorted internal phases include emulsions, water-reducible acrylic and polyester resins, urethane dispersions, some latexes in which the outer layer of the latex particles and layers adsorbed on them are swollen by water, and some pigment dispersions with comparatively thick adsorbed layers of polymer swollen with solvent. See Mewis and Vermant (2000) for discussion of the effect of adsorbed layers on particles on rheology.

While there may be other factors involved, many so-called thixotropic agents act by creating a swollen dispersed phase that can be distorted. For example, very small particle size SiO_2 adsorbs a layer of polymer swollen by solvent that is thick compared to the pigment and is distortable in a shear field. The degree of distortion increases with shear stress as well as with the time of shearing up to a limit. When shearing is stopped or decreased, the polymer layers recover their equilibrium shape and viscosity increases. Lightly cross-linked polymer gel particles are also used as thixotropic agents; the particles swell with solvent, forming a distortable dispersed phase. The shear thinning behavior depends on the particle size, concentration, and internal viscosity of the dispersed phase. Smaller particles lead to higher shear rates for shear thinning. Shear thinning decreases with decreasing concentration and increasing internal viscosity of particles.

The viscosity of dispersions is also affected by particle–particle interactions. If clusters of particles form when stirring of a dispersion is stopped, the viscosity of the dispersion increases; if these clusters separate again when shear is exerted, the viscosity drops. Examples of such shear thinning systems are flocculated pigment dispersions and flocculated latexes. Another example is the so-called gelation—really flocculation—induced by water in coatings containing treated clay dispersions (Jones et al., 1992). When clusters of particles form, continuous phase is trapped in the clusters, resulting in high V_i at low shear rates. At high shear rates, when the clusters break up, the value of V_i is reduced to that of individual particles of the dispersion without trapped continuous phase. As V_i increases, viscosity increases and vice versa.

One can also consider the dispersion from the point of view that V_i of the primary particles remains constant and that aggregation effects can be accounted for by changes in K_E and ϕ (see Eq. 3.10). An example is shown in Figure 3.14 (Hill and Wicks, 1982). The vertical axis is the ratio of the viscosity of the dispersion to the viscosity of the external phase (η/η_e). The viscosity of the dispersion increases rapidly as the number of particles n in aggregates increases.

Polymer solutions containing dispersed phases are complex physical systems whose flow properties continue to be investigated. In this discussion, we have used the Mooney equation; alternative treatments such as those of Krieger and Dougherty (1959) and Russel and Sperry (1994) are also useful. Goodwin and Hughes (1997) include a detailed analysis of effects of particle interactions on dispersion rheology.

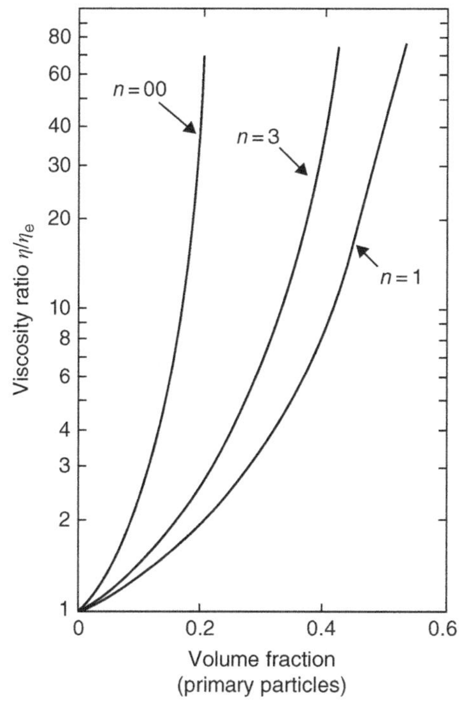

Figure 3.14 Effect of cluster formation on viscosity. Source: Hill and Wicks (1982). Reproduced with permission of Elsevier.

3.5.1 Thickeners for Latex Coatings

Latex paints require a thickener to control rheological behavior. Viscosity of the paint as a function of shear rate particularly affects ease of brushing, film thickness, leveling, sagging, and settling. Very commonly, combinations of water-soluble thickeners and *associative thickeners* are employed to balance these properties. The following discussion applies, to an extent, to all paints having binders that are dispersed or emulsified in water, for example, polyurethane dispersions.

Historically the water-soluble polymer hydroxyethylcellulose (HEC) was widely used as the only thickener and was very effective in controlling sagging and settling. Furthermore, by increasing the viscosity of the aqueous phase, HEC reduced penetration of water into porous substrates such as wood; this penetration could cause a rapid increase in viscosity just after application, reducing leveling. However, HEC posed several problems. High MW HEC is an efficient thickener but leads to increased extensional viscosity (see Section 3.6.3), which in turn leads to severe spattering when paint is applied by a roller. Therefore, it was replaced by larger amounts of less efficient medium to low MW HEC. Spattering was reduced, but cost increased and excessive shear thinning made it difficult to apply the paint as thickly as desired. Also, viscosity of HEC-thickened paints recovers very rapidly after shearing, which interferes with proper leveling. The resulting ridges and valleys were clearly visible in latex paint films of several decades ago.

Progress in minimizing these problems has been made with *associative thickeners*. Such thickeners are a subclass of *associative polymers, which are* water-soluble polymers that form supermolecular structures in solution (Winnik and Yekta, 1997). Such polymers have many end uses and are being widely studied (Lara-Ceniceros et al., 2014). The ones useful in coatings are often water-soluble polymers having two or more long-chain nonpolar hydrocarbon groups spaced along the backbone or located at the ends of the polymer chains. Examples include hydrophobically modified ethoxylated polyurethanes (HEUR), styrene maleic anhydride terpolymers (SMAT), hydrophobically modified alkali-swellable emulsions (HASE), and hydrophobically modified ethoxylated aminoplast (HEAT) polymers. An example of a HASE thickener is a copolymer of MAA, EA, and modified TMI in which one of the NCO groups is prereacted with an ethoxylated stearyl alcohol (Wu et al., 2002).

HEUR thickeners seem to be the most widely studied in coatings. Synthesis of a representative HEUR thickener starts by reacting a polyethylene oxide diol ($\bar{M}_w$ 8000–10000) with excess diisocyanate. The isocyanate ends then are capped with hydrophobic groups such as C_{12} alkyl or nonylphenol (Van Dyk et al., 2014; Nan et al., 2015). In simplified model systems consisting of a latex and such an HEUR, Van Dyk confirmed that the principle mode of thickening is the formation of a transient network formed when hydrophobic groups reversibly adsorb on latex particle surfaces, essentially causing bridging flocculation that is readily broken by shear forces. Associations of the thickener molecules themselves are of secondary importance at the low concentrations used in practical paints. In fully pigmented paints the situation is much more complicated, as described in the succeeding text.

Use of associative thickeners permits formulation of latex paints that exhibit less shear thinning so that viscosity at high shear rates can be higher; as a result, thicker wet films can usually be applied (Fernando et al., 1986). The thickness of the wet film promotes leveling, since the rate of leveling depends on wet film thickness, as discussed in Section 24.2. Low shear viscosity can also be maintained at an appropriate level, as with HEC, for controlling sagging and settling. To achieve the desired viscosity profile across a wide range of shear rates, combinations with HEC can increase the mid-shear viscosity. Good formulations including associative thickeners not only level better than HEC-thickened formulations but also give somewhat higher gloss (Hall et al., 1986). A formula that combines HEC with an associative thickener is described in Section 32.1.

The mechanisms of thickening in fully formulated latex paints may be quite complex. They likely include associations with pigment particles as well as with latex particles. Reynolds (1992) reviewed possible mechanisms of thickening by associative thickeners and factors involved in their use in formulating latex paints. He emphasized that greater care and skill is required in formulating with associative thickeners than with conventional water-soluble thickeners. The results obtained can be very dependent on the combination of the particular latex and thickener and on the amounts and types of surfactant present in the formulation (Chen et al., 1997). Santos et al. (2016) updated research on mechanisms of thickening. Compatibility is particularly important in selecting the surfactant to be used in tint colors for use with paints thickened with associative thickeners to avoid loss of viscosity after tinting.

Associative thickeners are also effective in polyurethane dispersion formulations (Nan et al., 2015) and probably in other paint formulations that include dispersions or emulsions of resins in water.

In latex and emulsion paints, associative thickeners interact with the pigments and the latex, potentially displacing some of the surfactant. They may interact with other paint additives as well. There are potential complications with paints that are tinted at the paint store. The patent literature reflects widespread ongoing research aimed at improving associative thickener technology. Suppliers periodically introduce new associative thickeners, claiming various improvements.

3.6 OTHER MODES OF FLOW

Although flow from application of shear stress is the most common type of flow encountered in making and using coatings, other modes of flow are sometimes involved.

3.6.1 Turbulent Flow

Turbulent flow occurs at very high rates of shear or in irregularly shaped containers and pipes. At low shear rates, flow occurs in a laminar fashion, as illustrated in Figure 3.1. However, as shear rate increases, a critical point is reached at which flow suddenly becomes chaotic. Laminar flow is disrupted; swirling eddies and vortices occur; and flow changes to turbulent flow. Even with Newtonian fluids, viscosity increases more than proportionally with shear rates above this critical level.

3.6.2 Normal Force Flow

When Newtonian fluids are stirred with a rotary stirrer, the liquid level becomes low in the center near the shaft of the stirrer and high on the walls of the vessel as a result of centrifugal force. This is shown schematically in Figure 3.15a.

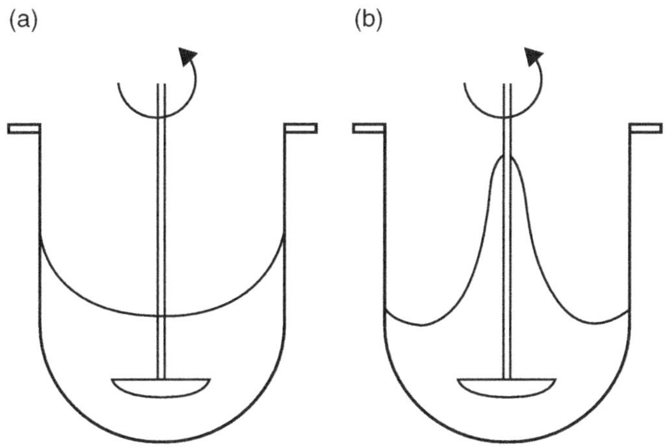

Figure 3.15 (a) Conventional compared to (b) normal force direction flow of liquids on stirring.

However, some liquids climb the shaft of the stirrer, as shown in Figure 3.15b, rather than the sides of the vessel. Such flow is normal (perpendicular) to the plane of force. This *normal force flow* behavior is typical of viscoelastic materials and of systems that are starting to gel. In the early stages of cooking resins, the flow pattern is as shown in Figure 3.15a; but if premature cross-linking leads to the onset of gelation, the flow pattern can change abruptly to that shown in Figure 3.15b. If this occurs, heating should be discontinued, and the reaction mixture should be dumped out as quickly and as safely as is possible before the gel becomes intractable.

Normal force flow effects have been observed in the handling and application of some coatings whose flow is viscoelastic. Under stress, the flow of these coatings has both elastic and viscous components (Aidun and Trianafillopoulos, 2012). It seems logical to speculate that a significant extent of elasticity could negatively effect atomization in spraying, film splitting in roller coating, and leveling. To study normal force, flow requires specialized, expensive instrumentation, such as an oscillatory plate rheometer. Instruments capable of good measurements have only recently become available.

3.6.3 Extensional Flow

Another mode of flow encountered in some methods of coating application is *extensional flow*, also called *elongational flow*. Extensional flow occurs when fluid deformation is the result of stretching. It can be visualized by stretching a rubber band. Various types of stretching are possible. In spin coating, for example, extension occurs in two dimensions. The extensional flow of greatest importance in most other coating processes is uniaxial—that is, in one direction. In uniaxial flow, the viscosity is properly called *uniaxial extensional viscosity*; we simply use the term *extensional viscosity*, but it should be remembered that there are several types of extensional viscosity.

Extensional viscosity is a measure of the resistance of a liquid to stretching forces. The difference between extensional flow and shear flow was first observed in fiber drawing. When the fiber material passes through the spinneret, the mode is shear flow. However, as the fiber is pulled after leaving the spinneret, there is no further shearing action; rather, the fiber is extended. The flow is extensional flow, and the resistance to flow is extensional viscosity. The symbol used for extensional viscosity is η_e. In the case of Newtonian fluids, $\eta_e/\eta = 3$.

Extensional flow is encountered when applying coatings by direct roll coating (Section 23.4). The material to be coated is passed through the nip between two rollers, one of which is covered with a layer of coating. Prior to entering the nip, the coating is subject to ambient pressure. As the coating goes through the nip, pressure rises and, as it exits the nip, pressure begins to drop and can drop below ambient pressure. The pressure drop can cause volatilization of solvents and/or release of dissolved air within the coating, thus causing cavitation. The cavities lead to the formation of ligaments of coating that gets elongated under the extensional deformation in the nip exit region. If the extensional viscosity is relatively low, the film splits quickly, leaving a ridged film. However, with higher extensional viscosity, fibers grow, and some of the longer fibers split in two places, resulting in formation of droplets, which are thrown into the air. This process is called *misting* in industrial coatings or *spattering* in architectural paints. Figure 3.16 shows an extreme case of fiber development in roll coating of paint with an impractically high extensional viscosity (Glass, 1978). Soules et al. (1988) discuss the relationship of variables and extensional viscosity effects in roll coating.

A large fraction of latex flat wall paints is applied by a roller. During roller application, latex paints *spatter*. Paints with high extensional viscosity are likely to spatter severely (Massouda, 1985). Extensional viscosity increases when

Figure 3.16 Fiber development in roll coating a high extensional viscosity paint. Source: Goodwin and Hughes (1997). Reproduced with permission of American Chemical Society.

high MW water-soluble polymers having very flexible backbones, such as high MW HEC, are used as thickeners in a latex paint (Glass, 1978). As a ligament of the paint is extended, the soluble thickener molecules line up in the direction of the extension, leading to increased resistance to the extension. Spattering can be minimized by using lower MW water-soluble thickeners with rigid segments in the polymer backbone, such as low MW HEC. It is more common in contemporary technology to address the spattering problem with associative thickeners, often in combination with HEC, as explained in Section 3.5.1.

Extensional flow can also be a critical factor in spray application. If, for example, a solution of a thermoplastic acrylic resin with $\bar{M}_w$ above about 100000 is sprayed, the solution emerges from the spray gun orifice as fibers rather than droplets. As the fiber extends, the mode of flow is extensional. The behavior is called cobwebbing. While cobwebbing is undesirable when painting a car, it can give a desirable effect for applying decoration to a Christmas tree or making cotton candy. Soules et al. (1991) discuss other possible extensional viscosity phenomena in spray application. Airless spray of latex paints having high extensional viscosity results in large particle size droplets and correspondingly poor film appearance (Fernando et al., 2000). This paper also provides a review of various aspects of extensional viscosity effects in coatings.

Measuring extensional (elongational) viscosity of liquids has been difficult, but devices are being developed that make it more practical. Extensional testers are commercially available. The McKinley laboratory at MIT has published extensively in the field. For example, Sharma et al. (2015) compare extensional viscosity measurement with three types of instruments: a capillary breakup rheometer, a forced jet extensional rheometer, and a cross-slot extensional rheometer. They show that hydrophobically modified cellulose derivatives (a class of associative thickeners) perform well in applications that require jetting (inkjet printers) or spraying.

A screen viscometer is a simple device that involves flow of a solution through a stack of screens (Prud'homme et al., 2005). ASTM developed a method for testing extensional spray characteristics of agricultural chemical dispersions (E 2108-04), which employs a screen viscometer. Such a device might be adaptable for study of paints.

GENERAL REFERENCES

Eley, R. R., Rheology and Viscometry in Koleske, J., Ed., *Paint Testing Manual: Gardner-Sward Handbook*, 14th ed., ASTM, Philadelphia, **1995**, pp 333–368.

Mezger, T. G., *The Rheology Handbook*, Vincentz, Hanover, **2014**.

Patton, T. C., *Paint Flow and Pigment Dispersion*, 2nd ed., Wiley-Interscience, New York, **1979**.

Reynolds, P. A., The Rheology of Coatings in Marrion, A., Ed., *The Chemistry and Physics of Coatings*, Royal Society of Chemistry, London, **1994**.

Schoff, C. K., *Rheology*, Federation of Societies for Coatings Technology, Blue Bell, PA, **1991**.

Schweizer, P. M.; Kistler, S. F., Eds., *Liquid Film Coating Scientific Principles and their Technological Implications*, Springer, Dordrecht, **2012**.

Tadros, T. F., *Rheology of Dispersions, Principles and Applications*, Wiley-VCH, Weinheim, **2010**.

REFERENCES

Aidun, C. K.; Trianafillopoulos, N., High-Speed Blade Coating in Schweizer, P. M.; Kistler, S. F., Eds., *Liquid Film Coating: Scientific Principles and their Technological Implications*, Springer, Dordrecht, **2012**, pp 661–666.

Allcock, H. P.; Lampe, F. W., *Contemporary Polymer Chemistry*, 2nd ed., Prentice Hall, Englewood Cliffs, **1990**.

Anwari, F., et al., *J. Coat. Technol.*, **1989**, *61(774)*, 41.

Armstrong, M. J., et al., *J. Rheol.*, **2016**, *60(3)*, 433–450.

Boggs, L. J., et al., *J. Coat Technol.*, **1996**, *68(855)*, 63.

Chen, M., et al., *J. Coat. Technol.*, **1997**, *69(867)*, 73.

Eiseman, M. J., *J. Coat. Technol.*, **1995**, *67(840)*, 47.

Eley, R. R.; Schwartz, L. W., *J. Coat. Technol.*, **2002**, *74(932)*, 43.

Erickson J. R., *J. Coat. Technol.*, **1976** *48(620)*, 58.

Erickson, J. R.; Garner, A. W., *ACS Org. Coat. Plast. Chem. Prepr.*, **1977**, *37(1)*, 447.

Fernando, R. H., et al., *J. Oil Colour Chem., Assoc.*, **1986**, *69*, 263.

Fernando, R. H., et al., *Prog. Org. Coat.*, **2000**, *40*, 35.

Glass, J. E., *J. Coat. Technol.*, **1978**, *50(641)*, 56.

Goodwin, J. W.; Hughes, R. W., Particle Interactions and Dispersion Rheology in Glass, J. E., Ed., *Technology for Waterborne Coatings*, ACS Symposium Series 663, American Chemical Society, Washington, DC, **1997**, pp 94–125.

Hall, J. E., et al., *J. Coat. Technol.*, **1986**, *58(738)*, 65.

Hare, C. H., *J. Protective Coat. Linings*, **2001**, April, 79.

Haseebuddin, S., et al., *Prog. Org. Coat.*, **1997**, *30*, 25.

Hester, R. D.; Squire, Jr., D. R., *J. Coat. Technol.*, **1997**, *69(864)*, 109.

Hill, L. W.; Wicks, Jr., Z. W., *Prog. Org. Coat.*, **1982**, *10*, 55.

Jones, F. N., *J. Coat. Technol.*, **1996**, *68(852)*, 25.

Jones, D. A. R., et al., *J. Colloid Interface Sci.*, **1992**, *150(1)*, 84.

Krieger, I. M.; Dougherty, T. J., *Trans. Soc. Rheol.*, **1959**, *III*, 137.

Kruithof, K. J. H.; van den Haak, H. J. W., *J. Coat. Technol.*, **1990**, *62(790)*, 47.

Lara-Ceniceros, T. E., et al., *J. Polym. Res.*, **2014**, *21*, 511.

Massouda, D. B., *J. Coat. Technol.*, **1985**, *57(722)*, 27.

Mewis, J.; Vermant, J., *Prog. Org. Coat.*, **2000**, *40*, 111.

Mezger, T. G., *The Rheology Handbook*, Vincentz, Hanover, **2014**.

Mooney, M., *J. Colloid Sci.*, **1951**, *6*, 162.

Nan, G.; Zhuo, Z.; Quingzhi, D., *J. Nanomater.*, **2015**, doi:10.1155/2015/137646

Nielsen, L. E., *Polymer Rheology*, Marcel Dekker, New York, **1977**, pp 56–61.

Patton, T. C., *Paint Flow and Pigment Dispersion*, 2nd ed., Wiley-Interscience, New York, **1979**.

Prud'homme, R. K., et al., *Polym. Mater. Sci. Eng.*, **2005**, *92*, 241.

Reynolds, P. A., *Prog. Org. Coat.*, **1992**, *20*, 393.

Rocklin, A. L.; Edwards, G. D., *J. Coat. Technol.*, **1976**, *48(620)*, 68.

Russel, W. B.; Sperry, P. R., *Prog. Org. Coat.*, **1994**, *23*, 305.

Santos, F. A., et al., *Proceedings of the American Coatings Conference*, Indianapolis, IN, April, , **2016**, paper 9.2.

Schoff, C. K., Rheological Measurements in Mark, H. F.; Bikales, N. M.; Overberger, C. G.; Menges, G.; Kroschwitz, J. I. Eds., *Encyclopedia of Polymer Science and Engineering*, 2nd ed., John Wiley & Sons, Inc., New York, **1988**, Vol. 14, pp 454–540.

Schoff, C. K., *Rheology*, Federation of Societies for Coatings Technology, Blue Bell, PA, **1991**.

Schoff, C. K., Concentration Dependence of the Viscosity of Dilute Polymer Solutions in Branderup, J., et al., Eds., *Polymer Handbook*, 4th ed., John Wiley & Sons, Inc., New York, **1999**, VII/265.

Sharma, V., et al., *Soft Matter*, **2015**, *11*, 16, 3251–3270. Available as an MIT open access article.

Sherwin, M. A., et al., *J. Coat. Technol.*, **1981**, *53(683)*, 35.

Soules, D. A., et al., *J. Rheol.*, **1988**, *32*, 181.

Soules, D. A., et al. Dynamic Uniaxial Extensional Viscosity in Glass, J. E., Ed., *Polymers as Rheology Modifiers*, American Chemical Society, Washington, DC, **1991**, pp 322–332.

Toussaint, A.; Szigetvari, I., *J. Coat. Technol.*, **1987**, *59(750)*, 49.

Van Dyk, A., et al., *Proceedings of the American Coatings Conference*, Atlanta, GA, April 7–9, **2014**, paper 5.3.

Wicks, Jr. Z. W.; Fitzgerald, G., *J. Coat. Technol.*, **1985**, *57(730)*, 45.

Wicks, Jr., Z. W., et al., *J. Coat. Technol.*, **1985**, *57(725)*, 51.

Winnik, M. A., Yekta, A., *Curr. Opin. Colloid Interface Sci.*, **1997**, *2*, 424–436.

Wright, A. J., *Eur. Coat. J.*, **1996**, *32*, 696.

Wu, W., et al., *Proceedings of the Waterborne High-Solids Powder Coating Symposium*, New Orleans, LA, **2002**, pp 343–355.

Chapter 4

Mechanical Properties

4.1 INTRODUCTION

While the mechanical properties of structural materials, such as high strength steel or fiber-reinforced polymers, are obviously paramount to their usefulness, the mechanical properties and requirements of coatings may seem less important. However, coatings face an assortment of mechanical stresses and insults throughout their life and must resist these stresses in order to fulfill their desired function. The coating on the outside of an automobile should withstand being hit by a piece of flying gravel without film rupture. The coating on the outside of a beer can must be able to withstand abrasion when cans rub against each other during shipment in a railroad car. The coating on outdoor wood should not crack when the wood expands and contracts as a result of changing temperatures during winter or due to swelling and shrinkage resulting from changes in moisture content of the wood. The coating on aluminum siding must be flexible enough to withstand fabrication of the siding and must resist scratching during installation on a house. In addition, many coatings must also withstand the effects of weather (Chapter 5), retain adhesion (Chapter 6), and protect metals from corrosion (Chapter 7).

4.2 BASIC MECHANICAL PROPERTIES

Before the mechanical behavior of coatings can be studied, the basics of mechanical behavior that are common to most materials need to be addressed. At the most fundamental level, the crux of mechanical performance is the relationship between applied forces and the dimensional changes that take place in a material in response to those forces. For example, when a person pulls on a rubber band, it elongates. The change in length is proportional (over a reasonable range) to the amount of force applied. While the forces and changes in length are easy to understand, to be useful those terms need to be normalized to the initial size of the object being deformed.

Materials can deform in different modes, but tensile deformation is the most easily visualized. The previously mentioned example of pulling a rubber band is an example of tensile deformation. In this mode, the forces applied to a material cause it to lengthen, and the lengthening occurs along the same axis to which the force is being applied. Typically the force is applied perpendicular to the surface of the material (Figure 4.1). Compressive deformation arises when the forces are applied such that the material shrinks in the direction of the applied forces.

The force, F, applied to a material is typically converted to a tensile stress, σ, by dividing the force by the appropriate cross-sectional area, A:

$$\sigma = \frac{F}{A} \tag{4.1}$$

In the SI system, forces are measured in newtons, N, and the area in m^2, and therefore, the units of stress are $N\ m^{-2}$, which is termed a Pascal, Pa. The change in dimension that occurs when the stress is applied is termed the strain, ε. The strain is defined as

$$\varepsilon = \frac{\Delta L}{L_0} \tag{4.2}$$

where L_0 is the initial length of the material and ΔL is the change in length in response to the applied stress. Strain is

Organic Coatings: Science and Technology, Fourth Edition. Frank N. Jones, Mark E. Nichols, and Socrates Peter Pappas.

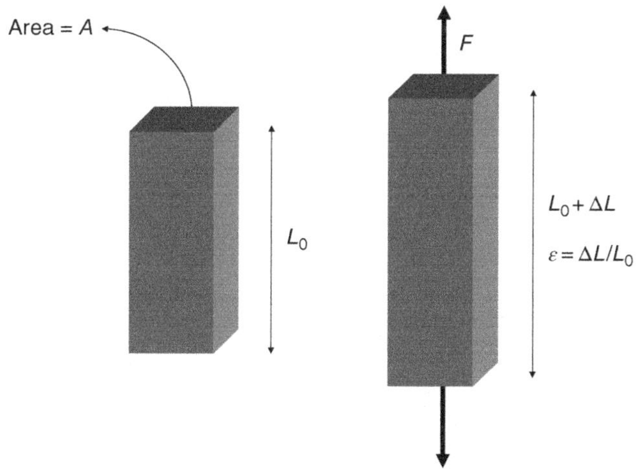

Figure 4.1 Tensile force, F, applied to a material of cross-sectional area, A, leads to an extension and a resultant strain, ε.

Table 4.1 Modulus of Various Engineering Materials

Material	Modulus (GPa)
Rubber	0.001–0.1
High density polyethylene	0.8
Epoxy	3.0
Magnesium	45
Aluminum	70
Steel	205
Carbon fiber	200–700

a dimensionless quantity. When the tensile stress and strain are small, they are related via simple proportionality:

$$\sigma = E\varepsilon \tag{4.3}$$

This equation is known as Hook's law and the proportionality constant, E, is known as Young's modulus. E is also often referred to as the tensile modulus, or simply the modulus, and is a measure of the stiffness of a material. Like stress, the units of modulus are Pa. The modulus of a variety of materials is given in Table 4.1. The modulus is an inherent property of a material. It is determined by the material's composition and process history.

While tensile deformation is most easily visualized, another type of mechanical deformation, shear deformation, is quite common. Shear deformation occurs when a force, T, is applied parallel to the surface of a material (Figure 4.2).

The shear stress, τ, is defined as the force divided by the area over which the shear force is applied. The shear strain, γ, is the angle by which the material deforms. As in tensile deformation, the shear stress and strain are related

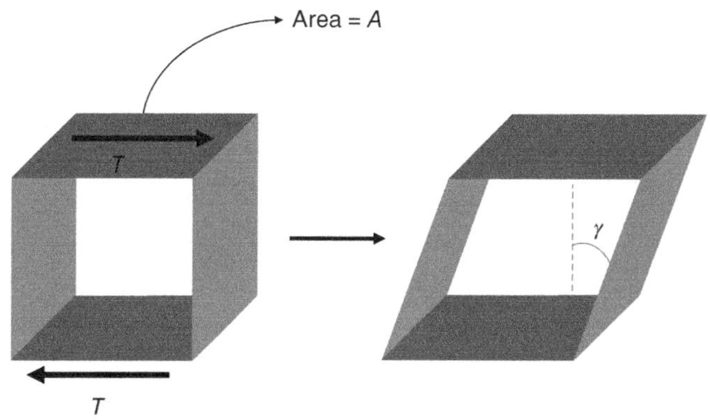

Figure 4.2 Shear deformation of a material. A force, T, parallel to a surface causes a shape change, quantified by the angular change, γ, the shear strain.

to one another via a proportionality constant termed the shear modulus, G:

$$\tau = G\gamma \tag{4.4}$$

The shear modulus is a measure of a material's resistance to shear deformation. G and E are not independent quantities. They are related to each other by Eq. 4.5:

$$E = 2G(1+\nu) \tag{4.5}$$

where ν is Poisson's ratio. When a material grows in one dimension, such as during tensile deformation, it typically contracts in another dimension. With reference to the stretching of a rubber band, the thickness of the rubber band decreases as the length increases. The ratio of the thinning to the lengthening is termed Poisson's ratio. This can be written mathematically as

$$\nu = -\frac{\varepsilon_x}{\varepsilon_y} \tag{4.6}$$

where ε_x and ε_y are the strains parallel and perpendicular to the applied tensile stress (Figure 4.3). Note that the negative sign is required, as the material contracts perpendicular to the applied stress.

To assess the mechanical properties of a material, the simplest experiment is one where a material is stretched and its change in dimensions is recorded as a function of the load (stress) applied. In its simplest form this is a tensile test, and the resulting data is typically plotted as a stress–strain curve. A stress–strain curve for a typical material is shown in Figure 4.4. A number of the curve's features are important. First, the initial slope of the curve is the modulus, E, of the material. This initial portion of the curve is usually fairly linear for small strains, such that the modulus

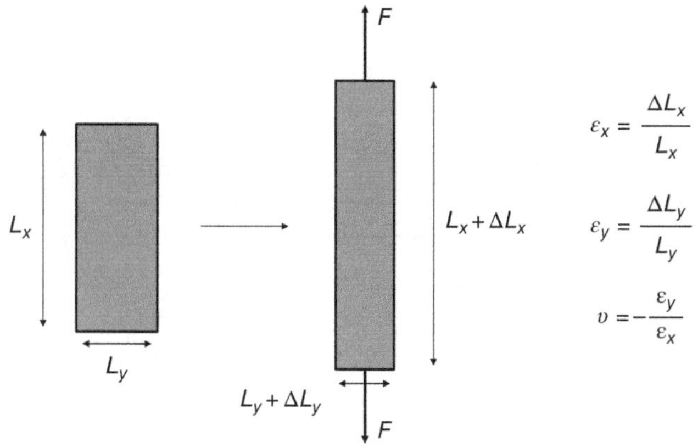

Figure 4.3 Two-dimensional Poisson's effects in materials. When stretched in one direction, the material will contract in the other direction(s).

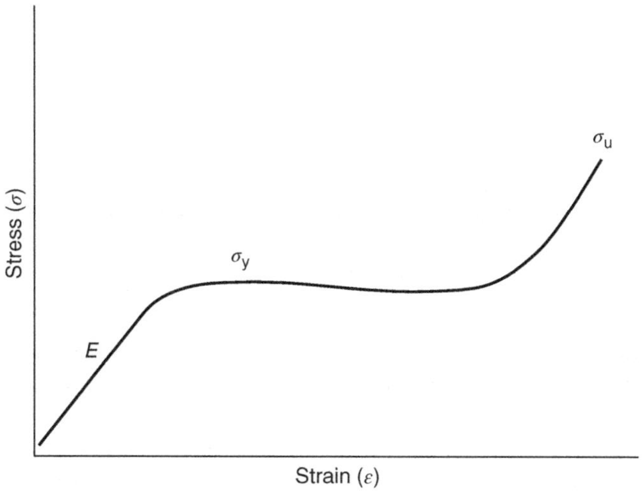

Figure 4.4 Typical stress–strain curve for a ductile polymer. The initial slope, E, is the modulus, the local maximum stress is the yield stress, σ_y, and the ultimate breaking stress is σ_u.

can be unambiguously calculated. The end point of the curve is where the material fails. All materials have a finite breaking strength, even though there are many modes by which failure can eventually occur. The stress at breaking is termed the tensile strength but is also often referred to as the ultimate strength (σ_u) or stress, or the rupture strength. The strain at failure is called the ultimate strain, breaking strain, and somewhat incorrectly the elongation at break.

Stress–strain curves can take on various shapes depending on a material's composition, the testing temperature, or the rate of deformation. Under some conditions, a local maximum in the stress may be observed in some materials. This is often called the yield strength (σ_y) or yield stress, as this point can be loosely correlated to the onset of permanent deformation, which is also termed plastic deformation. If the stress placed on a material is

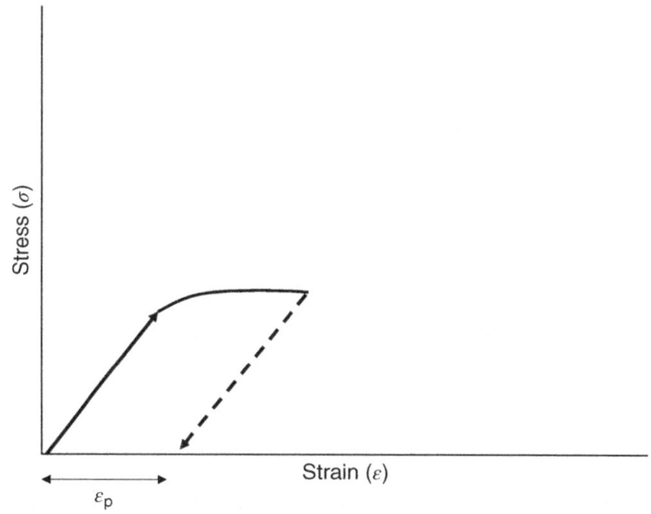

Figure 4.5 Loading and unloading paths for a polymer, which is plastically deformed and then unloaded. The plastic strain is denoted as ε_p.

relatively small, the deformation is recoverable; and upon unloading, the material returns to its original dimensions. When this occurs, the material is said to behave elastically. However, if larger stresses are placed upon a material, some of the deformation can become permanent. This is referred to as plastic deformation. When the load is removed, the material will recover its elastic deformation but will retain a permanent shape change due to the plastic deformation. Imagine a paper clip that can be elastically bent quite easily. However, if one applies more force to the paper clip, it can be permanently deformed into a new shape. It has been plastically deformed. This process can be explained graphically in Figure 4.5, where the stress–strain curve for loading and unloading has been drawn for a material that plastically deforms.

These fundamental concepts—stress, strain, modulus, tension, compression, and shear—form the basis for a discussion about the mechanical properties of any material, including coatings.

4.2.1 Glass Transition Temperature (T_g)

T_g is introduced in Section 2.2.1; aspects of the subject that are relevant to mechanical properties are reviewed here. While the mechanical behavior of all materials can be affected by the material's temperature, the mechanical behavior of polymers is remarkably influenced by the proximity of the material to its glass transition temperature, T_g. Above T_g polymer molecules possess sufficient thermal energy such that segments along the polymer chain can rotate freely with respect to the chain axis. This results in

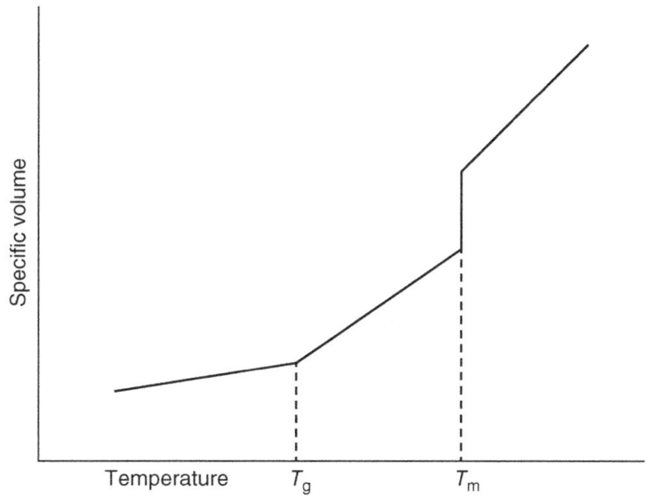

Figure 4.6 Specific volume of a semicrystalline polymer as a function of temperature. Note the step change at the melting temperature, T_m, and the slope change at the glass transition temperature, T_g. Source: From Nichols and Hill (2010). Reproduced with permission of American Coatings Association.

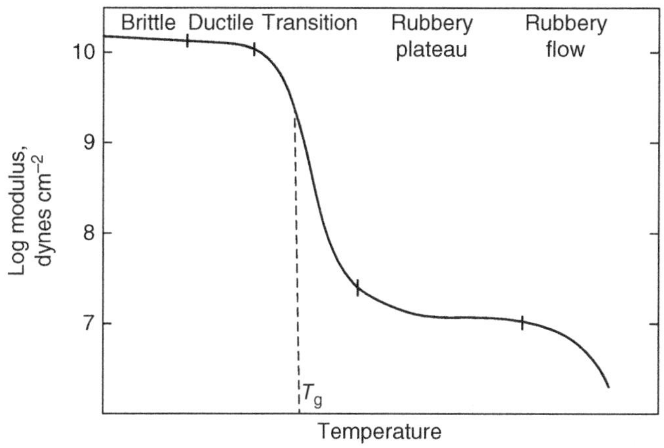

Figure 4.7 Modulus as a function of temperature for a thermoplastic polymer. Source: From Nichols and Hill (2010). Reproduced with permission of American Coatings Association.

the polymer behaving much more like a viscous fluid and much less like an elastic solid than it does when its temperature is below T_g. Thus, mechanical properties such as the modulus and strength are much higher below the glass transition temperature than above.

T_g is not a first-order thermodynamic transition like the melting temperature but is a more complex transition that is difficult to succinctly describe. As shown in Figure 4.6, at the glass transition temperature, the specific volume versus temperature curve of a polymer undergoes a change in slope. Above T_g the curve has a large slope, and below T_g the slope becomes smaller. This is due to the relative ease with which the polymer segments can move around. The transition is not as sharp as is illustrated in Figure 4.6 but gradually changes over several degrees.

Also, unlike the melting temperature, T_g is a function of the means by which it is probed and by the rate of measurement. A good rule of thumb is that the T_g of a polymer increases by 5–7°C for every factor of 10 increase in the heating rate or deformation rate (Nichols and Hill, 2010). Thus, the T_g of an acrylic copolymer might be measured as 30°C when tested in a differential scanning calorimeter (DSC) in which the sample is heated at 1°C min^{-1}. However, when heated at 10°C min^{-1}, the T_g would be measured as approximately 35°C. This is not an instrumental artifact, but is instead indicative of a kinetic component to the T_g. One can imagine that at very slow heating rates, or at very long times, the effective glass transition temperature can be many degrees lower than that observed on normal time scales. The same can be observed when the T_g is probed via mechanical means, which shall be discussed shortly. Note that, in a DSC, the T_g is recognized by a step change in the heat capacity and not via an exothermic or endothermic peak that would be indicative of melting or crystallization, which are first-order thermodynamic transitions.

The measured mechanical properties of a polymer are strongly dependent upon the relationship of the measuring temperature to the T_g of the polymer. Below T_g, polymers are in a glass-like state, and molecular motions are hindered. Only at very long time scales will the polymer be observed to flow. Above T_g, the polymer behaves much more like a viscous fluid and molecular motions can readily occur. Thus, the stiffness above T_g is at least one and as many as three orders of magnitude lower than that of the same polymer below T_g. (Almost all unfilled glassy polymers have a modulus of ~3 GPa below T_g.)

The modulus versus temperature behavior of a glassy polymer is shown in Figure 4.7. These plots are usually shown with the vertical axis on a log scale, as the changes in the modulus are quite large.

4.2.2 Viscoelasticity

As discussed earlier in Section 1.2, the binder in organic coatings is typically a high molecular weight (MW) polymer, which surrounds the pigments and other additives in a coating formulation. The long chains of such polymers result in their mechanical properties being highly *viscoelastic*, which signifies that the material's mechanical properties are a combination of elastic solid-like behavior and viscous liquid-like behavior. In practice, it follows that the mechanical behavior of polymers and coatings is strongly dependent upon their temperature and the rate of deformation. In contrast, metals and ionic solids display essentially elastic and plastic behavior throughout their entire range of use temperatures.

Viscoelastic effects are most easily identified by examining the creep and stress relaxation behavior of polymers in comparison with elastic materials such as a metal. In a creep experiment, a mechanical load is placed upon a material, and the load is held constant as a function of time. This is most easily imagined by hanging a fixed weight from the end of a material specimen that is rigidly clamped to a stationary object at the other end. The stress and strain history as a function of time is shown for both a polymer and metal in Figure 4.8. For both materials, the stress increases instantaneously when the load is applied and then stays constant throughout the experiment. In the case of the metal, the strain also increases instantaneously after the stress is applied and then remains constant throughout the experiment. This is true for metals and ionic solids unless the experiment is performed at a temperature approaching approximately 50% of the melting temperature. However, for the polymer or organic coating, the strain continues to increase as time progresses, which is a direct result of the long-chain nature of polymers. Unlike atomic solids, where the atoms can respond quickly to the application of forces, the long molecules in a polymer can only slowly change shape in response to an applied load, and, as long as that load is applied, the molecules will continue to move until they come to a new equilibrium position.

Many theories have been proposed to describe the mechanism by which polymers rearrange during flow or the application of stress. At this time, theories based on constrained tubes appear to most accurately predict polymer molecule motions. These theories propose that the long-chain molecules move via *reptation*, a motion that mimics that of a snake (reptile) (De Gennes, 1979). The chain molecules can slide down the length of an imaginary tube that surrounds the polymer chain and follows its contours. As the motion is highly constrained and only allowed along the axis of the tube, the bulk polymer can only respond slowly to the application of loads. The instantaneous strain is a result of intramolecular movements, whereas the slower intermolecular movement and disentanglement of the chains themselves give rise to the time-dependent increase in strain during a creep test.

The viscoelastic nature of coatings is also apparent during stress relaxation (Figure 4.9). During a stress relaxation experiment, a fixed strain is rapidly placed upon a material and then held constant for a given time. Imagine stretching a material to a new length and then holding the material at that new length for a prescribed time. For metals, the stress increases instantaneously when the strain is applied and then remains constant throughout the remainder of the test. However, for a polymer, the stress increases instantaneously but then *decreases* as time progresses. This again is due to the viscoelastic nature of polymers, which is brought on by their long-chain nature. When the strain is applied, the molecules can only slightly rearrange themselves on the time scale of the strain application. Thus, the stress response is quite high. However, as time progresses, the molecules have time to rearrange (reptate) into lower energy states, which reduces the stress in the material.

The time-dependent behavior of coatings during a creep or stress relaxation experiment is fundamental to

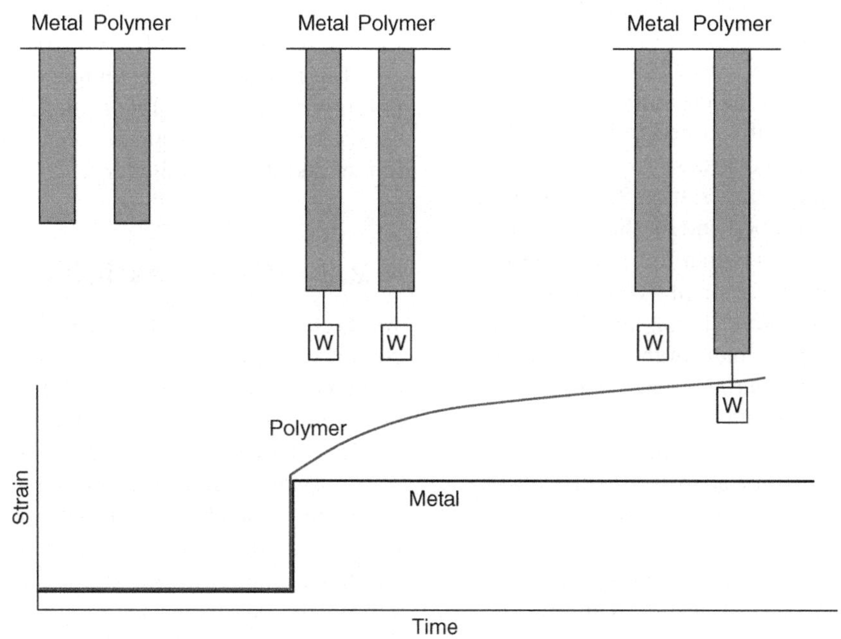

Figure 4.8 Creep behavior of a polymer and a metal. Note continuous extension of polymer over time. Source: From Nichols and Hill (2010). Reproduced with permission of American Coatings Association.

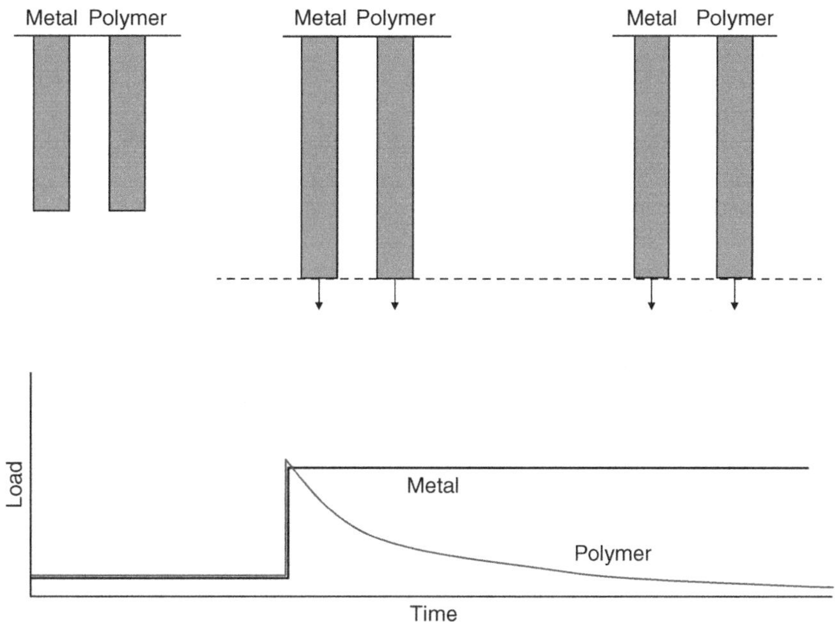

Figure 4.9 Stress relaxation of a metal and polymer. Note the decrease in stress as time progresses for the polymer.

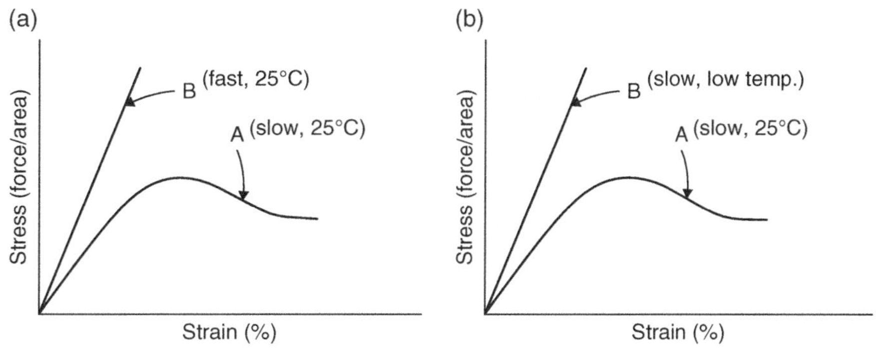

Figure 4.10 Effects of (a) rate of application of stress and (b) temperature on stress–strain response.

understanding their viscoelastic behavior. However, a more practical repercussion of viscoelasticity is the temperature-dependent nature of a coating's mechanical properties. As the temperature increases, the polymer molecules can reptate more rapidly, as the additional thermal energy increases the vibrational motion of the molecules.

The strong temperature dependence of the mechanical behavior is pervasive in all coating properties. This effect can be easily observed in simple stress–strain tests, as shown in Figure 4.10. Here, a number of changes can be observed as the testing temperature is increased. First, the initial slope of the curve decreases as the temperature increases. Thus, the coating becomes more compliant (less stiff) as the test temperature is increased. Second, the strain to break increases as the temperature increases. The higher temperature allows the polymer molecules to rearrange into more preferable conformations as the temperature increases, which allows bulk polymer to support larger strains. Finally, some coatings may show a pronounced yield point (local maximum stress) as the temperature increases. This local yielding is poorly understood from a theoretical standpoint but is accompanied by localized plastic (non-recoverable) strain and macroscopic necking (thinning) of the test specimen.

The final manifestation of viscoelasticity that needs to be addressed is the correspondence between temperature and deformation rate. As explained previously, a finite time is required for polymer molecules to rearrange and respond to an applied load. However, given enough time, they will eventually move to accommodate the load. One could imagine that, if the deformation was applied slowly enough, the polymer would have the ability to respond on the same time scale as the load was applied. Thus, there would be little stress response to the applied deformation. Conversely, if a deformation is applied rapidly, the polymer molecules would have even less time to reconfigure themselves and the stress response would be larger. This is exactly the response that is observed when we change the temperature of a polymer. At high temperatures, the material can respond quickly to a displacement. The stresses reached are not particularly high, and the elongation at break can be large as the polymer molecules can flow past one another. The response is viscous and fluid-like. At low temperatures, the material displays a strong elastic response.

The correspondence between rate and temperature is often referred to as time–temperature superposition, as it allows one to predict the response of a polymer to different temperatures by characterizing the polymer's response to different deformation rates and vice versa. This is extremely useful, as the characterization of materials at very high deformation rates is experimentally challenging, while at very low deformation rates it is not practical owing to the time required to collect data. An example of time–temperature superposition is shown in Figure 4.11, where a series of stress relaxation curves generated at different temperatures are combined together to form a complete stress relaxation curve that covers many orders of magnitude of time. To do so, the individual curves are shifted along the horizontal axis until they line up. This shifting can be described mathematically via a variety of theories. However, the most commonly applied theory is that of Williams, Landel, and Ferry and is termed the WLF theory. This approach attempts to predict the horizontal shift factors, a_T, used to superpose data from one temperature to another. Typically, the curves are shifted with respect to T_g, but other temperatures can be chosen as well. Formally, the WLF equation can be written as

$$\log\left(a_T\right) = \frac{-C_1\left(T - T_g\right)}{C_2 + T - T_g} \tag{4.7}$$

where C_1 and C_2 are constants that were originally believed to be universal, but have subsequently been shown to vary over a small range for most polymers, and T is the temperature at which an individual data set is acquired.

4.2.3 Dynamic Mechanical Behavior

The testing and properties outlined previously are typically termed "static" mechanical properties, which signifies that the properties are measured under constant or monotonically increasing stress (or strain), such as in a simple tensile test. However, much can be learned about a polymer or coating by subjecting it to a dynamic deformation. This is most commonly accomplished by applying a time-varying sinusoidal strain to a sample and measuring its time-dependent stress response. This type of testing is termed dynamic mechanical analysis (DMA) or dynamic mechanical thermal analysis (DMTA).

The stress and strain signals for a typical DMA test are shown in Figure 4.12. Here the strain is dynamically applied and the resultant stress response is measured. The time lag, δ, between the applied strain and the resultant stress is also measured and is related to the viscoelastic properties of the coating. This time lag is often termed the *phase lag*. As the deformations are quite small in DMA testing and only the recoverable stress–strain response is

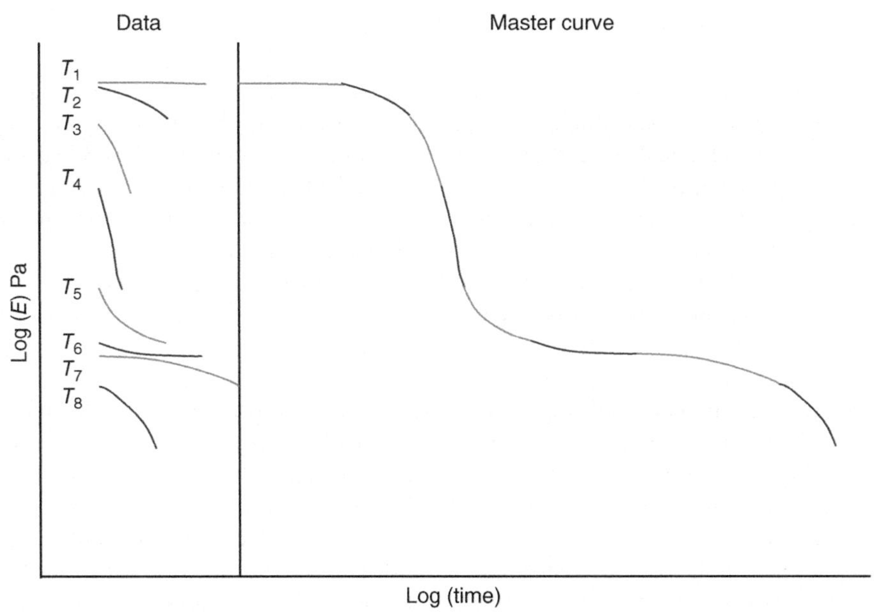

Figure 4.11 Master curve generation via the use of time–temperature superposition of creep curves taken at varying temperatures and shifted along the time axis. Source: From Nichols and Hill (2010). Reproduced with permission of American Coatings Association.

probed (no plastic deformation), the most frequently quantified property is the modulus of the coating. Because there are both elastic and viscous components to the modulus, it is broken down into two components, the storage modulus, E', and the loss modulus, E''. These are related to each other and δ by the following equations:

$$\tan\delta = \frac{E''}{E'} \tag{4.8}$$

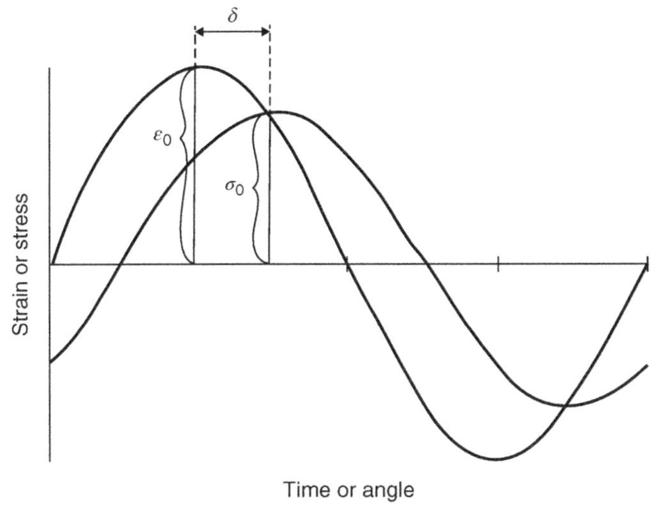

Storage modulus = $E' = \dfrac{\sigma_0 \cos\delta}{\varepsilon_0}$

Loss modulus = $E'' = \dfrac{\sigma_0 \sin\delta}{\varepsilon_0}$

Loss tangent = $\dfrac{E''}{E'}$ = $\tan\delta$

Figure 4.12 Stress response for a viscoelastic material tested with a dynamic sinusoidal strain input, such as in a DMA experiment. Source: From Nichols and Hill (2010). Reproduced with permission of American Coatings Association.

Since the strain is typically applied sinusoidally in DMA testing, the phase lag is often quantified by the angular difference between the applied stress and applied load. When the phase lag is zero, the loss modulus is zero and the material behaves in a perfectly elastic manner. When the phase lag is 90°, the material behaves in a completely viscous manner and the storage modulus is zero. Except when the temperature is near T_g, the storage modulus is typically much larger than the loss modulus. Depending on the specific instrument used and the geometry employed, DMA testing can be done in either tensile or shear modes, and the corresponding shear modulus data, G' and G'', can be collected.

In a typical DMA experiment, a sample is subjected to deformation at a fixed frequency, 2 Hz for example, and the temperature is slowly increased while the viscoelastic response is measured. The output of such an experiment is shown in Figure 4.13. Note the similarity of the E' curve to the stress relaxation curve shown in Figure 4.11. Both show a similar transition from stiff, rigid behavior to flexible, compliant behavior as the temperature is increased through T_g. In fact, DMA testing is often used as the most conclusive means for measuring T_g, with T_g often being defined as the maximum in the tan δ curve. Other authors prefer to define T_g as the peak of the loss modulus plot. As can be seen, there is a substantial difference between the two. The peak of the loss modulus plot is usually nearer to

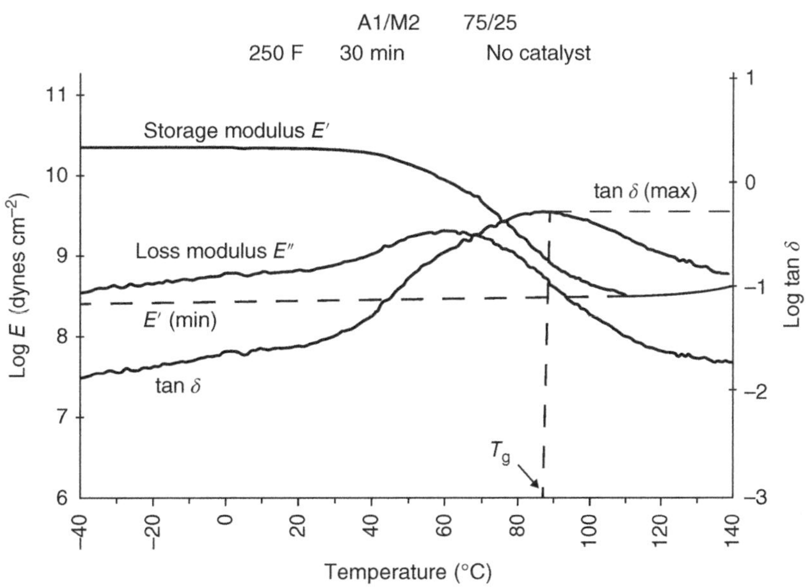

Figure 4.13 Storage modulus and tan δ for a polyester–melamine clear coat. The maximum in tan δ is taken as T_g. The minimum in the E' can be used for cross-link density calculations. Source: From Nichols and Hill (2010). Reproduced with permission of American Coatings Association.

the T_g determined using DSC. As previously noted, T_g is dependent on the rate of heating in DSC (Section 2.1.2) and on the rate of application of stress—the frequency of oscillation—as well as the rate of heating in a dynamic test. This dependence on experimental technique might lead one to wonder if T_g means anything. It does, but one must always be careful to compare T_g values determined in the same way. Alternatively, DMA experiments can be run at a fixed temperature and a frequency sweep can be made. This is the most common means used to generate data for the master curves shown in Figure 4.11.

Much useful information can be extracted from DMA data in addition to the simple temperature/frequency dependence of the modulus. The modulus in the rubbery plateau region (denoted E'_{min} in Figure 4.13) of the curve (well above T_g) can be used to estimate the cross-link density (XLD) of the coating. For rubbery materials, the XLD can be shown to be related to the shear modulus in Eq. 4.9, where R is the gas constant, T is the absolute temperature, and ν_e is the number of moles of *elastically effective network chains* per volume in the film (Hill, 1997):

$$\nu_e = \frac{G'}{RT} \quad \text{or} \quad \nu_e = \frac{E'}{3RT} \tag{4.9}$$

An elastically effective network chain is one that is connected at both ends to the network at different junction points—short cyclical chains and dangling ends are not elastically effective. As the XLD is closely related to many coating properties, its measurement can be highly informative to the formulator. Care must be taken during these measurements, however, as pigments in the coating will inflate the value of the modulus, particularly above T_g. Thus these measurements are best made on unpigmented films. Non-cross-linked coatings do not typically show much of a plateau region as the molecules in a thermoplastic are free to flow until all stress is relieved. Thus, the notion of XLD is not applicable to these polymers.

Note that ν_e divided by the film density provides the moles of network chains per mass. The inverse of ν_e, grams per mole of network chains, corresponds to the average MW of network chains, frequently called average MW between cross-links $\bar{M}_c$. Commonly, $\bar{M}_c$ is erroneously defined and used as the MW per branch point; therefore one must be careful when reading any paper dealing with MW of network chains and XLD. XLD can also be calculated and correlated with measurements by swelling. Correlation is good, although it is necessary to assume an interaction parameter from the extent of swelling of a film by solvent. While cross-linked films do not dissolve in solvent, solvent does dissolve into a cross-linked film. As cross-links get closer together—that is, as XLD increases—the extent of swelling decreases.

In addition to the XLD, the uniformity of cross-linking can be assessed via the width of the tan δ peak. A narrow XLD distribution is reflected in a narrow tan δ peak width, while a broad distribution of XLD in a sample is reflected in a wider tan δ peak. In the extreme case where there may be two phases present in a polymer, two tan δ peaks may be observed, which are indicative of the T_g of each individual phase. Thus, DMA can provide clues as to the morphology of a polymer as well as its viscoelastic performance. Finally, while the primary peaks observed in DMA data are due to the glass transition temperatures, other lower temperature peaks are often observed as well. These low temperature peaks, often termed secondary relaxations, can arise from specific subsegment motions in the polymer that become active at certain temperatures. These relaxations are often labeled as the β, γ, δ relaxations as one moves farther below the glass transition temperature (the α relaxation). In plastics, it is well established that tough impact-resistant materials generally have low temperature loss peaks; polymers made from bisphenol A (epoxies and polycarbonates) are common examples. It is reasonable to speculate that coatings with low temperature loss peaks may have good impact resistance (if adhesion is good), but the relationship is not well documented in the literature.

XLD can also be determined by nuclear magnetic resonance (NMR) imaging. This technique has the advantage that one can determine XLD as a function of depth through a film (Hellgren et al., 2001). For example, as expected, the top surface of air dry films of alkyds with only cobalt driers could be seen to have cross-linked selectively. Also, the poor cure of the surface of UV curing latex films could be seen to be air inhibited.

Equation 4.9 can also be used to predict the storage modulus above T_g from the XLD. In a system with stoichiometric amounts of two reactants whose functional groups react completely, one can estimate the XLD from the equivalent weights and the average functionality. If the reactant mixture contains molecules of several different functionalities, calculation becomes more difficult. A more general approach is provided by the Scanlan equation, Eq. 4.10 (Hill, 1992):

$$\nu_e = \frac{3}{2}C_3 + \frac{4}{2}C_4 + \frac{5}{2}C_5 \cdots \tag{4.10}$$

The C values are the concentrations of reactants with functionality of 3–5 (or more), expressed in units of moles per cubic centimeter of final cured film. The volume of the final film depends on density of the cured film and loss of volatile by-products of the reaction. Equation 4.10 does not include a term for difunctional reactants because these reactants do not create junction points in a network; they only extend chains. Recent modifications permit consideration of incomplete conversion (Hill, 1997). Although the Scanlan equation is convenient for stoichiometric reactions, it does not apply to other cases. For nonstoichiometric mixtures and/or incomplete conversion, Miller–Macosko equations are useful general equations. Bauer (1988) selected the Miller–Macosko equations most useful for coatings, gave examples of their applications, and provided a computer program.

Properties are affected by the extent to which cross-linking has proceeded to completion. Incomplete reaction leads to lower XLD and, hence, lower storage modulus above T_g. The extent of reaction can be followed by determining storage modulus as a function of reaction time (Skrovanek, 1990). As cross-linking continues, storage modulus increases until a terminal value is reached.

Thus, one can, at least in theory, design a cross-linked network to have a desired storage modulus above T_g by selecting an appropriate ratio of reactants of appropriate functionality. By proper selection of the structures between cross-links and XLD, one can design the T_g of the cross-linked network.

Intermolecular hydrogen bonding can provide desirable film properties associated with covalent cross-linking. For example, the strong hydrogen bonds between urethane groups lead to high abrasion resistance in urethane coatings (Chapter 12). When subjected to stress, the hydrogen bonds dissociate and relieve the stress without breaking chemical bonds. When the stress is released, hydrogen bonds reform, restoring the original properties.

A related phenomenon results from the presence of mesogenic groups on resins, which form liquid crystal domains when the film is cured. When stress is applied the crystal structure can be disrupted, and, when the stress is relieved, crystal domains can reform. The initial examples of such a coating system were prepared by cross-linking a mesogenic diol with hexamethoxymethylmelamine (HMMM) (Dimian and Jones, 1987). A variety of resin classes were investigated including epoxy/amine, polyester, alkyd, and acrylic resins having side chains from *p*-hydroxybenzoic acid. Non-cross-linked acrylic lacquers, class I melamine-formaldehyde (MF) cross-linked thermosetting coatings, and waterborne coatings, which contained mesogenic groups, were also investigated. Appropriate levels of mesogenic groups provided both high impact resistance and film hardness, which are generally difficult to achieve in combination (Dimian and Jones, 1987).

Yoshida et al. (2005) investigated MF resins in which HMMM is reacted with a mesogen such as the hydroxytriphenyl ether of 1,6-hexanediol to replace part of the methoxy groups of the HMMM. The resins, which showed liquid crystal domains, were tested as coil coating vehicles. They provided both high flexibility and hardness. A UV cure coating formulated with an acrylic ester of a mesogenic alcohol also provided both hardness and flexibility (Yoshida and Kakuchi, 2005). The high cost of these coatings has limited their commercialization.

Understanding relationships between composition and basic mechanical properties of films can provide a basis for more intelligent formulation. Most coating formulators were educated as chemists, not as engineers, and few have had formal education on mechanical properties.

In many cases, a coated metal object is subjected to mechanical forces either to make a product, as in forming bottle caps or metal siding, or in use, as when a piece of gravel strikes the surface of a car with sufficient force to deform the steel substrate. To avoid film cracking during such distensions, the elongation at break must be greater than the extension of the film under the conditions of fabrication or distortion.

The use of DMA has proven to be a valuable tool for studying cross-linking of hydroxy-functional resins with MF resins and their subsequent performance. As the stoichiometric ratio of methoxymethyl groups from the MF cross-linker to hydroxyl groups of the resin was raised from values less than 1, the storage modulus above T_g of the films increased up to the point that the ratio became one, corresponding to complete cure. As discussed in Section 11.3.2, these results show that all of the functional groups on the MF resin can react with hydroxyl groups and that the reaction is not limited by steric hindrance, as previously thought. When the amount of MF cross-linker was further increased so that excess methoxymethyl groups were present, storage modulus above T_g increased at higher temperatures during dynamic mechanical testing. This behavior has been explained on the basis that the excess methoxymethyl groups can undergo self-condensation reactions during the testing. The self-condensation reaction is relatively slow and was incomplete during the baking cycle used in preparing the film; hence, the reaction continued at the higher temperatures used in DMA, leading to the higher storage modulus. Self-condensation during baking of coatings also occurs when excess MF resin is used; then, the extent of self-condensation increases as baking time and temperature increase.

Pigmentation affects the mechanical properties of films. In many cases, as the pigment volume concentration (PVC) of films is increased up to the critical PVC, the tensile strength of the films increases (Section 22.2.1). A DMA study of aluminum pigmented base coats showed that the storage modulus (E') increased tenfold as the pigmentation increased in five steps from 0 to 0.13 PVC (Schlesing et al., 2004). It is also possible that imperfections resulting from some types of pigmentation may lead to crack propagation. However, in most cases, the storage modulus above T_g is strongly increased by the presence of pigment. For pigmented coatings, one cannot expect a direct proportional relationship between storage modulus (E') above T_g and XLD; however, if the pigment content is constant in a series of films varying in the polymeric portion, the relative values of E' should still be indicative of relative XLDs for the pigmented films. In general, pigmentation lowers, broadens, and shifts tan delta to higher temperature and increases

E' up to critical pigment volume concentration (CPVC) above which it decreases. Tensile strength also generally increases with PVC and then drops sharply above CPVC. However, if there is poor bonding between the pigment particles and the resin, as is the case with $CaCO_3$ in an acrylic binder, the tensile strength drops steadily as the PVC increases (Perera, 2004). See Section 22.2.1 for further discussion.

Nano-pigments (Section 20.5) can have substantial effects on mechanical properties. Nano-pigments are those having small particle size, defined by some authors as less than 100 nm; but more commonly the particle size is below 25 nm. The term is new but nano-pigments have been used for many years in coatings. For example, high strength channel blacks have particle sizes in the range of 5–15 nm. A review paper on nano-material technology applications in coatings is available (Baer et al., 2003).

The enhancement of properties caused by the addition of nano-sized pigments is mostly related to their high surface area. Researchers have postulated that immobilization of the polymer binder near the surface of the pigments gives rise to increased strength and stiffness, which are manifested in improved scratch resistance, permeability, and corrosion resistance, as well as other properties. In some cases, the films have two T_gs, one for the bulk binder and one for the binder immobilized near the pigment surface. Nano-pigments have been shown to increase E' and in some cases T_g of clear coat films (Schlesing et al., 2004). When the particle size of the nano-pigments is smaller than the wavelength of light, coatings become transparent (Perera, 2004). Mar-resistant clear coats are prepared using ball mills to disperse nano(aluminum silicate) in an acrylic resin with a commercial dispersing agent (Vanier et al., 2005).

The effect of heterogeneity in coating films has been increasingly recognized as an important factor with regard to mechanical properties (Kivilevich, 2004). Most obviously, as PVC of a film increases, the elastic modulus increases owing to the reinforcing effect of pigments. At the same PVC, the modulus of a coating will increase more for coatings with nonspherical pigments compared with spherical ones. This is due to the improved stress transfer from the binder to the pigment when the pigment has an aspect ratio greater than 1. Thus, the reinforcing effect is greatest for fibrous pigments followed by platy and then spherical pigments. If the pigment is increased above the CPVC, the tensile strength drops precipitously (Chapter 22).

One can significantly affect a coating's mechanical performance by incorporating elastomeric particles in a coating. The use of a combination of low T_g and high T_g components gives films having higher blocking resistance and longer elongation.

Mechanical properties of coatings are generally more complex than those of most plastics. One reason is that coatings are used as thin films on a substrate. Interactions with the substrate affect the mechanical properties of thin films. The substrate can limit the extent of deformation that occurs. During impact the substrate can act as an energy sink to dissipate the energy so that there is less effect on the coating film. Adhesion can have a profound effect on ability to withstand fabrication. If adhesion is good, fabrication and impact resistance of the films are almost always superior. Stress within the film can act to pull the film off the substrate. Coatings have been known to pop off post-formed metal bottle caps when a jar is sitting on a supermarket shelf. Stress within films can also arise during the last stages of solvent loss and/or cross-linking of films (Perera and Schutyser, 1994). Solvent loss and cross-linking result in shrinkage. If this shrinkage occurs when the temperature is near the T_g of the film, the resulting internal stresses may persist indefinitely. (See Section 6.2 for discussion of the effects of internal stresses on adhesion.)

Film thickness is also a factor in the ability of a coating film to withstand fabrication without cracking. Thin films can be used for deeper draws than thick films (see Section 4.3). In making coated exterior siding, the hardness of the film can be increased without encountering cracking by limiting the film thickness. Of course, thinner films of pigmented coatings provide poorer hiding; a common compromise in this case is in the range of 20–25 µm film thickness. Two-piece fish cans are coated as flat sheets with a relatively highly cross-linked phenolic coating that is quite brittle in order to minimize swelling with fish oil. Such cans can only be successfully formed without cracking only if the film thickness is 5 µm or less.

For pre-coated substrates subjected to deformation, an important variable can be the timing of fabrication or flexing after curing a coating. It is common for coatings to become less flexible as time goes on. Particularly in air dry coatings, some solvent may be retained in films. Since most coil coatings have T_g values near or slightly above room temperature, solvent loss may be very slow (Section 18.3.4). Solvents generally act as plasticizers, so as solvent is lost, T_g and storage modulus increase resulting in less flexible coatings. In cross-linkable coatings, if the cross-linking reaction was not complete, the reaction may continue slowly, increasing XLD and storage modulus and decreasing flexibility. Continued cross-linking is particularly likely to occur in air dry coatings, since reaction rates are likely to become mobility rate controlled, and, hence, the last part of the reaction is slow. Reactions during the use life of the coating, especially during exterior exposure (Section 5.1), can result in embrittlement. DMA of a variety of clear coats after Florida exposure and QUV testing show the changes in mechanical properties resulting from exposure (Hill et al., 1994). Effects of exposure on mechanical properties will be further discussed in Chapter 5.

Hardening of baked cross-linked coatings over time is also commonly observed. Although in some cases further volatile loss or continued cross-linking may be responsible, another possible factor is *densification*. If a polymer is heated above its T_g and then cooled rapidly (quenched), the density is commonly found to be lower than if the sample had been cooled slowly. As glassy polymers are not at thermodynamic equilibrium, the polymer will slowly proceed toward equilibrium over time. Rapid cooling "freezes" the structure of the polymer at a state farther from equilibrium than slow cooling, and thus the density is reduced compared with the ideal equilibrium state. On storage, the molecules in quenched films slowly move causing densification. Since these changes in properties with aging are not chemical, the process is called *physical aging*. Physical aging leads to embrittlement in most polymers. Thus, cracking is more likely during fabrication when the material has been aged for a significant time below T_g. This phenomenon has been widely observed in plastics and is increasingly being recognized in coatings. It may occur when coatings are baked on metal at high temperatures and then cooled rapidly after emerging from the oven. It has been suggested that densification may be a common cause of embrittlement during aging of baked coating films. Greidanus (1988) has studied physical aging at 30°C of polyester/MF films that had been baked at 180°C and then quenched to 30°C. There was a small, but reproducible, increase in modulus with time at 30°C. The aging rate (i.e., the rate of increase of modulus) decreased with time. If the sample was heated again at 180°C and again quenched to 30°C, the modulus returned to its lower value and underwent physical aging again. Perera (2003) has shown that relative humidity (RH) can also affect physical aging; samples stored at high humidity age more quickly that those stored at 0% RH because of the plasticizing effect of water relaxing the polymer so that it can rearrange to its "normal state." The previous reference also gives an extensive review of the field of physical aging. Further work is needed, but it is evident that physical aging can be an important phenomenon.

When coatings are baked industrially, further complications arise. Oven temperatures can vary not only the air temperature in the oven as a whole but also within the oven. The rate of heating of a coating can depend on the substrate thickness. For example, the temperature of the coating on the sheet metal roof of a car increases faster than the temperature of the coating on joints, where the metal thickness is greater. To achieve the desired properties, some minimum time at a temperature is required, but overbaking can lead to excessive cross-linking. There is a *cure window* for any baked coating; within this range of time and temperature, satisfactory properties are obtained. As discussed in Section 11.3, high solids acrylic melamine coatings have narrower cure windows than their conventional solids predecessors. The effects of some of the variables in thermal history have been modeled (Dickie et al., 1997).

4.3 FRACTURE MECHANICS

Many of the mechanical failures that coatings experience in the field are due to the propagation of cracks, either within the coating or at an interface between the coating and substrate. The propagation of cracks can be quantified by applying the principles of fracture mechanics. This discipline is used to describe the stress state at or near a crack tip and can be applied to calculate the stresses and strains in a material (or at an interface) that will cause cracks to grow (propagate). The propagation of cracks can be quite common in coatings. In particular, after outdoor exposure, many coatings can embrittle or delaminate from the substrate. Examples of these failures would be the cracking of an automotive clear coat after weathering or the cracking of an architectural paint on a wood substrate after many years of outdoor exposure.

The stresses that drive crack propagation are mainly due to the constrained nature of coatings. Stresses develop in a coating on metal because organic coatings typically expand more than metal substrates when the temperature is raised. On wood, variations in humidity cause uneven expansion and contraction of the wood. In addition, densification due to physical aging or continued cure can lead to greater residual stresses in the coating film, which can promote cracking. In the presence of a preexisting flaw, these stresses are often large enough to cause cracks to propagate. Some work suggests that use of expanding monomers during polymerization may offset some of the stress buildup during weathering and may be a fruitful strategy to employ to reduce long-term cracking (Bailey, 1973).

The failure of clear coats on automobiles has been studied extensively using the fracture mechanics approach. Such failures are especially encountered after weathering, which induces stresses in the coating and also embrittles the coating. The embrittlement is largely driven by UV radiation-induced photooxidation of the coating. This process can be mitigated by the use of UV stabilizers (see Chapter 5), but it can never be completely eliminated. Automotive clear coats typically possess high cross-link densities to combat both scratches and to provide solvent resistance. However, high cross-link densities are generally detrimental to cracking resistance, as little plastic deformation can take place in such materials and residual stresses tend to be large due to the high extent of reaction.

The *toughness* (opposite of brittleness) of any material can be quantified by measuring its *fracture energy*, G_c, which is the energy required to propagate a crack in a material. For bulk materials, many geometries and techniques exist to measure this material property. For coatings, however, there are significant challenges to assessing the toughness of a coating either as a free film or when it is attached to a substrate. Note that some coatings scientists refer to the toughness of a material as the area under the stress–strain curve, which is common in the textile or fiber science field as well. Others refer to toughness without specifying exactly what the term means. While the area under the stress–strain curve may be germane to some coating properties, it is not the same as the toughness term that is used to characterize the fracture resistance.

As the measurement of any coating property is preferably measured when the coating is adhering to the intended substrate, a method has been developed that quantifies the fracture energy of relatively brittle coatings when they are strongly adhering to their substrate or to another coating layer. By measuring the strain at which cracking occurs during a tensile test on the substrate/coating coupon, the fracture energy (G_c) of the coating can be calculated by Eq. 4.11:

$$G_c = \frac{\varepsilon^2 h E_f \pi g(\alpha,\beta)}{2} \tag{4.11}$$

in which ε is the strain at which cracking occurs, E_f is the modulus of the coating divided by $(1-\nu^2)$, $g(\alpha,\beta)$ is a parameter related to the mismatch in modulus between the coating and substrate, and h is the coating thickness (Nichols et al., 1998). For organic coatings on metallic substrates, $g(\alpha,\beta)$ is approximately 0.72. Experimentally, it has been shown that the fracture energy of many coatings decreases as weathering progresses, corresponding to embrittlement over time. During service, the driving force for cracking arises from the stresses detailed previously. The driving force (G) can be calculated via Eq. 4.12:

$$G = \frac{Z\sigma^2 h}{E_f} \tag{4.12}$$

where Z is a parameter related to the geometry of cracking (Z=3.951 for film cracking) and σ is the stress in the film (Hutchinson and Suo, 1991). When $G > G_c$, cracking will occur. These two relationships allow one to calculate failure envelopes for stress levels and coating thicknesses, as both the driving force for cracking and the propensity to crack are directly proportional to the thickness of the coating.

While this approach works well for relatively brittle films, when significant plastic deformation accompanies the propagation of a crack, this analysis is not valid for more ductile coatings. For those materials, such as for soft or elastomeric coatings, another method has been established, which requires the preparation of free films. A tensile test is performed on a free film having a preexisting crack of known size. The energy required to drive the crack forward can be separated from the energy used to plastically deform the coating. This technique is referred to as the method of essential work. The total work is

$$w_{\text{total}} = w_e l t + b w_p l^2 t \tag{4.13}$$

where w_e is the essential work of fracture, l is the ligament length (the uncracked length of the film), t is the film thickness, β is a shape factor, and w_p is the plastic work (Ryntz et al., 2000). By plotting the total work of fracture versus the ligament length for a number of different specimens, the essential work of fracture can be calculated as the y-intercept of the graph. This method has been shown to work well for low bake coatings used to coat thermoplastic olefin (TPO) fascias on automobiles (Ryntz and Britz, 2002).

4.4 ABRASION, SCRATCH, AND MAR RESISTANCE

Much confusion can arise when discussing the abrasion, scratch, and mar resistance of coatings. Abrasion is the wearing away of a surface, where significant material removal occurs over a relatively large surface area. Scratches are more localized phenomena where a thin but relatively deep defect occurs when another sharp object is translated over the surface of a coating while a normal force is applied. The scratching of a car's coating with a key is a good example. Marring results from surface deformations where the coating is locally plastically deformed, but where the coating does not fracture or become removed from the surface. All three phenomena fall within the field of *tribology*, the science of surfaces in sliding contact. Terminology is not standardized, and terms such as *buffing*, *gouging*, and *wearing* are also used with meanings that sometimes overlap *abrasion*, *scratching*, and *marring*.

4.4.1 Abrasion Resistance

One might suppose that hard materials are less likely to fail by abrasion than soft materials. In some cases, this supposition is true, but in many other cases, softer materials are more abrasion resistant; for example, rubber tires resist abrasion far better than steel tires.

Evans (1969) studied the mechanical properties of a series of floor coatings having known actual wear life. He determined tensile strength at break, elongation at break, and work to break. His data are given in Table 4.2, in which the coatings are listed in order of increasing wear life. One might suppose that higher tensile strength would provide higher abrasion resistance; the data show the reverse. (It should not be assumed from this limited data that abrasion

Table 4.2 Mechanical Properties of Floor Coatings

Floor coating	Tensile strength (psi)	Elongation at break (%)	Work to break (in.-lb in^{-3})	Taber[a] (rev mil^{-1})
Hard epoxy	9000	8	380	48×10^3
Medium epoxy	4700	19	600	33×10^3
Soft epoxy	1100	95	800	23×10^3
Urethane elastomer	280	480	2000	36×10^3

[a] See Section 4.6.3.4 for a discussion of these results.

resistance is always inversely related to tensile strength.) Elongation-at-break values properly ranked the wear performance of the coatings. Evans concluded that work to break, which he defined as the area under the stress–strain curve, best represented the relative wear lives.

However, in studies on another series of coatings, Evans and Fogel (1977) determined that work to break did not always correlate with abrasion resistance determined by loss of gloss in a ball mill abrasion tester when the stress–strain tests were carried out at ambient temperatures. They reasoned that the strain rate of their instrument was too low relative to the stress application in the test. Using a time–temperature superposition relationship, they calculated that, at an accessible strain rate, the tests should be carried out at −10°C in order to achieve the desired high rate of stress application that was inaccessible by the instrumentation at ambient temperature. The resulting work-to-break values did correlate with abrasion resistance for urethane films having a T_g equal to or greater than −10°C.

Urethane coatings often exhibit superior abrasion resistance together with good solvent resistance. This combination of properties may result from the presence of intersegment hydrogen bonds in addition to the covalent bonds. At low levels of stress, hydrogen bonds act like cross-links, reducing swelling on exposure to solvent. At higher levels of stress, the hydrogen bonds can dissociate, permitting the molecules to extend without rupturing covalent bonds. When the stress is released, the molecules relax and new hydrogen bonds form. Urethanes are used as wear layers for flooring, as well as top coats in aerospace applications, where this combination of properties is desirable.

Factors in addition to work to break are involved in abrasion resistance. The coefficient of friction of the coating can be an important variable. For example, abrasion of the coating on the exterior of beer cans during shipment can be minimized by incorporating a small amount of incompatible wax or fluorosurfactant in the coating. When the two coated surfaces rub against each other, the incompatible additive reduces surface tension and the coefficient of friction, so that transmission of shear force from one surface to the other is minimized and abrasion is reduced.

Another variable is surface contact area. Incorporation of a small amount of a small particle size SiO_2 pigment in a thin silicone coating applied to plastic eyeglass lenses reduces abrasion, adding to the effect of the low surface tension of the silicone surface. The pigment particles reduce contact area, permitting the glasses to slide more easily over a surface. Another example of the same principle is the incorporation of a small amount of coarse SiO_2 inert pigment in wall paints to reduce *burnishing*. If a wall paint without such a pigment is frequently rubbed, as around a light switch, it abrades to a smoother glossier surface—that is, it burnishes. The coarse inert pigment reduces burnishing by reducing contact area.

An approach that has been used for many years in resin-bonded pigment print colors on textiles is to incorporate rubber latex in the print paste. The latex particles are not soluble in the resin and remain as individual particles in the resin together with the pigment particles. The abrasion resistance is markedly improved by the latex addition. Similar work is now being done to improve the abrasion resistance of continuous coatings. Presumably, the relatively soft rubber particles act to dissipate stresses on the film, minimizing the occurrence of film rupture by reducing stress concentration. Glass microspheres have been shown to increase abrasion resistance of epoxy coatings by damping the energy released by impacts on the surface (Kotnarowska, 1997). Lee (1984) reviewed abrasion resistance as one type of wear in a broad approach to fracture and surface energetics of polymer wear.

4.4.2 Scratch and Mar Resistance

The difference between scratches and mars is not definitive. One person's scratches may be another person's mars. Both failures are significant problems with automotive coatings and are one of the chief complaints of vehicle owners. Scratching is typically associated with the fracture of the coating in the area of the defect, while mars are associated with permanent plastic deformation but no fracture. Topographic views of the two defects are given in Figure 4.14. When the normal forces are low, the plastic deformation merely moves the material from the area of

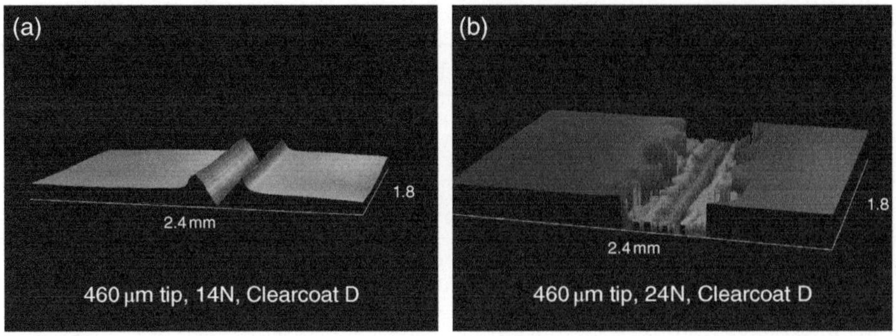

Figure 4.14 Non-contacting profilometer view of (a) mar made by a 14N normal force and (b) fracture scratch made by a 24N normal force in an automotive clear coat.

the furrow to the shoulder area. This is enough to cause local light scattering. As these defects are small, one often needs the lighting and observation angles to be just right in order to observe the defects. When the normal forces are higher, the material fractures alongside and within the scratch. Material removal often follows, resulting from the loss of coating adhesion to the underlying surface.

The physics of scratching and marring is complex. Different authors use different terms to describe the phenomena involved. The influence of viscoelasticity is strong. Different coatings vary widely in their responses; and the same coating may respond quite differently as the force or rate of marring stress changes. A further complication is that mars in some coatings can slowly heal by reflow (creep) (Seubert et al., 2012).

Efforts are underway to relate mar resistance to the chemical structure of coatings, but relatively few systematic studies have been published. In general, MF cross-linked acrylic clear coats are more resistant to marring than isocyanate cross-linked coatings (urethanes). However, MF cross-linked coatings have poorer environmental etch resistance, which is a concern to automotive manufacturers and consumers. MF cross-linked polycarbamates are an exception, combining etch and mar resistance (Section 11.3.4). Since urethanes generally have superior abrasion resistance, it is surprising that they have inferior mar resistance; possibly, this might be explained by differences between surface and bulk properties. A study of marring of clear coats by a scanning probe microscope indicated that an acrylic polyurethane had a thin layer of deformable plastic material on its surface, while an acrylic/MF clear coat had a layer of elastic material (Jones et al., 1998). Osterhold and Wagner (2002) review mar test methods and compare them by subjective estimations of performance. Methods include two simulated car wash tests, a crockmeter test originally developed for evaluating abrasion of colored textiles, micro-indentation hardness tests, scrub resistance tests, a Rota–Hub test, a collate sanding machine test, micro-scratch resistance, a nano-scratch tester, and a DMA. Yaneff et al. (2001) compare a variety of scratch tests on cross-linked trimethoxypropylsilane-functional clear coats for plastics. New instruments, *nanomechanical analyzers* (nano-indenters), that can characterize near-surface mechanical properties are now available and show some promise in improving our understanding of scratch and mar behavior (Lin et al., 2000; Schulz et al., 2001). Some workers have observed that the size of the asperity that causes a scratch is a factor. Accordingly, it is not clear that nano-sized scratches correctly mimic the behavior of real-world scratches made with larger asperities (Seubert and Nichols, 2007).

The improvement of scratch and mar performance of coatings is difficult. One can, in principle, improve the mar resistance (plastic deformation resistance) of a coating by increasing its yield strength. This can be done through the incorporation of fillers (not practical for clear coatings) or by an increase in XLD. Nanosilica and other nanoparticles have been added to automotive clear coats to improve their scratch resistance and to aid in rheology control. The latter method can be very successful, as has been shown in numerous studies. However, if the XLD becomes too high, the coating embrittles and the resistance to fracture scratches diminishes. Thus, it is difficult to simultaneously mitigate both failure modes. Courter (1997) proposes that maximum mar resistance is obtained with coatings that possess as high a yield stress as possible without being brittle. The reasoning is that high yield stress minimizes plastic flow and, thereby, avoidance of brittleness minimizes fracture. Courter's paper provides a good review of attempts to relate bulk mechanical properties of coatings to their scratch and mar resistance, but these studies have not led to a broadly applicable theory.

Film flexibility is an important factor influencing fracture resistance. Accordingly, use of 4-hydroxybutyl acrylate instead of 2-hydroxyethyl acrylate in an acrylic resin cross-linked with MF resin gave improved scratch behavior, as did use of a polyol-modified hexamethylene diisocyanate isocyanurate instead of isophorone diisocyanate isocyanurate in cross-linked urethane coatings (Gregorovich and

Hazan, 1994). Further improvement was obtained using silicone-modified acrylic resins, presumably due to a decrease in the friction coefficient.

Hara et al. (2000) have reported a study on the scratch resistance of 49 different clear top coats, subjected to a simulated car wash test. High correlation between measurements and visual evaluation was obtained. Determination of dynamic mechanical properties showed that the best performance was associated with high storage modulus at a relaxation time of one second together with minimum strain to reduce plastic deformation of the films.

A further problem related to mar resistance is *metal marking*. When a metal edge is rubbed across a coating, a black line is sometimes left on the surface of the coating where metal has rubbed off. A common test for metal marking resistance is to draw a coin across the surface of a coating and observe if a dark streak is left. Coating lore indicates that a Canadian nickel is the best kind of coin to use. Metal marking usually occurs with relatively hard coatings. The problem can be reduced or eliminated by lowering the surface tension of the coating, thereby reducing the coefficient of friction; then, the metal slips over the surface. Additives can be incorporated to increase slip. Modified polysiloxanes have been reported to be particularly effective (Fink et al., 1990). Care must be exercised in selecting the particular grade of silicone additive and the amount of the additive used so as to minimize marring, scratching, and metal marking without causing other defects, such as crawling.

While some improvements in the scratch resistance of coatings are possible, completely removing the susceptibility of organic coatings to scratches is a fool's errand. Most of the materials that can potentially cause a scratch are inorganics such as dirt (oxides and silicates) or metals. A simple examination of the Mohs hardness scale (Table 4.3) indicates that, aside from talc, coatings will be scratched by the other materials if enough force is applied.

Table 4.3 Mohs Hardness Values for a Variety of Materials

Material	Hardness
Talc	1
Most polymers	1–2
Gypsum	2
Steel	4–5
Apatite	5
Glass	5.5
Quartz	7
Diamond	10

4.5 MEASUREMENT OF MECHANICAL PROPERTIES

Many of the test methods used to probe the mechanical properties of coatings require free films. Two major disadvantages of using free films are that (1) the interaction of the film with the substrate can have major effects on some film properties and (2) free films are sometimes difficult to prepare and handle. Test results are generally more reproducible with thick free films than with thin ones; however, the results with thick films may not be applicable to thin films. Preparation of thin unsupported films can be difficult. In some cases, it is possible to make a film by drawing down a coating on a release paper with a wire-wound bar. Release papers are coated with low surface tension materials to minimize adhesion; but if the surface tension of the release paper is lower than the surface tension of the coating being applied, there is the possibility of crawling; that is, the coating tries to minimize surface free energy by drawing up into a ball (see Section 24.4 for discussion of crawling). One tries to find a release paper with a low enough surface tension so that adhesion is poor but high enough so that crawling does not occur. A generally more effective method is to apply the coatings to tin-plated steel panels. After curing, one end of the panel is placed in a shallow pool of mercury. Mercury creeps under the coating, forming an amalgam with the tin, and the film releases from the panel. Mercury vapor is toxic and care must be taken to minimize the hazard. The safety regulations of some laboratories forbid such use of mercury. After the film is released from the substrate, a specimen is cut from it. Cutting free films may result in nicks or cracks along the edge of the film. When subjected to stress, cracked films commonly tear easily starting at the imperfection, leading to meaningless results. Handling films having T_g above room temperature is especially difficult; they tend to be brittle and easily broken. Gently heating the films before cutting with a die is often helpful.

One must be careful about changes that may occur during storage of the films prior to testing, such as loss of residual solvent, chemical changes, or physical aging. Test results will correspondingly change with time. Storage conditions can be critical. Most films absorb some water from the atmosphere. If the T_g is near room temperature and especially if the film has groups such as urethanes that hydrogen bond strongly with water, the T_g and film properties can be strongly affected by the humidity conditions in storage, since water acts as a plasticizer. Comparisons should be done with samples that have been stored at the same temperature and humidity. In actual use, films encounter a variety of humidity conditions and, hence, exhibit a range of properties.

Several types of instruments are available for determining mechanical properties. Electromechanical (screw-drive) load frames are used for tensile (non-dynamic) experiments. The free sample is mounted between two jaws of the tester; care must be taken to ensure that the film is in line with the direction of pull. The instrument can be operated utilizing a range of rates of jaw separation, but even the highest rates are slow compared with the rates of stress application found in many real situations. This problem can be partially overcome by running the tests at low temperatures. This method has the advantage that stress can be increased until the film fails, making it possible to determine the tensile strength, tensile modulus, elongation at break, and work to break. However, one cannot separate the viscous and elastic components of the mechanical properties.

A thermal mechanical analyzer (TMA) is a penetrometer that measures indentation versus time and temperature. An advantage over most tensile instruments is that a TMA includes a furnace and temperature programmer so that heating, cooling, and isothermal operations are possible. TMAs can be used with films on a substrate. TMA can be used to measure *softening point*, which is related to the extent of cure of cross-linking films. Figure 4.15 shows a plot of probe penetration as a function of temperature for an undercured and a well-cured 25 μm acrylic coil coating. The softening points for the two samples are marked on the graph. The softening point is related to, but not identical to, T_g and is frequently used as an index of flexibility (Škrovanik and Schöff, 1988).

Various dynamic mechanical analyzers (DMA) are also available. The most versatile are those in which the free sample is subjected to an oscillating strain by attachment under tension to a fixed clamp on one end and a vibrating clamp on the other. Oscillating stresses are imparted to the sample. A range of frequencies can be used, and properties can be determined over a wide range of temperatures. Typical instruments are set up in line with a computer that analyzes the data and provides storage and loss modulus, as well as tan delta data that can then be analyzed via accompanying software.

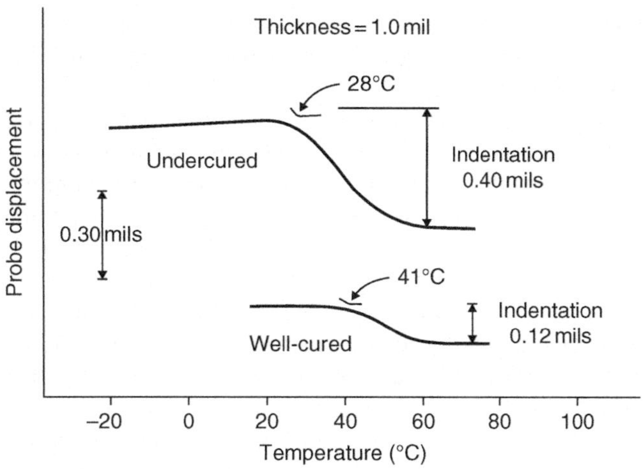

Figure 4.15 TMA plot of probe displacement against temperature for an undercured and well-cured acrylic coil coating. Source: From Skrovanik and Schöff (1988).

4.6 TESTS OF COATINGS ON SUBSTRATES

A variety of test methods has been established to characterize the properties of coatings. In general, these methods do not permit calculation of the basic mechanical properties, but rather provide information about some combination of properties of the coating. Hill (1998) refers to the relationship between fundamental coating properties and more practical coating properties as property–property relationships. There are two categories of practical coating tests: one type can be appropriate for prediction of actual performance, and the second type is only suitable for quality control. There are very real needs for both types of tests. But, too commonly, paint tests, which may be appropriate for quality control, are used to predict performance, even though the results do not permit performance predictions. Three broad classes of coating tests are field exposure tests, laboratory simulation tests, and empirical tests.

4.6.1 Field Exposure Tests

The only reliable way of knowing how a coating will perform is to use it and see how it performs. The next best approach is to use the coating in field applications on a smaller scale and under especially stringent conditions that may accelerate possible failure. The more limited the test and the greater the degree of acceleration, the less reliable are the predictions, but carefully designed and analyzed tests of this type can be very useful. There are many examples; we cite a few to illustrate the principles.

Highway marking paints can be tested by the application of stripes across the lanes of traffic instead of parallel to the traffic flow. In this way, exposure of the paint to wear is greater, and many paints can be tested and compared on a short length of highway, where they will receive the same amount of traffic wear. Controls with known performance are tested alongside new coatings. Tests should be done at different times of the year because the effects of hot sun, snow plows, salt application, and so forth must be considered. Tests should be set up on different types of highway materials, such as concrete and asphalt. Floor paints can be similarly evaluated in busy hallways. Test automobiles painted with new coatings are driven on *torture tracks* with stretches of gravel, through water, under different climate conditions, and so on. Sample packs of canned goods are made; the linings are examined for failure and the contents evaluated for flavor after various lengths of storage.

4.6.2 Laboratory Simulation Tests

Many tests have been developed to simulate use conditions in the laboratory. The value of these tests depends on how well use conditions are simulated and upon thorough validation procedures. An important key to the value of any test for performance prediction is the simultaneous evaluation of standards with known performances that cover the range from poor to excellent performance. It is not enough to use only the extremes of standards. While such information may be a first step for checking the possible applicability of a test, performance prediction requires more than two standard data points. Enough replica tests must be run to decide how many repeat tests must be run to give results within desired confidence limits. Chemists commonly think of standard deviations, but these are only 67% confidence limits; the odds are 33% that the proper value is outside of the standard deviation range. See Martin et al. (2008) for a comprehensive discussion of problems in predicting service life.

An example of a well-validated method is the laboratory shaker test devised to simulate the abrasion of can coatings when six packs of beer are shipped by railcar (Vandermeerche, 1981). Six packs are loaded onto a shaker designed to simulate the pressures, rate of shaking, range of motion, etc. actually encountered in rail shipments. The test was validated with cans of known field performance. The times to reach different degrees of abrasion failure were established. In unpublished work, in laboratories of several coatings suppliers and can manufacturers, the results were compared with known performance and with the results of standard abrasion tests. In contrast to the shaker test, none of the standard laboratory abrasion tests gave satisfactory predictions.

The automobile industry uses *gravelometers* to evaluate resistance of coatings to chipping when struck by flying gravel. Pieces of standard gravel or shot are propelled at the coated surface by compressed air under standard conditions, usually at low temperature. The tests have been standardized by comparison to a range of actual results and have been found to give reasonably good predictions of actual performance. A more sophisticated instrument, a precision paint collider, that permits variations in angle and velocity of impact and temperature has been described (Ramamurthy et al., 1999). This test was used in studies of impact failure of various coatings on automotive bumpers made with various plastic polymers. In some cases, cohesive failure occurred in the upper layers of the plastic, and, in other cases, the coating delaminated from the plastic.

Several laboratory devices are available that approximately reproduce stamping or other forming operations to test the ability of coated metal to withstand fabrication. Individual companies design such tests to match the conditions of their factory's forming operations as closely as possible.

Commonly, such simulation tests check only one or a few of the performance requirements, so they must be used together with other tests to predict overall performance. For example, the shaker test for beer cans obviously can give no information on the important ability of the coating to withstand the pasteurization procedure, but separate simulation tests are available for testing pasteurization performance.

In most cases, simulation tests are designed for use in performance prediction rather than quality control. Generally, the sample preparation and testing time are too long for checking whether production batches are equal to the standard.

4.6.3 Empirical Tests

A range of empirical tests is used for coatings. In some cases, the results can be used as part of the data to predict performance, particularly when comparing formulations that are very similar to standards with known performance. In most cases, empirical tests are more appropriate for quality control and are frequently required in product specifications, which is appropriate when empirical tests are used solely for quality control. But quality control tests established for one coating should be revalidated if they are to be applied to a new type of coating. Commonly, there is a considerable range of error, and many replicates should be run.

ASTM annually publishes books describing such empirical tests; most of the tests of importance to the coatings field are in Volumes 06.01, 06.02, and 06.03: *Paint—Tests for Formulated Products and Applied Coatings.* While the books are published annually, most of the methods are not changed, although they are supposed to be reviewed every 4 years. Each method has a number, such as ASTM D-2794-93 (1999). The "93 (1999)" signifies that the test was approved in 1993, as well as reevaluated and reapproved in 1999. If one finds a reference to a test as D-1876-71 and then refers to a newer 1997 ASTM book and finds it listed as D-1876-88, it means that the test was last reviewed, and possibly revised, in 1988. In general, one should use the new test method. Sometimes, a method will be designated as D-459a. The "a" signifies that there was a minor rewriting that did not change the basic method. A too common assumption is that, because a test method is given an ASTM designation, it can be used not just for quality control but also to predict performance. This is frequently not the case. However, following ASTM procedures does provide the best chance for obtaining comparable test results. Some ASTM methods include precision statements, usually based on repeatability and reproducibility studies involving different laboratories. The precision statements should not be ignored; many people believe the tests are more precise than they proved to be in ASTM round-robin tests. In Europe, tests established by the German Institute for Standardization (Deutsches Institut für Normung (DIN)) are widely used.

An essential reference book is the *Paint and Coating Testing Manual* (*Gardner-Sward Handbook*), now in its 16th edition (Koleske, 2012). (H. A. Gardner, who started the series, was perhaps the leading coatings scientist in the early twentieth century.) This handbook provides descriptions of a wide range of test methods and summaries of each major class of properties, as well as background information and comparisons of the utility of various tests. Nichols and Hill (2010) provide an informative, brief discussion of the more important tests in their monograph on mechanical properties.

4.6.3.1 Hardness

The measurement of hardness of a viscoelastic material is not as straightforward as it might appear. In some tests the units of hardness, force per unit area ($N\,m^{-2}$), are the same as those of modulus. When interpreting hardness and modulus data, it is important to know how the force was applied (tensile, shear, bending, or compressive), the rate of application of stress, and the temperature. Since the area changes during deformation, it is also important to determine the area before, during, and after deformation. As reviewed by Guevin (1995), three types of empirical hardness tests are used for coatings: indentation, scratch, and pendulum tests.

One indentation test is run with a Tukon Indentation Tester (ASTM D-1474M-13). An indenter having a diamond-shaped tip is pressed into the film with a fixed weight for a fixed time. The indenter is raised, and the indentation left in the film is measured with a calibrated microscope. The results are expressed in Knoop hardness numbers (KHN), which are related to the weight divided by the area of the indentation. Results are affected by film thickness; thin films on hard substrates give higher values than thicker films of the same coating. Meaningful results are obtained only with high T_g films. Intermediate T_g materials may have partial recovery of the indentation in the time needed to move the sample under the microscope and make the measurement. Low T_g films give considerable response variation, and the indenter may leave no indentation at all on a rubbery material. Accordingly, a careless tester may conclude that a rubbery material is very hard, even though it is obviously soft. The Tukon method is most appropriate for baked coatings, which are more likely to have a T_g above the testing temperature.

A widely used scratch test is the pencil hardness test (ASTM D-3363-05). Pencils with hardnesses varying from 6B (softest) to 9H (hardest) are available. The "lead"—actually graphite and clay—in the pencil is not sharpened as for writing but is squared off by rubbing perpendicularly on abrasive paper. For the test, the pencil is held at a 45° angle to the panel and pushed forward with a pressure just below that, which will break the lead. Hardness is reported as the grade of pencil that does not cause any marring of the surface. Experienced testers can reproduce results to +1 hardness grade. Probably, the test reflects some combination of modulus, tensile strength, and adhesion.

A pendulum test, particularly used in the United States, is the Sward rocker (ASTM 2134-93 (2012)). The rocker is a circular device made up of two rings joined with a glass level gauge and is weighted off center. The circumference of the rings rests on the panel. The rocker is rolled to a marked angle and released. The number of rocks (swings) required to dampen the motion to a smaller fixed angle is determined. The rocker is calibrated to give 50 rocks (a hardness reading of 100) on polished plate glass. Hard coatings give higher readings than soft coatings (but <100). Dampening is caused by rolling friction as well as by mechanical loss. The results are dependent on film thickness and surface smoothness. The Sward rocker is probably most useful for following the increase of hardness of a coated panel during drying of an ambient cure coating. It has only limited validity for comparing hardness of different coatings.

In Europe, Konig and Persoz pendulums are used. In these tests, a pendulum makes contact with a coated panel through two steel balls. As the pendulum swings back and forth through a small angle, movement of the balls requires some deformation of the coating near the surface. Test results are reported as the time (in seconds) for the swing to be dampened from a higher to a lower angle (from the perpendicular)—from 6 to 3° in the case of the Konig pendulum and from 12 to 4° in the Persoz test. Usually, harder coatings give longer times. However, soft, rubbery coatings may also give longer times. Based on the reasonable assumption that the main contribution to dampening the pendulum is absorption of mechanical energy by the film, these apparently conflicting results can be explained in terms of the loss modulus. As shown in Figure 4.13, loss modulus values are low in both the regions below and well above T_g and are highest in the transition region. Low loss modulus could account for longer dampening times for both soft, rubbery films, with T_g values below ambient temperature, and hard films, with T_g values well above ambient temperature. This hypothesis predicts that dampening times for coatings in the transition region at ambient temperature may be very sensitive to temperature, since loss modulus goes through a maximum in this region.

4.6.3.2 Formability and Flexibility

One flexibility test is a *mandrel bend test*, in which a coated panel is bent around a rod or cone (the mandrel). The panel is bent with the coated side away from the mandrel. Any crack in the coating on the bend is reported as a failure. In the cylindrical mandrel test, a series of mandrels having different diameters is used, and the smallest diameter that permits a bend without failure is reported as the mandrel flexibility. In a *conical mandrel test*, one end of the panel is clamped, and a lever is used to bend the panel over a cone. The distance from the small end of the cone to the end of a crack is reported. This distance, which is proportional to the radius of curvature of the mandrel at that point, can be used to estimate the elongation at break. Thicker films crack more easily than thinner films because the elongation at the same distance along the mandrel is greater. The bent edge should be inspected with a magnifying glass for hairline cracks, and the panels should also be inspected the next day because sometimes cracks appear later. If adhesion is poor, the film comes loose from the panel. The severity of the test can be increased by testing at low temperature (by putting the panel and tester in a freezer before testing). The severity of the test is also affected by the rate of bending.

Another formability test, widely used for testing coil coatings, is the *T-bend test*. The coated panel is bent back on itself with the coating side out. If there is no crack at the edge, the result is reported as 0T. The zero signifies that there is no additional layer of metal inside the bend. If the coating cracks, the panel is bent back on itself again. Repeated bends back over the original bend are made until the coating does not crack (see Figure 4.16). The radius of curvature becomes greater as the number of bends increases. The results are reported as 0T, 1T, 2T, 3T, and so forth, counting the number of thicknesses of metal inside the bend. The severity of the test is affected by temperature and rate of bending, and the panels should be reexamined after some time has elapsed.

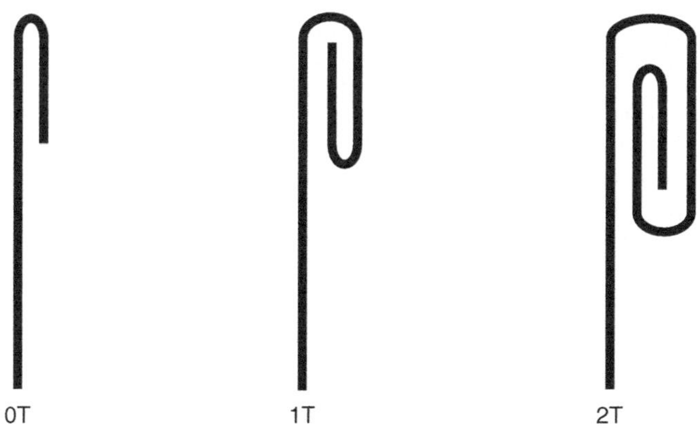

Figure 4.16 T-bend test for coil coated metals.

4.6.3.3 Impact Resistance

Impact tests evaluate the ability of a coating to withstand extension without cracking when the deformation is applied rapidly (ASTM D-2794-09 (2014)). A weight is dropped down a guide tube onto a hemispherical indenter that rests on the coated panel. An opening opposite the indenter in the base support on which the panel rests permits deformation of the panel. If the coated side is up—that is, the coating is directly hit by the indenter—the test is called a *direct impact test*. If the back of the panel is up, the test is called a *reverse impact test*. The weight is dropped from greater and greater heights until the coating on the panel cracks. In the United States, the results are reported in in-lbs—that is, the number of inches the weight falls times its weight. In the most common apparatus, the maximum is 160 in-lbs. Generally, reverse impact tests, in which the coating is extended, are more severe than direct impact tests in which the coating is compressed. The thickness, mechanical properties, and surface of the substrate substantially affect the results. Comparison of results on different substrates is meaningless. Different lots of test panels of the same type can have subtle differences in their surfaces that can affect test results. The test panels should be aged for at least one day before testing and should be inspected immediately after the test as well as on the next day.

4.6.3.4 Abrasion Resistance

In the *Taber Abraser* test, abrasive wheels roll on a panel, creating a circular wear path. The test is continued until the coating is worn through. Results are reported as the number of rotations required to wear through 1 mil (25 μm) of coating. The results often do not correlate with field use tests. For example, referring to the four floor coatings described in Table 4.2, the "hard epoxy" coating had the poorest abrasion resistance in actual field use but exhibited the highest rating in the Taber Abraser test. Another example, in which correlation was poor is the beer can abrasion test mentioned previously. On the other hand, some authors report that Taber tests correlate with visual observation of abrasion of clear coats on cars by automatic car washes (Betz and Bartelt, 1993), although others disagree. Generally, softer coatings tend to give poorer Taber Abraser results, probably because the abrasion disks rotate at a constant speed and, therefore, impart more energy to softer coatings. However, very soft coatings may clog the abrasive on the wheel and give spurious results.

Another abrasion test is the *falling sand test*, in which sand falls out of a hopper through a tube onto a coated panel held at a 45° angle to the stream of sand. The result is reported in liters of sand required to wear through a unit thickness of coating. A more sophisticated version of the falling sand test has been reported (Rutherford et al., 1997).

4.6.3.5 Solvent Resistance

Solvent resistance is not a mechanical property, but is included here because it is one of the properties that must be balanced with mechanical properties for many applications. It is also appropriate to mention here because resistance to swelling of cross-linked films is related to XLD, which affects many mechanical properties.

The most common test of solvent resistance is the methyl ethyl ketone (MEK) double rub test (ASTM D-4752-10 (2015)). This test can be done by rubbing a tissue soaked with MEK on the panel, but it is more conveniently done with a felt tip marker pen filled with solvent. The test can be mechanized so that there is one stroke back and forth (one double rub) on the film each second, and a timer can be used as a counter. A soft thermoplastic coating will rub off with very few rubs. In the case of thermosetting coatings, the number of rubs that the coating withstands increases as the degree of reaction increases. The test is sensitive to the development of XLD in the low range but insensitive to changes as the XLD gets higher. Usually, the test is stopped after 200 double rubs, and therefore, a series of highly cross-linked coatings may all be reported to give 200+ double rub resistance, even though there may be significant differences in the extent of cross-linking. At higher cross-link densities, DMA or solvent swelling can be used to determine the extent of cross-linking, as discussed in Section 4.2.

Another solvent resistance test is to expose the coating to the solvent for a certain period of time—say, 15 min—and then carry out a pencil hardness test on the exposed area. For example, a test for aircraft top coats specifies that the coating shall not lose more than two pencil hardness units after exposure to hydraulic fluid for 15 min.

GENERAL REFERENCES

Koleske, J. V. Ed., *Paint and Coating Testing Manual (Gardner-Sward Handbook)*, 15th ed., ASTM, Philadelphia, **2012**.

Mark, J. E., Ed., *Physical Properties of Polymers Handbook*, American Institute of Physics, Woodbury, NY, **1996**. (Particularly, Brostow, W., Kubat, J., and Kubat, M. M., *Mechanical Properties*, pp. 313–334.)

Nichols, M.; Hill, L. W., *Mechanical Properties of Coatings*, 2nd ed., American Coatings Association, Washington, DC, **2010**.

Rubenstein, M.; Colby, R. H., *Polymer Physics*, Oxford University Press, New York, **2004**.

Sperling, L. H., *Introduction to Physical Polymer Science*, 3rd ed., Wiley-Interscience, New York, **2001**.

REFERENCES

Baer, D. R., et al., *Prog. Org. Coat.*, **2003**, *47(3–4)* 342–356.

Bailey, W. J., *J. Elastomers Plast.*, **1973**, *5(3)*, 142–152.

Bauer, D. R., *J. Coat. Technol.*, **1988**, *60(758)*, 53.

Betz, P.; Bartelt, A.. *Prog. Org. Coat.*, **1993**, *22(1)*, 27–37.

Courter, J. L., *J. Coat. Technol.*, **1997**, *69(866)*, 57–63.

De Gennes, P.-G., *Scaling Concepts in Polymer Physics*, Cornell University Press, Ithaca, **1979**.

Dickie, R. A., et al., *Prog. Org. Coat.*, **1997**, *31(3)*, 209–216.

Dimian, A.; Jones, F., *Polym. Mater. Sci. Eng.*, **1987**, *56*, 640–644.

Evans, R. M., in Meyers, R. R.; Long, J. S., Eds., *Treatise on Coatings*, Marcel Dekker, New York, **1969**, Vol. 2.

Evans, R. M.; Fogel, J., *J. Coat. Technol.*, **1977**, *47(639)*, 50.

Fink, F., et al., *J. Coat. Technol.*, **1990**, *62(791)*, 47–56.

Gregorovich, B. V.; Hazan, I., *Prog. Org. Coat.*, **1994**, *24(1–4)*, 131–146.

Greidanus, P., FATIPEC Congress Book, Vol. *I*, **1988**, p 485.

Guevin, P., *J. Coat. Technol.*, **1995**, *67(840)* 61–65.

Hara, Y., et al., *Prog. Org. Coat.*, **2000**, *40(1–4)*, 39–47.

Hellgren, A. C., et al., *Prog. Org. Coat.*, **2001**, *43(1–3)*, 85–98.

Hill, L. W., *J. Coat. Technol.*, **1992**, *64(808)*, 28–42.

Hill, L. W., *Prog. Org. Coat.*, **1997**, *31*, 235–243.

Hill, L. W., *Polym. Mater. Sci. Eng.*, **1998**, *79*, 405.

Hill, L., et al., *Prog. Org. Coat.*, **1994**, *24(1)*, 147–173.

Hutchinson, J. W.; Suo, Z., *Adv. Appl. Mech.*, **1991**, *29*, 63–191.

Jones, F. N., et al., *Prog. Org. Coat.*, **1998**, *34(1–4)*, 119–129.

Kivilevich, A., *J. Coat.Technol.*, **2004**, *1(4)*, 38–48.

Kotnarowska, D., *Prog. Org. Coat.*, **1997**, *31(4)*, 325–330.

Lee, L. H., *Polym. Mater. Sci. Eng.*, **1984**, *50*, 65.

Lin, L., et al., *Prog. Org. Coat.*, **2000**, *40(1–4)*, 85–91.

Martin, J. W., et al., *Service Life Prediction of Polymeric Materials*, Springer, New York, **2008**.

Nichols, M. E.; Hill, L. W., *Mechanical Prorpeties of Coatings*, 2nd ed., American Coatings Association, Washington, DC, **2010**.

Nichols, M., et al., *Polym. Degrad. Stab.*, **1998**, *60(2)*, 291–299.

Osterhold, M.; Wagner, G., *Prog. Org. Coat.*, **2002**, *45(4)*, 365–371.

Perera, D. Y., *Prog. Org. Coat.*, **2003**, *47(1)*, 61–76.

Perera, D. Y., *Prog. Org. Coat.*, **2004**, *50(4)*, 247–262.

Perera, D.; Schutyser, P., FATIPEC Congress Book, Vol. *I*, **1994**, 25.

Ramamurthy, A. C., et al., *Wear*, **1999**, *225–229*, 936–948.

Rutherford, K. L., et al., *Wear*, **1997**, *203–204*, 325–334.

Ryntz, R. A.; Britz, D., *J. Coat. Technol.*, **2002**, *74(925)*, 77–81.

Ryntz, R. A., et al., *J. Coat. Technol.*, **2000**, *72(904)*, 47–53.

Schlesing, W., et al., *Prog. Org. Coat.*, **2004**, *49(3)*, 197–208.

Schulz, U., et al., *Prog. Org. Coat.*, **2001**, *42(1–2)*, 38–48.

Seubert, C. M.; Nichols, M. E. *J. Coat. Technol. Res.*, **2007**, *4(1)*, 21–30.

Seubert, C., et al., *Coatings*, **2012**, *2(4)*, 221–234.

Skrovanek, D. J., *Prog. Org. Coat.*, **1990**, *18(1)*, 89–101.

Skrovanik, D. J.; Schöff, C. K., *Prog. Org. Coat.*, **1988**, *16(2)*, 135–163.

Vandermeerche, G. A., *Closed Loop*, **1981**, April 3.

Vanier, N. R., et al., Google Patents, US6916368 B2, **2005**.

Yaneff, P. V. et al., Proceedings of the Waterborne, High Solids, and Powder Coatings Symposium, New Orleans, LA, **2001**.

Yoshida, K.; Kakuchi, T., *Prog. Org. Coat.*, **2005**, *52(3)*, 165–172.

Yoshida, K., et al., *Prog. Org. Coat.*, **2005**, *52(3)*, 227–237.

Chapter 5

Exterior Durability

Exterior durability of coatings refers to their resistance against the deleterious effects of outdoor exposure, which include changes of modulus, loss of strength, embrittlement, discoloration, loss of adhesion, chalking, loss of gloss, and environmental etching. Thus, both aesthetic and functional properties are involved. The terms *outdoor durability* and *weatherability* are also used. Corrosion protection is another aspect of durability that will be discussed in Chapter 7.

The most common chemical processes leading to degradation of coatings are photoinitiated oxidation and hydrolysis resulting from exposure to sunlight, air, and water. These processes can sometimes work together, such that the overall rate of degradation is greater than the sum of the individual components. Both photooxidation and hydrolysis are thermally activated processes, whose rates increase as the temperature is increased. Hydrolytic degradation may be enhanced by exposure to acid, as from acid rain or from acids that reside in the coating, for example, acid catalysts used to enhance the rate of cross-linking. Other atmospheric degradants include ozone and oxides of nitrogen and sulfur, but these play a minor role compared to photooxidation and hydrolysis.

When the previously mentioned chemical reactions proceed to a large enough extent, their repercussions are easily recognized and classified as physical failure. The definition of failure is governed by the type of coating and its end use. For example, the cracking or flaking of an exterior architectural paint would typically be considered a weathering-induced failure. The noticeable change in color of an exterior automotive paint would also be considered a weathering-induced failure. While no exterior coating can protect for an infinite amount of time, most coatings do have an intended *service life*. An entire branch of durability science is devoted to *service life prediction* (SLP), which we will discuss later in this chapter.

5.1 PHOTOINITIATED OXIDATIVE DEGRADATION

The first tenet of photochemistry is that a photon must be absorbed for photochemistry to be initiated. Thus, the minimization of absorption by the binder is of critical importance in reducing the rate of photooxidation in a coating. Accordingly, coatings formulated for exterior durability should exclude or minimize resin components that absorb UV radiation at wavelengths longer than 295 nm, as well as those that are readily oxidized. The thin layer of ozone in the upper atmosphere absorbs all radiation below 295 nm, and thus, the effect of radiation below that wavelength on coatings is unimportant. While many polymers appear to be transparent above 295 nm, subtle impurities and even end groups left over from polymerization can have a dramatic effect on the photooxidation rate of the binder (Gerlock et al., 1988).

Photoinitiated oxidation of polymers by a chain reaction is outlined in Scheme 5.1. Absorption of a photon by a polymer (P) or other coating component produces highly energetic photoexcited states (P^*) that undergo bond cleavage to yield free radicals (P·). Free radicals undergo a chain reaction with O_2 (autoxidation), leading to polymer degradation. Hydroperoxides (POOH) and peroxides (POOP) are unstable products of photoinitiated oxidation; they dissociate with sunlight and moderate heat to yield alkoxy (PO·) and hydroxyl (HO·) radicals, so the degradation reactions are autocatalytic. These radicals are highly reactive toward

Organic Coatings: Science and Technology, Fourth Edition. Frank N. Jones, Mark E. Nichols, and Socrates Peter Pappas.

hydrogen abstraction and yield polymer radicals (P·), which enter into the propagation stage of polymer degradation. Tertiary alkoxy radicals dissociate into ketones and a lower molecular weight polymer radical (P′·), resulting in scission of the polymer. As shown in Scheme 5.1, chain propagation leading to oxidative degradation proceeds by hydrogen abstraction from the polymer autocatalytically.

To achieve good exterior durability, functional groups in a coating that are vulnerable to hydrogen abstraction should be avoided or minimized. A general ordering of common <u>CH</u> groups having decreasing vulnerability to free radical abstraction is provided below. Accordingly, the groups at the bottom of the list are least susceptible to oxidative degradation:

$$\text{Amines}: -\underline{\text{CH}}_2-\text{NR}_2 \cong \text{diallyl}: -\text{CH}=\text{CH}-\underline{\text{CH}}_2-\text{CH}=\text{CH}- \cong \text{allyl ether}: -\text{CH}=\text{CH}-\underline{\text{CH}}_2-\text{O}- >$$
$$\text{Ethers, alcohols}: -\underline{\text{CH}}_2-\text{O}-\text{R}(\text{H}) \cong \text{urethanes}: -\underline{\text{CH}}_2-\text{NH}-\text{CO}-\text{OR} \cong$$
$$\text{Allyl}: \text{CH}=\text{CH}-\underline{\text{CH}}_2- \cong \text{benzyl}: \text{Ph}-\underline{\text{CH}}_2- >$$
$$\text{Esters}: -\underline{\text{CH}}_2-\text{CO}-\text{O}-\underline{\text{CH}}_2- \cong \text{tertiary alkyl}: \text{R}_3\underline{\text{CH}} >$$
$$\text{Secondary alkyl}: -\underline{\text{CH}}_2- > \text{primary alkyl}: \underline{\text{CH}}_3 >>$$
$$\text{Methylsiloxanes}: -\text{Si}(\underline{\text{CH}}_3)_2-\text{O}-$$

Initiation:

$$\text{Polymer (P)} \xrightarrow{\text{Sunlight}} \text{P}^* \qquad (5.1)$$

$$\text{P}^* \longrightarrow \text{free radicals (P}\bullet) \qquad (5.2)$$

Propagation:

$$\text{P}\bullet + \text{O}_2 \longrightarrow \text{POO}\bullet \qquad (5.3)$$

$$\text{POO}\bullet + \text{polymer (P—H)} \longrightarrow \text{POOH} + \text{P}\bullet \qquad (5.4)$$

Chain termination:

$$2\text{POO}\bullet \longrightarrow \text{POOP} + \text{O}_2$$

$$2\text{P}\bullet \longrightarrow \text{P—P or disproportionation products}$$

$$\text{POO}\bullet + \text{P}\bullet \longrightarrow \text{POOP or disproportionation products}$$

$$2\text{POO}\bullet \longrightarrow \text{ketones (aldehydes) + alcohols}$$

Autocatalysis:

$$\text{POOH(P)} \xrightarrow{\text{Sunlight}} \text{PO}\bullet + \bullet\text{OH}(\bullet\text{OP})$$

$$\text{PO}\bullet(\bullet\text{OH}) + \text{polymer (P—H)} \longrightarrow \text{POH} + (\text{H}_2\text{O}) + \text{P}\bullet$$

Polymer scission:

$$\text{PO}\bullet \longrightarrow \text{ketones} + \text{P}\bullet$$

Scheme 5.1

The order is based on literature bond dissociation energies and well-established activating effects of heteroatoms and aryl/vinyl groups; neighboring groups, steric hindrance, and statistical effects should also be considered. For example, poly(vinylidene fluoride), $-(\text{CH}_2-\text{CF}_2)_n-$, is known to weather substantially better than other polymers having $-\text{CH}_2-$ groups, which probably reflects the electronegativity effect of the neighboring F groups, discussed in the following text, as well as the statistical reduction of $-\text{CH}_2-$ groups. As another example, linolenic esters, having 1,4,7-triene groups, are more readily oxidized than linoleic esters, having 1,4-diene groups, which may be attributed to a neighboring group effect of the third ene group in linolenates, as well as to the statistical effect of two highly activated $-\text{CH}_2-$ groups between double bonds in linolenates compared with one in linoleates. In general, H-abstraction from tertiary alkyl groups is disfavored statistically and is expected to be particularly sensitive to steric hindrance effects.

Enhancement of H-abstraction by electron-donating N and O groups and retardation by strongly electron-withdrawing F groups can be attributed to polar contributions to the transition state for H-abstraction. Since most H-abstracting agents are electronegative, for example, peroxy or alkoxy radicals, the transition state will be stabilized by partial negative charge on the abstracting agent and, correspondingly, partial positive charge on the C from which the H is abstracted, as shown, where RX· represents the H-abstracting agent. Accordingly, the energy of the transition state will be lowered by electron-donating groups, which stabilize the partial positive charge, and raised by electron-withdrawing groups, which destabilize the partial positive charge.

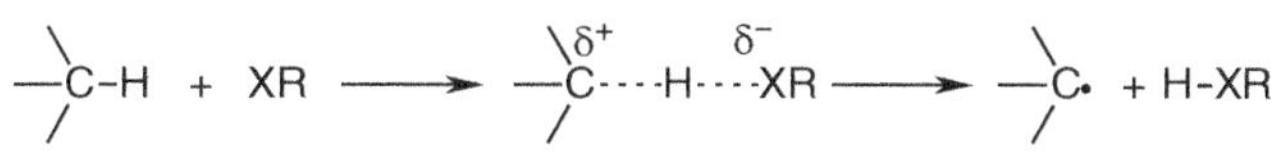

None of these functional groups absorb sunlight directly. In the absence of absorbing aromatic groups, absorption of sunlight occurs primarily by inadvertently present peroxide and ketone groups, both of which absorb UV above 295 nm. Essentially, all resins used in coatings contain traces of hydroperoxides, as do most organic substances. Peroxides, ketones, and aldehydes are also formed in photoinitiated oxidation (Scheme 5.1). Photolysis of peroxides and ketones is shown in Schemes 5.1 and 5.2, respectively. Photooxidation of aldehydes and ketones, as shown in Scheme 5.2, yields peracids, which are strong organic oxidants. Peracids may play a significant role in oxidative degradation.

Methyl-substituted silicones and silicone-modified resins (Chapter 16) exhibit high photooxidative stability, generally in proportion to the silicone content. The excellent exterior durability of fluorinated resins (Chapter 17) may be attributed, at least in part, to the absence (or reduced level) of C–H groups.

Aromatic groups having directly attached heteroatoms, as found in aromatic urethanes (Ar–NH–CO–OR) and bisphenol A (BPA) epoxies (Ar–O–R), absorb UV above 295 nm and undergo direct photocleavage to yield free radicals that can participate in oxidative degradation. BPA epoxies also have readily abstractable hydrogens positioned on carbon directly attached to ether and alcohol oxygen. Accordingly, BPA epoxies are expected to be susceptible to oxidative degradation, as evidenced by studies on model compounds and via practical formulation experience (Timpe et al., 1985). Weatherable polyurethanes must be formulated with aliphatic isocyanates to achieve acceptable exterior durability.

Coatings prepared from aromatic isocyanates yellow badly after only short exposures to UV radiation. Coatings based on BPA epoxies generally *chalk* rapidly on exposure outdoors. Chalking is a somewhat generic term that refers to the roughening and whitening of a coating's surface. It is most often due to the exposure of pigments when the binder is degraded and removed from the surface (see Section 5.2.4). However, the term is also applied to weathered unpigmented clear coats, when the surface of the clear coat erodes and microcracks leaving the appearance of a chalked surface, even though the mechanism is somewhat different. Since ketones absorb UV, they should be avoided. When acrylic resins are polymerized in ketone solvents such as methyl amyl ketone, ketone groups can be incorporated in the resin by chain transfer (Gerlock et al., 1988); it is preferable to use polymerization solvents such as esters or toluene to avoid introduction of ketone groups. The rate of photooxidation of binders polymerized with certain initiators, such as *t*-butyl perbenzoate and cumene hydroperoxide, can be twice as high as the same binder polymerized with an initiator such as AIBN, which will not leave end groups that promote photooxidation.

Highly chlorinated resins such as vinyl chloride copolymers, vinylidene chloride copolymers, and chlorinated rubber degrade by autocatalytic dehydrochlorination on exposure to either heat or UV. They must be formulated with stabilizers, as discussed in Section 5.3.

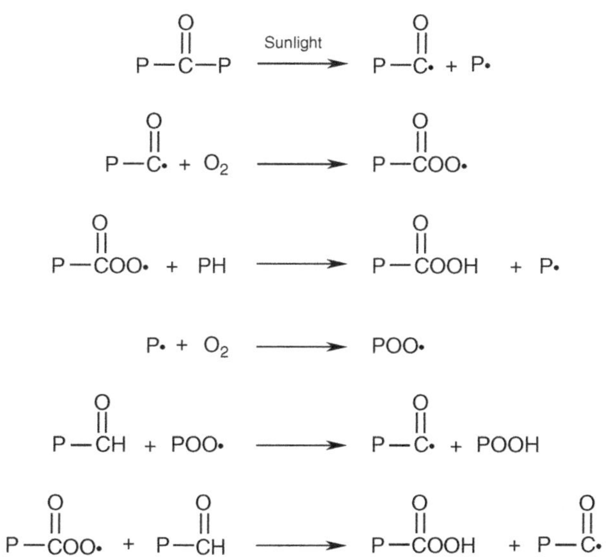

Scheme 5.2 Oxidation of aldehydes and ketones (peracid formation).

UV absorber (A):

$$A \xrightarrow{\text{Sunlight}} A^*$$

$$A^* \longrightarrow A + \text{heat}$$

$$\text{Overall: radiation} \longrightarrow \text{heat}$$

Excited state quencher (Q):

$$P \xrightarrow{\text{Sunlight}} P^*$$

$$P^* + Q \longrightarrow P + Q^*$$

$$Q^* \longrightarrow Q + \text{heat}$$

$$\text{Overall: radiation} \longrightarrow \text{heat}$$

Scheme 5.3

$$\text{POOH} + \text{Co}^{2+} \longrightarrow \text{POO}\bullet + \text{H}^+ + \text{Co}^{3+}$$

$$\text{POOH} + \text{Co}^{3+} \longrightarrow \text{PO}\bullet + \text{HO}^- + \text{Co}^{2+}$$

$$\text{Overall: 2POOH} \longrightarrow \text{POO}\bullet + \text{PO}\bullet + \text{H}_2\text{O}$$

Scheme 5.4

$$\text{R}_2\text{N—R}' + \text{O}_2 \xrightarrow{\text{UV}} \text{R}_2\text{NO}\bullet + \text{other products}$$

$$\text{R}_2\text{NO}\bullet + \text{P}\bullet \longrightarrow \text{R}_2\text{NOH} + \text{R}_2\text{NOP} + \text{P(minus H)} \quad (5.8)$$

$$\text{R}_2\text{NOH(P)} + \text{POO}\bullet \longrightarrow \text{R}_2\text{NO}\bullet + \text{POOH(P)} \quad (5.9)$$

Scheme 5.5

5.2 PHOTOSTABILIZATION

A major step toward formulation of a coating with exterior durability is to start with binders that resist photooxidation and hydrolysis. Substantial further improvement can often be achieved by use of photostabilizers. Defenses against photoinitiated oxidative degradation include UV absorbers (UVAs) to reduce UV absorption by the polymer (Scheme 5.1, Eq. 5.1), excited state quenchers to compete with bond cleavage of P^* (Scheme 5.1, Eq. 5.2), antioxidants to reduce oxidative degradation (Scheme 5.1, Eqs. 5.3 and 5.4), and hindered amine light stabilizers (HALS) (Scheme 5.5). Review articles and books on photostabilization and thermal stabilization of coatings, including degradative pathways, are available (Wypych, 2013).

5.2.1 UV Absorbers and Excited State Quenchers

Important characteristics of UV stabilizers—both absorbers (A) and quenchers (Q)—are photostability and chemical and physical permanence. Photostability requires that the photoexcited stabilizer (A^* or Q^*) can return to the ground state by converting the UV energy into thermal energy, as shown in Scheme 5.3. This process generally occurs by reversible intramolecular hydrogen transfer or *E–Z* (*cis–trans*) isomerization of double bonds.

An important feature of excited state quenchers is their effective quenching volume, within which quenching of photoexcited polymers occurs efficiently. Effective quenching volume depends on the mechanism of energy transfer. These considerations have been discussed and related to the prospects for stabilizing aromatic polymers (Pappas and Winslow, 1981). Stabilization by UVAs requires strong absorption in the wavelength region in which the polymer and/or trace impurities also absorb. Since quenchers function by energy transfer from excited state polymers and/or impurities, stabilization requires spectral overlap in the wavelength region in which the quencher absorbs and the excited state polymer and/or impurities emit radiation. Stabilizers, which function as UVAs in some coatings, may perform as excited state quenchers in other coatings. The most effective stabilizers probably function by both roles.

One cannot eliminate UV absorption by the resin by adding a UVA, which can only reduce absorption by the binder to slow the rate of photodegradation reactions. Since absorption increases as the path length increases and incident radiation is attenuated in the lower layers by absorption in the upper layers, UVAs are most effective in protecting the lower parts of a film or substrate—for example, a base coat, wood, or plastic under a clear top coat containing an absorber—and least effective in protecting the layer at the air interface. It follows that the thickness of a clear top coat, which contains a UVA can be an important variable affecting the protection of base coats or plastics under them, since thicker films transmit less radiation. In contrast to a UVA, a quencher does not absorb incident radiation. A photodegradable polymer or impurity does absorb incident radiation; but the ratio of bimolecular quenching to photodegradation of a photoexcited polymer (P^*) or impurity (I^*) is independent of the concentration of P^* or I^* and, therefore, independent of absorbed radiation throughout the coating. The extent of quenching is determined by the effectiveness of the quencher and its concentration. A critical consideration in the design and selection of UV stabilizers is their absorption spectra. In general terms, one would like to have very high absorption of UV radiation, from 295 to 380 nm. To avoid color effects by the stabilizer, ideally, there would be no absorption above 380 nm.

Substituted 2-hydroxybenzophenones, 2-(2-hydroxyphenyl)-2*H*-benzotriazoles, 2-(2-hydroxyphenyl)-4,6-phenyl-1,3,5-triazines, benzylidenemalonates, and oxalanilides are classes of UV stabilizers. Specific members of these classes may act as UVAs, as excited state quenchers, or as both, but they are frequently called "UV absorbers" or "UVAs" in the coatings literature.

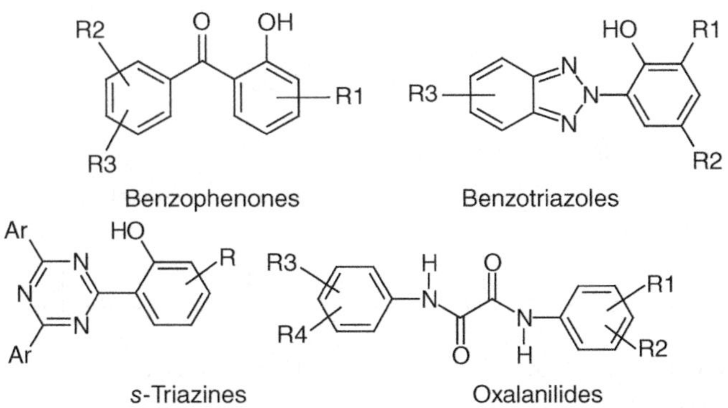

These UV stabilizers convert UV energy into heat by intramolecular hydrogen transfer or *cis–trans* isomerization. For example, UV absorption or excited state quenching by a 2-hydroxyphenyl-substituted stabilizer yields a photoexcited state that converts the excess electronic energy into chemical energy by undergoing intramolecular hydrogen transfer to yield an unstable intermediate. The unstable intermediate spontaneously undergoes reverse hydrogen transfer to regenerate the UV stabilizer with conversion of the chemical energy into heat. The process is illustrated with 2-hydroxybenzophenone.

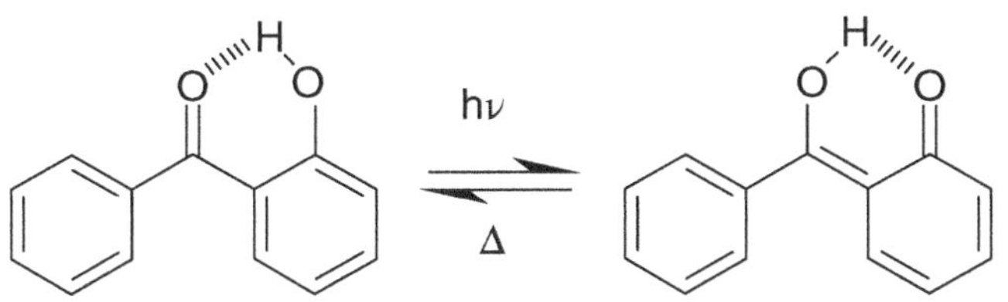

The UV stabilizer must be soluble in the coating film. (Pigments that serve as UVAs, discussed in Section 5.2.4, are insoluble.) Several grades of the various stabilizers are available with different substituents on the aromatic rings that provide for solubility in different polymer systems.

Commonly, the stabilizer is added to the top coat of a multicoat system. However, especially in baking systems, migration may result in the stabilizer being distributed through the whole coating, reducing the concentration in the top coat. This effect has been demonstrated by analyzing sections through a film of clear coat–base coat automotive finishes (Bohnke et al., 1991). In one combination of coatings, in which UV stabilizer was added only to the clear coat, section analysis indicated that the stabilizer content was essentially uniform throughout the whole film, both clear coat and base coat. In a second case, with a different type of base coat, a major fraction of the stabilizer remained in the clear coat. In coatings for plastics, it has been shown that stabilizers can migrate into the plastic (Haacke et al., 1996). Even mild temperatures can be sufficient to cause UVAs to migrate between layers in cured films (Peters et al., 2007). A potential solution is to add stabilizer to the lower coats.

A critical requirement of a UV stabilizer is permanence. Loss of stabilizer can result by two mechanisms. There can be physical loss by vaporization, leaching, or migration, and/or there can be loss by chemical or photochemical degradation of the stabilizer. The latter dominates for most coatings. If a UV stabilizer has even a small vapor pressure, it slowly volatilizes from the surface over the long-term period for which durability is desired. More importantly, the UVA can be volatilized during baking in those coatings that undergo thermal cure. This loss during baking can result in substantially lower concentration of UVA in the coating than was intended by the formulator.

Most hydroxyphenyltriazines have very low vapor pressures. They exhibit the least loss during baking and often the best photo-permanence followed by benzotriazoles, benzophenones, and oxalanilides. Longer-term physical permanence may be achieved by using oligomeric photostabilizers. Polymer-bound stabilizers can also be used, assuring physical permanence (Cliff et al., 2005).

More important than physical loss, organic UVAs are chemically consumed during the weathering process. The UVAs do not absorb light and disassociate, but are instead subject to attack by free radicals produced in the binder. Those UVAs most resistant to free radical attack tend to have the best permanence in coatings. Coatings with high degradation rates lead to shorter UVA lifetimes. Valet (1997) provides an extensive discussion of the complex factors involved in UVA permanence.

Because UVAs are chemically consumed during weathering, after long-term exterior exposure, a gradient is often observed in the concentration of UVA in a coating. As the UVA cannot protect the very top of the coating, degradation often is highest there and the rate of attack on the UVAs is therefore greatest at the surface of the coating (Gerlock et al., 2001b; Smith et al., 2001). While both UVAs and HALS have some mobility in unweathered coatings, the increase in T_g of the binder that typically occurs during weathering significantly retards mobility of these stabilizers as weathering progresses. Thus, significant gradients in their concentrations can be observed after only a few years of outdoor exposure. As mentioned previously, gradients can also be observed in the opposite direction if the UVA has a higher solubility in an underlying layer or in the substrate than it has in the coating. Such migration is expected to reduce stabilization, in particular, can lead to unexpectedly poor durability if the coating is used on a substrate on which it was not originally tested.

Related to the gradients in UVA concentration with enhanced depletion generally at the surface of a coating owing to consumption during weathering, the extent of degradation in a coating is also typically a function of depth into the coating. As UVAs and pigments restrict penetration of UV light into the coating, degradation is often greatest near the surface. Recent work from both simulation and experiments has shown that oxygen diffusion, long thought to be unimportant in coating degradation, can also lead to a gradient in degradation through the thickness of a coating (Kiil, 2012; Adema et al., 2015).

5.2.2 Antioxidants

Antioxidants may be classified into two groups: preventive and chain-breaking antioxidants. Preventive antioxidants include peroxide decomposers, which reduce hydroperoxides to alcohols and become oxidized into harmless products. Examples of peroxide decomposers are sulfides and phosphites, as shown in Eqs. 5.5 and 5.6 for dilauryl thiodipropionate (LTDP) and triphenyl phosphite. The initial products, sulfoxides and phosphates, respectively, may undergo further reactions:

$$\mathrm{POOH} + \mathrm{S}\left(\mathrm{CH_2CH_2CO} - \mathrm{OC_{12}H_{25}}\right)_2 \rightarrow \mathrm{POH} + \mathrm{O} = \mathrm{S}\left(\mathrm{CH_2CH_2CO} - \mathrm{OC_{12}H_{25}}\right)_2 \qquad (5.5)$$

$$\mathrm{POOH} + \left(\mathrm{PhO}\right)_3\mathrm{P} \rightarrow \mathrm{POH} + \left(\mathrm{PhO}\right)_3\mathrm{P} = \mathrm{O} \qquad (5.6)$$

Metal complexing agents are a second type of preventive antioxidant. They complex with transition metal ions that may be present as contaminants and, thereby, prevent the metal ions from catalyzing the undesired conversion of hydroperoxides into peroxy and alkoxy radicals by redox reactions, as exemplified in Scheme 5.4. The resulting peroxy and alkoxy radicals are undesirable because they promote oxidative degradation, as shown in Scheme 5.1. For oxidative cure of drying oils and alkyds (Section 14.2.2), transition metal ions are deliberately added as catalysts (driers) to accelerate cure. Thus, the reactions in Scheme 5.4 are involved in both oxidative cross-linking and oxidative degradation. Accordingly, drier concentrations should be minimized to reduce subsequent degradation during exterior exposure.

The tetrafunctional bidentate imine derived from *o*-hydroxybenzaldehyde (salicylaldehyde) and tetraaminomethylmethane is reported to complex a large number of transition metal ions effectively, including Co, Cu, Fe, Mn, and Ni (represented by M^{n+} in complex **1**) (Shelton and Hawkins, 1972).

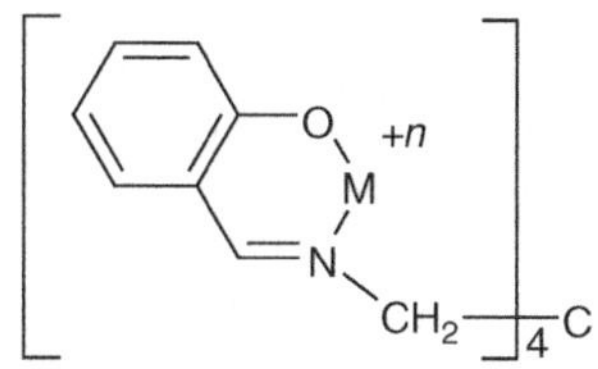

(Ar = 1,2-disubstituted benzene)

1

Chain-breaking antioxidants function by interfering directly with chain propagation steps of autoxidation, shown in Scheme 5.1. An example is a hindered phenol (2,2′-methylene-bis(4-methyl-6-*tert*-butylphenol)), which reacts with peroxy radicals in competition with hydrogen abstraction from a polymer (PH) to yield a resonance-stabilized, less reactive phenoxy radical, as shown in Eq. 5.7.

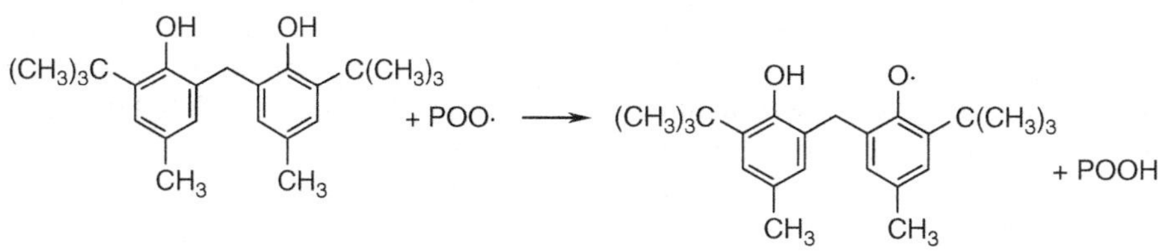

(5.7)

However, undesirable hydroperoxides are also generated in Eq. 5.7. This provides a basis for *synergistic stabilization* of polymers by a combination of chain-breaking phenolic antioxidants and peroxide decomposers. Synergistic stabilization signifies that the combination of stabilizers is more effective than the additive effect of each stabilizer by itself.

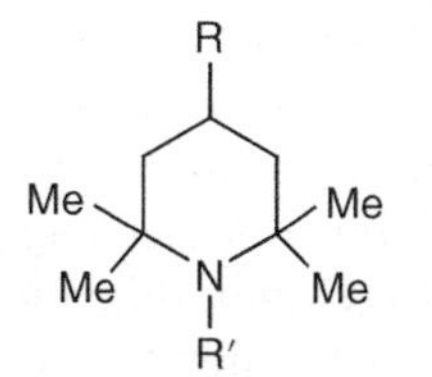

(R′ = H, alkyl, alkanoyl, alkoxy)

2

5.2.3 Hindered Amine Light Stabilizers

HALS are amines having two methyl groups on each of the two alpha carbons; most are derivatives of 2,2,6,6-tetramethylpiperidine, as shown in general formula **2**. They are reported to function both as chain-breaking antioxidants (Carlsson et al., 1984) and as transition metal complexing agents (Fairgrieve and MacCallum, 1984). The former function appears to be more important. Note that the 2,2,6,6-methyl groups prevent oxidation of the ring carbons attached to the nitrogen.

HALS derivatives undergo photooxidative conversion into nitroxyl radicals ($R_2NO\cdot$) that react with carbon-centered radicals by disproportionation and combination to yield corresponding hydroxylamines and ethers, respectively, as shown in Eq. 5.8 in Scheme 5.5. The hydroxylamines and ethers, in turn, react with peroxy radicals to regenerate nitroxyl radicals, shown in Eq. 5.9 in Scheme 5.5. In this manner, HALS derivatives interfere with propagation steps involving both carbon-centered and peroxy radicals in autoxidation (Scheme 5.1). In contrast to nitroxyl radicals, hindered phenols do not react directly with carbon-centered radicals. Oxygen is required as a co-stabilizer to convert carbon-centered radicals into peroxy radicals, with which hindered phenols can react, as shown in Eq. 5.7. Nitroxyl radicals can also react with alkoxy radicals to reduce the rate of photooxidation.

To be effective, HALS derivatives must undergo rapid oxidation to form nitroxyl free radicals on exterior exposure of the coating. In a photoexposed film, only a small fraction (about 1%) is present as the nitroxyl radical, the major storage components being the corresponding hydroxylamine (R_2NOH) and ethers (R_2NOP). Continued stabilization requires the presence of the nitroxyl radical, and its disappearance is followed shortly by rapid polymer degradation. Probably, the ultimate demise of HALS occurs, at least in part, by oxidation of the nitroxyl radical, accompanied by opening of the piperidine ring. Transition metal ions and peracids are potential oxidants for this process. New research has shown through *ab initio* calculations that HALS may function through the abstraction of a hydrogen from the alkoxyamine via a peroxyl radical (Gryn'ova et al., 2012).

A variety of HALS compounds is available. The "R" in general formula **2** is often a diester group that joins two piperidine rings; this increases the molecular weight, decreasing volatility. The first commercial HALS compounds, still used to a degree, had R′=H. Later versions with R′=alkyl exhibit better long-term stability. Both of these types are basic and interfere with acid-catalyzed cross-linking reactions, such as those involving melamine-formaldehyde (MF) resins. Alkanoyl HALS compounds with R′=C(=O)R″ are not basic, but their initial reaction to form nitroxyl free radicals is slower (Section 5.6.2) (Valet, 1997). More recently, hydroxylamine ethers (R′=OR″) have gained wide acceptance. The octyl ether provides a HALS with low basicity that converts rapidly to nitroxyl free radicals (Bauer, 1994). HALS compounds, especially with R′=H, can accelerate degradation of polycarbonate plastics, which is thought to occur via base-catalyzed hydrolysis.

Combinations of UVAs and HALS compounds can act synergistically (Gerlock et al., 1985). The UVA reduces the rate of generation of radicals, whereas the HALS compound reduces the rate of oxidative degradation by the radicals. A further factor is that UVAs are inefficient at protecting the outer surface of a film; in contrast, HALS compounds can effectively scavenge free radicals at the surface. As with UVAs, gradients can be observed in coatings containing HALS after only a few years of exterior exposure, as discussed in Section 5.6.1 (Gerlock et al., 2001c).

For automobile fascia use, a base coat and a clear coat are applied over a primed TPO (polypropylene–EPDM rubber blend). For example, an acrylic–MF base coat and an acrylic polyurethane clear coat could be applied. The UVA and HALS compound are included in the clear coat to assure exterior durability. However, in contrast to the same coating applied over steel, it has been found that the UVA and HALS can migrate through the coatings and into a TPO plastic resulting in poorer durability than over steel. To minimize this problem, the use of UVAs and HALS comprising hydroxyl groups that will react with the isocyanate in the clear coat to bind the stabilizers to the polymer chains has been studied. With the stabilizers bonded, their retention in the clear coat is much higher and durability is improved (Yaneff et al., 2004). This migration problem is even more severe in reaction injection molded (RIM) polyurethane materials that display an even greater affinity for stabilizer additives than TPO. With an acrylic–MF base coat and a 2K urethane clear coat, an extensive study also showed marked improvement in performance when using similar reactive UVAs and HALS compared with unreactive analogs (Cliff et al., 2005).

Wood discolors rapidly on outdoor exposure to UV, as well as visible light, and more slowly indoors. A major reason for the outdoor effect is photoreactivity of lignin in the wood. Hayoz et al. (2003) recommend a two-coat system to protect wood against discoloration. The base coat contained the nitroxyl derivative of 4-hydroxy-2,2,6,6-tetramethylpiperidine to trap radicals formed by the radiation. While, as noted earlier, in most cases it is desirable to place HALS in the top coat, in this case it was most important that HALS be present in the coating in direct contact with the lignin containing wood. A mixed bis-, tris-, and tetra-*para*-substituted tris-resorcinol-triazine was used as the UVA in the top coat to screen the radiation. A review includes much of the current thinking on the mechanisms of wood degradation and protection (Teaca et al., 2013).

The *N*-2,2,6,6-pentamethylhydroxypiperidinyl diester of 4-methoxyphenylmethylenemalonic acid combines the UVA function of a benzylidene malonic ester with the antioxidant properties of HALS derivatives. Furthermore, on outdoor exposure of films, this bifunctional UVA/HALS stabilizer reacts with MF cross-linked acrylic and 2K urethane acrylic coatings, thereby preventing loss by volatilization or leaching. Ávár and Bechtold (1999) reported that combining the bifunctional stabilizer with UV stabilizer, *o*-hydroxyphenyl-*s*-triazine, provides excellent stabilization over long periods of time. A comparison of three types of UV stabilizers with the bifunctional stabilizer has shown that photostability diminishes more rapidly in the order: *o*-hydroxyphenyl-*s*-triazines > *o*-hydroxyphenyltriazenes > oxanilides.

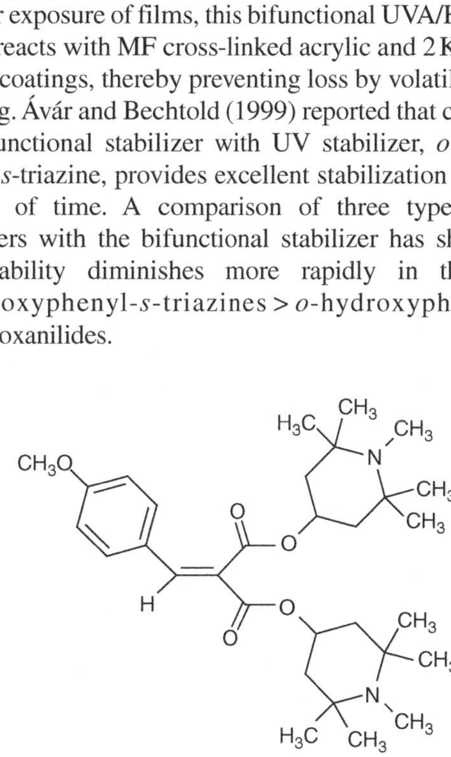

5.2.4 Pigmentation Effects

Many pigments absorb UV radiation. The strongest UVA known is fine particle size carbon black. Many carbon blacks have structures comprising multiple aromatic rings and, in some cases, phenol groups on the pigment surface. In the latter case, such black pigments are both UVAs and antioxidants. Enhanced exterior durability is obtained with carbon black pigmented coatings. When Henry Ford said, "The customer can have any color car he wants, as long as it is black," he was not just being crotchety. Black was by far the most durable paint available at the time, in addition to being the fastest drying color.

Other pigments absorb UV radiation to varying degrees. For example, in 50 μm coatings pigmented with fine particle size, transparent red iron oxide pigments absorb virtually all radiation below about 420 nm (Sharrock, 1990). This strong absorption is particularly useful in stains for use over wood, since the pigmented transparent coating protects the wood from photodegradation.

The formulator must also consider the interaction between pigments and organic stabilizers in the coating. Haacke and coworkers (Haacke et al., 1999) have shown that UVAs can be physisorbed onto the surface of various pigments including carbon black and organic pigments, thereby reducing their effective concentration in the binder and resulting in unexpectedly poor performance of certain colors.

Both nano-ZnO and nano-CeO_2 have been studied as inorganic permanent UVAs for many types of coatings and stains. While effective, even these nano-sized pigments can have a deleterious effect on the appearance of coatings and have therefore not gained significant market share compared with organic UVAs (Lima et al., 2009; Saadat-Monfared et al., 2012).

Rutile TiO_2 white pigments absorb UV strongly. Absorption is a function not only of wavelength and concentration but also of particle size of the pigment (Stamatakis et al., 1990). The optimum particle size for absorption of 300 nm radiation by rutile TiO_2 is 50 nm, which increases to 120 nm for 400 nm radiation. These sizes are smaller than the optimum particle size of 190 nm for hiding (Section 19.2.3). Rutile TiO_2 with an average particle size of 230 nm still absorbs UV strongly. Anatase TiO_2 also absorbs UV strongly, although not as strongly as rutile TiO_2 in the near UV. Thus, TiO_2, especially rutile TiO_2, functions as a UVA in coatings. However, TiO_2 can accelerate photodegradation of films on exterior exposure, causing chalking of coatings—that is, degradation of the organic binder and exposure of unbound pigment particles on the film surface that rub off easily, like chalk on a blackboard. Degradation of the binder is enhanced by interaction of photoexcited TiO_2 with oxygen and water to yield oxidants, as shown in Scheme 5.6 (Voelz et al., 1981).

$$TiO_2 + \text{light} \longrightarrow TiO_2^{*}(e/p)$$

$$TiO_2^{*}(e/p) + O_2 \longrightarrow TiO_2(p) + O_2{\cdot}^{-}$$

$$TiO_2(p) + H_2O \longrightarrow TiO_2 + H^{+} + HO{\cdot}$$

$$H^{+} + O_2{\cdot}^{-} \longrightarrow HOO{\cdot}$$

$$2HOO{\cdot} \longrightarrow H_2O_2 + O_2$$

$$TiO_2^{*}(e/p) + H_2O_2 \longrightarrow TiO_2 + 2HO{\cdot}$$

Scheme 5.6

Photoexcitation of TiO_2 results in promoting a low energy valence band electron into the higher energy conduction band, creating a separated electron (e)/hole(p) pair, signified by $TiO_2^{*}(e/p)$ in Scheme 5.6. Electron capture by O_2 (reduction) and hole capture by H_2O (oxidation) result in regeneration of ground-state TiO_2 and lead to hydroperoxy and hydroxy radicals that can participate in oxidative degradation, as shown in Scheme 5.1. Anatase TiO_2 is more photochemically active in promoting oxidative degradation than rutile TiO_2, so much so that anatase is generally not used in outdoor coatings except in rare cases where chalking is desired (see Section 20.1.1).

The photoactivity of TiO_2 pigments is reduced by coating the pigment particles with a thin layer of silica, alumina, or zirconia (Section 20.1.1) to form a barrier layer against the redox reactions. In this manner, rutile TiO_2 can be used in appropriate coatings that maintain excellent durability for many years of exterior exposure. Watson et al. (2012) have developed a laboratory test and database to compare the photoreactivity of various grades of TiO_2. Various stabilizing additives, including HALS, have also been reported (Braun, 1990).

Chalking of exterior house paints, which are predominantly pigmented with TiO_2, eventually leads to complete erosion of the film in the very long term. Chalking also leads to reduction in gloss, since the film becomes rougher. However, loss of gloss does not necessarily correlate with ease of chalking (Braun and Cobranchi, 1995). It was shown that initial gloss loss in some TiO_2 pigmented coatings resulted from film shrinkage, which, in some cases, was greater with more resistant grades of TiO_2. In paints containing both TiO_2 and color pigments, chalking results in color changes as a result of the gloss loss; the higher surface reflectance of the low gloss films gives an appearance of weaker colors.

The formulation of coatings for exterior use can be complicated by pigment–binder interactions. In some cases, a pigment shows excellent color retention after exterior exposure when formulated with one class of resins, but poor durability when formulated with another class. For example, thioindigo maroon had excellent color retention when used in nitrocellulose lacquers, but poor color stability when used in acrylic lacquers. While one can use experience with other systems as an initial basis in selecting pigments for a new class of resins, field tests are required to assure that each combination of pigments and resins is suitable.

5.3 DEGRADATION OF CHLORINATED RESINS

Highly chlorinated resins undergo dehydrochlorination on exposure to either heat or UV. The reactions are autocatalytic. As the number of conjugated double bonds in series increases, the polymer progressively discolors, finally becoming black. The resulting highly unsaturated polymer undergoes autoxidation, resulting in cross-linking and embrittlement. For poly(vinyl chloride), the ultimate products are conjugated polyenes.

The mechanism of degradation of poly(vinyl chloride) has been extensively studied. Dehydrochlorination is promoted when chlorine is situated on a tertiary carbon or on an allylic carbon. It has been proposed that at least one major weak point results from addition of a vinyl chloride monomer in a head-to-head fashion to the growing polymer chain, as in Eq. 5.10 in Scheme 5.7, followed by chain transfer of Cl to monomer as in Eq. 5.5 (Starnes, 1985; Georgiev et al., 1990). The resulting allylic chloride is highly susceptible to dehydrochlorination, which generates a new allylic chloride with two conjugated double bonds (Eq. 5.12). Progressive dehydrochlorination is favored because the growing number of conjugated double bonds increases the lability of allylic chlorides down the chain.

A variety of stabilizing agents is used. Since dehydrochlorination is catalyzed by hydrogen chloride, HCl traps, such as epoxy compounds and basic pigments, are useful. Dienophiles that can undergo Diels–Alder reactions can also act as stabilizers; Diels–Alder addition reactions reduce the chain of conjugated double bonds. Dibutyltin diesters are effective stabilizers. It is proposed that the activated chlorine atoms are interchanged with ester groups of the tin compounds to form the more stable ester-substituted polymer molecules. Dibutyltin maleate is a particularly effective stabilizer, since it acts both as an ester interchange compound and as a dienophile. Barium, cadmium, and strontium soaps act as stabilizers. Choice of stabilizer combinations can be system specific, especially depending on whether stabilization is needed against heat, UV, or both. In the case of UV stabilization, UVAs can further enhance stability.

5.4 HYDROLYTIC DEGRADATION

For coatings with good exterior durability, the rate of hydrolysis is generally much lower than the rate of photooxidation, as durable coatings are difficult to formulate if

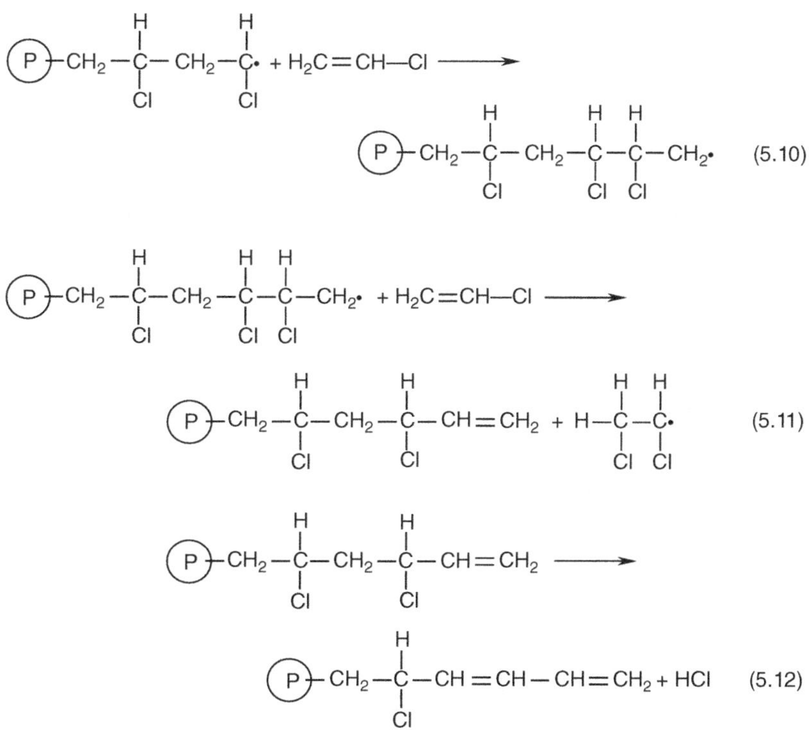

Scheme 5.7

they have even a moderate rate of hydrolysis. Photooxidation can be mitigated via stabilizer additives, but hydrolysis is difficult to moderate via additives. A general ordering of functional groups subject to hydrolysis is esters > carbonates > ureas > urethanes ≫ ethers, with esters being the most vulnerable and ethers the least. However, activated ethers, such as in etherified MF cross-linkers (Chapter 11), and etherified resole phenolics (Section 13.6), as well as their cross-linked products with hydroxy-functional resins, are more reactive than ureas and urethanes. The tendency of each type of group to hydrolyze can be reduced by steric hindrance—for example, by placement of alkyl groups in the vicinity of the susceptible groups, such as esters (Section 10.1.1). The alkyl groups may also reduce hydrolysis by decreasing hydrophilicity, thereby reducing rates of hydrolysis. Studies have shown that the lower the water solubility of the diacid or diol used to make a polyester, the greater the resistance to hydrolysis (Jones and McCarthy, 1995). Rates of hydrolysis are also influenced by neighboring groups. For example, *o*-phthalate half esters are more readily hydrolyzed under acidic conditions (often encountered during exterior exposure) than isophthalate half esters, in which the groups are meta (Section 10.1.2). "Superdurable" polyester resins have become commonplace; at least part of the reason for their enhanced durability is the use of polyols and perhaps polyacids that resist hydrolysis (Chapter 10).

Hydrolysis of polyesters results in backbone degradation. On the other hand, the backbones of (meth)acrylic resins are completely resistant to hydrolysis, since the backbone linkages are carbon–carbon bonds. Acrylate—and particularly methacrylate—ester side groups are very resistant to hydrolysis, owing to the steric effects of the acrylic backbone. Carbonate esters are less susceptible to hydrolysis than carboxylic esters.

Hydrolysis of MF cross-linked hydroxy-functional resins is enhanced by residual acid catalysts (generally sulfonic acids) used to catalyze cross-linking (Section 11.3.1). Curing temperatures can be reduced by increasing the concentration of sulfonic acid in the coating, but the sulfonic acid remaining in the cured film enhances susceptibility to hydrolysis. An apparently ideal solution would be to use transient, or fugitive, acid catalysts that either leave the film or become neutralized after cure. Nguyen et al. (2003) studied degradation of acrylic–MF coatings at 50°C and different humidity levels by Fourier transform infrared spectroscopy (FTIR) and atomic force microscopy (AFM). Formation of $–NH_2$ and –COOH groups was observed, and material was lost. It is interesting that the degradation was not uniform; pits form in the surface, deepen, and grow during exposure. Other workers have also found via simulation that the degradation of coatings can occur in a heterogeneous fashion, leading to void formation as degradation progresses. The heterogeneity may arise owing to local differences in cross-link density formed during curing of the polymer network (Zee et al., 2015).

Base coat–clear coat finishes for automobiles are subject to *environmental etching*. In warm, moist climates with acidic rain or dew, as in Jacksonville, FL, small, unsightly spots appear in the clear coat surface, sometimes within days. The spots are uneven, shallow depressions in the clear coat surface. These pits result from hydrolytic degradation of the resin in the area of a droplet of water containing a significant acid concentration. As the water evaporates, the concentration of acid increases, which increases the rate of hydrolysis. While testing by exposure of panels for 14 weeks in Jacksonville has been widely used to evaluate the acid etch resistance of coatings, considerable variability has been observed and has been attributed to large variations in rain, dew, sunshine, and acidity that occur on a yearly basis (Schulz et al., 2000). Therefore, an artificial weathering test is now recommended. A recent laboratory test (ASTM D7356) that has been shown to effectively mimic the performance of coatings in Jacksonville has recently been developed and includes UV, acid, temperature, and moisture exposure.

Several factors have been identified regarding the observed differences in resistance to environmental etching (Betz and Bartelt, 1993; Gregorovich and Hazan, 1994). Since urethane linkages are more resistant to acid hydrolysis than the activated ether cross-links obtained with MF cross-linked hydroxy-functional resins, urethane-polyol clear coats are generally less susceptible to environmental etching than many MF-polyol clear coats. When MF resins are used to cross-link carbamate-functional resins (Section 8.2.2), good resistance to environmental etching can be combined with mar resistance, as the cross-links formed are acid etch-resistant urethane cross-links. Temperature and T_g are also important, as is surface tension of the clear coat. A variety of approaches has been undertaken to further elucidate and minimize the problem; see Section 30.1.2 for further discussion.

Bauer et al. (1988) have reported that acrylic-urethane coatings are more effectively stabilized by HALS than are acrylic–melamine coatings. This finding may reflect, at least in part, the greater susceptibility of acrylic–melamines to degrade by hydrolysis, which HALS derivatives do not protect against. Both urethane and melamine coatings undergo oxidative degradation. CH groups adjacent to urethane nitrogen groups are activated toward H-abstraction by free radicals, as are CH groups adjacent to nitrogen and ether oxygen groups in melamines. However, evidence has been presented that hydroperoxide levels are significantly lower in melamine than in urethane cross-linked coatings, indicative of lower rates of free radical formation. The results have been attributed to the ability of melamines to decompose hydroperoxides (Mielewski et al., 1991), but may also be explained by lower susceptibility to oxidation of CH groups on melamine.

Bauer et al. (1988) have also reported that hydrolytic degradation of acrylic–urethane as well as acrylic–melamine coatings is accelerated by UV exposure. These results may arise from the increase of hydrophilic groups, such as hydroperoxides, alcohols, ketones, and carboxylic acids, resulting from photooxidation, which increase the hydrophilicity of the coating. Photooxidation may also occur at specific sites to generate groups that are more susceptible to hydrolysis. Bauer (1986) has also reported that photodegradation of acrylic–melamine coatings is accelerated in high humidity. This result was attributed to formation of formaldehyde (from hydrolysis of the coating), which, in turn, was oxidized into performic acid, a strong oxidant, capable of promoting photodegradation.

Silicon-containing coatings, which are highly resistant to photodegradation, are vulnerable to hydrolysis at cross-linked sites where silicon is attached to three oxygens (Brown, 1972; Hsiao et al., 1975; Pappas and Just, 1980). Apparently, the electronegative oxygens facilitate nucleophilic attack at Si by water (Section 16.1.2). The reaction is reversible, so the cross-links can hydrolyze and reform. If a silicone-modified acrylic or polyester coating is exposed to water over long periods or is used in a climate with very high humidity, the coating can get softer. To mitigate this problem, some MF resin is commonly included as a supplemental cross-linker in the formulation. Apparently, the MF–acrylic–polyester cross-links are more hydrolytically stable than the bonds between the silicone resin and the polyester or acrylic.

5.5 OTHER MODES OF FAILURE ON EXTERIOR EXPOSURE

Much of this chapter has focused on the changes in chemical composition that occur during the exterior exposure of coatings. Importantly, these chemical changes lead to changes in the physical appearance or performance of the coating: yellowing, color change, gloss loss, cracking, peeling, or chalking. Furthermore, environmental and climatic interactions of the coatings often determine the final physical manifestation of the chemical composition changes.

Cracking, for example, occurs when tensile stresses in the coating overcome the coating's fracture resistance. As discussed in Chapter 4, as weathering progresses, the fracture energy (toughness) of most coatings decreases and the stresses in the coating increase. The rate at which the coating embrittles is closely related to the rate at which the chemical composition changes. For automotive coatings and silicone hard coats (Nichols and Peters, 2002), the rate of photooxidation has been directly linked to the rate of embrittlement. Coatings in which chemical composition changed more slowly during weathering also embrittled more slowly (Nichols et al., 1999; Nichols and Peters, 2002).

Microclimates can also play a role in cracking failures. "Denver cracking" has been observed in dry, high altitude climates where the annual UV dose is high. The relatively high rate of photooxidation can lead to cracking that is typically not observed in warmer climates, simply because the stresses in the coating are particularly high during the cold, dry winter months (Nichols and Frey, 2014).

When paint is applied to wood, it must be able to withstand the elongation that results from the uneven expansion of wood grain when it absorbs moisture; otherwise, *grain cracking*—that is, cracking parallel to the grain—occurs. This failure mode can occur with interior coatings; however, it is more likely to take place with exterior coatings, particularly with alkyds or drying oils that embrittle with exterior exposure. As a result of greater exterior durability and extensibility, acrylic latex paint films seldom fail in this way.

Delamination, or peeling, has also been linked to changes in chemical composition. When an interface between coating layers or between a coating and a substrate begins to chemically degrade, the adhesive strength invariable decreases. Through the use of both infrared (IR) spectroscopy and time-of-flight secondary ion mass spectrometry (TOF-SIMS), the extent of chemical change at an interface has been directly correlated with delamination failure during exterior weathering. Those coatings that showed little chemical change at the interface also showed superior resistance to delamination after weathering (Gerlock et al., 1999).

The exposure time before failure occurs is determined by a complex interaction between the environmental loads (UV, temperature, moisture) and the response of the particular coating to each of these loads. For example, the same coatings are known to lose gloss more quickly when exposed in south Florida than in the Phoenix, Arizona area. The annual UV dose in Phoenix is approximately 10% higher than that in south Florida; and measurements have shown that the surface of paint systems exposed in Phoenix are *more* degraded than those exposed in Florida. However, coatings lose gloss by erosion of material from the surface of the coating. The rough surface left behind scatters light away from the specular direction, which is perceived as a loss in gloss. The main mechanism for surface erosion is the removal of degraded, loosely bound material by liquid water (rain or dew). The annual rainfall in south Florida is 10 times that of Phoenix. Accordingly, the loss of gloss is greater in Florida because most of the degraded material in the Florida exposed coatings is washed off the surface of the coating during rain events (Nichols and Misovski, 2009).

Blistering of exterior oil-based house paints on wood siding results in loss of adhesion. The blistering results from accumulation of water in the wood beneath the paint layer. The vapor pressure of the water increases with heating by the sun, and blisters form to relieve the pressure.

Since latex paints generally have higher moisture vapor permeability than oil-based paints, the water vapor can pass through a latex paint film, relieving the pressure before blisters grow. However, the high moisture vapor permeability of latex paint films can lead to failures of other types. For example, if calcium carbonate fillers are used in an exterior latex paint, *frosting* can occur. Water and carbon dioxide permeate into the film, dissolving calcium carbonate by forming soluble calcium bicarbonate, a solution of which can diffuse out of the film. At the surface of the film, the equilibrium changes direction, and the calcium bicarbonate is converted back to a deposit of calcium carbonate.

Dirt retention (dirt pickup) can be a difficult problem with latex house paints. Latex paints must be designed to coalesce at relatively low application temperatures. At warmer temperatures, soot and dirt particles that land on the paint surface may stick tenaciously and not be washed off by rain. Gilbert (2016) discusses the problems of formulation and testing. There can be wide variations in dirt retention among paints with similar hardness and sheen; the problem is addressed by testing each component of the coating. Dirt retention changes with exposure, requiring tests on specimens that have been exposed outdoors or in accelerated test devices. The problem is particularly severe in areas where soft coal is burned. As would be expected, dirt pickup is less for paints formulated with higher T_g polymers (Smith and Wagner, 1996; Wagner and Baumstark, 2002). It was also shown that the more hydrophobic styrene/acrylic latex paints picked up less dirt than acrylic latex paints having the same T_g (Gilbert, 2016).

Growth of fungi (mold and mildew) can lead to blotchy, dark deposits on the surface of coatings. Retardation of mildew growth is discussed in Section 32.1.

Often, one needs to be a detective to determine the origin of a paint failure. For example, there were suddenly numerous reports of ugly dirt spots on houses in Bismarck, North Dakota. It turned out that the flight pattern of airplanes taking off from the Bismarck airport had been changed. Oily droplets from plane emissions were landing on the paint, softening its surface and making it a good adhesive for dirt. The wide range of events that can occur outdoors complicates the problem of predicting coating performance based on laboratory testing.

5.6 TESTING FOR EXTERIOR DURABILITY

Ultimately, the most predictive method for determining the longevity of coatings is to apply the coating to the substrate of interest and expose the coated substrate in the environment/location in which it will be used. However, this is an impractical method for testing coatings for a number of reasons. First, it is economically unfeasible, to develop products based on real-time weathering data, because the feedback loop on performance of the coating would take about as many years for one cycle as the desired lifetime of the coating. Second, coatings are rarely used in only one environment/location. Vehicles are painted with the same coatings whether the customer lives in Alaska or Florida. The coating must perform adequately in both locations.

Thus, data on the long-term weathering performance of coatings must be generated via tests that accelerate the rate of degradation in order to facilitate formulation changes and incremental improvements. This data can be generated via three different approaches, which will be described in the following text. In addition, the tools used to probe the response of a coating to weathering can dramatically increase the amount of information gleaned from weathering exposures, which will be discussed.

5.6.1 Natural Weathering

The exposure of coatings in the outdoor environment, which at first does not seem like an accelerated exposure, does in fact increase the rate of weathering over exposure in the average location. South Florida has become the *de facto* standard for natural outdoor exposure, as this location provides high UV exposure (>300 MJ m^{-2} (295–385 nm)), warm temperatures (average ambient 24°C), high humidity (average 75%), and the highest average annual rainfall in the continental United States (sometimes in excess of 1 m). For many coatings systems, this combination of climatic conditions corresponds to a worst-case scenario, and thus, coatings exposed in all other locations in the United States and most places in the world will weather more slowly than in south Florida. For example, the rate of weathering in the northeast of the United States is on the order of half that observed for most coatings in south Florida. There are certainly exceptions to this rule, such as Denver cracking previously mentioned, but to a first approximation, south Florida is a good place to weather coatings and will provide acceleration over exposure in most other locations. Note that we differentiate between long-term weathering testing, which traditionally takes place in south Florida, and acid etch testing, which has traditionally taken place in Jacksonville, FL, due to very local environmental factors that were unique to that location.

The details of the exposure also affect the natural weathering performance. When coatings are exposed in the northern hemisphere (e.g., south Florida), the coating panels are typically oriented south, so that the sun faces the coated surface during the day. An important exception is when one tests the mold or mildew resistance of architectural coatings, where the panels are purposely exposed facing north so that the mold is allowed to grow in the shade. In addition to facing south, the coatings can be exposed at

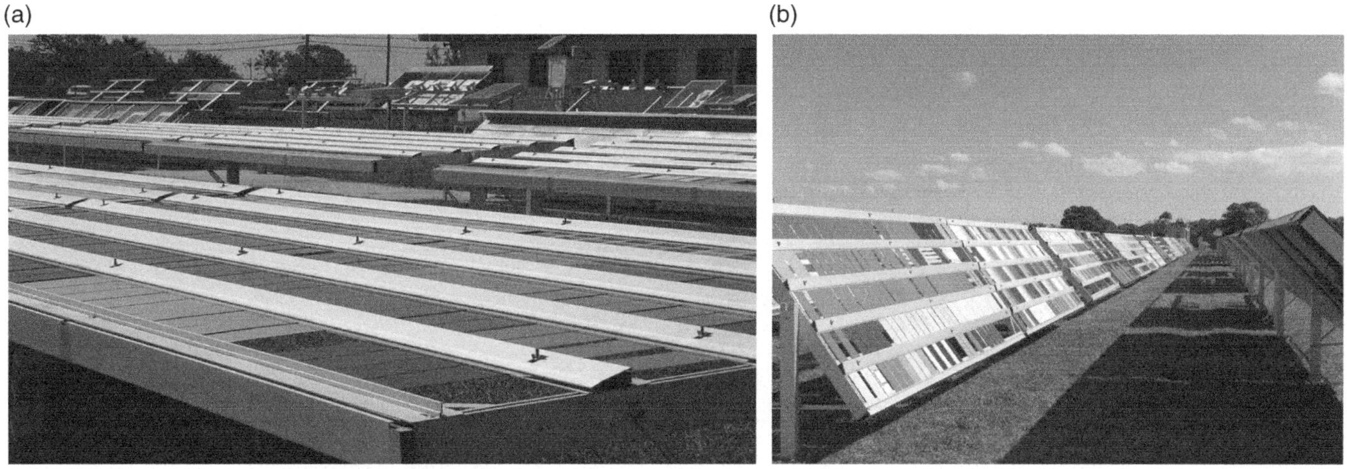

Figure 5.1 Panel exposure racks in south Florida. (a) 45° exposure, (b) 5° exposure. Source: Photo Courtesy of Atlas Materials Testing LLC.

various elevation angles. Automotive coatings are typically exposed at 5° from horizontal. This allows for high UV doses and also allows some water to pool on the surface during rain and dew events. The maximum UV exposure is always obtained when the coatings are exposed "at latitude," meaning exposure at an angle equal to the latitude of exposure. For example, panels exposed in south Florida (25°N) would receive the maximum solar radiation if exposed facing south at 25° from horizontal. For coil coatings, a major supplier exposes at 90° for sidewall panels and 45° for roofing panels (Pilcher, 2001). Architectural and industrial coatings are typically exposed at 45°. It is important to maintain records of the weather throughout the exposure period. Modern practice is to use spectral radiometers to record the cumulative amount of radiant energy striking the test specimen. Since different colors absorb different amounts of IR radiation, the maximum temperatures reached by coatings during exposure depends on their color; this difference can affect exposure test results. Johnson and McIntyre (1996) provide an analysis of various test methods. Hicks and Crewdson, in Koleske (2012), provide a review of different means of exposure and testing of the exposed panels.

Test specimens are examined periodically to compare their appearance before and after exposure. Usually, at least part of the coating surface is cleaned for the comparison. Ease of cleaning, change in gloss, change in color, degree of chalking, and any gross film failures are reported. Changes in color are particularly difficult to assess, since change in gloss or chalking changes the color even if there has been no change in the color of the components of the coating. The effects of changes in gloss and chalking on the color of the components can be minimized by cleaning the panels and then coating part of each panel with a thin layer of clear gloss lacquer. The lacquer layer minimizes the effect of differences in surface reflection on the color.

While most exposures are done with panels on racks without any backing behind the panels (see Figure 5.1), panels can be exposed with backing, which slightly increases the temperature of the panel and therefore may increase the rate of degradation. Whether this is a meaningful simulation depends on the end use of the coating. Some automotive components are tested in glass boxes to simulate the interior of vehicles. These components are obviously not exposed to the rain, but do experience high thermal and solar loads due to the strong heating of the vehicle interiors when parked in sunny climates. Both windshield glass, which can contain UVAs in the poly(vinyl butyral) inner layer, and side glass can be used as the glass cover.

Natural weathering exposures are certainly done in other locations. Phoenix, AZ, provides approximately 10% more solar radiation and an order of magnitude less water exposure than south Florida. For some coating failures driven strictly by UV, this location can be harsher than south Florida. Climatic conditions similar to south Florida can be found in northern Australia, so that coatings can be exposed to endless summer conditions by physically moving the exposure location between Florida and Australia to take advantage of the summer months in each location. Other exposure locations can be used to study different failure modes, such as seashore locations for corrosion performance, which will be discussed in a subsequent chapter.

5.6.2 Accelerated Outdoor Exposure

Weathering service providers have developed Fresnel mirror-based devices that focus the sun's radiation on test panels. These devices are engineered such that the intensity of the sunlight directed at the coating system is several times that of the ambient radiation. This exposure also guarantees that the radiation is of the proper wavelength so as not to distort the weathering chemistry. The mirrors on the

Figure 5.2 Fresnel reflector accelerated outdoor weathering device. Source: Photo Courtesy of Atlas Materials Testing LLC.

device are computer controlled to track the sun in the sky, thus assuring that the maximum intensity of sunlight will always be reflected onto the coating (Figure 5.2). Fresnel reflector devices with trade names such as *EMMAQUA* (Equatorial Mount with Mirrors for Acceleration plus water), *FRECKLE*, and *Sun-10* are operated in the Arizona desert (Verma and Crewdson, 1994). These devices enhance the intensity of sunlight on the panel surfaces by a factor of 8 over direct exposure and are reported to accelerate degradation rates 4–16 times the rate for non-accelerated exposure (see Hicks and Crewdson in Koleske, 2012). Panels are cooled by blowing air both over and under the samples. Since rainfall and humidity in the desert are low, the test facilities permit periodic spraying of water on the panel surfaces. In addition, research has shown that spraying water directly onto the samples does not emulate the amount of water absorbed by the same coating systems exposed in south Florida. Thus, the physical manifestations of failure may be different than those observed during natural weathering (Nichols and Misovski, 2009).

Fresnel reflector weathering devices record actual exposure energies per unit area, expressed in $MJ\,m^{-2}$. For comparison, the results can be approximately translated into Arizona or Florida exposure times based on relative solar intensity—that is, energy per unit time. Comparisons based on intensity of exposure to a 10 nm band of UV around 313 nm give results that correlate most closely with field exposures. An important conclusion from accelerated outdoor exposure results is that if the spectral power distribution (SPD) is correct, as it is in these Fresnel reflector devices, the degradation chemistry can be accelerated without altering the chemical compositional changes that occur during natural weathering (Gerlock et al., 2003). This same principle applies, advantageously, to exposure in accelerated weathering devices as well, where the light intensity can be increased by moving samples closer to the artificial light source or by increasing the power of the lamp.

As the accumulation of durability data is often the rate-limiting step in the coating development cycle, greater acceleration is very desirable. The ultimate acceleration that one can achieve without altering the chemistry is an open question. Ultra-accelerated outdoor weathering is an area currently under investigation, including the potential use of cold mirror technology to concentrate only the UV portion of the spectrum on samples while filtering the visible and IR (Hardcastle, 2015).

5.6.3 Accelerated Laboratory Weathering Devices

The need to accelerate the rate of weathering has led to the development of laboratory instruments that simulate the effects of outdoor exposures. Typically, these devices attempt to apply light, heat, humidity, and liquid water in some combination to mimic outdoor conditions. Many different specifications have been written on the various combinations of light, water, and heat to best simulate the outdoor environment. However, none provide a 100% faithful reproduction of natural weathering.

To accurately reproduce outdoor exposure in an accelerated weathering device, the influence of each of the weathering variables must be considered. As with accelerated outdoor weathering, an overarching requirement should be that the chemical changes, which occur in the accelerated weathering device, should be equivalent to those that occur during natural outdoor exposure. Otherwise, the accelerated test could be quite misleading. Many cases of false positives and false negatives have been found using various accelerated weathering tests. Some automotive base coat–clear coats containing alkanoyl HALS (Section 5.2.3) performed very well in accelerated devices that used a fluorescent UV light source but had poor resistance to delamination in the field. It was shown that this type of HALS was converted into the active nitroxyl radical more quickly by the short wavelength UV source in the fluorescent weathering device than by sunlight. One must be careful of blindly trusting accelerated weathering data (Bauer et al., 1990). False negatives can lead to missed business opportunities. False positives can lead to costly warranty claims.

An overriding concern in the validity of an accelerated weathering test is the light source used in the test chamber. Both xenon arc and fluorescent tube sources are used commercially. In the future, LED sources may become common. The original accelerated weathering devices used carbon arc light sources, but those have been almost completely removed from service. A comparison of the SPD of natural sunlight and xenon arc light sources filtered by various glass filters is shown in Figure 5.3.

An ideal accelerated weathering device would emit radiation that exactly matches the terrestrial solar spectrum. No such light source exists, as the radiation produced by nuclear fusion in the sun and filtered by the Earth's atmosphere cannot be exactly reproduced via any simple device (Adema et al., 2016). However, xenon arc lamps filtered by specific glass filters can provide a very good match to sunlight. Conventionally, xenon arc weathering has been filtered by a combination of two filters, one made from quartz and the other from borosilicate glass, commonly referred to as *quartz/boro weathering*. This combination of filters produces a SPD that mimics sunlight above 300 nm but allows some wavelengths down to 280 nm to reach the sample, corresponding to exposure below the solar *cut-on*. Exposure to these higher energy wavelengths below 300 nm, which are blocked from the Earth's surface by the ozone layer, can cause unrealistic weathering chemistry; therefore the quartz/boro filter combination should be avoided. A better match to sunlight can be obtained by utilizing a *boro/boro* combination of filters, which provide a reasonably close match to sunlight. But even this combination of filters allows some unnatural wavelengths of light to reach the specimen, which has been shown to alter the weathering chemistry in some paint systems (Gerlock et al., 2003). More recently a new set of filters, commercially available as RightLight or Daylight-F, provide an almost exact match to solar spectrum. These filters have been shown to reproduce the correct weathering chemistry in paint systems that are highly sensitive to the light source (Nichols et al., 2013). Recently a new weathering specification, ASTM D7869, has been written around these filters and has been shown to reproduce the vast majority of failure mechanisms observed during natural weathering.

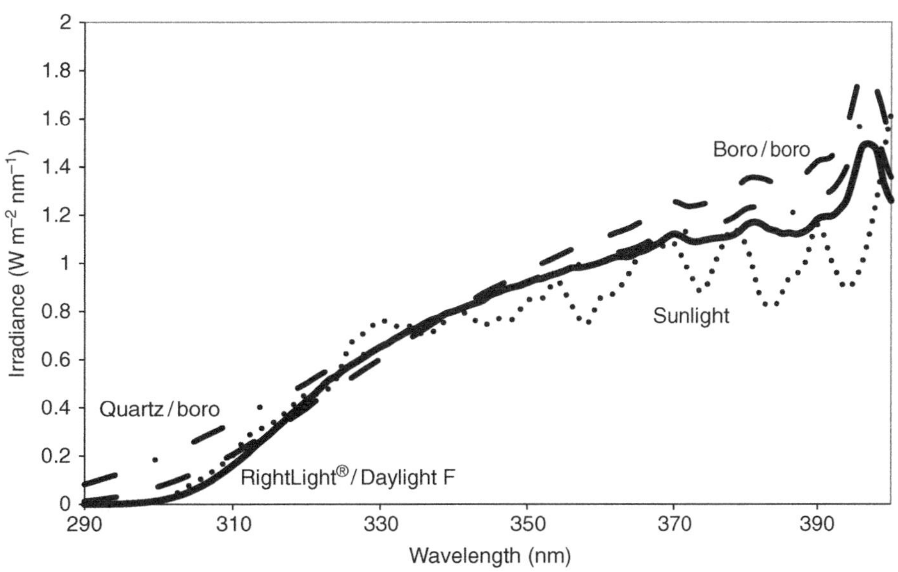

Figure 5.3 Spectral power distribution (SPD) of sunlight and xenon arc light filtered by typical filter combinations. All SPDs normalized to 0.8 W m^{-2} at 340 nm.

In addition to providing an accurate reproduction of solar exposure, the weathering device needs to provide similar water exposure to that experienced by painted panels tested outdoors. As discussed earlier, water exposure can vary greatly depending on the location around the globe. However, as south Florida is used as the most common weathering location, water exposure during accelerated weathering can be compared with that observed in south Florida, where the role of water during paint weathering is quite complex (Baukh et al., 2010; Lee et al., 2011; Hinderliter and Sapper, 2015). Aside from being wet by rain, the coatings are often saturated with water at night, as dew forms on the panels shortly after sunset. During the night, water diffuses into the paint system, which can aid in the transport of other small molecules into the paint system. The paint also becomes plasticized, resulting in swelling-induced mechanical stresses. Degraded material can also be washed from the surface. After sunrise, the system rapidly heats up and the water diffuses out of the paint system.

Some accelerated weathering protocols provide for alternating dark (lamp off) and light (lamp on) cycles; and most prescribed a liquid water spray during the light cycle. However, the coating is typically quite hot during the light cycle and little water can diffuse into the paint system. Newer methods, such as ASTM D7869, provide for much longer dark cycles during which water is allowed to penetrate deep into the paint system, thereby providing better correspondence with real-world failure mechanisms.

The temperature distribution during accelerated weathering is critical, as most of the degradation mechanisms that occur during weathering are thermally activated processes. The rate of degradation can be increased by increasing the temperature, but the chemistry can be altered if the temperature becomes too high. Most accelerated weathering protocols have settled on sample temperatures in the range of 60–70°C. The air temperature in the chamber is often less than this, as the samples are also heated by the intense radiation exposure. During natural weathering, the temperature of dark-colored coatings has been shown to be similarly elevated during the peak sunlight portions of the day in the summer. Clearly the temperature of the coatings both outdoors and in weathering devices is strongly dependent upon their color.

The previous discussion pertained to accelerated weathering devices that use a xenon arc light source. Most of the same considerations can be applied to weathering devices that use fluorescent tubes to provide radiation. These devices are typically simpler to operate and less expensive to buy. However, their simplicity reduces controllability and fidelity with respect to their ability to mimic outdoor exposure variables.

The SPD of radiation from fluorescent devices is determined by the composition of the phosphor that coats the inside of the fluorescent tubes. The most common types are UV-B, with a peak intensity at 313 nm, and UV-A, with a peak intensity at 340 nm. The UV-B bulbs emit radiation that extends well below the solar cut-on and have been shown repeatedly to distort weathering chemistry (Bauer et al., 1990). UV-A bulbs emit radiation that more closely corresponds to terrestrial UV but emit little visible or IR radiation. Accordingly, there is little temperature correspondence with natural weathering.

5.6.4 Analysis of Coating Changes during Weathering

Gloss and color change are the most common response variables that are measured either during natural or accelerated weathering tests. These two properties are important early indicators of the durability of any particular coating, as changes in the appearance of a coating are often the first sign of degradation during extended exposure.

Whereas measurement of gloss or color may be a good indicator of longer-term durability, these measurements do not provide information on the underlying chemical changes that are occurring and continue to occur. Such chemical information is expected to be more informative than physical manifestations in predicting long-term weathering performance and, perhaps more importantly, may provide insights into strategies for extending the long-term performance of the coating.

A number of possible chemical markers could be targeted for quantification by various analytical chemistry techniques. Most photooxidation and hydrolysis reactions produce hydroxyl and carboxylic acid functionalities in the binder. Their concentration can be determined in a straightforward way by measuring the intensity of the appropriate absorption peaks in the IR spectrum. IR spectra can be collected in transmission on free films, on powders incorporated into salt crystals, on the surface of a coating via attenuated total reflection (ATR) accessories, and from total paint systems via photoacoustic (PAS) FTIR. Growth of the carbonyl band has been used extensively to study the rate of oxidation of coating binders and other polymers (Bauer et al., 1991). Although this method offers a sensitive measure of degradation, the results are somewhat ambiguous, as some degradation processes can both form and break carbonyl groups. A more representative measure of degradation is the size of the broad peak between 2900 and 3600 cm^{-1} due to –OH, –NH, and –COOH groups in the polymer. As these groups are commonly produced when all coatings degrade, one can use the concentration of these groups to quantify the extent of weathering. Researchers have successfully used this method to study the degradation kinetics in automotive coatings and epoxy coatings (Gerlock et al., 2001a).

Since photooxidation is a free radical process, the presence of radicals can be quantified via electron spin resonance (ESR) spectroscopy. Gerlock et al. (1985) have correlated the rate of disappearance of stable nitroxyl radicals with loss of gloss in long-term Florida exposure. In this technique, nitroxyl radical precursors are incorporated into TiO_2 pigmented acrylic–melamine coatings subjected to ambient or accelerated short-term UV exposure. These studies allowed calculation of photoinitiation rates (PR) of free radical formation, which were found to correlate with rates of gloss loss (GLR) by the relationship: GLR $\propto$ $(PR)^{1/2}$. This proportional relationship of GLR with $(PR)^{1/2}$ is consistent with a free radical process and results from termination by second-order radical–radical reactions. PR, determined by this method, have also been used to evaluate experimental conditions for the synthesis of acrylic polyols by free radical polymerization, including the effect of initiator, temperature, and solvent on the projected exterior durability of the resulting acrylic–melamine coating (Gerlock et al., 1987).

Okamoto et al. (1986) have correlated the rate of buildup of phenoxy radicals from UV irradiation of acrylic–melamine coatings, containing BPA epoxy resins, in an ESR spectrometer cavity with cracking of corresponding coating films exposed in a QUV weathering device. The latter technique was used to evaluate UV stabilizers. The nitroxyl early detection method has also been used to investigate the synergistic stabilizing effect of a benzotriazole UV stabilizer and a HALS derivative in acrylic–urethane coatings. The UV stabilizer reduces the PR of free radical formation, whereas the HALS derivative reduces the propagation rate by lowering the concentration of free radicals (Gerlock et al., 2001a).

A somewhat more direct application of ESR for determining the photostability of coating films is to UV irradiate films well below T_g at a temperature of 140 K where free radicals are stable (Sommer et al., 1991). The method is appropriate for evaluating the effectiveness of stabilizers, such as HALS compounds, by comparing radical concentrations with and without stabilizer. Gerlock et al. (1985) determined that useful comparisons can be made in 3 h.

An alternative approach to utilizing expensive ESR instruments is to cryogenically grind exposed films to a fine powder and measure the buildup of hydroperoxides by titration with periodate. By this method and assuming the relationship GLR $\propto$ $(PR)^{1/2}$, previously determined by ESR, Gerlock et al. (1985) established a linear relationship between photooxidation rates and hydroperoxide concentrations in acrylic–urethane and acrylic–MF coatings.

The long-term performance of many paint systems is contingent upon the presence of the stabilizer additives (UVA and HALS) that are added to their formulation. Thus, the effectiveness and longevity of those stabilizers is critical to the coatings' long-term durability. As discussed earlier, organic UVAs do not last indefinitely in coatings exposed outdoors. UV spectroscopy can be used to measure the concentration of these additives as a function of weathering time to determine their depletion rate (Pickett and Moore, 1995; Smith et al., 2001). Samples, obtained by scraping and extraction into solution, can be analyzed by microtomy and transmission UV spectroscopy. Complete paint systems can be analyzed by micro-UV spectroscopy, as well as by TOF-SIMS instrumentation, which can determine UVA content as well as map the distribution of UVAs in coatings.

The concentration of HALS is more difficult to quantify by simple spectroscopic techniques. However, a peracid oxidation technique has been developed that allows for the quantitative determination of active HALS in a coating (Kucherov et al., 2000). In combination with microtomy, the depth-dependent concentration of HALS can be determined as a function of weathering time for a coating. As with UVAs, mentioned earlier, TOF-SIMS instruments are now capable of directly imaging and mapping the concentration of HALS as well (Nichols and Kaberline, 2013). Adamsons (2013) has recently reviewed the use of various analytical techniques.

Gebhard et al. (2005) have provided a study of various exposure tests and analytical tools that give early predictions of gloss retention of seven pigmented latex paints with known exterior durability. The most useful results were obtained with X-ray photoelectron spectroscopy (XPS). XPS was used to analyze the surface of films for titanium, aluminum, and silicon in relationship with polymeric carbon before and after exterior exposure. (The aluminum and silicon were present from the surface treatment of the TiO_2 in the paint.) Exposure leads to the degradation of polymer from the film surface, which increases the ratio of the metals to the polymeric carbon. Initially surfactant bloomed to the surface and interfered with the analysis but was absent after 2 weeks' exposure. The results obtained after 12 weeks' Florida exposure (following loss of the surfactant) allowed good predictions of gloss loss after 96 weeks of Florida exposure.

In addition to the analytical chemistry techniques detailed earlier, the physical properties of coatings can be measured as a function of weathering time to assess the rate of degradation and the manifestations of the chemical changes on the physical properties of the coatings. Dynamic mechanical thermal analysis (DMTA or DMA) has been successfully used to track changes in the modulus of coatings as weathering progresses. In particular, the modulus in the rubber plateau of cross-linked coatings has been used to assess changes in cross-link density during weathering (Hill et al., 1994; Mitra et al., 2014). Chemorheological techniques have also been used to assess cross-link density changes (Nichols et al., 1997).

As discussed in Chapter 4, many coatings embrittle as weathering progresses. Those changes can be quantified by measuring the fracture energy or essential work to fracture of coatings. Embrittlement can be related to both changes in cross-link density and the buildup of internal stresses due to chemical and physical aging. Many coatings embrittled at a rapid rate early in their lifetime and then more slowly as weathering progresses. The use of HALS, in particular, has been shown to greatly reduce the rate of embrittlement (Nichols and Gerlock, 2000).

AFM has been recommended for studying mechanical and chemical changes in the surface of films over time (Gu et al., 2004). Along these lines, various authors have shown that surface roughness increases as does binder heterogeneity during weathering.

5.7 SERVICE LIFE PREDICTION

Researchers at the US National Institute of Standards and Technology (NIST) have proposed a new methodology for assessing the weatherability of coatings (Martin et al., 2008). This approach is based on the SLP paradigm, which has been successfully applied to microelectronics and other modern materials/systems. The approach emphasizes an understanding of the mechanisms of degradation and how these mechanisms respond to environmental loads, such as temperature, moisture, and UV radiation. *Reciprocity*, *additivity*, and the precise measurement of the response of a coating and its constituents to these loads form the basis of this approach. Reciprocity signifies that a coating's response to light intensity is proportional to the magnitude of the input. If a coating degrades twice as fast when it is exposed to light at twice the intensity, then the coating conforms to the law of reciprocity. In that case, the product of the intensity and time (i.e., dose) is what controls the extent of degradation. Additivity signifies that the damage induced by radiation of a given wavelength can be summed with the damage induced by other wavelengths to give the total amount of degradation. Practical application of these principles is contingent upon very precise measurement of all the exposure variables and coating responses.

NIST has collected much data using its Simulated Photodegradation via High Energy Radiant Exposure (SPHERE) instrument. With this instrument, well-controlled exposure conditions including light intensity and SPD, temperature, and humidity have been used to expose a number of different coating materials. Degradation mechanisms have been deduced and lifetimes predicted for a few materials (Chin et al., 2005). Clearly this is an area that merits further research. Summaries of this approach can be found in a series of conference proceedings (Bauer and Martin, 1999; Martin and Bauer, 2002; Martin et al., 2009).

GENERAL REFERENCES

Bauer, D. R.; Martin, J. W., *Service Life Prediction of Organic Coatings—Methodologies and Metrologies*, American Chemical Society, Washington, DC, **2001**.

Valet, A., *Light Stabilizers for Paints*, translated by Welling, M. S., Vincentz, Hannover, **1997**.

Wypych, G., *Handbook of Material Weathering*, 5th ed., ChemTec, Toronto, **2013**.

REFERENCES

Adamsons, K., *J. Coat. Technol. Res.*, **2013**, *9*, 745–756.

Adema, K. N., et al., *Phys. Chem. Chem. Phys.*, **2015**, *17(30)*, 19962–19976.

Adema, K. N. S., et al., *Polym. Deg. Stab.*, **2016**, *123*, 121–130.

Ávár, L.; Bechtold, K., *Prog. Org. Coat.*, **1999**, *35(1–4)*, 11–17.

Bauer, D. R., *Prog. Org. Coat.*, **1986**, *14(3)*, 193–218.

Bauer, D. R., *J. Coat. Technol.*, **1994**, *66(835)*, 57–65.

Bauer, D. R.; Martin, J. W. E., *Service Life Prediction of Organic Coatings: A Systems Approach*, American Chemical Society, Washington, DC, **1999**.

Bauer, D. R., et al., *Ind. Eng. Chem. Res.*, **1988**, *27(1)*, 65–70.

Bauer, D. R., et al., *Polym. Deg. Stab.*, **1990**, *28(1)*, 39–51.

Bauer, D. R., et al., *Ind. Eng. Chem. Res.*, **1991**, *30(11)*, 2482–2487.

Baukh, V., et al., *Macromolecules*, **2010**, *43(8)*, 3882–3889.

Betz, P.; Bartelt, A., *Prog. Org. Coat.*, **1993**, *22(1)*, 27–37.

Bohnke, H., et al., *J. Coat. Technol.*, **1991**, *63(799)*, 53–60.

Braun, J. H., *J. Coat. Technol.*, **1990**, *62(785)*, 37–42.

Braun, J. H.; Cobranchi, D., *J. Coat. Technol.*, **1995**, *67(851)*, 55–62.

Brown, L. H., in Meyers, R. R.; Long, J. S., Eds., *Treatise on Coatings*, Marcel Dekker, New York, **1972**, Vol. *1*, pp. 536–563.

Carlsson, D. J., et al., *Die Makromol. Chem.*, **1984**, *8(Suppl. 8)*, 79–88.

Chin, J., et al., *J. Coat. Technol. Res.*, **2005**, *2(7)*, 499–508.

Cliff, N., et al., *J. Coat. Technol. Res.*, **2005**, *2(5)*, 371–387.

Fairgrieve, S. P.; MacCallum, J. R., *Polym. Deg. Stab.*, **1984**, *8(2)*, 107–121.

Gebhard, M. S. et al., *Waterborne High Solids Powder Coating Symposium*, New Orleans, **2005**.

Georgiev, G., et al., *J. Macromol. Sci. Chem.*, **1990**, *27(8)*, 987–997.

Gerlock, J., et al., *J. Coat. Technol.*, **1985**, *57(722)*, 37–46.

Gerlock, J., et al., *Prog. Org. Coat.*, **1987**, *15(3)*, 197–208.

Gerlock, J. L., et al., *Macromolecules*, **1988**, *21*, 1604–1607.

Gerlock, J. L., et al., *Polym. Deg. Stab.*, **1999**, *65(1)*, 37–45.

Gerlock, J., et al., A Brief Review of Paint Weathering Research at Ford in Martin, J.; Bauer, D. R., Eds., *Service Life Prediction Methodologies and Metrologies*, ACS Symposium Series 805, American Chemical Society, Washington, DC, **2001a**, pp 212–249.

Gerlock, J. L., et al., *J. Coat. Technol.*, **2001b**, *73(918)*, 45.

Gerlock, J. L., et al., *Polym. Deg. Stab.*, **2001c**, *73(2)*, 201–210.

Gerlock, J. L., et al., *J. Coat. Technol.*, **2003**, *75(936)*, 35–45.

Gilbert, J. A., *Coat. World*, **2016**, *21(9)*, 117–122.

Gregorovich, B.:, Hazan, I., *Prog. Org. Coat.*, **1994**, *24(1)*, 131–146.

Gryn'ova, G., et al., *J. Am. Chem. Soc.*, **2012**, *134*, 12979–12988.

Gu, X., et al., *J. Coat. Technol. Res.*, **2004**, *1(3)*, 191–200.

Haacke, G., et al., *J. Coat. Technol.*, **1996**, *68(855)*, 57–62.

Haacke, G., et al., *J. Coat. Technol.*, **1999**, *71(1)*, 87–94.

Hardcastle, H. K., Ultra-Accelerated Weathering II: Considerations for Accelerated Data-Based Weathering Service Life Prediction in White, C. C., et al., Eds. *Service Life Prediction of Exterior Plastics: Vision for the Future*, Springer International Publishing, Basel, **2015**, pp 165–184.

Hayoz, P., et al., *Prog. Org. Coat.*, **2003**, *48(2)*, 297–309.
Hill, L., et al., *Prog. Org. Coat.*, **1994**, *24(1)*, 147–173.
Hinderliter, B. R.; Sapper, E. D., *J. Coat. Technol. Res.*, **2015**, *12(3)*, 477–487.
Hsiao, Y. C., et al., *J. Appl. Polym. Sci.*, **1975**, *19(10)*, 2817–2820.
Johnson, B.; McIntyre, R., *Prog. Org. Coat.*, **1996**, *27(1)*, 95–106.
Jones, T. E.; McCarthy, J. *J. Coat. Technol.*, **1995**, *67(844)*, 57–65.
Kiil, S., *J. Coat. Technol. Res.*, **2012**, *9(4)*, 375–398.
Koleske, J. V. Ed., *Paint and Coatings Testing Manual*, ASTM, Conshohocken, **2012**.
Kucherov, A. V., et al., *Polym. Deg. Stab.*, **2000**, *69(1)*, 1–9.
Lee, S., et al., *J. Mater. Res.*, **2011**, *18(09)*, 2268–2275.
Lima, J. F. d., et al., *Appl. Surf. Sci.*, **2009**, *255(22)*, 9006–9009.
Martin, J. W.; Bauer, D. R. Eds., *Service Life Prediction Methodologies and Metrologies*, ACS Symposium Series 805, American Chemical Society, Washington, DC, **2002**.
Martin, J. W., Ryntz, R. A., Chin, J.; Dickie, R., Eds., *Service Life Prediction of Polymeric Materials: Global Perspectives*. Springer Science & Business Media, New York, **2008**.
Martin, J. W., et al., *Service Life Prediction of Polymeric Materials*, Springer, New York, **2009**.
Mielewski, D., et al., *Polym. Deg. Stab.*, **1991**, *33(1)*, 93–104.
Mitra, S., et al., *Prog. Org. Coat.*, **2014**, *77(11)*, 1816–1825.
Nguyen, T., et al., *J. Coat. Technol.*, **2003**, *75(941)*, 37–50.
Nichols, M. E.; Gerlock, J., *Polym. Deg. Stab.*, **2000**, *69(2)*, 197–207.
Nichols, M. E.; Kaberline, S. L., *J. Coat. Technol. Res.*, **2013**, *10(3)*, 427–432.
Nichols, M. E.; Misovski, T., The Influence of Water on the Weathering of Automotive Paint Systems in Martin, J. W. et al. Eds., *Service Life Prediction of Polymeric Materials: Global Perspectives*, Springer, New York, **2009**, pp 295–308.
Nichols, M. E.; Peters, C. A., *Polym. Deg. Stab.*, **2002**, *75(3)*, 439–446.
Nichols, M., et al., *Polym. Deg. Stab.*, **1997**, *56(1)*, 81–91.
Nichols, M. E., et al., *Prog. Org. Coat.*, **1999**, *35(1–4)*, 153–159.
Nichols, M., et al., *J. Coat. Technol. Res.*, **2013**, *10(2)*, 153–173.
Nichols, M. E.; Frey, J. R., *J. Coat. Technol. Res.*, **2014**, *12(1)*, 49–61.
Okamoto, S., et al., *XVII FATIPEC Congress*, **1986**, pp 239–255.
Pappas, S. P.; Just, R. L., *J. Polym. Sci. Polym. Chem.*, **1980**, *18(2)*, 527–531.
Pappas, S. P.; Winslow, F. H., *Photodegradation and Photostabilization of Coatings*, American Chemical Society, Washington, DC, **1981**.
Peters, C., et al., *Prog. Org. Coat.*, **2007**, *58(4)*, 272–281.
Pickett, J. E.; Moore, J. E., *Die Angew. Makromel. Chem.*, **1995**, *232(1)*, 229–238.
Pilcher, G., *J. Coat. Technol.*, **2001**, *73(921)*, 135–143.
Saadat-Monfared, A., et al., *Coll. Surf. A*, **2012**, *408*, 64–70.
Schulz, U., et al., *Prog. Org. Coat.*, **2000**, *40(1)*, 151–165.
Sharrock, R. F., *J. Coat. Technol.*, **1990**, *62(789)*, 125–130.
Shelton, J. R.; Hawkins, W. L., Environmental Deterioration of Polymers in Hawkins, W. L., Ed., *Polymer Stabilization*, Wiley-Interscience, New York, **1972**, p 110.
Smith, A.; Wagner, O., *J. Coat. Technol.*, **1996**, *68(862)*, 37–42.
Smith, C. A., et al., *Polym. Deg. Stab.*, **2001**, *72*, 89–97.
Sommer, A., et al., *Prog. Org. Coat.*, **1991**, *19(1)*, 79–87.
Stamatakis, P., et al., *J. Coat. Technol.*, **1990**, *62(789)*, 95–98.
Starnes, W., *Pure Appl. Chem.*, **1985**, *57(7)*,1001–1008.
Teaca, C.-A et al., *BioResources*, **2013**, *8(1)*, 1478–1507.
Timpe, H.-J., et al., *Polym. Photochem.*, **1985**, *6(1)*, 41–58.
Valet, A., *Light Stabilizers for Paints*, Vincentz Verlag, Hannover, **1997**.
Verma, M.; Crewdson, L. F., A study of the color change of automotive coatings subjected to accelerated and natural SAE weathering tetsts for exterior materials, SAE Technical Paper, #940856, **1994**.
Voelz, H. G. et al., in Pappas, S. P.; Winslow, F. H., Eds., *Photodegradation and Photostabilization of Coatings*, ACS Symposium Series 151, American Chemical Society, Washington, DC, **1981**, pp 163–182.
Wagner, O.; Baumstark, R., *Macromol. Symp.*, **2002**, *187*, 447–458.
Watson, S., et al., *J. Coat. Technol. Res.*, **2012**, *9(4)*, 443–451.
Wypych, G., *Handbook of Material Weathering*, 5th ed., ChemTec, Toronto, **2013**.
Yaneff, P. V., et al., *J. Coat. Technol. Res.*, **2004**, *1(3)*, 201–212.
Zee, M., et al., *Prog. Org. Coat.*, **2015**, *83*, 55–63.

Chapter 6

Adhesion

By definition, a coating must adhere to a substrate. Without an underlying material, a film is not a coating. From this fact, the importance of adhesion is obvious. If the bond between the coating and substrate is weak or deteriorates, the coating ceases to be attached to the substrate, and the coating has failed. Therefore, adhesion is perhaps the most critical property of a coating. Yet adhesion is almost certainly the least well understood and the most difficult property of a coating to quantify. The reasons behind this are multifold and will be explored in this chapter, in addition to what we do understand or can infer about adhesion and how it relates to coating formulation and performance.

One difficulty in dealing with adhesion is defining what adhesion means. In most cases, a coatings formulator thinks of adhesion in terms of the question: how hard is it to remove the coating? But a physical chemist would think in terms of the work required to separate two interfaces that are adhering. These perspectives can be different considerations; the latter is only one aspect of the former. Removal of a coating requires breaking or cutting through the coating, pushing the coating out of the way, and separating the coating from the substrate. Consider as an extreme example, adhesion of a plastic-covered wire used in electrical connections. Such a covering must have good "adhesion" to stay on the wire to protect against short circuits and shocks. However, to attach the wire to a fixture, one needs to make a cut through the covering and easily slip the coating off the metal. It is desirable to have minimal interactive forces between the plastic and the copper, but with considerable toughness within the coating so that accidental removal is unlikely. Complications such as this make coating adhesion difficult to quantify or even specify.

6.1 MECHANISMS OF ADHESION

Many different mechanisms have been proposed to explain the adhesion of one surface to another. In most cases, multiple mechanisms exist for any substrate–coating combination. The individual mechanisms will be discussed in Sections 6.2–6.5. An overriding theme, however, is that most surfaces must be thoughtfully prepared for coating. Surfaces rarely are ready to be painted without some type of cleaning, mechanical abrasion, chemical conversion, or other deliberate action, which is meant to improve the adhesion of the coating to the surface. Such surface preparations almost always result in improved chemical or physical interactions between the coating and substrate and provide the foundation for good adhesion, which is often paramount in achieving good corrosion performance and long-term durability.

6.1.1 Surface Mechanical Effects on Adhesion

Resistance to separation of coating and substrate can be enhanced by roughening the surface to be coated. Roughening increases the contact area of the coating with the surface. In theory, adhesive strength depends on the contact area between the coating and the surface, while the stresses that may cause failure do not. Consider the schematic representations in Figure 6.1. With a very smooth interface between coating and substrate, as shown in sketch A, the only forces holding the substrate and coating together are the interfacial attractive forces per unit of geometric area. With a rough surface on a microscopic scale, as represented in sketch B, the contact area is substantially

Organic Coatings: Science and Technology, Fourth Edition. Frank N. Jones, Mark E. Nichols, and Socrates Peter Pappas.

Figure 6.1 Geometries of surface interactions between a coating and a substrate. (a) Smooth interface between a coating and the substrate; (b) rough surface on a microscopic scale; and (c) rough surface with incomplete penetration of a coating.

increased. Another factor may be important: in some places, there can be undercuts in the substrate; to pull the coating off the substrate, one would either have to break the substrate or break the coating to separate them. This *lock and key* mechanism is analogous to a dovetail joint that holds two pieces of wood together.

The effect of roughening is controversial. For example, Abbott (2015) questions the importance of surface roughness. His view is that roughening often does not increase the surface area sufficiently to make a major difference and, further, that the surface profile rarely allows the lock and key effect to be significant. However, if surface roughness were proven unimportant, it would come as a great surprise to decades of people with experience in the coatings field.

Experience shows that better adhesion is generally obtained if the surface of a substrate is roughened before coating; however, as can be seen in sketch C, surface roughness can be a disadvantage. If the coating does not completely penetrate into the microscopic pores and crevices in the surface, dovetail effects are not realized, and the actual interfacial contact area is reduced and can, in extreme cases, be smaller than the geometric area. Furthermore, when water permeates through the film to the substrate, there will be areas of contact of water with uncovered substrate, which can be a major problem, especially if the coating is to protect steel against corrosion.

The scale of the roughness of surfaces can vary from macroscopic to microscopic to submicroscopic. It is important to consider the situation on a microscopic and submicroscopic scale. The situation is analogous to penetration of a liquid into a capillary on the microscale, although possibly not on the nanoscale. Equation 6.1 shows the variables affecting distance of penetration L (cm) into a capillary of radius r (cm) in time t (s), where γ is the surface tension (mN m^{-1}), θ is contact angle, and η is viscosity (Pa·s):

$$L = 2.24\left(\frac{\gamma r t}{\eta}(\cos\theta)\right)^{\frac{1}{2}} \qquad (6.1)$$

Rate of penetration, L/t, increases with increasing surface tension of the coating. However, there is an upper limit to this surface tension effect because the rate is strongly affected by contact angle. The rate is fastest when cosine of the contact angle is 1—that is, when the contact angle is zero. The cosine can only be 1 if the surface tension of the liquid coating is less than that of the solid substrate. The radius of the capillary is a variable of the substrate, not of the coating.

Accordingly, the variable over which the formulator has greatest control is viscosity (η). The critical viscosity is that of the continuous (external) phase of the coating, not the bulk viscosity of the coating, which includes the internal phase of pigment and polymer particles that are too large to effectively participate in penetrating microscopic and submicroscopic surface irregularities. The lower the viscosity of the external phase, the more rapid the penetration. Since in most cases, viscosity of the vehicle increases after application; it is important to maintain low viscosity for a long enough time for penetration to approach completion. Since viscosity of resin solutions increases with molecular weight (MW), one would expect that lower MW resins would provide superior adhesion after cross-linking, everything else being equal. This hypothesis has been confirmed in the case of epoxy resin coatings on steel (Sheih and Massingill, 1990). Another possible advantage of low MW resins is that a larger proportion of their molecules may be able to penetrate into smaller crevices than is possible for higher MW molecules. Coatings having low viscosity external phases, slow evaporating solvents, and relatively slow cross-linking rates have been found, in general, to provide better adhesion. In general, baked coatings provide better adhesion than do air dry coatings. (The term *air dry* is widely used but is potentially confusing. It generally means that the film is formed at ambient temperature and does not necessarily mean that oxygen is required for cross-linking.) When a coated article enters an oven, its temperature increases, the viscosity of the external phase decreases, and penetration into surface irregularities becomes easier. This is only one of several possible explanations for the advantages of baked coatings when adhesion is critical.

While roughening, such as sanding or sand blasting, of the surface typically improves coating adhesion, the practitioner must be aware that debris or surface contaminants left behind after roughening can significantly decrease the adhesion of the subsequently applied coating. Loose or lightly bound particles will interact with the coating but will not themselves be adherent to the substrate, thereby allowing the coating to be easily removed from the substrate.

6.1.2 Relationships between Wetting and Adhesion

Wetting is a major, and perhaps limiting, factor in adhesion. Intermolecular contact between a substrate and a coating requires that the coating spreads spontaneously over the substrate surface. If not, there cannot be interactions, and hence, no contribution to adhesion. Zisman (1972) extensively studied the relationships between wetting and adhesion. A liquid spreads spontaneously on a substrate if the surface tension of the liquid is lower than the surface free energy of the solid. (Surface free energies of solids have the same dimensions as surface tension, discussed in Section 24.1). If a liquid has a sufficiently low surface tension, it spreads spontaneously on the substrate and has a contact angle of 0°. If the surface tension of a liquid is significantly higher than the surface free energy of the solid substrate, a drop of the liquid stays as a drop on the surface of the solid—the contact angle is 180°. At intermediate surface tensions, there are intermediate contact angles. A schematic drawing of a drop of liquid with an intermediate surface tension is shown in Figure 6.2.

The relationship between contact angle θ and the surface free energy of a planar substrate γ_S, the surface tension of the liquid γ_L, and the interfacial tension between the solid and the liquid γ_{SL} is given in Eq. 6.2:

$$\cos\theta = \frac{\gamma_S - \gamma_{SL}}{\gamma_L} \qquad (6.2)$$

Good adhesion is promoted by a contact angle that approaches 0°. Experimental determination of contact angles of complex systems like coatings, especially on rough substrates with heterogeneous compositions, is difficult (Yekta-Fard and Ponter, 1992). In general, it is sufficient to conceptualize the relationships by saying that adhesion requires that the liquid have a lower surface tension than the surface free energy of the substrate to be coated. From a practical standpoint, it is useful to do cruder, but easier, experiments. One can apply a drop of coating on a substrate, put the sample in an atmosphere saturated with the solvents in the coating, and watch the spreading. If the droplet of coating stays as a small ball, spreading is poor and poor adhesion is expected. If the drop spreads out to a thin, wide circle, the coating meets at least one criterion for good adhesion. It is also useful to carry out a second type of experiment in which a coating is mechanically spread on a substrate, under a solvent saturated atmosphere and then observed. Generally, a liquid that spreads spontaneously in the first experiment will remain spread out in the second experiment. But sometimes, a liquid that spreads on the surface will draw up into droplets, or at least pull away in sections from the substrate upon standing without solvent evaporation.

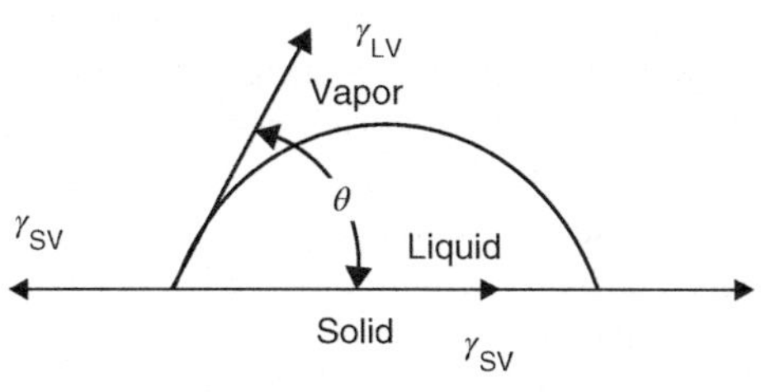

Figure 6.2 Contact angle.

Consider, for example, the application of *n*-octyl alcohol to a clean steel surface. The surface tension of *n*-octyl alcohol is lower than the surface free energy of steel, and it spontaneously spreads on steel. However, if one spreads out a film of *n*-octyl alcohol on steel, the film draws up to form droplets on the surface of the steel. The low surface tension of the *n*-octyl alcohol results from the linear hydrocarbon chain; however, after spreading on the polar surface of the steel, the hydroxyl groups of the *n*-octyl alcohol molecules interact with the surface so that a monolayer of oriented *n*-octyl alcohol molecules forms on the surface. This produces a new surface, aliphatic hydrocarbon, which has a low surface tension, lower than the surface tension of *n*-octyl alcohol. The *n*-octyl alcohol above the monolayer surface *dewets*. The behavior of *n*-octyl alcohol illustrates a principle important in formulating coatings: one must be careful about using additives with single polar groups and long hydrocarbon chains in coatings that are applied directly on metals. An example is the poor adhesion to steel that can result from use of dodecylbenzenesulfonic acid as a catalyst (Section 11.3.1). Adhesion of latex films can be affected by a layer of surfactant forming at the interface between the coating and the substrate (Charmeau et al., 1996).

6.1.3 Chemical Interactions between Coatings and Surfaces

While mechanical interlocking and good wetting are necessary to achieve good adhesion, some form of chemical interaction between the substrate and coating is also required. The strongest interaction would be covalent bonds between the substrate and coating binder. These can form if bonding sites on the surface of the substrate, for example, hydroxyl groups on an oxidized metal surface, react with appropriate functional groups in the binder.

Less strong hydrogen bonds can also form between coating and substrate due to interactions between H-bond donating and accepting groups such as carboxylic acid (strongly H-donating), amine (strongly H-accepting), hydroxyl, urethane, amide, and phosphate (the latter four being both H-donating and accepting). One might assume that a large number of such substituents on a molecule would be desirable. However, it is known from adsorption studies that if there are large numbers of polar groups, the adsorbed layer can be very thin at equilibrium. The principle can be illustrated by considering a polymer molecule with an aliphatic backbone chain with polar groups on every other carbon atom. At equilibrium, adsorption of adjacent polar groups is favored sterically, resulting in a thin adsorbed layer with the polar groups on the steel surface and mainly hydrocarbon groups exposed to the rest of the coating. The interactions between the rest of the coating and the hydrocarbon groups are expected to be weak, resulting in a weak boundary layer and poor cohesion. If a smaller number of H-bond donating groups are scattered along a resin chain, adsorption of resin molecules may occur with loops and tails sticking up from the surface so that some of the polar groups are adsorbed on the surface and some are on the loops and tails, where they can interact with the rest of the coating. On those parts of the resin molecule in the loops and tails, there can be groups to H-bond with molecules in the coating, as well as functional groups, which can react with a cross-linker in the coating.

BPA epoxy resins (Sections 13.1.1 and 15.8) and their derivatives commonly provide excellent adhesion to steel. These resins have hydroxyl groups and ether groups along the chain, which can interact with both the steel surface and other molecules in the coating. The cross-linkers employed with these resins often introduce additional hydrogen bonding groups, such as amines. It may also be important that the epoxy resin backbone consists of alternating flexible 1,3-glycidyl ether and rigid bisphenol A groups. It seems logical that such a combination could provide the flexibility necessary to permit multiple adsorptions of hydroxyl groups on the surface of the steel, along with the rigidity to prevent adsorption of all of the hydroxyl groups. The remaining hydroxyl groups can participate in cross-linking reactions or hydrogen bonding with the rest of the coating. Massingill et al. (1990) and Sheih and Massingill (1990) discuss the effects of variations in epoxy resin composition on adhesion.

In general, the results of many studies of adsorption of polymer molecules on metal surfaces are consistent with the picture given previously. However, many of these studies involve adsorption from dilute solutions. Observations are made over relatively long time intervals, permitting equilibrium conditions to develop. With polydisperse MW adsorbents, low MW species are adsorbed first, but, at equilibrium, they are displaced by higher MW molecules, which have larger numbers of polar groups. The resin is in a relatively concentrated solution, and the solvent evaporates in a relatively short time. There may not be time for equilibrium to be established. Depending on the coating, those groups that happened to be near the surface when the film was applied might remain there and could lead to poor adhesion, even if the same resins could provide good adhesion given the opportunity for appropriate orientation and equilibration to occur. Such a scenario is compatible with the improved adhesion using slow evaporating solvents, which can also permit a more complete penetration into surface crevices. Perhaps another reason for baking coatings commonly leading to improved adhesion is the greater opportunity for the orientation of molecules at the substrate–coating interface at the higher temperature. There is need for further research on interactions and orientation at the interface of coatings and substrates.

6.2 MECHANICAL STRESSES AND ADHESION

Coatings adhere to a substrate via the various mechanisms previously discussed. However, even the most weakly adhering coating will not spontaneously delaminate from a surface. The application of stress is required to remove the coating. In particular, peeling stresses are required. These stresses have a component normal to the surface and are responsible for a coating becoming detached from the surface. Tensile stresses in the coating cannot result in peeling. However, defects and edges in coatings can macroscopically modify tensile stresses such that peeling stresses arise in those locations.

While most stresses in a coating arise from external loads, internal stresses are often introduced during film formation or curing. As coatings, by definition, are constrained on a substrate, stresses arise during these processes. In thermosetting coatings, cross-linking reactions lead to the formation of covalent bonds that are shorter than the distance between two molecules before they react. When such reactions occur at temperatures near the T_g of the film, stresses result from the inability of the coating to undergo shrinkage. As the rate of cross-linking increases, stresses also tend to increase, since less time is available for polymer relaxation to occur. An extreme example is UV curing of acrylated resins by free radical polymerization that occurs in a fraction of a second at ambient temperatures (Section 29.8). Shrinkage, measured by thermomechanical analysis (TMA), has been shown to lag significantly behind polymerization (Kloosterboer, 1988). The high rates of polymerization, together with the relatively large shrinkage that accompanies polymerization of double bonds, contribute to the generally observed poor adhesion of UV cure acrylated resins to smooth metal surfaces. Heating after UV curing relaxes the cross-link network and often improves adhesion.

In lacquers, internal stresses arise as solvent evaporates from the coating and film formation progresses. Typically, meaningful stresses only arise once film formation has taken place and enough solvent has evaporated so that the coating is below its T_g. Above T_g, enough chain mobility exists to relieve most stresses.

6.3 ADHESION TO METAL SURFACES

The metal and the surface characteristics of the metal can have major effects on adhesion. Perfetti (1994) provides a review of metal surface characteristics, cleaning, and treatments. The surface tension of a clean metal surface (usually metal oxide) is higher than that of any potential coating. However, metal surfaces are frequently contaminated with oil, and such surfaces can have very low surface tensions. Whenever possible, it is desirable to clean the surface of the metal before applying a coating. Sometimes, the metal is wiped with rags wet with solvent. While this can remove some of the contaminants, they are often merely smeared around the surface. A more effective method is *vapor degreasing*, in which the object is hung from a conveyor that carries it above a tank above a boiling chlorinated solvent. The cold steel surface acts as a condenser, condensing solvent on the surface of the steel, where it dissolves oils. The solution drips off, removing the oil. The solvent is purified by distillation for reuse. Surfactant solutions are also used for cleaning oil from metals (Perfetti, 1994). Care must be exercised in selecting surfactants and in thoroughly rinsing the surface after cleaning. It is possible for some surfactants to adsorb on the surface, creating a hydrocarbon layer on the surface of the metal.

Perfetti (1994) also discusses abrasive particle blasting of steel. The surface of the steel, including rust, is removed, leaving a rough surface. Sandblasting is widely used for steel structures like bridges and tanks (Section 33.1), but it leaves the steel surface too rough for products such as automobiles and appliances. Sandblasting is effective, but there are environmental and health hazards to workers from inhalation of silica dust and old paint debris. Especially when the old paint contains lead pigments, expensive dust containment is required. The effect of different types of abrasive particles on the corrosion performance of steel has been investigated (Elsner et al., 2003). Other dry, abrasive materials, such as steel grit or water-soluble abrasives—including sodium bicarbonate and salt—can be used to replace sand; in cryogenic cleaning by blasting and in vacuum blasting, dry ice pellets and aluminum oxide can be used, respectively. For cleaning softer metals, such as aluminum, plastic pellet blasting is employed. Ultra-high pressure hydroblasting at pressures above 175 MPa (25 000 psi) is very effective in removing oil and surface contaminants, such as salts.

Formulations are tested on laboratory panels, but the surfaces of these panels are not the same as those of the product to which the coating will be applied. Furthermore, commercial test panels vary—for example, the side of the panel next to the wrapping of a package of panels was shown to have a different surface analysis than inner panels from the same package (Skerry et al., 1990). Washing the panels with warm water and rinsing with acetone before coating generally improved adhesion. When possible, laboratory tests should also be carried out on the ultimate substrate, or at least on sample pieces of metal to be used in production, simulating, as closely as possible, factory cleaning and treating procedures.

When excellent adhesion to aluminum substrates is required, caustic or acidic etching is often performed. These methods remove the contaminants and the native aluminum oxide layer. In its place a more controlled aluminum oxide layer quickly replaces the native oxide. Often a combination of alkaline cleaning and acidic "desmutting" is used to provide the most robust surface preparation for aluminum that is to be coil coated.

6.3.1 Conversion Coating and Pretreatment of Metal Substrates

When good adhesion, corrosion protection, and a relatively smooth surface are required, it is common to chemically treat the surface of the metal. These treatments are called *conversion coatings* or *chemical pretreatments*. In the past, pretreatments based on hexavalent chrome were common and provided excellent corrosion protection on both ferrous and nonferrous substrates. However, Cr^{6+} compounds are highly carcinogenic, and their use has been banned in most countries in most industries. An important exception is in civilian and military aviation, where these compounds are still used, as no replacement pretreatments have been found to give the required longevity of protection. Chrome-free pretreatments are an active area of research.

A variety of phosphate-based conversion coatings is used for steel substrates. One such coating involves spray or immersion application of a phosphoric acid-based "iron phosphate" solution. This method provides a mild etch of the steel surface and precipitation of a ferrous–ferric phosphate monolayer. Adhesion of coatings is markedly improved; however, corrosion protection is only slightly enhanced. More enhancements are often possible by using a zinc phosphate conversion coating. The steel object is immersed in a bath of zinc acid phosphate solution, and a coprecipitate of zinc and ferric phosphates is formed on the steel surface, as shown in Scheme 6.1. Both adhesion and corrosion protection are enhanced. The coprecipitate forms a mesh of crystals that adhere tightly to the surface, increasing the surface area on a microscopic scale. Depending on zinc concentration in the treatment bath, different crystals can be deposited. At relatively high zinc concentrations, the crystals are predominantly hydrated zinc phosphate,

$Zn_3(PO_4)_2{\cdot}4H_2O$, called hopeite. Under zinc-starved conditions, the crystals have been identified as phosphophyllite, $Zn_2Fe(PO_4)_2{\cdot}4H_2O$ (Dyett, 1989). The performance of the conversion coating is dependent on the uniformity and degree of surface treatment. Zinc phosphate coatings are generally applied in the range of 1.5–4.5 $g\,m^{-2}$. A variety of other phosphate coatings are also used. (See Perfetti (1994) and Sections 7.4.2, 27.1, 27.2, and 30.1.1 for further discussion of conversion coating.) Modern zinc phosphate coatings are more correctly termed tri-cationic zinc phosphate coatings, as nickel and manganese cations are also present in the bath to provide enhanced deposition as well as improved performance. The ratio of these cations must be closely controlled to optimize the performance of the pretreatment.

A schematic of modern automotive zinc phosphate systems is shown in Figure 6.3. The process starts with both spray and immersion cleaning steps, where alkaline cleaners plus surfactants are used to remove stamping lubricants, weld balls, and other dirt from the automobile body. The vehicle is then thoroughly rinsed by city water and then by deioniozed (DI) water before the rinse conditioner is applied to the surface of the vehicle via immersion or spray. The rinse conditioner consists of a dispersion of nano-sized TiO_2 or zinc phosphate. These small particles act as nuclei for the zinc phosphate coating that is applied in the next step. After the phosphate has reacted, the vehicle is rinsed. A final step often involves a "post-rinse" or sealing of the zinc phosphate surface. The post-rinse can apply either an organic layer, a silane, or a thin film of ZrO_2. The latter is produced by immersing the body in a dilute solution of hexafluorozirconic acid (H_2ZrF_6), which reacts with the surface to form an amorphous thin (~50 nm) film of zirconium oxide to seal the gaps between zinc phosphate crystals.

$$HPO_4^{2-} \rightleftharpoons H^+ + PO_4^{3-}$$
$$2H^+ + Fe \longrightarrow Fe^{2+} + H_2$$
$$Fe^{2+} + [Ox] \longrightarrow Fe^{3+}$$
$$3Zn^{2+} + 2PO_4^{3-} \longrightarrow Zn_3(PO_4)_2 \downarrow$$
$$Fe^{3+} + PO_4^{3-} \longrightarrow FePO_4 \downarrow$$

Scheme 6.1

The reactions shown in Scheme 6.1 are straightforward. Achieving the high rates of reaction required to permit minimum dwell times for treatment is more difficult. Proprietary formulations reduce times to the order of minutes or seconds. Treatment quality is dependent on time, temperature, and pH; these and other variables must be closely controlled to assure that the desired type and dimensions of crystals are formed. The treated surface must be thoroughly rinsed to remove any soluble salts, since these salts could lead to blister formation when water vapor permeates through a coating film applied over the soluble salts; rinsing also removes loosely adhering crystals.

The mechanism of action of the phosphate crystal layer is not fully understood. One factor is that the coating penetrates into the crystal mesh, giving a mechanical attachment (Section 6.2) to the crystals attached to the surface. Furthermore, the interfacial area for interaction is greater than for a relatively smooth steel surface. It is also possible that hydrogen bond interactions between these crystals and the resin molecules are stronger (i.e., less readily displaced by water) than those between the steel surface and the resin molecules.

Zinc phosphate is typically not used as a surface pretreatment for aluminum due to the abundant amount of

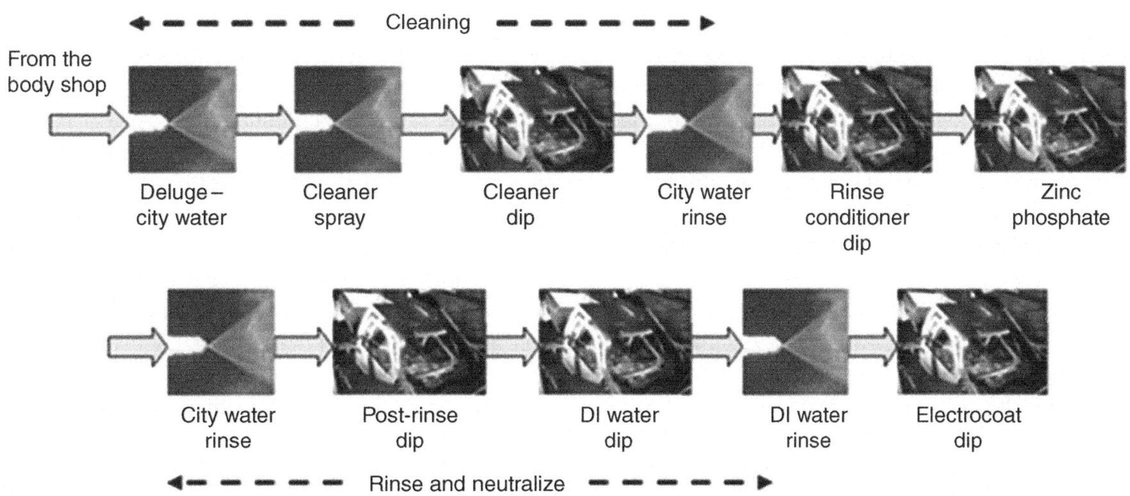

Figure 6.3 Layout of a modern automotive zinc phosphate system. Source: Nichols and Tardiff (2016). Reproduced with permission of Springer.

sludge produced by etching the aluminum surface. In the past chromated pretreatments were used and provided excellent adhesion and corrosion protection. However, as such pretreatments are being phased out, new pretreatments are taking their place. Thin film pretreatments based on mixed metal oxides, such as ZrO_2, Ce_2O_3, and mixed Ti–Zr oxides, are used on coil-coated aluminum. Silane-based pretreatments (Section 16.2) have also been shown to be effective adhesion promoters for aluminum substrates. Both industry and the US Navy have investigated the use of trivalent chromium compounds but have found them to be generally inferior to hexavalent chrome and other thin film pretreatments. All of these aluminum pretreatments are aimed at the removal of the native thin aluminum oxide–aluminum hydroxide surface that forms spontaneously on all aluminum exposed to air. Acidic or alkaline cleaning removes this oxide–hydroxide surface, and in its place, an engineered surface of the previously mentioned compounds is grown, consisting of a thin, dense, coherent layer of the native aluminum oxide with the chosen metal oxide earlier. Many new pretreatments for aluminum are based on sol–gel technology (Section 16.3.1). A recent review provides additional details (Figueira et al., 2015).

To provide greater protection against corrosion, steel coated with zinc is widely used in construction and on automobiles. Several types of zinc-coated steel are used; the best known is galvanized steel. There can be large variations in adhesion depending on the condition of the zinc layer of the galvanized steel. If zinc-coated steel has been exposed to rain or high humidity before coating, there may have been some degree of surface oxidation, leading to formation of a combination of ZnO, $Zn(OH)_2$, and $ZnCO_3$; all of these are basic and somewhat soluble in water. Therefore, it is important to use saponification-resistant resins in primers for galvanized steel. Resins such as alkyds, which tend to saponify, are likely to give poor adhesion in service.

Adhesion to stainless steel, because of its smoothness and lack of oxide and hydroxide groups on the surface, is difficult. In some cases, roughening of the surface provides a basis for anchoring; in others, pretreatment with an electrochemically produced flash of chromium–chromium oxide is required (Lori et al., 1996) but has lost favor due to the previously described toxicity of chromium containing pretreatment baths.

Plasma treatment is being investigated for cleaning and treating metallic surfaces (Lin et al., 1997). Initially, the metal is cleaned by plasma discharge, and then trimethylsilane is introduced into the plasma chamber leading to polymerization of a thin layer of polymer bonded to the surface. Laboratory tests have demonstrated superior adhesion of electrodeposition coatings (Chapter 27) to such a surface as compared to a conventionally pretreated galvanized steel substrate. Previously, plasma surface treatment was restricted to high vacuum chambers. However, recent advances in atmospheric pressure plasma treatment have opened this surface preparation technique to higher volume and less expensive applications, including those where the surface is simply activated by the plasma plume and those where the plasma polymerization of silane monomers applies a thin coating to the surface.

6.4 CHARACTERIZATION OF SURFACES

As mentioned previously, good adhesion starts with good cleaning and surface preparation. To assess those two factors, analytical surface science techniques can be particularly helpful in quantifying the surface chemistry of substrates and in determining the root cause of poor adhesion.

While optical microscopy can be useful for identifying some defects and contamination issues, the size of many potential adhesion-disrupting species is smaller than can be resolved with light microscopy. Scanning electron microscopy (SEM) can be useful for observing smaller contaminants on the surface of a substrate or for observing the quality of metal pretreatment. The size and morphology of zinc phosphate crystals are routinely monitored by SEM to ensure that the pretreatment process is performing properly. Unduly large or imperfectly shaped crystals are a sure sign that the chemical composition of one of the pretreatment bath stages is out of tolerance.

SEM coupled with energy-dispersive X-ray analysis (EDX) can map the elemental composition of a surface on a submicron scale. The elemental composition of a steel surface pretreated with zinc phosphate is shown in Figure 6.4. Note that different regions in the micrograph show different elemental compositions. This is typical of the type of data that can be generated with EDX and can be very useful in assessing the composition of the surface of inorganic substrates.

To determine elemental composition more quantitatively, the use of Auger electron spectroscopy (AES), particularly when it is coupled with sputter depth profiling, can provide improved elemental mapping of surfaces (Simko, 2011). Sputter depth profiling allows one to use ions to ablate away the surface at a known rate. Thus, the composition as a function of depth into the substrate can be determined. This can be particularly useful when trying to determine if a thin layer of a contaminant or oxide is confined to the surface or if the substrate is of uniform composition through its thickness. Sputter depth profiling is also particularly well suited for determining the thickness of oxide layers or thin film pretreatment layers, where EDS does not provide sufficient resolution.

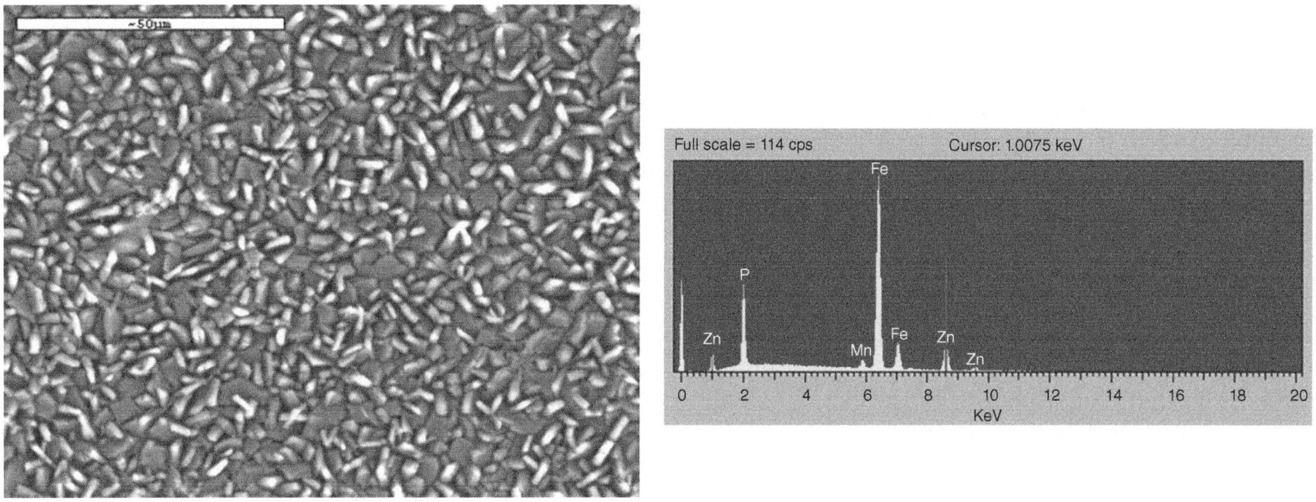

Figure 6.4 Left: SEM micrograph of zinc phosphate surface. Right: EDX elemental composition spectra of the zinc phosphate surface. Note Zn, Mn, and P peaks from the conversion coating and the Fe peak from the underlying cold rolled steel. Source: Reproduced with kind permission of Steven J. Simko, Ford Motor Company.

On all substrates, X-ray photoelectron spectroscopy (XPS) can be used to determine not only composition but also chemical state (bonding) information on the very top (~1–2 nm) of a surface (Haack, 2011). As very thin layers of organic materials that have not been cleaned from the substrate prior to painting can cause severe adhesion issues, XPS is often used to investigate poor adhesion. Because XPS gives not only elemental but also bonding information, it can be used to help identify particular organic compounds. For instance, XPS spectra differentiate between aliphatic carbon and carbon associated with carboxyl, ether, or ketone groups. This chemical structure information can also be used to determine if particular surface treatments, such as flame or plasma treatments, have sufficiently activated the surface for painting. Like AES, modern XPS instruments can depth profile surface via ion sputtering. In addition, mechanical microtoming and angle-resolved XPS can be used to quantify composition as a function of depth.

To assess surface composition with the best spatial resolution, time-of-flight secondary ion mass spectrometry (TOF-SIMS) is the preferred analytical instrument (Kaberline et al., 2011). TOF-SIMS can analyze surface composition at the monolayer thickness level with lateral resolution of 150 nm or less. This allows mapping of the chemical composition of surfaces at a very fine scale. While no bonding information is generated, as in XPS, the molecular fragments produced by the ion beam in SIMS can be matched to known ion masses and, therefore, compositions may be inferred. In addition, if contaminants on the surface are suspected, the mass "fingerprint" of the suspected contaminant can be compared to that of the contaminated surface to determine if a match is present. Quantification of the relative amount of a species on the surface is difficult with TOF-SIMS—unlike XPS, AES, or EDX—but its sensitivity is unrivaled.

For many surface quality assessments, traditional infrared spectroscopy is useful. While it does not have the surface sensitivity of the techniques discussed previously, sample preparation is simpler, high vacuum is not required, and the fingerprints of most coating constituents are well known. When used with an attenuated total reflectance (ATR) crystal, the first few microns of a surface can be examined for chemical composition information, provided the species of interest is present at more than a few percent of the total composition (Adamsons, 2012).

6.5 ORGANIC CHEMICAL TREATMENT OF SUBSTRATES TO ENHANCE ADHESION

Stronger interactions with a substrate surface should be possible by forming covalent bonds compared to the more readily displaced H-bonds. One such approach is the use of reactive silanes (Section 16.2), which are very effective in enhancing adhesion of coatings to glass and are now used more frequently to improve the adhesion of coatings to metals as well (Plueddemann, 1983). A variety of reactive silanes is available; most have a trialkoxysilyl group attached to a hydrocarbon chain, the other end of which has a functional group such as amine, mercaptan, epoxy, vinyl, and so forth. The alkoxysilyl group can react with hydroxyl groups on the surface of the substrate and with other alkoxysilyl groups after hydrolysis, so the surface of the substrate becomes covalently bonded to a series of hydrocarbon tails having reactive groups that can cross-link with the coating being applied.

For enhancing adhesion of a two package epoxy-amine coating to glass substrates, one could add 3-aminopropyltrimethoxysilane to the amine package. After application, the trimethoxysilyl group can react with silanol groups on the surface of the substrate to generate siloxane bonds, as shown in the first step of Scheme 6.2. The trimethoxysilyl groups can also react with water to produce silanol groups that can in turn react with remaining silyl methoxy groups to generate polysiloxane groups at the surface, as shown in the second step in Scheme 6.2. The terminal amine groups can react with epoxy groups in the resin so that the coating is multiply bonded to the surface, as shown in the third step in Scheme 6.2.

When water vapor penetrates through the coating to the substrate–coating interface, hydrolysis of some of the interfacial Si–O bonds occurs. However, with multiple interfacial bonds, some of the bonds remain intact and prevent the coating from delaminating. Furthermore, hydrolysis is reversible, so the hydrolyzed bonds can reform. The equilibrium of the reversible reaction strongly favors the interfacial Si–O bonds. The use of silanes has dramatically improved the coating scientist's ability to enhance the adhesion of coatings to substrates, especially to glass, to which robust adhesion and humidity resistance has traditionally been difficult to achieve.

Reactive silanes have also been added to coatings with the objective of improving adhesion to steel and other metal surfaces (Plueddemann, 1983). Adhesion can be enhanced by the addition of reactive silanes to the coating binder or by an additional pretreatment step before the organic coating is applied.

In theory, the trialkoxysilyl group could react with hydroxyl groups attached to iron or aluminum oxides on the metal surfaces. The use of silanes as pretreatments is a rapidly growing field, as the use of zinc and iron phosphate pretreatments face serious environmental and energy use deficiencies. In particular, the use of reactive silanes as pretreatments for aluminum shows great promise and commercial growth. In general, these systems are applied during spray, dip, or roll coating applications via an aqueous solution. The stability, extent of hydrolysis, and reactivity are controlled by the pH, temperature, and composition of the pretreatment. Electrical impedance studies have shown that these silanes can form strong, diffusion-resistant barriers to corrosion (Trabelsi et al., 2005; Naderi et al., 2013). The ability of the silane to then bond with the subsequently applied coating gives rise to the superior adhesion and corrosion protection properties of these coatings. While their performance has not yet achieved that of hexavalent chrome pretreatments, silane pretreatments with the addition of corrosion inhibiting additives provide good corrosion performance, as well as a substantially smaller environmental footprint than phosphate pretreatments. Several extensive reviews are available in the journal and patent literature (Van Ooij et al., 2005; Chico et al., 2012).

6.6 COVALENT BONDING TO GLASS AND METAL SUBSTRATES

Another approach to achieving chemical bonding to steel is the use of resins containing groups that can form coordination complexes with ferric compounds. For example, one can make resins with acetoacetic ester substituents (Section 17.6). Such esters are highly enolized and can coordinate with metal ions, including ferric salts. Preliminary reports indicate improvement in adhesion and corrosion protection (Del Rector et al., 1996). Because of the potential hydrolysis of acetoacetic esters, evaluation over relatively long time intervals will be required to assess their commercial utility.

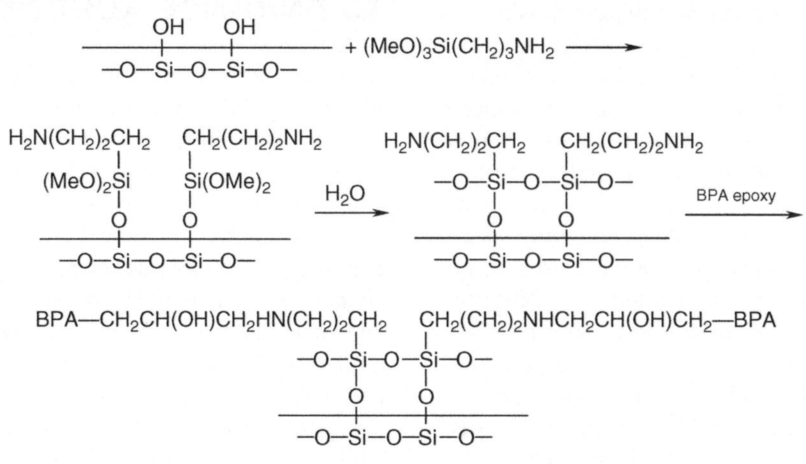

Scheme 6.2

6.7 ADHESION TO PLASTICS AND TO COATINGS

In contrast to clean steel and other metals, there is commonly a problem wetting the surface of plastic substrates with a coating. Wetting of many plastics is difficult, owing to their low surface free energies, as well as to the presence of mold release agents on molded plastic parts. Mold release agents should be avoided, if possible, when subsequent painting is expected. If essential, release agents should be selected that are relatively easily removed from the molded part, and care should be exercised to remove all traces. Power washers are the standard practice to remove all traces of mold release compounds and other contaminants.

Even after cleaning, the surface free energies of some plastics are lower than the surface tensions of many coatings, resulting in a contact angle >0°, which prevents spreading. The use of dyne pens or, more appropriately, computer-controlled contact angle measurement systems, together with image analysis, is the standard means of measuring the surface energy of substrates.

Attainment of satisfactory adhesion to polyolefins, which generally have low surface free energy, usually requires treatment of the surface to increase the surface free energy (Section 31.2.2). This can be done by oxidation of the surface to generate polar groups such as hydroxyl, carboxylic acid, and ketone groups. The presence of these groups not only increases surface free energy so that wetting is possible with a wider range of coating materials but also provides hydrogen bond acceptor and donor groups for interaction with complementary groups on coating resin molecules. A variety of processes can be used to treat the surface (Lane and Hourston, 1993; Ryntz, 2005). The surfaces of films, flat sheets, and cylindrical objects can be oxidized by flame treatment with gas burners using air–gas ratios such that the flames are oxidizing. Oxidation can also be accomplished by subjecting the surface to a corona discharge atmosphere; the ions and free radicals generated in the air by the electron emission serve to oxidize the surface of the plastic. While useful, both of these technologies are prone to workplace regulation, as open flames can be hazardous in certain environments, and the use of corona discharge systems produces harmful ozone.

Atmospheric pressure plasma systems (Section 31.2.2) are gaining popularity and provide the same type of surface modification as both corona and flame treatments. In this process a plasma is created by a high voltage source while air (or other gas) is forced through a nozzle. The resulting plasma jet contains sufficient energy to both clean and chemically modify the surface of most plastics. The process operates at essentially ambient temperature; both the composition and power of the plasma can be easily modified by altering the gas composition and flow rate. While the size of the individual plasma jets is small (on the order of a few centimeters), the jets can be grouped together to treat larger areas. The surface composition of these treated surfaces has been shown to be stable for several weeks, which allows reasonable time for shipping and painting.

Adhesion to untreated polyolefins has traditionally been enhanced by applying a thin *tie coat* of a low solids solution of a chlorinated polyolefin or chlorinated rubber. The mechanism of enhancement is thought to involve swelling of the near surface of the substrate by solvents in the tie coat, followed by diffusion of the chlorinated polyolefin binder into the swollen substrate. For thermoplastic olefin (TPO) substrates, the resin is thought to strongly diffuse into the EPDM rubber domains. This same mechanism of adhesion enhancement can take place in other polymeric substrate–coating systems provided that the substrate is coated above its T_g. Much of this enhancement is due to the solvents in the coating. The solvent swells the plastic, lowering its T_g and facilitating penetration of coating resin molecules into the surface of the plastic. The solvents should evaporate slowly to permit time for penetration to occur. Fast-evaporating solvents, like acetone, can cause *crazing* of the surface of high T_g thermoplastics, like polystyrene and poly(methyl methacrylate). Crazing is the development of large numbers of minute surface cracks; for further discussion of crazing, coatings for plastics, and surface treatments, see Section 31.2.2 and Ryntz (1994).

Both theoretical and experimental studies show that the molecules at the surface of a polymer are more mobile than those in the bulk material (Garbassi et al., 1998). One theoretical study indicated that the layer of mobile molecule segments is about 2 nm in thickness (Mansfield and Theodorou, 1991). Because polymer surfaces are dynamic, they adjust to the environment. Reduction of surface free energy favors migration of polar groups from a polymer–air interface into the bulk of the polymer. However, factors such as T_g, crystallinity and composition of the altered surface may prevent or substantially retard this migration.

Adhesion to other coatings, commonly called *intercoat adhesion*, is another example of adhesion to polymeric substrates. The same principles apply. The surface tension of the coating being applied must be lower than the surface free energy of the substrate coating to permit wetting. The presence of polar groups in both coatings permits hydrogen bonding; in the case of thermosetting coatings, covalent bonding enhances intercoat adhesion. It has been found empirically that the presence of relatively small amounts of amine groups on resins commonly enhances intercoat adhesion. Such comonomers as 2-(*N*,*N*-dimethylamino) ethyl methacrylate and 2-aziridinylethyl methacrylate have been used to make acrylic resins with enhanced intercoat adhesion.

As with metals, curing temperatures above T_g increases the probability of satisfactory adhesion. Compatible resins in the substrate coating and top coat also promote adhesion. Using solvents in the top coat that can swell the substrate coating is a commonly used technique for enhancing intercoat adhesion. Top coats generally adhere more readily to coatings having lower cross-link density, which are more swollen by solvents than are coatings with high cross-link density. Sometimes, one can undercure the primer, resulting in a lower cross-link density when the top coat is applied. Cure of the primer is completed when the top coat is cured. A common cause of intercoat adhesion failure is an unsatisfactorily high cross-link density in the coating layer that is subsequently coated.

Adhesion to high gloss coatings is difficult to achieve because of their surface smoothness. Glossy coatings that have undergone excessive cross-linking on aging are particularly difficult surfaces on which to apply an adherent coating. Sanding to increase surface roughness may be necessary to achieve intercoat adhesion, but thorough cleaning prior to application of a second coating is imperative. One reason for formulating primers with low gloss is that their rougher surfaces enhance adhesion. When possible, increasing the pigment loading of a primer above critical pigment volume concentration (CPVC) facilitates adhesion of a top coat (Chapter 23). Above CPVC, the dry film contains pores. When a top coat is applied, vehicle from the top coat can penetrate into the pores in the primer, providing a mechanical anchor to promote intercoat adhesion. Care must be exercised not to formulate the PVC too much higher than CPVC, in which case so much vehicle may be drained away from the top coat that the increase in PVC of the top coat will lead to a loss of gloss.

An essential requirement of many industrial coatings is *recoat adhesion*—that is, the ability of a coating to adhere to itself well enough that flawed or damaged objects can be repainted without extensive preparation. Even well-run paint shops do not perfectly paint all their objects without defects. When the first time through (FTT) is 98%, 2% of the objects will have to get recoated. Minimizing the work required to prepare the painted substrate for repainting is highly desirable. For high gloss coatings having high cross-link density, this objective can be challenging. In addition, many coatings contain small levels of additives that bloom to the surface during cure to reduce cratering or to improve wetting. These surface-active compounds can make it extraordinarily difficult to achieve good recoatability.

6.8 TESTING FOR ADHESION

As excellent adhesion is one of the central foundations upon which a superior coating system is founded, there is a strong need for quantitative, reproducible coating adhesion tests. Unfortunately, coating adhesion is notoriously difficult to quantify. As is well known, poor adhesion is easy to measure, as the forces needed to cause delamination are low and the geometrical and mechanical considerations are relatively easy to contain. However, strong adhesion is notoriously difficult to measure. When the adhesive strength approaches the cohesive strength of either the substrate or the coating, quantification of that adhesive strength becomes even more challenging. For this reason, the coatings industry has traditionally relied on qualitative measurements of coating adhesion.

At the most basic level, formulators can evaluate adhesion by determining how easily a penknife can scrape a coating from a substrate. By comparing the resistance of a new coating–substrate combination to combinations with known field performance, the formulator has some basis for performance prediction. While a penknife in the hand of an experienced person can be a valuable tool, it has major disadvantages as a test method. The experience is not easily transferred from one person to another; even the technique for the test is not easily transferred. Also, there is no good way to assign numerical values to the results. Thus, it is not useful for developing hypotheses based on small changes in adhesion with changes in composition.

Probably, the most widely used specification test is the *cross hatch adhesion* test. Using a device with 6 or 11 sharp blades, a scratch mark pattern is made across the sample, followed by a second pattern cut perpendicular to the first. A strip of pressure-sensitive adhesive tape is pressed over the pattern of squares and pulled off. Adhesion is assessed qualitatively on a scale of 5 to 0 by comparing the result to a set of photographs, ranging from trace removal along the incisions to removal of most of the area. The test is often made after subjecting the samples to extended water soaks or high humidity conditioning for several days. This water exposure almost always reduces the adhesion of the coating to the substrate, as water molecules are adsorbed onto the surface of the substrate, displacing coating–substrate bonds.

The cross hatch adhesion test is subject to many sources of error, one being the rate at which the cuts are made. If the cuts are made slowly, they are likely to be even. However, if the cuts are made rapidly, it is possible that there will be cracks proceeding out from the sides of the cuts, owing to more brittle behavior of coatings at higher rates of application of stress. Other important variables are the adhesive tape; the pressure with which it is applied; the angle and rate at which the tape is pulled off the surface; bending, if any, of the substrate during the test; and the surface of the coating to which the tape is applied. Some additives that appear to improve adhesion may actually only improve the test results by decreasing adhesion of the tape to the coating. The test may be useful for distinguishing between samples having poor adhesion and those having fairly good adhesion, but not much in distinguishing among higher levels of adhesion. See Koleske (2012) for a more detailed discussion of the variables affecting the test.

A similar test can be performed by using an X-scribe on the painted panel as opposed to the cross hatch pattern; it is widely used on architectural coatings. Results are qualitatively similar to the cross hatch test. Unfortunately, cross hatch–tape pull tests are *the* most common method of coating adhesion evaluation.

More satisfactory test methods are available for the evaluation of the adhesive strength of adhesives. However, few of these methods are applicable to coatings. Many investigators have worked on a variety of different methods in attempts to devise meaningful tests for evaluating the adhesion of coatings (Bullett and Prosser, 1972; Koleske, 2012). The central problem that confounds most coating adhesion tests is the low cohesive strength of the coatings themselves. All adhesion tests require a stress to be placed upon the coating–substrate interface. Of necessity that stress is almost universally transmitted through the coating itself. However, in the case of good adhesion, the coating will often rupture before enough stress is placed upon the coating to allow the interface to fail. Thus, tests such as peel tests that are often done on adhesives are not generally applicable to coatings, unless their adhesion is quite poor.

As most coatings are quite sensitive to flaws and thus will fail from tiny imperfections near their surface, some success in measuring coating adhesion has been found by constraining the coating layer under a stronger, stiffer material. One such test is the direct pull off test, which goes by various other names. A rod (also called a dolly) is fastened perpendicular to the upper surface of the coated sample with an adhesive. The panel is fastened to a support with a perpendicular rod on its back so that the two perpendicular rods are lined up exactly opposite each other. The assembly is put into the jaws of a mechanical testing machine (Section 4.5), and the tensile force required to pull the coating off the substrate is recorded. Since the procedure is subject to considerable experimental error, multiple determinations must be made. Experienced operators can achieve precisions of +15%. The adhesive must bond the rod to the coating surface more strongly than the coating is bonded to the substrate. It is also essential that the adhesive does not perturb the coating–substrate interface by penetrating into the coating. Cyanoacrylate adhesives are generally satisfactory. The rods must be aligned exactly with each other and perpendicular to the coating. If the rod is at even a slight angle to the surface, stress is concentrated on only part of the substrate–coating interface, and less force is required to break the bond. Sometimes, the weakest component is the substrate, which may be desirable for advertising purposes but does not provide a measure of the adhesive strength.

One must use caution in interpreting the results even when the sample appears to have failed adhesively at the substrate–coating interface. Sometimes, when no coating can be seen on the substrate surface after the test, there is a monolayer (or thin layer) of material from the coating left on the substrate surface. In this event, failure was not at the substrate surface but between the material adsorbed on the surface and the rest of the coating. Surface analysis (Section 6.4) is useful in determining the locale of failure and the identity of the adsorbed material. Fairly often, there is a combination of adhesive and cohesive failure. A possible explanation of such failures is that there was a fracture failure starting at some imperfection within the film and the initial crack propagated down to the interface. Tensile values from samples that fail in this way cannot be compared to the tensile values of samples that failed adhesively.

The direct pull test does not distinguish between the difficulty of breaking through a coating film and of shoving it out of the way, as mentioned in the beginning of this chapter. In spite of all the difficulties, direct pull tests are the most useful, available methods. Instruments have been devised for direct pull tests under field conditions. The method is quite widely used for quality control in high performance maintenance and marine coatings. Serious disadvantages for use on actual products are that the test is destructive and the tested area must be repainted. Koleske (2012) provides more detailed discussion of the effect of variables on the test results.

In addition to direct pull of tests, other constrained test geometries have been evaluated and some show promise. Blister tests have been developed to measure the strength of adhesively bonded materials (Islam and Tong, 2015). In these tests a small, circular defect is made at the interface between the substrate and adhesive (or coating). This can be done by placing a small amount of low surface energy material, such as a mold release, on the surface of the substrate. The surface is then coated. In the area where the defect was placed, a small hole is drilled through the substrate, but not through the coating. Pressure is then applied to the hole from the back side, typically with compressed air. The deflection of the surface is measured as a function of the air pressure, typically with a linear variable displacement transducer (LVDT). When the defect begins to propagate, the displacement of the surface, the modulus of the coating, the thickness of the coating, and the size of the blister can be used to calculate the energy required to propagate the defect. As with most other adhesion tests, the coating often fails before the defect propagates. To address this problem, inverted blister tests can be performed, where the pressure is applied through a hole in the coating and the thin substrate is pressurized. These tests are more successful for coatings but have not been widely adopted (Fernando and Kinloch, 1990).

Recent work on constraining the coating below a stiff upper layer has shown promise in preventing coating rupture and in mechanically driving the crack to the interface. This approach has been successfully used to measure the adhesive fracture energy of inorganic and sol–gel coatings on plastics and metals and holds promise if the technique can be found to work for more traditional organic coatings, whose adhesive strength is typically greater than that of inorganic coatings. Chen et al. (2014) provide an excellent review of the various mechanical means to evaluate coating adhesion.

Meth et al. (1998) have developed a unique method, which was used to measure the adhesive fracture energy of clear coats applied to automotive base coats. This test relies on a visible light laser being pulsed through the clear coat. As the clear coat is essentially transparent, the laser's energy is absorbed by the colored base coat. As the base coat was heated by the laser, some material was vaporized, which caused a blister to form between the base coat and clear coat. The size of the blister was directly related to the fracture energy of the clear coat–base coat interface. Other than the sophisticated equipment needed to perform the test, the main drawbacks are that its applicability is limited only to clear coats over color coats and that the thickness of the clear coat must be known very accurately, as the formula for calculating the fracture energy depends on the thickness of the clear coat to the fourth power.

GENERAL REFERENCES

Abbot, S., *Adhesion Science*, DEStech Publications, Lancaster, **2015**.

Baghdachi, J. A., *Adhesion Aspects of Polymeric Coatings*, Federation of Societies for Coatings Technology, Blue Bell, PA, **1996**.

Hartshorn, S. R., *Structural Adhesives: Chemistry and Technology*, Plenum, New York, **1986**.

Mittal, K. L., *Adhesion Aspects of Polymeric Coatings*, Plenum, New York, **1983**.

Nelson, G. L., Adhesion in Koleske, J. V., Ed., *Paint and Coatings Testing Manual*, 15th ed., ASTM, Philadelphia, **2012**.

REFERENCES

Abbott, S., *Adhesion Science: Principles and Practice*, DEStech Publications, Inc, Lancaster, **2015**.

Adamsons, K., *J. Coat. Technol. Res.*, **2012**, *9(6)*, 745–756.

Bullett, T.; Prosser, J., *Prog. Org. Coat.*, **1972**, *1(1)*, 45–71.

Charmeau, J., et al., *Prog. Org. Coat.*, **1996**, *27(1)*, 87–93.

Chen, Z.; Zhou, K.; Lu, X.; Lam, Y. C., *Acta. Mech.*, **2014**, *225*, 431–452.

Chico, B., et al., *J. Coat. Technol. Res.*, **2012**, *9(1)*, 3–13.

Del Rector, F., et al., *J. Coat. Technol.*, **1996**, *61(771)*, 31–37.

Dyett, M., *J. Oil Colour Chem. Assoc.*, **1989**, *72(4)*, 132–138.

Elsner, C. I., et al., *Prog. Org. Coat.*, **2003**, *48(1)*, 50–62.

Fernando, M.; Kinloch, A., *Int. J. Adhes. Adhesiv.*, **1990**, *10(2)*, 69–76.

Figueira, R., et al., *J. Coat. Technol. Res.*, **2015**, *12(1)*, 1–35.

Garbassi, F., et al., **1998**, *Polymer Surfaces: From Physics to Technology*, John Wiley & Sons Ltd, Chichester.

Haack, L. P., *Coat. Technol.*, **2011**, *8(2)*, 42–51.

Islam, M. S.; Tong, L., *Int. J. Adhes. Adhesiv.*, **2015**, *62*, 107–123.

Kaberline, S. L., et al., *Coat. Technol.*, **2011**, *8(4)*, 34–43.

Kloosterboer, J. G., *Electronic Applications*, Springer, Berlin, **1988**, pp. 1–61.

Koleske, J. V., **2012**, *Paint and Coatings Testing Manual*, ASTM, Conshohocken.

Lane, J.; Hourston, D., *Prog. Org. Coat.*, **1993**, *21(4)*, 269–284.

Lin, T., et al., *Prog. Org. Coat.*, **1997**, *31(4)*, 351–361.

Lori, L., et al., *Prog. Org. Coat.*, **1996**, *27(1)*, 17–23.

Mansfield, K. F.; Theodorou, D. N., *Macromolecules*, **1991**, *24(23)*, 6283–6294.

Massingill, J., et al., *J. Coat. Technol.*, **1990**, *62(781)*, 31–39.

Meth, J., et al., *J. Adhes.*, **1998**, *68(1–2)*, 117–142.

Naderi, R., et al., *Surf. Coat. Technol.*, **2013**, *224*, 93–100.

Nichols, M.; Tardiff, J., *Active Protective Coatings*, Springer, Dordrecht, **2016**, pp. 373–384.

Perfetti, B. M., *Metal Surface Characteristics Affecting Organic Coatings*, Federation of Societies for Coatings Technology, Blue Bell, PA, **1994**.

Plueddemann, E. P., *Prog. Org. Coat.*, **1983**, *11(3)*, 297–308.

Ryntz, R. A., *Painting of Plastics*, Federation of Societies for Coatings Technology, Blue Bell, PA, **1994**.

Ryntz, R. A., *J. Coat. Technol. Res.*, **2005**, *2(5)*, 351–360.

Sheih, P.; Massingill, J., *J. Coat. Technol.*, **1990**, *62(781)*, 25–30.

Simko, S. J., *Coat. Technol.*, **2011**, *8(3)*, 52–58.

Skerry, B. S., et al., *J. Coat. Technol.*, **1990**, *62(788)*, 55.

Trabelsi, W., et al., *Prog. Org. Coat.*, **2005**, *54(4)*, 276–284.

Van Ooij, W., et al., *Tsinghua Sci. Technol.*, **2005**, *10(6)*, 639–664.

Yekta-Fard, M.; Ponter, A., *J. Adhes. Sci. Technol.*, **1992**, *6(2)*, 253–277.

Zisman, W.A., *J. Coat. Technol.*, **1972**, *44(564)*, 42.

Chapter 7

Corrosion Protection by Coatings

Corrosion is the process by which metals are electrochemically transformed to metal oxides, in many cases transforming those metals back to the minerals or ores from which they were refined. The monetary toll of corrosion is frequently estimated at 3% of the US gross domestic product (GDP). The validity of this number is difficult to ascertain, but informed technologists know that enormous worldwide resources are devoted to preventing corrosion and mitigating its effects once it has begun.

The application of an organic coating to a metal is the main approach toward preventing corrosion; accordingly, we will focus on the mechanisms of corrosion and its prevention by organic coatings in this chapter. Specific types of coatings for corrosion control are discussed in Chapters 27, 30, and 33. It should be noted that the use of inorganic coatings to control corrosion is also a robust technology. For example, the use of a zinc layer on top of steel (galvanized steel) effectively enhances the corrosion performance of the underlying steel. However, the use of inorganic coatings is beyond the scope of this book.

7.1 CORROSION BASICS

As the field of corrosion science has its own nomenclature that can be confusing and daunting to the novice, a simple background on corrosion science is first presented.

At its simplest level, the corrosion of all metals is the oxidation of the elemental metal to its metal oxide. Subsequent reactions often occur, and the chemical pathway by which the oxidation takes place can depend strongly on the pH, temperature, environment, and other factors, but at its heart, the corrosion process is an oxidation–reduction reaction. During the corrosion process two complementary reactions occur at distinct locations, the anode and the cathode. The anode and cathode must be linked by a conductive pathway, typically provided by an aqueous environment that will allow both electron and mass transport. At the anode the metal (M) is ionized and gives up electrons to the surrounding media (usually water):

$$\mathrm{M} \rightarrow \mathrm{M}^{n+} + n\mathrm{e}^-$$

In the corrosion process, the metal is being consumed or transformed to its oxide at the anode. At the corresponding cathode, water, electrons, and oxygen combine to form hydroxide anions:

$$\mathrm{O_2 + H_2O + 4e \rightarrow 4OH^-}$$

Sufficient oxygen is typically dissolved in water to provide a ready source for corrosion; under normal atmospheric conditions the equilibrium concentration of oxygen in water is approximately 6 ml l^{-1}. As the positively charged metal and the negatively charged hydroxyl ions react readily, the first corrosion product is formed:

$$\mathrm{M^+ + OH^- \rightarrow MOH}$$

In some cases the metal hydroxide is stable, but often it further reacts with oxygen and water to form the metal oxide:

$$\mathrm{MOH + O_2 \rightarrow MO + H_2O}$$

Of course the details are much more complex than this, but the main points are that metal dissolution takes place at the anode and, in the presence of water and oxygen, metal oxides form, leading to a weakening of the metal and in many cases unsightly appearance.

Organic Coatings: Science and Technology, Fourth Edition. Frank N. Jones, Mark E. Nichols, and Socrates Peter Pappas.

7.2 CORROSION OF UNCOATED STEEL

As steel is the metal alloy that is most commonly used in society, we will begin our discussion of corrosion prevention with the mechanisms by which steel corrodes.

For steel, which is an alloy composed mainly of iron and other minor constituents such as carbon, aluminum, titanium, silicon, and other elements, the main electrochemical reaction is the oxidation of iron. Because steel has a heterogeneous microstructure, the composition of the surface varies from location to location; as a result, some areas are anodic relative to other areas that act as the cathode. Stresses and morphological structure of the metal surface can also be factors for setting up anode–cathode pairs. Cold-rolled steel has more internal stresses than hot-rolled steel and is generally more susceptible to corrosion; but cold-rolled steel is widely used because it is stronger. Internal stresses can also be created during fabrication or by the impact of a piece of gravel on an auto body, for example.

Like most metals, steel adsorbs on its surface a thin layer of water, which dissolves traces of soluble salts, thus providing the electrolyte for electrochemical reactions. In the absence of oxygen, ferrous ions are the primary anodic corrosion products of steel. In the absence of oxygen, hydrogen is formed at the cathode, which stops further dissolution of iron by cathodic polarization, except when the pH of the electrolyte is low:

$$\text{Anode}: Fe \rightarrow Fe^{2+} + 2e^-$$

$$\text{Cathode}: 2H^+ + 2e^- \rightarrow H_2$$

However, in the presence of oxygen, which is typical, depolarization of the cathode takes place, hydroxide ions are formed at the cathodes, and dissolution of iron continues:

$$O_2 + H_2O + 4e^- \rightarrow 4OH^-$$

$$2Fe + O_2 + 2H_2O \rightarrow 2Fe^{2+} + 4OH^-$$

The rate of corrosion of steel depends on the concentration of oxygen dissolved in the water at the steel surface, as shown in Figure 7.1 (Revie and Uhlig, 2011). At low concentrations, the rate increases with increasing dissolved oxygen concentration. At high concentrations, the rate declines because of passivation (Section 7.3.1).

Corrosion can occur at a significant rate only if there is a complete electrical circuit. The rate of corrosion depends on the conductivity of the water at the steel surface. Dissolved salts increase conductivity, which is one reason why the presence of salts increases the rate of corrosion of steel. Effects of salts on corrosion rates are complex; the reader is referred to Revie and Uhlig (2011) or other general texts on corrosion for detailed discussions. The relationship between NaCl concentration and corrosion rate is shown in Figure 7.2. The dashed vertical line in the figure indicates the salt concentration in seawater. At higher salt contents, the rate of corrosion decreases since solubility of oxygen decreases as NaCl concentration increases.

The rate of corrosion also depends on pH, as shown in Figure 7.3. Since iron dissolves in strong acid even without electrochemical action, it is not surprising that corrosion is most rapid at low pH. The corrosion rate is nearly independent of pH between about 4 and 10. In this pH region, the initial corrosion causes a layer of ferrous hydroxide to precipitate near the anode. Subsequently, the rate is controlled by the rate of oxygen diffusion through this layer. Underneath, the surface of the iron is in contact with an alkaline solution having a pH of about 9.5. When the environmental pH is above 10, the increasing alkalinity raises the pH at the iron surface. The corrosion rate then decreases because of passivation (Section 7.5.2).

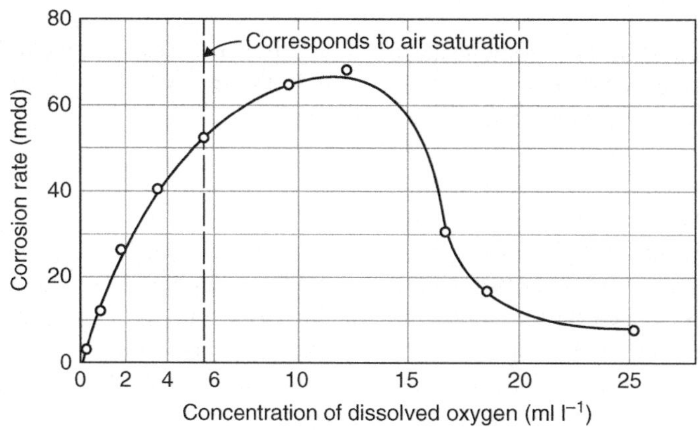

Figure 7.1 Effect of oxygen concentration on corrosion of mild steel in slowly moving distilled water, 48 h test, 25°C. Source: Revie (2011). Reproduced with permission of John Wiley & Sons.

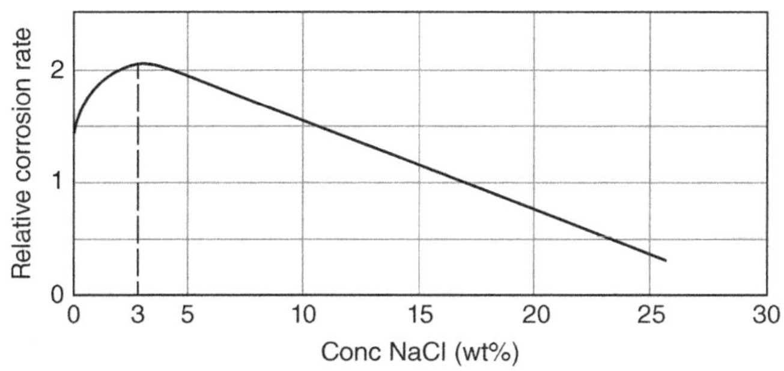

Figure 7.2 Effect of sodium chloride on corrosion of iron in aerated solutions at room temperature (composite of data from several investigations). Source: Revie (2011). Reproduced with permission of John Wiley & Sons.

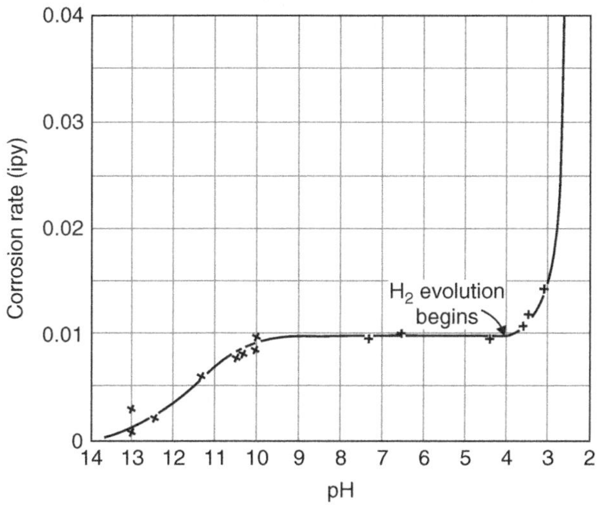

Figure 7.3 Effect of pH on corrosion of iron in aerated soft water at room temperature. Source: Revie (2011). Reproduced with permission of John Wiley & Sons.

Corrosion rate also depends on temperature, as shown in Figure 7.4. The reactions proceed more rapidly at higher temperatures, as indicated by the increase in corrosion rate in a closed system. However, the solubility of oxygen in water decreases as temperature increases, so in an open system, where the oxygen can escape, the rate of corrosion goes through a maximum at some intermediate temperature. The temperature at which the corrosion rate maximizes is system dependent.

7.3 CORROSION PROTECTION OF METALS

This section deals with strategies for controlling electrochemical corrosion without the use of organic coatings in order to introduce principles for corrosion protection using organic coatings (Sections 7.4 and 7.5). One can suppress

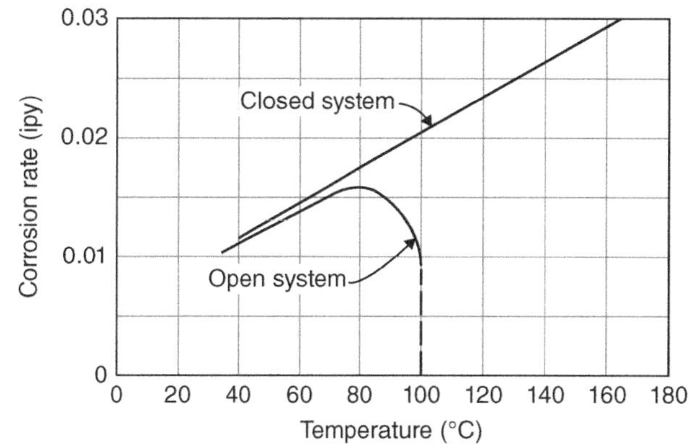

Figure 7.4 Effect of temperature on corrosion of iron in water containing dissolved oxygen. Source: Revie (2011). Reproduced with permission of John Wiley & Sons.

the anodic reaction, as discussed in Section 7.3.1; one can suppress the cathodic reaction, as discussed in Section 7.3.2; and one can prevent water, oxygen, and corrosion stimulants from contacting the surface, as discussed in Section 7.3.3.

7.3.1 Passivation: Anodic Protection

As shown in Figure 7.1, increasing oxygen concentration up to about $12\,\mathrm{ml\,l^{-1}}$ increases the corrosion rate because it acts to depolarize the cathode, as discussed in Section 7.2. At higher concentrations, more oxygen reaches the surface than is reduced by the cathodic reaction and corrosion is retarded. The mechanism of retardation has not been fully elucidated. According to one theory, if the oxygen concentration near the anode is sufficient, ferrous (Fe^{2+}) ions are oxidized to ferric (Fe^{3+}) ions soon after they are formed at the anodic surfaces. Since ferric hydroxide is less soluble in water than ferrous hydroxide, a barrier of hydrated ferric oxide forms over the anodic areas. Suppression of corrosion by retarding the anodic reaction is called passivation. The iron is said to be passivated.

The critical oxygen concentration for passivation depends on conditions. It increases with dissolved salt concentration and with temperature; and it decreases with increases in pH and velocity of water flow over the surface. At about pH 10, the critical oxygen concentration reaches the value for air-saturated water ($6\,\mathrm{ml\,l^{-1}}$) and is still lower at higher pH. As a result, iron is passivated against corrosion by the oxygen in air at sufficiently high pH values. It is impractical to control corrosion by oxygen passivation below about pH 10, since the concentrations needed are in excess of those dissolved in water in equilibrium with air. However, a variety of oxidizing agents can act as passivators. Chromate, nitrite, molybdate, plumbate, and tungstate salts are examples. As with oxygen, a critical concentration of these oxidizing agents is needed to achieve passivation; and lower concentrations may promote corrosion by cathodic depolarization. The reactions with chromate salts have been most extensively studied. Partially hydrated mixed ferric and chromic oxides are deposited on the surface, where they presumably act as a barrier to halt the anodic reaction.

Certain nonoxidizing salts, such as alkali metal salts of boric, carbonic, phosphoric, and benzoic acids, also act as passivating agents, which may result from their basicity. By increasing pH, they may reduce the critical oxygen concentration for passivation below the level reached in equilibrium with air. Alternatively, it has been suggested that the anions of these salts may combine with ferrous or ferric ions to precipitate complex salts of low solubility to form a barrier coating at the anode. Possibly, both mechanisms operate to some extent.

A newer approach to passivation is applying a film of electrically conductive polymer to a steel surface to protect it from corrosion. Polyphenyleneamine, commonly called polyaniline (PANI) and commercially available, may be effective by forming a tight, very thin metal oxide passivating layer on the surface of the metal. PANI powders are insoluble in all solvents, non-fusible, and difficult to disperse because of high surface tension, although dispersions in a variety of vehicles are available. Reviews of the effects of conducting polymer coatings on metals are available (Sitaram et al., 1997; Spinks et al., 2002). Electropolymerization and deposition of PANI from a phosphate buffer solution onto stainless steel increases the corrosion resistance of the steel (Moraes et al., 2003). Steel coated with polythiophene by electropolymerization also provides corrosion protection (Kousik et al., 2001). Another review of conducting polymer coatings emphasizes protection of aluminum with poly[2,5-bis(*N*-methyl-*N*-propylamino)phenylenevinylene] (Zarras, 1999). While this area of corrosion protection is an active area of research, its commercial significance is still minor owing to the processing and cost issues associated with conductive polymer technology.

Aluminum is higher in the electromotive series than iron and is more easily oxidized. Thus, one would expect aluminum to corrode more easily than steel or other iron alloys. However, aluminum generally corrodes more slowly than steel. A freshly exposed surface of aluminum oxidizes quickly to form a dense, coherent layer of aluminum oxide; that is, aluminum is passivated by oxygen at concentrations in equilibrium with air. While aluminum corrodes slowly at neutral pH, it corrodes rapidly at both low and high pH.

Unlike the corrosion of steel, corrosion of aluminum is generally localized and often cosmetic in nature. A particular type of aluminum corrosion, *filiform corrosion*, occurs on painted aluminum surfaces (Figure 7.5). This type of corrosion often occurs when chloride ions are present. Small threadlike corrosion tracks are observed to propagate from a defect. These threads typically follow the rolling direction in aluminum sheet alloys. The threads are localized corrosion cells where the propagating head of the filament is the anode and the cathode lies behind the head. The reaction products of the corrosion reaction are higher in specific volume than the aluminum metal and cause localized delamination of the paint at the head. As corrosion proceeds, the filament forms due to the ongoing delamination of the paint at the growing head, leaving a trail of delamination in its wake. Filiform corrosion can also occur on coated steel.

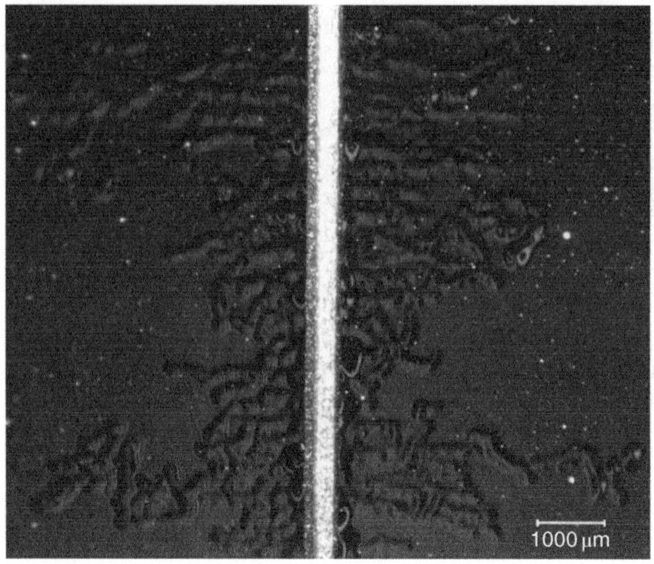

Figure 7.5 Filiform corrosion on a painted aluminum panel after a cyclic corrosion test.

7.3.2 Cathodic Protection

Consider connecting steel to a direct current source or to the positive pole of a battery, wherein a carbon electrode is connected to the negative pole. If both electrodes are immersed in salt water, the steel does not corrode. The impressed electrical potential makes the entire steel surface cathodic relative to the carbon anode. The result is electrolysis of water, rather than corrosion of steel. This is an example of cathodic protection.

A related method is to connect the steel electrically to a piece of metal higher in the electromotive series than iron—for example, magnesium, aluminum, or zinc. When a block of one of these metals, connected to steel, is immersed in an electrolyte, the more active metal is the anode in the circuit, and all corrosion takes place there. The more active metal is called a sacrificial anode. This method is often used to protect pipelines, the steel hulls of ships, and hot water heater tanks. The sacrificial anode is gradually used up and must be replaced periodically. Zinc and magnesium are generally the preferred sacrificial metals. Aluminum is often ineffective because a barrier layer of aluminum oxide forms on its surface (Section 7.3.1). However, aluminum is appropriate for ships that operate entirely in the ocean, since it corrodes readily in salt water.

Another method of cathodic protection is coating steel with zinc to make galvanized steel. The steel is protected in two ways: zinc functions as a sacrificial anode and also acts as a barrier preventing water and oxygen from reaching the steel surface. Since zinc is easily oxidized, it is passivated by oxygen at concentrations below $6\,ml\,l^{-1}$. If the surface of a galvanized sheet is damaged and bare steel and zinc are exposed, the zinc corrodes, but not the steel. After exposure to the atmosphere, the surface of the zinc becomes coated with a mixture of zinc hydroxide and zinc carbonate. Both are somewhat soluble in water and strongly basic.

7.3.3 Barrier Protection and Inhibition

For steel to corrode, oxygen and water must be in direct molecular contact with the surface of the steel. Barriers that can prevent oxygen and water from reaching the surface prevent corrosion. The zinc layer on galvanized steel acts as a barrier. It may even be considered that a layer of irreversibly adsorbed small molecules can act as a barrier. It is often incorrectly assumed that the tin coating on steel in tin cans acts electrochemically similarly to zinc in galvanized steel. However, tin is lower in the electromotive series than iron, so iron is the anode and tin the cathode. Before a can is opened, the tin coating is intact and acts as a barrier so that no water or oxygen reaches the steel. After a can has been opened, the cut bare edges expose both steel and tin to water and oxygen, and the steel corrodes relatively rapidly.

Many organic compounds are corrosion inhibitors for steel. Most are polar substances that tend to adsorb on high energy surfaces (Leidheiser, 1981). Amines are particularly widely used. Clean steel wrapped in paper impregnated with a volatile amine or the amine salt of a weak acid is protected against corrosion. Amines are also added to boiler water to minimize corrosion. The reason for their effectiveness is not clear. Amines may act as inhibitors because they are basic and neutralize acids. Alternatively, or in addition, amines may form a strongly adsorbed layer on the steel by hydrogen bonding or salt formation with acidic sites on its surface, in which the layer may act as a barrier to prevent oxygen and water from reaching the surface of the steel. However, the latter mechanism is feasible only in the presence of an aqueous medium containing the inhibitor.

Some aromatic compounds are also effective corrosion inhibitors in aqueous solutions. Benzotriazoles have been shown to be effective corrosion inhibitors. Certain natural products and extracts can also be corrosion inhibitors at low concentrations. Research in this area is ongoing, as the need to find nontoxic and/or naturally sourced inhibitors is increasing.

7.4 CORROSION PROTECTION BY INTACT COATINGS

Organic coatings can be effective barriers to protect steel, but only if the coating covers the entire substrate surface and remains intact in service. When it is anticipated that there will not be complete coverage of the substrate or that the layer will be ruptured in service, alternative strategies that can suppress electrochemical reactions involved in corrosion may be preferable; such strategies are discussed in Section 7.5. It is seldom effective to utilize both strategies in the same coating layer; one or the other is usually chosen. However, Sorensen et al. (2009) report that it is sometimes possible to combine both strategies.

7.4.1 Critical Factors

Until about 1950, it was generally believed that coatings protect steel by acting as a barrier, which excluded water and oxygen from the steel surface. Then it was reported by Mayne (1952) that the rate of permeation of water and oxygen entering the paint films is substantially greater than the rate of their consumption in the corrosion of uncoated steel. Accordingly, Mayne concluded that barrier action could not explain the effectiveness of coatings. He proposed that electrical conductivity of coating layers is the variable that controls the degree of corrosion protection. Presumably, coatings with high conductivity would give poor protection, as compared with coatings having lower conductivity. It was confirmed experimentally that coatings having very high conductivity afforded poor corrosion protection. However, little correlation between conductivity and protection has been found in comparison with layers having relatively low conductivity. Furthermore, high conductivity layers may fail because of their high water permeability. Nevertheless, some investigators continue to believe that conductivity of organic coatings is at least a factor in corrosion protection (Mayne, 1976; Leidheiser, 1979).

Current understanding of the protection of steel against corrosion by intact films is based, to a significant degree, on the work of Funke (1979, 1983, 1985, 1987). He found that an important factor not given sufficient emphasis in earlier work was the adhesion of a coating to steel in the presence of water. Funke proposed that water permeating through an intact film could displace areas of the coating from steel. In such cases, the film exhibits poor wet adhesion. Water and oxygen dissolved in the water would then be in direct contact with the steel surface; hence, corrosion would start. As corrosion proceeds, ferrous and hydroxide ions are generated, leading to the formation of an osmotic cell under the coating film. The resulting osmotic pressure would provide a sufficient force to remove more coating from the substrate, since osmotic pressure can be expected to range between 2500 and 3000 kPa, whereas the resistance of organic coatings to delamination is substantially lower, ranging from 6 to 40 kPa (Funke, 1979). Thus, blisters form and expand, exposing more unprotected steel surface. Martin et al. (1990) proposed that blisters can also grow by a non-osmotic mechanism, dependent on the modulus of the coating layer. At a sufficiently high modulus, blister growth is minimized owing to the smaller strains in the coating layer (Chuang et al., 1999).

In either osmotic or non-osmotic mechanisms, the key to maintaining corrosion protection by a barrier coating is sufficient adhesion to resist displacement forces. Both mechanisms predict that if the coating covers the entire surface of the steel on a microscopic, as well as macroscopic scale, and if strong wet adhesion is achieved at all areas of the interface, the coating would indefinitely protect steel against corrosion. However, it is difficult to achieve both of these requirements in the application of coatings. For example, owing to nano- or micro-roughness of the metal surface, all of its cavities may not be filled by binder molecules for steric reasons. These unprotected cavities, while small, are large enough to allow a local accumulation of water molecules that provides a medium for the action of corrosion cells (Funke, 1996). Accordingly, Funke (1985) found that in addition to wet adhesion, low water and oxygen permeability help increase corrosion protection. In any case, if wet adhesion is poor, corrosion protection is also poor. However, if the adhesion is good, a low rate of water and oxygen permeation may delay loss of adhesion long enough so that there is adequate corrosion protection for many practical conditions.

7.4.2 Adhesion for Corrosion Protection

Chapter 6 deals broadly with adhesion, but wet adhesion, which is especially critical to corrosion protection, is reviewed here. Good dry adhesion must be taken as a given for achieving corrosion protection. If there is no coating left on the substrate, it cannot protect the underlying metal. It had not been so obvious, however, that good wet adhesion is required. Good wet adhesion requires that the adsorbed layer of the coating will not desorb when water permeates through the film.

The first step to obtain good wet adhesion is to clean the steel surface before painting, especially to remove any oils and salts. Application of phosphate conversion coatings provides further advantages, in particular for steel substrates (Section 6.3.1). Various types of steel and coated steel may require different cleaning and treatment methods (Perfetti, 1994). In the past, a chromic acid solution was often used as a "sealing coat" after the phosphate conversion coating was applied. However, hexavalent chrome is now known to be a carcinogen and its use has been severely restricted. Only aviation and some military applications are allowed to utilize hexavalent chrome in Western nations, and the use in those applications is decreasing rapidly as equivalent performance is being demonstrated by other conversion coating technologies. Nickel compounds, which are often used in tri-cationic zinc phosphate conversion coatings, are also under scrutiny for possible toxicological harm and may face elimination in Europe and other locations in the near future.

A number of different sealing or post-rinse chemistries are under active investigation to replace hexavalent chromium chemistry. The use of organosilane compounds has been shown to be an effective strategy for sealing zinc phosphate. ZrO_2 chemistry or mixed ZrO_2/TiO_2 chemistry has also been shown to be effective. Zirconium oxide deposition is achieved by the use of an aqueous solution of hexafluorozirconic acid, which results in the deposition of an amorphous layer if ZrO_2 on the metal or pretreatment surface. Nair and Subbaiyan (1993) have reported that the addition of polyethyleneimine to a Ca—Zn phosphating treatment bath gives a satisfactory conversion treatment without a chromate rinse. Thin film pretreatments based on ZrO_2 and/or silanes are slowly displacing traditional zinc phosphate in the pretreatment of automobile bodies due to a reduced environmental footprint with equivalent in-field performance.

For many applications, aluminum does not need to be treated for corrosion control owing to the coherent aluminum oxide surface of the aluminum. But if exposure to salt or to low or high pH is expected, the surface must be treated before applying a coating. Again, chromate surface treatments were the norm but have been replaced in most industries (other than those mentioned previously that have been given grandfather exceptions until equivalent performance can be demonstrated). Equivalent performance on civil or commercial aircraft is difficult to demonstrate, as specifications can require 40-year corrosion protection on areas of an aircraft that are not amenable to inspection. Thus, proof of performance can take an enormous time to generate. Similar technologies, which have been evaluated for the surface treatment of steel, have been examined for the surface treatment of aluminum. ZrO_2 and ZrO_2/TiO_2 surface treatments have received the most commercial attention and are available for use in many industries including coil coating and automotive. Silane pretreatments are intriguing and have demonstrated efficacy by themselves or in concert with other additives. The use of trivalent chrome has shown some promise in corrosion protection, but has not been widely accepted. Results with cerium, lanthanum, vanadium, and other metal salts have also been mixed. A review of potential uses of silane treatments for steel, galvanized steel, and aluminum is available (Child and Van Ooij, 1999).

After cleaning and treating, the surface should not be touched and should be coated as soon as possible. Fingerprints leave oil and salt on the surface, which promote the formation of fine blisters after exposure to high humidity. A rusty handprint was once observed on a ship after only one ocean and lake passage (Leidheiser, 1981). It is critical to avoid having any salt on the metal surface when the coating is applied near the ocean.

It is also critical to achieve as complete penetration as possible into the micropores and irregularities in the surface of the steel to prevent pockets that could fill with water and oxygen (Section 7.4.1). An important factor for achieving penetration is that the viscosity of the external phase of liquid coatings be as low as possible and remain low long enough to permit complete penetration (Section 6.1). For this purpose, it is desirable to use solvents that evaporate slowly, coatings that cross-link slowly, and, when possible, baking primers. Macromolecules may be large compared with the size of small crevices, so lower molecular weight components may give better protection.

Wet adhesion requires that the coating is adsorbed strongly on the surface of the metal and is not displaced by water that permeates through the coating. Empirically, it is found that wet adhesion is enhanced by having several adsorbing groups scattered along the resin chain, with parts of the resin backbone being flexible enough to permit relatively easy orientation and other parts rigid enough to assure that there are loops and tails sticking up from the surface for interaction with the rest of the coating. Enhanced corrosion resistance of baking primers may also be attributed to greater opportunity for orientation of resin molecules at the steel interface at the higher baking temperatures. If the coating T_g is lower than the temperature at which corrosion protection is needed, wet adhesion may be improved, since desorption will be minimized. However, as discussed in Section 7.4.3, a higher T_g is preferable for reducing oxygen and water permeability. Amine groups are particularly effective polar substituents for promoting wet adhesion. Perhaps water is less likely to displace amines than other groups from the surface. Phosphate groups also promote wet adhesion. For example, epoxy phosphates have been used to enhance the adhesion of epoxy coatings on steel (Massingill et al., 1990). Phosphate-functional (meth)acrylic esters as comonomers increase the wet adhesion of solventborne and latex coatings (Yang et al., 2005).

Saponification resistance is another important factor in wet adhesion (Holubka et al., 1980; Holubka and Dickie, 1984). Corrosion generates hydroxide ions at the cathode, raising pH levels as high as 14, thereby enhancing the prospect of ester saponification. If the ester groups are in the backbone of a binder, saponification will result in polymer degradation near the interface, thereby reducing wet adhesion. Epoxy–phenolic primers are an example of high bake primers that are completely resistant to hydrolysis. In some epoxy–amine primers, there are no hydrolyzable groups. Amine-terminated polyamides, which are widely used in air dry primers to react with epoxy resins, have amide groups in the backbone that can hydrolyze. However, amides are more resistant to base-catalyzed hydrolysis than esters. Alkyd resins are used when only moderate corrosion protection is required and low cost is important. The superior corrosion protection of epoxy ester (Section 15.8) primers is often attributed to greater resistance to saponification of epoxy esters. Cathodic electrocoat primers (Chapter 27), which are based on aromatic epoxy compounds cross-linked by aromatic blocked isocyanates, also provide excellent wet adhesion, as they are quite resistant to saponification and are excellent barriers to both water and ions.

Water-soluble components that may remain in barrier primer layers should be avoided, because they can lead to blister formation. For example, zinc oxide is generally an undesirable pigment to use in primers. Its surface interacts with water and carbon dioxide to form zinc hydroxide and zinc carbonate, which are somewhat soluble in water and can lead to osmotic blistering. Insoluble pigments having adsorbed water-soluble contaminants should also be avoided. Passivating pigments, discussed in Section 7.5.2, cannot function unless they are somewhat soluble in water; their presence in coating layers at the metal surface can, therefore, lead to blistering. Funke showed that hydrophilic solvents, which become immiscible in the drying film as other solvents evaporate, can be retained as a separate phase and lead to blister formation (Funke, 1979).

7.4.3 Factors Affecting Oxygen and Water Permeability

Many factors affect permeability of coating layers to water and oxygen (Thomas, 1991). Water and oxygen can permeate, to at least some extent, through any amorphous polymer film, even though the film has no imperfections such as cracks or pores. Small molecules travel through the film by jumping from free volume hole to free volume hole. Since free volume increases as temperature increases above T_g, this factor points to designing coatings having a T_g above the temperature at which corrosion protection is desired. Since cross-linking reactions become slow as the

T_g of the cross-linking polymer approaches the temperature at which the reaction occurs and become very slow at $T<T_g$, air dry coatings cannot have T_g values much above ambient temperatures. Furthermore, since wet adhesion is improved by having a lower T_g (Section 7.4.2), compromises may be necessary. Permeability is also affected by the solubility of oxygen and water in a film. The variation in water solubility can be large. Permeability of oxygen and water through coating layers is reviewed in Sangaj and Malshe (2004).

Oxygen dissolves to only a small degree in resins used in coatings; but there are many other factors that affect permeability of films to oxygen. Higher density cross-linking and low polarity are two factors that lead to lower oxygen diffusion. Lateral side chains increase free volume and increase permeability. The proximity of ambient T to T_g is a major factor, which promotes the generally superior corrosion protection of baked coatings.

Permeability of coatings to water is affected by the same factors that affect permeability to oxygen; but there are other factors. Water is more soluble than oxygen in many coating films, and dissolved water increases permeability. Water can act as a plasticizer for coatings such as epoxy–amines and polyurethanes, and the swelling caused by water increases internal stress that can lead to delamination (Negele and Funke, 1996). Internal stress increases when a coating is cycled through wet and dry stages, as coatings will generally expand upon absorption of moisture and most metal substrates are dimensionally stable with respect to moisture.

Salt groups on a polymer increase the solubility of water in coatings. Accordingly, it is difficult to formulate highly corrosion-resistant air dry, water-reducible coatings that contain amine salts of carboxylic acids. Although to a lesser degree than salts, resins that contain polyethylene oxide segments are likely to have higher water permeabilities. On the other hand, water has low solubility in halogenated polymers; hence vinyl chloride and vinylidene chloride copolymers and chlorinated rubber are commonly used in formulating top coats for corrosion resistance. Fluorinated polymers have low permeabilities and are not easily wet by water. Accordingly, hydroxy-functional poly(vinylidene fluoride) cross-linked with polyisocyanates is reported to provide good corrosion protection even as a single coat (Barbucci et al., 1996). Although silicones have low surface tension and are used as water repellents, films of silicone resins are highly permeable to water probably due to their very low T_g and cross-link density.

Pigmentation can have significant effects on water and oxygen permeability. Oxygen and water molecules cannot pass through pigment particles; therefore, permeability decreases as pigment volume concentration (PVC) increases. However, if the PVC exceeds critical pigment volume concentration (CPVC), there are voids in the film, in which case passage of water and oxygen through the film is facilitated. (PVC and CPVC are discussed in Chapter 22.) Some pigments have high polarity surfaces that adsorb water, in which case water can displace polymer adsorbed on such surfaces. Furthermore, water permeability is expected to increase with increasing content of such pigments. As explained in the previous section, pigments should be insoluble and as devoid as possible of water-soluble impurities. In theory, the use of hydrophilic pigment dispersants should be avoided, or at least minimized.

Pigments with platelet-shaped particles can reduce permeability rates as much as fivefold when they are aligned parallel to the coating surface (Funke, 1983; Biegańska et al., 1988). Oxygen and water vapor cannot pass through the pigment particles and must circumvent them to pass through the coating. The alignment is critical, since permeability may increase if the platelets are not aligned, especially if the film thickness is small relative to the size of the platelets. A factor favoring alignment is shrinkage during solvent evaporation. (See Section 30.1.4 for discussion of additional factors.) Mica, talc, micaceous iron oxide, glass flakes, clay, and metal flakes are examples of such pigments. Aluminum flake is widely used; stainless steel and nickel platelets, while more expensive, have greater resistance to extremes of pH. When appearance permits, use of leafing aluminum pigment in the top coat is particularly effective (Section 20.2.5). Leafing aluminum is surface-treated to reduce its surface free energy. As a result, the platelets come to the surface during film formation, creating an almost continuous barrier. In formulating coatings with leafing aluminum, it is necessary to avoid resins and solvents that displace the surface treatment from the flakes.

The use of exfoliated clay to make nanocomposite coatings is increasing. If clay platelets can be properly dispersed and stabilized, the thin (<1 nm) layers that make up the clay platelets can become exfoliated, that is, individually dispersed at the atomic level. Such nanocomposites show a remarkable increase in their resistance to diffusion of small molecules (Tomić et al., 2016). The key to good dispersion is the exchange of cations within the clay galleries with other cations to expand and stabilize the exfoliated clay sheets. Many approaches have been investigated to achieve this end. The most successful have used organic cations in montmorillonite clay. In many cases the resistance to water and oxygen permeation is increased severalfold.

A Monte Carlo simulation model of the effect of several variables on diffusion through pigmented coatings has been devised (Bentz and Nguyen, 1990). The model indicates, as would be expected, that finely dispersed, lamellar pigment particles at a concentration near, but below, CPVC provide the best barrier performance.

Because the role of water transport is so important to corrosion performance, computer simulations of the permeation of water through a paint system have been devised. These simulations are particularly useful when trying to understand the fidelity of accelerated corrosion tests and how those tests compare to actual performance (Hinderliter and Sapper, 2015).

There are advantages in applying multiple layers of coatings. The primer can be designed for optimal wet adhesion and penetration into the substrate surface, without particular concern about other properties. Intermediate and top coat(s) can be designed for minimum permeability and other required properties. As long as the top coat is providing barrier properties, the primer film does not need to be thicker than needed to assure coverage of the entire surface. Funke (1987) has reported good results with a primer thickness of 0.2 μm and even as low as 10 nm thickness, utilizing a wet adhesion-promoting polymer. An advantage of applying multiple thin coats is to decrease the probability that any area of the substrate will be uncoated.

Film thickness affects the time necessary for permeation through coatings. Thicker films are expected to delay arrival of water and oxygen at the interface, but are not expected to affect the equilibrium condition. If the equilibrium condition was the only factor, corrosion protection afforded by intact films would be expected to be essentially independent of film thickness. However, film thickness also affects the mechanical performance of films. Obviously, erosion losses would take longer to expose bare metal as film thickness increases. On the other hand, the probability of cracking, resulting from bending, increases as film thickness increases so that there may be an optimum film thickness for the maintenance of an intact film. In certain air dry, heavy-duty maintenance coatings, there is an optimal film thickness, dependent on the coating, that provides a more than proportional increase in corrosion protection relative to thinner films. Commonly, this film thickness is as much as 400 μm or more. Funke (2005, private communication) suggests that below certain coating thicknesses, there may be microscopic defects extending down through the film to the substrate. The film may look intact, but there may be defects that are large compared to the free volume holes through which permeation occurs in fully intact films. A potential source of such defects is cracks resulting from shrinkage of films during the late stage of solvent loss from a coating, particularly when T_g of the solvent-free system is near ambient temperature. Funke suggests that if the layer is thick enough, such defects may not reach the substrate, hence substantially reducing passage of water and oxygen to the substrate interface. This hypothesis is consistent with the general observation that greater protection is achieved by applying more coats to attain the same film thickness. In line with this proposal, the use of barrier platelet pigments permits a reduction in the required film thickness without loss of protection. The platelets may minimize the probability of defects propagating through the film to the substrate. Such defects are less likely to occur in baked films, which may be another factor for the generally superior corrosion protection afforded by baked films, even though thinner films are used.

7.5 CORROSION PROTECTION BY NONINTACT FILMS

There are situations in which it is not possible to assure full coverage of the entire metal surface as required for barrier coatings. Furthermore, coatings that were intact initially may be damaged during their service lives, even when designed to minimize the probability of mechanical failure. In such cases, it is generally desirable to design coatings primarily to suppress electrochemical reactions rather than to enhance their barrier properties.

7.5.1 Minimizing Growth of Imperfections: Cathodic Delamination

If there are gouges through the film down to bare metal, water and oxygen reach the metal and corrosion starts. If wet adhesion of the primer to the metal is not adequate, water creeps under the coating, and the coating delaminates from the metal over a wider and wider area. Poor hydrolytic stability can be expected to exacerbate the situation. This mode of failure is called cathodic delamination. Control of cathodic delamination requires wet adhesion and saponification resistance. It has also been shown that blisters are likely to develop under a film near the location of a gouge (Leidheiser, 1981; Funke, 1983).

When wet adhesion varies on a local scale, filiform corrosion can occur (Bautista, 1996), which is characterized by the development of random tracks of thin threads of corrosion under the film, which never cross another track as shown previously in Figure 7.5. Formation of these threads often starts from the edge of a scratch. As introduced in Section 7.3.1, oxygen permeates through the film at the head of the thread, and cathodic delamination occurs. The head grows following the path of the poorest wet adhesion and often follows the rolling direction of the metal. Behind the head, oxygen is consumed by oxidation of ferrous ions and ferric hydroxide precipitates, passivating the area, explaining why threads never cross. Since the ion concentration decreases, osmotic pressure drops, and the thread collapses, but it leaves a discernable rust track. Filiform corrosion can be difficult to see through pigmented films. A good example of filiform corrosion can be observed on aluminum. Certain commercial aircraft liveries call for the "unpainted" aluminum look. If left unpainted, the bare aluminum will generally not corrode owing to self-passivation by its stabile oxide. However, if the same aluminum is coated with a clear coat, filiform corrosion will occur by the mechanism outlined previously. Similar behavior can be seen in unpainted aluminum campers/trailers, which can last for decades.

7.5.2 Primers with Passivating Pigments

Passivating pigments promote the formation of a barrier layer over anodic areas, thereby passivating the surface (Section 7.3.1). To be effective, such pigments must be water soluble, but only to a limited extent. If the solubility is too high, the pigment will leach out of the coating film too rapidly, thereby reducing the available time for corrosion inhibition. For the pigment to be effective, the binder must permit diffusion of water to partially dissolve the pigment. Therefore, the use of passivating pigments may lead to blistering after exposure to humid conditions. Accordingly, such pigments are most useful in applications wherein protecting the steel substrate after film rupture has occurred outweighs minimizing the probability of blistering. Such pigments are also useful when it is not possible to remove all surface contamination (blistering will probably occur anyway) or when it is not possible to achieve complete coverage of the steel by the coating.

Red lead pigment, Pb_3O_4 containing 2–15% PbO, was used in oil primers as a passivating pigment starting in the mid-nineteenth century for air dry application over rusty, oily steel. Although not fully understood, the mechanisms of action presumably include oxidation of ferrous ions to ferric ions followed by coprecipitation of mixed iron–lead salts or oxides. The somewhat soluble PbO raises the pH and neutralizes any fatty acids formed over time by hydrolysis of the drying oil. However, toxic hazards of red lead have resulted in widespread prohibition of its use.

The utility of chromate pigments for passivation of steel and aluminum is well established. Leidheiser (1981) has proposed various mechanisms to explain their effectiveness, all of which require that the chromate ions be in aqueous solution. Like all passivators, chromate ions accelerate corrosion at low concentrations. The critical minimum concentration for passivation at 25°C is approximately 10^{-3} mol $CrO_4^{2-}l^{-1}$. The critical concentration increases with increasing temperature and increasing NaCl concentration. Sodium dichromate is an effective passivating agent but a poor passivating pigment owing to its high solubility in water (3.3 mol $CrO_4^{2-}l^{-1}$). It would rapidly leach out of a film and probably cause massive blistering. At the other extreme, lead chromate (chrome yellow) is so insoluble (5×10^{-7} mol $CrO_4^{2-}l^{-1}$) that it has no electrochemical action.

So-called zinc chromates have been widely used as passivating pigments, although their use is rapidly declining as explained in the following paragraph. The terminology is poor, since zinc chromate itself is too insoluble and could promote corrosion rather than passivate. Zinc yellow pigment ($K_2CrO_4\cdot3ZnCrO_4\cdot Zn(OH)_2\cdot2H_2O$) has an appropriate solubility (1.1×10^{-2} mol $CrO_4^{2-}l^{-1}$ at 25°C) and has been widely used in primers. (Alternative ways of describing the same composition are $Zn_4K_2Cr_4O_{20}H_6$ and ($4ZnO\cdot K_2O\cdot4CrO_3\cdot3H_2O$).) Zinc tetroxychromate ($ZnCrO_4\cdot4Zn(OH)_2$ also written as $Zn_5CrO_{12}H_8$) has a solubility lower than desirable (2×10^{-4} mol $CrO_4^{2-}l^{-1}$). Strontium chromate ($SrCrO_4$) has an appropriate solubility in water (5×10^{-3} mol $CrO_4^{2-}l^{-1}$) and has been used in waterborne primers, especially latex primers, in which the more soluble zinc yellow can cause problems of package stability.

It has been established that zinc chromates and other soluble chromates (hexavalent chrome) are carcinogenic to humans. They must be handled with appropriate caution. In some countries, their use has been prohibited and worldwide prohibition is probable in the future. Enormous efforts have been undertaken to develop less hazardous passivating pigments (Smith, 1988). Almost all industries in the developed world have banned hexavalent chrome-containing coatings. The exception is the commercial and military aerospace and naval sectors, where hexavalent chrome is still allowed until comparable coating performance can be demonstrated using alternative nontoxic pigments.

The search for performance equivalent to chrome has been tried. In some cases, a formulation, optimized for one pigment, is compared with a formulation in which only the pigment is changed. However, the formulation may not be optimal for the substituted pigment. (A common example is the substitution of one pigment for another on an equal weight basis rather than formulating to the same ratio of PVC to CPVC; the results can be very misleading, since primer performance is quite sensitive to the PVC/CPVC ratio, as explained in Chapter 22.) Much of the published data is based on comparing corrosion resistance in salt fog chamber tests (or other laboratory tests) rather than in actual field experience. One of the many problems with accelerated tests is the difficulty in simulating the rate at which the pigment is leached from a film, which is an important factor in its performance. As discussed in Section 7.6, there is no laboratory test available that provides reliable predictions of field performance.

Non-chromate inhibitive pigments include salts of phosphates, molybdates, nitrates, borates, and silicates (Sørensen et al., 2009). As with chromates, their water solubility must be appreciable so they can be effective but low enough to minimize leaching. Sorensen attributes their effectiveness to the formation of passivating layers, but other mechanisms may also be at work. Basic zinc and zinc–calcium molybdates act as passivating agents in the presence of oxygen, apparently leading to precipitation of a ferric molybdic oxide barrier layer on the anodic areas. Barium metaborate, the salt of a strong base and a weak acid, may passivate by increasing the pH, thus lowering the critical concentration of oxygen required for passivation. The pigment grade is coated with silicon dioxide to reduce its solubility in water, but the solubility may still be too high for use in long-term exposure conditions. Zinc phosphate, $Zn_3(PO_4)_2{\cdot}2H_2O$, may act by forming barrier precipitates on the anodic areas. There is considerable difference of opinion as to its effectiveness. Calcium and barium phosphosilicates and borosilicates are being used increasingly; they may be passivators and/or they may act by increasing pH. Calcium tripolyphosphate has also been recommended (Vetere et al., 2001).

Del Amo et al. (2002) evaluated a series of passivating pigments in a waterborne epoxy coating. Zinc iron phosphate, zinc aluminum phosphate, zinc molybdenum phosphate, and basic zinc phosphate all performed well in laboratory tests; basic zinc phosphate was slightly poorer. Field performance remains to be demonstrated. Bethencourt et al. (2003) have reported that such dual metal phosphate pigments provide passivation superior to zinc chromate under acidic conditions but inferior to zinc chromate under alkaline conditions. Galliano and Landolt (2002) have shown that zinc aluminum phosphate is an effective corrosion inhibitor for waterborne epoxy–amine coatings. Zubielewicz and Gnot (2004) have reported a study of corrosion protection of steel by zinc phosphate, calcium zinc phosphate, zinc ferrite, and calcium-exchanged silica, each in a styrene/acrylic latex and a fatty acid-modified polyurethane dispersion. The corrosion resistance was evaluated by a series of tests including EIS and Prohesion. The polyurethane vehicle gave better resistance than the latex, and the best combination was zinc ferrite with the polyurethane; the next best was the calcium zinc phosphate.

Some spinel pigments, which are metal oxides with a mixture of two, preferably more, different cations in the crystalline lattice, offer the promise of corrosion inhibition with low toxicity but are as yet unproven in large-scale use (Sørensen et al., 2009).

Previously mentioned are all inorganic pigments; a wider range of potential oxidizing agents and bases would seem to be possible using organic pigments. An example of a commercially available organic pigment is the zinc salt of 5-nitroisophthalic acid, which is reported to be as effective as zinc yellow at lower pigment levels. However, since any passivating pigment will be lost by leaching over time, it seems doubtful that a substantially lower pigment content could provide an equal lifetime. The zinc salt of 2-benzothiazoylthiosuccinic acid has also been recommended as a passivating agent. A recurrent problem with organic passivators is that many perform extremely well initially but lose effectiveness with time by either leaching or evaporation.

Sinko (2001) reviewed the subject of passivating pigments and concluded that whereas no other practical inorganic pigment can match or surpass the performance of strontium chromate, hybrid inorganic/organic pigments can equal or excel its performance. He also provides manufacture procedures for inorganic/organic anticorrosion pigments such as $Zn(NCN)_2$/Zn(2-mercaptobenzothiazole). He discloses that the zinc salt of 2,5-dimercapto-1,3,4-trithiazole outperforms strontium chromate in protecting aluminum and that the triisocyanuric acid salt of dicyclohexylamine with strontium-doped amorphous silica outperforms strontium chromate on galvanized steel (Sinko, 2002).

Braig et al. (2000) report that the morpholine salt of toluylpropionic acid, the zirconium complex of toluylpropionic acid, metal salts of benzothiazoylthiosuccinic acid, and the tridecylammonium salt of benzothiazoylthiosuccinic acid impart corrosion inhibition and, further, that hindered amine light stabilizers (HALS) and UV absorbers improve corrosion protection. They further report that benzothiazoylthiosuccinic acid and calcium-exchanged silica enhance corrosion resistance of powder coatings and that the morpholine salt of toluylpropionic acid improves corrosion resistance of 2K epoxy–amine primers.

Recently Visser et al. (2015) reported that the use of lithium salts, in particular lithium oxalate and lithium carbonate, has shown great promise in inhibiting the corrosion of aluminum and steel. The mechanism by which this occurs on aluminum is still under investigation, but the appropriate solubility of lithium in the electrolyte leads to a passivation of the surface by forming a mixed hydroxide layer of lithium/aluminum hydroxide. Primers based on these pigments have been approved for use in aerospace applications, where previously only hexavalent chrome primers were used.

7.5.3 Cathodic Protection by Zinc-Rich Primers

The use of zinc-rich primers represents another approach to protecting steel with nonintact coatings as further discussed in Section 33.1.2. These primers were initially designed to replace the protection given by galvanized steel but have been found to provide excellent protection to steel structures after fabrication (Mayne and Evans, 1944). Such primers contain high levels of powdered zinc—over 84% by weight is usual. On a volume basis, the zinc content exceeds CPVC to assure good electrical contact between the zinc particles and with the steel. Furthermore, when PVC is greater than CPVC, the film is porous, permitting water to enter, completing the electrical circuit. CPVC of zinc powder varies, depending primarily on particle shape and particle size distribution; values on the order of 67% have been reported (Leclercq, 1990). The zinc serves as a sacrificial anode; and zinc hydroxide is generated in the pores. The structure of the zinc particles can affect performance. Smaller particle size spherical zinc provides better results than larger particle size. Lamellar zinc particles impart better mechanical properties to films, as well as satisfactory corrosion protection with lower weight fraction than required with spherical zinc particles (Kalendová, 2003).

Vehicles for zinc-rich primers must be saponification resistant, which eliminates the use of alkali-sensitive resins, such as alkyds. Both organic and inorganic binders are widely used. Among organic binders, epoxies are the most commonly used, and polyurethanes are gaining favor. The principal inorganic vehicles are oligomers derived from tetraethyl orthosilicate, produced by controlled partial hydrolysis with a small amount of water (Section 16.3). Ethyl or isopropyl alcohol is used as the principal solvent, since an alcohol helps maintain package stability. After application, the alcohol evaporates, and water from the air completes the hydrolysis of the oligomer to yield a film of polysilicic acid partially converted to zinc salts. Crosslinking is affected by relative humidity (RH); properties can be adversely affected if the RH is low at the time of application.

Properly formulated and applied, zinc-rich primers are very effective in protecting steel against corrosion. Their useful lifetime is not completely limited by the amount of zinc present, as one might first assume. Although the zinc metal is gradually consumed by the electrochemical reaction, the primer continues to protect the steel for an extended period if the coating remains intact. Possibly, the partially hydrated zinc oxide, formed in the initial stages of electrochemical corrosion, fills the pores and, together with the remaining zinc, acts as a barrier coating (Feliu et al., 1989). It is also possible that zinc hydroxide raises the pH to the level at which oxygen can passivate the steel.

Zinc is expensive, especially on a volume basis. Early attempts to replace even 10% of the zinc with low cost inert pigment caused a serious decrease in performance, presumably due to decrease in metal-to-metal contact, even though the PVC was above CPVC. A relatively conductive inert pigment, iron phosphide (Fe_2P), has shown promise. In ethyl silicate-based coatings, up to 25% of the zinc can be replaced with Fe_2P (Fawcett et al., 1984). However, with epoxy/polyamide coatings, replacement of part of the Zn with Fe_2P leads to a reduction in protection (Feliu et al., 1991).

Zinc-rich primers are frequently top coated to minimize corrosion of the zinc, protect against physical damage, and improve appearance. Formulation and application of top coats require care. If the vehicle of the top coat penetrates into the pores in the primer film, conductivity of the primer may be substantially reduced, rendering it ineffective. See Section 33.1.2 for further discussion. Zinc-rich primers can be formulated so that the painted objects can be welded. This has been a major advantage in the automotive industry, which has allowed coil-applied zinc-rich primers to be used in several automotive applications. As their welding performance is typically superior to that of Fe_2P coatings, zinc-rich coatings are preferred.

Waterborne inorganic zinc-rich primers have been developed in response to the need to reduce VOC. Parashar et al. (2001) have reviewed the chemistry of waterborne silicate binders. The binder is a combination of potassium, sodium, and lithium silicates with a dispersion of colloidal silica (Montes, 1993). The ratio of silicate to SiO_2 is critical; if there is insufficient base, phase separation can occur. Commonly a ratio of 1–4 or 5 is appropriate. The ratio can be changed by adding amorphous silica. Drying is accelerated by high temperature but impaired by high humidity. Anodic inhibition of the metal surface with amines before application reduces problems encountered in drying at high humidity. Hydroxy-functional silicones promote water compatibility as does addition of methyltrimethoxysilane. It is critical to use zinc dust that has not reacted with CO_2 and water to give a surface of zinc carbonate and zinc hydroxide (Szokolik, 1995). Environmental carbon dioxide also decreases pH, which can destabilize the silicate solution. Excellent performance on oil and gas production facilities in marine environments has been reported (Montes, 2001). Organic zinc-rich primers based on waterborne epoxy vehicles are also available.

Analogous to zinc-rich primers, which are typically used to protect steel, magnesium-rich primers can be used to protect aluminum. The magnesium serves as a sacrificial anode with respect to the aluminum. Development of these primers was spearheaded by Bierwagen (Nanna and Bierwagen, 2004; Battocchi et al., 2006). As the Mg is oxidized to MgO, a significant volume change takes place that changes the appearance of these coatings, even when used as primers, which can be problematic. In addition, because CO_2 can act as a buffer to the Mg to MgO reaction, the concentration of CO_2 in the atmosphere, particularly during accelerated corrosion testing, must be within normal limits or erroneous results can be obtained.

7.5.4 Smart Corrosion Control Coatings

Recently, "smart" coatings that sense the presence of incipient corrosion have been developed. The sensors are pH-sensitive molecules incorporated into the binder (Augustyniak and Ming, 2011). During corrosion over aluminum substrates, the locally lowered pH causes the sensor molecules to fluoresce well before corrosion is visible to the naked eye. Coatings of these types may prove to be beneficial for structures such as infrastructure and aircraft where inspection is common but difficult.

Other smart coatings (Chapter 34) have been developed that respond to mechanical damage in the coating. When the coating is breached, small microcapsules are ruptured. The contents of the capsules can often heal the crack. Alternatively, the capsules can contain corrosion-inhibiting substances, such as organic corrosion inhibitors, which can passivate the newly exposed metal and slow the progress of corrosion. The use of smart coatings such as these is an exciting area of research and holds promise for significant advances in the future.

7.6 EVALUATION AND TESTING

There have been many attempts to develop laboratory tests to predict corrosion protection by coatings, and these efforts continue. However, there is no laboratory test available that can be used to predict corrosion protection performance of a new coating system. This unfortunate situation is an enormous obstacle to research and development of new coatings but must be recognized and accommodated.

The barriers to robust, accurate corrosion testing are many. First, corrosion occurs in many different environments. Crevice corrosion, where the supply of reactants may be severely constrained, is very different from open atmospheric corrosion, which is very different than marine corrosion. Thus, one should not expect any one corrosion test to accurately mimic the corrosion performance of a coating/substrate system in all use cases. Second, corrosion typically proceeds by a very complex sequence of chemical reactions, many of which are not well understood. Thus, control of the reaction conditions by any laboratory method or at any outdoor exposure site is likely to accelerate some reactions at the expense of others. Accelerated corrosion testing is often done by introducing an artificial defect, typically a scribe, in the coating, thereby providing a place for corrosion to "start." In practice, however, corrosion often starts from defects that are much smaller than the macroscopic defects used in corrosion testing. Thus, the incubation period for corrosion in the field may differ substantially from that found during corrosion testing. Finally, the effect of processing conditions on corrosion performance is underappreciated. When one tests the corrosion performance of a coating/substrate combination, the coating technologist will likely take pains to ensure that the specimens are properly prepared—the substrate is well cleaned, the coating is applied at the appropriate thickness, and the coating is cured under the proper conditions. Commercially, however, coatings are often applied under nonideal conditions, and the corrosion performance is, therefore, inferior to that which was advertised.

Use testing is still the only reliable approach—that is, to apply a coating system and then observe its condition over years of actual use. The major suppliers and end users of coatings for such applications as bridges, ships, chemical plants, and automobiles have collected data, which correlate performance with composition of different coating systems over many years. These data provide a basis for selection of current coating systems for particular applications. They also provide insights for formulating new coatings to improve chances of success. While this data is useful to understand current performance and to make educated inferences about small formulation or process changes, the time scale of data collection renders this data less useful for most product development cycles. Computers have become almost essential in evaluating use testing data; Milmo (2016) describes a "big data" approach that may eventually provide valuable insights for the marine industry.

Simulated tests or accelerated tests are the next most reliable methods for predicting performance. One common approach is to expose laboratory-prepared panels on test fences in inland south Florida or on beaches in south Florida or North Carolina. The difficulties in developing tests to simulate corrosion in marine environments have been discussed (Lee and Money, 1984). Test conditions must simulate actual use conditions as closely as possible. For example, exposure at higher temperatures may accelerate corrosion reactions; however, oxygen and water permeability can be affected by $(T-T_g)$. If use temperatures are below T_g, but tests are run above T_g, no correlation should be expected.

Since wet adhesion is so critical to corrosion protection, techniques for studying wet adhesion can be very useful.

Electrochemical impedance spectroscopy (EIS) is widely used to study coatings on steel and aluminum. Its importance is growing. There are a number of good review papers that discuss its use and more importantly the interpretation of EIS data (Cano et al., 2009). In short, impedance is the apparent opposition to flow of an alternating electrical current and is the inverse of apparent capacitance. When a coating layer begins to delaminate, there is an increase in apparent capacitance. The rate of increase of capacitance is proportional to the area delaminated by wet adhesion loss. High performance systems show slow rates of increase of capacitance, so tests must be continued for long time periods. EIS is very sensitive for detecting defects but cannot determine whether the defect is characteristic of the coating system or a consequence of poor application. EIS can be very useful in comparing various coating formulations but, because of the wide variation in actual exposure conditions, cannot reliably predict field performance. A review of evaluation of coatings by electrochemical, physicochemical, and physical measurements is available (Sekine et al., 2002).

In North America, the most widely used test method for corrosion resistance is the salt spray (fog) test (ASTM Method B117-95). Coated steel panels are scribed (cut) through the coating in a standardized fashion, exposing bare steel, and hung in a chamber where they are exposed to a mist of 5% salt solution at 100% RH at 35°C. Periodically, the nonscribed areas are examined for blistering, and the scribe is examined to see how far from the scribe mark the coating has been undercut or has lost adhesion. However, it has been demonstrated, repeatedly, that there is little, if any, correlation between results from salt spray tests and actual performance of coatings in use (Mazia, 1977; Funke, 1979; Wyvill, 1982; Athey et al., 1985). A well-known example of salt spray test (B117-95) failure is its prediction that Zn-coated steel corrodes faster than uncoated steel, which is clearly untrue. Despite its poor correlation to field results, the test continues to be used in many industries, as paint companies and paint users both have extensive databases and histories with the test. In Europe, a variety of ISO tests are employed, summarized by Sorensen et al. (2009).

Many factors are probably involved in the unreliability of the salt spray test. One major flaw is that the test conditions are constant—no attempt is made to replicate the changes in conditions (warm/cool, wet/dry) that occur during most field exposures. Furthermore, outdoor exposure can have a significant effect on coating properties, and environmental factors such as acid rain vary substantially from location to location. The application of the scribe mark can be an important variable; narrow cuts generally affect corrosion less than broader ones. Also, if the scribe mark is cut rapidly, there may be chattering of cracks out from the main cut, whereas slow cutting may lead to a smooth cut. A passivating pigment with high solubility might be very effective in a laboratory test but may provide protection for only a limited time under field conditions, owing to loss of passivating pigment by leaching or evaporation.

Since with intact films it is common for the first failure to be blister formation, humidity resistance tests are also widely used (ASTM Method D2247-94). The unscribed face of a panel is exposed to 100% RH at 38°C, while the back of the panel is exposed to room temperature. Thus, water continuously condenses on the coating surface. This humidity test is a more severe test for blistering than the salt fog test because pure water on the film generates higher osmotic pressures with osmotic cells under the film than the salt solution used in the salt fog test. It is common to run the test at 60°C "because it is a more severe test." The pitfalls of this approach are obvious in view of the previous discussion of the importance of $(T-T_g)$. Humidity tests do not provide a prediction of the period of corrosion protection, but may provide useful comparisons of wet adhesion. Funke (2005, private communication) recommends testing for wet adhesion by scribing panels after various exposure times in a humidity chamber, immediately followed by applying pressure-sensitive tape across the scribe mark, and then pulling the tape off the panel. A peel adhesion test for wet adhesion has been described (McKnight et al., 1995). Wet adhesion can also be checked after storing panels in water (Hemmelrath and Funke, 1988).

It is often observed that alternating high and low humidity causes faster blistering than continuous exposure to high humidity. A possible explanation is that intermediate corrosion products form colloidal membranes, causing polarization and temporary inhibition of corrosion. The membranes are not stable enough to survive drying out and aging. Another factor may be the increase in internal stress that has been reported to occur by cycling through wet and dry periods (Negele and Funke, 1996). Many humidity cycling tests have been described, commonly involving repeated immersion in warm water and removal for several hours. In some industries, such tests have become accepted methods of screening coatings, although their predictive value is questionable. Simply correlating them with salt fog tests proves nothing.

A testing regimen called Prohesion (a trademark of BP Chemicals) has been reported to correlate better with actual performance than the standard salt spray test (Timmins, 1979). The procedure combines care in selection of substrates that will reflect real products, use of thin films, emphasis on adhesion checks, and a modified salt mist exposure procedure. Instead of 5% NaCl solution, a solution of 0.4% ammonium sulfate and 0.05% NaCl is used. Scribed panels are sprayed with the mixed salt solution cycling over 24 h, six 3 h periods alternating with six 1 h drying periods using ambient air. During these cycles, water can penetrate through the film to a greater extent than in salt fog chamber testing, in which the humidity is always at 100% and the 5% salt solution minimizes water penetration because of reverse osmotic pressure. In some laboratories, a QUV exposure cycle (Section 5.6.3) is included in the cycling regime (Skerry and Simpson, 1991).

Most of the major automotive companies have developed their own cyclic corrosion tests that they consider correlate well with full vehicle corrosion testing and actual field performance. These tests have been mainly developed to predict the corrosion response of cold-rolled and galvanized steel and have been formulated to predict the performance of vehicles in harsh corrosion environments (coastal marine or deicing salt regions). A large study showed that while a limited number of cyclic corrosion tests were appropriate for predicting field corrosion of vehicles, none of the cyclic tests were appropriate for predicting the corrosion of coated aluminum substrates owing to the neutral salt solutions used. Accordingly, modification of these tests for nonferrous substrates is under investigation.

Neither salt fog nor humidity tests have good reproducibility. It is common for differences between duplicate panels to be larger than differences between panels with different coatings. Precision can be improved by testing five to eight replicate panels of each coating system. (Unfortunately, decisions are commonly based on the results from testing two or three panels.) A further problem of evaluating panels for corrosion protection is the difficulty of detecting small blisters and rust areas underneath a pigmented coating film without removing the film. Infrared thermography has been recommended as a nondestructive testing procedure (McKnight and Martin, 1989). The use of a robotically controlled scribing tool, as opposed to scribing by hand, has improved the reproducibility of accelerated corrosion tests.

Other corrosion tests or laboratory assessments have been used with varying degrees of success. A great deal of effort has been expended on electrical conductivity tests of paint films and electrochemical tests of coated panels. (See Murray (1997) for an extensive review.) A variety of cathodic disbonding tests specifically for testing of pipeline coatings has been established by ASTM: G-8-96, G-42-96, and G-80-88 (reapproved 1992). In these tests, a hole is made through the coating and the exposed metal is made the cathode of a cell in water with dissolved salts at a basic pH. Disbonding (loss of adhesion) as a function of time is followed. While there is considerable variability inherent in such tests and their utility for predicting field performance is doubtful, useful guidance may be obtained for improving wet adhesion, not only for pipeline coatings but also more broadly. For a discussion of research on cathodic delamination, including investigation of the migration of cations through or under coating films, see Parks and Leidheiser (1984).

In general, the state of corrosion testing is akin to the state of weathering testing of coatings 30 years ago. Everyone knows that the current test methods, while useful, are far from being predictive under most use scenarios. The range of chemical reactions that occur in the corrosion process are now reasonably understood. However, mechanistically correct accelerated corrosion tests are lacking. It will take a concerted effort of academia, government, and industry to correctly produce and then propagate new test methods that accurately reproduce field performance before corrosion testing can become more useful and predictive of actual field performance. In addition, simulation tools must be developed, which are based on the correct chemical reactions and environmental conditions. These simulations would be extremely beneficial to those trying to predict the corrosion performance of components and coatings intended for extended use.

GENERAL REFERENCES

Dickie, R. A.; Floyd, F. L., Eds., *Polymeric Materials for Corrosion Control*, ACS Symposium Series, *322*, American Chemical Society, Washington, DC, **1986**.

Munger, C. C., *Corrosion Prevention by Protective Coatings*, National Association of Corrosion Engineers, Houston, **1997**.

Revie, R. W., Ed., *Uhlig's Corrosion Handbook*, 3rd ed., John Wiley & Sons, Inc., New York, **2011**.

Sorensen, P. A., et al., *J. Coat. Technol. Res.*, **2009**, *6(2)*, 135–176.

REFERENCES

Athey, R., et al., *J. Coat. Technol.*, **1985**, *57(726)*, 71.

Augustyniak, A.; Ming, W., *Prog. Org. Coat.*, **2011**, *71(4)*, 406–412.

Barbucci, A., et al., *Prog. Org. Coat.*, **1996**, *29(1)*, 7–11.

Battocchi, D., et al., *Corros. Sci.*, **2006**, *48(8)*, 2226–2240.

Bautista, A., *Prog. Org. Coat.*, **1996**, *28(1)*, 49–58.

Bentz, D.; Nguyen, T., *J. Coat. Technol.*, **1990**, *62(783)*, 57–63.

Bethencourt, M., et al., *Prog. Org. Coat.*, **2003**, *46(4)*, 280–287.

Biegańska, B., et al., *Prog. Org. Coat.*, **1988**, *16(3)*, 219–229.

Braig, A., et al., *Proceedings of Waterborne High Solids and Powder Coating Symposium*, New Orleans, LA, **2000**, pp 360–372.

Cano, E., et al., *J. Solid State Electrochem.*, **2009**, *14(3)*, 381–391.

Child, T. F.; Van Ooij, W. J., *Trans. Inst. Met. Finish.*, **1999**, *77(2)*, 64–70.

Chuang, T.-J., et al., *J. Coat. Technol.*, **1999**, *71(895)*, 75–85.

Del Amo, B., et al., *Prog. Org. Coat.*, **2002**, *45(4)*, 389–397.

Fawcett, N., et al., *J. Coat. Technol.*, **1984**, *56(714)*, 49–52.

Feliu, S., et al., *J. Coat. Technol.*, **1989**, *61(775)*, 63.

Feliu, S., Jr., et al., *J. Coat. Technol.*, **1991**, *63(793)*, 31–34.

Funke, W., *J. Oil Colour Chem. Assoc.*, **1979**, *62(2)*, 63–67.

Funke, W., *J. Coat. Technol.*, **1983**, 131–138.

Funke, W., *J. Oil Colour Chem. Assoc.*, **1985**, *68(9)*, 229–232.

Funke, W., *Farbe Lack*, **1987**, *93(9)*, 721–722.

Funke, W., *Prog. Org. Coat.*, **1996**, *28(1)*, 3–7.

Galliano, F.; Landolt, D., *Prog. Org. Coat.*, **2002**, *44(3)*, 217–225.

Hemmelrath, M.; Funke, W., *FATIPEC Congress Book*, **1988**, Vol. *IV*, p 137.

Hinderliter, B., Sapper, E., *J. Coat. Tech. Res.*, **2015**, *12(3)*, 477–487.

Holubka, J.; Dickie, R., *J. Coat. Technol.*, **1984**, *56(714)*, 43–46.

Holubka, J., et al., *J. Coat. Technol.*, **1980**, *52(670)*, 63–68.

Kalendová, A., *Prog. Org. Coat.*, **2003**, *46(4)*, 324–332.

Kousik, G., et al., *Prog. Org. Coat.*, **2001**, *43(4)*, 286–291.

Leclercq, M., *Mater. Technol.*, **1990**, (March), 57.

Lee, T.; Money, K., *Mater. Perform.*, **1984**, *23*, 28.

Leidheiser, H., *Prog. Org. Coat.*, **1979**, *7(1)*, 79–104.

Leidheiser, H., Jr., *J. Coat. Technol.*, **1981**, *53(678)*, 29–39.

Martin, J., et al., *J. Coat. Technol.*, **1990**, *62(790)*, 25–33.

Massingill, J., et al., *J. Coat. Technol.*, **1990**, *62(781)*, 31–39.

Mayne, J., *Off. Dig.*, **1952**, *24*, 127–136.

Mayne, J., in Shreir, L. L., Ed., *Corrosion*, Butterworth, Boston, **1976**, Vol. 2, pp 15:24–15:37.

Mayne, J.; Evans, U. R., *Chem. Ind.*, **1944**, *63*, 109.

Mazia, J., *Met. Finish.*, **1977**, *75(5)*, 77–81.

McKnight, M.; Martin, J., *J. Coat. Technol.*, **1989**, *61(775)*, 57–62.

McKnight, M. E., et al., *J. Prot. Coat. Linings*, **1995**, *12*, 82–82.

Milmo, S., *Coat. World*, **2016**, *21(11)*, 26–27.

Montes, E., *J. Coat. Technol.*, **1993**, *65(821)*, 79.

Montes, E., *Proceedings of Waterborne High Solids and Powder Coating Symposium*, New Orleans, LA, **2001**, pp 211–216.

Moraes, S. R., et al., *Prog. Org. Coat.*, **2003**, *48(1)*, 28–33.

Murray, J. N., *Prog. Org. Coat.*, **1997**, *31(3)*, 255–264.

Nair, U.; Subbaiyan, M., *J. Coat. Technol.*, **1993**, *65*(*819*), 59.
Nanna, M. E.; Bierwagen, G. P., *JCT Res.*, **2004**, *1*(2), 69–80.
Negele, O.; Funke, W., *Prog. Org. Coat.*, **1996**, *28*(*4*), 285–289.
Parashar, G., et al., *Prog. Org. Coat.*, **2001**, *42*(*1*), 1–14.
Parks, J.; Leidheiser, H., Jr., *FATIPEC Congress Book*, **1984**, Vol. *II*, p 317.
Perfetti, B. M., *Metal Surface Characteristics Affecting Organic Coatings*, Federation of Societies for Coatings Technology, Blue Bell, PA, **1994**.
Revie, R. W.; Uhlig, H. H., *Uhlig's Corrosion Handbook*, John Wiley & Sons, Inc., New York, **2011**.
Sangaj, N. S.; Malshe, V., *Prog. Org. Coat.*, **2004**, *50*(*1*), 28–39.
Sekine, I., et al., *Prog. Org. Coat.*, **2002**, *45*(*1*), 1–13.
Sinko, J., *Prog. Org. Coat.*, **2001**, *42*(*3*), 267–282.
Sinko, J., Google Patents, US7662241 B2 (**2002**).
Sitaram, S. P., et al., *J. Coat. Technol.*, **1997**, *69*(*866*), 65–69.
Skerry, B.; Simpson, C., *NACE Corrosion '97*, Cincinnati, OH, **1991**.
Smith, A., *Inorganic Primer Pigments*, Federation of Societies for Coatings Technology, Blue Bell, PA, **1988**.
Sørensen, P. A., et al., *J. Coat. Technol. Res.*, **2009**, *6*(2), 135–176.
Spinks, G. M., et al., *J. Solid State Electrochem.*, **2002**, *6*(2), 85–100.
Szokolik, A., *J. Prot. Coat. Linings*, **1995**, *12*(*5*), 56.
Thomas, N. L., *Prog. Org. Coat.*, **1991**, *19*(2), 101–121.
Timmins, F., *J. Oil Colour Chem. Assoc.*, **1979**, *62*(*4*), 131–135.
Tomić, M. D., et al., *J. Coat. Technol. Res.*, **2016**, *13*(*3*), 439–456.
Vetere, V., et al., *J. Coat. Technol.*, **2001**, *73*(*917*), 57–63.
Visser, P., et al., Faraday Discuss., **2015**, *180*, 511–526.
Wyvill, R., *Met. Finish.*, **1982**, *80*(*1*), 21–26.
Yang, H. S., et al., *J. Coat. Technol.*, **2005**, *2*(*13*), 44.
Zarras, P., et al., Semiconducting Polymers, ACS Symposium Series 735, American Chemical Society, Washington, DC, **1999**, pp 280–292.
Zubielewicz, M.; Gnot, W., *Prog. Org. Coat.*, **2004**, *49*(*4*), 358–371.

Chapter 8

Acrylic Resins

Acrylic polymers are used as the primary binder in a wide variety of industrial and consumer coatings. Their major advantages are photostability and resistance to hydrolysis. *Acrylic resins*, including thermoplastic, thermosetting, water-reducible, and non-aqueous dispersion acrylics, are prepared by chain-growth polymerization (Section 2.2.1) primarily of (meth)acrylic ester monomers in solution. In contrast, *acrylic latexes*, discussed in Chapter 9, are uniquely prepared by emulsion polymerization of (meth) acrylic monomers.

8.1 THERMOPLASTIC ACRYLIC RESINS

Solventborne coatings based on thermoplastic acrylic (TPA) resins have many excellent properties, especially exterior durability, but their use has declined because they require large amounts of solvent to reduce viscosity that is low enough for application. In North America, such coatings are called *lacquers*. In Europe, the term *lacquer* is sometimes applied to thermosetting as well as thermoplastic coatings. Lacquers based on TPA polymers were used for automotive top coats from the 1950s to the 1970s. Among their desirable features were brilliant metallic colors, made possible by the high solvent levels, which permitted orientation of aluminum flake pigment in the films parallel to the surface (Section 30.1.2). Acrylic lacquers are still used to a degree in refinishing automobiles (Section 33.3) and for other specialized coatings, but as volatile organic compound (VOC) regulations become more restrictive, their use is declining.

TPAs are copolymers of various (meth)acrylic monomers and, commonly, styrene (S). Monomers are selected on the basis of cost and their effect on properties, notably, outdoor durability and T_g. For automotive top coats, T_g must be higher than 70°C. Methyl methacrylate (MMA), S, and *n*-butyl acrylate (*n*BA) are often used. The acrylic copolymers are used with plasticizers. Control of molecular weight (MW) in the synthesis is critical. Film strength increases with MW, although above $\bar{M}_w$ of about 90 000, the change in properties with increasing MW is small. The upper end of MW is limited because solutions of acrylic polymers having $\bar{M}_w$ greater than about 100 000 exhibit *cobwebbing* on spraying, emerging as threads rather than atomizing to small droplets from the orifice of a spray gun. Viscosity of solutions increases with increasing MW, requiring that the solids content to decrease in order to allow the coating to be spray-applied. Viscosity is particularly affected by the very high MW fractions; therefore, it is critical to minimize this fraction by controlling MW distribution $\left(\bar{M}_w / \bar{M}_n\right)$ to a narrow range (Section 2.2.1). Commercial TPAs have $\bar{M}_w$ of 80 000–90 000 and $\bar{M}_w / \bar{M}_n$ of 2.1–2.3. The solids of TPA lacquers at application viscosity are in the range of 11–13 nonvolatile volume (NVV).

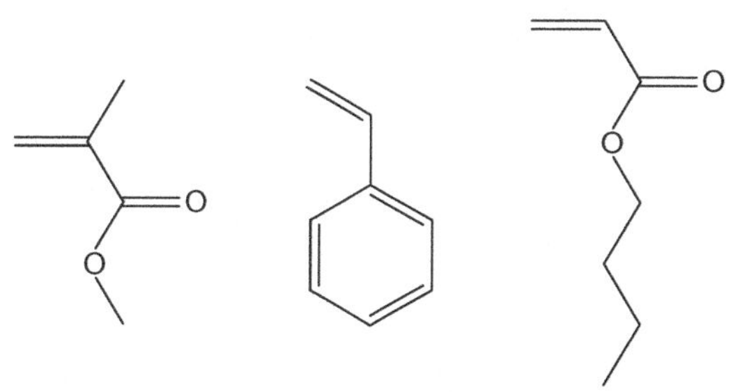

MMA   S   *n*BA

Organic Coatings: Science and Technology, Fourth Edition. Frank N. Jones, Mark E. Nichols, and Socrates Peter Pappas.

8.2 THERMOSETTING ACRYLIC RESINS

The low solids of thermoplastic lacquers can be increased by using lower MW thermosetting acrylic (TSA) resins, which are designed to react chemically after application. The post-reaction leads to a cross-linked polymer network, which is ideally one interconnected molecule with a very high MW, potentially resulting in good film properties (Section 2.2). The cross-linked film is not soluble in any solvent. The term *thermosetting resins* implies that the resins cross-link by themselves and require heat to cross-link; however, most TSA resins bear functional groups that are reacted with a different functional polymer or cross-linker, sometimes at ambient temperature.

8.2.1 Hydroxy-Functional Acrylic Resins

Hydroxy-functional acrylic resins are copolymers of non-functional monomers, often MMA, S, and BA, together with a hydroxy-functional monomer. They are cross-linked with amino resins (Chapter 11), with polyisocyanates (Chapter 12), with reactive silanes (Section 16.2), or with mixtures of cross-linker functionalities. Solventborne hydroxy-functional TSAs for conventional solids coatings were first developed in the 1950s. For conventional (low) solids formulations, their $\bar{M}_n$ values are usually 10000–20000 with $\bar{M}_w/\bar{M}_n$ of 2.3–3.3. TSAs are prepared by free radical-initiated chain-growth polymerization under monomer-starved conditions, as described in Section 2.2.1. Synthesis of conventional TSAs is easier to control than that of TPAs, since the application and film properties of TSAs is not as critically dependent on their MW. Owing to their lower MW, one can use less expensive aromatic hydrocarbon solvents for the polymerization medium. However, the lower MW of TSA polymers results in more end groups per unit weight. Therefore, the structure of the end groups can have greater effects on performance than in the case of TPA polymers. There are a large number of reactions that occur during conventional free radical polymerization that can lead to various end groups (Chung and Solomon, 1992; Moad et al., 2002; Moad and Solomon, 2006). Azo initiators are used predominantly, since they undergo relatively few side reactions and form end groups with minimal photochemical reactivity. Certain peroxy initiators also meet these requirements, but others introduce end groups that reduce exterior durability. Chain transfer to solvent can also result in end groups that reduce exterior durability. For example, terminal groups resulting from chain transfer to a ketone solvent, such as methyl *n*-amyl ketone (MAK), can have a deleterious effect (Gerlock et al., 1988).

Hydroxyl groups are often introduced by using a hydroxy-functional comonomer such as 2-hydroxyethyl-methacrylate (HEMA). If a lower T_g reactive monomer is desired, 2-hydroxyethyl acrylate (HEA) can be used, but its greater toxic hazard increases handling costs. Rather, it is often preferable to reduce T_g by increasing the proportion of low T_g nonreactive monomers, such as *n*-butyl acrylate. Commercial grades of HEMA (and HEA) contain appreciable amounts of diester, for example, ethylene glycol dimethacrylate (EGDMA) in the case of HEMA. Separation of the mono- and diesters is difficult since their boiling points are similar. When a relatively small fraction of hydroxy-functional monomer is used, the branching that results from the presence of correspondingly small amounts of diester may not be harmful. However, the branching resulting from larger amounts of diester can result in undesirable higher MW, broader MW distribution, and even gelation.

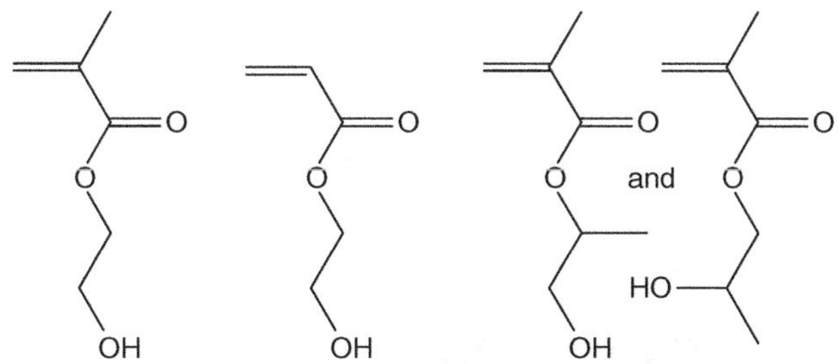

HEMA HEA HPMA

2-Hydroxypropyl methacrylate (HPMA) can also be used, which costs less than HEMA and generally has a lower diester content. HPMA is a mixture of isomers in which the secondary alcohol predominates. Since secondary alcohols are less reactive than primary alcohols, TSAs with HPMA substituted for HEMA require higher baking temperatures, longer baking times, or more catalyst to cure to a given cross-link density (XLD) with MF resins. For example, baking temperatures may have to be increased by 10–20°C. Isocyanate cross-linkers probably also react slower with HPMA-containing TSAs, but there is the accompanying advantage of longer pot life for two-package systems. In theory, the use of HPMA in place of HEMA could reduce resistance to photooxidation as a result of the hydrogen on tertiary carbon, which is more easily abstractable.

8.2.1.1 Acrylic Resins for Conventional Coatings

An example of a hydroxy-functional TSA *for conventional coatings, in which low VOC is not required*, is an MMA/S/BA/HEMA/MAA copolymer with a weight ratio of

50 : 15 : 20 : 14 : 1, corresponding to a mole ratio of 54.3 : 15 .6 : 16.9 : 11.7 : 1.5. Typically, $\bar{M}_w$ and $\bar{M}_n$ are about 35 000 and 15 000, respectively $\left(\bar{M}_w / \bar{M}_n = 2.3\right)$. With the given ratio of comonomers, the $\bar{P}_w$ is about 320 and the $\bar{P}_n$ about 140. For stoichiometric comparisons, one uses the number average $\bar{M}_n$. This resin has a hydroxyl equivalent weight of slightly over 900 and a number average functionality $\bar{f}_n$ of about 16 hydroxyl groups per polymer molecule. The small amount of carboxylic acid-functional monomer (methacrylic acid (MAA)) is introduced to reduce the *likelihood* of pigment flocculation (Section 21.1) in a liquid coating.

The nonfunctional monomers (MMA, S, and BA, in this case) are selected on the basis of their effect on T_g, exterior durability, and cost. This monomer combination provides good exterior durability with a relatively high T_g at a moderate cost, as is appropriate for an automotive top coat. This resin also has a relatively high level of hydroxy-functional monomer, appropriate when relatively high XLD films are desired. For applications requiring more flexible films, such as coil coatings or exterior can coatings, one would use a lower HEMA content. By adjusting T_g and functionality, TSAs can be designed for a wide range of end uses. Countless variations can be prepared.

An alternative method for introducing hydroxy functionality is to react a carboxylic acid-functional copolymer made using methacrylic (or acrylic) acid (MAA or AA) with an oxirane such as propylene oxide. Raw material cost is lower with propylene oxide than with HPMA, but exacting process control is required. A related approach is to react an AA or MAA copolymer with a glycidyl ester such as glycidyl versatate (Ryan and Walt, 1994). Versatic acid is a mixture of carboxylic acids having about 11 carbons and highly branched structures. Rapid cross-linking hydroxy-functional acrylic resins for use in automotive refinish coatings have also been prepared by prereacting glycidyl versatate with MAA and copolymerizing the resulting monomer with BA, S, and MMA. Rapid cure depends on reacting the MAA and glycidyl versatate under conditions that maximize the formation of primary hydroxy groups (Petit et al., 2001). Acrylic resins for high solids coatings can be made by polymerizing BA, BMA, S, AA, and HEMA in the presence of glycidyl versatate. During the polymerization, reaction of glycidyl versatate with a carboxylic acid group results in the formation of an additional hydroxy group and a long alkyl side chain. The T_g of the resins is lower than conventional acrylics having similar MW and hence permits higher solids coatings. To further increase the solids, incorporation of the reaction product of trimethylolpropane (TMP) with glycidyl versatate as a reactive diluent is also used. Polyisocyanates are used as cross-linkers (Slinckx et al., 2000). Use of *t*-butyl methacrylate (BMA) as a comonomer has been reported to provide films having better appearance. Refinish clear coats with VOC of 3.5 lb/gal have been formulated (Ball et al., 2002).

Thermosetting resins and cross-linkers should be selected or designed for use as a system. The appropriate composition of hydroxy-functional TSAs depends on the cross-linker that is used. A TSA resin optimized for a TSA–MF coating would probably not yield optimum properties with a polyisocyanate cross-linker. Furthermore, depending on the particular polyisocyanate used, the T_g of the final cross-linked films could be higher or lower than what would be obtained by cross-linking the same TSA with an MF resin. TSA resins designed for polyisocyanate cross-linking tend to have lower T_g and lower $\bar{f}_n$ than TSAs for cross-linking with MF resins, since intermolecular hydrogen bonding between urethane groups contributes to the covalent cross-linking and provides the further advantage of being reversible (Section 12.4). Another difference is that resins to be cross-linked with MF resins can have carboxyl functionality, possibly aiding pigment dispersion (but those designed for polyisocyanate cross-linking should not).

8.2.1.2 Acrylic Resins for High Solids Coatings

Beginning in the 1970s, major efforts were launched to prepare high solids solution acrylic resins that satisfy lower VOC emission requirements. It may seem that this problem is simple; why not just reduce $\bar{M}_n$ from 15 000 to 1 500? Actually, the problem is complex; as MW is decreased, greater care must be exercised in every aspect of polymerization, coating formulation, and application. In polymerization, the amount of non- or monofunctional resin must be kept to a very low fraction. Due to the larger number of chain ends, the effect of chain transfer to solvents, such as ketones, which can introduce photoreactive end groups, is potentially more serious. More care is needed in establishing the stoichiometric ratio of cross-linker, and, as noted in Section 25.2.2, the cure window becomes narrower. Controlling film defects during application becomes more difficult as the solids content becomes higher (Chapter 24). It is particularly difficult to avoid sagging of spray-applied baking coatings.

An example of a high solids TSA resin has the following composition: S/MMA/BA/HEA; weight ratio 15 : 15 : 40 : 30; $\bar{M}_w$, 5200; $\bar{P}_w$, 54; $\bar{M}_n$, 2300; $\bar{P}_n$, 20; $\bar{P}_w / \bar{P}_n$=2.7; equivalent weight, 400; $\bar{f}_n$=5.7 (Hill and Kozlowski, 1987). This resin was prepared at 65 NVW in methyl amyl ketone (MAK) for studying MF cross-linking of fairly high solids coatings. Compared with the conventional TSA described earlier in this chapter, the $\bar{M}_n$ has been reduced by a factor of 6.5, whereas $\bar{f}_n$ is only reduced by a factor of 2.8. The higher content of hydroxy-functional monomer in the high solids resin is required to attain about the same XLD in the final film, because the low MW resin must undergo more reactions. Although not reported, the application solids of coatings formulated with this resin would be around 45 NVV. This solids content corresponds to that used in high solids metallic automotive top coats. However, the solids are still too low to meet EPA regulations for some current end uses, and even stricter regulations are anticipated.

As explained in Section 2.2.1, conventional free radical polymerization inherently limits the extent to which MW of a TSA can be reduced. Satisfactory performance requires that a high fraction of the molecules have at least two hydroxyl groups, but this becomes statistically less probable as MW is reduced. The problem can be illustrated by comparing the conventional TSA ($\bar{M}_n = 15\,000$, $\bar{P}_n = 140$), described in Section 8.2.1.1, with a hypothetical high solids TSA having the same monomer ratio with $\bar{M}_n = 1070$ ($\bar{P}_n = 10$), which is about the level needed to formulate with VOCs around 300 g l^{-1}. The conventional resin has an $\bar{f}_n$ of 16 hydroxyl groups. While individual molecules have more or less than 16 hydroxyl groups, statistically the number of molecules with fewer than two hydroxyl groups is very low, and virtually all the molecules are capable of cross-linking. In contrast, $\bar{f}_n$ of the high solids TSA would only be about 1.2, and a very large fraction of the molecules could not be cross-linked. Molecules with no hydroxyl groups would either volatilize or remain in the film as plasticizers to the detriment of film properties. Molecules with one hydroxyl group would terminate cross-linking reactions, leaving *loose* or *dangling chain ends* in the coating. The theory of elasticity predicts that dangling chain ends seriously diminish the mechanical properties of a network. Experimental experience confirms that loose ends have a substantial effect on initial film properties (Spinelli, 1982; Nakamichi and Shibato, 1986). For example, each percent by weight of monohydroxy oligomer in a TSA may reduce T_g of the TSA-MF enamel by about 1 °C (Nakamichi and Shibato, 1986).

The problem results from both MW and functional group distribution. Figure 2.2 shows the distribution of molecules of differing degrees of polymerization of a series of resins having $\bar{P}_n = 12$ and $\bar{P}_w / \bar{P}_n = 1.07$, 1.5, and 3.0. The theoretical minimum polydispersity that can be obtained by anionic polymerization is 1.07, and the theoretical minimum achievable by common free radical initiated polymerization processes is 1.5. It is difficult to approach the theoretical minimum in practice; $\left(\bar{P}_w / \bar{P}_n\right) = 2.5$ is typical of reasonably well-controlled free radical polymerizations. The other factor involved is sequence length distribution, which is the distribution of sequences of different lengths of nonfunctional monomers separating functional monomers. If these sequence lengths are short compared with the chain length of the low MW parts of the resin, multiple functional groups will be incorporated in the molecules. But, if the sequence lengths of nonfunctional monomers are long in comparison with chain lengths, molecules with one or no functional monomer units can be expected.

Statistical methods have been used to calculate the proportions of nonfunctional molecules that would be formed during random copolymerization of monomer mixtures having differing monomer ratios for different MWs and MW distributions (Hill and Wicks, 1982; O'Driscoll, 1983). Due to the assumptions involved, the calculations are approximate. An example of such results for copolymers of S/BA/HEA (30:50:20 wt%) having varying $\bar{P}_n$ is given in Table 8.1 (O'Driscoll, 1983). For $\bar{P}_n = 9.5\left(\bar{M}_n = 1125\right)$ about 36% of all molecules, corresponding to 13% of the weight, are nonfunctional. (The mol% and wt% differ because of the MW distribution; low MW molecules are more likely to be nonfunctional than the higher MW ones.) The weight fraction of monofunctional oligomer is more difficult to calculate but would be expected to exceed that of nonfunctional oligomer. With a higher HEA content (30%) and larger $\bar{P}_n$ of 20, the proportion of monofunctional oligomer molecules would be minimized. But, if one targeted a volume solids of about 70 NVV, a $\bar{P}_n$ of about 10 and a narrow MW distribution would be required. The calculations show that the functional monomer content would have to be very high, so high that either the XLD of the cross-linked film would be too high for acceptable film properties or the coating would have to be formulated with a relatively small amount of cross-linker, leaving a high proportion of unreacted hydroxyl groups in the film, almost certainly causing water sensitivity and probably other problems.

Viscosity of acrylic resins can be reduced by changing the proportions of comonomers to reduce T_g. However, this approach also reduces T_g of the cross-linked film, other factors being equal. The effect of reduced T_g on film properties can be offset to an extent by increasing XLD, but if carried too far, this approach leads to an unsatisfactory combination of properties. Methacrylate esters of relatively rigid alcohols, such as 3,3,5-trimethylcyclohexyl (Kruithof and Van den Haak, 1990) and isobornyl (Wright, 1996) methacrylates, having a compact shape can combine relatively low viscosity and high T_g. Accordingly, such esters can be used as partial replacements for MMA and S.

Synthesis of acrylic resins having low but usable $\bar{M}_n$ and a low $\bar{M}_w / \bar{M}_n$ has been a long-standing objective. Various approaches are discussed in the following text.

Table 8.1 Percentage of Nonfunctional Molecules Statistically Predicted for S/BA/HEA (30:50:20 by wt%) Copolymers

$\bar{P}_n$	$\bar{M}_n$	Mol%	Nonfunctional (wt%)
36.8	4357	15	1.8
19.2	2273	24	5.8
9.5	1125	36	12.8

Source: Kruithof and van den Haak (1990).

$\bar{M}_n$ can be reduced by using chain transfer agents, but some chain transfer agents can have an adverse effect on exterior durability (Spinelli, 1982). Use of a chain transfer agent with a functional group is an interesting approach. For example, when 2-mercaptoethanol is used as a chain transfer agent, initiation by the resulting mercaptide radical also introduces a hydroxyl group on the polymer end, thereby reducing the fractions of non- and monofunctional molecules in the resin (Gray, 1985). Improved film properties are obtained, but the unpleasant odor of mercaptoethanol causes handling problems and may leave residual odor in the resin.

High levels of initiator with close control of temperature and concentrations during polymerization can minimize polydispersity. It appears that with high initiator concentration, polydispersity can be narrower than theory predicts. Initiator choice can be critical. Benzoyl peroxide tends to broaden distributions owing to chain transfer to polymer resulting from hydrogen abstraction. Less branching occurs with azo initiators such as azobisisobutyronitrile (AIBN); AIBN is also generally preferred over benzoyl peroxide because exterior durability of films is superior. Initiators based on *t*-amyl peroxides, such as ethyl 3,3-di(*t*-amylperoxy)butyrate, are reported to give narrow distributions (Kamath and Sargent, 1987). Other workers recommend initiators derived from *t*-butyl peroxide, such as *t*-butyl peracetate (Myers, 1995). A comparison of the use of azobis(2-methylbutyronitrile (ABMBN) versus *t*-butyl peroctanoate showed that the azo initiator gave narrower MW distributions (Ziemer et al., 2002).

8.2.1.3 High-Temperature Acrylic Polymerization

An important way to reduce $\bar{M}_n$ and $\bar{M}_w/\bar{M}_n$ is to carry out a free radical polymerization at elevated temperatures in pressure reactors. Acrylic resins produced by such processes are components of many automotive coatings (Wang and Hutchinson, 2008). Polymerization in an autoclave is a batch process. In addition, the pressure reactor can be fitted for injection of the ingredients at controlled rates, thereby enabling versatile semi-continuous batch processes.

Numerous patent citations describe polymerization of acrylic monomers in pressure reactors. For example, Brand and Morgan (1985) described numerous experiments with temperatures ranging from 190 to 270°C and brief residence in the batch reactor. Results varied, but in some cases $\bar{M}_w/\bar{M}_n$ well below 2.0 was achieved.

Studies by Burja et al. (2015) provide a recent example. They compared polymerization of S, HPMA, and AA at 140°C at atmospheric pressure with polymerization of the same monomers under pressure at 170°C. The 140°C product had $\bar{M}_w$ of 12 000 and $\bar{M}_n$ and an $\bar{M}_w/\bar{M}_n$ of 3.0, while the high pressure products had $\bar{M}_w$ of 4 000–5 500 and $\bar{M}_w/\bar{M}_n$ 1.9–2.3. The pressure processed materials had lower viscosities and, when cross-linked with a diisocyanate, cured rapidly to form films with improved anticorrosive properties.

While details are proprietary, several commercial resins have characteristics that are consistent with their preparation utilizing pressure processes. For example, a resin with $\bar{M}_n$ of 1300 and $\bar{M}_w/\bar{M}_n = 1.7$ can be formulated with class I MF resins in white, pigmented coatings at about 77 NVW. The resin provided hard, chemically resistant films with many good properties, although the films tended to be brittle.

Kinetics of high temperature polymerization depart significantly from the standard chain-growth kinetics described in Chapter 2. If the temperature is high enough, addition polymerization can become reversible (*depolymerization*), affecting the kinetics. Computational studies by Srinivasan et al. (2010) predict that high temperature polymerizations are accompanied by side reactions such as *backbiting* (chain transfer within the same polymer chain). Backbiting leads to the formation of branched polymers without increasing $\bar{M}_w/\bar{M}_n$. Branched polymers may have lower viscosities in high solids coatings than unbranched polymers of the same $\bar{M}_n$ and $\bar{M}_w/\bar{M}_n$. In some cases, high temperature polymerization of acrylates can be self initiated.

8.2.1.4 Controlled Radical Psolymerization (CRP)

As discussed in Secion 2.2.1.1, CRP enables the synthesis of acrylic polymers with various topologies, including block and highly branched structures. $\bar{M}_w/\bar{M}_n$ can be narrow, and functional groups can be introduced with controlled locations within the polymer (Krol and Chmielarz, 2014; Boyer et al., 2016). Such processes afford exciting possibilities for coatings. Very high solids acrylic formulations with good properties are feasible and are beginning to have a commercial impact. Previously, the cost of the products had apparently limited their usage in coatings to specialties such as pigment dispersants. Much of the very extensive research on CRP has been aimed at applications where cost is much less of a factor than it is with coatings. Low cost CRP processes for acrylics remain an enticing research target.

8.2.1.5 Other Considerations

In general, acrylic-based coatings do not provide good adhesion to metals and are often used over primers. However, acrylic coatings prepared with 1–4% of phosphate-functional (meth)acrylic esters are reported to provide excellent adhesion to a broad range of metals (Yang et al., 2005).

In coatings formulation it is quite common to blend acrylic resins with other generic types. A common method for increasing solids is to blend acrylic polyols with lower viscosity polyols, such as polyesters. Some reduction in VOC can be achieved in this way. There is risk that film properties, such as exterior durability and chemical resistance, may be reduced, but properties may be adequate for many uses. Highly weather-resistant coatings of this type have been reported (Okada, 1997).

8.2.2 Acrylics Having Other Functional Groups

Carbamate-functional acrylics are important components of certain automotive clear coats. They can be cross-linked with class I MF resins to provide films having better environmental etch resistance than MF cross-linked hydroxy-functional acrylics while retaining the advantage of mar resistance (Green, 2001) (Section 11.3.4). Carbamate-functional acrylic resins can be prepared by reacting isocyanato-functional acrylic resins with hydroxypropyl carbamate (Rehfuss and St. Aubin, 1994, 1997), by reacting acrylic resins with urea (Rehfuss and St. Aubin, 1997) and by ester interchange between a hydroxy-functional acrylic and the carbamate from propylene glycol monomethyl ether (Barancyk et al., 1995). Ohrbohm et al. (2000) disclosed using the combination of a carbamate-functional acrylic resin, a carbamate-functional urethane and MF resins. Carbamate-functional acrylics can be blended with hydroxyl-functional resins to reduce costs while retaining satisfactory levels of mar resistance.

Trialkoxysilyl-functional acrylics can be prepared by using a trialkoxysilylalkyl methacrylate as a comonomer (Chen et al., 1997). They provide high performance clear coats by moisture cure in the air or by ether interchange with hydroxyl groups (Section 16.2). Trialkoxysilyl-functional acrylics have also been synthesized by copolymerization of vinyltrimethoxysilane and butyl acrylate. Films of the resin cure by acid-catalyzed reaction with atmospheric water by forming Si—O—Si bonds during polymerization (Douskey et al., 2002). Other copolymers and blends of acrylics with silanes are gaining importance (Chapter 16).

Carboxylic acid-functional acrylic resins, prepared using acrylic acid or methacrylic acid as comonomers, can be cross-linked with epoxy resins (Section 13.3.2). This chemistry is also the basis for commercial automotive clear coats. Note that close quality control of acrylic acid monomer is required.

Acrylics with amide groups can be used in several ways. Acrylic copolymers, containing acrylamide, can be cross-linked with MF resins by generating the —NH—CH_2—NH—CO— linkage; a higher curing temperature is required than for hydroxy-functional resins. Alkoxymethyl derivatives of acrylamide copolymers can be made by reacting the amide groups with formaldehyde, followed by etherification, analogous to the preparation of amino resins, as described in Section 11.4.4 (Christson and Hart, 1961; Kelly et al., 1963). These resins can be used as cross-linkers for hydroxy-functional TSAs. Alternatively, one can use alkoxymethylacrylamides as comonomers. For example, *N*-(isobutoxymethyl)methacrylamide has been used as a comonomer together with hydroxy-functional monomers to prepare copolymers that can be cured with hydroxyl- or acid-functional resins at low temperatures (Rietberg and Van Roon, 1991). Acrylic copolymers, containing acrylamide and HEMA, can also be reacted with formaldehyde and then with an alcohol to make a self cross-linkable resin.

Epoxy-functional acrylics are made using glycidyl methacrylate (GMA) as a comonomer (Section 13.3.2). Such resins are used in clear coats for vehicles and have been used in powder clear coats (Ooka and Ozawa, 1994; Kenny et al., 1996; Agawa and Dumain, 1997) (Section 28.1.4). The epoxy-functional resins can be cross-linked with dicarboxylic acids, such as dodecanoic acid (Ooka and Ozawa, 1994; Kenny et al., 1996) or with carboxylic acid-functional acrylic resins (Agawa and Dumain, 1997). Coatings have also been prepared from mixtures of GMA acrylic copolymers, which contain 4-hydroxybutyl acrylate (HBA), together with carboxylic acid-functional acrylic copolymers, prepared using the half methyl ester of maleic anhydride as a comonomer. During curing, the half ester cyclizes to reform anhydride groups. The copolymers cross-link by reaction of the anhydride groups with the hydroxy groups from the HBA and by reaction of the resulting COOH groups with epoxy groups on the GMA copolymer. The cured films are reported to exhibit excellent mar and acid resistance as automotive clear coats (Fushimi et al., 2001). Three alternative systems based on GMA acrylic copolymers have been proposed. Firstly, a GMA copolymer can be cross-linked with carboxy-functional polyesters. Secondly, copolymers using both GMA and AA as comonomers undergo self cross-linking. Thirdly, an acrylic resin containing both GMA *t*-butyl acrylate monomers undergoes self cross-linking following thermal elimination of isobutylene from the *t*-butyl ester groups, which generates free COOH groups that react with the epoxy groups on GMA. The coatings are reported to exhibit excellent mar and acid resistance and to be useful, not only in high solids solventborne systems but also in powder coatings and, with a nonionic emulsifier, in waterborne coatings (Flosbach and Schubert, 2001).

Isocyanato-functional acrylics can be prepared by copolymerizing meta-isopropenyl dimethylbenzylisocyanate (m-TMI) with acrylates (Section 12.3.2) (Alexanian, 1994). They can be cross-linked with polyols and hydroxy-functional acrylic resins.

8.3 WATER-REDUCIBLE THERMOSETTING ACRYLIC RESINS

Another approach to reducing VOC emissions from TSA coatings is to make TSA resins that can be diluted (reduced) with water. Such resins are sometimes called "water-soluble resins"; but this terminology is misleading since the resins do not form solutions in water. Rather, solutions of amine salts of these resins in organic solvents can be diluted with water to form reasonably stable dispersions of polymer aggregates swollen by solvent and water. To minimize confusion, we have chosen to use the following terminology: *waterborne* is used broadly for all coatings with aqueous media, including latex coatings; *water-reducible* is used only for waterborne coatings based on resins having hydrophilic groups in most or all molecules, which excludes latexes; and *water-soluble* is limited specifically to substances that are truly soluble in water.

A typical water-reducible acrylic resin is a copolymer of MMA/BA/HEMA/AA, in a weight ratio of 60:22.2:10:7.8, prepared by free radical initiated polymerization using an azo initiator (Watson and Wicks, 1983). The polymerization is carried out at high solids (generally 70 NVW or higher) in a water-miscible organic solvent. Glycol ethers, such as 1-(*n*-propoxy)-2-propanol or 2-butoxyethanol, and butyl alcohol are the most widely used solvents. Except for the solvent and the higher proportion of AA, such resins are similar to conventional solids TSAs, having $\bar{M}_w$ and $\bar{M}_n$ of about 35 000 and 15 000, respectively, as described in Section 8.2.1.1.

Water-reducible acrylic resins typically have acid numbers of 40–60. (Acid number is determined by titration and is defined as mg of KOH required to neutralize 1 g of resin solids. Equivalent weight equals 56 100/acid number.) The resin is stored as a concentrated solution. The first step in preparing a coating is to add an amine, such as 2-(dimethylamino)ethanol (DMAE). As discussed later, less than the theoretical amount of amine required to neutralize all of the carboxylic acid groups is commonly used. The ratio used is called the *extent (of) neutralization* (EN). For example, if 75% of the theoretical amount were used, the EN would be 75. Other coating components (pigments, MF resin, sulfonic acid catalyst) are dispersed or dissolved in this solution, and the coating is diluted with water prior to application.

The change in viscosity with dilution of water-reducible resins is abnormal. An example is given in Figure 8.1, showing log viscosity as a function of concentration for a model resin (Hill and Richards, 1979). The height of the peak in the dilution curve depends on the particular resin and formulation, and the systems are highly shear thinning (see Section 3.2) in the peak region. Another abnormality of water-reducible resins is that their pH is larger than 7 (commonly 8.5–9.5), even though less than the theoretical amount of amine necessary to neutralize the carboxylic acid is used. For comparison purposes, a dilution curve for the same resin using an organic solvent (*t*-butyl alcohol in this case) is included in the figure, as is a typical dilution curve of a latex.

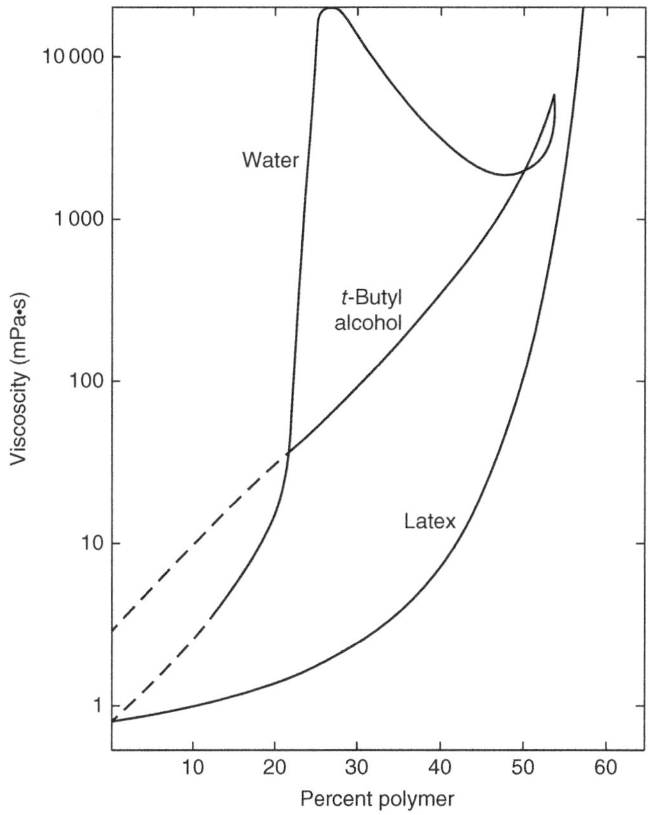

Figure 8.1 Viscosity dependence on concentration for a 10 mol% acrylic acid copolymer, 75 EN with DMAE, dissolved in *t*-butyl alcohol and then diluted with water. Also shown are curves for dilution of the same resin with *t*-butyl alcohol and a typical viscosity–solids dilution curve for a latex, with water. Source: Hill and Richards (1979). Reproduced with permission of American Coatings Association.

The morphology of water-reducible TSAs has been studied fairly extensively (Hill and Richards, 1979; Wicks et al., 1982; Armat et al., 1996). In Figure 8.1, the water-reducible resin is a 54% solution of a 90:10 BMA/AA copolymer in *t*-butyl alcohol at 75 EN using DMAE. The solvent dilution curve of log viscosity versus concentration is fairly linear, which is typical for most resin solutions in good solvents. The water dilution curve shows the abnormal response that is typical for water-reducible resins. During dilution with water, two kinds of changes are occurring simultaneously: the concentration is reduced as is the ratio of solvent to water.

In the first stages of dilution, viscosity drops more rapidly with water than it would with solvent. It is hypothesized that the viscosity is high before dilution because of association of ion pairs on different molecules. Water associates strongly with ion pairs, separating most of the intermolecular ion pairs and causing rapid reduction in viscosity. However, with further water dilution, the viscosity levels off and then increases, passing through a maximum, followed by a very steep decrease in viscosity. At application viscosity of about 0.1 Pa·s, such systems typically have solids in the range of 20–30 NVW. Thus, the application solids of coatings made with such resins is low; however, since a major part of the volatile components is water, VOC emissions are still low.

The viscosity changes on dilution with water are explained as follows: The amine salt of the resin is soluble in the *t*-butyl alcohol and in *t*-butyl alcohol–water solutions in which the ratio of solvent to water is high. But, as additional water is added, the ratio of solvent/water decreases to the point where some of the molecules are no longer soluble in the mixed solvent. These molecules do not precipitate in a separate macrophase, but the nonpolar segments of various molecules associate with each other to form aggregates. The predominantly nonpolar parts of the molecules are in the interior of the aggregates, and the highly polar carboxylic acid salt groups are on their periphery. As dilution continues, more molecules aggregate. Since the solvent is soluble in the resin, some solvent dissolves in the aggregates, swelling their volume. Also, water associates with the salt groups, and some dissolves in the solvent inside the aggregates, further swelling them. As the aggregates form, the system changes from a solution to a dispersion of aggregates in a continuous phase. As the number and volume of aggregates grow, the volume fraction of internal phase increases and the aggregates become more closely packed, leading to the increase in viscosity. (See Section 3.5 for discussion of the factors controlling the viscosity of liquids containing dispersed phases.) At the maximum viscosity, the material is predominantly a dispersion of highly swollen aggregates in water containing some solvent. The high viscosity can lead to difficulty in stirring. On still further dilution, the viscosity drops rapidly. This drop results from two factors. There is a dilution effect, that is, the decrease in volume fraction internal phase. However, the drop is steeper than would result from this effect alone. The balance of the decrease is attributed to a decrease in the swelling of the aggregates. The partitioning of solvent and water between the aggregates and the continuous phase changes throughout the dilution process. As more water is added, more solvent partitions into the continuous phase, resulting in the shrinking of the swollen aggregates.

The abnormal pH effect noted earlier can be explained in a way consistent with this picture of the morphology of the system (Hill and Wicks, 1980). When one neutralizes a simple carboxylic acid, like acetic acid, with 75% of the theoretical equivalent amount of an amine like DMAE, the pH is about 5.5. However, the situation with the carboxylic acid groups on the polymer chain is different. The AA groups are relatively "randomly" spaced along the polymer chain. In some cases, there are carboxylic acid groups near each other; in other cases, single AA groups are separated from each other by several hydrophobic ester monomer units. As aggregation occurs during dilution, many of the carboxylic acid groups are near the surface, where all, or almost all, are neutralized by the water-soluble amine. However, geometric factors require that a proportion of carboxylic acid groups reside in the interior of the aggregates; those most widely separated from other carboxylic acid groups by hydrophobic monomer residues are presumably most likely to be "hidden" in the interior of aggregates. DMAE partitions among the continuous (water/solvent) phase, the surface region of the aggregates, and the interior of the aggregates. Since DMAE and DMAE salts are highly soluble in water, it is expected that DMAE concentrates in the first two regions, leaving a fraction of carboxylic acid groups unneutralized in the interior of the aggregates. Even when there is only 75% of the amine necessary to neutralize all of the carboxylic acid groups, there is more than enough to neutralize those carboxylic acid groups at or near the surface of the aggregates, so that the remaining amine in the continuous phase results in a basic pH reading. Since the interaction is between a weak base and a weak acid, the change of pH with amine addition is slow. Due to this insensitivity, pH is not appropriate as a quality control specification.

Many variables affect the morphology of these systems. The shape of the viscosity–concentration (dilution) curve varies. In some cases, there is a very high viscosity peak—higher than the viscosity of the original undiluted material; in other cases, there may be only a shoulder in the dilution curve. Dependence of viscosity on MW depends on solvent structure and the ratio of solvent to water, in addition to concentration (Wicks et al., 1982). The effect of MW on viscosity is different at different stages of dilution. Log viscosity of an all-solvent solution varies with approximately the square root of the MW. Viscosity in the peak area of the dilution curve is very dependent on MW. As MW increases, the viscosity of the interior of the aggregate particles also increases, making them more difficult to distort; hence, the viscosity during the dilution process increases as MW increases. Very high peak viscosity leads to difficulty in dilution, in which case, it may be necessary to limit the MW so that the coating can be thoroughly agitated throughout the dilution cycle with the available mixing equipment. The viscosity of systems diluted with water to application viscosity is independent of MW. This advantage permits the application of coatings, using resins having an MW in the same range as those used in conventional solventborne TSA coatings but with VOC contents substantially lower and comparable with relatively high solids coatings.

Hill and Richards (1979) studied the dilution behavior of model TSAs, in which the mole percent of AA in a series of BA/AA copolymers varied from 10 to 50%. With 50% AA, the salts exhibit a viscosity dependence on dilution behavior approaching that of solvent systems. As carboxylic acid content is reduced, the abnormal rheological properties become more pronounced. This is an important reason that resins of this type are designed with the lowest carboxylic acid content (acid numbers 40–60) that still provides a stable dispersion at application viscosity. Low acid numbers favor formation of relatively unswollen aggregates, keeping viscosity down, but if the acid numbers are too low, the system separates into macrophases instead of forming a stable microphase dispersion.

The required acid content is lower for resins having increasing hydroxyl group content. While salts of carboxylic acid groups are much more hydrophilic than hydroxyl groups, hydroxyl groups are sufficiently hydrophilic to promote solubility in water–solvent blends, as indicated by the water solubility of the homopolymer of 2-hydroxyethyl acrylate. Therefore, the minimum required acid content decreases as the content of hydroxyl groups increases.

Large-scale production of water-reducible coatings by diluting the resin solution with water requires considerable care because of the complex rheology described earlier in this section. In some cases, it may be easier to add the resin solution to the water (Schoff, 2013).

Various types and amounts of amine can be used (Wicks and Chen, 1978; Hill and Richards, 1979; Ferrell et al., 1995). The lower the amine content, the lower the viscosity of the fully diluted systems, corresponding to higher solids at a fixed application viscosity. Figure 8.2 shows viscosity response to dilution as a function of EN of a 90 : 10 BMA/AA resin with DMAE. For any resin–amine combination, there is a minimum amount of amine required to give a stable dispersion at application viscosity, that is, to prevent macrophase separation. In the example shown in Figure 8.2, the 50 EN sample could not be fully diluted without macrophase separation. In a similar experiment using an 80 : 20 BMA/AA resin, 50 EN with DMAE gave a stable dispersion even when diluted until the viscosity was below 0.1 Pa·s. Viscosity at application viscosity changes rapidly with concentration, and it is possible to overshoot a desired reduction. The viscosity is also sensitive to amine content, so if a coating has been reduced with too much water, resulting in too low a viscosity, the viscosity frequently can be brought back up by addition of small amounts of amine.

While there may be some effect of base strength (less amine may be required with increasing base strength), the principal variable seems to be the water solubility of the amine. The amount of amine required for stable dispersions increases in the order DMAE < triethylamine (TEA) < tripropylamine. Hydroxyl-substituted amines are

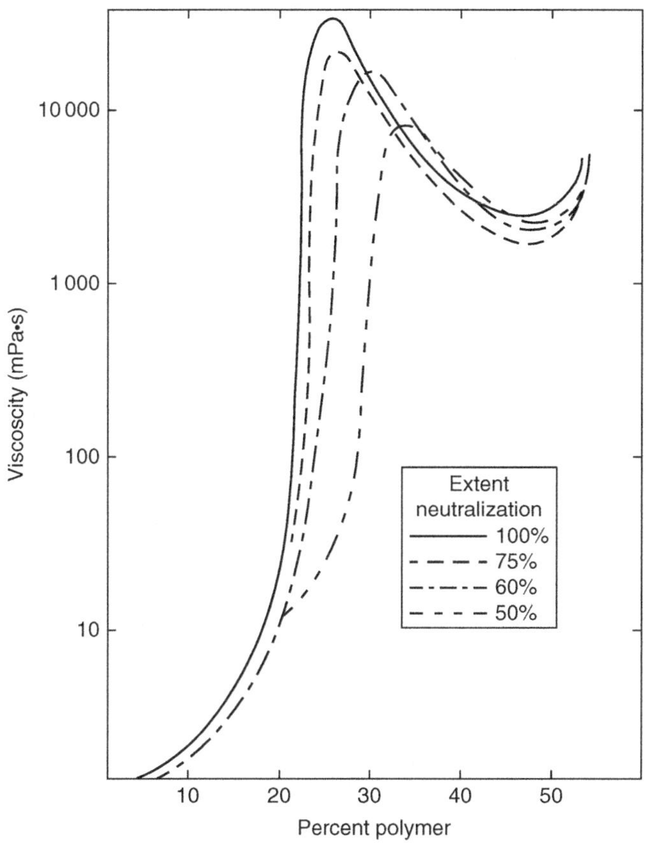

Figure 8.2 Viscosity as a function of weight percent resin when neutralized with varying levels of DMAE. The resin has 10 mol% of acrylic acid; dilution started at 54 wt% solids. Source: Hill and Richards (1979). Reproduced with permission of American Coatings Association.

the most widely used, but morpholine derivatives such as *N*-ethylmorpholine (NEM) are also effective. Potential toxic hazards of the amines should be carefully assessed.

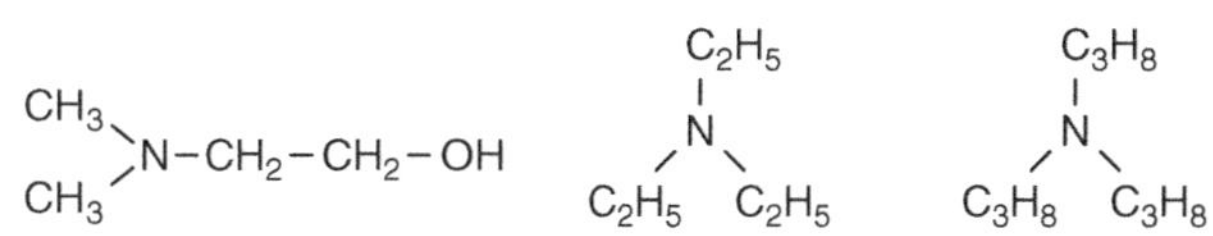

DMAE TEA TPA

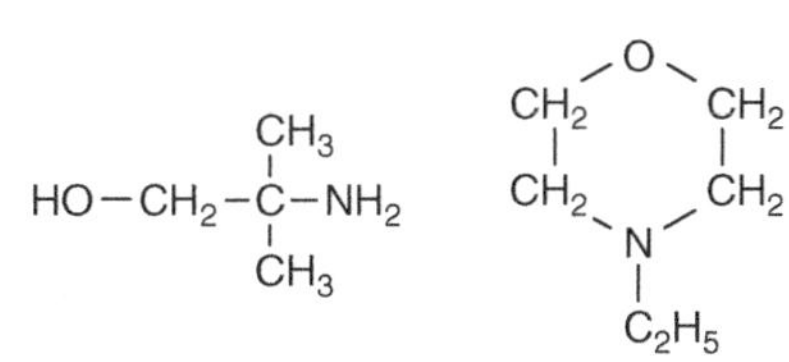

AMP NEM

In selecting amines, it is important to consider not only the effect on dispersion stability but also the effect on package stability and curing of the coating (Hill and Wicks,

1980). The amine improves the package stability by minimizing reactions of MF resins during storage. If class II MF resins are to be used, the amine must be tertiary. If, as is more common, class I MF resins are used, primary or secondary amines can also be used. Another consideration is the effect of amines on VOC. In North America, AMP is exempt from VOC regulations because of its low photochemical reactivity. The other amines are counted as VOCs. Cost of the amines is also a factor, as they are relatively expensive.

Melamine-formaldehyde resins (Chapter 11) are most commonly used as cross-linkers for TSA coatings. Class I and class II monomeric, methylolated MF resins are miscible in these water-reducible systems. They cross-link the TSA by reacting with hydroxyl and carboxyl groups. The reaction with OH groups is more facile, and the resultant ether bonds are more stable to hydrolysis than the ester bonds resulting from cross-linking with COOH groups. Thus, while the XLD of the final film depends on the total functionality of COOH and OH groups, it is generally desirable to attain as much of the cross-linking as possible through OH groups and to adjust XLD by varying the $\overline{f}_n$ of OH. A further advantage of using the lowest possible level of COOH is that less amine is required.

2K waterborne urethane coatings that use polyisocyanates as cross-linkers for water-reducible acrylics are discussed in Section 12.7.3.

Emulsion copolymerization (Chapter 9) of GMA is feasible. The resulting latexes can be cross-linked with polyamines, polyacids, and so on. Self-cross-linking latexes can be prepared by copolymerizing GMA, acid-functional monomers such as MAA, and nonreactive monomers. Many of the chemistries described in Section 8.2 could, in principle, be employed.

REFERENCES

Agawa, T.; Dumain, E. D., *Proceedings of the International Waterborne, High-Solids, and Powder Coatings Symposium*, New Orleans, LA, **1997**, p 342.

Alexanian, V., DE patent 68918315 D1 (**1994**).

Armat, R., et al., *J. Appl. Polym. Sci.*, **1996**, *60*, 1927.

Ball, P. A., et al., *Proceedings of the International Waterborne, High-Solids, and Powder Coatings Symposium*, New Orleans, LA, **2002**, pp 119–124.

Barancyk, S. V., et al., International patent application, WO 95/29947 (**1995**).

Boyer, C., et al., *Chem. Rev.*, **2016**, *116(4)*, 1803–1949.

Brand, J. A.; Morgan, L. W., US patent 4,546,160 (**1985**).

Burja, K., et al., *Prog. Org. Coat.*, **2015**, *78*, 275–286.

Chen, M. J., et al., *J. Coat. Technol.*, **1997**, *69(870)*, 43.

Christson, R. M.; Hart D. P., *Indust. Eng. Chem.*, **1961**, *53(6)*, 459.

Chung, R. P.-T.; Solomon, D. H., *Prog. Org. Coat.*, **1992**, *21*, 227.

Douskey, M. C., et al., *Prog. Org. Coat.*, **2002**, *45*, 145.

Ferrell, P. E., et al., *J. Coat Technol.*, **1995**, *67(851)*, 63.

Flosbach, C.; Schubert, W., *Prog. Org. Coat.*, **2001**, *43*, 123.

Fushimi, H., et al., *Prog. Org. Coat.*, **2001**, *42*, 159.

Gerlock, J. L., et al., *Macromolecules*, **1988**, *21*, 1604.

Gray, R. A., *J. Coat. Technol.*, **1985**, *57(728)*, 83.

Green, M. L., *J. Coat. Technol.*, **2001**, *73(918)*, 55.

Hill, L. W.; Kozlowski, K., *J. Coat. Technol.*, **1987**, *59(751)*, 63.

Hill, L. W.; Richards, B. M., *J. Coat. Technol.*, **1979**, *51(654)*, 59.

Hill, L. W.; Wicks, Z. W., Jr., *Prog. Org. Coat.*, **1980**, *8*, 161.

Hill, L.W.; Wicks, Z. W., Jr., *Prog. Org. Coat.*, **1982**, *10*, 55.

Kamath, V. R.; Sargent, J. D., Jr., *J. Coat. Technol.*, **1987**, *59(746)*, 51.

Kelly, D. P., et al., *J. Appl. Polym. Sci.*, **1963**, *7(6)*, 1991.

Kenny, J. C., et al., *J. Coat. Technol.*, **1996**, *68(855)*, 35.

Krol, P.; Chmielarz, P., *Prog. Org. Coat.*, **2014**, *77*, 913–948.

Kruithof, K. J. H.; van den Haak, H. J. W., *J. Coat. Technol.*, **1990**, *62(790)*, 47.

Moad, G.; Solomon, D. H., *The Chemistry of Radical Polymerization*, 2nd ed., Elsevier, Amsterdam, **2006**.

Moad, G., et al., *Macromol. Symp.*, **2002**, *182*, 65–80.

Myers, G. G., *J. Coat. Technol.*, **1995**, *67(841)*, 31.

Nakamichi, T.; Shibato, K., *J. Jpn. Soc. Colour Mater.*, **1986**, *59*, 592.

Nylund, J. E.; Pruskowski, S., *Resin Rev.*, **1989**, *39*, 17.

O'Driscoll, K., *J. Coat. Technol.*, **1983**, *57(705)*, 57.

Ohrbohm, W. H., et al., US patent 6,165,618 (**2000**).

Okada, K., *Proceedings of the International Conference on Organic Coating Science and Technology*, Athens, Greece, **1997**, Vol. *23*, p 337.

Ooka, M.; Ozawa, H., *Prog. Org. Coat.*, **1994**, *23*, 325.

Petit, H., et al., *Prog. Org. Coat.*, **2001**, *43*, 41.

Rehfuss, J. W.; St. Aubin, D. L., US patent 5,356,669 (**1994**).

Rehfuss, J. W.; St. Aubin, D. L., US patent 5,605,965 (**1997**).

Rietberg, J.; Van Roon, J. C., US patent 5,011,740 (**1991**).

Ryan, R. W.; Walt, M. B., *Proceedings of the International Waterborne, High-Solids, and Powder Coatings Symposium*, New Orleans, LA, **1994**, p 786.

Schoff, C. K., *CoatingsTech*, **2013**, 10(6, June), 48.

Slinckx, M., et al., *Prog. Org. Coat.*, **2000**, *38*, 163.

Spinelli, H. J., *Org. Coat. Appl. Polym. Sci.*, **1982**, *47*, 529.

Srinivasan, S., et al., *J. Phys. Chem. A* **2010**, *114*, 7975–7983.

Wang W.; Hutchinson R. A., *Macromol. React. Eng.*, **2008**, *2*, 199–214.

Watson, B. C.; Wicks, Z. W., Jr., *J. Coat. Technol.*, **1983**, *55(698)*, 59.

Wicks, Z. W., Jr.; Chen, G. F., *J. Coat. Technol.*, **1978**, *50(638)*, 39.

Wicks, Z. W., Jr., et al., *J. Coat. Technol.*, **1982**, *54(688)*, 57.

Wright, A. J., *Eur. Coat. J.*, **1996**, 696.

Yang, H. S., et al., *J. Coat. Technol.*, **2005**, *2(13)*, 44.

Ziemer, P. D., et al., *Proceedings of the International Waterborne, High-Solids, and Powder Coatings Symposium*, New Orleans, LA, **2002**, pp 99–107.

Chapter 9

Latexes

A *latex* is a dispersion of polymer particles in water. Most synthetic latexes are made by a free radical-initiated chain-growth polymerization process in which the monomers are emulsified in water at the start; hence, the process is called *emulsion polymerization*. It is not surprising that latex paints are sometimes misnamed *emulsion paints*, but this terminology is best avoided, since it can cause confusion with systems that are true emulsions. Aqueous dispersions of polymers are also prepared by other methods; for example, polyurethane dispersions in water are prepared by step-growth polymerization. These materials (discussed in Section 12.7.1) are usually called *aqueous dispersions* rather than latexes. Alkyd resin emulsions/dispersions prepared by emulsification processes are often called alkyd latexes (Chapter 15). To further confuse the terminology, some authors call latexes *colloidal dispersions* or *polymer colloids*.

Molecular weights (MWs) of polymers prepared by emulsion polymerization are generally high; commonly $\bar{M}_w$ is 1 000 000 or higher. However, unlike solution polymers, the MW of the polymer in latex particles does not affect the viscosity of the latex. Instead, latex viscosity is governed by the viscosity of the medium in which the polymer particles are dispersed (the continuous phase), by the volume fraction of particles, and by their packing factor (see discussion of the Mooney equation in Section 3.5). This lack of dependence of viscosity on MW makes it possible to formulate latex coatings at much higher solids than solutions of polymers having high MW. Latexes are used as the principal vehicle in a large majority of architectural coatings in the United States. A growing part of the OEM and special purpose coatings markets is latex based. Most latex paints form films by coalescence at ambient temperatures (Section 2.3.3). The durability of the films is often superior to films formed from drying oil and alkyd paints; VOC emissions are often lower; but the inability to achieve very high gloss is a significant limitation.

9.1 EMULSION POLYMERIZATION

Emulsion polymerization is carried out in water using monomer(s), surfactant(s), and a water-soluble initiator. The process is easily varied. Many of the same monomers are used as in solution polymerization; and the reactions are broadly similar at the molecular level. However, the physical circumstances of polymerization are different, which affects the properties of the latex. Van Herk (2013) provides a useful overview of the subject.

There have been many studies of the mechanisms of emulsion polymerization; however, there are so many variables that no general theory has been developed that can predict the results obtained in all emulsion polymerizations. Monomer structure, solubility, and concentration; surfactant structure and concentration; initiator concentration and rate of radical generation; presence and concentration of added electrolytes; and temperature are some of the critical variables. Reactor design and the rate of agitation can affect the results as well. A qualitative discussion of proposed mechanisms is provided in Section 9.1.2. The reader is referred to the general references provided at the end of the chapter for a more extensive discussion and proposed equations for predicting the number of particles and particle sizes of latexes.

Organic Coatings: Science and Technology, Fourth Edition. Frank N. Jones, Mark E. Nichols, and Socrates Peter Pappas.

Many early laboratory studies were carried out by a small-scale *batch process* in which all of the ingredients were placed into a sealed bottle that was then shaken in a temperature-controlled water bath. This process is often called the *pop-bottle process*. Many papers describing the mechanism of emulsion polymerization are based on such studies. However, batch processes cannot be used on a large scale because the heat evolved by the exothermic polymerization process would be uncontrollable. Instead, commercial latexes are produced by a *semicontinuous batch process*, using apparatus such as shown schematically in Figure 9.1 (Hoy, 1979). Monomers and initiators are added in proportions and rates such that rapid polymerization occurs. In this way, the monomer concentration at any time is low; and the polymerization is said to be done under *monomer-starved conditions.* This facilitates temperature control. Furthermore, the composition of copolymers formed under monomer-starved conditions approximately equals the composition of the monomer feed, regardless of the relative reactivity of the various monomers. This procedure also permits changing monomer composition during the course of a polymerization, as discussed in Section 9.1.3. It is common to start the polymerization in the presence of a seed latex (Section 9.1.2). Jones et al. (2005) describe a laboratory-scale process involving a seed latex.

The products obtained from batch and semicontinuous emulsion polymerizations of the same monomers are often quite different. The differences are magnified when copolymers are prepared by the two methods (Sections 9.1.2 and 9.3). Care must be exercised in reading the literature since conclusions derived from studies on the batch process are often misleading when applied to the semicontinuous process, and vice versa. Since many variables affect the composition and properties of the latex produced, it is vital to plan laboratory procedures to simulate eventual production operations as closely as possible to minimize problems in scaling up to production equipment.

9.1.1 Raw Materials for Emulsion Polymerization

9.1.1.1 Monomers

Many monomers have been converted to latexes by emulsion polymerization. The main requirements are that the monomers can undergo free radical chain polymerization and that they react with water at considerably slower rates, if at all. Monomers having limited solubility in water are most useful. Monomers having very high or very low solubility in water pose some difficulties but can be accommodated in copolymerization. The two major classes of latexes used in coatings are based on acrylic and methacrylic esters (Section 9.2) and on vinyl esters (Section 9.3). Styrene is a common comonomer for acrylics. The earliest latexes for coatings were copolymers of styrene with butadiene; they are obsolescent in architectural coatings but remain important coatings for paper. Various functional monomers are also used. Methacrylic acid (MAA) and acrylic acid (AA) improve colloidal stability, affect flow properties, improve adhesion, and have carboxylic acid groups to serve as cross-linking sites. Hydroxyethyl acrylate (HEA) and hydroxyethylmethacrylate (HEMA) provide hydroxyl groups for cross-linking with amino resins (Chapter 11) and polyisocyanates (Chapter 12). Monomers with ketone or aldehyde functionality provide sites for cross-linking with dihydrazides (Sections 17.6 and 17.11). Monomers

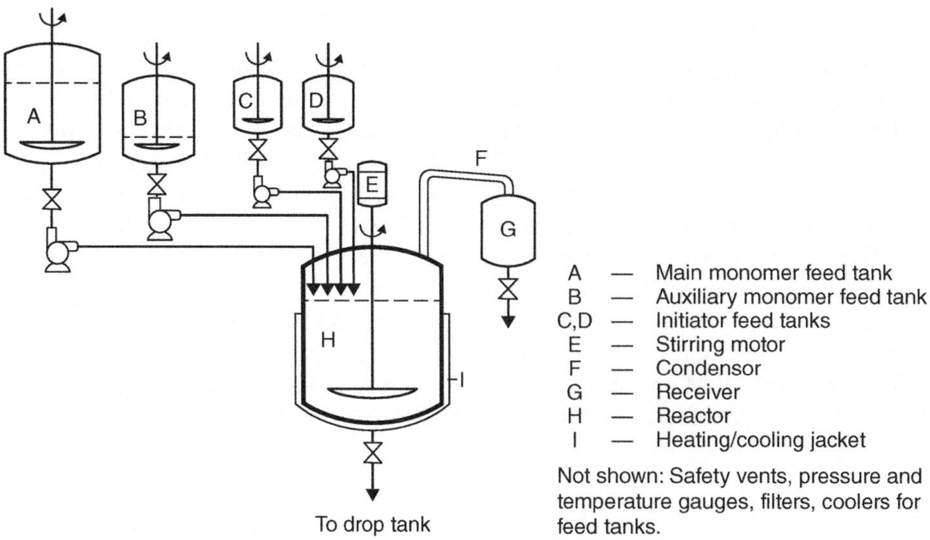

Figure 9.1 Schematic diagram of a semicontinuous batch process production unit. Source: Hoy (1979). Reproduced with permission of American Coatings Association.

have been developed that promote wet adhesion to painted surfaces; methacrylamidoethylethyleneurea is a common example. Phosphate-functional (meth)acrylate ester comonomers can promote adhesion to metals (Yang et al., 2005). Vinylidene chloride/acrylic ester copolymer latexes provide films with exceptionally low water permeability.

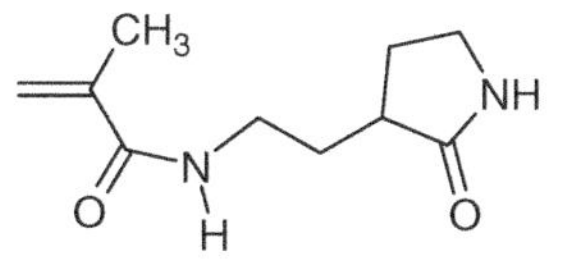

Methacrylamidoethylethyleneurea

9.1.1.2 Initiators

The principal initiators used in emulsion polymerization are soluble in water. The most common initiators are persulfate salts, especially ammonium persulfate (more correctly, ammonium peroxydisulfate). Persulfates cleave thermally in the water phase to sulfate anion radicals (Eq. 9.1) that initiate polymerization, as shown with ethyl acrylate (EA) in Eq. 9.2. Note that the terminal group is a sulfate half ester anion.

$$^-O_3S-O-O-SO_3^- \longrightarrow {}^-O_3S-O\bullet \qquad (9.1)$$

$$^-O_3S-O\bullet + H_2C{=}C(CH_3)(CO_2CH_2CH_3) \longrightarrow {}^-O_3S-O-CH_2-\overset{\bullet}{C}(CH_3)(CO_2CH_2CH_3) \qquad (9.2)$$

The sulfate anion radical can also abstract hydrogen from water, leading to the formation of bisulfate and a hydroxy free radical (HO$^\bullet$). While this reaction is not favored thermodynamically, it occurs to some extent, especially when monomer concentrations are low. Initiation by HO$^\bullet$ introduces a terminal hydroxyl group. Bisulfate lowers the pH, often requiring addition of a buffering agent.

$$^-O_3S-O\bullet + H_2O \longrightarrow HSO_4^- + HO\bullet$$

The half-life of persulfate is such that a 0.01 M solution of $(NH_4)_2S_2O_8$ at pH 10 produces 8.4×10^{12} radicals $ml^{-1}\,s^{-1}$ at 50°C and 2.5×10^{15} radicals $ml^{-1}\,s^{-1}$ at 90°C. For rapid polymerization at lower temperatures, radical production can be accelerated by reducing agents. For example, mixtures of ferrous, thiosulfate ($S_2O_3^{2-}$), and persulfate salts react faster than persulfate alone as a result of a sequence of oxidation and reduction reactions. Processes using this type of initiation are called *redox emulsion polymerizations*.

Using redox systems, polymerization can be initiated at room temperature; the exothermic polymerization reaction warms the reaction mixture to the desired temperature (often 50–80°C). Cooling is required to prevent overheating. Sometimes, redox initiation is used in the first stage of polymerization, and the latter stages are continued using thermal initiation. A variety of reducing agents have been used. Sodium formaldehyde sulfoxylate was widely used, but generates formaldehyde, and has been largely replaced. Ascorbic acid, isoascorbic acid, and sodium erythorbate are used, but tend to promote yellowing. A mixture of reducing agents, consisting of 2-hydroxy-2-sulfinatoacetic acid disodium salt, sodium sulfite, and 2-hydroxy-2-sulfanatoacetic acid disodium salt, has also been introduced that is reported to be very effective as a reducing agent without yellowing (Rothmann et al., 2001).

It is common to add a second, more lipophilic "chaser" initiator in the late stages of the semicontinuous process to facilitate conversion of monomer to high levels, preferably greater than 99%. An initiator such as *t*-butyl hydroperoxide, which is more soluble in the polymer particles than in water, is more effective than ammonium persulfate in the late stages, when most of the unreacted monomer is dissolved in the polymer particles. Even when an auxiliary initiator is used, latexes still contain some unreacted monomer. *t*-Butyl hydroperoxide has also been recommended as the sole initiator together with sodium formaldehyde sulfoxylate instead of potassium persulfate; higher conversion of monomers is obtained and the problem of residual persulfate ion is eliminated (Anderson and Brouwer, 1996).

9.1.1.3 Surfactants

Surfactants are an important component in emulsion polymerizations. A variety of both anionic and nonionic surfactants is used. One example of many available *anionic surfactants* is sodium lauryl sulfate (SLS), which is equivalent to sodium dodecyl sulfate (SDS). Both are often incorrectly called sulfonates.

$$CH_3CH_2CH_2CH_2CH_2CH_2CH_2CH_2CH_2CH_2CH_2CH_2OSO_3^-Na^+$$

Among *nonionic surfactants*, nonylphenylethoxylates (NPEs) have been popular. They are a member of the class of alkyl phenol ethoxylates (APEs), which have been major components of cleaning products such as laundry detergents. However, APEs (including NPEs) are ecotoxic and may be hazardous to humans. An EU directive essentially bans their use, and they are restricted in the United States. Fernandez et al. (2005) reviewed the biodegradation of NPEs to toxic products and proposed alternatives. Ethoxylates of $C_{12/14}$ mixed primary and secondary aliphatic alcohols are effective surfactants, which are biodegradable and do not require special labeling. They are available with different lengths of the ethoxyl chain, since the most suitable chain length can vary depending on the monomers to be polymerized. For example, preparation of acrylic latexes can be initiated with a seed latex, made using SLS as the surfactant, followed by adding the nonionic surfactant with an average of 40 ethoxyl groups (Fernandez et al., 2005). Santos et al. (2015) described unseeded and seeded emulsion polymerizations using APE-free surfactants.

$$CH_3(CH_2)_{11-13}O(CH_2CH_2O)_{40}H$$

As explained in the following text in this section, mixtures of nonionic and anionic surfactants usually give the best results. Nonionic types excel at controlling particle size and at stabilizing the latex. Nonionics also improve stability, especially when divalent cations are present in the paints; they also favorably influence rheology and enhance freeze–thaw stability.

Many cationic surfactants are available, but their use with latexes is limited to uncommon situations where low (3–4) pH is desired.

Surfactants generally have limited solubility in water. Above that solubility limit, the *critical micelle concentration* (CMC), nonpolar ends of surfactant molecules associate with each other in clusters called *micelles*. Micelles are submicroscopic aggregates typically containing 30–100 surfactant molecules in which the hydrophobic parts are oriented toward the center and the hydrophilic parts oriented outward in contact with water. The CMCs of different surfactants vary over a wide range from about 10^{-7} to 10^{-3} gl^{-1}. Surfactants play a critical role in stabilizing the dispersion of polymer particles and preventing coagulation of the latex on standing. If the particles approach each other closely, van der Waals forces tend to hold them together and the particles *flocculate*. As noted in Section 3.5, flocculation of a dispersed phase system leads to changes in flow properties; viscosity increases and flow becomes shear thinning. Lynn and Bory (1997) reviewed surfactants.

There are two general mechanisms for stabilization of dispersions; one is by *charge repulsion* in which the surfaces of the particles have an excess of one electrostatic charge, most commonly negative. For example, a latex can be stabilized by an anionic surfactant adsorbed on the surface of the polymer particles. The surfactant molecules orient with the long hydrophobic hydrocarbon tails in the polymer and the hydrophilic salt groups on the periphery associated with the water. As a result, the surface of the particles is covered with anions; each anion has an associated cation. This surface layer of cations is called a *Stern layer*; it is somewhat rigid and behaves as if it were physically part of the particle. Its presence induces a second diffuse layer of anions to surround the particle, giving it a negatively charged so-called double layer. When two particles approach each other, their diffused, negatively charged outer layers electrostatically repel each other. The stability of such dispersions can be affected by the addition of soluble salts. The stability is particularly sensitive to the presence of multivalent ions with the opposite charge of the stabilizing charge; hence deionized water is generally required. Charge repulsion is enhanced by copolymerizing small amounts of acidic monomers in the latexes. The acid groups form salts that tend to populate the surface and add to the Stern layer.

The second general mechanism for stabilization is by repulsion resulting from the outer layers of the particle. Three terms are widely used in the literature to characterize this type of stabilization: *steric*, *entropic*, and *osmotic* repulsion, which may be described as follows. If the outer surface of the particles is hydrophilic, water is absorbed and swells the surface. If this swollen layer is thick enough, the particles are not able to approach each other closely enough to flocculate; hence the term *steric repulsion*.

If a nonionic surfactant, for example, a surfactant with a hydrophobic nonpolar segment and a relatively long chain of repeating ethylene ether groups as a hydrophilic segment, is adsorbed on the surface of a latex particle, the hydrophilic ether groups will be on the outer surface. The ether groups hydrogen bond with water, and more water is absorbed in the layer. The absorbed layer can exist in a very large number of conformations, and water molecules can move in and out of the layer. When two particles approach each other, the layers are compressed, leading to a reduction in the number of conformations that molecules in the layers can assume. As a result, the system becomes less random, corresponding to a reduction in entropy. Resistance to the reduction in entropy leads to repulsion, hence the term *entropic repulsion*.

Some authors prefer to focus on the reduction in the amount of water in the layer if compression occurs. There is a tendency for the water to return to the equilibrium concentration in the layer; some consider this to be analogous to osmosis and call the result *osmotic repulsion*. Other authors prefer to avoid the controversy over terms by simply calling the repulsion, *steric repulsion*, which includes both entropic and enthalpic factors. We favor *entropic repulsion*, because the principal factor is considered to be entropy. In industrial practice, stabilization of the dispersion of polymer particles in a latex is found to be most effective when charge repulsion and entropic repulsion are combined.

Surfactants are a major factor controlling the particle size and particle size distribution of latexes (Section 9.1.2). In polymerizations where only one surfactant is used, anionic surfactant levels are usually 0.5–2 wt% based on polymer; they cost less than nonionic surfactants, and they reduce particle size more efficiently on a weight basis. On the other hand, nonionic surfactants, used at levels of 2–6 wt%, are more effective in stabilizing the latex against coagulation during freeze–thaw cycling, making it less sensitive to coagulation by salts (especially salts of polyvalent cations), less sensitive to changes in pH, and sometimes less likely to stabilize foam. In general terms, anionic surfactants stabilize primarily by charge repulsion, and nonionic surfactants stabilize primarily by entropic repulsion. The two surfactant types impart different rheological characteristics to the latex. Anionic surfactants lead to the formation of essentially rigid particles. Such latexes can have low viscosity at relatively high solids. On the other hand, with nonionic surfactants, the thicker, swollen layers on the surface of entropically stabilized particles lead to lower polymer solids or higher viscosity at a given solids. The surface layer of an entropically stabilized latex particle is not rigid and, hence, can be distorted when shear stress is applied, imparting shear-thinning characteristics.

As noted previously in this section, an emulsion polymerization recipe often contains both anionic and nonionic surfactants. More explicit generalizations about the use of surfactants in emulsion polymerization are imprudent because there are exceptions to almost any statement that can be made.

Surfactants impart some water sensitivity to films from latex paints. This sensitivity is reflected, for example, in a tendency of house paints to *waterspot* if rained on soon after application, as well as in the somewhat limited corrosion resistance of most latex coatings on steel. Selection of surfactant combinations that minimize such shortcomings is an ongoing subject of research. So-called "soap-free" latexes have been developed in which emulsion polymerization is carried out without use of conventional surfactants. Such approaches may reduce the water sensitivity associated with surfactants. Aslamazova (1995) provides an extensive review of emulsifier-free latexes. The most widely used approach is to incorporate a hydrophilic comonomer, such as MAA to form a Stern layer at the particle surface.

Another approach to the problem of water sensitivity is to use polymerizable surfactants (El-Aasser et al., 2001). Holmberg (1992) reported that a surfactant, comprised of allyl alcohol, butylene oxide, ethylene oxide, and a terminal sulfonate salt group, was completely incorporated into vinyl acetate (VAc)/butyl acrylate (BA) latexes during the polymerization. Presumably, a hydrogen atom, abstracted from the surfactant, yields a free radical that initiates the growth of another polymer chain, thereby covalently bonding the surfactant to a polymer molecule, which becomes incorporated into the latex. Alkenyl-functional surfactants in latex preparation have been reported to copolymerize with the monomers during polymerization providing latexes having superior shear stability and freeze–thaw resistance. Paints prepared with such latexes overcome the problems associated with the use of conventional surfactants, such as surfactant blooming (Grade et al., 2002). Tamer et al. (2016) described polymerizable nonionic surfactants, which contain polymerizable acrylamide groups.

Additional entropic stabilization can be achieved by including in the polymerization recipe a small amount of water-soluble polymer (sometimes called a *protective colloid*) that can readily form graft copolymers. A polymer commonly used for this purpose is poly(vinyl alcohol) (PVA), which has many abstractable hydrogens. During polymerization, initiator radical ions abstract a hydrogen atom from PVA to form a free radical on the PVA chain. A graft of the monomers being polymerized grows on the PVA chain. As the graft chain becomes longer, it becomes hydrophobic and associates with other polymer molecules in particles, carrying the PVA part of the molecule to the particle surface. The PVA segments associate with water to form an entropic stabilizing layer analogous to that formed by nonionic surfactants. Such layers can be thicker than the layer from adsorbed surfactant molecules and are effective in promoting stability. Water sensitivity of films may be reduced because less water-soluble surfactant is needed. Nguyen and Rudin (1986) have shown that grafting also occurs with hydroxyethyl cellulose (HEC). The resulting latex exhibits thixotropy and is more shear thinning than a latex prepared with no water-soluble polymer. While monomer–HEC grafting improves latex stability, it favors large particle size, broad particle size distribution, and low film clarity (Craig, 1989).

9.1.1.4 Water Quality: Other Additives

Variations in water quality can cause a variety of problems, especially when anionic surfactants are used. Accordingly, deionized water is used in emulsion polymerization and is usually used in latex paint production. Water is somewhat soluble in some polymers and acts as a plasticizer, reducing film formation temperature. Other ingredients sometimes used in emulsion polymerization processes include thickeners, which control viscosity, as well as buffers, which protect sensitive monomers from hydrolysis and sensitive surfactants from deactivation.

9.1.2 Emulsion Polymerization Variables

Many quantitative theories for the mechanisms of emulsion polymerization have been proposed, but none has proven to be applicable to all systems because of the wide range of variables involved. Since none of the theoretical models have proven to be generally applicable, the discussion here is limited to qualitative discussion of important considerations.

At the outset, the monomers are emulsified in water with surfactant molecules oriented at the surface of the emulsion droplets, stabilizing the emulsion. There is also surfactant dissolved in the water and excess surfactant present in micelles. In most systems of interest for coatings, at least one of the monomers is somewhat soluble in water, so there are monomer molecules in solution as well as in the emulsion droplets. For example, the solubilities of methyl methacrylate (MMA) and EA are approximately 1% in water. When initiator is added, it is also dissolved in the water.

There are three potential sites for initiation and initial propagation of the polymerization: (1) in the monomer-swollen micelles (*micellar initiation*), (2) in the aqueous phase (so-called homogeneous nucleation), and (3) in the emulsion droplets. All three of these modes may occur in any given polymerization, but the proportion of the three modes is system dependent. The predicted number of particles and particle size varies in each of the three modes (or combinations of them). However, in qualitative terms, particle size decreases and number of particles increase as surfactant concentration increases. Monomer solubility in water, surfactant composition, and emulsion droplet size may be particularly important variables.

When at least one of the monomers is somewhat soluble in water, it is commonly assumed that initiation occurs predominantly in the water phase. The initiating radical reacts with a monomer molecule, forming a new free radical that is soluble in water; these free radicals can react further with dissolved monomer molecules to form oligomers having terminal free radicals. As the MW of the growing oligomer chain increases, its solubility in water decreases; and one of three events may occur: (1) the growing chain may enter a micelle, where propagation continues, as monomer molecules also enter the micelle; (2) surfactant from the solution may be adsorbed on the surface of the growing oligomer molecule to form a new micelle; or (3) the growing chain can enter a monomer droplet. In any case, a surfactant-stabilized polymer particle containing a single free radical is formed. Such particles rapidly imbibe monomer, and propagation continues within the particles.

When all of the monomers are low in solubility, initiation in micelles may predominate. Again, the monomer in the micelle polymerizes rapidly to form a surfactant-stabilized polymer particle containing a single free radical.

Termination (by combination or disproportionation, as described in Section 2.2.1) can occur only after a second free radical enters a particle. After termination, the surfactant-stabilized polymer particle is temporarily inactive because it no longer contains a free radical. More monomer molecules then migrate into the inactive polymer particle. When a third free radical (IMMMMM•) enters the monomer-swollen particle, further initiation of polymerization occurs. Chain growth continues until a fourth free radical enters the particle and causes termination. Dissolution of monomer from the monomer droplets replenishes the monomer in solution. The process is repeated over and over, leading to the formation of additional polymer molecules in the particle. (This model does not apply when the T_g of the polymer is higher than the polymerization temperature, which is usually not the case.)

As the polymer particles grow, surface expansion requires additional surfactant. This need is satisfied by adsorption of more surfactant molecules from solution, which, in turn, is replenished by dissolution of micelles in which polymerization has not occurred. This results in the growth of the particles that acquired a polymer molecule early on and disappearance of micelles that did not. Polymer particle growth continues until unreacted monomer or initiator is used up; one uses sufficient initiator to assure high conversion of monomer. After the early stages of the process, much of the polymerization occurs within a fixed number of polymer particles. The number of particles per unit volume depends on the concentration of micelles in the early stages. If the concentration of micelles is high, the number of particles is high, and hence, the particle size at the end of the polymerization is small. At the same surfactant concentration, a smaller particle-sized latex is formed with surfactants having a low CMC than with those having a higher CMC.

In the homogeneous nucleation mechanism, surfactant is absorbed on the surface of the growing oligomer chains. As the number of particles and their size increase, the amount of surfactant becomes insufficient to stabilize them. Particles coalesce to form fewer larger particles until the surfactant concentration is just high enough to stabilize the particles.

Conditions for the third mechanism, initiation in the droplets of monomer emulsion, are usually unfavorable. (Exceptions are *miniemulsion polymerization* and *suspension polymerization*, discussed in the following text.) In most cases, the size of the droplets of monomer emulsion is relatively large, and their surface area is small relative to the surface area of the far more numerous micelles, so that the probability of an initiating radical or a growing oligomer radical entering an emulsion droplet is small. However, if the rate of agitation is very high and/or a high surfactant concentration is used, emulsion droplet size becomes smaller, and the probability of propagation in emulsion droplets increases. If a water-insoluble initiator were used in the early stages of polymerization, most of the polymerization would occur in the emulsion droplets, resulting in large, unstable particles. This method is the basis for a different polymerization method, called *suspension polymerization*.

In batch processes, the number of particles and, hence, their particle size are also affected by the rate of initiation (a function of the type of initiator, its concentration, and the temperature). The faster the rate of formation of initiating radicals, the larger their number and the smaller the size of the particles, everything else being equal.

Latexes are frequently prepared starting with a *seed latex*, that is, a small particle size latex (Gardon, 1997). Monomer and initiator are added slowly, and the polymerization occurs primarily in the seed particles, resulting in a nearly constant number of particles as polymerization proceeds. The seed latex can be prepared as the first step in a polymerization; or, commonly, a large batch of seed latex is produced and subdivided for subsequent use in many batches. The seed latex, which contains 10–20% solids, is diluted to 3–10% solids. Increasing the number of seed particles, with other factors being equal, yields a latex of smaller average particle size. Thus, controlling the number of seed particles provides a convenient way to control particle size in the finished latex. This practice can improve batch-to-batch reproducibility and is particularly useful in sequential polymerizations (see following text).

The $\bar{M}_w$ obtained by both batch and semicontinuous processes is generally very high, commonly greater than 1 000 000. In many cases, there is a broad distribution of MWs. High $\bar{M}_w$ results from several factors. First, during polymerization, the fraction of polymer particles containing more than one free radical at a given instant is low. As a result, the rate of termination reactions within the particle is reduced, and a single growing chain may consume most of the monomer present in the particle before it terminates. Second, the viscosity inside the particles is very high, reducing the mobility of the polymer chains. This situation favors diffusion of more mobile monomer molecules adding to the growing chain and increasing the MW, relative to reaction of two less mobile growing chains, resulting in termination. Third, hydrogen atoms on polymer chains can be abstracted, leading to chain transfer to polymer and formation of branched chain molecules. The environment within the polymer particle favors chain transfer to polymer because the concentration of free radicals is low (often, one per particle), while the concentration of polymer approaches 100%. Chain transfer to polymer increases $\bar{M}_w$ and $\bar{M}_w/\bar{M}_n$ because it is statistically probable that when growth of short chains is ended by chain transfer, the new chain will grow as a branch on a polymer molecule that already has a high MW. Sometimes a chain transfer agent is added to the monomer mixture to deliberately reduce MW, and sometimes a very small amount of difunctional monomer is added to increase it.

El-Aasser et al. (1983) compared batch and semicontinuous homopolymerization and copolymerization of VAc and BA, $\bar{M}_w/\bar{M}_n$ for batch homopolymers and copolymers was broad, ranging from 15 to 21; and the semicontinuous process resulted in even broader distributions, ranging from 9 to 175. The semicontinuous process commonly afforded bimodal distributions with considerable amounts of relatively low MW polymer. Semicontinuous poly(vinyl acetate) homopolymer had the broadest distribution; GPC indicated substantial fractions of material with M below 660 and above 1 700 000.

Particle size distribution is an important variable in formulation of latex paints. Particle size distribution is commonly measured by light scattering and can also be measured by hydrodynamic chromatography, ultracentrifuge, or transmission electron microscopy. It may be expressed as number average diameter $\left(\bar{D}_n\right)$ and volume average diameter $\left(\bar{D}_v\right)$. The ratio $\bar{D}_v/\bar{D}_n$ is a convenient index of particle size distribution. A broad or bipolar distribution is often advantageous because the packing factor is affected by particle size distribution. The volume fraction of latex particles that can be present at a standard viscosity is highest with broad particle size distribution (see Section 3.5).

The particle size of a latex is determined by the number of particles relative to the amount of monomer. The number of particles is governed primarily by the amount and type of surfactant(s) and initiator(s). Anionic surfactants tend to produce more particles than nonionic surfactants when used on an equal weight basis, resulting in smaller particles. The specific anionic surfactant used also affects the particle size. For example, use of NP-$(OCH_2CH_2)_{20}OSO_3^-\ NH_4^+$ (NP stands for the nonpolar part of the molecule) produced an acrylic latex having a particle diameter of 190 nm, whereas the diameter was 100 nm when the less hydrophilic NP-$(OCH_2CH_2)_9OSO_3^-\ NH_4^+$ was used (Alahapperuma and Glass, 1992). In general, particle size distribution is broader with anionic surfactants and more nearly approaches monodisperse particle size with nonionic surfactants. Changing the ratio of nonionic surfactant (nonylphenol ethoxylate consisting of about 40 mol of ethylene oxide) to anionic surfactant (sodium dodecylbenzene sulfonate) affects particle size and particle size distribution in the polymerization of a VAc/dibutyl maleate copolymer. The lowest ratio gave the smallest particle size and the broadest distribution of particle size (El-Aasser et al., 1983).

Particle size is also affected by agitation intensity and turbulence during polymerization. High turbulence leads to larger particle size thereby influencing rheological properties. This effect increases the problem in the scaling up of laboratory synthesis to production, requiring the design of lab equipment to approximate the agitation to be encountered in plant reactors. Oprea and Dodita (2001) studied the relationship of agitation and particle size during polymerization of styrene/acrylic latexes.

Emulsion polymerization usually produces latexes with essentially spherical particles. During the 1980s, latexes were commercialized in which groups of a few particles are fused together into nonspherical *lobed particles*

(Chou et al., 1987). At equal concentration, the viscosity of nonspherical latexes is higher, owing to their smaller packing factor relative to spherical latexes. Paints formulated with a lobed latex are shear thinning and result in a higher high-shear viscosity at the same low shear viscosity than do conventional latexes. The higher viscosity at shear rates experienced during application leads to thicker films with lower levels of thickener, thereby reducing the cost. Adhesion of paints to chalky surfaces using lobed latexes may also be improved relative to the use of conventional latexes.

As reviewed by van Herk (2013), feasibility of controlled radical polymerization (CRP) (Chapter 2) in emulsion polymerization has been demonstrated in numerous studies.

9.1.2.1 Miniemulsion, Microemulsion, and Suspension Polymerizations

Conventional emulsion polymerization starts with relatively large (10–100 µm) monomer droplets. Miniemulsion polymerizations start with much smaller (50–500 nm) droplets (Sudol and El-Aasser, 1997; Schork et al., 2005; Mittal, 2010). The small droplet sizes cause a major change in nucleation and polymer growth mechanisms. Most of the surfactant resides on droplet surfaces, so the population of micelles is low. Therefore, instead of the micellar nucleation mechanism discussed previously, much of the nucleation occurs in polymer droplets or at their surfaces. To the extent this happens, the polymers can be quite different than those made by conventional emulsion polymerization. The products are dispersions of very small polymer particles with relatively homogeneous compositions. Monomers come into less contact with water, suggesting that miniemulsion polymerization be used for water-sensitive monomers (Section 9.4). Microemulsion polymerizations (Chow and Gan, 2005) use even smaller droplets. The literature describes a huge variety of small particle dispersions made by these methods.

Suspension polymerization is a method for making solid acrylic resins in particle form. Relatively large monomer droplets containing initiator are suspended in water and heated to induce polymerization.

9.1.3 Sequential Polymerization

When latexes are prepared by semicontinuous polymerization under monomer-starved conditions, the composition of the monomer feed can be changed during the course of the polymerization. Depending on a number of variables, the resulting latex particles can have a variety of morphologies. In some cases, latexes having *core–shell* morphology can be made by changing from the initial monomer feed composition to a different feed part way through the process. Many such core–shell latexes display two T_g's after coalescence. In some cases the composition of the polymer in the center of each particle (the core) reflects the initial monomer feed composition, while the composition near the surface (the shell) reflects the second monomer feed composition. In other cases, an *inverse core–shell* morphology can be obtained in which the composition of the core polymer corresponds to that of the second monomer feed and the shell polymer is derived from the initial monomer feed. In yet other cases, a homogeneous composition is obtained in spite of a change in monomer feed composition. In still other cases, it is possible to make a gradient morphology by continually changing the monomer feed composition during the polymerization. There is an extensive literature on sequential polymerization. In particular, Padget (1994) reviewed the literature applicable to latexes of interest in coatings and discussed the variables involved.

Several factors affect the morphology of particles obtained in sequential polymerization. Formation of outer layers having the lowest interfacial tension with the water phase, is favored; thus, when different monomer compositions are fed into the polymerizing medium, the one containing the most polar monomers is likely to end up on the surface of the shell, regardless of its place in the sequence. Particle morphology can be affected by the amount of free monomer present during the polymerization, since free monomer acts as a plasticizer and facilitates inversion. Thus, formation of core–shell morphology tends to be favored under highly monomer-starved conditions. Presence of di- or trifunctional cross-linking monomers in at least one of the monomer combinations increases the probability of phase separated particle morphology. In latexes designed for coatings, generally only limited amounts of polyfunctional monomers can be used because of the need for coalescence to achieve good film formation. But, it is a common practice to use a low level of cross-linking monomer in the first composition in order to increase the likelihood that it will remain in the core. Examples of applications of sequential polymerization are given in Section 9.2.

Core–shell latexes having inorganic nanoparticle cores have been described. For example, Arai et al. (2016) prepared and studied MMA/BA/MAA latexes. First 20–30 nm silica nanoparticles were dispersed in a polymerizable nonionic surfactant and polymerized in the presence of a small amount of monomer. Then the rest of the monomer was added using starved feed semi-batch polymerization as in seeded core–shell methods. Some of the products formed transparent "nanocomposite" films with enhanced stiffness and toughness relative to coatings in which the nanosilica was blended into an ordinary latex made from the same monomers.

9.2 ACRYLIC LATEXES

Acrylic latexes are widely used for exterior paints because of their resistance to photodegradation. They are also more resistant to hydrolysis and saponification than are VAc latexes. These properties are critical for exterior paints and also enhance the usefulness of acrylic latex paints for alkaline substrates such as masonry and galvanized metal and for applications in which there is exposure to high humidity. Acrylic and styrene–acrylic latexes are being used increasingly for industrial maintenance coatings. Acrylic latexes are also generating increasing interest for wood finishes and for OEM automotive applications. For example, waterborne base coats, partially comprised of latexes, are being developed for OEM automotive applications. Hybrid acrylate/urethane dispersions have also become increasingly important (Section 12.7.2).

A critical decision in designing a latex is monomer selection. An important consideration is to select a monomer combination that produces a copolymer having the appropriate T_g. Monomer cost is also important, as the paint business is highly price competitive. For lower priced paints, adequate film properties are attained by copolymerizing acrylics with low cost monomers such as styrene or vinyl esters. The T_g must be low enough to permit coalescence of the latex at the lowest anticipated application temperature, yet high enough to assure adequate film hardness and toughness. T_g in the vicinity of 5–10°C is common for exterior house paints, which may be applied at temperatures as low as 2°C. The *minimum film forming temperature* (MFFT or MFT) (Section 2.3.3) of latexes is related to T_g, but is influenced by other factors such as particle size, phase separation within the latex particles, and plasticizing effects of water and surfactants. The MFFT tends to be somewhat lower than T_g. T_g and MFFT values in the desired range are usually attained by copolymerizing monomers whose homopolymers have much higher and much lower T_g values than the target values.

As a high T_g comonomer, MMA imparts excellent exterior durability and hydrolytic stability at moderate cost. Styrene is often substituted in part or completely for MMA because it costs less and provides a similar T_g effect. Such products are sometimes called styrene/acrylic latexes but usually just acrylic latexes. Styrene imparts excellent hydrolytic stability, but its effect on overall exterior durability is not clear-cut. Styrene homopolymer degrades relatively rapidly outdoors. However, in a copolymer, some MMA can be replaced by styrene without substantially reducing exterior durability. The amount of styrene that can be substituted apparently varies, depending on the other comonomers and probably on other variables, such as process conditions. For example, processes that would tend to produce blocks of homopolystyrene are suspect. It is safest to test each composition thoroughly outdoors before use. It is interesting that styrene/acrylics may be used more in northern Europe than in southern Europe, where weather exposure is more severe. Other properties may also be affected. For example, it cannot be assumed that the mechanical properties of the copolymer will be equal when styrene is substituted for MMA, even though its T_g is about the same. Polystyrene has a significantly higher brittle–ductile transition temperature than poly(MMA) even though it has a slightly lower T_g (Section 4.2).

Acrylic esters are generally used as the low T_g monomer(s), the choice of which is affected by cost and durability considerations. If the cost of the common acrylic esters is intermediate between that of MMA and of styrene, it is more economical to use EA (homopolymer $T_g = -24$°C) than BA (homopolymer $T_g = -54$°C) in styrene-free latexes in order to get as high proportion of EA in the composition as possible. However, with styrene/acrylic latexes, BA may be more economical when styrene is less expensive than EA or BA. Presumably, the different acrylic esters impart different film properties at a given T_g, but published data are inadequate to evaluate the effects. Recent decades have indicated an industry trend toward use of BA in preference to EA (Learner, 2001). While published data are scarce, one can speculate that BA has demonstrated better durability in proprietary tests. One can also speculate that it imparts better resistance to hydrolysis.

An example of a laboratory semicontinuous batch process for preparation of a 40:59:1 MMA/EA/MAA copolymer latex is provided in Table 9.1 (Anonymous, undated). Note from Table 9.1 that the measured MFFT of the latex is 9°C, which is somewhat lower than the T_g of 17°C estimated by the Fox equation for a 40:60 MMA/EA copolymer. Also, note that changing to a 50:50 MMA/EA ratio increases MFFT to 21°C, while the Fox equation predicts a T_g value of 28°C. These figures provide a general picture for how T_g and MFFT relate to composition. Many exemplary procedures can be obtained on the Internet from supplier literature. Jones et al. (2005) provide very detailed directions for laboratory-scale seeded emulsion polymerization with a more contemporary composition.

Commercial suppliers of latexes provide little information as to the composition of the latexes they offer. While a simple acrylic latex, prepared as described in Table 9.1, could be used to make a house paint, it would not be competitive. Commercial production involves many process nuances that are held confidential by the latex producing companies. It is difficult to replicate commercial latexes, even though the monomer and surfactant compositions can be precisely analyzed.

Acrylic acid, or the somewhat less soluble MAA, is commonly used in preparation of acrylic latexes, usually on the order of 1–2 wt% of the monomer charge. The carboxyl functionality incorporated with these monomers enhances mechanical stability of the wet latex paint, reducing the

Table 9.1 Laboratory Procedure for Preparation of a MMA/EA/MAA Copolymer Latex

Materials

1 l	Deionized water
96 g	Triton X-200 (sodium salt of an alkylaryl polyether sulfate ester) (Union Carbide Co.)
320 g	Methyl methacrylate (10 ppm MEHQ stabilizer) (MEHQ is the monomethylether of hydroquinone)
480 g	Ethyl acrylate (15 ppm MEHQ stabilizer)
8 g	Methacrylic acid (100 ppm MEHQ stabilizer)
1.6 g	Ammonium persulfate

Procedure

Prepare an emulsion of all reactants in 800 ml of water. Place 200 ml of the emulsion and 200 ml of water in a 3-l flask fitted with an inert gas inlet tube, a thermometer, a stirrer, an addition funnel, and a reflux condenser. Heat in a 92°C water bath while stirring until the internal temperature reaches 82°C. The mixture will begin to reflux, and its temperature will rise to about 90°C within a few minutes, indicating rapid polymerization. When refluxing subsides, add the remaining emulsion continuously over a period of 1.5 h. Heat to maintain refluxing; the internal temperature will be 88–94°C. After monomer addition is complete, heat to 97°C to complete conversion of the monomer. Cool and strain.

Properties

NVW 42.9 (calculated 43.1); pH 2.7; viscosity (Brookfield), 11.5 mPa-s; MFFT 9°C; film hardness (Tukon), 1.2 KHN. A similarly prepared 50 : 50 MMA/EA copolymer latex (no MAA) had similar properties, except that MFFT was 21°C and film hardness was 6.2 KHN.

Source: From De Bruyn et al. (2002).

amount of surfactant needed. The effects of acid monomers on stability and viscosity are maximized when they are incorporated in the latter part of the monomer feed and the polymerization medium is acidic (Hoy, 1979; Emelie et al., 1988). The viscosity of a latex made in this way depends on pH. As ammonia is added, there is little change in viscosity until the pH reaches about 7. Above pH 7, viscosity increases steeply until the pH reaches 9 or 10; at still higher pH levels, viscosity decreases. The pH of acrylic latex paints is usually adjusted to about 9, where there is a substantial viscosity effect: not only is the viscosity high, but the paint also exhibits shear thinning, which is often a desirable application characteristic (Sections 3.5 and 32.1).

Hoy (1979) describes a latex that shows this effect; it is a MMA/EA/BA/AA (40 : 52 : 6 : 2 by weight) copolymer, in which the AA was added late in the process. The T_g of this copolymer was 15°C. Neutralization with ammonia to pH 9 caused the surface layer of the particles to expand, so the diameter of the particles was about 1.8 times that at pH 7. The corresponding increase in volume was almost sixfold. The expansion resulted from the association of water with the highly polar salt groups in the polymer near the particle surface. The presence of the expanded layer at pH 9 increased viscosity at low shear rates. However, because this layer is easily distorted by shear forces, the latex is shear thinning.

In the same paper, Hoy describes the effects of using sequential polymerization during semicontinuous batch polymerization. In one experiment, an MMA/BA ratio of 40 : 60 was used, with 2 wt% of AA added late in the process. When the 40 : 60 ratio was maintained throughout monomer addition, the product, called a *uniform feed* latex, had a T_g of 20°C. A second *staged feed* latex of the same overall composition was made by a process designed to yield core–shell particles (Section 9.1.3). In the first half of the monomer addition, the MMA/BA ratio was 70 : 30; in the second half, it was 10 : 90. The resulting latex exhibited two T_g's, one at 60°C and the other at −10°C. Because the shell of this latex has a low T_g, its MFFT is substantially lower than that of the uniform feed latex. However, coalesced films core–shell particles are cloudy. The composition of the cores and the shells is sufficiently different that the polymers are not miscible with one another. Because the refractive index of the two polymers is different, light passing through the film is scattered, making the film cloudy. Since the T_g of the continuous phase is low, the blocking resistance of a film of the staged feed latex is inferior to a film of the uniform feed latex. Apparently the high T_g cores remain uncoalesced.

Hoy (1979) also describes a third latex having the same monomer composition made by a process called *linear power feed addition*, using the apparatus sketched in Figure 9.2. In this process there are two stirred tanks for monomer mixtures. The near tank was loaded with half of the monomer charge at an MMA/BA ratio of 70 : 30, and the far tank was loaded with the other half at a 10 : 90

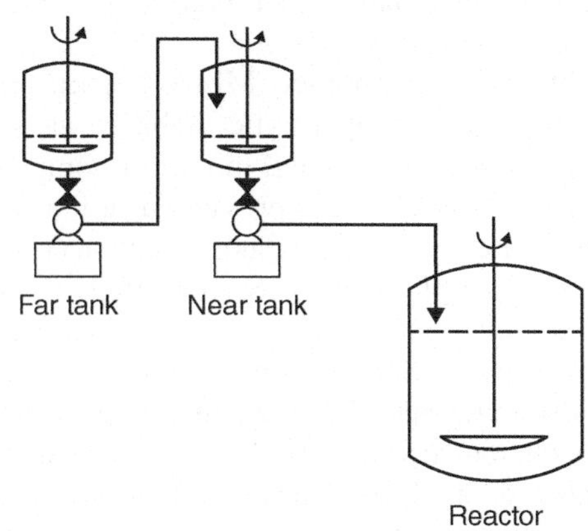

Figure 9.2 Schematic representation of an apparatus for semicontinuous latex polymerization using linear power feed addition. Source: Hoy (1979). Reproduced with permission of American Coatings Association.

ratio. During polymerization, monomer was pumped from the near tank into the reactor in the usual way. At the same time, monomer was being pumped at the same rate from the far tank to the near tank. Again, 2 wt% of AA was added in the late stage of the process. In this procedure, the monomer feed starts at 70 : 30 MMA/BA and ends at 10 : 90 but changes continuously through the process. Each latex particle presumably contains high T_g polymer at its center and low T_g polymer (with COOH groups) near its surface. It differs from the usual core–shell latex in that there is a linear gradient of composition and T_g from the center to the surface. Films cast from this latex have equal clarity and the same T_g, about 20°C, as those cast from the uniform feed latex. Desirably, the glass transition is much broader, the MFFT is lower and the blocking resistance temperature is higher for the linear power feed latex.

In North America, exterior house paints for large surfaces are usually formulated at low gloss; the formulations are heavily pigmented. Block resistance requirements are modest, and the pigment helps improve this property; T_g values about 3–10°C are adequate. Latex paint films are more permeable to water than films from oil or alkyd paints, which is an advantage when applied to wood surfaces because it reduces blistering, but a disadvantage when applied over metal.

Applications such as gloss trim and door paints, as well as interior trim paints (for kitchen cabinets, windowsills, etc.) have different requirements; to be glossy, they must be formulated at lower pigment levels, yet block resistance requirements are higher. This situation presents a challenge to designers of latexes. Mercurio et al. (1982) found that a latex with T_g of about 55°C is required to achieve adequate block resistance. Coalescence of acrylic latexes having this relatively high T_g can be attained by using a small particle size latex with substantial amounts of carefully selected coalescing solvent(s). The coalescing solvent increases VOC to 250 g of VOC/L of paint as applied, excluding water, which satisfies VOC regulations in some areas but not in others (Chapter 18).

Although the VOC of latex paints is usually lower than that of solventborne paints, there is increasing pressure to reduce VOC further. Reducing the amount of coalescing solvent required is the most evident approach to this end. Sequencing the composition of the monomer feed during polymerization so that the last part of the monomer feed has a higher fraction of monomers, which impart a low T_g, as in the power feed approach discussed previously, permits reduction of coalescing solvent (Section 9.1.3). Addition of AA in the later part of the comonomer feed may also permit reduction of coalescing solvent. Water associates with the neutralized carboxylic acid salt, plasticizing the surface of the particles, promoting coalescence, and reducing the need for coalescing solvent. Another method is to use cross-linkable latexes having a lower T_g; cross-linking after film formation increases block resistance to offset the effect of the lower T_g (Sections 9.4 and 2.3.3). It is also possible to design a coalescing solvent that cross-links after film formation. Another approach for eliminating the need for coalescing solvent is to use blends of high and low T_g latexes (Winnik and Feng, 1996; Eckersley and Helmer, 1997). When such a film dries, the high T_g latex does not coalesce but is dispersed in the continuous phase from the low T_g polymer. The hard particles act to reinforce the low T_g polymer film increasing the modulus of the film so that its block resistance is superior to that of a film using only the low T_g latex. It **is** critical that the ratio of the two latexes be such that there is sufficient low T_g latex to completely enclose the high T_g latex particles.

Various workers have studied core–shell latexes in which pigments instead of high T_g latex are encapsulated inside low T_g latexes. An example is described in Section 9.1.3 (Arai et al., 2016) using silica nanoparticles; enhanced film properties are reported.

A similar approach is to make nano polymer/nano clay composite latexes (Lorah et al., 2004). The latexes are made with relatively low T_g acrylic monomers polymerized with clay dispersed in the water in which the emulsion polymerization is carried out. Films prepared from the latex have unusual toughness without compromising film formation. They also exhibit low tack and dirt pickup in zero VOC paints.

A different use for acrylic emulsion polymerization is for the preparation of *microgel* particles. (Today it would be fashionable to call them nanogels.) For example, Nair (1992) prepared acrylic latexes, using a small fraction of divinylbenzene as a comonomer, by emulsion polymerization using a water-soluble azo initiator, 4,4′-azobis(cyanovaleric acid). Cross-linking within each growing particle limits accessibility of monomer and thus limits the size of the particles. The dried latex polymer readily dispersed in aliphatic hydrocarbon to yield a dispersion of an acrylic microgel. Among other applications, acrylic microgels are used as additives to control sagging of high solids coatings (Section 24.3).

9.3 VINYL ESTER LATEXES

VAc is less expensive than (meth)acrylic ester monomers. However, VAc latexes are inferior to acrylic latexes in both photochemical stability and resistance to hydrolysis. When an acetate group on a PVAc chain hydrolyzes, the liberated acetic acid catalyzes hydrolysis of more acetate groups, and the hydroxyl group may exert a neighboring group (anchimeric) effect that promotes hydrolysis of adjacent acetate groups. Hence, PVAc latexes find their principal use in interior coatings that do not have to withstand high humidity exposure or frequent wetting. One such use, flat wall paints, is the largest volume type of paint sold in North America and probably elsewhere. The T_g of PVAc homopolymer is 29°C, too high for film formation under ambient conditions. Thus, one must either formulate PVAc with a plasticizer or copolymerize VAc with a monomer that reduces T_g. A widely used comonomer is *n*-BA, although other comonomers, such as 2-ethylhexyl acrylate and di-*n*-butyl maleate, have been used. Longer chain vinyl ester also serve as comonomers as discussed in the following text; the resultant latexes have much better durability and hydrolytic stability than PVAc latexes.

Copolymerization of VAc with BA poses potential problems because of reactivity ratios; specifically, the rate of reaction of a free radical on a terminal BA group with another molecule of BA is greater than its rate of reaction with VAc; furthermore, the rate of reaction of a free radical on a terminal VAc group is much greater with BA than with VAc. Thus, in a batch polymerization, a 50:50 mixture of BA and VAc will produce a copolymer very rich in BA at the outset and very rich in VAc in the late stages. This problem is overcome by a semicontinuous batch process, wherein the monomer mixture is added at a rate equal to the polymerization rate. A steady state is reached in which the concentration of unreacted VAc is low, but the concentration of BA is even lower. Under these monomer-starved conditions, a relatively uniform copolymer forms with a composition similar to the ratio of monomers being fed. Such copolymers are made commercially on a large scale. An extensive study of the copolymerization of VAc with BA by batch and semicontinuous batch processes, as well as characterization of the copolymers, has been reported (El-Aasser et al., 1983; Misra et al., 1983) and is discussed in part in Section 9.1. Since the reactions rates are relatively low, the rate of monomer addition must be quite slow. If the initial charge is high in the least reactive monomer, the addition can be made faster while still producing reasonably uniform copolymerization. Mathematical simulations can predict the effects of different addition modes (Arzamendi and Asua, 1989).

VAc is more soluble in water than acrylic esters, as are the initial oligomeric free radicals formed from VAc than those from oligomeric radicals of acrylic esters at equal chain length. This factor increases the probability of termination of VAc-rich oligomers while still in aqueous solution, leading to low MW fractions. On the other hand, chain transfer to polymer is more facile with polyvinyl acetate, leading to high MW and broad MW distribution. The high water solubility of VAc also increases the need for effective initiator systems to remove residual monomer. Vandezande et al. (1997) review VAc latexes, including the effect of different initiator combinations.

An essential component of the procedure for polymerization of VAc is a buffer, such as sodium bicarbonate, to maintain the pH near 7. A buffer is essential because VAc monomer hydrolyzes at appreciable rates under either acidic or basic conditions. The hydrolysis is irreversible because the products are acetaldehyde and acetic acid. At pH less than 7, hydrolysis is autocatalytic because acetic acid catalyzes further reaction. Furthermore, acetaldehyde is oxidized by persulfate, which consumes an initiator and generates still more acetic acid to catalyze hydrolysis.

$$CH_3CO_2CH{=}CH_2 + H_2O \leftarrow CH_3CO_2H + CH_3CHO$$

Copolymer latexes of VAc and (meth)acrylic monomers have been designed for dual-purpose use in both exterior and interior flat paints. The objective is to save cost in two ways—lower raw material cost for exterior paints and lower inventory and storage costs by basing all production on a single resin. This approach has been commercially successful, although the latexes undoubtedly represent a compromise between the optimal resins for exterior and interior use. The amount of VAc in an acrylic latex can be increased by sequential polymerization to form inverse core–shell latexes, lowering the cost with minimal loss of properties (Vandezande and Rudin, 1994). For example, a seed latex of MMA/BA/MAA made with persulfate initiator was further reacted with VAc using azobisisobutyrylnitrile as initiator. Since the initial polymer has polar acrylate salt groups, the acrylate polymer becomes the shell with a polyvinyl acetate core.

Copolymer latexes of the vinyl ester of a C_{10} branched acid (vinyl *neo*-decanoate, registered as Vinyl Versatate®) with VAc are used in both interior and exterior paints. These copolymers have better hydrolytic stability than conventional VAc copolymers (Bassett, 2001). A combination of ethoxylated undecyl alcohol, a cellulose ether, and sodium vinyl sulfonate can be used as surfactant/protective colloid in preparing the latexes. Heldmann et al. (1999) studied the effect of changes in chain length of the ethylene oxide units on the properties of latexes and corresponding paints. Increasing degree of ethoxylation resulted in smaller particle size latex particles and higher viscosity. In flat paints, optimum properties are obtained with 2–3% emulsifier concentration and 17–28 degrees of ethoxylation. In gloss and semigloss paints, gloss improves with an increase of emulsifier content up to about 4% where it levels off. The gloss improves as the degree of ethoxylation increases to 17. Above 4% concentration and 17 ethoxyl units, blocking increases dramatically. Similar results were obtained with VAc/ethylene latexes.

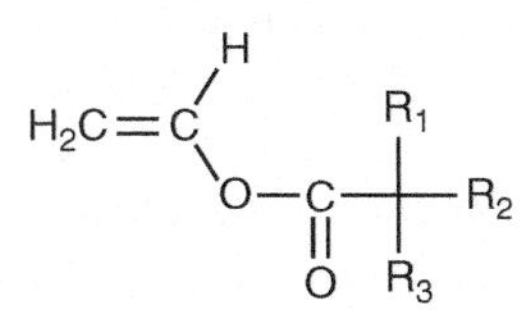

Vinyl *neo*-decanoate
(R_1, R_2, and R_3 are alkyl groups totaling to C_8H_{19})

Vanaken et al. (2014) describe VAc/vinyl *neo*-decanoate latexes that are sufficiently weatherable to be candidates for roof coatings in Brazil. They propose an "umbrella effect" in which the bulky C-10 groups protect adjacent VAc residues from hydrolysis.

Emulsion copolymerization of VAc with vinyl *neo*-decanoate can be problematic because of very low water solubility of the latter. Satisfactory processes have been developed, and the mechanism of copolymerization is reasonably well understood (De Bruyn et al., 2002).

Webster and Crain (2002) and Prior et al. (1996) discuss the use of a variety of vinyl esters, including vinyl pivalate and vinyl 2-ethylhexanoate, in VAc copolymer latexes. Such monomers also yield polymers that are more hydrophobic than VAc homopolymers and that have superior hydrolytic stability and scrub resistance.

Copolymers of ethylene and VAc are made by free radical emulsion polymerization under high pressure. Choi et al. (2015) describe a typical process, which is somewhat involved. The copolymers are called both EVA and VAE latexes. Such polymers have several large volume non-coating applications and are finding increasing usage in coatings, especially interior flat wall paints. The proportions of ethylene and VAc are quite variable; ratios of 10:90 to 20:80 appear common. Paints made with EVA latexes form films readily without large amounts of coalescing solvent and are said to have excellent scrub resistance. Other comonomers are often included. Inclusion of a second hydrophobic comonomer may enhance water, block, and scrub resistance of the paints. An example of such a terpolymer latex is VAc/ethylene/vinyl *neo*-decanoate (Avramidis et al., 2001). Formulations having very low VOC are feasible.

9.4 THERMOSETTING LATEXES

While the majority of latexes are thermoplastic, there are applications for thermosetting latexes, as reviewed by Taylor and Winnik (2004). The review emphasizes experimental results related to the importance of the rate of interdiffusion and cross-linking in films of thermosetting latexes. In order to develop good properties, significant interdiffusion must occur prior to extensive cross-linking. Film properties depend on the relative rate of interdiffusion between polymer particles and the rate of cross-linking. If the rate of cross-linking is fast in comparison to the rate of diffusion, cross-linking within the particles will interfere with film formation. If the rate of cross-linking is too slow, achievement of desirable film properties may take excessive time. A balance between the two is needed. In some cases, the functional groups on a latex particle are on the particle surface and an external cross-linker is utilized to cross-link the particles. In other cases, the functional groups are on the same polymer and interdiffusion is required to bind the particles together. Thermosetting latexes are introduced in Section 2.3.3, and the cross-linking chemistries are described in Chapters 11–14 and 16–17.

9.4.1 One-Package Thermosetting Latex Coatings That Require Baking for Cure

Many chemistries can be adapted for this type of coating. In general, the rate of interdiffusion is controlled by $(T-T_g)$ and the chain length of the latex. Thus, thermosetting latexes are designed having lower MW, which both reduces T_g and shortens chain length. Straight chains diffuse more rapidly than branched chains. Taylor and Winnik (2004) describe a copolymer of isobutoxymethylacrylamide (IBMAA) with MMA and BA having 2% IBMAA and $\bar{M}_w = 200000$. The MW was reduced by using dodecyl mercaptan as a chain transfer agent. IBMAA groups react with each other by an acid catalyzed reaction to form –CO–NH–CH_2–O–CH_2–NH–CO– cross-links. It was found that interdiffusion increases more rapidly with temperature than does cross-linking. At 80°C with 0.5% pTSA, the cross-linking occurred in 40 min.

9.4.2 Two-Package (2 K) Thermosetting Latex Coatings That Do Not Require Baking

There are many examples in this category that, in most cases, are useful for industrial applications. In general, a lower T_g latex can be used in such coatings, which permits coalescence without addition of a coalescing solvent, and/or at lower film formation temperatures. The rates of physical film formation must be faster than the chemical cross-linking reactions, but the latter must be fast enough to be practical.

A variety of cross-linkers for hydroxy-functional and carboxylic acid-functional latexes can be used. Hydroxy-functional polymers are readily prepared by using hydroxyethyl (meth)acrylate as a comonomer. The resultant latexes can be formulated with urea-formaldehyde (UF) or melamine-formaldehyde (MF) resins as cross-linkers. UF resins tend to evolve higher levels of formaldehyde from hydrolysis than MF resins. MF resins are soluble in BA/HEMA/MAA latex polymers so that they act as plasticizers to promote interdiffusion (Taylor and Winnik, 2004). As temperature increases, interdiffusion rates increase faster than the rate of cross-linking until many of the potential reactions have occurred, but at some point the cross-link density (XLD) reaches a level where interdiffusion ceases. The rate of cross-linking can be increased by the addition of a catalyst such as pTSA. Although this permits shorter cure times, the pot life of the coating decreases and the residual acid in the final film can lead to reduced resistance of the films to hydrolysis since acid catalysts also catalyze hydrolysis.

An approach for assuring distribution of the MF resin throughout the polymer particles is to dissolve the MF resin in the monomer mixture before polymerization (Huang and Jones, 1996). Premature cross-linking is minimized by controlling pH at or above 5. Films cross-link by the addition of catalyst; and the pot life is 1–2 days.

Acetoacetate-functional (Section 17.6) latexes can be cross-linked with polyamines, but the short pot life adversely affects coalescence (Geurink et al., 1996).

Carbodiimides (Section 17.8) can be used as cross-linkers (Taylor and Bassett, 1997). A water-emulsifiable polycarbodiimide that is terminated with a mixture of the methyl ether of diethylene glycol and the methyl ether of a polyethylene glycol is used as a cross-linker for COOH-functional latexes. The design of the latex markedly affected properties. The best properties were obtained with a BA/MMA/MAA latex that was prepared with the core containing 2% MAA and the shell 1% MAA (Porzio et al., 2003). Suppliers claim recent improvements in this technology.

Another possibility is to combine two latexes having different functional groups. For example, a copolymer of *t*-butylcarbodiimidoethyl methacrylate and 2-ethylhexyl methacrylate (EHMA) was mixed with a copolymer of MA, EHMA, MAA. The carbodiimide group and COOH group react to form *N*-acylurea groups. This graft copolymer is said to increase the solubility of one latex in the other. The reaction occurs at ambient temperature but improved properties were obtained at 60°C, because inter-diffusion increases more rapidly with temperature than the cross-linking reaction (Taylor and Winnik, 2004).

Polyfunctional aziridines (Section 17.7), such as the propylene imine addition product with pentaerythritol triacrylate, serve as cross-linkers for carboxylic acid-functional latexes (Pollano, 1996). Pot lives of 48–72 h are reported. These cross-linkers are reliably effective and can be used in relatively small quantities because of their low equivalent weights. They are widely used despite possible toxic hazards.

Carboxylic acid-functional latexes can also be cross-linked with epoxysilanes, such as β-(3,4-epoxycyclohexyl) ethyltriethoxysilane (Chen et al., 1997). Addition of an emulsion of this cross-linker to the latex increased solvent resistance and hardness, especially when baked 10 min at 116°C. Latexes having high COOH content and equivalent amounts of epoxysilane gave the largest improvements. Package stability is reported to be at least 1 year. Cross-linking can be catalyzed—for example, using 1-(2-trimethylsilyl)propyl)-1*H*-imidazole.

Various water-dispersable polyisocyanates can be effective cross-linkers. For example, *m*-isopropenyl-α,α-dimethylbenzyl isocyanate (TMI) (Section 12.3.2) reacts slowly with water and can be used to prepare thermosetting latexes (Inaba et al., 1994).

Yet another way to prepare 2 K ambient temperature curing latexes is to formulate hybrids of thermoplastic latexes with thermosetting resins. Various chemistries can be employed. Fu et al. (2014) report mixing a low MW epoxy resin with an acrylic latex. The epoxy is absorbed into the latex particles. When a reactive water-soluble amine cross-linker is added and a film is cast, the epoxy initially acts as a coalescing solvent followed by cross-linking with the amine. If the acrylic has a low T_g, it acts as an impact modifier. Very good film properties and low VOC formulations are possible.

9.4.3 One-Package Thermosetting Latex Coatings That Do Not Require Baking

Development of practical technologies for thermosetting latexes that cross-link at room temperature and are storage stable for the long times required for architectural paints has been challenging. One successful approach is to use adipic dihydrazide (ADH) as a cross-linker for keto- or aldehyde-functional latexes (Section 17.11). One key to stability is virtually complete segregation of the reactive groups—ADH in the water phase and the aldehyde or ketone in the polymer particles. The alkaline pH of the latex medium may also contribute to stability by suppressing the acid catalyzed reaction. Kessel et al. (2008) described synthesis of a diacetone acrylamide (DAAM) functional monomer and extensively studied its cross-linking with ADH. Model compound studies by these authors indicated that the cross-links are imines. Zhang et al. (2012) and Pi et al. (2015) made further refinements by cross-linking core–shell latexes containing DAAM with ADH. Pi et al. concluded that fastest cure and highest XLD were achieved when DAAM was in the particle shells. Good adhesion to polyethylene films was also noted. The effect of DAAM location on long-term stability was not reported.

Another approach is to cross-link by autoxidation. Latexes having allylic substitution cross-link on exposure to air after application (Collins et al., 1997; Monaghan, 1997). Hybrid alkyd/acrylic latexes have been prepared by dissolving an oxidizing alkyd (Section 15.1) in the monomers used in emulsion polymerization, yielding a latex with alkyd grafted on the acrylic backbone (Nabuurs et al., 1996; Van Hamersfeld et al., 1999). Elrebii et al. (2015) prepared hybrid alkyd/acrylic latexes by melt condensation reactions of a COOH-functional acrylic with a soya fatty acid alkyd. The products can be dispersed in water with ammonia; no solvent is needed. The dispersions can be formulated into pigmented paints that cross-link by autoxidation.

Silane chemistry (Chapter 16) provides additional ways to address the problem. Stable thermosetting latexes can be prepared using triisobutoxysilylpropyl methacrylate as a comonomer (Chen et al., 1997). The isobutoxy derivative, in contrast to the ethoxy derivative, has sufficient hydrolytic stability to permit emulsion polymerization at 80–90°C without having unacceptable reaction with water. The resultant latex has a package stability of more than a year, yet cross-links after application in a week using organotin catalysts. This unusual combination of storage stability and reactivity may signify that the trialkoxysilyl groups hydrolyze during storage, but in the presence of large amounts of water, do not cross-link until the water evaporates after the coating is applied. Guyer and Gai (2006) described acrylic latexes containing vinyltriethoxylsilane as a comonomer that provided films having increased solvent resistance and exterior durability, as well as improved heat and flow resistance, relative to a similar latex without the silane comonomeer. The silane modified latexes were also reported to be package stable if buffered with $NaCO_3$. Bergman et al. (2016) synthesized latexes incorporating three (meth)acrylate monomers that provided improved adhesion and film properties; but long-term stability of the products was not reported.

Miniemulsion polymerization (Section 9.1.2.1) appears to be an attractive route to package stable cross-linkable latexes because water-sensitive monomers can be encapsulated in the dispersed particles with minimal exposure to water. This possibility has been suggested by several authors (Marcu et al., 2003; Rodriguez et al., 2008) but does not appear to have been fully exploited.

Very fast-drying latex traffic striping paints are described in Section 33.4.

GENERAL REFERENCES

El-Aasser, M. S.; Lovell, P. A., Eds., *Emulsion Polymerization and Emulsion Polymers, John* Wiley & Sons, Inc., New York, **1997**.
Gilbert, R. O., *Emulsion Polymerization, A Mechanistic Approach*, Academic Press, New York, **1995**.
van Herk, A. M., *Chemistry and Technology of Emulsion Polymerization*, John Wiley & Sons, Inc., New York, **2013**.

REFERENCES

Alahapperuma, K.; Glass, J. E., *Prog. Org. Coat.*, **1992**, *21*, 53.
Anderson, L. L.; Brouwer, W. M., *J. Coat. Technol.*, **1996**, *58(855)*, 75.
Anonymous, Emulsion Polymerization of Acrylic Monomers, *Tech, Bull.* CM-104 A/cf, Rohm & Haas Co. (Now Dow Chemical Co. Midland, MI) Philadelphia, undated.
Arai, K., et al., *Prog. Org. Coat.* **2016**, *93*, 119–117.
Arzamendi, G.; Asua, J. M., *J. Appl. Polym Sci.*, **1989**, *38*, 2019.
Aslamazova, R., *Prog. Org. Coat.*, **1995**, *25*, 109.
Avramidis, K. S., et al., US patent 6,339,447 (**2001**).
Bassett, D. R., *J. Coat. Technol.*, **2001**, *73(912)*, 43.
Bergman, S. D., et al., *Proceedings of the American Coatings Conference*, Indianapolis, IN, Paper 16.3, **2016**.
Chen, M. J., et al., *J. Coat. Technol.*, **1997**, *69(875)*, 49.
Choi, H. C.; Kim, J. B.; Lee, W. J., Patent publication WO2015147352 A1 (**2015**).
Chou, C.-S., et al., *J. Coat. Technol.*, **1987**, *59(755)*, 93.
Chow, P. Y.; Gan, L. M., in Okubo, M., Ed., *Miniemulsion Polymerization*, Advances in Polymer Science 175, **2005**, pp 257–298.
Collins, M. J., et al., *Polym. Mater. Sci. Eng.*, **1997**, *76*, 172.
Craig, D. R., *J. Coat. Technol.*, **1989**, *61(779)*, 49.
De Bruyn, H., et al., *Macromolecules*, **2002**, *35*, 8371.
Eckersley, S. A.; Helmer, B. J., *J. Coat. Technol.*, **1997**, *69(864)*, 97.
El-Aasser, M. S., et al., *J. Polym. Sci.: Polym. Chem. Ed.*, **1983**, *21*, 2363.
El-Aasser, M., et al., *J. Coat. Technol.* **2001**, *73(920)*, 51.
Elrebii, M.; Kamoun, A.; Boufi, S., *Prog. Org. Coat.*, **2015**, *87*, 222–231.
Emelie, B., et al., *Makromol. Chem.*, **1988**, *189*, 1879.
Fernandez, A. M., et al., *Prog. Org. Coat.*, **2005**. *53*, 246.
Fu, Z. et al., *Proceedings of the American Coatings Conference*, Atlanta, GA, Paper 5.1, **2014**.
Gardon, J. L., A perspective on Resins for Aqueous Coatings in Glass, J. E., Ed., *Technology for Waterborne Coatings*, ACS Symposium Series 663, American Chemical Society, Washington, DC, **1997**, p 27.
Geurink, P. J. A., et al., *Prog. Org. Coat.*, **1996**, *27*, 73.
Grade, J., et al., *Proceedings of the Waterborne High-Solids Powder Coatings Symposium*, New Orleans, LA, **2002**, pp 145–157.
Guyer, K. L.; Gai, W. J., *Proceedings of the Waterborne High-Solids Powder Coatings Symposium*, New Orleans, LA, **2006**, pp 85–96.
Heldmann, C., et al., *Prog. Org. Coat.*, **1999**, *35*, 69.
Holmberg, K., *Prog. Org. Coat.*, **1992**, *20*, 325.
Hoy, K. L., *J. Coat. Technol.*, **1979**, *51(651)*, 27.
Huang, Y.; Jones, F. N., *Prog. Org. Coat.*, **1996**, *28*, 133.
Inaba, Y., et al., *J. Coat. Technol.*, **1994**, *66(833)*, 63.
Jones, F. N., et al., *Prog. Org. Coat*, **2005**, *52*, 9.
Kessel, N.; Illsley, D. R.; Keddie, J. L., *J. Coat. Technol. Res.*, **2008**, *5*, 285–297.
Learner, T., *Stud. Conserv.*, **2001**, *46*, 225–241.
Lorah, D. P., et al., *Proceedings of the Waterborne High-Solids Powder Coatings Symposium*, New Orleans, LA, **2004**, Paper No. 19.
Lynn, Jr., J. L.; Bory, B. H., in Kirk, R. E.; Othmer, D. F., Eds., *Encyclopedia of Chemical Technology*, 4th ed., Vol. *23*, John Wiley & Sons, Inc., New York, **1997**, pp 478–541.
Marcu, I., et al., *Macromolecules*, **2003**, *36(2)*, 328–332.
Mercurio, A., et al., *J. Oil Colour Chem. Assoc.*, **1982**, *65*, 227.
Misra, S. C., et al., *J. Poly. Sci.; Polym. Chem. Ed.*, **1983**, *21*, 2383.
Mittal, K., *Miniemulsion Polymerization Technology*, John Wiley & Sons, New York, **2010**.
Monaghan, G., *Polym. Mater. Sci. Eng.*, **1997**. *76*, 178.
Nabuurs, T., et al., *Prog. Org. Coat.* **1996**, *27*, 163.
Nair, M., *Prog. Org. Coat.*, **1992**, *20*, 53.
Nguyen, B. D.; Rudin, A., *J. Coat. Technol.*, **1986**, *58(736)*, 53.
Oprea, S.; Dodita, T., *Prog. Org. Coat.*, **2001**, *42*, 194.
Padget, J. C., *J. Coat. Technol.*, **1994**, *66(839)*, 89.
Pi, P. et al., *Prog. Org. Coat.*, **2015**, *81*, 66–71.
Pollano, G., *Polym. Mater. Sci. Eng.*, **1996**, *77*, 73.
Porzio, R. S., et al., *Proceedings of the Waterborne High-Solids Powder Coatings Symposium*, New Orleans, LA, **2003**, pp 129–143.
Prior, R. A., et al., *Prog. Org. Coat.*, **1996**, *29*, 209.
Rodriguez, R., et al., *Macromolecules*, **2008**, *41*(22), 8537–8546.

Rothmann, H., et al., *Proceedings of the Waterborne High-Solids Powder Coatings Symposium*, New Orleans, LA, **2001**, pp 465–479.
Santos, J. P., et al., *Paint & Coatings Industry*, **2015**, No. 6.
Schork, F. J., et al., Miniemulsion Polymerization in Okubo, M., Ed., *Miniemulsion Polymerization*, Advances in Polymer Science, 175, **2005**, pp 129–258.
Sudol, E. D.; El-Aasser, M. S., Miniemulsion Polymerization in Lovell, P. A.; El-Aasser, M. S., Eds., *Emulsion Polymerization and Emulsion Polymers*, John Wiley & Sons, West Sussex, **1997**, p 699.
Tamer, Y.; Yamak, H. B.; Yildirim, H., *J. Surfactant. Deterg.*, **2016**, *19*(*2*), 405–412.
Taylor, J. W.; Bassett, D. W., in Glass, J. E., Ed., *Technology for Waterborne Coatings*, American Chemical Society, Washington, DC, **1997**, p 137.
Taylor, J. W.; Winnik, M. A., *JCT Res.*, **2004**, *1*(*3*), 163.
Vanaken, D., et al., *Proceedings of the American Coatings Conference*, **2014**, Atlanta, GA, Paper 9.2.
Vandezande, G. A.; Rudin, A., *J. Coat. Technol.*, **1994**, *66*(*828*), 99.
Vandezande, G. A., et al., Vinyl Acetate Polymerization in Lovell, P. A.; El-Aasser, M. S., Eds., *Emulsion Polymerization and Emulsion Polymers*, John Wiley & Sons, Inc., New York, **1997**, pp 563–587.
Van Hamersfeld, E. M. S., et al., *Prog. Org. Coat.*, **1999**, *35*, 235.
van Herk, A. M., *Chemistry and Technology of Emulsion Polymerization*, 2nd ed., John Wiley & Sons, Inc., New York, **2013**.
Webster, D. C.; Crain, A. L., *Prog. Org. Coat*, **2002**, *45*, 43.
Winnik, M. A.; Feng, J., *J. Coat. Technol.*, **1996**, *68*(*852*), 39.
Yang, H. S., et al., *J. Coat. Technol.*, **2005**, *2*(*13*), 44.
Zhang, X., et al., *J. Appl. Polym. Sci.*, **2012**, *123*(*3*), 1822–1832.

Chapter 10

Polyester Resins

In the fields of fibers and plastics, *polyester* refers to high molecular weight (MW), partially crystalline, linear, thermoplastic polymeric esters of a short-chain diol and terephthalic acid (TPA). In coatings, however, polyester describes a different sort of material. Most coating polyesters have relatively low MWs and are amorphous and branched; and they must be cross-linked to form useful coating films. Furthermore, the term is applied only to certain polyesters, that is, those prepared from polyols and polybasic acids. Alkyd resins are also polyesters in the chemical sense of the word, but are not called polyesters in coatings. As discussed in Chapter 15, alkyds are prepared from polyols and dibasic acids, together with monobasic acids, usually derived from vegetable oils. Polyesters for coatings are sometimes called *oil-free polyesters* or *oil-free alkyds* to distinguish them from alkyds. They may also be called *saturated polyester resins* to distinguish them from the unsaturated resins commonly used in reinforced plastics and occasionally in coatings (Section 17.3).

Polyesters are made by step-growth polymerization; the general principles are discussed in Section 2.2.2. The relatively low MW polyesters for coatings are usually prepared from mixtures of diols, triols, and dibasic acids. Most commonly, the polyesters are hydroxy-terminated, requiring that excess polyol is used. *Hydroxy-terminated polyesters* are most commonly cross-linked with melamine-formaldehyde (MF) resins (Chapter 11) or polyisocyanates (Chapter 12). *Carboxylic acid-terminated polyesters* are also prepared and, generally, cross-linked with epoxy resins (Chapter 13), MF resins, or 2-hydroxyalkylamides (Chapter 17). *Water-reducible polyesters, having* both terminal hydroxyl and carboxylic acid groups, are also synthesized and are usually cross-linked with MF resins (Section 10.5).

Branched polyesters are made from monomer mixtures that include one or more monomers having functionality $F>2$. Kinetic analysis of such polymerizations is complex, but a few generalizations can be pointed out. As the proportion of monomer having $F>2$ increases, $\bar{M}_n$, $\bar{M}_w/\bar{M}_n$, and number average functionality $\bar{f}_n$ all increase; and the average $\bar{F}$ must be controlled to avoid gelation at high extents of reaction.

In general terms, better adhesion to metal substrates and better impact resistance are obtained with coatings based on thermosetting polyesters than on thermosetting acrylics (TSAs) (Section 8.2). On the other hand, coatings, based on TSAs, have superior water resistance and exterior durability when compared with conventional polyesters. In general, the differences can be attributed to the presence of the ester linkages in the backbone of polyesters—linkages that impart flexibility and promote adhesion but are vulnerable to hydrolysis. Accordingly, polyesters tend to be used for one-coat coatings on metal, whereas TSAs tend to be used in applications for which exterior durability and moisture resistance are particularly important, often with primers that assure excellent adhesion of the coating system to the metal. The superior exterior durability of TSAs results partly from the greater effectiveness of UV stabilizers in acrylics than in polyesters. Many exceptions to these generalizations can be found (Section 5.2.1). Coatings made with *superdurable polyester resins* (next section) are said to approach or even equal the exterior durability of those made with acrylics (Pilcher, 2012).

In keeping with a growing trend, polyester resins are increasingly blended with other generic resin types, often with hydroxy-functional acrylics. At best, the blends combine the attractive properties of the two types of resins. As MW of resins is reduced for high solids coatings (Section 10.2), the likelihood of finding compatible combinations increases. Blends of acrylics with "superdurable"

Organic Coatings: Science and Technology, Fourth Edition. Frank N. Jones, Mark E. Nichols, and Socrates Peter Pappas.

polyesters have been proposed and probably adopted for demanding applications such as automotive clear coats.

10.1 HYDROXY-TERMINATED POLYESTER RESINS FOR CONVENTIONAL SOLIDS COATINGS

Most hydroxy-terminated polyesters are made by co-esterifying four types of monomers: two polyols (a diol and a triol) and two diacids (an aliphatic dibasic acid or its anhydride and an aromatic dicarboxylic acid or its anhydride). Tribasic and cycloaliphatic acids are used to some extent. The ratio of moles of dibasic acid to polyol must be less than 1 to provide terminal hydroxyl groups and avoid gelation. MW is controlled by this ratio; the smaller the ratio, the lower the MW. The MW distribution, $\bar{M}_n$, and $\bar{f}_n$ are both controlled by the diol to triol ratio; $\bar{f}_n$ is critical because it affects the potential cross-link density after cure. With the same mole ratio of dibasic acid to total polyol, an increase in triol component increases the average number of hydroxyl groups per molecule, decreases hydroxy equivalent weight, broadens MW distribution, and increases cross-link density of a fully cross-linked film. The ratio of aromatic to aliphatic dibasic acids is the principal factor controlling the T_g of the resin; but differences in polyol structure also affect T_g. Mixtures of polyols and diacids also reduce the tendency of the resins to crystallize.

In commercial processes, mixtures of the monomers are esterified at 220–240°C while water is removed. Catalysts include organotin compounds, titanium orthoesters, and zinc acetate. Strong protonic acids also catalyze esterification but can cause side reactions and discoloration. The esterification is continued until a high conversion has been reached. Acid numbers (Section 8.3) of polyester resins are generally 5–10 for cross-linking with MF resins and less than 2 for cross-linking with polyisocyanates. Polyesters made from polyfunctional reactants can cross-link during polymerization, causing gelation. Since the probability of gelation increases with conversion and high conversions are generally desirable, an equation relating formula composition, extent of reaction, and probability of gelation would be useful. Progress has been made in predicting gelation in simple systems in which the reactivity of all hydroxyl groups and carboxylic acid groups is equal; but even with these "simple" systems, the equations have modest predictive value, because they do not take into consideration the formation of cyclic esters. In the case of polyesters used in coatings, variable reactivity of the functional groups and side reactions further complicate the situation. Textbook gelation equations are of even less value, as discussed by Odian (2004). If the mole ratio of dibasic acid to polyol exceeds, or even closely approaches 1, gelation may occur before esterification is essentially complete. This is seldom a problem in making hydroxy-functional polyesters for coatings, but it can become critical in the synthesis of alkyds (Section 15.1).

Misev (1989) provided equations that permit calculation of ratios of polyols and polybasic acids for preparing polyesters having a desired $\bar{M}_n$ and $\bar{f}_n$ at a desired extent of reaction of the acid groups. The equations are complex and require iterative computer solution but can be simplified by limiting the choice of raw materials, which is generally done, to diols, triols, and dibasic acids, and if essentially all of the acid groups are esterified. The latter assumption is reasonable, since, in commercial practice, the polyesters are cooked until about 98–99.5% of the carboxylic acid groups are esterified. In these simplified equations N_{p2} represents the number of moles of diol; N_{p3}, the number of moles of triol; and N_a, the number of moles of dibasic acid. M_{p2} represents the MW of the diol; M_{p3}, the triol; and M_a, the dibasic acid. If more than one dibasic acid (or other component) is used, then additional terms would be added to the equation for each additional component. Using these equations, one can calculate the amounts of diol, triol, and dibasic acid that should be used in an initial laboratory cook to obtain a desired $\bar{M}_n$ and $\bar{f}_n$.

In making hydroxy-terminated polyesters, the number of moles of polyol is one more than that of dibasic acid, as shown in Eqs. 10.1 and 10.2:

$$N_{p2} + N_{p3} = N_a + 1 \text{ or} \tag{10.1}$$

$$N_a = N_{p2} + N_{p3} - 1 \tag{10.2}$$

The relationship between the average functionality and the moles of reactants is expressed by Eq. 10.3:

$$\bar{f}_n = 2N_{p2} + 3N_{p3} - 2N_a \tag{10.3}$$

Substituting the value of N_a from Eq. 10.2 into Eq. 10.3 gives Eq. 10.4, which can be rearranged to Eq. 10.5:

$$\bar{f}_n = N_{p3} + 2, \text{or} \tag{10.4}$$

$$N_{p3} = \bar{f}_n - 2 \tag{10.5}$$

During esterification, 2 mol of water (total MW = 36) is eliminated for each mole of dibasic acid, resulting in Eq. 10.6 for $\bar{M}_n$. If an anhydride is used, only 1 mol of water (M = 18) is eliminated for each mole of anhydride, and the appropriate adjustment in Eq. 10.6 is needed:

$$\bar{M}_n = N_{p2}M_{p2} + N_{p3}M_{p3} + N_a\left(M_a - 36\right) \tag{10.6}$$

Substituting N_a and N_{p3} in Eq. 10.6 with their corresponding values from Eqs. 10.2 and 10.5, respectively, gives Eq. 10.7, which relates $\bar{M}_n$ to $\bar{f}_n$:

$$\bar{M}_n = N_{p2}M_{p2} + (\bar{f}_n - 2)M_{p3} + (N_{p2} + \bar{f}_n - 3)(M_a - 36) \quad (10.7)$$

A wide range of polyesters has been used commercially. In conventional polyester–MF coatings, polyesters having $\bar{M}_n$ on the order of 2000–6000, $\bar{M}_w/\bar{M}_n$ on the order of 2.5–4, and $\bar{f}_n$ of 4–10 hydroxyl groups per molecule are usual. Table 10.1 gives an example of an initial laboratory formulation calculated for a polyester having an $\bar{M}_n$ of 5000 and an $\bar{f}_n$ of 5 using Eqs. 10.1–10.7. Neopentyl glycol is NPG, trimethylolpropane is TMP, isophthalic acid is IPA, and adipic acid is AA. Structural formulas for these compounds are given in Sections 10.1.1 and 10.1.2.

A formulation such as that in Table 10.1 does not take into consideration all of the complexities of polyesterification; thus, it can only be expected to provide a starting point for experimentation. A common complication is partial loss of monomer from the reactor by volatilization. The diol is usually the most volatile component, so it is generally necessary to use excess diol. The amount of excess diol depends upon the particular reactor and conditions, the efficiency of separating water and diol, the rate of flow of inert gas, the temperature of reaction, and so forth. A formula established in the laboratory usually has to be adjusted when production is scaled up in a pilot plant, adjusted again when it is scaled up in a production reactor, and adjusted yet again when production is shifted from one reactor to another.

Esterification is a reversible reaction, and the formation of high yields of polyester requires removal of water. A few percent of a reflux solvent, such as xylene, may be added to accelerate removal of water and to assist in return of volatilized diol to the reactor. Alternatively, an inert gas purge can be used in the later stages of the process to help remove the last of the water. Ester groups that form during the process may hydrolyze or transesterify and re-form many times, leading to a mixture of kinetic and thermodynamic control in the formation of the final products. Another complication is that polyols can undergo self-condensation to form polyethers, which produces water and reduces the number of hydroxyl groups available for esterification.

Table 10.1 Starting Formulation for a Conventional Polyester Resin

Raw material	Weight (g)	Moles	Equivalents
NPG	2050	19	38
TMP	270	3	9
IPA	1740	10.5	21
AA	1530	10.5	21

10.1.1 Selection of Polyols

Polyols are selected on the basis of cost, rate of esterification, stability during high temperature processing (minimal decomposition and discoloration), ease of separation from water during processing, viscosity of esters at equal MW and functionality, effect on T_g, rate of cross-linking with MF resins (or other cross-linkers), and hydrolytic stability of their esters. Obviously, compromises are necessary. The following polyols are representative of the many available:

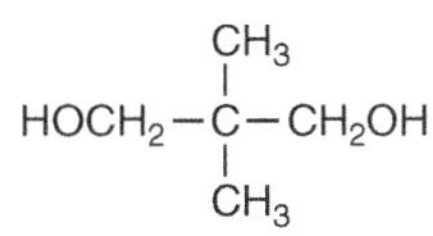

Neopentyl glycol (NPG)

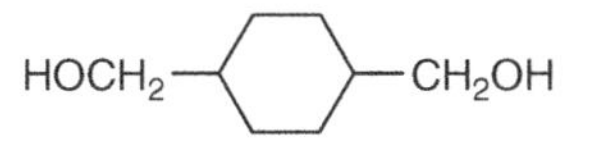

Cyclohexanedimethanol (CHDM)

Hexanediol

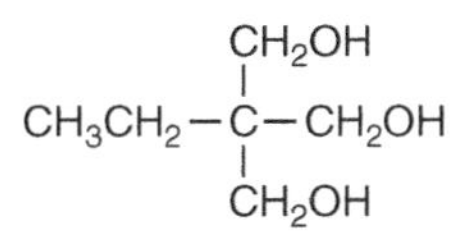

Trimethylolpropane (TMP)

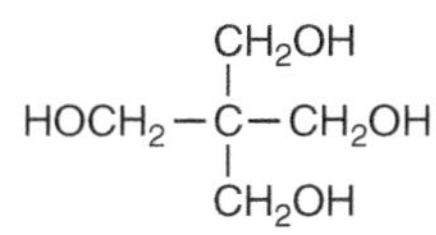

Pentaerythritol (PE)

The most widely used diol is probably NPG; and the most widely used triol is TMP. Cyclohexanedimethanol (CHDM) esterifies significantly more rapidly than NPG, which esterifies more rapidly than 2,2,4-trimethyl-1,3-pentanediol (TMPD) (Van Sickle et al., 1997). Generally, but not always, the slower the rate of esterification, the greater the resistance of the ester to hydrolysis. The hydrolytic stability of esters of NPG and TMP is better than those from less hindered glycols, such as ethylene glycol or propylene glycol. NPG (bp 213°C) is volatile at esterification temperatures, so well-designed fractionating and condensing equipment are needed to permit the removal of water with a minimum loss of NPG during processing at 220–240°C. Most diols start decomposing in the presence of strong acids at temperatures above about 200°C, so strong acid esterification catalysts should be avoided. Organotin compounds and titanium orthoesters are examples of appropriate catalysts.

While the kinetics and mechanisms of esterification and hydrolysis reactions have been extensively studied from a fundamental point of view, the literature contains relatively little basic data dealing with polyols of most interest in coatings resins. Turpin (1975) proposed an empirical *steric factor* based on Newman's "rule of six," which may be useful in predicting the relative hydrolytic stability of esters. The steric factor is based on the number of substituents in the 6 and 7 positions of an ester starting with the carbonyl oxygen as 1:

$$\text{Steric factor} = 4(\text{No. 6 position atoms}) + (\text{No. 7 position atoms})$$

For example, an NPG ester has a steric factor of 21, as shown.

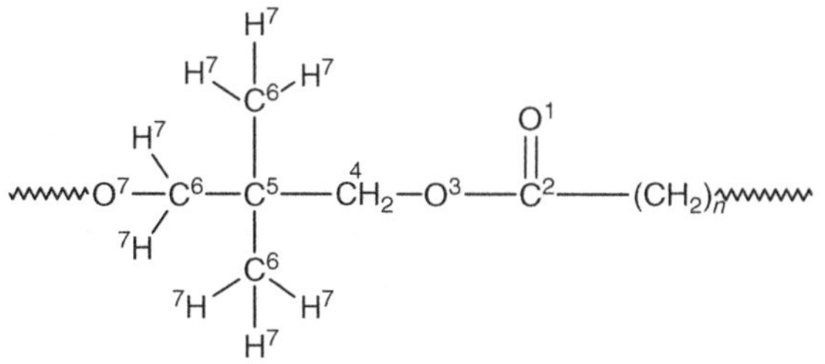

3 No. 6 atoms x 4 = 12
9 No. 7 atoms = 9
Total = 21

The value of 21 for NPG esters is the same for TMP esters, but for ethylene glycol and 1,6-hexanediol esters, the steric factors are only 13 and 15, respectively. Turpin's steric factor seems to correlate reasonably well with experience and with the limited published data for esters of acyclic alcohols. Turpin also pointed out that an *anchimeric*, or *neighboring group*, effect may be involved. For example, the location of the hydroxyl group in a half ester of ethylene glycol may promote the hydrolysis of its neighboring ester through a cyclic transition state and contribute to the poorer hydrolytic stability of ethylene glycol esters as compared with those of 1,6-hexanediol (Turpin, 1975). TMPD is also frequently used. The corresponding polyesters have good exterior durability. The ability of the ester groups in coating films to resist hydrolysis (as well as UV) is a crucial factor in exterior durability. Esters of cycloaliphatic diols, such as CHDM, seem more resistant to hydrolysis than predicted by Turpin's steric factor. Golob et al. (1990) tested films of coatings made with a series of polyesters from several polyols. Coatings based on CHDM polyesters, cross-linked with MF resins, showed the best test results with respect to properties related to hydrolytic stability. In contrast to general behavior, CHDM also esterifies more rapidly than NPG (Van Sickle et al., 1997). A liquid 1 : 1 mixture of (*cis, trans*) 1,3-cyclohexanedimethanol and (*cis, trans*) 1,4-cyclohexanedimethanol also provides good overall coating properties (Argyropoulos et al., 2003). The hydrophobicity of the cyclohexane ring may contribute to enhanced hydrolytic resistance of CHDM.

Linear or branched polyols can be produced by ring-opening reaction of caprolactone with diols or triols and subsequent polymerization (Union Carbide Corp., 1986). A variety of such materials is available.

The search for polyesters that have superior hydrolysis resistance (for outdoor durability and as alternatives for BPA in epoxy resins) has generated interest in other polyols. One promising candidate is 2,2,4,4-tetramethyl-1,3-cyclobutanediol (TMCD). Feng (2016) showed that polyesters based on this monomer, cross-linked with MFs and PUs, can give exceptional results in xenon arc and other accelerated weathering tests. While Feng does not identify specific resin compositions, Marsh et al. (2013) provide examples of compositions.

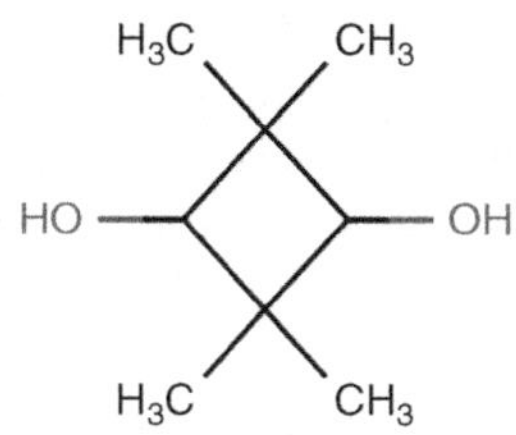

TMCD

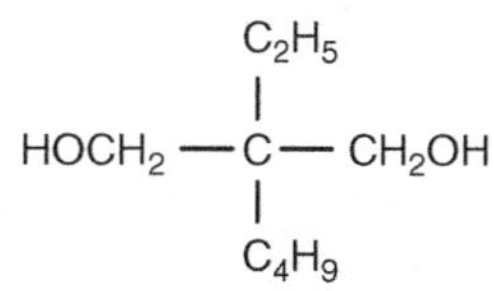

BEPD

Another diol associated with excellent outdoor durability is 2-butyl-2-ethyl-1,3-propanediol (BEPD). It is speculated that the alkyl groups provide both hydrophobicity and steric hindrance of the hydroxyl groups, which reduce vulnerability to hydrolysis relative to NPG. Applications of BEPD are described in Sections 10.5 and 10.6.

10.1.2 Selection of Polyacids

Most polyesters for coatings are made using a mixture of aromatic and aliphatic diacids. The ratio of aromatic to aliphatic acid is a major factor that controls the T_g of the resins and properties of the ultimate coatings.

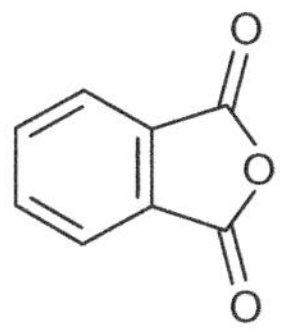

Phthalic anhydride (PA)

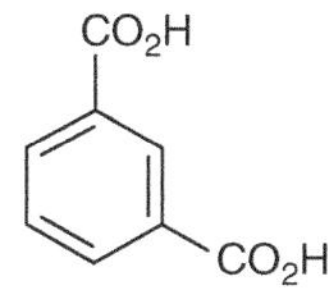

Isophthalic acid (IPA)

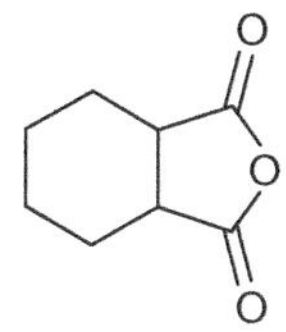

Hexahydrophthalic anhydride (HHPA)

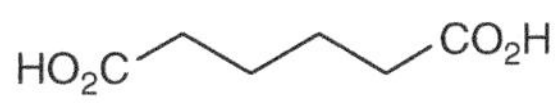

Adipic acid (AA)

Azelaic acid (AzA)

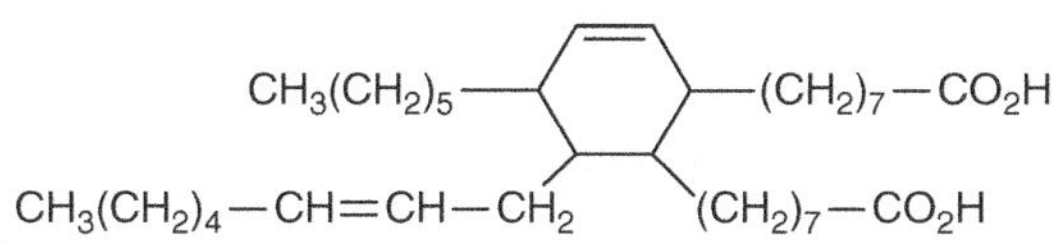

Dimer fatty acid (one of isomer mixture)

Aromatic acid esters hydrolyze more slowly than aliphatic esters unless there is an anchimeric effect. A processing cost advantage of phthalic anhydride (PA) results from its relatively low melting point (131 °C), allowing it to be handled as a molten liquid. PA is readily soluble in the reaction mixture, and its reactive anhydride structure rapidly forms monoesters at about 160°C. IPA (mp > 300°C) is more difficult to process because it dissolves more slowly in the reaction medium and has lower reactivity. Due to longer time at high temperatures, volatilization losses and acid-catalyzed side reactions (e.g., etherification or dehydration of polyols) are more serious problems with IPA than with PA (Section 15.5.2).

Nevertheless, IPA is the predominant aromatic acid used (for polyesters but not for alkyds) even though PA can be processed at lower cost. The preference for IPA is based on superior exterior durability of coatings made with IPA polyesters, which has been attributed to greater hydrolytic stability of isophthalic esters. Half esters of (ortho)phthalic acid are more easily hydrolyzed than half esters of IPA in the pH range of 4–8 (Bender et al., 1958). This difference results from anchimeric-assisted hydrolysis by the ortho-carboxylic acid group. During exterior exposure, resistance to hydrolysis in contact with rain water, which has a pH in the range of 4–6, is most important. Under these conditions, IPA polyesters are more resistant than phthalic acid polyesters. On the other hand, mono- and diesters of IPA hydrolyze more rapidly than those of phthalic acid under alkaline conditions (Cambon and Jullien, 1973; Jones and McCarthy, 1995).

Patents are readily available and free on the Internet. They can be excellent sources of information about a particular technology. *Disclosures* in patents often furnish perspective on the status of the technology and its shortcomings. *Examples* provide specifics on implementation. For example, Marsh et al. (2013) provide a detailed description of a procedure for preparing conventional polyester resins from NPG, TMP, IPA, and AA. The reaction mixture is heated at 160°C to dissolve the IPA and then at 230°C to drive esterification toward completion. Half of the TMP is withheld until late in the process. Butylstannoic acid is the catalyst. The patent continues with examples that describe the invention, which is highly weatherable polyesters based on TMCD (Section 10.1). US patent law requires that the best method of practicing the invention be disclosed at the time of filing. Of course, more advanced methods may have been found after filing.

AA is probably the most widely used aliphatic dibasic acid. Succinic and glutaric acids form esters having inferior hydrolytic stability, presumably due to the anchimeric effect (Turpin, 1975). Longer-chain acids such as azelaic and sebacic acids may provide somewhat better hydrolytic stability and greater reduction of T_g (on a molar basis) as compared with AA but are more expensive. Dimer acids, which are predominantly C_{36} diacids derived from dimerization of drying oil fatty acids, are also widely used. Fatty acid dimerization is discussed in Section 14.3.1. Dimer acids are relatively inexpensive and are available in various grades. The high quality grades of dimer acids normally used in polyesters have low levels of monobasic or tribasic acid contaminants and have been hydrogenated to saturate the double bonds in order to reduce oxidative degradation. Polyesters, prepared from mixtures of dimer acid and IPA, impart an excellent balance of properties to MF cross-linked coil coatings, including high resistance to marring combined with excellent resistance to cracking during fabrication of coated metal.

Cycloaliphatic dibasic acids or anhydrides can also be used. Hexahydrophthalic anhydride (HHPA) may provide exterior durability comparable with IPA. Aside from being aliphatic, HHPA half esters may be less susceptible to anchimeric-assisted hydrolysis than are phthalic acid half esters possibly owing to isomerization to the *E* (*trans*) isomers during processing. 1,4-Cyclohexanedicarboxylic acid (CHDA) provides a good balance of hardness, flexibility, and stain resistance. CHDA esters are much more stable to hydrolysis at pH 8–9 than IPA esters, but somewhat less stable to hydrolysis at pH 4–5 (Cambon and Jullien, 1973; Jones and McCarthy, 1995).

A statistically designed evaluation of the factors affecting the exterior durability of IPA/AA polyester–MF coatings showed that the major factor was the IPA/AA ratio. The polyols were NPG and TMP. Higher IPA content polyesters provided coatings with improved durability (Heidt et al., 2000).

10.2 POLYESTER RESINS FOR HIGH SOLIDS COATINGS

Low viscosity polyesters are required for reduction of VOC emissions; viscosity of concentrated polyester solutions depends on several variables. MW and MW distribution are two important factors. The number of functional groups per molecule also affects the viscosity; increasing the number of hydroxyl groups (and to an even greater extent, the number of carboxylic acid groups) increases the probability of intermolecular hydrogen bonding, in turn increasing viscosity. The hydrogen-bond effect can be minimized, but not eliminated, by using hydrogen-bond acceptor solvents such as ketones. The utilization of higher T_g resins also increases viscosity at a given concentration. For this reason, reducing the ratio of aromatic to aliphatic dibasic acids and use of acyclic polyols rather than cyclic polyols such as CHDM (Golob et al., 1990) result in lower viscosities; but reduction of T_g must usually be compensated by increasing cross-link density, and there is a lower limit of T_g below which desired film properties cannot be attained.

Polyesters have significant advantages over TSAs in high solids coatings. In contrast to the difficulty of making low MW TSAs having at least two hydroxyl groups on essentially all molecules, the synthesis of polyesters with two or more hydroxyl groups on essentially all of the molecules is relatively easy. With the low mole ratios of dibasic acids to polyols (2 : 3 is typical) required to make low MW resins, the probability of cyclization reactions to form nonfunctional materials is small. Virtually all of the final reactions result in terminal hydroxyl groups. While lowest viscosity is obtained with difunctional resins, such resins also require the most careful formulation and curing to achieve good film properties. If a hydroxyl is left unreacted on a conventional polyester having an average of 5 or so hydroxyl groups per molecule, any adverse effect on film properties would probably be relatively small. However, failing to react all of the hydroxyl groups on a resin with only two hydroxyl groups per molecule would magnify the effect. A compromise is required; commonly, an $\bar{f}_n$ of between two and three hydroxyl groups per molecule is used. Inclusion of trifunctional polyol increases f_n.

The major factor controlling M_n of a polyester is the mole ratio of dibasic acid to polyol. A ratio on the order of 2 : 3 is typical. Loss of polyol during production of a polyester results in a higher M_n than predicted from the starting ratio. Condensers are designed to remove water rapidly while returning a high fraction of NPG (or other polyol), but none is completely efficient. It is necessary to add some extra glycol to compensate for losses. Since production equipment and conditions vary, the final decision on the mole ratio to load must be based on experience with the particular reactor and with the particular formula. The amount of glycol being lost can be estimated by checking the refractive index of the water being removed from the reaction. In addition to evaporative loss of polyols, hydroxyl groups can also be consumed by etherification. Selected organotin and organotitanate catalysts minimize this problem.

Some low MW components of a resin may volatilize when a thin film is baked. Such loss has to be counted as part of the VOC emissions, together with solvent loss. Belote and Blount (1981) studied this volatility both in the absence of cross-linking agent and in its presence. They made a series of model resins from NPG and a 1 : 1 mole mixture of AA and IPA; the polyacid to polyol ratio was varied from 1 : 2 to 1 : 1.15. The presence of the cross-linker (class I MF resin) reduced volatile loss. They concluded that polyester resins having an $\bar{M}_n$ of about 800–1000 was optimum for achieving lowest VOC for resins having normal MW distributions. At $\bar{M}_n$ of about 600, so much low MW fraction volatilized that total VOC was higher than at $\bar{M}_n$ of 800, *even though the amount of solvent required was lower*. A broad MW distribution not only increases volatile losses because of the low MW material but also gives a higher viscosity because of the disproportionate effect on viscosity by the high MW fraction. Thus, lowest VOC results from having as narrow MW distribution as possible.

Unsymmetrical diols provide a means to reduce viscosity. An example is 2,2,4-trimethyl-1,3-pentanediol (TMPD), which has one primary hydroxyl group and one sterically hindered secondary hydroxyl group. Use of this diol in combination with a two-stage addition of the triol TMP during synthesis of a polyester gives lower viscosity and improved film properties (Hood et al., 1986). The unsymmetrical diol presumably narrows MW distribution because of the differential reactivity of the two hydroxyl groups. A drawback is the need for somewhat higher baking temperatures for coatings, presumably because much

Table 10.2 High Solids Polyester Formulations (Equivalents)

Raw material	Resin 1	Resin 2	Resin 3
TMPD	11.96	11.96	12.24
TMP	1.72	0.86	0.88
IPA	4.56	4.56	4.38
AA	4.56	4.56	4.38
Second stage			
TMP	—	0.86	0.88
Resin $\bar{M}_n$	1500	1075	1000
Coating NVW	75.6	76.5	77.6

of the hydroxyl functionality of the resin is the less reactive hydroxyl group (Table 10.2).

Very low $\bar{M}_n$ oligomeric diols are often used as *reactive diluents*. A polyester diol with a narrow MW distribution and a reported $\bar{M}_n$ of 425 is an example (Calbo, 1986). This resin is presumably manufactured by preparing a polyester with excess CHDM and then distilling off the low MW fraction with a vacuum thin-film evaporator. Low viscosity polyester diols and triols prepared by reacting caprolactone with a diol or a combination of diol and triol are also available. Very low $\bar{M}_n$ resins seldom provide satisfactory coatings when used by themselves, but they are useful in blends to increase the solids content of somewhat higher MW polyester or acrylic based coatings.

Matrix-assisted laser desorption/ionization time-of-flight mass spectrometry (MALDI-ToF-MS) has made it possible to study the complex mixtures of molecules in coatings polymers in unprecedented detail. For example, Willemse et al. (2005) studied polyesters designed for very high solids and solventless liquid formulations. One such polyester was made from 1,4-butanediol, AA, and IPA in a 1 : 0.375 : 0.375 mol ratio with an organotin catalyst. Figure 10.1a shows an overall MALDI spectrum of the product. Each cluster of peaks represents molecules with a specific degree of polymerization. Figure 10.1b shows an expanded spectrum focused on molecules composed of seven diol and six diacid monomers. The numbers designate the numbers of AA and IPA residues in the molecules; for example, the peak designated 4-2 with a mass of 1369.57 Da contains four AA and two IPA residues. Here the peaks are clusters because of the presence of molecules containing isotopes (e.g., ^2H, ^{18}O). Further expansion makes it possible to identify small peaks from cyclic structures and from molecules with one acid and one hydroxyl end group. A significant finding of this study was that the compositions of molecules within this polyester are close to random as calculated by Bernoullian statistics. Sequences of monomers within molecules cannot be determined by MALDI-ToF-MS alone.

In another study, Arnould et al. (2002) used MALDI-ToF-MS, tandem mass spectroscopy (MS/MS), and gel permeation chromatography (GPC) to study a higher MW polyester. It was made from NPG, TMP, TPA, and AA in a 50 : 1 : 45 : 5 ratio, a composition similar to powder coating resins (Section 10.6). The polymer was fractionated by GPC to facilitate study of higher MW fractions. Only low levels of cyclic products were detected. The MS/MS spectra suggested that, within individual molecules, certain specific sequences of monomers are favored.

Dendritic and *hyperbranched* polyesters have been studied for decades. In coatings they offer the potential for low solution viscosity and high functionality. Hyperbranched polyesters are readily synthesized by various methods that include more than the usual levels of tri- or tetrafunctional reactants. Dhevi et al. (2014) have reported their use in commercial coatings. Limitless variations are possible, including high solids, waterborne (Singh et al., 2016), radiation curable, and powder coatings.

Bio-based polyester polyols are attracting considerable research attention. Often they are based on esters of unsaturated fatty acids that have been epoxidized and ring opened to form hydroxyl groups. For example, Nelson et al. (2013) prepared polyols from epoxidized sucrose soyate that was ring opened with methanol or ethanol. These entirely bio-based polyols had about 10 hydroxyl groups per molecule. They were cross-linked with MF resins to give coatings with attractive properties.

Polycarbonate polyols are related to polyesters in that the diacid is, nominally, H_2CO_3. Polycarbonates are treated separately in Chapter 12 and in Section 17.9.

10.3 CARBOXYLIC ACID-TERMINATED POLYESTER RESINS

Carboxylic acid-terminated polyesters are copolymers of a diol and a molar excess of dibasic acid. A modest amount of a tri- or polyfunctional monomer, most commonly trimellitic anhydride (TMA), is almost always included to increase $\bar{f}_n$. With high ratios of high melting point diacids like IPA, it is usually preferable to use a two-step process. The slow dissolution and high melting point of IPA prevent clean reaction when it is used in excess in a single step. First, one prepares a hydroxy-functional polyester with the IPA and then reacts the terminal hydroxyl groups with a lower melting acid or anhydride. Careful control is needed to assure reproducible products, since varying degrees of transesterification occur in the second stage. Since the final products are governed by both kinetic and thermodynamic factors, resin properties are affected by reaction time as well as other process variables. Carboxylic acid-functional polyesters can be cross-linked with MF resins but in most

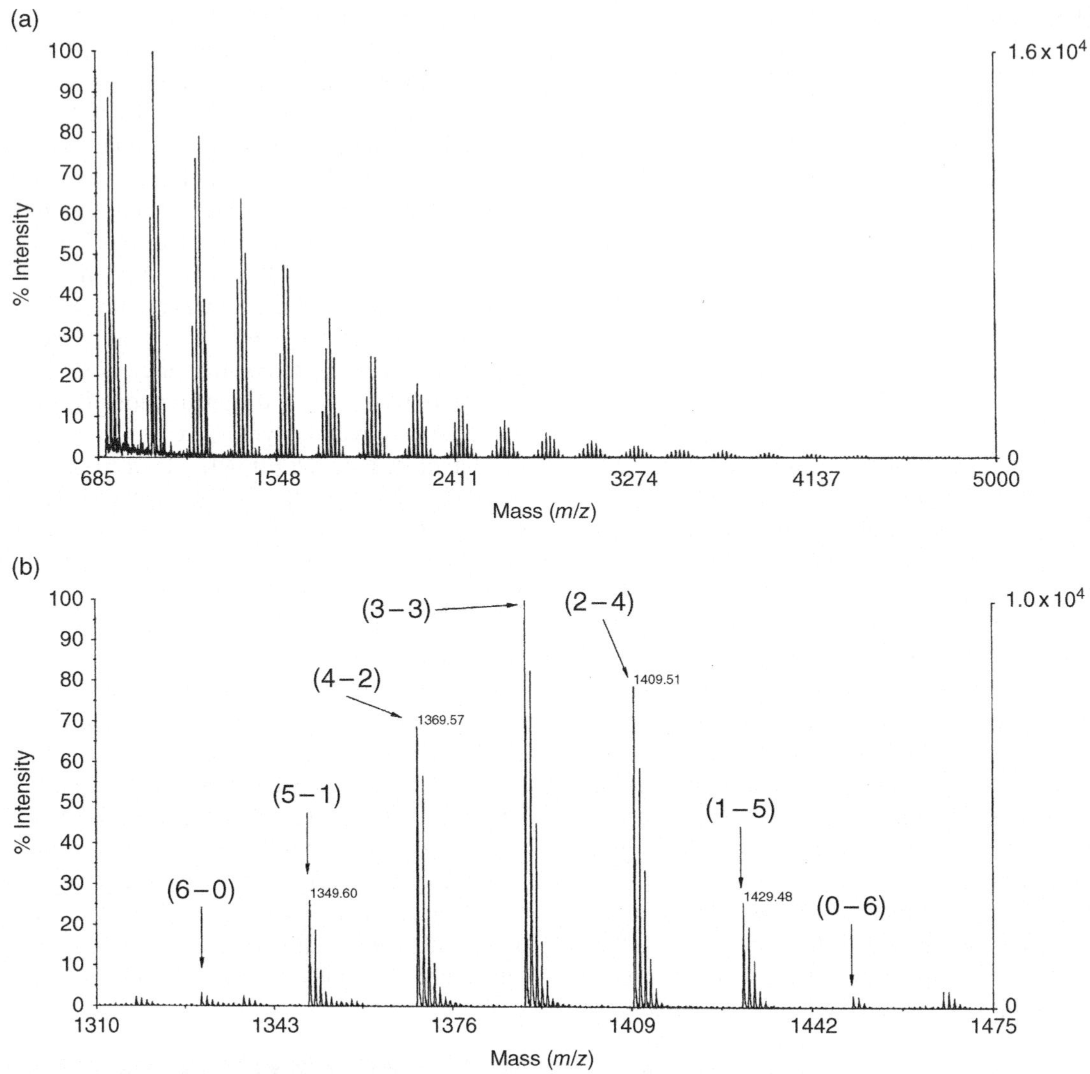

Figure 10.1 (a) MALDI-ToF-MS spectrum of a polyester resin made from 1,4-butanediol, adipic acid (AA), and isophthalic acid (IPA) in a 1 : 0.375 : 0.375 mol ratio. (b) Expanded spectrum focused on molecules composed of seven diol and six diacid monomer units. Source: Willemse et al. (2005). Reproduced with permission of American Chemical Society.

cases are cross-linked with epoxy resins or 2-hydroxyalkylamides, as discussed in Sections 13.3.2 and 17.5, respectively.

10.4 CARBAMATE-FUNCTIONAL POLYESTER RESINS

Carbamate-functional resins cross-linked with MF resins have been shown to give clear coats with good mar and acid resistance, exterior durability, and hydrolytic resistance. An example is an oligomer prepared by reacting citric acid with glycidyl neodecanoate, followed by transesterification with methyl carbamate (Green, 2001).

10.5 WATER-REDUCIBLE POLYESTER RESINS

Polyesters are also used in waterborne coatings having reduced VOC. Most such polyesters have both hydroxyls and carboxylic acids as terminal groups. As with other water-reducible resins, acid numbers in the range of 35–60 are required to obtain amine salt solutions in solvent that can be diluted with water to provide reasonably stable dispersions of aggregates of resin molecules swollen with water and solvent. These resins show abnormal viscosity dilution curves similar to those described in Section 8.3 for water-reducible acrylic resins.

In theory, one could prepare such water-reducible polyesters by simply stopping the esterification of a

Table 10.3 Water-Reducible Polyester Formulation

Material	Weight (g)	Mole ratio	Equivalent ratio
NPG	685	1.0	1.0
Adipic acid	192	0.2	0.2
TMA	84	0.067	0.1
IPA (85 : 15 IPA to TPA)	655	0.6	0.6
React at 235°C until the acid number is 16–18; cool to 180°C, and then add			
TMA	84	0.067	0.1
React at 180°C until the acid number is 40–45; cool to 160°C, and thin to 80% solids with diethylene glycol monobutyl ether			

combination of diol, triol, AA, and IPA at the desired acid number; however, in practice, it is almost impossible to do so reproducibly. Rather, it is necessary to utilize combinations of polybasic acids that have sufficiently different reactivity in order to control the ratio of unreacted hydroxyl and carboxyl groups. A widely used method has been to prepare a hydroxy-functional resin, followed by adding sufficient TMA to esterify a fraction of the hydroxyl groups, thereby introducing two carboxyl groups at each reacted site. Advantage is taken of the higher reactivity of the anhydride group of TMA at 180°C relative to carboxylic acid groups. Table 10.3 gives an example of a formulation for preparing a water-reducible polyester (Anon, 1984).

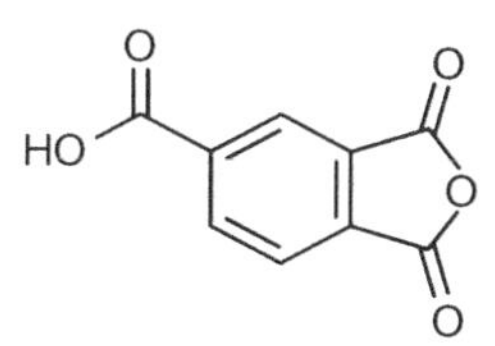

Trimellitic anhydride (TMA)

This approach has been used on a large scale but has disadvantages. Most seriously, the ester linkages in the resin are vulnerable to hydrolysis during storage of the formulated coating. The ester group of partially esterified TMA is particularly sensitive to hydrolysis owing to anchimeric assistance by the adjacent carboxylic acid group. Hydrolysis of the TMA partial ester results in removal of the solubilizing carboxyl groups and, thereby, destabilizes the resin dispersion. The film properties may also be adversely affected. In addition, the use of primary alcohols as a solvent, as recommended in Table 10.3, has been found to be undesirable. Primary alcohols have been found to esterify carboxylic acid groups and transesterify ester groups during thinning of the newly synthesized polyester at 160°C and at a slow, but appreciable, rate, during storage of the resin. This problem can be minimized by using a secondary alcohol as solvent, along with careful control of process temperatures. CHDA esters are more stable to hydrolysis under basic conditions than IPA (Jones and McCarthy, 1995). Water-reducible polyesters are used in applications for which long-term storage stability and hydrolytic stability are not required, such as industrial coatings with a fast turnover.

An alternative approach that provides somewhat better, but still limited, hydrolytic stability is to use 2,2-dimethylolpropionic acid as one of the diol components. The carboxylic acid group of this monomer is highly hindered by being located on a tertiary carbon. The resulting differential reactivity makes it possible to esterify the hydroxyl groups while leaving many of the acid groups unreacted. While too hindered to esterify readily, the acid groups still readily form salts. Polyesters of this type often serve as building blocks for polyurethane dispersions (Section 12.7.1).

Hydrolytic stability is also affected by the choice of polyol. In addition to the steric effect discussed earlier, it has been shown that polyols having low water solubility form polyesters that are more stable to hydrolysis under basic conditions than polyols with higher water solubility, presumably because the polymers are more hydrophobic. For example, polyesters made with 2-butyl-2-ethyl-1,3-propanediol (BEPD) (Section 10.1.1) are more stable to hydrolysis than NPG esters.

Another possible problem with water-reducible polyesters results from intramolecular reaction of terminal hydroxyl and carboxylic acid groups to form low MW nonfunctional cyclic polyesters. When coatings are baked, small amounts of the cyclic esters can volatilize out of the coating and gradually accumulate in cool spots in the oven. Eventually, sufficient resin can accumulate and drip down on products going through the ovens, marring their finish. Since the amounts are small, dripping may only start after weeks or months of operation of the coating line.

Water-thinnable polyester coatings have been formulated with low MW oligomeric hydroxy-terminated polyesters (Jones, 1996). Up to about 20% of water dissolves in a polyester–class I MF resin binder, reducing the viscosity by about half. This permits making solvent-free coatings. Further work is needed, but the idea is appealing not only because cosolvent is unnecessary, which reduces the VOC, but also because no amine is needed, which is expected to reduce the problem of hydrolysis.

10.6 POLYESTER RESINS FOR POWDER COATINGS

Chapter 28 is devoted to powder coatings. Polyester resins for powder coating are brittle solids having a relatively high T_g (50–60°C), in order that the powder coating does not sinter (partially fuse) during storage. These requirements are met with TPA and NPG as the principal monomers. Resins in which much of the TPA is replaced with IPA may provide coatings having superior outdoor weatherability (Merck, 2001). Smaller amounts of other monomers are added to increase $\bar{f}_n$ and reduce T_g to the desired level. Use of BEPD (Section 10.1.1) to reduce T_g may also improve outdoor durability. The relatively high T_g makes it possible to prepare hard, tough films having relatively low cross-link density. Widely used commercial products are amorphous, rather than crystalline. Both hydroxy- and carboxy-terminated polyesters are important. The former are most commonly cross-linked with blocked isocyanates (Section 12.5) and the latter with epoxy resins (Chapter 13). Other cross-linkers include 2-hydroxyalkylamides (Section 17.5) and tetramethoxymethylglycoluril (Section 11.4.3).

CHDA has been suggested as a replacement for IPA (Johnson and Sade, 1993). CHDA polyesters have lower T_g and lower melt viscosities; if the T_g is too low, thereby compromising storage stability, NPG can be fully or partially replaced with hydrogenated bisphenol A to improve storage stability. Processes involving direct esterification of TPA are used, but since TPA has a very high melting point, it is common to make the hydroxy-terminated polyester by transesterification of dimethyl terephthalate. An appropriate transesterification catalyst is tetraisopropyl titanate. If desired, the hydroxy-terminated TPA polyester can then be reacted with other polyacids to form a carboxylic acid-terminated product in an additional stage.

Recycled polyethylene terephthalate (PET) has been evaluated as a raw material for making polyesters for powder coatings. Kawamura et al. (2002) dissolved scrap PET in a mixture of NPG and TMP with a catalyst at 200°C. After cooling to 180°C, IPA was added and the reaction mixture was heated to 240°C until the required acid number was reached. Film properties of powder coated films were very similar to those obtained with a polyester prepared from IPA, ethylene glycol, NPG, and TMP.

REFERENCES

Anon., Amoco Chemicals Corp., Tech. Bull. TMA109e, **1984**.
Argyropoulos, J., et al., *Proceedings of the Waterborne High-Solids Powder Coatings Symposium*, New Orleans, LA, **2003**, pp 107–113.
Arnould, M. A., et al., *Prog. Org. Coat.*, **2002**, *45*, 305.
Belote, S. N.; Blount, W. W., *J. Coat. Technol.*, **1981**, *53(681)*, 33.
Bender, M. L., et al., *J. Am. Chem. Soc.*, **1958**, *80*, 5384.
Calbo, L. J., *Proceedings of the Waterborne Higher-Solids Coatings Symposium*, New Orleans, LA, **1986**, p 3.
Cambon, A.; Jullien, R., *Bull. Soc. Chim. France*, **1973**, *2003*.
Dhevi, N., et al., *Polym. Bull.*, **2014**, *71(10)*, 2671–2693.
Feng, L., *Proceedings of the American Coatings Association Conference*, Paper 13.6, Indianapolis, IN, **2016**.
Golob, D. J., et al., *Polym. Mater. Sci. Eng.*, **1990**, *63*, 826.
Green, M. L., *J. Coat. Technol.*, **2001**, *73(918)*, 55.
Heidt, P. C., et al., *Proceedings of the Waterbortne High-Solids Powder Coatings Symposium*, New Orleans, LA, **2000**, pp 295–307.
Hood, J. D., et al., *Proceedings of the Waterborne Higher-Solids Coatings Symposium*, New Orleans, LA, **1986**, p 14.
Johnson L. K.; Sade, W. T., *J. Coat. Technol.*, **1993**, *65(826)*, 19.
Jones, F. N., *J. Coat. Technol.*, **1996**, *68(852)*, 25,
Jones, T. E.; McCarthy, J. M., *J. Coat. Technol.*, **1995**, *67(844)*, 57.
Kawamura, C., et al., *Prog. Org. Coat.*, **2002**, *45*, 185.
Marsh, S. J., et al., Patent no. 2 US 8,524,834 B2 (45), date of patent: September 3, **2013**.
Merck, Y. *Surf. Coat. Intl. [B]: Coat. Trans.* **2001**, *84(B3)*, 231.
Misev, T. A., *J. Coat. Technol.*, **1989**, *61(772)*, 49.
Nelson, T. J., et al., *J. Coat. Technol. Res.*, **2013**, *10(6)*, 757–767.
Odian, G. W., *Principles of Polymerization*, 4th ed., Wiley, New York, **2004**, pp 40–62.
Pilcher, G. W., *CoatingsTech*, **2012**, *9(10)*, 26–38.
Singh, A. P., et al., *JCT Res.*, **2016**, *13(1)*, 41–51.
Turpin, E. T., *J. Paint Technol.*, **1975**, *47(602)*, 40.
Union Carbide Corp., Specialty Polymers and Composites, *Tech. Bull., TONE Polyols*, **1986**.
Van Sickle, D. E., et al., *Polym. Mater. Sci. Eng.*, **1997**, *76*, 288.
Willemse, R. X. E., et al., *Macromolecules*, **2005**, *38*, 6877.

Chapter 11

Amino Resins

Amino resins, also called *aminoplast resins*, are major cross-linking agents for baked thermosetting coatings. Amino resins based on melamine and urea also have a multitude of other uses, not described here. The amino resins most common in coatings are derived from melamine (2,4,6-triamino-1,3,5-triazine), urea, benzoguanamine, and glycoluril, as well as from copolymers of (meth)acrylamide (in the succeeding text).

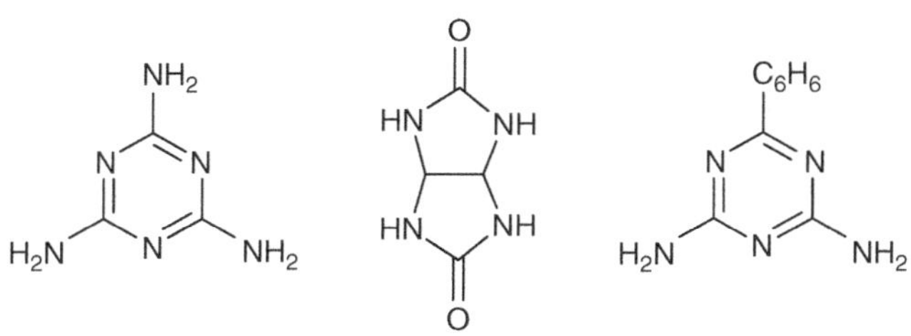

Melamine Glycoluril Benzoguanamine

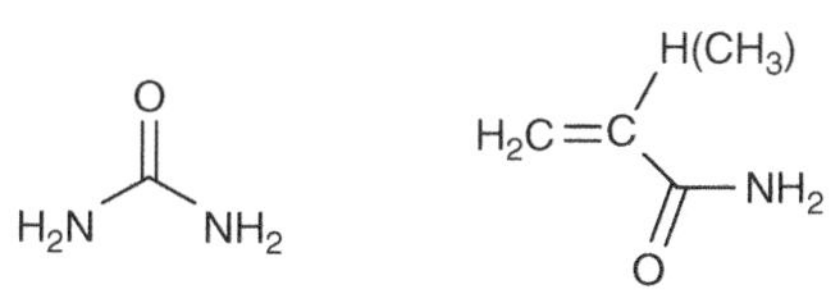

Urea (Meth)acrylamide

The five previously mentioned compounds and copolymers are all weak bases. They are converted into amino resins for coatings by reaction with formaldehyde ($H_2C{=}O$) and, subsequently, with alcohols (ROH) to yield *activated ethers* with the general structure $>NCH_2OR$. Such ethers are activated toward nucleophilic substitution by the neighboring N and are much more reactive than aliphatic ethers. When the nucleophile is the alcohol group of a polyol (POH), transetherification can occur, as shown in Eq. 11.1, resulting in the formation of a cross-linked polymer. The reaction is catalyzed by acid. Carboxylic acids, urethanes, and phenols having an unsubstituted ortho position also react, as shown in Eqs. 11.2, 11.3, and 11.4.

$$>N{-}CH_2{-}OR + P{-}OH \rightleftharpoons >N{-}CH_2{-}OP + ROH \quad (11.1)$$

$$>N{-}CH_2{-}OR + P{-}CO_2H \rightleftharpoons PCO_2{-}CH_2{-}N< + ROH \quad (11.2)$$

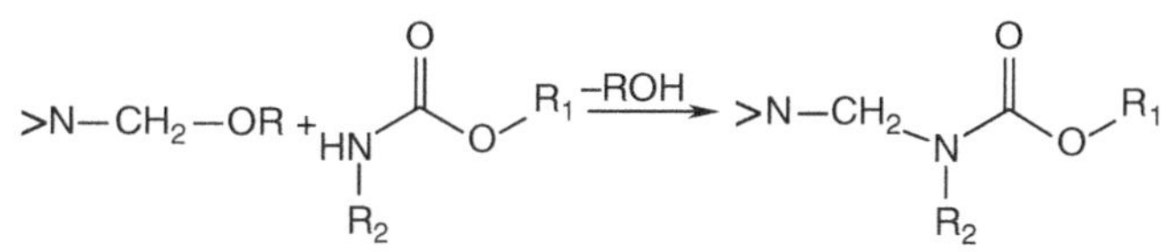

(11.3)

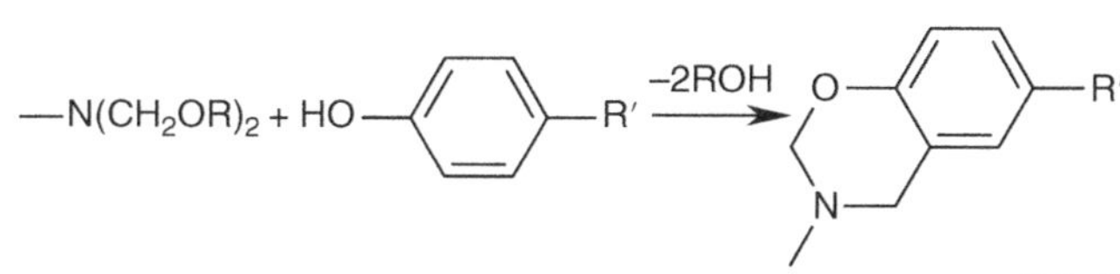

(11.4)

The discussion in the succeeding text will focus on amino resins derived from melamine. Other types of amino resins are discussed in less detail in Section 11.4. Many, but not all, of the same considerations apply to the other types.

11.1 SYNTHESIS OF MELAMINE–FORMALDEHYDE RESINS

The first step in the synthesis of melamine–formaldehyde (MF) resins is methylolation, the reaction of melamine with formaldehyde under basic conditions. With excess

Organic Coatings: Science and Technology, Fourth Edition. Frank N. Jones, Mark E. Nichols, and Socrates Peter Pappas.

formaldehyde, the reaction can be driven to form a mixture that contains substantial amounts of hexamethylolmelamine (**1** where R is H). With less than the stoichiometric 6 mol of formaldehyde per mol of melamine, a mixture of partially methylolated derivatives is formed, including symmetrical trimethylolmelamine (TMMM) (**2** where R is H). Methylolmelamine resins of this type are often used to fabricate laminates and plastics but are rarely used in coatings.

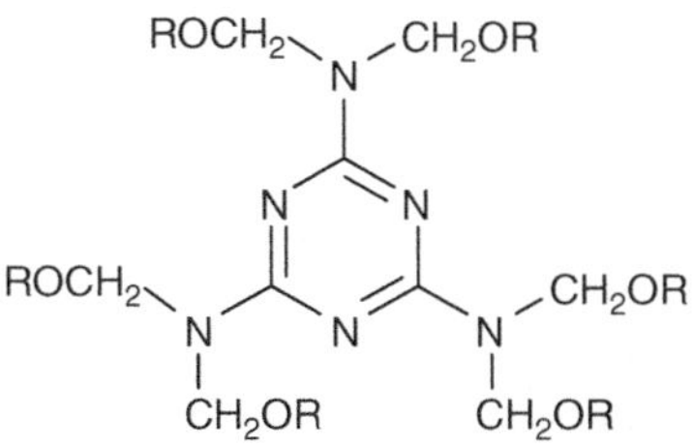

(1)

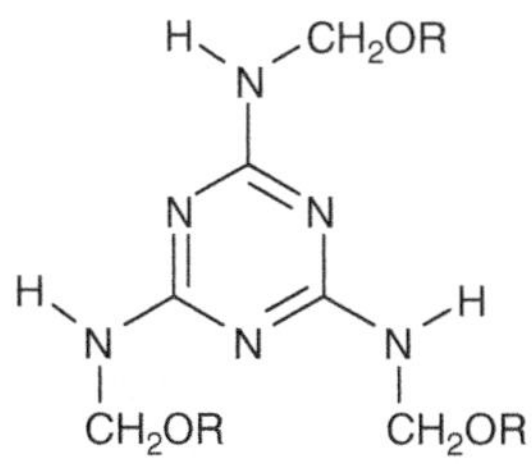

(2)

For coatings, the second step is acid-catalyzed etherification of methylolmelamines with an alcohol, such as methyl or butyl alcohol. Complete etherification of methylolmelamines (**1**) and (**2**), where R is H, yields alkoxymethyl derivatives (**1**) and (**2**), where R is alkyl. Many commercial MF resins are only partially etherified. In addition to monomeric species, commercial MF resins contain oligomeric species in which triazine rings are linked by methylene ($>NCH_2N<$) and dimethylene ether bridges ($>NCH_2OCH_2N<$).

11.1.1 The Methylolation Reaction

A probable mechanism for base-catalyzed reaction of melamine with formaldehyde is outlined in Scheme 11.1, where $-NH_2$ represents the melamine amino groups. The first step involves nucleophilic attack by the amino group on the electrophilic C of formaldehyde, which is facilitated by the removal of a proton from N by the base (B^-). This is followed by proton transfer from the resulting B—H to the negative oxygen atom, which yields the methylolated product and regenerates the base catalyst; both steps are reversible.

Early studies of the kinetics of the reaction indicated that the presence of one methylol group on N deactivates the group for a second reaction by a factor of 0.5–0.6. On the other hand, substitution on one amino group has less effect on the reactivity of the other amino groups. These kinetic factors favor formation of the symmetrical TMMM (**2**), where R is H. However, the preference is not strong enough to overcome the kinetic and thermodynamic tendencies to produce mixtures of products. Thus, at equilibrium, the reaction of 6 mol of formaldehyde with 1 mol of melamine yields a mixture of products including all levels of methylolation and free formaldehyde (Santer, 1984). A side reaction is methylolation on oxygen, as shown in Scheme 11.1.

11.1.2 The Etherification Reaction

Following methylolation, the base catalyst is neutralized with excess acid and the appropriate alcohol is added. The acid-catalyzed reversible reaction leads to formation of alkoxymethyl groups. Nitric acid is commonly used as a catalyst because nitrate salt by-products are relatively easily removed.

There is disagreement in the literature about the mechanism of the etherification reaction, as well as of the closely related transetherification reaction, shown in Eq. 11.1. The contending mechanisms for these substitution reactions are S_N1 and S_N2, both of which are outlined in Scheme 11.2, where R is H and R′OH is methyl or butyl alcohol. Note that the conjugate base (A^-) is omitted from the intermediate steps.

The distinguishing feature is whether the protonated methylol group dissociates to an intermediate resonance-stabilized carbocation, as in Eq. 11.5, before reacting with the alcohol, characteristic of the S_N1 mechanism, or reacts directly with the alcohol, as in Eq. 11.6, characteristic of the S_N2 mechanism. The controversy about the mechanism in MF resin synthesis also applies to the transetherification reaction involved in cross-linking MF resins with polyols, where R is commonly methyl or butyl and R′OH is the polyol. Experimental evidence bearing on this controversy

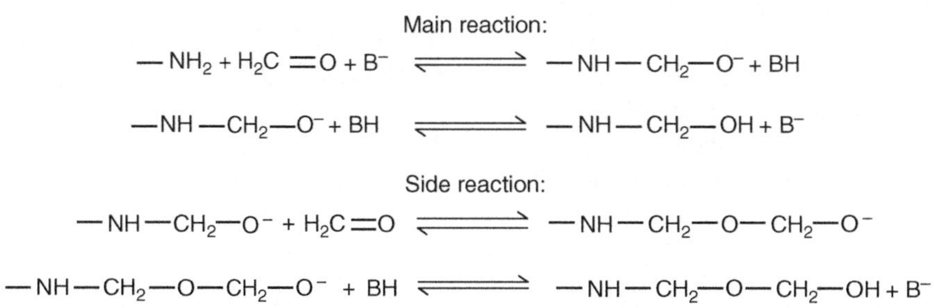

Scheme 11.1 Chemistry of methylolation reactions.

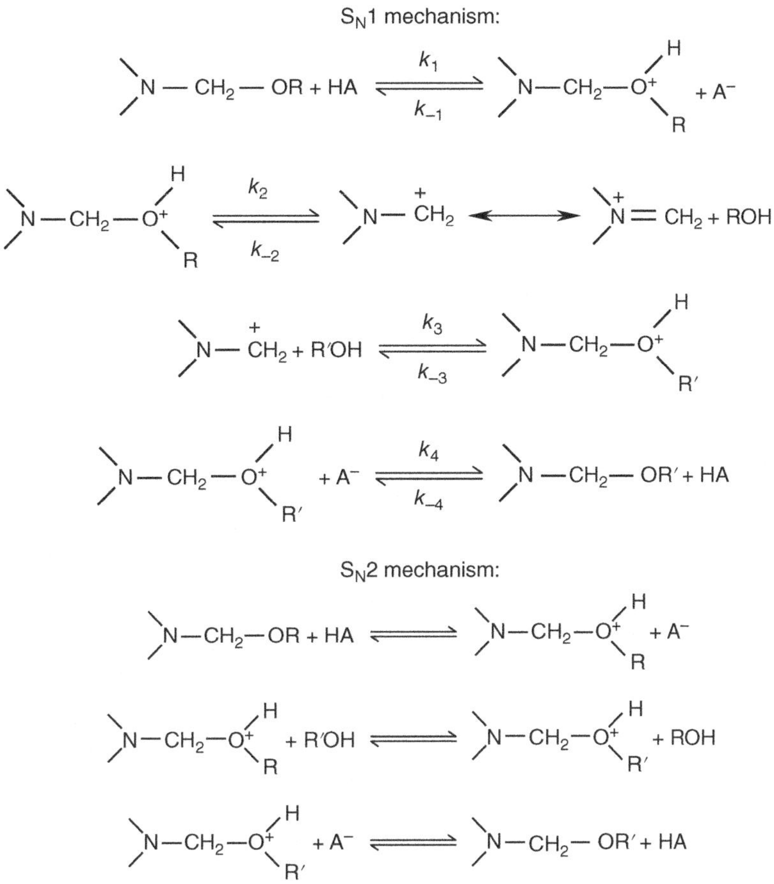

(11.5)

(11.6)

Scheme 11.2 Possible mechanisms of etherification reactions of methylolated amino groups containing no N–H moieties, including doubly methylolated amino groups.

has been acquired primarily in studies of coatings and is discussed in Section 11.3.2. Based on the available evidence, we favor the S_N1 mechanism.

Etherification of methylol groups on a singly substituted N (i.e., a N that also bears a H atom) is proposed to follow the mechanism outlined in Scheme 11.3, where R is H and R′OH is methyl or butyl alcohol.

The distinguishing feature is the formation of the neutral imine intermediate, shown in Eq. 11.7, which is possible due to the presence of the N—H group. Imines form by elimination of water catalyzed by relatively weak acids, such as carboxylic acids. Complexation of the methylolmelamine by an acid (A—H) is sufficient to yield the imine, either in a concerted push–pull mechanism, as shown in Scheme 11.3, or a stepwise mechanism. Complexation of the reactive imine by acid is also sufficient to effect subsequent addition of alcohol, resulting in overall substitution of water by alcohol. This reaction is also reversible. Since catalysis in Scheme 11.3 involves complexation, rather than protonation by the acid, the rate is dependent on the nature of H—A, as well as its concentration. Indeed, both the more positive H end and the more negative A end of H—A participate by facilitating breaking of both the C—O and the N—H bonds, respectively. By convention, this is called *general acid catalysis*. The mechanism in Scheme 11.3 also applies to cross-linking of polyols by MF resins having N—H groups by transetherification, where R is methyl or butyl and R′OH represents the polyol.

In contrast, elimination of water from methylolmelamines lacking N—H groups, either by the S_N1 or S_N2 mechanism (shown in Scheme 11.2), requires protonation of the methylol group, necessitating a strong ionizing acid, such as nitric, sulfuric, or sulfonic acid. Such catalysis, requiring protonation, is called *specific acid catalysis*, signifying that the rate is dependent only on the concentration of protons (H^+) and is independent of the conjugate base (A^-).

11.1.3 Self-Condensation Reactions

Self-condensation refers to reactions that form bridges between triazine rings, resulting in dimers, trimers, higher oligomers, and, ultimately, cross-linked polymer. The extent of such reactions depends on process factors including pH,

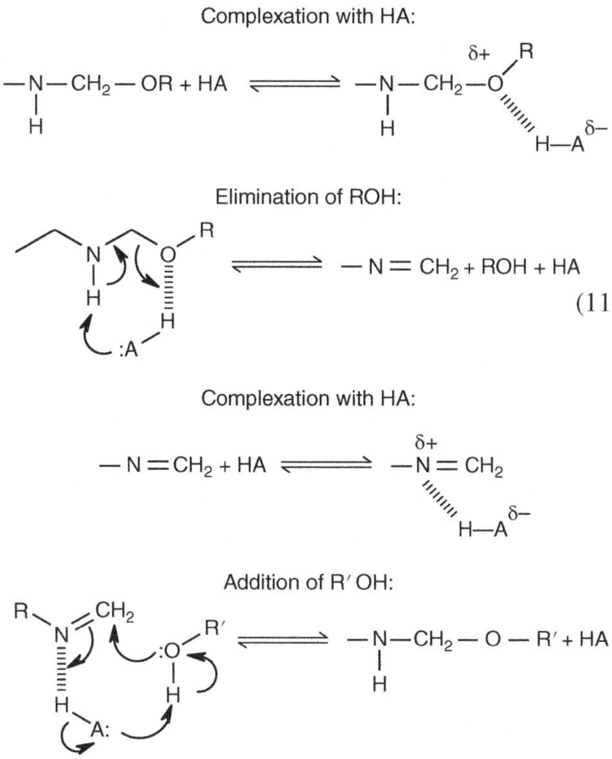

(11.7)

Scheme 11.3 Possible mechanism of etherification of singly methylolated amino groups.

ratio of reactants, reaction temperature, rate of removal of water, and probably others. Two types of linkages occur between triazines: methylene bridges (>NCH_2N<) and dimethylene ether bridges (>NCH_2OCH_2N<) (Chang, 1996; Subrayan and Jones, 1996; Lavric et al., 2015). Acid catalysis favors dimethylene ether bridges, while base catalysis favors methylene bridges (Nastke et al., 1986).

11.2 TYPES OF MF RESINS

Various MF resins are made commercially, which differ by the ratio of functional groups, the alcohol used for etherification, and the average degree of polymerization $\overline{P}$. To facilitate discussion, we will employ a fairly common convention in which MF resins are classified into two broad groups. Class I resins are made with relatively high ratios of formaldehyde to melamine resulting in most of the nitrogens having two alkoxymethyl substituents. Class II resins are made with lower ratios of formaldehyde; thus, a substantial fraction of the nitrogens have only one substituent.

From about 1940 to the 1950s, the predominant melamine resins were class II types in which the alcohol was usually *n*-butyl or isobutyl alcohol. Such resins have ample bridging so that $\overline{P}$ is three or more. They are economical and offer a wide latitude in usage; that is, coatings do not require exacting control of the formulation to produce acceptable application characteristics and film properties. The class II MF resins were used, primarily, as cross-linkers in alkyd formulations, in which they are readily miscible.

Class I MF resins were commercialized in the 1950s. Common etherification alcohols are methanol, *n*-butanol, isobutyl alcohol, or mixtures. Methylated class I resins are more compatible than butylated resins with certain coreactant resins and, in some cases, provide tougher films. Introduction of waterborne and high solids coatings during the 1970s accelerated the shift toward methylated class I resins. These methylated MF resins are frequently called *HMMM resins* for *hexamethoxymethylmelamine*, but the terminology *high HMMM resin* is more appropriate. Methylated MF resins are more miscible with the water, solvent, and resin blends used in waterborne coatings; and class I resins having low $\overline{P}$ impart lower viscosity to high solids coatings. These generalities are not universal; some class II resins are used in low solvent coatings.

Many possible structures are formed during synthesis. Hill and Kozlowski (1987) reported that a commercial methylated MF class I resin contains 62% monomers (i.e., one triazine ring), 23% dimers, and 15% trimers and higher oligomers. The equivalent weight of this resin is 80 g equiv.$^{-1}$, resulting from the presence of dimers, trimers, and higher oligomers, as compared with 65 g equiv.$^{-1}$ for pure HMMM. In estimating equivalent weight, it is necessary to make an assumption as to whether >NCH_2N< and/or >NCH_2OCH_2N< bridging groups react to form cross-links. (In this calculation it is assumed that they do not, which may not be strictly true.)

Van Dijk et al. (1980) published chromatograms of a similar high HMMM resin (Figure 11.1). As expected, –$N(CH_2OCH_3)_2$ groups predominate, but incompletely formylated and incompletely etherified groups are also present.

Characterization of MF resins continues to challenge analytical chemists. Lavric et al. (2015) reviewed previous efforts and employed advanced HPLC, MS, NMR, and Fourier transform infrared (FTIR) methods to study a high HMMM resin. They detected 57 species and were able to accurately measure the mass of a dozen monomers and dimers. They confirmed the presence of both >NCH_2N< and >NCH_2OCH_2N< bridges in self-condensation products.

A broad range of class I resins is available, having different degrees of polymerization $\overline{P}$ and different extents of formylation and etherification. In addition to methylated derivatives, butylated, as well as mixed methyl/*n*-butyl, methyl/isobutyl, methyl/ethyl, and methyl/isooctyl, derivatives are also available. Class I resins generally have lower $\overline{P}$ and, therefore, lower viscosity than class II resins at the same percent solids. The viscosity is also reduced by the presence of fewer polar groups. At a given $\overline{P}$, butyl ethers

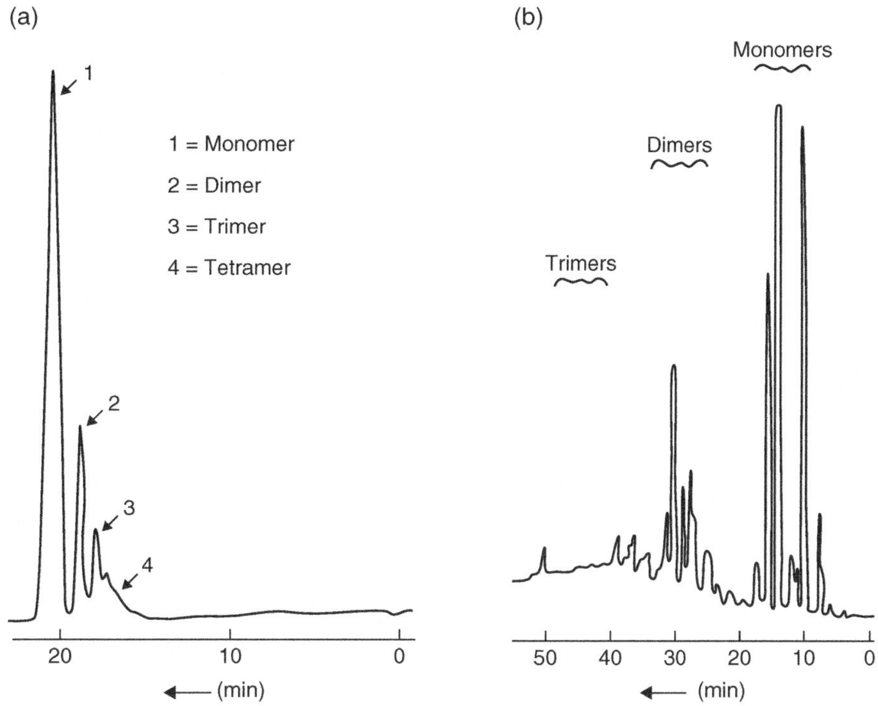

Figure 11.1 High performance liquid chromatograms (HPLC) of a typical class I high HMMM resin. (a) An SEC chromatogram and (b) a gradient HPLC chromatogram. Source: Reproduced with kind permission of Joop H. Van Dijk, Akzo Corporate Research.

exhibit even lower viscosities, attributable to lower T_g values, than the corresponding methyl ethers.

An even broader range of class II resins is available because the ratio of MF and of alcohol/formaldehyde can be varied over a wider range. It is harder to suppress bridge-forming reactions during synthesis of class II resins, and very low $\bar{P}$ resins are difficult to produce. The predominant reactive group present in class II resins is —NHCH$_2$OR. Hence, they are frequently called *high NH resins*. They also contain —NH$_2$, >NCH$_2$OH, and —N(CH$_2$OR)$_2$ groups. To minimize viscosity and maximize reactivity, resin producers strive to reduce $\bar{P}$ and maximize the amount of symmetrical TMMM (**2**), where R is Me. This is possible to a degree because methylolation of —NH$_2$ groups, yielding —NHCH$_2$OH, is kinetically favored over methylolation of NHCH$_2$OH, yielding —N(CH$_2$OH)$_2$. Owing to advances in HPLC analysis techniques, it is possible to follow the effects of small changes in process conditions on composition. Continuing refinement of processes has permitted commercial production of resins with as high as 50% TMMM. Various alcohols are used to make different resin grades.

Amino resins usually contain small amounts of unreacted formaldehyde that may be vaporized during application and baking of the coatings. Additional amounts of formaldehyde vapor may be formed during storage of formulated coatings and/or during the baking process. These small amounts of formaldehyde vapor pose a potential hazard to workers who are exposed daily. In 2016 the European Union tightened regulation of products containing more than 0.1% of free formaldehyde. Formaldehyde exposure levels are strictly regulated in the United States and were tightened in some areas in 2012. Current technology appears adequate to satisfy these regulations. While the regulations already appear stringent, it is possible that allowable exposure levels will be reduced further, and intensive research on ways to reduce formaldehyde emissions has been underway. Besides improving ventilation, approaches include changing resins, changing formulations, incinerating oven exhausts, and introducing formaldehyde scavengers such as hydrazides. MF resin producers have optimized production processes for many key resins to yield low-formaldehyde (<0.1 wt%) grades.

11.3 MF–POLYOL REACTIONS IN COATINGS

MF resins are used to cross-link coreactant resins having hydroxyl, carboxylic acid, urethane (carbamate), phenol, and/or amide groups. Acrylic (Chapters 8 and 9), polyester (Chapter 10), alkyd (Chapter 15), epoxy (Chapter 13), and polyurethane (Chapter 12) resins are the most important classes of coreactant resins. It is widespread practice to use combinations of cross-linkers, for example, MF resins and other types of amino resins, together with (blocked) polyisocyanates (Chapter 12), with phenolic resins (Section 13.6), or with resins bearing reactive siloxanes (Section 16.2).

Polyols (hydroxy-functional acrylics, polyesters, etc.) are the resins most commonly cross-linked by MF resins. The hydroxyl groups of polyols react with MF resins to form ether cross-links either by transetherification with the activated alkoxymethyl groups or by etherification of the methylol groups. The reaction with phenols, as in Eq. 11.4, has the advantage that a C—C bond is formed so that the product is stable to hydrolysis. Strong acid catalysts, such as sulfonic acids, are used for class I resins, and weak acid catalysts, such as carboxylic acids, for class II resins. The reactions are reversible, but are driven toward cross-linking by volatilization of the monofunctional alcohol or water produced, as shown in Eq. 11.8, where R is alkyl or H and POH is the polyol.

$$>N-CH_2-OR + (P)\sim\sim OH \underset{}{\overset{H^+}{\rightleftharpoons}} >N-CH_2-O\sim\sim(P) + R-OH \quad (11.8)$$

Rates of reaction with hydroxyl groups depend on the structure of the polyol and the MF resin, the type and amount of catalyst, and the temperature. The rate at which MF resins cross-link polyols increases with increasing volatility of the alcohol, according to the following order: methyl>ethyl>*n*-butyl. These rates are probably influenced by the rate of diffusion of the alcohol from the reactive site and by the rate of evaporation from the film.

During acid-catalyzed cross-linking with polyols, *self-condensation reactions* of MF resins also occur, including the formation of methylene bridges and dimethylene ether bridges. Reactions of MF resins with polyol (or another coreactant) have been termed *co-condensation reactions* (Blank, 1979). Both self- and co-condensation reactions contribute to the structure of the cross-linked polymer network and to its film properties.

A widely believed, but mistaken, conclusion is that steric hindrance limits the number of active ether groups on high HMMM resins that can participate in cross-linking of films to three, or at most, four of the six groups. This conclusion was apparently based on the need to use an excess over the stoichiometric amount of MF class I resin relative to polyol to achieve desired film properties at baking schedules, such as 30 min at 120°C, using around 0.3 wt% (on the total weight of polymers) of *p*-toluenesulfonic acid (pTSA). However, Hill and Kozlowski (1987) provide strong evidence in support of an essentially complete reaction of $-NCH_2OCH_3$ groups with certain polyols, showing that all $-NCH_2OCH_3$ groups of high HMMM resins can co-condense if there are enough hydroxyl groups present for them to react with. Their conclusions were based on studies of changes in mechanical properties, as discussed in Section 4.2. These results further substantiate the hypothesis that the excess MF resin used in most formulations is needed not to complete the co-condensation cross-linking but to permit some degree of cross-linking by self-condensation that may be necessary to achieve the desired film properties.

Factors in addition to cure response must be considered when selecting the MF resin. For example, viscosity at a given percent solids content can be reduced by substituting HMMM types by class I *mixed ether resins*. The alcohols in mixed ether resins are frequently 1 : 6 to 1 : 3 *n*-butyl or isobutyl alcohol to methyl alcohol. Only modest reductions of VOC are attained because mixed ether resins contain a higher weight fraction of volatile by-products; butyl groups represent a higher weight fraction of resins than do methyl groups. Perhaps, a greater advantage of mixed ether resins is that they impart lower surface tension to coatings than high HMMM types (Albrecht, 1986). High surface tension is often associated with development of film defects such as crawling, cratering, poor leveling (Chapter 24), and poor intercoat adhesion (Section 6.5).

Chu and Jones (1993) and Jones et al. (1994) have reported the rate of development of solvent resistance and film hardness when a fractionated low NH class I resin was used to cross-link polyester resins made with cyclohexanedimethanol (CHDM), neopentyl glycol (NPG), and 1,6-hexanediol. Film properties developed most rapidly with CHDM polyesters, followed in order by NPG polyesters (a close second), and hexanediol polyesters. It was also found that film properties generally developed more rapidly with acrylic polyols than with polyester polyols.

With strong acid catalysis, the apparent rate at which class I resins react with most polyols by co-condensation is faster than self-condensation. However, with class II resins, the apparent rates of co-condensation and self-condensation are similar. The relatively high levels of $-NHCH_2OR$ and $>NCH_2OH$ groups in class II resins promote self-condensation reactions. The methylol groups can self-react analogously to reaction with a polyol; and the $-NHCH_2OR$ groups can eliminate alcohol to yield reactive melamine imine ($-N=CH_2$) groups, which can form dimethylene ether bridges by reaction with $-NCH_2OH$ groups and methylene bridges by reaction with >NH groups. Based on the time and temperature required to achieve hard films, class I and II resins have also been classified as high and low cure temperature resins, respectively, which reflects, at least in part, the facility of self-condensation reactions by class II resins.

As noted previously, class I resins are used in coating formulations, wherein there is an excess of alkoxymethyl groups over polyol hydroxyl groups so that final development of film properties depends on the extent of self-condensation as well as on co-condensation reactions. In some cases, the use of excess MF resin is dictated by its lower cost as compared with some polyols. Self-condensation cross-links may also improve film properties. The amounts of MF resin and catalyst are determined empirically and are optimized for a specified baking time and temperature. If the use conditions differ from those for which the formulation was designed, the extent of self-condensation and co-condensation will differ from the optimum levels. Coating properties such as hardness, adhesion, exterior durability, and impact resistance may be affected. Many formulations are designed to achieve optimum properties when co-condensation is nearly complete and self-condensation is partly complete.

For high solids coatings in which the hydroxy equivalent weight and the average functionality of the polyol are lower in comparison with higher molecular weight polyols, used in conventional coatings, the film properties are particularly sensitive to variations in cure temperature and time. These considerations have been addressed in terms of a *cure window*, corresponding to the range of cure times and temperatures that provides films of acceptable properties (Bauer and Dickie, 1982). The cure window was shown to be smaller for high solids MF–polyol coatings, especially when class II resins were used.

11.3.1 Catalysis of MF–Polyol Reactions

With strong acid catalysts, commonly aryl sulfonic acids, usually in the range of 0.5–1 wt% of MF resin, reactions of MF class I resins with polyols give cured films in 10–30 min at 110–130°C. Co-condensation of MF class I resins with polyols can be catalyzed by weak acids, such as carboxylic acids; however, with weak acids elevated cure temperatures, generally greater than 140°C, are required (Lazzara, 1984). In the case of class II resins, carboxylic acids are more effective in catalyzing co-condensation in accordance with the mechanism provided in Scheme 11.3. Since many polyol resins also contain carboxylic acid groups, which are present to promote adhesion and facilitate pigment dispersion, no added catalyst may be needed for cross-linking with class II resins. At the high temperatures (air temperatures as high as 375°C) used in coil coating ovens, strong acid catalysts are needed even with class II resins, owing to the short cure times.

Cure time and/or temperature can be reduced by increasing the catalyst concentration. However, storage (or package) stability is reduced by this approach, since the reaction at ambient temperature is also catalyzed by acid. Another danger in increasing acid catalyst concentration to reduce cure time and/or temperature relates to the durability of the cured coating. In addition to catalyzing transetherification (and etherification) reactions of MF resins with polyols, acid residues catalyze hydrolysis of cross-links in cured coatings. The course of the hydrolysis reaction can be followed by reference to Schemes 11.2 and 11.3, where $>N{-}CH_2{-}OR$ represents a cross-linked polymer and R′OH represents water. The hydrolysis reaction cleaves cross-linked bonds and generates methylol groups that, at least in part, eliminate formaldehyde as shown in Eq. 11.9 (Blank, 1979).

$$>N{-}CH_2OH \rightleftharpoons\ >N{-}H + H_2C{=}O \qquad (11.9)$$

Free acids such as pTSA catalyze cross-linking under ambient conditions at a rate that may increase the viscosity of a liquid coating above the range for application in less than 6 months, which is within the desired storage period. One-package systems commonly use latent acid catalysts, also called *acid precursors* and *blocked acids*. Salts of aryl sulfonic acids with volatile tertiary amines (R_3N) are commonly used. Aryl sulfonic acids are strong acids (the pK_a of pTSA = −6), whereas their tertiary amine salts are weak acids (pK_a = 8–10) and are much less active as catalysts. As shown in Eq. 11.10, equilibrium exists between pTSA amine salt and the protonated MF resin.

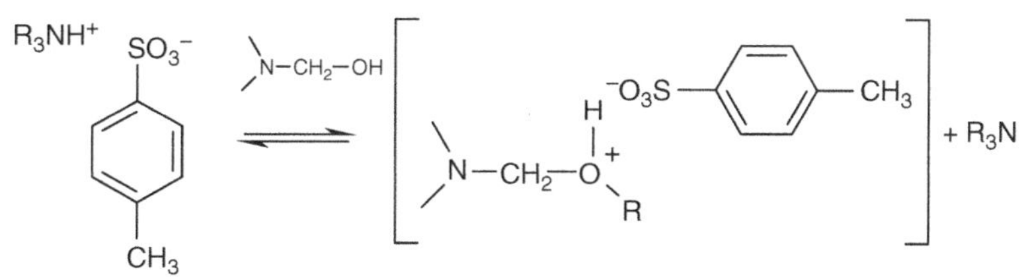

(11.10)

While the equilibrium strongly favors the left side, it is shifted to the right by volatilization of the amine, thereby increasing the concentration of protonated MF resin, the initial activated intermediate in the transetherification reaction (Scheme 11.2). Generally, storage stability of a coating with blocked catalyst approaches the stability of an uncatalyzed coating. In some cases, the cure rate with blocked acid approaches that of a coating catalyzed with the free acid; while in others, the cure rate is somewhat reduced (Mijs et al., 1983). The balance of storage stability and cure rate is reported to be particularly favorable for *N*-benzyl-*N,N*-dimethylanilinium sulfonates (Morimoto and Nakano, 1994). In this case, the MF resin is probably activated by O-alkylation with the benzyl cation, since quaternary anilinium sulfonates have no available proton.

The acid strength of the medium can be no stronger than that of protonated melamine resin, which exerts a leveling effect on acid strength, just as the strongest acid in water is the hydronium ion. Apparently, pTSA (TsOH) is almost completely ionized by MF resins, which explains why even stronger acids, such as hexafluorophosphoric acid, exhibit similar catalytic activity. On the other hand, pTSA is more effective for class I MF resins than weaker acids, such as carboxylic acids or butylphosphoric acid. The latter acid is sometimes used in the hope of enhancing adhesion to metal. With pTSA, the rate of co-condensation of an HMMM resin and acrylic polyol is reported to be proportional to the square root of the acid concentration (Bauer, 1986).

While the acid strength of a strong acid catalyst makes little difference in cure rates, the choice of acid (or blocked acid) catalyst can lead to differences in film properties of cured coatings. The most widely used catalyst is pTSA. The grade of pTSA is important, since some grades contain significant quantities of sulfuric acid, which can lead to pronounced yellowing of films during baking. Excess pTSA can also bloom to the surface of cured coatings. Water resistance of films can be improved by the use of a more hydrophobic sulfonic acid, such as dinonylnaphthalene disulfonic acid (DNNDSA), which is reported to produce films that are less likely to blister when exposed to high humidity conditions (Calbo, 1980). DNNDSA is particularly effective for coatings applied directly to metal, since such coatings are prone to blistering on humidity exposure with pTSA.

p-Dodecylbenzenesulfonic acid (DDBSA) provides similar catalytic activity and film properties to those of DNNDSA when used in coatings applied over a prime coat. However, when a DDBSA-catalyzed coating is used directly on steel, adhesion of the coating is likely to suffer. It seems probable that the sulfonic acid group on DDBSA is strongly adsorbed on the steel surface, resulting in a layer of dodecyl groups on the surface. The low surface tension of the long hydrocarbon dodecyl groups may cause dewetting by the rest of the coating or cause the formation of a weak boundary layer that reduces adhesion. The better adhesion with DNNDSA-catalyzed coatings may be related to the presence of two sulfonic acid groups or weaker surfactant properties.

The activity of acid (or blocked acid) catalysts can be affected by pigmentation of the coating. Some grades of titanium dioxide pigments lead to loss of catalyst activity with storage time. The loss is related to the composition of surface treatments on the TiO_2. Silicon dioxide-treated TiO_2 is preferable to aluminum oxide (alumina)-treated TiO_2 in this respect, probably owing to the basicity of alumina, which tends to neutralize the acid. When alumina-treated TiO_2 is required for other reasons, a higher concentration of acid catalyst may be necessary to counteract the effect.

In waterborne MF–polyol coatings, the acid catalyst is also neutralized by the amine utilized to neutralize the solubilizing carboxylic acid groups in the polyol (see Section 8.3 for discussion). It was also shown that commonly utilized hydroxy-functional amines, such as *N,N*-dimethylethanolamine (DMEA), may participate in transetherification of MF resins and/or transesterification of ester groups in the polyol during storage or initial stages of curing. In this event, volatilization of the amine may be retarded or prevented during cure, thereby retarding cure response (Wicks and Chen, 1979). It was also shown that 2-amino-2-methylpropanol (AMP) enhances cure since the primary amine can react with an acrylic resin to form amides, reducing the overall basicity. Formation of an amide with AMP coupled with transesterification by the accompanying hydroxyl group results in a cross-link. Furthermore, AMP has been shown to react with formaldehyde to form an oxazolidine, also reducing basicity and cure inhibition relative to DMEA (Ferrell et al., 1995). Dynamic mechanical analysis (DMA) also supports reduced inhibition as well as auxiliary cross-linking by AMP (Hill et al., 1996). In addition to these cure advantages, AMP is exempt from VOC regulations in the United States and Canada (Section 18.9.1).

11.3.2 Kinetics and Mechanism of MF–Polyol Co-condensation

Many studies have been directed at elucidating the mechanism(s) of reactions between MF resins and polyols. Until recently, most of these studies were based on following the time necessary to achieve some degree of film hardness. However, both co- and self-condensation reactions contribute to hardness. Therefore, conclusions from such studies applied specifically to the co-condensation reaction between polyol groups, and MF resins must be viewed with caution. Furthermore, many studies implicitly assume that the film structure is homogeneous when, in fact, there may be compositional gradients and other heterogeneities in the film, as discussed in the succeeding text.

A misconception resulting from using hardness results to study kinetics of class 1 resins is that the co-condensation reaction was thought to be much slower than it was later found to be. Disappearance of polyol hydroxyl groups, as followed by FTIR, showed that the reaction is more rapid than the development of film properties (Lazzara, 1984; Nakamichi, 1986; Yamamoto et al., 1988). Relatively slow development of film properties is characteristic of step-growth polymerization processes and may also reflect the slower rate of self-condensation reactions, which contribute to the achievement of optimal cross-link density.

Methods have been developed for measuring the extent of co-condensation directly. Infrared (IR) and FTIR spectroscopy have been used to follow changes in functional group concentrations using class I MF resins as a function of time and temperature (Bauer, 1986). In these studies, particular care must be taken in interpreting results because of band overlap in the hydroxy region. The rate of evolution of volatile reaction products, including formaldehyde and methyl alcohol, has been followed by gas chromatography (Blank, 1979; Lazzara, 1984) and by thermogravimetric analysis (Mijs et al., 1983). Major contributions have been made by the application of DMA to MF–polyol cross-linking coatings (Hill and Kozlowski, 1986, 1987); also see Section 4.2.3. Yamamoto et al. (1988) employed oscillating plate rheometry.

In class I resins, it has been observed that reactivity is reduced by the presence of imino (>NH) groups in the resin. Fractionation to remove essentially all of these groups from a commercial HMMM resin yielded a surprising increase in reactivity with hydroxy-functional resins, enabling lower cure temperatures (Jones et al., 1994; Hill and Lee, 1999). It is speculated that the relatively basic $-NHCH_2OCH_3$ groups present in most commercial HMMM resins tie up a portion of the catalyst.

The mechanism of co-condensation of MF class I resins with polyols remains controversial with experimental evidence being interpreted in support of (or against) the S_N1 or S_N2 mechanisms, provided in Scheme 11.2. However, the two mechanisms are not mutually exclusive and may occur simultaneously. Owing to the expectation that the activation parameters, E_a and A, are larger for the S_N1 reaction (Section 2.3.2 and Figure 2.6), the S_N1 mechanism is favored with increasing temperature.

The novel use of chiral monofunctional alcohols provided kinetic evidence for equal rates of reactivity of primary and secondary alcohols with a high HMMM resin, but higher reactivity in the reverse reaction with secondary alcohols (Meijer, 1986). These findings support the S_N1 mechanism and are consistent with the general order of reactivity in coatings of primary > secondary alcohols, since the curing reactions occur under reversible conditions. However, the utilization of an oversimplified rate expression may compromise these conclusions.

Using an HMMM–acrylic composition and following the co-condensation reaction by FTIR, Bauer (1986) provided support for the S_N1 mechanism. He also interpreted the role of methyl alcohol in the kinetic order of the reaction and the rate dependence on polyol in support of the S_N1 mechanism. A key point is that the rate of co-condensation by the S_N1 mechanism is expected to be dependent on the concentration of polyol if the methyl alcohol, formed from HMMM, competes with the polyol for the intermediate carbocation; refer to Scheme 11.2. Water also competes with the polyol for this carbocation. This valid interpretation of the S_N1 mechanism is an important consideration, since dependence of the rate on the polyol had mistakenly been interpreted as evidence against the S_N1 mechanism. Competition for the intermediate carbocation by methyl alcohol is expected to gain relative importance as the reaction proceeds because the methyl alcohol concentration increases, at least temporarily, whereas the polyol concentration decreases with increasing conversion. Thus, kinetic studies based on hardness or other film properties, which tend to develop only at high conversions, are expected to exhibit strong dependence on the concentration and nature of the polyol. Such results had generally been interpreted in favor of the S_N2 mechanism but are equally consistent with the S_N1 mechanism.

A further complication is the strong possibility that cross-linking with MF resins leads to a gradient of compositions and properties within films. While relatively few films have been studied in this way, most studies detected a gradient. For example, Haacke et al. (1995) used a microtome to separate layers in acrylic/MF clear coats and found that different layers had different compositions. Furthermore, Haacke et al. found that T_g was as much as 15°C higher at the surface than within in the film and that cross-link density was almost twice as high near the surface. They attributed the gradient to the effect of the differing extent of escaping by-product alcohol on the cross-linking equilibrium reactions at different levels in the film. There is the additional possibility that catalyst concentrations may not be uniform within the film, especially when catalysts blocked with volatile amines are used. Hiurayama and Urban (1992) observed melamine enrichment at the film/air interface under certain circumstances. Other studies have suggested that there may be a very thin layer of material at the surface of the coating that is quite different in composition and properties than the bulk (Jones et al., 1998). These scattered reports suggest that to fully understand the performance of amino resin cross-linked coatings (and perhaps of other types as well), it may be necessary to appreciate the effects of composition and property gradients within the films.

It has also been speculated that MF cross-linked films may be heterogeneous, owing to the growth of nano clusters of self-condensed melamine resin during the curing process. Conceivably, growth of such clusters may be partly responsible for embrittlement caused by excessive baking.

11.3.3 Package Stability Considerations

Package stability of coatings containing MF resins is affected by factors other than catalyst levels, discussed in Section 11.3.1. Formulations with class II resins generally exhibit poorer package stability than class I due to the preponderance of >NH and >NCH_2OH groups, which participate in weak acid-catalyzed reactions that occur slowly during storage at room temperature. Stability is somewhat improved by the addition of small quantities of tertiary amines. Primary or secondary amines, which react with formaldehyde, should not be used with class II resins. Free formaldehyde is in equilibrium with melamine methylol groups in these resins. Reaction of the formaldehyde with primary or secondary amines displaces the equilibrium reaction in favor of formaldehyde formation (i.e., demethylolation), which reduces the methylol group functionality of the MF resins.

Cross-linking and viscosity buildup occur both when the MF resin self-condenses and when it reacts with the polyol. Accordingly, an important approach for increasing package stability is to use as much monofunctional alcohol as possible in formulations containing either class of MF resins. The presence of monofunctional alcohol in the formulation extends the storage stability since its reaction with the MF resin does not lead to cross-linking. Generally, it is desirable to utilize the same alcohol that is used to synthesize the MF resin. If a different alcohol is used, undesirable changes may occur. For example, if *n*-butyl alcohol is used in the solvent with a methoxymethylmelamine resin, the cure response gradually becomes slower as the proportion of butyl ether increases. The viscosity of a coating may decrease during storage because butoxymethylmelamine is less viscous (it has a lower T_g) than the corresponding methoxymethylmelamine. Viscosity decrease may also result from cleavage of dimethylene ether bridges by excess alcohol.

11.3.4 MF Resin Reactions with Carboxylic Acids, Urethanes, Carbamates, and Malonate-Blocked Isocyanates

Carboxylic acid-functional resins react with MF resins to form the corresponding ester derivatives (Eq. 11.2); the reaction is slower than transetherification with hydroxyl groups. Carboxylic acid groups are present in substantial levels in waterborne hydroxy-functional resins to enhance dispersibility, as well as in some solventborne polyols, at low levels, to enhance the stability of pigment dispersions. When significant levels of carboxylic acid groups are present, it is desirable to cure under conditions in which the carboxylic acid groups react, since residual acid groups increase water sensitivity of cured films.

As shown in Eq. 11.3, MF resins react with urethane groups (—OC(=O)NH—), also called secondary carbamates. Polyurethanes, containing these groups and no hydroxyl groups, can be cross-linked with MF resins at temperatures only slightly higher than those used to cross-link polyols (Higginbottom et al., 1999). Accordingly, when hydroxy-terminated urethanes are used with MF resins, both groups are involved in cross-linking. Both groups also react with MF resins when polyurethane polyols are used as reactive diluents (Hill, 1997). A variety of MF resins can cross-link secondary carbamates; class I resins are more reactive than class II resins (Higginbottom et al., 1999). Secondary carbamates are nearly as reactive as hydroxyl groups despite their greater steric hindrance. A possible explanation for this surprising result is that the carbamate reaction is irreversible (Blank et al., 1997). Thus, the forward reaction of carbamate groups might be substantially slower than the forward reaction of hydroxyl groups, but the overall rate could be more nearly equivalent if the reverse reaction does not occur with carbamates.

The practice of cross-linking polyols with mixtures of MF and polyisocyanate resins (Chapter 12) has become increasingly common. Formulators should consider the strong possibility that some of the carbamate formed in the isocyanate/hydroxyl reaction will react further with MF resin.

Polymers having pendant —OC(=O)NH_2 structures (primary carbamates) are less hindered than secondary carbamates and would be expected to cross-link more rapidly. Conclusive evidence for this expectation is lacking, but support has been disclosed in patent examples, notably that polymers having primary carbamate groups cure with class I MF resins under the same conditions as polymers with primary —OH groups (Rehfuss and Ohrbom, 1994; Rehfuss and St. Aubin, 1994). Primary carbamate-functional resins can be made by reacting isocyanate-terminated resins (Rehfuss and Ohrbom, 1994) or —N=C=O functional acrylic resins (Rehfuss and St. Aubin, 1994) with hydroxypropyl carbamate. The isocyanate-based resins can be cross-linked with class I MF resins, but the cross-link density is too high for some applications. Functionality can be reduced by blocking half of the functional groups of class I resins with butyl carbamate (Rehfuss and St. Aubin, 1994). Automotive clear coats (Section 30.1.5) made from such combinations combine the usual high mar resistance of MF–polyol cross-linked coatings with enhanced resistance to environmental etching. Such clear coats play an important role in automotive finishing. Many refinements of the technology have been described, for example, by Jhaveri (2015).

He and Blank (1998) showed that class I MF resins react with adducts derived from the addition of malonic ester derivatives to isocyanates, at rates similar to their reactions with —OH groups. Their NMR studies of a model compound confirmed that environmentally resistant carbon–carbon cross-links are irreversibly formed in the process, indicating that MF reacts at the highly activated CH group of the adduct, which is attached to three carbonyl groups. Accordingly, a coating made from a malonic ester derivative of a triisocyanate and a class I MF resin exhibited superior environmental etch resistance. Malonic ester derivatives of isocyanates can also function as cross-linkers for hydroxy-functional resins by transesterification, as discussed in Section 12.5.2.

11.4 OTHER AMINO RESINS

Other amino resins are used on a smaller scale in coatings. Structures of precursor compounds are provided in the beginning of this chapter. While the chemistry of these resins is similar to that of MF resins, there are differences, particularly as a result of differences in basicity of the resins (Parekh, 1979).

11.4.1 Urea–Formaldehyde Resins

Substantial amounts of urea–formaldehyde (UF) resins are used in coatings because of their low cost and high reactivity. Urea reacts with formaldehyde to form methylol derivatives. The first and second formaldehyde units add easily, but trimethylol and tetramethylol derivatives of urea form more slowly, if at all. Methylolated ureas can be etherified by reaction with alcohols analogously to methylolated melamines. Self-condensation reactions also occur under both acidic and basic conditions. Various etherified UF resins are made utilizing different ratios of formaldehyde to urea and different alcohols for etherification.

In general, UF resins are the most economical amino resins, as well as the most reactive. In some circumstances, they are so reactive that they are used as two-package coating systems. With sufficient acid catalyst, coatings formulated with UF resins and polyols can cure at ambient or mildly elevated temperatures. However, such coatings have poor exterior durability and relatively poor resistance to hydrolysis, probably because the cross-links are relatively reactive. UF resins are used in coatings for temperature-sensitive substrates such as wood furniture, paneling, and cabinetry. In such applications, low temperature baking is essential, and corrosion resistance (related to the hydrolytic stability of the cross-links) is unimportant. Blends of UF and MF resins can combine fast initial cure, low cost, and acceptable film properties (Vaughan and Jacquin, 1999).

11.4.2 Benzoguanamine–Formaldehyde Resins

A range of resins analogous to MF resins can be prepared from benzoguanamine. As with melamine, benzoguanamine may be methoxymethylated and/or butoxymethylated to various extents. The average functionality is lower because there are only two $-NH_2$ groups per molecule. Etherified benzoguanamine–formaldehyde (BF) resins yield cross-linked films with greater resistance to alkali and to alkaline detergents, such as sodium tripolyphosphate, compared with MF resins. They also impart excellent toughness. Exterior durability, however, of BF-based coatings is poorer than MF-based coatings. Thus, applications for BF resins include coatings for washing machines and dishwashers, in which resistance to alkaline detergents and toughness is more important than exterior durability. The reduced exterior durability of BF-based coatings probably reflects lower photostability arising from the presence of the phenyl group on the triazine ring of benzoguanamine, although evidence supporting this reasonable hypothesis does not appear to be available. Aside from the hydrophobicity of the phenyl groups, a convincing explanation for enhanced alkali resistance is not intuitively forthcoming.

11.4.3 Glycoluril–Formaldehyde Resins

Glycoluril reacts with formaldehyde to give tetramethylolglycoluril (TMGU) (Parekh, 1979). In the pH range of 4–8, aqueous solutions of TMGU show an equilibrium level of about 3.6 methylol groups per glycoluril, with 0.4 parts free formaldehyde. Above pH 8.5, demethylolation is more favored. For example, at pH 11, there are approximately equal parts of methylol groups and free formaldehyde. Below pH 3, self-condensation occurs to form glycoluril dimers with methylene ether bridges, as shown in Eq. 11.11, where $>N-CH_2-OH$ represents methylolated glycoluril groups.

$$>N-CH_2-OH \underset{}{\overset{pH<4}{\rightleftharpoons}} >N-CH_2-O-CH_2-N< + H_2O \tag{11.11}$$

This behavior is unlike that of MF or UF resins. While TMGU does not self-condense above pH 4, methylolated melamines and ureas self-condense at pH 7 and higher.

TMGU reacts with alcohols in the presence of strong acid catalyst to form tetraalkoxymethylglycoluril (GF) resins. Tetramethoxymethylglycoluril is a relatively high melting solid and is used as a cross-linker in powder coatings (Section 28.1.3). In solution coatings, (dimethoxymethyldiethoxymethyl)glycoluril and tetrabutoxymethylglycoluril are used, since they are liquids and are more readily handled. The mixed methyl/ethyl ether is water soluble.

Relative to other amino resins, coatings based on GF resins exhibit greater flexibility at a similar cross-link density. Hence, GF resins are used in applications in which flexibility is particularly important, such as coil coatings and can coatings. GF resins possess the further advantage that less formaldehyde is evolved during cure, as compared with MF resins. Also, the glycoluril cross-linked polyols are more resistant to hydrolysis under acidic conditions than MF cross-linked polyols and are reported to be UV resistant (Parekh, 1979). The higher cost of GF resins limits their use.

11.4.4 Poly(meth)acrylamide–Formaldehyde Resins

Acrylic copolymers of *N*-isobutoxymethylacrylamide can be prepared by two routes: (1) synthesis of *N*-isobutoxymethylacrylamide monomer (from a stepwise reaction of acrylamide with formaldehyde and isobutyl alcohol), followed by copolymerization with acrylate monomers, and (2) copolymerization of acrylamide, followed by stepwise reaction with formaldehyde and isobutyl alcohol. Analogous reactions can be carried out with other alcohols and with methacrylamide.

Such alkoxymethyl(meth)acrylamide resins are used in coil coatings, where, with proper design, they are weather resistant and more flexible than coatings based on MF/acrylic polyol resins. One can speculate that their flexibility is enhanced by the absence of cross-linked clusters of self-condensed MF resins with their rigid triazine rings.

REFERENCES

Albrecht, N., *Proceedings of the 13th International Waterborne and Higher-Solids Coatings Symposium*, New Orleans, LA, **1986**, p 200.

Bauer, D. R., *Prog. Org. Coat.*, **1986**, *14*, 193.

Bauer, D. R.; Dickie, R. A., *J. Coat. Technol.*, **1982**, *54(685)*, 57.

Blank, W. J., *J. Coat. Technol.*, **1979**, *51(656)*, 61.

Blank, W. J., et al., *Polym. Mater. Sci. Eng.*, **1997**, *77*, 391.

Calbo, L. J., *J. Coat. Technol.*, **1980**, *52(660)*, 75.

Chang, T. T., *Prog. Org. Coat.*, **1996**, *45*, 211.

Chu, G.; Jones, F. N., *J. Coat. Technol.*, **1993**, *65(819)*, 43.

van Dijk, J. H., et al., *FATIPEC Congress Book*, **1980**, Vol. *II*, p 326.

Ferrell, P. E., et al., *J. Coat. Technol.*, **1995**, *67(851)*, 63.

Haacke, G., et al., *J. Coat. Technol.*, **1995**, *67(843)*, 29.

He, Z. A.; Blank, W. J., *Proceedings of the 25th International Waterborne, High-Solids, and Powder Coatings Symposium*, New Orleans, LA, **1998**, p 21.

Higginbottom, H. P., et al., *J. Coat. Technol.*, **1999**, *71(894)*, 49.

Hill, L. W., *Polym. Mater. Sci. Eng.*, **1997**, *77*, 387.

Hill, L. W.; Kozlowski K., *Proceedings of the International Symposium on Coatings Science and Technology*, Athens, **1986**, p 129.

Hill, L. W.; Kozlowski, K., *J. Coat. Technol.*, **1987**, *51(751)*, 63.

Hill, L.W.; Lee, S.-B., *J. Coat. Technol.*, **1999**, *71(897)*, 127.

Hill, L. W., et al., Effect of Amine Solubilizer Structure on Cured Film Properties of Water-Reducible Thermoset Systems in Provder, T. A., et al., Eds., *Film Formation in Waterborne Coatings*, ACS Symposium Series 648, American Chemical Society, Washington, DC, **1996**.

Hiurayama, T.; Urban, M. W., *Prog. Org. Coat.*, **1992**, *20*, 81–96.

Jhaveri, S. B., Patent Publication WOA1 (**2015**).

Jones, F. N., et al., *Prog. Org. Coat.*, **1994**, *24*, 189.

Jones, F. N., et al., *Prog. Org. Coat.*, **1998**, *34*, 119.

Lavric, S., et al., *Prog. Org. Coat.*, **2015**, *81*, 27–34.

Lazzara, M. G., *J. Coat. Technol.*, **1984**, *56(710)*, 19.

Meijer, E. W., *J. Polym. Sci. A Polym. Chem.*, **1986**, *24*, 2199.

Mijs, W. J., et al., *J. Coat. Technol.*, **1983**, *55(697)*, 45.

Morimoto, T.; Nakano, S., *J. Coat. Technol.*, **1994**, *66(833)*, 75.

Nakamichi, T., *Prog. Org. Coat.*, **1986**, *14*, 23.

Nastke, R., et al., *J. Macromol. Sci.*, **1986**, *A23*, 579.

Parekh, G. G., *J. Coat. Technol.*, **1979**, *51(658)*, 101.

Rehfuss, J. W.; Ohrbom, W. H., US patent 5,373,069 (**1994**).

Rehfuss, J. W.; St. Aubin, D. L., US patent 5,356,669 (**1994**).

Santer, J. O., *Prog. Org. Coat.*, **1984**, *12*, 309.

Subrayan, R. P.; Jones, F. N., *J. Appl. Polym. Sci.*, **1996**, *62*, 1237.

Vaughan, G. D.; Jacquin, J. D., *Proc. Int. Coat. Exhib.*, **1999**, *288*.

Wicks, Z. W., Jr.; Chen, G. F., *J. Coat. Technol.*, **1979**, *50(638)*, 39.

Yamamoto, T., et al., *J. Coat. Technol.*, **1988**, *60(762)*, 51.

Chapter 12

Polyurethanes and Polyisocyanates

Polyurethanes are polymers containing urethane (—NH—CO—O—) linkages; *carbamate* is a synonym for urethane. Urethanes are usually formed by reaction of an alcohol with an isocyanate, but they can also be made by other methods. Commonly, urethanes made by a non-isocyanate route are called carbamates. To add to the confusing terminology, the terms *urethane* and *polyurethane* are applied to almost any binder derived from isocyanates even though only part, if any, of the reaction products are urethanes.

Polyisocyanates are versatile cross-linkers for other resins, and they are building blocks for resins tailored for thousands of specialized applications, large and small. Here we will focus on the basic chemistry, the common polyisocyanates, as well as solventborne and waterborne formulations. Urethane-containing resins and cross-linkers for alkyd, powder, and electrodeposition coatings are described in Chapters 15, 28, and 29, respectively. Potential non-isocyanate alternatives to polyurethanes are described in Section 17.10. In our terminology diisocyanates are called polyisocyanates for convenience.

Urethane groups form intermolecular hydrogen bonds between polymer molecules, which may be acyclic and/or cyclic:

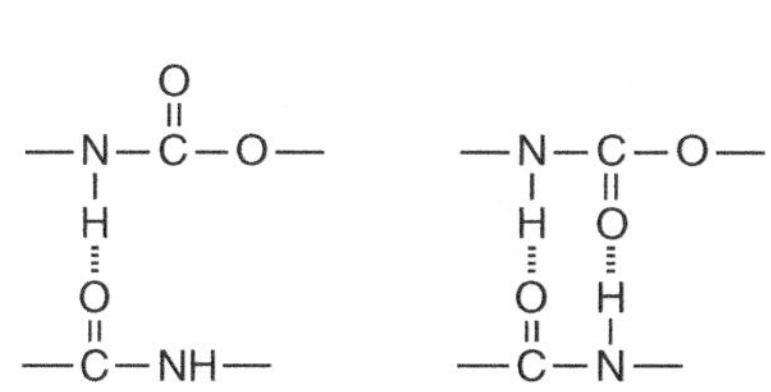

Under mechanical stresses, energy (about 20–25 kJ mol^{-1}) may be absorbed by separation of acyclic H-bonds, which can reform (probably in different positions) when the stress is removed. Energy absorption by this reversible bond-breaking/reforming process reduces the likelihood of irreversible breaking of covalent bonds leading to degradation. In addition, the urethanes can be designed with "soft" and "hard" segments, that is, segments having low T_g and high T_g, respectively. When stress is applied, the soft segments can extend between the hard segment anchors. This factor combined with the H-bonding permits the preparation of hard, yet extensible coatings, for example, 2H hardness and 300% extension. These properties permit design of polyurethanes that are abrasion resistant while still resisting swelling with solvents. Polyurethanes tend to absorb water; absorption is facilitated by H-bonding with water, and it plasticizes the coatings.

The isocyanate group is highly reactive, so polyisocyanates can be used to make coatings that cure at ambient temperature or at moderately elevated temperatures. Coatings based on aliphatic polyisocyanates exhibit exceptional exterior durability when stabilized with hindered amine light stabilizers (Section 5.2.3). Resistance of urethane coatings to environmental etching is superior to that of many melamine–formaldehyde (MF) cross-linked coatings (Nordstrom and Dervan, 1993).

The principal limitations of isocyanates are cost and toxicity (particularly of relatively low molecular weight (MW) compounds). Any cross-linker that reacts with hydroxyl, amine, and/or carboxylic acid groups near room temperature is likely to be toxic, since the body contains

Organic Coatings: Science and Technology, Fourth Edition. Frank N. Jones, Mark E. Nichols, and Socrates Peter Pappas.

proteins and other materials having such substituents. The important question is not toxicity as much as toxic hazard. Since higher MW reduces vapor pressure and permeability through body membranes, toxic hazard decreases, in general, as MW increases. With adequate ventilation and use of air masks and protective clothing, as commonly done in chemical factories and resin manufacturing plants, even relatively low MW isocyanates can be handled safely. Most low MW diisocyanates are sensitizers, however, which can result in an allergic reaction on prolonged exposure. The exposure leading to sensitization varies over a wide time period for individuals and may occur only after years of repeated exposure for some. The most common symptoms are hives and asthma. In extreme cases sensitized people cannot be in the same room, or even the same building, in which isocyanates are handled. Isocyanate suppliers provide safety instructions for handling and using isocyanates. Coating applicators, especially amateur users, are less likely than professional chemists to have appropriate facilities or even to follow safety instructions. To minimize the hazard, most isocyanates used in coatings are oligomeric or polymeric derivatives. The competence of the user to handle toxic materials should be taken into consideration in establishing formulations that contain any highly reactive cross-linker.

12.1 REACTIONS OF ISOCYANATES

Isocyanates react with any active hydrogen compound. Alcohols and phenols react with isocyanates to form urethanes, as in the following reaction, in which R and R′ can be aromatic or aliphatic groups. The reaction is reversible at elevated temperatures.

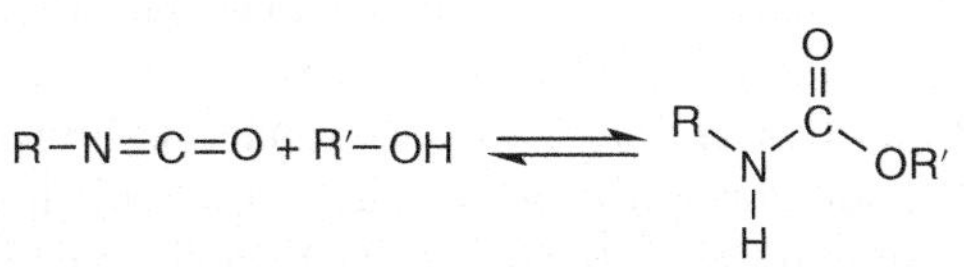

In general, rates of urethane formation decrease in the following order: primary alcohols > secondary alcohols > 2-alkoxyethanols > 1-alkoxy-2-propanols > tertiary alcohols. As a broad rule, ease of reversion is the inverse of reactivity. Urethanes from tertiary alcohols are relatively unstable and may decompose with heat to give alkenes, carbon dioxide, and amines, rather than alcohols and isocyanates.

Urethanes can react further with isocyanates to form *allophanates*. This reaction is much slower than the reaction of isocyanate with alcohol but can be accelerated with catalysts such as choline 2-ethylhexanoate. The allophanate structure has excellent resistance to hydrolysis.

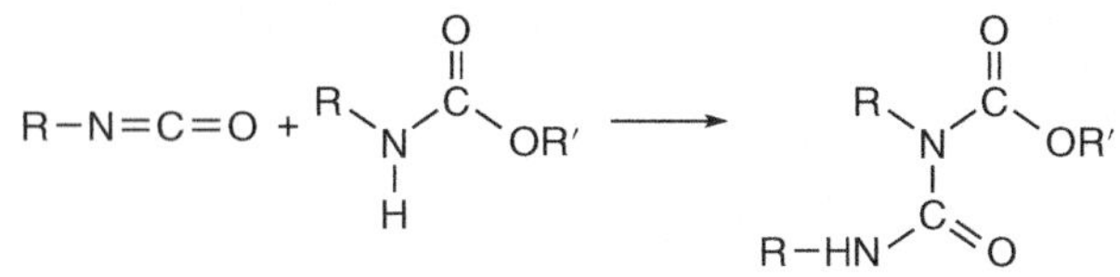

Isocyanates react rapidly with primary and secondary amines to form ureas. The reaction is much faster than the reaction of isocyanates with alcohols. For many coating applications, reactions are too rapid; however, hindered secondary amines have been developed that react sufficiently slowly to permit use in two-package (2K) coatings (Wicks and Yeske, 1997), as further discussed in Section 12.4.1.

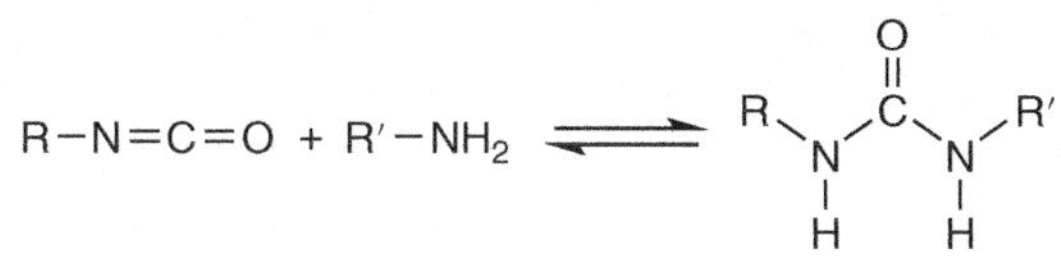

Isocyanates react with ureas to form biurets. Biuret formation is slower than urethane formation but faster than allophanate formation.

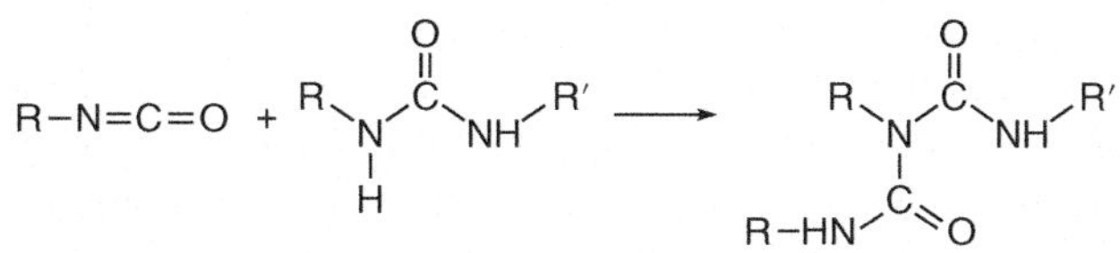

Isocyanates react with water to form unstable carbamic acids, which dissociate into carbon dioxide and an amine. The amine is so much more reactive than water that it reacts with a second isocyanate to form a urea even in the presence of excess water. In most circumstances, the reactivity of water with isocyanates is somewhat slower than that of secondary alcohols but more rapid than that of ureas.

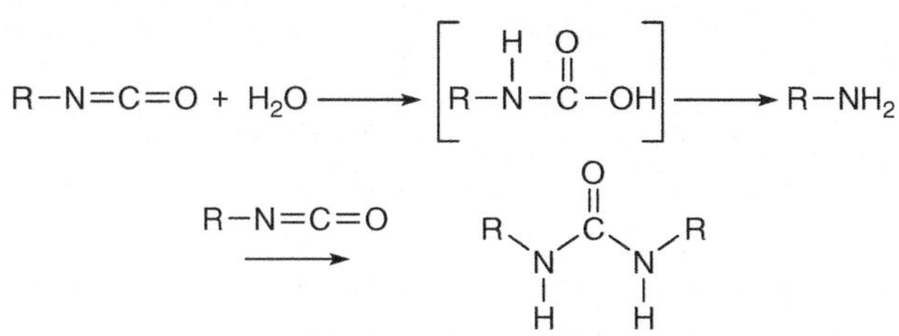

Carboxylic acids react relatively slowly with isocyanates to form amides and CO_2. Hindered carboxylic acid groups, such as in 2,2-dimethylolpropionic acid (DMPA) react very slowly; accordingly, DMPA is commonly used to incorporate COOH groups into resins for polyurethanes.

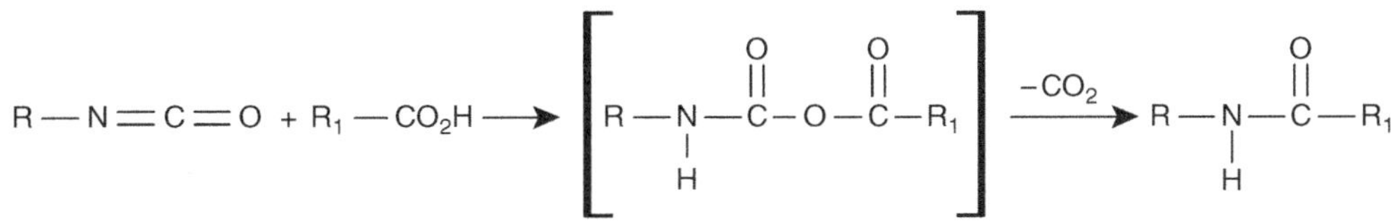

Isocyanates also react with thiols to form the sulfur analogues of urethanes. As a thiol is a stronger nucleophile than an alcohol, the thiol–isocyanate reaction is more rapid, particularly when catalyzed by a base. Coatings based on this chemistry have been developed for the automotive refinish market (Section 33.3; Dogan et al., 2006).

Isocyanates also react with each other to form dimers (uretdiones) and trimers (isocyanurates). Formation of uretdiones is catalyzed by phosphines. Trimerization to form aliphatic isocyanurates is catalyzed by quaternary ammonium compounds; trimerization of aromatic isocyanates is catalyzed by tertiary amines. Uretdiones decompose thermally to regenerate isocyanates and are used as blocked isocyanates (Section 12.5). Isocyanurates are stable and are extensively used as building blocks for multifunctional isocyanates (Section 12.3). Another useful cross-linker is a stable asymmetric trimer containing an imine group (structure in Section 12.3.2). Formation of asymmetric trimers is catalyzed by trialkyl phosphines or by quaternary phosphonium fluorides (Richter et al., 2000).

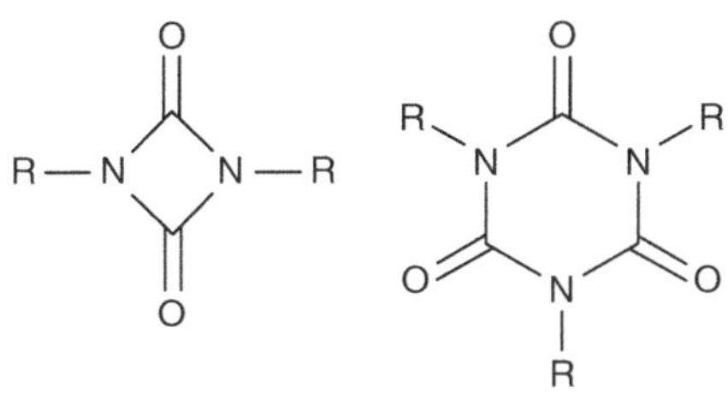

Uretdione Isocyanurate

12.2 KINETICS OF REACTIONS OF ISOCYANATES WITH ALCOHOLS

Although polyisocyanates are quite reactive, catalysts are generally employed both when they are used as cross-linkers and as building blocks. One reason is to achieve the desired reaction rates. Another, as shown below, is that uncatalyzed reactions often follow third-order kinetics and become very slow in the late stages unless they are *autocatalyzed* (catalyzed by the reaction products). Yet another reason is to increase selectivity, promoting the desired reaction at the expense of undesired side reactions.

The mechanisms and catalysis of the reaction of isocyanates with alcohols are complex. Kinetic studies can be revealing, but one must use caution when evaluating reaction rate data from the literature, especially in interpreting tables of rate constants compiled from different sources. Original literature should be consulted to make sure that comparisons are based on studies done in the same solvents and initial concentrations, since reaction rates depend on these variables. Also, rates of urethane formation are commonly determined by following disappearance of isocyanate. If other products, such as allophanates and isocyanurates, are also formed, isocyanate disappearance does not correspond to urethane formation. Further, many of the publications do not provide the number of replicates, if any, or statistical confidence limits for the data.

Formulators often compare rates of cross-linking by determining the methyl ethyl ketone (MEK) double rub resistance of films as a function of time, and/or temperature, and/or catalyst. Usually the test is stopped when the number of MEK double rubs exceeds 200 without marring of the surface. The test is sensitive to cross-link density in the relatively early stages of the reaction, but insensitive in assessing cure at high conversion for highly cross-linked coatings, for which there are better methods. Higginbottom et al. (1998) demonstrated that a 200+ MEK double rub test does not indicate full cure, or even correlate with the time to reach full cure, as determined by dynamic mechanical analysis (DMA, Section 4.2.3). In other words, a film that reaches 200+ double rub resistance faster than another film will not necessarily reach full cross-linking as rapidly as another that develops double rub resistance more slowly. In addition to DMA analysis, swelling by a solvent is a convenient method for measuring cross-link density at high conversions. Low conversion rates can be compared by measuring the time required for solutions of a triol and diisocyanates to gel or by observing rates of change in NMR spectra.

12.2.1 Noncatalyzed Reactions

One might expect urethane formation to follow second-order kinetics, in which the rate is proportional to the concentration of each reactant, as shown in Eq. 12.1. Indeed, the reaction of *n*-butanol with phenyl isocyanate in the dipolar aprotic solvent acetonitrile has been reported to follow second-order kinetics (Schwetlick et al., 1994).

$$\text{rate} = k\left[\text{R}-\text{N}=\text{C}=\text{O}\right]\left[\text{R}'-\text{OH}\right] \qquad (12.1)$$

Other kinetic studies show that the situation is more complex, and often the kinetic order changes as the reaction progresses. In at least some nonpolar or low polarity solvents, the reaction follows third-order kinetics, at least in its early stages: first order in isocyanate and second order in alcohol concentration. Since it seems reasonable to assume that reactions will be occurring in low polarity media in most coating applications, rate in Eq. 12.2 would apply to such cases. The implication of Eq. 12.2 is that two molecules of alcohol are involved in the reaction with one molecule of isocyanate:

$$\text{rate} = k[\text{R}-\text{NCO}][\text{R}'-\text{OH}]^2 \tag{12.2}$$

Such results can be reasonably interpreted in terms of the mechanism provided in Scheme 12.1. Reaction of isocyanate and alcohol (at a rate proportional to k_1) produces a zwitterionic reactive intermediate (**RI**), which can revert to starting materials (k_{-1}) or proceed to product, urethane (k_2). Product formation requires proton transfer from oxygen to nitrogen. The second molecule of alcohol may facilitate this transfer by way of a six-membered cyclic activated complex (**A**).

Equation 12.3 is a rate expression that takes into account the sequence of reactions in Scheme 12.1:

$$\text{rate} = k_1[\text{RNCO}][\text{R'OH}]\frac{k_2[\text{R'OH}]}{k_2[\text{R'OH}]+k_{-1}} \tag{12.3}$$

If the term k_2[R′OH] is substantially greater than k_{-1}, Eq. 12.3 simplifies to Eq. 12.1. This situation is favored when the alcohol concentration is high, such as at the start of the reaction. If, on the other hand, the term k_{-1} is substantially larger than k_2[R′OH], Eq. 12.3 simplifies to Eq. 12.2, second-order dependence on alcohol concentration is observed, and the overall rate is third order. The latter circumstance is more likely, since alcohol is consumed during the reaction, reducing [R′OH]. As a result, the reaction rate diminishes rapidly as the reaction proceeds (since the rate is proportional to the square of a small, diminishing alcohol concentration). The mechanism in Scheme 12.1 also provides an explanation for the strong solvent effects observed. Rates decrease as the hydrogen-bond acceptor potential of solvents increase in the order: aliphatic hydrocarbons, aromatic hydrocarbons, esters and ketones, ethers, and glycol diethers. Rates in aliphatic hydrocarbons can be two orders of magnitude faster than in glycol diethers. H-bonding of the zwitterionic **RI** and/or the alcohol with solvent may reduce the rate by reducing the concentration of cyclic activated complex **A**. The mechanism in Scheme 12.1 is also consistent with the observed higher reactivity of aromatic isocyanates. When R is an aromatic group, the negative charge on nitrogen in **RI** is delocalized in the π-electron system of the aromatic ring, resulting in a lower energy (by resonance stabilization) for the **RI** and a faster rate of urethane formation relative to aliphatic isocyanates.

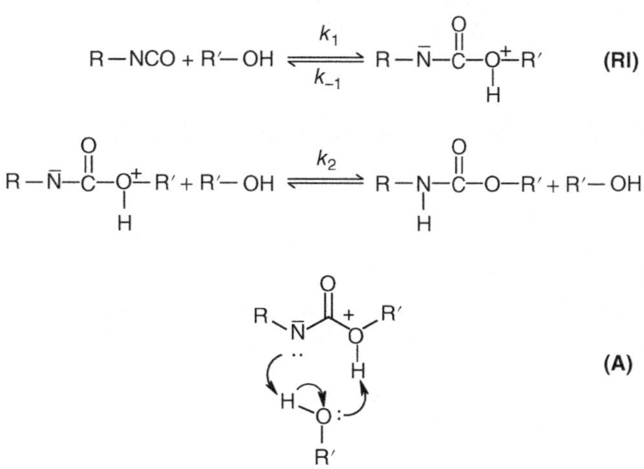

Scheme 12.1 Possible mechanism for uncatalyzed reaction of an isocyanate with an alcohol.

Rate constants reported in the literature for reactions of alcohols with isocyanates are commonly apparent initial second-order rate constants. Initial rate constants provide a useful way of comparing reactivities without becoming involved in the complexities of the kinetics, provided comparisons are made in the same solvent and the same initial concentrations of alcohols and isocyanates. Unfortunately, rate constants are sometimes compared that were determined in different solvents and/or at different initial concentrations. Such comparisons can be misleading.

The kinetics of the isocyanate–alcohol reaction are even more complex than discussed thus far, since the reaction is autocatalyzed to some extent by the urethane being formed. Another cyclic transition state **B** could be involved, in which a molecule of urethane facilitates proton transfer from oxygen to nitrogen in the zwitterionic **RI**.

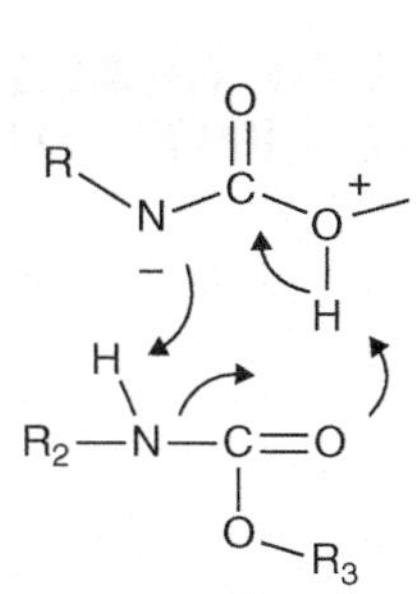

B

The autocatalyzed reaction would follow third-order kinetics but would be first order in alcohol. The decline in alcohol concentration would be offset by the increase in urethane concentration as the reaction proceeds if autocatalysis by the urethane is effective and if mobility is sufficient for it to reach the reaction site in the proper orientation. The apparent rate constant would change through the course of the reaction depending on the rate constant for the reaction involving activated complex **A** in comparison with that for complex **B**. Sato (1960) studied the reaction of various isocyanates with methyl alcohol in di-*n*-butyl ether. In the absence of added catalyst, his results can be expressed by Eq. 12.4, where a and b are the initial concentrations of isocyanate and methyl alcohol, x is the concentration of product (urethane), and k_2 represents the rate constant of the autocatalytic reaction:

$$\frac{dx}{dt} = k_1(a-x)(b-x)^2 + k_2(a-x)(b-x) \quad (12.4)$$

Assuming that there are no side reactions, $(a-x)$ in Eq. 12.4 equals [R—N=C=O] in Eq. 12.2, and $(b-x)$ equals [R′OH]. In most cases studied by Sato, k_2 was larger than k_1, but in a few cases, they were comparable. When k_2 is substantially larger, the second term in Eq. 12.4 dominates in the later stages of the reaction (as the urethane concentration builds up). Sato's studies were done in di-*n*-butyl ether; it would be of interest to know what changes in the rate constants would result from changes in solvent polarity. Sato reported that autocatalysis is more important for aliphatic than aromatic isocyanates, although the rate constants for autocatalysis were similar. This results from the higher reactivity of aromatic isocyanates with alcohols (k_1), which reduces the relative importance of autocatalysis.

12.2.2 Catalysts

Reactions of isocyanates with alcohols are catalyzed by a variety of compounds, including bases (tertiary amines, alkoxides, carboxylates), metal salts and chelates, organometallic compounds, acids, and urethanes. Most primary aliphatic amines react so rapidly with isocyanates at room temperature that their reaction rates and effect of catalysts have not been determined. Catalysts for the reactions of secondary amines with isocyanates include carboxylic acids and water.

The most widely used catalysts in coatings have been tertiary amines, commonly diazabicyclo[2.2.2]octane (DABCO[1]), and organotin(IV) compounds, most commonly, dibutyltin dilaurate (DBTDL). However, allowable amounts of DBTDL and certain other tin catalysts were strictly limited in Europe starting in 2015. Alternatives are discussed below in this section.

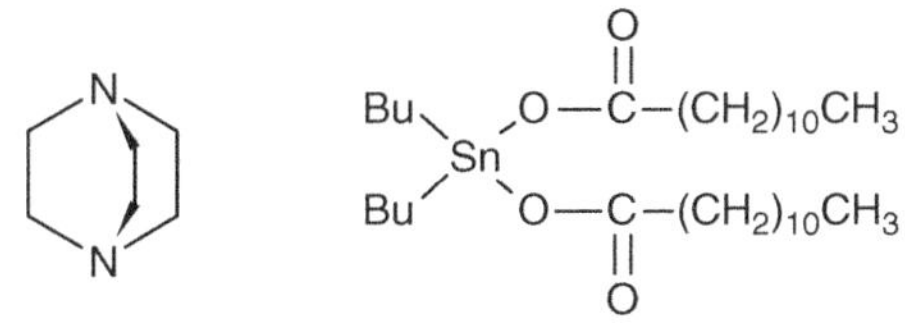

DABCO DBTDL

The mechanisms by which these and other catalysts operate are controversial. A reasonable explanation for catalysis by amines is that they facilitate proton transfer from the alcohol to the isocyanate (and similarly to alcohols and urethanes). Proton removal from the alcohol may occur during reaction with the isocyanate, thereby avoiding formation of the positive charge on oxygen and lowering the energy of the reactive intermediate, which may proceed to product by a proton transfer from the protonated amine, as shown in Scheme 12.2. Proton removal at an earlier stage by amines than by alcohols and urethanes is reasonable, owing to the greater basicity of amines.

Sato (1960) included catalysts in the study described in Section 12.2.1. For triethylamine-catalyzed reactions of isocyanates with methyl alcohol in dibutyl ether, Sato's data fit Eq. 12.5, where k_3 and (cat) represent the rate constant of the catalyzed reaction and the concentration of catalyst, respectively:

$$\frac{dx}{dt} = k_1(a-x)(b-x)^2 + k_2(a-x)(b-x) + k_3(\text{cat})(a-x)(b-x) \quad (12.5)$$

If the catalyst is effective, k_3 is larger than k_1 or k_2, and when the catalyst concentration is sufficient, the rate is governed by the third term and is first order in [R′OH] and in [R—N=C=O]. This is generally observed in amine-catalyzed reactions. The mechanism in Scheme 12.2 is consistent with the third term of rate Eq. 12.5. Amine basicity is not the dominant factor since DABCO, a weaker base, is a more active catalyst than triethylamine. The nitrogen electron pairs are more readily accessible with DABCO than triethylamine, which could account for DABCO's higher catalytic activity. The possibility, that both nitrogen basicity and electron accessibility are important factors, is supported by the even higher catalytic activity (relative to DABCO) of 1-azabicyclo[2.2.2] octane (quinuclidine), which combines both the nitrogen accessibility of DABCO and the high basicity of triethylamine.

1 DABCO is Air Products general trademark for a series of different catalysts, including diazabicyclo[2.2.2]octane that is DABCO L-33. However, the acronym DABCO has become widely used just for diazabicyclo[2.2.2]octane.

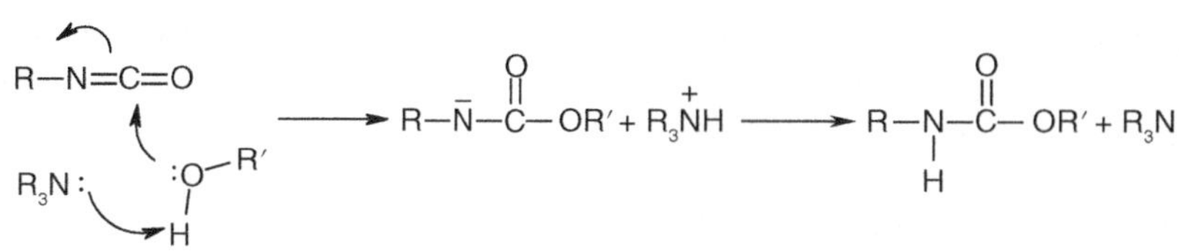

Scheme 12.2 Possible mechanism for amine-catalyzed reaction of an isocyanate with an alcohol.

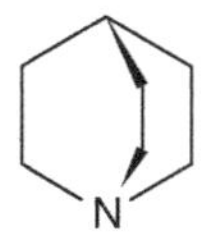

Quinuclidine

Catalysis of urethane formation by 1-8-diazabicyclo[5.4.0]undec-7-ene (DBU) is reported to result from the formation of the ammonium ion and an H-bond type complex with the polyol (Hira et al., 1983).

Wang et al. (2013) studied the reaction of 1,3-propanediol with phenyl isocyanate in dry butyl acetate solvent. They compared two catalysts, DABCO and tetramethyl ethylene diamine (TMEDA). These authors confirmed that DABCO is a far more effective catalyst even though TMEDA is much more electronegative (and therefore a stronger base). They attribute the effectiveness of DABCO to its lower steric hindrance, which outweighs electronegativity. The reaction follows second-order kinetics, with the expected deviations late in the process. Wang's results are consistent with those of Sato: however, rather than the 3-membered (troika) participation in Scheme 12.2, Wang et al. (2013) propose a three-step mechanism, favored by entropic considerations, in which DABCO interacts separately with the (R′OH) and the (R—N=C=O) groups.

Comparisons are complicated because catalysts, to varying degrees, also catalyze allophanate formation and trimerization of aromatic isocyanates to form isocyanurates. For example, Wong and Frisch (1986) determined that reaction of phenyl isocyanate with *n*-butyl alcohol (at 50°C in acetonitrile), catalyzed by pentamethyldipropylenetriamine (PMPTA), yielded 30% urethane, while 70% of the isocyanate was converted into triphenylisocyanurate. On the other hand, when DABCO was used as a catalyst, the urethane was the principal product, together with a small amount of an allophanate. Possibly, urethane formation is favored by sterically accessible amines (e.g., DABCO), and isocyanurate formation is less sensitive to this factor. Schwetlick and Noack (1995) measured rate constants for the formation of urethanes, allophanates, and isocyanurates with a variety of catalysts and cocatalysts. The zinc complex of 2,4-pentanedione [$Zn(AcAc)_2$], tin octanoate, and quaternary ammonium compounds, such as tetramethylammonium octanoate, specifically catalyzed allophanate formation.

Acids also catalyze urethane formation, perhaps by protonating the isocyanate group. Acetic acid is a more effective catalyst than a carboxylic acid-functional polyether polyester (Berlin et al., 1993). Stronger acids are even more effective (Nordstrom et al., 1997). Phenyl acid phosphate is less effective as a catalyst than DBTDL at temperatures below 100°C but is more effective at 130°C. Blocking the acids by forming salts with volatile amines extends pot life without reducing the cure rate at 130°C.

Catalysis by organometallic compounds of many types has been reported. Of the many metal derivative catalysts, DBTDL is the most widely used in coatings. It is soluble in a wide range of solvents, comparatively low in cost, colorless, and, in general, highly effective at levels on the order of 0.05 wt%. DBTDL promotes urethane formation without promoting allophanate formation (Yilgor and McGrath, 1985) or trimerization (Wong and Frisch, 1986). Being hydrophobic, it favors urethane formation over hydrolysis of the isocyanate when moisture is present. Dimethyltin diacetate (DMTDA) is usually a somewhat more effective catalyst than DBTDL and is particularly useful with sterically hindered isocyanates. Whereas aromatic isocyanates are more reactive than aliphatic isocyanates in uncatalyzed or amine-catalyzed reactions with alcohols, the reactivity of aliphatics and aromatics can be roughly equal with DBTDL catalysis.

Many mechanisms have been proposed for the catalytic activity of tin compounds, but none has been universally accepted. A plausible proposal is based on studies of the reaction of phenyl isocyanate with excess methyl alcohol using dibutyltin diacetate (DBTDA) as catalyst (Van der Weij, 1981; Lou et al., 1997). The reaction rate is first order in isocyanate concentration and half order in both alcohol and catalyst concentration. Restating the kinetic equation in the same form used thus far gives Eq. 12.6:

$$\frac{dx}{dt} = k_3 (\text{cat})^{1/2} (a-x)(b-x)^{1/2} \qquad (12.6)$$

Based on kinetic results and the observation that the rate was suppressed by addition of acid, the mechanism in Scheme 12.3 was suggested. The mechanism involves sequential complexation of alcohol (with loss of H^+) and isocyanate to the tin. The proposed proton loss is consistent

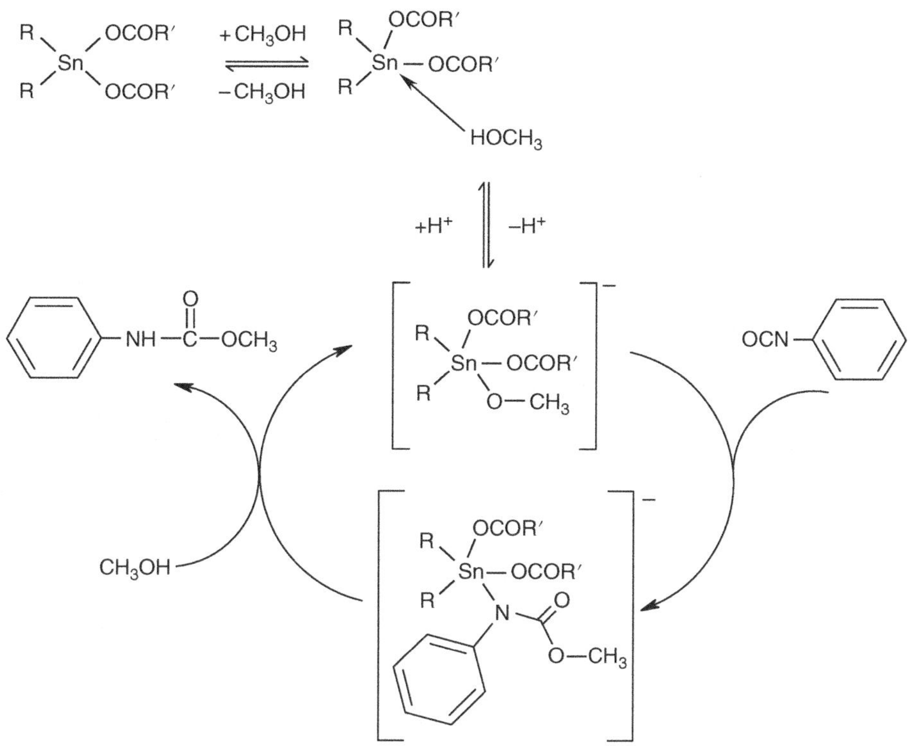

Scheme 12.3 Suggested mechanism for tin-salt-catalyzed reaction of an isocyanate with an alcohol.

with the observed reduction in activity of tin catalysts in the presence of carboxylic acids. Presumably, addition of H^+ favors decomplexation of alcohol and reversion to starting materials. This proposal is also consistent with the observation that tin compounds are not effective catalysts for reactions of isocyanates with amines.

Coordination of the isocyanate with tin is consistent with the observed roughly equivalent reactivity of aliphatic and aromatic isocyanates with tin catalysts, since stabilization of the reactive intermediate by electron delocalization with the aromatic ring is eliminated as a major factor. Furthermore, tin activates both the alcohol and isocyanate, whereas amine catalysts probably activate only the alcohol by facilitating proton removal.

The rate dependence on alcohol concentration has important implications on the pot life versus curing schedule with isocyanates. If the dependence changes from first to second order (characteristic of the uncatalyzed reaction), the rate will diminish rapidly as the reaction nears completion. With less sensitive half-order dependence, as reported for tin catalysis, the rate at high concentrations (during storage) will be lower and will not slow down as much as the reaction proceeds to completion, thereby favoring both a longer pot life and shorter cure time (or lower cure temperature). The first-order dependence with amine catalysis results in an intermediate situation.

An alternative mechanism that has gained favor is via alcoholization of the organotin ester to the organotin alkoxide:

$$R-OH+R'_2Sn(O_2CR'')_2 \rightleftharpoons R'_2Sn(OR)(O_2CR'')+R''CO_2H$$

$$R-OH+R'_2Sn(OR)(O_2CR'') \rightleftharpoons R'_2Sn(OR)_2+R''CO_2H$$

Here, dibutyltin di(2-ethylhexanoate) interacts with an alcohol to form tin alkoxides (mono or di), which then react with the isocyanate to form complexes which in turn react with the alcohol to give the urethane (Draye et al., 1999). The effect of solvent on the catalysis of the reaction of phenyl isocyanate with cyclopentanol in different solvents indicates that the principal reaction pathway is via the dialkyltin monoxide monoester. Kinetic and computational studies of urethane formation by Devendre et al. (2015) led to the conclusion that the dominant catalyst *in polar media* is an organotin alkoxide.

Protected catalysts for storage stable moisture curing coatings are prepared by reacting DBTDL with *p*-toluenesulfonylisocyanate (Richter et al., 1991). After the coating is applied, the catalyst is readily hydrolyzed with water or alcoholized with an alcohol.

Metal chelates of 2,4-pentanedione (acetylacetone, AcAc) also serve as catalysts. It has been proposed that AcAc metal derivatives such as $Zr(AcAc)_4$ activate the

Table 12.1 Tensile Modulus (psi) (% Elongation to Break)

Catalyst	Initial	1 Week	4 Weeks
DMTDC	2450 (270)	1900 (300)	800 (330)
DBTDL	1900 (270)	450 (400)	200 (436)
DBTM	1400 (280)	450 (330)	250 (400)

hydroxyl group followed by insertion of isocyanate (Blank et al., 1999). This catalyst is selective, significantly favoring reaction with hydroxyl groups over reaction with water, so it is particularly useful in 2K waterborne urethane coatings.

Many metal salts have been shown to be catalysts. There is a wide variation in the extent of catalysis of the NCO/OH reaction depending on the diisocyanate and catalyst. Stannous octanoate and lead octanoate are particularly effective although less so than DBTDL.

Numerous studies have been aimed at replacing tin and DABCO catalysts because of (perhaps exaggerated) concerns about their toxicity. Tin catalysts were strictly limited in Europe starting in 2015. Commercial *tin-free catalysts* include soluble carboxylic acid salts of zinc and bismuth (such as bismuth isooctoate) and a variety of substituted imidazolium salts (Schaefer et al., 2015). Belmokaddem et al. (2016) propose pentafluorophenyl phosphonic acid as a tin-free catalyst.

Another concern is that residual catalyst may adversely affect durability of the coatings. Increased catalyst concentration not only increases reaction rate but can also decrease hydrolytic stability of films after application and curing if the coreactants are vulnerable to hydrolysis. The effect is most evident in polyester/polyurethane films. Table 12.1 shows the effect on modulus (a measure of cross-link density) from exposure of films of H_{12}MDI/polyestertriol to 95% relative humidity at 70°C for different catalysts, DMTDC, DBTDL, and dibutyltin methoxide (DBTM) (Squiller and Rosthauser, 1987). The observed reductions in modulus seem reasonable, since hydrolysis is expected to cleave the urethane cross-link and produce the corresponding amine, alcohol, and CO_2. Similarly, amines catalyze not only the formation of urethanes but also the thermal cleavage of the urethane cross-links. For example, Okumoto et al. (1995) reported substantial reduction in the physical properties of polyurethane elastomers when exposed to temperatures >105°C for 400h, which was attributed to residual DABCO catalyst.

12.2.3 Interrelationships in Catalysis

Combinations of DABCO and DBTDL often act synergistically; that is, the effect of the combination is greater than the sum of the individual effects of the two catalysts. Based on mechanistic studies, Bechara (1981) suggested that DABCO and DBTDL form a highly active complex that facilitates alcoholysis of the tin carboxylates to a highly active tin alkoxide adduct. Sequentially, the tin alkoxide adduct adds isocyanate across the tin alkoxide bond to form the corresponding tin carbamate adduct, which reacts with alcohol to form the urethane and regenerate the tin alkoxide adduct.

As described in Chapter 10, organotin and organotitanate catalysts are often used in the synthesis of polyester polyols. Since residues of these catalysts can catalyze reactions of the polyols with polyisocyanates, their presence may lead to the false impression that polyester polyols are inherently more reactive than acrylic polyols.

In order to increase pot life without significantly reducing cure rate, it is common to add volatile inhibitors to 2K polyurethane coatings. Pot life of DBTDL catalyzed coatings can be extended by addition of small amounts of volatile carboxylic acids, such as formic or acetic acids; after application the acids evaporate and there is little, if any, effect on cure rate (Draye and Tondeur, 1999). Pot life can also be extended by addition of 2,4-pentanedione (acetylacetone) with minimal effect on dry times. Amine salts can extend the pot life of 2K coatings when strong acids are used as catalysts.

12.3 ISOCYANATES USED IN COATINGS

Both aromatic and aliphatic isocyanates are used in coatings; the former are less expensive and the latter provide films having better color retention and exterior durability.

Isoscyanates are manufactured by reaction of primary amines (RNH_2) with phosgene ($Cl_2C{=}O$). Do not try it at home—phosgene is a toxic and corrosive gas. The process has challenged engineers for at least 70 years, as evidenced by a long chain of process patents. Non-phosgene processes are being sought (Takamatsu et al., 2015).

12.3.1 Aromatic Isocyanates

The aromatic isocyanates most widely used in coatings are based on MDI and TDI. MDI is available in several grades: bis(4-isocyanatophenyl)methane, a mixture of 55% of the 2,4′ isomer and 45% of the 4,4′isomer, and several oligomeric (frequently called polymeric) grades having longer chains of methylene phenyl groups, as shown in the following structures. MDI is also used as a prepolymer with polyether polyols. Its volatility (particularly of the oligomeric grades) is sufficiently low to reduce toxic hazard, especially as compared to that of toluene diisocyanate (TDI).

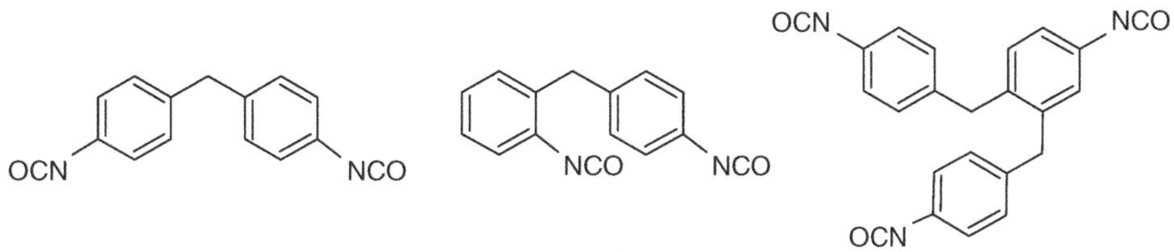

4,4′- and 2,4′- MDI isomers and an example of an oligomer

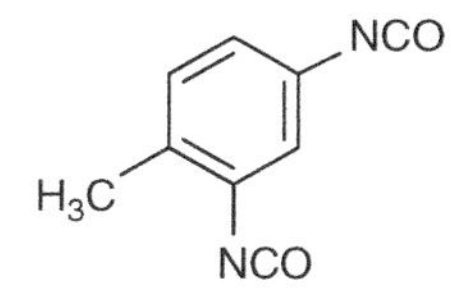

2,4-TDI

The most common grade of commercial TDI consists of a mixture of about 80% 2,4- and 20% 2,6-diisocyanate isomers. Nearly pure 2,4-TDI is also available at a premium price. Due to toxic hazards, TDI is not used as such in final coating formulations. For coatings that require unreacted isocyanate groups, TDI is converted into derivatives of higher MW and higher functionality. Higher MW reduces the toxic hazard, and the higher functionality forms solvent resistant films more rapidly.

2,4-TDI has the advantage of different reactivity of the *ortho*- and the *para*-isocyanate groups with alcohols. This differential in reactivity enables the synthesis of isocyanurates and prepolymers having narrower MW distribution than is possible with diisocyanates in which the isocyanate groups are equally reactive. At 40°C, the *para*-isocyanate group of TDI is about seven times more reactive than the *ortho*-group. Furthermore, whichever isocyanate group reacts first, the second group is less reactive than the first. After complete reaction of the *para*-isocyanate, the remaining *ortho*- group is 20 times less reactive than the *para*-isocyanate on a second TDI. The difference in reactivity decreases with increasing temperature, resulting in similar reactivities above 100°C. Thus, for maximum selectivity a prepolymer should be prepared at low temperature, albeit at the expense of slower reaction.

Catalysts can be used, but can also change the reaction selectivity. Furthermore, the catalyst remains in the product and, therefore, sets a lower limit on the amount of catalyst in the final formulation. In a study of several catalysts on the ratio of *para*-to-*ortho* urethane formation in the reaction of TDI with *n*-butanol in benzene at 20°C, the lowest ratio (2.8 : 1) was obtained with $SnCl_4$; the highest ratio (9.9 : 1) was obtained with tin octanoate, DBTDL, and triisopentylphosphine, whereas the ratio was 11.5 : 1 in the uncatalyzed reaction (Korzyuk and Zharkov, 1981). Palyutkin and Zharkov (1985) have also shown that amine catalysts tend to equalize the reactivity of the ortho- and para-NCO groups of TDI.

Almost any polyhydroxy compound can be reacted with excess TDI to make a prepolymer. Low MW hydroxy-terminated polyesters or mixtures of diols and triols are commonly used. For safety, the levels of unreacted TDI in the prepolymer must be very low. Low levels of TDI can be assured by using a less than 2 : 1 ratio of NCO : OH and by pushing the reaction to completion; but chain extension (caused by reaction of both NCO groups of some TDI molecules) increases the MW of the product. An alternative process, used when low MW prepolymers are required, is to react the polyhydroxy compound, often trimethylolpropane (TMP), with a large excess of 2,4-TDI, followed by removing excess TDI using a vacuum, wiped film evaporator. By this method, low MW products suitable for high solids coatings are obtained with very low levels of free TDI and minimal chain extension.

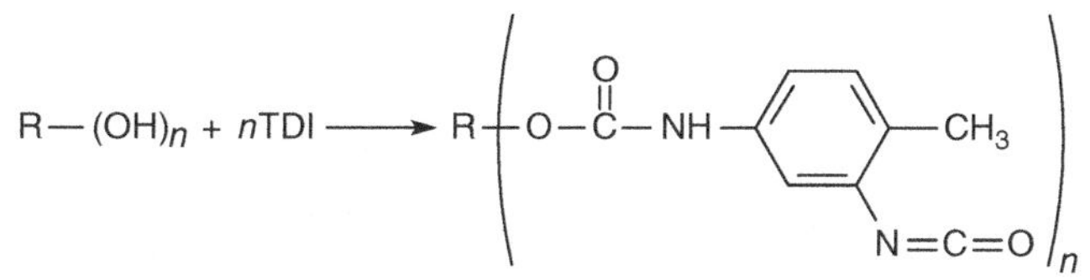

Idealized TDI Prepolymer Formation

The isocyanurate, prepared by trimerizing TDI (Section 12.1), has a lower toxic hazard than monomeric TDI. The trimerization reaction occurs almost exclusively with the *para*-isocyanate group.

12.3.2 Aliphatic Isocyanates

Aliphatic isocyanates are more expensive than the common aromatics but offer excellent outdoor durability and color stability.

The principal aliphatic isocyanates used are 1,6-hexamethylene diisocyanate (HDI), isophorone diisocyanate (IPDI), bis(4-isocyanatocyclohexyl)methane (H_{12}MDI), 1,3-xylylenediisocyanate (XDI), tetramethyl-*m*-xylidene diisocyanate (TMXDI), *m*-isopropenyl-α,α-dimethylbenzylisocyanate (TMI), and 2,2,5-trimethylhexane diisocyanate (TMHDI). Diisocyanates are usually converted to derivatives before use in coatings to increase functionality and reduce toxic hazard.

$O{=}C{=}N{-}(CH_2)_6{-}N{=}C{=}O$

HDI

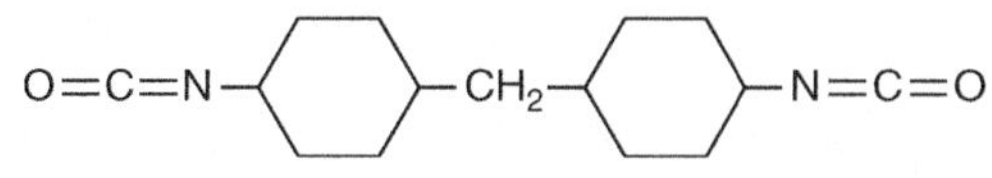

$H_{12}MDI$

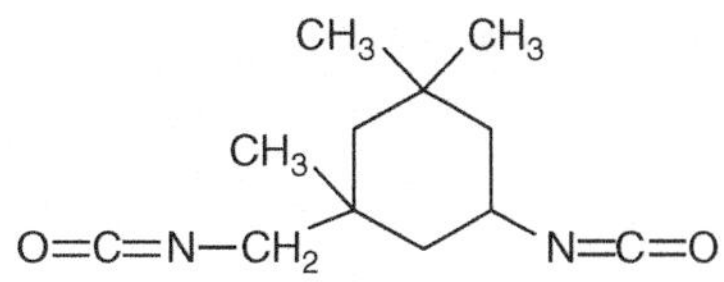

IPDI

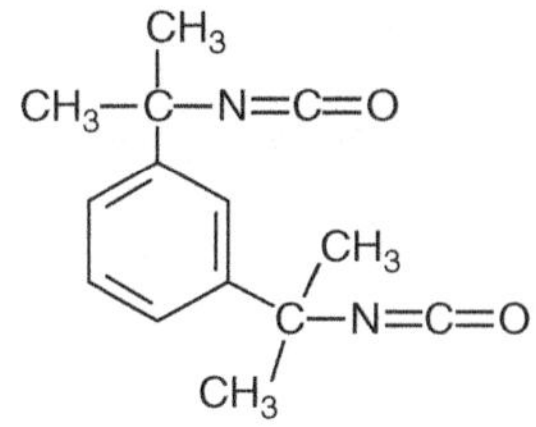

TMXDI

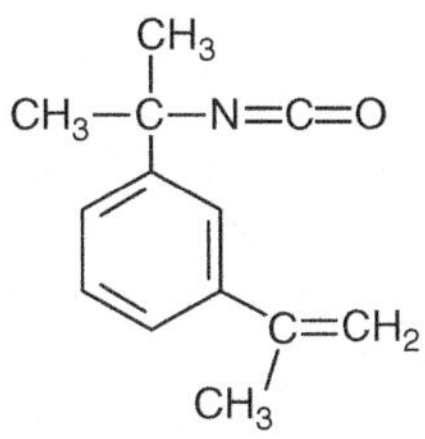

TMI

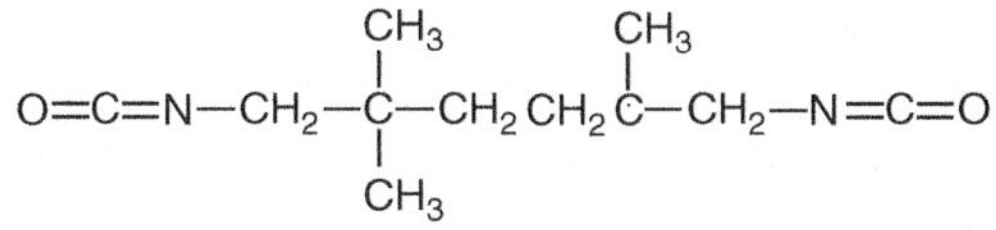

TMHDI

HDI is especially hazardous and is handled on a large scale only in chemical plants. The first less hazardous derivative was a biuret, which can be made by reacting HDI with a small amount of water and removing excess HDI. The structure of HDI biuret shown below is idealized; commercial products contain varying fractions of oligomeric biurets. The presence of oligomeric biurets increases the average functionality above 3. These polyfunctional isocyanates provide coatings with good color retention and weather resistance. The viscosity of an early commercial product was about 11.5 Pa·s at 20°C. Grades with lower average MWs (and average functionalities nearer to 3) are available with viscosities as low as 1.4 Pa·s.

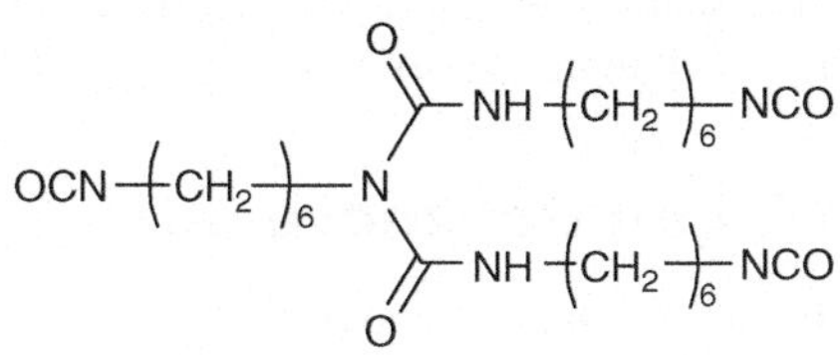

HDI biuret (and oligomers)

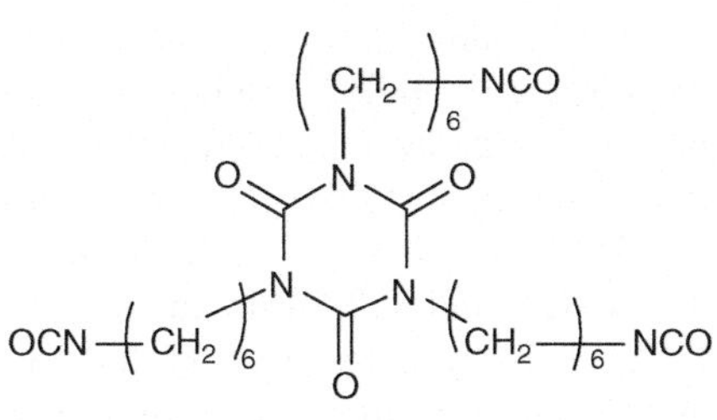

HDI isocyanurate (and oligomers)

HDI isocyanurates are used on a larger scale. The isocyanurate provides coatings with greater thermal stability and longer-term exterior durability than HDI biuret. Commercial products contain oligomeric material and average functionality is above 3. Grades having lower oligomer content and viscosities as low as 1 Pa·s are available.

Silica nanoparticles can be surface-functionalized with —N=C=O groups and dispersed in HDI isocyanurate, offering numerous possibilities for development (Maganty et al., 2016). Also see Sections 12.6 and 16.2.

Using ammonium fluoride as a catalyst, an isomeric HDI-iminooxadiazenedione trimer (called an unsymmetrical trimer of HDI) is formed in an approximately 50–50 mixture with HDI isocyanurate. The unsymmetrical trimer has the advantage of lower viscosity (1 Pa·s) than the corresponding HDI isocyanurate (3 Pa·s) (Richter et al., 1998). Unsymmetrical trimer having a viscosity of 0.7 Pa·s has been made available and is useful in very high solids coatings.

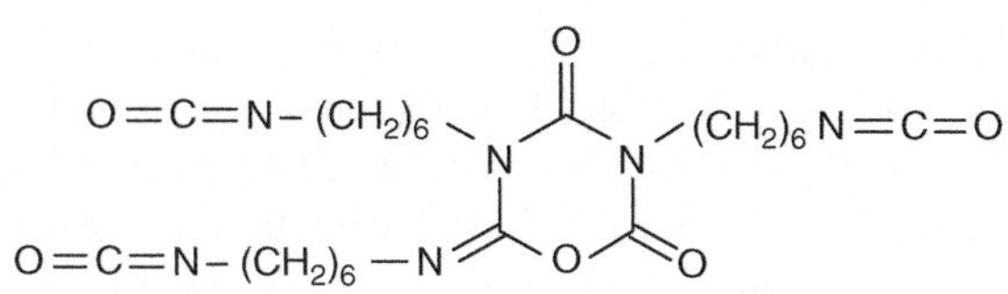

HDI unsymmetrical trimer

Very low viscosity grades (<100 mPa·s) of the uretdione dimer of HDI are also available that permit formulation of very low VOC coatings (Wojcik, 1994), which can cure under ambient conditions. (The uretdione structure is shown in Sections 12.1 and 12.5.2).

Allophanate derivatives of HDI and IPDI are also utilized in coatings. They are prepared by reacting an alcohol or diol with excess isocyanate, followed by removing unreacted diisocyanate with a wiped film evaporator (Potter and Slack, 1992). The structure of an allophanate made from a monoalcohol and HDI is shown below. The properties can be varied by using different alcohols to make compounds having different R substituents. For example, the cetyl alcohol monourethane from HDI yields an allophanate diisocyanate that is soluble in aliphatic hydrocarbons. Allophanates having higher functionality are made by reacting a glycol with excess diisocyanate.

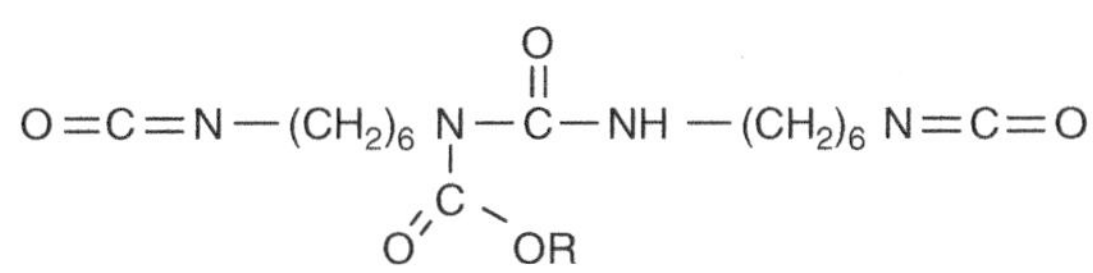

HDI allophanate

Commercial IPDI is a mixture of *Z*-(*cis*) and *E*-(*trans*) isomers in about a 75 : 25 ratio. The isomers are difficult to separate. The two different types of N=C=O groups, primary and secondary, exhibit different selectivity depending on the reaction conditions. With DBTDL catalysis, the secondary NCO group of both *Z* and *E* isomers is more reactive than the primary NCO group, and selectivity is greater with *sec*-butyl alcohol than with *n*-butyl alcohol (Ono et al., 1985; Hatada et al., 1990; Lomoelder et al., 1997). On the other hand, DABCO favors the reaction of the primary isocyanate group of IPDI in contrast to DBTDL as well as to other amines. In all cases, as expected, selectivity decreases with increasing temperature. High selectivity is particularly important in making IPDI prepolymers when low MW and narrow distribution of MW are desired. The aforementioned studies indicate that optimal conditions for synthesis of IPDI prepolymers are temperatures between 40 and 60°C with DBTDL catalysis.

Isocyanurate derivatives of IPDI, analogous to the HDI isocyanurates discussed previously, are widely used cross-linkers. The rigidity of IPDI affords films with higher T_g. By blending IPDI and HDI isocyanurates, formulators can dial in the desired T_g for a particular application.

Bis(4-isocyanatocyclohexyl)methane) ($H_{12}MDI$) is less volatile than HDI and IPDI and is sometimes used as a free diisocyanate in coatings to be applied by roll coating, but not by spray coating. $H_{12}MDI$ is a mixture of stereoisomers and has lower reactivity than HDI or IPDI.

TMXDI and TMI (structures above) have aromatic rings but impart color retention and exterior durability similar to aliphatic isocyanates. The exterior durability of polyurethane coatings, derived from these isocyanates probably results from the absence of isocyanate groups directly substituted on the aromatic ring, in contrast to the aromatic isocyanates. Durability of TMXDI and TMI coatings is also attributable to the absence of abstractable hydrogens on the carbons adjacent to nitrogen. Since the isocyanate group is on a tertiary carbon, the reactivity is lower than that of less sterically hindered aliphatic isocyanates. This difference can be offset by using higher catalyst levels and sterically accessible tin catalysts such as DMTDA instead of DBTDL where permitted. TMXDI is pre-reacted with TMP and offered as a low MW prepolymer, essentially free of unreacted TMXDI. TMI is used as a comonomer with acrylic esters to make 2000–4000 MW copolymers having 40–50 mol% TMI; thus each molecule has several isocyanate groups (Alexanian et al., 1993).

There is growing interest in bio-based polyurethanes. A review by Noreen et al. (2016) cites 41 examples of bio-based polyurethanes; many are prepolymers based on castor oil (which has hydroxy groups) or other vegetable oil derivatives that have been chemically modified to contain hydroxy groups. Narute and Palanisamy (2016) provide an example based on a polyol made by epoxidizing cottonseed oil and hydrolyzing the product. Li et al. (2015) describe a different approach. Totally bio-based poly(urethane urea) dispersions are synthesized using two commercially available diisocyanates, one prepared from tall oil fatty acid and the other from the ethyl ester of L-lysine (a naturally occurring amino acid having two amine groups). The two isocyanates are blended to achieve the desired T_g. Of course, partly biobased alkyd resins have been cross-linked with polyisocyanates for decades (Section 12.4).

12.4 TWO-PACKAGE (2K) SOLVENT-BORNE URETHANE COATINGS

A large volume of urethane coatings is formulated into 2Ks that are mixed just before application. One package contains the polyol (or other coreactant), pigments, solvents, catalyst(s), and additives; the other contains the polyisocyanate and moisture-free solvents. The catalyst may also be in a separate third package in order that cure rate can be adjusted for variations in ambient conditions. While the major reaction is formation of urethane cross-links, some urea cross-links result from the reaction of atmospheric water with the isocyanate. 2K urethane coatings can be formulated for ambient temperature cure or baking coatings.

In many applications, the 2K components are mixed at the start of a work shift. Viscosity starts to increase as soon as they are mixed, but the coatings must remain sprayable for 4 h (until lunch) or for 8 h after mixing, denoted as a 4-h or 8-h *pot life*. The pot life can be considerably shorter for factory application when special equipment is used to mix the components just before spraying. If very fast cure at relatively low temperatures is needed, reactive coreactants and/or high catalyst levels are used, in which case the pot life is very short. Such formulations can be applied using special spray equipment in which the two packages are fed to a spray gun by separate proportioning pumps and mixed inside the gun just before spraying or impingement mixed during spraying. Such spray equipment is expensive, and care is required to assure that the proper ratios are fed and thoroughly mixed.

An important variable in 2K polyisocyanate/polyol coatings is the ratio of NCO/OH on an equivalent basis. This ratio is commonly called *the index*. In ambient cure systems, it is often found that a ratio of the order of 1.1 : 1 usually gives better film performance than a 1 : 1 ratio. A probable reason is that part of the NCO reacts with water from solvent, pigment, or air to give urea cross-links. Since two hydroxyls remain unreacted for each water molecule that reacts, owing to the intermediary formation of an amine that preferentially reacts with a second NCO group (Section 12.1), use of excess NCO minimizes residual hydroxyl groups. Solvent resistance is also improved. Since polyisocyanates are typically lower in viscosity than the polyol, excess NCO results in lower VOC (Wicks and Yeske, 1997). Some finishes, notably for aircraft, are formulated with NCO/OH ratios as high as 2 : 1. Atmospheric moisture reacts with the surplus NCO groups to form urea cross-links to complete the cure. The high mobility of water together with the high reactivity of amine groups, resulting from the water/isocyanate reaction, may allow the development of higher T_g values than possible by the hydroxyl/isocyanate reaction itself. The reduced hydroxyl levels also provide longer pot life.

Since the rate of reaction of isocyanates generally decreases in the following order: primary alcohols > secondary alcohols > 2-hydroxyethyl ether alcohols > 2-hydroxypropyl ether alcohols, the required excess of NCO over OH tends to increase in that order.

Many suitable hydroxy-functional coreactants are available; hydroxy-terminated polyester and hydroxy-substituted acrylic resins are most common. In general terms, polyesters permit higher solids and provide films with greater solvent resistance and better adhesion to metals. Acrylics dry faster and provide better exterior durability, resulting from superior hydrolytic and photochemical stability. Furthermore, acrylics can be prepared having higher equivalent weight than that of polyesters, resulting in the need for less of the more expensive isocyanate, which may result in lower total cost.

Zhang et al. (2013) provide detailed studies of polyester polyols, made from the common monomers described in Chapter 10, and resulting polyurethane films. Their findings are in harmony with the principles described in Chapter 10 and provide detailed analytical data, as well as valuable insights for designing such polyesters. Of particular interest is the finding that wide angle X-ray analysis of the polyurethane films found no crystallinity but suggested short-range ordered structures.

Weatherability of 2K polyurethane coatings made with polyester polyols can be enhanced by utilizing hydrolysis-resistant polyols based on BEPD and TMCD (Section 10.1.1).

Other resins besides polyester and acrylic polyols, are also used in 2K coatings. Alkyd resins with low and medium oil length have unreacted hydroxyl groups, and their rate of dry can be accelerated by adding a polyisocyanate, such as the isocyanurate trimer of IPDI, just prior to application. Nitrocellulose (shipped wet with plasticizer instead of ethyl or isopropyl alcohol) is used in formulating cross-linking furniture lacquers with isocyanates. Bisphenol A (BPA) epoxy resins are also cross-linked with isocyanates through their hydroxyl groups. Hydroxy-terminated polyethers are widely used in urethane foams, and also find some use in coatings, often in combination with other polyols, although the resulting coatings show high moisture vapor permeability and relatively poor exterior durability.

Mobility constraints. As cross-linking proceeds, T_g of the film increases. If the T_g of a partially reacted system is near the temperature at which the cross-linking must occur, the rate of urethane formation becomes mobility controlled and cure rate is reduced. If the T_g of the fully reacted system is significantly above the cure temperature, the reaction virtually stops before the cross-linking reaction has gone to completion (Fiori and Dexter, 1986). Since in many ambient cure coatings, a T_g somewhat above the curing temperature is desired, and selection of a combination of polyisocyanate and polyol that provides an appropriate final T_g is critical. A polyisocyanate with relatively flexible aliphatic chains (such as HDI isocyanurate) can be used with a relatively high T_g acrylic or polyester. On the other hand, a polyisocyanate that yields relatively rigid cross-link segments (such as the TMP/α,α-tetramethylxylyl diisocyanate (TMXDI) prepolymer) requires a lower T_g acrylic or polyester. Mobility control of reactions is more likely to be encountered in ambient temperature cure coatings than in baked coatings. T_g can be dialed in by using mixtures of rigid and flexible isocyanurates. Along these lines, Wang et al. (2015) have described hybrid isocyanurates prepared from two different aliphatic diisocyanates.

2K coatings always involve a compromise between pot life and cure time (Section 2.3.2). However, several formulating variables are available to increase pot life with little effect on cure time. Reducing concentrations of reactive groups increases pot life, but results in higher VOCs. Supplementing the isocyanate/polyol reaction with moisture cure increases pot life, but may require humidity control after application. Since isocyanate–alcohol reactions proceed most slowly in media having high levels of H-bond accepting groups, strong H-bond accepting solvents (ketones and esters in polyurethane grades) should be selected to the extent possible. After application, as the solvent evaporates, the medium becomes less H-bond accepting and the reaction rate increases differentially.

Where permitted, organotin catalysts are generally preferable to amine catalysts (Section 12.2.2), because reaction rates often depend on alcohol concentration to the one-half power with organotin catalysts and to the first power with amine catalysts. Since the effectiveness of tin catalysts is reduced by carboxylic acids, one can add a volatile acid such as acetic or formic acid to the formulation. The acid inhibits reaction during the pot life stage but evaporates during application so that the inhibiting effect disappears. Similarly, 2,4-pentadione increases pot life by chelating with tin compounds and evaporating after application. Tin- and amine-free catalysts are described in Section 12.2.2.

Availability of lower viscosity isocyanate cross-linkers (Section 12.3.2) makes possible formulation of higher solids 2K coatings. However, one must consider the viscosity of not only the cross-linker but also the polyol, as well as the combination of polyol and cross-linker. If a lower viscosity cross-linker also has a lower equivalent weight, the weight ratio of cross-linker to polyol must be reduced, and the viscosity of the polyol becomes more important.

Reactive diluents are useful for increasing solids. For example, polytrimethylene carbonate polyols (Section 17.9) have been recommended as reactive diluents for acrylic polyols in coatings cross-linked with HDI isocyanurate (Zhou et al., 2000). Best results were obtained with relatively low MW triols made by reacting trimethylene carbonate with TMP. Reduction of VOC was accompanied with increasing film flexibility, without reduction of hardness, weather resistance, and mechanical properties.

Polyaspartic esters (Wicks and Yeske, 1997) are hindered amines that can cross-link triisocyanates in high solids coatings. They are manufactured by Michael addition reaction of readily available maleate esters and common diamines. The coatings are often called 2K polyurethane coatings although the cross-links are dominantly ureas, not urethanes. High solids coatings made from polyaspartic esters and polyisocyanates have proven to have excellent outdoor durability and resistance to mechanical damage. Major applications include auto refinish and maintenance coatings (Chapter 33). They are also candidates for use in OEM clear coats.

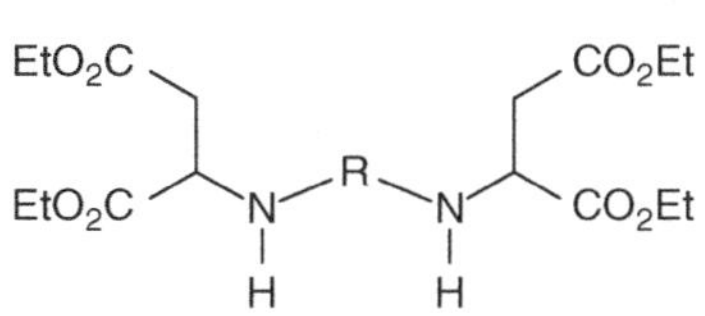

Polyaspartic esters; R is shown in Table 12.2

Table 12.2 Viscosities (100% Solids, mPad·s at 23°C) and Gel Times of Substituted Polyaspartic Acid Ethyl Esters

R	Viscosity	Gel time
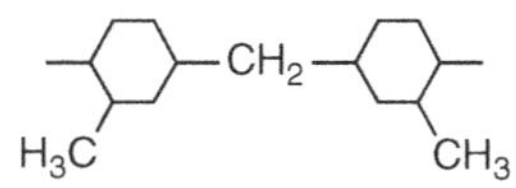	1500	>24 h
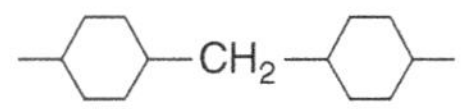	1200	2–3 h
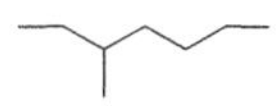	150	<5 min

As shown in Table 12.2, the gel time of polyisocyanate/polyaspartic ester combinations depends on the structure of the polyaspartic ester. Surprisingly, the pot life of these combinations is increased by addition of DBTDL (Wicks and Yeske, 1997). On the other hand, carboxylic acids and water accelerate reaction rates. By proper selection of isocyanate and aspartate and adjustment of inhibitors and catalysts, it is possible to formulate very high solids 2K coatings having a reasonable pot life and fast curing (Squiller et al., 2014).

Ketamines and aldimines are cross-linkers that also form predominately urea cross-links. Their complex chemistry is introduced in the third edition of this book and described by Bock and Halpaap (1987) and by Wicks and Yeske (1997). Polyaldimines offer the advantage of very low viscosity.

California regulations allow VOC of only $100\,g\,l^{-1}$ for certain 2K industrial maintenance coatings. Seipke et al. (2016) investigated formulation methods to satisfy this limit, taking advantage of exempt solvents (Section 18.9.1). In their formulations, the polyol is an 80/20 blend of acrylics and polyesters, the polyisocyanate is HDI isocyanurate, and the solvents are acetone (exempt but very fast evaporating) and *p*-chlorobenzotrifluoride (exempt and fairly fast evaporating), together with slow evaporating non-exempt "tail solvents."

12.4.1 2K Polyurea Coatings

As noted previously, most 2K polyurethanes contain some urea linkages, and those made with hindered amines are entirely ureas but are still called polyurethanes. The term *polyurea* coatings is commonly applied to those in which the final cross-linking reaction is between isocyanate and fast-reacting amine groups. The term 2K urea coatings is used in the maintenance field. Other applications of coatings, such as automotive refinish urethane clear coats, based on isocyanates and hindered amines, are called 2K urethane coatings (Section 12.4). Moisture-curable isocyanates cure by formation of urea bonds, but the coatings are seldom called polyureas.

The pot life of 2K urea coatings depends on the amine(s) and polyisocyanate(s) used. Primary aliphatic amines react too rapidly to be applied with most commercial equipment, but primary aromatic amines can be applied with special equipment and hindered amines can be used as 2K coatings. In applications such as roof coatings, floor coatings, traffic control paint, and thick-film sag-resistant maintenance coatings, rapid cure is very desirable (Chapter 33). This need has motivated the development of equipment that makes it possible to spray to apply coatings having pot lives as short as 1 s. 2K urea coatings, based on the reaction of MDI/polyoxypropylene prepolymers with aromatic amines, are used in maintenance coatings for which gloss and color retention are not critical. The coatings have the advantages that they cure rapidly, even at freezing temperatures, have zero VOC, and low temperature flexibility (Takas, 2004). Roofs and floors can be walked on within 30 s after they are coated with such polyureas. Adhesion to properly prepared metal and to concrete is obtained without the need of primer. Thick, tough coatings for pickup truck beds are made with aromatic isocyanates and amines. These typically black coatings cure in seconds at room temperature and protect the beds of pickup trucks from scrapes and abrasion. They provide an attractive alternative to drop-in liners made from polyethylene.

12.5 BLOCKED ISOCYANATES

Blocked isocyanates are used as cross-linkers in baked coatings. The general idea is to react an isocyanate with a *blocking group* that prevents appreciable reaction at storage temperatures but releases free isocyanate at elevated cure temperatures. Blocked isocyanates are of major importance in cationic electrodeposition coatings, powder coatings, coatings for electrical wire, and, increasingly, one-package (1K) automotive coatings. Common blocking agents include 2-ethylhexyl alcohol and 2-butoxyethanol in E-coats and ε-caprolactam, 2-butanone oxime (methyl ethyl ketone oxime, MEKO), and uretdiones in powder coatings. For detailed discussion of these topics, see Sections 27.2 and 28.1.3, respectively.

A series of extensive review articles on blocked isocyanates is provided by Wicks and Wicks (1975, 1981, 1999, 2001). While "blocked isocyanate" is the most common term, "masked isocyanate" and "capped isocyanate" are sometimes used. The isocyanate group can be "blocked" by any of several methods.

12.5.1 Principles of Blocking and Deblocking

There are two possible modes of reaction of blocked isocyanates with an alcohol: (1) dissociation of the blocked isocyanate to blocking group and free isocyanate, which subsequently reacts with the alcohol (i.e., elimination–addition) and (2) addition of the alcohol to the blocked isocyanate to yield a tetrahedral intermediate followed by elimination of the blocking agent (i.e., addition–elimination). Both pathways are shown in Scheme 12.4, in which B—H represents the blocking group. Evidence has been presented for both pathways with elimination–addition being favored by increasing temperature, in accordance with a consideration of the kinetic parameters, E_a and A, as discussed in Section 2.3.2 and illustrated in Figure 2.5.

At storage temperatures the equilibrium is essentially on the side of the blocked isocyanate, but as temperature increases, the rate of equilibration reactions increase, liberating isocyanate to cross-link with the coreactant. The temperature at which deblocking is rapid depends on the isocyanate and the blocking group. It also depends importantly on the relative reactivity and stability of the product of the reaction with the coreactant. Generally, if the final product is more stable than the blocked isocyanate, the reaction will proceed more rapidly and at a lower temperature than if the final product is less stable. The rate of cross-linking also depends on how effectively the coreactant reacts with the isocyanate in competition with the

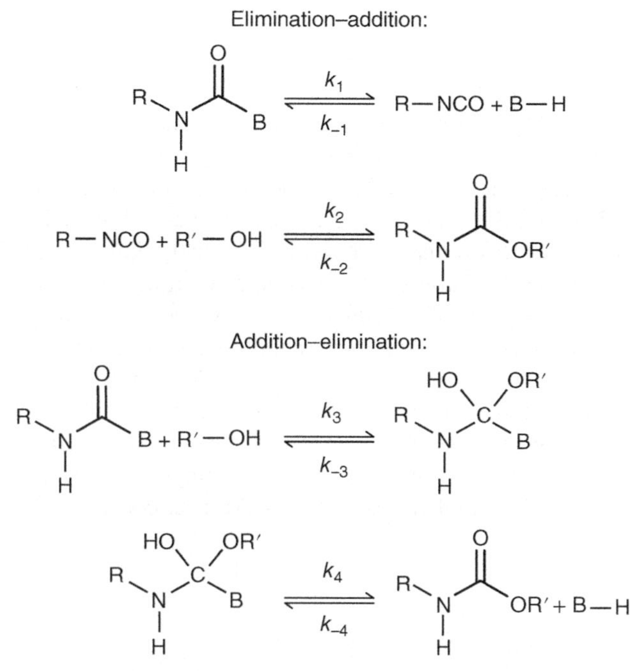

Scheme 12.4 Possible pathways for reaction of a blocked isocyanate with an alcohol.

blocking agent, as well as how rapidly the blocking agent evaporates from the film.

Kinetic studies of the reactions of blocked isocyanates are complicated by the fact that all, or almost all, the reactions are reversible and several side reactions are generally possible. Even the simplest case of heating a blocked isocyanate alone is complicated by possible side reactions of the isocyanates. At high temperatures, temporarily liberated isocyanate may trimerize or react with the original blocked isocyanate to form, for example, an allophanate or biuret. These in turn can also thermally decompose through different reaction pathways.

The curing schedule required for cross-linking of blocked polyisocyanates with polynucleophiles depends on many variables including

- Structures of the isocyanate, blocking agent, and nucleophile
- Relative rate of reaction of the nucleophile with the isocyanate compared with the reverse reaction rate of the isocyanate with the blocking agent
- The rates of diffusion and evaporation of the blocking agent
- Film thickness
- Polarity and hydrogen bonding potential of the reaction medium (solvents or coreactant)
- Concentrations of reactive groups
- Type and concentration of catalysts
- Extent of side reactions and whether they lead to cross-linking or termination
- Extent of cross-linking required to achieve optimum film properties.

Typically, these variables have to be evaluated for each application. A combination that leads to satisfactory cure for wire enamel in a few seconds at 265°C may not be appropriate for an automotive clear coat cured in 20 min at 140°C.

In view of the many variables, comparisons of published "deblocking temperatures" for different blocked isocyanates must be considered with caution, particularly if different methods were used to obtain the data. Furthermore, there is no threshold temperature for unblocking; rather what is measured is the temperature at which a specific analytical method detects some extent of the deblocking and/or cross-linking reaction.

Considering the many different blocking agents, isocyanates, catalysts, and reaction conditions, a study of these variables and their interactions would take a prohibitive amount of time and effort by conventional approaches to formulations even with statistical experimental design techniques. Such systems are candidates for high throughput experimental procedures carried out with combinatorial chemistry apparatus. Wicks (2002) discusses the background of such experiments and their potential applicability to the coatings field. Bach et al. (2002) evaluated various catalysts at 12 temperatures from 80 to 200°C, utilizing four isocyanates, five blocking agents, and an acrylic polyol. With 3,5-dimethylpyrazole as a blocking agent for all four isocyanates, DBTDL exhibited the greatest catalytic effect (curing at 110°C). With MEKO-blocked HDI isocyanurate, approximately equal results were obtained with DBTDL and titanium(IV) 2-ethylhexanoate (curing at 140°C). Unblocking of diethyl malonate-blocked HDI isocyanurate was not affected by catalysis (cure temperature 90°C). The 128 experiments required to obtain these results were completed in one day.

Of course, these unblocking temperatures depend on the method of analysis, heating rate, and other variables. Different analytical techniques are expected to give different unblocking temperatures for the same sample:

- Common methods for determining "deblocking temperature" follow some change in physical properties; for example, gel time—the time required for a mixture of blocked isocyanate and coreactant to gel at a given temperature (Katsamberis and Pappas, 1990). Another popular method is to determine the time required for a coating film to achieve solvent resistance sufficient to withstand 200 methyl ethyl ketone double rubs (Huang et al., 1995). But, as explained in the introduction to Section 12.2, cross-linking may continue after 200 MEK double rubs has been reached.
- Deblocking temperatures have been reported as the temperature at which a characteristic strong IR absorption band of isocyanates (near 2250 cm^{-1}) is first detected. The result depends on the rate at which the sample is heated, as faster rates of heating give higher apparent deblocking temperatures. Fourier transform infrared spectroscopy (FTIR) has been used to determine reaction kinetics from a single dynamic temperature scan (Carlson et al., 1984).
- Use of FTIR in combination with DMA is particularly effective in studying cure mechanisms by allowing the individual deblocking and cross-linking reactions to be observed during the curing process (Kuo and Provder, 1988). FTIR has also been used to determine the composition and rate of evolution of gases during cure; the data on blocking agent evolution can be used to calculate kinetic constants.
- Isothermal thermogravimetric analysis (TGA) has been used to determine kinetic parameters for deblocking reactions (Kordomenos et al., 1982). The extent of reaction is followed by tracking weight loss due to the release of the blocking agent.
- The changes in heat flow associated with deblocking as measured by differential scanning calorimetry (DSC) have been used to study the kinetics of the reaction (Provder et al., 1984).

- The chemical species can be tracked directly using solid-state NMR; solid-state ^{13}C-NMR can track the intensity of carbons within the blocking group that become significantly more intense after the deblocking reaction (Cholli et al., 1983).
- Since the reaction of amines with oxime-blocked isocyanates is first order, rate constants for deblocking of blocked isocyanates can be determined by following the rate of reaction with amines (Witzeman, 1996).
- Since the reaction of isocyanates with water leads to formation of CO_2, the deblocking temperature has been reported as the lowest temperature at which CO_2 is evolved when blocked isocyanates are heated in the presence of molecular sieves saturated with moisture (Mohanty and Krishnamurti, 1998).

Guo et al. (2016) provide a recent example of a kinetic study of deblocking using the TGA, DSC, and CO_2 methods.

Reaction rates and extents of reaction can be very dependent on volatilization of the blocking agent. Very different results were obtained by IR spectral analysis when a neat sample of MEKO-blocked HDI polyisocyanurate was heated at 140°C in open and closed cells (Wicks and Wicks, 1999). A related effect should be expected as a function of film thickness.

In general, blocked aromatic isocyanates deblock at lower temperatures than correspondingly blocked aliphatic isocyanates. This results from the π-conjugation of the aromatic ring with the isocyanate. Substitution of the aromatic ring with electron-withdrawing groups, such as Cl, NO_2, and COOR, increases deblocking rates, while electron donor groups, such as alkyl groups, decrease deblocking rates. There are also steric effects on deblocking rates. Blocked aliphatic isocyanates with the isocyanate group on a tertiary carbon (e.g., MEKO-blocked TMI) deblock at lower temperatures (Lucas and Wu, 1993). MEKO blocked on the ortho position of TDI deblocks 3.2 times faster than the less sterically crowded para isomer at 110°C; MEKO-blocked TMI deblocks 7.9 times faster than less sterically crowded MEKO-blocked cyclohexyl isocyanate at 110°C; and bis-isopropylketoxime-blocked TMI deblocks 5.5 times faster than MEKO-blocked TMI at 110°C (Pappas and Urruti, 1986).

Use of controlled radical polymerization (Section 2.2.1.1) to prepare precursors for blocked isocyanates opens up many opportunities for development.

12.5.2 Blocking Groups

The structures of blocking groups have a major effect on deblocking temperatures and cure rates of coatings.

Although generally requiring high deblocking temperatures, many alcohols have been used as blocking agents, particularly in applications where their stability is advantageous. Stability in waterborne coatings is a major reason 2-ethylhexyl alcohol is widely used in cationic electrodeposition primers (E-coats), where long-term tank stability is essential (Chapter 27). Lower curing temperatures are attainable with monoethers of ethylene glycol, such as 2-butoxyethyl alcohol (Hoppe-Hoeffler et al., 1991).

Phenol-blocked isocyanates deblock at lower temperatures than alcohol-blocked urethanes, in accordance with their slower rate of addition. Various alkyl phenols are also used.

Oximes deblock at lower temperatures than alcohols, phenols, and ε-caprolactam. They have been favored blocking groups for aliphatic polyisocyanates in solventborne coatings and were the first to see use in automotive top coats. Various structures are available, since almost any aldehyde or ketone can be reacted with hydroxylamine to form the oxime.

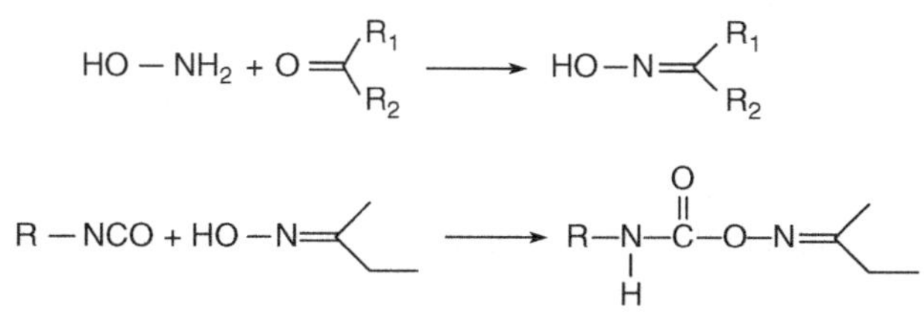

Among the advantages of the oxime groups is their high reactivity towards isocyanates, which allows the blocked products to be readily made without catalyst. The reaction rate of MEKO with cyclohexyl isocyanate was found to be between 50 and 75 times faster than octanol with the isocyanate (Witzeman, 1996). However, the high reactivity of MEKO is disadvantageous with regard to the deblocking reaction, because it competes more effectively than the polyol reactant for the intermediate isocyanate, thereby slowing the forward reaction; there is also concern about possible toxic hazards of MEKO.

ε-Caprolactam is a popular blocking agent, particularly in powder coatings (Chapter 28); it requires higher deblocking temperatures than oximes. 2-Pyrrolidone (also called pyrrolidinone) unblocks at even higher temperatures than ε-caprolactam (Regulski and Thomas, 1983), which probably reflects more steric crowding by the 7-membered lactam.

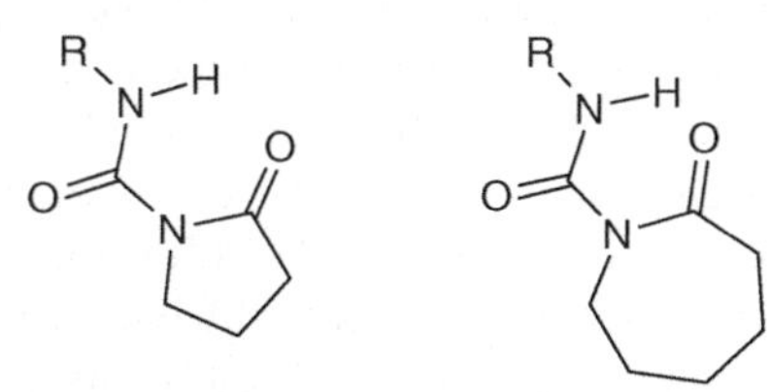

Isocyanates blocked with 2-pyrrolidone and ε-caprolactam

Volatile emissions depend on the MW of the blocking agent. In the case of IPDI-blocked ε-caprolactam, about 50% of the adduct's mass is lost by volatilization of the blocking group. Furthermore, caprolactam vapor condenses in cooler parts of the oven requiring extra maintenance.

Pyrazoles (Cooray and Spencer, 1998) and 1,2,4-triazoles (Keck et al., 1997) unblock at relatively low temperatures. As illustrated in the following, elimination of the pyrazole is probably facilitated by a cyclic proton transfer. Consistent with this hypothesis, the deblocking reaction is enhanced by increasing basicity of pyrazoles. In contrast to phenol-blocked isocyanates, cross-linking is catalyzed by DBTL, suggestive of an elimination–addition pathway. Unblocking is retarded by DABCO, which may interfere with the cyclic mechanism. 3,5-Dimethylpyrazole and 1,2,4-triazole also have the advantages of less yellowing than oximes.

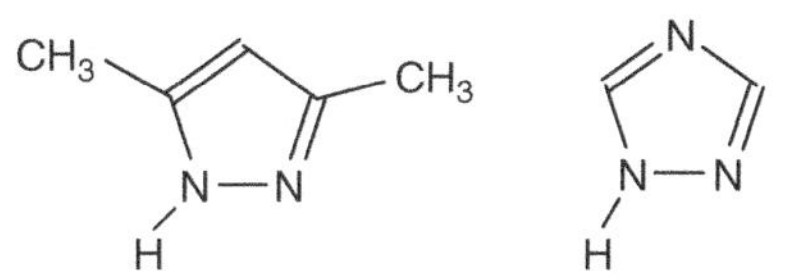

3,5-Dimethylpyrazole and a 1,2,4-triazole ring

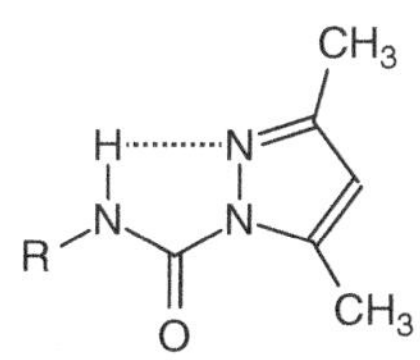

H-Bonded 3,5-dimethylpyrazole-blocked isocyanate

Secondary amines are useful blocking agents, but primary amines are not. Instead of unblocking, the urea from a primary amine and an isocyanate can cleave on the isocyanate or blocking group side of the central C=O group; cleavage on the isocyanate side releases a volatile, toxic monoisocyanate of the primary amine blocking group.

A study of a large number of hindered amines as blocking agents for powder coatings discloses that the most preferred amine is bis(2,2,6,6-tetramethyl-4-piperidinyl) amine, since it remains in the film to a large extent after baking and contributes to the UV stabilization of a coating (Gras and Wolf, 1997).

A large variety of active methylene compounds react with isocyanates by addition of the active methylene group to the carbon of the isocyanate group. The most useful active methylene compound for blocking is diethyl malonate (DEM). While classified as blocked isocyanates, DEM-blocked isocyanates do not unblock in the usual way to give free isocyanate. Instead, a model compound study showed that reaction of DEM-blocked cyclohexyl isocyanate with *n*-hexyl alcohol yields, primarily, transesterification products, together with other products (Scheme 12.5) (Wicks and Kostyk, 1977). It was proposed that the enol form dissociated to give ketenes, which formed the various products. Although cure temperatures are relatively low, hydrolytic stability is adequate to use malonate-blocked isocyanates in 1K waterborne coatings (Daude and Girard, 1996). The active hydrogen of DEM-blocked isocyanates also reacts with MF resins (Blank, et al., 1997).

There has been a trend to replace conventional blocked isocyanates in powder coatings with compounds that do not release a volatile blocking group on curing. Uretdione-blocked isocyanates are attractive in this respect. Because of their superior exterior durability, uretdiones from aliphatic isocyanates are of particular interest. Whereas attempts to make HDI uretdione result

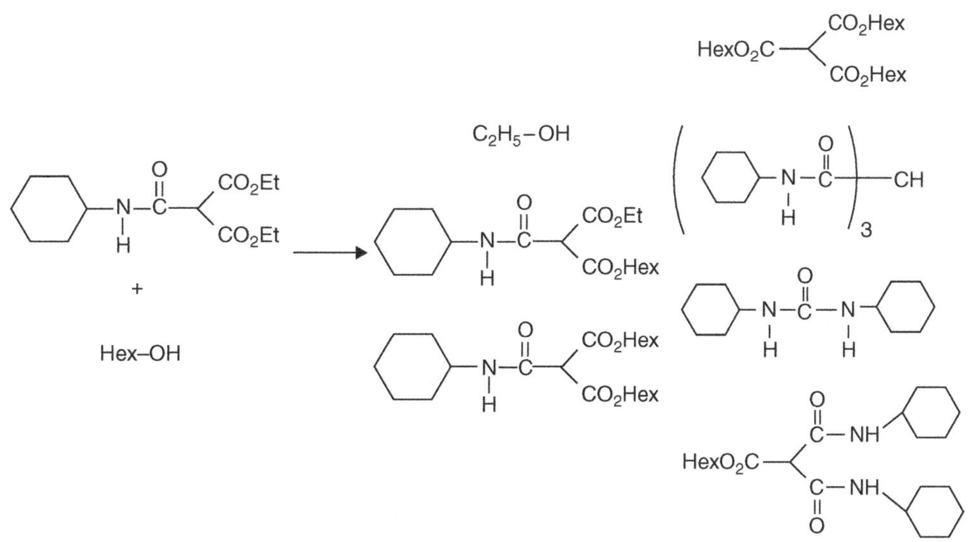

Scheme 12.5 Products of diethyl malonate-blocked cyclohexyl isocyanate reacting with hexyl alcohol.

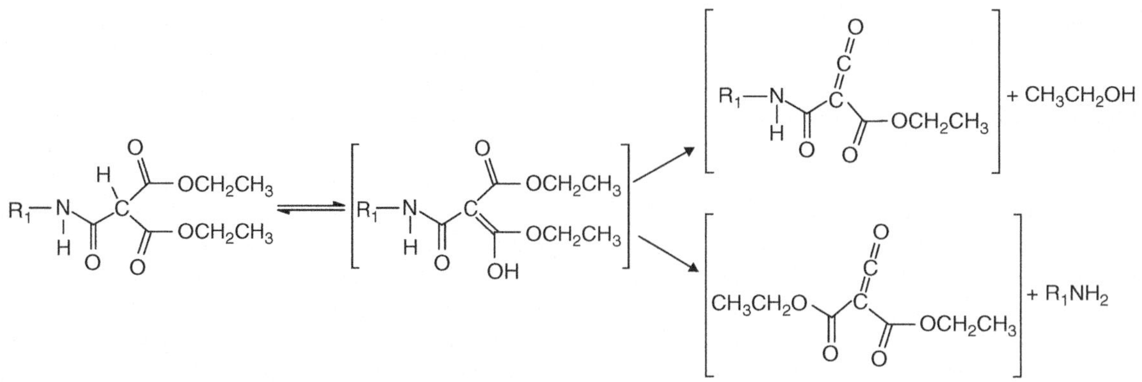

Scheme 12.6 Unblocking of a uretdione.

in a major conversion of the HDI to isocyanurate, IPDI uretdione can be prepared with only minor amounts of isocyanurate (Schmitt et al., 1998). The free isocyanate groups on the uretdione must also be reacted before use. Schmitt recommends reaction of the free isocyanate groups with a mixture of diols and triols to increase average functionality in the range of 2.3–2.5. Schmitt also shows that a mixture of IPDI and HDI can be dimerized into uretdiones. In the process some of the HDI also trimerizes to provide an overall functionality somewhat greater than 2. Unblocking of an IPDI uretdione is shown in Scheme 12.6:

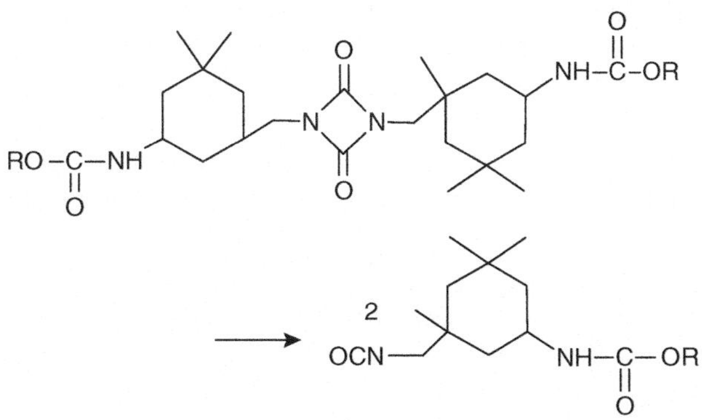

An alternative approach to 1K urethane coatings is to use encapsulated particles of isocyanates. These materials are surface treated to remain insoluble in the vehicle during storage but to dissolve in the coating during heating, thereby releasing free isocyanate that reacts with a hydroxy-functional polymer. One approach is to make a fine particle size dispersion of isocyanate in water, in which surface isocyanate groups react with water to form a urea-rich surface. The powder is filtered, dried, and dispersed in a coating (Kopp et al., 1994).

Another use for encapsulated isocyanates is in self-healing coatings (Wu and Baghdachi, 2015). Worldwide research on self healing has led to commercial products. Many approaches employ encapsulated isocyanates that remain intact in the coating film until they are broken by mechanical or other stress. For example, Huang and Yang (2011) describe a facile process for encapsulating an MDI prepolymer. 1,4-Butanediol is added to an oil-in-water emulsion of the prepolymer in order to polymerize a shell on the outside of the droplets, which encapsulates unreacted prepolymer inside. The capsules can be collected as a powder and dispersed into an epoxy coating to impart self-healing properties. This and other approaches to self-healing coatings are discussed in Section 34.3.

12.5.3 Catalysis of Blocked Isocyanate Coatings

Catalysts are often included in blocked isocyanate formulations but commonly without consideration of what reaction or reactions they are affecting. The same catalysts for 2K polyurethane formation are used with the blocked systems, though at higher levels. As indicated earlier, there are many reactions taking place and the catalyst may be involved in any one or more of the following:

- The deblocking reaction
- Reaction of the free isocyanate with the nucleophilic coreactant
- An addition–elimination reaction
- Side reactions

Carlson et al. (1987) studied the effect of DBTL on the deblocking of IPDI-based cross-linkers without a coreactant. DBTL showed no catalytic effect on the actual deblocking reaction but did, over a longer time, lead to a decrease in isocyanate concentration as a result of side reactions. As previously discussed, DBTL catalyzes the reaction of the unblocked isocyanate and a hydroxyl group, as well as allophanate formation.

Several catalysts were studied in the reaction of MEKO-blocked HDI isocyanurate, as well as other blocked isocyanates, with a hydroxy-functional acrylic resin (Blank et al., 1998). The films were intentionally undercured (20 min at 130°C) to emphasize the differences in catalytic activity. Of the catalysts tested, bismuth tris(2-ethyl hexanoate), cobalt bis(2-ethyl hexanoate), and $Ti(AcAc)_4$ were the most effective.

Catalysis of reactions of blocked isocyanate reactions with polyols in waterborne systems requires special considerations. It is well known that carboxylic acids inhibit catalysis by organotin compounds, but most waterborne systems include coreactants that are dispersed in water by amine salts of carboxylic acid groups. Blocked isocyanate anionic coatings are catalyzed by DBTDL, but to a lesser degree than the corresponding blocked 2K solventborne coatings; higher curing temperature (or longer time) is required to reach the same extent of cross-linking (Lomoelder and Reichel, 1998). These results may be attributed, at least in part, to the inhibition of DBTDL catalysis by carboxylic acids, which are present as amine salts in the anionic coating. Furthermore, DBTDL is relatively easily hydrolyzed and is, therefore a questionable catalyst for waterborne systems.

12.6 MOISTURE-CURABLE URETHANE COATINGS

Moisture-curable urethane coatings are solventborne 1K formulations that cross-link by reaction of isocyanate resins with atmospheric water. They can cure at temperatures as low as 0°C and are also called *moisture-cure* and occasionally *hydrocure*, *moisture-hardening*, or *moisture-tempered* coatings.

Such coatings are based primarily on isocyanate-terminated resins made from hydroxy-terminated polyesters or polyethers by reacting the terminal hydroxyl groups with excess polyisocyanate. To minimize the amount of unreacted diisocyanate, the resin is prepared utilizing an equivalent ratio of NCO to OH that is significantly less than 2 : 1. Ratios higher than 2 : 1 are used with oligomeric MDI since the toxicity of MDI is low and lower viscosity resins can be obtained.

Almost any type of polyol can be used as the backbone of moisture-curable resins although, if functionality is too high, it becomes difficult to avoid premature gelation. Polyether polyols are widely used as they are relatively inexpensive and the ether backbone is completely stable to hydrolysis. These resins tend to have low T_g, thereby enhancing flexibility; however, their exterior durability is relatively poor. Niesten et al. (2013) provide detailed procedures for preparing such coatings, including blends of MDI- and TDI-based polymers. Polyester polyols are also widely used, although the coatings may have limited saponification resistance. Depending on the components, resins having a broad range of T_g are available. Polycarbonate polyols provide resins having hydrolytic stability superior to polyester polyols. Moisture-curable block copolymers of polyols and siloxane segments are prepared by reacting excess isocyanate with a combination of polyols and hydroxy-terminated siloxanes (Kozakiewicz, 1996).

Hydrolytic resistance of coatings is affected by several variables, including cross-link density and free volume availability, as well as the functional groups present. Model compound studies suggest that substituted ureas hydrolyze somewhat more readily under both neutral and acidic conditions than analogous urethanes (Chapman, 1989). Ureas and urethanes from aliphatic isocyanates are somewhat more resistant to hydrolysis than those from aromatic isocyanates. Both urea and urethane groups are generally more stable to hydrolysis than unhindered ester groups.

Moisture-cure coatings are stable when stored in the absence of water; after application the isocyanates react with atmospheric water to form amines, which readily react with other isocyanates to form urea cross-links. Cure rates depend on the water content of the air; higher relative humidity is required at lower temperatures, since relative humidity increases with decreasing temperature for the same water content. Cure is rapid at high humidity and temperature; but the CO_2 released by the reaction of isocyanate with water can be trapped as bubbles, especially in thick films, and can cause pinholes and other defects (Gardner, 1996). Gardner also discusses other application considerations.

While most common moisture-curable urethanes are based on the direct reaction of NCO groups and water, there are several alternatives that avoid the CO_2 problem. Oxazolidines hydrolyze with water as shown below to yield free amine and hydroxyl groups. These in turn react with isocyanates to form urea and urethane linkages. Difunctional oxazolidines offer higher functionality (Howartl, 1996). They are, in effect, "blocked amines." Water is consumed but little or no CO_2 is formed, although there is a carbonyl byproduct.

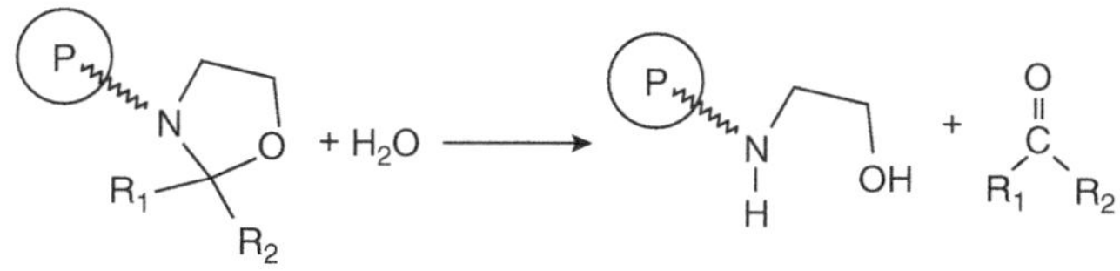

The isocyanate of a moisture-curable urethane can be reacted with 3-aminopropyltriethoxysilane to provide a resin having terminal triethoxysilyl groups that also moisture cure. An advantage of the triethoxysilyl resin is that part of the solvent can be ethyl alcohol. The ethyl alcohol permits a reasonable pot life in the presence of adventitious water, such as from solvents or pigments. Of course, if free isocyanate groups are present, the coating is unstable in the presence of either alcohol or water. The trialkoxysilyl approach also avoids formation of CO_2. See Section 16.2 for further discussion.

Many researchers have tried to enhance the film properties of moisture-curable and other polyurethances by incorporating nanoparticles into the binder matrix. Maganty et al. (2016) reviewed previous efforts and studied compositions consisting of a blend of (1) a polyurethane prepolymer and (2) a dispersion of —N=C=O functionalized silica nanoparticles (20 nm) in HDI isocyanurate. Coatings of the blend had significantly enhanced modulus, strength, ductility, and nanohardness in comparison to similar unpigmented compositions. Optimum properties were reached at about 9.75 wt% silica. The silica particles appeared well dispersed in the films.

12.7 WATERBORNE POLYURETHANE COATINGS

12.7.1 Polyurethane Dispersions

Polyurethane dispersions (PUDs) in water are a growing market. They are used in paper, textiles, and inks and as binders for organic coatings, probably their largest market. PUDs are frequently combined with acrylic latexes as blends or hybrids (Section 12.7.2).

PUD resins are linear, or lightly branched, relatively high MW polyurethane polymers dispersed in water. Such water dispersed materials are, by definition, latexes (Chapter 9), but in the coatings field, they are usually called PUDs, not latexes. Unlike most latexes, which are prepared by emulsion polymerization, PUDS are produced by step-growth polymerization processes followed by dilution with water, analogously to water-reducible acrylics (Chapter 8). Since urethane hydrogen bond strongly with water, the particles are swollen with water, plasticizing the polymer thereby permitting room-temperature film formation with higher T_g polymers than is possible with acrylic latexes (Noble, 1997).

There are many processes for producing PUDs, as reviewed by Kozakiewicz (2015). An example of an older process is outlined here, and some of the variations are then discussed. In the older example, a flexible polyester diol is dissolved in *N*-methylpyrrolidone (NMP, a water miscible solvent) and reacted with excess diisocyanate to form a diisocyanate prepolymer solution. The prepolymer is then reacted with DMPA and 1,4-butanediol. DMPA has two reactive hydroxyl groups and one carboxylic acid group that is unreactive because it is sterically hindered. The resulting polyurethane polymer has "soft" flexible polyester segments and "hard" segments that can associate to impart elasticity. The polyurethane also contains unreacted acid groups (from DMPA) along its backbone. A tertiary amine is added to "salt," and the acid groups and the solution are diluted with water to form a dispersion stabilized by the ionic amine salt groups. NMP is an especially desirable solvent for the process because it dissolves DMPA. However, NMP boils above 200°C, and it is not practical to remove it from the dispersion. PUDs made by this process typically contain 10–15% of NMP (Long-Susewitz, 2016), which functions as a coalescing solvent.

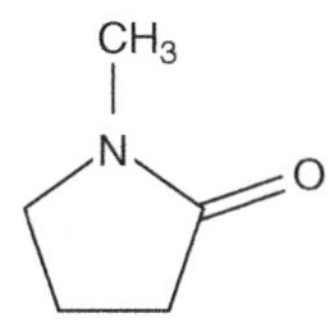

NMP

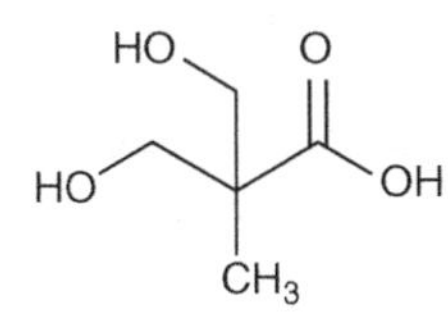
DMPA

Restrictions on the use of NMP because of toxicity concerns are being considered in North America and Europe. NMP also adds to the VOC of PUDs. Processes that replace NMP with acetone or other water miscible solvents have been developed. Low-boiling acetone can be distilled out of the product to yield very low VOC PUDs.

Numerous variations of the above process have been investigated, some of which are described in the following:

- The polyester diol may be linear or slightly branched. It may contain groups such as pendant polyethylene glycol oligomers and/or hindered acids to facilitate dispersion in water.
- The isocyanate prepolymer is often extended with diols such as 1,4-butane diol to form hard polyurethane segments for good elasticity, but it can also be extended with diamines or hydrazine to form hard polyurea segments (Satguru et al., 1994). Combinations of diol and diamine can be effective (Kozakiewicz, 2015; Peng et al., 2015).
- Polyester diols have the advantages of adhesion, toughness, abrasion resistance, exterior durability, and the ability to be designed to give a wide variety of mechanical properties. However, polyesters may suffer from limited hydrolytic stability. Hydrolysis-resistant polyester polyols (Chapter 10) and polycarbonates (Chapter 17) can provide hydrolytic stability, flexibility, and abrasion resistance but at higher cost. Polypropylene glycol has the advantage of imparting flexibility, hydrolytic stability, and low cost; but the films tend to have poor adhesion, poor exterior durability, and poor solvent resistance. Polytetramethylene glycol gives improved solvent resistance but at higher cost.
- Polyglycol monoethers are often used as nonionic, hydrophilic segments for stabilizing PUDs against

changes in pH and soluble ion concentration, and freeze–thaw cycles. They also reduce viscosity of the prepolymers. However, exterior durability of films can be compromised owing to the increased hydrophilicity.

- One can incorporate a small amount of trifunctional isocyanate into an isocyanate-terminated polymer, resulting in some branching on dispersion in water. Also one can chain extend with diethylene triamine to introduce branching or a low degree of cross-linking, but not to the extent that interferes with coalescence of the particles into a continuous film.
- Particle sizes of PUDs depend strongly on the level of DMPA in the polyester precursor. Peng et al. (2015) review the literature and provide detailed procedures for preparing PUDs having different particle sizes. The authors also provide a one-pot process by which PUDs with bimodal particle size distributions can be prepared. Bimodal distributions increase solids and improve packing of the particles during film formation.
- *Cross-linkable PUDs* are prepared by adding diethanolamine or ethanolamine to the isocyanate prepolymer dispersed in the water-solvent mixture. The predominant reaction of amine and isocyanate groups results in terminal hydroxyl groups, derived from the alcoholamines, which can be cross-linked with MF resins or polyisocyanates. Polyunsaturated PUDs are also being used (Section 15.7.2). Aldehyde and ketone functional PUD's can be cross-linked with dihydrazides (Section 17.11).
- Film properties of PUDs may be enhanced by modification with silanes, which may be carried out by chain extending the diisocyanate prepolymer with a diaminosiloxane oligomer.

Lower viscosity PUDs can be made using a *two-step process*. First a linear isocyanate-terminated polyurethane is made from a diisocyanate, a diol, and DMPA, and then in the second step, the terminal isocyanate groups are reacted with a triol (Hegedus and Kloiber, 1996). Satguru et al. (1994) describe a variety of other techniques for making PUDs; the patent literature provides numerous examples.

Various 1K waterborne coatings based on PUDs have been developed. One example is *soft-touch* (*soft feel*) *coatings*, which are used to coat plastic interior parts of cars and trucks. The soft-touch (*haptic*) effect can be achieved by using relatively low T_g binders heavily loaded with micron size pigments. The soft-touch effect of PUDs is attributed at least in part to partial coalescence of micron-sized PUD particles, forming a rough surface (Swaans et al., 2016). UV cure and 2K solventborne polyurethanes soft-touch coating formulations are also available.

Various routes to bio-based PUD's have been demonstrated, for example, by Li et al. (2015) and by Mannari (2015).

12.7.2 Acrylic/Polyurethane Blends and Hybrid Dispersions

There have been three major approaches to preparing aqueous acrylic/urethane systems: (1) blending of PUDs and latexes (or water-reducible acrylic resins), (2) polymerizing vinyl monomers, such as acrylates, in PUDs to make hybrid acrylic/urethane resins, and (3) cross-linking acrylic/urethane systems.

When combined, aqueous PUDs and acrylic latexes offer several potential advantages compared with each of the separate types of polymers. In general terms, polyurethane-based polymers offer superior abrasion resistance due to intermolecular hydrogen bonding and often superior adhesion. Acrylic resins are lower in cost owing both to cost of raw materials and processing. On the other hand, many latexes contain surfactants that can bloom to the surface of films thereby reducing gloss, whereas PUDs can be made with surfactants attached to the polymer. PUDs have low minimum film formation temperature (MFFT) relative to their dry film T_g owing to the plasticizing effect of hydrogen bonding with water. Also, owing to hydrogen bonding of urethane groups, film properties are not as dependent on T_g as in acrylic latexes. Accordingly, low MFFT can be achieved without reducing hardness. For example, a paint made with an acid-functional acrylic latex and a PUD having T_g of −40°C has longer open time, permitting rebrushing to assure hiding and lapping (Gray and Lee, 2001). Coalescing solvents are required to achieve film formation with acrylic latexes at low temperatures, whereas they are not required with PUDs, although residual NMP in many PUDs probably serves as a coalescing solvent. Films of PUDs alone have higher moisture vapor transmission than films from many acrylic latexes, a desirable attribute for coatings on wood.

While blends of PUDs with relatively inexpensive latexes can offer significant cost reductions and sometimes improvement in some properties, improvement of latex properties is not directly proportional to the ratio of the two products. For example, blending equal parts of an acrylic latex and a PUD provides films having tensile strength of 2900 psi, compared with 2000 and 6400 psi for the individual films, respectively, corresponding to an average of 4200 psi. Similar nonlinear effects have been observed with other mechanical and chemical properties. It is suggested that inhomogeneity, resulting from separate acrylic and urethane domains, contributes to the less than hoped for improvement in properties. Such inhomogeneous areas may develop excessive internal strains and/or achieve incomplete coalescence, resulting in increased permeability and reduced cohesive strength (Hegedus and Kloiber, 1996).

Improved film properties can be obtained by polymerizing the acrylate esters in the urethane. An early approach was to use particles of the PUD as "seed" for polymerization of the acrylates. The effect of various reaction parameters on the rate of polymerization of MMA, BA, and a 50:50 mix of MMA and BA in an anionic polyester PUD, dissolved in NMP, has been reported (Kukanja et al., 2002).

Many other variations have been reported. For example, Jahns et al. (2016) outline the current technology and provide procedures for utilizing stabilized PUD particles as seeds for emulsion polymerization of acrylic esters. Long-Susewitz (2016) described coatings made from acrylic/PUD hybrids; these hybrids are NMP-free because the PUD is produced using acrylic monomers, rather than NMP, as the solvent.

12.7.2.1 Cross-Linked PUD/Acrylate Systems

Two general methods for preparing cross-linked acrylic/urethane dispersions are (1) reaction of the same or different functional group on each of the urethane and acrylic resins and (2) use of a separate cross-linker that can react with functional groups on each of the resins. Several compositions and applications follow:

- An autoxidizable coating is based on a water-reducible acrylic resin together with a PUD made from a soy alkyd with HDI, neutralized with NH_4OH, and dispersed in water (Schneider et al., 1997).
- A study of cross-linking blends of hydrazine-terminated PUDs and keto-functional acrylic latexes has been reported (Okamoto et al., 1996). Following evaporation of water from the film after application, cross-linking occurs by reaction of the hydrazine groups with keto groups to form hydrazone (imine) linkages (Section 17.11).
- External cross-linkers are used to react with functional groups on acrylic latexes and PUDs. For example, a chip-resistant automobile primer is formulated with an H_{12}MDI/HDI/polyester polyol PUD, a carboxy- and hydroxy-functional acrylic latex, 2-amino-2-methylpropanol (AMP, presumably to neutralize the carboxy groups), and an MF resin (Gessner and Kandow, 2002).

12.7.3 2K Waterborne Urethanes

Since isocyanates readily react with water, it was assumed for many years that they could not be used directly in waterborne coatings. By the late 1980s this assumption was in question. Development of 2K waterborne coatings began and has advanced to large-scale commercial adoption (Melchiors et al., 2000). Cross-linkers are predominantly polyisocyanates and to a lesser extent polycarbodiimides, polyaziridines, and dihydrazides (Chapter 17). See Wicks et al. (2002) for an extensive review of waterborne 2K urethane systems.

12.7.3.1 Components of Waterborne 2K Coatings

Two types of polyisocyanate cross-linkers are used: conventional polyisocyanates and polyisocyanates that have been modified to be more hydrophilic. In general terms, hydrophilically modified isocyanates are more readily incorporated into a coating; but unmodified isocyanates are less expensive and may provide films that have better exterior durability.

Modifying isocyanates so that they are more readily dispersible in water facilitates mixing of the two packages. An early approach was to prepare nonionic hydrophilically modified polyisocyanates by reacting a fraction of the NCO groups on a triisocyanate (such as HDI or IPDI isocyanurate) with a polyethylene glycol monoether. The products are readily stirred into an aqueous dispersion of a coreactant to form a heterogeneous dispersion, in which the polyisocyanate and coreactant are in separate dispersed particles. It is critical that the components be noncrystalline. The ease of mixing and stability of the dispersions increase as the length or amount of the polyether modifier is increased (Shaffer and Bui, 1998). Adequate water dispersibility without crystallization can be achieved using monoethers with $n>5$ but <10 (Laas et al., 1993).

This approach has been successful in waterborne coatings for indoor use, particularly for low VOC wood furniture coatings. However, reduced functionality of the original triisocyanate and enhanced water sensitivity by the polyethylene glycol segments are drawbacks.

Water sensitivity of films is partially reduced by reacting a hydrophilically modified polyisocyanate with an aminoalkyltrialkoxysilane. For example, HDI is reacted with the monomethyl ether of a polyethylene glycol ($n>10$) and an amine-functional silane made by reacting diethyl maleate with 3-aminopropyltrimethoxysilane. Primers and top coats formulated with the trialkoxysilylated isocyanate and water-reducible acrylic resins show substantial advantages in gloss retention and reduced blistering on water immersion than corresponding films without the silane (Hovestadt et al., 1998). Also see Chapter 16.

PUDs having polyester backbones and DMPA have been widely used as coreactants for waterborne 2K coatings, but they are subject to ester hydrolysis, which cleaves the polymer backbone. The hydrolysis is autocatalytic since carboxylic acids formed from ester hydrolysis catalyze the hydrolysis, as does DMPA. Polycarbonate diols are reported to be more resistant to hydrolysis than polyester diols (Section 17.9).

Acrylic latexes have also been used in 2K waterborne urethane coatings. Feng et al. (1999) reported a statistical study of the effect of several variables on performance of finishes for wood kitchen cabinets, based on hydroxy-functional acrylic latexes cross-linked with a nonionic hydrophilically modified polyisocyanate. The study demonstrated that performance is enhanced by high hydroxy content of the latex (hydroxyl number 52), small particle size of the latex, core–shell preparation of the latex, and low T_g (obtained by increasing level of coalescing solvent).

In addition to reacting with the hydroxyl-functional coreactant, isocyanates can react with water. Although water is less reactive than primary alcohols, the large excess of water favors its participation. Since the reaction with water also results in cross-links (urea rather than urethane), the problem can be minimized by use of a large excess of isocyanate; the so-called "indexing" is sometimes 2:1 isocyanate:hydroxyl or even higher. However, urea linkages can lead to turbidity in films owing to relatively low compatibility and crystallinity of urea compared with urethane linkages in the polymer matrix. The relatively high cost of the isocyanates is another consideration, as well as evolution of CO_2 from reaction with water, which may cause foaming or bubbling of applied films. Accordingly, it is generally desirable to minimize reaction with water so that indexing can be reduced.

The required excess of water is affected by the hydroxy-functionality, $\bar{f}_n$, of the coreactant. Relative reaction rates of NCO with hydroxyl groups and water depend on the catalyst. $Zr(AcAc)_4$ provides better selectivity than DBTDL (Blank et al., 1999). See Section 12.2.2 for further discussion on the effect of various catalysts on the relative rates of reaction of isocyanates with water and alcohols.

12.7.3.2 Mixing and Application Considerations

It is relatively easy to prepare 2K waterborne polyurethane coatings having good properties in the laboratory; however, large-scale production and application are more challenging. There are several potential problems. In some systems, it is difficult to assure that uniform stoichiometric ratios are obtained throughout the film. If reaction occurs to a significant extent before application, coalescence will be inhibited and, therefore, film properties may be poor. With solventborne 2K coatings, pot life can be determined by monitoring viscosity increases. However, in many 2K waterborne coatings, this method is not useful because, as reaction between NCO and OH takes place, viscosity inside the aggregate particles increases, but not the bulk viscosity, which is measured.

To minimize mixing problems, the two dispersions should start out with similar viscosities. Generally it is desirable to use as low viscosity isocyanates as possible. The viscosity of HDI isocyanurate with a functionality of 3.3 is 1.7 Pa·s at 28°C compared with 8.5 Pa·s for HDI biuret with the same functionality, making the isocyanurate easier to disperse (Reiff, 1993). Use of a proprietary water-reducible acrylic resin with conventional HDI-based isocyanates is reported to be satisfactory (Bassner and Hegedus, 1996). Results are also dependent on the process used to mix the two packages together. High, but not excessive, shear is required to obtain relatively uniform particle size with an average diameter of about 150 nm (Bui et al., 1997).

In mixing two dispersions (one containing isocyanate, the other containing coreactant), maximum physical stability of the dispersion will generally be reached with the smallest particle size. However, smaller particle size requires high intensity agitation to break up the initial particles, which increases the possibility of contact of isocyanate groups with water. The probability of new particles being formed that contain both reactants also increases, resulting in nonuniform films. In making dispersions of hydrophilically modified polyisocyanates in dispersions of water-reducible acrylic resins, Dvorchak (1997) has shown that a bimodal particle size distribution of small particles and larger particles form at low shear rates; at high shear rate there was a single broader small particle size distribution, which broadened together increasing particle size resulting from excessive shearing. Carrying out the high intensity mixing in line with application can minimize the problems.

Performance can be affected by the rate of loss of water from films after application. If water is lost slowly, more of the isocyanate will react with water. Thicker films tend to increase the relative reaction with water by decreasing the rate of water loss through the film, requiring higher indexing (Dewhurst et al., 1999). For ambient cure coatings, relative humidity affects the rate of water loss as well as surface cure. Above about 70% RH, gloss is reduced due to bubble formation caused by increased evolution of CO_2. FTIR ATR analysis of 2K waterborne films demonstrates that reaction of isocyanate with water to form urea groups increases near the surface as RH increases (Urban, 2000).

Baking coatings can generally be formulated with lower indexing than ambient cure coatings because of the faster removal of water from the films. The velocity of air flow over the surface of the wet coating is a significant variable. A large fraction of the water evaporates during flash off and initial baking, reducing the extent of isocyanate reaction with water, thereby minimizing CO_2 evolution, and permitting lower ratios of NCO/OH. In one system, indexing of 1.3–1.8 for ambient cure coatings could be reduced to 1.1–1.3 for heat cure coatings (Hegedus et al., 1998).

12.8 HYDROXY-TERMINATED POLYURETHANES

It is often desirable to incorporate the film property advantages of polyurethanes into conventional baking coatings. Polyisocyanates and isocyanate functional polymers can be reacted with diols and triols at an N=C=O/OH ratio of less than 1 to make hydroxy-terminated polyurethanes. These can be cross-linked with MF resins or with other cross-linkers that react with hydroxyl groups. Such polyols can be used alone or blended with hydroxy-functional polyesters, acrylics, and so on. Compared with polyesters, hydrolytic stability is improved and coatings provide some of the flexibility, toughness, and abrasion resistance associated with urethane coatings. However, in solventborne formulations, the solids at equal MW and viscosity are lower owing to intermolecular hydrogen bonding; and the residual urethane groups increase moisture absorption. Whereas MW reduction of acrylic resins, made by conventional free radical initiated polymerization, is limited by the problem of ensuring that at least two hydroxyl groups are present on each oligomer molecule (Section 8.2.1), hydroxy-terminated polyurethanes have two (or more) terminal hydroxyl groups, even at very low MW (as with polyesters), owing to step-growth polymerization. Furthermore, the urethane (carbamate) groups may react to some extent with amino resins (Section 11.3.4), thereby increasing cross-linking functionality.

REFERENCES

Alexanian, V., et al., US patent, 5,254,651 (**1993**).

Bach, H., et al., *Farbe Lack*, **2002**, *108(4)*, 30.

Bassner, S. L.; Hegedus, C. R., *J. Prot. Coat. Linings*, **1996**, *September*, 52.

Bechara, I. S., The Mechanism of Tin-Amine Synergism in the Catalysis of Isocyanate Reaction with Alcohols in *Urethane Chemistry Applications*, ACS Symposium Series, *172*, American Chemical Society, Washington, DC, **1981**, pp 393–402.

Belmokaddem, F.-Z., et al., *Des. Monomers Polym.*, **2016**, *19(4)*, 347–360.

Berlin, P. A., et al., *Kinet. Katal.*, **1993**, *34*, 640.

Blank, W. J., et al., *Polym. Mater. Sci. Eng.*, **1997**, *77*, 391.

Blank, W. J., et al., *Polym. Mater. Sci. Eng.*, **1998**, *79*, 399.

Blank, W. J., et al., *Prog. Org. Coat.*, **1999**, *35*, 19.

Bock, M.; Halpaap, R., *J. Coat. Technol.*, **1987**, *59(755)*, 131.

Bui, H., et al., *Eur. Coat. J.*, **1997**, *97*, 476.

Carlson, G. M., et al., Cure Kinetics Characterization of Blocked Isocyanate Coatings by FTIR and Thermal-Mechanical Analysis in Frisch, K. C.; Klempner, D., Eds., *Advances in Urethane Science and Technology*, Technomic, Lancaster, **1984**, Vol. *9*, p 47.

Carlson, G. M., et al., *Polym. Sci. Technol.*, **1987**, *36*, 197.

Chapman, T. M., *J. Polym. Sci. A Polym. Chem.*, **1989**, *27*, 1983.

Cholli, A., et al., *J. Appl. Polym. Sci.*, **1983**, *28*, 3497.

Cooray, B.; Spencer, R., *Paint Resin*, **1998**, *October*, 18.

Daude, G.; Girard, P., US patent 4,623,592 (**1996**).

Devendre, R.; Edmonds, N. R.; Sohnel, T., *RSC Adv.*, **2015**, *5*, 48935–48945.

Dewhurst, J. E., et al., *Polym. Mater. Sci. Eng.*, **1999**, *81*, 195.

Dogan, N., et al., *RadTech Rep.*, **2006**, *20*, 43.

Draye, A.-C.; Tondeur J.-J., *J. Mol. Catal. A Chem.*, **1999**, *140*, 31.

Draye, A.-C., et al., *Main Group Met. Chem.*, **1999**, *22*, 367.

Dvorchak, M. J., *J. Coat. Technol.*, **1997**, *69(866)*, 47.

Feng, S. X., et al., *J. Coat. Technol.*, **1999**, *71(899)*, 51.

Fiori, D. E.; Dexter, R. W., *Proceedings of the International Waterborne, High-Solids, and Powder Coatings Symposium*, New Orleans, LA, **1986**, p 186.

Gardner, G., *J. Prot. Coat. Linings*, **1996**, *February*, 81.

Gessner, M. A.; Kandow, T. P., US patent 6,437,036 (**2002**).

Gras, R.; Wolf, E., Canadian patent application 2,186,089 (**1997**).

Gray, R. T.; Lee, J., US patent 6,303,189 (**2001**).

Guo, S., et al., *Materials*, **2016**, *9*, 110.

Hatada, K., et al., *J. Polym. Sci. A Polym. Chem.*, **1990**, *28*, 3019.

Hegedus, C. R.; Kloiber, K. A., *J. Coat. Technol.*, **1996**, *68(860)*, 39.

Hegedus, C. R., et al., *Proceedings of the International Waterborne, High-Solids, and Powder Coatings Symposium*, New Orleans, LA, **1998**, p 391.

Higginbottom, H. P., et al., *Prog. Org. Coat.*, **1998**, *34*, 27.

Hira, Y., et al., *Mater. Sci. Eng.*, **1983**, *49*, 336.

Hoppe-Hoeffler, M., et al., Ger. Offen., DE 3,938,883 (**1991**).

Hovestadt, W., et al., US patent 5,854,338 (**1998**).

Howartl, G. A., *Proceedings of the International Waterborne, High-Solids, and Powder Coatings Symposium*, New Orleans, LA, **1996**.

Huang, M.; Yang, J., *J. Mater. Chem.*, **2011**, *21*, 11123.

Huang, Y., et al., *J. Coat. Technol.*, **1995**, *67(842)*, 33.

Jahns, E., et al., Patent publication WO2016016286 A1 (**2016**).

Katsamberis, D.; Pappas, S. P., *J. Appl. Polym. Sci.*, **1990**, *41*, 2059.

Keck, M. T., et al., US patent 5,688,598 (**1997**).

Kopp, R., et al., *Angew. Makromol. Chem.*, **1994**, *223*, 61.

Kordomenos, P. I., et al., *J. Coat. Technol.*, **1982**, *54(687)*, 43.

Korzyuk, E. L.; Zharkov, V. V., *Kinet. Katal.*, **1981**, *22*, 522.

Kozakiewicz, J., *Prog. Org. Coat.*, **1996**, *27*, 123.

Kozakiewicz, J., *Polimerey*, **2015**, *60(9)*, 525–535.

Kukanja, D., et al., *J. Polym. Sci.*, **2002**, *84*, 2639.

Kuo, C.; Provder, T., *Polym. Mater. Sci. Eng.*, **1988**, *59*, 474.

Laas, H.-J., et al., US patent 5,252,696 (**1993**).

Li, Y., et al, *Prog. Org. Coat.*, **2015**, *86*, 134–142.

Lomoelder, R.; Reichel, D., *FATIPEC Congress Book*, **1998**, Vol. *D*, p D-25.

Lomoelder, R., et al., *J. Coat. Technol.*, **1997**, *69(868)*, 51.

Long-Susewitz, J., *Proceedings of the American Coatings Association Congress*, Indianapolis, IN, April, **2016**, paper 10.3.

Lucas, H. R.; Wu, K.-J., *J. Coat. Technol.*, **1993**, *820*, 59.

Luo, S.-G., et al., *J. Appl. Polym. Sci.*, **1997**, *65*, 1217.

Maganty, S., et al., *Prog. Org. Coat.*, **2016**, *90*, 243–251.

Mannari, V. M., US patent publication US8952093 B2 (**2015**).

Melchiors, M., et al., *Prog. Org. Coat.*, **2000**, *40*, 59.

Mohanty, S.; Krishnamurti, N., *Eur. Polym. J.*, **1998**, *34*, 77.

Narute, P.; Palanisamy, A., *J. Coat. Technol. Res.*, **2016**, *13(1)*, 171–179.

Niesten, M., et al., Canadian patent publication CA2502406 C (**2013**).

Noble, K. L., *Prog. Org. Coat.*, **1997**, *32*, 131.

Nordstrom, J. D.; Dervan, A. H., *Proceedings of the International Waterborne, High-Solids, and Powder Coatings Symposium*, New Orleans, LA, **1993**, p 3.

Nordstrom, J. D., et al., *Proceedings of the International Waterborne, High-Solids, and Powder Coatings Symposium*, New Orleans, LA, **1997**, p 70.

Noreen, A., et al., *Prog. Org. Coat.*, **2016**, *91*, 25–32.

Okamoto, Y., et al., *Prog. Org. Coat.*, **1996**, *29*, 175.

Okumoto, T., et al., *Nippon Gomu Kyokaishi*, **1995**, *68(4)*, 244.

Ono, H.-K., et al., *J. Polym. Sci. C Polym. Lett.*, **1985**, *23*, 509.

Palyutkin, V. G.; Zharkov, G. M., *Kinet. Katal.*, **1985**, *26*, 476.

Pappas, S. P.; Urruti, E. H., *Proceedings of the International Waterborne, High-Solids, and Powder Coatings Symposium*, New Orleans, LA, **1986**, p 146.

Peng, S.-J., et al., *Prog. Org. Coat.* **2015**, *86*, 1–10.

Potter, T. A.; Slack, W. E., US patent 5,124,427 (**1992**).

Provder, T., et al., Cure Reaction Kinetics Characterization of Some Model Organic Coatings Systems by FT-IR and Thermal Mechanical Analysis in Johnson, J. F.; Gill, P. S., Eds., *Analytical Calorimetry*, Plenum Press, New York, **1984**, Vol. *5*, p 377.
Regulski, T.; Thomas, M. R., *Org. Coat. Appl. Polym. Sci. Proc.*, **1983**, *48*, 1003.
Reiff, H., US patent 5,258,452 (**1993**).
Richter, R., et al., US patent 5,045,226 (**1991**).
Richter, F., et al., US patent 5,717,091 (**1998**).
Richter, F., et al., US patent 6,090,939 A (**2000**).
Satguru, R., et al., *J. Coat. Technol.*, **1994**, *88(830)*, 47.
Sato, M., *J. Am. Chem. Soc.*, **1960**, *82*, 3893.
Schaefer, H. et al., US patent publication US2A1 (**2015**).
Schmitt, F., et al., *Prog. Org. Coat.*, **1998**, *34*, 227.
Schneider, V., et al., US patent 5,688,859 (**1997**).
Schwetlick, K.; Noack, R., *J. Chem. Soc. Perkin Trans.*, **1995**, *2*, 395.
Schwetlick, K., et al., *J. Chem. Soc. Perkin Trans.*, **1994**, *2(3)*, 599.
Seipke, C., et al., *Proceedings of the American Coatings Association Congress*, Indianapolis, IN, April, **2016**, paper 10.2.
Shaffer, M.; Bui, H., *Proceedings of the International Waterborne, High-Solids, and Powder Coatings Symposium*, New Orleans, LA, **1998**, p 93.
Squiller, E. P.; Rosthauser, J. W., *Mod. Paint Coat.*, **1987**, *June*, 26.
Squiller, E. P., et al., Patent publication WOA1 (**2014**).
Swaans, R., et al., *Proceedings of the American Coatings Association Congress*, Indianapolis, IN, April, **2016**, paper 10.6.
Takamatsu, K., et al., US patent publication US2A1 (**2015**).
Takas, T. P., *JCT Coat. Technol.*, **2004**, *1(5)*, 40.
Urban, M. W., *Prog. Org. Coat.*, **2000**, *40*, 195.
Van der Weij, F. W., *J. Polym. Sci. A Polym. Chem.*, **1981**, *19*, 381.
Wang, J., et al., *J. Coat. Technol. Res.*, **2013**, *10(6)*, 859–864.
Wang, G., et al., *J. Coat. Technol. Res.*, **2015**, *12(3)*, 543–553.
Wicks, Z. W., Jr., *Prog. Org. Coat.*, **1975**, *3*, 73.
Wicks, Z. W., Jr., *Prog. Org. Coat.*, **1981**, *9*, 3.
Wicks, D. A., *Proceedings of the International Waterborne, High-Solids, and Powder Coatings Symposium*, New Orleans, LA, **2002**, p 1.
Wicks, Z. W., Jr.; Kostyk, B. F., *J. Coat. Technol.*, **1977**, *49(634)*, 77.
Wicks, D. A.; Wicks, Z. W., Jr., *Prog. Org. Coat.*, **1999**, *36(3)*, 148–172; 2001, 41, 1–85.
Wicks, D. A.; Yeske, P. E., *Prog. Org. Coat.*, **1997**, *30*, 265.
Wicks, Z. W., Jr., et al., *Prog. Org. Coat.*, **2002**, *44*, 161.
Witzeman, J. S., *Prog. Org. Coat.*, **1996**, *27*, 269.
Wojcik, R. T., *Polym. Mater. Sci. Eng.*, **1994**, *70*, 114.
Wong, S. W.; Frisch, K. C., *J. Polym. Sci. A Polym. Chem.*, **1986**, *24*, 2867, 2877.
Wu, L.; Baghdachi, J., Eds. *Functional Polymer Coatings*, John Wiley & Sons, Inc., New York, **2015**.
Yilgor, I.; McGrath, J. E., *J. Appl. Polym. Sci.*, **1985**, *30*, 1733.
Zhang, J.; Tu, W.; Dai, Z., *J. Coat. Technol. Res.*, **2013**, *10(6)*, 887–895.
Zhou, L., et al., *Proceedings of the International Waterborne, High-Solids, and Powder Coatings Symposium*, New Orleans, LA, **2000**, pp 262–281.

Chapter 13

Epoxy and Phenolic Resins

Epoxy and phenolic resins are important classes of step-growth resins with a multitude of end uses. Over 40% of worldwide epoxy resin production is used in coatings. Epoxies have been adapted for all low VOC coating technologies. High solids and waterborne epoxies are described in this chapter. Epoxies in electrodeposition coatings, powder coatings, and radiation cure coatings are discussed mainly in Chapters 27–29.

13.1 EPOXY RESINS

Terminology of epoxies can be confusing. Epoxy groups (also called epoxides) are three-membered cyclic ethers. In IUPAC and Chemical Abstracts nomenclature, they are called *oxiranes*. Most commercially important epoxy resins are derived from (chloromethyl)oxirane, more commonly known as epichlorohydrin (ECH) or simply "epi." The resins generally contain oxiranylmethyl ethers or esters, usually called glycidyl ethers or esters, respectively. In addition to the resins discussed in this chapter, the so-called epoxy esters are discussed in Section 15.8 because of their close relationship to the chemistry of alkyds. Acrylated epoxy resins are discussed in Section 29.2.5.

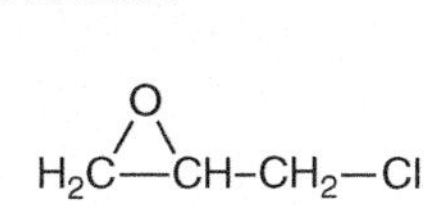

Epichlorohydrin (ECH)

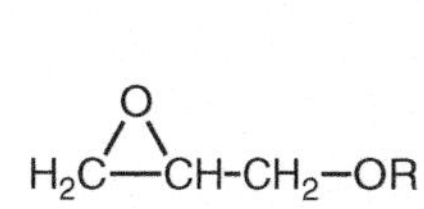

Glycidyl ether

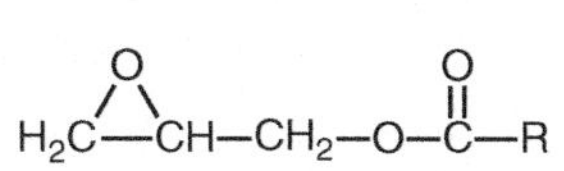

Glycidyl ester

Thousands of different glycidyl ethers and esters have been described, and many are commercial specialty products. The largest volume resins are made from ECH and bisphenol A (BPA). Pham and Marks (2012) describe many other glycidyl structures.

13.1.1 Bisphenol A Epoxy Resins

The first epoxy resins used in coatings were BPA epoxies made by reacting BPA with ECH. With many variations and refinements, they remain the workhorse resins for coatings because they impart exceptional toughness, adhesion, and corrosion resistance.

The chemistry of BPA epoxy production is detailed in Scheme 13.1. Under basic conditions, the initial reaction is the formation of a BPA anion (BPA^-), which attacks ECH and results in the formation of a new oxirane ring with elimination of chloride anion.

The initial product is the monoglycidyl ether of BPA (MGEBPA). Analogous reaction of the unreacted phenolic group of MGEBPA with NaOH and ECH gives the diglycidyl ether of BPA (DGEBPA), also called bisphenol A diglycidyl ether (BADGE). Molecular weight (MW) is controlled by the ratio of ECH to BPA. With a large excess of ECH, it is possible to make a resin that is dominantly DGEBPA, that is, where $n = 0$ in the general formula. The pure $n = 0$ compound is a crystalline solid, but the largest volume commercial grades are liquids with n values of 0.1–0.2 (the so-called standard liquid resin; it is commonly used in the familiar epoxy glues). A number of distinct resins within this n range are manufactured for various purposes.

Organic Coatings: Science and Technology, Fourth Edition. Frank N. Jones, Mark E. Nichols, and Socrates Peter Pappas.

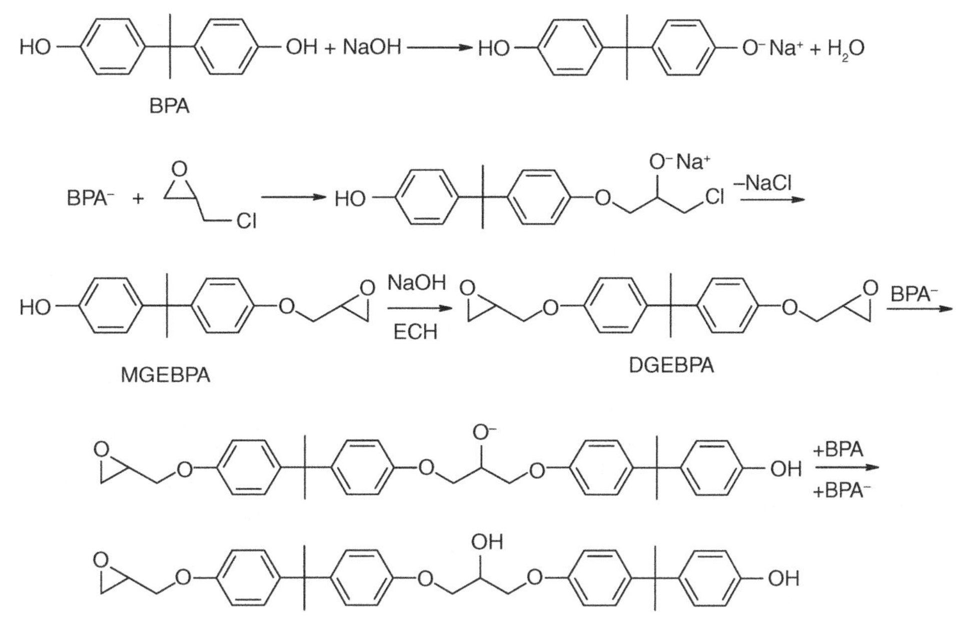

Scheme 13.1 Chemistry of the taffy process for synthesis of BPA epoxy resins.

The epoxy groups of MGEBPA and DGEBPA react with BPA⁻ to extend the chain, as shown for DGEBPA; these reactions introduce alcohol groups on the backbone. Continuation of these reactions results in linear polymers, since both the BPA and ECH are difunctional. BPA epoxy resins are made with excess ECH, so the end groups are glycidyl ethers. The polymers may be represented by the following general formula, where the molar ratio of ECH to BPA determines the average n value.

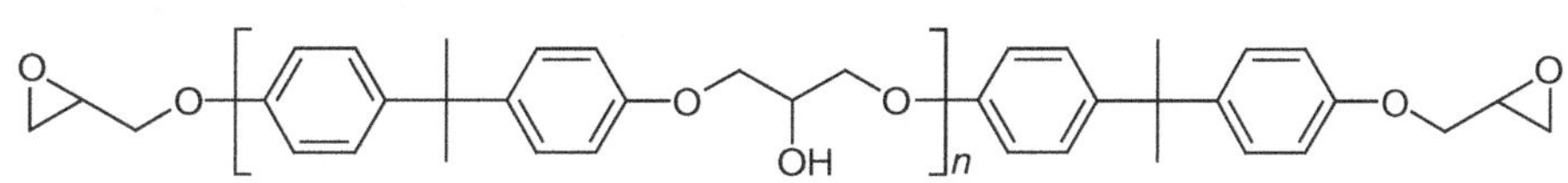

Bisphenol A epoxy resins; *n* can be zero or higher

The procedure shown in Scheme 13.1 for making BPA epoxy resins is called the *taffy process*. Stoichiometric amounts of NaOH are required, resulting in the formation of a large amount of NaCl, which must be removed from the resin by repeated washing with water. The washing step is relatively easy for the standard liquid resin due to its low viscosity. However, as the ratio of ECH to BPA is decreased, higher MW products are produced, the reaction mixture becomes highly viscous, and water washing to remove NaCl becomes difficult. Also, as MW and viscosity increase, the probability of branching increases. Historically, the taffy process was used to make high MW epoxy resins, but the processing difficulties now limit this process almost entirely to liquid resins and certain low MW solid resins.

Higher MW BPA epoxy resins are now usually made by reacting a standard liquid resin ($n = 0.1$–0.2) with BPA at high temperature in the presence of a catalyst, a method that is variously called the *advancement process*, the *upgrade process*, or the *fusion process*. The catalyzed reaction of BPA with both epoxide groups of the standard resin results in a higher MW resin with BPA end groups, which further react with standard resin to give epoxide end groups. Additional rounds of reactions with BPA and standard resin further advance the resin. The final M_n depends mainly on the ratio of standard resin to BPA. Purified grades of BPA are preferable. Such processes are carried out at higher temperature than the taffy process; hence, the viscosity is lower, mixing is better, and there is less branching. No NaCl is produced; therefore, the difficult salt removal step is unnecessary.

Traditionally, batch advancement/fusion/upgrade processes have been used. Large batches, however, pose the problem of how to rapidly cool a large hot mass of molten epoxy. Once the batch has reached the desired specifications, it is desirable to cool the entire mass quickly to obtain the most uniform possible product. As an alternative, continuous processes have been developed (Bochan et al., 2004) and are used in commercial production to some extent.

There can be variations in the resins obtained depending on the reaction conditions and the catalyst. Among many catalysts that have been used are imidazoles, ethyltriphenylphosphonium hydroxide, iminium salts, and alkyl phosphonium salts. Catalysts vary in their reactivity, selectivity, and tendency to promote formation of high MW molecules (Dante and Parry, 1969). Certain alkylphosphonium salt catalysts are designed to deactivate near the peak reaction temperature to improve stability of the products (Pham and Marks, 2012).

Resins made by the taffy process consist of oligomers with $n=0$, 1, 2, 3, 4, 5, and so on, whereas resins made by the advancement process have largely even number n value oligomers, that is, $n=0$, 2, 4, 6, and so on. The predominance of oligomers with even number n values in resins from the advancement process follows, starting with the DGEBPA. Each complete step in the polymerization attaches one mol of BPA and one mole of liquid resin to the growing polymer, so two n-units are added in each step. Roughly 10 wt% of odd n molecules result from the presence of about 10% of $n=1$ resin in the liquid resin.

Above an n value of about 1, the resins are amorphous solids at ambient temperature; T_g increases with MW. The resins may be said to have melting points, corresponding to the temperature of flow under specified test conditions. However, being amorphous, they do not melt in the sense that a crystalline solid does, so a preferable term is *softening point*, which is also used. Commercially available higher MW resins are often designated as types 1001, 1004, 1007, and 1009. Table 13.1 gives average n values, epoxy equivalent weights (EEW), and softening points for commercial BPA epoxy resins. EEW is often called weight per epoxide (WPE). EEW can be measured by titration or by nuclear magnetic resonance (NMR). As MW increases, EEW and average hydroxy functionality also increase. In some even higher MW ($n>60$) epoxy resins, the amount of epoxy groups present is so small that the resins are more appropriately polyfunctional alcohols, commonly called *phenoxy resins*.

The figures in Table 13.1 are subject to considerable variation. A given supplier will often offer subtypes and quote significantly different ranges for EEW (or WPE) for products within one of the types. For example, suppliers offer grades of type 4 resins specifically for powder coatings and other grades for other applications. Furthermore, resins classed as a particular type by different suppliers are likely to be measurably different.

According to legend, the designations type 1001, 1004, etc. originated as lab notebook page numbers of Sylvan Greenlee, one of the pioneers of epoxy resins. He had a good week.

Table 13.1 Broad Classifications of Commercial BPA epoxy resins

Resin type	Approx. n value	EEW ($g\,equiv.^{-1}$)	Softening point (°C)
Liquid	0.1–0.2	172–195	Liquid
Type 1001	2	450–560	70–85
Type 1004	5	800–950	95–110
Type 1007	15	1600–2500	120–140
Type 1009	25	2500–6000	145–160

Source: Pham and Marks (2012).

Although, theoretically, there should be two epoxy groups on each molecule, various other end groups are present to small extents. Varying amounts of these end groups can significantly affect performance of the resins. To minor degrees, unreacted phenol and 1,2-chlorohydrin terminal groups are present; additional non-epoxy end groups result from side reactions. For example, 1,2-glycol end groups can result from inadvertent hydrolysis of epoxy groups, as shown in Eq. 13.1, or of chlorohydrin groups.

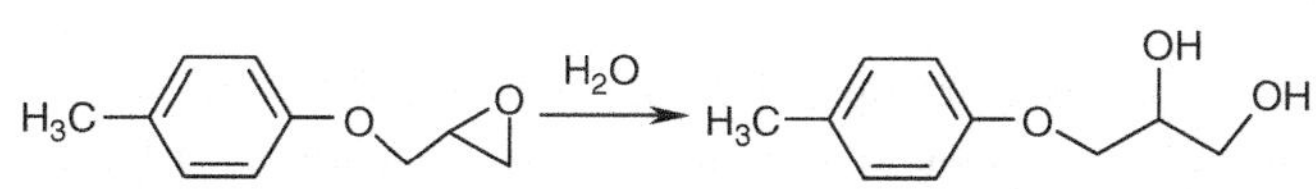

(13.1)

Instead of ring closing to give an epoxy group, the hydroxyl group of the 1,2-chlorohydrin can react with another ECH molecule to give the addition product shown in Eq. 13.2.

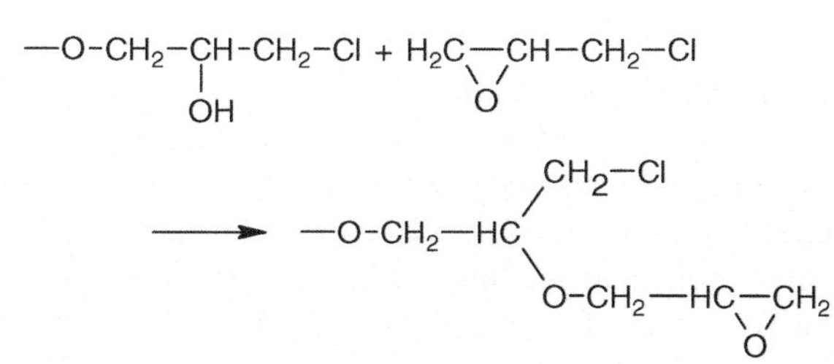

(13.2)

Another side reaction is ring opening by a phenoxide anion polymer end group with ECH at the less favored, more sterically hindered position, as shown in Eq. 13.3. The resulting 1,3-chlorohydrin derivative (after proton transfer) cannot ring close to give an oxirane and is relatively stable under the reaction conditions. This group is the main site of *non-hydrolyzable chlorine* that is commonly reported in specifications of commercial epoxy resins.

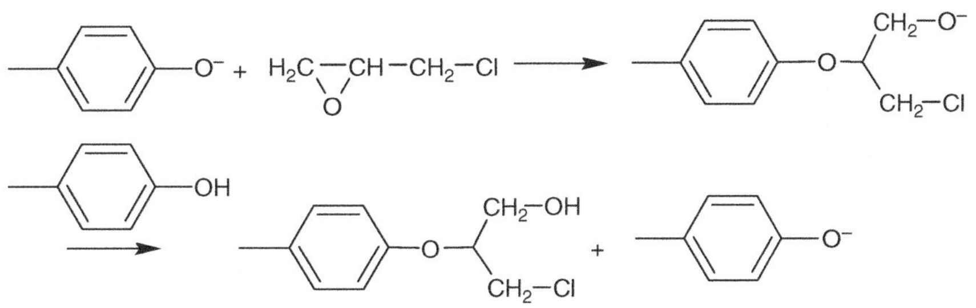

(13.3)

A hydroxyl group on the polymer backbone can also react with ECH to yield branched molecules with an epoxy functionality $(\bar{f}_n)$ of three, as shown in Eq. 13.4.

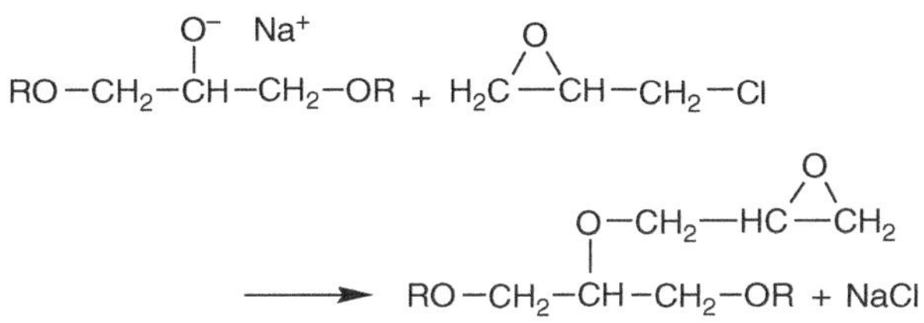

(13.4)

If there were no side reactions, the epoxy $\bar{f}_n$ of BPA epoxy resins would be 2.00. However, the net effect of side reactions is that commercial BPA epoxy resins generally have an $\bar{f}_n$ less than 2, commonly about 1.8–1.9. In some cases, this lower $\bar{f}_n$ can have an important effect on film properties. The presence of terminal glycol groups results in chain termination, thereby reducing MW and resin viscosity. The presence of small, controlled amounts of terminal glycol groups can have beneficial effects on adhesion (Sheih and Massingill, 1990). It is possible to increase the $\bar{f}_n$ above 1.8–1.9 by adding small amounts of trifunctional phenolic material (Section 13.6) to the BPA used in the process.

Development of high performance liquid chromatography (HPLC) and gel permeation chromatography (GPC, SEC) analytical techniques has permitted improvements in process development and control. Scheuing (1985) showed that HPLC can distinguish not only the individual oligomers ($n=0$, 1, 2, 3, etc.) but also the oligomers in which one or both ends have 1,2-dihydroxy groups or 1,3-chlorohydrin or can be separated (Eqs. 13.1 and 13.3). GPC is especially useful for higher MW resins because it can measure M_n, M_w, M_z, and other characteristics of each resin. Variations in these quantities can affect application and film properties. Analysis of the products resulting when minor process changes are made in reaction conditions permits establishing process parameters to meet relatively narrow product specifications. Tight process control is particularly critical for epoxy resins to be used in electronic applications and in powder coatings (Chapter 28).

BPA epoxy resins perform especially well in coatings applications in which excellent adhesion, electrical properties, toughness, and corrosion resistance are required. A limitation is poor exterior durability, primarily resulting from direct absorption of UV radiation by the aromatic ether groups, which ultimately leads to photooxidative degradation. Lower MW resins require less VOC and also give better chemical resistance when cross-linked with amines since the cross-link density of films is higher. However, when higher MW resins are used with cross-linkers that react with hydroxyl groups (e.g., phenolic or amino resins), chemical resistance and flexibility are better. Cross-linked higher MW resins also give films that dry more rapidly and have improved corrosion resistance.

BPA epoxy resins contain minute amounts of unreacted BPA. Studies in animals have suggested that larger amounts of BPA may cause a variety of health issues, including disruption of the endocrine system. After much study, the US Food and Drug Administration (FDA) concluded in 2014 that the very small amounts of BPA found in foods and beverages packed in cans with epoxy can linings are safe. Around the globe, BPA regulations vary. France effectively banned BPA epoxy resins for food contact uses in 2015. According to Gander (2016), alternatives are proving less satisfactory, prompting reconsideration of the ban. Even where not forbidden by regulation, some coatings customers are pressing epoxy coatings suppliers to develop BPA-free alternatives. It is a challenging problem because of the excellent properties of BPA coatings. For further discussion, see Section 30.3.1.

13.1.2 Other Epoxy Resins

Alternative epoxy resins are available. Major motivations for use of alternative resins are to reduce viscosity for high solids coatings, to increase $\bar{f}_n$ for greater cross-link density, to eliminate concerns about possible toxic hazards of BPA, and to improve weatherability. Alternative resins can be used by themselves or blended with BPA epoxy resins.

Bisphenol F (BPF) epoxy resins (from reaction of ECH with BPF in place of BPA) have the advantage of lower viscosities at the same n value. Standard liquid BPA epoxies have viscosities on the order of 12–14 Pa·s at 25 °C, whereas the viscosities of comparable BPF epoxies are 2.5–4.5 Pa·s. Most BPF resins are low MW liquids because low viscosity is their most valuable feature. These and other low viscosity resins are playing an increasing role in very low VOC and solventless epoxy coatings (Chapters 25 and 33).

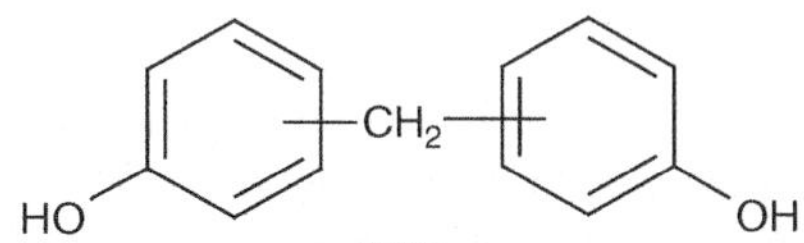

BPF

Bisphenol S ((BPS), 4,4′-sulfonyldiphenol) is used in certain food contact plastics as a substitute for BPA and has, at least experimentally, been incorporated into epoxy resins for coatings (Chen et al., 2016). The potential health effects of BPS are unclear.

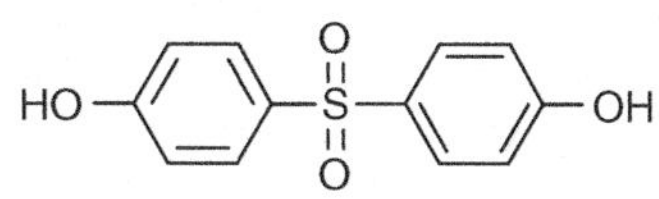

Epoxy resins having lower viscosity than BPA epoxy resins can also be made by substituting a flexible diol for BPA in the advancement process to make what might be called *copolymer epoxy resins*. Examples are advancement resins made from combinations of propylene or dipropylene glycol and BPA (Massingill et al., 1990). When cross-linked with phenolic resins, these resins form films that are more flexible than films from homopolymer BPA epoxy resins and also provide excellent adhesion. Lower T_g, lower viscosity, and more flexible epoxy resins can also be prepared from bisphenols with longer chain links between the two phenol rings, especially oligomeric ethylene and propylene glycol groups (Dubois and Sheih, 1992). Experimental resins to demonstrate the idea were prepared by reacting a series of such bisphenols with DGEBPA using ethyltriphenylphosphonium acetate/acetic acid in methyl alcohol as a catalyst.

Hydrogenated BPA epoxy resins have lower T_g and lower viscosity than BPA resins having the same n value. They also exhibit better exterior durability relative to both BPA and BPF resins, resulting from the absence of the UV-absorbing aromatic ether groups.

Brominated BPA epoxy resins are used in flame retardant coatings and in electronic circuit boards.

Epoxy resins are also prepared by reaction of ECH with novolac phenolic resins (Section 13.6.2). The resulting novolac epoxy resins are useful in applications in which more than two epoxy groups per molecule are desirable, especially in powder coatings where solid epoxy resins are needed. Epoxy resins derived from the reaction of o- or p-cresol-formaldehyde novolacs and ECH are available, having an epoxy $\bar{f}_n$ of 2.2–5.5. Cross-linked novolac epoxy resin coatings are more chemically resistant than BPA epoxy resins due to the higher cross-linked density resulting from higher functionality. A general structure of novolac epoxies is shown as follows.

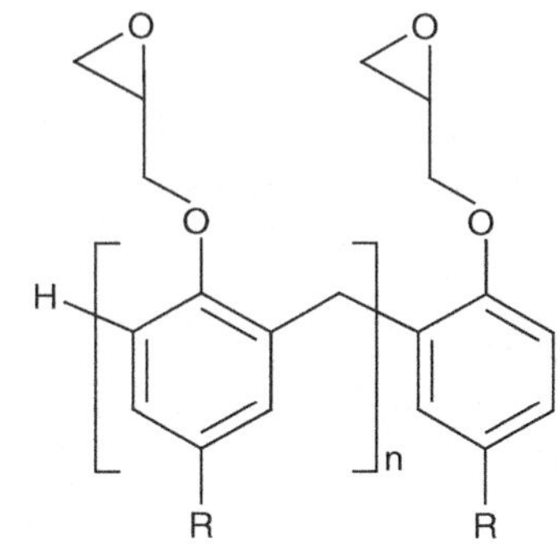

Other epoxy resins with higher functionality are made by reacting ECH with substrates having multiple phenolic groups or with aromatic diamines. For example, resins with a functionality of four are based on tetraglycidyl methylenedianiline. They are used mainly in composite materials.

Triglycidylisocyanurate (TGIC) is a solid trifunctional epoxy cross-linker used in powder coatings (Chapter 28). The presence of three functional groups gives higher cross-link density than obtained with BPA epoxy resins, and photochemical stability of the cured coatings is superior. However, there is concern that use of TGIC may present toxic hazards.

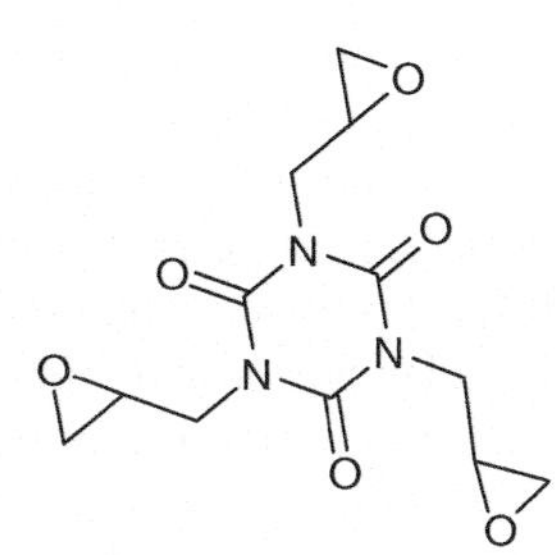

Triglycidylisocyanurate

In order to reduce VOC, reactive diluents are used. A coating formulated with the triglycidyl ether of TMP and a phenol-formaldehyde novolac resin showed improved resistance to methylene chloride, acetic acid, and sulfuric acid than the one formulated with a BPF resin (Kincaid and Schulte, 2001). Neopentyl glycol, butanediol, and cyclohexanedimethanol diglycidyl ethers are also used.

Monofunctional epoxies such as the glycidyl ether of n-butyl alcohol, o-cresol glycidyl ether, and the glycidyl ester of neodecanoic acid are used as reactive diluents in very high solids coatings. Such monofunctional epoxies reduce cross-link density, generally resulting in some loss of properties when used with BPA epoxy resins. They are more useful for reducing VOC of novolac epoxies, which have higher functionality as well as higher viscosity than BPA epoxies. Multifunctional aliphatic epoxies, including glycidated castor oil (with almost three hydroxyl groups per molecule), are also used as reactive diluents to reduce viscosity and film T_g without major impact on cross-link density.

Other epoxy products are made by the reaction of polyols, such as glycerol, sorbitol, and polyethylene or polypropylene glycols, with ECH and a base catalyst. Bio-based aliphatic epoxy products are produced similarly. Epoxidized soy and linseed oils are used in making acrylate derivatives for UV cure resins (Section 29.2.5) and thermal cationic cure resins (Section 13.3.5).

Also available are low MW cycloaliphatic diepoxy compounds such as 3,4-epoxycyclohexylmethyl-3′,4′-epoxycyclohexylcarboxylate (**1**), prepared by epoxidation of the corresponding dicyclohexene derivative, generally with peracetic acid. Another is diglycidyl 1,2-cyclohexanedicarboxylate (**2**). Such low MW epoxy derivatives are particularly useful as reactive diluents in cationic coatings (Sections 13.3.6 and 29.3.1). They can also be used as cross-linking agents for polyols (Section 13.3.3), carboxylic acids, and anhydrides (Section 13.3.2).

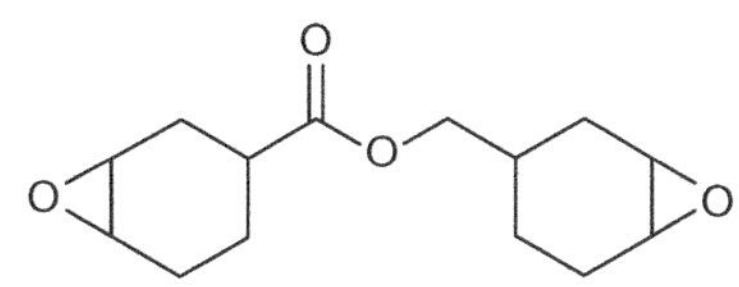

1

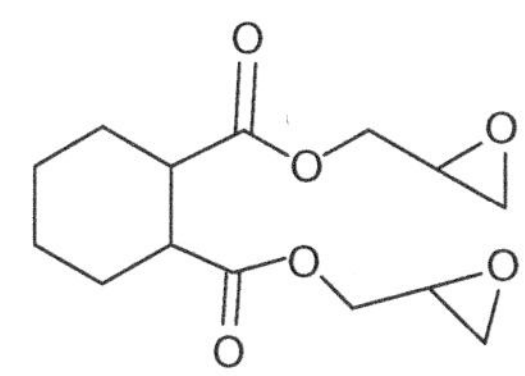

2

Amine-functional silanes such as APS are used to modify epoxy resins. One of many possibilities is to react a standard epoxy resin with APS to provide the following generalized structure (Source: Canadian patent publication CN103468095 A, 2013).

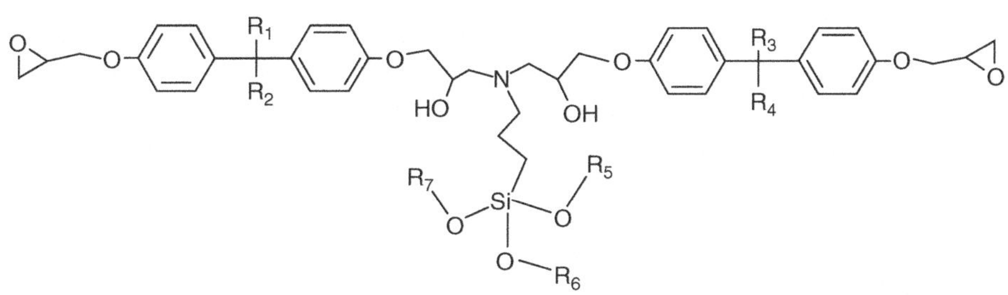

Such resins can be cross-linked with epoxy cross-linkers (Section 13.2) and can also cross-link by moisture cure (Chapter 16).

A widely used method to make epoxy-functional resins is by free radical copolymerization of acrylic esters together with glycidyl methacrylate (GMA) as a comonomer. Both solution and emulsion copolymerizations are feasible. By varying comonomers, GMA content, and MW, a range of materials can be made. Exterior durability and acid resistance can be excellent. GMA-modified acrylic resins are used in clear coats for automobiles and are candidates to replace BPA epoxy resins for food contact applications.

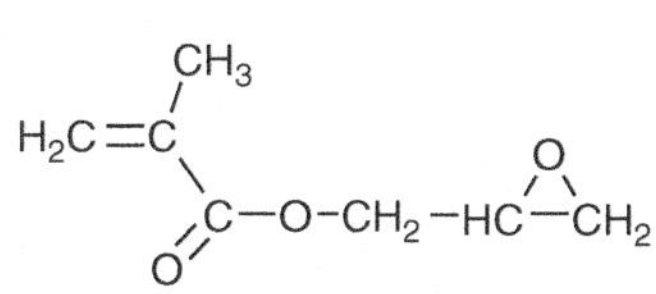

Glycidyl methacrylate (GMA)

13.2 AMINE CROSS-LINKED EPOXY RESINS

Epoxy groups react at ambient temperatures with primary amines to form secondary amines and with secondary amines to form tertiary amines. Tertiary amines react at higher temperatures to form quaternary ammonium compounds.

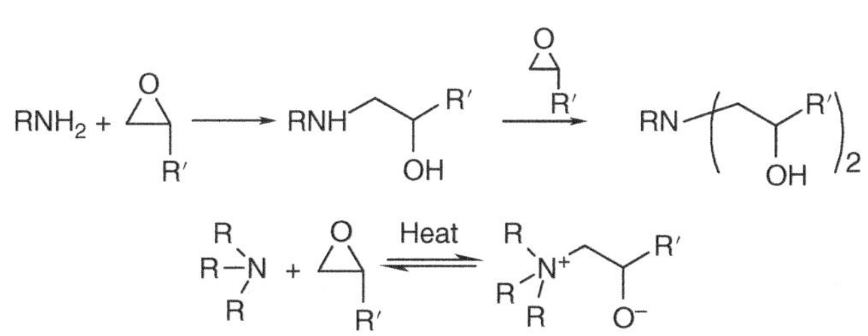

Reaction rates depend on epoxy and amine structure, concentration, catalysis, and media effects. Terminal epoxy groups, such as glycidyl ethers and esters, are more reactive

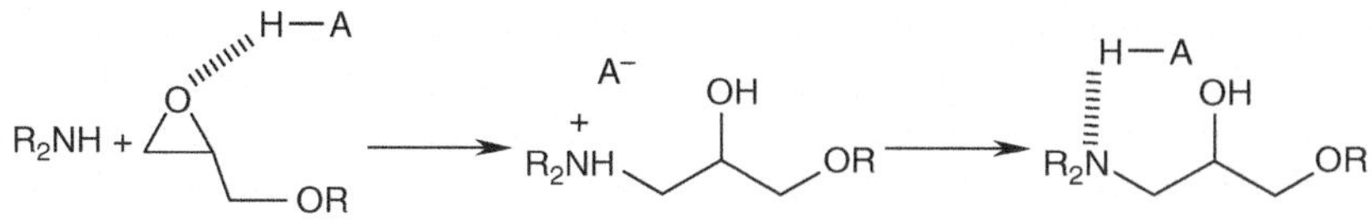

Scheme 13.2 Probable mechanism of catalysis by a weak acid of reaction of an amine with an epoxy resin.

than internal epoxy groups (e.g., cycloaliphatic diepoxide (**1**)), which are more sterically hindered.

Reactivity of amines tends to increase with base strength and decrease with steric crowding. The general order of reactivity, primary > secondary ≫ tertiary amines, can be attributed to steric effects, as well as to the absence of a transferable proton in the case of tertiary amines. Cycloaliphatic amines have reduced reactivity; the second reaction of such an amine is particularly slow. Aliphatic amines are more reactive than aromatic amines, which are less basic. The reaction is catalyzed by water, alcohols, tertiary amines, and weak acids (most notably by phenols), which promote ring opening by proton complexation with the epoxide oxygen. Strong acids are not effective catalysts for the reaction, since they protonate the amine in preference to the epoxide. Hydrogen-bond acceptor solvents tend to reduce reaction rates, probably by complexing with hydrogen donors in competition with the epoxy group.

Scheme 13.2 shows the attack on an epoxy group by the nonbonded electron pair of a secondary amine. Attack is primarily at the less sterically hindered terminal end of the epoxy group. Scheme 13.2 also shows catalysis by a weak acid (HA), which facilitates ring opening by H-complexation with the epoxide oxygen. Following H-transfer, the catalyst is regenerated by removal of the proton from the amine nitrogen by the conjugate base (A^-). The conjugate base could also participate in the catalysis by assisting the removal of the amine proton during the ring-opening reaction. Although we are unaware of experimental evidence directly supporting this hypothesis in epoxy–amine reactions, the general importance of concerted weak acid–weak base catalysis is well established (Jencks, 1969).

2,4,6-[Tris(dimethylaminomethyl)]phenol, which has both phenolic and tertiary amine groups, is an important catalyst for epoxy–amine reactions.

Epoxy–amine reactivity is too high at ambient temperatures to allow sufficient storage stability of a coating containing polyamine and polyepoxide in the same package; thus two-package (2K) coatings are required. With many aliphatic amines the pot life is limited to a few hours, and the coating generally takes about a week to cure at ambient temperature.

13.2.1 Pot Life and Cure Time Considerations

Epoxy–amine coatings are formulated to maximize pot life and minimize cure time. Many factors must be considered, including reactive group concentrations; structural effects of amine, epoxy, and solvents on reaction rates; equivalent and MWs; and $\bar{f}_n$ of the reactants. As the MW of a BPA epoxy resin is increased, the number of equivalents per liter of epoxy groups decreases; therefore, the reaction rate is slower. Furthermore, as the MW increases, viscosity increases. Thus, to formulate at the same viscosity, the amount of solvent must be increased, which decreases the concentration of both the amine and epoxy groups and lengthens pot life. The need to reduce VOC emissions forces the formulation of higher solids coatings, which have shorter pot lives and sometimes inferior film properties. See Section 2.3.2 for a discussion of kinetic limitations on achieving long-term one-package stability together with moderate cure temperatures, as well as ways to obviate these limitations.

An approach to the important problem of lengthening pot life without reducing cure time is to use blocked amine cross-linkers. Ketones react with primary amines to give ketimines, which do not readily react with epoxy groups. However, ketimines hydrolyze with water to release free amine plus ketone. Thus, the reverse reaction of ketimine formation unblocks the amine. Ketimine–epoxy moisture-curable coatings are stable in the paint can (assuming that care is taken to exclude moisture), but cure after application and exposure to ambient moisture. Most commonly, the ketimine of methyl ethyl ketone (MEK) is used. The high volatility of MEK from thin coating films minimizes reverse reaction with amine.

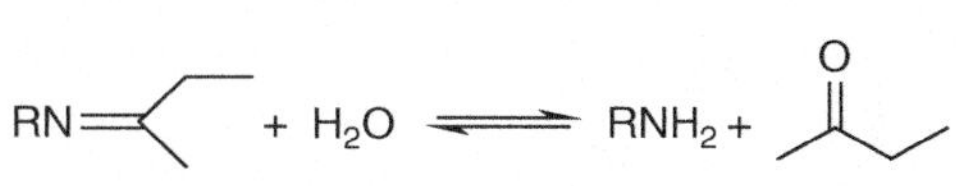

Ketimine–epoxy systems are stable indefinitely in the absence of water and should, in theory, permit formulation of one-package coatings. Nevertheless, they are most commonly used in long pot life two-package coatings, because of the difficulty of drying all of the components in a coating.

Many epoxy–amine coatings are pigmented, and pigment surfaces have a layer of water on them before they are dispersed. The water, which can hydrolyze the ketimine, remains in the coating. While removal of water from the surface of the pigment is possible, it adds to the cost. The amount of water is not high, usually less than 1% of the pigment weight, but the MW of water is so low that a small amount of water will hydrolyze a larger amount of ketimine. Solvents also contain water, and use of anhydrous solvents also increases costs. A further difficulty is that cure rate depends upon relative humidity as well as temperature.

McCarthy (2016) describes progress in overcoming the shortcomings of ketimine curing agents. Solventborne one-package epoxies, fairly stable for 12 weeks at 120°F and for a year at ambient temperature, were formulated. Salt spray and Prohesion test performance was essentially equal to conventional two-package epoxy–amine curing agent formulations. McCarthy refers to a patent application by Mower and Sheth (2010) describing stable formulations in which a siloxane moisture scavenger, such as vinyl trimethoxy silane (Chapter 16), is added to epoxy formulations employing ketimine-blocked amine curing agents. Also included in the formulations are hydrocarbon resins and sometimes an alkoxy silane resin. VOC is 2.6 lbs./gal (312 g l^{-1}).

Tertiary alcohol (e.g., *t*-butanol) blocked urethanes are relatively unstable and may thermally decompose to give alkenes, carbon dioxide, and amines. Advantage is taken of this reaction by using *t*-butyl alcohol blocked isocyanates as sources of amines to cross-link epoxy resins.

A commonly used cross-linker for epoxy powder coatings is dicyandiamide (DICY), a crystalline compound (mp 205°C) that also provides latency by insolubility. The complex reactions of epoxies with DICY have been the subject of numerous studies (Gilbert et al., 1991; Fedtke et al., 1993). Pham and Marks (2012) regard the multistep Gilbert mechanism as the most plausible.

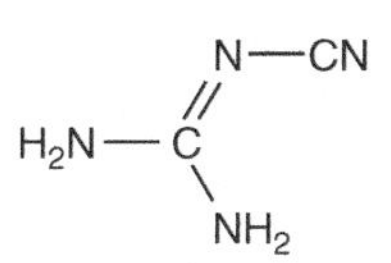

Dicyandiamide (DICY)

DICY requires curing temperatures near 200°C in epoxy formulations. Cure temperatures can be reduced by adding accelerators such as tertiary amines (e.g., benzyldimethylamine), various imidazoles (e.g., 2-methylimidazole), or substituted ureas. Imidazoles are probably the most common; they probably serve both as catalysts for DICY and as secondary cross-linkers.

Extensive research has aimed at practical ways to incorporate aminosiloxane cross-linkers and siloxane flexibilizers into epoxy coatings to reduce VOC, enhance flexibility, and improve weatherability without sacrificing adhesion (Geismann et al., 2014; Honaman and Witucki, 2014).

13.2.2 Toxicity and Stoichiometric Considerations

Other factors are involved in selecting amines besides reaction rates. Many amines are toxic. While they can be handled safely in a chemical plant, toxic hazards can arise if relatively inexperienced, careless, or uninformed personnel mix and apply 2K epoxy coatings incorporating some amines. For example, diethylenetriamine (DETA) is an efficient cross-linker for epoxies, but the handling hazard is high. In general, toxic hazards are reduced by increasing MW and reducing water solubility. As MW increases, volatility decreases, which reduces the chances of inhaling dangerous amounts of amine. Also, as water solubility decreases and MW increases, permeability through body membranes such as the skin decreases, generally reducing the toxic hazard. There are, of course, exceptions to these broad generalizations. Safety data sheets supplied by manufacturers should always be referenced for safe handling recommendations.

Other disadvantages of using low MW, highly functional amines are their low equivalent weights (EW) and viscosities. The EW of pure DETA is 21. If DETA were used with an epoxy resin having an EEW of about 500, the stoichiometric weight ratio of the two components would be about 25 : 1. This disparate amount would reduce the probability of accurately mixing stoichiometric amounts in two-package coatings, as well as increase the difficulty of uniform mixing.

One approach to designing amine cross-linkers with higher EW and lower toxic hazard is to make the so-called amine adducts. For example, standard liquid BPA epoxy ($n = 0.13$) may be reacted with an excess of a multifunctional amine such as DETA. Removal of excess amine provides the amine-terminated adduct, as shown in the following idealized structure. The low MW DETA is handled only in the chemical factory with proper precautions. Similar adducts, having a range of cure rates and pot lives, are prepared using a variety of amines. Epoxy–amine adducts contain no hydrolyzable groups, a potential advantage for corrosion resistance.

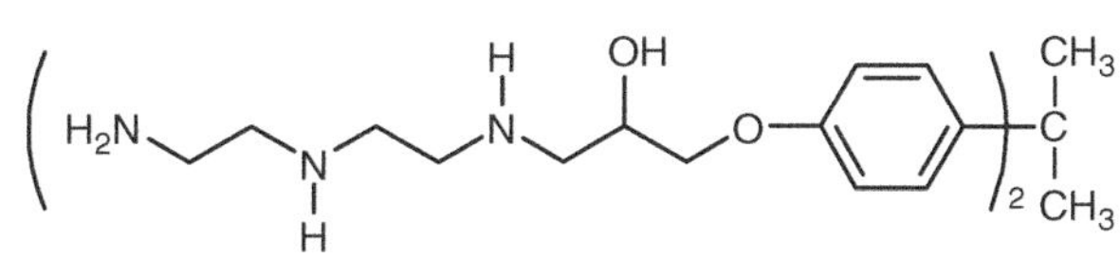

Epoxy amine adduct

Another approach is to react a multifunctional amine (e.g., DETA) with aliphatic mono- or dicarboxylic acids to form amine-terminated amides. Such products are called "polyamides" in the coatings industry, although the functional groups are amines. Dimer fatty acids are widely used; they are complex mixtures, predominately C_{36} dicarboxylic acids, made by acid-catalyzed dimerization of unsaturated C_{18} unsaturated fatty acids (Section 14.3.1). The reaction yields amine-functional "polyamides." An example is to react dimer acids with excess DETA. A mixture of products is obtained, including the simplest polyamide, shown in the following. The amine EW is the corresponding MW divided by the number of reactive NH groups (six in the polyamide shown). The amide groups do not react with epoxies, but may contribute to adhesion. A range of polyfunctional amines is used in making polyamides, including DETA, triethylenetetramine (TETA), aminoethylpiperazine, and aromatic diamines, such as *m*-phenylenediamine. The long aliphatic chains from the dimer acids improve flexibility, wetting, and adhesion as compared to use of an amine such as DETA.

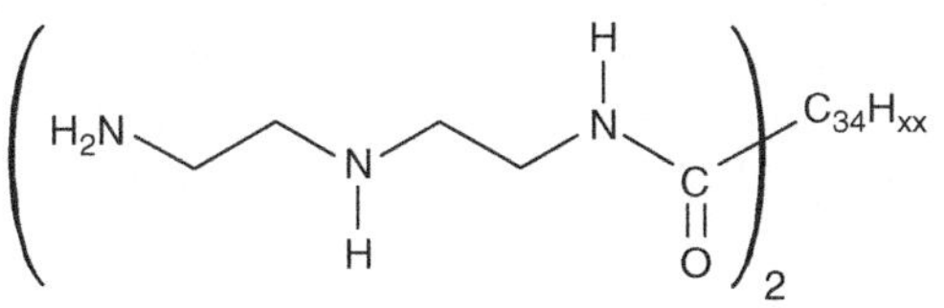

Polyamide

In the preparation of polyamides using an amine, such as DETA, the polyamide can react further by eliminating water and forming a terminal imidazoline group. The extent of this reaction varies, depending upon reaction conditions. The fraction of imidazoline groups affects solubility, compatibility, and pot life (and inversely reactivity). Imidazoline formation reduces average functionality and cross-link density of cured films (Brytus, 1986; Brady and Charlesworth, 1993); reduced functionality is desirable in some cases, as it improves flexibility of the film. Commercial grades of polyamides contain from 35 to more than 85% imidazoline functionality, a range from which the formulator may choose in designing coatings for a particular application. While amide groups can be hydrolyzed, they are less vulnerable than ester groups.

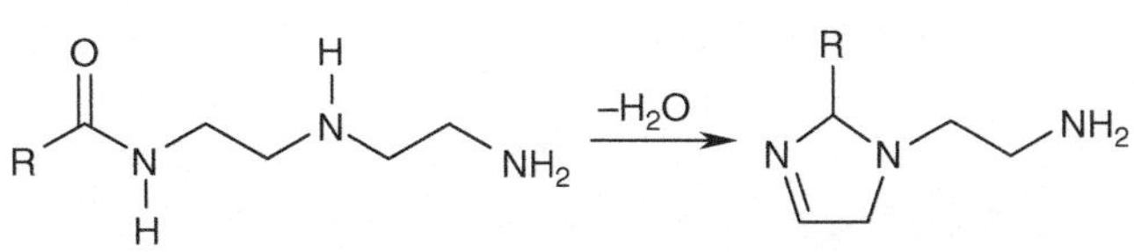

Imidazoline

A third approach is to form cross-linkers by reacting DETA with monocarboxylic acids, often a fatty acid. The products are frequently called *amidoamines*. Like polyamides, they contain varying amounts of imidazolines. The viscosity of amidoamines is lower than that of polyamides, an advantage in formulating high solids coatings.

"Polycyclamine" curing agents are a relatively recent development. They are oligomers containing several cyclohexylamine units, along with aromatic structures (Winter, 2011). Epoxy or epoxy/novolac formulations with such cross-linkers are said to cure relatively rapidly even at cool temperatures and to afford good protection in oil-field applications (Idlibi et al., 2016).

Conventional amine curing agents are unsatisfactory when it is necessary to cure an epoxy coating at temperatures below 5°C, as the cure rate is too slow. This problem has been addressed by using Mannich bases, prepared by reacting a methylolphenol with excess polyamine. The resulting aminomethylphenol has reduced amine functionality, but the presence of the phenolic hydroxyl, particularly, having *ortho*-aminomethyl groups, accelerates the epoxy–amine reaction.

13.2.3 Graininess and Blushing

BPA epoxy resins and polyamides are mutually soluble in the solvents used in epoxy–amine coatings, but most are not compatible in the absence of solvents. Thus, as solvent evaporates, phase separation can occur, resulting in a rough surface, called *graininess*. Graininess can be avoided by mixing the two packages 30 min to an hour prior to application. Partial reaction of the two components takes place during the waiting period. The elimination of graininess can be attributed to increased viscosity and/or the formation of reaction products that improve the compatibility of the mixture. When liquid standard resin is used, longer times are required after mixing to avoid phase separation. Proprietary amine cross-linkers are available that exhibit better compatibility, minimizing the problem (Tess, 1988).

Blushing or *scumming* is the appearance of a grayish, greasy deposit on the surface of films, usually accompanied by incomplete surface cure. Low temperature, high humidity conditions increase the probability of blushing. Blushing decreases gloss, increases yellowing, gives poor recoatability, and may interfere with intercoat adhesion. Blushing is said to result from formation of relatively stable carbamate salts of some amine groups on exposure to carbon dioxide and water vapor in the atmosphere (Tess, 1988).

$$RNH_2 + CO_2 + H_2O \longrightarrow RNH-CO_2H \xrightarrow{RNH_2} RNH-CO_2^- \ \overset{+}{H_3N}R$$

As with graininess, it is often possible to minimize blushing by mixing the epoxy and amine components 1 h or so before application. During the waiting period, some of the most reactive amine groups (those most likely to form carbamates) react with epoxy groups, so when the film is applied after this waiting period, visible blushing is less likely. If the waiting period is too long, viscosity increase is excessive. Amine adduct and Mannich base curing agents for epoxy coatings show little if any blushing.

13.2.4 T_g Considerations

It is critical to select a combination of epoxy resin and amine cross-linker that permits relatively complete reaction of the amine and epoxy groups at the temperature encountered during application. As polymerization and cross-linking proceed, T_g increases. As the T_g of a homogeneous network approaches the cure temperature (T), (T–T_g) and free volume decrease, and the reaction rate becomes limited by the mobility of the reactants, rather than by their reactivity. If the T_g approaches a value of about 40–50°C above the reaction temperature, the reaction essentially ceases and functional groups remain unreacted (Section 2.3.3). Unreacted functional groups cause adverse effects on mechanical properties and solvent resistance. If a coating that cures well when applied to an offshore oil rig near the equator in the Straits of Malacca is applied to the support the structure of an oil rig in the North Sea, where the water temperature even in summer does not exceed 4°C, adequate curing may not occur. A possible example of the effect of T_g is a report that after 7 days curing at 25°C, films from butyl ether-modified BPA epoxy resin (Section 13.1.2), cross-linked with an amine adduct, had better methyl alcohol resistance than films made with the same amine adduct and unmodified BPA epoxy resin (Payne and Puglisi, 1987; Bozzi and Helfand, 1990). There is no evident explanation for these results based on the chemical compositions. The superior methyl alcohol resistance may have resulted from a greater extent of reaction of the lower T_g butylated derivative before mobility limitation slowed the reaction.

13.2.5 Other Formulating Considerations

Solventborne epoxy–amine coatings can be formulated for application under water to oil rigs, bridges, etc. In such formulations the polyamine must be insoluble in water, and the solubility of water in the polyamine must be minimal. Water plasticizes epoxy–amine coatings, lowering T_g.

Epoxy–amine coatings tend to exhibit limited solvent resistance and are particularly susceptible to attack by acidic solvents, such as acetic acid. At least in part, this sensitivity results from diffusion of acetic acid into the film, followed by the formation of acetate salts with amine groups. The hydrophilic salt groups increase the solubility of water in the film, increasing water permeability, which lowers T_g by plasticizing the film, making it more susceptible to damage. This situation is exacerbated when cross-link density is low. BPA epoxy resins have an $\bar{f}_n$ of about 1.9, so even with highly functional amine components, cross-link density is limited, especially if there is deviation from stoichiometry. Since the amines are usually polyfunctional, it is generally best to formulate with a small excess (about 10%) of amine cross-linker to assure that the epoxy groups are fully reacted. The problem can be further alleviated by use of higher functionality resins such as novolac epoxies, which have $\bar{f}_n$ as high as 5. With the higher functionality and more viscous novolac epoxies, care must be exercised in selection of an amine cross-linker to ensure adequate reaction. Sometimes, blends of BPA and novolac epoxy resins are used.

Various methods can be used to combine epoxy resins with silanes, as discussed in Chapter 16. For example, addition of a coupling agent, [3-(2-aminoethylamino)propyl]trimethoxysilane, to BPA epoxy/polyamide coatings improves heat resistance, decreases color change with heat, increases hardness, and increases solvent resistance as a result of the additional cross-linking that occurs by hydrolysis and subsequent coupling of the silane group (Kiatkamjornwong and Yusabai, 2004).

Siloxane-modified epoxy resins are also reported to give superior coating properties. Formulation of a BPA epoxy resin, modified by reaction with hydroxyl-terminated dimethylsiloxane, and a polyamide cross-linker resulted in improved resistance to impact, mandrel bending, scratching, humidity, and NaOH. Due to the lower surface tension, films picked up much less dust (Ahmad et al., 2005).

Coatings in which cycloaliphatic epoxy resins are combined with amino silanes are reported to have excellent weatherability (Echeverria et al., 2016).

Witucki (2013) reports that a diamino-functional silane serves as a highly reactive chain extender or cross-linker for epoxy resins. Witucki's low VOC 2K formulation with a cycloaliphatic epoxy resin afforded good film properties including excellent gloss retention and chemical resistances.

Another factor to be considered when formulating epoxy–amine coatings is the effect of solvent composition on the coating. As noted earlier, hydrogen-bond acceptor solvents extend pot life. However, ketones and ester solvents should be avoided. Ketones form ketimines, and esters undergo aminolysis, especially with primary amines at room temperature, which reduces the concentration of reactive amine groups. An exception is *t*-butyl acetate, in which case amine loss is essentially as slow as when xylene is used (Cooper et al., 2001). In addition to its stability toward aminolysis, *t*-butyl acetate also exhibits negligible photochemical reactivity, resulting in its exemption from VOC regulations in the United States in 2005 and its absence on the HAP list.

Alcohols and water catalyze the reaction of amines with epoxy resins and also can react directly with epoxy groups, affecting the stability of the epoxy component package. Formulators generally judge package stability by changes in viscosity, but if monoalcohols are used as solvent, there could be a change in epoxy functionality without a significant change in viscosity. To the extent that epoxy groups react with an alcohol solvent, the potential for cross-linking is decreased. Since $\bar{f}_n$ of BPA epoxy resins is somewhat less than 2, further loss of functionality could cause the final film to have inferior properties. Apparently, in many systems, there is no difficulty. For example, some epoxy resins are sold as solutions in glycol monoethers. However, in other cases, a decrease in epoxy content has been observed after storage over a period of several months.

A similar situation can arise with water. As with alcohols, reaction of an epoxy group with water is slow, but epoxy groups are consumed. In some cases, in which the epoxy package is pigmented with TiO_2, it has been shown that epoxy content decreases with storage time, presumably as a result of the reaction of water from the surface of the TiO_2. This reaction, as well as that with alcohols, may be catalyzed by basic and/or acidic impurities or by alumina (basic) and silica (acidic), generally present as surface treatments on TiO_2. In most cases epoxy-package pigmented coatings have been used without problems. Apparently, the effects of both alcohols and water on package stability are system dependent. In view of such reports, it is recommended that package stability be checked periodically by epoxy group analysis rather than by relying solely on monitoring viscosity changes.

13.2.6 Waterborne Epoxy–Amine Systems

To reduce solvent content, waterborne epoxy–amine coatings have been widely investigated. Klippstein et al. (2012) reviewed this subject, emphasizing the patent literature.

An early approach was to make two-package coatings consisting of a liquid epoxy resin and a special waterborne curing agent (Zhang and Procopio, 2014). These coatings are sometimes called *type I waterborne epoxies*. Their application solids are similar to those of an organic solvent-soluble system, but the VOC content is lower, since a substantial fraction of the volatiles is water. Some loss of epoxy groups can be expected during the pot life of the mixed coating due to the high water content. Depending on formulation, such coatings tend to cure slowly. The films can be hard and chemical resistant, but they may be somewhat brittle because of their high cross-link density, owing to the use of low MW liquid epoxy resins.

Two-package *type II waterborne* epoxies comprise an aqueous dispersion of a relatively high MW epoxy with pigment and various additives, as Part A; Part B is an aqueous dispersion of polyamine curing agent (Zhang and Procopio, 2014). The curing agents are made from DETA or TETA, liquid epoxy resin, and nonionic surfactants. Film properties equivalent to solventborne epoxy–amine coatings are attainable. Proprietary "self-emulsifiable" epoxy resins and polyamides are available, which are made by prereacting a surfactant with the epoxy resin or polyamide; properties approaching those of solventborne coatings can be achieved (Wegmann, 1993).

Another approach to waterborne systems is the use of salts of resins having primary amine groups. The amine groups of a concentrated solution of the resin in organic solvents are neutralized with hydrochloric acid. When the solution is diluted with water, polymer aggregates, swollen with solvent and water, are formed in which the amine salt groups are on the outer periphery of the aggregates suspended in a continuous water phase. The behavior is analogous to that of water-reducible resins discussed in Section 8.3. When an organic solution of epoxy resin is mixed into the system, it enters the resin aggregates. The epoxy groups are, thus, kept separate from the amine hydrochloride salt groups, permitting a pot life of several days. After the coating is applied, followed by evaporation of the water and solvent, the amine hydrochloride salt and epoxy groups find themselves in the same phase. They react to yield a chlorohydrin and a free primary amine. Then, the amine can react twice with two additional epoxy groups. Since BPA epoxy resins have less than two epoxy groups per molecule and about one in three of these groups will be converted to chlorohydrins, it is desirable to blend novolac epoxy, having $\bar{f}_n$ up to 5, with the BPA epoxy. Then, when a third of the groups are converted to chlorohydrins, there is still an average of more than two epoxy groups per molecule to cross-link with amine groups.

Yet another approach is to use weakly acidic solvents, such as nitroalkanes (Albers, 1983; Lopez, 1989). Nitroalkanes form salts of amines, as shown in the following equation; the salt groups stabilize epoxy–amine emulsions and allow the system to be reduced with water. Following application, the nitroalkane solvent evaporates, shifting the acid–base equilibrium to the free amine, that is, to the left in the equation. Thus, the amine–nitroalkane combination functions as a transient emulsifying agent that stabilizes the emulsion during storage and, yet, is not present to adversely affect the final film properties. Conversion of amine groups to salts prolongs pot life of the mixed composition, since polar salt groups are oriented outward into the water phase, while the epoxy groups are in the interior of the emulsion particles.

$$R{-}NH_2 + R_2CH{-}NO_2 \rightleftharpoons R_2C{=}NO_2^- \ H_3\overset{+}{N}{-}R$$

Liu et al. (2016) describe a method for dispersing graphene (Section 20.5) uniformly into conventional 2K waterborne epoxy formulations. Graphene (3–10 layers thick) is predispersed in water and stabilized by sodium polyacrylate. Addition of the dispersion at 0.5 wt% graphene to the epoxy coating improved electrochemical impedance spectroscopy (EIS) performance (see Section 7.5) by an order of magnitude. Salt spray resistance was also substantially improved.

Resins used in automotive primers applied by cathodic electrodeposition are prepared by the reaction of epoxy and amine resins. The use of aqueous dispersions of carboxylate salts of such resins is discussed in Section 27.2. Waterborne epoxy acrylic graft copolymers and hybrid epoxy acrylic latexes are discussed in Section 13.4.

13.3 OTHER CROSS-LINKING AGENTS FOR EPOXY RESINS

A review by Massingill and Bauer (2000) provides a variety of epoxy cross-linking reactions, including discussion of their mechanisms.

13.3.1 Phenols

Epoxy resins, including BPA and novolac epoxies, can be cross-linked with phenolic resins. The reaction with phenols occurs predominantly at the less hindered CH_2 site of unsymmetrical epoxides (e.g., glycidyl derivatives), as shown. Presumably, steric hindrance also affects the reactivity of the phenol group. The uncatalyzed reaction is relatively slow but is generally acid catalyzed, commonly with pTSA or phosphoric acid.

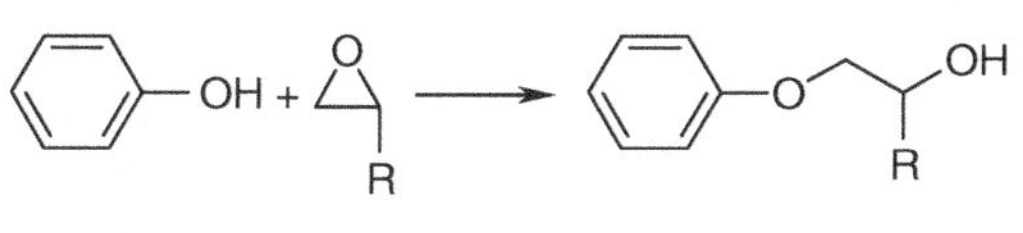

The phenolic hydroxyl groups of both resole and novolac phenolic resins (described in Sections 13.6.1 and 13.6.2) react with epoxy groups. In addition, the methylol groups of the resole phenolics undergo self-condensation and can also react with hydroxyl groups on the epoxy resin. Thus, cross-link density is higher with resole phenolics. Reaction also occurs with the hydroxy groups formed from the reaction of epoxies with phenols. The coatings require baking. Package stability is relatively limited but can be extended with etherified resole resins (Section 13.6.3). Increased solids and high functionality are reported using butoxymethylolated BPA as the phenolic resin (Payne and Puglisi, 1987; Bozzi and Helfand, 1990).

Unpigmented epoxy–phenolic coatings are used as linings for beverage cans and for some types of food cans as discussed in Section 13.1.1. Howe et al. (1998) and Wingender et al. (1998) described pioneering studies as to whether trace amounts of BPA are extracted in food or beverage cans from BPA epoxy containing can linings.

Pigmented epoxy–phenolic coatings are used as high performance primers. In both cases, the major advantages are adhesion to metals, even in the presence of water, and complete resistance to hydrolysis. In both of these applications, neither the discoloration that occurs on baking nor poor exterior durability is important.

Waterborne epoxy–phenolic emulsion coatings have been developed. Since conventional surfactants remain in the final film and reduce water resistance, water-soluble amine salts of methacrylic acid copolymers have been recommended as emulsifiers (Kojima et al., 1993). An example is the amine salt of methacrylic acid/methyl methacrylate/ethyl acrylate/styrene (40 : 20 : 20 : 20) copolymer.

13.3.2 Carboxylic Acids and Anhydrides

Carboxylic acids are among the most widely used cross-linkers for epoxies, as reviewed by Ooka and Ozawa (1994). The reaction of a carboxylic acid and an epoxy group yields a hydroxy ester. Ring opening occurs predominantly at the less hindered CH_2 carbon, although reaction at the more hindered CH–R site is significant. This chemistry has been adapted to produce highly weatherable coatings, including high solids clear coats for automobiles and trucks. Both one-package and two-package versions of such coatings are commercially important (Chapter 30).

The reaction has been shown to be second order in carboxylic acid. Most likely, one of the acid groups functions by nucleophilic attack at the CH_2 group, and the other functions as an electrophile and assists ring opening by complexation with the epoxy oxygen, as shown in Scheme 13.2 for the epoxy–amine reaction. The second-order dependence of reaction rate on acid concentration also results in a rapidly decreasing rate with conversion, exacerbating the difficulty of achieving high conversions:

$$\text{Rate} = k[\text{epoxy}][\text{RCOOH}]^2$$

The third-order dependence of the reaction (first order in epoxy and second order in carboxylic acid) also results in a small Arrhenius *A* value owing to the high molecular ordering of the epoxide and two carboxylic acid groups required in the transition state, reducing reactivity at all temperatures. GMA copolymers and cycloaliphatic epoxides, such as 3,4-epoxycyclohexylmethyl-3′,4′-epoxycyclohexylcarboxylate (**1**) and diglycidyl-1,2-cyclohexanedicarboxylate (**2**), react more rapidly than BPA epoxies. Tertiary amines catalyze the reaction of carboxylic acids with epoxies; triphenylphosphine is reported to be particularly effective. With triphenylphosphine catalysis and an excess of epoxy groups to carboxylic acid groups, coatings can be formulated that cross-link at 25°C (Shalati et al., 1990).

Merfeld et al. (2005) studied the effect of catalysts on curing temperature and film properties in TGIC/carboxylic acid-functional polyesters for potential use in powder coatings. Benzyltrimethylammonium chloride gave the best combination of catalyst effects: fastest cure at 120°C, slowest cure at 80–90°C, and least yellowing on overbake. The relatively slow cure at 80–90° is important for stability during extrusion.

Latent amine catalysts have been designed for powder coatings formulated with BPA epoxy resins and carboxylic acid-functional polyesters (Pappas et al., 1991). The catalysts are crystalline solid amic acids (e.g., the amic acid derived from the reaction of 3-methylphthalic anhydride and *N*,*N*-dimethylaminopropylamine). The amic acids are zwitterions, in which the tertiary amine is protonated, minimizing both their solubility in the powder coating and their catalytic activity. On heating, the amic acid melts and undergoes intramolecular cyclization to an imide having a free tertiary amine, which catalyzes the carboxylic acid–epoxy reaction.

Hydroxyl groups compete with carboxylic acid groups in reactions with epoxides. When carboxylic acids are used as cross-linking agents, part of the cross-links result from the reaction of epoxies with hydroxyl groups originally present on the epoxy resin or generated in the epoxy–carboxylic acid reaction. Esterification of carboxylic acid groups with hydroxyls may also occur.

Carboxylic acid-functional acrylics can be cross-linked with BPA epoxy resins, as well as with acrylic copolymers having pendant epoxy groups (Section 13.1.2). Self-cross-linking acrylics can be prepared by incorporating both (meth)acrylic acid and GMA in the same polymer. While the reaction rate of epoxy groups with carboxylic acid groups at storage temperatures is slow, it is not zero. This illustrates a drawback of self-cross-linking resins: the time of storage stability starts when the resin is made. On the other hand, when the functional groups are on two different resins, the time of storage stability does not start until the resins are mixed to make the liquid coating.

Cyclic anhydrides are also used as cross-linkers for epoxy resins (Ooka and Ozawa, 1994). Anhydrides react initially with the epoxy resin hydroxyl groups, yielding esters and carboxylic acids. The resulting carboxylic acid groups then react with epoxy groups, which generates a new hydroxyl group for further reaction. Epoxy groups can also react directly with anhydrides. Tertiary amines are generally used as catalysts; they probably function primarily by reacting with the epoxy to form a transient zwitterion, which then reacts with the anhydride.

13.3.3 Hydroxyl Groups

Although, when BPA epoxies react with resole phenolic resins, carboxylic acids, and anhydrides, reaction also occurs with hydroxyl groups, hydroxy-functional resins are not sufficiently reactive with BPA epoxy resins to be useful as cross-linkers in the absence of catalysts. However, with proper catalysis, cycloaliphatic epoxies serve as cross-linking agents for polyols in films baked at 120°C. This reactivity provides the basis for waterborne coatings that comprise caprolactone polyols and 3,4-epoxycyclohexylmethyl-3′,4′-epoxycyclohexylcarboxylate (**1**), together with diethylammonium triflate as a blocked catalyst (Eaton and Lamb, 1996). The active catalyst triflic acid is sufficiently strong that it is not consumed by direct addition to epoxide groups (see related discussion in Section 13.3.5).

Higher MW BPA epoxy resins can be cross-linked by reactions of their backbone hydroxyl groups. Both MF and UF amino resins are used; cross-linking occurs mainly between the activated ether groups of the MF or UF resin and the hydroxyl groups of the epoxy resin by transetherification (Chapter 11). Generally, amine salts or esters of pTSA or other sulfonic acids are used as latent catalyst. This type of cross-linking is important in container coatings (Sections 13.4 and 30.3).

Polyisocyanates also cross-link the hydroxyl groups of epoxy resins. Blocked isocyanates are generally preferred because they permit one-package stability (Section 12.5).

13.3.4 Mercaptans

Mercaptans (or thiols) (RSH) react with epoxies to yield sulfides. The reaction is strongly catalyzed by tertiary amines, which convert the mercaptan into the more highly reactive mercaptide anion RS^-. Following addition of RS^-, the intermediate alkoxide anion is protonated by the ammonium cation, thereby regenerating the catalyst, as shown.

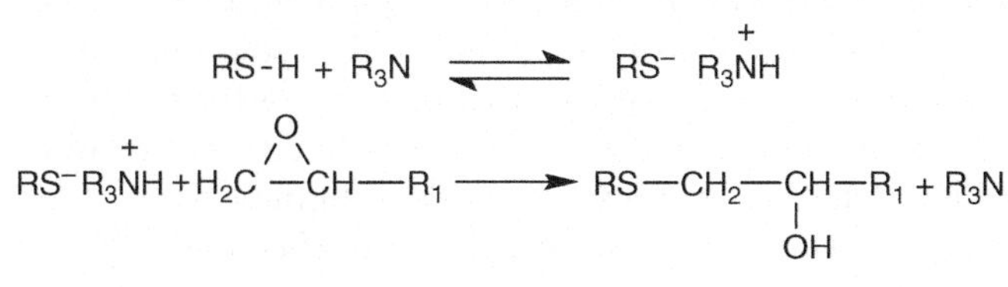

Reactivity of the mercaptide anion is sufficient for ambient temperature cure of 2K coatings. Polysulfide rubbers are low MW mercaptan-terminated polymers, which have been used to cross-link BPA epoxy resins in primers for aircraft. In general, unpleasant odors have been a drawback of mercaptan cross-linkers. Relatively low odor multifunctional mercaptans have been developed for thiol–ene radiation cure coatings (Section 29.2.5).

13.3.5 Homopolymerization

Epoxy groups undergo homopolymerization to polyethers in the presence of tertiary amines, Lewis acids, and very strong protic acids (super acids). Acid precursors are most commonly used as initiators. There are two types: (1) blocked or latent acids that undergo thermal decomposition to give the free acid and (2) photoinitiators that release acid on exposure to UV. UV-curable epoxy coatings are discussed in Section 29.3.1. The following equation shows initiation and the first step of polymerization.

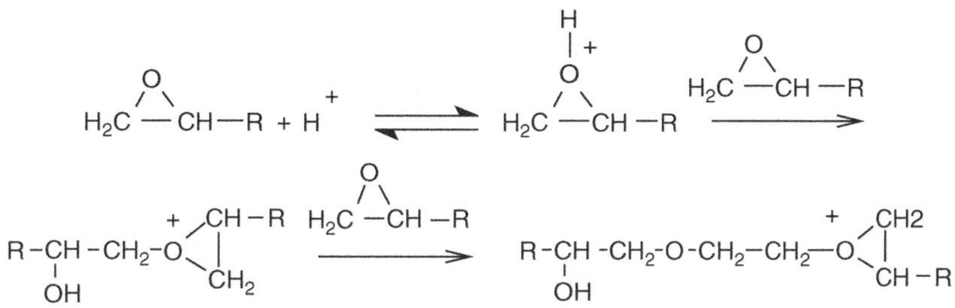

Only super acids are effective for homopolymerization of epoxies. Suitable super acids are trifluoromethylsulfonic acid (triflic acid) (F_3CSO_3H), perchloric acid ($HClO_4$), hexafluoroantimonic acid ($HSbF_6$), hexafluoroarsenic acid ($HAsF_6$), hexafluorophosphoric acid (HPF_6), and boron trifluoride (BF_3) etherate. Since strong acids have weak conjugate bases, the corresponding counterions are non-nucleophilic. Relatively strong acids, such as HCl and pTSA, are ineffective because the conjugate bases of such acids are nucleophilic enough to add to the protonated epoxy group, preventing addition of a second epoxy group, as required in the homopolymerization reaction. In the case of HCl, the result is to form a chlorohydrin, as shown in Eq. 13.5, rather than polymerization. Similarly, a sulfonate ester is formed in the case of pTSA.

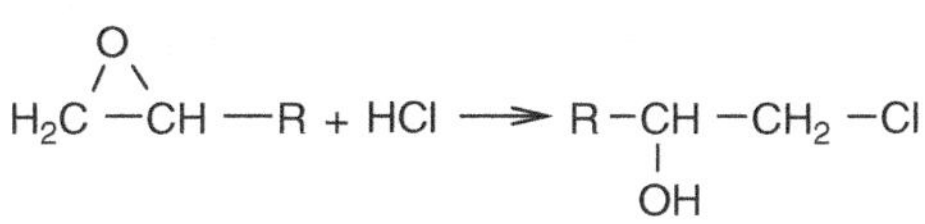

(13.5)

Homopolymerization can also be used for thermosetting coatings. Using α,α-dimethylbenzylpyridinium hexafluoroantimonate as a blocked catalyst, a GMA copolymer together with 3,4-epoxycyclohexylmethyl-3′,4′-epoxycyclohexylcarboxylate (**1**), as a reactive diluent, may be cured at 120° while retaining adequate pot life (Nakano and Endo, 1996). Cycloaliphatic epoxides such as (**1**) have been used together with polyols in thermal cationic cure coatings (Section 13.3.3). Part of the cross-linking is from homopolymerization and part from reaction with hydroxyl groups. Epoxidized linseed oil can be added to increase impact resistance (Eaton, 1997).

Walker et al. (2002) demonstrated that films cast from BPA epoxy resin emulsions in water can be cationically polymerized by perchloric acid to yield higher MW polyols although some epoxy groups remain. The polyols are formulated with MF resins to yield cross-linked films. The resins have the advantage that the amount of free BPA and $n=0$ BPA epoxy resin is lower than when conventional epoxy resins are used, thereby reducing the toxic hazard.

13.4 WATER-REDUCIBLE EPOXY/ACRYLIC GRAFT COPOLYMERS: EPOXY/ACRYLIC HYBRIDS

Epoxy resins are used on a large scale to make water dispersible acrylic graft copolymers (Woo et al., 1982; Woo and Toman, 1991). One way to prepare such graft copolymers is to react an acrylic copolymer, derived from ethyl acrylate, styrene, and methacrylic acid, with a BPA epoxy resin in a glycol ether solvent under conditions where only a fraction of the carboxyl groups react. A phenolic resin or a class I MF resin is added as a cross-linker, the acid groups are neutralized with an amine, and the system is diluted with water. The result is a dispersion that is used as a spray-applied coating for the interior linings of two-piece beverage cans.

A second method is to polymerize ethyl acrylate, styrene, and methacrylic acid using benzoyl peroxide as initiator in the presence of a BPA epoxy resin in a glycol ether solvent. This method is used in large volume production. The reaction is carried out at about 130°C, at which temperature both benzoyloxy and phenyl radicals are generated. These radicals can initiate polymerization as well as abstract hydrogen atoms from the epoxy resin (Woo and Toman, 1993). Abstraction of a hydrogen atom results in a free radical on the epoxy resin backbone, which serves as an initiating site for polymerization of the vinyl comonomers. Thus, a graft copolymer is formed with acrylic/styrene side chains substituted with carboxylic acid groups (from the methacrylic acid). The product is a mixture of epoxy/acrylic graft copolymer, nongrafted acrylic copolymer, and unreacted epoxy resin. The complex resin mixture is mixed with amine, cross-linker, and water as described in the previous paragraph to form a dispersion. Being a dispersion, the epoxy resin backbone can have a high MW, an advantage for coatings that must be highly flexible. Since the water-solubilizing groups are attached by C–C bonds rather than by ester groups, as in the first method, the resins are resistant to hydrolysis during storage. A latex may be blended into the epoxy/acrylic dispersions to reduce cost.

Many subsequent investigators have built on this concept. Examples are phosphated epoxy/acrylate dispersions for coil coating primers (addition of phosphoric acid enhances adhesion) (Section 13.5) (Yu et al., 2014) and waterborne epoxies combined with conductive polyaniline nanoparticles for enhanced corrosion protection (Chen and Liu, 2011; Jadhav et al., 2011).

Epoxy/acrylic hybrids can be made simply by blending acid-functional acrylic latexes with dispersions (or emulsions) of epoxy resins. This blend (Part A) is then combined with an amine-functional cross-linker (Part B) and the coating is applied. One use of such products is wall coatings. Leman (2016) compared the film properties of commercial coatings of this type and with polyamine-cured epoxies. Wide variations were observed. Leman goes on to describe a process in which an epoxy emulsion is present during emulsion polymerization of the acrylic. With appropriate selection of ingredients and conditions, the epoxy resin migrates into the latex particles during polymerization. In 2K formulations with polyamines, the epoxy migrates out of the latex particles during film formation, resulting in a cured epoxy matrix containing well-dispersed latex particles. The films have better adhesion and corrosion resistance than unmodified latexes. They also compare favorably with blended epoxy/acrylic hybrids.

13.5 EPOXY RESIN PHOSPHATE ESTERS

Phosphoric acid reacts with BPA epoxy resins to generate phosphate esters. Complex reactions occur; the predominant product is the monophosphoric acid ester of the primary alcohol together with other products. Most epoxy groups are hydrolyzed during the reaction to give the corresponding 1,2-diols (Massingill, 1991).

Low MW *epoxy phosphates* are used as adhesion promoters (Massingill and Whiteside, 1993). Higher MW epoxy resins can also be modified by reacting with minor amounts of aqueous phosphoric acid. Their use in epoxy–phenolic formulations provides coatings with improved adhesion and flexibility, without the need for a phosphoric acid catalyst, compared with corresponding unmodified epoxy–phenolic coatings (Massingill, 1991).

13.6 PHENOLIC RESINS

When phenol reacts with formaldehyde under appropriate conditions, a highly cross-linked solid is formed. It is Bakelite, the first major synthetic thermosetting plastic, invented by Baekeland in 1907. Cross-linking occurs because phenol has three reactive sites and formaldehyde has two. The phenolic resins used in coatings, however, are soluble molecules or oligomers. They are usually made from phenols that have one of the reactive sites blocked, reducing functionality to two. The product structures depend on the phenol(s) used, the stoichiometric ratio of phenol to formaldehyde, and the pH during the reaction. Phenolic resins are divided into two broad classes: *resole phenolics*, which are made using alkaline catalysts and high ratios of formaldehyde to phenol, and *novolac phenolics*, which are made using acid catalysts and low ratios of formaldehyde to phenol.

13.6.1 Resole Phenolic Resins

Under alkaline conditions, the initial reaction product of phenol and formaldehyde is a mixture of *ortho* and *para* methylolated phenols. The methylolated phenols are more reactive with formaldehyde than the unsubstituted phenol, resulting in the rapid formation of 2,4-dimethylolphenol and, subsequently, 2,4,6-trimethylolphenol; the latter is the predominant product, with a large excess of formaldehyde and a relatively short reaction time. With lower ratios (but still a molar excess) of formaldehyde to phenol and longer reaction times, formation of higher MW resole phenolic resins is favored. Polymerization occurs primarily by a methylol group on one phenol reacting at the *ortho* or *para* position of another phenol to form a methylene bridge connecting the two phenols. Dibenzyl ether bridges connecting two phenols also form by reaction of two methylol groups with each other. With excess formaldehyde, methylol groups are present on the terminal phenol groups of resole resins.

Such phenol-based resole resins cross-link on heating and are used in adhesive and plastics applications. However, they are not suitable for coatings applications, primarily because their cross-link density is higher than appropriate for any coating. Furthermore, the package stability of the resins is limited. Resole phenolics useful in coatings applications are made from monosubstituted phenols or from mixtures of monosubstituted phenols with phenol. The use of substituted phenols reduces the potential cross-link density. There are two broad categories of such resins: (1) those that are soluble in alcohol and other low MW oxygenated solvents, commonly called *alcohol-soluble, heat-reactive phenolics*, and (2) those that are soluble in vegetable oils and are called *oil-soluble, heat-reactive phenolics*.

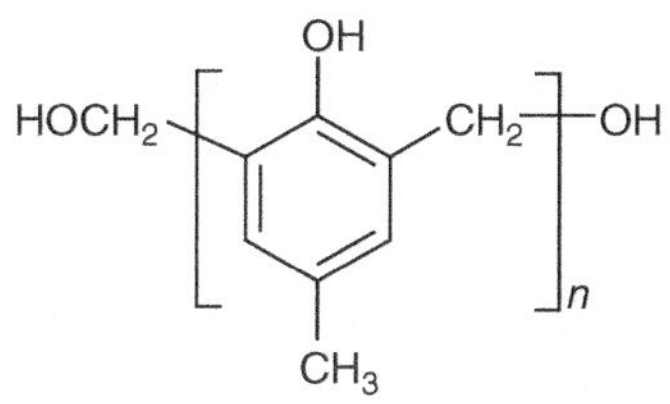

Resole phenolic resin (from *p*-cresol)
(idealized structure)

Alcohol-soluble, heat-reactive resole resins are prepared by reacting phenol, *o*- or *p*-cresol, and formaldehyde in the presence of a base catalyst at less than 60°C while removing water under vacuum. The catalyst is neutralized, alcohol is added, and the salt resulting from catalyst neutralization is removed by filtration. Potential cross-link density is controlled by the ratio of phenol to cresol; MW is controlled by the ratio of formaldehyde to phenols and by reaction time.

Such resole phenolic resins are used in interior can coatings and tank linings. In order to cure in short times, baking and an acid catalyst are required. To enhance flexibility and adhesion, they are commonly blended with low MW poly(vinyl butyral) as a plasticizer. The films are resistant to swelling by oils, such as encountered in canned fish, and are completely resistant to hydrolysis. These resins and other heat-reactive phenolics discolor during baking due to quinone methide formation, which restricts their usage to applications for which development of a yellow-brown color is permissible.

Resole resins are also blended with epoxy resins in thermosetting coatings for applications such as primers and can coatings (Section 13.3.1). The absence of hydrolyzable bonds and generally excellent adhesive properties are their chief advantages.

13.6.2 Novolac Phenolic Resins

Novolac phenolics of interest for coatings are made with acid catalysts and *o*- or *p*-substituted phenols. MW is controlled by the molar ratio of phenol to formaldehyde, which is always greater than 1. In contrast to resole phenolics, the terminal phenol groups are not metholylated, as shown in the idealized structure.

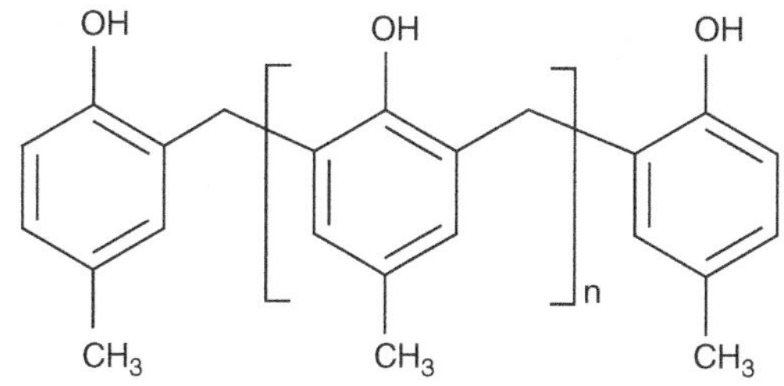

Novolac phenolic resin

Three types of novolac resins are used in coatings:

1. Alcohol-soluble, nonheat-reactive, low MW phenolics are derived from *o*- or *p*-cresol. An important use for these resins is in the preparation of novolac epoxy resins by reaction with ECH (Section 13.1.2).
2. Oil-soluble, nonheat-reactive novolac phenolic resins are made using a low ratio of formaldehyde, an acid catalyst, and a substituted phenol (e.g., *p*-phenylphenol, *p-t*-butylphenol, or *p*-nonylphenol). The resins are used together with drying oils, particularly tung oil or tung/linseed oil mixtures, to make varnishes. Such varnishes are still used to a modest extent as marine spar varnish owing to their reputation for durability in the do-it-yourself marine yacht market. The durability of phenolic varnishes may well result, at least in part, from the antioxidant activity of the phenolic groups (see Section 5.2.2 for a discussion of phenolic antioxidants).
3. Rosin-modified phenolic resins are the principal type of modified phenolics still in use. Their use in coatings is limited to low cost varnishes. The largest volume use is in printing inks. For inks, the phenolic resin is prepared in the presence of rosin esters and/or zinc or calcium salts of rosin. The structures of the reaction products, which are high melting, hydrocarbon-soluble resins, are not completely known. An important example of their use is in publication gravure inks such as those used in mail-order catalogs and some magazines.

13.6.3 Ether Derivatives of Phenolic Resins

The package stability of alcohol-soluble resole resins (Structure 3) and their compatibility with epoxy resins can be improved by partial conversion of the methylol groups to ethers. The ether groups undergo exchange reactions with hydroxy groups of the epoxy in the presence of acid catalysts. Allyl ethers of phenolic resins have been used for many years to cross-link epoxy resins in interior can coatings.

Low MW *n*- and isobutyl ethers of phenolic resins are also used for cross-linking epoxy resins and other hydroxy-substituted resins, primarily by etherification and transetherification reactions (Gardner and Mallalieu, 1992). A typical resin has an average of 2.2 aromatic rings per molecule. Due to its low average MW (ca. 320), the resin has a moderate viscosity, as supplied in butyl alcohol solution. The reactive groups in the resins are primarily butoxymethyl groups, but there are also benzyloxy groups and some free methylol groups. Furthermore, the phenol groups can react with epoxy groups, although the reaction of sterically hindered phenol groups is probably slow. Acid catalysts, such as phosphoric or sulfonic acids, are generally used. Blocked acids can extend shelf life.

GENERAL REFERENCES

Dornbusch, M.; Christ, U.; Rasing, R., *Epoxy Resins*, Vincentz, Hannover, **2016**.

Ellis, B., Ed., *Chemistry and Technology of Epoxy Resins*, Blackie Academic & Professional, London, **1993**.

Massingill, Jr., J. L.; Bauer, R. S., Epoxy Resins, in Craver, C. D.; Carraher, Jr., C. E., Eds., *Applied Polymer Science 21st Century*, Elsevier, Amsterdam, **2000a**, pp 393–424.

May, C. A., Ed., *Epoxy Resins—Chemistry and Technology*, Marcel Dekker, New York and Basel, **1988**.

Pham, Ha Q.; Marks, M. J., Epoxy Resins in Elvers, B., *Ullmann's Encyclopedia of Industrial Chemistry*, Wiley-VCH, Weinheim, **2012a**, Vol. *13*, pp 112–244.

REFERENCES

Ahmad, S., et al., *Prog. Org. Coat.*, **2005**, *54*, 248.

Albers, R., *Proceedings of the Waterborne Higher-Solids Coatings Symposium*, New Orleans, LA, **1983**, pp 130–143; US patent 4,352,898 (1982).

Bochan, A.; Mallen, T. R.; Lucarelli, M. A., US patent 6,803,004 (**2004**).

Bozzi, E. G.; Helfand, D., *FSCT Symposium*, Louisville, KY, May **1990**.

Brady, Jr. R. F.; Charlesworth, J. M., *J. Coat. Technol.*, **1993**, *65(816)*, 81.

Brytus, V., *J. Coat. Technol.*, **1986**, *58(740)*, 45.

Chen, F.; Liu, P., *ACS Appl. Mater. Interfaces*, **2011**, *3(7)*, 2694–2702.

Chen, Y., et al., Exposure to the BPA-Substitute Bisphenol S Causes Unique Alterations of Germline Function, *PLoS Genet.*, **2016**.

Cooper, C., et al., *J. Coat. Technol.*, **2001**, *73(922)*, 19.

Dante, M. F.; Parry, H. L., US patent 3,477,990 (**1969**).

Dubois, R. A.; Sheih, P. S., *J. Coat. Technol.*, **1992**, *64(808)*, 51.

Eaton, R. F., *Polym. Mater. Sci. Eng.*, **1997**, *77*, 381.

Eaton, R. F.; Lamb, K. T., *J. Coat. Technol.*, **1996**, *68(860)*, 49.

Echeverria, M., et al., *Prog. Org. Coat.* **2016**, *92*, 29–43.

Fedtke, M., et al., *Polym. Bull.*, **1993**, *31*, 429.

Gander, P., Bisphenol A-free can coating in limbo, **2016**, from http://www.foodpackagingforum.org/news/bpa-free-can-coatings-progressing-slowly.

Gardner, K. J.; Mallalieu, G. T. X., US patent 5157080 A (**1992**).

Geismann, C.; Kumar, V.; Kondos, C., US patent Appl. 2A1 (**2014**).

Gilbert, M. D.; et al., *Macromolecules*, **1991**, *24*, 360.

Honaman, L. A.; Witucki, G. L., **2014**, from http://www.dowcorning.com (accessed April 19, 2017).

Howe, S. R., et al., *J. Coat. Technol.*, **1998**, *70(877)*, 69.

Idlibi, Y., et al., *J. Prot. Coat. Linings*, **2016**, November.

Jadhav, R. S., et al., *Polym. Adv. Technol.*, **2011**, *22(12)*, 1620–1627.

Jencks, W. P., *Catalysis in Chemistry and Enzymology*, McGraw-Hill, New York, **1969**, pp 199–211.

Kiatkamjornwong, S.; Yusabai, W., *Surface Coat. Int. Part B: Coat. Trans.*, **2004**, *87-B3*, 149.

Kincaid, D. S.; Schulte, J. A., *Proceedings of the Waterborne Higher-Solids Coatings Symposium*, New Orleans, LA, **2001**, pp 127–141.

Klippstein, A.; Cook, M.; Monaghan, S., *Polymer Science: A Comprehensive Reference, Vol. 10, Polymers for a Sustainable Environment and Green Energy*, Elsevier, Amsterdam, **2012**, pp 519–539.

Kojima, S., et al., *J. Coat. Technol.*, **1993**, *65(818)*, 25.

Leman, A. A., Acrylic-Epoxy Hybrid (AEH) Coatings for Commercial and Institutional Wall Applications, *Proceedings of the International Conference on Coatings*, Indianapolis, IN, **2016**.

Liu, S., et al., *J. Mater Sci. Technol.*, **2016**, *32*, 425–431.

Lopez, J. A., US patent 4,816,502 (**1989**).

Massingill, J. L., *J. Coat. Technol.*, **1991**, *63(797)*, 47.

Massingill, Jr., J. L.; Bauer, R. S., Epoxy Resins, in Craver, C. D.; Carraher, Jr., C. E., Eds., *Applied Polymer Science 21st Century*, Elsevier, Amsterdam, **2000b**, pp 393–424.

Massingill, J. L.; Whiteside, R. C., *J. Coat. Technol.*, **1993**, *65(824)*, 65–71.

Massingill, J. L., et al., *J. Coat. Technol.*, **1990**, *62(781)*, 31.

McCarthy, J., *J. Prot. Coat. Linings*, **2016**, July.

Merfeld, G., et al., *Prog. Org. Coat.*, **2005**, *52*, 98.

Mower, N. R.; Sheth, K., US patent application 2A1 (**2010**).

Nakano, S.; Endo, T., *Prog. Org. Coat.*, **1996**, *28*, 143.

Ooka, M.; Ozawa, H., *Prog. Org. Coat.*, **1994**, *23*, 325.

Pappas, S. P., et al., *J. Coat. Technol.*, **1991**, *63(796)*, 39.

Payne, K. L.; Puglisi, J. S., *J. Coat. Technol.*, **1987**, *59(752)*, 117.

Pham, Ha Q.; Marks, M. J., Epoxy Resins in Elvers, B., *Ullmann's Encyclopedia of Industrial Chemistry*, Wiley-VCH, Weinheim, **2012b**, Vol. *13*, pp 112–244.

Scheuing, D. R., *J. Coat. Technol.*, **1985**, *57(723)*, 47.

Shalati, M. D., et al., *Proceedings of the International Conference on Coatings and Science Technology*, Athens, **1990**, p 525.

Sheih, P. S.; Massingill, J. L., *J. Coat. Technol.*, **1990**, *62(781)*, 25.

Tess, R. W., Epoxy Resin Coatings in May, C. A., Ed., *Epoxy Resins—Chemistry and Technology*, Marcel Dekker, New York and Basel, **1988**, p 743.

Walker, F. H., et al., *Prog. Org. Coat.*, **2002**, *45*, 291.

Wegmann, A., *J. Coat. Technol.*, **1993**, *65(827)*, 27.

Wingender, R. J., et al., *J. Coat. Technol.*, **1998**, *70(877)*, 75.

Winter, M., *NACE International Corrosion Conference*, Houston, TX, **2011**, paper 11040.

Witucki, G., Practical Applications of Polysiloxane Coatings, **2013**, from http://webcache.googleusercontent.com/search?q=cache:bBAx5UwndPcJ:http://docslide.net/documents/practical-applications-of-polysiloxane-coatings-gerald-l-witucki-august-2013.html%2BPractical+Applications+of+Polysiloxane+Coatings&hl=en-IN&gbv=2&ct=clnk (accessed April 20, 2017).

Woo, J. T. K.; Toman, A., *Polym. Mater. Sci. Eng.*, **1991**, *65*, 323.

Woo, J. T. K.; Toman, A., *Prog. Org. Coat.*, **1993**, *21*, 371.

Woo, J. T. K., et al., *J. Coat. Technol.*, **1982**, *54(689)*, 41.

Yu, J.; Pan, H.; Zhou, X., *J. Coat. Technol. Res.*, **2014**, *11(3)*, 361–369.

Zhang, Y.; Procopio, L., *Proceedings of the American Coatings Conference*, **2014**, Atlanta, GA Paper 14.5.

Chapter 14

Drying Oils

Oils and fats are *triglycerides*—triesters of glycerol and fatty acids. They are also called *lipids*. *Drying oils* are certain triglycerides that are capable of reacting with oxygen in air to form cross-linked films. The ability of triglycerides to cross-link in air results from the presence of multiple double bonds in their fatty acids. Drying oils are renewable resources because they are obtained from seeds of plants and from fish.

Galen described drying oil varnishes about 1800 years ago (Eastlake, 1848), as did several medieval authors (see Orna, 2013). However, few, if any, specimens of paints or varnishes from before 1400 AD have survived. In the nineteenth and early twentieth centuries, binders of most paints were drying oils. During the twentieth century, most drying oil binders were replaced by synthetic binders owing to their superior performance. Drying oils still have certain applications, for example, in specialized anticorrosive paints and in artists' paints.

Today, drying oils are raw materials for important binders such as alkyd, epoxy ester, and uralkyd resins (Chapter 15). Understanding drying oil chemistry is a foundation for understanding such resins, which can be considered to be synthetic drying oils.

14.1 COMPOSITIONS OF NATURAL OILS

Esters of many different fatty acids occur in nature. Fatty acids having 18 carbon atoms are common. Structures of some of the fatty acids that are important in coatings are provided in the following text. The numbers in parentheses designate the position of the first carbon of each of the double bonds, which is followed by the letters c or t, corresponding to *cis or trans* orientation.

$$\begin{aligned}
&\text{Stearic acid:}\\
&CH_3(CH_2)_{16}COOH\\
&\text{Palmitic acid:}\\
&CH_3(CH_2)_{14}COOH\\
&\text{Oleic acid:}\\
&CH_3(CH_2)_7CH=CH(CH_2)_7COOH(9c)\\
&\text{Linoleic acid:}\\
&CH_3(CH_2)_4CH=CHCH_2CH\\
&\qquad\qquad =CH(CH_2)_7COOH(9c12c)\\
&\text{Linolenic acid:}\\
&CH_3CH_2CH=CHCH_2CH=CHCH_2CH\\
&\qquad\qquad =CH(CH_2)_7COOH(9c12c15c)\\
&\text{Pinolenic acid:}\\
&CH_3(CH_2)_4CH=CHCH_2CH=CH(CH_2)_2\\
&\qquad\qquad CH=CH(CH_2)_3COOH(5c9c12c)\\
&\text{Ricinoleic acid:}\\
&CH_3(CH_2)_5CH(OH)CH_2CH=CH(CH_2)_7COOH\\
&\alpha-\text{Eleostearic acid:}\\
&CH_3(CH_2)_3CH=CHCH=CHCH\\
&\qquad\qquad =CH(CH_2)_7COOH(9t11c13t)
\end{aligned}$$

The oils are mixtures of triglycerides with hundreds of possible structures. Separation of the different but closely similar triglyceride molecules present in an oil borders on the impossible. Therefore, oils are characterized by their fatty acid content. First the oil is transesterified to form methyl esters of the individual fatty acids. The resulting mixture of methyl esters can then be accurately analyzed

Organic Coatings: Science and Technology, Fourth Edition. Frank N. Jones, Mark E. Nichols, and Socrates Peter Pappas.

by high performance liquid chromatography (HPLC) or by gas chromatography (GC) (Ackman, 1972; Khan and Scheinmann, 1977; King et al., 1982). A gas chromatograph linked to a mass spectrometer (GC-MS) is a powerful tool for this analysis (Izzo, 2010). Typical fatty acid contents of selected oils are given in Table 14.1. Compositions of the oils vary, sometimes quite widely, with variations in plant strain, climate, soil, and other growth conditions. Differences in sunflower oils are especially large, as illustrated in Table 14.1 for sunflower oils from Minnesota and Texas. In general, oils derived from seeds grown in colder climates have higher unsaturation to reduce their freezing points. Industrial oils and fatty acids are marketed in a variety of grades reflecting the degree to which the raw oils have been purified (refined) by the producer.

Included in Table 14.1 are *tall oil fatty acids* (TOFAs). They are not derived from drying oils, but are important coatings raw materials. TOFAs are mixtures of fatty acids obtained as a by-product of the sulfate pulping process for making paper. The word *tall* is the Swedish word for pine. TOFAs have a range of compositions, especially between acids of North American and European origins as shown. In some TOFAs the fatty acid mixture is similar to that of soybean oil fatty acid.

Most drying oils are obtained by crushing plant seeds to liberate the liquid oil. They contain triglycerides of a wide spectrum of highly unsaturated fatty acids, including C_{18}–C_{26} fatty acids having up to five nonconjugated double bonds, each separated by single methylene groups. Drying oils can also be obtained from fish.

Table 14.1 Typical Fatty Acid Compositions of Selected Oils

	Fatty acid				
Oil	Saturated[a]	Oleic	Linoleic	Linolenic	Other
Linseed	10	22	16	52	
Safflower	11	13	75	1	
Soybean	15	25	51	9	
Sunflower, MN	13	26	61	Trace	
Sunflower, TX	11	51	38	Trace	
Tung	5	8	43		80[b]
Tall oil fatty acids[c]	8	46	41[d]	3	2[e]
Tall oil fatty acids[f]	2.5	30	45	1	14[g]
Castor	3	7	3		87[h]
Coconut	91	7	2		

[a] Saturated fatty acids are mainly mixtures of stearic (C_{18}) and palmitic (C_{16}) acids; coconut oil also contains C_8, C_{10}, C_{12}, and C_{14} saturated fatty acids.
[b] α-Eleostearic acid.
[c] North American origin.
[d] Linoleic plus geometric and conjugated isomers.
[e] Rosin.
[f] European origin.
[g] Pinolenic acid.
[h] Ricinoleic acid.

14.2 AUTOXIDATION AND CROSS-LINKING

In nonconjugated oils the most reactive sites for autoxidation are the diallylic methylene groups ($-CH=CH\mathbf{CH_2}CH=CH-$) of the 1,4-diene structures. For conjugated oils the methylene groups adjacent to the 1,3-diene ($-\mathbf{CH_2}-CH=CH-CH=CH-\mathbf{CH_2}-$) structures are the most reactive. The reactivity of each oil is roughly governed by the abundance of these structures. Methylene groups attached to isolated double bonds are relatively unreactive. Early authors defined and classified oils based on their *iodine values*, that is, grams of iodine required to saturate the double bonds of 100 g of the oil, as follows: drying oils have an iodine value greater than 140; semidrying oils have an iodine value in the range of 125–140; and nondrying oils have an iodine value less than 125. However, although iodine values can serve as satisfactory quality control specifications, they can be quite misleading in predicting reactivity of a drying oil because the test does not discriminate between reactive and relatively unreactive unsaturated oils.

14.2.1 Nonconjugated Drying Oils

Nonconjugated drying oils have double bonds separated by methylene groups, in contrast to conjugated drying oils, such as tung oil, which are considered in the next section. Linoleic and linolenic acids have one and two diallylic methylene groups ($-CH=CH\mathbf{CH_2}CH=CH-$) per molecule, respectively. These highlighted methylene groups are the initial reactive groups for drying, and the average number of diallylic methylene groups per molecule corresponds to *functionality*, f_n. If f_n is greater than about 2.2, the oil is a drying oil, having sufficient reactive groups to form a cross-linked network. If the number is moderately below 2.2, the oil is a semidrying oil; it will cure to some extent but leave a weak, sticky film (Rheineck and Austin, 1968). Based on the data in Table 14.1, $\overline{f}_n$ of the linseed oil is 3.6; it is a drying oil. The $\overline{f}_n$ of soybean oil is about 2.07; it is a semidrying oil. The higher the $\overline{f}_n$ of a drying oil, the more rapidly a solvent-resistant, cross-linked film forms on exposure to air, other factors being equal.

The diallylic methylene groups, activated by their allylic relationship to two double bonds ($-CH=CH\mathbf{CH_2}CH=CH-$), are much more reactive than methylene groups allylic to only one double bond ($-CH_2CH=CH\mathbf{CH_2}CH_2-$). The relative rates of autoxidation of synthetic triolein (glyceryl trioleate), trilinolein, and trilinolenein are 1 : 120 : 330, reflecting the corresponding number of diallylic methylene groups in the three triglycerides of 0, 3, and 6, respectively (Chipault et al., 1951). The corresponding theoretical iodine values of the triglycerides are 86, 173, and 262, which are in the same order of reactivity, but the relative order of values (1 : 2 : 3) does not come close to reflecting the magnitude of the relative reactivity values. Accordingly, drying capacity is related to the average number of diallylic methylene groups per molecule, but not to the average number of double bonds per molecule, as measured by the iodine number.

Functionality considerations apply to synthetic drying oils (Chapter 15) as well as to natural oils. Since diallylic methylene groups are the initial reactive sites, it is convenient to relate the average number of such groups per molecule to the number average functionality $\bar{f}_n$ of the triglyceride or synthetic drying oil. This is important, because it is possible to make synthetic drying oils from semidrying oils such as plentiful soybean oil. For example, a synthetic drying oil with an average of six soybean oil fatty acid residues per molecule has an $\bar{f}_n$ twice that of the corresponding soybean oil triglyceride, thereby enhancing film formation, as discussed in Section 14.3.4.

The reactions taking place during drying are complex. However, the advent of analytical instrumentation provided tools to address the problem (Hartshorn, 1982; Falla, 1992; Muizebelt et al., 1998). Cross-linked films form from linseed oil in the following stages: (1) an induction period during which naturally present antioxidants (mainly tocopherols such as vitamin E) are consumed, (2) a period of rapid oxygen uptake with a weight gain of about 10% (FTIR shows an increase in hydroperoxides and appearance of conjugated dienes during this stage), and (3) a complex sequence of reactions in which hydroperoxides are consumed and a cross-linked film is formed. Hartshorn (1982) found that steps 1, 2, and 3 were far along in 4, 10, and 50 h, respectively, when catalyzed by a *drier* (Section 14.2.2). Sands (2011) reported that white pigmented linseed oil artists' paints gain over 15% in weight during the curing process.

Cleavage reactions to form low molecular weight by-products occur during the later stages of film formation. Slow continuing cleavage and cross-linking reactions through the lifetime of the film lead to embrittlement, discoloration, and formation of volatile by-products. Oils comprising significant quantities of fatty acids having three double bonds separated by methylene groups, such as linolenic acid, discolor to a particularly marked degree.

The following scheme illustrates some of the many reactions that occur during cross-linking. Initially, naturally present hydroperoxides decompose to form free radicals:

$$\text{ROOH} \longrightarrow \text{RO}\cdot + \text{HO}\cdot$$

$$\text{HO}\cdot\ (\text{or RO}\cdot) + -\text{CH}{=}\text{CH}{-}\text{CH}_2{-}\text{CH}{=}\text{CH}- \longrightarrow$$

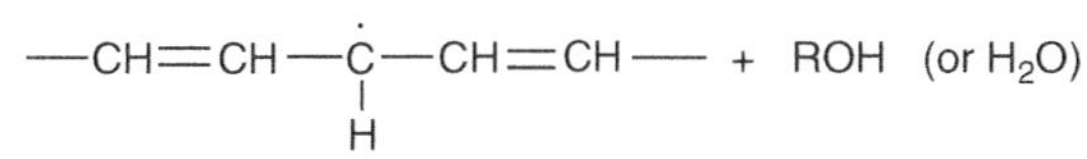

1

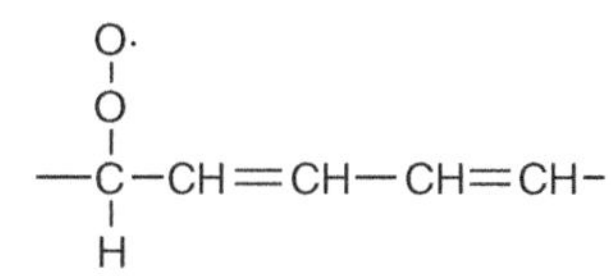

2

At first, these highly reactive free radicals react mainly with antioxidants that are naturally present to stabilize oils within the seeds. But, as the antioxidant is consumed, the free radicals react with other compounds. The diallylic methylene hydrogens are particularly susceptible to abstraction, yielding the highly delocalized, resonance-stabilized free radical (**1**), which reacts with oxygen to give predominantly a conjugated peroxy free radical, such as (**2**).

The peroxy free radicals can abstract hydrogens from other diallylic methylene groups to form additional hydroperoxides and regenerate free radicals such as (**1**). Thus, a chain reaction is established, resulting in autoxidation. At least part of the cross-linking occurs by radical–radical combination reactions forming C—C, ether, and peroxide bonds. These reactions correspond to termination by combination in free radical chain-growth polymerization (Section 2.2.1).

$$\text{R}\cdot + \text{R}\cdot \rightarrow \text{R}-\text{R}$$
$$\text{RO}\cdot + \text{R}\cdot \rightarrow \text{R}-\text{O}-\text{R}$$
$$\text{RO}\cdot + \text{RO}\cdot \rightarrow \text{RO}-\text{OR}$$

Cross-linking reactions

Reactions analogous to the addition step in chain-growth polymerization could also produce cross-links. For example, free radicals may add to conjugated double bonds, either initially present or formed from oxidation of diallylic groups to give (**3**).

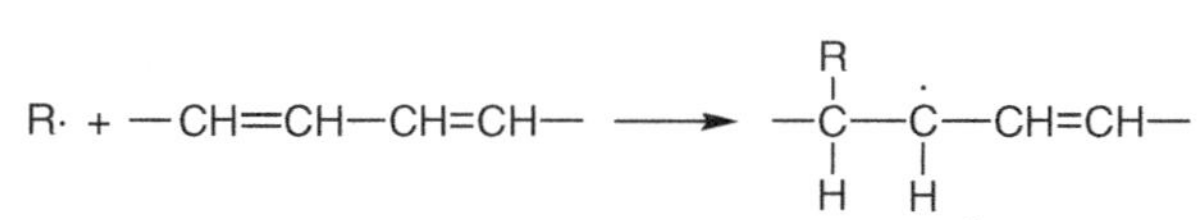

3

Such addition reactions could yield C—C or C—O linkages depending on the structure of the free radical. Subsequently, free radical (**3**) can rearrange, add oxygen to form a peroxy free radical, abstract a hydrogen from a diallylic methylene group, combine with another free radical, or add to a conjugated double bond. Studies of the catalyzed (Section 14.2.2) reactions of ethyl linoleate with oxygen by ^{1}H and ^{13}C NMR indicated that the predominant cross-linking reactions are those that form ether and peroxy cross-links (Falla, 1992; Muizebelt et al., 1998). Falla's FTIR and FT Raman analyses of drying linseed oil lead to the conclusion that only C—O—C and C—C bonds are formed. Muizebelt's MS studies showed that about 5% of the cross-links are new C—C bonds. Mallegol et al. (1999, 2000) detected substantial levels of epoxy groups in the reaction mixture, rising to a maximum in about 5 days and virtually disappearing in 100 days. As the film ages, epoxy groups may react with carboxyl groups in the film to form esters.

Rearrangement and cleavage of hydroperoxides to form aldehydes and ketones, among other products, lead to low molecular weight by-products. It has also been shown that low molecular weight carboxylic acids are released from linseed oil films as they age (van den Berg et al., 2001). The characteristic odor of oil and alkyd paints during drying is attributable to such volatile by-products, as well as to the odor of organic solvents.

Undesirable odor has been a factor motivating replacement of oil and alkyds in paints with latexes, particularly for indoor applications. The reactions leading to these odors have been extensively studied in connection with flavor changes of vegetable cooking oils (Frankel, 1980). Aldehydes have been shown to be major by-products from the catalyzed autoxidation of methyl oleate, linoleate, and linolenate, as well as from curing of drying oil-modified alkyd resins (Frankel, 1980; Hartshorn, 1982). Hancock et al. (1989) found that C_9 acid esters remain in the nonvolatile reaction mixture. Mallegol et al. (2000) proposed that the formation of aldehydes, ketones, and carboxylic acids in the aging of films of linseed oil results from photooxidation.

Dried films turn yellow or brown with aging, a process called *yellowing*. Linseed oil films yellow significantly, which has generally been attributed to the roughly 50% levels of linolenic acid that have three double bonds. The yellow color bleaches significantly when exposed to light; hence, yellowing is most severe when films are covered, such as by a picture hanging on a wall. The reactions leading to yellowing are complex and are not fully understood. One cause of yellowing is the incorporation of nitrogen compounds in the film. Color is markedly increased by exposure to ammonia, a common ingredient in household cleaners. Robey and Rybicka (1962) proposed that ammonia reacts with 1,4-diketones formed in autoxidation to yield pyrroles, which oxidize to yield highly colored products. Mallegol et al. (2001) observed that yellowing of poppy seed oil is almost as severe as that of linseed oil; yet poppy seed oil contains only few fatty acids having three double bonds. Based on their studies, the authors proposed that yellowing results from contaminants in the oils, but specific contaminants were not identified. Yellowing appears to be a multi-faceted process.

14.2.2 Catalysis of Autoxidation and Cross-Linking

The rate at which uncatalyzed nonconjugated drying oils react with oxygen is slow. Centuries ago, it was found that certain oil-soluble metal salts (*driers, siccatives*) catalyze drying. Common driers are oil-soluble cobalt, manganese, lead, zirconium, and calcium salts of 2-ethylhexanoic acid or naphthenic acids. Salts and complexes of many other metals, including rare earths, are also effective. In spite of many studies, the mechanisms by which driers work are not fully understood. Soucek et al. (2012) reviewed driers, and Hage et al. (2016) reviewed cobalt, iron, and manganese driers.

Oil-soluble cobalt and manganese salts, which catalyze drying that starts near the film surface, are called *top driers* or *surface driers*. Lead and zirconium salts catalyze drying throughout the film and are called *through driers*. Calcium salts show little, if any, activity alone but function as cocatalysts and reduce the amount of active driers needed. The surface-drying catalysis by cobalt is reported to accelerate the reaction of oxygen with methylene groups to form hydroperoxides, whereas manganese salts are reported, primarily, to accelerate decomposition of hydroperoxides to free radicals (Verkholantsev, 2000). Understanding of the drying process was advanced by nuclear magnetic resonance (NMR) depth profiling using alkyd resins (Erich et al., 2006). With cobalt catalysts a sharp cross-linking front was observed, starting at the exposed surface of the film and progressing into the film. No such front was observed with manganese catalysts. It was surmised that cobalt catalysts are far more effective driers, the rate of which is limited only by oxygen transport within the drying film.

Through driers do not function well in the absence of top driers. Therefore, combinations of metal salts are almost always used. Mixtures of lead with cobalt and/or manganese are particularly effective, but, owing to lead's toxicity, lead driers are no longer used in consumer paints in North America, in Europe, and in some other countries. Alternative driers include combinations of cobalt and/or manganese with zirconium, frequently including calcium. Co/Zr and Co/Zr/Ca combinations can be very effective (Mallegol et al., 2002; Soucek et al., 2012). Zirconium may inhibit the deactivation of cobalt by carboxylic acids; and cobalt by itself may primarily catalyze oxidation rather than polymerization. One study indicated that zirconium is more effective than lead (Meneghetti et al., 1998). Hein (1999) studied driers for high solids alkyds and recommended a combination of cobalt and neodymium salts, together with 2,2′-bipyridyl complexing agent. Chelating agents such as 1,10-phenanthroline enhance the activity of cobalt and manganese driers. A probable catalytic redox mechanism for cobalt is shown:

$$Co^{2+} + ROOH \rightarrow RO\cdot + OH^- + Co^{3+}$$
$$Co^{3+} + ROOH \rightarrow ROO\cdot + H^+ + Co^{2+}$$

Catalyzed decomposition of hydroperoxides

The net result is the formation of water and free radicals. Note that cobalt cycles between the two oxidation states. The activity of through driers has not been adequately explained.

It is suspected that cobalt salts may be carcinogenic and/or genotoxic. For this reason they may eventually be banned. Many European manufacturers have already replaced them (Najdusak and Goi, 2016). Furthermore, their long-term stability in waterborne paints is limited. Soucek et al. (2012), de Boer et al. (2013), and Najdusak and Goi (2016) reviewed ongoing efforts to find replacements for cobalt; possibilities include alternative metal driers. Iron and manganese are effective but tend to cause discoloration. A tetranuclear cluster (Mn_4O_2(2-ethylhexanoate)$_6$(bipyridine)$_2$) and a complex of manganese acetylacetonate/2,2′-bipyridine were found to be very effective (Warzeska et al., 2002; Oyman et al., 2005). Nonmetal and biomimetic catalysts are also being studied (Soucek et al., 2012). An iron(II)-bispidon catalyst is said to be as effective as cobalt at levels so low that discoloration by iron is negligible (de Boer et al., 2013). Pirs et al. (2015) compared the iron(II)-bispidon drier with a cobalt drier in high solids alkyd coatings. Drying times were comparable, but reaction kinetics was different; the iron drier also produced more uniform films, suggesting that it functions both as a surface and a through drier. The iron(II)-bispidon drier has the further advantage of good stability in aqueous formulations. Bispidon is a generic term for a family of complex ligands (Hage et al., 2016).

Calcium ions do not undergo redox reactions and may promote drying by preferentially adsorbing on pigment surfaces, thereby minimizing adsorption of active driers.

The optimum types and amounts of driers are system specific. Solventborne, waterborne, and high solids coatings each require different drier combinations. The driers should be utilized at the lowest effective level, since they catalyze not only drying but also the reactions that cause post-drying embrittlement, discoloration, and bond cleavage.

Pigments can strongly influence drying rates. Lead pigments (now mostly obsolete) often accelerate drying, while other pigments may retard it, possibly by adsorbing part of the drier.

Skinning occurs when liquid paints prepared from drying oils or oxidizing alkyds are exposed to the atmosphere, especially when driers have been added. Thus, antiskinning additives are usually required. An effective antiskinning agent is methyl ethyl ketone oxime (MEKO) (Bielman, 2000). MEKO forms an inactive complex with cobalt ions, $(Co(MEKO)_{1-8})^{3+}$, which is in equilibrium with the components. When the paint is applied, MEKO evaporates relatively rapidly, reversing the equilibrium and releasing the active cobalt catalyst. The MEKO is added to the coating just prior to filling the containers. Addition of 0.2% of MEKO prevents skinning for more than 250 days of storage but may result in a modest increase in cure time, depending on the time involved in reversing the equilibrium reaction by evaporation of MEKO. There are also regulatory concerns about MEKO; potential replacements are being developed (Najdusak and Goi, 2016).

14.2.3 Conjugated Drying Oils

Oils containing conjugated double bonds, such as tung oil (China wood oil), dry more rapidly than any nonconjugated drying oil. Drying of the ethyl ester of artificially conjugated linoleic acid has been studied (Muizebelt et al., 2000). The initial cross-links were determined to be peroxide links that, over time, became ether cross-links. Higher oligomers were formed than under comparable conditions with the nonconjugated ethyl linoleate.

In general, the water and alkali resistance of films derived from conjugated oils is superior, presumably because more of the cross-links are stable ether bonds. However, since the α-eleostearic acid in tung oil has three double bonds, discoloration on baking and aging is severe.

14.3 SYNTHETIC AND MODIFIED DRYING OILS

For at least 600 years, artists have improved the quality of their paints by pretreating their oils, for example, by exposing them to the sun, water, and air for several weeks or months (Meyer and Sheehan, 1991). Such treatments bleach the oil, increase its viscosity, and hasten its drying time. De Viguerie et al. (2016) reviewed historic practices, which often involved lead siccatives. Historic methods are impractical on an industrial scale. Today the processes discussed in Section 14.3.1 are utilized. Further chemical modifications are discussed in Chapter 15.

14.3.1 Heat Bodied Oils, Blown Oils, and Dimer Acids

Both nonconjugated and conjugated drying oils can be thermally polymerized by heating in an inert atmosphere to form *bodied oils*. The process increases viscosity and improves application and performance characteristics of paints. Process temperatures may be as high as 300–320°C for nonconjugated oils and 225–240°C for conjugated oils, although the reactions occur at an appreciable rate at somewhat lower temperatures. At least in part, bodying may result from thermal decomposition of hydroperoxides, always present in natural oils, to yield free radicals. The process increases molecular weight and perhaps causes limited formation of gel particles. It has also been shown that thermal rearrangement to conjugated systems occurs, followed by Diels–Alder reactions, which lead to the formation of dimers (Wheeler and White, 1967). Since tung oil has a high concentration of conjugated double bonds, it undergoes thermal polymerization more rapidly than nonconjugated oils (e.g., linseed oil). One must be careful to control the heating of tung oil to prevent complete gelation.

Viscosity can also be increased by passing air through drying oils at relatively moderate temperatures, 140–150°C, to produce *blown oils*. Presumably, reactions similar to those involved in air drying cause autoxidative oligomerization of the oil. Blown oils can be shear thinning (Gueler et al., 2004), suggesting the presence of gel particles.

Polyunsaturated fatty acids also dimerize or oligomerize by heat treatment using acid catalysis. For example, the fraction of fatty acids in TOFA that have 1,3-diene structures can be dimerized or oligomerized. Unreacted monobasic acids are then removed by distillation. The products, called *dimer acids*, are predominantly C_{36} dicarboxylic acids, with small fractions of monocarboxylic acid and C_{54} trimer acids. In some uses, it is desirable to eliminate residual double bonds by hydrogenation. Dimer acids are used to make polyesters (Section 10.1.2) and polyamides (Section 13.2.2).

14.3.2 Varnishes

The apparent drying rate of oils can be accelerated by dissolving a solid resin in the oil and diluting with a hydrocarbon solvent. Such solutions are called *varnishes*. The solid resin increases the T_g of the solvent-free film. The higher T_g results in more rapid hardening of the films, even though the rate of cross-linking does not increase. Thus, varnish films harden faster than films made from the corresponding oils, but the time required for the films to become solvent resistant is not shortened. Almost any high "melting" thermoplastic resin soluble in drying oil will serve the purpose. The higher the "melt point" of the resin, the greater the effect on T_g. Naturally occurring resins, such as congo, copal, dammar, and kauri resins; synthetic resins, such as ester gum (glyceryl esters of rosin); phenolic resins (Section 13.6); and coumarone–indene resins have been used.

In traditional varnish manufacture, the drying oil (usually linseed oil, tung oil, or a mixture of the two) and the resin are cooked together to a high temperature to obtain a homogeneous solution. In the past, small amounts of PbO (litharge) were added as a catalyst, but this practice ended where strict limits on lead content of varnishes is in effect. The varnish is then thinned with hydrocarbon solvents to achieve application viscosity. During cooking, some dimerization or oligomerization of the drying oil occurs; in some cases, reaction between the oil and resin has been demonstrated. Such varnishes were common for centuries, but during the twentieth century, they were gradually replaced by a variety of other products, especially alkyds, epoxy esters, and uralkyds. The term *varnish* has come to be used more generally for transparent coatings, even though few of them today are varnishes in the historical meaning of the word. Traditional *spar varnishes* are still favored for protecting wood in marine environments.

14.3.3 Synthetic Conjugated Oils

Tung oil dries rapidly, but is expensive and its films discolor rapidly, presumably due to the presence of three conjugated double bonds. This presumption led to efforts to synthesize *conjugated oils*, containing fatty acids having only two conjugated double bonds. One approach is to dehydrate castor oil using acid catalysis. A major component (87%) of the fatty acid content of castor oil triglycerides is ricinoleic acid, 12-hydroxy-(*Z*)-9-octadecenoic acid, which dehydrates to mixed geometric isomers of 9,11-conjugated and 9,12-nonconjugated fatty acid esters. *Dehydrated castor oil* dries relatively rapidly at room temperature but, as the film ages, the surface becomes tacky. This *after tack* has been attributed to the presence of geometric isomers, formed during the dehydration, which do not dry. Alternatively, aging may result in cleavage reactions that generate oily products. Dehydrated castor oil and its fatty acids are mainly used to prepare alkyds and epoxy esters for baking coatings, which do not exhibit after tack.

Nonconjugated oils can be partially isomerized to conjugated oils by heating with a variety of catalysts, such as alkaline hydroxides. A similar process can be used to partially conjugate double bonds of TOFAs. Synthesis of conjugated fatty acids by treatment of oils at high temperature with aqueous alkali hydroxides accomplishes isomerization and saponification simultaneously (Bradley and Richardson, 1942). The principal use of such conjugated oils and fatty acids has been to make alkyds and epoxy esters.

14.3.4 Esters of Higher Functionality Polyols

When oil-derived fatty acids are esterified with polyols having more than three hydroxyl groups per molecule, the number of cross-linking sites per molecule increases relative to the corresponding natural triglyceride oil. Whereas soybean oil is a semidrying oil, since $\bar{f}_n$ is 2.07, the pentaerythritol (PE) tetraester of soybean fatty acids is a drying oil, since $\bar{f}_n$ of 2.76. The PE tetraester of linseed fatty acids has an $\bar{f}_n$ of about 5 and affords dry, solvent-resistant films more rapidly than linseed oil. Still faster drying rates can be achieved by a reaction with still higher functional polyols, such as di- and tripentaerythritol. Alkyds, epoxy esters, and uralkyds (Chapter 15) made with fatty acids from such higher functional oils can be considered as higher functionality synthetic drying oils.

14.3.5 Maleated Oils

Both conjugated and nonconjugated oils and fatty acids react with maleic anhydride to form adducts. Conjugated oils, such as dehydrated castor oil, react at moderate temperatures by a Diels–Alder reaction:

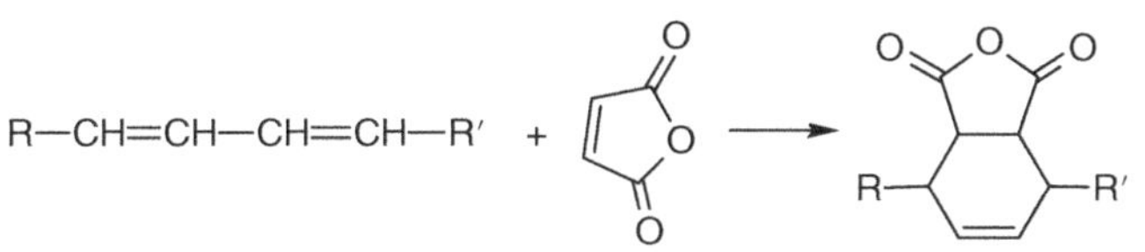

Nonconjugated oils, such as soybean and linseed oils, require higher temperatures (>100°C) and form a variety of adduct structures. Model compound studies using methyl linoleate demonstrate that maleic anhydride undergoes an ene reaction to give succinyl anhydride adducts. The ene reaction often results in conjugated bonds, as shown in general structure (**4**); subsequent Diels–Alder reaction with a second maleic anhydride gives a dianhydride, as shown in general structure (**5**) (Rheineck and Khoe, 1969).

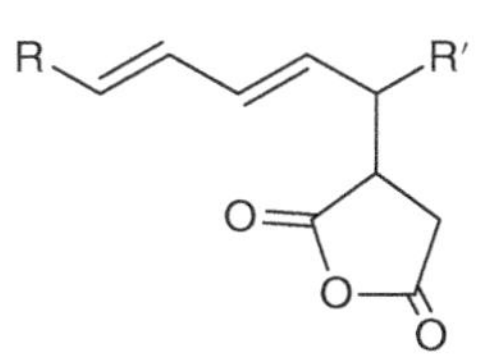

4

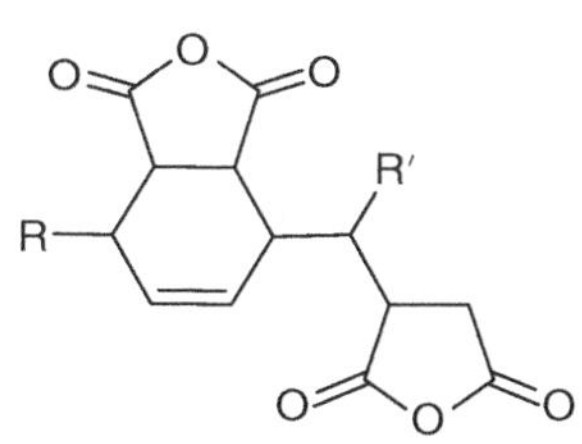

5

The products of these reactions, termed *maleated oils*, or sometimes, *maleinized oils*, react with polyols to provide moderately higher molecular weight derivatives that dry faster than the unmodified drying oils. For example, maleated soybean oil, esterified with glycerol, dries at a rate comparable with that of a bodied linseed oil having a similar viscosity.

Maleated oils can be made water reducible by hydrolysis with aqueous ammonia to convert the anhydride groups to the ammonium salts of the diacid. This process is used to make water-reducible alkyds and epoxy esters (Sections 15.3 and 15.8). Maleated oils and epoxy esters were important vehicles for early anionic electrodeposition coatings (Section 27.1), but this technology is now obsolescent.

14.3.6 Vinyl-Modified Oils

Both conjugated and nonconjugated drying oils react in the presence of a free radical initiator with unsaturated monomers, such as styrene, vinyl toluene, and (meth)acrylic esters. High degrees of chain transfer result in the formation of a variety of structures, including low molecular weight polymers of the monomer or mixture of monomers, short-chain graft copolymers, and dimerized drying oil molecules. The reaction of drying oils with such monomers is not commercially important, but the same principle is used to make modified alkyds (Section 15.6).

Linseed oil modified with cyclopentadiene has found fairly sizeable commercial use. This product is made by heating a mixture of linseed oil and dicyclopentadiene above 170°C under pressure. At this temperature monomeric cyclopentadiene is liberated at an appreciable rate by the reverse Diels–Alder reaction, and the resulting cyclopentadiene reacts with linseed oil. The product is inexpensive and dries faster than linseed oil. However, odor and dark color limit its applications.

GENERAL REFERENCES

Fox, F. L., *Oils for Organic Coatings*, Federation of Societies for Coatings Technology, Blue Bell, PA, **1965**.

Rheineck, A. E.; Austin, R. O., Drying Oils in Myers, R. R.; Long, J. S., Eds., *Treatise on Coatings*, Marcel Dekker, New York, **1968**, Vol. *I, No. 2*, pp 181–248.

Wicks, Z. W., Jr., Drying Oils in Kirk, R. K.; Othmer, D. F., Eds., *Kirk–Othmer Encyclopedia of Chemical Technology*, 5th ed., John Wiley & Sons, Inc., New York, **2003**.

REFERENCES

Ackman, R. G., *Prog. Chem. Fats Other Lipids*, **1972**, *12*, 165.

Bielman, J., Antiskinning Agents in Bielman, J. H., Ed., *Additives for Coatings*, Wiley-VCH, Weinheim, **2000**.

van den Berg, J. D. J., et al., *Prog. Org. Coat.*, **2001**, *41*, 143.

de Boer, J. W., et al., *Eur. J. Inorg. Chem.*, **2013**, July, 3581.

Bradley, T. F.; Richardson, G. H., *Ind. Eng. Chem.*, **1942**, *34*, 237.

Chipault, J. R., et al., *Off. Dig.*, **1951**, *23*, 740.

Eastlake, C. L., *(London) Quarterly Review American Edition*, Leanord Scott & Co., New York, **1848**, p 213.

Erich, S. J. F., et al., *Prog. Org. Coat.*, **2006**, *55(2)*, 105.
Falla, N. A. R., *J. Coat. Technol.*, **1992**, *64(815)*, 55.
Frankel, E. N., *Prog. Lipid Res.*, **1980**, *19*, 1.
Gueler, O. K., et al., *Prog. Org. Coat.*, **2004**, *51*, 365.
Hage, R.; de Boer, J. W.; Maaijen, K., *Inorganics*, **2016**, *4(2)*, 11.
Hancock, R. A., et al., *Prog. Org. Coat.*, **1989**, *17*, 321, 337.
Hartshorn, J. H., *J. Coat. Technol.*, **1982**, *54(687)*, 53.
Hein, R. W., *J. Coat. Technol.*, **1999**, *71(898)*, 21.
Izzo, F. C., *20th Century Artists' Oil Paints: A Chemical-Physical Survey*, Doctoral Dissertation, University of Venice, **2010**.
Khan, G. R.; Scheinmann, F., *Prog. Chem. Fats Other Lipids*, **1977**, *15*, 343.
King, J. W., et al., *J. Liq. Chromatogr.*, **1982**, *5*, 275.
Mallegol, J., et al., *J. Am. Oil Chem. Soc.*, **1999**, *76*, 967; 2000, *77*, 249.
Mallegol, J., et al., *J. Am. Oil Chem. Soc.*, **2000**, *77*, 257.
Mallegol, J., et al., *Stud. Conserv.*, **2001**, *46*, 121.
Mallegol, J., et al., *J. Coat. Technol.*, **2002**, *74*, 113.
Meneghetti, S. M. P., et al., *Prog. Org. Coat.*, **1998**, *33*, 219.
Meyer, R.; Sheehan, S., *The Artist's Handbook of Materials and Techniques*, 5th ed., Viking, New York, **1991**, pp 173–174.
Muizebelt, W. J., et al., *J. Coat. Technol.*, **1998**, *70(876)*, 83.
Muizebelt, W. J., et al., *Prog. Org. Coat.*, **2000**, *40*, 121.
Najdusak, R.; Goi, F., *Coat. World*, **2016**, June, 8.
Orna, M. V., *The Chemical History of Color*, Springer, New York, **2013**, pp 54ff.
Oyman, Z. O., et al., *Polymer*, **2005**, *46*, 1731.
Pirs, B., et al., *J. Coat. Technol. Res.*, **2015**, *12(6)*, 965–974.
Rheineck, A. E.; Austin, R. O., Drying Oils in Myers, R. R.; Long, J. S., Eds., *Treatise on Coatings*, Marcel Dekker, New York, **1968**, Vol. *1, No. 2*, pp 181–248.
Rheineck, A. E.; Khoe, T. H., *Fette Seifen Anstrichm.*, **1969**, *71*, 644.
Robey, T. L. T.; Rybicka, S. M., *Paint Res. Sta. Tech. Paper 217*, **1962**, *13(1)*, 2.
Sands, S., *Just Paint, No. 25*, **2011**, Golden Artist Colors, from http://www.goldenpaints.com (accessed April 22, 2017).
Soucek, M. D.; Kattab, T.; Wu, J., *Prog. Org. Coat.*, **2012**, *73*, 435–454.
Verkholantsev, V., *Eur. Coat. J.*, **2000**, *12(1–2)*, 120.
de Viguerie, L., et al., *Prog. Org. Coat.*, **2016**, *93*, 46–60.
Warzeska, S. T., et al., *Prog. Org. Coat.*, **2002**, *44*, 243.
Wheeler, D. H.; White, J., *J. Am. Chem. Soc.*, **1967**, *44*, 298.

Chapter 15

Alkyd Resins

While no longer the largest volume vehicles in coatings, alkyds still are of major importance. On the order of 1M metric tons of alkyd resins are used worldwide each year. Alkyd consumption in Europe, Middle East, and Africa was projected at 305 000 metric tons in 2013 (Hofland, 2012). During 2016–2024, worldwide alkyd resin production is projected to grow at an annual rate of 5.3% (Diamond, 2016). Presumably, driving forces for growth include the maturation of alkyd emulsion technology (Section 15.3.2) and increasing demand for bio-based coatings.

Alkyds are prepared primarily from polyols, dibasic acids, and fatty acids. Other types of monomers are sometimes included. Alkyds are polyesters, but in the coatings field the term polyester is reserved for "oil-free polyesters," that is, free of fatty acids. Such polyesters are discussed in Chapter 10. The term *alkyd* is derived from *al*cohol and ac*id*. Alkyds tend to provide coatings that are relatively easy to apply, exhibit fewer film defects during application (Chapter 24), and generally provide good adhesion. Many cross-link by reaction with air, which is free. However, durability of alkyd films, especially outdoors, tends to be poorer than films from acrylics, polyesters, and aliphatic polyurethanes, although better than BPA epoxies and aromatic polyurethanes.

Development of alkyd paints began about 1910 as the General Electric Co. was seeking improved wire insulation. At that time, however, basic polymer science was undeveloped and progress was very slow (Lanson, 1975). In 1927 Henry Reichhold became the first merchant supplier of alkyds (Danneman and Chu, 2016). An important advance was elucidation of the relationship of branched polymers with gelation (Kienle and Hovey, 1929). He and others used this knowledge to develop alkyd technology, which gradually displaced vegetable oils and became dominant paint resins from roughly the 1940s into the 1960s. In turn, alkyds have been gradually supplanted by acrylics, vinyls, polyesters, polyurethanes, and epoxies, all of which are derived almost entirely from fossil fuels. Currently, there is renewed interest in alkyds. Reasons include the evolution of technologies that enable very low VOC alkyd coatings (Section 15.3.2), as well as mounting interest in substituting bio-based materials for those derived from fossil fuels. Many alkyds are well over 50% bio-based.

There are many types of alkyds. One means of classification is *oxidizing* and *nonoxidizing* types. Oxidizing alkyds cross-link by the same mechanism as drying oils, as discussed in Chapter 14. Nonoxidizing alkyds are used as polymeric plasticizers or as hydroxy-functional resins, which are cross-linked by melamine-formaldehyde (MF) resins, urea-formaldehyde (UF) resins, or isocyanate cross-linkers.

A second classification, adapted from terminology used to classify varnishes, is based on the proportion of monobasic fatty acids utilized in their preparation. Varnishes with high ratios of oil to resin were called long oil varnishes; those with a lower ratio, medium oil varnishes; and those with an even lower ratio, short oil varnishes. Oil length of an alkyd is calculated by dividing the weight of "oil" in the final alkyd by the total weight of the alkyd solids, expressed as a percentage, shown in the first equation as follows. The 1.04 factor in the second equation converts the weight of fatty acids to the corresponding weight of triglyceride oil. Alkyds with oil lengths greater than 60 are *long oil alkyds*; those with oil lengths from 40 to 60, *medium oil alkyds*; and those with oil lengths less than 40, *short oil alkyds*. There is variation in the literature in the dividing lines between these classes:

$$\text{Oil length} = \frac{\text{weight of "oil"}}{\text{weight of alkyd} - \text{water evolved}} \times 100$$

Organic Coatings: Science and Technology, Fourth Edition. Frank N. Jones, Mark E. Nichols, and Socrates Peter Pappas.

$$\text{Oil length} = \frac{1.04 \times \text{weight of fatty acids}}{\text{weight of alkyd} - \text{water evolved}} \times 100$$

A third classification designates whether alkyd resins are designed for conventional solids, high solids or waterborne paints. Hofland (2012) projected that alkyd consumption in Europe, the Middle East, and Africa is about equally divided among the three types.

Yet another classification distinguishes *unmodified* from *modified alkyds*. Modified alkyds contain other monomers in addition to polyols, polybasic acids, and fatty acids. Examples are *styrenated alkyds* (Section 15.6) and *silicone alkyds* (Section 16.1.2).

15.1 OXIDIZING ALKYDS

Oxidizing alkyds are polyesters of one or more polyols, one or more dibasic acids, and fatty acids from one or more drying or semidrying oils. Accordingly, they can be considered as synthetic drying oils. The most commonly used polyol and dibasic acid are glycerol and phthalic anhydride (PA), respectively. Fatty acids made by hydrolysis of soybean oil provide a good example. Let us consider a simple, idealized example of an alkyd prepared from 1 mol of PA, 2 mol of glycerol, and 4 mol of soybean fatty acids. Using the fatty acid composition data for soybean oil from Table 14.1, we can calculate that this alkyd has an $\bar{f}_n$ of 2.76 (i.e., it would have an average of 2.76 activated diallylic **—CH$_2$—** groups (—CH=CH**CH$_2$**CH=CH—) per molecule) and, therefore, would dry to a solid film (see Sections 14.2.1 and 14.3.4 for a discussion of $\bar{f}_n$). The alkyd would form a solvent resistant film in about the same time as a pentaerythritol (PE) ester of soybean fatty acids (Section 14.3.4) since they have the same $\bar{f}_n$. However, the alkyd would form a tack-free film faster because the rigid aromatic rings from PA increase the T_g of the film.

If the mole ratio of PA to glycerol were 2:3, corresponding to an equivalent ratio of 4:9, 5 mol of soybean fatty acid could be esterified to yield an alkyd with an $\bar{f}_n$ of 3.45. This 2:3:5 mole ratio alkyd would cross-link more rapidly than the 1:2:4 mole ratio alkyd and would also form tack-free films even faster because the ratio of aromatic rings to long aliphatic chains would be 2:5 (40%) instead of 1:4 (25%). Although perhaps not intuitive, these simple relationships show that increasing the molar ratio of PA to glycerol increases the proportion of hydroxyl groups available for reaction with fatty acid as well as the ratio of PA groups to fatty acid. Accordingly, as the ratio of PA to glycerol is increased further, the average functionality for autoxidation increases and the T_g after solvent evaporation increases, as well, owing to the increasing ratio of aromatic groups to long aliphatic chains. Both factors promote faster film drying.

A theoretical alkyd prepared from 1 mol each of glycerol, PA, and fatty acid would have an oil length of about 60. However, if one were to try to prepare such an alkyd, the resin would gel prior to complete reaction. Gelation would result from reaction of a sufficient number of trifunctional glycerol molecules with three difunctional PA molecules to form cross-linked polymer molecules, swollen with partially reacted components. Gelation can be avoided by using a sufficient excess of glycerol to reduce the extent of cross-linking during alkyd synthesis. When the reaction is carried to near completion with excess glycerol, there are few unreacted carboxylic acid groups, but many unreacted hydroxyl groups.

Alkyds are manufactured in batch step-growth polymerization processes (Section 15.5) usually at temperatures above 200°C. It is often desired to adjust monomer compositions and process conditions so that the polymer approaches but does not reach gelation at the end of the process. If the alkyd gels in the kettle, a very expensive cleanup job is required. There have been many attempts, none fully successful, to calculate the ratios of functional groups and the extent of reaction that can be reached without encountering gelation. The problem is complex. Reactivity of the hydroxyl groups can be different; for example, glycerol contains both primary and secondary alcohol groups. Under esterification conditions, polyol molecules can self-condense to form ethers and, in some cases, can dehydrate to form volatile aldehydes (Section 15.5.2). Reactivity of the carboxylic acids also varies. The rate of formation of the first ester from a cyclic anhydride is more rapid than formation of the second ester. Aliphatic acids esterify more rapidly than aromatic acids. Polyunsaturated fatty acids and their esters can dimerize or oligomerize to form cross-links. Of the many early papers in the field, Blackinton (1967) recognized the complexities best, including the extent of formation of cyclic compounds by intramolecular esterification reactions, in addition to the previously described complexities. The method (Section 10.1) of Misev (1989) has been applied to calculating alkyd formulas.

In practice, alkyd resin formulators have found that the **mole** ratio of dibasic acid to polyol should be less than 1 to avoid gelation. How much less? It depends on many variables. Composition variables are discussed in Sections 15.1.1 through 15.1.3; the effect of variables in reaction conditions is discussed in Section 15.5.

For medium oil alkyds, the ratio of dibasic acid to polyol is not generally changed much for alkyds having an oil length of about 60, but the fatty acid content is reduced to the extent desired. This practice results in a larger excess of hydroxyl groups in the final alkyd. A common generalization is that as the oil length of an oxidizing alkyd is reduced below 60, the drying time decreases to a minimum at an oil length of about 50. However, this conventional wisdom must be viewed cautiously. The ratio of aromatic rings to aliphatic chains continues to increase after the solvent evaporates from the film tending to increase the T_g and shorten the time to form a tack-free film. However, at the same molecular weight (MW), the number of fatty acid ester groups per molecule decreases as the oil length decreases below 60, since more hydroxyl groups remain unesterified. Therefore, the time required to achieve sufficient cross-linking for solvent resistance increases.

Long oil alkyds are soluble in aliphatic hydrocarbon solvents. As the oil length decreases, mixtures of aliphatic and aromatic solvents are required, and oil lengths below about 50 require aromatic or oxygenated solvents, which are more expensive than aliphatics. The viscosity of solutions of long oil alkyds, especially of those having oil lengths below 65, is higher in aliphatic than in aromatic solvents; analogously, in medium oil alkyds, which require mixtures of aliphatic and aromatic solvents, viscosity decreases as the proportion of aromatic solvents increases. Formerly and to some extent still today, it was considered desirable to use a solvent mixture that afforded the highest possible viscosity; then, at application viscosity, the solids were lower and the raw material cost per unit volume of liquid paint was less. Accordingly, alkyds were designed to have high dilutability with aliphatic solvents. This was false economy, but it was a common practice. Increasingly, the emphasis is on reducing VOC, so the question becomes how to design alkyds having low solvent requirements rather than high dilutability potential. Furthermore, since aromatic solvents are on the HAP list, oxygenated solvents are often preferred for high solids alkyds, as discussed in Section 15.2.

15.1.1 Monobasic Acid Selection

Drying alkyds can be made using fatty acids from semidrying oils, since the $\bar{f}_n$ can be well above 2.2. For alkyds manufactured directly from vegetable oils (by the monoglyceride process; Section 15.5.1), soybean oil is used predominantly. Soybean oil is economical and supplies are dependable because it is a large-scale agricultural commodity; alkyd production utilizes only a few percent of the world supply. For alkyds made by the fatty acid process (Section 15.5.1), tall oil fatty acids (TOFAs) are, at times, more economical than soybean fatty acids. Both soybean oil and TOFA contain roughly 40–60% linoleic acid and significant amounts of linolenic acid (Table 14.1). White coatings containing linolenic acid esters gradually turn yellow. Premium cost "nonyellowing" alkyds are made with safflower or sunflower oils, which are high in linoleic acid, but contain very little linolenic acid.

Applications, in which fast oxidative drying and high cross-link density are important, require alkyds made with drying oils. The rate of oxidative cross-linking is affected by the functionality of the drying oils used. At the same oil length and MW, the time required to achieve a specific degree of cross-linking decreases as the average number of diallylic methylene groups $(\bar{f}_n)$ increases. Linseed long oil alkyds therefore cross-link more rapidly than soybean long oil alkyds, although their color and color retention are poorer than that of soybean alkyds, probably owing to the large fraction of esters of fatty acids having three double bonds in linseed alkyds. The cross-linking effect is especially large in very long oil alkyds having oil lengths well above 60% where $\bar{f}_n$ is very high even with soybean oil and the effect of further increase in functionality by using linseed oil is small. The effect is less noticeable in soybean alkyds with oil lengths around 60%.

Tung oil-based alkyds dry still faster, because of the high proportion of esters with three conjugated double bonds. Tung oil alkyds also exhibit a high degree of yellowing. Dehydrated castor alkyds have fairly good color retention but dry more slowly, since they contain only a small proportion of esters of fatty acids with three double bonds; they are used primarily in baking coatings.

As discussed in Section 14.3.1, drying oils and drying oil fatty acids undergo dimerization at elevated temperatures. Dimerization occurs concurrently with esterification during alkyd synthesis and generates difunctional acids, increasing the mole ratio of dibasic acids to polyol. The rate of dimerization is faster with drying oils having a higher average number of diallylic methylene groups per molecule and with those having conjugated double bonds. Thus, the MW and, therefore, the viscosity of an alkyd made with the same ratio of reactants depend on the fatty acid composition. The higher the degree of unsaturation, the higher the viscosity owing to the greater extent of dimerization. Accordingly, linseed alkyds have higher viscosities than soybean alkyds made with the same monomer ratios under the same conditions. The effect is particularly marked with tung oil, in which case the unsaturation is conjugated. It is difficult to prepare alkyds, using only tung oil, because of the high risk of gelation. Mixed linseed–tung alkyds are easier to produce and still have high oxidative cross-linking functionality.

A critical factor involved in the choice of fatty acid is cost. Drying oils are agricultural products and, hence, tend to be volatile in price. Depending upon relative prices, one drying oil is often substituted for another in certain alkyds. By adjusting for functionality differences, substitutions can frequently be made without significant changes in properties.

Benzoic acid is sometimes added to esterify excess hydroxyl groups remaining in the preparation of medium oil alkyds. The benzoic acid increases the ratio of aromatic to aliphatic chains in the alkyd, thus contributing to a higher T_g of the solvent-free alkyd and more rapid formation of a tack-free film. At the same time, the reduction in the free hydroxyl content may somewhat reduce water sensitivity of the dried films. Rosin can also be used in the same fashion. Although rosin is not an aromatic acid, its polynuclear ring structures impart sufficient rigidity to increase T_g and reduce tack-free time. However, if a critical requirement in drying is rapid development of solvent resistance, which depends on cross-linking, such benzoic acid and rosin modifications do not serve that purpose. Frequently, benzoic acid-modified alkyds are called *chain-stopped* alkyds, which implies that benzoic acid stops chain growth. This is not the case; the benzoic acid simply esterifies hydroxyl groups that would not have been esterified if the benzoic acid were absent. The effect on degree of polymerization is negligible.

15.1.2 Polyol Selection

Glycerol is the most widely used polyol because it is present in naturally occurring oils from which alkyds are commonly synthesized (Section 15.5.1). The next most widely used polyol is PE. The tetrafunctionality of PE must be taken into account when replacing trifunctional glycerol with PE in order to avoid gelation. If the substitution is made on a mole basis, rather than an equivalent basis, chances for gelation are minimized. As mentioned earlier, the ratio of moles of dibasic acid to polyol should be less than 1, and generally, a slightly lower mole ratio is required with PE than with glycerol. At the same mole ratio of dibasic acid to polyol, more moles of fatty acid can be esterified with PE. Hence, in long oil alkyds, the average functionality for cross-linking is higher, and the time to achieve a given degree of solvent resistance is shorter for a PE alkyd as compared with a glycerol alkyd. Due to this difference, one must be careful in comparing oil lengths of glycerol and PE alkyds.

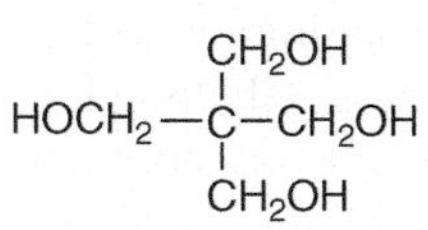

Pentaerythritol (PE)

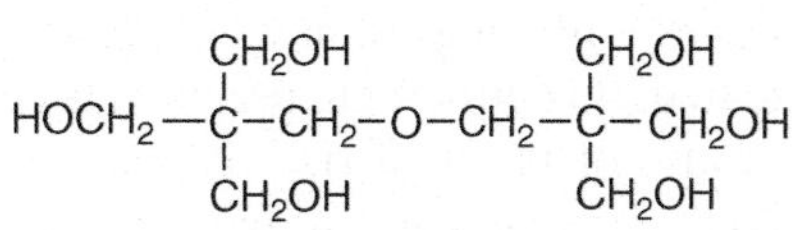

Dipentaerythritol

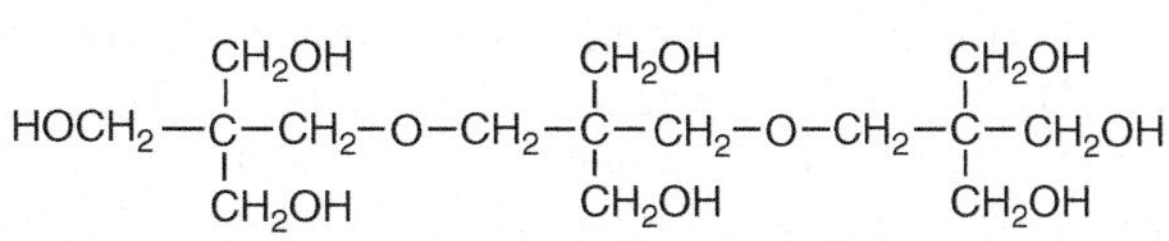

Tripentaerythritol

When PE is synthesized, di- and triPE are by-products and are present in commercial PE. Consequently, care must be exercised in changing sources of PE, since the amount of the higher polyols may differ. Because of the very high functionality, diPE and triPE ($F=6$ and 8, respectively) are useful in making fast drying low MW alkyds (Section 15.2).

Styrene and allyl alcohol can copolymerize to form a high-functionality polyol for alkyds. For example, it can be used in combination with PE to make long oil alkyds having relatively low viscosities and good hardness and humidity resistance (Porreau and Smyth, 2004).

To reduce cost, it is sometimes possible to substitute mixtures of PE and ethylene or propylene glycol for glycerol. A 1 : 1 mole ratio of tetra- and difunctional polyols affords an average functionality of three, corresponding to glycerol. The corresponding alkyds can be expected to be similar, but not identical.

Trimethylolethane (TME) and trimethylolpropane (TMP) are also useful triols. With TMP the rate of esterification is slower than with glycerol. Although all of TMP's alcohol groups are primary, they are somewhat sterically hindered by the neopentyl structure (Kangas and Jones, 1987). However, TMP-based alkyds have a narrower MW distribution and a somewhat lower viscosity than a comparable glycerol-based alkyd. A kinetic study demonstrated that esterification of one or two of the hydroxyl groups of TMP has little effect on the rate constant for esterification of the third hydroxyl group (Bacaloglu et al., 1988). It can be speculated that PE behaves similarly.

Glycerol is bio-based, but part of the supply is derived from fossil fuels. PE is usually fossil fuel based, although it can be made from plant-based ethanol or methanol. Promising bio-based polyols include sucrose (Nelson and Webster, 2013) and sorbitol (Yin et al., 2014). Sorbitol is hexafunctional and inexpensive, but it imparts a brown color to resins made by conventional high temperature alkyd processes. Nonconventional processes are an inviting research target.

15.1.3 Dibasic Acid Selection

Dibasic acids used to prepare alkyds are usually aromatic. Their rigid aromatic rings increase T_g of the resin. Cycloaliphatic anhydrides, such as hexahydrophthalic anhydride, are also used. While they are not as rigid as aromatic rings, the cycloaliphatic rings also increase T_g.

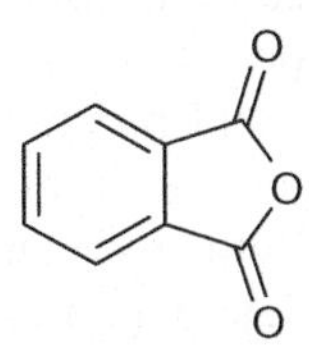

Phthalic anhydride
PA

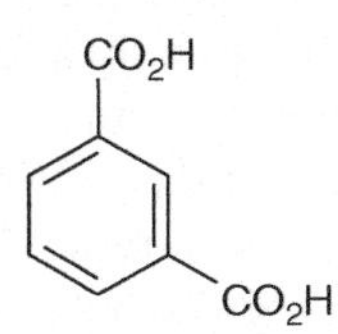

Isophthalic acid
IPA

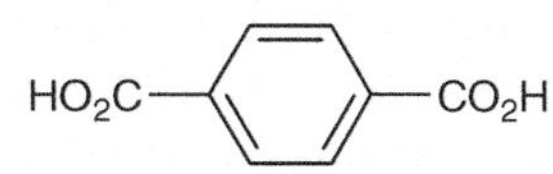

Terephthalic acid
TPA

By far the most widely used dibasic acid is PA. It has the advantage that the first esterification reaction proceeds rapidly by opening the anhydride ring. The amount of water evolved is lower, which also reduces reaction time. The relatively low melting point (the pure compound melts at 131°C) is desirable, since the crystals melt and dissolve readily in the reaction mixture. In large-scale manufacturing, molten PA is used, which reduces packaging, shipping, and handling costs.

The next most widely used dibasic acid is isophthalic acid (IPA). As discussed in Section 10.1.2, esters of IPA are more resistant to hydrolysis than those of PA in the pH range of 4–8, the most important range for exterior durability. On the other hand, under strongly alkaline conditions, esters of phthalic acid are more resistant to hydrolysis than isophthalic esters. The raw material cost for IPA is not particularly different from PA (even after adjusting for the extra mole of water that is lost), but the manufacturing cost is higher. Owing to its high melting point (330°C), IPA requires higher temperatures and longer times to dissolve in the reaction mixture than PA; hence, more dimerization of fatty acids occurs with IPA resulting in higher viscosity. The longer time at higher temperature also leads to greater extents of side reactions of the polyol components, which are discussed in Section 15.5.2 (Brown et al., 1961). Thus, when substituting IPA for PA, one must use a lower mole ratio of IPA to polyol in order to make an alkyd of similar viscosity.

Terephthalic acid is not useful in making alkyds because it dissolves very slowly even at the highest feasible process temperatures. However, corresponding esters, usually dimethyl terephthalate, can be used in the synthesis of alkyds by transesterification. Most notable is the use of post-consumer poly(ethylene terephthalate) (PET). Kawamura et al. (2002) described a process in which PE, a small amount of ethylene glycol, soy fatty acid, and dibutyltin oxide catalyst are heated until molten; recycled PET is added and the temperature raised to 240°C until the viscosity and acid number meet specifications. Properties of films from this resin and a comparable standard alkyd were essentially equal. Guclu and Orbay (2009) recycle PET by a more complex process that yields alkyds having better film properties than a standard alkyd.

15.2 HIGH SOLIDS OXIDIZING ALKYDS

The need to minimize VOC emissions has led to efforts to increase solids content of alkyd resin coatings. Solids can be increased by a change of solvents. Aliphatic (and to a lesser degree, aromatic) hydrocarbon solvents promote intermolecular hydrogen bonding, especially between carboxylic acids, as well as between hydroxyl groups, thereby increasing viscosity. Use of a hydrogen-bond acceptor solvent, such as a ketone or an ester, or hydrogen-bond acceptor–donor solvent such as an alcohol, even as a portion of the total solvent, significantly reduces viscosity at equal solids. Utilization of such oxygen solvents in place of solvents such as xylene (which is on the hazardous air pollutants (HAP) list) is particularly desirable.

An additional approach to increasing solids is to decrease MW, which is easily accomplished by decreasing the dibasic acid to polyol ratio and utilizing longer oil length alkyds. However, reducing VOC, significantly, by this route results in an alkyd having lower functionality for cross-linking and a lower ratio of aromatic to aliphatic chains. Both changes increase the time for drying and may compromise ultimate film properties. The effect of longer oil length on functionality can be minimized by using drying oils having higher average functionality. However, oils containing linolenic or α-eleostearic acid tend to discolor. Alternatively, one can use sunflower or safflower oils, which have a higher linoleic acid content and less linolenic acid than soybean oil. Proprietary fatty acids comprising 78% linoleic acid are commercially available. Increasing the concentration of driers (Section 14.2.2) not only accelerates drying but also accelerates yellowing and embrittlement.

Solids can be increased by making resins with narrower MW distributions. For example, adding a transesterification catalyst near the end of the alkyd cook results in more uniform MW and a lower viscosity product. Kangas and Jones (1987) utilized dicyclohexylcarbodiimide, as the transesterification reactant, to synthesize and evaluate model alkyds having very narrow MW distribution. Under the low reaction temperature conditions, dimerization by reactions of double bonds and self-etherification of polyols was negligible. The model alkyds dried more rapidly, but their film properties, especially impact resistance, were inferior to those obtained with control resins having the usual broad MW distribution. This problem is further discussed at the end of Section 15.5.

Solutions of alkyds prepared from dendritic or hyperbranched polyols generally have lower viscosities than comparable conventional alkyds (Ikladious et al., 2015). For example, alkyd resins made from a hyperbranched polyester polyol and TOFA are less viscous than comparable conventional alkyds and are reported to yield films having excellent properties (Murillo et al., 2010, 2011). Dendritic precursors can be monodisperse, but hyperbranched precursors, while polydisperse, are less expensive.

Another approach to high solids alkyds is the use of *reactive diluents* in place of part of the solvent. The idea is to have a component of lower MW and much lower viscosity than the alkyd resin, which reacts with the alkyd during drying and, thereby, does not contribute to VOC emissions. Zabel et al. (1999) list key properties of a reactive diluent: low viscosity, compatibility with alkyds, low volatility (bp > 300°C), nontoxicity, low color, economic replacement of solvent, and, importantly, ability to participate in oxidative cure.

Some reactive diluents contain allyl ethers as the cross-linking site, but there is concern that allyl ethers may emit hazardous vapors during drying. Proposed alternatives to allyl ethers include:

- Melamine derivatives that comprise acrylic and drying oil fatty acid amide groups (Strazik et al., 1981). Such reactive diluents contain high levels of $>NCH_2NHCOCH{=}CH_2$ and $>NCH_2NHCOC_{17}H_x$ moieties and promote fast drying.
- A combination of 2,7-octadienyl maleate and fumarate and 2-(2,7-octadienyloxy)succinate has been reported to be particularly effective (Zabel et al., 1999).
- Tetra-(2,7-octadienyl) titanate both increases coating solids and improves film properties (Alidedeoglu et al., 2011).
- A low viscosity polymer made from diPE, drying oil fatty acids, and an aliphatic diisocyanate (Bracken, 2000).
- Epoxidized soybean oil can be used as a reactive diluent for high solids alkyds using a combination of cobalt and neodymium driers (Kuang et al., 1999).

Formulation of high performance high solids alkyd coatings is challenging. Even with special drier combinations, tack-free times are long because such coatings dry mainly by cross-linking, in contrast to partial physical drying of conventional alkyds. They tend to sag and to pull away from edges and corners. Film physical properties tend to be weak relative to conventional alkyds, perhaps because more cross-linking is needed to produce equal properties. Hofland (2012) lists expedients that can mitigate some of the problems. He reported that the lower limit of VOC attainable with practical high solids coatings is about 160–200 $g\,l^{-1}$. Modest further progress can be expected. Lower VOC targets can possibly be achieved by using *t*-butyl acetate as a solvent in jurisdictions where it is exempt from VOC regulations. Pirs et al. (2015) compared cobalt and iron (bispidon) driers (Section 14.2.2) in long oil high solids alkyd formulations, finding that the iron drier afforded more uniform, less permeable films.

A potential approach toward upgrading the film physical properties of high solids alkyd paints is to disperse various nano-pigments in the coating (Dederichs, 2013).

15.3 WATERBORNE OXIDIZING ALKYDS

15.3.1 Water-Reducible Alkyds

One way to prepare waterborne alkyds is to:

1. Dissolve the resins having acid numbers in the range of 35–50 in water-miscible solvents such as glycol ethers. At this stage, the solution is about 75% solids.
2. Neutralize the acid groups with ammonia or an amine.
3. Dilute the resulting solution with water to form a dispersion of solvent swollen aggregates in water.

Dilution behavior of such resins is similar to that of the water-reducible acrylics discussed in Section 8.3. Note that the use of primary alcohol solvents must be avoided because they can more readily transesterify with the alkyd during resin production and storage, leading to reduction in MW and $\bar{f}_n$ (Bouboulis, 1982). The water-miscible solvents remain in the dispersion, adding to VOC in the final paint.

Hydrolytic instability can be a problem with water-reducible alkyds. If the carboxylic acid groups are half esters of phthalic or trimellitic acid, hydrolytic stability will probably be inadequate for paints that require a shelf life of more than a few months. Owing to the anchimeric effect of the neighboring carboxylic acid group, such esters are readily hydrolyzed. As hydrolysis occurs, the solubilizing acid salt is detached from the resin molecules, resulting in instability of the aqueous dispersion. A more satisfactory way to introduce free carboxylic acid groups is by reacting a completed alkyd with maleic anhydride. Part of the maleic anhydride adds to the unsaturated fatty acid esters in Diels–Alder and/or ene reactions, as discussed in Section 14.3.5. The anhydride groups are then hydrolyzed with amine and water to give the desired carboxylate salt groups, which are attached to resin molecules by C—C bonds rather than by C—O ester bonds, in which case hydrolysis results in the loss of solubilizing groups. There is still a hydrolytic stability problem with the alkyd backbone, but hydrolysis does not result in destabilization of the dispersion.

Hydrolytic stability can be improved by grafting carboxy-functional acrylic resins to the alkyd resin to make *alkyd/acrylic hybrids* (see Section 15.3.2).

After the film is applied, water, solvent, and amine evaporate; and the film cross-links by autoxidation. Since there are a fairly large number of residual carboxylic acid groups remaining in the cross-linked binder, the water resistance and particularly the alkali resistance of the films are reduced but are still satisfactory for some applications. *Early water resistance* can be a problem if, for example, a freshly painted surface is rained on before all the amine has evaporated from the film. Commonly, ammonia is used as the neutralizing amine because it is assumed that ammonia volatilizes faster than any other amine. However, if the T_g of the alkyd film is sufficiently high before all of the amine has volatilized, loss of amine becomes controlled by diffusion rate rather than by volatility. An important factor that controls the rate of diffusion of amine through the carboxylic acid-functional film is its basicity. A less basic amine, such as morpholine, may leave the film faster than ammonia even though its volatility is considerably lower.

Superior resistance to hydrolysis during storage can be achieved by reacting the hydroxyl groups of an alkyd with isophorone diisocyanate (IPDI) and dimethylolpropionic acid (DMPA) to form an aqueous polyurethane dispersion (PUD) (Hofland, 2012). See Section 15.7.2 for further discussion of autoxidizable PUDs.

15.3.2 Alkyd Emulsions

Alkyd emulsion technology has been in commercial use since the 1930s (Cheetham and Pearce, 1943; Osterberg et al., 1994; Osterberg and Bergenstahl, 1996), and it continues to advance. This technology offers the potential for very low VOC bio-based coatings, a major stimulus for research on emulsion preparation and formulation of emulsified alkyd paints. Hofland (2012) proposed that "cosolvent and amine-free alkyd emulsions can be made from most alkyds providing the resin viscosity is not too high and sufficient shear forces are applied to achieve a low particle size that ensures storage stability."

Molten alkyds are emulsified in water at 80–95°C. The emulsions can be prepared with little, if any, volatile solvent, although emulsifying agents are usually required. Effective emulsifiers include certain anionic and nonionic surfactants and polyacrylic acid. Such surfactants impart some degree of water sensitivity to the films; nonionic surfactants are theorized to be the least problematic. Stability of the early emulsions was questionable, especially if the process does not achieve the required small particle size. Reactive emulsifiers that copolymerize with the alkyd during oxidative film formation may mitigate this problem (Palmer, 2014).

Kan et al. (2014) provide examples of a process for emulsifying alkyd resins; they add a surfactant to stabilize the emulsion, and they employ a rotor–stator mixer to reduce particle sizes to 18–26 μm. Arendt and Kim (2014) describe a solvent-free alkyd emulsion that contains no surfactant, although it may contain an acrylic copolymer that serves the purpose. Film properties (notably ASTM B 117 salt spray resistance) of very low VOC coatings made with this emulsion were comparable to coatings from solventborne alkyds and better than conventional waterborne alkyds.

Earlier pigmented coatings formulated with alkyd emulsions required longer dry times, perhaps because of absorption of cobalt drier on pigment surfaces and/or because cobalt hydroxide partly precipitates during storage. Possible solutions to the problem are use of different driers, for example:

- A combination of cobalt neodecanoate and 2,2′-bipyridyl (Weissenborn and Motiejauskaite, 2000).
- Iron(II)-bisphidon driers (de Boer et al., 2013). Also see Section 14.2.2.

Vogel et al. (2015) described mechanical emulsification of alkyds in a two-stage continuous process. Most alkyds can be emulsified if their viscosities are not overly high. The emulsions have narrow particle size distributions between 100 and 300 nm; and solids content can be 50 wt% or more. Relatively little surfactant is required, sometimes none. Vogel describes near-zero VOC formulations whose film properties—notably gloss, adhesion, and salt spray resistance—are generally better than commercial waterborne alkyds and roughly equivalent to solventborne alkyd formulations. Optimized formulations utilizing combinations of Co, Zr, Mn, and Fe driers showed some improvement in reducing dry times. Long-term stability tests were not described.

For decades it has been common practice to add a few percent of an alkyd–surfactant blend to latex paints to improve adhesion to chalky surfaces (Section 32.1), as well as to metals (see Section 33.1.3). For this purpose, it is particularly important to use alkyds that are as resistant as possible to hydrolysis. Consumers expect paints to be storable for many years.

More recently, bio-based architectural paints have appeared in the US market. They are presumably acrylic latexes blended with high percentages of emulsified alkyd or copolymerized to form *hybrid alkyd/acrylic latexes*. Grafting (meth)acrylic polymer chains onto alkyds is discussed in Section 15.6. Hybrid alkyd/acrylic latexes are attractive binders for emulsion paints, as exemplified by the following reports:

- (Meth)acrylated fatty acid macromonomers can be used to synthesize water-reducible hybrids (Zuchert and Biemann, 1993; Weger, 1990). The carboxy groups cannot be hydrolyzed; and it is theorized that the acrylic chains form a shell around the alkyd, partly protecting it from hydrolysis.
- Alkyd grafted acrylic latexes can be prepared by emulsion polymerization of the (meth)acrylate monomers containing a dissolved oxidizing alkyd (Nabuurs et al., 1996; Guyot et al., 2007).
- Hybrid oil/alkyd/acrylic latexes have been prepared by emulsion polymerization. For example, hydroperoxidized sunflower oil was used as the initiator to polymerize a combination of sunflower oil, a long oil alkyd, and ethyl methacrylate. Films formed from the latex underwent the fast dry, expected for a latex, followed by autoxidation to give cross-linked films (van Hamersfeld et al., 1999).
- Elribii et al. (2015) prepared water-dispersible alkyd/acrylic hybrid resins by simply melt blending alkyds with acid-functional acrylic polymers. They prepared and studied dispersions of the products, finding that particle size decreases as acid content of the acrylic increases.

Formulation of practical waterborne paints from alkyd emulsions presents numerous challenges. The commercial success of these paints shows that formulators are meeting the challenges for many applications. Nonionic associative thickeners have been recommended to control rheology. Modification of a waterborne alkyd coating with nano-alumina improves its electrochemical impedance spectroscopy (EIS) performance, indicative of improved corrosion resistance (Dhoke and Khanna, 2012).

15.4 NONOXIDIZING ALKYDS

Certain low MW short–medium and short oil alkyds are compatible with such polymers as nitrocellulose and thermoplastic polyacrylates. Accordingly, such alkyds can be used as plasticizers for these polymers. They have the advantage over monomeric plasticizers (e.g., dibutyl or dioctyl phthalate) in that they do not volatilize appreciably when films are baked. It is generally not desirable to use oxidizing alkyds, which would cross-link and cause yellowing and embrittlement of the films, especially on exterior exposure. Therefore, nondrying oil fatty acids (or oils) are used in the preparation of alkyds for such applications. For exterior acrylic lacquers, pelargonic acid (n-$C_8H_{17}COOH$)-based alkyds combine excellent resistance to photodegradation and good compatibility with the thermoplastic acrylic resins. Castor oil-based alkyds are particularly appropriate plasticizers for nitrocellulose lacquers for interior applications, since the hydroxyl groups on the ricinoleic acid of castor oil promote compatibility.

All alkyds, particularly short and short–medium oil alkyds, contain excess hydroxyl groups to avoid gelation. These hydroxyl groups can be cross-linked with MF resins or with polyisocyanates. In some cases, relatively small amounts of butylated MF resin are used to supplement the cross-linking during baking of medium oil oxidizing alkyds. Such coatings provide somewhat better durability and faster curing than alkyd resins alone, with little increase in cost. Furthermore, there is the important advantage of the alkyd component imparting relative freedom from film defects, common to alkyd coatings (Chapter 24). However, high levels of unsaturation remaining in the cured films impart a tendency for discoloration on overbake, as well as for discoloration, loss of gloss, and embrittlement on exterior exposure. These difficulties can be reduced by using nondrying oils having minimal levels of unsaturated fatty acids. Coconut oil is a common example. Performance can be further enhanced by hydrogenation of the low levels of unsaturated acids present in the nondrying oils.

Since IPA esters are more stable to hydrolysis in the pH range of 4–8 than phthalate esters, the highest performance exterior alkyd-MF enamels use nonoxidizing IPA alkyds. For many applications, alkyd-MF coatings have been replaced with acrylic–MF or polyester–MF coatings to improve the overall balance of film properties. However, this trend may be stalling or reversing as demand for bio-based coatings increases.

Polyamide-modified alkyd resins are used as thixotropic agents to increase the low shear viscosity of alkyd paints. For example, high solids thixotropic alkyds based on polyamides made with aromatic diamines have been developed, which provide superior performance in high solids alkyd coatings (Bakker et al., 2001).

15.5 SYNTHETIC PROCEDURES FOR ALKYD RESINS

Various synthetic procedures, each with many variations, are used to produce alkyd resins. Anonymous (1962), Kaska and Lesek (1991), and Wicks (2002) reviewed manufacturing procedures. Alkyds can be made directly from oils or by using free fatty acids as raw materials.

15.5.1 Synthesis from Oils or Fatty Acids

Monoglyceride Process. In the case of glycerol alkyds, it would be absurd to saponify an oil to obtain fatty acids and glycerol and then re-esterify the same substances in a different combination. Rather, the oil is first transesterified with sufficient glycerol to give the total desired glycerol content, including the glycerol in the oil. Since PA (the mostly used diacid) is not soluble in the oil, but is soluble in glycerol, transesterification of the oil with glycerol must be carried out as a separate step before the PA is added; otherwise glycerol phthalate gel particles would form early in the process. This two-stage procedure is often called the *monoglyceride process*. The first step, transesterification, is carried out at 230–250°C in the presence of a catalyst, many of which are effective. Prior to the strict regulation of lead in coatings, litharge (PbO) was widely used; the residual lead also acted as a drier. Examples of catalysts now used in the United States are tetraisopropyltitanate, lithium hydroxide, and lithium ricinoleate. It is also common to transesterify the oil with PE, which is less expensive than glycerol, to obtain mixed partial esters. The process is run under an inert atmosphere such as CO_2 or N_2 to minimize discoloration and dimerization of drying oils.

Although the process is called the monoglyceride process, the transesterification reaction with glycerol results in a mixture of unreacted glycerol, monoglycerides, diglycerides, and unconverted drying oil. Using PE, the "monoglyceride" is even more complex. The composition depends on the ratio of glycerol (or PE) to oil as well as on the catalyst, time, and temperature. In general, the reaction is not taken to equilibrium. The viscosity and properties of the final alkyd can be affected by the extent of reaction prior to PA addition. While many tests have been devised to evaluate the extent of transesterification, none is very general because the starting ratio of glycerol to oil varies over a considerable range, depending on the oil length of the alkyd being made. (In calculating the mole ratio of dibasic acid to polyol, the glycerol already present in the oil must also be included.) The operator must decide when to start the second stage by adding PA to the reactor. A useful empirical test is to follow the solubility of molten PA during the first stage. This test has the advantage that it is directly related to a major requirement that must be met.

The second stage, esterification of the "monoglyceride" with PA, is carried out at a temperature of 220–255°C. Both stages are carried out in an inert atmosphere.

Fatty Acid Process. It is often desirable to prepare an alkyd utilizing a polyol other than glycerol, commonly PE. In this case, fatty acids must be used instead of oils. The saponification process can be performed in a single step with reduced time in the reactor. Any drying, semidrying, or nondrying oil can be saponified to yield fatty acids, but the cost of separating fatty acids from the reaction mixture increases the cost of the alkyd. A more economical alternative is to use TOFAs, which have the advantage that they are produced as fatty acids. TOFA compositions are somewhat similar to those of soybean fatty acids (Table 14.1), although there may be significant differences in TOFAs from different regions. Specially refined tall oils having higher linoleic acid content are available, as are other grades that have been treated with alkaline catalysts to isomerize the double bonds partially to conjugated structures. Generally, in the fatty acid process, the polyol, fatty acids, dibasic acid, and catalyst are all added at the start of the reaction, and the esterification of both aliphatic and aromatic acids is carried out simultaneously in the range of 220–255°C.

15.5.2 Process Variations

Since esterification is a reversible reaction, an important factor affecting the rate of esterification is the rate of removal of water from the reactor. Most alkyds are produced using a reflux solvent, such as xylene, to promote the removal of water by azeotroping. The amount of xylene, generally less than 5%, is determined empirically, dependent on the reactor. A sufficient quantity is required for vigorous reflux, but not so much as to cause flooding of the condenser, since the reaction is run at a temperature far above the boiling point of xylene. Some of the xylene is distilled off along with the water; water is separated and xylene is returned to the reactor. Methyl isobutyl ketone (MIBK) has also been reported as a reflux solvent. MIBK reduces the viscosity more than xylene. In either case, the solvent also reduces accumulation of sublimed solid monomers, mainly PA, in the reflux condenser; and its vapor serves as an inert atmosphere, reducing the amount of inert gas needed.

Reaction time is affected by reaction temperature. Figure 15.1 shows the relationship among temperature, time, and viscosity in preparing an alkyd. Figure 15.2 shows the relationship of the variables with acid number (i.e., milligrams of KOH required to titrate the acid in 1 g resin solids). There are economic advantages to short reaction times. Operating costs are reduced; and shorter times permit more batches of alkyd to be produced in a year,

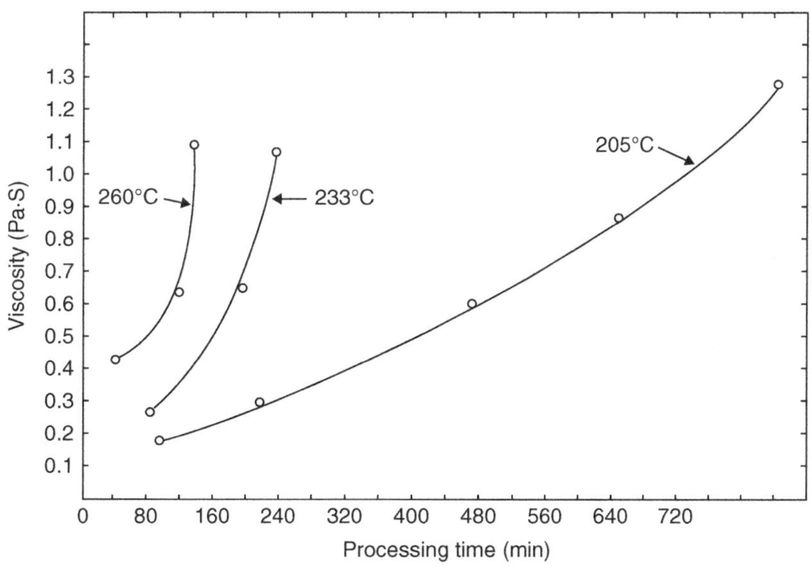

Figure 15.1 Effect of temperature and processing time on viscosity. Source: Reproduced with permission of Monsanto Chemical Co.

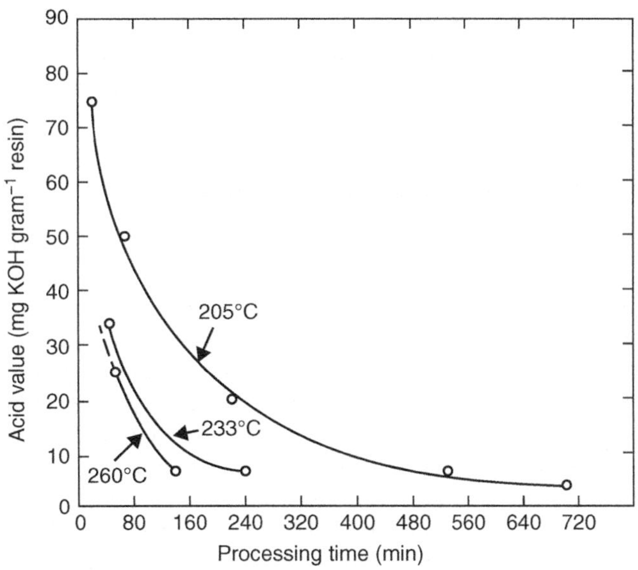

Figure 15.2 Effect of temperature and processing time on acid value of a typical medium oil linseed alkyd. Source: Reproduced with permission of Monsanto Chemical Co.

increasing capacity without capital investment in more reactors. Therefore, it is desirable to operate at as high a temperature as possible without risking gelation.

A critical aspect of alkyd production is deciding when the reaction is completed. Disappearance of carboxylic acid is followed by titration; and increase in MW is followed by viscosity. These determinations require time during which the reaction is continuing. After it is decided that the extent of reaction is sufficient, the reaction mixture is "dropped" into a larger tank containing solvent (for solventborne and water-reducible alkyds), a procedure that also requires time while the reaction is continuing. Accordingly, the decision to start dropping the batch must be made so that the acid number and viscosity of the batch will be on target after the subsequent procedures of sampling, determination of acid number and viscosity, and discharging of the reactor. The time for these determinations becomes the rate-controlling step in production. If they can be done rapidly enough, the reaction can be carried out at 240°C or even higher without overshooting the target acid number and viscosity. On the other hand, if the control tests are done slowly, it may be necessary to run the reaction at a lower temperature, for example, 220°C, which may require 2 h or more of additional reaction time (Figures 15.1 and 15.2). Automatic titration instruments permit rapid determination of acid number, so the usual rate-limiting step is viscosity determination. While attempts have been made to use viscosity of the resin at reaction temperature to monitor change in MW, the dependence of viscosity on MW at the high reaction temperatures is not sensitive enough to be very useful. The viscosity must be determined at some lower standard temperature. Since viscosity depends strongly on solution concentration and temperature, these variables must be carefully controlled.

In manufacture of emulsified alkyds, the molten alkyd must be cooled as quickly as possible to about 100°C. Surfactant is usually added, and the warm resin is mixed with hot (but not boiling) water in a high shear apparatus, such as a rotor–stator disperser to assure small droplet size of the resin (Kan et al., 2014). Lower emulsification temperatures may be possible with a specially adapted extruder (Vogel et al., 2015).

In alkyd production, viscosity is commonly determined using Gardner bubble tubes as discussed in Section 3.3.4. The cook is continued until the viscosity is high enough so that by the time the resin batch is dropped into the solvent and the batch cooled, its viscosity will match the value in the specification. This requires that discharge of the reactor must be started when the test sample is at some lower viscosity. It is not possible to generalize how much lower the viscosity should be; it depends on the specific alkyd composition, the temperature at which the reaction is being run, the time required to do the determination, the time required to empty the reactor, and so on. Viscosities can be determined more rapidly using a cone and plate viscometer (Section 3.3.2) than with bubble tubes. Furthermore, the very small sample required for a cone and plate viscometer can be cooled and equilibrated at the measurement temperature more quickly.

One of the many variables that affect the acid number and viscosity of alkyds is the ratio of reactants: the closer the ratio of moles of dibasic acid to moles of polyol approaches 1, the higher the MW of the backbone of the resin, and the greater the likelihood of gelation. A useful rule of thumb for a starting point is a mole ratio of 0.95. The final ratio is determined by adjustments such that the combination of acid number and solution viscosity match the desired levels. The greater the ratio of hydroxyl groups to carboxylic acid groups, the faster the acid groups are reduced to a low level. The degree of completion of the reaction is an important factor controlling the viscosity, as well as the acid number. It is usually desirable to have a low acid number, typically in the range of 5–10 mg of KOH per gram of alkyd. Water-reducible alkyds have acid numbers of 35–50.

The composition of the fatty acids is a major factor affecting the viscosity; and composition of an oil or grade of TOFA can be expected to vary somewhat from lot to lot. Dimerization and oligomerization of the unsaturated fatty acids occur in the same temperature range at which the esterification is carried out (Section 14.3.1). Fatty acids with conjugated double bonds dimerize more rapidly than those with nonconjugated bonds; and dimerization rates increase with the level of unsaturation. At the same ratio of phthalic to polyol to fatty acids, the viscosity of alkyds having the same acid number and solution concentration will increase in the order soybean < linseed < tung.

Some volatilization of polyol, PA, and fatty acids out of the reactor will occur, depending on the design of the reactor, the rate of reflux of the azeotroping solvent, the rate of inert gas flow, and the reaction temperature, among other variables. The amount and ratio of these losses affect the viscosity at the standard acid number. The exact ratio of reactants must be established in the reactor that is actually used for synthesis. Since gelation can occur if the ratio of dibasic acid to polyol is overly high, it is prudent not to add all of the PA into the reactor in the beginning of fatty acid processes. If the viscosity is too low when the acid number is nearing the standard value, more PA can easily be added. The amount of PA held back can be reduced as experience is gained cooking a particular alkyd in a particular reactor.

Other side reactions can affect the viscosity–acid number relationship. Glycerol and other polyols form ethers to some degree during the reaction. Glycerol can also form toxic acrolein by successive dehydrations.

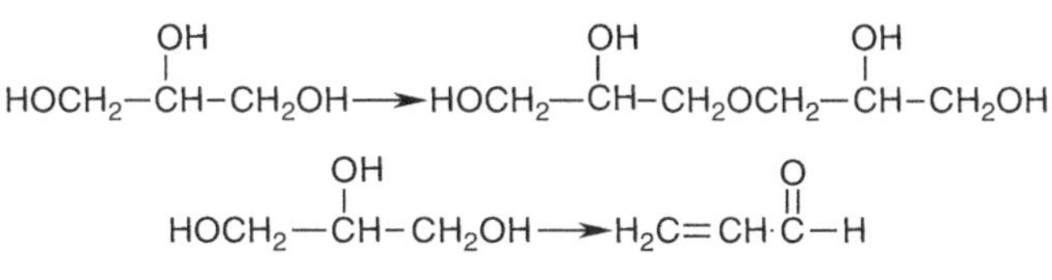

When these reactions occur, the mole ratio of dibasic acid to polyol increases and the number of hydroxyl groups decreases; therefore, at the same acid number, the MW will be higher. Excessive viscosity and even gelation can result. Ether formation is catalyzed by strong protonic acids; therefore, it is desirable to avoid them as catalysts for the esterification. Monobutyltin oxide has been used as an esterification catalyst; presumably, it does not significantly catalyze ether formation. As noted earlier, PE and TMP seem less vulnerable than glycerol to undesirable side reactions such as ether formation; and glycerol is the only polyol that can decompose to form acrolein. A hydroxyl group on one end of a growing polyester chain can react with a carboxylic acid group on another end of the same molecule, leading to ring formation. Transesterification of chain linkages can have the same result. Since cyclization reactions reduce chain length, as well as the proportion of polar groups, their net effect is to reduce viscosity.

Many alkyd resins have broad, irregular MW distributions. Figure 15.3 shows an SEC trace of a contemporary alkyd; it is reminiscent of the best performing alkyd described by Kumanotani et al. (1984) as described in Section 2.1.1 and Figure 2.2b. Even modest changes in reaction conditions can cause large differences in MW distribution, which can have significant effects on final film properties. In many alkyds, very small gel particles, *microgels*, are formed, which play an important role in

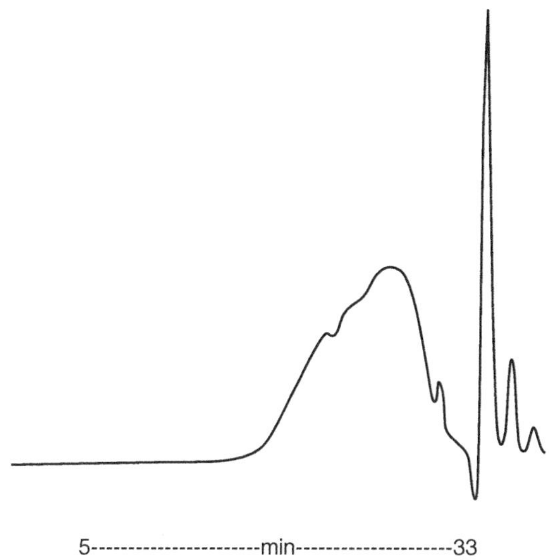

Figure 15.3 Size exclusion chromatogram of an alkyd resin. Note the very broad molecular weight distribution and the irregularities in the high molecular weight portion. Source: Reproduced with permission, Courtesy of Agilent Technologies, Inc.

determining the mechanical properties of final films (Kumanotani et al., 1984). Process changes that may make the alkyd more uniform may be undesirable. For example, allowing glycerolysis to approach equilibrium before addition of PA and using transesterification catalysts in the final stages of esterification both favor narrower MW distributions and lower viscosities, but films made from the more uniform alkyds may exhibit inferior mechanical properties.

15.6 MODIFIED ALKYDS

Oxidizing alkyds can be modified by reaction with vinyl monomers. The most widely used monomers are styrene, vinyl toluene and methyl methacrylate, and methacrylic acid for waterborne hybrid alkyds (Section 15.3.2). Essentially any vinyl monomer can be reacted in the presence of an alkyd to give a modified alkyd. Methyl methacrylate imparts better heat and weather resistance than styrene but at higher cost. Wang (2013) used advanced nuclear magnetic resonance (NMR) techniques to confirm that grafting of acrylates to alkyds having $-CH{=}CHCH_2CH{=}CH-$ structures occurs by abstraction of a doubly allylic hydrogen followed by growth of polyacrylate at that site. Alkyds without double bonds are grafted at the glyceride unit, presumably more slowly.

In making styrenated alkyds, an oxidizing alkyd is prepared in the usual way; then styrene and a free radical initiator, such as benzoyl peroxide, are added. The resulting free radical chain process leads to a variety of reactions, including formation of low MW homopolymers of styrene, grafting of polystyrene onto the alkyd, and dimerization of alkyd molecules. The reaction is generally carried out at about 130°C, which favors decomposition of benzoyl peroxide to form phenyl free radicals; phenyl radicals have a strong tendency to abstract hydrogen, which favors grafting. The ratio of alkyd to styrene can be varied over a wide range and is often about 50% alkyd and 50% styrene. The ratio of aromatic rings to aliphatic chains is greatly increased, and as a result, the T_g of styrenated alkyds is higher and tack-free time is shorter. Styrenated alkyds form a "dry" film in 1 h or less versus 4–6 h for the corresponding non-styrenated alkyds. However, the average functionality for oxidative cross-linking is reduced, not just by dilution with styrene but also because a portion of activated methylene groups are consumed in the styrenation. As a result, the time required to develop solvent resistance is longer than for the counterpart alkyd. Styrenated alkyds are very attractive for applications in which fast drying and low cost are desirable, but not for applications in which the time required for cross-linking is more critical.

Styrenated alkyd vehicles are often used for air dry primers. In these applications, the top coat must be applied almost immediately or not until the film has had ample time to cross-link. During the intermediate time interval, application of top coat is likely to cause nonuniform swelling of the primer, leading to what is called *lifting* of the primer. Lifting results from the development of wrinkled areas in the surface of the dried film. End users who are accustomed to using alkyd primers, which do not give a hard film until a significant degree of cross-linking has occurred, are particularly likely to encounter problems of lifting if they switch to styrenated alkyd primers.

15.7 URALKYDS AND OTHER AUTOXIDIZABLE URETHANES

Diisocyanates are used to make a variety of resins that share similar characteristics with alkyds, as discussed in this section. For a description of the isocyanates, see Section 12.3.

15.7.1 Uralkyds

Uralkyds are also called *oil-modified urethanes*. Traditionally they were solventborne; but waterborne versions having less VOC are also available. Uralkyds are used primarily as the vehicle for so-called urethane varnishes for contractors and the do-it-yourself market. Uralkyd-based varnishes provide abrasion and hydrolytic resistance greatly superior to that obtained with conventional varnishes or alkyd resins. Many so-called varnishes sold to the consumer today are based on uralkyds (see Section 14.3.2 for the historical definition of a varnish). They are used as transparent coatings for furniture, woodwork, gymnasium floors, and hardwood floors, applications in which good abrasion, mar, scuff, chemical, and saponification resistance are important. They are also widely used in maintenance paints.

Uralkyds are prepared by transesterification of a drying oil with a polyol such as glycerol or PE to make a "monoglyceride," followed by reaction with a diisocyanate. The NCO/OH ratio is somewhat less than 1. To further assure that no NCO groups remain unreacted, methyl alcohol (or other low MW alcohol) is often added at the end of the process. For lower cost, one can use a combination of PA and diisocyanate.

Uralkyds prepared from aromatic diisocyanates, primarily TDI and, less often, MDI, will tend to discolor with time, even indoors; those prepared from aliphatic diisocyanates have much better color retention but are more expensive. For example, a uralkyd made by reacting aliphatic H_{12}MDI with the polyol made transesterification of linseed oil with PE had improved color retention compared with the corresponding uralkyd made using TDI (Mountfield et al., 1967). An extensive review paper on isocyanates is available (Wicks and Wicks, 2005).

Narayan (1998) compared the properties of uralkyds, alkyds, and 2K urethane coatings. Four uralkyds, two prepared from linseed oil and either TDI or MDI and two prepared similarly from safflower oil, were compared with linseed and safflower alkyds. All of the alkyds and uralkyds had an oil length of 60%. They were further compared with two 2K polyester/MDI resin systems. Primers and pigmented top coats were formulated with each resin. The cohesive strengths of the uralkyd paint films were greater than those of the alkyd films and approached those of the 2K coatings. Of the uralkyd films, the highest cohesive strength was obtained using the linseed oil/TDI resin. Scratch resistance and tensile strength tests showed the same ranking of the films made from the eight resins. Salt spray resistance and humidity resistance of the uralkyd coatings were superior to those of the alkyd coatings but inferior to the 2K urethane coatings. Particularly notable, the chemical and solvent resistance of the uralkyd films was markedly better than the alkyd films, although somewhat inferior to the 2K films.

Studies on the permeability rate of water into uralkyd and alkyd films indicated that the alkyd films are more permeable than the uralkyd films (Xu et al., 2002). Permeability of uralkyds can be reduced by incorporation of surface-treated montmorillonite clay (Kowalczyk, 2014).

An advantage of uralkyds as compared with alkyds is ease of manufacture. The time for the reaction of isocyanate groups is substantially shorter and the reaction temperature is lower, on the order of 70°C, as compared with

220–245°C for the esterification of alkyds. This results not only in lower production costs and higher equipment utilization but also in less dimerization of the unsaturated chains, thereby reducing viscosity at the same molar ratios.

VOC of traditional uralkyd varnishes has been over 500 g l^{-1}. This level can be lowered by reducing the MW of the uralkyd. In this case, performance can approximate that of higher VOC uralkyds by careful balancing of driers and accelerators. Another approach is to use the exempt solvents such as 4-chlorobenzotrifluoride (PCBTF) (Section 18.1) to replace part of the mineral spirits solvent. Higher solids varnishes are advantageous in that two coats are sufficient rather than the usual three; however, great care must be taken to avoid obtaining too thick films.

As with other types of coatings, VOC regulations stimulated development of waterborne uralkyds. Every major approach used to develop waterborne alkyds has been investigated for uralkyds. Water-reducible uralkyds having VOCs of less than 250 g l^{-1} are on the market. Autoxidizable acrylic/urethane hybrid dispersions (Section 15.7.2) permit preparation of high quality products at a higher cost. 2K waterborne urethanes have many advantages for the industrial and contractor markets, but are not suitable for the do-it-yourself market (Caldwell, 2004). See Section 12.7.3.

15.7.2 Autoxidizable Polyurethane Dispersions

Water-reducible PUDs (Section 12.7) can be made autoxidizable by introducing appropriate fatty acids. Sometimes the products are called *fatty acid-modified polyurethane dispersions* (FAPUDs). Patel et al. (2010) described a synthetic process for FAPUDs and studied the effects of compositional variations on film properties. This rather complicated process may be summarized as follows:

1. Linoleic acid, PA, and various polyols are combined to make an oligomeric alkyd having high OH functionality.

The resulting alkyd is reacted with excess IPDI, DMPA, and additional polyol in *N*-methylpyrrolidone (NMP) solvent to form a polymer comprising —COOH, —NCO, and fatty acid functional groups.

2. The resulting polymer is neutralized with an amine and then dispersed in water.
3. Further reaction of the —NCO groups with water (Section 12.6) increases the molecular weight and fatty acid functionality of the polymers in the dispersed particles. The resulting high $\bar{f}_n$ of the fatty acid enhances the potential for oxidative cross-linking.

An extensive study of films made from such FAPUDs indicated that excellent coating properties are attainable by optimizing the compositional variables. A drawback of the aforementioned process is that NMP remains in the dispersion, adding VOC. Variations in the process permit removal of the solvent (Hofland, 2012).

Liu et al. (1999) described FAPUDs having excellent hydrolytic stability. The films also have the expected abrasion resistance. Drying speed, color, and cost depend on the drying oil and diisocyanate used.

Other examples include a FAPUD made from a dehydrated castor oil (Prantl et al., 1994) and a PUD for wood flooring coatings prepared from castor oil and soy oil (Irle et al., 2003). The coatings are said to resist solvents, abrasion, and heel marking.

In acrylic/urethane hybrid dispersions, part of the expensive urethane is replaced with relatively inexpensive acrylic. Coogan and Damery (2003) provide an example of such a hybrid. A drying oil fatty acid amide of diethanolamine is reacted with DMPA, cyclohexanedimethanol, a monomethyl ether of polyethylene glycol, and TDI in NMP. Methyl methacrylate is added; the product is diluted with dipropylene glycol monomethyl ether, neutralized with dimethylaminoethanol (DMAE), and then diluted with water. After dilution with water, the methacrylate monomer is polymerized; the resultant PUD gives films that cross-link by autoxidation.

15.8 EPOXY ESTERS

BPA epoxy resins (Section 13.1.1) can be reacted with fatty acids to produce what are commonly called *epoxy esters*. Drying or semidrying fatty acids are used so that the products cross-link by autoxidation. During resin synthesis the epoxy groups undergo a ring-opening reaction with carboxylic acids to generate an ester and a hydroxyl group (Section 13.3.2). These hydroxyl groups, as well as the hydroxyl groups originally present on the epoxy resin, can further esterify with fatty acids.

Epoxy esters are generally made by starting with a low MW epoxy resin (i.e., the standard liquid resin, $n = 0.13$) and extending with BPA by the advancement process (Section 13.1.1) to the desired MW; off-specification epoxy resin is often used to reduce cost. The fatty acids are added to the molten hot resin, and the esterification reaction is continued until the acid number is low, usually less than 7 mg of KOH per gram of resin. In the esterification reaction with fatty acids, the average number of sites for reaction is the n value, corresponding to the number of hydroxyl groups on the resin, plus twice the number of epoxy groups. The resin synthesis is carried out at high temperatures (220–240°C). The rate of esterification slows as the concentration of hydroxyl groups diminishes, and side reactions occur, especially dimerization of the drying oil fatty acids (or their esters). Accordingly, it is not practical to esterify more than about 90% of the potential hydroxyl groups, including those from ring opening the epoxy groups.

TOFAs are commonly used because of their low cost. Coatings based on linseed fatty acids cross-link faster owing to higher average functionality. However, their viscosity is higher because of the greater extent of dimerization during esterification; and their cost is higher. For still faster cross-linking, part of the linseed fatty acids can be replaced with tung oil fatty acids, but this further increases viscosity and cost. The color of epoxy esters from linseed and linseed–tung fatty acids is darker than from tall oil esters. Dehydrated castor oil fatty acids afford faster curing epoxy esters for baked coatings. As with alkyds, the rate of formation of a dry film from epoxy esters depends on two factors: the average number of diallylic methylene groups $\left(\bar{f}_n\right)$ and the ratio of aromatic rings to long aliphatic chains. The $\bar{f}_n$ can be maximized by using higher MW BPA epoxy resin and by using sufficient fatty acid to react with a large fraction of the epoxy and hydroxyl groups. The ratio of aromatic rings to fatty acids can be maximized by using high MW epoxy resin and esterifying a smaller fraction of epoxy and hydroxyl groups.

Epoxy esters are used in coatings in which adhesion to metal is important. While the reasons are not completely understood, it is common for epoxy coatings, including epoxy esters, to have good adhesion to metals and to retain adhesion after exposure of the coated metal to high humidity, a critical factor in corrosion protection (Chapter 7). A distinct advantage of epoxy esters over alkyd resins is their greater resistance to hydrolysis and saponification. The backbone of alkyds is held together with esters from PA and the polyol, whereas in epoxy esters, the backbone is held together with C—C and ether bonds. Of course, the fatty acids are bonded to the backbone with ester groups in both cases, but the fraction of polymer bonds in a dry film subject to hydrolysis is substantially lower in the case of epoxy esters. On the other hand, exterior durability of epoxy ester coatings is poor, as is the case for all films made with BPA epoxy resins. Owing to these advantages and disadvantages, the major uses for epoxy ester resins are as primers for metal, for can coatings, and for crowns (bottle caps), in which the important requirements are adhesion, flexibility, and hydrolytic stability. In baking primers, it is sometimes desirable to add a small amount of MF resin into the formulation to supplement the oxidative cross-linking by MF cross-linking of free hydroxyl groups on the epoxy ester.

Epoxy ester resins with good exterior durability (better than alkyds) can be prepared by reacting fatty acids with epoxy-functional acrylic copolymers (e.g., copolymers containing glycidyl methacrylate; Section 13.1.2). The product is an acrylic resin having multiple fatty acid ester side chains. By appropriate selection of acrylate ester comonomers and MW, the T_g of the resin can be designed so that a tack-free film is obtained by solvent evaporation, followed by oxidative cross-linking. For applications such as repainting an automobile at ambient temperatures, catalysis by metal salt driers may not be necessary. The rate of cross-linking is slower without driers, but exterior durability is improved.

Epoxy esters can also be made water reducible, for example, by reacting maleic anhydride with epoxy esters prepared from dehydrated castor oil fatty acids to generate adducts having anhydride groups. Subsequent addition of a tertiary amine, such as 2-(dimethylamino)ethanol, in water results in ring opening of the anhydride to give amine salts. Like other water-reducible resins (Section 8.3), these resins are not soluble in water, but form a dispersion of resin aggregates swollen with water and solvent in an aqueous continuous phase. The hydrolytic stability of these epoxy esters is better than corresponding alkyds and is sufficient for use in anionic electrodeposition primers (Section 27.1), now largely replaced by cationic primers. Water-reducible epoxy esters are still used in spray-applied baking primers and primer-surfacers. They are also used in dip coating primers in which nonflammability is an advantage.

GENERAL REFERENCES

Anonymous, *The Chemistry and Processing of Alkyd Resins*, Monsanto Chemical Co. (now Solutia, Inc.), St. Louis, **1962a**.

Wicks, Z. W., Jr., Alkyd Resins in Mark, H. F.; Bikales, N. M., Eds., *Encyclopedia of Polymer Science and Technology*, 3rd ed., John Wiley & Sons, Inc., New York, **2002a**, pp 318–340.

REFERENCES

Alidedeoglu, A. H., et al., *JCT Res.*, **2011**, *8(1)*, 45.

Anonymous, *The Chemistry and Processing of Alkyd Resins*, Monsanto Chemical Co. (now Solutia, Inc.), St. Louis, **1962b**.

Arendt, J.; Kim, J., *Proceedings of the American Coatings Conference*, **2014**, April 7–9, Atlanta, GA.

Bacaloglu, R., et al., *Angew. Makromol. Chem.*, **1988**, *164*, 1.

Bakker, P. J., et al., *Proceedings of the 28th International Waterborne Higher-Solids and Power Coatings Symposium*, New Orleans, LA, **2001**, pp 439–453.

Blackinton, R. J., *J. Paint Technol.*, **1967**, *39(513)*, 606.

de Boer, J. W., et al., *Eur. J. Inorg. Chem.*, **2013**, *2013(21)*, 3581.

Bouboulis, C. J., *Proceedings of the International Waterborne Higher-Solids and Power Coatings Symposium*, New Orleans, LA, **1982**, p 18.

Bracken, J., US patent 6,075,088 (**2000**).

Brown, R., et al., *Off. Dig.*, **1961**, *33*, 539.

Caldwell, R. A., *JCT Coat. Tech.*, **2004**, *2(3)*, 30.

Cheetham, H. C.; Pearce, W. T., Alkyd Emulsion Paints in Mattiello, J. J., Ed., *Protective and Decorative Coatings*, John Wiley & Sons, Inc., New York, **1943**, p *488*.

Coogan, R. G.; Damery, S., US patent 6,548,588 (**2003**).

Danneman, J.; Chu, J., *CoatingsTech*, **2016**, *13(9)*, 30–39.

Dederichs, F., *Adv. Coat. Surf. Technol.*, **2013**, September, 8.

Dhoke, S. K.; Khanna, A. S., *Prog. Org. Coat.*, **2012**, *74*, 92–99.
Diamond, C., *Coat. World*, **2016**, *21(11)*, 31–34.
Elribii, M., et al., *Prog. Org. Coat.*, **2015**, *87*, 222–231.
Guclu, G.; Orbay, M., *Prog. Org. Coat.*, **2009**, *65(3)*, 362.
Guyot, A.; Landfester, K.; Schork, F. S.; Wang, C., *Prog. Poly. Sci.*, **2007**, *32*, 1439–1461.
van Hamersfeld, E. M. S., et al., *Prog. Org. Coat.*, **1999**, *35*, 235.
Hofland, A., *Prog. Org. Coat.*, **2012**, *73*, 274–282.
Ikladious, N. E., et al., *Prog. Org. Coat.*, **2015**, *89*, 252–259.
Irle, C., et al., US patent 6,559,225 (**2003**).
Kan, C. S., et al., US patent 8,709,607 (**2014**).
Kangas, S. L.; Jones, F. N., *J. Coat. Technol.*, **1987**, *59(744)*, 89.
Kaska, J.; Lesek, F., *Prog. Org. Coat.*, **1991**, *19*, 283.
Kawamura, C., et al., *Prog. Org. Coat.*, **2002**, *45*, 185.
Kienle, R. H.; Hovey, A. G., *J. Am. Chem. Soc.*, **1929**, *51*, 509.
Kowalczyk, K. J., *Coat. Technol. Res.*, **2014**, *11*, 421.
Kuang, Z., et al., *Proceedings of the International Waterborne Higher–Solids and Power Coatings Symposium*, New Orleans, LA, **1999**, pp 126–141.
Kumanotani, J., et al., *Adv. Org. Coat. Sci. Technol. Ser.*, **1984**, *6*, 35.
Lanson, H. J., Chemistry and Technology of Alkyd Resin in Craver, J. K.; Tess, R. W., Eds., *Applied Polymer Science*, American Chemical Society, Washington, DC, **1975**, pp 531–547.
Liu, W., et al., *Proceedings of the International Waterborne Higher–Solids and Power Coatings Symposium*, New Orleans, LA, **1999**, p 202.
Misev, T. A., *J. Coat. Technol.*, **1989**, *61(772)*, 49.
Mountfield, B. A., et al., UK patent GB 1,059,936 (**1967**).
Murillo, E. A.; Vallejo, P. P.; Lopez, B. L., *Prog. Org. Coat.*, **2010**, *69*, 235–240.
Murillo, E. A., Vallejo, P. P.; Lopez, B. L., *J. Appl. Polym. Sci.*, **2011**, *120(6)*, 3151–3158.
Nabuurs, T., et al., *Prog. Org. Coat.*, **1996**, *27*, 163.
Narayan, R., et al., *Paint India*, **1998**, September, 116.
Nelson, T. J.; Webster, D. C., *JCT Res.*, **2013**, *10(4)*, 515–525.
Osterberg, G.; Bergenstahl, B., *J. Coat. Technol.*, **1996**, *68(858)*, 39.
Osterberg, G., et al., *Prog. Org. Coat.*, **1994**, *24*, 281.
Palmer, C., *Proceedings of the American Coatings Conference*, **2014**, April 7–9, Atlanta, GA.
Patel, A., et al., *Prog. Org. Coat.* **2010**, *67*, 255.
Pirs, B., et al., *J. Coat. Technol. Res.*, **2015**, *12(6)*, 965–974.
Porreau, D. R.; Smyth, S. E., *JCT Coat. Technol.*, **2004**, *1(2)*, 40.
Prantl, B., et al., US patent 5,319,052 (**1994**).
Strazik, W. F., et al., US patent 4,293,461 (**1981**).
Vogel, E., et al., *American Coatings Conference*, Indianapolis, IN, April 11–13, **2015**.
Wang, Q., *Investigation of Acrylated Alkyds*, Thesis, University of Akron, **2013**, from https://etd.ohiolink.edu/rws_etd/document/get/akron1366047898/inline (accessed April 22, 2017).
Weger, W., *Fitture e Vernici*, **1990**, *B66(9)*, 25.
Weissenborn, P. K.; Motiejauskaite, A., *Prog. Org. Coat.*, **2000**, *40*, 253.
Wicks, Z. W., Jr., Alkyd Resins in *Encyclopedia of Polymer Science and Technology*, 3rd ed., John Wiley & Sons, Inc., New York, **2002b**, pp 318–340.
Wicks, D. A.; Wicks, Z. W., Jr., *Prog. Org. Coat.*, **2005**, *54*. 141.
Xu, Y., et al., *Prog. Org. Coat.*, **2002**, *45*, 331.
Yin, X., et al., *Prog. Org. Coat.*, **2014**, *77*, 674–678.
Zabel, K. H., et al., *Prog. Org. Coat.*, **1999**, *35*, 255.
Zuchert, B.; Biemann, H., *Farg Lack Scand.*, **1993**, *(2)*, 9.

Chapter 16

Silicon Derivatives

Three classes of silicon derivatives are used in organic coatings: silicones, reactive silanes, and orthosilicates. Silicones are versatile polymer building blocks, cross-linkers, polymer modifiers, and additives. Unmodified silicones are used as coating binders and as additives. Reactive silanes are polymer building blocks (especially for silane-terminated polymers) and coupling agents. Orthosilicates are included for their use in organic/inorganic coatings. As a very broad generality, silicon-containing coatings are expensive, but the cost is often justified when their performance, especially durability, is superior. Thus, silicon-containing resins are well established in automotive clear coats and are growing in industrial maintenance and marine paints.

16.1 SILICONES

Silicones, more properly called polysiloxanes, are polymers with backbones consisting of $-[Si(R)_2-O]-$ repeating units. Polysiloxanes are prepared from chlorosilane monomers. The principal commercial monomers are the following methyl- and phenyl-substituted silanes.

Me_3SiCl	Trimethylchlorosilane
Ph_2SiCl_2	Diphenyldichlorosilane
$PhSi(Me)Cl_2$	Phenylmethyldichlorosilane
Me_2SiCl_2	Dimethyldichlorosilane
$PhSiCl_3$	Phenyltrichlorosilane
$MeSiCl_3$	Methyltrichlorosilane

Chlorosilanes react with water to form silanols that in turn can condense to form siloxanes. For example, dimethyldichlorosilane reacts with water to form a polysiloxane, often represented as a linear polysiloxane. The following example, polydimethylsiloxane (PDMS), enjoys a central position in the silicone industry.

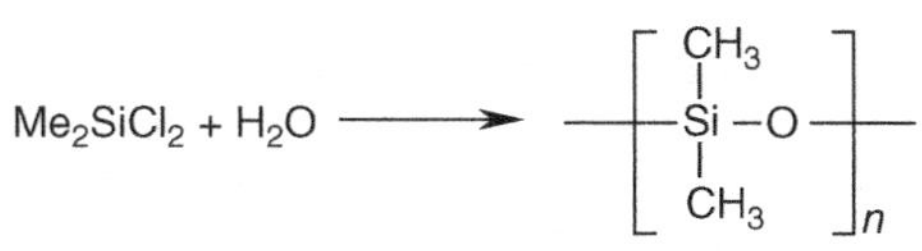

PDMS

Structures of the polymers can be more complex than represented by this linear structure. If chain branching is desired, small amounts of $MeSiCl_3$ can be added to the aforementioned reaction to form branch sites. Siloxanes are more apt than hydrocarbons to form large ring cyclic structures, such as the 8-membered ring $(Me_2SiO)_4$. Such cyclic compounds are intermediates in the polymerization and are present to varying degrees in the finished products. Chain branching is not the only possible outcome with trichlorosilanes. They can react with water to form three-dimensional cluster compounds (polyhedral oligomeric silsesquioxanes, commonly abbreviated POSS). Certain POSS structures can also be obtained at reasonable cost by pyrolysis of oat or rice hulls. The $Si_8O_{12}R_8$ structure shown in the following is widely studied.

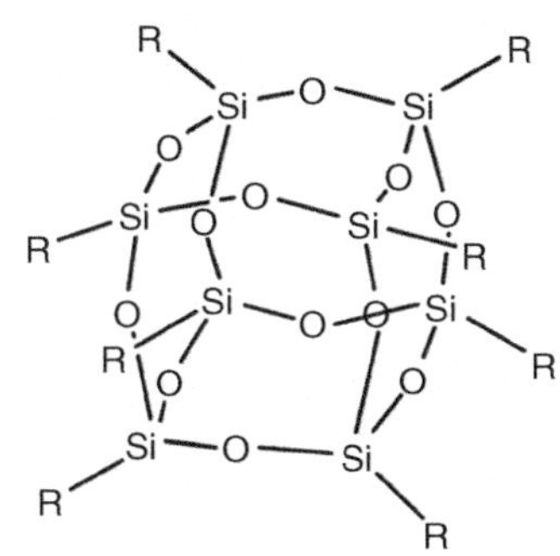

An octahedral POSS

Organic Coatings: Science and Technology, Fourth Edition. Frank N. Jones, Mark E. Nichols, and Socrates Peter Pappas.

Owen (1990) lists four characteristics of PDMS that account for its singular properties:

1. Low intermolecular forces between methyl groups
2. Unique flexibility of the siloxane backbone
3. High energy of the siloxane bond
4. Partial ionic nature of the siloxane bond

To this list we can add (5) virtual transparency to terrestrial UV.

These characteristics combine to explain the attributes of PDMS, including its low surface tension, its unique flexibility, its thermal stability, and its exceptional weatherability. On the other hand, the partial ionic nature of the siloxane bond also causes vulnerability to hydrolysis at low or high pH. As we shall see, PDMS can be combined in various ways with carbon-based polymers to impart some of the desirable attributes.

Oligomers are made using monochlorosilanes as chain-terminating groups; molecular weight (MW) is reduced in proportion to the monochlorosilane–dichlorosilane ratio. The oligomeric products are called silicone fluids or silicone oils. As discussed in Chapter 23, small amounts of silicone fluids are used as additives in coatings. PDMS fluids reduce surface tension because the ease of rotation around the Si—O—Si bonds leads to rapid orientation of methyl groups at the surface. The methyl groups provide a surface having very low surface tension. Low MW silicone fluids can be used as VOC-exempt solvents.

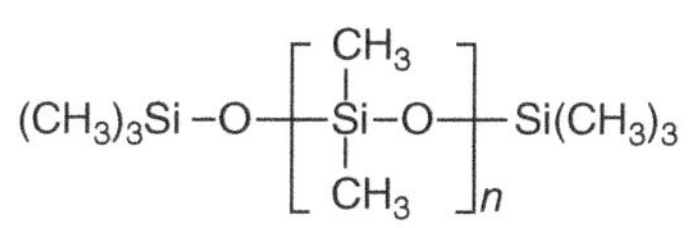

A silicone fluid

PDMS have limited compatibility with many coating resins. Chemically modified silicone fluids with broader ranges of compatibility have been described (Fink et al., 1990). Examples are polysiloxane–polyether block copolymers; polysiloxane–polyether copolymers, which also contain some longer alkyl chains on the siloxane groups; and PDMS modified with esters as well as with aryl-substituted alkyl groups. Siloxanes having a polyether terminal group or side chain have been reported to be excellent surface tension additives to prevent film defects in waterborne coatings. They can be prepared by reacting allyl-terminated polyethers with PDMS having Si—H bonds (Spiegelhauer, 2002).

16.1.1 Silicone Rubbers and Resins

Silicone rubbers are cross-linked polysiloxanes, and *silicone resins* are usually branched polysiloxanes. In both cases, trichlorosilanes are used to provide chain branching and, ultimately, cross-linking to form silicone rubbers. While silicone rubbers are expensive, they have unique and useful properties; they exhibit excellent resistance to oxidation accompanied by retention of flexibility at low temperatures. They have a low T_g as a result of the ease of rotation around the Si—O—Si bonds.

Polymerization of a mixture of mono-, di-, and trichlorosilanes, carried out by hydrolyzing a solution of the monomers in a water-immiscible solvent, results in a silicone resin having some unreacted hydroxyl groups. The silicone resin is separated from the acidic water phase, and residual acid is removed by extraction. A stable solution of a silicone resin is obtained if the last water is removed from the reaction mixture at relatively low temperatures in the absence of catalyst. The $\bar{M}_n$ of such resins is generally in the range 700–5000. Different copolymerization processes yield different proportions of linear, branched, cyclic, and POSS structures; the composition affects properties. Typical functionality is three to four end groups per molecule.

Such silicone resins can be baked to form cross-linked films. A typical cure schedule is 1 h at 225°C, but the time and temperature can be reduced with catalysts, for example, zinc octanoate. Since the cross-linking reaction is reversible, silicone films are sensitive to water, especially under basic conditions. Ammonia and amines are especially destructive to such films. Evidence has been presented that aminolysis occurs selectively at Si groups having three oxygens (i.e., at cross-linking sites) probably resulting from the greater electrophilicity of such Si groups with three electronegative oxygen atoms (Hsiao et al., 1975; Pappas and Just, 1980).

The MW and viscosity of silicone resins depend on the proportion of monochloro-, dichloro-, and trichlorosilane monomers as well as on the process. The ultimate crosslink density depends on the fraction of trichlorosilane in the recipe and on the fraction of this monomer that forms branched, rather than cyclic or POSS structures, during the process.

Most silicone resins for coatings are copolymers of methyl- and phenyl-substituted monomers; other alkyl monomers are sometimes included. Properties depend on the phenyl/methyl ratio, as summarized in Table 16.1 (Finzel and Vincent, 1996). The rate of the cross-linking

Table 16.1 Properties of High-Methyl versus High-Phenyl Silicone Resins

High methyl	High phenyl
Lower weight loss in curing	Greater weight loss in curing
Faster cure rate	Longer package stability
Higher UV stability	Higher heat stability
Lower temperature flexibility	Greater oxidation resistance

Source: Adapted from Finzel and Vincent (1996).

reaction is faster with high-methyl-substituted silicone resins. Consequently, a larger amount of material is lost during heat curing of high-phenyl resins since low MW components evaporate in the oven to a larger degree before polymerizing. The reactivity differences lead to faster cure rate of the high-methyl-substituted silicones and longer package stability of the high-phenyl-substituted silicones. Better gloss retention during exterior exposure is obtained with high methyl relative to high-phenyl-substituted silicone resins owing to UV absorptivity of phenyl groups. The exterior durability of well-formulated silicone coatings is superior to that of acrylic–MF, polyester–MF, or urethane coatings and approaches that of highly fluorinated polymers (Chapter 17).

While high-methyl silicones are more weatherable, high-phenyl silicones are superior to high-methyl silicones for applications requiring high-temperature resistance. In this respect they are superior to other organic coatings except certain fluoropolymers. While inferior to that of high-phenyl silicones, the thermal stability of high-methyl silicones is much greater than that of silicones having longer alkyl groups. The half-lives at 250°C, based on weight loss studies, are reported to exceed 100 000 h for phenyl silicone films compared with over 10 000 h for methyl silicone (PDMS) films and only 2 h for propyl silicone films (Finzel and Vincent, 1996). The useful life of a methyl silicone film at 350°C is given as 1000 h; for comparison a "polyester" film (presumably cross-linked with a MF resin) has a useful life of 1000 h at only 223°C.

The large difference in thermal stability of methyl and propyl silicones is attributable to differences in bond dissociation energies. Methyl silicone is held together entirely by strong bonds with dissociation energies above 100 kcal mol^{-1} (Si—O about 110, Si—C about 104, and methyl C—H about 105), whereas C—C bonds in propyl groups have dissociation energies of about 86 kcal mol^{-1}.

When silicones are thermally decomposed, the ultimate product is silica (poly SiO_2) that, though brittle, can serve as a temperature-resistant coating binder. For example, chimney paints are made from silicone resins pigmented with aluminum flake for use at over 500°C. At the high service temperature, the organic substituents burn off, leaving behind a film of the aluminum pigment in a matrix of silicon dioxide—essentially glass—containing some aluminum silicate. The brittle film continues to provide protection if it is not mechanically damaged. Finzel and Vincent (1996) discuss coatings having varying heat resistance.

Coatings that contain methyl silicone resins exhibit low temperature flexibility superior to those containing phenyl silicones as well as to most other organic coatings. Methyl silicones also possess the unusual combination of being repellant to liquid water but permeable to water vapor.

16.1.2 Modified Silicone Resins

Terminal —OCH_3 groups on silicone resins can be hydrolyzed to terminal —OH groups with base or acid catalysis (Osterholtz and Pohl, 1992). The products are called *silanols*. High solids silanols are available that cure either directly using zinc octanoate catalysts or by cross-linking with trialkoxysilanes using a titanate catalyst (Finzel, 1992). Florio and Ravichandran (2016) describe other catalysts, as well as the reactions involved in self-condensation of silanol groups.

Silicone resins can be emulsified in water with the help of surfactants. Mixtures of emulsified methyl silicone resins, an amino-functional polysiloxane emulsion, and a styrene/acrylic latex are used to formulate exterior house paints having outstanding exterior durability (Mangio et al., 2002). Silicone resins designed for this purpose are available (Jiang et al., 2016).

Silicone resins can be modified with a variety of functional groups. For example, Greene (2002) has reported that epoxy-functional resins can be prepared by reacting methylphenyldimethoxysilane, phenyltrimethoxysilane, and dimethyldimethoxysilane with water and a KOH catalyst, followed by reacting the resulting hydroxy-functional silicone resin with 3-glycidyloxypropyltriethoxysilane. Thermosetting coatings were formulated with the resulting epoxy silicone by utilizing a diaminodialkoxysilane cross-linker. Epoxy-functional silicone resins cross-linked with carboxylic acid-functional acrylic resins are used to formulate marine finishes that have excellent durability (Fransehn et al., 2004). Florio and Ravichandran (2016) list silicone resins modified with oxime, acetoxy, amine, and epoxy groups.

Octahedral POSS ($Si_8O_{12}R_8$—shown in Section 16.1) can be synthesized having as many as eight R groups bearing a reactive group, such as —OH, —NH_2, epoxy, and other groups. They have been called octopus monomers. Thus, the functional POSS derivatives are unique polymer building blocks and cross-linkers. For example, Markevicius et al. (2012) described coatings based on an aliphatic polyurethane/aliphatic diol/POSS binder with promising corrosion resistance on aluminum. As another example, Yari and Mohseni (2015) added up to 9% of hydroxyl-functional POSS to acrylic–melamine clear coats, which enhanced the rate of cross-linking, as well as the cross-link density, even when the hydroxyl content of the formulations was held constant.

16.1.3 Silicone-Modified Resins

The high cost of pure silicone coatings and the long, high temperature cures can be reduced by combining silicones with other coating resins. The earliest approach was simply to blend a silicone resin with an alkyd resin in the reactor at the end of the alkyd cook. While some covalent bonds between silicone resin and alkyd may form by this procedure, probably most of the silicone resin simply dissolves in the alkyd. Exterior durability of silicone-modified alkyd coatings made in this way is significantly better than unmodified alkyds. The improvement in durability is roughly proportional to the amount of added silicone resin; 30% by weight of silicone resin is a common degree of modification. Silicone resins designed for this purpose may contain higher alkyl as well as methyl and phenyl groups to improve compatibility. Alkyd coatings modified predominantly with phenyl silicone resins are more faster air drying but more thermoplastic and soluble than those modified predominantly with methyl silicones. These results can be attributed to the higher rigidity of the aromatic rings, which leads to a "solid" film at an earlier stage of drying, resulting in less cross-linking than occurs with methyl silicone modification.

Further improvements in exterior durability are obtained by adding the silicone intermediate at the onset of alkyd synthesis. Such intermediates contain functional groups that readily coreact with free hydroxyl groups of the alkyd. Silicone-modified alkyds prepared in this way are said to impart better durability than alkyds in which the silicone resin is cold blended. Such modified alkyds are used mainly in outdoor air dry coatings, in which the application process is expensive (e.g., in a top coat for steel petroleum storage tanks). Longer coating service life more than offsets the extra cost of the silicone addition.

Silicone-modified polyester and acrylic resins are also made by reacting silicone intermediates with the hydroxy-functional resins. Silicone intermediates having silanol (Si—OH) groups can undergo co-condensation with hydroxyl groups on the resin as well as self-condensation with other Si—OH groups. The ratio of the two reactions is controlled by catalyst choice. Effective catalysts for promoting the desirable co-condensation between the resin and the silicone intermediate are titanates, such as tetraisopropyl- or tetraisobutyltitanate. In model compound studies, tetraisopropyltitanate was shown to favor co-condensation over self-condensation by 3.4 : 1 as compared with a 0.23 : 1 ratio with no catalyst. Co-condensation between a hydroxy-functional silicone intermediate and acrylic polyol is shown in the following text. Self-condensation between silicone intermediates can also occur resulting in Si—O—Si cross-links. Excessive self- or co-condensation of a polyfunctional silicone intermediate and a polyfunctional resin can result in gelation.

The reaction is easier to control when the reactive functional groups on the silicone intermediates are silylmethoxy (Si—OMe) rather than silanol (Si—OH) groups. Finzel and Vincent (1996) have described a methoxylated silicone intermediate that has a weight average MW of 470 and an equivalent weight of 155. An idealized structure of the intermediate is

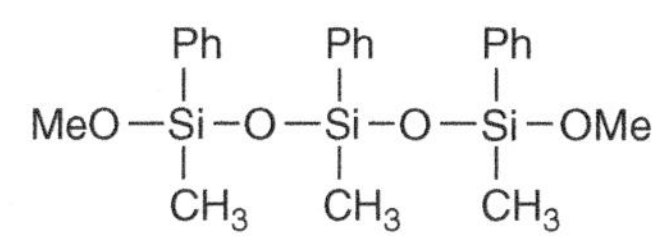

A solution of the polyester or acrylic is heated with the desired amount of silicone intermediate and a titanate catalyst at 140°C until a predetermined viscosity is reached. The reaction is relatively slow at 140°C; higher temperatures reduce reaction time and hence process cost but increase the risk of overshooting the desired viscosity as well as the risk of gelation. Typical silicone-modified polyesters and acrylics have 30–50 wt% silicone content on a solids basis. Modification with less than about 25% silicone provides little improvement in exterior durability. Exterior durability is enhanced with increasing silicone content above 30%, but cost also increases.

The same reactions occur during cross-linking after the coating is applied. Usually zinc octanoate is used as the catalyst for cross-linking the coating, since titanate esters are hydrolyzed by the adventitious water brought into the coating by pigments. The principal applications are in baking coatings, especially in coil coatings for metal siding (Section 30.4), in which case a typical cure schedule is 90 s at peak metal temperatures of 300°C. Such coatings tend to soften when exposed to high humidity for prolonged periods. The softening is called reversion, as it has been attributed to reversible hydrolysis of the cross-links, predominantly at sites where three oxygen atoms are bound to a single silicon atom, as discussed previously. The film hardens again if the ambient humidity decreases, but the film is vulnerable to physical damage while it is soft. To minimize this problem a small amount of MF resin can be added as a secondary cross-linker.

Waterborne silicone-modified resins can be prepared from water-reducible acrylic and polyester resins. Also, acrylic latexes prepared with hydroxyethyl (meth)acrylate as a comonomer can be modified with silicone intermediates. Silicone-modified epoxy resins have also been made

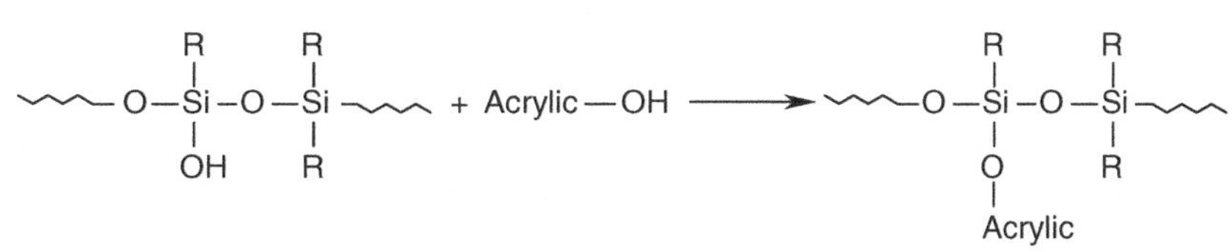

(Ogarev and Selector, 1992). Finzel (1992) provides examples of formulations and preparation of various silicone-modified resins.

16.2 REACTIVE SILANES

Reactive silane is the term used for compounds having a trialkoxysilyl group and an alkyl group bearing a reactive substituent. Reactive silanes were originally developed to improve adhesion of plastic to glass in fiber-reinforced plastics. Dozens, perhaps hundreds, are commercially available. Among reactive silanes that are important in coatings are 3-methacryloxypropyltrimethoxysilane (MPS), 3-aminopropyltriethoxysilane (APS), *N*-(2-aminoethyl)-3-aminopropyltrimethoxysilane, and 3-glycidyloxypropyltrimethoxysilane. Utilization of these and other reactive silanes in coatings have been reviewed (Plueddemann, 1991; Ogarev and Selector, 1992; Witucki, 1993; Mittal, 2007).

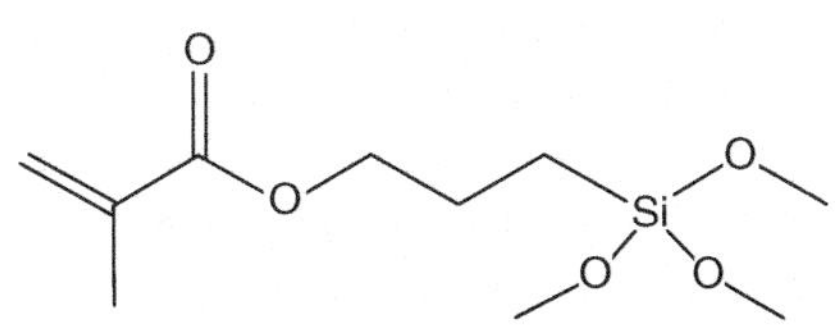

MPS

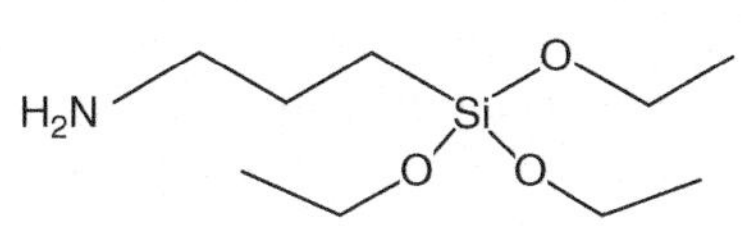

APS

An important use of reactive silanes is to promote adhesion of coatings to glass and, increasingly, to metals having oxide surfaces as described in Section 6.8. In this context they are called *coupling agents.* In an oversimplified view of a complex situation, the trialkoxysilane end of the reactive silane molecule forms covalent bonds to the glass or metal oxide surface, and the reactive group on the other end forms covalent bonds to the coating binder, bonding the coating to the surface. The reactive group is matched with the functionality of the binder.

Reactive silanes are sometimes supplied in water in partly hydrolyzed form; in this form they also function as adhesion promoters. Although described as aqueous solutions, such compositions may actually contain sols, which are colloidal dispersions (Section 16.3.1).

Beyond their value as coupling agents, reactive silanes can also be used to modify resins. In one important example, resins comprising multiple trialkoxysilyl groups can serve as binders for moisture cure coatings. For example, isocyanate-terminated urethanes prepared by reacting polyols and MDI are endcapped with 3-aminotrimethoxysilane to prepare a resin used in moisture-curable coatings (Baghdachi et al., 2002). Such coatings cross-link to a polymer network after application and exposure to humid air. Multifunctional isocyanates may be used to prepare resins that provide increased cross-link density. An advantage of the triethoxysilyl resin over the isocyanate resin is that part of the solvent can be ethyl alcohol. The ethyl alcohol extends the pot life in the presence of water, avoiding the need for expensive moisture-free pigments. The trialkoxysilyl approach also avoids formation of CO_2, which can lead to film imperfections, such as pinholing, in moisture cure urethanes. After the alcohol evaporates, the reaction with water proceeds rapidly at ambient temperature.

Another example of urethane modification with reactive silanes is provided by Noh et al. (2012). An IPDI trimer (Chapter 12) is blocked partly with a reactive silane and partly with a conventional blocking agent. Addition of this material to an acrylic–melamine clear coat enhanced the scratch resistance of the coating. Many other examples can be found in the literature.

Amine functional silanes such as APS are used to modify epoxy resins. One of many possibilities is to react a standard epoxy resin with APS to give the following generalized structure (Source: Min et al. (2013)):

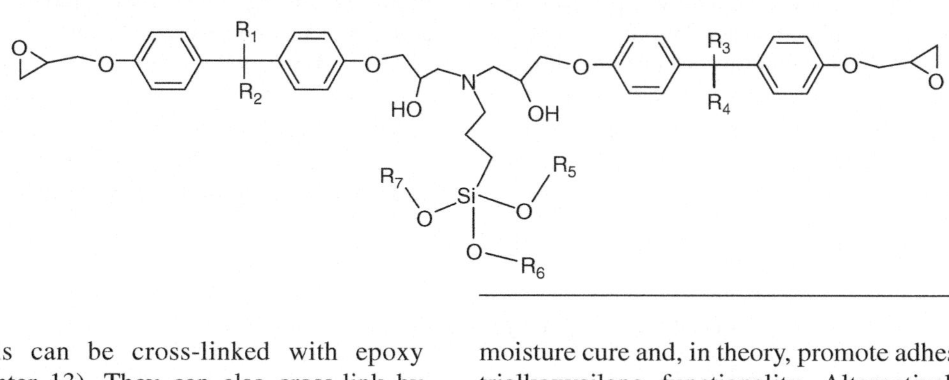

Such materials can be cross-linked with epoxy cross-linkers (Chapter 13). They can also cross-link by moisture cure and, in theory, promote adhesion through the trialkoxysilane functionality. Alternatively, Yuan et al.

(2016) utilized APS to cross-link a commercial silicone–epoxy resin with a —OH functional siloxane-modified polyacrylate to make an anticorrosive coating for aircraft-grade aluminum.

There have been many reports that coatings based on silane-containing cycloaliphatic epoxies afford excellent weatherability (Echeverria et al., 2016). Reactive silanes can play a role. For example, Witucki (2013) reports that a diamino-functional silane serves as a highly reactive chain extender or cross-linker for epoxy resins. A low VOC 2K formulation with a cycloaliphatic epoxy resin (Section 13.1.2) afforded good film properties including excellent gloss retention and chemical resistances. In this application, cycloaliphatic epoxy resins that contain ester groups may be problematic.

Trialkoxysilyl-functional acrylic resins can be made by copolymerizing a trialkoxysilylalkyl methacrylate such as MPS with other acrylic monomers (Ooka and Ozawa, 1993; Furukawa et al., 1994; Nordstrom, 1995). Coatings based on such resins cure on exposure to atmospheric moisture; the reaction is catalyzed with organotin compounds or organic acids. The resulting coatings are reported to exhibit excellent exterior durability, resistance to environmental etching and marring, and adhesion to aluminum. They are used in automobile OEM and refinish clear coats. Trimethoxypropylsilyl-functional acrylic resins are also used in automotive clear coats for plastics. Methoxysilyl groups hydrolyze to silanol groups at room temperature, followed by self-condensation during baking to form siloxane cross-links. The scratch and mar resistance before and after weathering is superior to MF cross-linked coatings (Yaneff et al., 2002).

Automotive clear coats are being made by combining trialkoxysilylalkyl-functional and hydroxy-functional acrylic resins with MF resins or blocked isocyanates (Nordstrom, 1995). Detailed studies of the cross-linking reactions have not been reported, but one can speculate that they include hydrolysis and condensation of the trialkoxysilyl groups to form siloxane cross-links, reaction of hydroxyl groups with MF resins or blocked isocyanates, and transetherification of the hydroxyl groups with the trialkoxysilyl groups. The resulting coatings are reported to have excellent exterior durability as well as high resistance to environmental etching and marring. Another advantage is that formaldehyde emissions are reduced because lower levels of MF resin are used.

Trialkoxysilylated acrylic and vinyl acetate latexes can be prepared using 3-methacryloxypropyltriisobutoxysilane as a comonomer in emulsion polymerization (Chen et al., 1997). The more hindered isobutoxysilyl derivatives exhibit higher resistance to hydrolysis during polymerization and storage. Thus, the latexes are storage stable and do not cross-link until water evaporates from a film. Possibly the alkoxysilyl groups hydrolyze to some extent during storage, but the silanols formed do not condense until after a film is formed because of the large excess of water present. Coatings derived from such resins are reported to have superior adhesion as well as high chemical, solvent, and mar resistance.

A third important use for reactive silanes is to treat the surfaces of pigments for improved dispersibility. This method appears to be especially useful with nano-pigments. Ribeiro et al. (2014) review various methods to treat nanosilica particles with silane coupling agents to enhance their dispersibility. They go on to describe preparation of core shell latex particles in which the silica nanoparticles are surface modified by MPS (structure as shown previously) and then used as a seed for emulsion polymerization (Chapter 9). Nearly uniform distribution of the silica particles within a coating film was achieved.

16.3 ORTHOSILICATES

Tetraethyl orthosilicate ($Si(OEt)_4$ or TEOS) is used in making coating binders for zinc-rich primers for corrosion protection of steel (Sections 7.4.3 and 33.1.2). When exposed to water vapor in the atmosphere, $Si(OEt)_4$ hydrolyzes, with acid or base catalysis, to form a complex polysilicic acid network binder.

The primers are prepared by partial polymerization of an ethyl alcohol solution of $Si(OEt)_4$ by adding a small amount of water—just enough to increase the MW sufficiently to give the desired viscosity, followed by dispersing zinc powder pigment into the resulting oligomer solution. The ethyl alcohol solvent retards polymerization by adventitious water in the coating formulation during storage. When the coating is applied and the alcohol evaporates, water is absorbed from the atmosphere, and the cross-linking reaction continues to completion at ambient temperatures. The cross-linked binder is predominantly polysilicic acid with zinc salts incorporated in the matrix. The salts form by reaction with the zinc hydroxide/carbonate always present on the surface of zinc metal pigments. Ferric ions from the steel surface may also be incorporated in the matrix. Parashar et al. (2001) review the chemistry and properties of zinc-rich primers based on $Si(OEt)_4$. For waterborne zinc-rich primers, aqueous solutions of potassium, sodium, and/or lithium silicate are binders.

16.3.1 Sol–Gel Coatings

Sol–gel technology (Brinker and Scherer, 1990) involves hydrolysis and condensation of metal alkoxides to produce a *sol*, a colloidal dispersion of metal oxide polymers or particles in water, having diameters of 1–1000 nm. The most widely studied alkoxide is $Si(OEt)_4$, but alkoxides of many other metals, including Ti, Al, and Zr, are also important. As described in an excellent open-access review (Danks et al., 2016), the sol can form a gel by various mechanisms including further hydrolysis/condensation or involvement of organic molecules such as citric acid. Subsequent removal of water yields inorganic materials in a variety of physical forms. Sol–gel technology is important for fabricating ceramics, inorganic coatings, and nanoparticles. However, most inorganic coatings are outside the scope of this book.

Sol–gel technology has been adapted to prepare organic/inorganic coatings. They are commonly called *sol–gel coatings*, *hybrid sol–gel coatings*, or *ormosils*. Wang and Bierwagen (2009) as well as Figueira et al. (2016) reviewed hybrid sol–gel anticorrosive coatings. There is potential for excellent barrier properties, but technical and economic barriers have prevented widespread adoption. A typical approach is to prepare separate sols from 3-glycidoxypropyltrimethoxysilane (a reactive silane; Section 16.2) and tetra-*n*-propoxyzirconium, combine the sols, and apply the blend by dipping or other means. Diamine cross-linkers can be added to react with the epoxy (glycidoxy) groups. Anticorrosive pigments or nano-pigments can also be added. In one example, Alvarez et al. (2016) found that anticorrosive performance was enhanced by adding 0.5% of a functionalized ZnO nano-pigment. This and other recent publications in this area make extensive use of analytical methods, such as electrochemical impedance spectroscopy (EIS) and scanning electron microscopy (SEM) with energy-dispersive X-ray (EDS) capabilities to examine the morphology and composition of sound and damaged areas of the coatings (see Section 6.7). These methods were used to demonstrate that ZnO reduces corrosion rates by forming a coprecipitate of Zn and other metal oxides that help protect damaged areas (Alvarez et al., 2016).

Extensive efforts are underway to find sol–gel technology that can replace chrome(VI) pigmented primers for aluminum. Substantial progress suggests that practical products may eventually result.

Sol–gel methods are also extended to other types of coatings. For example, a sol–gel, prepared from the condensate of ethylsilane and tetramethyltetravinylcyclotetrasiloxane, can be combined with TEOS to provide a sol–gel clear coat. Application over an automotive base coat provides a finish having better resistance to automatic car washing and sulfuric acid spot tests than the control without the sol–gel coating (Hofacher et al., 2002). Another example is the combination of sol–gel and waterborne polyurethane technologies (Section 12.7) (Albert, 2016). The opportunities for innovation are almost limitless.

GENERAL REFERENCES

Mittal, K.L., Ed., *Silanes and Other Coupling Agents*, Leiden, Boston, **2007**, Vol. *4*.
Plueddemann, E. P., *Silane Coupling Agents*, 2nd ed., Springer Science + Business Media, New York, **1991**; reprinted in 2013.
Ziegler, J. M.; Fearon, F. W. G., Eds., *Silicon-Based Polymer Science A Comprehensive Resource*, American Chemical Society, Washington, DC, **1990**.

REFERENCES

Albert, P., European patent Publ. EP2987836 A1 (**2016**).
Alvarez, D., et al., *Prog. Org. Coat.*, **2016**, *96*, 3–12.
Baghdachi, J., et al., *J. Coat. Technol.*, **2002**, *74(932)*, 81.
Brinker, C. J.; Scherer, G. W., *Sol-Gel Science*, Academic Press, Boston, **1990**.
Chen, M. J., et al., *J. Coat. Technol.*, **1997**, *69(870)*, 43.
Danks, A. E.; Hall, S. R.; Schnepp, Z., *Mater. Horiz.* **2016**, *3*, 91–112.
Echeverria, M., et al., *Prog. Org. Coat.*, **2016**, *92*, 29–43.
Figueira, R. B., et al., *Coatings*, **2016**, *6*, 12.
Fink, F., et al., *J. Coat. Technol.*, **1990**, *62(791)*, 47.
Finzel, W. A., *J. Coat. Technol.*, **1992**, *64(809)*, 47.
Finzel, W. A.; Vincent, H. L., *Silicones in Coatings*, Federation of Societies for Coatings Technology, Blue Bell, PA, **1996**.
Florio, J.; Ravichandran, R., *CoatingsTech* **2016**, February, 46–57.
Fransehn, P., et al., *Proceedings of the Waterborne High-Solids, and Powder Coatings Symposium*, New Orleans, LA, **2004**, Paper No. 7.
Furukawa, H., et al., *Prog. Org. Coat.*, **1994**, *24*, 81.
Greene, J. D., US patent 6,344,520 (**2002**).
Hofacher, S., et al., *Prog. Org. Coat.*, **2002**, *45*, 159.
Hsiao, Y.-C., et al., *J. Appl. Polym. Sci.*, **1975**, *19*, 2817.
Jiang, P., et al., *Proceedings of the American Coatings Conference*, Indianapolis, Indiana, April 11–13, **2016**; Paper 12.1.
Mangio, R., et al., *Proceedings of the Waterborne High-Solids and Powder Coatings Symposium*, New Orleans, LA, **2002**, pp 41–56; Paint and Coatings Industry, 2001, January, 50.
Markevicius, G., et al., *Prog. Org. Coat.*, **2012**, *75(4)*, 319–327.
Min, Q., et al., Canadian patent publication CN103468095 A (**2013**).
Mittal, K.L., Ed., *Silanes and Other Coupling Agents*, Leiden, Boston, **2007**, Vol. *4*.
Noh, S. M., et al., *Prog. Org. Coat.*, **2012**, *74*, 192–203.
Nordstrom, J. D., *Proceedings of the Waterborne, Higher-Solids, and Powder Coatings Symposium*, New Orleans, LA, **1995**, p 192.
Ogarev, V. A.; Selector, S. L., *Prog. Org. Coat.*, **1992**, *21*, 135.
Ooka, M.; Ozawa, H., *Prog. Org. Coat.*, **1993**, *23*, 325.
Osterholtz, F. D.; Pohl, E. R., *J. Adhes. Sci. Technol.*, **1992**, *6(1)*, 127–149.
Owen, M. J., Siloxane Surface Activity in Ziegler, J. M.; Fearon, F. W. G., Eds., *Silicon-Based Polymer Science A Comprehensive Resource*, American Chemical Society, Washington, DC, **1990**, pp 705–739.
Pappas, S.P.; Just, R.L., *J. Polymer Sci. [A]*, **1980**, *18*, 527–531.
Parashar, G., et al., *Prog. Org. Coat.*, **2001**, *42*, 1.
Ribeiro, T.; Baleizao, C.; Farinha, J. P. S., *Materials*, **2014**, *7*, 3881–3900.
Spiegelhauer, S., *Proceedings of the Waterborne High-Solids, and Powder Coatings Symposium*, New Orleans, LA, **2002**, pp 161–170.
Wang, D.; Bierwagen, G.P., *Prog. Org. Coat.*, **2009**, *64(4)*, 327–338.
Witucki, G. L., *J. Coat. Technol.*, **1993**, *65(822)*, 57.
Witucki, G. L., **2013**, from http://www.paintsquare.com/education/branding_images/gwwebinar.ppt (accessed April 21, 2017).
Yaneff, P. V., et al., *J. Coat. Technol.*, **2002**, *74(933)*, 135.
Yari, H.; Mohseni, M., *Prog. Org. Coat.*, **2015**, *87*, 129–137.
Yuan, X., et al., *J. Coat. Technol. Res.*, **2016**, *13(1)*, 123–132.

Chapter 17

Other Resins and Cross-Linkers

Chapters 8–16 describe the types of resins and cross-linkers used in the largest volumes in coatings. In this chapter other commercially important materials and selected promising developments are discussed.

17.1 HALOGENATED POLYMERS

Halogenated polymers have desirable properties such as low water vapor permeability. Some are used in top coats for corrosion protection. Others serve as binders for adhesion promoting primers for plastics. Certain fluorinated polymers offer exceptional weatherability and thermal stability. Coatings based on halogenated resins are applied as solutions, dispersions, latexes, or powder coatings.

17.1.1 Solution-Grade Thermoplastic Vinyl Chloride Copolymers

Vinyl chloride resins for coatings are copolymers of vinyl chloride with comonomers such as vinyl acetate or vinyl isobutyl ether. Comonomers are needed to reduce the T_g (about 80°C) and crystallinity of poly(vinyl chloride) (PVC) homopolymer. Additional comonomers may be added to impart properties such as adhesion and cross-linkability. The vinyl chloride/comonomer weight ratio can range from 90/10 to 60/40. The vinyl chloride imparts water resistance. Vinyl chloride copolymer resins are produced by suspension polymerization or emulsion polymerization. Applications of the resins include industrial coatings, heat-sealable coatings, and inks.

Vinyl chloride copolymers require stabilization to prevent thermal and photochemical degradation. As discussed in Section 5.3, the polymers undergo dehydrochlorination in an autocatalytic chain reaction. Stabilizing agents include organotin esters such as dibutyltin dilaurate; barium, cadmium, and strontium soaps; maleates; and oxirane (epoxy) compounds.

An example of a soluble vinyl chloride resin is a terpolymer of vinyl chloride, vinyl acetate, and maleic acid having a weight ratio of 86 : 13 : 1 (mole ratio 81 : 17 : 1) and an $\bar{M}_w$ of about 75 000. Such polymers have been used as interior coatings for beverage cans. The maleic acid promotes adhesion. Other solution vinyl resins are made with hydroxy functionality. Use of solution vinyl resins has declined because of the low solids required for thin film application, about 10–12 nonvolatile volume (NVV) solids.

17.1.2 Vinyl Chloride Dispersion Copolymers

Dispersion grade vinyl chloride copolymers with high MWs can be used in high solids coatings. They are prepared as particles with an average diameter of a few micrometers by suspension polymerization. In this process, a solution of the monomer/initiator mixture is added with vigorous stirring to a hot dilute solution of poly(vinyl alcohol) (PVA) in water. The monomers are dispersed into small droplets, inside of which the polymerization is initiated and completed, resulting in the formation of particles. Monomer-soluble initiators are used in suspension polymerization in contrast to the water-soluble initiators used in emulsion polymerization. The PVA stabilizes the suspension by minimizing coalescence of the particles while the temperature of the reaction is above the T_g. Since the PVA has many easily abstractable hydrogens, graft side chains grow on the VC copolymer chains. After polymerization is complete, the polymer particles are separated by filtration, washed, and dried. The molecular weight (MW) of

Organic Coatings: Science and Technology, Fourth Edition. Frank N. Jones, Mark E. Nichols, and Socrates Peter Pappas.

polymers made by suspension polymerization depends primarily on initiator structure and on the combination of reaction temperature and initiator concentration. Commonly, $\bar{M}_w$ is on the order of 100000.

In plastics, dispersion-grade vinyl chloride copolymers are used as *plastisols*. Plastisols are polymer particles dispersed in a plasticizer. Since the T_g of the copolymer is well above room temperature and is partially crystalline, it does not dissolve in the plasticizer at room temperature at an appreciable rate. When the plastisol is heated above the copolymer T_g and above the melting point of its crystalline regions, the copolymer dissolves in the plasticizer and the particles coalesce to a molten state. When cooled, the product is a plastic consisting of a homogeneous solution of copolymer and plasticizer. In coatings, the viscosity of plastisols is generally too high for application, but viscosity but can be reduced by addition of solvent. Solvents must be chosen that dissolve the plasticizer without significantly swelling the copolymer particles. Application viscosities can generally be achieved with 20% solvent or less. Such coatings are properly called *organosols* but frequently called plastisols even though there is volatile solvent present.

17.1.3 Chlorinated Rubber, Chlorinated Ethylene/Vinyl Acetate Copolymers, and Chlorinated Polyolefins

Chlorinated rubber is used in top coats for heavy-duty maintenance and marine paints because of the barrier properties resulting from its low water permeability. Many paints for masonry swimming pools are based on chlorinated rubber, which is also used in *tie coats* (primers) on polyolefin plastics. Chlorinated rubber is compatible with some alkyd resins and imparts fast drying qualities to alkyd-based traffic paints. Like PVC, chlorinated rubber dehydrochlorinates and requires stabilizers similar to those used with PVC. Some metal salts, especially of iron, tend to promote degradation of chlorinated rubber when applied over rusty steel (Morcillo et al., 1993).

Chlorinated rubber is prepared from synthetic polyisoprene, polybutadiene, or natural rubber. The natural rubber is milled to reduce its MW. A solution of milled rubber in CCl_4 is reacted with chlorine or some other chlorinating reactant. The reactions are complex: addition to the double bonds, substitution, and cyclization reactions all occur. To eliminate most of the double bonds from the rubber, the final product contains 65–68 wt% chlorine. Various MW grades are manufactured. Polymer strength increases with MW, but the solution viscosity also increases. The formulator makes an appropriate compromise between coating solids and film properties for a particular application.

Chlorinated ethylene/vinyl acetate copolymers have been developed, which can be used to replace chlorinated rubber in at least some applications. Since the initial resin does not contain double bonds, the chlorine content need not be as high as in chlorinated rubber; typically it is 52–58 wt%. Higher chlorine content improves barrier properties but reduces flexibility and impact resistance. Performance is considered to be comparable to chlorinated rubber with the advantage of better storage stability.

Chlorinated polyolefins (CPOs) are an important class of adhesion promoters in primers for plastics, such as polypropylene, and "thermoplastic polyolefin" (TPO) (Section 31.2.2). TPO is used in large volume by the auto industry in applications, such as bumpers. The predominant CPOs for this purpose are CPOs modified with maleic anhydride (MAnh).

17.1.4 Fluorinated Polymers

Fluorine-containing polymers find use in hundreds, perhaps thousands, of coating applications. They vary in structure and fluorine content. As a rule to which there are exceptions, the higher the fluorine content, the more exceptional the properties but the more expensive the coatings are to make and apply.

Polytetrafluoroethylene (PTFE) [$-(CF_2CF_2)_n-$] has the greatest exterior durability and heat resistance of any polymer used in organic coatings. However, PTFE is insoluble in all solvents, and its fusion temperature is so high that coating uses are limited to applications in which the substrate can withstand high temperatures (Munekata, 1988). For example, aqueous dispersions of PTFE are used for coating the interior of chemical processing equipment and cookware (Batzar, 1989). After application, the polymer particles are sintered at temperatures as high as 425°C. PTFE has such a low surface free energy that it is not wet by either water or oils, thereby providing a "non-stick" cooking surface; it also has excellent resistance to chemicals and moisture and a very low coefficient of friction.

Adhesion of PTFE is, however, problematic. Usually the metal surface to be coated must be roughened so that "dovetail joints" ("lock and key joints") can be formed when the coating and substrate are baked (Section 6.1). Even then it is difficult to obtain excellent long-lived adhesion. This problem has led to the development of other highly fluorinated polymers that are more tractable. Fluorinated ethylene propylene (FEP) and copolymers of TFE with perfluoropropyl vinyl ether (PFA) are examples. Coatings based on these and other highly fluorinated polymers are also classed as PTFE coatings. They find use in food and chemical processing, medical, military, industrial, electrical, and other applications as well as in nonstick pans. The global market for such coatings is growing and is estimated to reach $1B by 2019.

Perfluorocarboxylic acids (PFCAs), particularly perfluorooctanoic acid (PFOA), were formerly used in the production of PTFE but are being phased out. These acids do not biodegrade and are thought to present persistent environmental and probably human health hazards. Most nonstick pans are advertised as PFOA-free. Schlummer et al. (2015) detected PFCAs when PFOA-free pans were overheated, although the amounts appeared too small to present immediate health hazards. Environmental contamination by PFCAs remains a concern, as these acids are accumulating in fish and animals, especially in the arctic (Prevedouros et al., 2006; Wang et al., 2014).

Having a somewhat lower fusion temperature than PTFE, poly(vinylidene fluoride) (PVDF) [$-(CH_2CF_2)_n-$] can be used in a broader range applications. For example, PVDF is used in coil coatings as a plastisol-like dispersion in a solution of acrylic resin (Gaske, 1987; Munekata, 1988). The temperature required for fusion of the films is 230–250°C. The exterior durability is outstanding but the cost is high. Copolymers of vinylidene fluoride (VDF) are also finding use in powder coatings.

Hybrid VDF/acrylic copolymer latexes are made by polymerizing acrylic monomers in the presence of a PVDF latex. Such products can be film formers at ambient temperature. Hydroxy-functional acrylic monomers are included in order that the films can be cross-linked with aliphatic polyisocyanates, which can be used in waterborne coatings (Section 12.7.3). Excellent film properties and exterior durability are reported (Beaugendre et al., 2014).

Vinyl fluoride (VF) ($CH_2{=}CHF$) polymers and copolymers are reported to have excellent resistance to sunlight degradation, chemical attack, water absorption, and solvents. Additionally, they have a high solar energy transmittance rate (Ebnesajjad, 2013). While more highly fluorinated polymers may be superior in some of these respects, VF polymers and copolymers are very weatherable. Among their uses are paints for metal building panels and for construction of solar panels. VF polymers can be synthesized by bulk, suspension, and emulsion polymerization.

Halofluoroethylene ($CF_2{=}CFX$)/vinyl ether copolymers (usually X = Cl) have been used in steel building panels and to a limited extent in clear coats for automobiles (Munekata, 1988). Vinyl ethers and $CF_2{=}CFX$ form alternating polymers. Functional groups can be introduced by copolymerizing hydroxy- and/or carboxy-substituted vinyl ether comonomers. The T_g is controlled by the ratio of fluorinated monomer to vinyl ether monomers and by the chain length of the alkyl group on the vinyl ether. Copolymers with hydroxyl groups can be cross-linked with melamine–formaldehyde (MF) resins or polyisocyanates. Masuda et al. (2015) compared polyurethane coatings based on fluorinated polymers of this type with conventional polyurethanes and alkyds after 6 years on bridges in Vietnam and Okinawa. The fluorinated polyurethanes had superior gloss retention; chemical impedance measurements indicated strong and similar corrosion protection by the fluoropolymer and the polyurethane, perhaps owing to the epoxy primers used with both.

(Meth)acrylic monomers that have fluorinated side chains, for example, 2,2,2-trifluoroethyl methacrylate and perfluoroalkyl acrylates, are available. Polymers containing such monomers tend to have very low surface energies and have been studied for applications such as antifouling and antigraffiti coatings. Perfluoroalkyl acrylate esters have been copolymerized with hydroxyethyl methacrylate (HEMA) to form solvent-soluble resins. Munekata (1988) reported superior gloss retention of poly(perfluoroalkyl acrylates) relative to that of acrylic coatings of that era on exterior exposure. However, the monomers are expensive, and fluorinated alcohols in side chains can be oxidized to PFCAs, which raise environmental concerns as discussed earlier in this section.

17.2 CELLULOSE DERIVATIVES

Cellulose is a naturally occurring polymer of repeating anhydroglucose units. It is abundant because it is the main building block of plant structures. Although cellulose itself is insoluble in water and organic solvents, water-soluble derivatives can be made by reactions of the hydroxyl groups. Other derivatives are soluble in organic solvents.

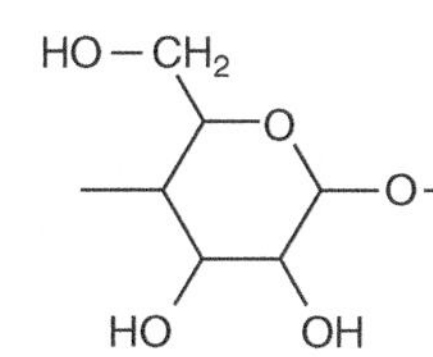

Anhydroglucose unit

Many solvent-soluble cellulose derivatives have been prepared. Those of most importance in coatings are nitrocellulose (NC) and cellulose acetobutyrate (CAB); ethyl cellulose, cellulose acetate, and cellulose propionate have also been used to some extent. Water-soluble cellulose derivatives are used to modify the flow properties of latex paints (Chapter 32).

17.2.1 Nitrocellulose

NC is another example of poor nomenclature since it is not a nitro compound but rather a nitrate ester. Three types of NC are available with differing extents of esterification.

Table 17.1 Composition of Nitrocellulose Types

Type	Nitrogen	$-ONO_2$ groups (%) per anhydroglucose unit
SS (spirit soluble)	10.7–11.2	1.9–2.0
RS (regular solubility)	11.8–12.3	2.15–2.25
Explosive	12.3–13.5	2.25–2.5

Their compositions are given in Table 17.1. One of the isomeric dinitroanhydroglucose units of NC is

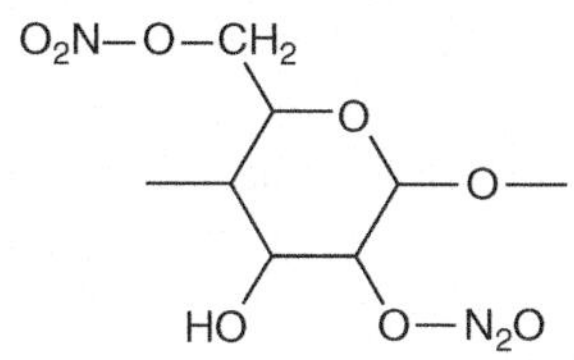

Dinitroanhydrocellulose

The SS type is alcohol soluble and is a binder for flexographic printing inks, but has few uses in coatings. The RS type is soluble in esters and ketones and is generally used in coatings. The explosive type is included in Table 17.1 to emphasize the hazards of NC. The RS type is not explosive but is highly flammable. To reduce the handling hazards, NC is shipped wet with a nonsolvent. In most cases the RS type is shipped at 70 NVW with the balance being ethyl or isopropyl alcohol. While RS-type NC is not soluble in alcohol, it is soluble in solvent mixtures containing ketones, esters, and alcohols (Section 18.2).

NC is made by reacting cellulose with nitric acid. Small amounts of water are present and some hydrolysis of the cellulose occurs, reducing MW. A range of MW grades is available. Higher MW enhances polymer strength but lowers the solids of coatings that can be made. Viscosity of NC solutions is determined using a falling ball viscometer in which the time for a ball to fall a specified distance through an NC solution of standard concentration is measured. The various MW grades are designated as ¼-, ½-, 1-s, and so forth. The times are proportional to the time in the falling ball viscosity test. Longer times indicate higher MW.

NC was used in lacquers for automobile top coats prior to development of alkyd enamels and acrylic lacquers (Chapter 30). Exterior durability of acrylic lacquers is far superior. NC-based refinish lacquers continued to be used for some years for repairing automobiles because they polished more easily and provided a deeper gloss appearance than room-temperature dried acrylic lacquers. NC lacquer primers continue to be used to some degree in refinishing automobiles.

The principal remaining use of NC lacquers in coatings is in wood finishing (Section 31.1). These lacquers have relatively low solids but continue to be used to a significant extent because they enhance the appearance of wood grain to a greater extent than any other coating. Other desirable characteristics for NC lacquers for wood finishing are rapid drying, which permits handling and shipment soon after finishing and easy damage repair, since the films are thermoplastic. Increasingly stringent VOC emission regulations are likely to further limit the consumption of NC.

Other applications of NC varnishes include inks, finger- and toe-nail varnishes, and leather coatings. The ink industry is probably the largest consumer of NC, where its fast drying and good adhesion are paramount.

Use of NC lacquers has declined in North America and Europe because of their high VOC, but in other parts of the world, NC use is still substantial. Similarly, automobile refinish NC lacquers continue to be used.

17.2.2 Cellulose Acetobutyrate

A variety of organic esters of cellulose are used in fibers, films, plastics, and, to a degree, coatings. The predominant cellulose ester in coatings is a mixed acetate, butyrate ester (CAB). CAB lacquers are used to some extent; they exhibit lighter color and better color retention together with reduced handling risk relative to NC lacquers. The principal application has been as a component of acrylic automotive coatings. CABs help to control flow of the wet paint and particularly to facilitate orientation of aluminum flakes parallel to the surface of the film during application and film formation. Orientation of the flakes in the final film enhances the appearance of metallic finishes (McBane, 1987). See Chapter 30.

Several types of CAB are available with differing MWs and differing contents of acetate, butyrate, and unesterified hydroxyl groups. The composition affects the solubility and compatibility of the polymer. Higher proportions of butyrate ester allow larger amounts of aromatic hydrocarbon in the solvent mixture. Also, T_g tends to decrease as the butyrate content increases. Compatibility with acrylic resins is dependent on both the CAB and the acrylic polymer. An example of a CAB grade used in acrylic lacquers has an average of 2.2 acetate groups, 0.6 butyrate groups, and 0.2 unreacted hydroxyl groups per anhydroglucose unit. This corresponds to about 30% acetyl content, 17% butyryl content, and 1.5% hydroxyl content, based on the total weight of polymer. Solubility in exempt solvents (Section 18.2) is a consideration in North America.

17.3 UNSATURATED POLYESTER RESINS

Unsaturated polyester resins usually include maleic ahnhydride (Manh) or fumaric acid in the polymerization formula as the source of unsaturation together with polyols and diacids. Glass-reinforced plastics, commonly called "fiberglass," are usually based on styrene solutions of such resins. The solutions are cross-linked by free radical polymerization to form the plastic. These resins are used on a smaller scale in some coating applications.

A variety of diacids and diols can form unsaturated polyesters; the most common are phthalic anhydride (PA), MAnh, and propylene glycol (or a mixture of propylene glycol and propylene oxide). These raw materials typically have the lowest cost; in addition, the initial esterification reaction can be carried out at moderate temperatures. During polymerization MAnh reacts with the diols to form maleic esters, which partially isomerize to give a mixture of maleate (*Z*-(*cis*) configuration) and fumarate (*E*-(*trans*) configuration) groups in the polymer backbone. The fraction that isomerizes depends on the reaction conditions and may also depend on the other components of the polyester. Reaction conditions during esterification must be carefully controlled to ensure that the extent of isomerization is reproducible from batch to batch. Fumarate groups are more reactive than maleates during cross-linking. Accordingly, use of fumaric acid instead of MAnh results in a higher fumarate diester content; such polyesters generally afford harder plastics than those made with MAnh. More flexible plastics can be obtained by substituting aliphatic diacid (e.g., adipic acid) for part of the PA, which reduces T_g. The properties of the cured materials are also affected by the ratio of polyester to styrene, typically 70 : 30.

Dai et al. (2015) demonstrated synthesis of fully bio-based unsaturated polyester resins and prepared waterborne coatings from them. Wojcieszak et al. (2015) reviewed synthesis of bio-based MAnh. However, petrochemical-based unsaturated polyester resins are, at present, lower in cost, so that adoption of bio-based counterparts is likely only when customer preferences outweigh cost considerations.

The resin/styrene solution is cross-linked using free radical initiators. The product is a complex mixture derived from copolymerization of styrene with the maleate or fumarate double bonds, graft copolymerization, and homopolymerization of styrene. Cross-link density is affected by the ratio of MA/PA in the resin, commonly 15–40 mol% of MAnh. Peroxide initiators, such as benzoyl peroxide, are used at temperatures of 70–100°C. For applications that require cross-linking at ambient temperatures, a more active initiator, such as methyl ethyl ketone (MEK) peroxide, and promoters, such as a mixture of dimethylaniline and cobalt naphthenate, are added to the resin/styrene solution just before application. Such combinations react rapidly at room temperature to generate free radicals, which initiate cross-linking. MEK peroxide is a complex mixture in which the major components contain both peroxide and hydroperoxide groups (Sheppard, 1988) as shown. The cobalt salt acts as a redox catalyst for the decomposition of the peroxide and hydroperoxide groups into free radicals; dimethylaniline further promotes that reaction.

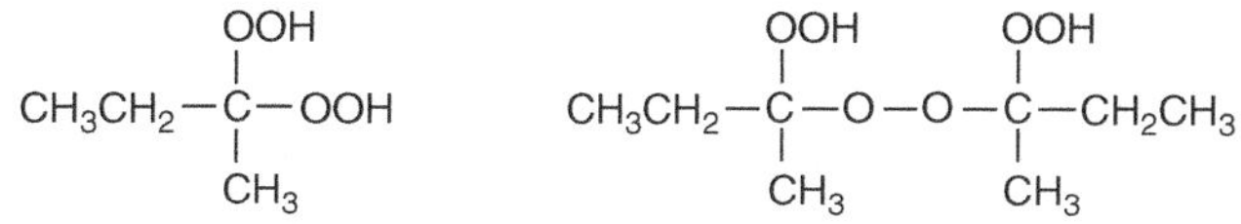

Major components of MEK hydroperoxide

Applications include *gel coats*, which are pigmented unsaturated polyester/styrene coatings (with initiator and promoter); they are sprayed on the inside of a mold surface (Section 31.2.1). The exposed (back) surface of the gel coat is then sprayed with glass fiber-loaded unsaturated polyester/styrene compound, covered with plastic film and cross-linked. After the part is taken from the mold, the front surface is the gel coat that cured in the absence of oxygen. Many glass-reinforced plastic objects ranging from prefabricated shower stalls to boat hulls are made this way. In the case of boat hulls, hydrolytic stability is particularly important in order that the surface retains gloss after outdoor exposure and immersion in water. Unsaturated polyesters made using neopentyl glycol, MA, and isophthalic acid provide better gloss retention than those made from propylene glycol and PA, but cost is higher.

The free radical curing reactions are oxygen inhibited. This is not a problem for many plastics applications since the reaction is commonly carried out in a mold, so the resin surface is not exposed to air. In most coating applications, the top surface is exposed to the air and will remain sticky after the polymerization is complete below the surface. A further problem is the volatility of styrene. These problems can be minimized by incorporating an insoluble semicrystalline paraffin wax in the formulation. After the coating is applied, the low surface tension wax particles preferentially come to the surface. The wax layer reduces the rate of styrene loss and also reduces oxygen concentration near the surface of the polymerizing coating, thereby minimizing the extent of surface stickiness. The wax layer, however, results in a relatively uneven low gloss surface, suitable for some applications but not for others.

Gel coats can be pigmented to achieve various colors and appearance effects. Fumed and/or precipitated silica are often added to control rheology (Toth and Romaine, 2009).

Oxygen inhibition can be minimized using coreactants having allyl ether groups in the preparation of the unsaturated polyesters (Traencknerm and Pohl, 1982). A free radical abstracts a hydrogen atom from the methylene group activated by both the adjacent double bond and the ether oxygen. The resulting free radicals from this reaction then react with oxygen to form peroxy free radicals, which in turn abstract hydrogen from the activated methylene groups to form hydroperoxides. The reactions consume some of the oxygen at the surface and generate new hydroperoxide groups in a chain reaction, thereby reducing air inhibition.

Use of allyl ether reactions has been extended to waterborne unsaturated polyester resins. A low MW ester is prepared by reacting 2 mol of MAnh with 1 mol of a mixture of low MW diols and polyalkylene glycols. The resulting partial ester is further esterified with 2 mol of trimethylolpropane diallyl ether (Dvorchakm and Riberi, 1992). Emulsification in water is promoted by the polyalkylene glycol ester segments. Films can be cured either with hydroperoxide/cobalt initiators or by photoinitiators and UV radiation.

Early UV cure coatings were based on styrene solutions of unsaturated polyesters, which have been largely replaced by more highly reactive acrylate-functional compositions—see Section 17.4. Unsaturated polyesters can also be cross-linked by Michael addition reactions—see Section 17.10.

17.4 (METH)ACRYLATED OLIGOMERS

Thermosetting resins can be made with (meth)acrylate double bonds as cross-linkable groups. (Meth)acrylated polyols, epoxies, polyurethanes, and siloxanes are available. Such resins are used in radiation-cured coatings (Section 29.2.5). They can also be used in ambient cure and at somewhat elevated temperatures (force-dry) with free radical initiators. Both acrylate and methacrylate polymerizations are air-inhibited, particularly methacrylate systems, which generally require higher cure temperatures to minimize the effect of oxygen inhibition.

Acrylated oligomers can also be cross-linked with polyfunctional primary amines by a Michael addition reaction as described in Section 17.10. Furthermore, they can be cross-linked by a Michael reaction with acetoacetylated resins and their enamine derivatives as discussed in Sections 17.6 and 17.10.

17.5 2-HYDROXYALKYLAMIDE CROSS-LINKERS

Esterification of most alcohols with carboxylic acids is too slow for practical use as a cross-linking reaction. However, 2-hydroxyalkylamides undergo esterification reactions more rapidly than simple alcohols. Polyfunctional 2-hydroxyalkylamides (e.g., the tetrafunctional hydroxyalkylamide derived from aminolysis of dimethyl adipate with diethanolamine (structure in the following text) can serve as cross-linkers for carboxylic acid-functional acrylic or polyester resins (Lomax and Swift, 1978). Similar cross-linkers are made from diisopropanylamine.

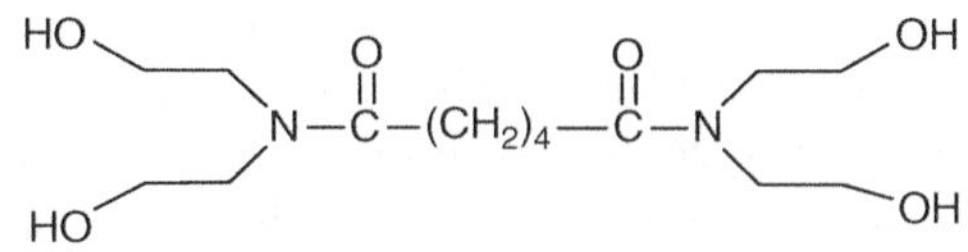

Tetra-*N*,*N*,*N*′,*N*′-(2-hydroxyethyl)adipamide

The properties of coatings obtained by cross-linking carboxylic acid-functional acrylic resins with 2-hydroxyalkylamides compare favorably with those obtained using MF resins as cross-linkers with the same resins. An advantage relative to MF cross-linkers is the absence of formaldehyde, which is emitted in low concentrations when MF-based coatings are baked. A disadvantage is that higher baking temperatures are required. Such cross-linkers are well suited for powder coatings (Chapter 28).

2-Hydroxyalkylamides are soluble in both water and common coating solvents and, hence, are useful as cross-linkers in either waterborne or solventborne coatings. Tetra-*N*,*N*,*N*′,*N*′-(2-hydroxyethyl)adipamide is a solid well suited for use in powder coatings (Mercurio, 1990). See Section 28.1.3. However, discoloration can occur on baking (Stanssens et al., 1993), presumably due to the formation of some free amine groups by thermal rearrangement of the hydroxylalkylamides. Proprietary hydroxyalkylamide cross-linkers with less tendency to discolor are available.

The esterification reaction of 2-hydroxyalkylamides exhibits unique features compared with general esterification reactions. 2-Hydroxyalkylamides having secondary hydroxyls esterify more rapidly than the corresponding primary alcohol derivative; the reaction is not catalyzed by acid; and aromatic carboxylic acids esterify faster than aliphatic carboxylic acids. All of these characteristics are the opposite of the behavior of other alcohols. In addition to the faster reaction of 2-hydroxyalkylamides with aromatic acids, the saponification resistance of the products is greater than of the corresponding esters from aliphatic acids. Mechanistic studies of the 2-hydroxyalkylamide/carboxylic acid esterification reaction, as well as related reactions, have been reported to support the intermediate formation of highly reactive oxazolinium carboxylates (Wicks et al., 1985). Direct evidence for formation of oxazolinium intermediates has also been obtained (Stanssens et al., 1993).

2-Hydroxyalkylamides also react with epoxies. At elevated temperatures, the amide rearranges to the aminoester and the resulting amine group then reacts with the epoxy (Jones and Lin, 1984).

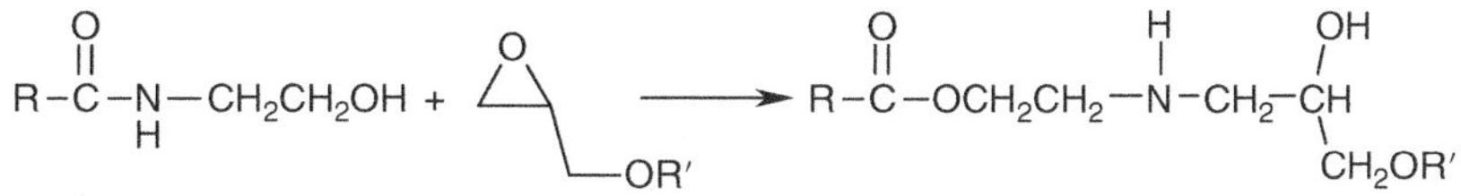

17.6 ACETOACETATE CROSS-LINKING SYSTEMS

The chemistry of β-keto esters, such as acetoacetic esters, has been widely studied. Acetoacetic esters are weak acids and exist to a significant degree in the tautomeric enol form.

They undergo a wide range of reactions, some of which can be adapted for coatings purposes. Several methods are available to prepare resins bearing acetoacetate groups. Acetoacetoxy-functional acrylic resins can be made by copolymerizing acetoacetoxyethyl methacrylate (AAEM) with other acrylate monomers (Rector et al., 1989; Witzeman et al., 1990; Li and Graham, 1993). One can also use AAEM to make acrylic latexes (Geurnik et al., 1996), although the long-term stability of acetoacetic ester polymers in aqueous media has been questioned.

There are alternative synthetic routes to acetoacetoxy-functional resins. Hydroxy-functional resins can be reacted with diketene or transesterified with methyl acetoacetate to form acetoacetylated resins. Replacement of the hydroxyl group with the less polar acetoacetate group leads to a reduction in viscosity, permitting higher solids of solvent-borne coatings.

A variety of coreactants with acetoacetylated resins have been investigated. MF resins react with acetoacetate groups in the presence of an acid catalyst, although somewhat less rapidly than with hydroxyl groups (Rector et al., 1989). Film properties of the resulting coatings using acetoacetylated resins are reported to be comparable with those made from hydroxy-functional resins. There are indications of improved wet adhesion, perhaps resulting from chelating interactions with metal surfaces.

Isocyanates also react with acetoacetate groups; the cure rate is slower than with hydroxyl groups; as a result, pot life of 2K coatings is longer. Film properties are reported to be similar to those of conventional polyurethanes. In other examples of 2K coatings, based on MDI, the solids at equal viscosity were about 10% higher with an acetoacetylated resin compared with the original hydroxy resin and excellent film properties were obtained (Narayan and Raju, 2002).

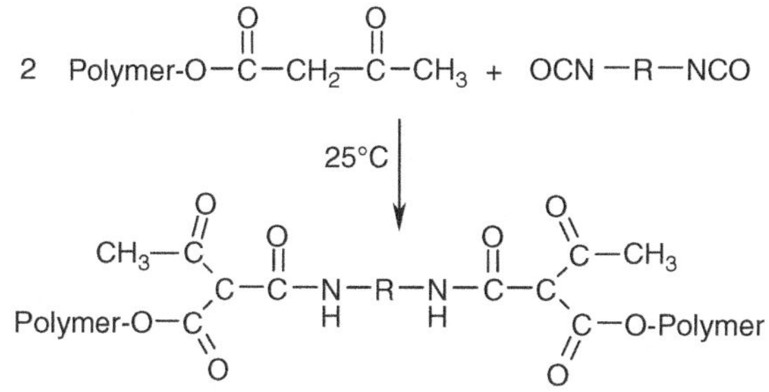

Amines react rapidly with acetoacetic esters. Thus, polyamines have been reported as cross-linkers for acetoacetate-functional latexes (Geurnik et al., 1996). Amines have also been reported to be excellent cross-linkers for acetoacetate-functional water-reducible acrylic resins. Although reactions are too rapid under many conditions to have sufficient pot life as two-package (2K) coatings, it has been found that, if the resin is neutralized with ammonium hydroxide and the solids of the vehicle are restricted to 15–20%, stability is improved. For example, a copolymer of AAEM, MAA, BA, and MMA, neutralized with an equal equivalent of NH_4OH at 18% solids and blended with 1,6-hexamethylenediamine, is storage stable for a year. It was proposed that NH_4OH reacts with a portion of the acetoacetoxy groups to form the ketimine–enamine equilibrium, thereby acting as a blocking group. After application, ammonia evaporates and the resin is cross-linked by the hexamethylenediamine (Esser et al., 1999).

In contrast to the situation with latexes and water-reducible resins, the acetoacetate–amine reaction is so fast in solution at ambient temperatures that pot life is very limited. This problem can be minimized by blocking the amine groups with ketones (Section 13.2.1). The resultant ketimine hydrolyzes in the presence of water, permitting the cross-linking reaction to proceed (Clemens and Rector, 1989). The cross-links are tautomeric ketimine/eneamine groups that are thought to interact strongly with metal surfaces (Zabel et al., 1998). A ketimine/acetoacetate cross-linking primer is reported to provide excellent adhesion and corrosion resistance when applied to an aircraft grade aluminum alloy with a chromate-free pretreatment.

Ketimine Enamine

Ketimine/eneamine tautomers can also serve as Michael reactant cross-linkers with polyacrylates(Schubert, 1993).

Cross-linking of acetoacetates by Michaels addition reactions (Section 17.10) and by dihydrazides (Section 17.11) are growing in importance.

17.7 POLYAZIRIDINE CROSS-LINKERS

Aziridine, the nitrogen three-membered ring counterpart of oxirane, and its derivatives have been studied for many years; in some cases trifunctional aziridines, often called *polyaziridines*, have been used as cross-linkers. The common name of aziridine is ethyleneimine. Ethyleneimine is highly toxic and may be carcinogenic. Propyleneimine is somewhat less toxic. Polyaziridines are skin irritants and some individuals may become sensitized. Mutagenicity of polyaziridines is controversial; however, dilution by coating vehicles reduces possible toxic effects (Pollano, 1997). Ethyleneimine is even more reactive with acids than is ethylene oxide. In the presence of relatively strong acids, ethyleneimine polymerizes very rapidly to yield poly(ethyleneimine), $(-CH_2CH_2NH-)_n$.

Among the many reactions of aziridines, one of the greatest interests in coatings applications is the reaction of a multifunctional aziridine with a polyfunctional carboxylic acid to form 2-aminoester cross-links. Some 2-aminoesters can spontaneously rearrange to the corresponding 2-hydroxyamides, but this reaction does not break the cross-link.

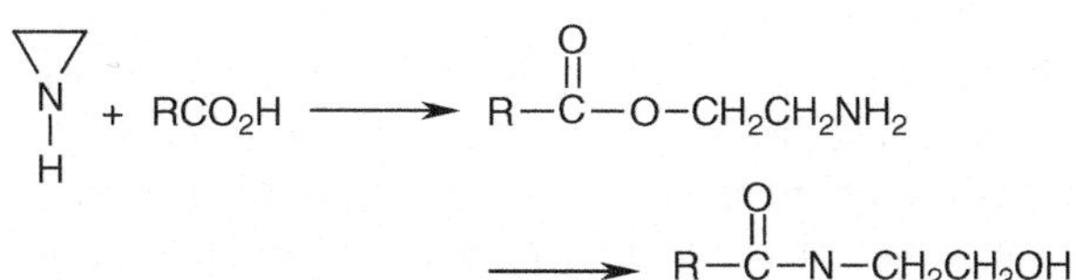

An example of a trifunctional aziridine cross-linker is the Michael addition product of 3 mol of aziridine to 1 mol of trimethylolpropane triacrylate.

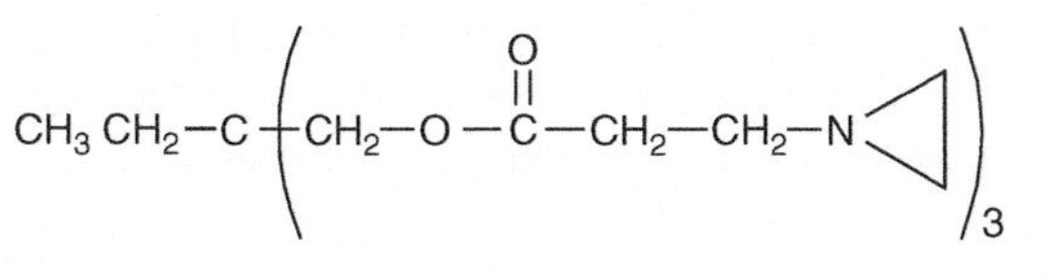

Roesler and Danielmeier (2004) reviewed *tris*-3-(1-aziridino)propionates and their uses. Trimethylolpropane *tris*-[3-(2-methyl-1-aziridino)propionate], trimethylolpropane *tris*-[3-(1-aziridino)propionate], and pentaerythritol *tris*-[3-(1-aziridino)propionate] were evaluated. Note that the 2-methyl-1-aziridino analog is based on propyleneimine, rather than ethyleneimine. Some commercial products are propyleneimine derivatives, perhaps because of health concerns.

Carboxy groups react by opening the aziridine ring, giving a secondary amine as a cross-link. Subsequent rearrangement of the structure is possible. The reactions are acid catalyzed. The main use of such multifunctional aziridines is to cross-link with carboxylic acid groups on latexes and waterborne polyurethanes. They are particularly useful for cross-linking COOH-functional latexes in coatings for factory-coated wood cabinetry and flooring, where a modest degree of cross-linking improves film properties over those attainable with un-cross-linked latexes. These are 2K coatings—the aziridine is added shortly before application. Aziridine groups react much faster with carboxylic acids than with water, but the reaction rate with water is such that that pot lives are 48–72 h (Pollano, 1997).

Multifunctional aziridines can also cross-link polyurethane dispersions (PUDs). Solvent swelling tests on cured films show that cross-linking of PUDs is more complete with aziridines than with carbodiimides (Section 17.8) or MF resins. (See Section 12.7.1 for further discussion.)

In view of the potential toxic hazards, manufacturer's recommendations for safe handling should be carefully followed.

17.8 POLYCARBODIIMIDE CROSS-LINKERS

Carbodiimides react with carboxylic acids to form *N*-acylureas significantly faster than their reaction with water, which allows them to be used as cross-linkers in waterborne coatings, including latexes. The product of the reaction with a carboxylic acid is an *N*-acylurea.

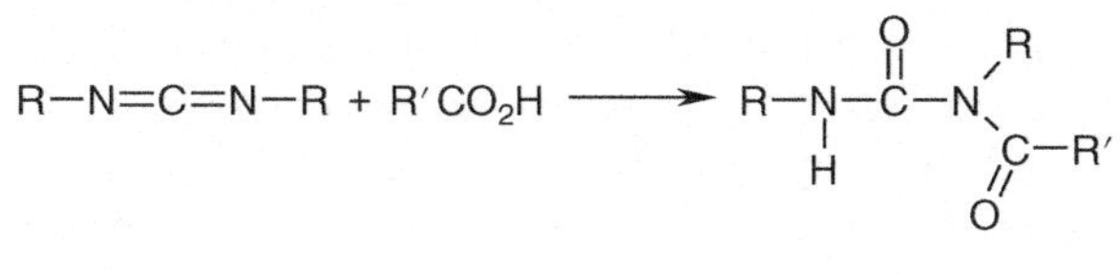

Multifunctional carbodiimides, which can be made from corresponding isocyanates, are available to cross-link carboxylic acid-functional resins including latexes, PUDs, and blends of the two (Taylor and Basset, 1997). Cross-linking occurs within several days at ambient temperature and much faster with heat. In one study, with latexes, curing conditions ranged from 60°C to 127°C for 5–30 min; better film properties are obtained at higher temperatures (Taylor and Basset, 1997), suggesting that film properties depend not only on the extent of chemical cross-linking but also on temperature-induced physical film forming effects. New carbodiimide cross-linkers periodically appear on the market.

17.9 POLYCARBONATES

Polycarbonates are a subclass of polyesters (Chapter 10) in that the diacid monomer is, nominally, H_2CO_3. As with polyesters, the polycarbonates used in coatings are quite different from the familiar thermoplastic polycarbonate plastics. Polycarbonates for coatings are generally low in MW and are hydroxy-functional for subsequent chain extension or cross-linking.

Webster (2003) reviewed the preparation and reactions of polycarbonate resins. Polycarbonate polyol resins can be produced by ester interchange reactions of diols or polyols with carbonates, such as dimethyl carbonate (Kainz et al., 2014), by ring-opening polymerization of cyclic carbonates (Hult et al., 2005), or by reaction of epoxy resins with CO_2 (Wang et al., 2014; Willkomen et al., 2014). Pressure reactors are often required.

As with other resins, a range of properties can be achieved by varying structure, branching, functionality, and MW. Polycarbonate polyols can be cross-linked with any of the common cross-linkers for hydroxy-functional resins. For example, polyurethanes made from polycarbonate polyols (sometimes blended with polyether polyols) and multifunctional isocyanates in a 2-pack reaction were reported to have excellent wear resistance, mechanical properties, and hydrolytic stability (Anon., 2014). When polycarbonate diols made with blends of 1,4-cyclohexanedimethanol (CHDM) and butyl ethyl propane diol (BEPD) were cross-linked with aliphatic polyisocyanates (Chapter 12) or with HMMM (Chapter 11), coatings with excellent mechanical properties resulted and their resistance to hydrolysis was superior to that of conventional polyesters (Zhou and Johnson, 2012).

Polycarbonate diols can also be chain extended with diisocyanates to make thermoplastic resins. An example is a resin formed by reaction of a polycarbonate and butane diol with an aliphatic diisocyanate. Coatings made from such resins can have strong mechanical properties and the excellent abrasion resistance required of linings for the insides of pipes (Jofre-Reche et al., 2016).

H. Zhou et al. (2014) report that waterborne PUDs made from polycarbonate resins and aliphatic isocyanates have attractive properties in applications requiring moisture resistance, flexibility, and weatherability. Performance was enhanced by addition of 3-aminopropyltriethoxysilane.

L. Zhou et al. (2000) recommended polytrimethylene carbonate polyols as reactive diluents for acrylic polyols in coatings cross-linked with HDI isocyanurate (Section 12.3.2). Best results were obtained with relatively low MW triols made by reacting trimethylene carbonate with TMP. VOC was reduced, film flexibility was increased, and hardness, weather resistance, and mechanical properties were maintained.

17.10 NON-ISOCYANATE TWO-PACKAGE BINDERS

As discussed in Chapter 12, 2K polyurethanes provide excellent film properties even when cured at ambient temperature. There has always been concern about the hazards to workers applying these coatings because of possible exposure to unreacted isocyanates. While such hazards can be mitigated with proper procedures, the concern seems to be growing, and regulations that would severely restrict 2K polyurethanes have been proposed in California and elsewhere. This situation has motivated research on 2K binder chemistries that contain no unreacted isocyanates. Two examples of commercially offered 2K systems are described here.

17.10.1 Carbamate–Aldehyde Chemistry

As described in Section 11.3.4, carbamates (ROC(O)NHR′) can be cross-linked with amino resins. Primary carbamates (R′=H) also react with aldehydes to form imines. With acid catalysts, this reaction can proceed rapidly at room temperature. Thus, resins with multiple primary carbamate groups can be cross-linked by multifunctional aldehydes, for example, glutaraldehyde, at temperatures as low as 0°C (Argyyropoulos et al., 2016). When the polycarbamates are made from polyisocyanates, coatings with urethane-like properties including good weatherability can be formulated. Worker exposure to isocyanates is eliminated except in the resin factory, where exposure can be reliably contained. The formaldehyde by-product of cross-linking carbamates with amino resins is also eliminated.

Argyropoulos et al. (2016) describe implementation of technology using alkyd resin carbamates cross-linked with cycloaliphatic dialdehydes to make topcoats and primers.

17.10.2 Michael Addition Chemistry

(Meth)acrylated oligomers and polymers (Section 17.4) can be cross-linked with polyfunctional primary amines by a Michael addition reaction, as illustrated in the following text for monofunctional reactants. This reaction is very fast, so early practical systems are based on blocked amines, commonly ketimines (Noomen, 1989). As described in Section 13.2.1, when a coating film is exposed to atmospheric moisture, the ketimine reacts with water to release the free primary amine.

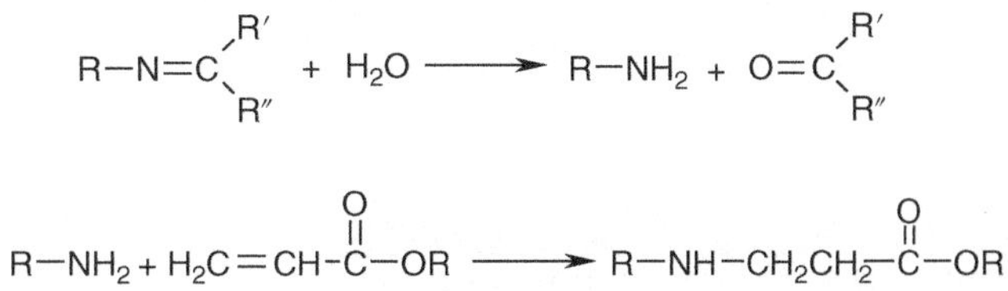

(Meth)acrylated oligomers (Section 17.4), acrylated acrylic polymers, and unsaturated polyester resins (Section 17.3) can also be cross-linked by a Michael reaction with acetoacetylated resins and their enamine derivatives (Section 17.6). Acrylated acrylics can be made by reaction of acrylic acid with glycidyl methacrylate copolymers. The reaction is illustrated in the following equation for a diacrylate.

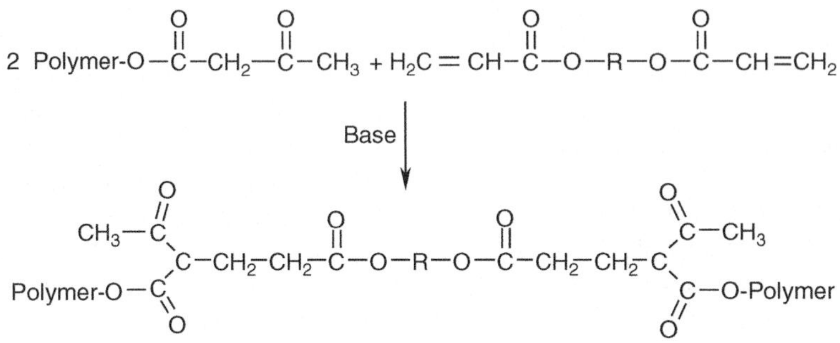

The catalysts are often strong organic bases such as benzotriazines or tetramethylguanidine (Geurnik et al., 1996). With such catalysts, these reactions are generally too fast for practical coatings as pot life is too short. Pot life can be extended by utilizing amine salts of volatile acids, such as the formic acid salt of 1,8-diazabicyclo[5.4.0]undec-7-ene (Li and Graham, 1993). Noomen et al. (1991) and Noomen (1997) applied this chemistry to unsaturated polyester resins (Section 17.3) and acrylated acrylic polymers. They also demonstrated various ways to extend pot life.

Brinkhuis et al. (2013, 2014) further refined this chemistry using combinations of basic catalysts and acid inhibitors that permit the final user to manipulate the pot life and cure rate to suit one's requirements. A feature of this technology includes carbonate salts in the formulation that form CO_2, which evaporates quickly after application.

17.11 DIHYDRAZIDES

The structure of hydrazides is analogous to that of amides except that they are made from hydrazine ($H_2N{-}NH_2$) rather than amines. Hydrazides react at elevated temperatures with epoxy groups and at ambient temperatures with isocyanates, ketones, and aldehydes.

Dihydrazides are gaining importance as cross-linkers, and several are commercially available. The most common is adipic dihydrazide (ADH):

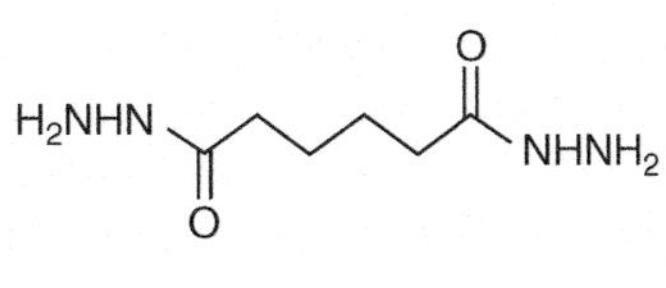

ADH has been used for decades to cross-link epoxy powder coatings (Chapter 28), especially for the steel pipelines.

ADH now finds applications as a cross-linker in latex coatings. A small amount of a ketone or aldehyde monomer, such as diacetone acrylamide, is copolymerized into the latex. ADH, which is readily soluble in water, is added to the coating formula. Thus, in the liquid coating the ketone and the hydrazide are physically separated, especially if a core-shell latex is employed (Zhang et al., 2012). Furthermore, the latex coating contains a volatile base (e.g., ammonia) that impedes the acid-catalyzed reaction. During film formation, water and the volatile base evaporate, ADH diffuses into the polymer, and cross-linking occurs at ambient temperature. The cross-links are imides formed by reaction of the ketone and the hydrazide groups. Kessel et al. (2008) illuminated the chemistry of cross-linking and showed that the rates of evaporation, deformation, coalescence (Section 2.3.3), and cross-linker diffusion, as well as the cross-linking chemical reaction, may all affect the structure and, hence, the properties of the coating film.

Hydrazides react readily with formaldehyde and are employed as formaldehyde scavengers.

REFERENCES

Anon., *Coatings World*, December 25, **2014**.

Argyyropoulos, J. N., et al., US patent publication US20160289386 A1 (**2016**).

Batzar, K., *Proceedings of the 15th International Conference in Organic Coatings Science and Technology*, Athens, Greece, **1989**, p 13.

Beaugendre, A.; Skilton, R. W.; Wood, K., *CoatingsTech*, **2014**, July, 34–39.

Brinkhuis, R. H. G., et al., Crosslinkable composition crosslinkable with a latent base catalyst, EP2556107 A1 (**2013**).

Brinkhuis, R. H. G., et al., Composition crosslinkable by a real Michael addition (rma) reaction, WOA1 (**2014**).

Clemens, R. J.; Rector, F. D., *J. Coat. Technol.*, **1989**, *61(770)*, 83.

Dai, J., et al., *Prog. Org. Coat.* **2015**, *78*, 49–54.

Dvorchakm M. J.; Riberi, B. H., *J. Coat. Technol.*, **1992**, *64(808)*, 43.

Ebnesajjad, S., *Polyvinyl Fluoride: Technology and Applications of PVF*, Elsevier, Amsterdam, **2013**, p 6.

Esser, R. J., et al., *Prog. Org. Coat.*, **1999**, *36*, 45.

Gaske, J. E., *Coil Coatings*, Federation of Societies for Coatings Technology, Blue Bell, PA, **1987**.

Geurnik, P. J. A., et al., *Prog. Org. Coat.*, **1996**, *27*, 73.

Hult, A., et al., *Prog. Org. Coat.*, **2005**, *54*, 269.

Jofre-Reche, J. A., et al., *Proceedings of the American Coatings Conference*, Indianapolis, IN, **2016**; paper 10.4.

Jones, F. N.; Lin, I.-C., *J. Appl. Polym. Sci.*, **1984**, *29*, 3213.

Kainz, B., et al., US patent application US2A1 (**2014**).

Kessel, N.; Illsley, D. R.; Keddie, J. L., *J. Coat. Technol. Res.*, **2008**, *5*, 285–297.

Li, T.; Graham, J. C., *J. Coat. Technol.*, **1993**, *65(821)*, 864.

Lomax, J.; Swift, G. F., *J. Coat. Technol.*, **1978**, *50(643)*, 49.

Masuda, K., et al., *Proceedings of the CORCON Conference*, **2015**, Chennai, Paper CL-01.

McBane, B. N., *Automotive Coatings*, Federation of Societies for Coatings Technology, Blue Bell, PA, **1987**.

Mercurio, A., *Proceedings of 16th International Conference in Organic Coating Science and Technology*, Athens, Greece, **1990**, p 235.

Morcillo, M., et al., *Prog. Org. Coat.*, **1993**, *21*, 315.

Munekata, S., *Prog. Org. Coat.*, **1988**, *16*, 113.

Narayan, R.; Raju, K. V. S. N., *Prog. Org. Coat.*, **2002**, *45*, 59.

Noomen, A., *Prog. Org. Coat.*, **1989**, *17*, 27.

Noomen, A., *Prog. Org. Coat.*, **1997**, *32*, 137.

Noomen, A., et al., European patent EP0448154 A1 (**1991**).

Pollano, G., *Polym. Mater. Sci. Eng.*, **1997**, *77*, 383.

Prevedouros, K., et al., *Environ. Sci. Technol.*, **2006**, *40(1)*, 32–44.

Rector, F. D., et al., *J. Coat. Technol.*, **1989**, *61(771)*, 31.

Roesler, R. R.; Danielmeier, K., *Prog. Org. Coat.*, **2004**, *50*, 1.

Schlummer, M., et al., *Chemosphere*, **2015**, *129*, 46–53.

Schubert, W., *Prog. Org. Coat.*, **1993**, *22*, 357.

Sheppard, C. S., Peroxy Compounds in Mark, H.; Bikales, N., Eds., *Encyclopedia of Polymer Science and Engineering*, 2nd ed., John Wiley & Sons, Inc., New York, **1988**, Vol. *11*, pp 1–21.

Stanssens, D., et al., *Prog. Org. Coat.*, **1993**, *22*, 379.

Taylor, J. W.; Basset, D. R., The Application of Carbodiimide Chemistry to Coatings in Glass, J. E., Ed., *Technology for Waterborne Coatings*, American Chemical Society, Washington, DC, **1997**, p 137.

Toth, J.; Romaine, M., Canadian patent publication CN101348686 A (**2009**).

Traencknerm, H.-J.; Pohl, H. U., *Angew. Makromol. Chem.*, **1982**, *108*, 61.

Wang, Z., et al., *Environ. Int.*, **2014**, *70*, 62–75.

Webster, D. C., *Prog. Org. Coat.*, **2003**, *47*, 77.

Wicks, Z. W., Jr., et al., *J. Coat. Technol.*, **1985**, *57(726)*, 51.

Willkomen, W., et al., *Proceedings of the American Coatings Conference*, Atlanta, GA, **2014**, paper 6.2.

Witzeman, J. S., et al., *J. Coat. Technol.*, **1990**, *62*(*789*), 101.

Wojcieszak, R., et al., *Sustain. Chem. Process.*, **2015**, *3*, 9.

Zabel, K. H., et al., *Prog. Org. Coat.*, **1998**, *34*, 236.

Zhang, X., et al., *J. Appl. Polym. Sci.*, **2012**, *123*(*3*), 1822–1832.

Zhou, J.; Johnson, A., US patent application publication 2012/0041143 (**2012**).

Zhou, L., et al., *Proceedings of the Waterborne High-Solids and Powder Coatings Symposium*, New Orleans, LA, **2000**, pp 262–281.

Zhou, H., et al., *Prog. Org. Coat.*, **2014**, *77*, 1073–1078.

Chapter 18

Solvents

Most liquid coatings contain volatile material that evaporates during application and film formation. The volatile components reduce viscosity for application and control viscosity changes during application and film formation. In some cases, the volatile material must be a solvent for the resins in a formulation; in other cases, nonsolvent volatile components are desired. Often, any volatile organic material is called a solvent whether it dissolves the resin or not.

IHS (2013) reports that worldwide consumption of solvents in 2012 was 28 million metric tons of which 40% was in coatings. Well over half of solvent consumption was in Asia.

Too often, formulators do not consider the effects that solvent choice can have on coating performance. The choice of volatile components affects popping, sagging, and leveling and can affect adhesion, corrosion protection, and exterior durability.

Solvent usage is limited in North America, in Europe, and in parts of Asia by various regulatory schemes. In the United States, most solvents are classified by the US Environmental Protection Agency (EPA) as photochemically reactive *volatile organic compounds* (VOCs). Since the 1960s their use has been regulated to reduce air pollution. Because some solvents are less harmful to the atmosphere than others, a growing trend is to classify and regulate solvents according to their *maximum incremental reactivity* (MIR) and *photochemical ozone creation potential* (POCP). In 1990, the US Congress listed certain common solvents as *hazardous air pollutants* (HAPs), further restricting their usage. VOCs, MIRs, POCP, HAPs and their regulation are discussed in Sections 18.7, 18.8, and 18.9.

18.1 SOLVENT COMPOSITION

Various organic compounds and mixtures are used as solvents. They can be classified in three broad categories: weak hydrogen bonding, hydrogen-bond acceptor, and hydrogen-bond donor–acceptor. Solvent suppliers publish tables listing key properties of their solvents; they are readily available on the Internet.

Weak hydrogen-bonding solvents are petroleum-distillate mixtures of aliphatic and aromatic hydrocarbons, called *naphthas*. They vary in volatility and in aromatic content. Commercial *aliphatic naphthas* are mixtures of straight chain, branched chain, and alicyclic hydrocarbons with fairly low aromatic hydrocarbon content. The more volatile fractions are called varnish makers and painters (VM&P) naphthas. Less volatile fractions are called *mineral spirits*; special grades with reduced aromatic content and, hence, less odor are available. An advantage of aliphatic naphthas is low cost, especially on a volume basis since their densities as well as their price per unit weight are low.

Aromatic hydrocarbon solvents are more expensive than aliphatics, but they dissolve a broader range of resins. Toluene, xylene, and naphthas having high aromatic content have been used on a large scale. Commercial xylene is a mixture of isomeric xylenes (dimethylbenzenes) and ethyl benzene. Higher boiling mixtures rich in aromatic hydrocarbons (*high flash aromatic naphthas*) are predominantly mixed alkyl and dialkyl benzenes with three to five carbons in the substituent groups. Use of benzene itself is prohibited because it is carcinogenic. Toluene and xylene have been used on a large scale; however, they are listed in the United States as HAPs (Section 18.8).

Organic Coatings: Science and Technology, Fourth Edition. Frank N. Jones, Mark E. Nichols, and Socrates Peter Pappas.

Chlorinated solvents, in general, are no longer used in coatings primarily because most of them lead to ozone depletion in the stratosphere. They are also potent greenhouse gasses. An exception is *p*-chlorobenzotrifluoride (PCBTF), a VOC-exempt solvent discussed in Section 18.9.

Oxygenated solvents, such as esters, ester ethers, and ketones, are hydrogen-bond acceptors. Ketones (e.g., methyl ethyl ketone (MEK) 2-butanone and methyl isobutyl ketone (MIBK)) are generally less expensive than esters with similar evaporation rates. The cost differential is particularly marked when the solvents are purchased by weight, since ketones have lower densities. Important considerations in solvent selection are whether they are on the list of HAPs or are exempt from VOC regulations (Section 18.9.2). MEK and MIBK, formerly listed as HAPs in the United States, were removed from the HAP list in 2015. Concern about MEK persists, and additional regulation is being considered.

Esters and ester ethers such as 2-butoxy ethyl acetate are often preferred over slower evaporating ketones, like isophorone and methyl *n*-amyl ketone, because of their lower odors. Several esters (e.g., *t*-butyl acetate (TBAC) and propylene carbonate (PC)) are not HAPs and are exempt as VOCs. Esters should not be used as solvents for resins bearing primary or secondary amine groups owing to possible aminolysis of ester groups. Sterically hindered TBAC is an exception (Cooper et al., 2001).

Oxygenated solvents with active hydrogen groups, such as alcohols and alcohol ethers, are strong hydrogen-bond donor–acceptors. The most widely used volatile alcohols are methyl, ethyl, isopropyl, *n*-butyl, *sec*-butyl, and isobutyl alcohols. Many latex paints contain a slow evaporating, water-soluble solvent, such as propylene glycol, that dissolves mostly in the water phase and serves as antifreeze (Section 32.1). In water-reducible acrylic and polyester resin coatings, ether alcohols with low volatility, such as 1-propoxypropan-2-ol and 2-butoxyethanol, are used. The latter solvent was delisted as an HAP in 2004.

Nitroparaffins, such as 2-nitropropane, are highly polar hydrogen-bond accepting solvents. Their high polarity results in increased electrical conductivity, which is useful for adjusting solvent combinations in electrostatic spraying (Section 23.2.3).

For sustainability, solvents that are (or could be) made from renewable resources are of increasing interest. These include methanol, ethanol, ethyl lactate, glycerol, and 2-methyl tetrahydrofuran.

In addition to their effects on the atmosphere, solvents have "cradle-to-grave" impacts throughout their production and use. Solvents can potentially impact the environment, human health and safety, and aquatic life. Attempts to account for their total impact are underway in other industries. The American Chemical Society Green Chemistry Institute evaluated about 25 characteristics of solvents used in pharmaceutical processes (Alfonsi et al., 2008; Constable, 2013), balancing their environmental impacts against their performance. Preferred solvents for the pharmaceutical industry include 1- and 2-propanol, 1- and *t*-butanol, ethyl acetate, and MEK. MEK is, however, restricted in California. Similar analysis for the solvents in formulated products is underway and can be expected to influence solvent selection in the future.

18.2 SOLUBILITY

In the early days of the paint industry, there was a little problem in selecting solvents, as almost all resins dissolved in hydrocarbons. An exception was shellac, which dissolved in ethyl alcohol. Early in the twentieth century, when nitrocellulose lacquers were introduced, solvent selection became a greater challenge. RS-type nitrocellulose (Section 17.2.1) is soluble in esters and ketones but not in hydrocarbons or alcohols. However, RS nitrocellulose does dissolve in a mixture of ketones or esters with hydrocarbons, which reduces solvent cost. If ethyl alcohol is included in the solvent mixture, a higher ratio of hydrocarbons can be used, further reducing cost. In those days, esters and ketones were classified as *true solvents*, hydrocarbons as *diluents*, and alcohols as *latent solvents*.

To control the drying rate of nitrocellulose lacquers under various conditions, mixtures of solvents are required. At a minimum, the solvent blends contain two esters or ketones, two hydrocarbons, and an alcohol. Solvent mixtures must be selected to maintain solubility throughout solvent evaporation from films. If the slowest evaporating solvent were a hydrocarbon, nitrocellulose would precipitate before all the solvent was gone, resulting in an uneven film with poor appearance and physical properties. As many as 10 solvents are blended to control drying and solubility through the drying cycle at the lowest volume cost.

18.2.1 Solubility Parameters

After 1930, more and more types of resins were adopted for coatings, and empirical selection of solvents and solvent mixtures became more difficult. The general rule of "like dissolves like" was broadened by the experience with nitrocellulose that mixtures of weak hydrogen-bonding hydrocarbons with strong hydrogen-bonding alcohols give solvency similar to medium hydrogen-bonding esters and ketones. Burrell (1955) initiated studies to develop a more scientific basis for selecting solvents and formulating solvent mixtures. He applied the theories of Hildebrand (1916), who analyzed miscibility of small molecule organic compounds in terms of thermodynamics. Hildebrand showed that the tendency of a pair of chemicals to mix spontaneously can be described in terms of Gibb's free energy equation:

$$\Delta G_{\mathrm{m}} = \Delta H_{\mathrm{m}} - T\Delta S_{\mathrm{m}}$$

Miscibility is thermodynamically favored when the change in free energy of mixing, ΔG_{m}, is negative. Hildebrand pointed out that the change in entropy ΔS_{m} is usually positive, since a solution is usually less ordered after mixing than before. (There are exceptions when strong interactions increase the entropic order of mixtures relative to the separate materials, resulting in a negative entropy change. But, in most cases, ΔS_{m} is positive and the entropy change favors mixing.) Hildebrand focused on ΔH_{m}, the enthalpy of mixing, since it usually determines the sign of ΔG_{m}. Change in ΔH_{m} is in turn related to the change in the energy of mixing ΔE_{m}, as shown, where R is the gas constant and T is temperature (Kelvin):

$$\Delta E_{\mathrm{m}} = \Delta H_{\mathrm{m}} - RT$$

Liquids have intermolecular attractive forces strong enough to hold the molecules together; otherwise, they would be gases. The forces can be measured by determining the energy needed to vaporize a liquid at that temperature. Results are expressed in terms of the molar energy of vaporization divided by the molar volume V. Hildebrand reasoned that the energy required to separate molecules during mixing is related to cohesive energy density $\Delta E_{\mathrm{v}}/V$. He expressed the change in ΔE_{m} of ideal solvents by the following equation, where V_{m} is the average molar volume of the mixture and ϕ_1 and ϕ_2 are the volume fractions of two components:

$$\Delta E_{\mathrm{m}} = V_{\mathrm{m}}\phi_1\phi_2\left[\left(\frac{\Delta E_{\mathrm{v}}}{V}\right)_1^{1/2} - \left(\frac{\Delta E_{\mathrm{v}}}{V}\right)_2^{1/2}\right]^2$$

The square root of cohesive energy density of a solvent was defined as the *solubility parameter* δ. Two solvents are miscible if the differences in solubility parameters approach zero. Then, the ΔE_{m} of the pair approaches zero, the ΔH_{m} is small, and the free energy of mixing is controlled by ΔS_{m}. When ΔS_{m} is positive, the solvents are miscible. ΔE_{v} and V_{m} vary with temperature; therefore, δ varies with temperature. Most tables of solubility parameters provide the values at 25°C, although commonly, no temperature is specified.

There is risk of confusion about units. The older units of solubility parameter were $(\mathrm{cal\,cm^{-3}})^{1/2}$, sometimes designated as Hildebrand's, h. In the SI system, the proper units are $(\mathrm{MPa})^{1/2}$; 1 $(\mathrm{cal\,cm^{-3}})^{1/2} = 0.488(\mathrm{MPa})^{1/2}$. The SI units have not been widely adopted and $(\mathrm{cal\,cm^{-3}})^{1/2}$ units are still used, although frequently not stated.

Table 18.1 Small's Molar Attraction Constants, $(\mathrm{MPa})^{1/2}$ $(\mathrm{cm^3\,mol^{-1}})$ at 25°C

Hydrocarbon group	*G*	Other groups	*G*
$-CH_3$	284	O (ethers)	236
$-CH_2$	270	O (oxiranes)	361
–CH–	176	Cl	420
=CH–	249	CO (ketones)	539
$=CH_2$	259	COO (esters)	668
Phenyl	1400	OH	463
Phenylene	1370		

When data are not available for energy of vaporization, solubility parameters can be estimated by using empirical equations relating boiling points, vapor pressure data, or surface tension data (Patton, 1979b). They can also be estimated by summation of Small's molar attraction constants, G, using the following equations where ρ is density and M is molecular weight (MW):

$$\delta = \left(\frac{\rho}{M}\right)\Sigma G = \left(\frac{1}{V}\right)\Sigma G$$

Selected Small's constants are listed in SI units in Table 18.1. As with solubility parameters, the units for Small's constants are commonly not specified and are frequently the old units. Hoy (1970) used Small's constants to calculate solubility parameters of 640 compounds.

Both Small's constants and solubility parameters of H-bond donor–acceptor molecules vary with the environment. The value for the alcohol hydroxyl group included in Table 18.1 depends on other groups in the solvent with which the hydroxyl group might hydrogen bond and on the polarity of other components in mixed solutions. Water is the most extreme case. While specific values of a solubility parameter of water are sometimes encountered, the published values vary widely, since they are very dependent on the medium. Their usefulness is limited to comparing systems having similar compositions.

Burrell's pioneering work on application of two-dimensional solubility parameters to coatings is described in previous editions of this book and in his publications (Burrell, 1955). His method achieved a modest degree of success, but it has been supplanted by three-dimensional approaches.

18.2.2 Three-Dimensional Solubility Parameters

Other investigators sought to improve Hildebrandt's two-dimensional approach with three-dimensional systems. The system of Hansen (1967, 1970, 1995) came to be most widely accepted. Hansen reasoned that since there are three types of interactive forces between molecules, there should be three types of solubility parameters: dispersion, δ_d; polar, δ_p; and hydrogen bond, δ_h. The total solubility parameter was arbitrarily set equal to the square root of the sum of the squares of the partial solubility parameters:

$$\delta = \left(\delta_d^2 + \delta_p^2 + \delta_h^2\right)^{1/2}$$

For mixed solvents, one can calculate a weighted average of the three partial solubility parameters:

$$\delta_{d(\text{blend})} = (\phi\delta_d)_1 + (\phi\delta_d)_2 + \cdots + (\phi\delta_d)_n$$
$$\delta_{p(\text{blend})} = (\phi\delta_p)_1 + (\phi\delta_p)_2 + \cdots + (\phi\delta_p)_n$$
$$\delta_{h(\text{blend})} = (\phi\delta_h)_1 + (\phi\delta_h)_2 + \cdots + (\phi\delta_h)_n$$

Three-dimensional solubility parameters can be determined or calculated by a variety of methods. Although values are commonly given as three figures, it should not be inferred that the third figure is significant. Table 18.2 provides solubility parameters for representative solvents selected from Hansen's tables and the *Polymer Handbook* (Grulke, 1999).

Hansen (1995) determined three-dimensional solubility parameters for a group of resins by experimentally testing solubilities of 34 polymers in 90 solvents. Previously, Hoy (1969) had calculated three-dimensional solubility parameters of resins by a less laborious method analogous to the use of Small's constants. The values determined or calculated by various methods are frequently not the same; tables of (presumably) self-consistent values have been published and are available in computer databases.

Table 18.2 Three-Dimensional Solubility Parameters $(\text{MPa})^{1/2}$

Solvent	Total	δ_d	δ_p	δ_h
n-Hexane	14.9	14.9	0	0
Toluene	18.2	18.0	1.4	2.0
o-Xylene	18.0	17.0	1.4	3.1
Acetone	19.9	15.5	10.4	7.0
Methyl ethyl ketone	19.0	16.0	9.0	5.1
Methyl isobutyl ketone	17.0	15.3	6.1	4.1
Isophorone	19.8	16.6	8.2	7.4
Ethyl acetate	18.2	15.8	5.3	7.2
Isobutyl acetate	16.8	15.1	3.7	6.3
n-Butyl acetate	17.4	15.8	3.7	6.3
Methyl alcohol	29.7	15.1	12.3	22.3
Ethyl alcohol	26.6	15.8	8.8	19.4
Isopropyl alcohol	23.5	16.4	6.1	16.4
n-Butyl alcohol	23.1	16.0	5.7	15.8
2-Butoxyethanol	20.9	16.0	5.1	12.3

Three-dimensional solubility parameters are based on thermodynamic laws, but several assumptions and arbitrary choices are involved in their derivation and use. Assertions that they are "theoretically sound" can be misleading (Patton, 1979c). They are best regarded as an empirical method that has proven useful for finding alternative solvent mixtures having similar solvency characteristics.

Attempts have been made to apply solubility parameters to other problems, but the theoretical foundations are shaky and results have been erratic. For example, use of solubility parameters to predict solubility of one polymer in another (*compatibility*) frequently gives erroneous predictions.

Even three-dimensional solubility parameters are an oversimplification of the complex factors involved in solubility. The difficulties result from at least two interrelated factors. First, they neglect entropy changes and, second, the H-bond solubility parameter combines the effects of donors and acceptors. In particular, entropy changes are likely to be significant in H-bonding systems. Hansen (1995) has suggested that increased weight should be ascribed to the hydrogen-bond parameter in some cases.

The difficulties increase as polymer MW increases. As molecules get larger, interaction between solvent and polymer molecules must become greater to overcome intermolecular polymer–polymer interactions. Intramolecular interactions may also play a role. Estimates of solubility parameters of polymers by use of group attraction constants do not take MW into consideration.

The solubility behavior of polymers seems to fly in the face of one's experience with the solubility of smaller molecules. Small molecules have upper limits of solubility. For example, 36.1 g of NaCl dissolves in 100 g of water at 25°C. In general, there is no upper limit to the solubility of a polymer. If a small amount is entirely soluble, it is quite safe to conclude that any larger amount will be soluble in that solvent. On the other hand, it is common for there to be a lower limit of solubility. A high concentration of polymer may be soluble in some solvent(s), but on dilution, part of the polymer may precipitate. This phenomenon can be used to fractionate polymers by dilution. The fractions precipitating first are generally the highest MW components. As dilution is continued further, progressively lower MW fractions are precipitated. In some cases, the fractionation is based on the polarity of parts of the resin. Many alkyds are soluble at high concentrations in aliphatic solvents but partially precipitate on dilution. The first fractions precipitating are high MW molecules with a larger than average number of hydroxyl and/or carboxylic acid groups. To understand this phenomenon, one can think of the situation in reverse: some solvent is soluble in the polymer (resin), and all of the polymer is soluble in the combination of polymer and solvent. As more solvent is added, solvency of the system changes, and parts of the polymer are not soluble in the more dilute polymer solution. The precipitate is highly swollen with solvent.

Hoy (1969) showed that glycol ethers can change apparent polarity by assuming the polar nature of their surrounding environment. In polar solvents, these ethers behave as polar solutes. However, in nonpolar solvents, they undergo intramolecular bonding or intermolecular hydrogen bonding to form less polar dimers. In view of this consideration, it is not surprising that solubility predictions often fail, since the resins are often polydisperse, polyfunctional materials with multiple H-bond donor and acceptor sites.

It is likely that solubility parameters will continue to play an important role in paint formulation and reformulation. Tighter VOC regulations are being considered in much of the world (Section 18.9.2); and there is growing emphasis on regulating solvent toxicity (Section 18.7). Fortunately, solvent suppliers provide formulators access to up-to-date data and computerized reformulation starting points.

18.2.3 Other Solubility Theories

Numerous refinements of solubility parameters have been proposed. Some have involved adding a fourth parameter, for example, by dividing δ_h into hydrogen-bond donor and acceptor terms. Huyskens and Haulait-Pirson (1985) proposed equations that reflect changes in entropy and attempt to account for the differences resulting from both hydrogen-bond acceptor and donor groups. They provide somewhat better predictions; but only relatively simple cases of poly(vinyl acetate), poly(methyl methacrylate), and poly(ethyl methacrylate) were studied. They explain the shortcomings of the three-dimensional system but fall short of giving a broadly applicable alternative.

Studies involving cooperative work in several laboratories led to the development of what a technical news story called a "universal solubility equation." The researchers themselves "believe this to be somewhat of an overstatement" (Kamlet et al., 1986). The complexity of polymer solubility can be seen from the fact that this universal equation has 13 factors in a 5-term equation and still does not account for several important situations. Among the limitations, critical to coatings, the equation has thus far been applied mainly to monofunctional solutes.

18.2.4 Practical Considerations

The need for solvent changes is frequent, motivated by new toxic hazard information, changes in relative cost, new regulations, temporary shortages, and so on. These are the situations for which solubility parameters are most useful. What can formulators do? While three-dimensional solubility parameters are imperfect, they can be very valuable. One can use solvent computer programs with data banks that include solubility parameters, evaporation rates, cost, density, and so on. These programs permit calculation of alternative solvent mixtures with solubility characteristics that do not change substantially during evaporation while minimizing cost per unit used to calculate alternative solvent blends on a customer service basis. The result of such a calculation is given in Table 18.3 in which solubility parameter units are provided in units of $(MPa)^{1/2}$ (ARCO Chemical Co., 1987). Solvent blend 1 contained 2-ethoxyethyl acetate that, because of health concerns, had to be removed from the formulation. Blends 2 and 3 were calculated to have nearly equal average solubility parameters and evaporation rates. Blend 2 was unsatisfactory because the slowest evaporating solvent, xylene, was not a solvent for the resin. Blend 3 was a better starting point since the slowest evaporating solvent, 1-methoxy-2-propyl acetate, was a true solvent for the resin. Relative evaporation rates are also given in Table 18.3. Relative evaporation rates of mixed solvents must be viewed with caution (Section 18.3.3).

A second type of problem faced by formulators is to predict solvents and solvent combinations for a new resin. Solubility parameters may be useful, but in most cases, the old idea that "like dissolves like" is just as effective for finding a starting point, and it takes less time. The problem of selecting solvents for new resins has become easier because, in order to increase solids, lower MW resins are being used. A wider range of solvents dissolves a resin as its MW becomes lower.

Table 18.3 Solvent Blends, Weight Percents

Component	1	2	3
MEK	9.9	6.1	14.7
MIBK	29.7	32.8	19.7
Xylene	21.9	25.2	24.3
Toluene	20.1	18.4	18.2
n-Butyl alcohol	13.5	17.5	17.3
2-Ethoxyethyl acetate	4.9	—	—
1-Methoxy-2-propyl acetate	—	—	5.7
Properties			
Solubility parameter, δ^a	19	19	19
δ_d	18	18	18
δ_p	5.7	5.7	5.3
δ_h	5.7	5.1	5.3
Relative evaporation rate	1.16	1.17	1.14
Cost^b	0.70	0.68	0.68

^{a} $(MPa)^{1/2}$.

^{b} At the time of study, in dollars per kilogram.

18.3 SOLVENT EVAPORATION RATES

During application and film formation, the volatile material evaporates out of the coating. The rate at which evaporation occurs affects not only the time required to convert a coating to a dry film but also the appearance and physical properties of the final film. Like so many topics in the coatings field, evaporation rate appears simple but turns out to be complex.

18.3.1 Evaporation of Single Solvents

The rate of evaporation of a solvent is affected by four variables: temperature, vapor pressure, surface to volume ratio, and rate of airflow over the surface. The rate of evaporation of water is also affected by relative humidity (RH).

The important temperature is the temperature at and near the surface. Initially, it is that of the surrounding air. However, as the surface cools owing to solvent evaporation, the air near the surface is cooled. Depending on the circumstances, thermal diffusion within the liquid may occur rapidly, in which case the surface temperature does not fall much during evaporation, or slowly, resulting in a sharp drop in surface temperature. The cooling effect is largest for those situations in which the solvent evaporates most rapidly. Also, the higher the heat of vaporization of the solvent, the greater the temperature drop if all other variables are equal.

The important vapor pressure is that of the solvent at the temperature(s) at which evaporation occurs. A common error is to assume that boiling points (the temperatures at which solvents have a vapor pressure of 1 atm (101.3×10^2 kPa)) are proportional to vapor pressures at other temperatures. However, boiling points are poor indicators of vapor pressure. For example, at 25°C *n*-butyl acetate (bp 126°C) evaporates more rapidly than *n*-butyl alcohol (bp 118°C).

The ratio of surface area to volume is important because solvent evaporation occurs at the solvent–air interface. If 10 g of solvent are spread out over an area of 100 cm^2, it evaporates more rapidly than if the surface area is 1 cm^2. Accordingly, the concentration of the resin solution and the viscosity increase more slowly during solvent evaporation from thicker films having the same surface area. When a coating is applied by a spray gun, it is atomized into small particles as it emerges from the orifice of the gun. The ratio of surface area to volume is very high, as is the rate of solvent loss. Thus, a major fraction of solvent evaporates after the spray droplets leave the orifice of the spray gun and before they arrive at the surface to which the coating is being applied.

The rate of airflow over the surface is important because the rate of evaporation depends on the partial pressure of the solvent vapor in the air at the air–solvent interface. If vaporized solvent molecules are not carried away from the surface quickly, the partial pressure of solvent builds up and evaporation is suppressed. Airflow rates vary substantially, depending on the application method and, therefore, the solvents used in a coating must be selected for the particular application conditions. For example, spraying a coating with an air spray gun results in significantly more loss of solvent than when the coating is sprayed from an airless gun because of the greater airflow over the surface of the droplets by the former method (Sections 23.2.1 and 23.2.2). The rate of solvent loss from a freshly coated surface depends on the rate of airflow through the spray booth. If the same coating is applied to the outside and the inside of a pipe, solvent evaporates more rapidly from the outside unless the inside is ventilated. Airflow results in nonuniform evaporation from coated objects; solvent evaporates more rapidly near the edges of a coated panel than from its center.

RH has little effect on the evaporation rates of most solvents; however, it has a major effect on the evaporation rate of water. The higher the RH, the more slowly water evaporates when all other conditions are equal. As RH approaches 100%, the rate of evaporation of pure water approaches zero. The evaporation rate of water from solutions can be negative at high humidity. RH decreases as air is warmed; sometimes it is feasible to partly compensate for the effects of humidity on drying by modestly increasing the air temperature, since warming the air both increases the vapor pressure of the water and decreases the RH.

18.3.2 Relative Evaporation Rates

As introduced in Section 18.2, formulation of nitrocellulose lacquers requires complex solvent combinations. An important criterion involved in solvent selection is evaporation rate. From the standpoint of solvent selection, vapor pressure is an important variable that affects evaporation rate. However, it is difficult to judge how much more rapidly one solvent will evaporate than another from vapor pressure data. Consequently, procedures for measuring absolute evaporation rates were developed, and, because of the difficulties of controlling air and heat flows in different types of measuring devices, *relative evaporation rates* were determined. Rates of solvent evaporation were related to the evaporation rate of *n*-butyl acetate, which at the time was a standard nitrocellulose solvent. Relative evaporation rate E is determined by Eq. 18.1, where t_{90} is the time for 90 wt% of a sample to evaporate in a given type of apparatus under controlled conditions and t_{90} of *n*-butyl acetate is defined as 1. Note that t_{90} values are inversely related to evaporation rate:

$$E = \frac{t_{90}(n\text{-butyl acetate})}{t_{90}(\text{test solvent})} \tag{18.1}$$

Some authors express E as a percentage figure, corresponding to 100% for *n*-butyl acetate. In either case, the higher the E value, the faster the evaporation rate of the test solvent. One should be sure which reference point is being used; there is at least one paper in the literature in which an author used data from two different tables without realizing that he was mixing data that differed by a factor of 100; his conclusions were absurd.

Determination of relative evaporation rates requires measurement under carefully controlled, standardized conditions. The Shell thin-film evaporometer (ASTM D3539) is an example. A top loading balance is placed in an enclosure designed to minimize variation in airflow over the surface of the balance pan. A flow of 25°C air with a RH of less than 5% at a rate of 21 l min^{-1} is maintained. A sample of 0.70 ml solvent is dispensed onto a piece of filter paper on the balance pan, and the time required to lose 90% of the sample weight is determined. Rates determined this way are volume-based relative evaporation rates. While volume-based evaporation rates are the most common in the literature, some experimental procedures use a fixed weight, rather than a fixed volume of solvent. Care must be taken not to compare data obtained with the different methods. One should also avoid comparing relative evaporation rates determined using different instruments because different evaporation conditions can change not only the absolute evaporation rates but also the relative evaporation rates.

Rocklin (1976) compared the relative evaporation rates of 66 solvents measured by the standard procedure (evaporation from filter paper) with rates measured by evaporation directly from the flat aluminum pan of the balance. Table 18.4 gives his data for several solvents. Note that both sets of data are relative to *n*-butyl acetate. As can be seen by comparing the E_{paper}/E_{metal} ratios, significant differences exist between relative evaporation rates of the same solvents for evaporation from filter paper and from a smooth metal surface in the same instrument at the same settings.

The most notable differences between E_{paper} and E_{metal} occur with faster evaporating solvents and with water and alcohols. In the case of evaporation from filter paper, the ratio of surface area to volume is much higher than in the case of the evaporation from a metal surface. Accordingly, solvents evaporate more rapidly from filter paper, initially, resulting in a sharp drop in temperature, thereby decreasing vapor pressure, which slows subsequent evaporation. In the case of evaporation from a metal surface, the surface area is much smaller and the thermal conductivity is much higher. Both factors minimize the temperature drop, leading to less reduction in evaporation rate. The difference is greater with fast evaporating solvents, such as *n*-pentane and acetone, than with *n*-butyl acetate, which evaporates more slowly, resulting in their low ratios of E_{paper} to E_{metal}. The experimental temperature was 25°C, but that is the air temperature. What controls the rates of evaporation and, hence, the relative rates is the actual temperature(s) of the surfaces.

Table 18.4 Volume-Based Relative Evaporation Rates at 25°C

Solvent	E_{paper}	E_{metal}	Ratio
n-Pentane	12	38	0.32
Acetone	5.7	10	0.55
Ethyl acetate	4.0	6.0	0.67
Methyl ethyl ketone	3.9	5.3	0.74
n-Heptane	3.6	4.3	0.83
Toluene	2.0	2.1	0.92
Ethyl alcohol	1.7	2.6	0.65
Methyl isobutyl ketone	1.7	1.7	1.0
Isobutyl acetate	1.5	1.5	1.0
n-Butyl acetate	1	1	1
sec-Butyl alcohol	0.93	1.2	0.81
m-Xylene	0.71	0.71	1.0
n-Butyl alcohol	0.44	0.48	0.92
2-Ethoxyethanol	0.37	0.38	0.98
Water	0.31	0.56	0.56
Methyl *n*-amyl ketone	0.34	0.35	0.96
2-Ethoxyethyl acetate	0.20	0.19	1.1
n-Decane	0.18	0.16	1.1
2-Butoxyethanol	0.077	0.073	1.1
Isophorone	0.023	0.026	1.0
Diethylene glycol monoethyl ether	0.013	0.014	0.99

The E_{paper} to E_{metal} ratios are also low in the cases of water and alcohols. This results from the greater extent of hydrogen-bond interaction of the hydroxyl groups with the very large surface area of cellulose of the paper as compared with the smooth aluminum surface. This retards evaporation of water and alcohols relative to *n*-butyl acetate. Note that relative to *n*-butyl acetate, *sec*-butyl alcohol evaporates more rapidly from the smooth metal surface and more slowly from the filter paper.

When formulating baking coatings for spray application, it is common to use a mixture of fast and very slow evaporating solvents. A significant fraction of the fast-evaporating solvent evaporates before the spray droplets reach the object being coated, raising viscosity and reducing the tendency of the coating to *sag*, while the slow evaporating solvent keeps the viscosity low enough to promote *leveling* and to minimize the probability of *popping* when the coated object is put into a baking oven. (See Chapter 23 for further discussion of sagging, leveling, and popping.) In selecting slow evaporating solvents, formulators have generally used tables of relative evaporation rates at 25°C and boiling points, which are available for a large number of solvents. Jackson (1986) published evaporation rate data for a limited number of slow evaporating solvents over the range of 75–150°C. The rates were determined using a thermogravimetric analysis (TGA) isothermally at a series of furnace temperatures. The rate relationships can be quite different at different temperatures. For example, at 25°C, the evaporation rate for a commercial mixture of the dimethyl esters of succinic, glutaric, and adipic acids is about five times slower than for isophorone. However, at 150°C, their evaporation rates are approximately equal.

Which evaporation rate data are "correct"? All of them are "correct"; they depend on the particular circumstances under which they were determined. However, we do not apply coatings to filter paper (or to aluminum balance pans or to TGA pans) nor do we dry them in a Shell evaporometer (or a TGA furnace). The rates at which solvents evaporate from coatings depend on the particular situation as discussed in Section 18.3.4. First, we need to consider evaporation from mixtures of solvents.

18.3.3 Evaporation of Mixed Solvents

Compared with pure solvents, evaporation of solvent blends adds further complications. In *ideal* homogeneous solutions, vapor pressure is governed by Raoult's law, which predicts that the vapor pressure P_i of the ith component of the solution is reduced from the vapor pressure of the pure liquid P_i^0 in proportion to its mole fraction x_i:

$$P_i = x_i P_i^0$$

Since vapor pressure of each solvent in the blend is different, the composition of the solvents that evaporate is different than that of the initial solvent blend. Therefore, the partial vapor pressures change continuously as solvent evaporates from a mixture. Raoult's law provides a good approximation for many combinations of solvents, especially when structures are similar and intermolecular interactions are minimal. However, many other solvent mixtures are *nonideal* as a result of interaction effects. These effects change as ratios of solvents change. The vapor pressure P_{total} of any mixture of miscible solvents can be calculated by the following equation, where χ represents an empirical adjustment factor for the interaction effects, usually given the more dignified-sounding title of an activity coefficient:

$$P_{\text{total}} = P_1 + P_2 + \cdots + P_i = \chi_1 P_1^0 x_1 + \chi_2 P_2^0 x_2 + \cdots + \chi_i P_i^0 x_i$$

Activity coefficients for many solvents have been evaluated and incorporated into computer programs that can calculate partial vapor pressures of each solvent in a mixture throughout its evaporation. One such program is called UNIFAC (Skjold-Jorgenson et al., 1979). Results of such calculations are often presented as the partial vapor pressures at each 10% interval through the evaporation. Several assumptions are involved in calculating vapor pressures, and, in order to relate vapor pressure to evaporation rate, it is assumed that all other factors affecting evaporation rate (temperature, surface to volume ratio, and airflow) are fixed. Thus, the results of such calculations are only approximations. However, in view of the uncertainties of relative evaporation rates, a high level of accuracy is not needed. Evaporation of solvent blends is discussed in an extensive review paper (Yoshida, 1972).

Relative evaporation rates of mixed solvent compositions can be determined experimentally in an evaporometer. The relative evaporation rate of a mixed solvent (E_T) can also be calculated from volume fractions c, activity coefficients a, and relative evaporation rates E of the individual solvents (Patton, 1979c, p 340):

$$E_T = (caE)_1 + (caE)_2 + \cdots + (caE)_n$$

Such calculated E_T values are of dubious significance: the composition changes over time, so the value of E_T changes. The difference between experimental and calculated relative evaporation rates is especially large when comparing a solvent mixture with a narrow range of E values with another with a wide range of E values. For example, one could calculate E_T for a mixture of two solvents with high and low E values that would be equal to the E value of a single solvent with an intermediate E value. However, it is evident that the mixed solvent would actually require a longer time for evaporation than the single solvent since, after the fast evaporating solvent was gone, the slow evaporating solvent would evaporate more slowly than the single intermediate E value solvent.

This description makes it sound almost impossible to formulate a satisfactory solvent mixture for a coating. However, formulation does not start in a vacuum of knowledge. Experience with other formulations that proved useful under somewhat similar circumstances provides guidance. Using such formulations together with any suitable table of relative evaporation rates, one can make a reasonable first stab with a new formulation. The coating is then applied under the particular circumstances and adjustments made as needed. The final adjustment is usually made by someone in the user's factory, experienced with production spray guns and operating conditions. The use of relative evaporation rate tables assists development of a formulation that is in the ball park of what is needed. Because final in-factory adjustments are needed anyway, it may not matter much which relative evaporation rate table was used.

Evaporation rates of water–organic solvent mixtures require special consideration for at least four reasons. First, strong interactions often cause deviations from Raoult's law. Second, RH affects the evaporation rate of water but has little or no effect on the evaporation rates of organic solvents. Third, azeotropic effects may occur. Fourth, the heat capacity and heat of vaporization of water are unusually high.

The humidity effect is illustrated by the fact that the relative evaporation rate (E) of water at 0–5% RH and an air temperature of 25°C is 0.31; but, at 100% RH, E is 0. If a solution of 2-butoxyethanol (E=0.077) in water evaporates at low RH, water evaporates more rapidly, and the remaining solution becomes enriched in 2-butoxyethanol. At high RH, 2-butoxyethanol evaporates more rapidly, and the remaining solution becomes enriched in water. At some intermediate RH, the relative evaporation rates of water and 2-butoxyethanol are equal. When an aqueous solution evaporates at this RH, the composition of the remaining solution is constant. This RH has been called the *critical relative humidity* (CRH) (Dillon, 1977). The CRH for 2-butoxyethanol solutions in water is estimated at 80%. If the relative evaporation rate of the solvent were larger than that of water, even at 0–5% RH, there would be no CRH, since the solvent would evaporate more rapidly than water at any RH. At the other end of the scale, if the relative evaporation rate of the solvent were very low, the CRH would approach 100%.

Azeotropic behavior is likely to occur with water solutions. Rocklin (1986) has studied the role of azeotropy in speeding up water/solvent evaporation in humid air. He reported development of a computer model, the AQUEVAP program that permits calculation of the fastest evaporating water–solvent blends at various RH values. For example, at 40% RH, the time required for 90% evaporation of a 20 wt% solution of 2-butoxyethanol in water is 1820 s compared with 2290 for water alone. Thus, co-evaporation of water and 2-butoxyethanol accelerates evaporation.

The high heat capacity and heat of vaporization of water can also affect the evaporation rates of water and water–solvent blends in an oven. For example, when room temperature samples were placed in a TGA furnace at 150°C, the times for 99% weight loss of 2-butoxyethanol (bp 171°C), water, and a 26 : 74 blend of 2-butoxyethanol/water were 2, 2.6, and 2.5 min, respectively (Watson and Wicks, 1983). Since the air temperature in the furnace was 150°C, the samples took some time to heat up. The higher heat of vaporization of water (2260 J g^{-1} at its boiling point) compared with 2-butoxyethanol (373 J g^{-1} at its boiling point) slowed the rate of heating of the water and water–solvent blend enough to more than offset the expected evaporation rates based on boiling points or E values.

18.3.4 Evaporation of Solvents from Coating Films

Except in high solids coatings (Section 18.3.5), the resin or other coating components have little effect on initial rate of solvent evaporation when solventborne coating films are applied. Initial rates of evaporation of solvent from resin solutions are, within experimental error, the same as or close to the rates of evaporation of the solvents alone under the same conditions. Such observations are not inconsistent with Raoult's law, which predicts that dissolved resins will have little effect on vapor pressure because of their high MWs. However, as solvent loss from a coating continues, a stage is reached at which the rate of evaporation slows sharply. As the viscosity of the remaining coating increases, availability of free volume decreases and the rate of solvent loss becomes dependent on the rate of diffusion of solvent through the film to the surface, rather than on the rate of evaporation from the surface. The solids level at which the transition from evaporation rate control to diffusion rate control occurs varies widely but is often in the 40–60 NVV range.

Hansen (1970) described the situation as first stage and second-stage losses of solvent. In the first stage, the rate is governed by the factors that govern evaporation of solvent mixtures: vapor pressure, surface temperature, airflow over the surface, and surface to volume ratio. At this stage, there is a first-power dependence on wet film thickness. After a transitional stage, evaporation slows and the rate of solvent loss becomes dependent on the rate of diffusion of the solvent molecules through the film. During this stage, the rate of diffusion depends on the square of film thickness. In line with the changes in viscosity, the first stage is sometimes called the *wet stage* and the second stage the *dry stage*. An example of a plot of weight of solution remaining as a function of time for evaporation of MIBK from a soluble vinyl chloride copolymer resin solution is shown in Figure 18.1 (Newman and Nunn, 1975).

During the second stage, diffusion rate is primarily controlled by free volume availability. That is, the solvent molecules move through the film by jumping from one free volume hole to another. The most important factor controlling free volume availability is (T–T_g). If solvent evaporation is occurring at a temperature well above the T_g of the solvent-free resin, diffusion rate will not restrict the rate of solvent evaporation at any stage in drying. As the T_g of the

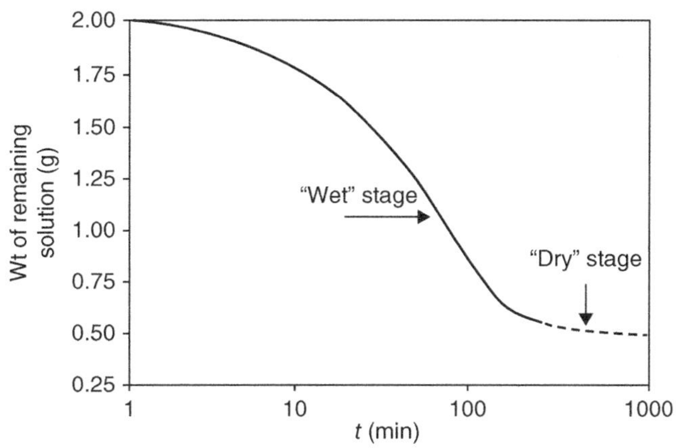

Figure 18.1 Two-stage release of solvent; MIBK in Vinylite® VYHH at 23°C, initially at 20 wt% polymer. Source: Newman and Nunn (1975). Reproduced with permission of Elsevier.

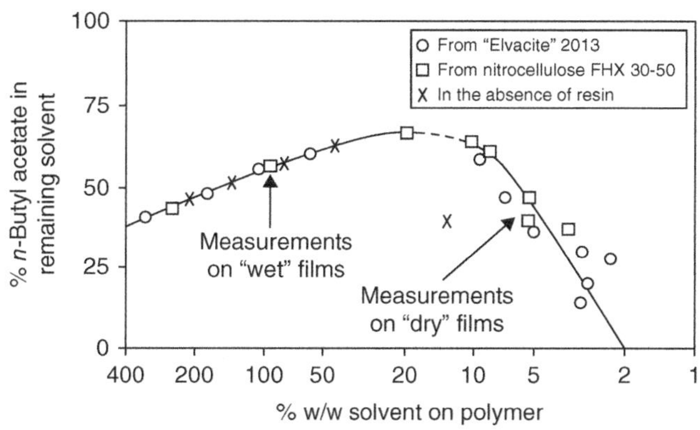

Figure 18.2 Changes in remaining solvent concentration during wet and dry stages of solvent evaporation at 23°C from films of an acrylic resin (Elvacite® 2013), from nitrocellulose, and from resin-free solvent, each initially in 60 : 40 IBAc to BAc. Source: Newman and Nunn (1975). Reproduced with permission of Elsevier.

resin approaches the temperature at which drying is occurring, solvent evaporation will become increasingly diffusion rate controlled as $(T–T_g)$ becomes smaller. No experimental data are available as to the values of $(T–T_g)$ when diffusion rate control becomes important; such values are probably system dependent.

When the second stage is reached, the rate of evaporation becomes dependent on how fast the solvent molecules get to the surface, not on their vapor pressure. As solvent loss continues, concentration of the remaining resin solution increases and, therefore, T_g increases and the rate of diffusion slows further. If the T_g of the resin increases to levels well above the temperature of the film, the rate of solvent loss will, in time, approach zero. Years after films have been formed, there will still be residual solvent left in the film. If solvent must be essentially completely removed in a reasonable time, the film must be baked at a temperature above the T_g of the solvent-free resin. It is a common mistake for novice polymer chemists to think that the rate of solvent loss from polymer samples can be increased by drying under a vacuum at temperatures below T_g. When the rate of solvent loss has reached a stage at which it is independent of vapor pressure of the solvent, it is also independent of atmospheric pressure above the sample.

Solvent evaporation is affected by the addition of a plasticizer. The concentration of solvent at which solvent loss becomes diffusion controlled decreases as plasticizer concentration increases. The amount of solvent retained after some time interval decreases as plasticizer concentration increases. Perhaps, less obviously, in the case of mixed-solvent systems, the ratio of slow to fast evaporating solvent retained in the film increases as plasticizer concentration increases.

There have been many attempts to quantify the diffusion of solvents from coating films during solvent evaporation, but success has been limited (Hansen, 1967; Waggoner and Blum, 1989). Some of the problems are discussed by Lasky et al. (1988) who suggest possible approaches to modeling the diffusion through scaling analysis. As a broad generalization, small molecules tend to evaporate faster than large molecules in both the first and second stages, because small molecules tend to be more volatile in the first stage and to diffuse faster in the second stage. However, other factors are involved. One is molecular configuration. Since the molecules must jump from hole to hole, the smaller the size of the solvent molecule, the greater its chance of finding sufficiently large holes. Even though its relative evaporation rate is higher, cyclohexane is retained in films to a greater degree than toluene because cyclohexane has a higher molar volume.

Figure 18.2 shows the results of an experiment illustrating the effect of cross-sectional size. Nitrocellulose and acrylic lacquers were made at 20 wt% solids in 60 : 40 isobutyl acetate (IBAc) and *n*-butyl acetate (BAc), and the ratios of the two solvents remaining in the films were monitored. Like other branched–linear isomeric pairs, the relative evaporation rate of branched IBAc is higher than that of linear BAc. Thus, the ratio of IBAc to BAc remaining in the film decreased during the first, wet stage of solvent loss, reaching a minimum of about 35 : 65. During the second dry stage, the linear compound diffused more rapidly; accordingly, in contrast to the first stage, the ratio of IBAc to BAc remaining in the film increased and approached 90 : 10 (Newman and Nunn, 1975). Equations have been developed that model effect of solvent size on diffusion based on free volume of polymers (Vrentas et al., 1996).

In air dry coatings, the T_g sometimes increases to a level above ambient temperature, while significant amounts of solvent remain in the film. When this happens, solvent evaporation becomes extremely slow, and detectable amounts of volatile solvents may remain in the film for years. Residual solvent can affect film properties such as corrosion and moisture resistance adversely (Section 7.3.1). Thus, selection of solvents that have minimal detrimental effects may be a factor in optimizing long-term film performance.

18.3.5 Evaporation of Solvents from High Solids Coatings

Generally, it is more difficult to control sagging of spray-applied high solids coatings than their lower solids counterparts (see Sections 24.3 and 25.2.2). While other factors may be involved, there is considerable evidence that high solids formulations lose much less solvent than conventional coatings while atomized droplets are traveling between a spray gun and the object being coated (Wu, 1978; Bauer and Briggs, 1984). As a result, there is less increase in viscosity and, therefore, a greater tendency to sag. Explanations for this decreased solvent loss do not appear to be adequate. One factor is the colligative effect of the lower mole fractions of solvent(s) in a high solids coating. A sample calculation illustrates this difference. If the vehicle of a conventional coating consists of 30% by weight solids solution of a resin having $\bar{M}_n$ of 15 000 in 70% of a solvent mixture with $\bar{M}_n$ of 100, the mole fraction of solvent is 0.997. However, if a high solids coating has 70% solids of a resin having $\bar{M}_n$ of 1000 and 30% of a solvent mixture with $\bar{M}_n$ of 100, the mole fraction of solvent is 0.811. While the lower mole fraction is consistent with slower solvent evaporation from the high solids coating, the magnitude of the difference does not seem large enough to account for the large changes in solvent loss that have been reported.

Hill and Wicks (1982) have suggested that high solids coatings may undergo transition from first-stage to second-stage solvent loss with relatively little solvent loss as compared with conventional coatings. The reasoning was that T_g of the more concentrated solution in high solids coatings might be expected to increase more rapidly with increasing concentration and, hence, attain the second stage of free volume control after less solvent loss relative to conventional coatings. Consistent with this hypothesis, Ellis (1983) found that high solids polyester coatings were formulated at concentrations that exceeded the transition concentration where solvent loss rate becomes diffusion controlled. He also found that the transition points occur at higher solids with linear molecules such as *n*-octane versus isooctane and *n*-butyl acetate as compared with IBAc. Further research is warranted owing to the potential severity of sagging problems in the application of high solids coatings.

18.3.6 Volatile Loss from Waterborne Coatings

For solventborne coatings, the formulator has available variety of solvents having a wide range of properties and volatility. But for waterborne coatings the major volatile component is water, which comes with only one vapor pressure–temperature curve, one heat of vaporization, and so forth. Furthermore, the RH at the time of application and drying of the coatings can have a major effect on rates of volatile losses from waterborne coatings. Formulators use limited levels of organic solvents to modify evaporation rates, but tightening regulations restrict this approach in many cases.

A major class of waterborne coatings is based on water-reducible resins, which are not soluble in water (Section 8.3). A salt of the resin, dissolved in an alcohol or an ether alcohol, is diluted with water. During dilution, the resin forms aggregates, swollen with solvent and water, that are dispersed in a continuous aqueous phase, which also contains solvent. The relative evaporation rate of water is affected by the RH; accordingly, with miscible solvents like 2-butoxyethanol, there is a CRH (Section 18.3.6). It has been found that the CRH is different for evaporation from coatings than from water–solvent blends without resin. For example, CRH is 65% for 10.6 wt% (based on volatile components) 2-butoxyethanol in a coating, in contrast to 80% calculated for the same concentration of a 2-butoxyethanol-water blend without resin (Brandenburger and Hill, 1979). A possible explanation for the different CRH values is that in coatings the concentration of 2-butoxyethanol in the continuous water phase is below the average, and the solvent–water ratio inside the aggregates is above the average. This type of distribution and its effect on the composition of vapor above a model aggregate system using a more volatile solvent, *t*-butyl alcohol, has been demonstrated (Wicks et al., 1982). Such effects can be critical in controlling sagging and, particularly, popping of such waterborne coatings (Sections 24.3, 24.7, and 26.1).

The other major class of waterborne coatings uses latexes as the principal binder. Evaporation of water from a drying latex paint film is controlled by temperature, humidity, evaporative cooling, and rate of airflow over the surface (Croll, 1987), and it resembles first-stage drying throughout most of the process. After most of the water has left, evaporation slows as a result of coalescence of a surface layer through which water must diffuse. In latex paints to be applied by brush or roller, it is desirable to retard the development of a partially coalesced surface layer to permit lapping of wet paint on wet paint. This generally has required the presence of a slow evaporating solvent, such as propylene glycol. The presence of these solvents does not affect the initial rate of water loss but does slow the development of a surface skin (Sullivan, 1975). The presence of such water-soluble solvents also facilitates the loss of coalescing solvent. Use of both types of solvents is discussed further in Chapter 31 along with expedients to curtail solvent use.

18.4 VISCOSITY EFFECTS

Solvent selection can have a major effect on viscosity of resin solutions. Differences result from two factors: the viscosity of the solvent itself and the effect of solvent–resin interactions. The direct effect of solvent viscosity on solution viscosity is apparent from equations that relate viscosity and resin concentration. Even the simplest relationship, which is valid over only a narrow range of concentrations for solutions with viscosities in a range of about 0.1–10 Pa·s, shows this dependence on solvent viscosity, where K is a constant and (Conc) is a resin concentration term:

$$\ln \eta_{\text{solution}} = \ln \eta_{\text{solvent}} + K(\text{conc.})$$

This relationship yields a surprising prediction: A difference of viscosity of 0.2 mPa·s for two solvents (1.0 and 1.2 mPa·s) may appear trivial, but it can cause a difference of 2000 mPa·s in the viscosity of a 50 wt% solution of a resin (10 Pa·s vs. 12 Pa·s) (Patton, 1979a, p 109). The relationship between concentration and viscosity of resin solutions is discussed in more detail in Section 3.4.3.

One must be careful in comparing viscosities of solutions. Literature data make comparisons based on weight relationships, volume relationships, weight of solvent per volume of coating, and weight of resin per volume of solution. One must decide what type of comparison is appropriate for a particular purpose. Most commonly, weight ratios are used when considering the viscosity of resin solutions, but they can be misleading in relationship to air pollution regulations, which are based on weight of solvent per volume of coating. Comparing the viscosity of resin solutions with different solvents on a weight basis can lead to the wrong choice in trying to minimize weight of solvent per volume of coating. Since flow is generally related to volume considerations, use of volume fraction comparisons might seem desirable, but polymer solutions are seldom ideal, creating uncertainty about the significance of volume fraction values.

Sprinkle (1983) demonstrates the effect of solvent choice on viscosity of an acrylic solution containing 400 g l^{-1} of solution in various solvents (Table 18.5). For comparison, the densities and viscosities of the solvents alone are also given. The data in Table 18.5 illustrate the substantial effect that solvent selection can have on solution viscosity of a given polymer. The weight concentrations of these solutions are not constant; it would have been interesting to compare as well the viscosities of equal weight ratio solutions. The data suggest that solvents with low density and low viscosity tend to provide low solution viscosity, as predicted by theory. However, other factors are involved.

Table 18.5 Viscosities at 25°C of Solutions of a High Solids Acrylic; Solvent Concentration 400 g l^{-1} Solution

Solvent	Solvent viscosity (m Pa·s)	Solvent density (gMl^{-1})	Solution viscosity (m Pa·s)
Methyl propyl ketone	0.68	0.805	80
Methyl isobutyl ketone	0.55	0.802	110
Ethyl acetate	0.46	0.894	121
Methyl amyl ketone	0.77	0.814	147
n-Butyl acetate	0.71	0.883	202
Toluene	0.55	0.877	290
Isobutyl isobutyrate	0.83	0.851	367
Xylene	0.66	0.877	367

Source: Sprinkle (1983).

Another factor that is sometimes important is the effect of solvent–solvent interactions on viscosity of solvent mixtures. In general, these effects are small except for mixtures that include alcohols or water. When relatively moderate (generally <40%) amounts of alcohols are mixed with other solvents, they do not increase the viscosity of the solution proportionally to their relatively high neat viscosities (Rocklin and Edwards, 1976). This results from reduced multiple hydrogen-bond interactions among the alcohol molecules in the presence of other solvents. The effects are larger and less obviously predictable in the case of solutions of solvents in water.

Large viscosity effects can result from solvent–resin interactions. To distinguish these effects from the effects of solvent viscosity discussed previously, it is common to compare relative viscosities (η/η_s) of solutions. The effects of interactions are complex and not fully understood in even relatively simple systems. At least two factors play a role. First, most resins used in coatings have polar and hydrogen-bonding substituents, such as hydroxyl groups or carboxylic acid groups, which tend to associate with polar groups on other molecules, often substantially increasing viscosity. Solvents that prevent or minimize such interactions by interacting with the polar groups tend to reduce viscosity. Polar solvents with single hydrogen-bond acceptor sites are effective in minimizing relative viscosity; examples are ketones, ethers, and esters. Water can also serve to reduce viscosity of oligomers that are capable of hydrogen bonding. As much as 20% of water can dissolve in solvent-free coatings (depending on the formulation);

adding water, in the range of 5–15%, has been shown to reduce viscosity by 40–60% (Jones, 1996).

A second factor is the effect of a given solvent on the hydrodynamic volume of individual resin molecules with their closely associated solvent molecules. If the interaction between resin and solvent molecules is strong, the chains of the resin molecules become extended, and their hydrodynamic volume increases. For less strong interactions, the molecules contract, and the hydrodynamic volume becomes smaller. Relative viscosity tends to be directly related to hydrodynamic volume. However, if solvent–resin interaction is so weak that resin–resin interactions form clusters, relative viscosity and viscosity increase. In the majority of coating applications, and especially high solids coatings, which contain low MW resins, resin–resin interactions are the stronger of the two effects, although hydrodynamic volume effects can be significant. To counteract the effects of resin–resin interactions, most high solids formulations include hydrogen-bond acceptor solvents. The situation with mixed solvents is more complex. In a study involving a limited number of systems, Erickson and Garner (1977) found that relative viscosity was dominated by the effect of the solvent that interacted most strongly with the resin.

A word of caution: relative viscosities can be useful for understanding the factors affecting viscosity, but viscosities, rather than relative viscosities, are more important in applying coatings; a solvent that reduces relative viscosity may result in a comparatively high viscosity solution. Viscosity of resin solutions is discussed further in Section 3.4.

Solvent selection can also affect storage stability. Inclusion of alcohols in formulations containing melamine–formaldehyde resins can increase storage life by minimizing cross-linking during storage (Section 11.3.3). Polyesters can transesterify with primary alcohols during storage, which can be particularly problematic with water-reducible polyesters (Section 10.5). Alcohols can react slowly with epoxy resins during storage. H-bond acceptor solvents can extend the pot life of epoxy–amine coatings (Section 13.2.5). Reaction rates of isocyanates with hydroxyl groups are reduced in H-bond acceptor solvents, extending pot life (Section 12.4 and Hazel et al., 1997). Urethane grade esters and ketones that contain no more than traces of alcohols or water are generally used for two-package polyurethanes.

18.5 FLAMMABILITY

Fires and explosions of flammable coating solvents can inflict terrible burns. Tragic accidents have occurred in which people were killed or severely burned. Sadly, these accidents could have been prevented.

Most solvents used in coatings are flammable. Care should be exercised when working with solvents in the laboratory, in the coatings factory, and in end use environments. Flammability depends on structure and vapor pressure. Generally, there is both an upper and a lower level of vapor concentration that limits flammability or explosion. If the partial pressure of solvent vapor is low enough, not enough energy is released during burning to maintain the vapor–air mixture above the ignition temperature of the system. If the partial pressure of the solvent is high enough, there is insufficient oxygen for explosion or fire. Tables of upper and lower explosive limits of many solvents are available. A full container of solvent may present less of a fire risk than a recently emptied container; the vapor phase in the former might have the solvent at concentrations above the upper explosive limit, while the "empty" container may have a concentration in the explosive range.

The most common cause for fires in coating factories has been static electricity. Solvent flowing out of one tank and into another tank by gravity can pick up enough electrostatic charge to cause a spark, which can set off a fire or explosion. To avoid such charge accumulation, all tanks, pipes, and so forth used in handling solvents and solvent-containing mixtures should be electrically grounded at all times. A second cause is sparking of electric motors or faulty electrical connections. Factories and laboratories should be equipped with explosion proof electrical installations. One should never try to bypass these installations. And, of course, smoking is a potential ignition source.

There are two main types of flammability tests: *open cup* and *closed cup*, both measure a *flash point*, the minimum temperature at which solvent can be ignited by a hot wire. ASTM specifies standard conditions for both tests. Generally, open cup testers give results more appropriate for indicating degree of hazard of a mixture when exposed to air, as during a spill. The closed cup flash point more nearly describes the fire hazard of a liquid enclosed in a container. US Department of Transportation regulations for shipment of flammable liquids are based on closed cup tests. Transportation costs can be substantially affected by flash points of the material being shipped. A closed cup flash point is lower than an open cup flash point. Solvents used in architectural paints should have closed cup flash points over 38°C. Extensive tables of flash points of solvents are available. Different sources sometimes disagree on flash point values. Variations are not surprising for solvents such as naphthas and mineral spirits that are variable mixtures, but variations are also found with single component solvents such as *n*-butyl acetate. The ASTM methods are said to be accurate to +2.5°C. Reproducibility is worse at temperatures below 0°C.

Many coatings contain a mixture of solvents. It is safest to determine flash points of mixtures experimentally. Ellis (1976, p 45) discusses the factors affecting flash points, including molecular interactions in blends. Wu et al. (1988) report that good predictions of closed cup flash points for mixed solvents can be made by a UNIFAC computer program that requires only flash points and molecular structures of the pure components. Their best results were obtained using UNIFAC group interaction parameters derived from flash points of binary solvent mixtures. Calculation of useful approximations of flash points of mixed solvents using simpler equations has been reported (McGovern, 1992).

Risk of fire or explosion can be eliminated by meeting either of the two conditions. If solvent vapor concentrations in air are kept well outside of the concentration range within which ignition is possible or if all sources of ignition are eliminated, there can be no fire. Unfortunately, many accidents have occurred because neither of these conditions was satisfied. Severe burns and fatalities can result. Because it is difficult to be certain which one of these conditions can be met in all circumstances, prudent practice dictates that all possible steps should be taken to meet both conditions, providing redundancy. The importance of good ventilation cannot be overemphasized, especially since the vapors of solvents, particularly the relatively dense ones, may stratify in stagnant air. As a result, solvent vapor concentrations can be higher in the lower part of the workspace, perhaps high enough to explode even when the average concentration is below the lower explosive limit.

18.6 OTHER PHYSICAL PROPERTIES

Density can be an important variable, including having a major effect on cost. Most solvents are sold on a weight basis, but the critical factor and, correspondingly, critical cost in almost all cases in the coatings field is the cost per unit volume. Most US air pollution regulations (Section 18.9) are based on weight of solvent per unit volume of coating that also favors use of low density solvents in formulations.

Conductivity can affect solvent choice. As discussed in Section 23.2.3, use of electrostatic spray guns requires control of the conductivity of the coating. In general, solventborne formulations that have appreciable, but low, conductivity work best. The conductivity of hydrocarbon solvents is too low to permit pickup of adequate electrostatic charge. Alcohols, nitroparaffins, and small amounts of amines are common solvents or additives to increase conductivity to a desired range.

The relatively high conductivity of waterborne coatings can introduce complications in electrostatic application, such as the need to insulate the spray apparatus and relatively fast loss of charge from spray droplets. Additives can minimize the latter problem. For example, inclusion of glycol ethers improves sprayability of waterborne coatings, apparently by reducing surface conductivity of water. Presumably, the alkyl groups of the ethers orient quickly to the surface of the droplets.

Surface tension (Section 24.1) can be another important factor influencing solvent selection. Solvent can affect the surface tension of coatings, which, in turn, can have important effects on the flow behavior of coatings during application as discussed at length in Chapter 24. Solvent selection can also be a factor affecting the development of surface tension differentials across the surface of a drying film during application and film formation; these differentials can affect flow behavior (Hahn, 1971; Overdiep, 1986). Since surface tensions depend on temperature and concentration of resins in solution, solvent volatility can have a large effect on the development of surface tension differentials. For a coating to wet a substrate, the surface tension of the coating must be lower than that of the substrate. Solvent selection can be a factor particularly in aqueous systems. While surface tension of aqueous systems can be reduced by adding surfactant, it is often more desirable to accomplish the same purpose with a solvent such as 2-butoxyethanol. Solvent evaporates, while surfactant leaves residues in the final film that may be detrimental to properties such as adhesion and humidity resistance. Surfactants may also introduce different problems, as discussed in Chapter 24.

18.7 TOXIC HAZARDS

In considering the toxic risks of volatile solvents, the extent of exposure must be considered in combination with toxicity data. All solvents are toxic at some level of exposure. Obviously, one should avoid ingesting them. The hazard from skin contact can be controlled by wearing protective clothing. Generally, the greatest potential risk comes from inhalation.

Three general types of toxicity data are important. Acute toxicity indicates the level of intake in single doses that can be injurious or lethal; this kind of information can be particularly important in cases of accidental ingestion or spills. A second type of toxicity data concerns the level of exposure that is safe when people are going to be exposed 8 hours a day for long periods of time. This kind of data is used, for example, to set the upper concentration limit for a solvent in a spray booth. Third, is risk of exposure over periods of years to low levels of materials that can increase health risks such as cancer. When it is found, usually by animal tests, that a solvent may be carcinogenic, very low levels of permissible exposure are set. The levels are frequently too low to be controlled by economically feasible methods. The effect is that carcinogenic solvents are banned. For example, benzene has not been used in coatings for many years for this reason. While most solvents used in large quantities in coatings have been tested and are thought to be noncarcinogenic, it is prudent for the user to be aware of current knowledge of the materials he/she is using and to minimize inhalation and contact with all solvents.

Extensive tables of all three types of data are available. In formulating coatings, it is also necessary to consider the clientele that will use them. While coatings sold to retail consumers are carefully labeled to include cautionary information, one must assume that many people will not read the labels. When selling to a large corporation, it is reasonable to assume that the material safety data sheets will be read and appropriate practices will be established. But when selling to small customers such as automobile repair shops, one cannot assume that the customer will pay adequate attention to precautions. It may be ethical to sell a coating with a somewhat toxic solvent to one class of customer but unethical, although perhaps legal, to sell it to another. A common difficulty is to know what the level of exposure will be. Smith et al. (1987) describe an approach to assessing possible exposures when retail consumers apply coatings in a room.

Since 1976, chemicals in the United States have been regulated under the Toxic Substances Control Act (ToSCA or TSCA). This act was amended in 2016, establishing new, risk-based safety standards and clarifying procedures for enforcement by the EPA. Implementation is underway and will undoubtedly have an impact on coatings formulation (see https://www.epa.gov/assessing-and-managing-chemicals-under-tsca).

In 1990 a law passed by the US Congress identified 189 HAPs for which use was to be reduced (Brezinski in Koleske, 1995). Many solvents important in coatings were on the list. The list is revised occasionally, for example, 2-butoxy ethanol and MEK were delisted in 2004 and 2005. Implementation of the law by the US EPA is discussed in Section 18.9.2.

In 2007 the European Union (EU) launched an 11-year program of Registration, Evaluation, Authorization, and Restriction of Chemicals (REACH). Implementation is underway (Section 18.9.2).

Looking ahead, coatings formulators and producers can expect many challenges as regulation of solvent toxicity intensifies around the world. For example, in 2016 the US reformed its ToSCA; California's Department of Toxic Substances Control is planning aggressive regulations (DTSC, 2014); and EU's REACH regulations will likely be refined and become more restrictive.

18.8 ATMOSPHERIC PHOTOCHEMICAL EFFECTS

Since the 1950s, it has been generally realized that the presence of organic compounds in the atmosphere can lead to serious air pollution. Terminology can be confusing; in the older literature, such compounds are referred to as "hydrocarbons," meaning any organic compound, not just unsubstituted hydrocarbons. More recently, they are called VOCs. The terms *reactive organic gases* (ROG) and *reactive organic compounds* (ROC) are also used. California classifies solvents according to their MIR. Particularly in Europe, solvents are classified by their POCP.

Three end effects of VOC emissions into the atmosphere are important: formation of eye irritants, of particulates, and of toxic oxidants, especially ozone.

While all three factors are important, the most critical for coatings is the level of ozone in the *troposphere*, the layer of atmosphere closest to the Earth's surface. (In the upper atmosphere, ozone shields the Earth from destructive UV radiation.) While ozone is a naturally occurring component of the atmosphere, it is toxic to plants and animals. When you go into a forest in the mountains and smell the wonderful "fresh air," part of the odor is ozone. Pine and oak trees emit substantial quantities of *biogenic* VOC into the atmosphere, the UV level increases with altitude, and as a result, ozone generation is high. Plants and animals evolved in the presence of some ozone and can tolerate its presence up to a point. By some estimates, global VOC emission from biogenic sources is 10 times larger than man-made emissions. However, with the growth of VOC emissions from man-made (*anthropogenic*) sources, ozone levels on many days of the year in many parts of the world, especially in and around cities, have increased risks to human health, especially of susceptible individuals. The US EPA (2013) detailed current knowledge about ozone's effects in its *Integrated Science Assessment for Ozone and Related Photochemical Oxidants*.

Annual biogenic emissions of VOC in the United States are very roughly estimated at 40 million short tons, three times more than man-made emissions. However, most man-made emissions are concentrated in densely populated areas. The EPA reports that man-made VOC emissions in the United States peaked in 1970 about 33.7 million short tons per year and have been declining fairly steadily ever since, reaching about 16 million short tons per year by 2004 (Anon. (US EPA), 1900–1998) and about 12.3 million short tons per year by 2011. This 63% reduction was a major accomplishment, especially in view of the substantial increases in GDP, population, and vehicle miles traveled during this period. Since 1970 VOC emissions from coatings decreased roughly in proportion, accounting for 11–15% of total man-made emissions throughout the period. However, ozone levels were reduced by only 17% between 1990 and 2010. The coatings industry has done a lot to reduce VOC but will certainly be asked to do more.

From 1979 to 1997 the US National Ambient Air Quality Standard for ozone allowed no more than a 1-h period per year (averaged over 3 years) when the concentration of ozone in the air exceeded 0.12 ppm (US EPA,

2015). In 1997 the measurement protocol was changed and the standard was reduced to 0.08 ppm. If was further reduced to 0.075 ppm in 2008 and to 0.070 ppm in 2015 (US EPA, 2016). While there has been gradual improvement, by 2016 few US metropolitan areas had met the 2008 standard (US EPA *Green Book*, 2017), let alone the proposed 2015 standard. The most severe problems continue to be in large areas of California. EPA's ozone standards are controversial; estimates of the cost of implementation and of the benefits vary widely.

Photochemical reactions in the atmosphere are complex and depend on many variables. The amount and structure of VOCs are important, as are the concentrations of various nitrogen oxides. Full discussion of the reactions is beyond the scope of this text.

Dobson (1994), Seinfeld (1989), and Atkinson (1990, 2000) provide brief and more detailed reviews of the science. Probably, the principal pathways leading to the generation of ozone are by way of hydrogen abstraction from the VOC compounds. Some of the reactions that have been proposed to explain ozone generation are shown in Scheme 18.1, in which RH represents a VOC compound. Of course, this scheme is simplistic and does not reflect the many complexities. For details of this evolving subject, the reader is referred to the May 27, 2015 special issue of *Chemical Reviews*, which contains several lengthy reviews on atmospheric chemistry.

Note that nitrogen oxides (NO_x) participate in the ozone-forming reaction sequence shown in Scheme 18.1. A study by the National Research Council (Anon., 1991) concluded that, in many areas of the United States, substantial reductions of ozone can be accomplished only by reducing NO_x levels "in addition to or instead of reducing VOC." Coatings emit almost no NO_x; the main sources are electricity generation and transportation.

An important variable in the rate of ozone generated from VOCs is the ease of abstraction of hydrogens from ROG by free radicals, such as the hydroxyl free radicals shown in the first equation of Scheme 18.1. Atkinson (1990, 2000) published data on the rate constants for a wide variety of organic compounds. In general terms, compounds with hydrogens on carbon atoms alpha to amines or ethers, hydrogens on tertiary carbon atoms, allylic hydrogens, and benzylic hydrogens are examples of easily abstractable hydrogens (Section 5.1). POCP values are determined by rates of reaction of hydroxy free radicals with the various solvents.

$$RH + \cdot OH \longrightarrow H_2O + R\cdot$$
$$R\cdot + O_2 \longrightarrow ROO\cdot$$
$$ROO\cdot + NO \longrightarrow RO\cdot + NO_2$$
$$NO_2 + h\nu \longrightarrow NO + O$$
$$O_2 + O \longrightarrow O_3$$

Scheme 18.1 Simplified equations involved in formation of ozone in the atmosphere.

It was found that the effect of hydroxy free radicals on ozone production used in POCP depended on the ratio of NO_x to ROC. Therefore, Atkinson measured the propensity of different solvents to generate ozone by injecting them into a smog chamber filled with a standard ROG mixture and including NO_x levels appropriate to the area of interest. The incremental effect of the solvent on ozone generation is calculated by dividing the maximum amount of ozone generated by the amount of compound added. Table 18.6 gives the MIR values for some of the solvents tested. See Section 18.9.2 for discussion of aerosol regulations.

VOC emissions can be substantially affected by the *transfer efficiency* in spray-applying coatings. When a coating is sprayed, only a part of the coating is actually applied to the object being coated. Transfer efficiency represents the percentage of the coating used that is actually applied to the product. As the transfer efficiency increases, the VOC emissions decrease, since less coating is wasted. Transfer efficiency depends on many variables, particularly the type of spray equipment utilized (Section 23.2). In some cases, regulations have been established requiring certain kinds of spray equipment or setting a lower limit on transfer efficiency, such as 65%.

Besides reducing the VOC content of coatings and increasing transfer efficiency, two other broad approaches are used to minimize VOC emissions: solvent recovery and incineration. In some cases, it is feasible to recover the solvent evaporated from coatings. One can pass the airflowing out of a drying chamber through activated carbon beds where the solvent is adsorbed. When the carbon becomes saturated with solvent, it is heated to distill and recover the solvent. Alternatively, the solvent-laden air can be passed through condensers cooled with liquid nitrogen; as an added benefit, nitrogen that is vaporized is fed into the

Table 18.6 MIR Values of Selected Solvents, $m(g_{ozone}/g_{solvent})$

Solvent	MIR
Methyl acetate	0.07
PCBTF	0.11
t-Butyl acetate	0.22
Acetone	0.43
n-Butyl acetate	0.89
Odorless mineral spirits	0.91
2-Butanol	1.60
1-Methyl-2-propyl acetate	1.71
VM&P Naphtha	2.03
Toluene	3.97
Methyl isobutyl ketone	4.31
Mixed xylenes	7.37

drying chamber, reducing the oxygen content and, hence, permitting higher concentrations of solvent without exceeding the lower explosive limit. Solvent recovery is desirable when it is feasible, but feasibility is limited by low solvent concentration in the air stream needed to stay safely below the lower explosive limits. The effluent air from spray booths must also be below concentrations that could have toxic hazards; the solvent concentration of this effluence is generally too low to permit economic recovery of the solvent.

Minimizing VOC emissions by incineration involves heating the effluent solvent-laden air stream in the presence of a catalyst to a temperature high enough to burn the solvent. As with solvent recovery, this approach is economically feasible only when solvent concentrations are relatively high. Incineration has been found to be particularly applicable in coil coating (Section 30.4). In this case, most of the solvent is released in the baking oven. Part of the effluent air from the baking oven is recirculated back into the oven; the amount of air recirculation is limited so that the solvent content does not approach the lower explosive limit. The balance of the effluent air is fed to the gas burners that heat the oven. The solvent in the air is burned along with the gas; the fuel value of the solvent reduces the gas requirement. Thus, VOC emission is minimized and the fuel value of the solvent is used.

A high percentage of VOC emissions from auto and truck assembly plants arises from the coating operations. Between 1970 and 1995, VOC emissions from typical auto assembly plants were reduced by 70–80% by the combination of reduced VOC coatings, improved transfer efficiency, and incineration (Praschan, 1995). Progress continues: from 2002 to 2013 Toyota's 14 US manufacturing locations reduced VOC emissions by 63%.

18.9 REGULATION OF SOLVENT EMISSIONS FROM COATINGS

In Europe, Asia, and North America, emissions of VOC and HAP solvents into the atmosphere are limited by government regulations. Regulations are also being developed or considered in parts of South America and Africa. In addition, major efforts are underway to improve indoor air quality. Regulations change periodically, and vast amounts of information about the current situation is available on the Internet. Before describing the regulations, we will consider the problem of measuring VOC.

18.9.1 Determination of VOC

VOC of solventborne coatings containing neither exempt solvents nor water is defined simply by

$$\text{VOC} = (10)(100 - \text{NVW})(\text{density of liquid coating})$$

NVW is nonvolatile by weight, measured by heating a sample of coating at 110°C in a convection oven for 1 h. Density is expressed in g ml^{-1}. The factor of 10 is needed to adjust the VOC units to g l^{-1}. Pounds per gallon can be calculated by multiplying g l^{-1} by 120.

A more general definition is needed for coatings containing water and/or exempt solvents:

$$\text{VOC} = \frac{\begin{array}{l}(10)(\text{wt\% of all volatiles} - \\ \text{wt\% of water} - \text{wt\% of exempt solvents}) \\ (\text{density of coating})\end{array}}{\begin{array}{l}\text{liters of coating} - \text{liters of water} - \\ \text{liters of exempt solvents}\end{array}}$$

A convenient VOC calculator can be downloaded from the South Coast Air Quality Management District (SCAQMD) website.

The definition of VOC as weight/volume follows from the conventions that air pollution data are based on weights of pollutants and that liquid coatings are sold and applied on a volume basis. Regulatory limitations are based on VOC content of the coating "as applied." If, as is very common, the user is to add solvent to the coating after receiving it from the coating manufacturer, the VOC is based on the coating after this solvent has been added. The water terms in the VOC equation are there to exclude coatings having low solids that are diluted with water but still have high VOC contents relative to the dry film thickness that has to be applied. The exempt solvent term is mainly of interest in North America.

Exempt solvents. The early US EPA and California regulations treated almost all solvents useful in coatings as equally undesirable. It was soon realized that certain solvents have negligible photochemical reactivity in the atmosphere, and the EPA now exempts some such solvents from regulation—hence the exempt solvent terms in the above VOC equation. Useful exempt solvents are water, acetone, TBAC, CO_2, 4-chlorobenzotrifluoride (PCBTF), methyl acetate, dimethyl carbonate (DMC), PC, and 2-amino-2-methyl-1-propanol (AMP), a popular solubilizing amine for waterborne coatings. Many states accept the EPA definition, but about half of them review EPA's data before approving them. The exempt list can change. New solvents can be exempted if they are shown to have negligible reactivity, but new information could cause deletions of solvents from the list. CARB Rule 102 (www.arb.ca.gov/.../r102) lists solvents that are exempt in California; TBAC is exempt only in certain types of coatings. Rule 102 is amended fairly often.

Most of the exempt solvents evaporate too rapidly to be used as the entire solvent for most coatings. PCBTF evaporates slower but is more expensive; its solvency is similar to xylene. Ostrowski (2000) discusses formulation with PCBTF.

Several US states and Canada limit paint thinners and strippers to $50\,g\,l^{-1}$ or 3.0 wt% of VOC; to satisfy such regulations these products contain mainly exempt solvents.

A nagging problem: How to define the V in VOC? It might appear easy to calculate or measure the amount of VOC emitted by a coating, but it is not. In most cases, only approximations of potential VOC emissions can be calculated, even when the exact formula of a coating is known. There are many reasons for this. Solvent can be retained in films for very long periods of time. In latex paints, coalescing solvents are used that are only slowly released from the coating. In cross-linking coatings, volatile by-products may be generated by the reaction. For example, MF cross-linking leads to the evolution of a molecule of volatile alcohol for each co-condensation reaction, and in self-condensation reactions, there can be emission of alcohol, formaldehyde, and methylal. The amount released depends on curing conditions and the level of catalyst used. On the other hand, when slow evaporating glycol ether solvents are used in an MF cross-linking system, it is probable that some of the glycol ether transetherifies with the MF resin and is not emitted from the film. Amines used in "solubilizing" water-reducible coatings may volatilize to different extents, depending on the conditions and amine structure (Wicks and Chen, 1978; Hill et al., 1996). As noted previously, 2-amino-2-methyl-1-propanol (AMP) is exempt from regulation as a VOC in the United States and Canada.

With high solids two-package coatings, the amount of volatile material can be affected by many variables, most obviously by the time interval between mixing the two packages and application. Very high solids coatings contain low MW oligomers. Particularly when baked, some of the oligomer may volatilize.

In much of the EU, Canada, and some parts of Asia, any solvent having a boiling point above 250°C is not considered a VOC. While this definition has the great virtue of simplicity, it enables formulators to use coalescing solvents that at least partly evaporate and presumably contribute to ozone formation. The EU is considering raising its ceiling to 280°C. The US EPA, the California Air Resources Board (CARB), and the multistate Ozone Transport Commission (OTC) exempt solvents with boiling points above 216°C, vapor pressures less than 0.1 mm Hg, or more than 12 carbon atoms in many consumer products *but not in most paints* (www.arb.ca.gov/consprod/regs.pdf).

Calculations of VOC from the coatings formula are inexact; nor is experimental measurement of VOC in a coating straightforward. The amount of VOC released depends on the conditions under which the coating is used. Time, temperature, film thickness, airflow over the surface, and, in some cases, amount of cross-linking catalyst are among the variables that affect the results obtained. While it would seem that determination under the conditions of actual use would be most appropriate, this is not easy to do. For air dry coatings, the time required for the determination would be very long. For baking coatings, there can be differences in conditions for use of the same coating. It is generally agreed that it would be desirable to have a standard method for determining VOC. However, there is little agreement as to what that standard method should be; in view of the effect of application variables, it is doubtful whether a single appropriate standard method will ever be developed.

Vo and Morris (2012) illuminated the problem by comparing various solvents and fluids in four VOC test methods:

- US EPA Method 24—weight loss at 110°C in a forced air oven for 1 h. When water or exempt solvents are present, they must be analyzed separately or calculated based on the formula (see below). This method is a variant of ASTM D-2369-89. The European counterpart is ISO 11890-1.
- SCAQMD Method 313—a gas chromatography/mass spectroscopy (GC/MS) method in which all compounds that elute faster than a high boiling marker (methyl palmitate bp 332°C) are said to be VOCs unless they can be identified as exempt solvents. (ASTM Method D6886 is somewhat similar; in Europe the standard method uses GC, ISO 11890-2, but with a lower boiling (250°C) marker.)
- ASTM Method E-1868-10—a thermogravimetric analysis (TGA) method. It has the same limitations regarding water and exempt solvents as Method 24.
- Weight loss at ambient (20–30°C) temperature of 1-g samples in Petri dishes or aluminum pans over 6 months. These tests were performed on "neat" fluids; evaporation from paint films would almost surely be slower.

The authors documented numerous discrepancies among the results obtained by the four test methods. The ambient weight loss experiments led the authors to classify the materials studied as volatile (total weight loss within 6 months), semi-volatile (60–5% weight loss), and nonvolatile (negligible weight loss). The ambient method classified two popular coalescing solvents as volatile even though their boiling points are above 250°C, and they are not VOCs in the EU. Hydrocarbons in the C-15 to C-17 range, with boiling points of 270–302°C, were classed as semi-volatiles.

Arendt et al. (2014) also reviewed VOC test methodologies and applied them to analysis of selected coalescing solvents.

Weight loss methods for determination of VOC are subject to significant experimental errors, as shown by round-robin testing of ASTM D-2369-89. With waterborne coatings the errors are magnified by the need to determine water content; water analysis methods are surprisingly poor. A standard gas chromatography method, ASTM D-3792-99, is reproducible to only 5%. A rather complicated Karl Fischer procedure (ASTM D-4017-02) is reproducible to only 5.5%, an improvement over earlier versions of the procedure. A round-robin test of waterborne automotive coatings by ASTM D-3960-96 had a reproducibility of only 9.75%, with most of the poor reproducibility attributable to the water analysis methods. Note that, in the previous formula, as the ratio of water to regulated solvent increases, errors in water analysis are magnified in the calculation of VOC. This becomes a critical problem for very low VOC paints, such as those formulated to meet current VOC limits of $50\,g\,l^{-1}$. Efforts are underway to develop improved test methods.

As noted previously, an alternative to VOC analysis is to calculate VOC based on the formulation of the coating. This calculation requires knowledge of the solvent content of all coating components and assumptions about what fraction of the solvents are actually emitted and about how much additional VOC is produced by chemical reactions such as cross-linking. Even so, calculated VOC values may be more reliable than measured values in many cases, especially for waterborne coatings with very low VOC.

In the United States and probably elsewhere, paints are marketed as "zero VOC." Sometimes the basis for this claim is that no VOC is detected by US EPA Method 24. In 2017, parts of California approved "Method 313" a rather complex procedure for determination of VOC employing gas chromatography and mass spectrometry. It is expected that this method will provide more accurate measurements.

18.9.2 Regulations

Cogar (2015) provides a summary of VOC regulations in North America, Europe, and Asia. The regulations are complex and are significantly different in different countries and localities. Here we will give a few examples, using architectural paints as an example.

All states and localities within the United States are required to adhere to EPA regulations (Anon. US EPA, 2017). Localities are allowed to set more rigorous rules. California has enforced stricter limits for many years, and quite a few other states have followed suit. The complexity of the VOC regulatory scene in the United States is illustrated by a 2014 table, provided in Cogar (2015), which compares VOC limits by the US EPA for dozens of architectural and industrial maintenance (AIM) coatings with corresponding limits within 23 different jurisdictions in California. And this is just in one state. Additional rulemaking is underway, and the table will need frequent updating (Anon., 2016).

Here we will outline regulation of architectural paints as an example. VOCs for industrial and special purpose coatings are also highly regulated. Readers are referred to websites of large paint manufacturers for current information.

As of 2015, the US EPA allows architectural flat paints to contain $250\,g\,l^{-1}$ of VOC, while key areas in California allow only $50\,g\,l^{-1}$, which will probably be reduced further. The EPA allows $450\,g\,l^{-1}$ for "quick dry enamels," but most of California allows $250\,g\,l^{-1}$ and Canada allows $100\,g\,l^{-1}$. Non-flat and high gloss paints are permitted $380\,g\,l^{-1}$ by the EPA; in California the limits range from 250 to $50\,g\,l^{-1}$, with the lowest limits set by the SCAQMD.

The CARB and the SCAQMD were pioneers in regulating VOCs because ozone and smog problems were acute in California, especially in the Los Angeles area. Through their efforts, smog and ozone problems were substantially improved, especially in coastal areas. Parts of California, however, still have the worst ozone problems in the US California agencies continue to strictly regulate coatings as low VOC technology is judged feasible, and sometimes they adopt "technology forcing" rules. Their rules are sometimes viewed as models for other jurisdictions.

As of 2016 about half of the US states have adopted tighter VOC regulations than those required by the US EPA. For example, the OTC involving about 13 northeastern states developed two model rules specifying different levels of VOC reduction. Most of the member states have enacted the less strict of the model rules and are considering the more strict ones (www.otcair.org).

VOC limits in the EU are set by Council Directive 2004/42/EC, as subsequently augmented. For waterborne matte paints, the limits are $30–40\,g\,l^{-1}$. However, VOC is defined as organic compounds having boiling points below 250°C. Under this definition, certain high boiling coalescing solvents do not count as VOC, and the limits are more lenient than they appear.

In Asia the definitions of VOC are similar to those in Europe, with various differences. China allows VOC of $120–150\,g\,l^{-1}$ for wall paints, while Korea allows $40\,g\,l^{-1}$ (European Union, 2010). China is considering tighter restrictions, and it started to tax VOC in 2015. Japan sets very low limits, but they are voluntary. Many other Asian countries have voluntary limits ranging from 50 to $150\,g\,l^{-1}$.

Each automotive assembly plant is regulated by "local permitting," which varies considerably from place to place. Often regulations are based on total volatile emissions from a factory rather than individual coatings. This approach gives the automakers more flexibility than the approach of regulating individual VOCs; they can use a combination of low VOC coatings, incineration, and solvent recovery to attain the goal. In Europe industrial emissions are often regulated in terms of weight of carbon emitted/area of surface coated, usually expressed in $g\,m^{-2}$. This sensible method seems to be spreading.

In addition to the regulation of VOC, solvents are also regulated to reduce toxic hazards. As described in Section 18.7, the US EPA began regulating selected HAPs in 1990. HAPs on the 1990 list included MEK, MIBK, *n*-hexane, toluene, xylenes, methyl alcohol, ethylene glycol, and ethers of ethylene glycol—all of them important coating solvents at the time. Mandatory HAP limits are included in EPA's Unified Air Toxics Regulations, issued for all major categories of coatings users in 1995–1999. A revised method for analysis of HAPS in coatings has been published (US EPA, 2014).

The limit on HAP emissions at any facility has been set by the EPA at 10 tons per individual HAP per year and a total of 25 tons per year for all HAPs. Lower limits are set for individual chemicals thought to be more dangerous. Over the years, several solvents including TBAC, MEK, and 2-butoxy ethanol have been removed from the HAPs list. Also removed was caprolactam, not a solvent but important in powder coatings. Some US states developed their own lists that include solvents not on the national HAP list. The present HAP lists provide motivation for replacing listed solvents with solvents not on the list. Most mono- and diethers of ethylene glycol, diethylene glycol, and triethylene glycol are listed on the HAP list, but ethers of propylene glycol are not. Such substitutions may do little or nothing to reduce hazards or to improve air quality if the unlisted solvent is equally hazardous as the listed one.

Additional regulation of toxic solvents is in place or in process in Europe, where the 2007 REACH legislation requires extensive testing of all major chemicals. Full implementation is scheduled for 2018. The European Commission's reports on progress (Anon., 2017) describe considerable success. On the other hand, Bergkamp and Herbatscheck (2014) point out that the program establishes several regulatory regimes that are supposed to be complementary but often are not. For example, there is a regulatory structure for authorizing chemicals and a separate structure for restricting them. It can be perplexing for manufacturers to know which to follow. The authors illustrate the difficulties for the case of polar, aprotic solvents.

In establishing future regulations, there are differences of opinion as to whether all nonexempt solvents should be considered as equally undesirable in the atmosphere, as they usually are now. However, it may well be that replacing more reactive with less reactive solvents would be advantageous by allowing at least some opportunity for dissipation in the atmosphere to minimize the probability of local excess ozone concentrations. The present approach is enforceable but may not be optimal. Some steps in this direction have been taken. In Europe, some regulations are based on the POCP of individual solvents. Atkinson (1990) provides a list of POCP values and examples of reformulation of solvents to minimize POCP emissions. The POCP values have been calculated by computer modeling, using reaction kinetic data to assess changes in ozone formation and checked in smog chambers. Values for about 200 VOCs are available. Xylene and related aromatic hydrocarbons have particularly high POCP values.

Table 18.7 MIR Limits ($g\,l^{-1}$) for Selected Aerosol Coatings

Flat paints	1.21
Non-flat paints	1.40
Primers	1.11
Auto body primers	1.57
Automobile top coats	1.77
Auto bumpers and trim	1.75
Traffic paint	1.18
Marine spar varnish	0.87

The CARB has also begun to depart from the total VOC approach. CARB's regulations controlling solvents (and propellants) used in aerosol coatings are based on MIR values ($g_{ozone}/g_{solvent}$) (Table 18.6) rather than the total solvent. MIR values can then be used to calculate the MIR values of any mixture of solvents, which in turn permits the calculation of the MIR/gram of coating. Before regulation, the emissions of ROC from aerosol coatings in California were estimated to be 26.5 tons per day with the capacity to produce 56.3 tons of ozone per day. Table 18.7 provides the regulatory amounts of grams of ozone per gram of product for selected types of aerosol coatings. Other states are beginning to follow CARB's lead. CARB projects tighter limits in 2017.

Indoor air quality. VOC regulations are aimed at protecting the outdoor atmosphere in the regulated regions. Indoor air is also a concern (Cogar, 2015). A number of organizations promote use of paints that have minimal effect on indoor air. One example is the "Leadership in Energy and Environmental Design" (LEED) green building certification process. By using low emitting paints, builders and remodelers can gain points toward certification as green buildings. Protocols for measuring emissions from paints during the first month after application are available.

GENERAL REFERENCES

Ash, M.; Ash, I., *Handbook of Solvents*, 2nd ed., Synapse Information Resources, Endicott, **2003**.

Chemical Reviews special issue "Chemistry in Climate," American Chemical Society, Washington, DC, May 27, **2015**.

Ellis, W. H., *Solvents*, Federation of Societies for Coatings Technology, Blue Bell, **1986**.

Lagowski, J. J., *The Chemistry of Nonaqueous Solvents*, a six-volume treatise, Academic Press, New York/London, **1970–2012**.

Wypych, G., Ed., *Handbook of Solvents*, ChemTech/William Andrew, New York/Toronto, **2001**.

REFERENCES

Alfonsi, K., et al., *Green Chem.*, **2008**, *10*, 31–36.

Anon., Summary of Architectural Coating Rules in California as of December, **2016**, from http://www.arb.ca.gov/coatings/arch/rules/VOClimits (accessed April 21, 2017).

Anon., from http://www.ec.europa.eu>EuropeanCommission>Environm ent>Chemicals (accessed March 20, **2017**).

Anon. (National Research Council), *Rethinking the Ozone Problem in Urban and Regional Air Pollution*, National Academy Press, Washington, DC, **1991**.

Anon. (US EPA), *National Air Pollution Emission Trends*, **1900–1998**. *Air Emission Trends—Continued Progress Through 1994*. Detailed reports are accessible on the EPA web site, www.epa.gov and are updated periodically.

Anon. (US EPA), *Hazardous Air Pollutants Strategic Implementation Plan*, **2017**, from http://www.epa.gov/ttn/atw/index.html (accessed March 20, 2017).

ARCO Chemical Co., ARCOCOMP Computer Solvent Selector Program, Newton Square, **1987**.

Arendt, W.; Conner, M. M.; McBride, E., *Paint and Coatings Industry*, **2014**, May 1.

Atkinson, R., *Atmos. Environ.*, **1990**, *24A*, 1.

Atkinson, R., *Atmos. Environ.*, **2000**, *32*, 2063–2101.

Bauer, D. R.; Briggs, L. M., *J. Coat. Technol.*, **1984**, *56(716)*, 87.

Bergkamp, L.; Herbatschek, N., *Intl. Eur. Chem. Regul.*, **2014**, *23(2)*, 221–245.

Brandenburger, L. B.; Hill, L. W., *J. Coat. Technol.*, **1979**, *51(659)*, 57.

Brezinski, J. J.; Litton, R. K., Regulation of Volatile Organic Compound Emissions from Paints and Coatings in Koleske, J. V., Ed., *Paint and Coating Testing Manual*, ASTM, Philadelphia, **1995**, pp 3–14.

Burrell, H., *Off. Digest*, **1955**, *27*, 726.

Cogar, J., *Paint and Coating Industry*, **2015**, January 6. www.pcimag.com/events (accessed March 20, 2017).

Constable, D. J. C., Green Chemistry Innovation and Opportunity in the Pharmaceutical and Specialty Chemical Industries, **2013**. www.industrialgreenchem.com (accessed April 17, 2017).

Cooper, C., et al., *J. Coat. Technol.*, **2001**, *73(922)*, 19.

Croll, S. G., *J. Coat. Technol.*, **1987**, *59(751)*, 81.

Dillon, P. W., *J. Coat. Technol.*, **1977**, *49(634)*, 38.

Dobson, I. D., *Prog. Org. Coat.*, **1994**, *24*, 55.

DTSC.ca.gov, Priority Product Work Plan, **2014**.

Ellis, W. H., *J. Coat. Technol.*, **1976**, *48(614)*, 45.

Ellis, W. H., *J. Coat. Technol.*, **1983**, *53(696)*, 63.

Erickson, J. R.; Garner, A. W., *Org. Coat. Plast. Chem. Prepr.*, **1977**, *37(1)*, 447.

European Union, Directive 2010/75/EU of the European Parliament and of the Council, November 24, **2010**.

Grulke, E. A., Solubility Parameter Values in Brandrup, J., et al., *Polymer Handbook*, 4th ed., John Wiley & Sons, Inc., New York, **1999**, p VII/675.

Hahn, F. J., *J. Paint Technol.*, **1971**, *45(562)*, 58.

Hansen, C. M., *The Three Dimensional Solubility Parameter and Solvent Diffusion Coefficient*, Thesis, Danish Technical Press, Copenhagen, **1967**.

Hansen, C. M., *Ind. Eng. Chem. Prod. Res. Dev.*, **1970**, *9*, 282.

Hansen, C. M., Solubility Parameters in Koleske, J. V., Ed., *Paint & Coatings Testing Manual*, 14th ed., ASTM, Philadelphia, **1995**, pp 383–404.

Hazel, N., et al., *Proceedings of the Waterborne High-Solids Powder Coatings Symposium*, New Orleans, LA, **1997**, p 237.

Hildebrand, J., *J. Am. Chem. Soc.*, **1916**, *38*, 1452.

Hill, L. W.; Wicks, Jr., Z. W., *Prog. Org. Coat.*, **1982**, *10*, 55.

Hill, L. W., et al., Dynamic Mechanical Analysis of Property Development During Film Formation in Provder, T., et al., Eds., *Film Formation in Waterborne Coatings*, American Chemical Society, Washington, DC, **1996**, p 235.

Hoy, K. L., *Tables of Solubility Parameters*, Union Carbide Corp., Chemicals & Plastics, R & D Dept., Charleston, **1969**.

Hoy, K. L., *J. Paint Technol.*, **1970**, *42(541)*, 76.

Huyskens, P. L.; Haulait-Pirson, M. C., *J. Coat. Technol.*, **1985**, *57(724)*, 57.

Jackson, H. L., *J. Coat. Technol.*, **1986**, *58(741)*, 87.

Jones, F. N., *J. Coat. Technol.*, **1996**, *68(852)*, 25.

Kamlet, M. J., et al., *Chemtech*, **1986**, September, 566.

Lasky, R. C., et al., *Polymer*, **1988**, *29*, 673.

Linak, E.; Bizzari, S. N., Global Solvents: Opportunities for Greener Solvents, **2013**, from https://www.ihs.com (accessed April 21, 2017).

McGovern, J. L., *J. Coat. Technol.*, **1992**, *64(810)*, 33, 39.

Newman, D. J.; Nunn, C. J., *Prog. Org. Coat.*, **1975**, *3*, 221.

Ostrowski, P., *Proceedings of the Waterborne High-Solids Powder Coatings Symposium*, New Orleans, LA, **2000**, pp 448–459.

Overdiep, W. S., *Prog. Org. Coat.*, **1986**, *14*, 159.

Patton, T. C., *Paint Flow and Pigment Dispersion*, 2nd ed., Wiley-Interscience, New York, **1979a**, p 109.

Patton, T. C., *Paint Flow and Pigment Dispersion*, 2nd ed., Wiley-Interscience, New York, **1979b**, pp 306–310.

Patton, T. C., *Paint Flow and Pigment Dispersion*, 2nd ed., Wiley-Interscience, New York, **1979c**, p 340.

Praschan, E. A., *ASTM Standardization News*, **1995**, October, 24.

Rocklin, A. L., *J. Coat. Technol.*, **1976**, *48(622)*, 4.

Rocklin, A. L., *J. Coat. Technol.*, **1986**, *58(732)*, 61.

Rocklin, A. L.; Edwards, G. D., *J. Coat. Technol.*, **1976**, *48(620)*, 68.

Seinfeld, J. H., *Science*, **1989**, *243*, 745.

Skjold-Jorgenson, S., et al., *Ind. Eng. Chem. Prod. Res. Dev.*, **1979**, *18*, 714.

Smith, R. L., et al., *J. Coat. Technol.*, **1987**, *59(747)*, 21.

Sprinkle, Jr., G. F., *Modern Paint & Coatings*, **1983**, April, 44.

Sullivan, D. A., *J. Paint Technol.*, **1975**, *47(610)*, 60.

US EPA, EPA/600/R-10/076F, *Integrated Science Assessment for Ozone and Related Photochemical Oxidants*, Washington, DC, **2013**, www.epa.gov/ord (accessed March 20, 2017).

US EPA, *Method 311—HAPS in Paints & Coatings*, Technology Transfer Network, Emission Measurement Center, **2014**.

US EPA, *National Ambient Air Quality Standards (NAAQS) for Ozone*, **2015**, from http://www.epa.gov (accessed March 20, 2017).

US EPA, *Table of Historical Ozone National Ambient Air Quality Standards (NAAQS)*, **2016**, from http://www.epa.gov (accessed March 20, 2017).

US EPA, **2017**, from https://www.epa.gov/green-book (accessed April 17, 2017).

Vo, U-U. T.; Morris, M. P., *Non-Volatile, Semi-Volatile or Volatile: Redefining Volatile for Volatile Organic Compounds*, South Coast Air Quality Management District, Diamond Bar, **2012**, from http://www.aqmd.gov (accessed March 20, 2017).

Vrentas, J. S., et al., *Macromolecules*, **1996**, *29*, 3272.

Waggoner, R. A.; Blum, F. D., *J. Coat. Technol.*, **1989**, *61(768)*, 51.

Watson, B. C.; Wicks, Jr., Z. W., *J. Coat. Technol.*, **1983**, *55(698)*, 59.

Wicks, Jr., Z. W.; Chen, G. F., *J. Coat. Technol.*, **1978**, *50(638)*, 39.

Wicks, Jr., Z. W., et al., *J. Coat. Technol.*, **1982**, *54(688)*, 57.

Wu, S. H., *J. Appl. Polym. Sci.*, **1978**, *22*, 2769.

Wu, D. T., et al., *FATIPEC Congress Book*, Aachen, Vol. IV, **1988**, p 227.

Yoshida, T., *Prog. Org. Coat.*, **1972**, *1*, 72.

Chapter 19

Color and Appearance

Color and the interrelated topic of gloss are important to the decorative aspects of coatings and, sometimes, to their functional aspects. We have all experienced color since we were babies, but most people have little understanding of color. Many technical people think of color as an aspect of physics dealing with the distribution of visible light. While that is a factor, color is a psychophysical phenomenon. The difficulty of understanding color can be grasped by considering its most rigorous definition: color is that characteristic of light by which an observer may distinguish between two structure-free fields of view of the same size and shape. In effect, it says that color is what is left to distinguish between two objects when all the other variables are removed—not a very satisfying definition.

Color has three components: an observer, a light source, and an object. (The single exception is when the light source is the object being viewed.) There is no color on an uninhabited island. This is not just a semantic statement; color requires an observer. There is no color in the absence of light; in a completely darkened room, there is no color, not because you cannot see it, but because it is not there. There must be an object; if you look out the window of a spaceship without looking at a planet or star, there is no color—there is an observer, there is light, but there is no object.

Appearance of a coating is affected by gloss as well as color. The eye can resolve irregularities in surface smoothness of approximately 25 μm, depending on the distance from the object. Below that size, the general texture of the surface determines its level of gloss. If a surface is very smooth at a size scale on the order of the wavelength of light or less, the surface will be perceived as having a high gloss. If the surface is rough at a size scale on the order of the wavelength of the illuminating light or slightly greater, the surface will appear to have low gloss. In addition to the overall roughness, heterogeneities in the surface, such as brush marks or orange peel, can exist in both high and low gloss coatings. Adding to the complexity, color and gloss interact; changing either one changes the other.

19.1 LIGHT

Light is electromagnetic radiation to which our eyes are sensitive. The range of visible wavelengths varies somewhat among individuals, but in most cases it is 390 through 770 nm. The sensitivity of the eye varies as a function of wavelength (Figure 19.1) and has a maximum in the region of green light (Billmeyer and Saltzman, 1981). Figure 19.1 also shows the response of photomultiplier tubes and silicon photodiodes, two solid-state technologies used to detect light photons. The eye is relatively insensitive to the shorter and longer ends of the range. The response of our eyes depends on the distribution of wavelengths of light emitted by the light source. In the case of monochromatic light sources, the colors we see range from violet through blue, green, yellow, and red with increasing wavelength. We see different colors as the ratios of wavelengths in polychromatic light sources change. If we look at a light source with nearly equal content of all wavelengths, we see white.

Sunlight is considered the standard light, but sunlight varies depending on the time of day, latitude, season, cloudiness, and so forth. When work toward understanding color began, light from overcast north sky (in the northern hemisphere) was accepted as the standard. Based on many measurements of energy distribution, also called the *spectral power distribution* (SPD), a light source designated as D_{65} was adopted as a standard related to average daylight. The mathematical description of a light source is called an *illuminant*. A graph of standard illuminant D_{65} is

Organic Coatings: Science and Technology, Fourth Edition. Frank N. Jones, Mark E. Nichols, and Socrates Peter Pappas.

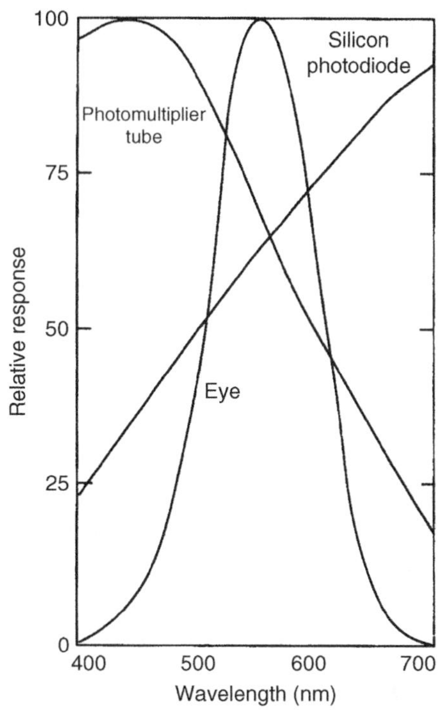

Figure 19.1 Sensitivity of the eye, photomultiplier tube, and silicon photodiode as a function of wavelength. Source: Billmeyer and Saltzman (1981). Reproduced with permission of Elsevier.

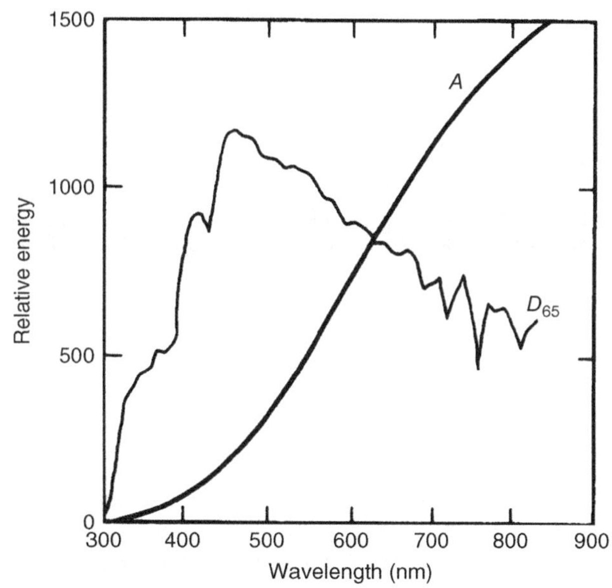

Figure 19.2 Relative spectral power distributions of CIE standard illuminants A and D65. Source: Billmeyer (1979). Reproduced with permission of Elsevier.

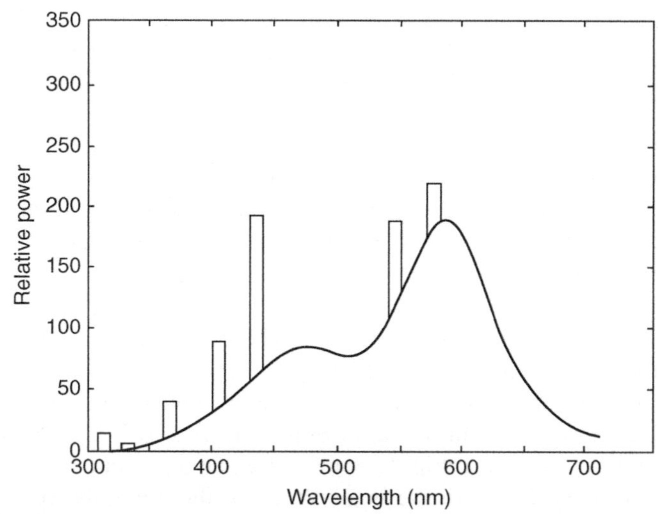

Figure 19.3 Spectral power distribution of a cool white fluorescent lamp (IES 1981). Source: Billmeyer and Saltzman (1981). Reproduced with permission of Elsevier.

shown in Figure 19.2 (Billmeyer, 1979). Tables giving energy distributions as a function of wavelength for various bandwidths from 1 to 20 nm wide are available (Wyszecki and Stiles, 1982).

The SPD from tungsten lights is different. Another standard light source, *A*, is a carefully specified tungsten light operated under specified conditions. A graph of illuminant *A* is also shown in Figure 19.2, and tables of its energy distributions are available. Fluorescent lights are another type of light source, and many types are available. As shown in Figure 19.3, they exhibit a continuous energy distribution with peaks at a series of wavelengths due to specific phosphors used in the manufacturing of the fluorescent bulbs. Even if the underlying continuous spectrum were the same as D_{65}, the peaks would lead to changes in colors when these lights are used as a light source in comparison with daylight.

Indoor lighting is undergoing a revolution, mandated by improved energy efficiency. Light-emitting diodes (LEDs) are being used more frequently as indoor light sources. The SPD of a variety of warm LEDs used in indoor lighting is shown in Figure 19.4 (Rich, 2016). LEDs themselves produce very narrow wavelength distributions of light. However, white light is produced by either mixing LEDs of various colors together within a single bulb or, more commonly, coating the glass surrounding the LED with a phosphor that emits light when excited by the blue LED inside the glass.

19.2 LIGHT–OBJECT INTERACTIONS

An important factor that affects color is the interaction between the light and the object being viewed.

19.2.1 Surface Reflection

When a light beam is directed at a surface, some light is reflected at the surface and some passes into the object. As shown in Figure 19.5, if the surface is optically smooth, light

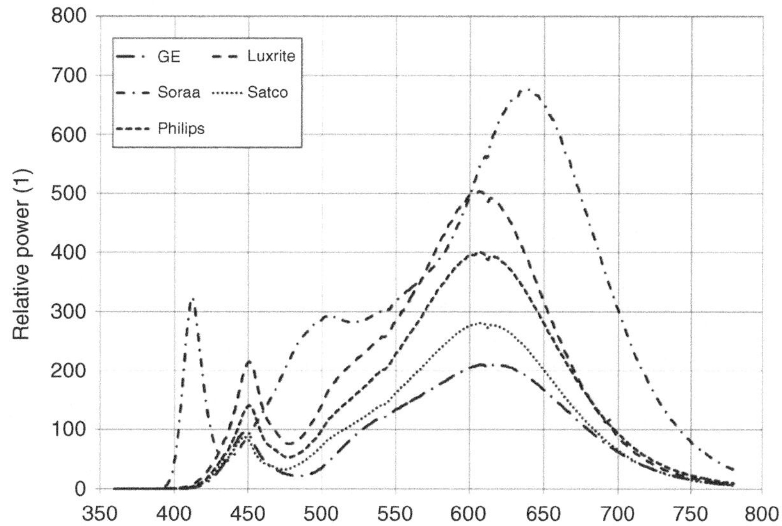

Figure 19.4 Spectral power distribution of a variety of white LED light sources. Source: Rich (2016). Reproduced with permission of American Coatings Association.

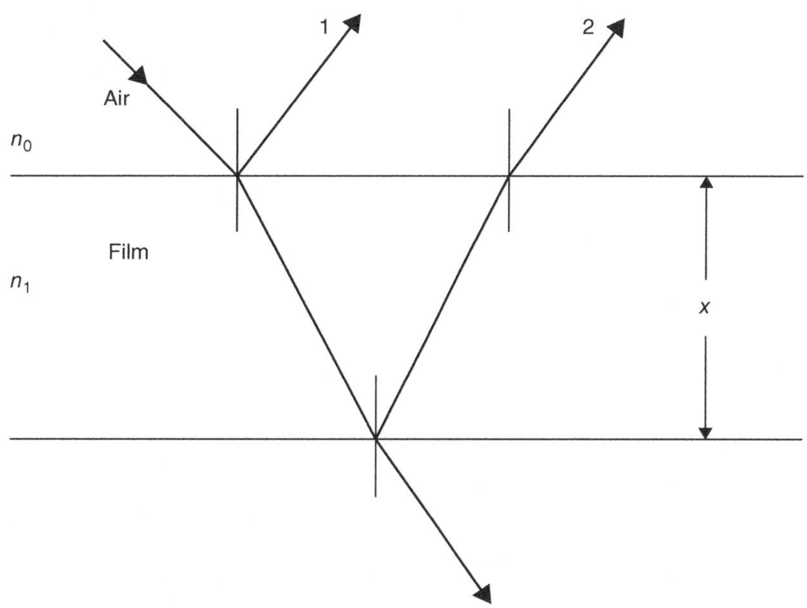

Figure 19.5 External and internal reflection and refraction of light by a nonabsorbing film (refractive index, n_1, thickness, x) with optically smooth parallel surfaces. Source: Greenstein (1988). Reproduced with permission of John Wiley & Sons.

is reflected at the same angle (r) as the angle of incidence (i) (Greenstein, 1988). This kind of reflectance is called *specular* (mirrorlike) *reflectance*. By convention, an angle of incidence normal to the surface is designated as 0° and the grazing angle is 90°. The fraction of light reflected (R) varies with angle of incidence and difference in refractive index (n) between the two phases. If there is no difference in index of refraction, no light is reflected at the interface; as the difference increases, the fraction of light reflected increases. For angles of incidence near 0°, the fraction reflected can be calculated using the following simplified version of the Fresnel equations given in both the general form, Eq. 19.1a, and where the first medium is air ($n_1 = 1$), Eq. 19.1b:

$$R = \left(\frac{n_2 - n_1}{n_2 + n_1}\right)^2 \tag{19.1a}$$

$$R = \left(\frac{n-1}{n+1}\right)^2 \tag{19.1b}$$

Most resins have refractive indexes of about 1.5; Eq. 19.1b shows that approximately 4% of incident light is reflected when the angle of incidence is near 0°. The dependence of reflection on angle of incidence is illustrated in Figure 19.6 (Judd and Wysecki, 1975); reflectance approaches 100% as the angle of incidence approaches 90°. The family of curves in Figure 19.6 illustrates the combined effects of refractive index difference and angle of incidence.

Light not reflected at the surface enters the object. When a light beam enters an object, it is *refracted*; that is, it is bent, also shown in Figure 19.5. The angle of refraction varies with the ratio of the refractive indexes of the two media and is described by Snell's law (Eq. 19.2):

$$\sin r = \frac{n_1}{n_2} \sin i \tag{19.2}$$

When light passes from air into a nonabsorbing polymeric film having parallel, optically smooth surfaces and an index of refraction of 1.5, the angle of refraction is smaller than the angle of incidence (Figure 19.5). If the angle of incidence is near 0°, the fraction reflected at the first surface is 0.04, and 0.96 is transmitted into the film. If there is no absorption, 0.96 reaches the second film/air interface. There, 0.96×0.96 of the original light is transmitted into the air on the other side of the film. The light transmitted into the air is also refracted so that the angle of refraction is equal to the original angle of incidence. There is also reflection at the second surface, and 0.04×0.96 is reflected back toward the first surface. Again, there is reflection of 4% and transmission of 96% at the back of the first surface. As a result, $(0.96)^2 \times (0.04)$ of the original light is transmitted out of the film at an angle equal to the original angle of incidence, and $(0.96) \times (0.04)^2$ is reflected back into the film. This bouncing back and forth continues, if there is no absorption, until the total transmission equals 0.92 and total reflectance equals 0.08.

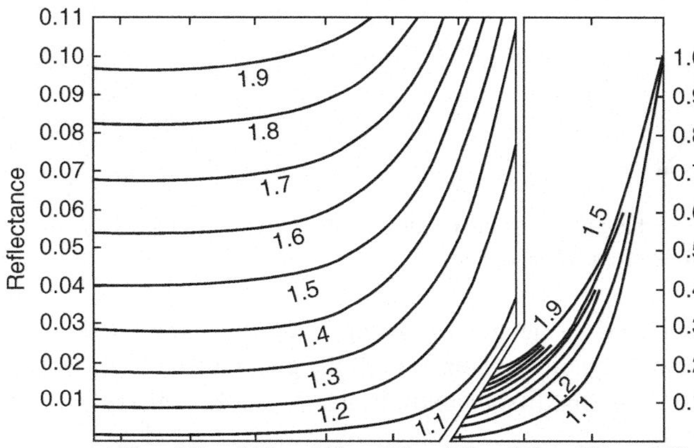

Figure 19.6 Fraction of light reflected from a smooth surface as a function of the angle of incidence, *i*, with various differences in refractive index. Source: Judd and Wysecki (1975). Reproduced with permission of Elsevier.

When a light beam passes from a medium of higher refractive index to one of lower refractive index, the angle of the beam increases. If the angle of incidence is high enough, all light is reflected back and none is transmitted out. If the refractive indexes are 1.5 and 1, the critical angle (at which all the light is reflected) is calculated to be 41.8°.

Total internal reflection is the basis for optical fibers. If light enters the end of a nonabsorbing fiber in which there are no bends, such that the angle of incidence with the inner surface of the fiber is less than its critical angle, all of the light will follow the fiber, even all the way across an ocean.

If the thickness of a film is small (60–250 nm), interference colors will be seen. The intensity of some wavelengths of reflected light is enhanced, and others are transmitted selectively. With a refractive index of 1.5 in air, at a film thickness of 66 nm, bluish white light is reflected and yellowish white light is transmitted; as thickness is increased, other colors are preferentially reflected. The colors seen are affected by the angle of incidence of the light. This phenomenon can easily be observed in a puddle of water that contains a small amount of oil and is illuminated by sunlight. The rainbow of colors is due to the small differences in the thickness of the oil on top of the water and the varying angle of incidence.

19.2.2 Absorption Effects

In almost all coatings, the color we observe is affected by differential absorption of various wavelengths of light. Colorants, dyes, pigments, and, to a degree, some resins absorb some wavelengths of light more strongly than others. These absorptions are controlled by the chemical structures of the colorants. First, we consider the absorption effects in transparent systems, that is, systems in which the colorant is in solution or is so finely divided that it does not significantly reflect light at the colorant–resin interface. The extent of absorption depends on chemical composition, wavelength, particle size, optical path length (film thickness), concentration, and media–colorant interactions.

Each colorant has an absorption spectrum that controls the absorption of various wavelengths of light. Chemists generally speak of molar absorptivities, ε, in units of $1\,\text{mol}^{-1}\,\text{path length}^{-1}$, corresponding to the absorbance per molar concentration of a substance. Physicists generally speak of absorption coefficients, K, in units of path $\text{length}^{-1}\,\text{mass}^{-1}$, corresponding to the absorbance per unit mass of the system.

The smaller the particle size, the greater the fraction of light absorbed by the same quantity of a colorant. The highest molar absorptivities are exhibited by individual molecules in solution. In the case of pigments, the smaller the particle size of a pigment, the greater the absorption. The longer the path followed by a beam of light through a medium containing absorbers, the greater the degree of absorption. In transmission of a beam of light at 0°, the optical path equals the film thickness. At any other angle, the optical path length is greater than the film thickness. If, in passing through a unit path length, one-half of the light of a particular wavelength is absorbed and one-half is transmitted (ignoring surface reflection), on passing through two units of path length, three-fourths is absorbed and one-fourth is transmitted. Mathematically this relationship is expressed by an exponential equation known as the Beer–Lambert law, which is shown in Eqs. 19.3a and 19.3b, in which X is path length, I is intensity of light transmitted, and I_0 is the original light intensity.

Chemists commonly use base 10 (Eq. 19.3a), and physicists base e (Eq. 19.3b):

$$\frac{I}{I_0} = 10^{-\varepsilon X} \tag{19.3a}$$

$$\frac{I}{I_0} = e^{-KX} \tag{19.3b}$$

Ideally, the same relationship holds when the concentration of the colorant in the medium is changed (Eq. 19.4):

$$\frac{I}{I_0} = 10^{-\varepsilon CX} \tag{19.4}$$

Equation 19.4 holds only over a limited range of concentrations; the width of the range is system dependent. When absorbers are in solution, there are less likely to be intermolecular interactions between molecules in dilute solutions than in more concentrated solutions. In the case of pigment dispersions, concentration effects are likely to be larger. Possible interactions with the media are another complication. In the case of solutions, a change in solvents can sometimes lead to association of molecules, effectively increasing particle size and hence reducing absorption. There can also be effects of hydrogen bonding between a soluble dye molecule and different solvents. This changes the structure, leading to a change in absorption spectrum. In dispersions of pigments, changes in the medium, such as dilution with solvent, can lead to flocculation (agglomeration) of pigment particles; this causes a larger particle size and, hence, reduced absorption.

Figure 19.7 shows transmission spectra of idealized reddish magentas. Spectra a and b result from transmission

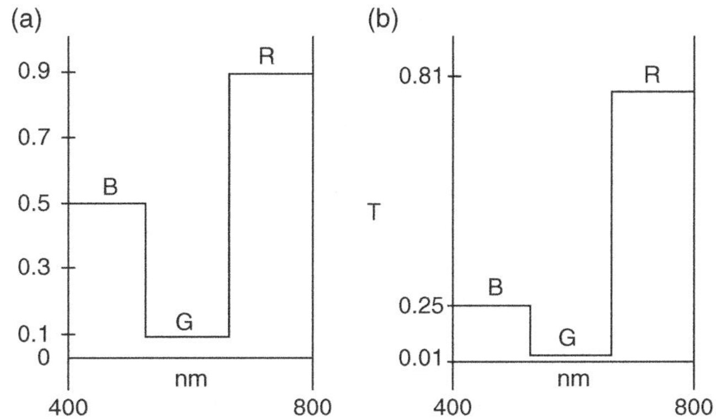

Figure 19.7 Transmission spectra of idealized magentas: (a) path length = x; (b) path length = $2x$.

of light through the same transparent coating of path length x and $2x$, respectively. (Surface reflection effects are ignored.) The fraction of light of all wavelengths transmitted through the thicker coating, shown in spectrum b, is less. But also the relative transmissions of the blue (B), green (G), and red (R) parts of the spectra are different. The color seen in case b would be a redder purple than in case a. The composition is the same, but the shade of the color is affected by path length. The same type of change takes place if the concentration is doubled at the same film thickness. This is an idealization of a real situation. Reddish magenta is one of the inks in four-color process printing that permits printing of multicolored pictures. The film thickness of magenta ink affects the color of the magenta print and, hence, the combined color picture.

19.2.3 Scattering

Scattering is another phenomenon that can occur during the passage of light through a film. If there are small particles dispersed in the film that have a different refractive index than the medium, light is reflected and diffracted at the interfaces between the particles and the medium. The physics involved in scattering is complex and beyond the scope of this brief presentation, but the results can be stated in a simplified manner. When a beam of light passes through a film containing nonabsorbing particles, the light is reflected in all directions, so it changes from a beam to diffuse illumination inside the film. Light that reaches the back of the top surface at angles greater than the critical angle is reflected back into the film; part of the light reaching this surface at angles less than the critical angle leaves the film. If the film is thick enough, no light can pass completely through it; all the light is reflected back out of the top surface. The reflection, however, is not just at the specular angle, but at all angles, that is, this light is diffusely reflected even though the top surface of the film is optically smooth. The extent of light scattering by nonabsorbing particles depends on the refractive index difference between particles and medium, particle size, film thickness, and particle concentration.

The greater the difference in refractive index, the greater the degree of light scattering. The degree of scattering increases steeply as the difference in refractive index increases (Figure 19.8). The degree of scattering is the same whether the particles have a higher or lower refractive index than the medium. For example, water droplets in air (fog) scatter light as efficiently as air droplets of the same size at the same concentration in water (foam). An ideal white pigment would absorb no light and have a very high refractive index, so the difference in refractive index from that of binders would be large. Rutile TiO_2 comes close to meeting these specifications; it has an average refractive index of 2.73, but it absorbs some light below about 420 nm. Another crystal type of TiO_2, anatase, absorbs less light, but its refractive index is lower, 2.55. The smaller index of refraction difference leads to less efficient scattering.

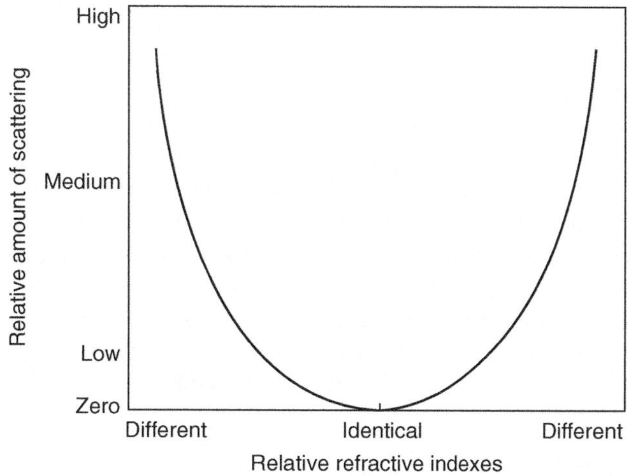

Figure 19.8 Scattering as a function of refractive index difference; particles have higher refractive indexes than the media on the right-hand part of the curve and lower values on the left. Source: Billmeyer and Saltzman (1981). Reproduced with permission of Elsevier.

Scattering is affected by particle size (Figure 19.9). For rutile TiO_2 in a resin with a refractive index of about 1.5, the scattering coefficient for 560 nm light increases with particle size to a maximum at 0.19 μm diameter and then drops off steeply (Mitton, 1973). Commercial TiO_2 has a range of particle sizes. Since efficiency drops off more rapidly on the small diameter size side of the maximum, TiO_2 pigments are produced with an average particle diameter somewhat over 0.2 μm. See Fields (1993) for calculations showing the effect of particle size distribution and particle agglomeration on scattering.

The particle diameter with maximum scattering depends on refractive index difference. For calcium carbonate ($n = 1.57$), the scattering coefficient goes through a maximum at about 1.7 μm (note the use of different scales). As expected from the small difference in refractive index, scattering by even optimum particle sizes is low.

The extent of light scattering is affected by film thickness. If no absorption occurs, the light that is not reflected back out of the top surface is transmitted, unless the film thickness is great enough so that all the light is reflected back. Scattering is also affected by concentration. At low concentrations of particles, such as rutile TiO_2, transmission through a film (neglecting surface reflection) follows a relationship (Eq. 19.5) similar to that for absorption:

$$\frac{I}{I_0} = e^{-ScX} \tag{19.5}$$

As concentration increases, however, scattering efficiency reaches as maximum then decreases. In the case of rutile TiO_2 pigmented coatings, the efficiency decreases so much that finally the fraction of light transmitted increases and the fraction reflected decreases. The result can be seen in a plot of scattering coefficient S as a function of pigment

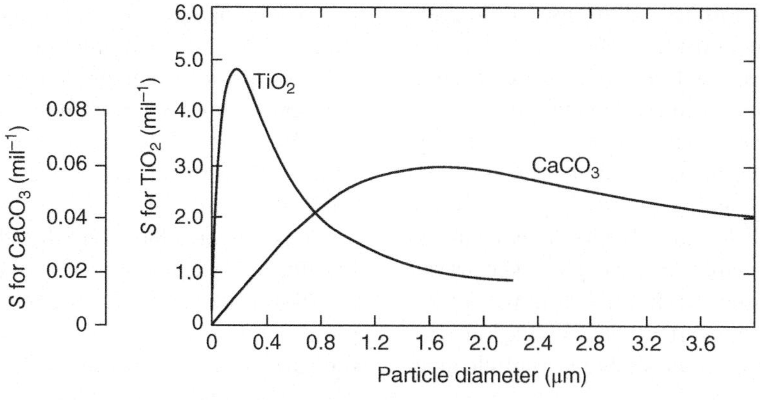

Figure 19.9 Scattering coefficients as a function of particle size for rutile TiO_2 and $CaCO_3$. Source: Mitton (1973). Reproduced with permission of John Wiley & Sons.

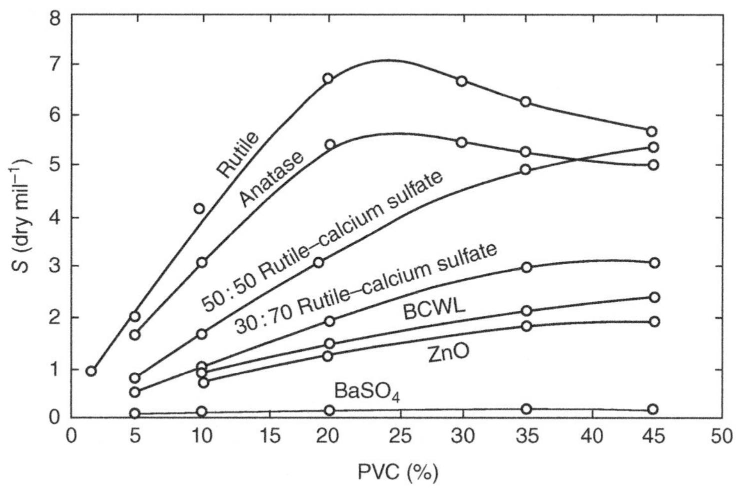

Figure 19.10 Experimental plots of scattering coefficients versus PVC for selected pigments in a dry acrylic lacquer. BCWL is basic carbonate of white lead. Source: Mitton (1973). Reproduced with permission of John Wiley & Sons.

volume concentration (PVC) of rutile TiO_2 in dry films (Sharma et al., 2005) (Figure 19.10). PVC is defined as the volume percent pigment in a *dry* film of coating. In commercial practice, the cost effectiveness of pigmenting with TiO_2 drops off sufficiently at PVCs above about 18% that it is not generally economically sound to use a higher PVC of TiO_2. The optimum value varies somewhat from system to system; it depends on how well the pigment is dispersed in the film and on the TiO_2 content of the particular TiO_2 pigment, which may vary from grade to grade by more than 10% (Section 20.1.1).

The drop-off in scattering effectiveness as the concentration of scatterers is increased is due to crowding effects. As the PVC of TiO_2, or any other scatterer, is increased, the distance between particles is decreased. When particles move very close together or touch, their effectiveness as scatterers decreases substantially, and the effect of increasing PVC decreases and eventually leads to decreasing opacity with increasing PVC. This trend dramatically reverses once the critical pigment volume concentration (CPVC) of the coating is reached and entrapped air begins to act as an additional scattering center.

19.2.4 Multiple Interaction Effects

The effects of surface reflection, absorption, and scattering are interdependent. Light sources generally are not narrow beams, but relatively broad beams or diffuse sources. Surfaces are not often optically smooth; in some cases, we strive for surface roughness, such as with matte coatings. In coatings, we seldom deal with pigments or pigment combinations that only absorb or only scatter light; commonly, both occur simultaneously. We are seldom interested in free films, but rather films on a substrate so that the reflectance at the bottom surface of a film is not the same as that at the top surface of the film. The eye does not distinguish light that is reflected from the top surface, from within the film, or from the bottom surface of a film. The combined light from all three sources is integrated when the eye responds.

Surface reflectance increases as the angle of incidence increases (Figure 19.6). If a panel of a high gloss (smooth surface) blue paint illuminated with a beam of white light is observed at any angle other than the specular angle, one sees a comparatively dark blue color, resulting from the diffuse reflectance back out of the film. If the angle of incidence approaches 90°, one sees a very light blue color at the specular angle because the observed light contains a high fraction of surface reflected light and only a little light from within the film. If the illumination is diffuse instead of a beam, the color is a lighter blue from most angles of observation because one sees a higher fraction of surface reflected light. If the blue paint is low gloss (rough surface) with exactly the same composition as in the previous example, a lighter blue color is seen under diffuse illumination at most viewing angles, as compared with the high gloss paint. This results from a greater ratio of surface reflected light reaching the eye. The color is darkened at most viewing angles by wetting the surface with water, since water fills in the surface roughness.

The effects of pigment concentration and film thickness on coatings that absorb or weakly scatter light without rescattering the primary scattered light can be modeled by simple equations. However, interactions resulting from both absorption and multiple scattering together are more complex, and equations that model reflection and transmission are more complex. Significant progress has been made in both analytical and computer models that accurately predict the scattering and appearance of simple pigmented films. The earliest successful model was the Kubelka–Munk model. Figure 19.11 gives a set of Kubelka–Munk equations that model reflectance R_1 and transmittance T_1 of a translucent or opaque film or slab of material of thickness X containing light absorbers and scatterers. The observer views the sample in air, while the absorbers and scatterers are embedded in a medium such as plastic or resin.

The Kubelka–Munk reflectance R and transmittance T of the film in the absence of a refractive index boundary include the effects of absorption K, scattering S, and film thickness X. The boundary reflections from both sides of the top surface r_1 and the inner and bottom sides of the film, r, as well as the corresponding boundary transmittances t_1 and t influence the reflectance R_1 and the transmittance T_1 seen by the observer viewing the sample in an air medium. Two boundary reflectances r_1 and r are required because light normally incident on the film from air has a reflectance value of 0.04, while diffuse light incident on the boundary from inside the resin has a reflectance value of 0.596 for a material of refractive index about 1.5. The differences between R_1 and R, as well as those between T_1 and T, in the case of typical plastics and resins are very significant. Other Kubelka–Munk equations are available that permit calculations when the reflectances at the inner side of the top and bottom sides of the film are different (Wicks and Kuhhirt, 1975).

Kubelka–Munk models are relatively simple owing to the assumptions about microstructure and pigment–light interactions. More sophisticated models based on additional radiation fluxes or more complex microstructures are now possible owing to improvements in computer power. The models of Torrence and He provide relatively accurate representations of surfaces that contain both specular and diffuse reflections. Those models are now being incorporated into computer rendering software to realistically depict painted surfaces in movies, video games, and web-based animations (He et al., 1991).

$$R_1 = r_1 + \frac{t_1 t[R(1-rR) + rT^2]}{(1-r_1R)^2 - r^2T^2}$$

$$T_1 = \frac{t_1 tT}{(1-r_1R)^2 - r^2T^2}$$

$$t_1 = 1 - r_1; \qquad t = 1 - r$$

$$R = \frac{\text{Sin h } bSX}{a \sinh bSX + b \cosh bSX}$$

$$T = \frac{b}{a \sinh bSX + b \cosh bSX}$$

$$a = 1 + \frac{K}{S}; \qquad b = (a^2 + 1)^{1/2}$$

For completely opaque slabs or films, simplified equations can be used.

$$\frac{K}{S} = \frac{(1 - R_\infty)^2}{2R_\infty}; \qquad R_\infty \text{ is the reflectance at complete hiding}$$

Surface reflection must be corrected for using Saunderson's equation:

$$R_1 = r_1 + \frac{t_1 tR_\infty}{1 - rR_\infty}$$

Assumptions

1. The equations apply to a single wavelength at a time.
2. K and S are the same throughout the film.
3. Pigment particles are randomly oriented.
4. The internal radiation flux is perfectly diffuse.
5. Edge effects are ignored.

Figure 19.11 Kubelka–Munk equations and assumptions.

19.3 HIDING

Color is affected by light that reaches the substrate and is reflected back through the coating. It is common to compare the colors of a coating applied over a white substrate and one having black stripes. If the pattern of stripes is visible, the coating is said to exhibit poor *hiding*, also called *black and white hiding*. The difference results from the reflection of light reaching the white stripes compared with absorption of light reaching the black stripes. The effect is as if some black pigment was put in the coating above the black stripes. If all light entering the film is absorbed or scattered back out of the film prior to reaching the substrate, there is no effect of substrate on color; hiding is complete. Simplified Kubelka–Munk equations can be used for such opaque films (Figure 19.11).

Hiding is a complex phenomenon that is affected by many factors. Hiding increases as film thickness increases. Low hiding coatings require thicker films, which results in coverage of less area per unit volume and higher cost. Hiding increases as the efficiency of light scattering increases; that is, hiding is affected by the refractive index differences, particle sizes, and concentrations of scattering pigments present. Hiding increases as absorption increases. Carbon black pigments, which have high absorption coefficients for all wavelengths, are particularly effective. Colorants also increase hiding but not as much as black. Surface roughness increases hiding, since a larger part of the light is reflected at the top surface, which reduces differences of reflection from the substrate to which a coating is applied. TiO_2 provides the best hiding of all the commercially practical white pigments. As TiO_2 is expensive, paint manufacturers are highly motivated to use the minimum amount to achieve sufficient hiding. The use of alternative pigments or hollow polymeric spheres in architectural coatings is targeted at reducing the amount of TiO_2 used while maintaining adequate hiding.

An important factor affecting hiding, sometimes forgotten in testing coatings, is uniformity of film thickness. Application of coatings commonly results in nonuniform thickness. Coatings are generally formulated to level, that is, to flow after application to enhance uniformity (Section 24.2). Leveling is often incomplete, however, which can affect hiding. Consider a brush-applied coating having an average film thickness of 50 μm of dry coating that levels poorly. Brush marks may remain, which result in adjacent film thicknesses of 65 and 35 μm, for example. If the hiding at 50 μm is just adequate, hiding will be inferior at 35 μm. The difference in color is enhanced by the thin layers of coating being right next to the thick layers. The contrast results in a poorer rating in hiding by the uneven film than for a uniform 35 μm film of the same coating. Further complicating hiding is variation of substrates over which the coating is applied. As many DIY home painters have found, hiding of a white coating applied over a white surface may be rated as excellent, whereas hiding of the same coating over a dark surface may be unacceptable.

There are quality control tests that compare hiding of batches of the same or similar coatings, but no test is available that can provide an absolute measure of hiding (Cremer, 1981). The only way to establish covering power of a coating is to apply it to an appropriate large surface with a film thickness that gives adequate hiding and then calculate the actual coverage in square meter per liter or square feet per gallon.

In some cases, hiding is not desired; an example is a coating for metallized plastic Christmas tree ornaments. One wants transparent red, green, blue, and other color coatings. Transparent coatings, in general, require that there is no light scattering within the film; accordingly, the particle size of any colorant particles must be very small.

19.4 METALLIC AND INTERFERENCE COLORS

Metallic coatings are widely used on automobiles. They are formulated with transparent colorants together with *non-leafing* aluminum pigment (Section 20.2.5). These metallic coatings exhibit unusual shifts in color as a function of both illumination and viewing angle (Figure 19.12). When viewed from an angle near the normal and when the illuminating light is also near normal, the path length of light through the film is short because it is reflected back by the aluminum. In addition, much of the light is specularly reflected from the metal flakes, which are preferentially aligned parallel to the surface of the coating. Each flake acts like a small mirror that reflects most of the incoming light directly back to the observer's eye. On the

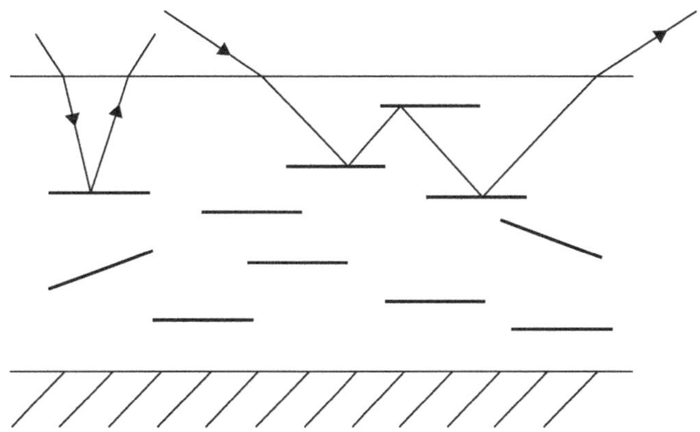

Figure 19.12 Idealized diagram of the reflection of light in a metallic coating.

other hand, from a greater viewing angle, much of the light is specularly reflected away from the observer's eye, so that only a small fraction of the light reaches the observer. In addition, the light that does reach the observer has typically traveled a longer path length, and the color is therefore darker. Accordingly, metallic coatings are lighter in color when viewed near the normal angle (the *face color*) and darker when viewed from a larger angle (the *flop color*). It is desirable to achieve a high degree of *color flop* or *travel*, that is, a large difference between face and flop colors. This requires a smooth surface (high gloss), a transparent film having very little light scattering from the resin or color pigment dispersion, and a parallel alignment of the aluminum flake with respect to the surface of the film. The degree of travel is often quantified via the *flop index*, as shown, where the various L^* values are the lightness of the coating at the specified angle (°) of the specular reflection (Streitberger and Dossel, 2008), (Eq. 19.6). See Section 30.1.4 for discussion of possible mechanisms for orientation of aluminum during film formation and further discussion of color effects in automotive coatings. The evaluation and quantification of colors with high travel is an ongoing area of research (Mirjalili et al., 2014):

$$\text{Flop index} = \frac{2.69\left(L^*_{15^\circ} - L^*_{110^\circ}\right)^{1.11}}{\left(L^*_{45^\circ}\right)^{0.86}} \tag{19.6}$$

Pigments that produce colors by interference, often called *effect pigments*, are also used in automotive coatings, sometimes in combination with metallic flake. Such pigments are described in Section 20.2.5 and, typically, are composed of mica or glass coated with various inorganic oxides to produce different hues, which also depend upon the angles of illumination and observation. Vacuum-deposited films can also provide high levels of color shift and are based on either diffraction or interference effects.

In addition to their effect on color and travel, coatings that contain flake-like pigments also produce sparkles or glints of light when the angles between the illumination source, the flake surface, and the observer are precisely aligned such that a strong specular reflection reaches the observer's eye. These sparkles are one of the hallmarks of metallic colors, although their characterization is difficult. Recent progress has been made using digital photographs of the metallic coating surface illuminated under specific conditions. The number and intensity of the sparkles are then thresholded via certain image analysis algorithms. Some instruments can measure both the sparkle and the graininess (salt-and-pepper) effect in metallic coatings.

19.5 THE OBSERVER

The third critical aspect of color is the observer; both the eye and the brain are involved. The retina of the eye has two types of photodetectors, rods and cones. These detectors are stimulated by light photons to send signals through the optic nerves to the brain. Rods are sensitive at low levels of illumination to all wavelengths of light and become insensitive as the level of illumination increases. Cones are insensitive at low levels of illumination but are sensitive at higher levels. The response of cones is wavelength dependent, having three ranges of sensitivity that overlap. One peaks in the blue region of the spectrum, another in the green, and the third in the red region. These overlapping ranges lead to complex signals to the brain, which integrates the signals so that we can see thousands of different colors. At low levels of illumination, only rods respond and vision is limited to shades of gray. At intermediate levels of illumination, both rods and cones respond; but one does not see bright colors, only grayish colors. At higher levels of illumination, the rods are inactive and we see a full range of colors from the response of cones.

The mechanism of color vision is complex, only partly understood, and beyond the scope of this text. There are variations from person to person in the degree of response to the three ranges of sensitivity. As a result, colors seen by different observers are not identical. Usually, the differences are small, but in some cases they are large. In extreme cases, some people are *color blind*. There are different types of color blindness, most commonly red-green color blindness, which affects up to 8% of males (Chan et al., 2014).

To specify color and predict responses to mixtures of colorants, a mathematical model of a *standard human observer* was established by an international committee of experts known by the acronym of its French name as CIE. Figure 19.13 shows a graph of three functions that model color vision. At any given wavelength, the standard observer has the same response to the ratio of the values of $\bar{x}$, $\bar{y}$, and $\bar{z}$ at a particular wavelength, as to monochromatic light of the same wavelength. Tables of *CIE color matching functions* are available as a function of wavelength with bandwidths of 1, 10, and 20 nm (Wyszecki and Stiles, 1982).

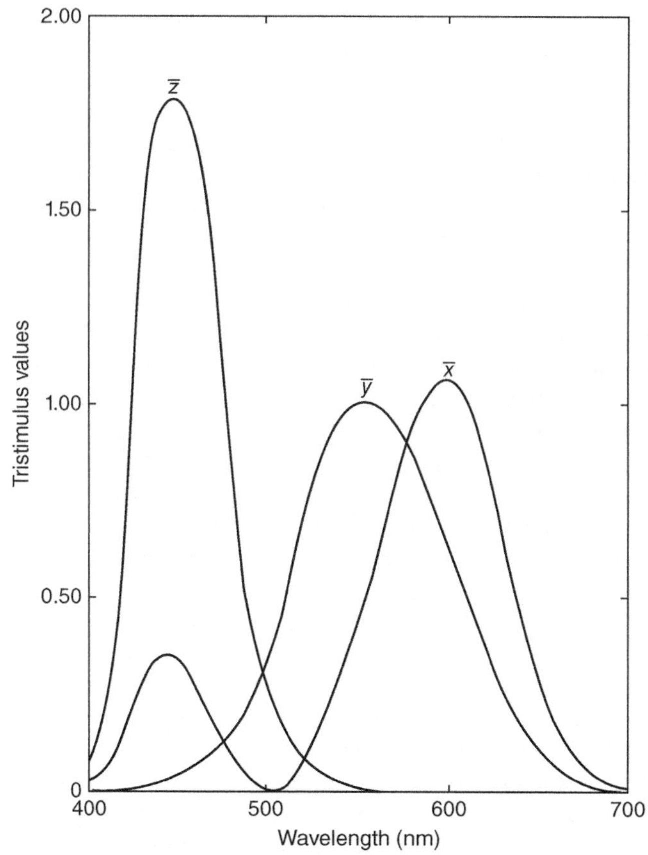

Figure 19.13 CIE color matching functions $\bar{x}$, $\bar{y}$, $\bar{z}$, for equal energy spectra. Source: Billmeyer and Saltzman (1981). Reproduced with permission of Elsevier.

Within the range of higher illumination, the eye adapts to changes in levels of illumination. For example, if an area of white is surrounded by black, the white appears whiter than if the black were not present. The eye adapts to the level of light reflected by the combined black and white and responds more to the white in the presence of the black. Similar effects occur when two strongly colored fields of view are adjacent to each other. Yellow surrounded by blue-green appears more orange than if the blue-green were absent. In general, if one looks at color chips when selecting a coating, the color is different than if that coating is applied to a large surface because of the effect of the surrounding area on the color. Many other effects result from such interactions between the eyes and combinations of colors on a surface. See Judd and Wysecki (1975) for further discussion.

19.6 INTERACTIONS OF LIGHT SOURCE, OBJECT, AND OBSERVER

Color depends on the interaction of the three factors: light source, object, and observer. If any factor changes, the color changes. If we observe an object under a light source with the energy distribution of illuminant *A* (Figure 19.2) and shift to a different illuminant, for example, illuminant *C* (similar to D_{65}), the color changes. Light source *A* has relatively lower emission in the blue end of the spectrum and relatively higher emission in the red end. The light reaching the eye when the object is illuminated with source *A* has more red light and less blue light than the light reflected from the same object illuminated with source *C*. The color is different. The situation is illustrated in spectral sets *a* and *b* in Figure 19.14, which shows the different responses as product spectra of light source × object × observer for an object with illuminants *A* and *C*. Rich (2016) provides an excellent discussion of the evolution of various light sources and the difficulties in matching colors under different light sources, including newer LED lights.

The only way to be sure that a pair of coatings will match under all illuminations is to use colorants with identical chemical composition and physical states. If the chemical composition and physical state of the colorants in two coatings are the same, their reflectance spectra are identical, and the coatings match under any light source unless there are differences in pigment distribution within the coatings. It is possible for two materials with different colorant compositions and, hence, different reflectance spectra to appear as the same color under a given light source. However, such a pair will not match under light sources with different SPDs. This phenomenon is called *metamerism*. In a spectral match, the two panels change color with a new light source, but it is the same change in both cases. In a metameric pair, the color is the same with one light source, and the colors of both panels also change when the light source is changed, but the extent of change is different between the two panels. This situation is illustrated in Figure 19.14, which shows responses of a second object with illuminants *C* (spectral set *c*) and *A* (spectral set *d*). The first and second objects (a metameric pair) are identical under illuminant *C*, but their colors are different under illuminant *A*.

19.7 COLOR SYSTEMS

The human eye can discriminate thousands of colors. However, it is difficult for a person to tell another person what colors he/she sees. A color description such as "light grayish blue-green" illustrates this difficulty. Many systems have been developed to permit definitive identification of colors. There are two types of systems: one that uses color samples in specific arrangements and one that identifies colors mathematically. All color systems require at least three dimensions to include all the possible colors.

The most widely used visual color system in the United States is the *Munsell Color System* in which carefully prepared and selected color chips are classified in a three-dimensional system. The dimensions of the Munsell system are called *hue*, *value*, and *chroma*. Hue refers to the dimension of color described as blue, blue-green, green, greenish yellow, yellow, yellowish red, red, reddish purple, purple, purplish blue, and back to blue. The value dimension characterizes the lightness of a color as compared to a series of gray samples, a *gray scale*. The value 0 is assigned to pure black and the value 10 to pure white. A light blue has a high grey scale value, while a dark blue of the same hue has a low value. The chroma dimension refers to the difference between a color and a gray of the same value and hue. A bright red has a high chroma, whereas a grayish red of the same hue and value has a low chroma.

The color chips in the Munsell system are prepared so that there are equal visual differences between all pairs of adjacent chips. The chips are systematically labeled; for example, a chip labeled G5/6 is a green with a value of 5 and a chroma of 6. One can look at a set of Munsell chips and know the color that someone meant when they made such a designation. There are two limitations to this statement. The light source must be specified. Chip G5/6 gives a different color under source *A* than under source D_{65}. Secondly, surface roughness affects color, so comparisons have to be made at equal gloss levels. Two sets of Munsell chips are available: one with high gloss and the other with low gloss. Significant errors can result in comparing either of these with other materials that are semigloss. In Europe, a different color chip system, the Natural Color System, is most widely used (Pierce and Marcus, 1994).

The mathematical color system is the *CIE Color System*. It is based on mathematical descriptions of light sources, objects, and a standard observer. Light sources are specified by their relative energy distributions, objects are specified by their reflectance (or transmission) spectra, and the observer is specified by the CIE standard human observer tables. For color analysis, the light reflected (or transmitted) from (or through) an object is measured with a spectrophotometer. Since in most cases the reflection is diffuse, it is essential to use a spectrophotometer with an integrating sphere so that all of the reflected light is sampled, not just that at some narrow angle. For the most accurate data, reflectance measurements are taken at each wavelength, and the values over the range of 380–770 nm would be used in the summation. For most purposes, the accuracy is sufficient using 16 measurements at 20 nm intervals from 400 to 700 nm.

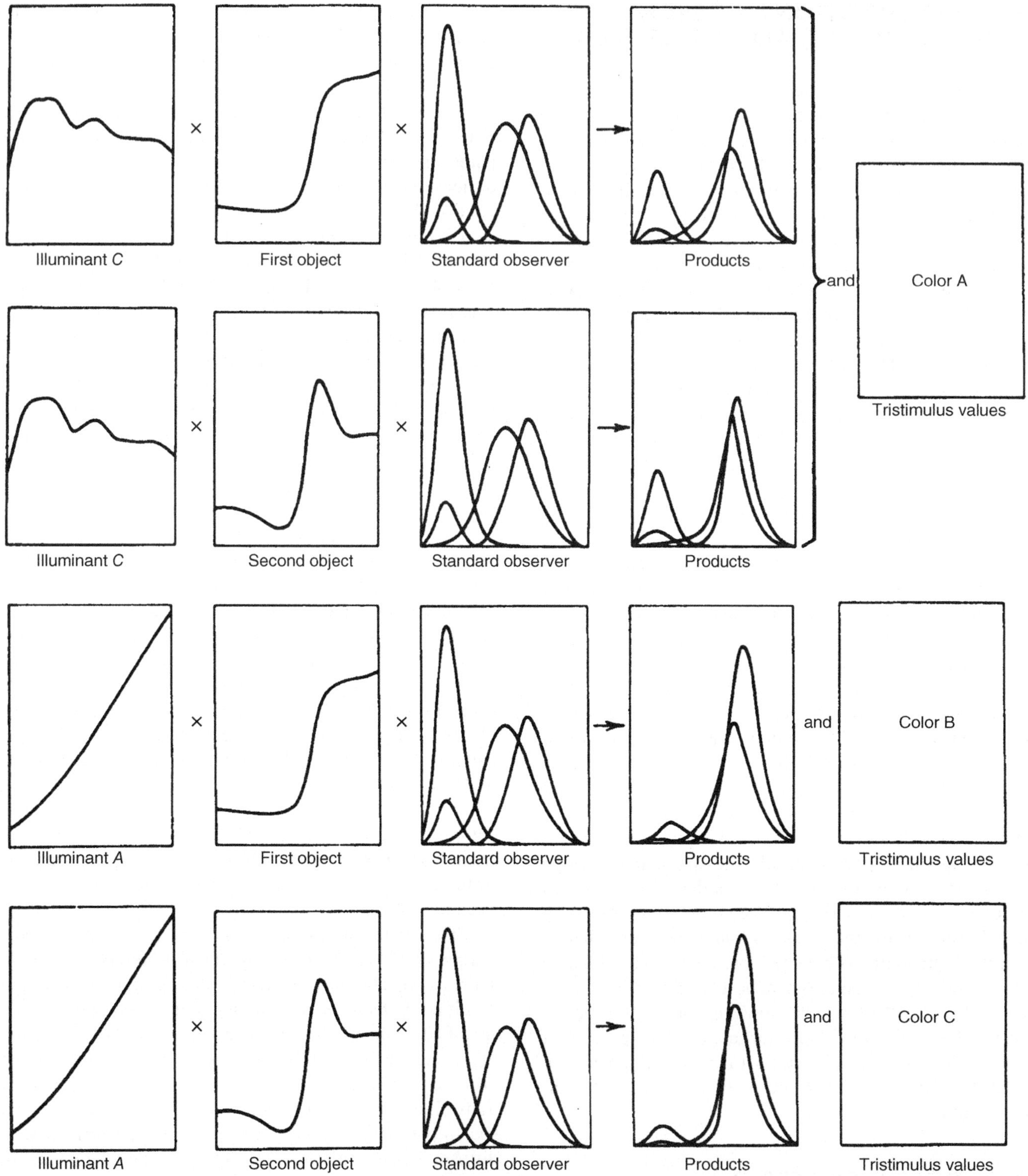

Figure 19.14 Light source effects and metamerism. Source: Billmeyer and Saltzman (1981). Reproduced with permission of Elsevier.

To identify the color resulting from the interaction of a light source, an object, and a standard observer, one uses the data for these three dimensions to calculate the *tristimulus values* X, Y, and Z using Eqs. 19.7a, 19.7b, and 19.7c:

$$X = \sum_{380}^{770} \bar{x}_\lambda E_\lambda R_\lambda \quad (19.7a)$$

$$Y = \sum_{380}^{770} \bar{y}_\lambda E_\lambda R_\lambda \quad (19.7b)$$

$$Z = \sum_{380}^{770} \bar{z}_\lambda E_\lambda R_\lambda \quad (19.7c)$$

Tristimulus values are different for the same object and the same observer when the energy distribution E of a different illuminant is used. This is as it should be, since we know that colors change as the light source changes. The tristimulus values uniquely and unequivocally define colors. For example, $X=14.13$, $Y=14.20$, and $Z=51.11$ is a definitive description of a color, but what color? Unfortunately, even experts often cannot say by looking at the numbers. This set of tristimulus values is for a blue, but few people could look at the values and tell you it is a blue, much less whether it is a grayish blue and approaches being a purplish blue, which it happens to be.

The X and Y tristimulus values can be converted to *chromaticity values* x and y by normalization as shown in Eq. 19.8:

$$x = \frac{X}{X+Y+Z} \quad y = \frac{Y}{X+Y+Z} \quad (19.8)$$

If one has a metameric pair of coated panels, the tristimulus values and the chromaticity values are the same with the light source under which the panels match. They are not the same, however, if calculated with the energy distributions from another light source. When the light source changes, the X, Y, and Z and the x and y values of both panels change, but the changes are to different degrees.

The chromaticity values of each wavelength of the spectrum can be calculated and plotted against each other to form the *CIE spectrum locus* (Figure 19.15). The ends of the locus are connected by a straight line called the *purple line*. There are no purples in the spectrum; in CIE color space, hues of purple lie along this line. As shown, the plot can be divided into color areas, so one can look at the x and y values and have a reasonable idea of the shade of a color. The third dimension rises vertically from the plot; it is the Y tristimulus axis, *luminance*; $Y=100$ (or 1 in some conventions) at the x, y values of the light source. The Y at the spectrum locus line approaches 0. At x, y points between the spectrum locus and the point of the source, Y is always less than 100. As the value of Y gets larger, the gamut of possible colors narrows (Figure 19.16).

If one draws a line from the point of the source through the point of a sample extended out to the spectrum locus, the wavelength at the intercept is called the *dominant wavelength* of the color. This dimension corresponds to the hue dimension in the Munsell system, but the scale is different. If this extrapolation intercepts the purple line, then the line is extrapolated in the opposite direction, and the intercept with the spectrum locus is called the *complementary dominant wavelength*. If one divides the distance from the source point to the sample point by the total distance from the source point to the spectrum locus (some workers express it as a percentage), one obtains the *purity*. Purity is the same dimension as chroma in the Munsell system, but the scale is different. The vertical Y dimension is a gray scale, as in the Munsell value dimension, but again the scale is different.

Figure 19.16 shows a topographical diagram of three-dimensional CIE color space with illuminant C. All real colors that can be seen with light source C fall inside this color space. In the Munsell system, the boundaries are limited not by reality of color, but by the color purity of the available pigments to make the reference chips. CIE color space is not visually uniform (Figures 19.15 and 19.16). For example, a small difference in x and y in the blue part of the color space represents a substantial difference in color, whereas in the green part of color space, the same difference in x and y values represents little difference in color. In this sense, Munsell color space is preferable, since the differences are visually uniform. However, color calculations of the types needed for instrumental color matching are feasible with CIE color space, but not with Munsell space. If the comparisons are to be visual, one commonly uses the Munsell system. If the comparisons are to be mathematical and include all possible color space, one uses the CIE system.

There have been many attempts to mathematically transform CIE color space to a visually equal color space; then differences could be used as meaningful measures for specification purposes. Some progress has been made, and color differences (ΔE) can be calculated using the CIE 1976 $L^*a^*b^*$ equations. L^* represents lightness, a^* redness/greenness, and b^* yellowness/blueness (Hardy, 1936), (Eqs. 19.9a, 19.9b, and 19.9c). These quantities are calculated from tristimulus values by the following equations:

$$L^* = 116\left(\frac{Y}{Y_n}\right) - 16 \quad (19.9a)$$

$$a^* = 500\left[f\left(\frac{X}{X_n}\right) - f\left(\frac{Y}{Y_n}\right)\right] \quad (19.9b)$$

$$b^* = 500\left[f\left(\frac{Y}{Y_n}\right) - f\left(\frac{Z}{Z_n}\right)\right] \quad (19.9c)$$

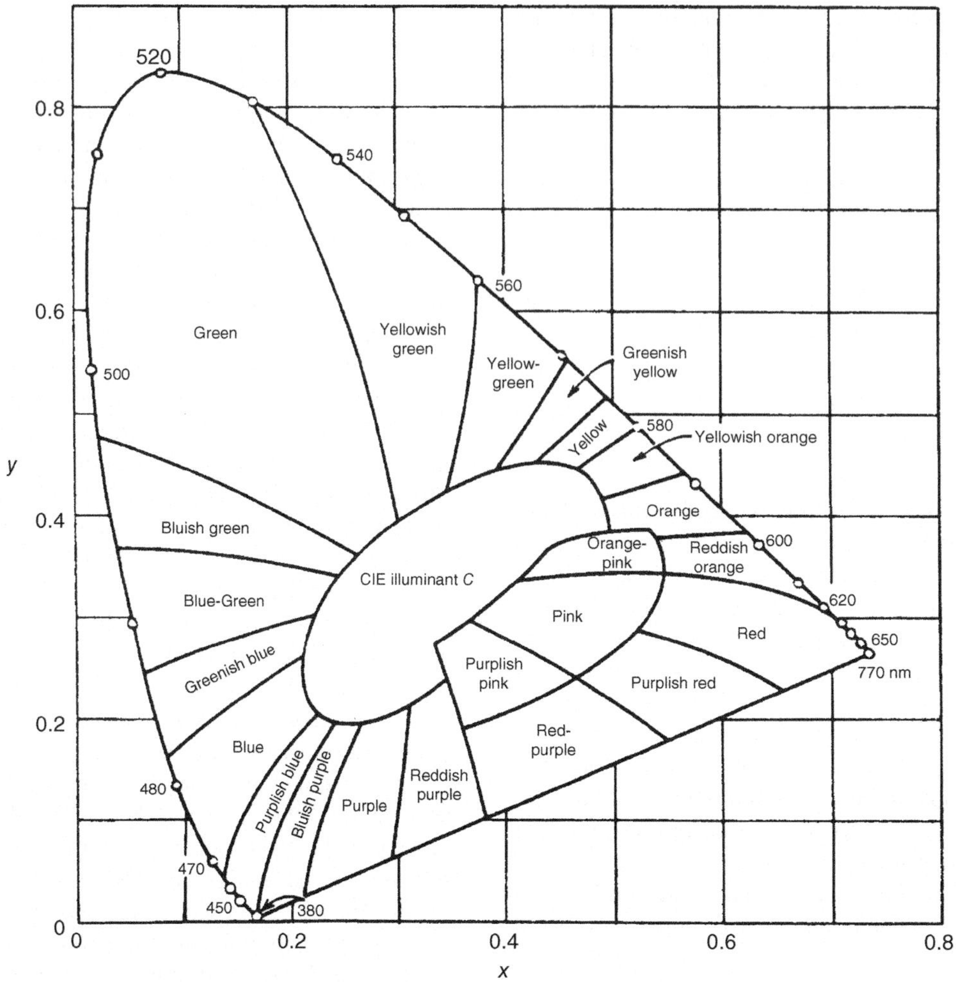

Figure 19.15 CIE chromaticity diagram showing the location of various hues. Source: Billmeyer and Saltzman (1981). Reproduced with permission of Elsevier.

In these equations, X, Y, and Z are the tristimulus values of the sample; X_n, Y_n, and Z_n are the tristimulus values of the light source; $f(Y/Y_n) - (Y/Y_n)^{1/3}$ for values of $(Y/Y_n) > 0.008856$; $f(Y/Y_n) = 7.787(Y/Y_n) + 16/116$ for values of (Y/Y_n) equal to or less than 0.008856; and the functions $f(X/X_n)$ and $f(Z/Z_n)$ are similarly defined.

The classical equation for CIELAB color difference is (ASTM 2016), (Eq. 19.10)

$$\Delta E\left(L^* a^* b^*\right) = \left[\left(\Delta L^*\right)^2 + \left(\Delta a^*\right)^2 + \left(\Delta b^*\right)^2\right]^{\frac{1}{2}} \quad (19.10)$$

These equations still do not represent fully uniform color space. If specifications are written for a line of colors specifying a fixed ± range for ΔE for the whole series, the requirements will be more stringent for some colors than for others. Even if color difference equations were available that were visually uniform, their use would still be limited for specifications. The color would be permitted to vary equally in any direction from the central standard. However, it is common for people to be more concerned about deviations in one direction in color space as compared with other directions. For example, there is commonly a greater tolerance for whites to be off in the blue direction than in the yellow direction. Most untrained observers are more attuned to small color differences than lightness differences.

CIE has developed a series of improvements over the CIELAB color difference formula, most recently "CIEDE2000" (CIE, 2001). Implementation of this system requires a 22-step calculation (Sharma et al., 2005), easy with computers but beyond the scope of this book. Automotive paints are often evaluated using ΔE_{CMC}, which is a further variation of the simple ΔE formula. ΔE_{CMC} correctly weights differences in chroma more than differences in lightness (AATCC, 2009).

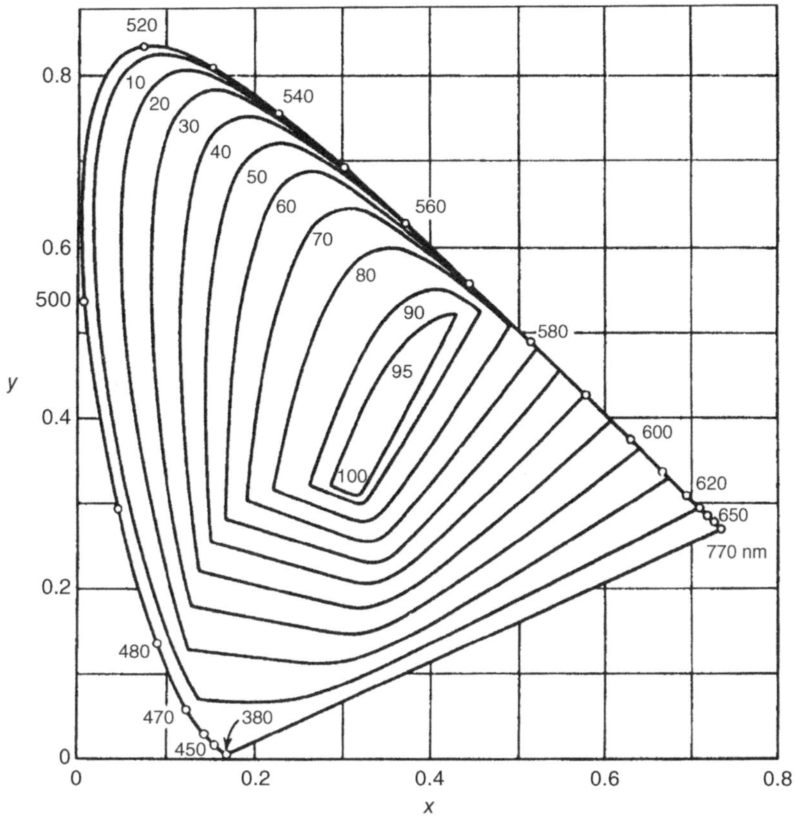

Figure 19.16 Topographical diagram of a three-dimensional color space with illuminant *C*. Source: Billmeyer and Saltzman (1981). Reproduced with permission of Elsevier.

19.8 COLOR MIXING

There are two types of color mixing: *additive* and *subtractive*. In additive mixing, the primary colors are red, green, and blue. Additive mixing is involved in theatrical stage use of overlapping colored spot lights, as well as in color television (TV), where three colors of dots (red, green, and blue) are projected near each other on the screen. The lights from nearby dots are "added" when we look at them to give colors, which depend on the ratio of the three colors in nearby dots. In additive color mixing, equal amounts of blue and green light give blue-green (*cyan*); similarly, blue and red light give purple (*magenta*); green and red light give *yellow*; equal addition of all colors gives white light. With appropriate light sources, all color lights can be displayed. Note that we see as yellow the combination of all wavelengths in the green, yellow, orange, and red segments of the spectrum.

In almost all cases, however, color mixing encountered in coatings is not additive but subtractive mixing. We use colorants that absorb (subtract) some of the light of some wavelengths from white light. If we add a second colored pigment to a coating containing colored pigment, we subtract more; we cannot add back any intensity of wavelengths by using a further colorant. The primary colors in subtractive mixing are cyan, magenta, and yellow. If we mix equal quantities of an ideal cyan and an ideal magenta, the result is blue. That is, the cyan absorbs red and the magenta absorbs green, so blue remains. Similarly, cyan and yellow yield green; magenta and yellow yield red; a mixture of equal amounts of all three ideal absorbing colorants absorbs all light and yields black. Idealized transmission (or reflectance) spectra of the primary colors, which illustrate their complementary nature, are provided in Figure 19.17.

19.9 COLOR MATCHING

A large fraction of pigmented coatings is color-matched. The customer chooses some color for a refrigerator, automobile, or another product. A coatings formulator is given a sample of a material with this color and asked to match the color using a coating formulation appropriate for the end use. After the formulator establishes the appropriate ratio of the appropriate pigments and after the customer approves the color match, the factory must color match batch after batch of the coating to meet this standard.

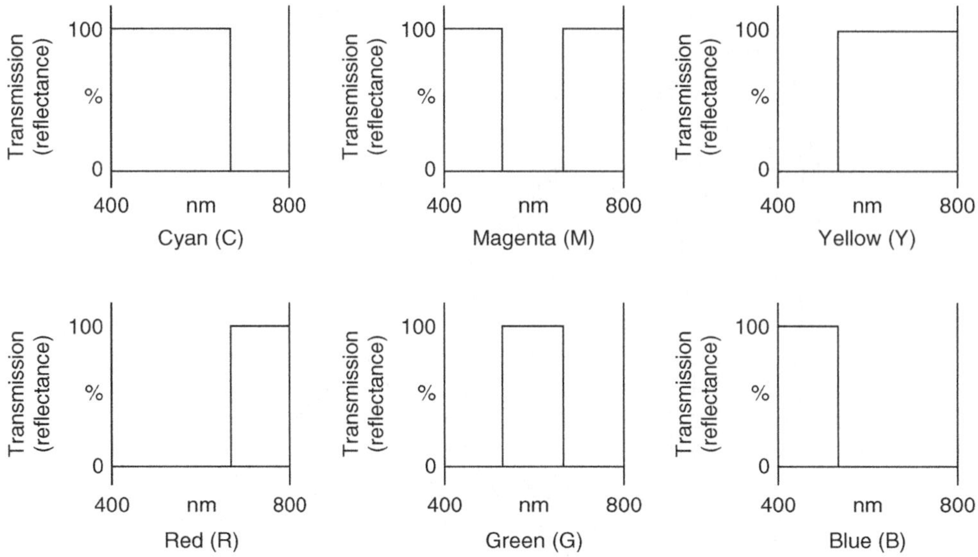

Figure 19.17 Transmission (or reflectance) spectra of cyan, magenta, and yellow colorants, together with their complementary colors.

Sometimes the formulator develops the formula first, and the color is accepted by the customer.

19.9.1 Information Requirements

Before starting the initial laboratory color match, the color matcher needs a substantial amount of information:

1. *Metamerism.* Is a spectral (nonmetameric) match possible? That is, can exactly the same colored pigments (including white and black) be used to establish the match as were used to make the customer's sample? If not, the customer must realize that any match will be metameric; that is, the colors will match under some light source, but not under others. For example, if the sample is a dyed fabric, the color of the textile cannot be exactly matched under all lights by any pigmented coating, since dye and pigment colorants cannot be identical. If, as another example, the customer has been using a particularly old formulation of a coating made with one or more lead-containing pigments and wishes to have a lead-free coating, only a metameric match is possible.
2. *Light sources.* If the match is to be metameric, the customer and the supplier must agree on the light source(s) under which the color is to be evaluated. If there is to be more than one light source, a decision should be made as to whether it is more desirable to have a close match under one light source without regard as to how far off the match might be under other light sources, or to have fair color matches under several light sources, but not a close match under any individual light source.
3. *Gloss and texture.* The color of a coating depends on its gloss and texture. Some of the light reaching the eye of an observer is reflected from the surface of the film and some from within the film. The color seen by the observer depends on the ratio of the two types of reflected light. At most angles of viewing, more light is reflected from the surface of a low gloss coating than from the surface of a high gloss coating. It is impossible to match the colors of a low gloss and high gloss coating at all angles of viewing. There must be agreement as to the gloss of the coating. If the gloss of the standard is different from the gloss desired for the new coating, the angles of illumination and viewing must be agreed on. It is impossible to make even a metameric match of the color of a fabric sample with a paint at all angles of viewing since both the colorants and the surface textures have to be different. When paint manufacturers advertise on TV that their paint stores can do this, they are engaging in misleading advertising.
4. *Color properties.* Colorants have to be chosen that permit formulation of a coating that can meet the performance requirements, examples being exterior durability, resistance to solvents, resistance to chemicals like acids and bases, resistance to heat, satisfying a toxicity regulation, and so on.
5. *Film thickness and substrate.* Since the coating may not completely hide the substrate, the color of the substrate can affect the color of the coating applied to it, and the extent of that effect will vary with the film thickness. This variable is particularly important in applications, such as can coatings and coil coatings for

which relatively thin coatings are applied (<25 μm in the case of can coatings). A thin coating that was color-matched over a gray primer will not match over a red primer. A coating setup for one coat application on aluminum will not match the color standard if it is applied over steel.

6. *Baking schedule.* Since the color of many resins and some pigments are affected by heating, particularly at high temperatures, the color of a coating can be affected by the time and temperature of baking. It is also necessary to specify what the color requirements are for overbaking.
7. *Cost.* The color matcher should know the cost limitations. There is no sense in making an excellent color match with pigments that are too expensive for a particular application.
8. *Tolerance.* How close a color match is needed? Coatings for exterior siding or automobile top coats require very close color matches. For many other applications, close matching is unnecessary, but some customers, in their ignorance, set tight tolerance limits. Overly tight tolerances raise cost without performance benefits. For coatings that will be produced over time with many repeat batches, the most appropriate way to set color tolerances is to have an agreed upon set of limit panels. For example, for a deep yellow coating, these would be the limits of greenness and redness, brightness and darkness, and high and low chroma. Since panels can change with age, spectrophotometric measurements should be made of the standard and the limit panels, and CIE tristimulus values calculated. The customer would then agree to accept any batch of coating having a color within this volume of color space. As mentioned earlier, while ΔE color differences, calculated from the 1976 CIELAB color difference equations, have been used to set color tolerances, they have limitations. It is undesirable to use them as a general statement of tolerance for a series of colors. Furthermore, they permit variation equally in any direction from the standard color, which is frequently undesirable.

19.9.2 Color Matching Procedures

There are two approaches to matching colors: visual and instrumental. In visual color matching, experienced color matchers examine the sample to be matched and from their experience select a combination of pigment dispersions that they think will permit matching the color. A formulation is mixed and the coating applied; since gloss affects color, first the gloss must be compared with the standard and adjusted if needed. The color matcher looks at the resulting color in comparison with the standard and decides what further addition of either one or more of the same pigment dispersions, or perhaps dispersions of different pigments, are needed to improve the color match.

Spectrophotometric curves of the original standard sample can be used in an analytical mode to help identify component colorants in the standard, thereby simplifying pigment selection for visual matching. Organic pigments can often be identified by examining the absorption spectra of solutions of colorants from a sample of coating (Kumar et al., 1985).

This process continues until a satisfactory combination of pigment dispersions has been selected. Color matching must be based on dry films. It can be useful to compare samples of wet coatings to estimate progress toward a match; but since color changes considerably during application, film formation, and drying, final decisions have to be based on dry films prepared under conditions approximating the way in which the coating will actually be used.

The visual color matcher not only has the responsibility of selecting the colorants and their ratios to make a color match but must also attempt to make the color match so that it will be as easy as possible to produce batch after batch efficiently in the factory. It is not economically feasible for pigment manufacturers to produce successive batches of color pigments that are exactly the same color and color strength. Furthermore, there will be batch-to-batch variation in dispersions made from these pigments. The formulation should be set up so that it can be adjusted in the factory to permit matching the color standard in spite of these variations. It is desirable to use four pigments (counting black and white, but not inerts) to make the original match. This provides the four degrees of freedom necessary to move in any direction in three-dimensional color space. Sometimes use of single pigments cannot be avoided, but it is undesirable especially in applications in which close color matches are needed. Sometimes, more than four pigments are needed, but this adds to the complications of production and should be avoided, if possible. The factory should not make any changes in the pigments in the formulation, which would result in a metameric match.

Visual color matching is a highly skilled craft requiring years of apprenticeship to master. It is still practiced and still important, but the coatings industry has largely moved to computerized instrumental color matching. Instrumental color databases, along with computer programs, can be used to select colorants and their ratio both for original color matches in the laboratory and to provide information as to the amount of the different pigment dispersions to be added in the factory to match production batches. Establishing such a program requires a major effort. The database must be carefully set up. The pigments must be made into dispersions; and multiple single color coatings with different concentrations of color pigment dispersion in a series of appropriate white coatings must be made and applied. The reflectance values must be measured at 16 wavelengths; for critical matches, values at 35 wavelengths may be needed. See Pierce and Marcus (1994) for further discussion of creation of a pigment database. Commercial paint companies use closely guarded algorithms to guide their color matching efforts.

Analysis of computer color matching is beyond the scope of this text. Pierce and Marcus (1994) and Rich (1995) review computer color matching and provide detailed discussion (also see the general references cited). Solid color coatings that provide complete hiding can be accurately color-matched by current computer color matching software. The basis for these programs is the Kubelka–Munk theory outlined in Figure 19.11. These programs attempt to achieve a tristimulus match between the sample and standard or object to be color-matched. The mathematics of this operation is complicated by the nonlinear nature of the relation between the colorant concentrations in the sample and the tristimulus values of the sample and standard.

Typical color measurements consist of either 16 or 35 reflectance measurements at a set of wavelengths in the visible region. To utilize this data for color matching, the reflectances must be corrected for boundary reflections by means of Saunderson's equation to give Kubelka–Munk reflectances, as provided in Figure 19.11 (Saunderson, 1942), (Eqs. 19.11a, 19.11b, and 19.11c). These reflectances can be used to calculate the ratio *K*/*S* for the sample to be matched at each wavelength. The *K* and *S* values are assumed to be linear functions of the pigment concentrations chosen to match the sample. In the case of a four-pigment match,

$$K = k_1c_1 + k_2c_2 + k_3c_3 + k_4c_4 \qquad (19.11a)$$

$$S = s_1c_1 + s_2c_2 + s_3c_3 + s_4c_4, \text{ and} \qquad (19.11b)$$

$$\left[\left(\frac{K}{S}\right)s_1 - k_1\right]c_1 + \left[\left(\frac{K}{S}\right)s_2 - k_2\right]c_2 + \cdots \left[\left(\frac{K}{S}\right)s_4 - k_4\right]c_4 = 0 \qquad (19.11c)$$

Values need to be determined for each of the 16 (35) wavelengths, since the measured *K*/*S* ratio and k_i and s_i values for the individual pigments are different at each wavelength. In addition, the pigment concentrations are usually expressed as percents to satisfy the equation

$$c_1 + c_2 + c_3 + c_4 = 100$$

It would seem straightforward to select three wavelengths to give three equations that combined with the pigment fractions would give four linear equations with four unknowns that on solution would give the required pigment concentrations to match the sample. Unfortunately, this is not the case in practice. Even using all 16 (35) *K*/*S* equations and using various estimates such as tristimulus weighted averaging or least squares solutions of the determined set of 17 (36) equations in four unknowns does not give, in most cases, a close tristimulus color match. Allen (1974) devised a method that is used by most commercial software to obtain a tristimulus match. Starting from an initial estimate of pigment concentrations obtained using the *K*/*S* ratios and tristimulus weighting, these concentrations are iteratively refined to approximate a tristimulus match.

The ΔX value is the difference between the measured *X* value (tristimulus value) of the object to be matched and the *X* value of the match calculated from the initial concentrations. The partial derivatives of *X* with respect to concentration are calculated from the initial match concentrations. The values for *Y* and *Z* are calculated in the same fashion. The details of these calculations can be found in the cited references. The resulting four equations are solved for the four unknown Δc_i values, and a new set of match concentrations are calculated by adding the Δc_i values to the initial estimate of c_i values used to set up these equations. The process is repeated until a tristimulus match or no further improvement in the match is obtained. Usually, the best tristimulus match with the four chosen pigments is obtained in less than three iterations.

If the four colorants chosen are close to or identical with the colorants in the object to be matched, a nonmetameric match results. If the degree of metamerism as judged by the variation in color difference between the object and match for two or more illuminants is unacceptable, combinations of different pigments can be tried until an acceptable match is attained.

Matching of metallic colors has been more difficult to computerize because the colors have to match at multiple "aspecular" angles, that is, angles with differences between the angle of illumination and the angle of viewing. The reflectances of the samples and all of the colorants in the database must be determined at multiple angles. Instruments are now available that measure at a variety of aspecular angles. The angle of illumination is fixed and the angle of viewing can be set at different angles. In the United States the accepted aspecular angles are 15°, 45°, and 110°. See ASTM E 2194-14: Standard Practice for Multiangle Measurement of Metal Flake Pigmented Materials.

Matching colors with pearlescent flakes, interference, and cholesteric pigments is further complicated by changes in hue as a function of angle of illumination and angle of viewing (Section 20.2.5). Instruments are now available that have three illuminating angles, 25°, 45°, and 75°, with viewing angles for each of the illuminating angles of +15° and −15°, but research continues, and more elaborate geometries may be needed. ASTM E2539-14 provides the standardized method for measurement.

Other information on the colorants, such as cost, properties, regulatory restrictions, and so forth, can be built into the database. The computer can then calculate a series of alternative colorant combinations selected for lowest cost, least degree of metamerism, excluding pigments with inadequate exterior durability, and so forth. The formulator can then choose the most appropriate combination for the particular color matching assignment. The formulator then makes up an experimental batch of paint, applies it to the appropriate substrate, bakes or dries it, and measures the reflectance values. The result is seldom a satisfactory color match on the first attempt. Reflectance values of the first hit are instrumentally compared with those of the standard, and the computer calculates the amounts of the various colorants needed so that the color will match. The number of hits required to make color matches has been found to be significantly less using computer color matching than the number required by even experienced visual color matchers.

The same program is used in production of factory batches of colored coatings. The colorants to be used have already been selected and must not be changed. The laboratory formulation is used for the initial mix, but a fraction of each of the colorants is held back. (Otherwise, if the factory batch of some pigment dispersion were stronger than standard, too much of that colorant would be put in.) The batch is mixed, a panel coated, and its reflectance measured. The data is then used to calculate what additions need to be made to the batch to obtain a color match. If necessary, the process is repeated. The savings in time and, hence, in cost using computer color matching can be large, but the cost of establishing and maintaining the database is substantial. Since the colors of pigment dispersions vary from batch to batch of pigment and of pigment dispersion, measurements usually have to be made of each batch for comparison with the standard database values. Computer programs can be written to correct for small differences in the colors from batch to batch. The matching of metallic colors, such as for automobile refinish coatings, is strictly done by spectrophotometric measurements and complex computer analysis that are proprietary to each refinish coating supplier.

19.9.3 Rendering of Color

All of the preceding discussion is focused on the appearance of physical samples of paint. However, increasingly paint colors are rendered on a computer screen so one can visually perceive how a certain color will appear in a room or on a vehicle before either the paint or the object is purchased. The display of a color on a computer screen is called *rendering the color.* When a color is rendered, its color is translated into a binary description of the color and then displayed on a computer screen, typically through the use of the red/green/blue (RGB) color coordinate system. As the RGB system is not commonly used to describe actual paint colors, the paint color must be converted to RGB values. This is not difficult, and many online tools allow for the conversion of colors from one color space, say, L^*, a^*, b^*, to RGB (easyrgb.com, 2008). The bit depth of the RGB scale must be defined. Eight-bit colors have $2^8=256$ possible values of each R, G, or B value. For an 8-bit color, R, G, B=0, 0, 0 is a completely black color, while 255, 255, 255 is completely white.

The fidelity of the color match between the rendered color and the physical object is often less than ideal. This is typically the result of both hardware and software limitations. The dynamic range of most computer monitors is not sufficient to display the variations in brightness that can be observed on an actual painted part. In addition, the subtleties of color are often difficult to render in a computationally efficient manner. Thus, altering and displaying colors in real time on a computer screen is challenging, but will continually improve as computational power increases.

For simple, solid colors effective rendering can be done using equations like the Kubelka–Munk equations. Thus, many DIY painters can now render their new paint colors by photographing a room in their home and "virtually" painting the room with a new color using simplified computer interfaces in paint stores or via the Internet. For more complex colors, more sophisticated models are required. The He–Torrance model has proven popular in the gaming and movie industries, as it is computationally efficient and produces relatively realistic images.

For highly *gonioapparent* colors, that is, colors whose appearance changes depending on the angle of both illumination and observation, the only sufficient way to fully describe the color is through the use of the bidirectional reflectance distribution function (BRDF), which fully describes the wavelength-dependent reflection of light from a material as a function of illumination direction, observation direction, and azimuthal angle. In principle, measurement of the BRDF gives a complete mathematical description of the reflectance properties of a coating. However, the BRDF can only be measured by a time-consuming process using an expensive goniospectrophotometer. Thus, the BRDF is often approximated with data from a multi-angle spectrophotometer. As little as 5 or over 20 combinations of illumination and observation angle have been used to approximate a complete BRDF. A two-dimensional slice through a full BRDF is shown in Figure 19.18. A *Lambertian* surface is perfectly *matte* (very low gloss) and is depicted as a semicircular function, while a metallic coating with large specular and near specular components is also shown. The effect of formulation variability on the BRDF is only beginning to be explored.

The use of computerized design tools to both design new colors and alter existing colors is also an area of rapid advances. Simple colors and surface textures have been so well modeled that their appearance and adjustment via

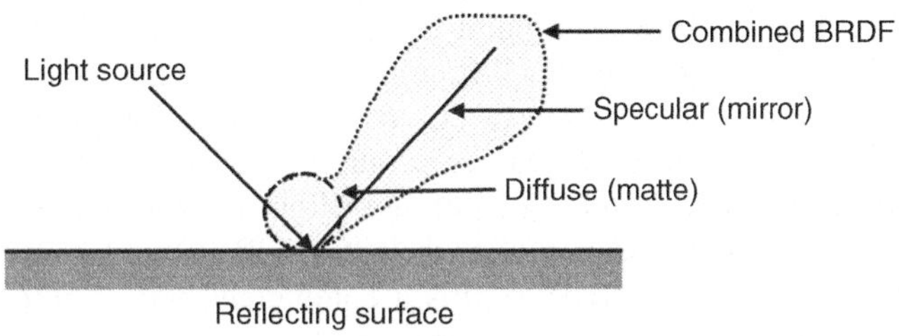

Figure 19.18 Two-dimensional slice through a BRDF. The diffuse, directionally diffuse, and specular reflections add to give a complex BRDF.

computer systems is now a reality. Orange peel and simple metallic colors with limited travel can be rendered and linked to formulation databases to find the closest match to the created color.

19.10 GLOSS

Gloss is a complex phenomenon. Individuals frequently disagree on the visual assessment of gloss and generally do not have in mind a clear definition of gloss, partly because there are several types of gloss. *Specular* (mirrorlike) *gloss* is the type most often considered in coatings. A high gloss surface reflects a large fraction of the light that is reflected from the surface at the specular angle, which corresponds to the angle of the incident light beam. Lower gloss surfaces reflect a larger fraction at nonspecular angles. When considering gloss, people visually compare the amount of light reflected at the specular angle with the amounts reflected at other angles. If the difference or contrast in reflection is high, gloss is said to be high. Note that gloss is *not* directly related to the fraction of light reflected from the surface, as seems to be commonly assumed. Contrarily, at most angles of illumination, surface reflection from a low gloss surface is actually greater than from a high gloss surface. The fraction of light reflected at a surface increases as the angle of illumination increases and surface reflection at the specular angle increases as the refractive index of an object increases (Figure 19.6).

If a surface is rough on a small scale, but still at a scale substantially greater than the wavelength of light, the angle of incidence of a beam of light is not the same as the geometric angle of the surface with the light beam. Light is reflected at specular angles between the light beam and individual rough facets of the surface. If a surface has many small facets oriented at all possible angles, a beam of light is reflected in all directions. As mentioned earlier, such a surface is a *diffuse or* Lambertian reflector and has a low gloss; it is called *matte* or *flat*. Flat is unfortunate terminology, since a perfectly smooth (flat) surface gives high gloss, whereas a microscopically rough surface gives what we call a flat surface. At intermediate surface roughness, the gloss is intermediate and is called *semigloss*, *eggshell*, *satin*, or a variety of other terms.

A phenomenon related to specular gloss is *distinctness-of-image* (DOI) gloss. A perfect specular reflector is a perfect mirror with an image that exactly mimics the original. If a surface has perfect diffuse reflection, no mirror image can be seen. At intermediate stages, the image is more and more blurred as the ratio of specular to diffuse reflection decreases. Furthermore, larger surface irregularities lead to distortion of the image. Commonly, one sees some degree of both blurring and distortion. DOI was commonly used as a metric for automotive finishes but is used less often today.

Sheen, as the term is used in the coatings industry, refers to reflection of light when a low gloss coating is viewed from near a grazing angle. A high gloss coating reflects a high fraction of light whose angle is near grazing. Reflection from most low gloss surfaces is low at a grazing angle. A low gloss coating is said to have a high sheen if there is significant reflection at a grazing angle. The effect is easy to see, but difficult to describe; there is not a glare (as when a glossy surface is illuminated and viewed at a grazing angle), but a "soft," relatively high reflection.

Luster is another type of gloss effect. (To further illustrate the problem of defining gloss, a dictionary definition of luster is gloss or sheen.) Luster is directional gloss. For example, some woven fabrics are much glossier when viewed parallel to the warp (lengthwise yarn) than parallel to the woof (transverse yarn). Such fabrics exhibit variations in gloss looking at folds in a draped fabric. If the contrasts are great, the fabric is said to have a high luster. Similar effects can sometimes be observed in textured coatings.

Haze can be considered as a form of gloss. When light enters a hazy film, it is scattered to some degree, causing some diffuse reflection to reach the eye of the observer. The result is similar to having some of the light reflected at nonspecular angles at the surface. The contrast between the fraction of light reflected at specular and nonspecular angles is reduced. In pigmented coatings, it may be difficult to distinguish visually between gloss reduction resulting from haze and from the scattering of light by a pigment. It is always desirable for a formulator to make a batch of coating with no pigment to check the clarity of a dried, cross-linked, pigment-free film. Haze can be quantified in nominally clear coatings by measuring the amount of light transmitted outside of a narrow angular range adjacent to the main beam of light being transmitted. Typically, light outside of 1.5° from the main beam will contribute to haze measured by commercial instruments.

Bloom gives an appearance similar to haze. If a liquid component of a coating film is not soluble in the resin binder, it can separate from the body of the film in small droplets. These can come to the surface, making it uneven, diffusing light beams striking the surface and reducing gloss. In contrast to haze, bloom can be wiped off a surface with a cloth dampened with solvent for the blooming material; commonly, bloom reappears.

Mottle is observed as a change in color or lightness over a reasonably large length scale, perhaps a few tens of centimeters. Mottle is typically due to application differences that result in spatially varying film thicknesses or pigment morphology. Inconsistent flake alignment in metallic coating can give rise to mottle on automobile bodies.

At the smallest scale, gloss is due to the interaction of light waves/photons with the surface. If the roughness of the surface is on a much larger scale than the wavelength of the light, geometric optics prevail and a surface can be considered to be composed of a joined series of facets, each of which gives rise to a specular reflection. When a surface's roughness is much smaller than the wavelength of light, the material appears to be optically smooth (high gloss, mirror-like), and little scattering occurs other than Rayleigh scattering. For the intermediate cases, where the roughness of the surface is on the order of the wavelength of light, no simple approximations exist to predict the gloss or scattering of a surface. Complex mathematics, which are beyond the scope of this book, can be used to simulate the scattering behavior of these surfaces. In many ways, scattering of this type is analogous to the scattering of microwaves from the surface of the ocean, where the wavelength of the radiation is similar to that of the surface waves of the ocean. As important as gloss is, it is still fundamentally not well understood or mathematically describable.

19.10.1 Variables in Specular Gloss

Some of the variables of specular gloss can be visualized by use of schematic diagrams based on hypothetical measurements by an idealized goniophotometer. A goniophotometer can direct a beam of light onto a surface at any angle of incidence and can measure the amount of light reflected at various angles away from the surface. A schematic representation of an idealized instrument is shown schematically in Figure 19.19. Figure 19.19a shows a cross section through the plane of the beam of light. Figure 19.19b shows a schematic view looking down on the instrument. The drawing shows that the only light detected by the photometer is that reflected in the plane of the beam of light. Also, there is no way to build an instrument that would measure light reflected at 90°. In this idealized instrument, the beam of light is narrow enough so that only a point on the surface is illuminated, and the photometer can detect a reflected beam with a beam width approaching zero.

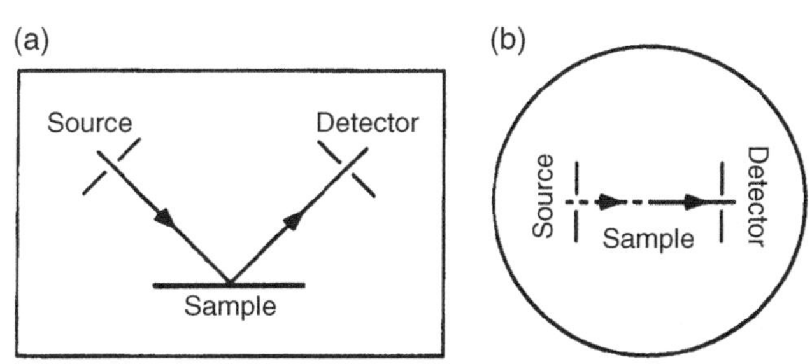

Figure 19.19 Simplified representation of a goniophotometer: (a) side view and (b) top view.

All light reflected from a perfect mirror is at the specular angle and in the same plane as the beam of light. From a perfect diffuse reflector, light is reflected at all angles in all planes. Line S_0 in Figure 19.20 shows photocurrent as a function of angle of viewing from a perfect diffuse reflector illuminated at 45°, as measured by an ideal goniophotometer (Zorll, 1972). At first glance, the results do not seem rational, but it must be recalled that the photometer can measure all the light reflected at 0°, but only a small fraction (approaching zero) of the light reflected in all planes at 90°.

Also shown in Figure 19.20 is the idealized response curve *u* for a semigloss white coating. Comparison of these two curves illustrates three important factors in considering specular gloss. The first factor is the *relative brightness* *A* of the sample compared with that of the diffuse reflector, in which *A* is the ratio of light intensities in the perpendicular direction u_0/S_0. Second is *height of gloss*, *h*, which is calculated from the *peak height*, *p*, by Eq. 19.12:

$$h = \frac{P - u_{45}}{S_{45}} = \frac{p}{S_{45}} - \frac{A}{100} \tag{19.12}$$

Third is the DOI aspect of gloss δ, as defined by Eq. 19.13, in which *P* is the total area under the sample curve, *U*

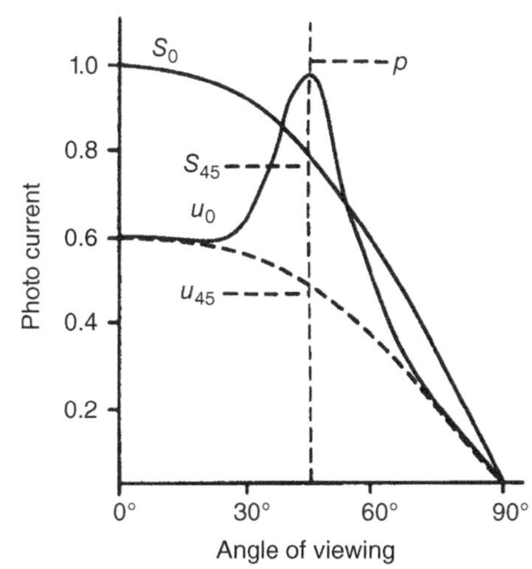

Figure 19.20 Schematic representation of the photocurrent reflected as a function of viewing angle. Source: Zorll (1972). Reproduced with permission of Elsevier.

is the area of diffuse background reflection from the sample, and S is the area under the standard diffuse reflector curve (Zroll, 1972):

$$\delta = \frac{h}{F} \quad F = \frac{P-U}{S} \tag{19.13}$$

If A is large, an observer will consider the gloss of a sample to be lower than that of a sample with a lower A. If h is small, the gloss of a panel will be lower than that of a sample with a higher h. It is instructive to compare a white gloss coating with a black gloss coating. A is larger and h is smaller for the white coating because the white coating scatters light that enters a film, resulting in a high extent of diffuse reflection. On the other hand, a black coating absorbs almost all the light entering its film, so there is little diffuse reflection from within the film. If the surfaces of the two films are equally smooth, surface reflection is equal, but the black coating will be seen as glossier even though it reflects less of the incident light. It is impossible to make a white coating with as high a gloss as that of a high gloss black one. Gloss of colored coatings is intermediate; the darker the color, the higher the potential gloss. In Section 19.2.4 we considered the effect of gloss on color; now we see that color also affects gloss. If the value of δ is high, the surface acts as a mirror, giving a clear image. As the value of δ decreases, there is more and more blurring of the specular image.

One must be careful when comparing these values with direct evaluation of gloss by an observer. As has already been stated, different observers rate the gloss of pairs of panels differently. Intensity of reflection (peak height) is not a major factor in an observer's evaluation of gloss; the principal factors are the contrast between reflection at the specular angle and the distinctness of image. Gloss is also affected by the distance between the observer and an object. An observer, who is close enough to visually resolve irregularities on the surface, will say that a surface is a rough, high gloss surface. On the other hand, one observing the same surface at a greater distance, which does not allow visual resolution of the irregularities, will say it is a smooth, low gloss surface. Recent studies with human observers have demonstrated that the perception of gloss for even achromatic colors is strongly influenced by lightness and that 60° gloss is most closely associated with visually perceived gloss (Mirjalili et al., 2014).

The major factor controlling gloss of coatings is pigmentation. As a film shrinks owing to loss of volatile components, pigment particles can cause irregularities in the surface. Roughness of the surface varies with the ratio of PVC to the CPVC in the dry film (Section 22.1). Braun (1991) discusses the effect of pigmentation on gloss as solvent evaporates.

In solventborne, high gloss coatings, the pigment concentration in the top micrometer or so of a dry film contains little, if any, pigment. This pigment-depleted layer results from the movements within a film as solvent evaporates. Initially, convection currents are set up within the film; and both resin solution and dispersed pigment particles move freely. As solvent evaporates, viscosity of the film increases, which decreases the movement of pigment particles. However, the movement of resin solution continues longer, which results in reduced pigment levels in the top surface. Accordingly, application of a clear top coat over a pigmented base coat increases gloss unless the clear coat itself contains matting agents.

As PVC increases, the amount of pigment at the surface increases, decreasing gloss. Particle size of the pigment affects gloss; if aggregates of pigment are not broken up in the dispersion process, gloss will be lower, since flocculated pigment systems have a lower CPVC. However, since large particles become immobilized before small ones, flocculated particles can lead to increased gloss in low PVC coatings. Braun and Fields (1994) and Simpson (1978) discuss effects of pigment particle size and thickness of clear top layers on specular gloss.

Low gloss together with a high degree of transparency is desirable in some coatings (e.g., furniture lacquer). This is accomplished using a small quantity of very fine particle size silicon dioxide as a pigment. The combination of small particle size and low refractive index difference results in minimal light scattering as long as concentration is low. When solvent evaporates from such a lacquer, the SiO_2 particles keep moving until the viscosity of the surface of the film becomes high. The result is a higher than average concentration of pigment in the top of the film, reducing gloss at relatively low PVC.

Poor leveling can reduce gloss. If the irregularities are large, as is commonly the case with brush marks, the surface is seen as a glossy, but wavy surface. However, if the irregularities are small, as is commonly the case with orange peel, gloss will be rated as lower. An irregular surface can result from not only poor leveling over a smooth surface but also applying a coating over a rough substrate, also known as *telegraphing*. It has been shown that smoothness and gloss are affected by the extent of roughness of a substrate, film thickness, and viscosity (Ngo et al., 1993). See Section 24.2 for further discussion of leveling.

Orange peel values are strongly dependent on application parameters, paint formulation, substrate quality, and drying/curing conditions. Minimum values are particularly important for automotive coatings, where a high quality surface finish is associated with overall vehicle quality. Quantification of orange peel has been aided by the development of instruments that measure both the amplitude and lateral dimensions of surface roughness. The roughness is often binned into five categories of increasing size scale:

Wa (0.1–0.3 mm), Wb (0.3–1.0 mm), Wc (1.0–3.0 mm), Wd (3.0–10 mm), and We (10–30 mm). A further parameter, dullness (du), is used to assess roughness on the size scale of less than 0.1 mm and is correlated with the inverse of the gloss value. The typical appearance spectrum is shown in Figure 19.21, where high levels of roughness are seen at the Wa and Wb sizes, which indicates poor short-range roughness values. While formulation can certainly affect the roughness of a coating, the Wa–We values are most often used to tune application parameters to minimize orange peel under certain operating constraints and with a given paint formulation.

Wrinkling affects gloss. If cross-linking occurs at the surface of a film prior to the lower layers, wrinkling is likely to occur (Section 24.6). When the lower layers subsequently cross-link, shrinkage occurs, causing the top of the film to pucker up in wrinkles. If the size of the wrinkles is small enough, gloss will be low, but when examined with a microscope, the surface can be seen to be an irregular glossy surface.

Since the fraction of light reflected at a surface increases as refractive index increases, gloss of high gloss coatings tends to increase as refractive index of a coating increases, since the contrast between reflection at the specular angle and other angles is greater. Refractive index differences between different binders are small, so while detectable, this effect is generally small compared with surface roughness (Zorll, 1972). Refractive indexes of pigment may also be a factor.

Latex coatings generally exhibit lower gloss than solventborne coatings at equal levels of pigmentation in the dry film (Section 32.3). There are several reasons for this:

1. Both resin and pigment are dispersed in latex coatings. Accordingly, there is not the same opportunity during drying for generation of a pigment-depleted thin layer at the top of the film as in a solvent coating. Latexes with smaller particle size provide somewhat higher gloss films, since larger size particles become immobilized sooner as a film dries.
2. Pigment-free dry films of many latexes are not transparent. Rather, they are hazy owing to the presence of dispersants and water-soluble polymers that are not completely soluble in the latex polymer, thereby reducing gloss.
3. Commonly, surfactants bloom to the surface of a latex coating, reducing gloss.
4. It is generally more difficult to achieve good leveling with latex coatings. Surface roughness such as brush marks reduce gloss. The appearance of brush marks depends on how close an observer is to the surface. If one is close enough to visually resolve the brush marks, the coating can look like a wavy, gloss surface. From a greater distance, the gloss appears to be low. Leveling has been improved by formulation with associative thickeners.

Gloss can change during the life of an applied film. In pigmented films, gloss loss proceeds as the binder degrades and is worn away, often by the action of liquid water. When pigment particles are exposed, they scatter light in aspecular

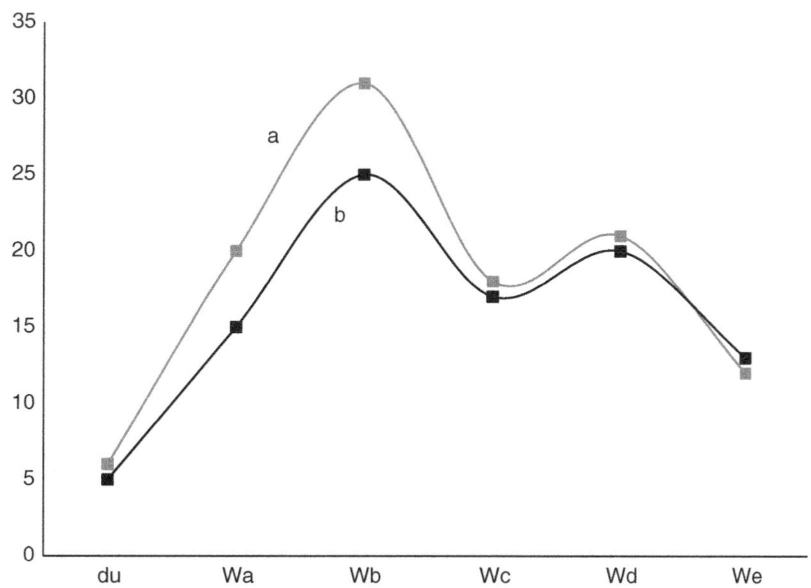

Figure 19.21 Surface roughness measurements from a Wavescan instrument showing higher roughness values in the low wavelength regions, Wa and Wb for coating (*a*), while appearance is equivalent to coating (*b*) at longer wavelengths.

directions, reducing gloss. As more and more pigment is exposed, the gloss continues to decrease. Eventually, the surface of the film may embrittle and then crack as the film expands and contracts. The absence of pigments in clear coats is the main reason for their superior gloss retention compared with pigmented coatings. The surface of a clear coat remains relatively smooth as it degrades and erodes, since there are no pigment particles to scatter light. Erosion of binder in pigmented films can proceed to a stage in which pigment particles on the surface easily rub off, a phenomenon called *chalking*. Chalky surfaces have dramatically lower gloss; and, therefore, the color becomes lighter. Loss of gloss can also result from loss of volatile components as a film ages, leading to film shrinkage and increased surface roughness. Braun and Cobranchi (1995) offer an excellent review of durability and gloss.

In some cases, the gloss of a low gloss surface increases with use. It is common for the gloss of flat wall paints to increase if rubbed, for example, near a light switch. This increase in gloss is called *burnishing*.

19.10.2 Gloss Measurement

No fully satisfactory method for measuring gloss is available, and no satisfactory rating scale for visual observation of gloss has been developed. While all people will agree regarding which film is glossier if the gloss difference is large, they frequently disagree if the difference is small. Even the same observer experiences difficulty in consistently ranking a series of panels if the differences in gloss are small. Differences are exacerbated unless the lighting and viewing conditions are well controlled.

Caution must be exercised in using instruments that measure gloss. In Section 19.10.1, we discussed an idealized goniophotometer. In practice, one has to accept less than this ideal. The beam diameter cannot actually approach zero because there must be sufficient light intensity reflected from all angles to give a measurable response on available photodetectors. In real instruments, there is a light source shining on a slit aperture at a fixed distance from a sample surface. The reflected light goes through another slit aperture at a fixed distance to a photodetector.

Specular gloss meters are widely used, but correspondence between meter readings and visual comparisons is limited, although improving (Leloup et al., 2016). Such instruments give significantly different readings with differences in intensity of reflected light, but observers are relatively insensitive to such differences. Furthermore, the aperture of the slit in a gloss meter is about 2°, whereas the limit of resolution of a human eye is about 0.0005° of arc (Braun, 1991). A gloss meter is, therefore, less sensitive to DOI than the eye (when the observer is close to the object). The distance between the aperture and a panel is fixed in a gloss meter, whereas a person can view a panel from any distance.

The most widely used *gloss meters*, also called *reflectometers*, are simplified goniophotometers in which one measures a response only at the specular angle. Those most commonly used in the coatings industry can make measurements when the angles of incidence and viewing are 20°, 60°, and 85°. A schematic drawing is shown in Figure 19.22. The first step for using a gloss meter is to calibrate the instrument with two standards: one with high gloss and the other with a lower gloss. If the second standard does not give the standard reading after the instrument is set with the first standard, something is wrong; most commonly, one or both of the standards are dirty or scratched. Other possible problems include panel misalignment, deterioration of the light source, or a malfunction of the photometer. One must use the standard that has been calibrated at the selected angle. Black and white standards are available. Reflection at the specular angle is not the same from a white and a black standard with equal surface roughness (Section 19.10.1). It is not feasible to have standards of all different colors; white standards are used for light colors and black standards for dark colors. (One should always report which set of standards is used.) Standards can be certified by NIST using a research goniophotometer described in Nadal and Thompson (2000). In contrast to gloss meters, such research instruments allow both angles of illumination and viewing to be varied independently; but they are large, expensive, and relatively difficult to maintain.

In normal practice, one first measures at 60°. If the reading obtained is over 70, readings should be made at 20° rather than at 60° since precision is higher nearer the midpoint of the meter reading. It is common to read low gloss panels at both 60° and 85°. Readings at 85° may have a relationship to sheen. It is essential to report the angle at

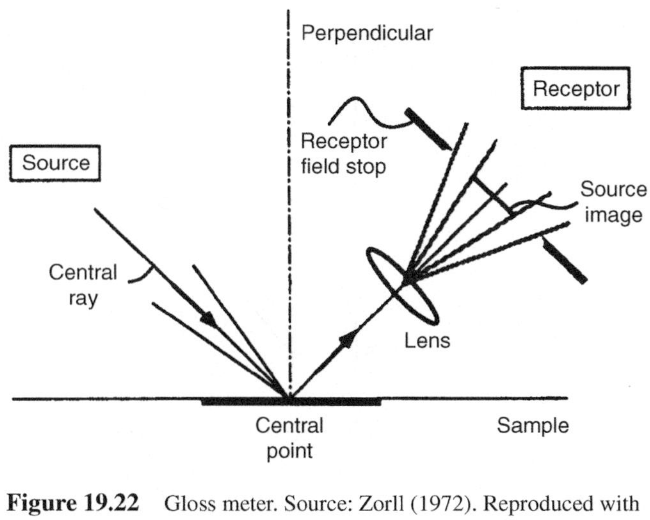

Figure 19.22 Gloss meter. Source: Zorll (1972). Reproduced with permission of Elsevier.

which readings are made. Multiple readings should be taken and analyzed statistically. Some gloss meters can perform this analysis. This reduces the probability that local surface irregularities or dirt particles are affecting the meter reading. Slight brush marks may not affect readings if the direction of the brush marks is parallel to the plane of incidence. If results are compared between two laboratories, it is essential to check the compatibility of their instruments. This is best done by measuring at least three black and three white standards on each instrument. Readings are reproducible on carefully calibrated instruments to ±3%, but it is more appropriate to express error in terms of units, such as ±2 gloss units (Huey, 1964). This is a high percent of error in the low gloss range.

Gloss meters provide numerical results between 0 and 100, but there is confusion as to what the numbers mean. Commonly, the values are expressed as percents. However, the values are *not* the percent of light reflected at the surface. They approximate the percent of light reflected at that angle compared with the reading that would be obtained if a perfectly smooth surface were measured. The values are more appropriately called gloss units or just meter readings. As noted previously, the total reflection from a black matte surface is much higher at most angles of illumination and viewing than that from a high gloss black surface. The point can be emphasized by comparing meter readings of the same panels at 20° and 60°. Meter readings are lower for the same panel when the setting is 20°, as compared with 60° (Figure 19.23) (Freier, 1967).

With sufficient care, specular gloss instruments can be used for quality control comparisons of lots of the same or very similar coatings and for following loss of gloss on aging. The instruments are not appropriate for specification purposes other than quality control. Rather, one must rely on standard visual panels. The customer selects three panels of each color, one with the desired gloss and the others representing the upper and lower acceptable limits.

Figure 19.23 also illustrates a procedure for following loss of gloss when a high gloss panel is exposed to exterior weathering. Initial readings are taken at 20°; then after some exposure, at both 20° and 60°, then for panels that have shown considerable loss of gloss only at 60°. Different scale units in the 20° and 60° readings (left scale) provide for a continuous slope at the transition point. The right scale designates gloss grade as observed visually.

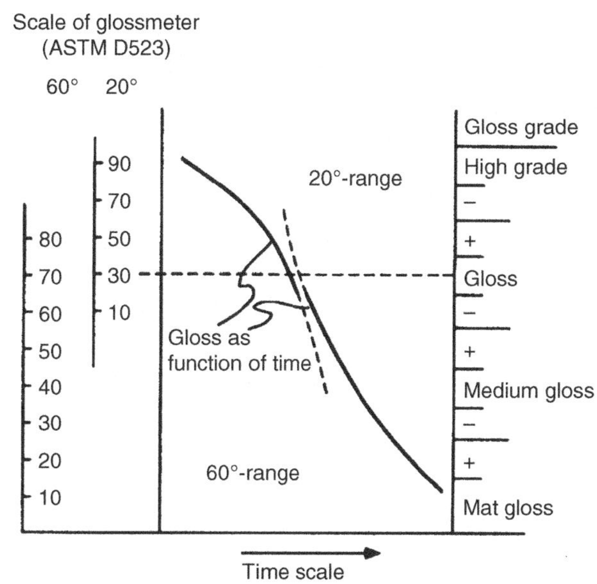

Figure 19.23 Transition of gloss readings. Source: Zorll (1972). Reproduced with permission of Elsevier.

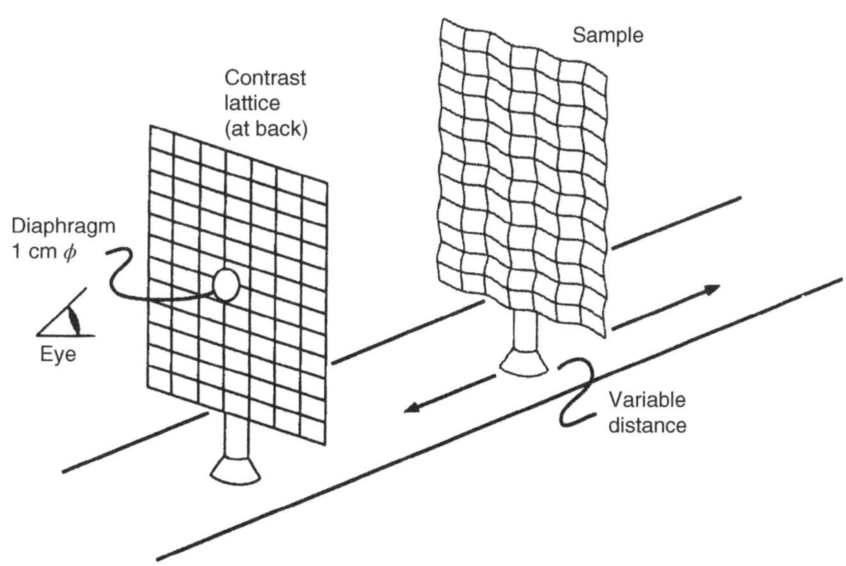

Figure 19.24 DOI meter. Source: Zorll (1972). Reproduced with permission of Elsevier.

DOI meters rely on using the sample as a mirror (Figure 19.24). The reflection of a grid on the surface of the panel is compared visually to a set of photographic standards ranging from a nearly perfect mirror reflection to a blurred image in which the grid cannot be detected. One reports the comparison of the degree of blurring and also a qualitative statement about distortion. Correspondence with visual assessment in the high gloss range is better than with specular gloss meters. Instruments are available that make comparisons based on optical density rather than on visual assessment.

GENERAL REFERENCES

ASTM, *Standards on Color and Appearance Measurement*, 3rd ed., ASTM, Philadelphia, **1991**.

Burns, R. S., *Billmeyer and Saltzman, Principles of Color Technology*, 3rd ed., John Wiley & Sons, Inc., New York, **2000**.

Hammond, H. K., III; Kigle-Boeckler, G., Gloss in Koleske, J. V., Ed., *Paint and Coating Testing Manual*, 14th ed., ASTM, Philadelphia, **1995**, pp 470–480.

Hunter, R. S.; Harold, R. W., *The Measurement of Appearance*, 2nd ed, John Wiley & Sons, Inc., New York, **1987**.

Kuehni, R. G., *Color, an Introduction to Practice and Principles*, 2nd ed., John Wiley & Sons, Inc., New York, **2004**.

Pierce, P. E.; Marcus, R. T., *Color and Appearance*, 2nd ed., Federation of Societies for Coatings Technology, Blue Bell, PA, **2003**.

Westland, S.; Ripamonti, C., *Computational Colour Science using MATLAB*, John Wiley & Sons, Inc., New York, **2004**.

Zorll, U., *Prog. Org. Coat.*, **1972**, *1*, 113.

REFERENCES

AATCC, Test Method 173-2009, AATCC, **2009**.

Allen, E., *J. Opt. Soc. Am.*, **1974**, *64*(*7*), 991–993.

ASTM D2244-16, *Standard Practice for Calculation of Color Tolerances and Color Differences from Instrumentally Measured Color Coordinates*, ASTM International, West Conshohocken, PA, **2016**, www.astm.org.

Billmeyer, F. W., *Encyclopedia of Chemical Technology*, John Wiley & Sons, Inc., New York, **1979**.

Billmeyer, F. W.; Saltzman, M., *Principles of Color Technology*, John Wiley & Sons, Inc., New York, **1981**.

Braun, J. H., *J. Coat. Technol.*, **1991**, *63*(*799*), 43–51.

Braun, J. H.; Cobranchi, D., *J. Coat. Technol.*, **1995**, *67*(*851*), 55–62.

Braun, J.; Fields, D., *J. Coat. Technol.*, **1994**, *66*(*828*), 93–98.

Chan, X. B. V., et al., *Asia Pac. Fam. Med.*, **2014**, *13*(*1*), 1.

CIE, Publ. 142-2001, Central Bureau of the CIE, Vienna, Austria, **2001**.

Cremer, M., *Prog. Org. Coat.*, **1981**, *9*(*3*), 241–279.

easyrgb.com, **2008**, from http://www.easyrgb.com/ (accessed April 22, 2017).

Fields, D. P., *Surf. Coat. Int.*, **1993**, *76*, 87.

Freier, H.-J., *Farbe Lack*, **1967**, *73*, 316.

Greenstein, L. M., in Lewis, P. A., Ed., *Pigment Handbook*, Wiley-Interscience, New York, **1988**, Vol. *1*, pp 829–858.

Hardy, A. C., *Handbook of colorimetry*. 1st ed., The Technology Press, MIT, **1936**.

He, X. D. T., et al., *Comput. Graph. Forum*, **1991**, *25*(*4*), 175–186.

Huey, S., *Off. Dig.*, **1964**, *36*, 344.

Judd, D. B.; Wysecki, G., *Color in Business, Science, and Industry*, John Wiley & Sons, Inc., New York, **1975**.

Kumar, R., et al., *J. Coat. Technol.*, **1985**, *57*(*720*), 49–54.

Leloup, F. B., et al., *J. Coat. Technol. Res.*, **2016**, *13*(*6*), 941–951.

Mirjalili, F., et al., *J. Coat. Technol. Res.*, **2014**, *11*(*6*), 853–864.

Mitton, P., in Patton, T. C., Ed., *Pigment Handbook*, John Wiley & Sons, Inc., New York, **1973**, Vol. *3*, pp 289–339.

Nadal, M. E.; Thompson, E. A., *J. Coat. Technol.*, **2000**, *72*(*911*), 61–66.

Ngo, P.-A. P., et al., *J. Coat. Technol.*, **1993**, *65*(*821*), 29–37.

Pierce, P. E.; Marcus, R. T., *Color and Appearance*, Federation of Societies for Coatings Technology, Blue Bell, PA, **1994**.

Rich, D., *J. Coat. Technol.*, **1995**, *67*(*840*), 53–60.

Rich, D. C., *J. Coat. Technol. Res.*, **2016**, *13*(*1*), 1–9.

Saunderson, J. L., *J. Opt. Soc. Am.*, **1942**, *32*, 727.

Sharma, G., et al., *Col. Res. Appl.*, **2005**, *30*(*1*), 21–30.

Simpson, L., *Prog. Org. Coat.*, **1978**, *6*(*1*), 1–30.

Streitberger, H.-J.; Dossel, K.-F., *Automotive Paints and Coatings*, John Wiley & Sons, Chichester, **2008**.

Wicks, Z. W., Jr.; Kuhhirt, W., *J. Paint Technol.*, **1975**, *47*(*610*), 49.

Wyszecki, G.; Stiles, W. S., *Color Science*, John Wiley & Sons, Inc., New York, **1982**.

Zorll, U., *Prog. Org. Coat.*, **1972**, *1*(*2*), 113–155.

Chapter 20

Pigments

Pigments are insoluble, fine particle size materials used in coatings for one or more of the five reasons: to provide color, to hide substrates, to modify the application properties of a coating, to modify the performance properties of films, and/or to reduce cost. Pigments are divided into four broad classes: white, color, inert, and functional pigments. Pigments are insoluble materials used as colloidal dispersions. Dyes are soluble colored substances; they are used only in a few specialized coatings such as stains for wood furniture (Section 31.1.1). Some color pigments are called *lakes* and others are called *toners*. The original meaning of lake was a dye that had been converted into a pigment by irreversible adsorption on an insoluble powder. The term lake is now sometimes used when a colored pigment is blended with an inert pigment. Toners are pigments that are essentially all-color pigment.

Particle size affects color strength, transparency or opacity, exterior durability, solvent resistance, and other properties of pigments. For any given pigment, the manufacturer selects the particle size, which most appropriately provides a combination of the desired properties, and designs the process to produce that average particle size consistently. In general, there will be a distribution of particle sizes and shapes. Manufacture of many pigments involves precipitation from water; the process conditions determine the particle size range. Many pigments are surface treated during or after precipitation. The precipitated pigment is filtered and the filter cake is dried. During drying, pigment particles become cemented together in *aggregates*. The coating manufacturer often receives pigment from the pigment manufacturer as a dry powder of aggregates and must make a stable dispersion in which the aggregates are broken up and the pigments are reconverted into their original particle size (Chapter 21).

Few entirely new pigment compositions have been commercialized in recent years, but progress has not stood still. Pigment manufacturers strive to improve existing compositions by refining their processes. Optimization of particle size distribution, purity, consistency, shade, and ease of dispersion reduces costs for their customers and in some cases may help customers improve product quality. Numerous patents describe coating of pigments with polymers. Specialized books on pigments include those by Hunger et al. (2014), Buxbaum and Pfaff (2005), Faulkner and Schwartz (2009), and Patton (1973).

20.1 WHITE PIGMENTS

A large fraction of coatings contains white pigment. White pigments are used not only in white coatings but also in a substantial fraction of other pigmented coatings to provide lighter colors than would be obtained using color pigments alone. Furthermore, many color pigments have poor hiding, and the white pigment provides a major part of the hiding power of the coating. As discussed in Section 19.2.3, the ideal white pigment absorbs no visible light and has a high scattering coefficient. Since a major factor controlling scattering efficiency is the difference in refractive index between the pigment and the binder, refractive index is a critical property of a white pigment.

20.1.1 Titanium Dioxide

The most important white pigment used in coatings is TiO_2. The worldwide market for TiO_2 pigments is projected to reach $17B USD by 2020, with about half used in coatings (Anonymous, 2015).

Organic Coatings: Science and Technology, Fourth Edition. Frank N. Jones, Mark E. Nichols, and Socrates Peter Pappas.

The technology of TiO_2 pigments is reviewed by Winkler (2013) and Braun (1995). Two different crystal types are used: *rutile* and *anatase*. Rutile predominates primarily because it provides significantly greater hiding power than anatase, owing to the higher average refractive index of rutile of 2.73, compared with 2.55 for anatase. On the other hand, although a coating pigmented only with rutile looks white, it is yellowish alongside a coating pigmented with anatase. As shown in Figure 20.1, rutile absorbs some violet light, whereas anatase absorbs almost no visible light.

The color and hiding of rutile TiO_2 coatings can be adjusted by tinting. If one adds a small amount of a dispersion of a violet pigment, such as carbazole violet, the resulting coating is less yellow. The violet pigment strongly absorbs wavelengths other than violet and reduces the difference in reflection of violet compared with the other wavelengths. The resulting coating is light gray, but unless there is a higher reflectance white pigment present for comparison, it is seen as white. At the same time, there is the added advantage that hiding power is increased due to the additional light absorption. Phthalocyanine blue can be used in the same way. It does not give as "white a white" as carbazole violet, but is less expensive. Surprisingly the most common tinting pigment used in white paints is carbon black. Carbon black is less expensive than either violet or blue pigments and provides a greater increment to hiding. Since carbon black absorbs all wavelengths of light, it does not reduce the yellowness of the white, but it provides a lower cost method of increasing hiding.

Being more photoreactive than rutile TiO_2, anatase-pigmented coating films fail more quickly on outdoor exposure by *chalking* than rutile-pigmented films. Chalking is the formation of loose particles on the exposed coating surface resulting from the erosion of binder by photodegradation (Section 5.2.4). Chalking is usually undesirable, although at one time *self-cleaning* white house paints were deliberately formulated with anatase TiO_2 so that they chalked readily. As the surface of the film erodes away, dirt accumulation is also removed so that the paint surface stays whiter.

In most exterior applications, chalking must be minimized, and rutile TiO_2 is chosen. However, rutile is still sufficiently photoreactive to reduce exterior durability. The problem is minimized by treating the surface of the TiO_2 particles during pigment manufacture (Braun, 1995). The most common surface treatments are silica (SiO_2) and/or alumina (Al_2O_3); other metal oxides are also used. A typical surface treatment for exterior grades of rutile is about 6 wt% silica and 4 wt% alumina. An ideal surface treatment would be a continuous shell on each TiO_2 particle to keep air away from the TiO_2 surface. Perfection has not been achieved, but the best chalk-resistant grades of rutile show little acceleration of degradation. Treated anatase pigments accelerate chalking more than treated rutiles but much less than the untreated anatase pigments. Since TiO_2, particularly rutile, strongly absorbs UV (Figure 20.1), it enhances exterior durability of coatings as a UV absorber, compensating to some extent for its photoreactivity.

Surface treatment affects other properties of the pigments besides chalking tendencies. Alumina treatments

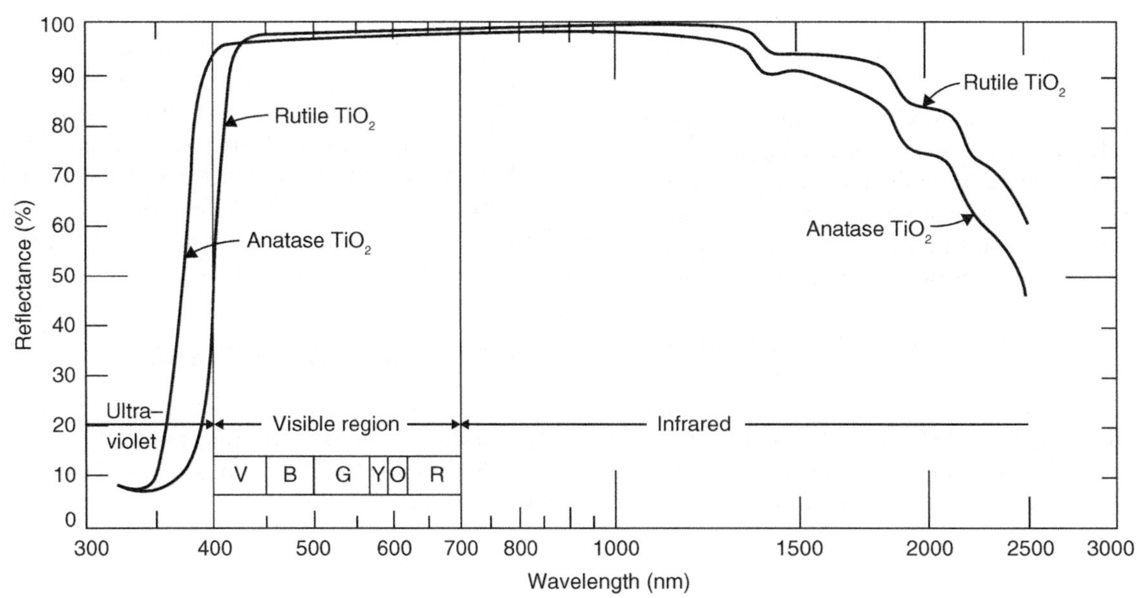

Figure 20.1 Reflectance spectra of rutile and anatase pigments in TiO_2 coatings. Source: Patton (1973). Reproduced with permission of John Wiley & Sons.

improve dispersibility in solventborne coatings. However, alumina can partly neutralize acid catalysts, such as those used with coatings containing Class I MF resins (Section 11.3.1), thereby reducing cure rates. Specially treated TiO_2 pigments permit production of more stable dispersions in waterborne coatings. A surface treatment consisting of small polymer particles that adhere strongly to TiO_2 is credited with improving hiding efficiency by improving TiO_2 particle spacing (Hook, 2014). Spacing is further discussed in Section 20.3.

TiO_2 dispersions in water, called *slurries*, are used in the manufacture of latex paints (Section 21.3) as an alternative to dry pigment. The slurries are manufactured by the TiO_2 producer and shipped in tank cars or tank trucks. The price on a pigment basis is about the same, but the slurry is a pumpable liquid, and aggregation of pigment particles is minimized, so there are substantial savings in material handling and dispersion costs.

The lower absorption of near-UV radiation by anatase TiO_2 is an advantage over rutile in white-pigmented UV cure coatings, since there is less interference with absorption of UV radiation by photoinitiators (Section 29.5).

Various TiO_2 pigments are offered by each manufacturer. Formulators need to be aware of the differences among the types. Coating failures and economic waste have resulted from the use of the wrong type of TiO_2 and from indiscriminate substitution of one type for another. Some types are designed for maximum exterior durability, some to provide maximum gloss, some are more easily dispersed in water systems, and others are more easily dispersed in solvent systems. The TiO_2 content of the various types of pigments varies from as high as 98% to as low as 75%. Generally, the hiding varies with TiO_2 content, so the amount of TiO_2 needed for the hiding varies with pigment type. Another consideration is the range of particle sizes; the broader the range, the larger the fraction of particles that provide less than optimum hiding.

Nanoparticle TiO_2 pigments are too small to appreciably scatter light, but they are finding other uses as discussed in Sections 20.5 and 22.2.1.1.

20.1.2 Other White Pigments

Opaque white pigment technology has a fascinating history (Steig, 1975). For centuries white lead (basic lead carbonate, $2PbCO_3{\cdot}Pb(OH)_2$) was the dominant white pigment in coatings. Its relatively low refractive index, 1.94, gives low hiding power, less than one-tenth that of rutile TiO_2, but it performed well in linseed oil paints, the dominant binder until the 1930s. Starting just before 1900, white lead was gradually supplanted by pigments with better hiding (see the following text), although substantial use continued into the 1940s.

White lead is slightly soluble in water and presents a toxic hazard. Many children have suffered from lead poisoning attributed to ingesting paint containing white lead paints that peel from the walls and woodwork in older buildings or paint dust containing white lead. As a result, the lead content of any paint sold to retail consumers through interstate commerce in the United States is limited to 0.06% of the dry weight. There are similar restrictions in Europe. However, white lead was obsolete well before these regulations took effect. Lead poisoning should not automatically be assumed to result from lead paint. There are other possible sources of lead exposure, including lead pipes in water systems and dirt contaminated with lead from auto exhausts during the era of leaded gasoline.

Zinc oxide, ZnO, was a popular white pigment from about 1900 to 1950, although its low refractive index, 2.02, made its hiding power only a bit better than white lead's. ZnO is still used in exterior house paints as a fungicide and in some can linings as a sulfide scavenger. Zinc oxide should not be used in primers because it is somewhat water soluble and can, therefore, induce blistering by osmotic pressure when water vapor penetrates through the top coats into the primer coat.

Other white pigments, important at one time, but now largely obsolete, are zinc sulfide (ZnS), with a refractive index of 2.37, leaded zinc oxide, and lithopone, a coprecipitate of ZnS with $BaSO_4$. During the late 1920s and 1930s, lithopone was the largest selling white pigment, although it was best suited for indoor paints. It had almost disappeared from the market by 1950 (Steig, 1975).

It was long known that TiO_2, with its high refractive index, could be a superior white pigment. It was a challenge, however, to manufacture good pigments from it. The first commercial TiO_2 appeared in 1916, but for the next 20 years or so, the evolving technology led to gradual increase in market share. Technical improvements continued, and by the 1940s white house paints were commonly pigmented with combinations of zinc, lead, and titanium oxides. Early TiO_2 pigments were composites of TiO_2 with barium or calcium sulfates. The first 100% rutile pigment was commercialized in 1939, but World War II delayed its implementation. Silica- and alumina-coated TiO_2 appeared in the 1950s. By the early 1970s, rutile TiO_2 "was in undisputed possession of the coatings market" (Steig, 1975). That is not the end of the story, however. Ever since, producers have improved their products by, for example, improving surface treatments and improving particle size distributions (Gao et al., 2014; Veronovski et al., 2014). Thus, TiO_2 pigment technology has been evolving for 100 years, and further evolution can be expected.

Another chapter in pigment history involves pigmentation by air. White flowers use this effect to scatter light. In some cases, cost reductions are possible by taking advantage of scattering by small air bubbles in coating films. The refractive index of air, 1.00, gives a refractive index difference of about 0.50 from that of a typical binder. While this difference is small compared with the difference of 1.2 between binder and rutile TiO_2, it still scatters light and adds to hiding. In coating films, a common way of incorporating air bubbles is by the use of such high levels of pigmentation after the film forms so that there is not enough resin to adsorb on all the pigment surfaces and fill all the interstices between the pigment particles. The resulting air voids increase hiding. Such coatings are said to be formulated at pigment volume levels above the *critical pigment volume concentration* (CPVC) (Chapter 22). In the case of rutile TiO_2, the very large refractive index difference of 2.7 between the TiO_2 and air is especially effective in increasing hiding. However, the voids make the film porous, reducing protective properties and increasing dirt pickup. In ceiling paints, advantage is taken of the added hiding, since requirements for other properties are not as important.

Tinting strengths of white coatings increase as the PVC of a series of colored coatings is increased beyond CPVC. The air voids present above CPVC increase light scattering so that a colored paint dries with a lighter color than one having the same amount of color pigment but with PVC below CPVC.

Particles containing air bubbles provide added hiding from the air/resin interface without imparting porosity. For example, high T_g latexes containing water trapped within the particles are available and are used in latex paints as a partial replacement for TiO_2 (Fasano, 1987). Such latexes are prepared by sequential emulsion polymerization (Herbst and Hunger, 2006). In the first step, a low T_g latex copolymer of acrylic acid and (meth)acrylic esters is prepared; these particles are swollen with water at alkaline pHs. The swollen particles are then encapsulated in shells of high T_g cross-linked polymer, such as a styrene/divinylbenzene copolymer. When the paint containing these particles is applied and the water evaporates, the conventional latex particles of the binder coalesce to form the film, but the high T_g latex particles do not coalesce. Water diffuses out of the particle cores, leaving air bubbles within the particles. The diameters (about 250–300 nm) of the air bubbles can be optimized for light scattering, and the shell around each bubble prevents undesirable crowding effects (Hook, 2014; Nuasaen and Tangboriboonrat, 2015).

20.2 COLOR PIGMENTS

A wide variety of color pigments is used in coatings. Lewis (1995) and Hunger et al. (2014) provide detailed discussions of their chemistry, properties, economics, and uses. First, it is appropriate to summarize the considerations involved in selecting color pigments for a particular coating application:

1. *Color*. The first criterion is the color of the pigment. Pigment suppliers provide technical bulletins that contain color chips showing the colors obtainable with each pigment. Generally, there are two or three color chips for each pigment. The *mass tone* is displayed by a color chip in which the pigment is used as the sole pigment in the coating. Then, there are one or two additional color chips showing the color obtained when the pigment is used together with TiO_2 in different ratios; these are called *tint colors*. In some cases, color chips showing combinations with effect pigments (Section 20.2.5) are also provided. The Society of Dyers and Colourists in the United Kingdom and the Association of Textile Chemists and Colorists in the United States developed and maintain a *colour (color) index* system in which each pigment or dye is assigned a code number. For example, a certain pigment might be designated C. I. Pigment Red 202 or simply PR 202. The P in the code stands for pigment, the R designates the hue, and the number is chronologically assigned.
2. *Color strength*. Some colorants are strong and others weak corresponding to the absorption coefficients and particle sizes of the pigments. It may well be more economical to use an expensive (i.e., high cost per unit volume), strong pigment than a lower cost weak pigment.
3. *Opacity or transparency*. Depending on the end use, it may be desirable to use a pigment that increases hiding by both scattering and absorbing radiation, or it may be important to select pigments that scatter little, if any, light in the coating film so that a transparent color can be obtained.
4. *Ease of dispersion*. If everything else were equal, one would select the most easily dispersed pigment. Many pigments are surface treated by the manufacturers to enhance their ease of dispersion (Hays, 1984, 1986, 1990). Many types of surface treatment have been used. As discussed previously in Section 20.1.1, TiO_2 particles are coated with thin layers of Si, Al, and other oxides; organic pigments can also be treated in this way (Bugnon, 1996). Molecules having structures similar to that of the pigment but with attached polar groups can adsorb on organic pigment surfaces to enhance dispersibility. Adsorbed polymer monolayers and polymeric particles (Hook, 2014) can also be effective. Pigment dispersion is discussed in Chapter 21.
5. *Exterior durability*. Photodegradation of color pigments leads to loss of color or change of hue. Pigment

manufacturers provide data that can be useful in screening pigments for exterior applications. However, there can be significant variation in exterior durability depending on the combination of pigment and resin. Therefore, for critical applications, the durability of specific formulations must be determined. (Chapter 5 describes testing for exterior durability.) A related property is *light fastness*. This term usually refers to the ability of pigments to resist fading when used indoors. ASTM classifies pigments into five light fastness grades, I–V. Only those classed I or II are suitable for applications, such as artist paints, for which long-term color stability is critical. Unfortunately, the works of many artists, such as van Gogh, are fading because either knowingly or unwittingly they employed vulnerable pigments (Everts, 2016).

6. *Heat resistance*. Coatings that are baked or exposed to high service temperatures require heat-resistant pigments. A few pigments (e.g., yellow iron oxides) undergo chemical changes with heat and change color. A more common problem is that some organic pigments slowly sublime at elevated temperatures.
7. *Chemical resistance*. Many coatings are exposed to chemicals, most often to acids and bases. For example, automobiles are exposed to acid rain, and laundry machines are exposed to alkaline detergents. The pigments must resist color change under those conditions.
8. *Water solubility*. In most applications, pigments with significant water solubility are best avoided because they leach out of a film, resulting in loss of properties. The presence of water-soluble substances in a primer can lead to blistering of the film when water permeates through it. In latex paints, water-soluble materials can reduce the stability of the dispersion of the latex particles. The stability of colloidal aqueous dispersions can be particularly affected by the presence of low concentrations of polyvalent salts. On the other hand, pigments that help protect against corrosion by passivation must be somewhat soluble in water to be effective (Sections 7.4.2 and 20.4).
9. *Solvent solubility*. Partial solubility of pigments in some solvents can be a problem. For example, if one is making a red coating for a bicycle that is to have a white stripe on the red coating, the red pigment must not dissolve in solvents used in the white striping coating. If the red pigment is somewhat soluble, the stripes will turn pink. Such pigments are said to *bleed*. There is risk that certain pigments that do not bleed when overcoated with solventborne coatings may bleed into waterborne overcoats because of the aggressive amines in waterborne formulations.
10. *Moisture content*. Most "dry" pigments adsorb small to moderate amounts of water. The adsorbed water can result in serious problems with water-reactive binders such as 2K polyisocyanates.
11. *Toxic and environmental hazards*. Filter face masks should always be worn when handling dry powder pigments; it is dangerous to inhale fine particle dusts, even if they are chemically inert. In most cases, there is little or no hazard once the pigment is incorporated into a coating. In some cases, however, toxicity in a coating film can be important. The hazards of lead-containing pigments are described in the previous text. Zinc chromate pigments are generally thought to be human carcinogens. In some countries, their use in coatings is prohibited except in aircraft coatings; regulations can be expected to become increasingly restrictive. Another factor that can affect pigment choice is the increasing cost of disposal of materials containing heavy metals.
12. *IR reflectance*. Inorganic pigments that reflect rather than absorb infrared radiation are available in a variety of colors, as discussed in Section 20.4. Combinations of color pigments that absorb all visible wavelengths result in black films; thus, reflective inorganic pigments can be formulated to make gray paints having significantly less IR absorption than grays made with absorptive or carbon black pigments.
13. *Cost*. Cost is always important, but as noted earlier, one cannot determine which pigment is the least expensive by just considering the price per unit weight. The critical question is: How much will the final coating cost on a volume basis? The answer depends on pigment color strength and density.

20.2.1 Yellow and Orange Pigments

Inorganic Yellows and Oranges. Iron oxide yellows, FeO(OH), are low chroma brownish-yellow pigments. They provide opaque films with good hiding and high exterior durability; chemical and solvent resistance is excellent. The pigments are generally easily dispersed and are comparatively inexpensive. When heated above 150°C, they gradually change color to a low chroma red because they dehydrate to form iron oxide red, Fe_2O_3. They can be stabilized against color change up to 200°C by encapsulating each particle in glass. Most iron oxide yellows are made synthetically, but some natural or ochre pigments are used. In some cases, particularly with the natural pigments, the presence of soluble iron and other metal salts can affect the stability of coatings. Extremely fine particle size iron oxide pigments are also available and are used when transparency is required. This effect of particle size on opacity is an excellent example of the relationship between light scattering and particle size (Section 19.2.3).

Chrome yellow pigments are bright, high chroma yellows. Medium chrome yellow pigments are predominantly lead chromate ($PbCrO_4$). Greenish yellow colorants, called primrose yellow and lemon yellow, are co-crystals of lead chromate with lead sulfate. Reddish yellows, that is, chrome oranges, are co-crystals of lead chromate with PbO. Still redder oranges, molybdate oranges, are co-crystals of lead chromate with lead molybdate ($PbMoO_4$) and lead sulfate. Chrome yellows are relatively low cost, although higher than iron oxides, partly because their density is higher. Chrome yellows slowly discolor on exterior exposure to form lower chroma yellows, but their exterior durability is adequate for many outdoor uses. They bleed but are heat resistant. Due to their lead content and concerns about chromate, chrome yellows are seldom used in the United States and Western Europe, although they are still used elsewhere. A major application was yellow traffic striping paint. Monoarylide yellows and bismuth vanadate are their predominant replacements.

Bismuth vanadates ($BiVO_4$) are yellow pigments with high brilliance. Durability is satisfactory for automotive paints (Endriss et al., 2009). Cadmium pigments, exemplified by CdS·ZnS (PY 35), are another durable pigment, popular for coloring plastics (Dunning et al., 2009). However, in 2014 the EU proposed banning cadmium pigments in Europe.

Titanium yellows are made by introducing other metal ions into the lattice of anatase TiO_2 crystals, followed by calcining to convert to the rutile crystal structure. Greenish yellow shades are based on introducing antimony and nickel; reddish yellow shade grades contain antimony and chromium. Resistance to exterior exposure, chemicals, heat, and solvents is excellent. However, only relatively weak yellow colors can be produced, thereby requiring greater amounts, resulting in high cost.

Organic Oranges and Yellows. Chemical structures of representative pigments are shown in Figure 20.2.

Diarylide yellows are bisazo pigments derived from 3,3′-dichlorobenzidene, for example, PY 13, as shown in Figure 20.2. These pigments have high color strengths and high chroma. The hue and photostability are controlled by the number, positions, and structure of substituent groups on the aromatic ring of the anilide portion of the molecule. Even the most photochemically stable diarylide yellow pigments fade on exterior exposure, especially when used in tints. On the other hand, their solvent, heat, and chemical resistances are excellent. Due to their high strength and low density, their cost is relatively low. They are used in interior coatings for which a bright yellow color is needed in tints and in such applications as coatings for pencils. Grades with excellent transparency can be made. Diarylide yellows are the major yellows used in printing inks, especially in yellow inks used in four-color process printing.

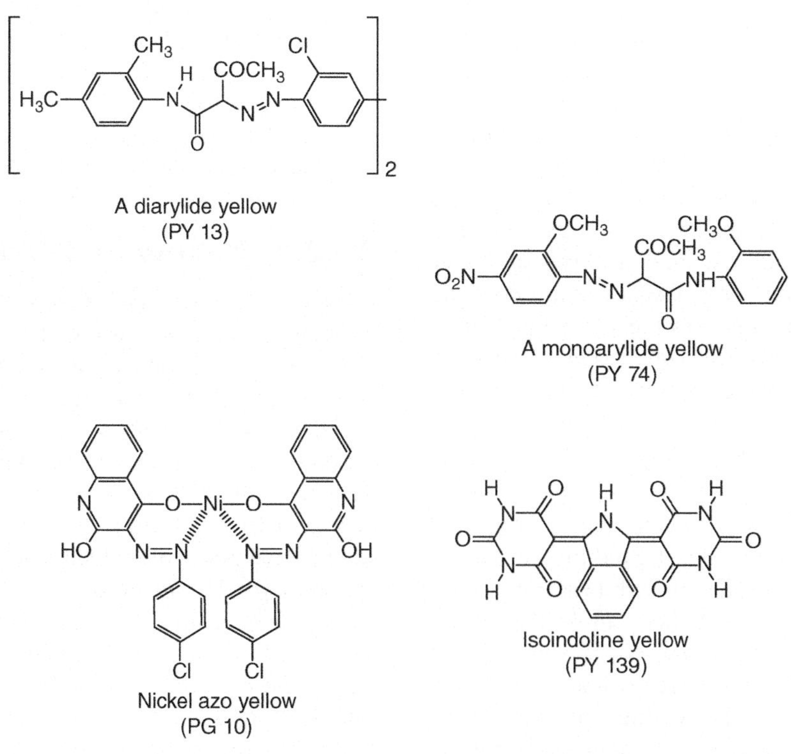

Figure 20.2 Examples of organic yellow pigments.

Monoarylide (monoazo) yellow pigments, such as PY 74, also have high chromas. They have poor bleed resistance and sublime when exposed to high temperatures. However, their light fastness is better than diarylide yellows, although still inferior to inorganic yellow pigments. Some opaque grades have sufficient light resistance for use in outdoor coatings. They are partially replacing chrome yellows in traffic paints (Section 33.4).

Nickel azo yellow is a relatively weak, very greenish yellow; in fact, as can be seen from the designation PG 10 in Figure 20.2, it is classified as a green. Nickel azo yellow exhibits excellent exterior durability and heat resistance. It gives transparent films and is used predominantly in automotive metallic coatings. Since bleeding into stripes has been reported in some applications, specific applications must be checked for bleeding.

Vat yellow pigments, as exemplified by isoindoline yellow PY 139, give transparency with excellent exterior durability, as well as heat and solvent resistance. However, these pigments are expensive and used only when their outstanding properties are required (e.g., in automotive metallic coatings).

Benzimidazolone orange pigments offer excellent light fastness and resistance to weather, heat, and solvent. They are used as replacements for molybdate orange.

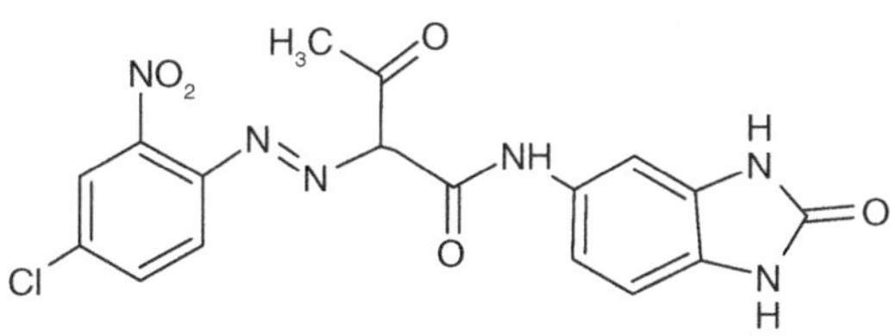

Benzimidazolone orange PO 36

20.2.2 Red Pigments

Inorganic Reds. Iron oxide (Fe_2O_3) gives the familiar "barn red" color. It is a low chroma red with excellent properties and low cost. In contrast to iron oxide yellows, iron oxide reds are thermally stable. When their particle size is optimal for scattering, they provide a high degree of hiding. An orange–red iron oxide is now available. There are also very fine particle grades available that provide transparent films. The excellent exterior durability makes the transparent grades suitable for use with aluminum in metallic automotive top coats. Transparent iron oxide is an excellent UV absorber and is used in transparent wood stains to protect the wood from photooxidation.

Organic Reds. Toluidine red pigment, PR 3, is a moderate cost, bright red azo pigment with high color strength, good exterior durability in deep colors, good chemical resistance, and adequate heat resistance to permit use in baking enamels. As shown in Figure 20.3, it is an azo derivative of β-naphthol. Toluidine red is soluble in some solvents and gives coatings that are likely to bleed.

Bleed resistance with azo pigments can be achieved by the presence of carboxylic acid salts. For example, 2-hydroxy-3-naphthoic acid (BON) can be coupled with diazo compounds. Permanent Red 2B is an example of such bleed resistant high chroma red azo pigments. It is available as the calcium, barium, or manganese salt. The somewhat higher cost manganese salt has better exterior durability than the calcium or barium salts. Azo pigments that have somewhat different shades are also available. This general class is the largest volume of organic red pigments used in coatings and inks. However, many are sensitive to bases and are not suitable for some latex paints.

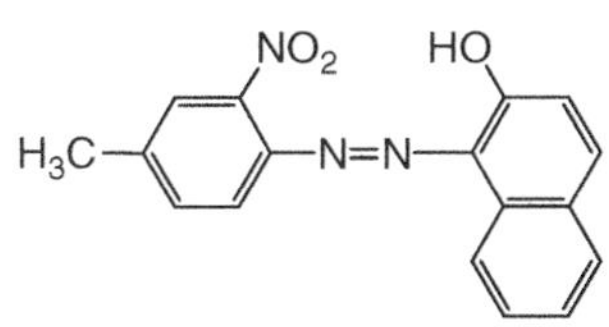

Toluidine red
(PR 3)

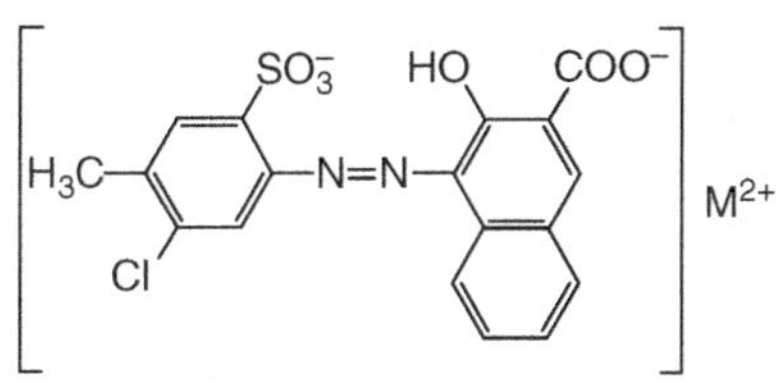

Permanent red 2B
Ca salt: PR 48 : 1
Ba salt: PR 48 : 2
Mn salt: PR 48 : 4

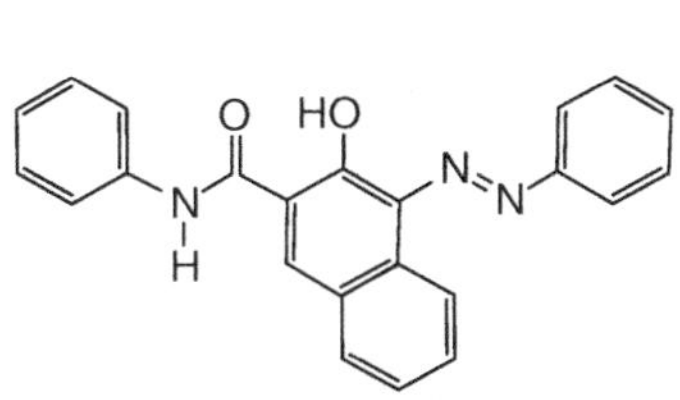

Generic structure of naphthol reds

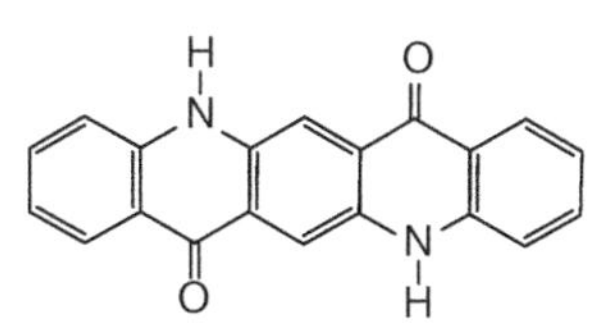

Ring structure of quinacridone reds

Figure 20.3 Examples of organic red pigments.

Naphthol reds are a large family of azo pigments with various ring substituents (Cl, OCH_3, NO_2, etc.). The generic structure is shown in Figure 20.3. Naphthol reds are more resistant to bases, soap, and acid than the permanent reds and also have fairly good exterior durability and solvent resistance.

Quinacridone pigments are heat and chemical resistant and also provide outstanding exterior durability, even in light shades. However, their cost is high. Depending on substitution and crystal form, a variety of orange, maroon, scarlet, magenta, and violet colors is available. Large particle size grades are used when opaque pigments are needed; and fine particle size grades are available for use in metallic automotive top coats. Surface treatments are applied to increase resistance of their dispersions to flocculation (Jaffe et al., 1994). Many other high performance red pigments are available (Faulkner and Schwartz, 2009; Hunger et al., 2014).

20.2.3 Blue and Green Pigments

Inorganic Blues and Greens. Iron blues are a family of coordination complexes involving Fe^{++}, Fe^{+++}, and CN^- ions. An example is ferric ammonium ferrocyanide, $FeNH_4Fe(CN)_6$, an intense reddish shade blue with fairly good properties. Historically, iron blues were important as one of the first widely used synthetic pigments. Since the 1930s, they have been increasingly supplanted by phthalocyanine blues, which have greater color strength.

Organic Blues and Greens. The principal blue and green pigments are copper phthalocyanine pigments, commonly called phthalo blue and phthalo green (Figure 20.4). They exhibit outstanding exterior durability, bleed and chemical resistance, are heat stable, and have high tinting strengths. The apparently complex molecule can be synthesized from phthalic anhydride, urea, and a copper salt. Although their cost per pound is fairly high, the high tinting strength and relatively low density translate into a moderate cost in use.

Phthalo blues are available commercially in three crystal forms: alpha, beta, and the seldom-used epsilon. The beta form has a relatively green shade of blue and is stable. The alpha form is redder, but not as stable; when some grades of the alpha form are used, there can be serious problems of change of color and strength during storage of coatings or during baking. More stable alpha form pigments are available, which incorporate various additives that stabilize the crystal form and minimize problems with flocculation of dispersions. Slightly chlorinated phthalocyanine blues have greener blue shades.

Phthalo greens are made by halogenating copper phthalocyanine to produce mixtures of isomers in which many of the 16 hydrogens of CPC have been replaced with chlorine or mixtures of chlorine and bromine. Phthalo greens containing 13–15 chlorines but no bromine have blue–green shades. Partial replacement of chlorine with bromine shifts the color toward a yellow–green shade. The yellowest shades have a high ratio of bromine to chlorine; an example is PG 36, which has on average about nine bromines and three chlorines.

20.2.4 Black Pigments

Most black pigments used in coatings are carbon blacks. They absorb UV radiation as well as light; with many binders, black colors are the most stable on exterior exposure. They also absorb IR radiation, causing the coatings to absorb heat. Carbon blacks are made by a variety of processes of partial combustion and/or cracking of petroleum products or natural gas. Depending on the process, the particle size and, therefore, the degree of *jetness* (intensity of blackness) vary. High color *channel blacks*, which are of

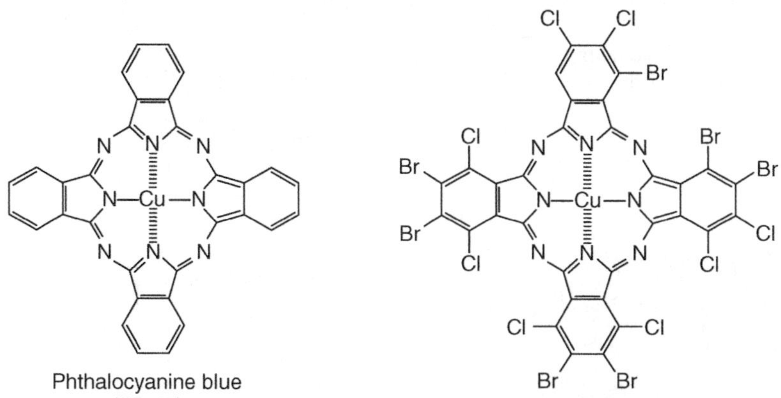

Phthalocyanine blue (PB 15)

Phthalocyanine green mixed isomers

Figure 20.4 Representative phthalocyanine pigments.

the smallest particle sizes, with diameters of 5–15 nm, have the highest jetness. They are used when intensely black, glossy coatings are desired. *Furnace blacks* have larger particle sizes, are lower in cost, and give less jet black coatings. Various grades are available with average diameters of 50–200 nm. *Lampblacks* have still larger particle size, on the order of 0.5 μm, and have lower color strengths than other carbon blacks. They are primarily used in making gray coatings and are preferable to high color blacks for this purpose; if a small excess of high color black is added by mistake, it may be necessary to increase the batch size 50% or more to make up for the mistake. The shade of gray obtained changes much more gradually with the addition of lampblack, making color matching easier.

While all carbon blacks are predominantly elemental carbon arranged in polynuclear six-member rings, the chemical structure of the surface varies depending on the raw material and process used in their manufacture. The surfaces are generally quite polar and in some cases are acidic. There are also variations in the porosity. Black coatings can present difficulties in formulating, especially when channel blacks are used. Due to their small particle size, the ratio of surface area to volume of the pigment particles is very high. The large surface area leads to adsorption of high ratios of resin on the pigment particles—commonly, many times their volume of polymer, substantially increasing their volume, thereby resulting in high viscosity at relatively low pigment loading levels. In addition, owing to the polarity of the surface and the large surface area, carbon black pigments can selectively adsorb polar additives such as catalysts from a coating formulation. For example, in drying oil and oxidizing alkyd formulations, the metal salt driers slowly adsorb on black pigments, retarding drying of the coating.

Acetylene black pigments increase the electrical conductivity of films. Conductivity is required for coatings such as primers for plastic parts that are to be coated by electrostatic spray (Section 31.2.2).

Kesler (2016) describes an organic black pigment (PBk 32) that absorbs less IR, which makes it suitable for high solar reflectance coatings.

20.2.5 Effect Pigments: Metallic, Interference, and Cholesteric Pigments

Effect pigments enable a variety of color effects such as colors that appear to change with the angles of viewing and illumination, pearl-like appearance, and an illusion of depth (Pfaff, 2008). Three types are considered here: metallic, interference, and cholesteric pigments.

The most important metallic pigments are aluminum flake pigments. They are produced by milling finely divided aluminum metal suspended in mineral spirits in steel ball mills to make thin flakes. A variety of particle size flakes is available. There are two major classes used in coatings: *leafing* and *non-leafing* pigments.

Leafing aluminum pigments are surface treated (e.g., with stearic acid) in order to decrease their surface tension. As a result, when a coating containing leafing aluminum is applied, flakes migrate to and orient near the surface. This gives a bright metallic appearance and acts as a barrier to permeation of oxygen and water vapor through the film. As a result of barrier development, leafing aluminum pigments are used in top coats for corrosion protection of steel structures (Section 7.3.3).

Non-leafing aluminum pigments have higher surface tension and do not migrate to the surface of the coating. They are used in metallic top coats for automobiles and trucks. As discussed in Sections 19.4 and 30.1, formulations using non-leafing aluminum are designed to maximize the fraction of flakes that are oriented parallel to the film surface. Films made with transparent color pigments and non-leafing aluminum change depth of color and shade with the angle of viewing. Non-leafing aluminum pigments are surface treated to minimize their sensitivity to acid in the environment.

The aluminum pigments used in water-reducible coatings are surface treated or encapsulated to minimize the reaction of water with the aluminum (which generates hydrogen). In early work it was necessary to mix in the aluminum paste just before application. Various treatments have been developed to permit formulation of premixed coatings (Muller, 1995); an example is treatment with a low acid number styrene/maleic acid copolymer.

Aluminum pigments are produced in solvent and sold as pastes in order to minimize problems of dusting and the danger of explosions. But such pastes introduce VOC into waterborne coatings. Aluminum pigments have been developed, which are pelletized with a melamine–formaldehyde surface coating and treated to reduce water sensitivity. This treatment enables the pigments to be handled easily and stirred into water/glycol ether, which in turn can be added to waterborne coatings with minimal increase of VOC (Ferguson, 2004).

Bronze, nickel, and stainless steel pigments, in the form of platelets similar to those of aluminum pigments, are also available but are used in lower volume. Bronze alloy flakes provide "gold" colors; depending on the alloy composition, shades from greenish yellow to reddish golds are available. Generally, they are surface treated to promote leafing. Due to the presence of copper, bronze alloy pigments change color on outdoor exposure to a blotchy, muddy green appearance.

Interference pigments are flake pigments that provide color by interference (Sections 19.2.1 and 19.4). Pearlescent pigments are platelets of mica having thin surface treatment layers of TiO_2 or Fe_2O_3, which serve to produce interference patterns of light striking the pigment surfaces (Greenstein, 1988; Maile et al., 2005). As a result, on some areas of the platelet surface, certain wavelengths of light are strongly reflected and others are transmitted; while on other areas, where the film thickness of the treatment layer is different, different wavelengths are reflected and transmitted. The result is a mother-of-pearl effect and an illusion of optical depth. These pigments are used in automotive coatings to give an effect related to that obtained with aluminum flake, but with the addition of hue changes with angle. They are sometimes combined with aluminum flake or other effect pigments to produce various optical effects.

Another type of pigment that gives colors as a result of interference effects is available (Droll, 1998). The pigment particles are thin (around 1 μm in thickness) flakes composed of a thin layer of reflective metal sandwiched between two layers of clear plastic, having very thin top layers of a semiopaque, very thin metal coating on each plastic surface. As a result of interference reflection (Section 19.4), certain wavelengths of light are reflected and others absorbed. The color obtained depends on the film thickness and the angle of illumination and viewing. As with other flake pigments, these flakes must be oriented parallel to the surface of the coating.

Cholesteric pigments are based on polymeric liquid crystal structures. One process for making these pigments starts with a combination of an acrylated nematic liquid crystal and an acrylated chiral compound that forms helical structures. The combination is dissolved in a vinyl monomer, such as divinylbenzene, together with a photoinitiator. The solution is coated on a polyethylene support, linear stress is applied to align the helices, and the combination is UV cure. The film is removed from the support, milled, and classified for particle size. The resulting flakes consist of a translucent off-white highly cross-linked powder. The helical structure causes optical interference, which changes color when viewed at different angles. By changing the ratio of chiral additive and liquid crystals, the wavelengths of light reflected and transmitted change. The flakes are sensitive to UV and must be used in a photostabilized composition. In waterborne automotive base coats, containing a UVA, and an adequate thickness of clear coat, containing a UVA and HALS, the flakes exhibit excellent durability (Nowak, 2004). Dramatic polychromatic effects, such as red to green, are attainable.

Maile et al. (2005) provide a review of these, as well as a variety of other effect pigments for various applications.

20.3 INERT PIGMENTS

Inert pigments absorb little, if any, visible light and have refractive indexes close enough to those of binders that they provide little light scattering in coatings. Several synonymous terms are used: *inert pigments*, *inerts*, *fillers*, and *extenders*. Commonly, but not always, these pigments are inexpensive and reduce the cost of a coating. The principal function of most inert pigments is to occupy volume in a film. Other functions are to adjust the rheological properties of fluid coatings and the gloss and mechanical properties of films. As discussed in Chapter 22, many film properties are controlled by the volume of pigment in the film. Solomon and Hawthorne (1983), Patton (1973), Lewis (1988), and Gyseau (2017) detail the multitude of inert pigments available.

For decades there has been controversy as to whether or not inert pigments having particle sizes smaller than TiO_2 can improve hiding by improving the spacing of TiO_2 particles within a film. Diebold (2011) and Hook (2014) resolved this question by showing that all inert pigments reduce the scattering efficiency of TiO_2 by crowding the particles, but the smallest inert particles have the least ill effect.

Calcium carbonate ($CaCO_3$) pigments are widely used inerts. The lowest cost grades are ground limestone or dolomite (calcium magnesium carbonate rock). Synthetic calcium carbonate pigments are more expensive but are whiter. Pigments are available with a variety of average particle sizes. Calcium carbonate reacts with acids (e.g., acid rain) and is problematic in exterior latex paints (Section 32.1). In the presence of acid, for example, water and carbon dioxide, which permeate through the film of a latex paint, can react with calcium carbonate to produce calcium bicarbonate. Calcium bicarbonate is water soluble and the solution can permeate back out of the film. On the film surface, the water evaporates and the reaction reverses, leaving a *frosting* of insoluble calcium carbonate deposited on the film. Frosting is especially noticeable on dark color paints.

A wide range of clays (aluminum silicates) is used as inert pigments. They are available in various particle size ranges. Cost is frequently related to whiteness. As mentioned in Section 20.4, bentonite and attapulgite clays are used to modify viscosity of coatings. Mica (aluminum potassium silicate) has a platelet structure and, when oriented parallel to the surface of the film, can be useful in reducing permeability of oxygen and water vapor.

In addition to limestone and dolomite, other pulverized minerals can serve as inert pigments. Examples are pulverized quartz and nepheline syenite, a complex mineral. Magnesium silicate minerals are also used as inerts. Talcs of various crystal structures affect the film strength of coatings differently. Some talcs are platey and reduce vapor permeability; others are fibrous and may be particularly effective in film reinforcing. Asbestos is a very fibrous magnesium silicate that is no longer used because it can cause lung cancer when fibers are inhaled.

Silicon dioxide pigments are an important group of inert pigments. Ground natural silicon dioxide is used in a variety of particle sizes. An unusual example of a SiO_2 pigment is *diatomaceous earth*, which is composed of fossil skeletons of diatoms. The large ratio of surface area to volume affects rheology and film properties (Section 22.2). Very fine particle size synthetic silicon dioxide pigments are used as *flatting pigments* (Sections 20.4 and 31.1.1). Nano-silicon dioxide is emerging as an important nano-pigment (Section 20.5).

Barytes (barium sulfate) has been used as an inert pigment, especially in automotive primer formulations and powder coatings. It is said to provide a "harder" primer. The density of barytes is $4.5\,g\,cc^{-1}$, roughly twice that of most other inert pigments. In at least some cases, the use of barytes is based on weight comparisons with other inert pigments, rather than on volume comparisons as should be done. While barytes is not expensive on a weight basis, it is more expensive than most other fillers on a volume basis. Note that barytes adds inexpensive weight to powder coatings, which are sold by weight.

While most inert pigments are inorganic minerals, organic materials can also be used as inert pigments. For example, powdered polypropylene is insoluble and acts as an inert pigment. High T_g latexes, such a polystyrene latex, can be used as inert pigments in latex paints. Synthetic fibers, such as aramid fibers, have been shown to be effective for increasing the mechanical strength of coating films.

20.4 FUNCTIONAL PIGMENTS

Functional pigments are used to modify the application characteristics, appearance, or film properties of coatings. Perhaps the most important are corrosion-inhibiting pigments (Smith, 1989). Complex zinc chromate pigments, strontium chromate, barium phosphosilicates, barium borosilicates, red lead, and zinc phosphate are used in primers to inhibit corrosion of steel by passivation of anodic areas. In contrast to other pigments, they must be at least slightly soluble in water in order to function. Because of toxic hazards, use of chromates and red lead are decreasing rapidly. Zinc metal pigment provides corrosion protection by cathodic action and is used in a class of primers known as zinc-rich primers. Discussion of anticorrosion pigments is found in Sections 7.4.3 and 33.1.2. Patton (1973) and Solomon and Hawthorne (1983) provide considerable information.

Flatting pigments are a type of functional pigments used to reduce gloss. While low gloss is often attained by formulating coatings with a high volume ratio of pigment in the dry film, this approach is not always desirable. For example, in lacquers for wood furniture, it is essential to have optimal transparency of the dry film so that the beauty of the wood grain is not concealed. Fine particle size silicon dioxide is widely used in such coatings as a flatting pigment. Silane treatment of such pigments is reported to improve transparency in liquid coatings, providing excellent flatting with smooth surfaces and having little influence on rheology (Christian, 2004). Powdered polypropylene has been used in a somewhat analogous fashion. As discussed in Section 19.10.1, desired gloss reduction can be achieved by solvent evaporation from the coating film, which sets up convection currents that concentrate fine particles at the surface of the dry film. The combination of low total pigmentation, small particle size, and small refractive index difference causes little reduction in transparency.

Some pigments are biocides. For example, zinc oxide is used as a fungicide. *Antifouling coatings* are used to minimize the growth of barnacles, algae, and other organisms on the outer hulls of ships, as detailed in Section 33.2. Cuprous oxide is currently a common ingredient of antifouling coatings. Organotin pigments and compounds are more effective, but they have been widely banned because of environmental concerns.

Finely divided silver is effective in antimicrobial coatings, such as coatings for kitchen appliances and hospital furniture. Other antimicrobial pigments have been developed, and rapid growth of antimicrobial coatings is anticipated, as further discussed in Section 34.5.

Zinc oxide is also used in linings for tin-plated cans used for packing vegetables, such as corn, that evolve hydrogen sulfide during cooking. ZnO reacts with the H_2S to form white zinc sulfide. This prevents black stains that result from the formation of tin sulfide by reaction of H_2S with the tin oxide layer on the surface of the tin-plated cans. The stains are harmless but worrisome.

Antimony oxide (Sb_2O_3) is a white pigment with a refractive index too low (2.18) and a cost too high to use simply for hiding. It is used in fire retardant coatings. When a combination of Sb_2O_3 and a chlorinated or brominated polymer is heated to high temperature, combustion byproducts are generated that suppress flame propagation.

Viscosity modifiers are another class of functional pigments. In general terms, these pigments increase the low shear viscosity of coatings either to inhibit pigment settling during storage of the coating and/or to reduce sagging after application of the coating. An important example is quaternary ammonium salt-treated bentonite clay, used in solventborne coatings (Memnetz et al., 1989). Another type of clay that is used in both solventborne and waterborne coatings is *attapulgite clay* (*fuller's earth*), a fibrous clay. The fibrils of this clay associate, providing an increase in viscosity that reverses with agitation. In waterborne coatings, it also acts by absorbing water to make swollen particles that distort in shape with stress.

Fine particle size silica has been used for many years as a viscosity modifier. Fumed silica (also called pyrogenic silicon dioxide) is a fine particle size SiO_2 that is useful for increasing the low shear viscosity of coatings. It is used in both solventborne and waterborne coatings to provide thixotropic coatings having reduced settling, as well as thixotropic viscosity that allows application of thicker films without sagging. Surface treatment of the SiO_2 with polydimethylsiloxane is in general most suitable for solventborne coatings. The hydrophilic untreated silica is generally most suitable for waterborne coatings (Ettlinger et al., 2000). The effects of such pigments on flow are discussed in Section 3.2.

Infrared reflecting (NIR) *pigments* for roofing and siding coatings reduce heat absorption and hence air conditioning costs. Thus, *cool roof* technology is an environmental goal. Aluminum and white pigments have generally been used, but brown, black, and brightly colored inorganic pigments having high IR reflectance are available (Detrie and Swiler, 2009; White, 2009; Decker et al., 2012). They are known as *complex inorganic color pigments*, as *mixed metal oxide pigments*, and as *ceramic colorants*. These pigments are generally fused mixtures of several metal oxides. Many have excellent outdoor durability, and, when used on building roofs and siding, their IR reflectance significantly reduces solar heating relative to conventional color pigments. Kesler (2016) describes several organic NIR pigments and proposes mathematical models that predict total solar reflectance of various formulations.

20.5 NANO-PIGMENTS

The literature and patents reflect intense worldwide efforts to produce and use nano-pigments in coatings, plastics, cosmetics, sunscreens, inks, photonics, microelectronics, and other applications.

"Nano-pigment" is a fairly new name for materials that have been around for a long time. They were used by the ancient Greeks and have been used for many decades in coatings. For example, high strength channel blacks and fumed silicas have particle diameters in the range of 5–15 nm. Certain organic pigments used in transparent and metallic coatings qualify as nano-pigments. Some clays can also be considered as nano-pigments, as they can be separated (*exfoliated*) into sheets on the order of 1 nm thick. Such sheets are impermeable to water and oxygen. Sheets of graphene are even thinner. Fernando (2004) reviewed nanomaterial applications in coatings; Fernando and Sung (2009) discuss various aspects of the subject. Prospects for wider application in coatings are enormous, as dozens of inorganic substances (including oxides of Si, Ti, Al, Fe, and Zn) and metals are now available in nanoparticle form, some with average diameters of 5 nm or less.

For purposes of risk assessment and regulation, it is important to define the size boundaries of nano-pigments. Many authors regard pigments having average particle diameters below 25 nm as nano-pigments. Others specify 50 nm as the upper limit. In commerce, particles with diameters of up to 100 nm are sometimes described as nano-pigments. The US National Nanotechnology Initiative (NNI) strategic plan defines nanomaterials (presumably including nano-pigments) as having diameters of "approximately 1–100 nm" in the smallest dimension (http//www.nano.gov/sites/default/files/pub_resources/2016-nni-strategic-plan.pdf). The European Commission Directorate General for Health and Consumers agrees (http://www.ec.europa.eu/health/scientific_committees/emerging/docs/scenihr_o_030.pdf). These groups and others are engaged in risk assessment of nanotechnology for the preparation of regulation (Naidu, 2014). Risk assessment requires answers to questions that are hard to answer. For example, if a safe substance is produced in the form of nanosized particles, what are the conditions, if any, under which it becomes unsafe? Inhalation of nanoparticle dusts should be avoided, but nano-pigments can be handled as dispersions in liquids. The broad 1–100 nm range might lead to the regulation of conventional pigments with average particle sizes above 100 nm if they contain appreciable levels of particles below 100 nm. Future regulation is anticipated (Bergeson, 2015), but the timing and scope are unpredictable. In 2015 the US EPA proposed a broad rule requiring that all production and use of nanoscale materials be reported. Coatings and other industries have objected to the proposed rule on the grounds that it is excessively burdensome and threatens innovation.

Applications of nano-pigments are discussed in Section 22.2.1.1. In summary, nano-pigments have potentially remarkable benefits in coatings with the prospect of improving a long list of properties in the right circumstances (Fernando, 2004; Fernando and Sung, 2009; Lin et al., 2010). Fajzulin et al. (2015) reviewed one notable application: inorganic nanoparticles as UV absorbers; and Nikolic et al. (2015) extensively reviewed nanofillers in wood coatings.

Production of nano-pigments can be difficult. As particle diameters are reduced, especially below 10 nm, physical properties of the materials change, causing various complications, many related to the very high surface area/mass ratios of the particles. The conventional pigment dispersion methods discussed in Chapter 21 are usually incapable of reducing particle diameters to the lower part of the nano range, but some successes have been reported (Perera, 2004). Section 21.4.4 discusses the variables involved in selecting mills, media, and media size; pin mills are recommended. Nano color pigments for automotive metallic coatings are prepared by grinding suspensions of the conventional color pigments in butyl acetate with a commercial dispersing agent using very small high density balls in the mill. The mill base is then used to make tinting pastes for color matching metallic colors (Vanier et al., 2005a).

Ball mills are used to disperse nano(aluminum silicate) in an acrylic resin with a commercial dispersing agent. The dispersion is used in making mar-resistant clear coats (Vanier et al., 2005b). Thermal plasma processes, sol–gel technology, and various spray processes have potential for production of nano-pigments.

Silica and other metal oxide nano-pigments can be surface treated to introduce reactive functional groups, such as —OH, $—NH_2$, and —N=C=O. Riberio et al. (2014) reviewed the use of reactive silanes (Section 16.2) for this purpose. Functionality facilitates incorporation of nano-pigments into composite materials. For examples, see Maganty et al. (2016) and Section 12.6.

As further discussed in Section 22.2.1.1, preparation of stable dispersions of nano-pigments can be difficult (Croll and Lindsay, 2012).

Various nano forms of carbon offer intriguing possibilities:

- *Graphene* is a single layer of carbon in the hexagonal form that composes graphite. It is the thinnest known material, very strong, impermeable, transparent, and an excellent electrical conductor (Scott, 2016), all properties of potential value in coatings. Graphene's ability to reinforce polymers is well established. Between 2005 and 2014, 26000 graphene-related patents were filed. Commercial graphene is usually not a single sheet of carbon, but typically contains stacks of 5–30 sheets. The surfaces can be oxidized to improve bonding of the sheets with polymers. In 2016 the price was less than $100 kg^{-1} and falling.
- *Carbon nanotubes* can also reinforce polymers. Shen (2016) found that they can increase conductivity and improve corrosion resistance of an epoxy hybrid coating. Furthermore, carbon nanotubes can be formed on surfaces using plasma-enhanced chemical vapor deposition (Sembukutiarachilage et al., 2012). The resulting structures can absorb up to 99.96% of incident light, creating a striking appearance sometimes called a "black hole." They are used to absorb stray light in astronomical equipment, and striking art works have been made.

GENERAL REFERENCES

Braun, J. H., *White Pigments*, Federation of Societies for Coatings Technology, Blue Bell, PA, **1995**.

Buxbaum, G.; Pfaff, G., *Industrial Inorganic Pigments*, 3rd ed., Vincentz, Hannover, **2005**.

Faulkner, E. B.; Schwartz, R. J., Eds., *High Performance Pigments*, 2nd ed., Wiley-VCH Verlag, Weinheum, **2009**.

Hunger, K., et al., *Industrial Organic Pigments*, 4th ed., Vincentz, Hannover, **2014**.

Patton, T. C., Ed., *Pigment Handbook*, Wiley-Interscience, New York, **1973**, Vol. *3*.

Winkler, J., *Titanium Dioxide*, 2nd ed., Vincentz, Hannover, **2013**.

REFERENCES

Anonymous, *Paint and Coatings Industry*, June 28, **2015**.

Bergeson, L., *ChemicalProcessing.com*, May, **2015**, 19.

Braun, J. H., *White Pigments*, Federation of Societies for Coatings Technology, Blue Bell, PA, **1995**.

Bugnon, P., *Prog. Org. Coat.*, **1996**, *29*, 39.

Buxbaum, G.; Pfaff, G., *Industrial Inorganic Pigments*, 3rd ed., Vincentz, Hannover, **2005**.

Christian, H.-D., *Proceedings of the International Waterborne, High-Solids, and Powder Coatings Symposium*, New Orleans, LA, **2004**, Paper No. 17.

Croll, S.; Lindsay, M., *Int. J. Nanotechnol.*, **2012**, *9(10/11/12)*, 982.

Decker, E. L., et al., Canadian patent CA 2677255 C (**2012**).

Detrie, T.; Swiler, D., Infrared Reflecting Complex Inorganic Colored Pigments in Faulkner, E. B.; Schwartz, R. J., Eds., *High Performance Pigments*, Wiley-VCH Verlag, Weinheum, **2009**, pp 467–488.

Diebold, M. P., *JCT Res.*, **2011**, *8(5)*, 541.

Droll, F. J., *Paint. Coat. Ind.*, **1998**, *14(2)*, 54.

Dunning, P., Cadmium Pigments in Faulkner, E. B.; Schwartz, R. J., Eds., *High Performance Pigments*, 2nd ed., Wiley-VCH Verlag, Weinheum, **2009**, pp 13–26.

Endriss, H., Bismuth Vanadates in Faulkner, E. B.; Schwartz, R. J., Eds., *High Performance Pigments*, 2nd ed., Wiley-VCH Verlag, Weinheum, **2009**, pp 7–12.

Ettlinger, M., et al., *Prog. Org. Coat.*, **2000**, *40*, 31.

Everts, S., *Chem. Eng. News*, **2016**, February 1, 32.

Fajzulin, I.; Zhu, X.; Moller, M., *JCT Res.*, **2015**, *12(4)*, 617–632.

Fasano, D. M., *J. Coat. Technol.*, **1987**, *59(752)*, 109.

Faulkner, E. B.; Schwartz, R. J., *High Performance Pigments*, Wiley-VCH Verlag, Weinheum, **2009**.

Ferguson, R. L., *JCT Coat. Technol.*, **2004**, *1(7)*, 42.

Fernando, R., *JCT Coat. Technol.*, **2004**, *1(5)*, 32.

Fernando, R. H.; Sung, L.-P., Eds., *Nanotechnology Applications in Coatings*, ACS Symposium Series no. 1008, Oxford University Press, Oxford, **2009**.

Gao, B., et al., *Ind. Eng. Chem. Res.*, **2014**, *53(1)*, 189–197.

Greenstein, L. M., Nacreous (Pearlescent) Pigments and Interference Pigments in Lewis, P. A., Ed., *Pigment Handbook*, 2nd ed., Wiley-Interscience, New York, **1988**, Vol. *I*, pp 829–858.

Gyseau, D., *Fillers for Paints*, 3rd ed., Vincentz, Hannover, **2017**.

Hays, B. G., Am. Inkmaker, **1984**, June, 28; 1986, October, 13; 1990, November, 28.

Herbst, W.; Hunger, K., *Industrial Organic Pigments: Production, Properties, Applications*, 3rd ed., John Wiley & Sons, Inc., New York, **2006**.

Hook, J. W., III, Modeling White Hiding Power Helps Deliver Decades of Innovation, *Proceedings of the American Coatings Conference*, Atlanta, GA, April 7–9, **2014**.

Hunger, K., et al., *Industrial Organic Pigments*, 4th ed., Vincentz, Hannover, **2014**.

Jaffe, E. E., et al., *J. Coat. Technol.*, **1994**, *66(832)*, 47.

Kesler, W., *CoatingsTech*, **2016**, *13(7)*, 36–44.

Lewis, P. A., Ed., *Pigment Handbook*, 2nd ed., Wiley-Interscience, New York, **1988**, Vol. *I*.

Lewis, P. A., *Organic Pigments*, Federation of Societies for Coatings Technology, Blue Bell, PA, **1995**.

Lin, F.; Yang, L.; Han, E., *JCT Res.*, **2010**, *7(3)*, 301–313.

Maganty, S., et al., *Prog. Org. Coat.*, **2016**, *90*, 243–251.

Maile, F. J.; Pfaff, G.; Reynders, P., *Prog. Org. Coat.*, **2005**, *54*, 150.

Memnetz, S. I., et al., *J. Coat. Technol.*, **1989**, *61(776)*, 47.

Muller, B., *J. Coat. Technol.*, **1995**, *67(846)*, 59.

Naidu, D., *Biotechnology and Nanotechnology Regulation*, LexisNexis, New Providence, **2014**.

Nikolic, M., et al., *JCT Res.*, **2015**, *12(3)*, 445–461.

Nowak, P. J., *Proceedings of the International Waterborne, High-Solids, and Powder Coatings Symposium*, New Orleans, LA, **2004**, Paper No. 13.

Nuasaen, S.; Tangboriboonrat, P., *Prog. Org. Coat.*, **2015**, *79*, 83–89.

Patton, T. C., Ed., *Pigment Handbook*, Wiley-Interscience, New York, **1973**, *3* Vols.

Perera, D. Y., *Prog. Org. Coat.*, **2004**, *50*, 247.

Pfaff, G., *Special Effect Pigments*, Vincentz, Hannover, **2008**.

Riberio, T.; Baleizao, C.; Farinha, J. P. S., *Materials*, **2014**, *7*, 3881–3900.

Scott, A., *Chem. Eng. News*, **2016**, April 11, 28–33.

Sembukutiarachilage, R. S.; Jensen, B. P.; Gaun, Y. C., Nanostructure Production Methods, EP1885909 (**2012**).

Shen, W., *Prog. Org. Coat.*, **2016**, *90*, 139–146.

Smith, A., *Inorganic Primer Pigments*, Federation of Societies for Coatings Technology, Blue Bell, PA, **1989**.

Solomon, D. H.; Hawthorne, D. G., *Chemistry of Pigments and Fillers*, Wiley-Interscience, New York, **1983**.

Steig, F. B., The Science and Technology of Opaque White Pigments in Coatings in Craver, J. K., Tess, R. W., Eds., *Applied Polymer Science*, Organic Coatings and Plastics Chemistry Division of the American Chemical Society, Washington, DC, **1975**, pp 246–254.

Vanier, N. R., et al., US patent 6,916,368 (**2005a**).

Vanier, N. R., et al., US patent 6,875,800 (**2005b**).

Veronovski, N.; Lesnik, M.; Verhovsek, D., *JCT Res.*, **2014**, *11(2)*, 255–264.

White, J., Complex Inorganic Pigments: An Overview in Faulkner, E. B.; Schwartz, R. J., Eds., *High Performance Pigments*, 2nd ed., Wiley-VCH Verlag, Weinheim, **2009**, pp 41–52.

Winkler, J., *Titanium Dioxide*, 2nd ed., Vincentz, Hannover, **2013**.

Chapter 21

Pigment Dispersion

Manufacture of pigmented liquid paints usually involves mixing liquid concentrates of dispersed pigments with the other ingredients. This chapter addresses production of these concentrates. When made at the paint factory, they are commonly called *mill bases*; when purchased from specialist companies or pigment suppliers, they are called *dispersions*, *colorants*, *slurries*, *pastes*, or *flushings*.

As described in Chapter 20, pigments are manufactured with a particle size distribution that provides an acceptable compromise of properties, but the particles often become cemented together into aggregates during processing. Breaking these aggregates and forming stable dispersions of optimally sized pigment particles are a critical process in the manufacture of coatings. Three stages are involved in making dispersions: (1) *wetting*, (2) *separation*, and (3) *stabilization*. Most authors agree that there are three stages, but some use different terms, sometimes with conflicting meanings. Be careful in reading papers to understand how an author is using the terms, especially in the older literature. Some authors add a fourth stage—*let down* into the final paint. The stabilized dispersion must be robust enough to survive this process.

21.1 DISPERSION IN ORGANIC MEDIA

This section concerns the dispersion of pigments in an organic solvent medium. Pigments are also dispersed in aqueous media. However, owing to the effects of the high surface tension and polarity of water, aqueous dispersion is treated separately in Section 21.3.

21.1.1 Wetting

Wetting refers to the displacement of air, and sometimes of water or other contaminants, from the surface of pigment particles and aggregates by the medium, which is usually a solution of polymer and additives in organic solvent. In some cases, the medium may be the coating vehicle itself. Wetting requires that the surface tension of the medium is lower than the surface free energy of the pigment. This requirement is met with organic media for all inorganic and most organic pigments. There can be important differences in the rate of wetting. When a dry pigment is added to a vehicle to make a mill base, the pigment tends to clump up in clusters of aggregates. For wetting to occur, the vehicle must penetrate through these clusters and into the pigment aggregates. The rate of wetting depends on the viscosity of the vehicle, the stirring rate, and the surface tension. (In aqueous media, wetting can be accelerated by addition of proprietary surfactant blends, some of which may also reduce foaming.)

21.1.2 Separation

Processes are designed to separate pigment aggregates into individual crystals without grinding crystals to smaller particle size; it is generally undesirable to decrease the crystal size, since the pigment manufacturing process was designed to produce an optimal particle size. Many different types of machinery are used to carry out the separation stage: important examples are described in Section 21.5. Dispersion machines function primarily by applying a shear stress to the aggregates suspended in the vehicle.

Organic Coatings: Science and Technology, Fourth Edition. Frank N. Jones, Mark E. Nichols, and Socrates Peter Pappas.

Compressive forces resulting from impact may play a secondary role. If the aggregates are easily separated, a comparatively small shear stress is sufficient. If aggregates require a relatively large force for separation, then machinery that can apply a higher shear stress is required. Pigment manufacturers have been increasingly successful in processing and surface treating pigments so that more and more aggregates are relatively easily separated.

Recall from Section 3.1 that shear stress is equal to shear rate times the viscosity of the mill base. The available shear rate for any particular dispersion equipment is set by machine design. The formulator selects appropriate dispersion machinery and formulates a mill base that can effectively transfer sufficient shear stress to the aggregates. For the fastest rate of separation of aggregates, the mill base should have as high a viscosity as the equipment can handle efficiently. Then, the highest shear stress is exerted on the pigment aggregates, and separation is accomplished in minimum time. Winkler and Dulog discussed the engineering theory and provided equations modeling the forces for separation (Winkler et al., 1987).

21.1.3 Stabilization

Wetting and separation are essential steps in making pigment dispersions but are seldom a problem in organic solvent systems. On the other hand, stabilization can frequently be challenging and is usually the key to making good pigment dispersions. If the dispersion is not stabilized, the pigment particles will be attracted to each other and undergo *flocculation*. Flocculation is a type of aggregation, but the aggregates formed are not cemented together like the aggregates in the dry pigment powder. Flocculation can be reversed by applying relatively low levels of shear stress.

Flocculation is usually undesirable. With light-scattering pigments, the larger particle size resulting from flocculation reduces scattering and, therefore, reduces hiding. With color pigments, the larger particle size reduces light absorption and, thus, color strength. Larger size floccules in the final film can reduce gloss. Flocculation of pigments, including inert pigments, can change critical pigment volume concentration (CPVC) and, thus, affect performance of the coating films (Section 22.2). Flocculated dispersions are shear thinning and have higher viscosities at low shear rates than well-stabilized dispersions. Flocculated dispersions do have the advantage that any settling forms soft pigment-bearing sediments that are easily stirred back to uniformity. However, settling problems can usually be minimized by addition of thixotropic additives such as treated clay or gels without the adverse results of flocculation.

There are two principal mechanisms for stabilization: *charge repulsion* and *entropic repulsion*. As discussed in Section 9.1.1, some authors prefer to use the terms *osmotic repulsion* or *steric repulsion* rather than entropic repulsion. Centeno et al. (2014) describe a third mechanism, *hard-core repulsion*, which sometimes comes into play.

In charge repulsion, particles with like electrostatic charges repel each other. Calcium and zinc soaps can sometimes stabilize pigment dispersions by absorbing on pigment surfaces and thereby introduce ionic charges that provide charge repulsion. Charge repulsion, which is more important in aqueous dispersions, is discussed in Section 21.3.

In most cases, entropic repulsion is the primary stabilizing mechanism in nonaqueous media. Entropic repulsion is a term used to describe the repelling effect of layers of adsorbed material on the surface of the particles of a dispersion that prevents the particles from getting close enough together for flocculation to occur. In many pigment dispersions in organic media, the adsorbed layer consists of polymers or large surfactant molecules swollen with solvent. The particles are in rapid, random (Brownian) motion. As they approach each other, their adsorbed layers become crowded; there is a reduction in the number of possible conformations of molecules of polymer or surfactant, together with associated solvent, in the adsorbed layers. This introduces more order, which constitutes a reduction of entropy. Reduction of entropy corresponds to an increase in energy, which is unfavorable and promotes repulsion owing to the more favorable, lower energy, less ordered state of the separated particles. Similarly, compression of the layers could lead to a more ordered system by squeezing out solvent. The accompanying reduction in entropy would again lead to repulsion. Only certain parts (*anchors*) of the polymer or surfactant are adsorbed with other parts sticking out into the solution (*loops* and *tails*).

Much of our understanding of entropic stabilization of pigment dispersions comes from the seminal work of Rehacek (1976). He devised an experimental technique to determine the thickness and composition of the adsorbed layer on the surface of a pigment dispersed in a polymeric resin solution. Details of his method are described in the third edition of this book (Wicks et al., 2007).

There is general agreement that, if the average adsorbed layer thickness of resin plus solvent is less than 9–10 nm and the dispersion stabilizer is a typical resin, the dispersion is not stable and flocculation occurs (Rehacek, 1976; Saarnak, 1979; Dulog and Schnitz, 1984). However, with monofunctional surfactants and with specially designed pigment dispersion stabilizers, the adsorbed layer can be thinner and still protect against flocculation. In one case, McKay (1980) has shown that an adsorbed layer thickness of 4.5 nm of surfactant and associated solvent was adequate. There are two credible explanations. First, while the resin may adsorb to form a combination of loops and tails of polymer extending into the medium, the surfactant or stabilizer can be structured to form only tails, which are more entropically effective. Second, the adsorbed layer of resin is probably nonuniform in thickness, while the special stabilizer layer can be more uniform, so it does not have to be as thick to provide stabilization.

For resin-stabilized dispersions, Rehacek showed that the resin and solvent compete for adsorption to the pigment. The resin–solvent balance depends both on the relative affinities of resin and solvent molecules for the pigment surface and on the concentration of resin. He observed instances in which concentrated resin solutions yield a stable dispersion, but dilution of the resin causes flocculation.

What controls the thickness of the adsorbed layer? In the case of a surfactant with its polar end adsorbed on the surface of a polar pigment, the length of the nonpolar aliphatic chain is the primary factor. In the case of resins with several adsorption sites, the largest single factor is probably molecular weight. For example, Saarnak (1979) showed that the adsorbed layer thickness on TiO_2 dispersed in a series of BPA epoxy resins in MEK, increased from 7 to 25 nm as the molecular weight of the epoxy resin increased. The lowest molecular weight resin provided an adsorbed layer thickness of 7 nm, which was insufficient to prevent flocculation, as predicted. Higher molecular weight epoxy resins gave stable dispersions.

Average adsorbed layer thickness is also affected by the pigment surface. Organic pigments are more likely to give significant differences in adsorbed layer thickness with different resin–solvent combinations than more polar inorganic pigments. However, even with inorganic pigments, significant differences can be encountered. For example, an alumina-treated TiO_2 forms a more stable dispersion than a silica-treated TiO_2 in the same long oil alkyd solution (Brisson and Haber, 1991). The authors proposed that the adsorbed layer is more compact on the silica-treated TiO_2. Commonly, the interactions between pigment surfaces and adsorbed molecules are considered to be hydrogen-bond interactions, but some authors prefer to interpret the interactions as acid–base interactions (Dasgupta, 1991; Lara and Schreiber, 1991).

The spacing and number of functional groups along the resin chain affect layer thickness. As an extreme example, consider a linear aliphatic chain resin with a polar group on every other carbon atom adsorbed on the surface of a polar pigment like TiO_2. At equilibrium, adsorption of tightly-bound single molecules on the pigment surface would be favored, resulting in a thin adsorbed layer. However, if the resin has only occasional polar groups along the chains, the longer segments between polar groups result in loops and tails of resin swollen with solvent, projecting out from the pigment surface, thereby providing a thicker adsorbed layer at equilibrium.

Effective layer thickness is also affected by solvent–resin interaction. If the loops and tails interact more strongly with the solvent, there are more solvent molecules in the layer, the average conformation of the resin is more extended, and the layer thickness is greater. Resin molecules that have multiple adsorbing groups have an advantage in competition with solvent molecules, but if the solvent interacts strongly with the pigment surface and the resin only interacts weakly, the more numerous solvent molecules will "win" the contest. For example, less polar toluene favors adsorption of macromolecules such as nitrocellulose, polyurethanes, and phenoxy resins on magnetic iron oxide pigment particles as compared with tetrahydrofuran (Dasgupta, 1991). Addition of solvent to a stable pigment dispersion can, in some circumstances, lead to flocculation. If the ratio of resin to solvent is just sufficient to allow adequate adsorption of resin to stabilize the dispersion, addition of more of the same solvent can shift the equilibrium, displacing part of the resin and reducing the average adsorbed layer thickness below the critical level for stabilization, resulting in flocculation. Furthermore, if a second solvent is a poor solvent for the resin, the loops and tails may collapse. In either case, the dispersion is said to have suffered *solvent shock*.

Centeno et al. (2014) applied computer modeling to assess the various factors involved in entropic stabilization. They developed equations to calculate the *potential mean force*, the total free energy of a many-bodied colloidial system. The equations isolate the various factors affecting colloidial stability. They can be used to optimize pigment/dispersant/solvent systems and to quantitatively predict the effects of system changes on stability.

For most conventional solventborne coatings, the resin used as binder in the coating can stabilize the pigment dispersion. Most conventional alkyds, polyesters, and thermosetting acrylics stabilize dispersions of many pigments. In the past, most dispersions were made in the resin to be used in the paint. If there was a problem, the two most common changes to give adequate stability were to increase the molecular weight or the number of polar groups, such as, hydroxyl, amide, and carboxylic acid, at least on that part of the resin to be used in the mill base. It has been shown that the higher molecular weight components of resins are selectively adsorbed (Lara and Schreiber, 1991).

Frequently, however, it is difficult to make stable pigment dispersions in the resins used in high solids coatings. Increasing the solids of organic solution coatings requires decreasing the molecular weight of the resins and reducing the number of functional groups per molecule. Reduced molecular weight results in thinner adsorbed layers of resin and associated solvent molecules. The reduced number of functional groups per resin molecule also decreases the probability of the adsorption of resin molecules, which favors the probability of solvent adsorption and hence a greater likelihood of flocculation. Remedies for this problem are discussed in the following section.

21.1.3.1 Pigment Dispersion Stabilizers for Nonaqueous Coatings

The trend in modern paint production is to add pigment dispersion stabilizers during the dispersion process. They may be used in combination with a primary resin for the coating or, in many cases, by themselves. This practice often produces superior dispersions for conventional solids coatings and is essential with many high solids coatings. Conventional monofunctional surfactants can be used for this purpose. However, if the concentration of surfactant is low, adsorption of the solvent is likely to be favored over most surfactant molecules owing to the larger number of solvent molecules, even though the surfactant may be more strongly adsorbed. This can be offset by increasing the concentration of surfactant molecules to shift the equilibrium in favor of surfactant, but doing so leaves excess surfactant in the final film, which tends to reduce performance properties such as adhesion to metal. Thus, in organic media, monofunctional surfactants are not generally a desirable choice for stabilization. A commonly used additive is *lecithin*, a naturally occurring choline ester of phosphoglycerides, which is quite strongly adsorbed on many pigments.

Surfactants have been designed that are so strongly adsorbed on a pigment that they are preferentially adsorbed even in the presence of a large amount of solvent. For example, phthalocyanine blue modified by covalently attaching aliphatic side chains has been used as a surfactant with phthalocyanine blue pigment; the phthalocyanine end of the surfactant molecule in effect joins the crystal structure of the surface of the pigment particles so that little, if any, is in solution (McKay, 1980). The average adsorbed layer thickness needed to protect against flocculation was shown to be about 4.5 nm. Specific, proprietary surfactants have been designed for other pigments. While effective, this approach finds only limited use because the surfactants are expensive and are designed to work only with specific pigments.

These difficulties have led to the design of special dispersing aids called *polymeric dispersants*, *A-B dispersants*, *hyperdispersants*, or *superdispersants* (Jakubauskas, 1986; Schofield, 1991; Sastry and Thakor, 2009). Jakubauskas describes the design parameters of such dispersants. He concluded that the most effective class of dispersant has a polar end with several functional groups (anchoring groups) and a less polar tail that is soluble in the medium and has sufficient length to provide for a surface layer that is at least 10 nm thick. Tails are more effective than loops for entropic stabilization. Examples of such dispersants include polycaprolactone polyol–polyethyleneimine block copolymers and polycaprolactone capped with toluene diisocyanate post-reacted with triethylenetetramine. Schofield (1991) reviewed other types of hyperdispersants, for example, low molecular weight polyesters from polyhydroxystearic acid. Various other block, comb, and brush copolymers are effective. Controlled free radical polymerization (CRP) (Section 2.2.1.1) is used to synthesize dispersants, as detailed in Section 21.3.1. In some cases, it is desirable to treat the surface of a pigment so that there are sites on the surface to interact with the anchors of the dispersant (Jaffe et al., 1994). Bio-based dispersion stabilizers are also available.

Another approach to stabilization is to covalently bond long chains to the surface of pigments. For example, the surface of SiO_2 pigments particles can be reacted with trialkoxysilanes for subsequent attachment of long alkyl chains (Hamann and Laible, 1978). In many cases, pigment manufacturers offer pigments that have been surface treated to provide stable dispersions even when the pigment is dispersed in solvent. Schroeder (1988) and Bugnon (1996) reviewed surface treatment of pigments.

A single dispersant that anchors firmly to all pigments and is compatible with all types of coatings is desirable. It could reduce paint production cost, as the production and inventory of multiple mill bases is a significant expense. This ideal is probably not attainable, but part of the potential savings can be realized by using broadly effective dispersants or combinations of dispersants that are compatible with a variety of types of coatings.

21.1.3.2 Other Considerations

High solids coatings present a further problem of adsorbed layer thickness on pigment surfaces. In conventional coatings, the volume of pigment in the wet coating of even highly pigmented coatings is relatively low, so differences in adsorbed layer thickness do not make a large difference in the viscosity of the wet coating (provided the dispersion is stable against flocculation). However, in high solids coatings, especially highly pigmented high solids coatings, the adsorbed layer thickness can have a significant effect on the viscosity. This problem is especially limiting in the formulation of high solids primers, where high pigment loadings near the CPVC (Chapter 22) are often desirable. The effect of layer thickness can be seen by considering a modification of the Mooney equation (Section 3.5) shown in Eq. 21.1. Here, the internal phase volume (V_i) is replaced by the sum of pigment (V_p) and adsorbed layer (V_a) volumes. Shear viscosity is designated by η and η_e, the viscosity of the continuous phase. Using Eq. 21.1, model calculations were made to illustrate the effect of adsorbed layer thickness on viscosity (Hill and Wicks, 1982):

$$\ln \eta = \ln \eta_e + \frac{2.5(V_p + V_a)}{1 - (V_p + V_a)/\phi} \qquad (21.1)$$

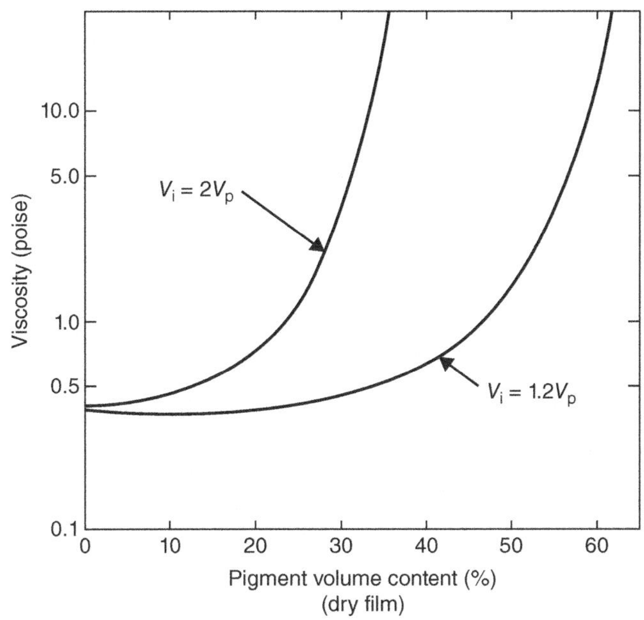

Figure 21.1 Calculations of the effect of PVC on the viscosity of two formulations, both at 70 NVV, but differing in thickness of the layer of adsorbed polymer solution. See text for assumptions. Source: Hill and Wicks (1982). Reproduced with permission of Elsevier.

Figure 21.1 shows the calculated dependence of the viscosity of coatings with 70 NVV as a function of the pigment volume concentration in dry films prepared from the coatings. The following principal assumptions were made in making the calculations: $V_i = 1.2V_p$ and $V_i = 2V_p$ (these values are equivalent to an 8 nm and a 25 nm adsorbed layer, respectively, on a 200 nm pigment); $\phi = 0.65$; solvent: $\rho = 0.8$ and $\eta = 0.4$ mPa·s; and oligomer: $\rho = 1.1$, and $\eta = 40$ mPa·s at 70 NVV, 4×10^5 mPa·s at 100 NVV. The viscosity dependence was assumed to follow Eq. 21.1, in which w_r is weight fraction of resin and η_s is the viscosity of the continuous phase:

$$\ln \eta = \ln \eta_s + \frac{w_r}{0963 + 0.763 w_r} \tag{21.2}$$

As shown in Figure 21.1, the possible level of pigmentation decreases sharply as the adsorbed layer becomes thicker.

In retail outlets, paints are usually inventoried as white paints and white tinting bases. Colors are made to order by adding color pigment dispersions ("colorants") to the white base paints. Since both waterborne and solventborne paints are tinted, it is desirable to have "universal colorants," dispersions that can be used in both types of paint. An example of a vehicle for such dispersions is a proprietary dispersing agent, water, propylene glycol, modified alkylpolyglycol ether surfactant, biocide, and antifoam additive (Bieleman, 2004).

21.2 FORMULATION OF NONAQUEOUS MILL BASES

As discussed in Section 21.1.2, the formulator must design a mill base for dispersing a pigment in the most appropriate dispersion equipment at optimum efficiency. Pigment dispersion machinery is the most expensive machinery in a coatings factory, both from the standpoints of capital and operating costs. It is, therefore, important to maximize the amount of pigment dispersed per unit time. Higher pigment loading increases production efficiency and occurs when the viscosity of the vehicle (solvent plus resin) used in the mill base is low. Low viscosity also facilitates wetting. A well-stabilized pigment dispersion exhibits Newtonian flow, and its viscosity follows the Mooney equation as shown in Eq. 21.3:

$$\ln \eta = \ln \eta_e + \frac{K_E V_i}{1 - (V_i / \phi)} \tag{21.3}$$

While mill bases usually contain a single pigment, there are circumstances in which it is advantageous to mill two or more pigments together. Such mill bases are called *composite grinds.*

The volume of pigment (internal phase) can be maximized by using the lowest possible viscosity (η_e) vehicle. Low viscosity, fast pigment wetting, and high pigment content can be obtained using only solvent. But solvent alone cannot stabilize a dispersion against flocculation; it is necessary to include resin (or hyperdispersant) in the mill base. For maximum pigment loading, it is desirable to use the minimum concentration of resin solution that provides stability. Many years ago, Fred Daniel devised a simpler and fast, although not very accurate, method to estimate the minimum effective amount of resin (Patton, 1979).

21.2.1 Daniel Flow Point Method

The *Daniel flow point method* is a simple tool for formulating mill bases efficiently, especially for dispersions to be made in ball and other media mills. This method provides an estimate of the most appropriate resin concentration to use with a particular pigment. One makes a series of solutions with different resin concentrations and adds varying amounts of each solution to a weighed amount of pigment. Each mixture is then dispersed by rubbing vigorously with a spatula on a flat glass plate to determine which dispersion has a viscosity just low enough to flow readily off the spatula. The spatula serves as both a dispersion machine and a crude viscometer. Since the clearance between the spatula and the glass plate is small, the shear rate is high; a spatula is a fairly good dispersing machine. One plots the volume

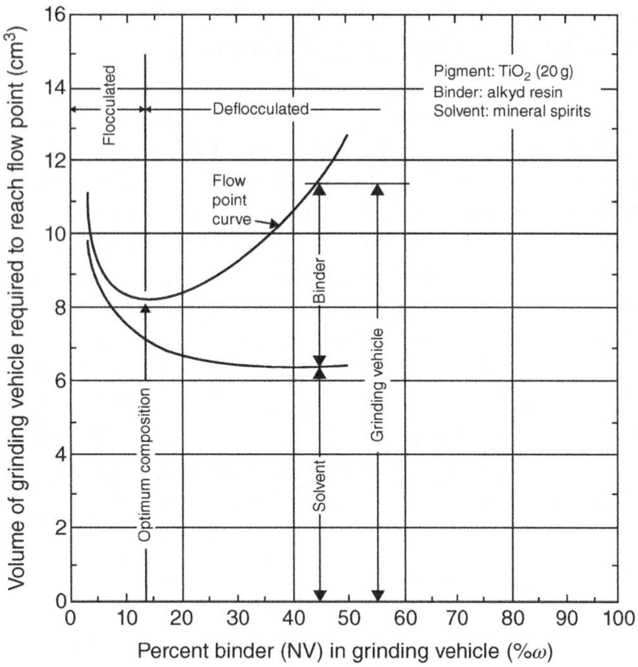

Figure 21.2 Daniel flow point plot: milliliters of solutions of an alkyd resin in mineral spirits per 20 g TiO_2 as a function of NVW resin in solution. Source: Patton (1979). Reproduced with permission of John Wiley & Sons.

of each solution needed for flow against the concentration of that solution. This plot is an isoviscosity plot; that is, the viscosity of each of the dispersions is approximately the same. (The actual low shear rate viscosity is ~10 Pa·s.) An example is shown in Figure 21.2; in addition to the volume of resin solution, the volume of solvent in the resin solution is also plotted.

Any resin–solvent–pigment combination that gives a stable dispersion exhibits a minimum in the curve, as shown in Figure 21.2. Let us consider the significance of the minimum. To the right of the minimum, the viscosity of the external phase is increasing because the concentration of resin is increasing. As can be seen from Eq. 21.3, if η_e increases and the viscosity stays constant, then V_i must decrease; that is, the amount of vehicle required per unit of pigment must increase. To the left of the minimum, the amount of vehicle per unit of pigment is also increasing, even though η_e is decreasing. In this region, the concentration of resin is insufficient to stabilize the dispersion. The pigment dispersion flocculates increasingly as the concentration of resin in the solutions decreases and the viscosity increases more and more steeply, so additional resin solution must be added to reach the isoviscosity level. The minimum point corresponds to the minimum concentration of that resin in the solvent that permits preparation of a stable dispersion of that pigment. Since the determination is not highly accurate, it is usual to start with experimental dispersions having a somewhat higher resin concentration. If one adds more solvent to a stable dispersion with near-the-minimum resin concentration, the dispersion will flocculate. In some cases, no minimum is found—the amount of resin solution required per unit of pigment continues to decrease as the concentration increases. The latter behavior signifies that a stable dispersion of the pigment cannot be made with that particular combination of resin and solvent.

21.2.2 Oil Absorption Values

In the early days of formulating coatings, it was observed that the weight of pigment that could be incorporated with linseed oil in a mill base varied greatly from pigment to pigment. To simplify formulating mill bases to approximately equal viscosity, *oil absorption values* for various pigments were determined. Linseed oil was slowly added to a weighed amount of pigment while rubbing with a spatula. Initially, balls of pigment wet with oil formed. When just enough oil had been added and worked into the pigment so that a coherent mass formed, the end point was reached. The oil absorption value is calculated as the number of pounds (grams) of linseed oil required to reach the end point with 100 lb (100 g) of pigment. At the end point, there is approximately enough linseed oil to adsorb on the surface of all the pigment particles and fill the interstices between close-packed particles.

Oil absorption values for different pigments vary over a wide range. The smaller the particle size, the higher the oil absorption. A small particle size pigment has more surface area; therefore, a larger amount of linseed oil is adsorbed on the surface. In some cases, such as some grades of carbon black, the pigment particles are porous. Some oil penetrates into these pores, increasing the amount of oil required and, hence, the oil absorption value. Diatomaceous earth (Section 20.3) has a very high surface area and, hence, very high oil absorption values. Pigment density also has a major effect. High density pigments require less weight of oil to adsorb on the surface of a unit weight of pigment and to fill the interstices; hence, they have lower oil absorptions. Other vehicles besides linseed oil give similar oil absorption values. Therefore, data obtained with linseed oil can be a starting point for formulating mill bases with any vehicle that gives stable dispersions.

The precision of oil absorption determinations by the spatula method is not high. Operators working in different laboratories commonly report values that vary by ±15%. Deviations may be reduced to ±2–3% for a single operator with experience. In spite of these error ranges, it is common to see oil absorption values having three supposedly significant figures. Improved accuracy and precision can be achieved by using a mixing rheometer (Section 3.3.2), such as a brabender plastometer for carrying out the determination (Hay, 1974). The mixer is loaded with a known amount of linseed oil; pigment is added slowly. The mixer imparts the necessary shear to separate pigment aggregates. The power required to turn the blades is recorded. As the amount of pigment is increased, the power requirement increases. When the oil absorption end point is passed, the dispersion mass breaks up into chunks, leading to erratic readings. The values obtained are more reproducible and are generally a few percent higher than those obtained by the spatula method.

Formulators can also take advantage of oil absorption values provided by pigment vendors. These oil absorption values provide a starting point for pigment content when formulating a mill base. Batch-to-batch variations can be expected. Oil adsorption values can also be used to make preliminary estimates of CPVC of films containing those pigments as discussed in Section 22.2.

21.3 DISPERSION IN AQUEOUS MEDIA

Dispersion of pigments in aqueous media involves the same stages as in organic media: wetting, separation, and stabilization. However, the unique properties of water add extra complications. First, the surface tension of water is high, so there is more likely to be a problem in wetting the surfaces of low-polarity pigment particles. Secondly, in some cases, water interacts strongly with the surface of pigments; in such cases the anchor groups on the stabilizers have to interact more strongly with the pigment surface to compete with water. Thirdly, since many applications of aqueous dispersions are in latex paints, the systems have to be designed so that stabilization of the latex dispersion and of the pigment dispersion does not adversely affect each other. Peck (2014) presents an overview of aqueous dispersion formulation and recommends a systematic approach.

Inorganic pigments, such as TiO_2, iron oxides, and most inert pigments, have highly polar surfaces, so there is little or no problem with wetting them with water. Most organic pigments, however, require the use of a surfactant to wet the surfaces. Some organic pigments are surface treated with adherent layers of inorganic oxides to provide a polar surface that is more easily wet by water (Bugnon, 1996). While the surfaces of inorganic pigments interact strongly with water, the adsorbed layer of water does not by itself stabilize against flocculation.

21.3.1 Stabilization of Aqueous Dispersions

In contrast to dispersions in organic media, stabilization by charge repulsion (Section 9.1.1) can be a major mechanism in aqueous media. The stability of the dispersions depends on pH, since pH affects surface charges. For any combination of pigment, dispersing agent, and water, there is a pH at which the surface charge is zero; this pH is called the *isoelectric point* (*iep*). At *iep*, there is no charge repulsion; above *iep*, the surface is negatively charged and below *iep*, it is positively charged. The stability of dispersions is at a minimum at $iep \pm 1$ pH unit (Morrison, 1985). The *iep* value for pigments varies, for example, 4.8 for kaolin clay and 9 for $CaCO_3$.

In addition to stabilization of the final dispersion, Peck (2014) shows that it is desirable to include additives for dynamic stabilization of the newly separated particles during the grind. Certain nonionic dispersants serve this purpose.

At least one supplier provides a tool on the Internet that calculates formulas for mill bases for a long list of pigments. The formulas include pigment, dispersion stabilizer, antifoam additive, and water. The supplier emphasizes that these are only starting point formulas to be adapted to specific circumstances.

Since latex paints were introduced in the 1950s, continuous efforts have been made to improve colorants. By the 1990s, introduction of waterborne automotive base coats (Section 30.1.2) increased the need for high quality aqueous pigment dispersions. More recently, a need has developed for solvent-free colorants that are effective in "zero VOC" paints. These needs motivated intensive research on pigment dispersants that continues today. An early goal was to find polymeric dispersants that are effective in aqueous media. The sodium salt of polyacrylic acid and certain polyacrylamide copolymers are effective in latex paints (Farrokhpay et al., 2006). A variety of acrylic copolymers made by conventional free radical polymerization proved to be useful. For example, copolymers of (1) (meth)acrylic acid, (2) monomers bearing one or more highly polar anchoring groups that can absorb on pigment surfaces, and (3) a mono-(meth)acrylic ester of polyethylene oxide are effective as are polyetheralkanolamines (Nguyen, 2007). All these copolymers are often used in combination with conventional surfactants.

Controlled free radical polymerization techniques have been studied in order to obtain polymeric dispersants that are more effective and more efficient than is possible from conventional polymerization. Early research on dispersants made by CRP was summarized by Perrier (2004). These techniques can be adapted to produce AB and ABA block copolymers, as well as graft, star, and hyperbranched copolymers with controlled structures. When made with appropriate adsorbing and stabilizing groups, all of these types of polymers can function as pigment dispersion stabilizers. See Section 2.2.1.1 for discussion of CRP methods.

One CRP technique, atom transfer radical polymerization (ATRP), permits synthesis of polymers having predictable molecular weights, low molecular weight distributions, and controllable block structures. For example, an AB block copolymer pigment dispersant is synthesized by (1) loading a reactor with toluene, then 1-naphthalene sulfonyl chloride, copper powder, 2,2′-bipyridyl, and glycidyl methacrylate, which are reacted at 80°C for 6 h to form the A block; (2) adding the methacrylate of polyethylene glycol monomethyl ether in methyl 2-hydroxypropyl ether and heating for an additional 4 h at 80°C to attach a hydrophilic B block; (3) reacting the glycidyl epoxy groups with undecanoic acid to make the A block highly hydrophobic; and (4) removing the copper complex by precipitation with magnesium silicate and filtration. The resulting AB block copolymer solution is reduced to 70% weight solids, dispersed in water, and the pH is adjusted to 8 with DMAE (White et al., 2002).

Another CRP technique, catalytic chain transfer polymerization (CCTP), differs from ATRP in that a conventional initiator is used in combination with a cobalt (or other metal) complex that mediates the reaction (Visscher and McIntyre, 2003). In one example, a graft copolymer dispersant is prepared using an azo initiator with a cobalt(II) complex as chain transfer agent. First, CCTP is used to prepare an acrylic macromonomer containing 20 wt% methacrylic acid. This macromomer is then copolymerized with a hydrophilic nonionic macromonomer and hydrophobic monomers. The result is a graft copolymer with a hydrophobic backbone and both anionic and nonionic hydrophilic side chains. When neutralized with AMP, the graft copolymer becomes an effective dispersant. In aqueous media, the hydrophobic backbone presumably anchors the polymer to pigment surfaces, and the hydrophilic grafts act as tails, providing entropic and charge stabilization. Base coats formulated with TiO_2, dispersed with the graft copolymer, provided higher gloss coatings after being clear coated (Visscher and McIntyre, 2003).

Another CRP technique is nitroxide mediated polymerization (NEM), in which hindered nitroxide compounds are used as mediating agents to prepare narrow molecular weight distribution block copolymers. For example, a block copolymer of butyl acrylate and dimethylaminoethyl acrylate, useful as a pigment dispersion stabilizer, is made by sequential bulk polymerization using piperidine nitroxyethers as mediators (Auschra et al., 2002).

Yet another technique is RAFT living radical polymerization. Saindane and Jagtap (2015) used this method to synthesize AB block copolymers of ethyl acrylate and acrylic acid with M_n of 5000–10000. These copolymers proved to be very effective dispersant stabilizers in white acrylic latex paints.

The possibilities are almost limitless, but the polymer structures must be designed correctly for good results. For example, an AB diblock copolymer of 2-(dimethylamino)ethyl methacrylate and *t*-butyl methacrylate, made by living ionic polymerization, is a very effective stabilizer; whereas a corresponding ABA triblock copolymer of the same components, having the amine-functional blocks on both ends, leads to particle bridging and severe flocculation (Creutz and Jerome, 2000).

A broad study on stabilization of aqueous TiO_2 dispersions by anionic and nonionic surfactants concluded that a high molecular weight nonionic copolymer provides the greatest resistance to flocculation both in the dispersion and during drying of a gloss latex paint film (Clayton, 1997). Since stabilization was primarily entropic rather than by charge repulsion, it was not affected by changes in pH. On the other hand, a study of stabilization of TiO_2 dispersions with sodium polyacrylate showed that both charge and entropic repulsion affected stability. The surface layer did not need to be as thick as in solvent-based dispersions. Even as little as an average 1 nm layer thickness combined with charge repulsion sufficed for stabilization. Adsorption on the TiO_2 surface was flat but there were loops and tails extending into the water phase. Higher molecular weight gave thicker films but adversely affected viscosity (Banash and Croll, 1999).

Surface treatments can have important effects on the stability of aqueous pigment dispersions. Specially treated TiO_2 pigments have been developed for waterborne coatings. Substantial differences in dispersion stability can result from differences in the composition and completeness of the surface treatment (Losoi, 1989). In some cases, the amount of surface treatment reduced the TiO_2 content by as much as 25%. Since hiding is related to the actual TiO_2 content, larger amounts of such highly treated pigments are needed to obtain equivalent hiding.

Slurries of up to 80 wt% TiO_2 in water are used on a large scale with substantial cost savings (Section 20.1.1). The dispersions are often stabilized with carboxylic acid-functional polymeric dispersants; pH is controlled by adding amine. The formulations are established to permit minimal flocculation in order to retard settling. Bactericide is added to inhibit microbial growth.

Aqueous dispersion of fine particle size pigments, such as transparent iron oxide, can be difficult for producing coatings without haze. It has been shown that good results can be obtained using dispersants prepared by copolymerizing vinyl ether-terminated polyethers with maleic anhydride. High color yield dispersions can be achieved that provide films having much less haze than with other dispersing agents. For special frosty appearance effects, dispersions of fine particle size "transparent" TiO_2 can be used (Silber et al., 2002).

Most latex paint formulations contain several pigments and several surfactants including anionic and nonionic surfactants. The *iep* of the various pigments are different, which complicates the problem of charge stabilization. Polymeric anionic surfactants (such as salts of acrylic copolymers in which acrylic acid and hydroxyethyl acrylate are used as comonomers) provide salt groups for strong adsorption on the polar surface of the pigment and hydroxyl groups for interaction with the aqueous phase; nonpolar intermediate segments add adsorbed layer thickness. Ammonia or amine salts are recommended because the base can escape from the film during drying providing films that are less sensitive to water than when sodium or potassium salts are used. Polymeric surfactants are less likely to lead to performance problems in the final films than monomeric surfactants. Since polyacrylates are not effective wetting agents, they are frequently combined with nonionic surfactants and/or an anionic surfactant. It is common to also add potassium tripolyphosphate, the basicity of which may assure that the pH is above the *iep* of all pigments. Note that potassium salt of tripolyphosphate is used in paints, rather than the sodium salt, which is used in laundry detergents. The potassium salt is less likely to deposit on the surface of a film as scum after being leached out of a dry paint film by water. Large paint producers develop proprietary surfactants for their major product lines. Some anionic surfactants can also act as fungicides.

Additives can sometimes affect dispersion stability. For example, a hydrocarbon solvent used in a latex paint as a defoamer was shown to cause flocculation (Smith, 1988). The paint formula had several surfactants. Smith found that the order in which the surfactants were added controlled whether or not flocculation occurred.

Organic pigments, unless surface treated, generally have surface free energies lower than the surface tension of water. Therefore, surfactants are needed to reduce the surface tension of the water to permit the wetting essential for pigment dispersion. Either anionic or nonionic surfactants can be used. In the case of anionic surfactants, the dominant mode of stabilization is probably charge repulsion. In the case of nonionic surfactants, entropic repulsion is probably predominant, since the relatively long polyether alcohol end is oriented out into the water and associated with multiple water molecules.

Wilker (1999) has reported that organic pigments dispersed in aqueous polyurethane dispersions (PUDs) show greater color strength than when dispersed in solventborne media. Depending on the pigment, the increase is between 10 and 30%. The effect is attributed to water's ability to dissolve salts in the pigment agglomerates; these salts are not dissolved by organic solvents and increase adhesion within the agglomerates.

There is reasonable understanding of the stabilization of single-pigment systems. However, paints usually contain several pigments, commonly with both polar and nonpolar surfaces, and one or more latexes (Chapter 32). Furthermore, the latex formulations often contain one or more water-soluble polymers and/or associative thickeners (Section 32.3) to adjust the rheological properties of the paint, and these can interact with dispersants and pigments. As a result, the selection of appropriate dispersing agent combinations is done at least partly on an empirical basis. High throughput methods can help formulators understand how various ingredients interact (Rosano, 2014). As in any situation in which empirical knowledge is the key to success, it is desirable to build a database of combinations that work and do not work to facilitate formulating with a new pigment combination or latex.

21.4 DISPERSION EQUIPMENT AND PROCESSES

A variety of equipment is used to make pigment dispersions. An important difference between the machines is the level of shear stress that can be exerted on the pigment aggregates. Easily separated pigment aggregates can be dispersed in low shear stress equipment, whereas strongly associated pigment aggregates require very high shear stress for separation. Discussion of all types of such machinery and their operation is beyond the scope of this text. Patton (1979) provides more discussion of some types. Winkler and Dulog (1987) deal more fundamentally with engineering aspects of some dispersion methods. We discuss a few important types with emphasis on their advantages and disadvantages. A formulator must become familiar with dispersion equipment in his or her company's factories in order to design formulations appropriate for production with the available machinery.

Often, an excellent source of up-to-date information about machine characteristics and operating procedures is the manufacturer of the machine.

From a processing point of view, there are three stages in making and using pigment dispersions: (1) premixing, that is, stirring the dry pigment into the vehicle and eliminating lumps; (2) imparting sufficient shear stress to separate the pigment aggregates in the presence of a dispersion stabilizer; and (3) *let down*, that is, combining the pigment dispersion with the balance of the ingredients to make a coating. Some machines can carry out only the second step, some two of the three, and others, all three.

Some pigment dispersions are made for specific batches of a coating and are let down with the other components to make the final coating. More often, pigment dispersions are made for use in several related types of coatings or as *tinting pastes* (*mill bases*) for color matching a variety of coatings. To minimize inventory of tinting pastes, one tries to select vehicle and stabilizer combinations that are compatible with a range of coatings made by that company.

21.4.1 High-Speed Disk (HSD) Dispersers

High-speed disk (HSD) dispersers consist of a shaft with a disk that rotates at high speed in a cylindrical tank. HSD dispersers are also called *high-speed impellers*, *dissolvers*, and *high-intensity mixers*. The disks have flanges around the perimeter that are bent at a sharp angle from the plane of the disk. A schematic drawing of a typical disk is shown in Figure 21.3. Increasingly, the disks are made from engineering plastics that provide greater abrasion resistance than steel. The shearing action takes place by the differential laminar flow rates streaming out from the edge of the disk, which typically rotates with a peripheral speed of 20–26 m s^{-1}. Sometimes, the speed of an HSD is expressed in revolutions per minute (rpm), but the important speed is the peripheral speed, which depends on the radius of the disk. The shear stress developed is relatively low, and therefore, HSD machines are appropriate only for relatively easily separated pigments. Obtaining predominantly laminar flow requires a minimum viscosity depending on the dimensions; it is usually somewhat over 3 Pa·s. The higher the viscosity, the greater the shear stress exerted on the aggregates and, therefore, the faster the process. The viscosity should be set such that the motor driving the shaft is run at peak power.

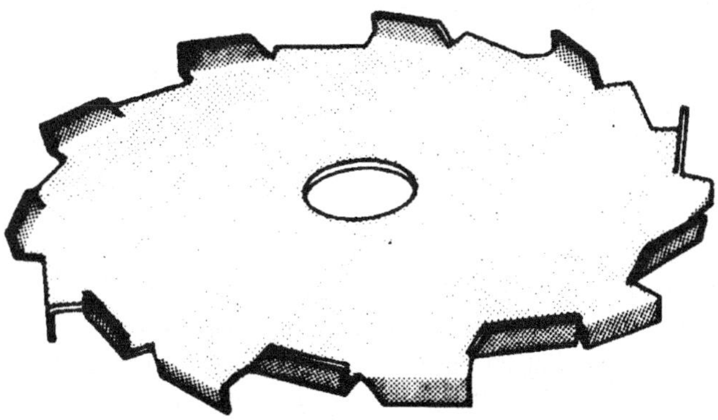

Figure 21.3 Schematic drawing of a high-speed impeller disk. Source: Patton (1979). Reproduced with permission of John Wiley & Sons.

For best efficiency, pigment loading should be maximized. In solvent-based coatings, this can be done by using a vehicle solution having the minimum concentration of resin that provides stabilization against flocculation. The Daniel flow point method can be used to estimate the resin–solvent ratio. Then, the ratio of this vehicle to the pigment is set to achieve a viscosity at high shear rate such that the motor draws peak amperage. Alternatively, pigment dispersion stabilizers (Section 21.1.3.1) can be added or used in place of the resin. In making latex paints, water-soluble polymers are commonly included in the pigment dispersion stage to increase viscosity during dispersion.

A diagram of flow in using an HSD is shown in Figure 21.4. Centrifugal force leads to flow up the sides of the tank. If the mill base is Newtonian and the dimensions and operating conditions are appropriate, the whole

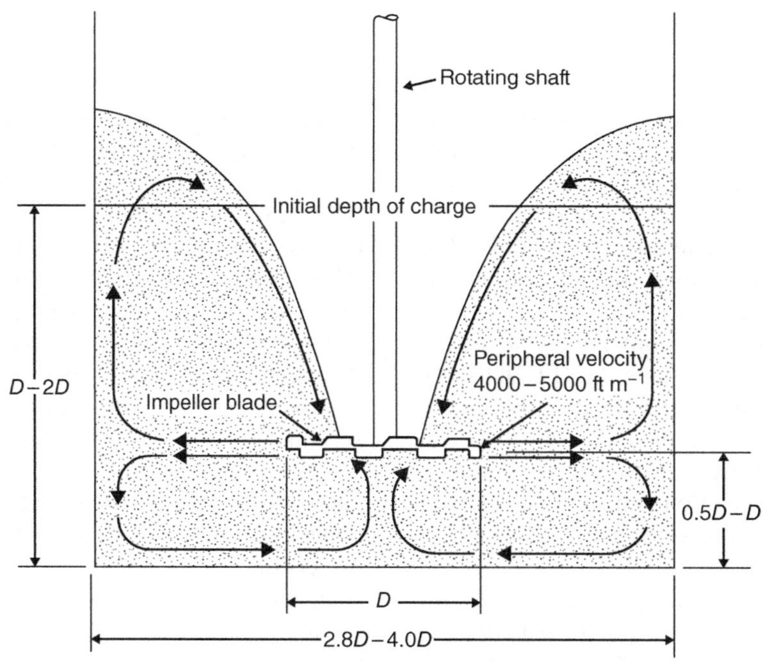

Figure 21.4 Diagram of a high-speed impeller disperser showing correct positioning of disk and optimum dimension ratios. Source: Patton (1979). Reproduced with permission of John Wiley & Sons.

charge becomes intimately mixed, and all portions of the mill base pass repeatedly through the zone of highest shear, near the edges of the disk. However, if the mill base is shear thinning, the viscosity is high at the upper edge of the material on the sides of the tank where shear rate is low, resulting in hang up and incomplete mixing. Some mill bases contain a pigment designed to make the final coating shear thinning. The resulting hang up problem can be minimized by dispersing all of the pigments except the one that gives the shear thinning effect. After the balance of the pigments has been separated, the pigment that causes shear thinning is slowly added. A desirable way to minimize hang up is to design the dispersing tank with a slow-speed scraping blade that travels around the upper inside of the tank while the HSD is spinning in the center of the tank.

HSD machines are commonly used for premix, dispersion, and let down operations. One initially loads the vehicle components, mixing at a low rpm, and then slowly loads the dry pigment near the shaft. After the pigment is loaded, the speed is turned up for the dispersion stage. In a properly formulated mill base, the dispersion stage requires about 15 min. The speed is reduced and the let down is carried out with the rest of the components in the formulation. When producing latex paints, the latex cannot be present during the dispersion stage, because most latexes coagulate when exposed to high shear. The latex is added in the let down stage at low speed.

Compared with other dispersion equipment, HSD machines generally have the lowest capital and operating costs. No separate premix is required, and the let down can be carried out in the same tank. Cleaning in order to change from one color to another is relatively easy. Solvent loss can be kept to a minimum by using a covered tank. The major limitation is that the shear stress imparted to pigment aggregates is relatively low, so the equipment can only be used for relatively easily separated pigments. Pigment manufacturers have made substantial progress in making pigments with easily separated aggregates. Laboratory HSDs are also available, and results can be correlated with production machines.

21.4.2 Rotor/Stator Mixers

Shear stress imparted to pigment aggregates can be increased by substituting rotor/stator mixers for HSD machines. These mixers are somewhat similar to HSD machines, but the rotating disk spins in close proximity to fixed elements. The mixers can be used for large batches or in line. Paterson (2002) compared these mixers with HSD machines. He reported that rotor/stator mixers can provide superior dispersions in aqueous media, including critical dispersions for automotive basecoats.

21.4.3 Ball Mills

A *ball*, or *pebble*, *mill* is a cylindrical container, mounted horizontally and partially filled with balls or pebbles. The mill base components are added to the mill, and it is rotated at a rate such that the balls are lifted up one side and then roll in a cascade to the lower side, as shown schematically in Figure 21.5. Intensive shear is imparted to pigment aggregates when the balls roll over each other with a relatively thin layer of mill base between them.

Two classes of such mills are used. In steel ball mills, the balls and lining of the cylindrical mill are steel. In pebble mills, the balls are ceramic and the mill lining is porcelain. In early days, the balls were large smooth pebbles, hence the term pebble mills. The term ball mill is used by some to mean either class of mill and by others to mean only steel ball mills. We use ball mill in the former sense and specify steel ball mill or pebble mill for each particular class. Steel balls have the advantage of higher density and, therefore, shearing is greater and milling times can be shorter. But their use is limited to dark-colored pigments and pigments where color is not critical (such as primers), because the balls wear and discolor light colors. Pebble mills are used when discoloration is undesirable.

Ball mills operate most efficiently when their diameter is large so that the length of the cascade is long. The efficiency of operation depends on the loading of the mill. A mill should be loaded about half full of balls; this gives the longest cascade. The mill base volume should be just over the volume needed to cover the balls when the mill is at rest. If excess mill base is used, the time required for satisfactory separation increases. Assuming the balls are spheres of equal diameter, the volume of balls is approximately 32%, and the volume of mill base is only a little over 18% in a properly loaded mill. If the balls or pebbles are not spheres and/or not uniform in size, there is even less space for mill base.

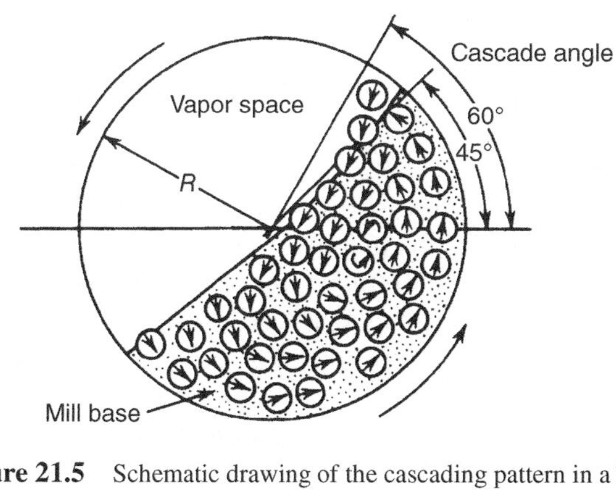

Figure 21.5 Schematic drawing of the cascading pattern in a ball mill. Source: Patton (1979). Reproduced with permission of John Wiley & Sons.

Operation of a ball mill requires care in setting the rpm of the mill. If the rpm is too low, the balls are not carried high enough. If it is too high, the balls are carried past the 60° angle shown in Figure 21.5, and some fall rather than cascade. This reduces shear, can lead to ball breakage, and is more likely to lead to breaking the ultimate crystals of the pigment rather than just shearing the aggregates. Experienced mill operators can judge whether the rpm setting is proper by the sound. Milling efficiency is also affected by the viscosity of the mill base. If the viscosity is too low, ball wear is high. If it is too high, the balls roll more slowly and efficiency is reduced. Optimum viscosity depends on density of the balls and on mill size. Viscosities in the neighborhood of 1 Pa·s are common.

The Daniel flow point method (Section 21.2.1) was originally developed to help formulate mill bases for ball mills. It permits loading the maximum amount of pigment at the viscosity required. Proper mill base formulating and use of proper volumes of balls and mill bases can make large differences in the time required for satisfactory separation. The time required depends on how easily the pigment aggregates are separated. The minimum time is usually on the order of 6–8 h. Even difficult pigments should require no more than 24 h. The requirement of substantially longer time indicates poor formulation or improper mill loading.

Ball mills can be used to disperse all but the most difficult to separate pigments. Although the capital cost of ball mills is relatively high, they last a long time and operating cost is low. No premixing is required, and the mills can run unattended. They are difficult to clean and, therefore, most appropriate for making batch after batch of the same dispersion. Another difficulty is that batch sizes are limited because the mills contain only about 18% by volume of mill base.

Laboratory ball mills are useful, but they do not correlate with production mills. This is because the efficiency of a ball mill is so dependent on its diameter. Production mills range from 1.25 to 2.5 m (4–8 ft) in diameter. Some laboratories use so-called jar mills. They are usually less than 30 cm (1 ft) in diameter and are often rolled at a much less than ideal rpm.

In the laboratory, *quickee mills* are much faster for making small batches of dispersions roughly similar to production dispersions in a ball mill. A steel container is filled a little over half full with 30 mm steel balls and enough mill bases to moderately cover the balls. The container is then shaken on a paint shaker of the type used in paint stores. Easily separated pigments require 5–10 min. Difficult-to-separate pigments may take up to an hour of shaking. Glass jars or vials partly filled with glass beads, sand, or ceramic media can also be used as quickee mills.

21.4.4 Media Mills

Media mills (also called *agitator mills*) were invented by S. Hochberg to get around batch size limitations of ball mills and to enable continuous production. In the earliest mills, the media was sand that had an average diameter of about 0.7 mm, and the mills were called *sand mills*. Sand has largely been replaced by small ceramic balls; in which case the mill may be called a *bead mill*. Small steel balls are also used, in which case, the mill is commonly called a *shot mill*, because the original steel balls were the shot used in shotgun shells. Shot mills are generally used to disperse carbon black or other dark pigments.

A schematic drawing of a vertical media mill is shown in Figure 21.6. A high-speed rotor with disks is mounted in a cylinder. The space between the rotor and the cylinder is partially filled with sand or other media. A premixed mill base is pumped into the bottom of the mill. It is exposed to shear as it flows up through the mill, passing between the rapidly moving particles, and out through a screen that retains the media in the mill. A typical vertical media mill operates at a peripheral rotor speed of 10 m s^{-1}. This impels the media particles at high speed so that, even though their size is small, the shear rate between them is high, and, of course, in passing through the mill, a pigment aggregate passes between many pairs of particles. Mill bases should be formulated using slightly higher resin concentrations than the minimum in the Daniel flow point curve. Viscosities range from 0.3 to 1.5 Pa·s. For more difficult-to-separate pigments, two or three passes through a mill (or a battery of mills in series) may be required. Mills have been designed that automatically recycle part of the mill base.

The mill shown in Figure 21.6 has a vertical rotor, but use of horizontal media mills is increasing. A variety of designs is available. Some have simple disks on the rotor and various rotor configurations are available.

Pin mills are also being increasingly used. Each pin is a cylinder made of abrasion-resistant steel, typically 1–2 mm in diameter and 20 mm in length. Pins are placed in opposite walls of a rotating cylinder so that the pins on one side rotate between pins on the other side. A pin mill runs at high speed, which enhances grinding and energy efficiency.

Way (2004) provides a discussion of the factors involved as the size of the media is reduced. The ultimate size of the dispersed pigment decreases as the size of the media decreases so that the smallest particle size media will provide the smallest particle size pigment dispersion in the shortest time. However, as the particle size of the media decreases, the difficulty of separating the media balls from the pigment dispersion increases. The factors that affect separation of the mill base from the media are as follows:

- Mill base viscosity. High viscosity increases the power necessary to rotate the shaft and may surpass the

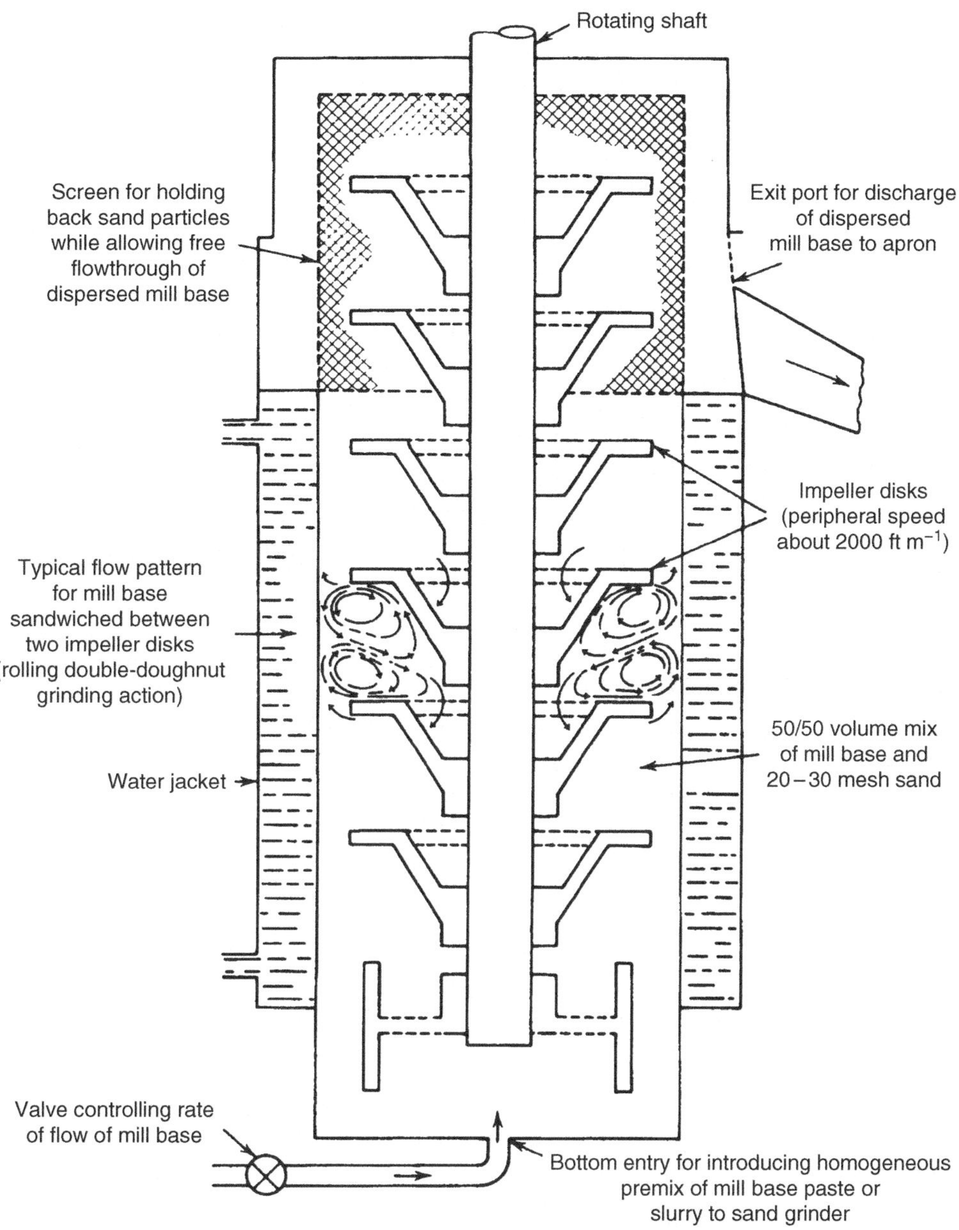

Figure 21.6 Schematic drawing of a media mill. Source: Patton (1979). Reproduced with permission of John Wiley & Sons.

design limit for rotating the shaft rapidly enough to disperse pigments. Also, as the viscosity increases, the media beads are carried toward the screen leading to screen blockage.

- Flow rate. If the flow velocity is higher than the separation velocity, the beads are carried to the screen.
- Mill base density. It becomes harder to separate the dispersion from the beads as density increases.
- Bead size and density. As bead size decreases, separation of the beads centrifugally becomes more difficult. As one changes the density of the media from glass balls (density = 2.6), to yttrium stabilized zirconium oxide (density = 6), to steel balls (density = 7.8), and to tungsten carbide (density = 14), separation becomes easier. However, high density media increase wear of the dispersing machine.
- Agitator speed. Higher speed gives better separation but there are practical limits.
- Design of the separation system. The size of the screen is critical. There must be sufficient screen area in order that the velocity of material passing through the screen is not so high that media balls will be carried there and clog the screen.
- Machine design. Horizontal pin mills are more efficient with small media balls than the traditional vertical mills discussed earlier in this section.

Nano-pigments are particularly difficult to disperse; pin mills with small media are recommended. Miranda (2011) states a rule of thumb that media diameter should be about 1000 times greater than the target diameter for the processed particles. He describes specially designed horizontal mills that can use 30-μm media to target particle sizes of 30 nm and below.

Media mills require premixing, and let down must be done separately. Batch size is flexible. Cleaning is not easy; generally, certain mills are reserved for similar colors to minimize cleaning. Media mills are effective for separating all but the most difficult pigment aggregates, provided that the aggregates are small compared with the media particles. Laboratory media mills are available that correlate well with production mills. Rather than pumping the premix in from the bottom of the mill, it is poured into the top of the laboratory mill. Horizontal laboratory mills are also available. The use of media mills and related equipment is probably second only to HSD machines.

21.4.5 Three Roll and Two Roll Mills

Three roll mills are still important in ink production, but their use in coatings manufacture has declined sharply as easily dispersed grades of pigments become available and media mill technology improves.

Two roll mills exert even greater shear rates than three roll mills, but their capital and operating costs are very high. They remain important in the rubber industry but are seldom needed to disperse modern coatings pigments. Formerly they were used to disperse high value pigments and certain carbon black pigments when the desired jetness was only attained with virtually complete separation. More easily dispersed jet black pigments are now available.

21.4.6 Extruders

Extruders are increasingly used for pigment dispersion. Pigment dispersion for almost all powder coatings, as well as some high viscosity liquid dispersions, is processed in extruders (Section 28.4.1). Extruders have one or two screws feeding through a cylinder and forcing the material out the end through a die. The barrel of the extruder can be operated over a wide temperature range. Extruders can be configured so that the shear imparted by the screw(s) is very high, and almost any pigment aggregate can be broken down in an extruder. Difficult-to-disperse pigments require longer residence times.

21.4.7 Ultrasound

Exposure of a liquid/pigment premix to low frequency ($f \approx 20$ kHz) ultrasound at intensities above the cavitation threshold generates alternately growing and imploding bubbles. The mechanical effect breaks pigment aggregates into individual particles. Van Eecke and Piens (2000) report that sonication times of the order of 15 s can yield tinting strength equal to conventional pigment dispersions.

21.4.8 Stir-in Pigments

Certain pigment manufacturers offer *stir-in pigments.* Essentially the pigment manufacturer performs the pigment dispersion step and converts the dispersion to dry powder or granules that contain, for example, a polyurethane dispersant (Johnson, 2014). The paint manufacturer disperses the dry dispersion into liquid paint by simple stirring.

21.5 EVALUATION OF DISPERSIONS

Assessment of *degree of dispersion* is a critical need for establishing original formulations and optimizing processing methods, as well as for quality control. Differences in degree of dispersion come from two factors: incompleteness of separation of the original aggregates into individual crystals and flocculation after separation. Frequently, the coatings industry does a poor job in this critical evaluation.

For white and colored pigments, the most effective evaluation method is by determination of *tinting strength* in comparison to a standard. For a white dispersion, one weighs out a small sample of the dispersion and mixes into it a small weighed amount of a standard color dispersion, for example, a blue. One also weighs out a standard white sample with the same ratio of the blue standard. After thorough mixing, one places a small amount of each tint mixture side by side on a piece of white paper and draws down both samples with a stiff flat-ended spatula or a drawdown bar so that the edges of the two samples touch each other. One can then compare the color of the two drawdowns. If the batch is a darker blue than the standard, the tinting strength of the batch of white is low, usually meaning that it is not equally dispersed. To test a color dispersion, say, blue, one carries out the same procedure but uses the standard white in both samples, together with either the standard blue or the batch of blue.

One can check for flocculation by rubbing some of the wet tint mix with a finger. If the color changes, the dispersion was flocculated. For example, if a mix of blue with white becomes bluer where it is rubbed, the blue pigment dispersion is flocculated. If the rubbed portion becomes a lighter blue, the white is flocculated. Flocculation can also be detected by examining the flow of the dispersion. Well-stabilized dispersions have Newtonian flow properties. If a dispersion is shear thinning (and does not contain a component designed to make it shear thinning), it is flocculated.

Transparency is a critical requirement for dispersions of color pigments that are used in applications such as automotive metallic coatings. Transparency can be tested by drawing down a standard and a batch side by side on a glass plate. The degree of haze or difference from the standard can be evaluated visually or measured instrumentally.

A further method of assessing pigment dispersion is by settling or centrifugation experiments. The rate of settling is governed by particle size and difference in density of the dispersed phase from the medium. A well-separated, well-stabilized dispersion settles slowly; but when settling is complete, the amount of sediment is small. A well-separated but poorly-stabilized dispersion settles quickly to a bulky sediment. The floccules settle more quickly because of their large size and they form a bulky sediment because continuous phase is trapped within the floccules. Flocculated sediment is readily stirred, or shaken back to a uniform suspension, in contrast to the sediment formed from the nonflocculated dispersion. The test can be accelerated with a centrifuge. Settling or centrifugation tests provide qualitative or semiquantitative information sufficient for development work and quality control purposes.

One can also examine the dispersion with a microscope. Brownian motion is observed in deflocculated dispersions, but flocculation stops the motion. One must use caution in preparing the samples for examination. In general, it is necessary to dilute the sample. If the sample is diluted with solvent, there is a possibility of solvent-induced flocculation. In such a case, one may report that the dispersion is flocculated when actually the flocculation was a result of sample preparation. In assessing, for example, TiO_2 dispersions with an optical microscope, only the largest particles are seen.

Electron microscope studies of the surfaces of etched dry coating films can be useful for assessing variations in dispersion (Brisson et al., 1991).

Laser light scattering is probably the most precise of measuring pigment particle size and particle size distribution (Van et al., 2001). Instruments are now available that make the measurements rapidly and reproducibly. Concentrated dispersions must be diluted to perform the measurements, and shear forces are applied as the measurement is in progress. These instruments, therefore, are better suited for measuring de-agglomeration than for measuring flocculation.

Balfour and Hird (1975) pioneered infrared backscattering to study TiO_2 dispersions, calling it the *flocculation gradient technique*. Backscattering can be used to measure the degree of pigment dispersion in both liquid coatings and dry films. The extent of scattering of 2500 nm infrared radiation by a film as a function of film thickness was measured. (Particles, generally, scatter longer wavelength radiation more effectively than they do visible light.) There must be a significant difference in the refractive index at 2500 nm between the pigment and the vehicle. A plot of backscatter against film thickness gives a straight line whose gradient increases with increasing flocculation. Dasgupta (1991) and Hall et al. (1989) provide examples of the use of infrared backscattering measurements.

A widely used method of testing for *fineness of grind* in the coatings industry is a drawdown gauge, commonly the *Hegman gauge*. Other similar gauges are available. A sketch of one such gauge is shown in Figure 21.7 (Patton, 1979). A sample of the dispersion is placed on the steel block before the zero reading and drawn down by a steel bar scraper. One then lifts the block up and quickly looks across the drawdown sample to see at which graduation one can start to see particles projecting or streaks caused by particles being dragged along. The lore is the higher the scale reading, the "better" the dispersion.

Unfortunately, the device is not capable of measuring degree of dispersion. A major objective is to determine whether the dispersion is flocculated. However, the gauge cannot detect flocculation at all, since the drawdown step breaks up any floccules. Furthermore, the particle sizes of properly dispersed pigments are small compared with the depth of the groove on the gauge. The gauge shown in Figure 21.7 ranges from 4 mil (100 μm) to 0 in steps of 0.5 mil (12.5 μm). TiO_2 pigment particles have an average size of about 0.23 μm, about two orders of magnitude smaller than the graduation steps of the gauge. Even aggregates of 10^4–10^5 particles could escape detection. Many color pigment particles are even smaller and carbon black particles can be as small as 5 nm. Some inert pigment particles are larger but are still an order of magnitude smaller than the groove depths. Obviously, the gauge cannot test whether or not all or most of the particles are less than the groove depths. Blakely (1972) showed that in TiO_2 dispersions only approximately 0.1% of the total pigmentation of a coating was responsible for an unacceptable fineness of grind rating.

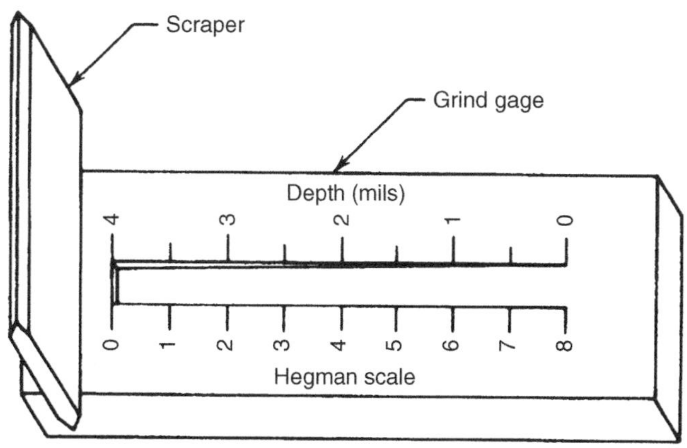

Figure 21.7 Sketch of a grind gauge and scraper for measuring "fineness of grind" of pigment dispersions. Source: Patton (1979). Reproduced with permission of John Wiley & Sons.

Why are such devices used? They do give some idea as to whether large aggregates are broken up, as well as whether dirt particles are present. The determinations are fast, taking about a half minute. A tinting strength determination by an experienced person requires 2 or 3 min, but after spending this time, one gains an assessment of the dispersion instead of a close to meaningless number. In the nineteenth century, when very coarse, difficult-to-separate pigments were common, the gauge might have had some value. There is no excuse for using it in the twenty-first century.

GENERAL REFERENCES

Parfitt, G. D., *Dispersions of Powders in Liquids*, 3rd ed., Applied Science Publishers, London, **1981**.

Patton, T. C., *Paint Flow and Pigment Dispersion*, 2nd ed., Wiley-Interscience, New York, **1979a**.

Winkler, J., *Dispersing Pigments and Fillers*, Vincentz Network, Hannover, **2012**.

REFERENCES

Auschra, C., et al., *Prog. Org. Coat.*, **2002**, *45*, 83.

Balfour, J. G.; Hird, M. J., *J. Oil Colour Chem. Assoc.*, **1975**, *58*, 331.

Banash, M. A., Croll, S. G., *Prog. Org. Coat.*, **1999**, *35*, 32.

Bieleman, J., *J. Oil Colour Chem. Assoc.*, **2004**, *87*(4), 173.

Blakely, R. R., *Proceedings of FATIPEC Congress*, **1972**, p 187.

Brisson, A.; Haber, A., *J. Coat. Technol.*, **1991**, *63*(794), 59.

Brisson, A., et al., *J. Coat. Technol.*, **1991**, *63*(801), 111.

Bugnon, P., *Prog. Org. Coat.*, **1996**, *29*, 39.

Centeno, R. C.; Perez, E.; Goicochea, A. G., *J. Coat. Technol. Res.*, **2014**, *11*(6), 1023–1031.

Clayton, J., *Surf. Coat. Int.*, **1997**, *94*, 414.

Creutz, S.; Jerome, R., *Prog. Org. Coat.*, **2000**, *40*, 21.

Dasgupta, S., *Prog. Org. Coat.*, **1991**, *19*, 123.

Dulog, L.; Schnitz, O., *Proceedings of FATIPEC Congress*, **1984**, Vol. *II*, p 409.

Farrokhpay, S.; Fornasiero, D.; Morris, G. E.; Self, P., *J. Coat. Technol. Res.*, **2006**, *3*(4), 275.

Hall, J. E., et al., *J. Coat. Technol.*, **1989**, *61*(770), 73.

Hamann, K.; Laible, R., *Proceedings of FATIPEC Congress*, **1978**, p 17.

Hay, T. K., *J. Paint Technol.*, **1974**, *46*(591), 44.

Hill, L. W.; Wicks, Z. W., Jr., *Prog. Org. Coat.*, **1982**, *10*, 55.

Jaffe, E. E., et al., *J. Coat. Technol.*, **1994**, *66*(832), 47.

Jakubauskas, H. L., *J. Coat. Technol.*, **1986**, *58*(736), 71.

Johnson, M. W., European patent EP2057234 B1 (**2014**).

Lara, J.; Schreiber, H. P., *J. Coat. Technol.*, **1991**, *63*(801), 81.

Losoi, T., *J. Coat. Technol.*, **1989**, *61*(776), 57.

McKay, R. B., *Proceedings of the International Conference Organic Coatings Science and Technology*, Athens, **1980**, p 499.

Miranda, S., *Powder and Bulk Engineering*, **2011**, November, from https://www.powderbulk.com (accessed April 18, 2017).

Morrison, W. H., Jr., *J. Coat. Technol.*, **1985**, *57*(721), 55.

Nguyen, D. T., *J. Coat. Technol. Res.*, **2007**, *4*(3), 2007.

Paterson, J., *Paint Coat. Ind.*, **2002**, February 58, from http://www.pcimag.com (accessed April 21, 2017).

Patton, T. C., *Paint Flow and Pigment Dispersion*, 2nd ed., Wiley-Interscience, New York, **1979b**.

Peck, K. M., *CoatingsTech*, **2014**, *11*(3), 50–57.

Perrier, S., *Surf. Coat. Int. Part B Coat. Trans.*, **2004**, *87*(B4), 235.

Rehacek, K., *Ind. Eng. Chem. Prod. Res. Dev.*, **1976**, *15*, 75.

Rosano, W., *Proceedings of the American Coatings Conference*, Atlanta, GA, April 7–9, **2014**.

Saarnak, A., *J. Oil Colour Chem. Assoc.*, **1979**, *62*, 455.

Saindane, P.; Jagtap, R. N., *Prog. Org. Coat.*, **2015**, *79*, 106–114.

Sastry, N. V.; Thakor, R. R., *J. Coat. Technol. Res.*, **2009**, *6*(1), 11.

Schofield, J. D., *J. Oil Colour Chem. Assoc.*, **1991**, *74*, 204.

Schroeder, J., *Prog. Org. Coat.*, **1988**, *16*, 3.

Silber, St., et al., *Prog. Org. Coat.*, **2002**, *45*, 259.

Smith, R. E., *J. Coat. Technol.*, **1988**, *60*(761), 61.

Van Eecke, M. C.; Piens, M., *Prog. Org. Coat.*, **2000**, *40*, 285

Van, S. T., et al., *J. Coat. Technol.*, **2001**, *73*(923), 61.

Visscher, K. B.; McIntyre, P. F., US patent 6,599,973 (**2003**).

Way, H. W., *J. Coat. Technol. Res.*, **2004**, *1*(1), 54.

White, D., et al., US patent 6,462,125 (**2002**).

Wicks, Z. W., Jr., et al., *Organic Coatings Science and Technology*, 3rd ed., John Wiley & Sons, Inc., New York, **2007**, pp 437–439.

Wilker, G., *Proceedings of the International Waterborne, High-Solids, and Powder Coatings Symposium*, New Orleans, LA, **1999**, pp 360–372.

Winkler, J.; Dulog, L., *J. Coat. Technol.*, **1987**, *59*(754), 55.

Winkler, J., et al., *J. Coat. Technol.*, **1987**, *59*(754), 35, 45.

Chapter 22

Effect of Pigments on Coating Properties

While color and hiding are major reasons for including pigments in coatings, there are many other effects that result from pigmentation. Here we address the effects of pigments on coating film properties. Most of these effects are related to the pigment volume concentration (PVC) and the critical pigment volume concentration (CPVC). Accordingly, we begin with these concepts. Extensive reviews of these concepts are available (Perera, 2004; Lestarquit, 2016). The latter review extends earlier work and presents a detailed analysis of "pigment/binder geometry" and its consequences for film properties. Pigmentation effects on the viscosity of liquid coatings are discussed in Sections 3.5 and 25.1.2.

22.1 PVC AND CPVC

Traditionally, coatings formulators have worked with weight relationships, but volume relationships are generally of more fundamental importance and practical significance. While there were a few previous isolated examples recognizing the importance of volume considerations in coatings, credit for full realization of this importance belongs to Asbeck and Van Loo (1949). They considered a series of performance variables as a function of PVC, that is, the volume percent of pigment in a *dry* film. (The term PVC should never be used to specify the volume of pigment in a wet coating. This has been done occasionally in the literature and has been responsible for serious misinterpretations.) PVC is expressed as a percentage value, or, less often, as pigment volume fraction, Φ_{pig} or V_{pig}.

Asbeck and Van Loo observed that, as PVC is increased in a series of formulations, many properties of films change abruptly at a particular value of PVC. They designated the PVC at which these changes occurred as the CPVC. They postulated that CPVC corresponds to the PVC where there is just sufficient binder to provide a completely adsorbed layer on the pigment surfaces and to fill all of the interstices between the particles in a close-packed system. Below CPVC, the pigment particles are not close-packed and binder occupies the "excess" volume in the film. Above CPVC, the pigment particles are closely packed, and there is insufficient binder to occupy all of the volume between the particles, resulting in voids in the film. Slightly above CPVC, the voids are air bubbles in the film, but as PVC increases further, the voids interconnect and film porosity increases sharply.

The distribution of pigment particles in the dry film is not uniform. (A random distribution of particles is not uniform.) Some parts of the film contain more than the average concentration of pigment. As a result, scattered voids will form in a film as PVC approaches CPVC (Bierwagen et al., 1999). Furthermore, such effects as flocculation and flow within the drying paint can result in nonrandom distributions of pigment that increase the probability of voids below CPVC. Different methods of measuring CPVC (Section 22.1.2) often give differing values.

22.1.1 Factors Controlling CPVC

There are large variations in CPVC, depending on the pigment or pigment combination in a coating. Pigment dispersion affects CPVC; coatings in which the pigment is flocculated exhibit lower CPVC than corresponding coatings with nonflocculated pigment. Coatings with flocculated pigment clusters are more likely to have regions where there are local high concentrations of pigment. Solvent containing resin can be trapped inside clusters of pigment particles. When the coating dries, the solvent diffuses out of the resin solution trapped in the floccules, leaving behind insufficient binder to fill the spaces. Asbeck

Organic Coatings: Science and Technology, Fourth Edition. Frank N. Jones, Mark E. Nichols, and Socrates Peter Pappas.

(1992) reported that CPVC decreased from 43 to 28% with increasing flocculation in one example. Asbeck suggests using the term *UCPVC*, *ultimate CPVC*, to designate CPVC with a nonflocculated pigment combination. It seems to us, better to recognize that flocculation is one of many variables that affect CPVC, especially since it may be difficult to determine UCPVC experimentally.

Pigment particle size is also a factor. Other factors being equal, smaller particles give lower the CPVC. A layer of binder can be adsorbed on the surface of the pigment strongly enough to increase the effective volume of the pigment as the coating dries. Because smaller pigment particles have a higher volume fraction of adsorbed binder, their effective volume during film formation is larger, and there is less room in the film for pigment particles. Bierwagen et al. (1999) propose a method for correcting CPVC calculations for different thickness of adsorbed layers based on the pigments' OA values. This method has proven reasonably successful. However, during film formation, some of the adsorbed material may be displaced from the pigment surface or incorporated into the binder phase. No method for direct measurement of the adsorbed binder thickness in dry films seems to be available. High molecular weight fractions of the binder probably adsorb preferentially on the pigment surfaces.

While small particles reduce CPVC, broad particle size distributions usually increase it. As discussed in Section 3.5, broader particle size distribution of spherical, dispersed-phase systems increases packing factor.

In low gloss coatings, the least expensive component of the dry film is inert pigment; to minimize cost, it is desirable to maximize inert pigment content. In reformulating to increase the level of inert pigments, the PVC-to-CPVC ratio should remain constant to minimize effects on film properties, as explained in Section 22.2.1. Other factors being equal, the lowest cost coatings, therefore, are those with the highest CPVC. Accordingly, cost is reduced by minimizing the amount of very fine particle size pigment, since CPVC decreases with decreasing particle size, and, at the same time, by maximizing particle size distribution, which increases CPVC. Compromises are needed, but broad particle size distribution is often advantageous.

CPVC of coatings with different pigments varies over a wide range, at least from 18 to 68%. One might expect CPVC to depend on binder composition. Since the composition affects the thickness of adsorbed layers, thicker adsorbed layers might be expected to lead to lower CPVC. Actually, CPVC with a given pigment or pigment combination is often essentially independent of the binder composition (except in latex paints, which are discussed in Section 22.2). Possibly, shrinkage forces encountered when films are formed press the particles against each other to the extent that only some minimal layer of binder is left between particles, regardless of the original adsorbed layer thickness. Possibly, the differences are too small to detect by the relatively imprecise methods for determining CPVC.

22.1.2 Determination of CPVC

CPVC has been determined by many different procedures (Bierwagen and Hay, 1975; Braunshausen et al., 1992; Wang et al., 2014). In many cases, the precision, that is, reproducibility of the value, is relatively poor; and the accuracy may be indeterminate. A further complication is that there is no single true value, because CPVC depends on the method of measurement. Bierwagen (1992) and Fishman et al. (1993) emphasized that in the formation of films with PVC at or near CPVC, local fluctuations within a film of volume fractions of binder and pigment are possible. Thus, while the average PVC might be less than CPVC, there could be parts of a film with PVC>CPVC. In view of these uncertainties, one must be careful in assessing the importance of small differences in PVC and CPVC values.

Many changes in film properties have been used as a means of determining CPVC; tinting strength is one of the most widely used. A series of white paints with increasing PVC are prepared and tinted with the same ratio of color to white pigment. Above CPVC, the white tinting strength of the coating increases due to the "white" air bubbles above CPVC. It is suggested that the values of CPVC obtained by this and other optical methods are somewhat low, due to nonuniform distribution of pigment, resulting in parts of a film being above CPVC, while others are below (Fishman et al., 1993). The technique is most easily applicable to white paints, but it can also be applied to colored paints.

Another method is to measure film density as a function of PVC. Since the density of most pigments is higher than those of binders and the density of air is lower than those of binders, density maximizes at CPVC. Owing to nonuniform distribution of pigment in the films, Fishman et al. (1993) suggested that the density method tends to give high CPVC values. The CPVC can be determined by filtering a coating and measuring the volume of the pigment filter cake. Asbeck (1992) recommends a specially designed filter that he calls a *CPVC cell*.

Wang et al. (2014) list numerous methods for determining CPVC and introduce a new one. Coating specimens with varying PVC are stained with a fluorescent dye, sliced at a shallow angle with a microtome, and examined with a fluorescence microscope. Fluorescent dye in pores (if there are any) in the coating are readily visible and can be analyzed by image software that measures the area of fluorescence. As predicted, a gradual onset of pore formation below CVPC was observed.

The CPVC for a pigment or pigment combination can be calculated from oil absorbtion (OA) values (Section 21.2.2) using Eq. 22.1. This equation is only valid if the OA value is based on a nonflocculated dispersion or is measured at a sufficiently high shear rate that any floccules are separated. The definitions of both OA and CPVC are based on close-packed systems with just sufficient binder to adsorb on the pigment surfaces and fill the interstices between the pigment particles. OA is expressed as grams of linseed oil per 100 g pigment; CPVC is expressed as milliliters of pigment per 100 ml of film; ρ is the density of the pigment(s); and 93.5 is 100 times the density of linseed oil, as both OA and CPVC are usually expressed as percentages.

$$\text{CPVC} = \frac{1}{\left(1 + \frac{(\text{OA})(\rho)}{93.5}\right)} \tag{22.1}$$

The significance of the interrelationship depends on the observation that OA and CPVC are approximately independent of the binder, provided the pigment particles are not flocculated. Since the accuracy of calculated CPVC depends on the accuracy of OA determinations, OA values determined by a mixing rheometer, such as a Brabender plastometer (Section 21.2.2), are preferable to values determined by the spatula rub up method (Hay, 1974). Huisman (1984) reported the effects of variation in procedures for determination of OA, including the use of other liquids besides linseed oil.

Since OA values are not always accurate, Asbeck (1992) recommended against using such calculations, but many workers have found the calculations useful. Although the CPVC for combinations of pigments can be calculated from OA values of the individual pigments, such calculations of are problematic. A reason is that differences in particle size distribution with pigment combinations affect the packing factor. Preferably, OA values for each combination of pigments can be determined experimentally. Nonetheless, a variety of equations has been developed to calculate CPVC from data on individual pigments (Braunshausen et al., 1992). The most successful equations use OA values, densities, and average particle sizes of the individual pigments (Bierwagen, 1972; Hegedus and Eng, 1988). The equations assume that the particles are spheres, a fair assumption for some pigments. Calculated values correspond reasonably well to experimentally determined CPVC values. Combinations of spherical and platelet pigments present additional difficulties (Lestarquit, 2016).

22.1.3 CPVC of Latex Coatings: LCPVC

There has been controversy about the applicability of the CPVC concept to latex paints. The current consensus is that the concept of CPVC is useful in latex paints but is affected by the latex as well as by latex coalescing agents.

In a study of the effect of PVC on hiding of latex paints, Anwari et al. (1990) concluded that CPVC is lower in latex paint than in solvent-based paint made with the same pigment composition. Patton recommended that the term *latex CPVC*, LCPVC, be used to distinguish from CPVC of solventborne paints. Although CPVC is approximately independent of the binder in solventborne paints, LCPVC varies with the latex and some other components of latex paints. LCPVC increases (1) as the particle size of the latex decreases, (2) as the T_g of the latex polymer decreases and (3) by addition of a coalescing agent. Since LCPVC is smaller than CPVC, the ratio of volume percent of binder in the film of a solvent-based paint, V_s, to that of a latex-based paint, V_l, is always less than 1 with the same pigment combination. This ratio, shown in Eq. 22.2, has been called the *binder index e* (Patton, 1979), also sometimes abbreviated as BI:

$$\text{BI} = e = \frac{V_s}{V_l} \tag{22.2}$$

The binder index of lattices can range from 0.4 to 0.9 (Lestarquit, 2016). Generally, this ratio is almost independent of the pigment combination. If one knows the binder index for a latex, one can estimate the LCPVC for paints made with that latex, which contains any pigment, using its CPVC value, calculated from OA. If the latex contains a mixture of pigments, CPVC can be calculated from OA, density, and particle size. As a caution, Bierwagen and Rich (1983) emphasize that experimental errors in determining CPVC of latex paints are even greater than with solventborne paints.

The difference between CPVCs of latex- and solvent-based paints can be understood from the difference in film formation. To illustrate, let us consider a highly idealized comparison. Let us assume that all the pigment particles in the solvent-based paint are spheres with the same diameter and that the CPVC is 50%. In the latex paint, we will use the same pigment together with spherical latex particles that have the same diameter as the pigment particles. In the solvent-based paint, there is a layer of resin swollen with solvent on the surface of all the pigment particles. Ideally, when a solventborne paint with PVC = CPVC is applied, the solvent evaporates and the resin-coated pigment particles arrange themselves in a close-packed order with binder filling in all the spaces between the pigment particles. At the same ratio of pigment to binder as in the solvent-based paint, the idealized latex paint would contain equal numbers of latex particles and pigment particles. When we apply a layer of the latex paint, the water evaporates and we get a closely packed system of particles. However, some of the particles are pigment and some are latex. There would not be a uniform arrangement of alternating latex and pigment particles in a three-dimensional lattice; rather, there would be a statistical distribution of particles. In some areas there would be clusters of pigment particles, and in other areas there would be clusters of latex particles. As the film forms, the latex particles coalesce, flowing to some extent around the pigment particles. However, the viscosity of the coalescing binder is high, and it is difficult for the polymer to penetrate into the center of clusters of pigment particles. As film formation proceeds, the water left inside the pigment clusters diffuses out of the film, leaving behind voids. Although the PVC of 50% in the solvent-based paint equaled CPVC, the same PVC of 50% in the latex paint results in a film with PVC>LCPVC.

The probability of having clusters of pigment particles can be decreased by increasing the number of latex particles until void-free films are obtained, but this necessarily reduces PVC. If the T_g of the polymer is lower, the distance that the latex can penetrate into clusters of pigment particles will usually increase. Thus, the LCPVC of paint made with lower T_g latex is higher, although still lower than the CPVC of a solvent-based paint. Analogously, reduction of viscosity of the polymer by coalescing agent increases LCPVC. Hypothetically, LCPVC should also increase if the temperature during film formation is elevated. Furthermore, it can be speculated that LCPVC should depend on time. Although the viscosity of the polymer is high and inhibits flow between the particles, the viscosity is not infinite, and perhaps in time, voids would fill.

If a smaller particle size latex is used with the same ratio of pigment to binder volume, the number of latex particles in our idealized paint would be larger than the number of pigment particles. Now the probability of clusters of pigment particles forming would be reduced. As a result, one would expect, as is found experimentally, that LCPVC increases with decreasing particle size of the latex. Particle size distribution of a latex affects its packing factor (Hoy and Peterson, 1992), which would presumably affect LCPVC. del Rio and Rudin (1996) reported a quantitative study of the effect of latex particle size on LCPVC using a series of monodisperse vinyl acetate/butyl acrylate latexes with TiO_2 pigment, demonstrating that LCPVC depends on the ratio of the number of latex and pigment particles as well as the ratio of their diameters. These authors also provided a review of the literature.

Not only does larger particle size latex decrease LCPVC, flocculation of the latex particles would be expected to decrease LCPVC. If one formulates a latex paint so that the PVC is slightly less than LCPVC with an unflocculated latex, but the latex flocculates, the PVC would be greater than LCPVC. A simulation program using particle size distributions of latex and pigment and a measure of the deformability of latex particles has been developed to predict the CPVC of simple latex paints (Nolan and Kavanaugh, 1995).

Floyd and Holsworth (1992) proposed a different point of view about LCPVC. They suggested that a phase inversion occurs at LCPVC in which air becomes the external phase and the pigment/latex phase becomes the internal phase. This proposal is consistent with the finding that latex films are somewhat porous at PVC values considerably below LCPVC. The porosity is caused by the clustering of pigment particles and of latex particles as described earlier in this section. When latex coatings dry there will be air pockets in the film; and, as the PVC increases, more and more of these air pockets join to make continuous passages of air through the film. As PVC reaches LCPVC, the rate of the increase of pore formation increases rapidly with increased PVC. The result of this behavior is that there is not an abrupt change of all properties at LCPVC. For example, scrub resistance shows a broad transition retaining significant scrub resistance above LCPVC. On the other hand, opacity increases quite abruptly at LCPVC. Since porosity increases at PVC well below LCPVC (at PVCs of the order of 20–30 PVC), barrier-related properties such as permeability and corrosion resistance of latex paints begin to change at these relatively low PVCs.

22.2 RELATIONSHIPS BETWEEN FILM PROPERTIES AND PVC

Changes in volume relationships almost always affect physical properties of pigmented films. Thus, there is a major advantage in formulating coatings on the basis of volume rather than weight relationships. For any particular application, there is a ratio of PVC to CPVC that is most appropriate for the combination of properties needed for that application. This ratio is often designated by the Greek letter Λ. Once this ratio has been established, changes in pigment combinations for that application should usually be made without changing Λ. This important concept is developed in detail by Bierwagen and Hay (1975). Bierwagen (1992) also emphasizes that one must exercise particular care when formulating with PVC near CPVC; in this region, relatively small changes in pigment ratios, packing, or flocculation can substantially affect film formation and film properties.

While accurate determinations of CPVC are desirable, even reasonable estimates can be useful, since they permit one to focus a series of experiments within a range of the desired PVC to CPVC ratio. Furthermore, the decision as to the proper pigment loading of a particular formulation should be based on actual experimental data, not on the theoretically best PVC value. Even without CPVC data, the concept is of value to the formulator. It is critical to recognize that performance properties vary with volume, not with weight relationships, and that CPVC increases with increasing particle size distribution. Qualitative use of OA and density values for individual pigments, together with the recognition that CPVC increases when mixtures of pigments with different particle size distributions are used, permit one to start formulating in a reasonable range of compositions. On the other hand, using weight relationships, one is working blindly.

22.2.1 Mechanical Properties

In the majority of cases, T_g of films increases with PVC; however, there are examples in which T_g decreases or is unchanged. It is common for there to be two T_gs, one resulting from the continuous phase of binder and the other from the adsorbed layer of resin on the pigment surfaces.

Tan δ is generally lower and broader, and the peak is shifted to a higher temperature for highly pigmented coatings as compared with the tan δ of the unpigmented binder. The broadening and temperature shift increase with increasing PVC/CPVC and are more pronounced with lamellar pigments in comparison with spherical pigments. Figure 22.1 schematically represents tan δ as a function of temperature for a pigmented coating as compared with the unpigmented binder (Perera, 2004).

Increase in pigmentation usually increases the elastic modulus especially in the glass transition and rubbery regions as seen in Figure 22.2 (Perera, 2004). This results from the replacement of binder with rigid pigment. In Figure 22.2, CPVC is located between PVC 45 and 55% as indicated by the lower E' in the glassy region of the coatings pigmented at PVC of 55 and 60%.

Tensile strength of films generally increases with PVC to a maximum near CPVC but then decreases above CPVC. Below CPVC, the pigment particles serve as reinforcing particles and increase the strength. It can be considered that polymer molecules adsorb on the surface of multiple pigment particles, providing the equivalent of cross-links.

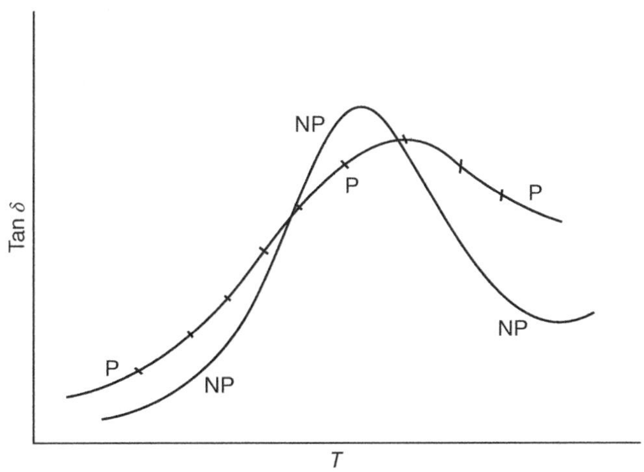

Figure 22.1 Schematic description of tan δ as a function of T for a pigmented coatings (P) and a non-pigmented (NP) coatings. Source: Perera (2004). Reproduced with permission of Elsevier.

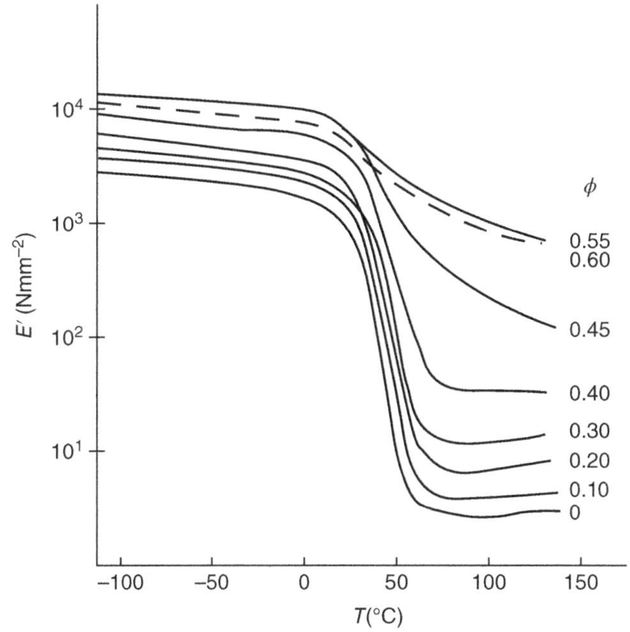

Figure 22.2 Storage elastic modulus (E') as a function of temperature (T) for a polyacrylate coating with different PVC (ϕ) of TiO_2. Source: Perera (2004). Reproduced with permission of Elsevier.

Accordingly, more force is required to break this physical network as pigment level increases. However, above CPVC, air voids weaken the film; abrasion and scrub resistance of films drop above CPVC.

The relative tensile strength of an acrylic binder with a series of pigments is shown in Figure 22.3. As expected, tensile strength of films pigmented with TiO_2 and barytes increases with PVC and then drops off above CPVC. However, if the bonding between pigment particles and the resin is poor as with $CaCO_3$, tensile strength of the films drops steadily as PVC increases. With microtalc, the platelet structure of the pigment provides especially high reinforcement at low PVC.

Thermal expansion coefficient is affected by pigmentation. As can be seen in Figure 22.4, the thermal expansion coefficient of a TiO_2 pigmented epoxy coating decreases as the PVC increases.

Internal stress in dried coatings, in which $T_g > T$, increases with PVC up to CPVC and then decreases. Accordingly, the probability of delamination or film cracking increases for such a system as a result of the stress.

22.2.1.1 Nano-pigment Effects

Nano-pigments are introduced in Section 20.5; their effects on mechanical properties are briefly discussed in Section 4.3. Compared with the effects of conventional pigments, nano-pigments, especially nanofillers, have huge potential to improve important coating properties.

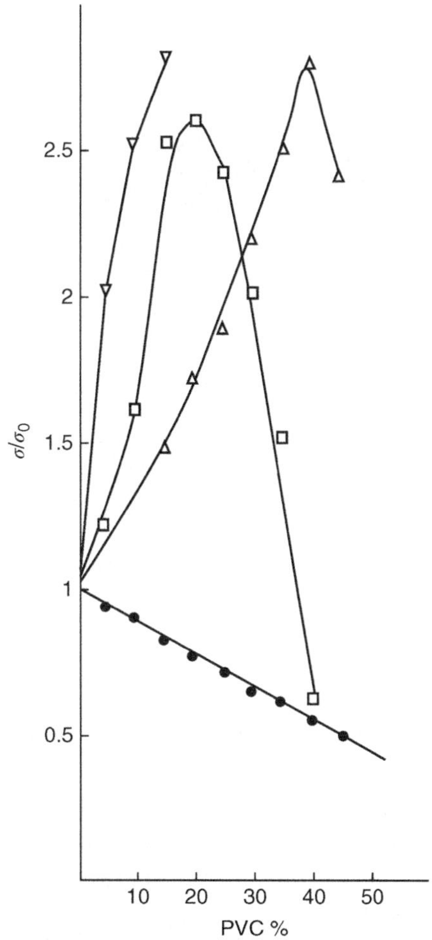

Figure 22.3 Relative tensile strengths (ratio of pigmented film strengths to strength of unpigmented film) dependence on PVC for an acrylic system containing (●) $CaCO_3$, (∇) microtalc, (Δ) TiO_2, and (□) barytes. Source: Perera (2004). Reproduced with permission of Elsevier.

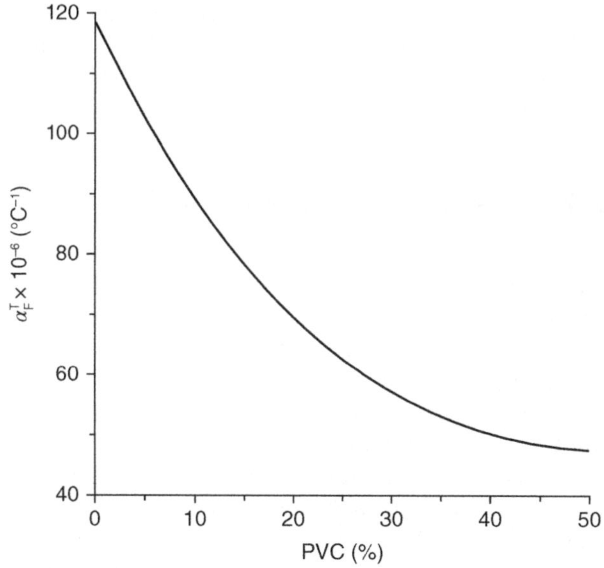

Figure 22.4 Thermal coefficient dependence on PVC at 21°C and 0% RH for TiO_2 pigmented epoxy coatings. Source: Perera (2004). Reproduced with permission of Elsevier.

Improvements in scratch, mar, abrasion, heat, radiation, block, and swelling resistance, decrease in water permeability, and increase in hardness, weatherability, modulus, and strain-to-failure are all possible with nano-pigments (Sung et al., 2008).

Some nano-pigments have been with us for a long time. Acrylic microgels could be considered nano-pigments; they have been used for decades for sag control of automotive topcoats. Nano-iron oxide reduces permeability of acrylic latex coatings, making them suitable for encapsulation of lead-containing paints. Nano-carbon black has a long history in jet black auto finishes. Nano-silver and nano-copper are well known to impart antibacterial properties to paints (Section 34.5).

The high surface areas of nanoparticles result in an increase in adsorption of resin on the pigment surfaces and, hence, a larger fraction of immobilized resin in the paint film. The immobilized resin can be detected by tapping atomic force microscopy (Gu et al., 2012). Often, the films have two T_gs detectable by differential scanning calorimetry. Based on an adsorbed layer of 10 nm, the fraction of resin adsorbed on 300 nm pigment particles has been calculated to be 3% and on 50 nm nano-particles to be 22% (Fernando, 2004). The percentage increases as nano-particles become even smaller. Silica and other nano-pigments with average diameters below 5 nm are available. If they are de-agglomerated and deflocculated, even a small weight fraction of such tiny particles is expected to substantially affect film properties.

Nano-pigments, however, have a strong tendency to flocculate and agglomerate (Croll and Lindsay, 2012). If there is substantial flocculation or agglomeration, the advantages of nano-pigments will not be realized. Various counter approaches have been utilized: surface treatments of the pigments (e.g., with selected siloxanes), encapsulation, *in situ* polymerization in the presence of the pigments, special pigment dispersants, ultrasonic dispersion (Perera, 2004), grafting polymers on the nanoparticle surface (Kumar and Jouault, 2013), and formation of nano-particles *in situ* with sol–gel technology (Section 16.3). For example, ball mills are used to disperse nano(aluminum silicate) in an acrylic resin with a commercial dispersing agent. The dispersion imparts mar resistance to clear coats (Vanier et al., 2005).

Nano-pigments are usually specified by composition and average particle size. Formulators should be aware that they may also vary in particle size distribution, shape, porosity, surface treatment, functionalization, and crystal form. Both anatase and rutile nano-TiO_2 are available, for example, nano-pigments are commercially available in dry form and also as dispersions in solvents, in resins, and in monomers for radiation curing.

Nano-pigment technology is well established in coatings (Fernando and Sung, 2009). They enhance scratch and mar resistance and perhaps weatherability of automotive clear coats. Similar enhancements of transparent coatings for plastics and wood are attainable. Nikolic et al. (2015) review published information about nanofillers in wood coatings, concluding that nano-sized TiO_2, ZnO, and CeO_2 are effective, long-lived UV absorbers. These authors also summarize research on mechanical property effects and on antibacterial and antifungal coatings (Section 34.5), which contain nanoparticles of Ag or Cu.

In addition to their established uses, nano-pigments show tantalizing promise for many other types of coatings, for example:

- Addition of just 1% of nanoparticles can substantially improve solvent and block resistance of waterborne 1K and 2K polyurethane wood coatings. Film formation of the 1K polyurethane was accelerated. The effect of the particles is said to be similar to that of a crosslinker (Burgard and Herold, 2014). The nanoparticles are probably surface-treated zinc oxide (Pilotek et al., 2009).
- Nano-ZnO is said to increase corrosion resistance of an electrodeposited polyurethane film by two orders of magnitude (Rashvand and Ranjbar, 2013).
- Nanocomposite uralkyd varnishes, containing 0.5 wt% of exfoliated montmorillonite clays, enhance the protection of wood (Kowalczyk, 2014).
- Epoxy/clay nanocomposite coatings significantly improved salt spray resistance (Tomic et al., 2013).
- TiO_2 nanoparticles can be surface-functionalized with reactive isocyanate groups (Li et al., 2012).
- Emulsion polymerization in the presence of 0.5–2.5% of 40-nm $BaSO_4$ nanoparticles affords films having well-dispersed nanoparticles and enhanced properties (Kulkarni et al., 2013).
- Polymer nanocomposites containing graphene (Kim et al., 2010) and carbon nanotubes (Moniruzzaman and Winey, 2006) have substantially altered properties. Only small amounts of the additives are needed if they are well dispersed within the polymer.
- A paste of CdS, CdSe, and TiO_2 nanoparticles was brushed onto conductive glass. After annealing, this assembly could photoconvert sunlight into electricity (Genovese et al., 2012). While this "solar paint" is far from a practical roof coating, it suggests a target for further research.

Of course, substantial improvements in one property may be accompanied by undesirable changes in other properties, so formulators must be judicious in the use of nanoparticle pigments and fillers. Deployment of nano-pigment technology is complex, and there are many unanswered questions. For example, will nano-pigment paint prove to have long-term package stability? This is especially a concern for waterborne paints containing nano-ZnO owing to the slight solubility ($0.005\,g\,l^{-1}$) of ZnO in water. How will nano-pigments behave during ultrafiltration of electrodeposition primers? Eventually such questions will be answered and the potential scope of nanocomposite coatings will be clarified. The field appears highly promising.

22.2.2 Effects of Film Porosity

Stain resistance decreases above CPVC, since staining liquids can penetrate into pores, leaving color behind that is difficult to remove. Porosity also affects other properties. If one applies a single layer of a coating, in which PVC > CPVC, to steel and exposes the panel to humidity, rapid rusting can occur, since the pores permit water and oxygen to get to the surface of the steel with little interference.

It is almost always desirable to formulate primers with a high PVC, since the rougher, low gloss surface gives better intercoat adhesion than a smooth, glossy surface. In most cases, primers for metals should have PVC that is high enough to roughen the surface and to minimize permeability of water and oxygen, but lower than CPVC, where permeability increases. Minimum permeability often occurs at roughly PVC/CPVC = 0.9.

In primers for nonmetals it is sometimes desirable to formulate the primer with PVC > CPVC. Adhesion of a top coat to such a primer is enhanced by mechanical interlocking resulting from penetration of vehicle from the top coat into pores of the primer. The primer PVC should be only enough higher than CPVC to assure good adhesion. If PVC is too high, too much of the top coat binder will be depleted by penetrating into the pores of the primer, thereby decreasing PVC of the top coat, affecting its properties. Such a primer is said to have poor *enamel hold out*.

Conventional zinc-rich primers (Section 33.1.2) require PVC > CPVC. The porosity permits water to enter the film establishing a conductive circuit with the steel surface. PVC may be less critical in zinc-rich primers incorporating conductive additives such as carbon black or polyaniline. Primers using 2K or moisture cure urethane coatings that have PVC > CPVC have the additional advantage that the porosity permits CO_2 to escape without causing blistering.

PVC also affects hiding; as pigmentation increases, hiding generally increases. Initially, hiding increases rapidly but then tends to level off. In the case of rutile TiO_2, hiding goes through a maximum, gradually decreases with further increase in PVC and then increases above CPVC. The decrease is caused by excessive crowding of the TiO_2 particles. The increase in hiding above CPVC results from air voids left in the film when PVC is above CPVC. Hiding by air voids is discussed in Section 20.1.2.

If one applies an alkyd-based coating or primer, in which PVC>CPVC, to a wood substrate, one is less likely to get blistering than with a similar coating, in which PVC<CPVC. When water gets into the wood behind the coating, it can escape through pores in an alkyd coating when PVC is above CPVC but not when PVC is below CPVC.

22.2.3 Effects on Curing and Film Formation

There are many examples of pigmentation effects on curing of thermosetting coatings. Titanium oxide coated with alumina inhibits the acid catalyzed curing of coatings, for example those which contain MF resins as the cross-linker. The basic alumina neutralizes the acid catalyst. Silica coated TiO_2 is acidic and does not inhibit the cure. In powder coatings, carboxy-functional resins and alumina-coated TiO_2 can interact, leading to higher yield values and higher viscosity to the detriments of leveling.

Curing of diallyl phthalate prepolymers, pigmented with silica, progresses well when catalyzed by dicumyl peroxide. However, kaolinite pigment virtually eliminates the curing because the acidic groups present on the clay surface decompose the catalyst. As another example, liquid UV cure coatings pigmented with natural iron oxide gel prematurely in a matter of hours because of contaminants in the pigment. Synthetic iron oxide pigmented coatings can be used without difficulty.

REFERENCES

Anwari, F., et al., *J. Coat. Technol.*, **1990**, *62*(786), 43.
Asbeck, W. K., *J. Coat. Technol.*, **1992**, *64*(806), 47.
Asbeck, W. K.; Van Loo, M., *Ind. Eng. Chem.*, **1949**, *41*, 1470.
Bierwagen, G. P., *J. Paint Technol.*, **1972**, *44*(574), 46.
Bierwagen, G. P., *J. Coat. Technol.*, **1992**, *64*(806), 71.
Bierwagen, G. P.; Hay, T. K., *Prog. Org. Coat.*, **1975**, *3*, 281.
Bierwagen, G. P.; Rich, D. C., *Prog. Org. Coat.*, **1983**, *11*, 339.
Bierwagen, G. P., et al., *Prog. Org. Coat.*, **1999**, *35*, 1–9.
Braunshausen, R. W., Jr., et al., *J. Coat. Technol.*, **1992**, *64*(810), 51.
Burgard, D.; Herold, M., *Eur. Coat. J.*, **2014**, March, 24–28.
Croll, S.; Lindsay, M., *Int. J. Nanotechnol.*, **2012**, *9*(10/11/12), 982.
Fernando, R., *J. Coat. Technol. Res.*, **2004**, *1*(5), 32.
Fernando, R. H.; Sung, L.-P., Eds., *Nanotechnology Applications in Coatings*, ACS Symposium Series 1008, Oxford University Press, Oxford, **2009**.
Fishman, R. S., et al., *Prog. Org. Coat.*, **1993**, *21*, 387.
Floyd, F. L.; Holsworth, R. M., *J. Coat. Technol.*, **1992**, *64*(808), 63.
Genovese, M. P.; Lightcap, I. V.; Kamet, P. V., *ACS Nano*, **2012**, *6*(1), 865–872.
Gu, X.; Chen, G., et al., *JCT Res.*, **2012**, *9*(3), 251.
Hay, T. K., *J. Paint Technol.*, **1974**, *46*(591), 44.
Hegedus, C. R.; Eng, A. T., *J. Coat Technol.*, **1988**, *60*(767), 77.
Hoy, K. L.; Peterson, R. H., *J. Coat. Technol.*, **1992**, *64*(806), 59.
Huisman, H. F., *J. Coat. Technol.*, **1984**, *44*(574), 46.
Kim, H.; Abdata, A.; Macosco, C., *Macromolecules*, **2010**, *43*(16), 6515–6530.
Kowalczyk, K., *JCT Res.*, **2014**, *11*(3), 421–430.
Kulkarni, R. D.; Ghosh, N., et al., *Polym. Compos.*, **2013**, *34*(10), 1670.
Kumar, S. K.; Jouault, N., *Macromolecules*, **2013**, *46*(9), 3199–3214.
Lestarquit, B., *Prog. Org. Coat.*, **2016**, *90*, 200–221.
Li, D.; Liao, B., et al., *Mater. Chem. Phys.*, **2012**, *135*(2.3), 1104.
Moniruzzaman, M.; Winey, K. J., *Macromolecules*, **2006**, *39*(16), 5194–5205.
Nikolic, M.; Lawther, J. M.; Sanadi, A. R., *J. Coat. Technol. Res.*, **2015**, *12*(3), 445–461.
Nolan, G. T.; Kavanaugh, P. E., *J. Coat. Technol.*, **1995**, *67*(850), 37.
Patton, T. C., *Paint Flow and Pigment Dispersion*, 2nd ed., Wiley-Interscience, New York, **1979**, p 192.
Perera, D. Y., *Prog. Org. Coat.*, **2004**, *50*, 247.
Pilotek, S.; Burgard, D., et al., *Proceedings of the International Waterborne, High-Solids, and Powder Coatings Symposium*, New Orleans, LA, February 18–20, **2009**.
Rashvand, M.; Ranjbar, Z., *Prog. Org. Coat.*, **2013**, *76*, 1413–1417.
del Rio, G.; Rudin, A., *Prog. Org. Coat.*, **1996**, *28*, 259.
Sung, L.-P.; Comer, J., et al., *JCT Res.*, **2008**, *5*(4), 419.
Tomic, M. D.; Dunjic, B., et al., *Prog. Org. Coat.*, **2013**, *77*, 518–527.
Vanier, N. R., et al., US patent 6,916,368 (**2005**).
Wang, J., et al., *Prog. Org. Coat.*, **2014**, *77*, 2147–2154.

Chapter 23

Application Methods

Many factors affect the choice of method to be used to apply a coating, including capital costs, operating costs, film thickness, appearance requirements, and structure of the object to be coated. Reduction of VOC emissions and improved efficiency are major driving forces for the improvement of application methods and equipment. Formulation of coatings and development of application methods are strongly linked. Application of electrodeposition, powder, and radiation-curable coatings is discussed in Chapters 27, 28, and 29, respectively. Development of these technologies required close collaboration among coatings scientists, formulators, and application equipment engineers. Further information and discussion of additional application methods can be found in the general references at the end of the chapter.

23.1 BRUSHES, PADS, AND HAND ROLLERS

Brushes, pads, and hand rollers are frequently used for the application of architectural paints. Although the same paints can usually be applied by a spray gun, few do-it-yourselfers use spray guns; on the other hand, professional painters use spray guns whenever possible to save time.

23.1.1 Brush and Pad Application

A variety of brushes is available: narrow and wide, long handled and short handled, nylon, polyester, and natural bristle, such as hog or oxtail bristles (Levinson, 1988). Natural bristles work well for solventborne paints, but not for waterborne. Nylon bristles are appropriate for waterborne paints but are swollen by some solvents. Polyester bristle brushes can be used with either. High quality brushes can be serviceable for years if they are properly cleaned after each use. Brushes have in common a large number of bristles that hold paint in the spaces between the bristles. When the paint is applied, pressure forces paint out from between the bristles. The forward motion of the brush splits the layer of paint so that part is applied to the surface and part remains on the brush. Open-cell polyurethane foam "brushes" are another option.

Characteristics of paint viscosity are critical when using brushes. Pickup of paint on the brush is controlled by paint viscosity at a relatively low shear rate, around 15–30 s^{-1}, corresponding to the shear rate of dipping and removing a brush from a paint can. If the low shear viscosity is too high, too much paint is brought out of the container with the brush; if it is too low, too little paint is on the brush. Ease of brushing requires a low viscosity. The viscosity at relatively high shear rate controls the ease of brushing; excessive viscosity leads to high "brush drag," which can be quite tiring (Schoff, 1991). Film thickness increases with increasing viscosity at high shear rate. For most applications, a viscosity at high shear rate of 0.1–0.3 Pa·s is appropriate. (See Chapter 32 for further discussion.) Solventborne paints are formulated with solvents having relatively slow evaporation rates to slow the increase in viscosity of the paint on the brush.

When paint is applied by brush, the surface of the wet film has furrows called *brush marks*. These furrows do not result from the individual bristles of the brush, as is evident from comparing the size and number of brush marks with the bristles. When polyurethane foam "brushes" are used, brush marks still result, even though the brush has no bristles. Brush marks result from splitting the wet film between the brush and the substrate as the paint is applied. Whenever layers of liquids are split, the surfaces are initially irregular. These irregularities are made into lines by the linear

Organic Coatings: Science and Technology, Fourth Edition. Frank N. Jones, Mark E. Nichols, and Socrates Peter Pappas.

movement of the brush. It is desirable to formulate a coating such that the brush marks flow out before the film dries. Low viscosity promotes leveling but increases the probability of sagging (Sections 24.2 and 24.3). Thixotropic flow properties (Section 3.2) are generally desirable because they delay the increase in viscosity after brushing, permitting a compromise between leveling and sagging. Formulation problems are particularly challenging in the case of latex gloss paints (Section 32.3.2).

Pad applicators are also used in the do-it-yourself market (Levinson, 1988). The most common type of pad consists of a sheet of nylon pile fabric attached to a foam pad that is attached to a flat plastic plate with a handle. For low viscosity coatings like stains and varnishes, a lambs-wool pad is used. Pads have a number of advantages compared to brushes. Pads hold more paint than a similar width brush and can apply paint up to twice as fast. In general, pad application leaves a smoother layer than brush application. Extension handles can be used with pads, reducing the need for moving ladders. Pads, especially refills, are less expensive than brushes. On the other hand, pads require use of trays, which results in some paint loss and solvent evaporation. Cleanup of pads is more difficult than brushes.

23.1.2 Hand Roller Application

Hand rollers are the fastest method of hand application for paints on walls and ceilings. A variety of rollers and roll coverings are available (Levinson, 1988). Rollers having built-in wells minimize the need for dipping in the tray; there are also power-filled rollers.

Viscosity requirements are similar with those for brush application. Wet film splitting also occurs when paint is applied by a roller. As the roller moves, the film is stretched; flow is in an extensional mode (Section 3.6.3) within the film. Due to imbalances of pressures, the film breaks at different times, leaving a ribbed surface. Some of the film extends into fibers still farther from the nip. As the roller moves, fibers are drawn longer and longer and eventually break. After breaking, the ends of the fibers are pulled back to the roller and the applied paint by surface tension to minimize the surface area of the applied coating, but if there is not sufficient time at low viscosity for leveling to occur, track marks are left in the coating. A further complication arises with longer fibers that may break in two places; as a result, droplets of loose paint fly off to land on the painter or the floor. This is called *spattering*. If spattering could be eliminated, painting speed could be increased, and time spent on masking and laying drop cloths could be reduced. However, all current formulations spatter, some much more than others. The phenomenon is not completely understood. Glass (1978), Fernando and Glass (1988), and Fernando et al. (2000) have pointed out that as the fibers are drawn out, extensional (elongational) flow, rather than shear flow, is involved (Section 3.6.3). It follows that extensional viscosity, rather than shear viscosity, affects development of fibers from roller application and, in turn, the roughness of the film and the degree of spattering. Spattering can occur with any kind of paint but is particularly difficult to minimize in latex paints (Section 32.2).

23.2 SPRAY APPLICATION

Spraying is a common method for applying paints and coatings in architectural and especially in industrial and maintenance applications. For skilled painters, spraying is much faster than application by brush or hand roller. Spraying is used on flat surfaces but is particularly applicable to coating irregularly shaped articles. Many different kinds of equipment are used for spraying; all atomize the liquid coating into droplets. Droplet size depends on the type of spray gun and coating; variables include air and fluid pressure, fluid flow, surface tension, viscosity, and, in the case of electrostatic application, voltage. The choice of spray system is affected by capital cost considerations, efficiency of paint utilization, labor costs, size, and shape of objects to be coated, among other variables. Coating formulations are adjusted for the particular spray equipment and conditions.

The principal disadvantage is the inefficiency of application, since only a fraction of the spray particles are deposited on the object being sprayed. One must mask work areas where coating is not desired, and there is a likelihood of contaminating the area with spray dust—small airborne particles of dry paint that do not reach the target. Some of the droplets approach the surface and *bounce back*, (an unfortunate term as they do not actually strike the surface) carried by eddy currents of air. The higher the pressure, the higher the forward velocity of the air and, as a result, the greater the percentage of bounce back. Some droplets miss the object being coated, called *overspray*. Some droplets may fall out of the spray pattern under the force of gravity. *Fall out* is higher with large spray patterns and longer distances between the spray gun and the surface being sprayed. The sum of all this waste determines *transfer efficiency*, defined as the percent of coating solids leaving the gun that is actually deposited on the coated object.

Spray dust can cause many problems. If spray dust lands on the surface of a wet coating, dirt contamination results. If the surface tension of the spray droplets is different than that of the wet coated surface, cratering (Section 24.4) may result. Minimizing spray dust by choice of application equipment, ventilation, and careful maintenance and cleaning of equipment is vital.

Transfer efficiency is an important cost factor, since high transfer efficiency corresponds to less coating used for the same surface area; cost and VOC emissions are reduced. Transfer efficiency is affected by many variables. The size and shape of the product being coated is a major variable. Transfer efficiency is low when spraying a chain link fence and high when spraying a large wall. Less obvious variables are conveyor line speed, localized airflows, and the way objects being coated are hung on a conveyor line. The spray method is a major factor in determining transfer efficiency, that is, whether the application is manual or automated; whether air, airless, or rotary spray guns are used; and whether an electrostatic system is used. For manual systems, the skill of the sprayers is very important. For automated systems, system design is critical. Some variables involved are distance between the gun and the surface, angle of the spray gun, stroke speed uniformity, extent and uniformity of overlapping, and precision of triggering.

To determine transfer efficiency of different spray methods, the EPA together with spray gun and coating manufacturers has developed a standard procedure for comparing *baseline transfer efficiency*, adopted by the ASTM as method D5009–96. Measurement of transfer efficiency is difficult because small changes in airflow can affect the results. Typical percentages are given in Table 23.1 (Adams, 1990); as noted previously in this section, actual efficiencies are affected by many variables besides the spray method.

Very reactive coatings with short pot lives, such as some two-package polyurethane and polyurea coatings, require dual feed spray guns. Most commonly, the two components are metered into a small, efficient mixing chamber just before they reach the spray gun orifice. The average dwell time in the mixing chamber is a fraction of a second, and the average residence time in the gun is not much longer. Frequent checking and maintenance are required to assure that the proper ratio of components is being applied. When spraying is interrupted, the equipment automatically flushes the mixing chamber and gun with solvent to prevent clogging. In other types of equipment, the components are sprayed from two orifices, in which case, mixing occurs after atomization so that the coating cannot gel in the equipment.

Table 23.1 Typical Baseline Transfer Efficiencies

Type of spray gun	Transfer efficiency (%)
Air	25
Airless	40
Air-assisted airless	50
High volume low pressure air	65
Electrostatic air	60–85
Electrostatic rotary	65–94

Source: Form Adams (1990).

23.2.1 Air Spray Guns

Air spray guns cause atomization of the coating by fine streams of compressed air. It is the oldest spray method and is still used. Figure 23.1 shows a cross section of an air spray gun. A stream of coating is driven through the nozzle orifice by relatively low pressure, 10–50 kPa (1.5–7 psi), or, in other types of guns, by the suction caused by the rapid airflow past the outside of the orifice. The stream of coating emerging from the orifice is atomized into small droplets by fine streams of compressed air at pressures of 250–500 kPa (35–70 psi). The degree of atomization is controlled by the following factors: (1) viscosity of the coating (the higher the viscosity at the high shear rate encountered going through the orifice, the larger the particle size), (2) air pressure (higher air pressure, smaller particle size), (3) diameter of the orifice (smaller orifice, smaller particles), (4) pressure forcing or pulling the coating through the orifice (higher pressure, smaller size), and (5) surface tension (lower surface tension, smaller particle size). The outer jets of compressed air shown in the figure adjust the shape of the stream of atomized particles emerging from the gun. If these jets were not present, a cross section of the stream would be roughly circular. Generally, an elliptical cross-sectional pattern permits more efficient application; the pattern is a flattened cone, often called a *fan*. Guns can be handheld or attached to robots.

Air spray guns are less expensive than other types of guns. Atomization can be finer than with other spray guns. The system is versatile, and any sprayable object can be sprayed with air spray. The level of control can be high if the operators are skilled. However, transfer efficiency is the lowest of all the spray methods (Table 23.1).

A substantial improvement in transfer efficiency of air spraying has been made by the use of *high volume, low pressure* (HVLP) air guns. These guns are designed to operate at lower air pressures, 20–70 kPa (3–10 psi), but with higher air volumes and large, unrestricted air passages to handle a large volume of air. Because of the low pressures, bounce back is reduced and transfer efficiencies of 65% or even higher can be achieved. HVLP guns are used increasingly in automotive repair shops, where California South Coast VOC regulations require 65% transfer efficiency and air pressures of 70 kPa or less. Electrostatic HVLP guns offer additional improvements in transfer efficiency. Alternatively, low volume, low pressure (LVLP) guns may be used. LVLP guns also operate at air pressures less than 70 kPa, but the air volume is reduced by mixing the air and coating inside the gun.

23.2.2 Airless Spray Guns

With airless spray guns, coating is forced out of an orifice at high pressure, 5–35 MPa. The coating emerges from the orifice as a "sheet." As the sheet extends in moving away

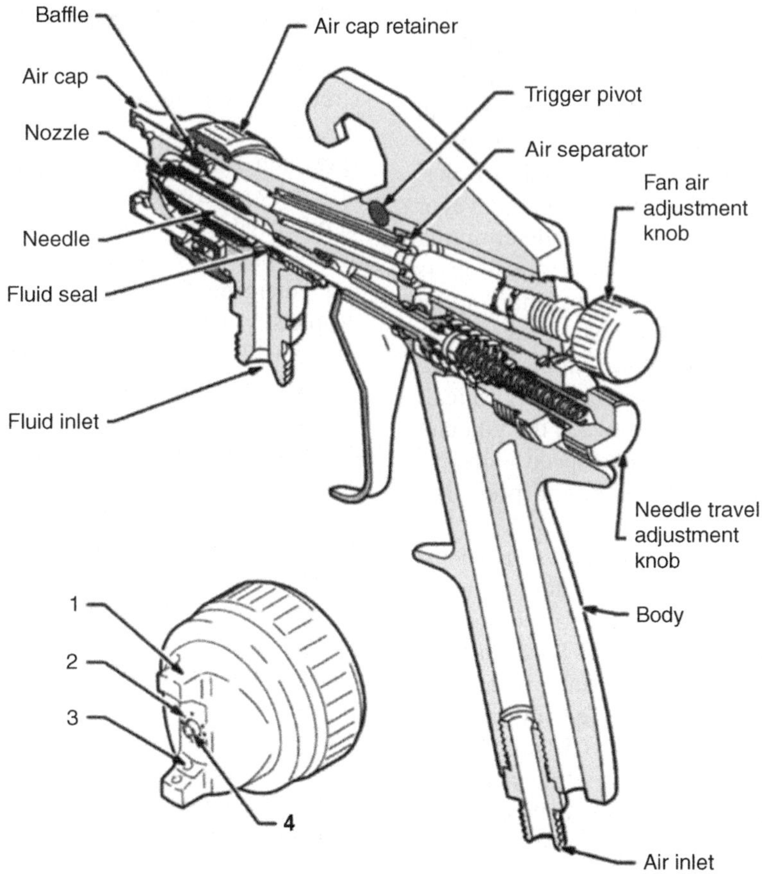

Figure 23.1 A schematic cross section of an air spray gun and a schematic of spray gun nozzle (Delta Spray™, Graco, Inc.): (a) Wings or horns; (b) Angular converging holes; (c) Side-port holes; and (d) Annular ring around the fluid tip. Source: Courtesy of Graco, Inc.

from the orifice, flow instabilities cause the formation of ligaments, followed by further disintegration into droplets (Lefebvre, 1989). Atomization is controlled by the relative velocity between the sheet and the contiguous air (higher relative velocity, smaller droplets), viscosity (higher viscosity, larger particle size), pressure (higher pressure, smaller particle size), and surface tension (lower surface tension, smaller particle size). The shape of the fan or pattern of the spray is controlled by the orifice size and shape. Air-assisted airless spray guns are also available; the atomization is airless, but there are external air jets to help shape the fan pattern, confining most of the smaller droplets within the spray pattern (Easton, 1983). Both handheld and robot airless guns are available.

The Fraunhofer Institute in Stuttgart has developed methods for measuring droplet size and momentum as the coating leaves the airless spray gun and for analyzing the resulting application characteristics with computational methods. For example, Ye et al. (2011) explore the effects of variables on airless spray, focusing on the field application of anticorrosive coatings under windy conditions. At a painting distance of 0.3 m, side winds of 5 and 10 m s^{-1} affect film thickness distribution but have little effect on transfer efficiency. At greater distances, both are affected. Use of an obstacle to deflect wind while maintaining the optimum distance is recommended.

The size of droplets from airless spray guns is larger than those from air spray guns, 70–150 μm compared with 20–50 μm (Nielsen et al., 1990). Optimized airless guns give a so-called fishtail spray; that is, there is a relatively sharp edge to the spray droplet fan with quite uniform droplet distribution within the fan (Ye et al., 2011). In contrast, the fan from air guns is *feathered* at the edge; that is, the number of droplets decreases at the edge of the fan, with some being quite widely spaced. As a result of these differences, one can generally achieve more uniform film thickness with air spray than with airless spray; air-assisted airless application gives intermediate results.

Coatings can be applied more rapidly by airless than by air guns, permitting more rapid production. However, as the application rate increases, the likelihood of applying excessive coating thickness also increases, particularly when coating objects with complex shapes. Excess coating thickness not only is wasteful but may also lead to sagging. Since a stream of compressed air does not accompany the particles and because the droplet size is generally larger, there is less solvent evaporation from the atomized particles from airless guns than from air guns. Solvents with higher relative evaporation rates are generally required in formulating coatings for airless spray application.

The absence of the air stream in airless guns reduces the problem of spraying into closed recesses of irregularly shaped objects. On the other hand, spraying into a recessed section that is open on the opposite end is easier with an air gun, since the air stream helps to carry the particles along. Airless spray equipment can cause problems with some waterborne coatings. Excess air is dissolved in the water under the high pressure and emerges as bubbles from the gun owing to the release of pressure. Entrapment of bubbles in the film can result in pinholing (Section 24.7).

With some latex coatings, droplet size from the gun is larger than would be expected from the shear viscosity of the coating. This phenomenon is associated with relatively high extensional viscosity. As with roller application, extensional viscosity influences spray behavior from airless guns. Latex paint formulations with 1.7% of an HEUR type of associative thickener gave the finest atomization (Fernando et al., 2000).

2K coatings can be applied with dual feed airless spray equipment, combining the advantages of airless spray and dual spray (Pucken, 2004). Such equipment has been developed to the point where it is possible to spray coatings that cure within a few seconds, enabling the use of 2K polyurea coatings (Section 12.4.1).

An aerosol coating container is a type of airless spray unit. A liquefied gas, commonly propane, supplies the pressure to force the coating out of the orifice. Since the pressure is relatively low, the viscosity of the coating must be low to achieve proper atomization.

23.2.3 Electrostatic Spraying

Transfer efficiency can be substantially higher with electrostatic spray units (Table 23.1). In the simplest case, a wire is built into the orifice of the spray gun. An electric charge on the order of 50–125 kV is impressed on the wire. At the fine end of the wire, an electric discharge ionizes the air. As atomized coating particles pass through this zone of ionized air, they pick up a negative charge. The object to be coated is electrically grounded. When the coating particles approach the grounded surface of the object, the differential in charge attracts the particles to the surface. Considering the example of a chain link fence, an increased fraction of coating deposits on the metal fence even to the extent that particles, which had passed through the holes, are attracted back to the back side of the fence. This *wrap around* effect permits coating both sides of the fence by spraying from only one side. There is a lower, but still fairly high, overspray loss even with electrostatic spray. In large-scale production most of the paint is precisely applied by computer-controlled robots. Utilizing robots, overspray losses can be reduced by over 50%. With objects like automobile bodies, transfer efficiencies above 80% are possible (Section 30.1.1).

In addition to electrostatic spray guns, other devices can be used for electrostatic spraying. Figures 23.2 and 23.3 show bell- and disk-type spray units, both of which are called rotary atomizers. In both cases, coating is pumped through a tube leading to the middle of the rapidly spinning bell or disk. The coating flows out to the edge of the disk or bell and is thrown off as filaments that quickly break up to form paint droplets. Martinez (2012) provides a detailed study of the bell process with excellent photos and illustrations. The droplets can be charged internally by direct contact with a charge impressed on the disk or bell or externally by passing them through ionized air. Droplet size is controlled by viscosity and the peripheral speed of the unit. Small (2–7 cm diameter) bells that spin at speeds up to 70 000 rpm permit application of coatings with viscosities as high as 1.5–2 Pa·s compared with 0.05–0.15 Pa·s with conventional speed disks (around 900 rpm) and bells as well as spray guns. The spray pattern is adjusted with *shaping air*. This ability to handle higher viscosity coatings with bells permits the use of less solvent. However, it has been reported that application of pigmented coatings by high speed electrostatic bells tends to give films with lower gloss (Tachi et al., 1990). It is suggested that the centrifuging effect leads to differences in pigment content among atomized particles, leading to the formation of uneven

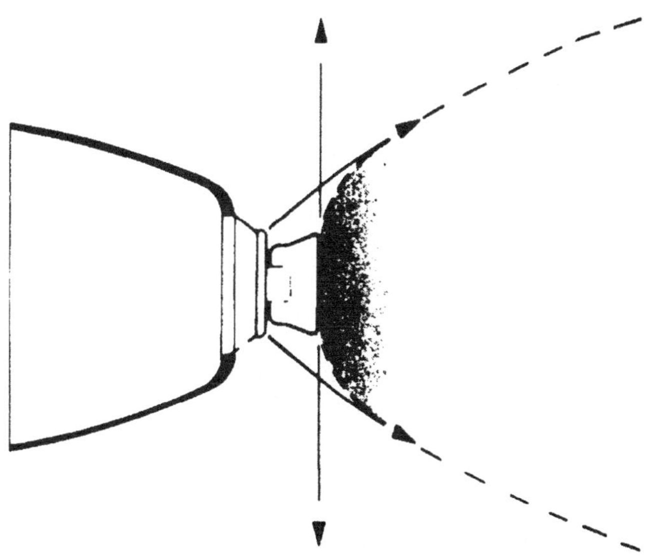

The bell: The geometry of the spray pattern is varied according to the object to be painted by adjusting an annular compressed air shroud

Figure 23.2 Bell electrostatic spray equipment. Source: Levinson (1988). Reproduced with permission of The Federation of Societies for Coatings Technology.

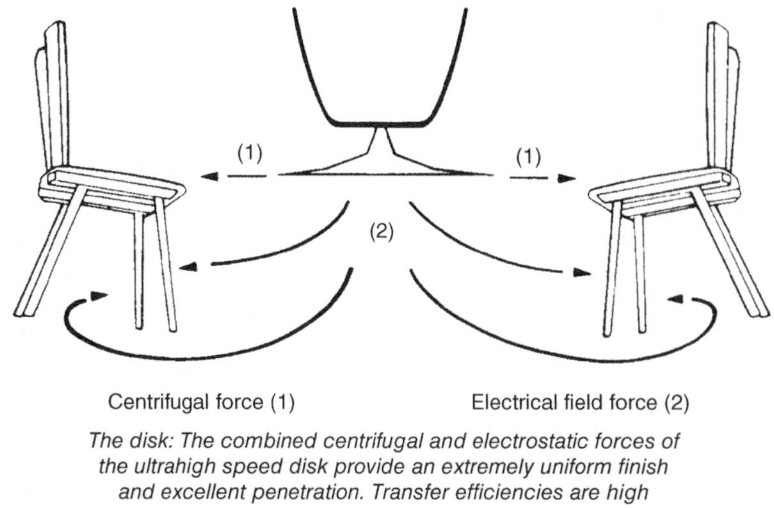

Figure 23.3 Disk electrostatic spray equipment. Source: Levinson (1988). Reproduced with permission of The Federation of Societies for Coatings Technology.

films. Uniformity of application and transfer efficiency can often be improved by compressed air shaping of the cloud of atomized coating, analogously to the effect of the shaping air in conventional air guns. Further efficiency improvements can result from impressing an electric potential on the object to be painted (Section 30.1.1).

Low overspray losses and good wrap around with electrostatic spray depend on charge pickup by the atomized particles, which is determined by conductivity of the coating. If conductivity is too low, the particles do not pick up sufficient charge from the ionized air. This is most likely to be a problem with coatings that have mainly hydrocarbon solvents, especially aliphatic hydrocarbons. Good results can be obtained by replacing a portion of the hydrocarbon solvents with nitroparaffin or alcohol solvents. For coatings with free carboxylic acid groups on the resin, addition of a small amount of a tertiary amine like triethylamine increases conductivity sufficiently to give good charging. If the conductivity is too high, there is increased danger of electrical shorting. Generally, conductivity is measured, but the results are expressed as resistivity; optimum resistivity varies with equipment and the coating operation, ranging from 0.05 to 20 MΩ. Newer gun designs permit operation with a wide range of conductivities.

Conductivity of waterborne coatings is higher than solventborne coatings; resistivities of the order of 0.01 MΩ are reported (Diana, 1992). Addition of slow evaporating water-miscible solvents with nonpolar ends, such as 2-butoxyethanol, somewhat reduces surface conductivity. However, spray equipment must be specially designed so that the coating line is electrically isolated; otherwise the charges will be dissipated, and the atomized droplets will not be charged. Furthermore, the hazard of shocks would be high. Although the whole application line can be electrically isolated, this approach is expensive and does not eliminate the hazard of electric shocks. Alternatively, just the atomizer and a short hose connected to a "voltage blocking device" can be charged (Konieczynski, 1995). Another alternative to electrical isolation is to fit the gun with an external probe that charges the atomized paint. These expedients permit effective charging of the atomized droplets and reduce the hazard of shocks at substantially lower cost than isolating the entire application line.

Electrostatic spray application is not without limitations. The substrate must be electrically conductive so that the necessary charge differential can be established by grounding the object to be sprayed. Coating of recessed areas can also be difficult even with airless electrostatic spray application due to the *Faraday cage effect*. A Faraday cage results from the pattern of field lines between the electrode on the gun and the grounded object. The strong electrical field induced by the difference in voltage establishes field lines that the atomized particles follow between the gun and the object. However, areas surrounded by grounded metal, such as inside corners of a steel box, are shielded from the electric fields by the metal; a Faraday cage is established and few particles enter such shielded areas.

Electrostatic spray is the most important method for applying powder coatings (Section 28.5.1).

23.2.4 Hot Spray

Since the viscosity of coatings must be relatively low, generally, 0.05–0.15 Pa·s, for proper atomization, the solids must usually be low, especially for lacquers prepared with relatively high molecular weight (MW) resins. An approach for increasing solids is *hot spray*, also called *temperature-conditioned spray*. A hot spraying unit is designed with a heat exchanger to heat the coating to temperatures from

38 to 65°C. The unit is designed so that the coating is recirculated when the gun is turned off, even temporarily. At the elevated temperature, the viscosity is reduced sufficiently to permit a significant increase in solids. While these systems were originally designed for use with quite low solids coatings like lacquers, they are also useful for high solids coatings. For example, it is reported that the solids of a top coat for appliances can be increased from 55 to 65% by using a heater in front of a disk electrostatic spray gun (Schoff, 1991). The viscosity of high solids coatings generally decreases more sharply with increasing temperature than conventional coatings, which is a desirable characteristic for hot spray application. A further advantage is that the temperature drops in the interval between leaving the spray gun orifice and arriving on the work, resulting in a correspondingly large viscosity increase. This characteristic is desirable for reducing sagging that can be serious with high solids coatings (Section 24.3). Another advantage is that temperature control eliminates the effects of variation in ambient temperature on viscosity of the coating.

The term *thermal coating* usually refers to processes for applying molten metal coatings, which are outside the scope of this book.

23.2.5 Supercritical Fluid Spray

Supercritical fluid spraying was intensively investigated in the 1990s (Nielsen et al., 1995; Senser et al., 1995) and commercial-scale units were developed. The method was envisaged as a way to lower VOC in coating furniture and other items. The supercritical fluid of choice is carbon dioxide, which has a critical temperature of 31.3°C and a critical pressure of 7.4 MPa. In the supercritical state, carbon dioxide exhibits solvency characteristics similar to hydrocarbons, but is not counted as VOC. The impact of this method on large-scale industrial processes seems to have been small, perhaps because of the high equipment cost and improvements in other low VOC technologies.

Recently, there has been renewed interest in supercritical fluid spray for a variety of specialized applications. For example, Ovaskainen et al. (2014) demonstrated that electrostatic supercritical solution spray of a poly(vinyl acetate)/poly(vinylpivalate)-based coating results in film surfaces having the micro- and nanoscale roughness required for superhydrophobicity (lotus effect). Such coatings have extreme water repellency (Section 34.1).

23.2.6 Formulation Considerations for Spray-Applied Coatings

Formulation of solvent mixtures for spray-applied coatings and effective use of spray guns requires taking into consideration the large ratio of surface area to volume of the atomized coating droplets and the flow of air over the surface of those droplets. As discussed in Section 18.3, these factors affect the rate of solvent evaporation. When lacquers are applied by spray, more than half the solvent in the coating may evaporate during the interval between the orifice of the spray gun and the surface of the object. If the solvent mixture is balanced properly and the spray gun is used properly, a relatively smooth, sag-free film can be obtained. Otherwise, sagging may occur when not enough solvent is evaporated or rough surfaces may result from dry spray. Low viscosity of the coating on the surface not only generally facilitates leveling but also increases sagging (Sections 24.2 and 24.3).

Proper control requires a careful balance of solvent evaporation rate with the particular spray equipment and procedure. The greater the distance from the spray gun orifice to the object, the greater the fraction of solvent lost. Good hand-spray technique requires that the spray gun be moved sideways at a near-constant distance from the target. Flexing the wrist is likely to result in nonuniform film thickness. In extreme cases of poor technique, both sagging and rough surfaces can occur on the same object.

The rate of solvent loss is affected by the degree of atomization. If the average particle size is smaller, the surface to volume ratio is higher and the extent of solvent loss is greater. The rate of solvent evaporation is also affected by airflow over the surface of the droplets. More solvent evaporates with air guns than with airless guns or spinning bells. The rate of airflow through the spray booth can affect the extent of solvent loss. Temperature in the spray booth can also be an important factor; during hot weather, it is common to change the solvent mixture to reduce the rate of evaporation. When formulating and testing a solvent mixture for a coating in the laboratory, the same type of spray gun at approximately the same distance from the work that will be encountered in the customer's factory should be used. A final adjustment must be made in the customer's plant to accommodate regular production conditions. If those conditions change, the solvent mixture may have to be changed. This is one reason solventborne industrial coatings are almost always shipped at higher concentrations than the customer will actually use. This permits modifications of both solvent levels and solvent composition (by changing the reducing solvent) to accommodate temperature changes in the spray booth and other variables.

Viscosity of the coating must be adjusted to obtain appropriate atomization for the spray gun being used. The critical viscosity for atomization corresponds to its value at the high shear rates of 10^3–$10^6\,s^{-1}$ (Schoff, 1991) that are encountered as the coating passes through the orifice of the gun. Coatings with high low shear viscosity can be sprayed successfully if they are shear thinning. The upper limitation on viscosity at low shear rate is the need to have a satisfactory flow rate of coating through the tubing to the spray gun, which varies from gun to gun. Architectural and maintenance paints, which must not sag significantly when thick films are applied to large wall expanses, are shear thinning. Often, the same coating used for brush application can be used for spray; in the case of latex paints, the viscosity at low shear rate may need to be reduced by diluting with a small amount of water, with some spray equipment.

Many conventional industrial solventborne coatings exhibit Newtonian flow; however, most high solids coatings and waterborne coatings are shear thinning. Viscosity for application is generally checked with an efflux cup (Section 3.3.5). However, efflux cups can be misleading because they do not detect shear thinning or thixotropic flow properties. The proper way to establish viscosity for production spraying is by using the production spray gun under the conditions of use. Then, having found the optimum formulation, one can establish an efflux cup time as the control standard for the particular coating, gun, and conditions of use.

Many water-reducible and high solids coatings are shear thinning after reduction to spray viscosity, which signifies that for proper atomization, the efflux cup time of the reduced waterborne or high solids coating will be longer than for most conventional solventborne coatings. Atomization is controlled by viscosity at high shear rate and efflux cup time is controlled by viscosity at a lower shear rate. Sagging of some spray-applied high solids coatings cannot be controlled by the adjustment of solvent evaporation rates. In such cases, the coating has to be formulated so that it is thixotropic (Section 24.3).

In spray applying waterborne coatings, air bubbles can be entrapped in the coating film. If they persist until the viscosity of the film surface becomes too high for the bubbles to escape, pinholing or popping will occur. The troublesome air voids are tiny (~10 μm). The problem can be particularly severe in airless spray application, since more air dissolves in the coating due to the high pressure. When the pressure is released as the coating leaves the spray gun, bubbles of air form in the droplets. While larger air bubbles can escape by rising to the surface and breaking before the viscosity at the surface increases, microform bubbles only dissipate by dissolution of the air in the coating followed by diffusion through the film (Gebhard and Scriven, 1994). In air spray, HVLP spray, and air-assisted airless spray, the problem can be minimized by using CO_2 as the driving gas instead of air. The improvement results from the higher solubility of CO_2 in a waterborne coating as compared with air.

23.2.7 Overspray Disposal

In industrial production, the advantages of decreased overspray do not only relate to savings in the cost of lost coating and lower VOC emissions. Overspray needs to be trapped so that it does not contaminate the surrounding area. This purpose is served by water-washed spray booths in which the wall behind the work being sprayed is a continuous waterfall that is recirculated. The overspray is collected as a sludge. While it is sometimes possible to rework the sludge into low-grade coatings, generally the sludge must be disposed in approved hazardous solid waste disposal landfills. The collection process and disposal of sludge result in substantial costs; increased transfer efficiency can reduce waste disposal cost.

Although water-wash spray booths work efficiently with solventborne coatings, the separation of sludge is less efficient when some waterborne coatings are sprayed. Poor coagulation makes separation more difficult and can limit recirculation of the water. Froth flotation methods adapted from ore recovery processes permit relatively rapid separation of the sludge (Fuchs et al., 1988). Separation of sludge from waterborne coatings can be improved by adding to the water tank an emulsion of a melamine–formaldehyde resin together with water-soluble cationic and/or nonionic acrylamide polymers, which promote detackification and coating flocculation, respectively (Kia et al., 1991).

An alternative approach uses dry filters to capture the overspray. A variety of filters is available. The filters gradually clog with paint and must periodically be replaced and sent to the landfill. Relatively inexpensive, effective filters have been developed.

23.3 DIP AND FLOW COATING

Dip coating can be an efficient procedure for applying coatings, offering both relatively low capital cost equipment and low labor requirements. The principle is simple: the object is dipped into a tank of coating and pulled out; excess coating drains back into the dip tank. In practice, satisfactory coating application by dipping is quite complex. While excess coating is draining off the object, a gradation of film thickness develops; the coating thickness at the top of the object is thinner than at the bottom. During draining, solvent is evaporating. If the object is withdrawn slowly enough and the solvent evaporates rapidly enough, film thickness approaching uniformity can be achieved on vertical flat panels. In actual production, the rate of withdrawal is usually faster than optimum for uniform film thickness in a compromise with production rate.

Solventborne dip coatings present severe flammability hazards. In addition, viscosity must be held constant by replacing any solvent that evaporates. Changes in viscosity result in changes in film thickness, which increases with increasing viscosity. Maintaining constant viscosity, as required for optimum film thickness, becomes more difficult as the volatility of the solvent is increased.

Successful use of dipping in production lines requires that the coating be very stable. Viscosity can increase not only by loss of solvent but also by premature cross-linking of coating components. A small fraction of the coating in the tank is removed each time an object is dipped; and fresh coating is added to make up for these removals so that the coating in the tank is a mixture of aged and fresh coating.

An advantage of dip coating is that all surfaces are coated, not just the outer surfaces accessible to spray. However, there are difficulties in dipping irregularly shaped objects. Coating may be held in pockets or depressions giving pools of coating that do not drain. To minimize this problem, the point(s) at which the objects hang on the hooks of the conveyor line that carries it into and out of the tank must be carefully designed and selected. Objects to be dip coated must be designed with drain holes that minimize pooling, but do not interfere with the performance or appearance of the product. Lower edges, especially lower corners, build up high film thickness. Buildup can be reduced by electrostatic *de-tearing*, that is, by passing the object over a highly charged electrode that causes a charge concentration at such points, causing drops to be pulled off.

Waterborne coatings are supplanting solventborne dip coatings in many applications. They reduce flammability hazards (and thus, insurance costs) as well as VOC emissions.

In many large-scale applications, dip coating has been replaced by electrodeposition and autophoretic coatings, which are also dipping processes (Chapter 27), and by powder coatings (Chapter 28). However, dip coating remains important in various small-scale applications (Faustini et al., 2014).

Flow coating and dip coating are related methods. Objects to be flow coated are carried on a conveyor through an enclosure in which streams of coating are squirted on the object from all sides. The excess coating runs off and is recirculated through the system. There is a gradation of film thickness from the top to the bottom of the object, but usually less than that from dip coating. Highly automated flow coating lines have been used for applying coatings to major appliances. The design permits more rapid line speeds than with conventional dip tanks. Again, where feasible, electrodeposition offers major advantages. Powder coatings (Chapter 28) are also effective for coating irregularly shaped objects.

23.4 ROLL COATING

Roll coating is widely used and is efficient but is only applicable to uniform flat, or cylindrical surfaces. Relatively slow evaporating solvents must be used to avoid viscosity buildup on the rolls of the coater. The coating must be stable, since the rate of turnover of coating through the system is relatively low. There are many types of roll coating procedures; the two most common are *direct roll coating* and *reverse roll coating*.

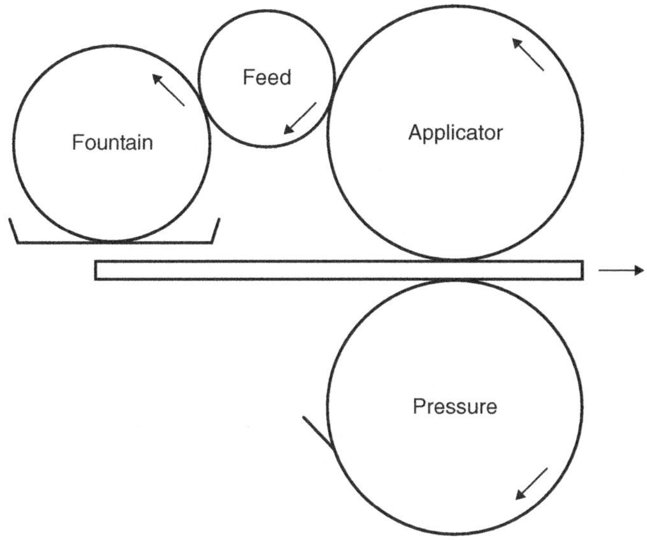

Figure 23.4 Direct roll coater. Source: Levinson (1988). Reproduced with permission of The Federation of Societies for Coatings Technology.

In direct roll coating, the stock passes between two rollers, an applicator roller and a backup roller, which are rotated in opposite directions. The rollers pull the material being coated between them as illustrated in Figure 23.4. Direct roll coating is used for coating sheet stock (especially for cans) and sometimes for coil stock. In direct roll coating, the applicator rollers are generally covered with a relatively hard polyurethane elastomer. Coating is fed to the applicator roll by a smaller feed, or doctor, roll that is, in turn, fed by a pickup roll. The pickup roll runs partially immersed in a tray, called a *fountain*, which contains the coating. Film thickness is controlled by the clearance between the feed and applicator rolls and by the viscosity.

Several variations on direct roll coating use different types of applicator rolls. The applicator rolls can have sections cut out so that they do not coat the entire substrate surface. The applicator roller can be engraved with small recessed cells over its whole surface. The cells are filled with coating, the surface is scraped clean with a doctor blade, and the coating remaining in the cells is transferred to the substrate being coated. Such so-called precision coaters apply a controlled amount of coating only to the areas of the substrate that contact the roller. In another variation, the applicator roller is a brush roller for applying thick coatings to relatively rough surfaces.

In direct roll coating, the wet layer of coating is split between the roller and the substrate as the coated substrate emerges from the nip between the rollers. At this point, the release of pressure results in cavitation. As the coated stock moves away from the nip, the walls between the cavities become filaments. Further along, filaments may break and fall back on the fresh coating. The result, almost invariably, is a ribbed surface, called *roll tracking*. The coating must be designed to level so that this tracked appearance becomes less obvious, before the coating stops flowing. To minimize solvent loss on the rolls and to keep the viscosity low to promote leveling, slow evaporating solvents are used, and, when possible, coatings are formulated so they exhibit Newtonian flow. The rate of shear on the coating as it passes through the nip depends on speed and gap distance, but is relatively high, on the order of 10^4–$10^5\ s^{-1}$. Accordingly, if the coating exhibits thixotropy, it is desirable to have the rate of recovery of the low shear viscosity be as slow as possible.

As the coating emerges from a nip, the filaments may grow quite long. When they finally break, at least some split in more than one place leading to drops being released, called *misting* (Figure 23.5). Extended filaments result when extensional viscosity (Section 3.6.3) is high (Fernando et al., 2000).

In reverse roll coating, the two rollers rotate in the same direction so that the material being coated must be pulled through the nip between the two rolls, as shown in Figure 23.6. It is generally not feasible to coat sheets by reverse roll coating, but the process is widely used for coating coil stock. Reverse roll coating has the advantage that the coating is applied by wiping rather than film splitting. A smoother film is formed, and problems of leveling are minimized. Section 30.4 describes coil coating, a major application for reverse roll coating.

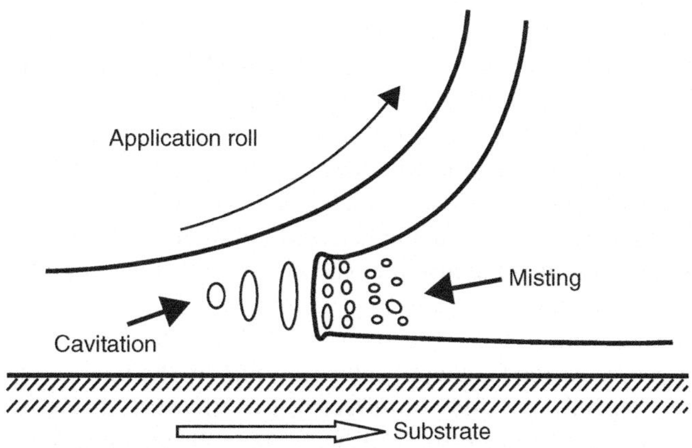

Figure 23.5 Cavitation and misting in roll application. Source: Fernando et al. (2000). Reproduced with permission of Elsevier.

23.5 CURTAIN COATING

Curtain coating is a cost-effective way to coat flat sheets such as wall panels, metal doors, and printed circuit boards. The coated sheets are baked or, increasingly, cured by UV radiation (Chapter 29). Coating is pumped through a slot in the coating head so that it flows out as a continuous curtain. The sheet to be coated is carried through the curtain on a conveyor. The curtain must be wider than the substrate being coated to avoid edge effects on the film thickness. A recirculating system returns the overflow from the sides and from between sheets to the coating reservoir. A schematic diagram is shown in Figure 23.7.

The width of the slot, the pumping pressure, the viscosity of the coating, and the rate of the passage of the substrate being coated control film thickness. The faster the line runs, the thinner the coating. Where applicable, curtain coating is an excellent method. No film splitting is involved, so the film that is laid down is essentially smooth. Film thickness can be very uniform. The coating must be very stable; solvent is added to make up for loss by evaporation. If a particle of low surface tension lands on the flowing curtain, surface tension differential driven flow can lead to a hole in the curtain that in turn leads to a gap in the coating being applied to the substrate (Section 24.4). As the surface tension of the coating is increased, particles in the air are more likely to have lower surface tension than the coating, and the probability of holes is increased.

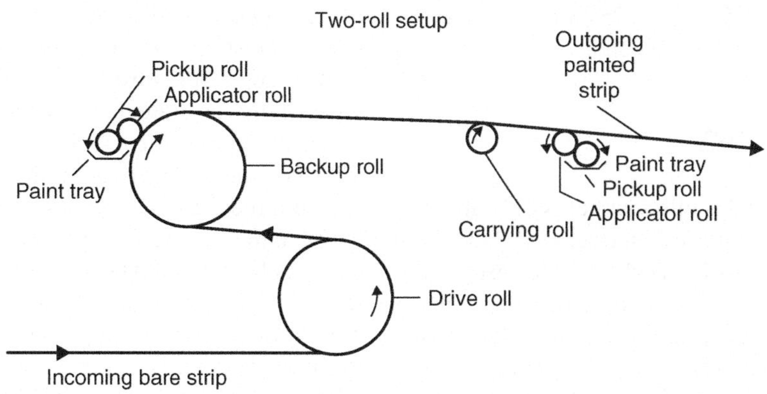

Figure 23.6 Reverse roll coater for coating both sides of coil stock. Source: Levinson (1988). Reproduced with permission of The Federation of Societies for Coatings Technology.

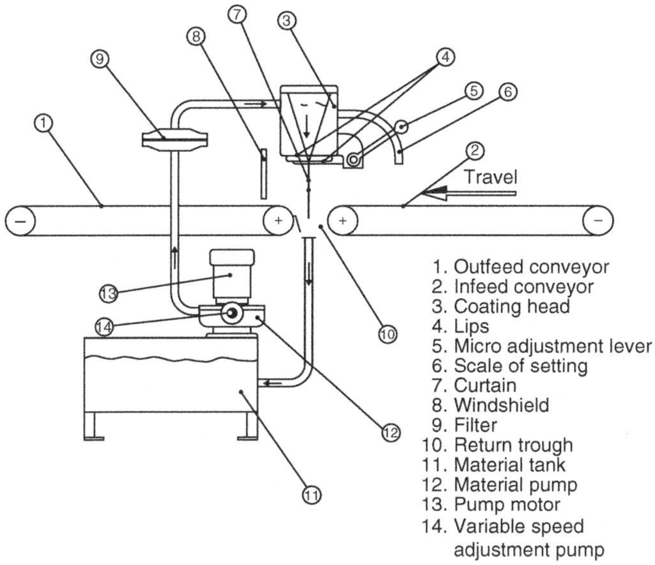

Figure 23.7 Schematic diagram of a curtain coater. Source: Levinson (1988). Reproduced with permission of The Federation of Societies for Coatings Technology.

GENERAL REFERENCES

Goldschmidt, A.; Streitberger, H.-J., *BASF Handbook on Basics of Coating Technology*, Vincentz, Hannover, **2003**, pp 437–617. Second edition, 2007.

Levinson, S. B., *Application of Paints and Coatings*, Federation of Societies for Coatings Technology, Blue Bell, PA, **1988**.

REFERENCES

Adams, J., *Spray Applications for Environmental Compliance*, FSCT Symposium, Louisville, KY, May **1990**.

Diana, M. J., *Products Finishing*, **1992**, July, 54.

Easton, M. G., *J. Oil Colour Chem. Assoc.*, **1983**, *66*, 366.

Faustini, M., et al., *ACS Appl. Mater. Interfaces*, **2014**, *6*(19), 17102–17110.

Fernando, R. H.; Glass, J. E., *J. Rheol.*, **1988**, *32*, 199.

Fernando, R. H., et al., *Prog. Org. Coat.*, **2000**, *40*, 35.

Fuchs, E. W., et al., *J. Coat. Technol.*, **1988**, *60*(767), 89.

Gebhard, M. S.; Scriven, L. E., *J. Coat. Technol.*, **1994**, *66*(830), 27.

Glass, J. E., *J. Coat. Technol.*, **1978**, *50*(641), 72.

Kia, S. F., et al., *J. Coat. Technol.*, **1991**, *63*(798), 55.

Konieczynski, R. D., *J. Coat. Technol.*, **1995**, *67*(847), 81.

Lefebvre, A. H., *Atomization and Sprays*, Hemisphere Publishing Corporation, New York, **1989**, pp 59–61.

Levinson, S. B., *Application of Paints and Coatings*, Federation of Societies for Coatings Technology, Blue Bell, PA, USA **1988**.

Martinez, L. A., *Automotive Rotary-Bell Spray Painting*, M.S. Degree Project, University of Gothenburg. Undated–about **2012**.

Nielsen, K. A., et al., *Polym. Mater. Sci. Eng.*, **1990**, *63*, 996.

Nielsen, K. A., et al., *Proceedings of the International Waterborne, High-Solids, and Powder Coatings Symposium*, New Orleans, LA, **1995**, p 151.

Ovaskainen, L., et al., *J. Supercritical Fluids*, **2014**, *95*, 610–617.

Pucken, W., *J. Prot. Coat. Linings*, **2004**, *21*(9), 51.

Schoff, C. K., *Rheology*, Federation of Societies for Coatings Technology, Blue Bell, PA, **1991**.

Senser, D. W., et al., *Proceedings of the International Waterborne, High-Solids, and Powder Coatings Symposium*, New Orleans, LA, **1995**, p 161.

Tachi, K., et al., *J. Coat. Technol.*, **1990**, *62*(791), 19.

Ye, Q., et al., *ILASS: Europe 2011, 24th European Conference on Liquid Atomization and Spray Systems*, Estoril, Portugal, September, **2011**.

Chapter 24

Film Defects

Obvious defects can result from incomplete coverage during application of a coating, resulting in thin spots or holes, often called *skips* or *holidays*. Many other kinds of defects and imperfections can develop in a film during or after application. In this chapter, we describe the most important defects and, to the extent possible, discuss possible remedies. Unfortunately, nomenclature for many defects is not uniform. LeSota (1995) provides definitions for coating terms including those for film defects. Readily available www pages also define many of the terms and provide illustrations of the defects (e.g., Schoff, C. K., www.paint.org/article/automotive-coatings-application-defects/). Surface tension is fundamental to many types of defects, so we start there.

24.1 SURFACE TENSION

Surface tension occurs because the forces at an interface of a liquid differ from those within the liquid owing to unsymmetrical force distributions on the surface molecules. The surface molecules possess higher free energy, defined as the energy per unit area required to remove the surface layer of molecules. The dimensions of surface tension are force exerted in the surface perpendicular to a line; SI units are newtons per meter or millinewtons per meter ($mN\,m^{-1}$). Older units, still commonly used, are dynes per centimeter (fortunately, $1\,mN\,m^{-1} = 1\,dyne\,cm^{-1}$). Similar surface orientation effects are present in solids, which have surface free energies, expressed in units of free energy per unit area, commonly millijoules per square meter ($mJ\,m^{-2}$), which are numerically and dimensionally equal to $mN\,m^{-1}$. Frequently, people refer to surface tension of solids; while not formally correct, errors seldom result because the values are identical. Owen (1996) provides an excellent discussion of surface and interfacial properties, including data on many polymers.

Surface forces work to decrease the surface free energy of liquids. Surface tension draws liquids into spheres, since a sphere encloses a minimum ratio of surface area to volume. In a spaceship, floating liquid droplets are spherical; on Earth, the force of gravity distorts the spheres. For the same reason, surface tension drives flow of an uneven liquid surface toward becoming a smooth surface. The smooth surface has less interfacial area with air than the rough surface; hence, there is a reduction in surface free energy as the surface becomes smoother.

Segments of molecules that minimize surface tension tend to orient at the surface. The lowest surface tension results from perfluoroalkyl groups at the surface, followed by methyl groups. Poly(dimethylsiloxane) has a low surface tension because the very flexible, easily rotatable backbone of siloxane bonds permits orientation of a large population of methyl groups at the surface. The surface tension of linear aliphatic hydrocarbons increases as chain length increases, reflecting the larger ratio of methylene to methyl groups oriented at the surface. Progressively higher surface tensions result from aliphatic chains, aromatic rings, esters and ketones, and alcohols. Dipolar groups, such as ketones, and H-bonding groups, such as alcohols, impart higher surface tension because the stabilizing intermolecular dipolar and H-bonding interactions are absent at the surface, resulting in higher surface energy. As the chain length of aliphatic esters, ketones, and alcohols increases, the surface tension increases because methylene groups impart higher surface tensions than methyl groups. Surface tension decreases as temperature increases. Oliveira et al. (2011) successfully employed van der Waals density gradient

Organic Coatings: Science and Technology, Fourth Edition. Frank N. Jones, Mark E. Nichols, and Socrates Peter Pappas.

theory to calculate the surface tensions of 37 esters as a function of temperature and to predict surface tensions of mixtures.

Water has the highest surface tension of the volatile components used in coatings. Its surface tension is $72\,mN\,m^{-1}$ at 25°C and $59\,mN\,m^{-1}$ at 100°C. Surface tensions of organic liquids at 25°C range from about $10\,mN\,m^{-1}$ for perfluorocarbons to about $50\,mN\,m^{-1}$ for highly polar materials such as glycerol.

Addition of small amounts of surfactants to water reduces surface tension because the hydrophobic parts of surfactants gather at the surface. Addition of poly(dimethylsiloxane) to an organic solvent solution reduces surface tension owing to orientation of the methyl groups at the surface.

A classical method for measuring surface tension of liquids is the Du Noüy ring tensiometer (ASTM D1331-14). The method involves measuring the force required to slowly pull a platinum–iridium wire ring out of the liquid. With hand-operated instruments, considerable finesse is required to obtain reproducible results. Electronically controlled instruments are available, which are considered to be easier to use and to provide more accurate results. For many paints the Wilhelmy plate method (also described in ASTM D1331-14) is preferred. The electronically controlled instruments can also perform Wilhelmy measurements.

Why is the Wilhelmy method preferred? Because the liquid is not deformed during this measurement. When a liquid is agitated, the molecules at the surface are mixed in with the rest of the liquid. When the agitation is stopped, reorientation to give the lowest surface tension occurs, but the equilibrium surface composition is not reestablished instantly. When coating films are applied, they are subjected to considerable agitation. The time to regain equilibrium after agitation has ceased varies widely, depending on the composition, viscosity, temperature, and probably other factors. As Bierwagen (1975, 1991) pointed out, the surface tension of importance in governing some aspects of coating behavior may not be the *equilibrium surface tension*, but rather a *dynamic surface tension* (Smith, 1983; Schwartz, 1992). ASTM D3825-09 provides a standardized method for measuring dynamic surface tension—the "fast bubble technique." DIN has a similar method. These methods have limitations, and research on improved methods continues, for example, see Yang et al. (2014). Improved instruments are becoming available and offer opportunities for research.

Qualitatively, one can say that equilibrium is established most rapidly when the coating contains small, flexible molecules and when there are large differences in the polarity of components in the system as in waterborne coatings. Reaching equilibrium takes longer when the molecules with the lowest potential surface tension groups are polymers. But if the polymers have moderate molecular weight and flexible backbones, they can apparently reach the surface relatively rapidly. Poly(dimethylsiloxane) mentioned previously is an example. Low molecular weight octyl acrylate copolymers are used as additives to reduce surface tension of coating films as they are forming. In waterborne coatings, Smith (1983) and Schwartz (1992) showed that surfactants differ in their rates of reaching equilibrium surface tension. Surface tension increases with decreasing temperature, and solvents generally have lower surface tensions than resins. Therefore, surface tension increases as solvent evaporates from a film of resin solution, owing to both the change in concentration and temperature.

Surface free energy of solids (also just called *surface energy*) is measured by contact angle experiments, as detailed in Section 34.1.

If two miscible liquids with different surface tensions are brought in contact with each other, the liquid of low surface tension flows to cover the liquid with higher surface tension, since the overall surface free energy is thereby reduced. Such flow is *surface tension differential-driven flow*, also called *surface tension gradient-driven flow*. This type of flow has been observed for millennia, but Carlo Marangoni, a nineteenth-century Italian physicist, is credited with providing a scientific understanding of the phenomenon (Scriven and Sternling, 1960).

An example of the Marangoni effect is the flow that occurs when a clean glass containing wine is tipped, wetting the side of the glass with the liquid, and then returned to an upright position. Liquid then flows up the side of the glass, forming a bead of greater film thickness along the upper edge of the wetted area. When the amount of liquid collecting in the bead becomes large, droplets flow back down the inside of the glass. These "tears of wine" (also called "legs") have been known since biblical times, but why does the phenomenon occur? The ethyl alcohol in wine has a higher relative evaporation rate and lower surface tension (about $22\,mN\,m^{-1}$) than water (about $72\,mN\,m^{-1}$); evaporation occurs most rapidly along the edge and exposed sides of the layer of wine remaining on the side of the glass. Preferential evaporation of the alcohol results in a higher concentration of water and, accordingly, higher surface tension along the edge and sides relative to the bulk of the liquid. To minimize surface free energy, the lower surface tension (higher alcohol concentration) liquid in the glass defies gravity and flows up the side of the glass to cover the higher surface tension liquid at the edge. Evaporation continues at the edge, reducing the alcohol concentration and leading to a continuation of the process. Eventually, gravity causes droplets of liquid to flow back down the sides of the glass resembling tears. The surface tension differential is also affected by the temperature change; as alcohol evaporates, temperature decreases, further increasing the differential in surface tension.

In summary, two types of flow result from surface tension effects. Surface tension-driven flows occur to minimize surface area of a liquid. Surface tension differential-driven flows occur to cover a liquid or other surface of higher surface tension with a liquid of lower surface tension.

24.2 LEVELING

Most methods of applying coatings initially lead to formation of a rough wet film. It is usually desirable for appearance and performance to have the irregularities level out. The most widely studied leveling problem has been leveling of brush marks. A person unacquainted with the field might first think that leveling results from gravitational effects. If gravity were a significant factor, paints applied to ceilings should level more poorly than paints applied to floors, but they do not. Based on studies of flow of mineral oil, Orchard (1962) presented evidence that the driving force for leveling is surface tension. He further established mathematical models for the variables that he proposed would control the rate of leveling. Orchard's treatment has been widely applied to flow of coatings, although coatings are more complex than mineral oil. Patton (1979, p 554) illustrated the Orchard model for an idealized cross section of a wet film exhibiting brush marks that follow a sine wave profile, as shown in Figure 24.1.

Patton provides several forms of the Orchard equation and illustrates their derivation. A convenient form relating the change in amplitude of the sine wave to time is shown in Eq. 24.1, where a_0 = initial amplitude (cm), a_t = amplitude at time t (cm), x = average coating thickness (cm), λ = wave length (cm), and γ = surface tension (mN m^{-1}), η = viscosity (Pa·s), t = time (s):

$$\ln\frac{a_0}{a_t} = \frac{5.3\gamma x^3}{\lambda^4}\cdot\frac{dt}{\eta} \tag{24.1}$$

Leveling is promoted by thicker films, smaller wavelengths, lower viscosity, and higher surface tension. Unfortunately, the formulator often has minimal control over most of the variables. High surface tension increases the rate of leveling; however, the formulator is limited in optimizing this factor, since high surface tension often causes other defects (Section 24.4). Thicker films promote leveling; however increasing film thickness increases coating cost and increases the probability of sagging (Section 24.3) as well as other defects. Thicker films also increase wavelength in brush application, which decreases leveling. The principal means of control remaining is viscosity.

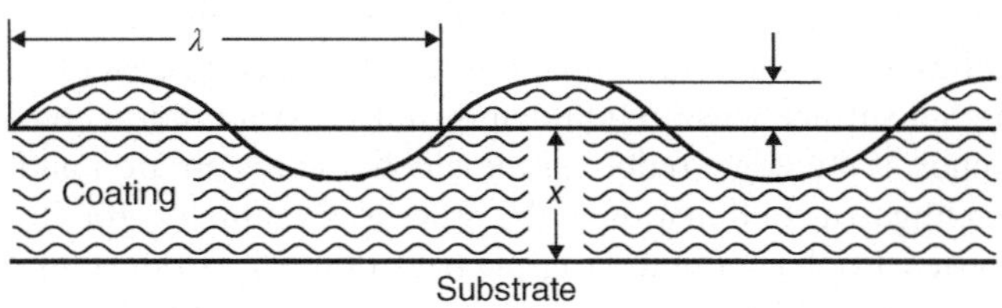

Figure 24.1 Schematic diagram of a cross section of brush marks. Source: Patton (1979). Reproduced with permission of John Wiley & Sons.

The Orchard model provides satisfactory correlation between experimental data and predictions when the liquid film has Newtonian flow properties and sufficiently low volatility such that viscosity does not change much during the experimental observations. With coatings, however, these conditions are usually not satisfied. Viscosity is usually changing during the time in which leveling occurs. As solvent evaporates, viscosity increases. Furthermore, if the system is thixotropic (Section 3.5), the viscosity is reduced by the high shear rate exerted during application and subsequently increases with time at the low shear rate involved in leveling. Another potential shortcoming in the Orchard treatment is the assumption of constant surface tension.

Overdiep (1978, 1986) devised methods for observing the location of the ridges and valleys during drying of two solventborne alkyd coatings. Surprisingly, he found that brush marks leveled to an essentially smooth film, but then ridges grew where there had been valleys and valleys formed where there had been ridges. Although uniform surface tension can and does cause ridges to level, it cannot cause a level film to form ridges, which would create more surface area. Overdiep also proposed that surface tension differential is the major driving force for emergence of the new ridges. As can be seen in Figure 24.1, the wet film thickness in the valleys of the brush marks is less than in the ridges. The concentration of resin is the same in both areas, so that evaporation of the same amount of solvent per unit area of surface results in more concentrated resin solutions in the valleys. Since surface tension of the resin is higher than that of the solvent, the surface tension of resin solution in the valleys becomes higher than in the ridges. In accordance with the Marangoni effect, coating flows from the ridges into the valleys. In other words, Overdiep proposed, and was able to demonstrate experimentally, that with volatile solvents, the primary driving force for leveling is not surface tension, but surface tension differential. In some cases, these differentials lead to overshooting of the smoothest stage and cause the growth of ridges. The extent of the flow driven by surface tension differential depends on the rate of evaporation of the solvent.

Kojima and Moriga (1995) studied solvent evaporation and leveling of water-reducible coatings and showed that the forces driving leveling depend on the solvent in the formulation. With a fast evaporating solvent, such as isobutyl alcohol, surface tension increased during drying promoting leveling by surface tension differential forces; but with a slow evaporating solvent such as ethylene glycol monohexyl ether, surface tension decreased, causing an adverse effect on leveling. Equations have been developed that model the drying process with the changes in surface tension differentials and in viscosity during solvent evaporation (Wilson, 1997).

Overdiep (1978) was particularly interested in leveling of coating films over a rough substrate. He reasoned that surface tension-driven flow might give the smoothest film. However, as illustrated in Figure 24.2a, this may be undesirable because protection in thin areas might be limited. On the other hand, as shown in Figure 24.2b, surface tension differential-driven flow would tend to yield equal film thickness, with the surface of the film following the roughness of the substrate rather than being level. Overdiep suggests it might be best to adjust a coating so that both types of flow are significant to achieve a compromise of reasonable film smoothness while avoiding regions where film thickness is too thin, as shown diagrammatically in Figure 24.2c. Such a compromise could be controlled by the solvent selection. Since leveling is predominately surface tension driven with solvents of low volatility and predominately surface tension differential driven with solvents of high volatility, the desired balance could be achievable by selecting solvents having intermediate volatility.

Neither the Orchard equation nor Overdiep's work takes into account the effect of thixotropy. Cohu and Magnin (1996) developed equations that predict the effect of thixotropy on leveling of non-evaporative coatings.

A frequent result of poor leveling is reduction in gloss. A study of the effect of leveling on gloss showed that three factors are important: extent of roughness of the substrate, film thickness, and viscosity (Ngo et al., 1993). The addition of fluorocarbon surfactant did not significantly change leveling in these experiments, which suggests that surface tension differentials did not develop in the drying of the films.

In spray application, surface roughness consists of bumps surrounded by valleys rather than ridges and valleys. Since the effect is somewhat reminiscent of the appearance of orange skins, it is called *orange peel*. Figure 24.3 shows a photograph of paint with an orange peel surface. Orange peel is not necessarily a defect. For example, it is often desired in coatings for appliances and metal cabinets because it hides small irregularities in the metal. The bumps are generally much larger than spray droplets. Most commonly, orange peel is encountered when spraying coatings that have solvents with high evaporation rates. It is commonly concluded that owing to fast evaporation, the viscosity of the coating on the substrate builds up too rapidly for leveling to occur; in some situations, that is probably the case. However, in the late 1940s, it was found that leveling of sprayed lacquer films could frequently be improved by addition of very small amounts of silicone fluid. Contrary to the common wisdom at the time—that leveling is favored by high surface tension (see Eq. 24.1)—here was a case in

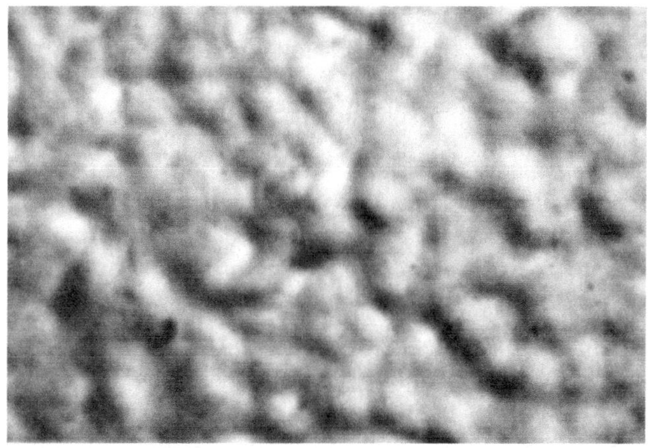

Figure 24.3 Typical orange peel pattern (15X). Source: Pierce and Schoff (1994). Reproduced with permission of The Federation of Societies for Coatings Technology.

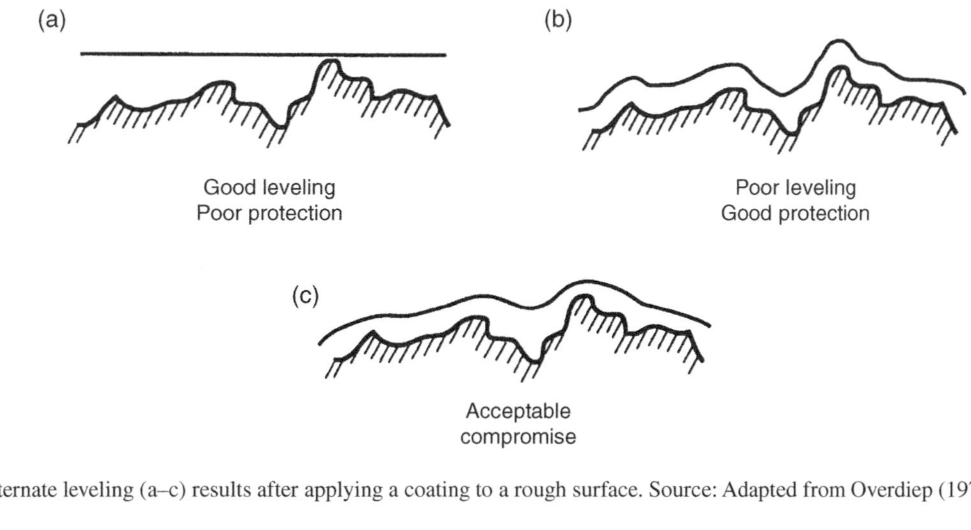

Figure 24.2 Alternate leveling (a–c) results after applying a coating to a rough surface. Source: Adapted from Overdiep (1978), with permission.

which adding a material known to reduce surface tension substantially improved leveling.

Hahn (1971) explained the phenomenon. When one sprays a lacquer, initially the surface is fairly smooth and, if you watch, orange peel grows (Figure 24.3). Hahn also proposed that the growth of orange peel results from surface tension differential-driven flow. He reasoned that the last atomized spray particles to arrive on the wet lacquer surface have traveled the longest distance between the spray gun and the surface; hence, they have lost more solvent and have a higher resin concentration and, therefore, a higher surface tension than the main bulk of the wet film. The lower surface tension wet lacquer flows up the sides of these last particles to minimize overall surface free energy; that is, surface tension differential-driven flow grows the orange peel. If one adds a surface tension depressant, such as a silicone fluid that orients very rapidly to the surface; the surface tension of the wet lacquer surface and the surface tension of the last atomized particles to arrive are almost uniformly low, so there is little differential to promote growth of orange peel. Octyl acrylate copolymer additives can also lower the overall low surface tension and minimize orange peel growth.

Electrostatically sprayed coatings are likely to show surface roughness greater than non-electrostatically sprayed coatings. It has been speculated that the greater surface roughness obtained with electrostatic spray may result, because the later arriving charged particles that deposit on a coated surface are more electrically insulated from ground. These later arrivals may retain their charges sufficiently long to repel each other and thereby reduce the opportunity for leveling. It has also been suggested that coatings applied by high-speed bell electrostatic spray guns may develop differentials in the pigment concentration within the spray droplets, resulting from the centrifugal forces (Tachi et al., 1990). Pigment concentration differentials can lead to rougher surfaces, reduction in gloss, and aluminum flake striping owing to differential flake concentrations.

Leveling problems are particularly severe with latex paints, especially those formulated entirely with cellulosic thickeners. Three factors combine to work against good leveling:

- Rapid increase in viscosity after application
- Low surface tension
- Low surface tension differentials

Compared with solventborne paints, latex paints with cellulosic thickeners exhibit a greater degree of shear thinning during spraying or brushing and more rapid recovery immediately afterward. The main cause of the difference is that latex paints have a higher dispersed phase content, so their viscosity increases more rapidly during evaporation of volatile materials than the viscosity of solventborne paints. It seems probable that the leveling is primarily surface tension driven. The surface tension of water is high, but surfactants that reduce surface tension are always present in latex paints. Furthermore, it is probable that this low surface tension is established rapidly after the agitation of application stops. Perhaps more importantly, the surface tension is uniformly low, since it is almost unchanged as water evaporates. Thus, the generally poor leveling of latex paints may result in part from the absence of surface tension differentials to promote leveling. Latex paints formulated partly or entirely with associative thickeners level better, at least in part, because their viscosity recovers more slowly after shearing during application. Leveling of latex paints is discussed further in Section 32.3.2.

The leveling of powder coatings is discussed in Section 28.3.

24.3 SAGGING AND DRIP MARKS

When a wet coating is applied to a vertical surface, the force of gravity can cause it to flow downward (*sagging*). Differences in film thickness at various places lead to differing degrees of sagging, sometimes resulting in curtains, or drapes, of coating. The variables that affect the volume of coating that has sagged (V_s) after time (t) are given by Patton in Eq. 24.2, in which x is initial film thickness (cm), ρ is density (g cm^{-1}), g is gravitational constant (s cm^{-2}), t is time (s), and η is viscosity (Pa·s) (Patton, 1979, p 572):

$$V_s = \frac{x^3 \rho g t}{300\eta} \tag{24.2}$$

Sagging increases with coating density and thickness; the driving force is gravity. Some latitude is available to the formulator in minimizing density and thickness, but the major variable available for controlling sagging is viscosity. The tendency to sag can be evaluated by observing the behavior of films applied under conditions simulating field use, and various tests have been developed. The most common test is a sag-index blade, a straight edge applicator blade with a series of 1/4 in gaps of different depths at 1/16 in intervals across the blade (Wu, 1978; Patton, 1979, p 578). A drawdown, which is a series of stripes of coating of increasing thickness, is made on a chart, which is immediately followed by placing the chart paper in a vertical position. Sag resistance is rated by observing the thickest stripe that does not sag down to the next stripe. For research purposes, Overdiep (1986) developed the sag balance, a more sophisticated method, which provides a numerical basis for evaluating the extent of sagging. Overdiep also developed equations for sagging that take into consideration the changes in viscosity after application. Lade et al. (2015)

describe a further refinement in which 25 μm spores are dusted on the wet coating surface and tracked with a microscope as the coating sags. Their results agree with theory for Newtonian coatings.

In spray-applied coatings dissolved in solvents, sagging can generally be minimized together with achieving adequate leveling by a combination of proper use of the spray gun (to avoid excessively thick areas) and control of the rate of evaporation of solvents. The formulators' goal is to control viscosity so that it is initially low for leveling and builds up before noticeable sagging occurs. In brush and hand roller-applied coatings, in which slow evaporating solvents are used, thixotropic systems permit leveling to occur before the viscosity recovers. Latex paints, which are almost always thixotropic, are less likely to exhibit sagging than solvent solution paints. This is the positive aspect of their relatively poor leveling, discussed earlier in the previous section.

Sagging can be a serious problem with high solids coatings, especially when spray-applied. Although other factors may be involved, a major cause of sagging of high solids coatings is that substantially less solvent is lost during spraying (i.e., after the droplets leave the gun and before they arrive at the substrate surface) (Wu, 1978; Bauer and Briggs, 1984). The lower solvent loss leads to less increase in viscosity of a high solids coating as compared with a conventional coating, resulting in a greater likelihood of sagging. The reasons for lower solvent loss have not been clearly established, but are probably related to the following considerations. First, perhaps because of their higher surface tension, high solids coatings atomize to give larger particle size droplets than conventional coatings. The lower ratio of surface area to volume would lead to lower solvent loss. However, one should be able to adjust spray equipment and conditions to obtain equivalent atomization. A second possibility is a colligative effect on solvent evaporation. Resins in high solids coatings are lower in molecular weight, and their concentrations are higher. For both of these reasons, the ratio of the number of solvent molecules to resin molecules is lower in the case of high solids coatings than for conventional coatings. This undoubtedly leads to a decrease in rate of solvent loss; see Section 18.3.5 for a model calculation. However, it seems doubtful that this difference could account entirely for the large differences in solvent loss, such as reported by Wu (1978). A third theory is that the onset of diffusion control of the rate of solvent loss, which substantially reduces the rate of solvent evaporation, occurs earlier with high solids than with conventional coatings; accordingly, solvent evaporation from spray droplets of high solids coatings is markedly reduced (Hill and Wicks, 1982; Ellis, 1983). At later stages of solvent evaporation from a film, the rate of diffusion of solvent molecules to the surface becomes the factor limiting the rate of evaporation of solvent. Solvents with linear rather than branched backbones are desirable since they can diffuse faster (Section 18.3.4).

Hot spraying can help control sagging (Section 23.2.4). When the coating cools on striking the object, the viscosity increase reduces sagging. High-speed electrostatic bell application permits application of coatings at higher viscosity, which also helps control sagging (Section 23.2.3).

It is not uncommon that sagging of high solids coatings cannot be adequately controlled by adjustment of solvent composition of the coating and application variables. An expedient is to impart thixotropy to the systems with additives. For example, dispersions of fine particle size silicon dioxide, precipitated silicon dioxide, bentonite clay treated with a quaternary ammonium compound, or polyamide gels can be added to impart thixotropy. One tries to formulate so that recovery to high viscosity is slow enough to permit reasonable leveling and rapid enough to control sagging. However, such agents increase the high shear viscosity somewhat and require higher solvent levels. They also tend to lower gloss and are frequently not effective at higher temperatures.

The problem of sagging in high solids automotive effect coatings can be particularly severe (Section 30.1.4). Even a small degree of sagging, which might not be noticeable in an opaque coating, is very evident in a metallic coating, since it affects the orientation of the flake pigments. Use of SiO_2 to impart thixotropy is undesirable, since even the small light scattering of SiO_2 is enough to reduce color flop in the coatings. Acrylic microgels have been developed that impart thixotropic flow by flocculation of the swollen gel particles (Maklouf and Porter, 1979; Backhouse, 1981; Wright et al., 1981). In the final film, the index of refraction of the polymer from the microgel is nearly identical with that of the cross-linked acrylic binder polymer so that light scattering does not interfere with color flop. The effect of the gel particles depends on interaction with the low molecular weight acrylic resin; Ishikura et al. (1988) discusses the rheological properties of these systems. There can also be an improvement in the strength of the final film when microgels are incorporated (Boggs et al., 1996). Microgels (which today might be called "nanogels") are sometimes added to coatings for this reason alone.

Another sagging problem that can be encountered with high solids coatings is *oven sagging*, also called *hot sag* (Hill and Wicks, 1982). The coating appears to be fine until it is put into the oven; then sagging occurs. Oven sagging results from the strong temperature dependence of the viscosity of high solids coatings. Compared with conventional coatings, there is a much steeper drop in viscosity as the coated product enters the hot oven, which favors sagging; and the coating continues to flow even after the solvent is gone. Initial stages of cure involve chain extension of the low MW resin. The viscosity does not jump until substantial cross-linking has occurred. Oven sagging can be somewhat controlled by zoning the oven. A lower temperature in the initial zone allows more time for solvent loss, and perhaps, some cross-linking, so viscosity increases as a result of the higher solids or higher molecular weight before the film is subjected to maximum bake temperature.

Water-reducible coatings are less likely to exhibit sagging problems than high solids coatings, but there are circumstances when they undergo delayed sagging. The viscosity of these coatings is very dependent on the ratio of water to solvent, as well as solids content (Section 8.3). As water and solvent evaporate, the residual water to solvent ratio can sometimes decrease, leading to lower viscosity in spite of the higher solids content; and sagging can result. Such behavior can depend on the relative humidity during the flash-off period after spraying. It has been found for a water-reducible acrylic enamel that sagging occurred above, but not below, a critical relative humidity (Brandenburger and Hill, 1979). (See Sections 18.3.3 and 18.3.6 for the definition and discussion of critical relative humidity.)

A problem related to sagging is drip marks. For example, when one applies a coating to a vertical surface in which nail heads have been countersunk, the coating often flows into the holes in the surface. The coating may sag, but, in addition, it will flow out of the nail head depression where the film thickness is greater, leaving a drip. Similar effects can occur at other places on the paint surface, such as dripping from boltholes, cutouts, etc., where film thickness tends to be greater.

Eley and Schwartz (2002) investigated sagging and drip marks in latex paints. They developed a computer model and correlated it with the actual performance of the paints. The model takes into account not only the effect of gravity, thixotropy, and film thickness but also the change in flow properties of the thixotropic system with time. It is pointed out that viscosity varies with shear stress during the flow of films rather than with shear rate as commonly thought. The model also accounts for nonuniform film thickness. As an example, the model was designed to predict the effect of a countersunk nail head filled with liquid coating on a vertical surface. The coating is thicker in the depression, and gravity causes it to flow out of the depression resulting in the formation of a drip mark.

24.4 CRAWLING, CRATERING, AND RELATED DEFECTS

If a coating having a relatively high surface tension is applied to a substrate having comparatively low surface free energy, the coating will not wet the substrate. The mechanical forces involved during application may spread the coating on the substrate surface, but, since the surface is not wetted, surface tension forces tend to draw the liquid coating toward a spherical shape, forming a bump. Meanwhile, solvent is evaporating, and, therefore, viscosity is increasing, so that the viscosity becomes high enough that flow essentially stops. The result is an uneven film thickness with areas having little, if any, coating adjoining areas of excessive film thickness. This behavior is commonly called *crawling* or *retraction*. Figure 24.4 shows an example of crawling. For waterborne coatings, crawling can depend on the rate of establishment of equilibrium surface tension with different surfactants (Schwartz, 1992).

Crawling can result from applying a coating to steel with oily contamination on the surface. It is especially common in coating plastics. In some cases, crawling results from failure to completely remove a mold release agent from a plastic molded part. Application of a top coat to a primer having a lower surface free energy can also lead to crawling. If a primer contains silicone fluids or fluorocarbon surfactants, there is likely to be crawling when a subsequent coat is applied. If one handles a primer surface with bare hands and then applies a relatively high surface tension top coat, it is likely that the top coat will draw away from the oils left behind in fingerprints. This type of crawling, which copies a pattern of low surface tension areas on the substrate, has been called *telegraphing*. However, this is only one of several phenomena called telegraphing.

Crawling can also result from the presence within the coating of surfactant-like molecules that can orient rapidly on a polar substrate surface. The polar group of the surfactant associates with the substrate and the long nonpolar end becomes the surface. If one adds excess silicone fluid to a coating to correct a problem like orange peel, small droplets of insoluble fractions of the poly(dimethylsiloxane) can migrate to the substrate surface and spread on it, leaving

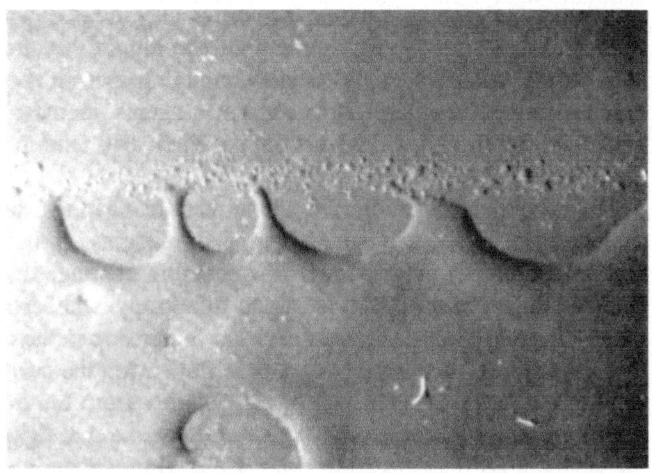

Figure 24.4 Crawling of a top coat applied over a low surface energy primer (7X). Source: Pierce and Schoff (1994). Reproduced with permission of The Federation of Societies for Coatings Technology.

a new substrate surface that the coating cannot wet, resulting in crawling and/or cratering. A little silicone fluid can solve some defect problems, but even a small excess can cause what might be a worse problem. Higher molecular weight fractions of poly(dimethylsiloxane) are reported to be insoluble in many coating formulations (Fink et al., 1990). Modified silicone fluids, such as polysiloxane–polyether block copolymers, have been developed that are compatible with a wider variety of coatings and are less likely to cause undesirable side effects. Many of the proprietary additives offered for sale are of this type. The effect of a series of additives on crawling and other film defects has been reported (Berndimaier et al., 1990).

High solids coatings often have higher surface tensions than conventional coatings. To achieve high solids, lower molecular weight resins with lower equivalent weights must be used, which requires that the concentration of polar functional groups such as hydroxyl groups is higher, and hence, surface tension is generally higher. Also, solvents that provide low viscosity coating solutions are likely to have relatively high surface tension. Since such solvents are desirable for high solids coatings, this further increases the likelihood of crawling problems with high solids coatings.

Cratering is the appearance of small round depressions in a coating's surface, often with a slightly raised crest. They often look somewhat like volcanic craters—hence the name. They are also called *fish eyes.* Schoff (1999) calls cratering "the nastiest, most frustrating defect of all." A schematic drawing and a photograph are shown in Figures 24.5 and 24.6, respectively. Cratering is sometimes mistaken for popping and vice versa; ways to differentiate these film defects are described in Section 24.7.

Cratering results from a small particle or droplet of low surface tension contaminant. It can be on the substrate, in the coating, or land on the wet surface of a freshly applied film. The particles or droplets can be very small. Some of the low surface tension material dissolves in the adjacent film, creating a localized surface tension differential. As a result of the Marangoni effect, this low surface tension part of

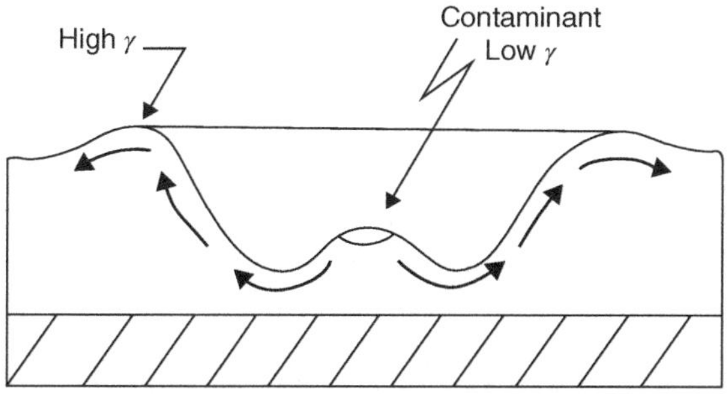

Figure 24.5 Schematic diagram of a crater. Source: Pierce and Schoff (1994). Reproduced with permission of The Federation of Societies for Coatings Technology.

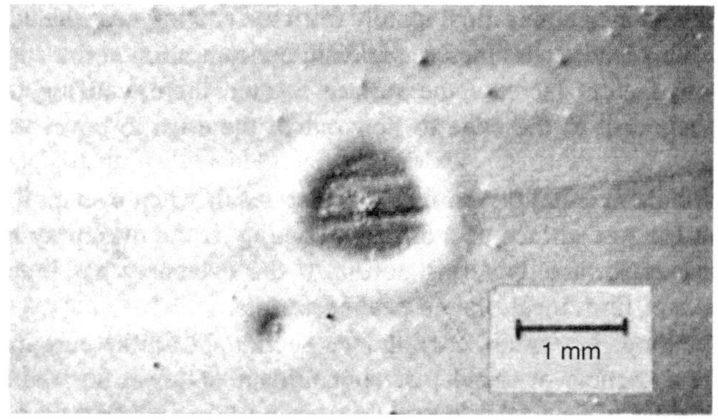

Figure 24.6 A typical crater. Source: Pierce and Schoff (1994). Reproduced with permission of The Federation of Societies for Coatings Technology.

the film flows away from the particle to try to cover the surrounding higher surface tension liquid coating. Furthermore, the flow is self-sustaining, since solvent evaporates as the flow occurs, which increases the differential in surface tension. However, loss of solvent causes an increase in viscosity, eventually freezing the crater in place.

Craters must be avoided in applying coatings. One approach is to minimize the probability of low surface tension contaminants arriving on the wet coating surface. For example, the spraying of lubricating oils or silicone fluids on or near the conveyor carrying freshly coated parts is an almost sure way to cause craters. Paint booth air must be filtered multiple times. Otherwise harmless dust particles may adsorb enough oily contaminants to cause cratering.

In a number of cases, particles from cosmetics worn by production workers have caused cratering of automotive coatings. In some factories where paint is applied, workers are forbidden to wear cosmetics of any kind. A sporadic cratering problem in an automotive assembly plant was, after much difficulty, traced to the lubricant sprayed once every two weeks into a soft drink vending machine. An author of this book has observed persistent cratering in a mirror factory, likely caused by the use of aerosol deodorants elsewhere in the factory. All aerosol spray cans are suspect. Purchasers of industrial coatings sometimes instruct their suppliers to avoid all silicone and perfluorocarbon fluids in formulating the coatings. They fear that airborne droplets of such coatings may cause cratering of other coatings.

It is possible in some cases to determine the nature and source of the contaminating particle causing craters. Schoff (1999) and Hare (2001) advocate use of low power (2x-60x or 10x-80x) optical microscopes for preliminary investigation of craters. If necessary, more powerful optical microscopes and scanning electron microscopy (SEM) can be applied to the problem. SEMs with an X-ray attachment (EDX) are especially valuable. Two other powerful tools for such analyses are laser microprobe mass analysis

(LAMMA) and time-of-flight secondary ion mass spectroscopy (ToF-SIMS). LAMMA is primarily useful in identifying inorganic materials and ToF-SIMS can identify organic contaminants; even different silicone oils can be distinguished. In the case of a crater through a clear coat down to a base coat, ToF-SIMS showed that the contaminant was a polydimethylsiloxane and the structure of the siloxane was determined. The source of the siloxane was traced to a resin provided by a new supplier. In another case, craters in a primer coat on a car body were shown by LAMMA to be from a skin cream used by an employee in the paint shop. In another case, a contaminating particle in an E-coat primer was shown by LAMMA to be a welding pearl from the copper-coated steel welding wire (Wolff et al., 2004). McDonough et al. (2006) describe a case study of identifying the culprit in cratering of an epoxy coating inside food cans. Many of the previously mentioned tools were used. The cause was traced to contaminated cooling tower water used to cool the lubricant/coolant fluid used in production. Even with these powerful tools, the cause cannot always be found, perhaps because the cratering agent has evaporated or dissolved in the film.

In most factories, the presence of contaminating particles cannot be entirely avoided; in that case, the formulator must design coatings that minimize the probability of cratering. Lower surface tension coatings are less likely to form craters, since fewer contaminating particles will have even lower surface tensions. Alkyd coatings have low surface tensions and seldom have cratering (or crawling) problems. In general, polyester coatings are more likely to crater than acrylic coatings, which tend to have lower surface tensions. High solids coatings, because of their generally higher surface tensions and low viscosities even after solvent loss, are more prone to cratering problems than conventional solids coatings. Some waterborne coatings are also vulnerable to cratering. Powder coatings are quite vulnerable to cratering, as discussed in Chapter 27.

Additives can be used to minimize cratering. Small amounts of silicone fluid generally eliminate cratering, but, as noted before, caution is required in selecting the amount and type of silicone fluid used in order to avoid crawling or recoat adhesion problems. Fluorosurfactants can be effective but, again, must be used carefully. Poly(2-ethylhexyl acrylate) (polyoctyl acrylate) or other hydrophobic polyacrylate additives usually reduce cratering. These additives provide a uniform surface tension, which promotes spreading of the low surface tension contaminant widely across the surface. If the entire surface has a uniform surface tension, there will be no surface tension differential-driven flow and no cratering.

Suppliers offer a variety of proprietary additives to control cratering, as well as for other purposes. Often these are dilute solutions of silicones, modified silicones, or combinations of additives. In recent years, new organically modified silicones have been introduced (e.g., see Penny and Berglund, 2014). However, it is necessary to test such additives in the specific paint formulation. Given all the variables involved, an additive that works well in one paint may not be effective in another. Some suppliers offer additive test kits with samples of several dozen additives and recommendations for their use. Comparisons of the effects of a range of additives on the control of defects, such as cratering, have been reported (Schnall, 1991; Waelde et al., 1994). Side effects, such as reduction of gloss and loss of adhesion to coatings applied over the treated coatings, are discussed. Modified silanes and siloxanes may be preferable to siloxanes themselves. For example, siloxanes modified with a polyether terminal group or side chain are reported to prevent film defects in waterborne coatings and to have little effect on recoatability. They are prepared by reacting allyl-terminated polyethers with poly(dimethylsiloxanes) having Si—H bonds (Spiegelhauer, 2002). In another example Seraj et al. (2014) report that a silane-modified epoxy resin sharply reduces cratering in a cathodic electrocoat.

There are many other examples of film defects that result from surface tension differential-driven flows. Tin plate sheets are coated by roller application, and the coated sheets are mounted on warm wickets that carry the sheets approximately vertically through an oven. In some cases, one can see a pattern of the wicket as a thin area on the final coated sheet. Heat transfer to the tin sheet is fastest where it is leaning against the warm metal wicket. The surface tension of the liquid coating on the opposite side decreases locally in the pattern of the wicket because of the higher temperature. The lower surface tension material flows toward the higher surface tension surrounding coating, leaving an area of thin coating in the shape of the wicket pattern, a defect that has also been called *telegraphing*. Similarly, on large plastic parts with thicker reinforced areas on the back of the part, telegraphing through to the top coat(s) can be encountered.

With the adoption of high solids clear coats to paint plastic auto body components, there has been an increasing problem with "bond line readout." Typically this occurs over the areas of SMC plastic where, on the other side, there are adhesive joints to the car structure. A series of model system studies showed that the bond line defects resulted from differential surface tension-driven flow. The areas over the bond line heat more slowly than the adjacent areas, setting up differentials in surface tension and flow away from this bond line (Blunk and Wilkes, 2001).

In spraying flat sheets, one can get an effect called *picture framing*, or *fat edge*. The coating is thickest at the edges and thinner than average adjacent to the edge. The contrast in hiding of the substrate can make the differences in film thickness very evident. Solvent evaporates most rapidly from the coating near the edge, where the airflow is greatest. This leads to a lower temperature and to an increase in resin concentration at the edge. Both factors increase the surface tension there, causing the lower surface tension coating adjacent to the edge to flow out to the edge to cover the higher surface tension coating.

Surface tension differential-driven flow can also result when overspray from spraying one coating lands on the wet surface of a different coating. If the overspray has lower surface tension than the wet surface, cratering occurs. If the overspray has higher surface tension, local orange peeling results.

In applying coatings by curtain coating (Section 23.5), the curtain of coating must remain intact. If a particle or droplet of contaminant of lower surface tension than the coating lands on the surface of the flowing curtain, surface tension differential-driven flow can cause a thin area in the curtain, leading to a hole in the curtain, which results in an uncoated area on the panel being coated. The problem is minimized by using coatings of the lowest possible surface tension. Since the curtain is flowing, dynamic surface tension is the important quantity (Bierwagen, 1991).

24.5 FLOATING AND FLOODING: HAMMER FINISHES

Two related defects result from uneven distribution of pigment in a film as it is drying: *floating* and *flooding*. Both phenomena are sometimes called floating. We follow the more common terminology in which floating describes a mottled effect and flooding is used when the surface color is uniform, but darker or lighter in color than should result from the pigment combination used. Floating results from horizontal separation of different pigments within the film, whereas flooding results from vertical separation.

Floating is most evident in coatings pigmented with at least two pigments. For example, a light blue gloss enamel panel can show a mottled pattern of darker blue lines on a lighter blue background. The pattern tends to be hexagonal, but seldom perfectly so. Alternatively, with a different light blue coating, the color pattern might be reversed: the lines could be light blue with a darker blue background. These effects result from pigment segregations that occur as a result of convection current flows driven by surface tension differentials while a film is drying. Rapid loss of solvent from a film during drying leads to considerable turbulence. Convection patterns are established whereby coating material flows up from lower layers of the film and circulates back down into the film. As the fresh material flows across the surface before it turns back down, solvent evaporates, concentration increases, temperature drops, and surface tension increases. The resultant surface tension differential sustains the convection current. The flow patterns are roughly circular, but as they expand, they encounter other flow patterns and the convection currents are compressed. If the system is quite regular, a pattern of hexagonal *Bénard cells* is established. The cells are named after a seventeenth-century French scientist who pointed out the commonness of hexagonal flow patterns in nature. As solvent evaporation continues, viscosity increases, and it becomes more difficult for the pigment particles to move. The smallest particle size, lowest density particles continue moving longest, and the largest, highest density particles stop moving sooner. A segregated pattern of floating results.

Floating is particularly likely to occur if one pigment is flocculated and the other is a nonflocculated dispersion of finer particle size. The fine particle size pigment keeps moving longest and is trapped where the convection current turns back into the film at the border between adjacent cells. The border between the cells has a higher concentration of the finer particle size material, whereas the center of the cells is more concentrated in the coarser pigment. If, in the example of the light blue coating, the white pigment is flocculated and the blue is not, one will find darker blue lines on a lighter blue background. If, on the other hand, the blue is flocculated and not the white, there will be lighter blue lines on a darker blue background. Figure 24.7 shows a schematic diagram of convection patterns in Bénard cell formation. Floating can also occur in single pigment coatings, resulting in uneven pigmentation near the surface, which reduces gloss.

Floating can be reduced by properly stabilizing pigment dispersions so that neither pigment is flocculated. However, floating can occur even without flocculation, resulting from the use of pigments with very different particle sizes and densities. An example is use of fine particle size high color carbon black with titanium dioxide to make a gray coating. Not only is the particle size of the TiO_2 many times that of the carbon black, but the TiO_2 also has about

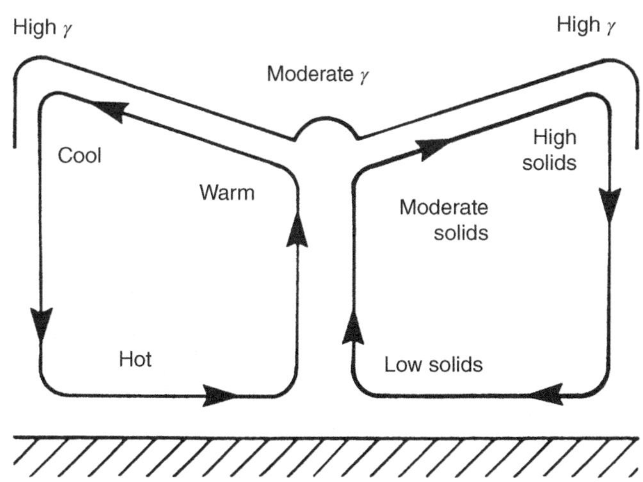

Figure 24.7 Schematic diagram of Bénard cell formation. Source: Pierce and Schoff (1994). Reproduced with permission of The Federation of Societies for Coatings Technology.

fourfold higher density. A much larger particle size, weaker black, such as lamp black, can be used to make a gray with a lower probability of floating.

Floating is reduced if slower evaporating solvents are used. Surface tension differentials are then less likely to be established, and Marangoni flow is limited. If a coating exhibits floating, the formulator should first reformulate to eliminate the problem; most commonly, selective pigment flocculation is the cause and should be corrected. Sometimes the flocculation is caused by the solvents used to reduce the coating to application viscosity, in which case solvent change eliminates the problem. In other cases, a change in the resin or dispersing agent used to stabilize the pigment dispersion is required (Section 21.1.3). A second approach is to use an additive. As with other flow phenomena driven by surface tension differentials, floating can be prevented by adding a silicone fluid. To avoid crawling, the minimum amount of silicone must be used, which can be facilitated by using a very dilute solution of the silicone fluid in solvent. Since slow evaporating solvents also help reduce floating, slow evaporating solvents are commonly used to make additive solutions for this purpose.

The term *flooding* is applied when the color of the surface is uniform, but different than should have been obtained from the pigment combination involved. For example, one might have a uniform gray coating, but a darker gray than that expected from the ratio of black to white pigments. The most troubling part of flooding is that the extent of flooding can vary with the conditions encountered during application, leading to different colors on articles coated with the same coating. Flooding results from surface enrichment by one or more of the pigments in the coating (Schnall, 1989; Pierce and Schoff, 1994). The stratification is thought to occur as a result of different rates of pigment settling within the film, which are caused by differences in pigment density and size or flocculation of one of the pigments. Flooding is accentuated by thick films, low vehicle viscosity, and low evaporation rate solvents—that is, any factor that tends to keep the film at low viscosity longer and allow more pigment settling. Flooding is easily detected by rubbing a section of a wet film: if the rubbed area has a different color, the coating is flooding. The remedies are to avoid flocculation and low density, fine particle size pigments and, if possible, to use faster evaporating solvents and higher viscosity vehicles. Schnall (1989) further discusses floating, flooding, and related color defects.

Floating and flooding can also occur in latex paints (Ying et al., 2004). Latex paints contain one or more surfactants, as well as inorganic and organic pigment dispersions. In some cases, the combinations are unstable and lead to flocculation of the paint. In other cases, there can be incompatibility that affects only one of the pigments, which leads to floating or flooding. When a latex paint is diluted with water, floating can result. This is attributed to a shift in the equilibrium between the dispersed particles and water, resulting in less stabilization of the pigment dispersions. Also, the viscosity decreases facilitating movement of particles; and increased surface tension may promote formation of Bénard cells. Expedients to reduce flooding are:

- Decreasing the particle size of the inert pigments
- Using a mixture of higher and lower molecular weight acrylic carboxylate dispersants
- Assuring compatibility among the thickeners and the dispersants

While floating is usually undesirable, ingenious coatings formulators have taken advantage of the problem by purposely inducing floating to make attractive coatings. The coatings are called *hammer finishes* because they look a bit like the pattern one would get by striking a piece of metal with a ball-peen hammer. Such coatings contain large particle size, non-leafing aluminum pigment, and dispersions of transparent, fine particle size pigments, commonly phthalocyanine blue. One way of getting a hammer effect is to spray a metallic blue coating and then spray a small amount of solvent on the wet film to create surface tension differentials. There are *self-hammer* coatings, formulated to give a hammer finish pattern without need for a spatter spray of solvent. Fast evaporating solvents are used with a resin, such as a styrenated alkyd resin, which dries rapidly. Hammer finishes were once used on a large scale to hide defects in cast iron components. Their use decreased as smooth plastic parts replaced metal castings.

24.6 WRINKLING: WRINKLE FINISHES

The term *wrinkling* refers to the surface of a coating that looks shriveled or wrinkled with many small hills or ridges and valleys. In some cases, a wrinkle pattern is so fine that, to the unaided eye, the film appears to have low gloss rather than to look wrinkled. However, under magnification, the surface can be seen to be glossy but wrinkled. In other cases, the wrinkle patterns are broad or bold and are readily visible to the naked eye. Wrinkling results when the surface of a film becomes high in viscosity, while the bottom of the film is still relatively fluid. This situation can result from rapid solvent loss from the surface, followed by later solvent loss from the lower layers. It can also result from more rapid cross-linking at the surface of the film than in the lower layers of the film. Subsequent solvent loss or cure in the lower layers results in shrinkage, which pulls the surface layer into a wrinkled pattern. Wrinkling is more apt to occur with thick films than with thin films.

Early examples of wrinkling were with drying oil films, especially if all or part of the oil was tung oil and cobalt salts were used as the only drier. Tung oil cross-links relatively rapidly when exposed to oxygen from air, and cobalt salts are active catalysts for the autoxidation reaction near the surface, but they are poor through driers (Section 14.2.2). These factors favor differential surface cure, which results in wrinkling of the surface layer. The wrinkle pattern can be fine or bold depending on the ratio of tung oil to other drying oils or, in the case of alkyd systems, the oil length of the alkyd, as well as to the ratio of cobalt salts to other driers, such as lead (or zirconium) salts that promote through dry.

While, in many cases, wrinkling is undesirable, formulators have turned the disadvantage into an advantage. For many years, *wrinkle finishes* were sold on a large scale for applications such as office equipment. Like hammer finishes, wrinkle finishes hid unevenness of cast metal parts. Their usage has also decreased as plastic molded parts have supplanted many metal castings.

Today, wrinkling is usually an undesired defect. It is most often encountered in improperly formulated or applied MF cross-linked coatings in which amines are used to neutralize acidity in the coating formulation and/or amine-blocked sulfonic acid catalysts are used for package stability. The probability of wrinkling in such coatings increases as the volatility of the amine increases. For example, triethylamine leads to wrinkling under conditions in which dimethylaminoethanol does not. Increasing catalyst concentration tends to increase wrinkling, as does greater film thickness (Wicks and Chen, 1978).

Another situation in which wrinkling can occur is in UV curing of pigmented acrylate coatings by free radical polymerization (Section 29.5). High concentrations of photoinitiator are required to compete with absorption by the pigment. Penetration of UV through the film is thereby reduced by absorption by the pigment as well as by the photoinitiator. The rapid cross-linking at the surface and slower cross-linking in the lower layers of the film may cause wrinkling. Wrinkling is likely to be more severe if the curing is done in an inert atmosphere rather than in air. In the latter case, the cure differential is reduced by oxygen inhibition of surface cure. UV curing by cationic polymerization, which is not air inhibited, is even more prone to surface wrinkling.

24.7 BUBBLING AND POPPING

Bubbles can occur near the surface of a film; *popping* is the formation of broken bubbles at the surface of a film that do not flow out. Both effects can occur when the surface viscosity increases to a high level, while volatile material remains in the lower parts of the film. They are most common in baked coatings. If the viscosity is very high at the surface, some of the bubbles of solvent rise to the surface, but do not break. Popping occurs when the viscosity at the surface of a film increases sufficiently that solvent bubbles break, but the coating does not flow out before further viscosity increase prevents leveling. If the "pops" are very small, the effect is sometimes called *pinholing*. There often is a mixture of open pops, bubbles, and pinholes in a surface.

In solventborne coatings, popping and bubbling result from rapid loss of solvent at the surface of a film during initial flash-off. The surface develops a high viscosity relative to that of the solvent-rich lower layers of the film. When the coated object is placed in an oven, solvent volatilizes in the lower layers of the film, creating bubbles that do not readily pass through the high viscosity surface. As the temperature increases further, the bubbles expand, finally bursting through the top layer, resulting in popping. The viscosity of the film meanwhile has increased to the extent that the coating cannot flow together to heal the eruptions.

Popping can also result from, or at least be worsened by, entrapment of air bubbles in a coating. If the surface of the film has a high viscosity, the bubbles may remain in the film until the coating goes into the oven. The air expands with higher oven temperatures; and the bubbles may burst through the surface. Air bubbles are especially likely to be entrapped during spray and hand roller application of waterborne coatings. Bubbling and popping can also result from solvent that remains in primer coats or base coats when the top coat is applied too soon.

Waterborne urethane coatings that have free isocyanate groups, such as moisture-curable urethanes (Section 12.6), and 2K waterborne urethane coatings (Section 12.7.3) can generate CO_2 bubbles from the reaction of isocyanate groups and water.

In coating plastics, solvents can dissolve in the plastic, causing bubbling or popping during bake. More commonly, air or moisture in voids in the plastic blow out into the coating, resulting in popping. This is usually called *gassing*, but it may be indistinguishable from solvent popping. Similarly, hydrogen, methane, or other volatiles trapped in or under the zinc layer in electrogalvanized steel can blow out and cause popping (*galvanized gassing*). Electrical discharge and hydrogen evolution during cationic electrodeposition at high voltage can produce pop-like pinholes and craters (Schoff, 2013). Another potential cause of popping is evolution of volatile by-products from cross-linking reactions, which occur after the surface viscosity has increased to the extent that the bubbles of volatile material cannot readily escape through the surface. Automotive waterborne base coat/solventborne clear coat combinations can exhibit popping if the base coat bake does not remove sufficient water. The clear coat may also pop if the dehydration bake is so hot that a porous base coat results. If solvent from the clear coat is imbibed by the base coat, it may later blow out through the partially cured clear coat.

The probability of popping increases with film thickness, since there is a greater chance of developing a differential in solvent content as film thickness increases. A means of evaluating the relative likelihood that a series of coatings will show popping is to determine the maximum film thickness of each coating that can be applied without popping when the films are prepared, flashed off, and baked under standardized conditions (Watson and Wicks, 1983). This thickness is called the *critical film thickness for popping*. Formulators strive to make it as high as possible. Popping can be minimized by spraying more slowly in more passes, by longer flash-off times before the object is put into the main oven, and by zoning the ovens so that the first stages are relatively low in temperature. The probability of popping can also be reduced by having a slow evaporating good solvent in the solvent mixture. This tends to keep the surface viscosity low enough for bubbles to pass through and heal before the viscosity at the surface becomes too high.

Popping can be particularly severe with water-reducible baking enamels, as shown in Table 24.1. The pairs of enamels were identical except that in one set, they were reduced for application with solvent and in the other set, with water (Watson and Wicks, 1983). As can be seen, the critical film thickness for popping was consistently lower for the water-reduced coatings. The data in Table 24.1 also illustrate another variable that affects the probability of popping. Critical film thickness for popping in these coatings decreases as the T_g of the acrylic resin in the coating increases. This is true in both the solvent-reduced and water-reduced coatings, but the effect is particularly large in the water-reduced compositions.

There are probably many reasons for the greater difficulty of controlling popping of water-reducible coatings. Solvents with different evaporation rates are available for adjusting formulations of solventborne coatings, but water has only one vapor pressure/temperature curve, and its curve is steeper than for any organic solvent. Water can be retained by forming relatively strong hydrogen bonds with polar groups on resin molecules at room temperature; these hydrogen bonds break at higher temperature, releasing the water. The heat of vaporization of water, $2260\,J\,g^{-1}$, is higher than that of organic solvents, $373\,J\,g^{-1}$ for 2-butoxyethanol as an example. This higher heat of vaporization slows the rate of temperature increase of films of water-reduced coatings in an oven, further increasing the probability of popping (Schnall, 1989; Section 18.3.6).

In contrast to increased probability of popping with higher T_g water-reducible coatings, popping is more likely to occur with lower T_g latex polymers. This contrasting behavior may reflect the higher likelihood of premature coalescence at the surface of a lower T_g latex occurring before the water has completely evaporated.

Water-reducible automotive base coats are feasible because relatively thin films are applied, and there are provisions for adequate flash-off. Thicker films of the same coatings would pop. Thin films also promote orientation of flake pigments.

Popping and cratering often appear similar but should be distinguished because the remedies are different. One clue is that popping tends to be worst where the film is thickest, while cratering is more or less independent of thickness. Another distinction is that pops are often accompanied by unbroken bubbles.

Table 24.1 Critical Film Thickness for Popping

	Critical dry film thickness (μm)	
Copolymer T_g (°C)	Water	Solvent
−28	50	120
−13	30	>70, <95
−8	20	>70, <95
14	10	55
32	5	25

24.8 FOAMING

During manufacture and application, a coating is agitated and mixed in air, creating the opportunity for foam formation. Foam can cause surface irregularities in the final film, and incorporation of foam in a liquid coating can lead to pinholing or popping. The problem can be particularly severe with waterborne coatings applied by spray, especially airless spray, or by hand rollers. Formation of foam involves the generation of a large amount of surface area, so that the lower the surface tension, the less energy required to generate a given amount of foam.

Foam bubbles in low viscosity pure liquids are not stable and break almost instantaneously in the absence of surfactants or other ingredients that stabilize the foam. Although pure water has a high surface tension and, therefore, might not be expected to generate foam bubbles easily, waterborne paints contain chemical species that can rapidly migrate to the surface of a bubble to stabilize it. For example, a surfactant not only reduces the surface tension of water, facilitating foam formation, but also migrates to the surface to give an oriented surface layer with a high viscosity, stabilizing the foam bubbles. In formulating a latex paint, an important criterion in selecting surfactants or water-soluble polymers as thickeners is their effect on foam stabilization (Bierwagen, 1975; Schwartz and Bogar, 1995). Acetylene glycol surfactants, such as 2,4,7,9-tetramethyl-5-decyne-4,7-diol alkoxylates, are reported to be effective surfactants that do not increase the viscosity of the surface of bubbles as much as surfactants such as alkylphenol ethoxylates (Heilin et al., 1994).

A variety of additives can be used to break foam bubbles. Most depend on creating surface tension differential-driven flow on the surface of bubbles. If the surface tension of a spot on the bubble surface can be lowered, liquid from that area will flow away to try to cover neighboring higher surface tension areas, weakening and eventually breaking the thin wall of the bubble. For example, poly(dimethylsiloxane) fluids are effective in breaking a variety of foams, since their surface tension is low compared with almost any foam surface. Of course, as in other uses of silicone fluids, a little may be fine, but a little extra can cause problems. Other low surface tension additives, such as poly(octyl acrylates), act as defoamers. Small particle size hydrophobic SiO_2 can also act as a defoamer and/or a carrier for active defoaming agents (Smith, 1988). Also, a small amount of immiscible hydrocarbon solvent will often reduce foaming of an aqueous coating. However, a hydrocarbon solvent used in a latex paint as a defoamer was shown to cause flocculation (Kozakiewicz et al., 1993). In this study the paint formula contained several surfactants; and the order in which the surfactants were added controlled whether or not flocculation occurred. Siloxanes with a polyether terminal group have been reported to prevent foaming in waterborne coatings, as well as to minimize entrapment of air bubbles (Spiegelhauer, 2002).

Przybyla et al. (2014) describe two methods for testing defoamers. These authors point out that very low and near-zero VOC waterborne coating formulations often require complex combinations of surface-active ingredients. This has complicated defoamer selection and made many conventional defoamers ineffective. They recommend defoamers that include polymeric additives. Penny and Berglund (2014) recommend an organically modified silicone defoamer; they also claim improved block resistance and other benefits.

Several companies sell lines of antifoam additives and offer test kits with small samples of their products. The formulator evaluates the products in a coating to find one that overcomes, or at least minimizes, the foaming problem. While it is possible to predict which additive will break a foam in a relatively simple system, such predictions are difficult for waterborne paints because of the variety of components that could potentially be at the foam interface. The combination of surfactants, wetting agents, water-soluble polymers, and antifoam agents can be critical. Empirical testing of different additives in each paint formula is still necessary (Przybyla et al., 2014). A variety of test methods have been used to compare the tendency of coatings to foam (Smith, 1988; Kozakiewicz et al., 1993; Heilin et al., 1994).

24.9 DIRT

Dirt has been called the most common defect of all. Schoff (2015) says, "The paint is usually blamed for dirt problems, but rarely is the culprit." In a factory, a wide variety of solid particles can land on the wet surface of a freshly applied film. If the particles contain low surface tension materials, they can cause cratering. If not, they become defects if large enough to be visible. Sanding dust, floor dirt, dust blown or tracked in from outdoors, fibers from wiping cloths or clothing worn by operators, and oven dirt are examples of such particles. Outdoors, dirt on the substrate and windblown dust are common problems. Even insects are sometimes fatally attracted to wet paint. Optical microscopy is invaluable for identifying the type of dirt, and FTIR spectroscopy can often identify the source. Prevention of dirt problems requires clean raw materials, clean paint, and a clean paint shop, preferably isolated from the rest of the factory. The air supply to spray booths and spray guns must be clean. Sanding should be minimized and sanding dust cleaned up before painting. Ovens should be carefully and frequently cleaned. Lint-free protective clothing and wiping cloths reduce lint contamination.

Even with precautions, some dust is almost inevitable. Therefore, it is a desirable characteristic of coatings that they dry/cure rapidly to a state where dust particles will not stick to their surfaces. This characteristic is especially important for coatings that are not baked. Simple *dust free time* measurements involve sprinkling a specified dust or lint on the surface at time intervals and then blowing it off. Fast drying is even more important for shops that do not or cannot take all the recommended precautions.

GENERAL REFERENCES

Heilen, W., *Additives for Waterborne Coatings*, Vincentz, Hannover, **2009**.

Koleske, J. V., *Paint and Coating Testing Manual*, 15th ed., ASTM, West Conshohocken, **2012**.

Pierce, P. E.; Schoff, C. K., *Coating Film Defects*, 2nd ed., Federation of Societies for Coatings Technology, Blue Bell, PA, **1994**.

Schoff, C. K., Coatings Clinic, on the last inside page of almost every issue of JCT CoatingsTech starting from 1(4) (**2004**). Publication of an archive of these articles in the American Coatings Association website is planned.

Weldon, D. G., *Failure Analysis of Paints and Coatings*, Revised Edition, John Wiley & Sons, Inc., Chichester, **2009**.

REFERENCES

Backhouse, A. J., US patent 4,268,547 (**1981**).

Bauer, D. R.; Briggs, L. H., *J. Coat. Technol.*, **1984**, *56*(716), 87.

Berndimaier, R., et al., *J. Coat. Technol.*, **1990**, *62*(790), 37.

Bierwagen, G. P., *Prog. Org. Coat.*, **1975**, *3*, 101.
Bierwagen, G. P., *Prog. Org. Coat.*, **1991**, *19*, 59.
Blunk, R. H. J.; Wilkes, J. O., *J. Coat. Technol.*, **2001**, *73*(918), 63.
Boggs, L. J., et al., *J. Coat. Technol.*, **1996**, *68*(855), 63.
Brandenburger, L. B.; Hill, L. W., *J. Coat. Technol.*, **1979**, *51*(659), 57.
Cohu, O.; Magnin, A., *Prog. Org. Coat.*, **1996**, *28*, 89.
Eley, R. P.; Schwartz, L. W., *J. Coat. Technol.*, **2002**, *74*(932), 43.
Ellis, W. H., *J. Coat. Technol.*, **1983**, *55*(696), 63.
Fink, F., et al., *J. Coat. Technol.*, **1990**, *62*(791), 47.
Hahn, F. J., *J. Paint Technol.*, **1971**, *43*(562), 58.
Hare, C. M., *J. Prot. Coat. Linings*, **2001**, *18*(12), 57.
Heilin, W., et al., *J. Coat. Technol.*, **1994**, *66*(829), 47.
Hill, L. W.; Wicks, Z. W., Jr., *Prog. Org. Coat.*, **1982**, *10*, 55.
Ishikura, S., et al., *Prog. Org. Coat.*, **1988**, *15*, 373.
Kojima, S.; Moriga, T., *Polym. Eng. Sci.*, **1995**, *35*, 1998.
Kozakiewicz, J., et al., *J. Coat. Technol.*, **1993**, *65*(824), 47.
Lade, R. K., Jr., et al., *Prog. Org. Coat.*, **2015**, *86*, 49–58.
LeSota, S., Ed., *Coatings Encyclopedic Dictionary*, Federation of Societies for Coatings Technology, Blue Bell, PA, **1995**.
Maklouf, J.; Porter, S. M., US patent 4,180,619 (**1979**).
McDonough, F.; Niemeyer, W.; Shuster, M., *Modern Microscopy*, **2006**, February 8.
Ngo, P.-A. P., et al., *J. Coat. Technol.*, **1993**, *65*(821), 29.
Oliveira, M. B.; Coutinho, J. A. P.; Queimada, A. J., *Fluid Phase Equilib.*, **2011**, *303*, 56–61.
Orchard, S. E., *Appl. Sci. Res.*, **1962**, *A11*, 451.
Overdiep, W. S., in Spalding, D. B., Ed., *Physicochemical Hydrodynamics*, V.G. Levich Festschrift, Advance Publications Ltd., London, **1978**, Vol. *II*, p 683.
Overdiep, W. S., *Prog. Org. Coat.*, **1986**, *14*, 159.
Owen, M. J., Surface and Interfacial Properties in *Physical Properties of Polymers Handbook*, American Institute of Physics, Woodbury, **1996**.
Patton, T. C., *Paint Flow and Pigment Dispersion*, 2nd ed., Wiley-Interscience, New York, **1979**, p 554.
Patton, T. C., *Paint Flow and Pigment Dispersion*, 2nd ed., Wiley-Interscience, New York, **1979**, p 572.
Patton, T. C., *Paint Flow and Pigment Dispersion*, 2nd ed., Wiley-Interscience, New York, **1979**, p 578.
Penny, T.; Berglund, B., *Proceedings of the American Coatings Conference*, Atlanta, GA, April, **2014**, Paper 9.4.
Pierce, P. E.; Schoff, C. K., *Coating Film Defects*, 2nd ed., Federation of Societies for Coatings Technology, Blue Bell, PA, **1994**.
Przybyla, D.; Peter, W.; Pirrung, F., *Proceedings of the American Coatings Conference*, Atlanta, GA, April, **2014**, Paper 9.6.
Schnall, M., *J. Coat. Technol.*, **1989**, *61*(773), 33.
Schnall, M., *J. Coat. Technol.*, **1991**, *63*(792), 95.
Schoff, C. K., *J. Coat. Technol.*, **1999**, *71*(888), 57.
Schoff, C. K., *JCT CoatingsTech*, **2013**, August.
Schoff, C. K., *JCT CoatingsTech*, **2015**, June, 56.
Schwartz, J., *J. Coat. Technol.*, **1992**, *64*(812), 65.
Schwartz, J.; Bogar, S. V., *J. Coat. Technol.*, **1995**, *67*(840), 21.
Scriven, L. E.; Sternling, C. V., *Nature*, **1960**, *187*(4733), 186.
Seraj, S.; Ranjbar, Z.; Jannesari, A., *Prog. Org. Coat.*, **2014**, *77*(11), 1735–1740.
Smith, R. E., *Ind. Eng. Chem. Prod. Res. Dev.*, **1983**, *22*, 67.
Smith, R. E., *J. Coat. Technol.*, **1988**, *60*(761), 61.
Spiegelhauer, S., *Proceedings of the International Waterborne, High-Solids, and Powder Coatings Symposium*, New Orleans, LA, **2002**, pp 161–170.
Tachi, K., et al., *J. Coat. Technol.*, **1990**, *62*(791), 19.
Waelde, L. R., et al., *J. Coat. Technol.*, **1994**, *66*(836), 107.
Watson, B. C.; Wicks, Z. W., Jr., *J. Coat. Technol.*, **1983**, *55*(698), 59.
Wicks, Z. W., Jr.; Chen, G. F., *J. Coat. Technol.*, **1978**, *50*(638), 39.
Wilson, S. K., *Surf. Coat. Int.*, **1997**, *80*, 162.
Wolff, U., et al., *Prog. Org. Coat.*, **2004**, *51*, 163.
Wright, H. J., et al., US patent 4,290,932 (**1981**).
Wu, S. H., *J. Appl. Polym. Sci.*, **1978**, *22*, 2769.
Yang, L., et al., *Phys. Fluids*, **2014**, *26*, 113103.
Ying, H., et al., *JCT Res.*, **2004**, *1*(3), 213.

Chapter 25

Solventborne and High Solids Coatings

Chapters 25 through 29 present the principal classes of coatings: solventborne, waterborne, electrodeposition, powder, and radiation cure coatings. Our intent in these chapters is to discuss the principles involved in each of these classes and to compare the various types of resins applicable to each group. In Chapters 30 through 33, end-use applications for these various types of coatings are discussed.

Historically, almost all organic coatings were solventborne or solvent-free paints. Exceptions were paints sold in powder form that were suspended in water by the user. Starting in the 1930s waterborne coatings gradually began to replace many solventborne coatings. By the 1970s, powder, electrodeposition, and radiation curable coatings were also gaining importance.

The original motivations for waterborne coatings were to reduce fire hazards and odor and to permit cleanup with water. Since the 1960s, a major driving force has been to reduce VOC emissions (Chapter 18). Despite predictions that solventborne coatings will wither, they accounted for 48% of the global market in 2014 (Challener, 2014). Further replacement of solventborne by other technologies is expected as tighter environmental regulations are imposed and other technologies advance. However, in many situations, solventborne coatings have advantages over waterborne and other alternatives, for example:

- Capital cost for application is generally lower, especially as compared with waterborne coatings for which stainless steel equipment is needed.
- Electrostatic spray installations for solventborne coatings are less expensive than that for waterborne coatings.
- Solvent evaporation rates do not depend on humidity, eliminating a variable that can bedevil waterborne coatings.
- There are fewer problems of air entrapment and popping.
- In certain applications, solventborne coatings offer superior film properties, especially where exceptional resistance to water and corrosion are required.
- Solvent incineration and solvent recovery processes can almost eliminate VOC. They are economically practical in certain situations, notably in coil coating. Furthermore, the technology of high solids solventborne coatings continues to advance, reducing VOCs while retaining the inherent advantages of solventborne.
- Technical improvements in high solids coatings and in-application equipment have made it possible to satisfy many VOC regulations.
- In the United States and Canada, certain solvents are exempt from VOC regulations on the basis that their photochemical reactivity is minimal (Section 18.9.1). Exempt solvents include acetone, *t*-butyl acetate, CO_2, 4-chlorobenzotrifluoride (PCBTF), methyl acetate, dimethyl carbonate (DMC), and propylene carbonate (PC). Evaporation rates of these solvents are fast, but they can be combined with slow evaporating solvents to formulate coatings that satisfy many current regulations. High solids coatings having excellent properties and VOCs of 200–300 g l^{-1} (1.7–2.5 lb gal^{-1}) or below are available with such formulations; and there is potential for further VOC reduction if required.
- There is considerable potential for introducing bio-based resins into solventborne coatings, and only part of this potential has been realized.
- In certain cases, cradle-to-grave analysis shows that solventborne coatings have less total environmental impact than the competing technologies (Section 30.1.4).

Organic Coatings: Science and Technology, Fourth Edition. Frank N. Jones, Mark E. Nichols, and Socrates Peter Pappas.

Research and development aimed at still higher solids coatings and, ultimately, at *solventless* (solvent-free) liquid coatings are bearing fruit. High solids coatings with VOC below 250 g l^{-1} (2.1 lb gal^{-1}) are commonplace industrial, marine, maintenance, and railroad coatings. "Solvent-free" liquid (almost 100% solids) coatings are available for specialized applications (Brand, 2016; Section 25.2.2). In view of present advantages and projected future improvements, it seems possible that the substitution of other technologies for solventborne coatings will continue, but at a slower rate than some experts project.

For some end uses, especially when cost is particularly important, a single coating is adequate; however, in many other end uses, performance requirements can only be met by applying at least two coats. When more than one coat is to be applied, it is often preferable to have a *primer* specifically designed to be the first coat and a different coating for the *top coat*. The primer is designed to adhere strongly to the substrate and to provide a surface to which the top coat adheres well. It is not necessary for the primer to meet all the other requirements that are critical for top coat performance, although primers often influence top coat properties. Commonly, primers are less expensive than top coats.

25.1 PRIMERS

The first consideration in formulating a primer is to achieve adequate adhesion to the substrate (Chapter 6). Ideally, the substrate will be clean and have a uniformly rough surface, but in many cases the primer must accommodate less-than-ideal situations. The surface tension of the primer must be lower than the surface free energy of the substrate, except for primers that penetrate into plastic substrates (Section 31.2.2). The viscosity of the continuous phase of the primer should be as low as possible to promote penetration of the vehicle into the pores and crevices on the surface of the substrate. Penetration is also promoted by use of slow evaporating solvents, use of slow cross-linking systems, and, whenever feasible, use of baking primers. The primer binder should have polar groups scattered along the backbone of the resin that can interact with the substrate surface. Presence of soluble salts (sometimes present in pigments) within the film should be minimal. Interaction between the binder and the substrate should be such that the coating will resist displacement by water when water molecules permeate through the coating to the interface (Section 7.4.2). For primers over metal substrates, as well as over alkaline surfaces such as masonry, saponification resistance of the primer is an important criterion in binder selection. Permeability to water vapor is undesirable for metal primers, but primers for wood products should be permeable enough to avoid blistering.

25.1.1 Binders for Primers

The binders in many primers for metal substrates are bisphenol A (BPA) epoxy resins and their derivatives (Sections 13.1.1 and 15.8). While a full explanation is not available, experience has shown that, in general, BPA epoxy resin-based coatings exhibit superior adhesion to clean metal. Addition of a small amount of an epoxy phosphate ester (Section 13.5) can further enhance adhesion, especially wet adhesion. For baked coatings, epoxy–phenolic coatings are particularly appropriate. For air dry coatings, epoxy–amine coatings are commonly selected. Both types have excellent wet adhesion and saponification resistance, critical for long-term corrosion protection. In general, baked epoxy–phenolics excel in these properties.

Primers based on alkyd (Chapter 15) and epoxy ester (Section 15.8) resins are also widely used. Material and application costs are less with epoxy coatings, but wet adhesion to metal is usually not as good, except possibly to metal contaminated with oily residues. Saponification resistance of most alkyd resins is limited, while that of epoxy esters is intermediate between alkyds and epoxy–amines.

The lowest cost vehicles used in metal primers are styrenated alkyds (Section 15.6). They are generally used in air dry primers. Higher T_g resulting from the large fraction of aromatic rings leads to primers that give dry-to-touch films, or even dry-to-handle films, rapidly. However, styrenated alkyds cross-link more slowly and with lower cross-link density than their nonstyrenated counterparts. As a result, the films develop solvent resistance slowly and have a wider "do not coat" time window within which they should not be top coated. If the primer is top coated within this time interval, *lifting* is likely to occur.

The surface of galvanized steel is a layer of zinc hydroxide, zinc oxide, and zinc carbonate; all are somewhat water-soluble, strong bases. Especially if the metal is not phosphate treated shortly before coating, the binder in the primer must be resistant to saponification. Alkyd primers lack the necessary saponification resistance. For air dry coatings, acrylic latex vehicles are more appropriate than alkyds.

Primers are also required for many surfaces other than metal substrates. In many architectural coating applications (Chapter 32), primers are desirable, although architectural paints that do not require primer are plentiful. Solventborne alkyd primers play an important role here.

Adhesion to polyolefin plastics can be difficult to achieve and can be enhanced by use of a primer based on a chlorinated polymer (Sections 6.5 and 31.2.2). Concrete and other masonry surfaces are alkaline and often require special surface treatments, most commonly washing with hydrochloric or phosphoric acid. The acid wash not only neutralizes the surface alkalinity but also etches the surface. Saponification resistant primers such as epoxy–amine or latex primers provide longest life. Concrete blocks have very porous surfaces, and penetration of solventborne paints into the surface requires a relatively large volume of coating for coverage. Latex paints offer better coverage (Section 32.1).

25.1.2 Pigmentation of Primers

Pigments play a vital role in corrosion protection, as covered in Chapter 7.

In other ways, selection of pigments and their amount is critical in primer formulation (Chapter 22). Pigmentation level affects adhesion of top coats to the primer film. Formulations with a high ratio of PVC to CPVC are low gloss coatings; the roughness and increased surface area of low gloss films provide improved intercoat adhesion. In some cases, it is desirable to formulate primers with PVC > CPVC. The resultant primer film is somewhat porous, permitting penetration of top coat vehicle into the pores, thereby promoting intercoat adhesion. Primers with PVC > CPVC can be sanded more easily and are less likely to clog sand paper than primers with PVC < CPVC. Since inert pigments are generally the least expensive components of the dry film, high PVC minimizes cost. However, the PVC should be only slightly higher than CPVC or the absorption of vehicle from the top coat may reduce gloss of the top coat.

Since primers almost always are low gloss coatings, inexperienced formulators may think it is not necessary to have good pigment dispersions for primers. It is not so. Since CPVC is affected by the degree of dispersion and particularly by the extent of flocculation, and since primers are usually formulated to be either slightly above or slightly below CPVC, pigment dispersion can be critical. If the pigment dispersion is not properly stabilized, CPVC decreases and the PVC/CPVC ratio becomes greater than intended.

If only a single color top coat is to be applied over the primer, it is usually desirable for the primer to have a similar color, since this minimizes the effect of the primer color on the final top coat color. In many cases, several different color top coats are applied over the same primer; then, it is usually desirable to use a light gray primer. Gray primers have better hiding than white primers as they contain a pigment that strongly absorbs light, usually lamp black. Light gray primers have relatively little effect on top coat color. Primers pigmented with red iron oxide provide good hiding at low cost but have a larger effect on the final top coat color.

In addition to adhesion, water and oxygen permeabilities are important factors in corrosion protection of metals by barrier coatings (Sections 7.3.3 and 33.1.1). Permeability can be strongly affected by pigments. Up to a PVC/ CPVC ratio of about 0.9, the higher the PVC of the film, the lower the gas and vapor permeability. To occupy the largest possible volume of the dry film with impermeable pigment, a combination of pigments that provide a high CPVC should be used in which PVC is close to CPVC. Platelet-shaped pigment particles tend to provide better barriers to oxygen and water permeability. Mica and micaceous iron oxide are widely used in primers for metals because of their platelet form. One must also use resin–pigment combinations in which the resin is strongly adsorbed on the pigment surface. If the pigment has a polar surface with a weakly adsorbed resin, water permeating through the film may displace the resin from the pigment surface, increasing water permeability.

Pigments for barrier primers should be completely insoluble in water. If, for example, zinc oxide were used as a pigment in a primer formulation, some of the zinc oxide would dissolve in water permeating through the film, resulting in the establishment of an osmotic cell that can cause blistering. Passivating pigments can be useful for protecting steel against corrosion when the coatings on the steel substrate have been ruptured (Section 7.5.2). However, passivating pigments must be somewhat soluble in water in order to passivate the steel. Furthermore, dissolving the pigments requires that the binder is able to swell with water to a degree. These characteristics mean that blistering is more likely with a primer that contains a passivating pigment. In most baking primers for OEM products, it is preferable not to use passivating pigments but to rely on barrier properties to provide corrosion protection. However, coatings for bridges, storage tanks, ships, and offshore drilling platforms cannot be baked and are exposed to physical damage that could rupture the film. Passivating pigments or zinc-rich primers are often preferred for such applications (Sections 33.1.2 and 33.1.3).

As reviewed by Serra et al. (2016), worldwide research is underway to develop primers based on sol–gel technology (Section 16.3.1). Often the candidates are epoxy/sol–gel hybrids. Such technology provides a method for generating silica nano-pigments within the coating film. These studies are often aimed at a particularly challenging problem: replacement of chromate pigments in primers for aluminum alloys (Section 30.5).

25.1.3 High Solids Primers

In conventional low solids primers, the effect of pigmentation on the viscosity of the coating is relatively small if the pigment is not flocculated. However, as solids are increased, the volume of dispersed phase (including both the volume of the pigment and of the adsorbed layer on the surfaces of the pigment particles) increases and becomes an important factor controlling the viscosity of the coating, thereby, limiting the solids at which a coating can be applied.

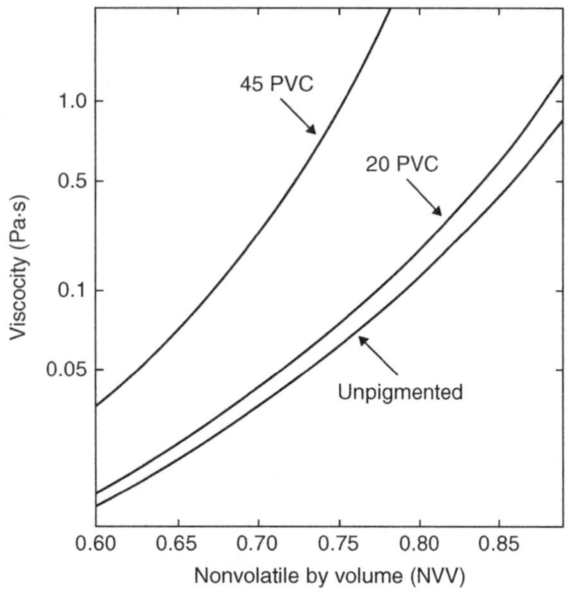

Figure 25.1 Effect of pigmentation on viscosity as a function of volume solids for an unpigmented coating and for two pigmented coatings based on the same binder with pigment loadings sufficient to give 20% PVC and 45% PVC in the dry films. See Hill and Wicks (1982) for the assumptions made in the calculations. Source: Hill and Wicks (1982). Reproduced with permission of Elsevier.

See Section 25.1.2 for factors affecting viscosity of high solids pigmented coatings.

Pigmentation levels such as encountered in low gloss coatings with PVC of 45 or higher substantially increase the viscosity of the wet coating, requiring reduction of solids for application. Figure 25.1, based on model calculations by Hill and Wicks (1982), shows plots of viscosity as a function of volume solids for three sets of calculations: one for an unpigmented coating, one for a 20 PVC (i.e., glossy) coating, and one for a 45 PVC (i.e., low gloss) coating.

The upper limit of solids for a primer having PVC near to or above CPVC is somewhere in the neighborhood of 60 NVV. In many cases, even 60 NVV is too low to satisfy VOC regulations. Exempt solvents (acetone, *t*-butyl acetate, CO_2, PCBTF, methyl acetate, DMC, and PC (Section 18.9.1)) offer a route to higher solids. Nonetheless, viscosity limitations tilted technology development toward waterborne, electrodeposition, and powder primers, to be discussed in the following chapters.

25.2 TOP COATS

Top coats include coatings applied over primers or directly on a substrate. In the former case, the primer provides adhesion to the substrate and a major part of the corrosion protection for coatings on metal. The top coat must adhere well to the primer and provide the desired appearance and other properties. A single coating layer must combine both functions. In general, it is preferable to use a primer/top coat system; however, a single coating is less expensive than multiple coat systems. Single coats are satisfactory on products that need little corrosion protection and for which the need to maintain adhesion in the presence of water is not critical. When appearance and exterior durability requirements are minimal, a primer with excellent corrosion protection properties may be used without a top coat, for example, inside ballast tanks of ships and for the interior of structural components of aircraft.

25.2.1 Binders for Top Coats

To an important degree, the properties of a top coat are controlled by the class of resin used as the principal binder in the top coat. The chemistry of these various binders is discussed in Chapters 8 through 17; in this chapter, the advantages and disadvantages of some classes of resins used in top coats, especially for OEM product coatings on metal, are compared. While the discussion focuses on single generic types of resins, it is becoming more and more common for formulators to use blends or hybrid resins.

25.2.1.1 Alkyds

From the late 1930s through the mid-1950s, alkyds (Chapter 15) were the dominant binders for coatings. While increasingly being replaced by other binders, alkyds, which are partly bio-based, are still used on a large scale. A major advantage of alkyds in many applications is lower cost. A second major advantage is that application of alkyd coatings is, in general, the most foolproof. Solventborne alkyd coatings are least subject to film defects of all the classes of coatings. This advantage results from the low surface tensions of most alkyd coatings. Therefore, problems, such as crawling, cratering, and other defects that result from surface tension driven flows or surface tension differential driven flows (Section 24.4), seldom occur. Pigment dispersions in alkyds are relatively easy to stabilize against flocculation. A further major advantage of alkyds is their ability to cross-link by autoxidation, which allows air dry or low temperature baking and avoids the need for cross-linking agents that may introduce potential toxic hazards. In essence, air is the cross-linker.

The major limitations of alkyds are comparatively poor color retention on baking, as well as limited exterior durability and saponification resistance. It is also difficult to achieve very high solids in alkyd solventborne coatings (Section 15.2); alkyd emulsions are an exception. A further problem can be generation of smoke in baking ovens.

Oxidizing alkyd binders are used for low cost coatings such as for steel shelving, machinery, wire coat hangers, and exterior coatings for steel drums. Individual applications may be relatively small, but there is a host of products for which the performance of these coatings is adequate. Oxidizing alkyds are also used for architectural gloss enamels (Section 32.3.1). With metal salt driers, oxidizing alkyd coating films air dry in a few hours, force dry at 60–80°C in an hour, or bake at 120–130°C in half an hour or less. Baking time is shortest with highly unsaturated drying oil derived alkyds having an oil length of about 60. Tung oil-modified alkyds give the fastest cross-linking, followed in turn by mixed tung–linseed oil alkyds and linseed oil modified alkyds. Cured films from such alkyds are quite yellow and turn yellow brown on overbaking. Generally, they are only used for dark color coatings. Since the cured films still contain substantial amounts of unsaturation and metal driers, the films embrittle with age. Better color and color retention can be obtained by using alkyds made with fatty acids or oils such as soy or tall oil fatty acids that are less unsaturated. These alkyds are also less expensive, but they cure more slowly than the more highly unsaturated alkyds.

Medium oil alkyd molecules have a substantial number of hydroxyl groups that can be cross-linked with a variety of cross-linkers. Urea–formaldehyde (UF) (Section 11.4.1) and melamine–formaldehyde (MF) (Chapter 11) cross-linkers, being economical, are the most widely used. Butylated, isobutylated, or octylated UF and MF resins are the most compatible. Generally, the more reactive, class II MF resins with high NH content are used (Section 11.2). The solvent system should contain some butyl alcohol to promote package stability and reduce viscosity as compared with straight hydrocarbon solvents. Usually, with class II cross-linkers no added catalyst is needed, since the residual unreacted carboxylic acid groups on the alkyd are sufficient for catalyzing the reaction with class II MF resins. Relatively weak acid catalysts such as alkyl phosphoric acids are alternatives. Color and speed of cure are affected by the type of fatty acid in the alkyd. Cross-linking results from the alkyd unsaturation as well as the MF resin reactions. Highly unsaturated fatty acids impart dark color, so that soy, tall oil, and dehydrated castor alkyds are used. The color, color retention, and resistance to embrittlement are better than obtained using the oxidizing alkyd without MF resin, since MF cross-linking minimizes or eliminates the need for metal driers, which catalyze the oxidation reactions.

The best color, color retention, and exterior durability are obtained using nonoxidizing, saturated fatty acid based alkyds. For example, solid color exterior top coats (monocoats) based on coconut oil alkyds and MF resins were popular automotive top coats at one time. Degradation due to hydrolysis under acidic conditions was minimized when the alkyds were made with isophthalic acid (or hexahydrophthalic anhydride) rather than phthalic anhydride. The exterior durability was good in opaque colors, but not as good as obtained with the best acrylic coatings. The coatings are considered to exhibit a greater appearance of depth than is obtained with acrylic–MF coatings. The cost of these coatings was similar to polyester–melamine coatings; and film defects during application were less likely.

The hydroxyl groups on alkyds can also be cross-linked at room temperature or under force dry conditions with polyisocyanates. IPDI isocyanurate prepolymers are often used for this purpose, since their color stability and exterior durability are superior to aromatic isocyanates, such as TDI derivatives (Section 12.3.2). In many cases, use of a polyisocyanate with an oxidizing alkyd permits sufficiently rapid development of tack free coatings so that the need for metal driers is reduced or eliminated which further improves exterior durability.

In the United States, the federal government mandates that agencies and contractors use bio-based coatings in certain applications (www.biopreferred.gov). Minimum bio-based content for interior solventborne alkyds, for example, is 67%, achievable with long oil alkyds. Use of bio-based coatings is also favored by various "green building" programs (Section 18.9.2) and by some manufacturers, for example, of furniture.

25.2.1.2 Polyesters

Polyester resins (Chapter 10) are one of the major classes of resins replacing alkyd resins in MF cross-linked baking enamels. Polyesters are also widely used in urethane coatings. Their cost is generally somewhat higher than oxidizing alkyds, but in some cases, less than nonoxidizing alkyds. Color, color retention, exterior durability, and resistance to embrittlement are better than obtained with most alkyds, but exterior durability and resistance to saponification have generally not been as good as with acrylics. However, development of "super durable" polyesters (Section 10.1.1) may have narrowed or closed the performance gap. Adhesion and impact resistance of polyester-based coatings without primers over clean, treated steel and aluminum substrates are comparable with alkyds and generally superior to acrylics. Polyester coatings generally have higher surface tensions than alkyd coatings and, hence, are more subject to crawling and surface tension differential driven flow defects, such as cratering. Most polyesters are hydroxy-terminated and are cross-linked with UF or MF resins or with isocyanates. Methylated or mixed methylated/butylated MF resins are usually used. Polyester/MF formulations are common auto-body primers ("chip-resistant primers") applied to many vehicles directly over the electrodeposition coating. For lower temperature cure or air dry coatings, aliphatic polyisocyanate cross-linkers can be used. A major advantage of polyesters over alkyds and most acrylic resins is the relative ease of preparing polyester resins suitable for very high solids and even solventless coatings (Section 25.2.2). Polyesters are also adaptable for use in powder coatings (Section 10.6, Chapter 28).

25.2.1.3 Acrylics

In general terms, the major advantages of acrylic binders (Chapter 8) are their low color, excellent color retention, resistance to embrittlement, and exterior durability at relatively modest cost. The best acrylics are quite photochemically stable and very hydrolytically stable. In general, their surface tensions are intermediate between alkyds and polyesters. As a result, the susceptibility of acrylic-based coatings to film defects is intermediate. Generally, they require primers for application to metals because their adhesion to metal surfaces is inferior to epoxy, alkyd, and polyesters coatings.

In the 1960s and 1970s, millions of new cars were monocoated with thermoplastic acrylic resins (Section 8.1). The principle monomer was methyl methacrylate, but the T_g was adjusted with comonomers or plasticizers to be a bit lower than that of pure PMMA. Cellulose acetate butyrate was added to improve orientation of metallic flake pigments. These coatings were durable in metallic formulations, were easily repairable, and gave excellent appearance. However, VOC was very high, and such coatings are now obsolete except for specialty situations.

Thermoplastic acrylics were replaced with thermosetting acrylics (TSAs) to reduce VOC emissions and enhance damage resistance (Section 8.2.1). Most commonly, TSAs are hydroxy-functional, sometimes with a minor amount of carboxylic acid functionality. They are cross-linked with either class I or class II MF resins; the choice usually depends on curing temperature requirements. They can also be cross-linked with polyfunctional isocyanates or with siloxanes, or with mixtures of cross-linkers. Aliphatic isocyanate cross-linkers are more expensive than MF resins and present greater toxic hazards, but they cure at lower temperatures, frequently have better environmental etch resistance, and, together with HALS stabilizers, usually provide somewhat greater exterior durability.

The major types of high solids clear coats for cars and trucks are (1) carbamate-functional acrylics cross-linked with MF resins, (2) epoxy-functional acrylics cross-linked with polyacids, and (3) hydroxyl-functional acrylics cross-linked with polyisocyanates or with some combination of siloxanes and MF resins (Section 30.1.5). These chemistries can be formulated at about 55% NVV.

Cross-linking by Michael addition reaction of acetoacetates with polyacrylate-functional resins is described in Sections 17.6 and 17.10. This chemistry has been developed as an alternative to two-package polyurethanes (Brinkhuis et al., 2011; Hendrikus et al., 2011).

25.2.1.4 Epoxies and Epoxy Esters

While the major uses for epoxy resins (Chapter 13) are in primers, significant volumes of BPA and novolac epoxy-based top coats are used in applications where exterior durability is not an issue. They are well suited to such purposes because adhesion of epoxy-based coatings in the presence of water or water vapor is generally superior to other types, and they are highly resistant to saponification. Applications such as beer and soft drink can linings (largely shifted to waterborne epoxies) and aircraft fuel tanks do not require exterior durability.

Weatherable, non-BPA epoxies have gained importance. Epoxy-functional acrylic resins cross-linked with polybasic acids provide films with excellent exterior durability (Section 13.1.2). Among other applications, they are widely used in high solids auto clear coats (Chapter 30).

Epoxy esters (Section 15.8) offer properties intermediate between alkyds and epoxies. An example of their use in top coats is in coating bottle caps and crowns, for which they exhibit the requisite combination of hardness, formability, adhesion, and resistance to water.

25.2.1.5 Urethanes

Polyisocyanate cross-linkers (Chapter 12) have been mentioned in connection with alkyds, polyesters, and acrylics. They are widely used as cross-linkers because of low temperature curing and abrasion resistance combined with resistance to swelling of cured films with solvent. Urethanes can also be backbone links in resins that have reactive groups other than isocyanates. Hydroxy-terminated urethane resins (Section 12.8) can be cross-linked with polyisocyanates or with MF resins. When MF resins are used, the coatings have the advantage of being one-package coatings without the concern for toxic hazards of free isocyanate. As with polyesters, there is no difficulty making low molecular weight resins where substantially all of the molecules have a minimum of two hydroxyl groups. Cost is higher than polyesters but the hydrolytic stability and, therefore, exterior durability can be superior. However, intermolecular hydrogen bonding of urethane groups leads to higher viscosity of solutions having equal concentration and equal molecular weight compared with polyester resins. Low molecular weight, hydroxy-functional urethanes can be blended with other generic types of hydroxyl-functional resins to modify coating properties.

Urethanes, especially those having primary $-O-CO-NH_2$ (carbamate) groups, react with MF resins to provide coatings having greater hydrolytic stability under acidic conditions than hydroxy-functional resins cross-linked with

MF (Section 11.3.4). Carbamate-functional acrylic and silicone resins, cross-linked with MF resins, are useful in clear automotive coatings as well as other applications.

High solids two-package coatings based on carbamate-functional resins with aldehyde cross-linkers, described in Section 17.10, are potential alternatives to two-package polyurethanes (Anderson et al., 2014).

25.2.1.6 Silicone and Fluorinated Resins

The highest resistance both to thermal degradation and photooxidation is obtained with silicones (Chapter 16) and fluorinated resins (Section 17.1.4). The cost of both is high, especially that of highly fluorinated resins. Silicone alkyds provide greater exterior durability than alkyds for air dry coatings. In a compromise to obtain excellent outdoor durability at an intermediate cost, silicone-modified polyesters and acrylics (Section 16.1.3) are relatively widely used. Fluorinated copolymer resins cross-linked with MF resin or aliphatic polyisocyanates have outstanding exterior durability and are used, for example, on skyscrapers where repainting would be very expensive.

Silicone-containing resins are playing an increasingly important role in applications such as automotive clear coats (Section 30.1.5) and marine coatings (Section 33.2). Patent activity on such resins is intense.

25.2.2 Formulating Solventborne Coatings for Low VOC

Although often asked for, there is no single definition of high solids coatings. For automotive clear coats, high solids corresponds to about 55 NVV. For a highly pigmented primer, high solids might be 50 NVV. For some high gloss pigmented coatings well above 75 NVV is possible, so that the answer is related, at least in part, to what is feasible. The situation is further complicated by the difficulty of exactly measuring the VOC of a coating (Section 18.9.1). For example, in some cases, solvents with functional groups may partially react with the cross-linker and, hence, not be evolved. On the other hand, there can be volatile byproducts of cross-linking that should be included in VOC. Furthermore, low molecular weight components may volatilize before cross-linking, the extent to which will vary with baking conditions.

A limitation on solids content is the increasing difficulty of achieving desired mechanical properties as molecular weight and average functionality per molecule $\bar{f}_n$ are decreased and molecular weight distribution is narrowed (as discussed in Section 8.2.1).

As one aims for still lower molecular weights and functionalities for still higher solids coatings, achieving high performance properties becomes even more difficult. For solution coatings with NVV of 70, the $\bar{M}_w/\bar{M}_n$ must be on the order of 2000/800 or less, with an average of little over two hydroxyl groups per molecule. It is critical that essentially all molecules have at least two functional groups per molecule. Molecules with only one functional group cannot cross-link and will leave dangling ends in the network; any molecules with no functional groups will be plasticizers and, if low enough in molecular weight, may partially volatilize in a baking oven. It is difficult to synthesize acrylic resins for very high solids coatings (Section 8.2.1). On the other hand, it is relatively easy to make polyurethane or polyester resins with $\bar{M}_w/\bar{M}_n$ of 2000/800, or even lower, wherein essentially all of the molecules have at least two hydroxyl groups (Sections 10.2 and 12.8). Polymerization of acrylics at high (160–180°C) temperatures offers a route to resins with somewhat improved characteristics for high solids coatings (Chapter 8).

These problems of conventional free radical polymerization can potentially be overcome by using living or controlled radical polymerization (CRP) methods (Sections 2.2.1.1 and 30.1.5). Among various CRP methods, atom transfer radical polymerization (ATRP, Section 8.2.1) shows considerable promise. Krol and Chmielarz (2014) reviewed ATRP, detailing the method's value for high solids acrylic coatings as well as for many other specialty coatings. As one of many examples, Ma and Barsotti (2008) used ATRP to selectively locate the hydroxyl groups in acrylic resins for high solids two-package coatings. Large-scale adoption of this remarkable technology has been hindered by its large cost relative to conventional polymerization methods.

In conventional solids coatings, the *cure window* is relatively large; that is, it makes little difference if the baking temperature, baking time, or catalyst amount is off by ±10%. In high solids coatings, the cure window is narrower (Bauer and Dickie, 1979). If there are a large number of hydroxyl groups on each resin molecule and 10% are not reacted, the change in properties may be small. If, however, there are only a little over two hydroxyl groups per average molecule and 10% are left unreacted, a significant fraction of the molecules will be connected to the network in only one place, resulting in a dangling end, with detrimental effects on film properties. The problem can be minimized by using cross-linkers with higher $\bar{f}_n$; at least then, the problem only arises from the polyester (or other resin), not from both. Class I MF resins, because of their greater average functionality compared with class II MF resins, generally offer broader cure windows (Section 11.3). The extent of self-condensation reactions of MF resins is particularly dependent on time, temperature, and catalyst concentration.

Because of the narrow cure window of high solids coatings, the applicator should be very careful in controlling the time and temperature in baking ovens according to the coating supplier's recommendations. The formulator must be more careful in checking the film properties when the temperature is about 10°C above and below the standard temperature. In making recommendations of cure cycles to a customer, one should use pieces of the customer's metal for establishing the baking schedule. The critical temperature for baking is that of the coating itself, not that of the air in the oven. Coatings applied to a heavy piece of metal heat up more slowly in an oven than coatings on light gauge sheet metal. On sheet metal, the coating over an area where the sheet metal has been welded to a supporting member heats up more slowly than on the rest of the surface. In conventional coatings, variations resulting from such differences are generally small; high solids coatings are more likely to be subject to variation in properties due to differences in coating temperature and time.

To increase solids, polydispersity of resins can be decreased; however, this may narrow the breadth of the T_g transition range (Section 4.2), which may adversely affect the mechanical properties of films (Kangas and Jones, 1987). An approach to overcoming this problem is to blend resins with differences in composition, but still similar enough to each other to be compatible; for example, polyester/acrylic blends using low molecular weight hydroxy-terminated polyesters to reduce the viscosity and thereby increase the solids (Hill and Kozlowski, 1987). Resins and cross-linkers for high solids coatings generally have $\bar{M}_n$ well below 5000. In contrast to high molecular weight polymers, the entropy of mixing of different low molecular weight resins is large enough to be a significant factor favoring compatibility (Section 18.2). The effect of molecular weight is illustrated by 50:50 blends of several acrylic and methacrylic homopolymers that are compatible, when their $\bar{M}_n$ values are less than 5000 but incompatible when $\bar{M}_n$ values are above 10000 (Cowie et al., 1992). The broader compatibility of low $\bar{M}_n$ resins permits formulation of high solids coatings based on mixtures of different generic type resins. Not all blends of low $\bar{M}_n$ resins are compatible, and there is the possibility that phase separation could occur as molecular weight increases in the early stages of cross-linking. Unpigmented films should be made and checked for transparency; even slight haziness suggests phase separation on a scale large enough to detract from appearance or properties.

Isocyanurates, unsymmetrical trimer, biuret, and allophanate polyisocyanates (Section 12.3.2) with lower molecular weights and, hence, lower viscosities are available. However, the lower viscosity of these polyisocyanates is only one factor affecting the VOC of coatings (Jorissen et al., 1992). In some cases, the lower viscosity cross-linker also has a lower equivalent weight, which requires that the weight ratio of low viscosity cross-linker to higher viscosity polyol is decreased. For this reason, the lowest VOC could be obtained using a somewhat higher equivalent weight polyisocyanate, even though it had a somewhat higher viscosity. Aldimines and hindered diamines permit preparation of very high solids clear top coats (Section 12.4).

Solvent free liquid coatings for certain applications are not new. In the 1960s P. I. Meli pioneered 100%-solids two-package epoxy coatings for the insides of storage tanks (Brand, 2016). Evolution of the technology has led to solvent-free coatings for other applications that require toughness and corrosion protection but not weatherability. Examples are the insides of ships, storage tanks, and railroad cars (Lotz, 2015). They are applied by airless spray. The near absence of volatile solvents is especially an advantage when the coatings are applied in enclosed spaces.

Polyesters afford a major opportunity for very low VOC coatings, since low molecular weight oligomers can be made such that all molecules have at least two reactive groups, usually hydroxyl groups. For example, 1,4-butanediol esters of mixed glutaric, adipic, and azelaic acids with $\bar{M}_n$ of 300 have a viscosity at 25°C of 250 Pa·s (Jones, 1996). Solvent-free coatings can be formulated with polyisocyanate cross-linkers. With MF resins, film hardness of baked films is too low; however, the coatings can be thinned with water (Section 18.4), permitting formulation with a variety of other components in solvent-free coatings. Such polyols are also effective reactive diluents to lower the VOC of polyurethane or acrylic formulations.

An example of a "zero" VOC coating that uses water as a diluent is an epoxy–hydroxyl system (Eaton and Lamb, 1996). Caprolactone polyols are used with 3,4-epoxycyclohexylmethyl-3′,4′-epoxycyclohexane carboxylate at a 2:1 epoxy to hydroxyl ratio and a triflic acid derivative as catalyst.

VOC reduction is being aided by application methods that permit use of higher viscosity coatings. Hot spray (Section 23.2.4), high-speed electrostatic disks (Section 23.2.3), and supercritical fluid spray (Section 23.2.5) are examples. Improvements in transfer efficiency are making a significant contribution to reducing VOC emissions (Sections 23.2.1 and 23.2.3).

As solids increase, it becomes more difficult to avoid pigment flocculation (Hochberg, 1982). The primary factor controlling stabilization of pigment dispersions is the thickness of the adsorbed layer on the surface of the pigment particles (Section 21.1.3). The very low molecular weight molecules in resins used in very high solids coatings are incapable of providing an adequate adsorbed layer. Furthermore, as the number of functional groups per molecule decreases, the solvent can compete more effectively for adsorption sites on the pigment surface promoting flocculation. Hyperdispersants (Section 21.1.3) have been developed for use in high solids coatings.

Another limiting factor for some high solids coatings is surface tension effects. Generally speaking, as resin molecular weight gets lower, more functionality must be added because the number of reactions required to achieve a polymeric network with the required cross-link density will be greater. In most coatings, the functional groups are highly polar, such as hydroxyl and carboxylic acid groups. Increased levels of such groups result in higher surface tensions. Furthermore, achieving high solids at a given viscosity with such resins generally requires using hydrogen-bond acceptor solvents rather than hydrocarbon solvents, which further increases surface tensions and the probability of film defects during application. When coating metal, the importance of clean surfaces increases as coating solids increases. When coating plastics, removal of mold release agents is more critical and a broader range of plastic materials have to be surface treated to avoid crawling and to achieve adhesion (Sections 6.7 and 31.2). Since the surface tension of a freshly applied high solids coating is generally higher than that of a conventional coating, a larger fraction of contaminating particles in the atmosphere have lower surface tensions than that of the wet high solids coating, thereby increasing the probability of cratering (Section 24.4).

Sagging is a greater problem with high solids than with conventional coatings (Bauer and Briggs, 1984). This results from the slower rate of evaporation of solvents from high solids coatings than from conventional coatings (Section 24.3). While the reasons for this difference have not been completely elucidated, the consequences pose significant problems. Sagging of spray-applied high solids coatings cannot be easily controlled by adjusting the evaporation rate of the solvent in the coating or by changing the distance between the spray gun and the substrate. Sagging can be minimized by using hot spray (Section 23.2.4) or supercritical fluid spray (Section 23.2.5). For many applications, thixotropic flow properties must be built into a high solids coating for spray application. Fine particle size SiO_2, bentonite clay pigments, zinc stearate, and polyamide gel thixotropes are additives that can be useful. Another alternative is to include a modest fraction of highly reactive cross-linker that will cause a quick increase in viscosity; mixtures of isocyanates and MF resins can be used (Teng et al., 1997). Roughly 30% of cars and light trucks in the United States are finished with high solids base coats. The problems of sagging and/or mottling in metallic basecoats is particularly challenging; microgels and nanoparticles are effective rheology modifiers that have little or no affect on polymer transparency, a requirement for metallic base coats and clear coats (Sections 24.3 and 30.1).

Even when sagging is not encountered during application, it can occur with high solids coatings during baking. The temperature dependence of viscosity is greater for high solids than for conventional coatings, as illustrated in Figure 25.2. Accordingly, *oven sagging* may occur when the increase in viscosity resulting from the cross-linking reactions is insufficient to counteract the decrease in viscosity of high solids coatings as temperature is increased during baking (Bauer and Briggs, 1984).

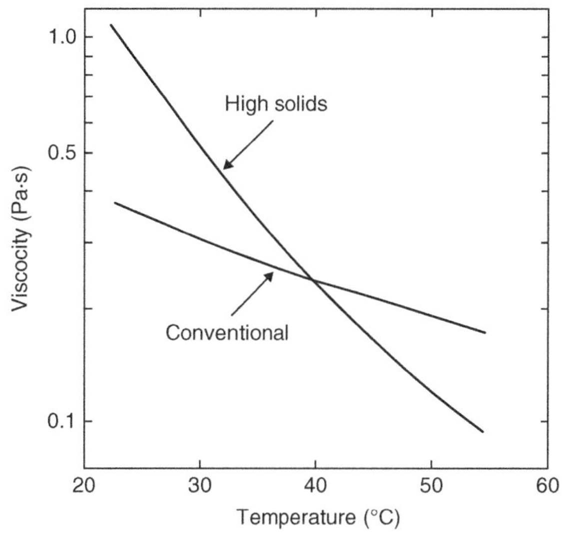

Figure 25.2 Viscosity as a function of temperature for a conventional and a high solids resin solution. The high solids solution is of a 1500 molecular weight polyester at 90% solids in ethylene glycol monoethyl ether acetate. The conventional solution is of a 20 000 molecular weight polyester at 25% solids in methyl ethyl ketone and ethylene glycol monoethyl ether acetate. Source: Hill and Wicks (1982). Reproduced with permission of Elsevier.

GENERAL REFERENCES

Hill, L. W.; Wicks, Jr., Z. W. *Prog. Org. Coat.*, **1982**, *10*, 55.

Mannari, V.; Chintankumar, J. P., *Understanding Coatings Raw Materials*, Vincentz, Hannover, **2015**.

REFERENCES

Anderson, J. R. et al., US patent 8, 653,174 B2 (**2014**).

Bauer, D. R.; Briggs, L. M., *J. Coat. Technol.* **1984**, *56*(716), 87.

Bauer, D. R.; Dickie, R. A., *J. Coat. Technol.*, **1979**, *54*(685), 101.

Brand, W., *Paint Square News*, **2016**, September 8. Available online from the Technology Publishing Network, from http://www.paintsquare.com (accessed April 18, 2017).

Brinkhuis, R. H. G., et al., WO2011124665 (**2011**).

Challener, C., *JCT CoatingsTech.*, **2014**, November/December, 26–33.

Cowie, J. M. G., et al., *Macromolecules*, **1992**, *25*, 3170.

Eaton, R. F.; Lamb, K. T., *J. Coat. Technol.*, **1996**, *68*(860), 49.

Hendrikus, R., et al., Crosslinkable composition crosslinkable with a latent base catalyst, WO 2011124665 A1 (**2011**).

Hill, L. W.; Kozlowski, K., *J. Coat. Technol.*, **1987**, *59*(751), 63.

Hill, L. W.; Wicks, Jr., Z. W. *Prog. Org. Coat.*, **1982**, *10*, 55.

Hochberg, S., *Proceedings of the Waterborne Higher-Solids Coatings Symposium*, New Orleans, LA, **1982**, p 143.

Jones, F. N., *J. Coat. Technol.*, **1996**, *68*(852), 25.

Jorissen, S. A., et al., *Proceedings of the Waterborne Higher-Solids Powder Coatings Symposium*, New Orleans, LA, **1992**, p 182.

Kangas, S. L.; Jones, F. N., *J. Coat. Technol.*, **1987**, *59*(744), 99.

Krol, P.; Chmielarz, P., *Prog. Org. Coat.*, **2014**, *77*, 913–948.

Lotz, A., *JCT CoatingsTech.*, **2015**, May, 26–33.

Ma, S.-H.; Barsotti, R. J., US patent application 2A1 (**2008**).

Serra, A.; Ramis, X.; Fernandez-Francos, X., *Coatings*, **2016**, *6*(1), 8.

Teng, G., et al., *Polym. Mater. Sci. Eng.*, **1997**, *76*, 117.

Chapter 26

Waterborne Coatings

Until about 1930, almost all architectural organic coatings except whitewash were solventborne or, in the case of certain vegetable oil-based coatings, solvent-free. Waterborne paints based on casein (an emulsifiable protein in milk) had been used by artists since ancient times, and they gained favor as interior wall paints during the 1930s, reaching a US volume of \$5M in 1940 (Radcliffe, 1943). Casein paints were gradually supplanted by paints based on casein-stabilized emulsions of linseed oil and/or alkyd resins owing to their better film properties. Such emulsions enjoyed considerable commercial success in the 1940s (Cheetham and Pearce, 1943). Starting about 1950, latex architectural paints (Chapters 9 and 32) were introduced. Latexes replaced earlier waterborne technologies and accelerated the replacement of solventborne coatings in architectural markets.

Initially, the driving forces for adoption of latexes were better performance, easier cleanup, and reduced fire hazards. Since 1970, the desire to reduce VOC emissions resulted in further gradual replacement of solventborne by waterborne coatings. This trend is expected to continue as technology improves and as restrictions on VOC become more stringent. One study projects a global growth rate for waterborne of 5.9% $year^{-1}$ in US dollars between 2014 and 2019. However, as pointed out in Chapter 25, solventborne coatings technology is not standing still, and the replacement rate may be slower than some experts project.

Many waterborne coatings contain some organic solvent. The solvents play a variety of important roles in resin manufacture, coating production, application, and film formation. Resin chemists and coatings formulators are learning how to reduce and, in some cases, eliminate solvents from waterborne coatings.

The largest classes of waterborne coatings are based on latex polymers (Section 26.2), followed by *water-reducible* coatings (Section 26.1), and the smaller but growing amounts of *emulsion coatings* (Section 26.3). The terminology is not uniform (Padget, 1994). We use the term water reducible for resins made in solvent and reduced with water to form a dispersion of resin in water. Other authors refer to these resins as *aqueous dispersion* resins or *water-soluble resins*, although they are not soluble in water. Some authors refer to latex coatings as emulsion coatings; we recommend against this practice to avoid confusion of latexes (dispersions of solid polymer particles in water) with coatings that really are emulsions (dispersions of liquids in water). Latex resins are usually prepared by emulsion polymerization. However, polyurethane latexes are an exception and are usually called *aqueous polyurethane dispersions (PUDs)*. There is a growing use of blends of the different types, for example, combinations of latex and water-reducible resins.

Water is also used as a cosolvent in solventborne coatings, which we call *water thinning* (Section 25.2.2). *Slurries* of powder coatings in water are described in Section 28.6.

Schoff (2013) says that waterborne coatings are "a different animal" from solventborne counterparts. He is right. Water has both advantages and disadvantages compared with organic solvents. On the positive side, water presents no toxic hazard, is odor-free, and is not flammable. Nonflammability reduces risks and insurance costs and permits use of less makeup air in baking ovens, reducing energy consumption in some cases. There are no emission or disposal problems directly attributable to the use of water. With some formulations, cleanup of personnel and equipment is easier with waterborne coatings; however, in other cases, cleanup is more difficult. Water costs less than organic solvents, but it does not necessarily follow that the cost of waterborne coatings is lower.

On the other hand, a wide range of solvents with different evaporation characteristics enables the formulator to

Organic Coatings: Science and Technology, Fourth Edition. Frank N. Jones, Mark E. Nichols, and Socrates Peter Pappas.

fine-tune evaporation rates from solventborne coatings, but there is only one kind of water. The heat capacity and heat of vaporization of water are high, resulting in high energy requirements for evaporation. With a given amount of energy available for evaporation, water evaporates more slowly than solvents with similar vapor pressures, although the vapor pressure of water increases relatively rapidly with increasing temperature. In contrast to organic solvents, evaporation of water is markedly affected by relative humidity (RH); variations in RH as waterborne coatings are applied can lead to major problems (Section 26.1).

The surface tension of water is higher than that of any organic solvent, so pure water does not wet most surfaces well. Surfactants are usually needed to reduce surface tension, which can also be reduced with solvents such as butyl alcohol or a butyl glycol ether. Surfactants are usually present in latexes, and water-reducible resins often contain surface-active molecules. The presence of surfactants tends to impart poorer water resistance to films; various approaches are being used to reduce or even eliminate the need for free surfactants. In water-reducible coatings, pigments can be dispersed in solvent solutions of the resins before addition of water.

Water increases corrosion of paint storage and application equipment such as storage tanks, coating lines, and ovens. Often, corrosion-resistant equipment must be used for applying waterborne coatings, increasing capital cost. For example, mild steel lines may have to be replaced with stainless steel. Since water is electrically conductive, special adaptations of electrostatic spray equipment are required, which also increase cost (Section 23.2.3).

Many waterborne coatings are vulnerable to foaming (Section 24.1), because they usually contain surfactants and/or surface-active polymer molecules. A wide assortment of antifoam additives is available, but general principles guiding their selection are in short supply. Trial-and-error experimentation for each specific formulation is often necessary.

The binders of waterborne coatings are heterogeneous, which adds complexities relative to solventborne coatings in which resin solutions are often homogeneous. Formulation of waterborne coatings is often more difficult. Overbeek (2010) points out that there is great opportunity in this complexity because potentially useful types of heterogeneity are more accessible in films cast from water. He extensively reviews heterogeneous binders, such as hybrid resins of acrylics together with alkyds, epoxies, urethanes, or siloxanes, and shows that the hybrids can solve practical problems. Since his review, published literature and patents indicate that extensive use is being made of hybrid and other heterogeneous binders.

Section 26.1 describes coatings based on water-reducible resins, which account for the majority of waterborne OEM product coatings. Water-reducible electrodeposition coatings involve such different formulation and application procedures that they are discussed separately in Chapter 27.

Latex binders (Chapter 9) are used in a majority of architectural coatings. While the general principles of latex coatings are provided in Section 26.2, detailed discussion of architectural coatings is postponed to Chapter 32. Increasing fractions of both OEM product coatings and special purpose coatings are latex based. The growing field of emulsion coatings is discussed in Section 26.3.

26.1 WATER-REDUCIBLE COATINGS

Most water-reducible coatings are produced as high solids solutions in water-miscible solvent(s). The resins have either carboxylic acid or amine groups that are at least partially neutralized with low molecular weight amines or acids, respectively. Pigment is usually dispersed in the partly neutralized resin solution, cross-linker is added along with additives such as catalysts, and the coating is diluted with water to application viscosity. The resin is neither soluble in water nor in the aqueous solution of the solvent that results. Instead, aggregates form in which the salt groups are oriented toward the water–particle interfaces and the low polarity parts of the resin molecules occupy the interior of the aggregates. The solvent partitions between the water phase and the aggregates. The aggregates are swollen by solvent, as well as by water, which associates with the salt groups and dissolves in the solvent. The cross-linker is dissolved in the resin–solvent aggregates, and the pigments usually reside inside the aggregates. The aggregates exist in a dynamic equilibrium, with solvent moving freely from one to another and probably with aggregates fusing and separating. All major types of resins can be adapted for use as water-reducible coatings, as exemplified by acrylics (Section 8.3).

When solutions of these neutralized resins are diluted with water, the change of viscosity with concentration is abnormal. Figure 8.1 shows a typical relationship between resin concentration and log viscosity when water-reducible acrylic resin solutions are diluted with water. As water is added, there is an initial rapid drop in viscosity. As more water is added, the viscosity plateaus and, frequently, with further addition of water, increases. Still further addition of water leads to a rapid drop in viscosity. The viscosity decreases rapidly as application solids are approached and is very dependent on the ratio of amine to carboxylic acid. If a small excess of water is inadvertently added, the viscosity may be too low. However, this problem can usually be remedied by adding a further small increment of amine, which increases the viscosity. This abnormal rheology can make the production of the coatings unforgiving. In some cases, the difficulties can be reduced by adding the formulated resin to the water rather than by adding the water to the resin (Schoff, 2013). At application viscosity the coatings are usually lower in solids compared with solvent-borne coatings.

While the viscosity of the organic solvent solution of resin is Newtonian, the viscosity of the water-diluted system in the region of the peak or plateau is highly shear thinning and becomes only slightly shear thinning upon dilution to application viscosity. As explained in Section 8.3, this behavior is consistent with the formation of swollen aggregates. In the case of carboxylic acid-functional resins neutralized with a low molecular weight amine, the pH is basic even though less than stoichiometric amounts of amine are frequently used. This abnormal pH behavior is also consistent with swollen aggregates, as explained in Section 8.3.

Selection of the amine for neutralization is an important formulation consideration (Section 8.3). Volatility and base strength are important. If there is insufficient amine, there will be macrophase separation of the coating. Therefore, it is desirable to select amines that efficiently stabilize the aggregate dispersion at low concentration. The principal factor controlling the required amine level appears to be water solubility of the amine. For example, the amount of tripropylamine required is substantially larger than the more soluble triethylamine (on an equivalent weight basis as well as a weight basis). Still more efficient are amino alcohols: *N*,*N*-dimethylaminoethanol (DMAE) has been the most widely used amine. Amines are expensive, and they generally add to VOC emissions, 2-amino-2-methyl-1-propanol (AMP) being an exception. AMP, an effective amino alcohol neutralizer, has been gaining in popularity because of its recent exemption from VOC regulations in North America. Somewhat less efficient, but still more effective than trialkylamines, are morpholine derivatives. Thus, the more water soluble the amine, the more effective in stabilizing the aggregate dispersion.

Most types of resins can be chemically modified to make them water reducible. The most widely used resins are acrylics, which contain both carboxylic acid and hydroxy functionality (Section 8.3). Water-reducible polyesters are used, but their hydrolytic stability is limited and low molecular weight cyclic oligomers can volatilize in baking ovens (Section 10.4). Since polyesters can be synthesized in the absence of solvent, it is possible to make solvent-free dispersions (Engelhardt, 1996). The resin is cast as a solid, pulverized, and stored as a solid until needed, delaying possible hydrolysis. When a coating is to be made, the powdered resin is stirred into a hot aqueous solution of dimethylaminoethanol to make a dispersion.

Water-reducible alkyds (Section 15.6) are desirable because they are partly bio-based. Their adoption has been slowed by technical problems such as package instability, moisture sensitivity of the films, and VOC, but solutions to these problems are being developed with new products, including:

- Reactive emulsifiers that can participate in oxidative cross-linking to reduce moisture sensitivity (Palmer, 2014).
- Various acrylic/alkyd blends and hybrids with superior package stability and good film properties; near-zero VOC is possible (Overbeek, 2010; Elrebii et al., 2015; Kim, 2015).
- Dispersions of short oil alkyds that solidify at ambient temperature to make zero VOC, hydrolytically stable binders for baked coatings (Ortiz et al., 2013).
- Emulsified alkyds—see Section 26.3.

Water-reducible epoxy esters (Section 15.8) and uralkyds (Section 15.7) have acceptable hydrolytic stability for many applications. Water-reducible urethanes have excellent saponification resistance and a minimum of cyclic oligomers (Section 12.7). Rosthauser and Nachtkamp (1987) reported solvent-free and low solvent water-reducible urethanes, a technology that is gaining importance. Urethane resins can give films with excellent properties but are generally more expensive than acrylic resins.

Coatings made with water-reducible resins have advantages and limitations. An important advantage follows from the high molecular weights that can be used, as high as that of resins used in conventional solution thermosetting coatings. This is possible because the viscosity at application dilution is almost independent of molecular weight. For example, one can use water-reducible acrylics with $\bar{M}_w/\bar{M}_n$ on the order of 35 000/15 000 and an average of around 10 hydroxyl groups and 5 carboxylic acid groups per molecule. Together with a class I melamine-formaldehyde (MF) cross-linker, film properties essentially equal to those of a conventional solution acrylic enamel can be achieved. The cure window is comparable with that of a conventional thermoset acrylic, and the problems related to molecular weight and functionality of high solids coatings are not encountered. A disadvantage is that the application solids are low. Typically, such coatings are applied by spraying, roll coating, or curtain coating at around 20–30 NVV. Low solids means that more wet film thickness has to be applied in order to achieve the same dry film thickness. On the other hand, low solids is an advantage in automotive metallic coatings, owing to better orientation of the aluminum pigment in the film.

Water-reducible resins for cross-linking with blocked isocyanates are widely available. A combination of blocked isocyanate and MF resin cross-linking gives coatings having both mar and etch resistance.

Before application, the viscosity of water-reducible resin coatings depends strongly on the ratio of water to solvent. During application and cure, evaporation rates are highly dependent on variations in humidity and temperature. If the humidity is above a critical level, water evaporates more slowly than even slow evaporating solvents, such as 2-butoxyethanol (Section 18.3.6). If the RH is over 70%, the rate of evaporation of water is very slow, and as RH approaches 100%, water may be absorbed rather than evaporated. In extreme cases, the viscosity of the coating during flash-off after application can decrease, rather than increase, as evaporation continues. The result is delayed sagging (Brandenburger and Hill, 1979).

In the mid-RH range, the effect can be reduced by a relatively modest increase in temperature, since RH decreases with temperature. At one time, certain automotive painting operations cooled the air to condense out some of the water and then rewarmed it, an expensive expedient. For factory-applied coatings, it is often desirable to formulate for application at a relatively high RH, say, 60%, since it is less expensive to increase RH than to decrease it.

As detailed in Section 24.7, popping during baking of water-reducible coatings is more difficult to control than with solventborne coatings (Watson and Wicks, 1983). The probability of popping increases as film thickness increases. Since the solids, as well as the rate of evaporation of water during spraying and flash-off, are lower for water reducible than for solventborne coatings, the wet film thickness required to apply the same solids is higher.

The greatest difficulties occur when spray applying thick coats. One cannot apply completely uniform films by spray. To assure that sufficient coating is applied on all areas, there will be some areas with substantially more than the average film thickness. Also, there is a greater chance of air entrapment with spray-applied coatings. If the air bubbles do not break, the air in the bubbles expands when the coating is baked, ultimately leading to popping. Air entrapment is particularly severe with airless spray, since at high pressure more air dissolves in the coating and is released as pressure decreases on leaving the gun (Section 23.2.6). Additives based on siloxanes with a polyether terminal group have been reported to minimize air entrapment in waterborne coatings (Spiegelhauer, 2002).

Thin coatings applied by roll coating or curtain coating seldom have a popping problem. The films are more uniform than spray-applied films, and there is no air entrapment problem. With reasonable care regarding flash-off times, curtain coating of relatively thick films can be accomplished without major problems. Water-reducible coatings are used on a large scale for such applications, including can coatings and panel coatings.

One might think that the high surface tension of water would cause serious problems with crawling and cratering. This is usually not the case, presumably because orientation of nonpolar segments of solvent near the surface is rapid, reducing surface tension. Surfactant-like molecules present in resins probably also reduce surface tension.

At one time it was widely assumed that waterborne urethane coatings having free isocyanate groups would not be feasible owing to the reactivity of isocyanates with water. However, two-package (2K) urethane coatings with adequate pot life for commercial use have been widely commercialized (Jacobs and Yu, 1993; Wicks et al., 2002; Section 12.7). For example, a water-dispersible, aliphatic, "hydrophilically modified" polyisocyanate is used in one package, and the second package contains a hydroxy-terminated, water-reducible polyurethane with carboxylic acid groups from 2.2-dimethylolpropionic acid. A 2:1 ratio of —N=C=O to —OH is used to offset partial reaction with water. Films cross-linked within a week at 25°C when the RH was 55% or lower. At high humidity, solvent resistance did not develop at 25°C, but warming to 31°C permitted cross-linking even at 80% RH. Acrylic polyols can also be used. Pot life problems can be minimized by use of appropriate mixing and application equipment.

Alternatively, low viscosity polyisocyanates, such as blends of HDI dimer and trimer, can cross-link aqueous PUDs (O'Connor et al., 1997). Waterborne polyurethanes are further discussed in Section 12.7.

Graft copolymer dispersions and semi-block copolymer dispersions, prepared by various catalytic chain transfer polymerization methods, have the advantage of narrower molecular weight distributions compared with water-reducible acrylic resins made by free radical polymerization (Huybrechts et al., 2000). Such methods enable synthesis of a huge array of potentially useful block copolymers. For example, Overbeek (2010) describes a copolymer comprising a block of acrylic acid and isobornyl acrylate (T_g 95°C) and a block of butyl methacrylate and butyl acrylate (T_g 0°C), which has good adhesion to polypropylene. Few other polymers adhere to polypropylene.

Waterborne primers are particularly important, since reducing VOC levels in solventborne primer formulations is particularly difficult (Section 25.1.3). Waterborne coatings are especially suited for dip coating applications (Section 23.3) because they eliminate the fire hazards of solventborne dip coatings. For example, maleated epoxy esters (Section 15.8) are used for spray or dip-applied primers for steel. All electrodeposition coatings (Chapter 27) are water reducible. Interior linings for beverage cans are water-reducible coatings based on grafting acrylates, styrene, and acrylic acid onto BPA epoxy (Section 13.4). Development of BPA-free can liners is discussed in Section 30.3.1. Waterborne sol-gel coatings are described in Section 16.3.

26.2 LATEX-BASED COATINGS

As detailed in Section 32.1, latexes are the major type of vehicle for architectural coatings. For certain household applications, such as flat wall paint, the advantages of latex paints over any solventborne paint are so large that solventborne paints are seldom marketed in North America.

Important advantages of interior latex paints include rapid drying, low solvent odor, absence of odor from oxidation by-products of drying oils and alkyds, easy cleanup, reduced fire hazard, and better long-term retention of mechanical properties. For exterior paints, a major advantage is superior exterior durability of high performance latex paints, relative to drying oil or alkyd paints. On wood siding, blistering is reduced, since the latex films are more permeable to water vapor. On the other hand, adhesion of latex paints to chalky surfaces is inferior to that of solventborne paints.

A further advantage of acrylic, styrene/acrylic, and styrene/butadiene latexes over alkyds is their excellent resistance to saponification. Paints with shelf lives of many years are attainable. Latex paints generally show better long-term adhesion to galvanized metal surfaces than alkyd paints. They generally show better performance over cement and concrete surfaces than alkyds, which tend to be hydrolyzed by the alkaline surfaces, although alkyds designed for use over masonry are available. Latex paints also give better coverage over porous cement surfaces (Section 32.1).

The polymers in liquid latex coatings exist in discrete particles and do not move from one latex particle to another, although solvent and surfactant molecules may do so. Latex coatings form films by coalescence of the polymer particles (Section 2.3.3) as the polymer particles interpenetrate. Coalescence can occur only if the film formation temperature is higher than the T_g of the polymer particles. While initial coalescence proceeds rapidly at temperatures just a few degrees above T_g, completion of coalescence is relatively slow unless the temperature is significantly higher than T_g. For most architectural paints, slow final coalescence is not a real problem, so the T_g need be only a few degrees below the film formation temperature. In baked industrial coatings, film formation should be essentially complete by the time the coated article comes out of the baking oven; therefore, baking temperatures have to be significantly above the T_g.

There are limitations to latex paints, particularly in regard to the minimum film forming temperature (MFFT), which allows proper coalescence for film formation. To obtain a final film with a high enough T_g to resist blocking, it is common to use *coalescing solvents* in the formulation. The coalescing solvent dissolves in the polymer particles, reducing the T_g, permitting film formation at a lower temperature. After film formation, the coalescing solvent slowly diffuses out of the film and evaporates. Even with the use of coalescing solvents, however, there are limitations in achieving the proper balance of MFFT and T_g required for good film formation (Section 32.1).

Latex paints with less organic solvent or, in some cases, no VOC have been developed and marketed. Technical approaches to latexes having no VOCs include (1) blends of latexes of different T_g and particle sizes (Winnik and Feng, 1996; Eckersley and Helmer, 1997) and (2) latexes with particles having a gradient T_g made by sequential polymerization (Sections 2.3.3 and 9.2) (Hoy, 1979).

Thermosetting latexes permit the use of a low T_g polymer, facilitating film formation at relatively low temperatures; subsequent cross-linking gives the needed block resistance and other properties (Section 9.4.1). Examples of cross-linkers are given in the following text.

Thermosetting latexes are also used in two-package coatings for industrial use; pot life has to be long enough to permit coalescence before a significant degree of cross-linking occurs. Various cross-linkers are effective in one-package latexes. Examples are polymers having allylic groups that permit autoxidative cross-linking, polymers having trialkoxysilyl groups, and copolymers with aldehyde or keto groups that can be cross-linked with dihydrazides. Also, hybrid alkyd/acrylic latexes have been prepared by dissolving an oxidizing alkyd resin in the monomers used in the emulsion polymerization.

As reviewed by Overbeek (2010), latex technology offers almost limitless opportunities for innovation. One promising method involves infusing low molecular weight cross-linkable materials into host thermoplastic latexes. The infused material can act as a coalescing solvent, lowering or eliminating VOC, and can form cross-links to enhance film properties. Examples of thermoplastic latexes infused with cross-linkable materials are:

- Acrylic epoxy hybrid dispersions in which liquid epoxy resins are infused into various latexes (Fu et al., 2014). The coatings are formulated with water-soluble epoxy cross-linkers. The infused epoxy can be modified with a monoamine to form polyhydroxy amino ethers, which can also be cross-linked.
- Latexes containing tetraethoxysilane (TEOS) (Chapter 16) within the particles. During film formation, TEOS accelerates coalescence and then reacts in a sol–gel process to form nanosilica particles, resulting in films with enhanced properties (Picarra et al., 2014).

Polyurethane latexes are usually called *aqueous* PUDs. They can be thermoplastic or cross-linked (Section 12.7.1). An important factor promoting their increasing usage is that films of higher T_g polymers can be formed at ambient temperatures than with acrylic and other conventional latexes, hence reducing or eliminating the need for coalescing solvents. The water acts as a plasticizer for the polyurethane, reducing the T_g of the polymer and permitting film formation (Satguru et al., 1994).

Bio-based (up to 97%) PUDs have been demonstrated by Li et al. (2015). The building blocks are diisocyanates derived from fatty acids and lysine and isosorbide, a diol derived from glucose. Good properties were reported.

Latexes, water-reducible acrylic resins, and water-reducible polyesters can be blended with PUDs. Acrylic monomers can also be polymerized in PUDs to make hybrid acrylic/urethane resins. There have been three major approaches to preparing aqueous acrylic/urethane systems:

- Blending of PUDs and latexes (or other water-reducible acrylic resins) (Section 12.7.2)
- Polymerizing acrylic monomers in PUDs (Section 12.7.2)
- Cross-linking urethane/vinyl systems (Section 12.7.2)

In broad terms, acrylate/PUD blends combine the advantages of both acrylic latexes and PUDs. Acrylic latexes are lower in cost and have excellent exterior durability, while PUDs provide higher abrasion resistance and often better adhesion. MFFTs of PUDs are lower than those of acrylic latexes having the same T_g, which permits reduction or even elimination of coalescing agents and hence reduction of VOC. Hybrid resin films provide higher tensile strength than acrylic films. A variety of cross-linkers can be used with acrylated PUDs having functional groups, which are of particular interest in automotive base coats.

Another limitation in the formulation of latex coatings is the difficulty of formulating high gloss latex paints (Sections 19.10.1 and 32.3). The main problem of formulating gloss coatings results from the nonuniform distribution of pigment and latex particles as the volatiles evaporate. Accordingly, it is difficult to obtain an upper surface of the film that is pigment-free or has a low pigment content, as with solventborne coatings. This problem can be minimized, but not eliminated, by using small particle size latexes. Gloss is also limited by the presence of surfactants in the dry film, which can lead to haze and blooming. Latexes with very low surfactant content or with polymerizable surfactants are being developed to minimize or eliminate this problem (Section 9.1.1).

Latex coatings tend to be excessively shear thinning. When the viscosity at high shear rate is appropriate for application, the viscosity at the low shear rates encountered in leveling tends to be high. This is one of several reasons that leveling of latex-based coatings tends to be poorer than that of solventborne coatings. Flat coatings with only fair leveling have acceptable appearance, but as gloss is increased, the unevenness of the films becomes obvious. Substantial progress has been made by using *associative thickeners* (Section 32.3).

Use of latexes has been more limited in OEM coatings for many reasons. The problems of evaporation of water on a conveyor line and in ovens, mentioned in the last section, are part of the problem. Popping can also be a problem with latex coatings. It is desirable to use a latex polymer with as high a T_g as possible to minimize coalescence of the latex particles at the surface of the film before the water has completely evaporated. Probably, the major limitation for industrial applications has been flow properties of latex-based coatings. In addition, uncross-linked latexes lack the robust resistance to solvents, water, and mechanical failure that are required for many OEM applications. Leveling requirements for many industrial coatings are more rigorous than for architectural coatings. In general, latex coatings exhibit relatively high degrees of shear thinning and, in many cases, thixotropy. In some cases, these flow characteristics result from flocculation of latex particles. Flocculation of latex increases low shear viscosity to a major degree; furthermore, it further exacerbates the gloss problem. Use of associative thickeners minimizes this problem as further discussed in Section 32.3.

Combinations of latexes with water-reducible resins are increasingly used. For example, automotive base coats often contain a mixture of water-reducible and latex acrylics. Frequently, these combinations have better flow properties than latex coatings.

An advantage of latexes is that their high molecular weights provide mechanical properties that are fully satisfactory for many applications without the need for cross-linking. However, cross-linking is generally required for good solvent resistance.

Methods for cross-linking latexes are detailed in Section 9.4. One is to copolymerize a small amount of diacetone acrylamide into the latex and then formulate the coating with adipic dihydrazide cross-linker (Section 17.11); the cross-links are hydrazones formed from the carbonyl group on diacetone acrylamide and the dihydrazide (Mestach et al., 2004). This approach is used in commercial architectural paints and probably in industrial coatings.

The viscosity of latexes is independent of molecular weight, so they can be applied at relatively high solids even though their molecular weight is high. In highly pigmented coatings, the fraction of internal phase volume becomes so large that the solids level has to be reduced; but the reduction is done primarily with water so that VOC emissions are minimal.

As pressure to reduce VOC emissions increases still further, it is anticipated that there will be substantial increases in the use of latexes, particularly of thermosetting and hybrid latexes, in industrial applications, including both OEM product coatings and special purpose coatings.

26.3 EMULSION COATINGS

Most waterborne coatings are based on water-reducible or latex binders. Nevertheless, *emulsion coatings*, defined here as a dispersion of a liquid resin in a liquid (water in this case), have found applications and have potential for wider use. Examples have been described where the dispersed particles are on the borderline between liquid and solid (e.g., by Elrebii et al., 2015). It is convenient to simply call them *dispersions*.

Commercially important emulsion coatings are two-package coatings in which one package is a liquid BPA (or novolac) epoxy resin solution and the second is an amine-terminated cross-linking agent (Section 13.2.2) containing a nonionic surfactant (Albers, 1982; Galgoci et al., 1997). The amine cross-linker package is diluted with water, and the epoxy resin solution package is added with vigorous stirring. The pot life is limited to a few hours, since the epoxy resin can react not only with the amine groups but also, slowly, with water. While pot life of solventborne epoxy–amine coatings is limited by viscosity increase, emulsion systems show little, if any, change in viscosity with age, since the viscosity is controlled by internal phase concentration not by molecular weight and because reaction with water does not lead to cross-linking. Instead, pot life is limited by the resulting poor film properties of applied coatings, including reduced gloss. Such emulsion epoxy paints are used where hard, easily cleaned wall coatings are needed, for example, in hospitals and food processing plants. The residual surfactant reduces the corrosion protective properties for application on metal surfaces. Emulsion coatings based on aliphatic epoxy resins and aliphatic carboxylic acid-functional resins are reported to give superior pot life and improved properties (Eslinger, 1995).

An approach to eliminating surfactants is the use of nitroethane as one of the solvents (Lopez, 1989; Wegmann, 1993). As discussed in Section 13.2.6, nitroethane forms a salt with an amine group of an amine-terminated polyamide, which acts as a surfactant. When the film dries, the nitroethane evaporates, leaving a less water-sensitive film. Such coatings are used in high performance applications such as aircraft primers (Section 33.4).

Variants of this technology use dispersions of solid epoxy resins with specially designed hardeners (Zhang and Procopio, 2014; Zirngast et al., 2014). Potential applications include concrete floor coatings and the industrial, railway, agriculture, and construction equipment sectors. The solid epoxies are apparently low melting enough to enable ambient temperature cure.

For baking coatings, acrylic copolymers such as amine salts of methacrylic acid/methyl methacrylate/ethyl acrylate/styrene (40 : 20 : 20 : 20) can be used as emulsifying agents for epoxy–phenolic coatings (Kojima et al., 1993). During baking, the carboxylic acid groups of the acrylic react with epoxy groups, incorporating the surfactant into the polymer structure, which avoids the ill effects of nonreactive surfactants on film properties.

Another example is nitrocellulose lacquers emulsified into water for use as top coats for wood furniture (Section 31.1; Winchester, 1991). Such emulsions have significantly lower VOC than solventborne lacquers, but longer drying times are required to achieve print resistance.

Alkyd emulsion paints and paints containing alkyd emulsions are growing commercially (Hofland, 1997; Barrios et al., 2014). Higher gloss can be obtained than with latex paints, and the problem of minimum temperature required for film formation experienced with latex paints is obviated. Very low VOC is attainable. Most film properties are likely to be similar to those of the corresponding alkyds, although high surfactant levels of 5–10% in the alkyd emulsions are common and contribute to water sensitivity. Water sensitivity is reduced by using an emulsifier having a polyunsaturated hydrocarbon chain to cross-link with the alkyd after application (McNamee et al., 2004) or with an emulsifier bearing cross-linkable allyl groups (Palmer, 2014).

Long oil alkyds with anionic or anionic/nonionic blend surfactants provide the most stable emulsions and the lowest VOCs. Drying is generally slower than with solventborne alkyds, perhaps because the driers are less effective. Flow properties of the formulations can be adjusted with HUER-type associative thickeners.

For decades, alkyds have been emulsified into latex paints to improve adhesion to chalky surfaces (Section 32.1). It is speculated that package stability is enhanced by partial infusion of the alkyd molecules into the latex particles, where they are somewhat shielded from hydrolysis.

Elrebii et al. (2015) have described acrylic/alkyd hybrid dispersions in which the binder is formed by melt condensation reactions of alkyds with COOH-functional acrylic prepolymers. When combined with water and an amine, the binders form dispersions with various morphologies and particle sizes. No solvent is needed, and good film properties are reported.

GENERAL REFERENCES

Glass, J. E., Ed., *Technology for Waterborne Coatings*, American Chemical Society, Washington, DC, **1997**.

Heilen, W., Ed., *Additives for Waterborne Coatings*, Vincentz, Hannover, **2009**.

Karsa, D. R.; Davies, W. D., Eds., *Waterborne Coatings and Additives*, The Royal Society of Chemistry, Cambridge, **1995**.

REFERENCES

Albers, R., *Proceedings of the International Waterborne, High-Solids, and Powder Coatings Symposium*, New Orleans, LA, 1983, p 130; US patent 4,352,898 (**1982**).

Barrios, S. P., et al., *Proceedings of the American Coatings Conference*, Atlanta, GA, April 9, **2014**, paper 13.1.

Brandenburger, L. B.; Hill, L. W., *J. Coat. Technol.*, **1979**, *51*(659), 57.

Cheetham, H. C.; Pearce, W. T., Alkyd Emulsion Paints in Mattiello, J. J., Ed., *Protective and Decorative Coatings*, **1943**, Vol. *III*, pp 488–494.

Eckersley, S. A.; Helmer, B. J., *J. Coat. Technol.*, **1997**, *69*(864), 97.

Elrebii, M.; Kamoun, A.; Boufi, S., *Prog. Org. Coat.*, **2015**, *87*, 222–231.

Engelhardt, R., *Proceedings of the International Waterborne, High-Solids, and Powder Coatings Symposium*, New Orleans, LA, **1996**, p 408.

Eslinger, D. R., *J. Coat. Technol.*, **1995**, *67*(850), 45.
Fu, Z., et al., *Proceedings of the American Coatings Conference*, Atlanta, GA, April 8, **2014**, paper 5.1.
Galgoci, E., et al., *Proceedings of the International Waterborne, High-Solids, and Powder Coatings Symposium*, New Orleans, LA, **1997**, p 106.
Hofland, A., Making Paint from Alkyd Emulsions in Glass, J. E., Ed., *Technology for Waterborne Coatings*, American Chemical Society, Washington, DC, **1997**, p 183.
Hoy, K. L., *J. Coat. Technol.*, **1979**, *51*(651), 27.
Huybrechts, J., et al., *Prog. Org. Coat.*, **2000**, *38*, 67.
Jacobs, P. B.; Yu, P. C., *J. Coat. Technol.*, **1993**, *65*(622), 45.
Kim, K.-J., WO2015077677 A1 (**2015**).
Kojima, S., et al., *J. Coat. Technol.*, **1993**, *65*(818), 25.
Li, Y.; Noordover, B. A. J.; van Benthem, R. A. T. M.; Konig, C. E., *Prog. Org. Coat.*, **2015**, *86*, 134–142.
Lopez, J. A., US patent 4,816,502 (**1989**).
McNamee, W., et al., *Proc. Waterborne High-Solids Powder Coat. Symp*, New Orleans, LA, **2004**, paper No. 27.
Mestach, D. E. P., et al., US patent 6,730,740 (**2004**).
O'Connor, J. M., et al., *Proceedings of the International Waterborne, High-Solids, and Powder Coatings Symposium*, New Orleans, LA, **1997**, p 458.
Ortiz, R.; Spilman, G.; Young, T. J.; Sandoval, R. W., WO2013056162 A1 (**2013**).
Overbeek, A., *J. Coat. Technol. Res.*, **2010**, *7*(1), 1–21.
Padget, J. C., *J. Coat. Technol.*, **1994**, *66*(839), 89.
Palmer, C., *Proceedings of the American Coatings Conference*, Atlanta, GA, April 9, **2014**, paper 13.2.
Picarra, S., et al., *Langmuir*, **2014**, *30*(41), 12345–12363.
Radcliffe, R. S., Casein Paints in Mattiello, J. J., Ed., *Protective and Decorative Coatings*, John Wiley & Sons, Inc., New York, **1943**, Vol. *III*, 461–480.
Rosthauser, J. W.; Nachtkamp, K., Waterborne Polyurethanes in Frisch, K. C.; Klempner, D., Eds., *Advances in Urethane Science and Technology*, Technomic Publishers, Westport, **1987**, Vol. *10*, p 121.
Satguru, R., et al., *J. Coat. Technol.*, **1994**, *66*(830), 47.
Schoff, C. K., *CoatingsTech*, January, **2013**, 48.
Spiegelhauer, S., *Proceedings of the American Coatings Conference*, New Orleans, LA, **2002**, pp 161–170.
Watson, B. C.; Wicks, Z. W., Jr., *J. Coat. Technol.*, **1983**, *55*(732), 61.
Wegmann, A., *J. Coat. Technol.*, **1993**, *65*(827), 27.
Wicks, Z. W., Jr., et al., *Prog. Org. Coat.*, **2002**, *44*, 161.
Winchester, C. M., *J. Coat. Technol.*, **1991**, *63*(803), 47.
Winnik, M. A.; Feng, J., *J. Coat. Technol.*, **1996**, *68*(852), 39.
Zhang, Y.; Procopio, L. J., *Proceedings of the American Coatings Conference*, Atlanta, GA, April 9, **2014**, paper 14.5.
Zirngast, M., et al., *Proceedings of the American Coatings Conference*, Atlanta, GA, April 9, **2014**, paper 14.4.

Chapter 27

Electrodeposition Coatings

Electrodeposition is an efficient and unique method of applying high performance coatings. Electrodeposition coatings are a major class of waterborne coatings, which are used not only in largest volume as primers but also in single- and two-coat systems. The general principle is relatively simple; development and initial commercialization were complex and difficult. Electrodeposition coating systems are *anionic* or *cationic*, alternatively *anodic* or *cathodic*, corresponding to deposition at the anode or cathode, respectively. E-coat, electrocoat, electropaint, ED, EDP, ELDEP, and ELPO are fairly common synonyms for electrodeposition coatings. Cationic E-coat is used worldwide for priming auto bodies. The first automotive body (a Ford Thunderbird) was E-coated in 1963, and general adoption of E-coat in the 1970s and 1980s led to a major improvement in the corrosion resistance of motor vehicles. Krylova (2001) provides a review of electrodeposition coating.

The object to be coated is immersed in a bath containing the waterborne electrocoat. The process of electrodeposition in the bath can be thought of as a four-step process. The first step is the *electrolysis* of water. In cathodic electrodeposition, water is reduced at the cathode, producing hydrogen gas and hydroxyl anions. This increases the local pH near the cathode. In anodic electrodeposition, water is oxidized at the anode, and hydrogen cations (or protons) are produced in addition to oxygen. The local pH is decreased near the anode.

The second step in electrodeposition is *electrophoresis*. All electrocoats are aqueous dispersions of positively (in cathodic E-coat) or negatively (in anodic E-coat) charged particles in water. During anodic electrodeposition, the negatively charged paint particles are attracted to the anode by electrostatic forces. In the case of cathodic electrocoat, the cathode is the substrate to which the positively charged paint particles are drawn. When the charged particles closely approach the surface to be coated, their charge is neutralized by the locally high concentration of either hydrogen cations or hydroxyl anions. The particles then fall out of dispersion and are deposited on the substrate, which is called *electrodeposition proper*, the third step of the process.

As the newly deposited film is a mixture of polymer particles and water, in order to achieve a dense film, the water must be removed from the film. This occurs during the fourth and final step, called *electroosmosis*. In this step water diffuses out of the film by the process of *syneresis*, resulting in a dense film. The object is then removed from the bath, rinsed, and baked.

There are many generic requirements placed upon the formulation of an electrocoat. In both cathodic and anodic electrocoat, the binder is almost always a thermoset, and the coatings are baked at relatively high temperatures in an oven to induce cross-linking. The coatings must be designed so that all coating components are attracted to the electrode at the same rate; otherwise, the bath composition will change with time. A vehicle, in which pigments can be dispersed and cross-linkers dissolved, must form a stable, electrically charged dispersion of aggregate particles when diluted with water. The pigment must be preferentially wet by the resin so that it does not migrate out of the resin aggregates. It is possible to electrodeposit polymer films from solutions of the salt of a polymer in water, but one cannot use a dissolved polymer as a binder for pigmented coatings as the resin, pigment, cross-linker, and other components would not deposit in the same ratio over time. Polymer films can be electrodeposited from a pigment-free latex, but not from pigmented latex coatings, since pigment particles and latex particles would be attracted to the electrode, that is, substrate, at different rates.

Organic Coatings: Science and Technology, Fourth Edition. Frank N. Jones, Mark E. Nichols, and Socrates Peter Pappas.

The coating is diluted to 10–20% solids with water. This relatively low solids is used for three reasons. When the coated object is brought out of the tank, it carries a layer of the bath liquid with it that must be rinsed off; losses are less and rinsing is easier if the solids are low. Furthermore, ultrafiltration (Section 27.4) is easier with lower solids. For coating products like automobile bodies and appliances, the electrodeposition tanks are very large—up to 500 000 l. The tank is kept full at all times; therefore, part of the capital investment is the cost of a tank full of coating—the investment is lower with 10–20% solids than it would be with 50% solids.

A critical requirement for electrodeposition coatings is indefinite stability after dilution. As the coating is applied, coating solids are removed from the bath, and they have to be continually replaced to maintain the same composition in the tank. Ideally, the tank would never be emptied. Some material from the original loading will be in the tank for long times and, therefore, must be very stable to hydrolysis, premature cross-linking, and mechanical agitation. The cross-linker must be stable in the diluted coating at a pH above 7 for anionic coatings and below 7 for cationic coatings. Stability to oxidation is critical, since air is continually mixed into the bath by agitation. If an oxidizing type vehicle is used, an antioxidant that volatilizes in the baking oven is an essential additive.

27.1 ANIONIC ELECTRODEPOSITION COATINGS

Resins used in anionic systems are substituted with carboxylic acid groups that have an acid number in the range of 50–80 mg KOH g^{-1} resin. The pigments and other components are dispersed in the resin, and the carboxylic acid groups are partially neutralized with an amine such as 2-(*N*,*N*-dimethylamino)ethanol. To load the tank, the coating is diluted to about 10% solids with water. The degree of substitution with salt groups is designed to be such that the resin is not soluble in water, but rather forms aggregates on dilution. The aggregates are stabilized as a dispersion in water with salt groups on the outer surface of the particles. Even with less than the theoretical amount of amine to neutralize the carboxylic acid groups, the pH is above 7 due to entrapment of unneutralized COOH groups in the center of aggregate particles (Section 8.3).

In early work on electrodeposition primers for automobiles, maleated linseed oil (Section 14.3.5) was used as a vehicle. The anhydride moiety was bonded to the linseed oil molecules by carbon–carbon bonds that do not hydrolyze, and hence, the coating was reasonably stabile. The anhydride was hydrolyzed to form the necessary COOH groups. However, adhesion to steel was relatively poor, and maleated linseed oil was soon replaced with maleated epoxy esters (Section 15.8), which had even better hydrolytic stability and provided superior adhesion to steel. The cross-linking obtained through the drying oil fatty acid esters in the epoxy ester was supplemented by using some melamine-formaldehyde (MF) resin as a cross-linker. Mixed methyl ethyl ether class I MF resins (Section 11.2) are most appropriate for electrodeposition. Such resins have sufficient solubility in water to permit easy incorporation and also are more soluble in the resin aggregates than in water, so a constant ratio deposits as the bath is used over long time periods.

Carboxylic acid-substituted resins made by reacting moderate molecular weight (MW) polybutadiene with maleic anhydride have also been used as vehicles for anionic primers. Since the backbone linkages are all carbon–carbon bonds, there is no problem of hydrolysis in the bath.

Maleated epoxy esters are not appropriate for many top coat applications because their color stability and chalking resistance are poor. Today, the most widely used resins for anionic electrodeposition top coats are acrylic copolymers made using acrylic (or methacrylic) acid and 2-hydroxyethyl methacrylate as comonomers. The carboxylic acid groups permit formation of stable aqueous dispersions having electronegatively charged particles; the hydroxyl groups, as well as the carboxylic acid groups, serve as sites for cross-linking with MF resin as described in Section 8.2.1.

The primary reaction occurring at the anode in anodic electrodeposition is electrolysis of water to yield protons. The protons neutralize carboxylate anions on the resin at the surface of the anode. When the salt groups are neutralized, the surface of the aggregates becomes destabilized and they coalesce on the metal substrate:

$$2H_2O \rightarrow 4H^+ + O_2 + 4e^-$$

$$RCOO^- + H^+ \rightarrow RCOOH$$

Not all the salt groups have to be neutralized for precipitation to occur. When the film is formed, some salt groups and, hence, ammonium ions are trapped within the film. Side reactions can also occur at the anode. Iron can dissolve to form ferrous ions that are oxidized to ferric ions. The ferric ions can form insoluble salts with the carboxylic acids on the resin, leading to reddish brown discoloration. Discoloration is not a problem for primers, but limits making light-colored or white electrodeposition top coats. Zinc phosphate conversion coatings (Section 6.3.1) generally reduce discoloration to acceptable levels. Anionic top coats can be used on aluminum, since aluminum salts do not affect the color of a coating. In fact, the protective oxide layer on aluminum is enhanced when it is the anode, so anionic E-coat is preferable to cationic for coating aluminum.

An important side reaction in anionic electrocoating of phosphate conversion treated steel is partial dissolution of the iron-zinc phosphate layer by hydrogen ions generated at the anode surface:

$$Zn_3(PO_4)_2 + 2H^+ \rightarrow 3Zn^{2+} + 2(HPO_4)^{2-}$$

This partial removal of conversion coating has two potentially deleterious effects. Damage to the phosphate coating can lead to poorer adhesion to the steel surface and less corrosion protection. Also, the soluble ion concentration in the bath is increased, which leads to higher conductivity of the water phase. As discussed later, maintaining constant, relatively low conductivity of the water phase is critical.

Electrolysis of water at the anode also generates oxygen. The oxygen can be generated at the metal surface after it has been coated, leading to film rupture as the bubbles of oxygen escape through the film, which can be observed as small pinholes in the coating. Deposition at high voltages usually results in greater film rupture.

At the cathode, hydrogen gas and hydroxide ions are generated. The hydroxide ions neutralize the ammonium counterions:

$$4H_2O + 4e \rightarrow 4OH^- + 2H_2$$

$$R_3NH^+ + {}^-OH \rightarrow R_3N + H_2O$$

The amines formed are water soluble, and only a fraction of the amine is removed when the coated article is taken out of the bath. This leads to accumulation of amine in the bath. Methods for control of amine concentration are discussed in Section 27.4. An advantage of electrodeposition as compared with other waterborne systems is that only a small amount of counterion is deposited on the substrate with the rest of the coating. Some anionic electrodeposition tanks have been run using potassium hydroxide as the neutralizing base.

27.2 CATIONIC ELECTRODEPOSITION COATINGS

Cationic coatings have positively charged aggregates that are attracted electrophoretically to the cathode. The resins have amine groups neutralized by a low MW water-soluble acid such as formic, acetic, or lactic acid. Coatings stable at a pH a little below 7 are preferred; at lower pH, stainless steel or other expensive corrosion-resistant piping and handling facilities would be required. Some suppliers recommend corrosion-resistant equipment in all cases. Commercial cationic electrodeposition tanks are operated within a narrow pH range, 5.8–6.2. The resins commonly used in the formulation of cationic E-coat automotive primers are based on BPA epoxy resins reacted with poly- or diamines, such as diethanolamine (DEA) to yield a modified BPA epoxy resin, which has amine and hydroxyl groups, followed by reaction of the amine groups with a low MW carboxylic acid to form ammonium salts, as shown in Scheme 27.1.

The vast majority of cationic E-coats are cross-linked with blocked isocyanates.

Half-blocked isocyanates, such as the reaction product of equimolar amounts of 2,4-TDI and an alcohol, such as 2-ethylhexyl alcohol (2EH-OH) or 2-butoxyethanol, can be reacted with hydroxyl groups on a modified BPA epoxy resin, as shown in the first equation of Scheme 27.2. The blocked isocyanate cross-linking agent is stable in the slightly acidic water system, whereas MF resins are not. The chemistry of blocked isocyanates is discussed in Section 12.5. Wicks and Wicks (2001) provide a review of blocked isocyanates.

During baking, the isocyanate unblocks and reacts with a hydroxyl group to form a urethane cross-link. However, significant deblocking of both isocyanate groups occurs during cure releasing free TDI, which can migrate into base coats and cause yellowing. The release of monomeric TDI can be suppressed by avoiding urethane linkages on both sides of the TDI. A patented method discloses the reaction of the second NCO groups of a half-blocked TDI with a bis-aspartate ester to form a bis-urea, as shown in the

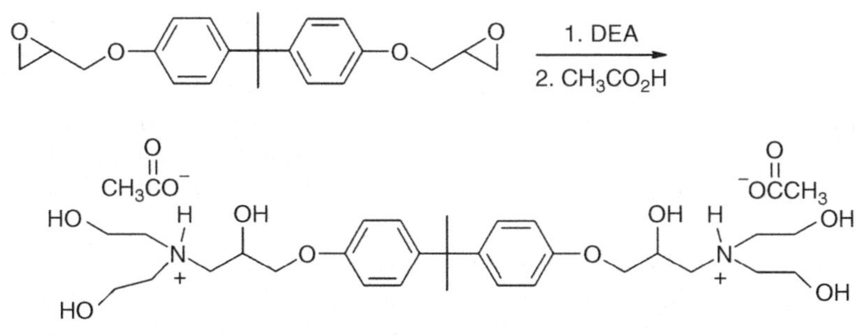

Scheme 27.1

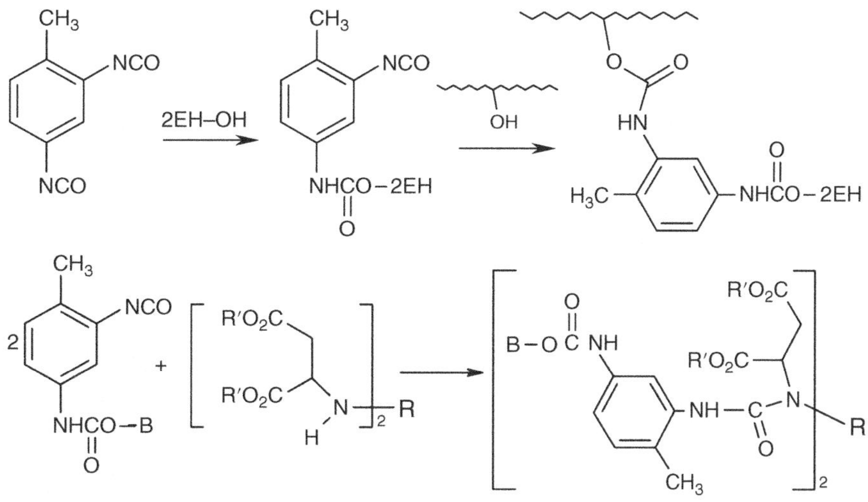

Scheme 27.2

second equation of Scheme 27.2. The urea groups cyclize to thermally stable hydantoins upon heating (Yonek et al., 1996). When the coating is baked, deblocking of the terminal blocking groups (B–OH) produces a diisocyanate cross-linker, which reacts with a hydroxyl-functional polymer, such as the modified BPA epoxy resin in Scheme 27.1.

Alternatively, the yellowing problem can be essentially eliminated by utilizing blocked aliphatic polyisocyanates, such as caprolactam-blocked IPDI/TMP prepolymer as the cross-linker (Debroy and Chung, 1988). In some cases, delamination on exterior exposure between the E-coat and top coats occurs; incorporation of HALS and a UV absorber in a 2-(2-butoxyethoxy)ethanol-blocked TDI primer has been reported to overcome this problem (Zwack and Eswarakrishnan, 1995).

MDI and polymeric MDI (Section 12.3.1) have the advantage of being less mobile than TDI and thus less likely to diffuse into base coats. E-coat dispersions are reported to be more stable using a polymeric MDI (12% 2,4′, 35% 4,4′, and 53% poly) blocked with methyl alcohol (or other alcohols) (Mauer et al., 1989). Coatings with enhanced flexibility can be obtained by partially reacting the MDI with polyether polyols prior to blocking. A cationic polyurethane dispersion (PUD), prepared from polymeric MDI, partially blocked with 2-(2-butoxyethoxy)ethanol, TMP, and *N,N*-dimethylaminopropylamine, neutralized with formic acid and dispersed in water, provides a solvent-free cross-linker for E-coats (Schafheutle et al., 1999).

The choice of the blocking groups is important, as it controls the bake window of the coating. In general, lower bake temperatures are preferred, as they use less energy and cross-linking can be achieved more quickly. However, the blocking–unblocking reaction is a reversible reaction. If the unblocking reaction is even slightly possible at the temperature of the electrocoat bath, the resin particles would slowly cross-link in the tank and the bath would not be stable. Thus, most cathodic electrocoats require bake temperatures of greater than 160°C. Research on new blocking agents and catalysts is an ongoing area of interest, as lower bake temperatures accompanied with acceptable bath stability would be a significant advancement for cathodic electrocoats.

Glycol ethers, such 2-butoxyethanol, are widely used blocking agent. The corresponding blocked isocyanates cross-link at lower temperatures than those blocked with non-ether alkyl alcohols, since the reverse reaction of the glycol ethers to reform the blocked isocyanates is slower. In addition, owing to lower volatility, glycol ethers remain in the film longer, promoting better flow and leveling. Glycol amides such as *N,N*-dibutylglycolamide have been disclosed as blocking agents, permitting lower temperature cure in E-coats as compared with 2-ethylhexanol (Moriarity, 1984).

Resins blocked with pendant morpholine-2,3-dione groups are reported to be effective blocking agents (Anderson and Gam, 2005). Methyl ethyl ketone oxime (MEKO)-blocked isocyanates are reported to provide primers that cure at temperatures as low as 80°C (Morikami et al., 2014). Such oxime systems were widely used in Japan; but, in the E-coat process, it has been shown that HCN is generated, leading to dangerous waste discharges (Nonomura et al., 1993; Modler and Nonomura, 1995). There is also the concern of premature of unblocking in the electrodeposition bath.

Considerable effort has been devoted to development of new catalysts for E-coats. There have been four driving forces: preparation of lead-free primers, reduction of curing temperatures, removal of tin from the formulation, and increased hydrolytic stability. Dibutyltin oxide was traditionally the most commonly used catalyst; it provides good stability, but it is insoluble and must be dispersed with the other pigments. Furthermore, it is not highly efficient and generally requires a lead compound as a cocatalyst. Trialkyltin compounds, such as bis(trioctyltin) oxide (TOTO), are claimed to be effective catalysts especially when used with ZnO (Bossert et al., 1999a). Bossert et al. (1999b) disclose various catalysts and particularly recommend TOTO, which permits curing at 150°C. It is claimed that thicker coatings, greater throw power, and better edge coverage can be obtained with TOTO. TOTO is a liquid that is readily incorporated into the coating, can be used to make clear coatings, and is low in toxicity.

Bismuth salts have been reported to be effective catalysts. Bismuth salts of lactic acid and dimethylolpropionic acid (DMPA) offer the advantage that the phosphate conversion coating does not need a chromate wash when the salts are used as catalysts for the E-coat primer (Kerlin and Hamacher, 1997). The effectiveness of a series of catalysts on the cure of a cationic E-coat in which the cross-linker was a glycol ether-blocked polymeric MDI concluded that bismuth tris(isostearate) together with isostearic acid provided the best cure response as well as superior resistance to hydrolysis (He et al., 2002). Wicks and Wicks (2001) provide an extensive review of catalysts in cationic E-coats. Since both tin and bismuth compounds are now preferably eliminated from electrocoat formulations, Peters et al. (2013) and Ravichandran et al. (2014) propose alternative organometallic compounds as effective electrocoat cure catalysts

While aromatic isocyanates have been used in most cationic E-coat primers, problems arise if UV or short wavelength visible radiation reaches the primer surface as a result of inadequate opacity of the subsequently applied coatings. Photodegradation of the primer can lead to discoloration and worse to delamination after outdoor exposure. Accordingly, primers have been developed using aliphatic isocyanates. However, since aliphatic isocyanates deblock at higher temperatures than aromatic isocyanates, they require higher cure temperatures/longer bake times.

Acid dissolution of the phosphate conversion coating that occurs in anionic deposition is not a problem in cationic E-coating. Depending on zinc concentration in the treatment bath, different crystals can be deposited (Section 6.4.1). At relatively high zinc concentrations, the crystals are predominantly hydrated zinc phosphate, $Zn_3(PO_4)_2{\cdot}4H_2O$, called hopeite. Under zinc-starved conditions, the crystals have been identified as phosphophyllite, $Zn_2Fe(PO_4)_2{\cdot}4H_2O$ (Dyett, 1989), which is the preferred treatment for E-coating steel. Accordingly, Zn–Ni, Sn–Mn, and Zn–Mn–Ni phosphate conversion treatments are used on zinc-coated steels for E-coats, since zinc phosphate by itself would give only hopeite (Schoff, 1990).

E-coats based on BPA epoxies and TDI or MDI are used only as primers. If used as top coats, they would exhibit poor color retention and exterior durability. For top coats, acrylic resins together with blocked aliphatic diisocyanates provide better color retention and exterior durability. 2-(*N*,*N*-Dimethylamino)ethyl methacrylate and hydroxyethyl methacrylate as comonomers provide the needed amine groups for salt formation and hydroxyl groups for cross-linking, respectively. The curing temperature required for alcohol-blocked aliphatic isocyanates to react with hydroxyl groups is relatively high. Oxime-blocked isocyanates cross-link at lower temperatures, but the bath stability may be limited. Alternatively, one can incorporate amine groups on acrylic resins by using glycidyl methacrylate as a comonomer, followed by reaction the pendant epoxy groups with primary amines. Alcohol-blocked aliphatic isocyanates can be used to cross-link the secondary amine groups to form urea cross-links, which require relatively high temperature bakes, but chemical stability in the tank is excellent, probably resulting from the essential irreversibility of the urea cross-links, compared with urethane cross-links. If photostable electrocoats are required, anodic chemistry is preferred as quite photostable acrylics can be more readily synthesized than their cationic counterparts.

Nevertheless, cationic E-coats have replaced anionic types as automotive primers primarily owing to superior corrosion protection. Amine-substituted cationic resin binders provide greater corrosion protection for steel resulting, perhaps, from strong interaction between the amine groups and the substrate surface, thereby enhancing wet adhesion, the most critical factor in corrosion protection (Section 7.4.2). Wet adhesion of lead-free coatings is especially critical for corrosion protection, since lead-containing compounds inhibited corrosion in earlier E-coats.

27.3 EFFECT OF VARIABLES ON ELECTRODEPOSITION

Deposition of coating is not instantaneous with application of current. Some time elapses before enough hydrogen ions (anionic E-coat) or hydroxide ions (cationic E-coat) are formed to neutralize enough charges on aggregate particles to cause precipitation. After this initial time interval, the rate of deposition is affected by the rate of electrophoresis of the aggregates. This rate is importantly affected by the impressed voltage; the higher the voltage, the faster the film deposition. Coatings are designed to coat in 2–3 min with 225–400 V; the high voltage is not needed to electrolyze the water, but rather to increase the driving force for electrophoretic attraction of particles to the electrode and for adequate coating of recessed areas.

The first areas covered are the edges of the metal, since current density is highest there. After edges, outer flat surfaces of the object are coated, followed by recessed and

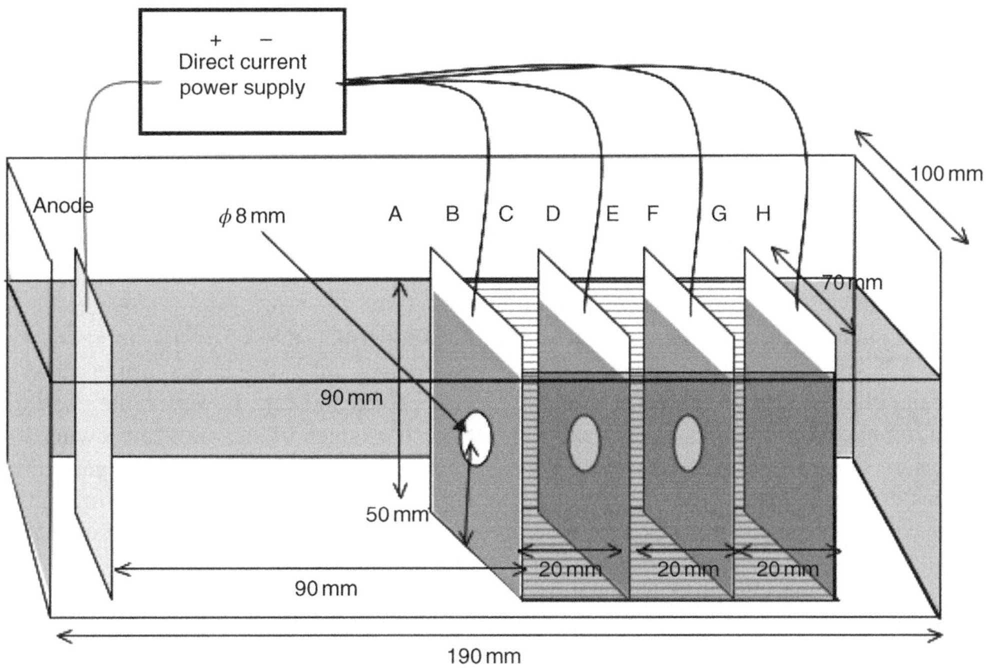

Figure 27.1 Schematic of Nagoya box used to determine E-coat throwing power. Current flows through holes in the bottom of the front panels. Sides of the panels are labeled A (facing the anode), B the back of the panel closest to the anode, and so forth.

enclosed areas. The further back in a recess, the later the area is coated. Particularly for corrosion protective primers, it is desirable to have the entire surface of the steel coated, so it becomes important to coat the furthest recessed areas in the 2–3 min dwell time in the tank. As film thickness increases, electrical resistance increases, reducing the rate of deposition at those sections coated first. There is a limiting film thickness beyond which deposition of further coating stops or, at least, becomes very slow.

The rate of deposition is also affected by the equivalent weight of the coating. The higher the equivalent weight, the greater the amount of coating precipitated by each hydrogen (or hydroxide) ion and, therefore, the faster the buildup of film thickness. On the other hand, it is critical to have a low enough equivalent weight so that there are sufficient polar salt groups to maintain the stability of the dispersion of aggregates in the coating bath. The rate is also affected by the amount of soluble low MW ions present in the bath. These ions are attracted to the electrodes in competition with the aggregate particles. Since they are small, they move more rapidly in the electrophoretic field. Thus, concentration of soluble salts must be low and maintained at essentially a constant level.

A coating that deposits in recessed areas quickly is said to have a high *throw power* (also called *throwing power* or just *throw*). Standard tests for throw power determine how far up into a pipe or an open-ended box the coating is applied at a standard voltage in a standard time. More recently the automotive industry has adopted the Nagoya box, which was first developed by Toyota, as a measure of throw. The experimental setup is shown in Figure 27.1. The box consists of a series of four panels placed in parallel with a fixed distance between them. The anode is placed in front of the first panel. At the bottom of each of the first three panels is a small hole to allow both current and fluid to pass among the panels. The faces of the panels are numbered from A to H, going from the first face to the last. The ratio of the film build on face A to face G is a good measure of the electrocoat's throw.

Throw power increases with higher impressed voltage and longer dwell times. However, if the voltage is increased to too high a level, there will be film rupture of the coating applied to the outer surfaces. At such high voltage levels, the current breaks through the film leading to local generation of gas under the film (oxygen in anionic electrodeposition and hydrogen in cationic electrodeposition) and bubbles of gas can blow out through the film, leaving film defects. It has been shown that, at higher voltages, electric discharges occur through the film during electrodeposition, causing visible sparks (Smith and Boyd, 1988). These sparks may also be responsible for film ruptures. Sparking is reported to occur at lower voltages (about 240 V) when the substrate is galvanized steel than when the substrate is steel (about 300 V or higher). As conductivity of the deposited film increases, film rupture tends to occur at lower voltages. Direct current electricity obtained by rectifying AC electricity has a relatively wide variation in voltage; the current is said to *ripple*. The effect of rippling can be to break through the film at lower (average) voltage (Vincent, 1990). Model equations have been developed to permit predictions of throw power and film build on complex shaped surfaces like an automobile (Boyd and Zwack, 1996). The prediction of the film build on the exterior is relatively straightforward. However, prediction of the film build in recessed areas requires extensive knowledge of the local electric current and a sufficient film growth model (Ellwood et al., 2009).

Throw power is affected by the conductivity of the bath; the higher the conductivity, the greater the throw power. However, there is a limitation; as the conductivity owing to the presence of soluble salts increases, the rate of electrophoresis of coating aggregates decreases. Increased numbers of salt groups on the resin increase conductivity throw power, but the rate of deposition then decreases. Entrapment of conductive material in the film increases with a corresponding increase in the likelihood of film rupture. A compromise on conductivity must be reached. Conductivities used are in the range of 1000–1800 microsiemens (μS); the older unit, mho, is still used; 1 μS = 1 μmho.

Film rupture, and hence throw power, is affected by variations in coating composition. If the viscosity of the aggregates precipitated on the surface is too high, complete coalescence is not achieved and a porous film results, resulting in high film conductivity and poor throw power. On the other hand, if the viscosity of the precipitated aggregates is too low, the films will be soft; if electrolysis takes place below the surface, bubbles will break through easily and film rupture will be severe. A compromise must be made between the extremes. The glass transition temperature of the resin is a controlling factor; correspondingly, the temperature at which the electrodeposition is carried out is important. Temperature of the bath must be controlled within a fairly narrow range, typically 32–35°C. Higher T_g resins can be used only if solvent is included in the formula. Many E-coats now use a small amount of solvent; no solvent is a desirable environmental goal. Solvents can affect the electrical conductivity of the deposited film, so care must be taken in their selection. Excess solvent leads to film rupture at lower voltage, which reduces throw power. Also, the partition coefficient must be such that almost all the solvent is dissolved in the aggregates with very little in the water, or the solvent concentration in the bath will build up over time. Substantial variations in smoothness of the primer surface can result from changes in E-coat primer composition (Gilbert, 1990).

Pigment concentration also affects coalescence. If the PVC is near or above CPVC and the amount of solvent, if any, is small, the deposited film will not coalesce. The amount of pigment also has a significant effect on leveling of the film. Since the film as applied has a low solvent concentration, its viscosity is very dependent on pigment concentration. Unless PVC is relatively low, less than half of CPVC for most coatings, leveling will be poor because of the high viscosity. Because of the low level of pigmentation as compared with conventional primers, the gloss of electrodeposited primers is higher, especially if the pigment content is reduced to low enough levels to permit good leveling.

If the PVC is lowered further, the viscosity of the coating after application and before cross-linking is lower. During baking, some leveling occurs prior to cross-linking owing to the temperature induced viscosity reduction. Subsequently, the viscosity increases as cross-linking progresses, after which no further flow can occur. If the viscosity is too low, however, the coating will flow away from the edges of the substrate owing to surface tension effects. The temperature of the edges increases first, reducing the surface tension of the coating in that area, resulting in flow over the adjoining higher surface tension area. This compromise between surface tension, viscosity, and cure kinetics is one of the fundamental formulation challenges in electrocoats, particularly for the cathodic E-coats used on automobiles. Smoother electrocoats are beneficial as this leads to less orange peel in the subsequently applied primer and top coat. However, the relatively low viscosity required for good smoothness may result in poorer edge coverage (called *high edge* in the industry). One measure of edge coverage is the ability of the electrocoat to coat a razor blade. The film thickness at the razor's edge after curing is a measure of its edge coverage.

In general, most pigments used in electrocoat are inert. The use of corrosion-inhibiting pigments is known in the patent literature; but since the removal of lead and hexavalent chrome pigments from commercial use, active corrosion-inhibiting pigments are rarely used commercially.

27.4 APPLICATION OF ELECTRODEPOSITION COATINGS

A schematic diagram of an E-coat system is shown in Figure 27.2. The object to be coated is hung from a conveyor and carried into a dip tank. Both to return coating carried out of the tank by the object and to avoid local accumulation of excess coating on the object, the object is rinsed with ultrafiltrate as it comes out of the tank. At the rinsing stage, the coating film has not been cross-linked, but is of sufficiently high viscosity to remain intact; only the bath liquid is washed from the surface of the coating. Concentration of coating in the bath must be kept constant by adding "makeup" coating continuously to replace coating solids removed by deposition. The heat exchanger maintains temperature within a narrow range.

The bath liquid is continuously recirculated through an ultrafiltration unit. Adoption of ultrafiltration was an essential factor in allowing large-scale cationic electrodeposition.

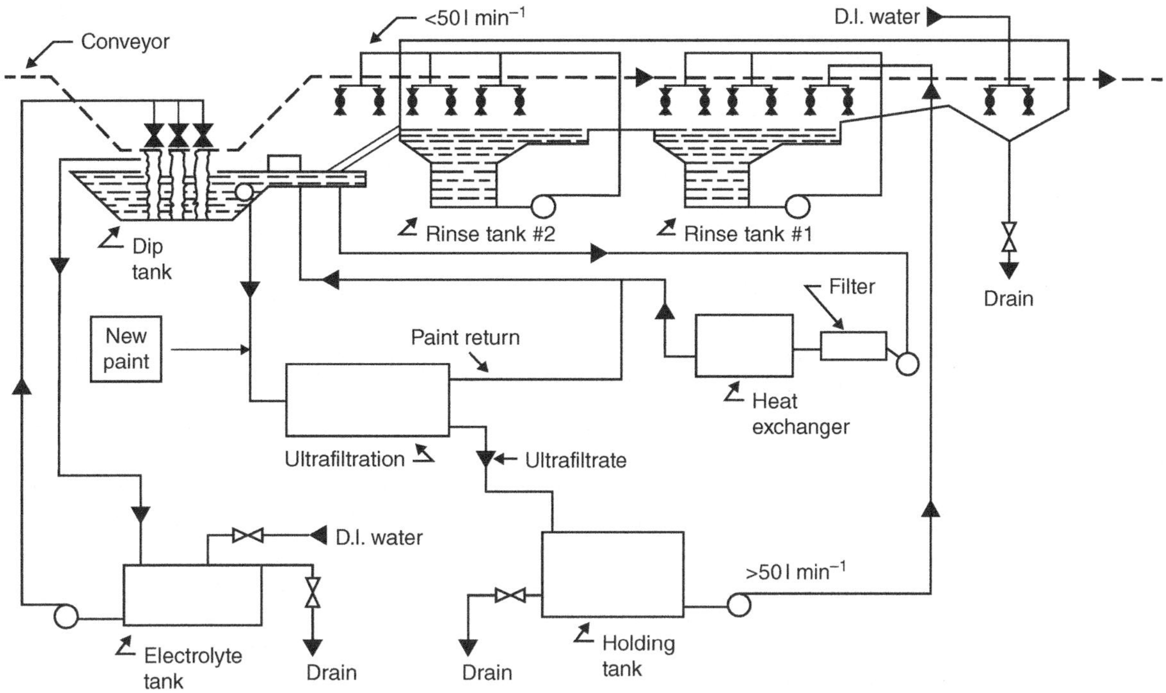

Figure 27.2 Cationic electrocoat deposition layout for an automotive E-coating line. Source: Wismer et al. (1982). Reproduced with permission of The Federation of Societies for Coating Technology.

The ultrafiltration membrane permits removal of excess water and water-soluble materials while not removing the aggregates containing the resin, pigment, and cross-linker. Ultrafiltration permits maintaining the concentration of soluble salts essentially constant so that there is constant conductivity.

Figure 27.2 also shows an electrolyte tank. This represents a system for controlling the concentration of solubilizing agent (acid for cationic coatings and amine for anionic ones). When coating is precipitated on the substrate surface, a corresponding amount of solubilizing agent is released. The concentration of solubilizing agent must be maintained at a constant level. A small amount of solubilizing agent is carried out of the bath with the coated product, and some is removed by ultrafiltration, but these losses of solubilizing agent are less than the amount being released by electrocoating. There are two ways of maintaining the balance. A sufficiently low level of solubilizing agent content is used in the makeup coating, and the remainder of the required level of solubilizing agent is the excess left in the bath. The other more effective approach is to have the counter electrode in a microporous polypropylene box. The pore size of the membrane must be such that the aggregate particles cannot pass through the membrane, but water and carboxylate (or ammonium) ions can easily go through. In some cases, it is necessary to use ion-selective membranes. The clear liquid in the box is recirculated to the electrolyte tank, where the concentration is monitored and corrected automatically. Maintenance of the proper level of solubilizing agent is critical. It is common lore that the pH of the bath must be kept constant. While this is true, the pH of these coatings is insensitive to the ratio of the weak acid and weak base (Section 8.3). Conductivity is a more important control criterion. The application lines can be highly automated with feedback control of rate of addition of makeup coating, solubilizing agent, and water.

27.5 ADVANTAGES AND DISADVANTAGES OF ELECTRODEPOSITION

Electrodeposition is used for applying coatings to a variety of products. Almost all new automobiles have electrodeposited primer coats. Many appliances are primed by electrodeposition. Aluminum extrusions, drapery fixtures, metal toy trucks, and steel furniture are a few of many examples of single-coat electrodeposition applications.

Development of electrodeposition in the 1960s and 1970s was a long and expensive process, and even today it is expensive to install a new system and get it running smoothly. Once it is running, electrodeposition can be a highly automated system with low labor requirements, especially as compared with spray application. A startling example of manpower savings has been reported in the use of cationic electrodeposition to apply an epoxy coating to air conditioners (Miranda, 1988). The former coating system of a flow-coated primer and a spray-applied acrylic top coat required 50 people, including those who did the required touch-up and repair. The E-coat operation requires only one operator. Further savings result from elimination of coating losses from overspray. Coating utilization in excess of 95% is reported. The economic advantage of the combination of these factors is large in assembly line operations. However, the capital cost of the automated line is high, limiting applicability of highly automated lines to large production operations. Simpler installations are used for applications, such as coating metal toys.

The solvent content of E-coats is relatively low, so VOC emissions are low and fire hazard is reduced. Another environmental advantage as compared with spray-applied coatings is the absence of overspray sludge; unless a poor job of formulating, coating production, bath maintenance, or control makes it necessary to dispose of 500 000 l of bad coating! Since the solids of the deposited film are high, only 3–5 min flash-off time is required before entering the oven, an advantage relative to spray-applied coatings.

Another advantage, assuming adequate throw power, is that complete coverage of surfaces is obtained. There can be differences in film thickness; for example, the recessed areas will generally have thinner deposits than the exposed face areas, but the entire surface will have some coating. Recessed and enclosed areas that cannot be coated by spray application can be coated by electrodeposition. Objects with many edges, such as drapery fixtures, can be coated better by electrodeposition than by any other way. The edges are uniformly coated, in contrast to excess coating at the bottom edges of objects coated by dipping.

Sagging can be experienced when the electrocoated part is heated in an oven. However, due to the relatively high viscosity of the coating immediately on application, severe sagging is less likely than during spray or conventional dip application methods. Also, one is less likely to encounter the relatively large differences in film thickness between the top and bottom of the dipped object commonly experienced with conventional dip coating.

Uniform film thickness can lead to a problem, especially with relatively highly pigmented primers: the applied electrocoating follows the surface contours of the metal closely, so a rough metal gives a rough primer surface. This is referred to as *telegraphing*. If there is minimal replication of metal scratches or a rough metal profile in the surface of the primer coat, the primer is said to show good *filling*. E-coat films are relatively thin, varying from 15 to 30 μm, depending on coating composition and application variables. When E-coat primers are sanded to smooth the surface and improve adhesion of a subsequent coating, only light "scuff sanding" is done. If necessary, heavier "spot sanding" will smooth out sags, runout, seeds, etc. and remove dirt, weld balls, and weld spatter. Spot sanding will often result in "sand-through" to expose metal. If so, the area must be patched with a "spot primer," a catalyzed spray primer that contains corrosion inhibitors.

There can be a problem achieving adequate adhesion of top coats and base coats to the surface of electrodeposited primers. Adhesion is affected by cross-link density of the E-coat; overbaking can lead to poorer intercoat adhesion. The gloss of electrodeposited primers is relatively high, and the smooth surface makes achieving intercoat adhesion more difficult. When the PVC is further reduced to promote leveling, the smoother surface further increases the difficulty of achieving intercoat adhesion. For auto bodies, a *primer-surfacer* (often just called a primer) is usually applied over the E-coat primer. If the solvents are carefully selected, penetration into the electrodeposited film promotes intercoat adhesion. The primer-surfacer can have a PVC > CPVC to enhance adhesion of top coat, and it can be sanded (except in the case of "wet on wet" application) to smooth the surface without reducing adhesion (see Section 30.1.1). The primer-surfacer also improves adhesion of sealants and adhesives and further protects the E-coat from UV degradation.

The substrate must be conductive; most applications have been for metal primers or for one-coat metal coatings. Spatially varying conductivity of the surface can lead to defects called *mapping* where the E-coat shows patches of slightly different appearance on the substrate. Changes in the local surface conductivity can be due to differences in the cleaning prior to pretreatment or locally different phosphate deposition times, resulting from heterogeneous spray coverage. Mapping is often contained by sanding the surface of the electrocoat but is better managed through a well-controlled pretreatment system.

Another limitation that electrodeposition shares with all other dipping system is the difficulty of formulation changeovers. If it is decided to change color of an automotive primer, what do you do with 500 000 l of coating? At least once when this problem was faced, the coating supplier made the new color primer compatible with the old one, so that the new coating could be introduced into the existing tank. Of course, for a period of time there was a slow change of primer color from the old color to the new color. This was not serious for a primer, but would not be acceptable for a top coat. Accordingly, electrodeposited top coats are limited to applications with long runs of the same color, such as agricultural equipment, garden tractors, and lawn mowers, so that a line can be dedicated to a particular color. In small installations, such as for coating toys, one coating bath can be pumped out of the electrodeposition tank into a holding tank, where it must be kept continuously agitated, and another color is pumped into the tank. But these are relatively small tanks; this procedure would not be economically feasible with large tanks.

27.6 AUTODEPOSITION COATINGS

Autodeposition coating, also called *Autophoretic®* coating, is, as the names imply, a process in which coatings are applied without external electricity, in contrast to anionic and cationic electrodeposition. A comparative study of the properties of steel coated by autophoresis and by cationic E-coat has been reported (Almeida et al., 2003). Autophoresis was carried out in a stabilized poly(vinylidene chloride) (PVDC) latex bath containing a weak acid (hydrofluoric acid), an oxidant (hydrogen peroxide), deionized water, FeF_3, surfactants, and additives. The PVDC latex is designed to be stable in the presence of Fe^{3+} ions but to be coagulated by Fe^{2+} ions. When a steel object is immersed in the coating bath, two reactions take place. The principal reaction is depicted in Eq. 27.1 and the minor reaction in Eq. 27.2:

$$Fe^0 + 2FeF_3 \rightarrow 3Fe^{2+} + 6F^- \qquad (27.1)$$

$$Fe^0 + 2HF \rightarrow Fe^{2+} + H_2(g) + 2F^- \qquad (27.2)$$

These reactions generate Fe^{2+} ions near the metal surface. These ions form an iron complex with the latex, destabilizing it so that the resulting coagulum deposits on the surface of the steel object. Some of the ferrous ions in the complex are oxidized to ferric ions by the hydrogen peroxide and hydrofluoric acid in the bath, as shown in Eq. 27.3. Thus, FeF_3 is regenerated:

$$2Fe^{2+} + H_2O_2 + 2HF \rightarrow 2Fe^{3+} + 2H_2O + 2F^- \qquad (27.3)$$

Initially, the deposited coating remains adherent and porous, allowing the acid to continue to penetrate to the substrate as the reactions continue. The first areas of the steel coated are the anodic areas of the steel such as edges. As the initial anodic areas are coated, other areas of the steel become anodic and coating continues. After application, the coating is baked at about 105°C to remove water and become cohesive.

The main advantage of autophoretic coatings is their unlimited throw, which allows any recessed area to be covered as long as it is exposed to the liquid coating. Autophoretic coatings for nonferrous substrates are still under development. Where the film properties are adequate, autophoresis has several advantages over E-coat: bake temperatures are lower, permitting application in the presence of heat-sensitive components, such as plastic materials. Only thorough cleaning of the steel is required, not phosphate coating, required for E-coat. Capital cost of application tanks and equipment is lower. The coatings have little or no VOC. The method is being used on a large scale commercially.

Almeida et al. report that, in the 2003 study, the corrosion protection by intact PVDC autophoretic films is comparable with a cathodic E-coat film in unscribed coatings. However, when the coatings are scribed, the cathodic E-coat is better, possibly because the passivating pretreatment is used with E-coat but not with autodeposition or perhaps due to superior adhesion. Bryden et al. (1998) disclose an acidic pretreatment that may improve this situation.

The films from autophoretic coating are rougher than cationic E-coats. The resistance to brake fluid of PVDC-based autophoretic coatings is also inferior. Thus, cationic E-coat is preferable for coating car bodies, and autophoretic coating uses are limited to applications where less corrosion protection and solvent resistance are required, for example, seat structures, fans, head light casings, and the like. The available coatings are usually black.

In addition to PVDC, a variety of other binders have been patented, including vinylidene chloride, acrylic acid, methyl acrylate copolymer latex (Hall, 2001), an acrylic latex (Roberto and Maxim, 1996), a BPA epoxy resin with an IPDI uretdione prepolymer as a blocked cross-linking agent, and a resole phenolic resin with an acrylonitrile/butadiene latex (Roberto and Maxim, 1996). Flexibilized BPA epoxy resins with caprolactam-blocked HDI isocyanurate are reported to provide coatings with excellent adhesion and corrosion protection without the need for a phosphate conversion coating (Bammel and Maxim, 2003). Epoxy autophoretic coatings are commercially available (Bryden et al., 1998).

GENERAL REFERENCE

Dini, J. W., *Electrodeposition: The Materials Science of Coatings and Substrates*, Noyes Publications, Norwich (**1993**).

REFERENCES

Almeida, E., et al., *Prog. Org. Coat.*, **2003**, *46*(1), 8–20.
Anderson, A. G.; Gam, A., US patent, US6908539 B2 (**2005**).
Bammel, B. D.; Maxim, M. A., *Waterborne, High Solids, and Powder Coatings Symposium*, New Orleans, LA, **2003**.
Bossert, E. C., et al., US patent, US 5859165 A (**1999a**).
Bossert, E. C., et al., US patent, US5902871 A (**1999b**).
Boyd, D. W.; Zwack, R. R., *Prog. Org. Coat.*, **1996**, *27*(1), 25–32.
Bryden, T. R., et al., US Patent application, US20080280046 A1 (**1998**).
Debroy, T. K.; Chung, D.-Y., US patent, US4755418 A (**1988**).
Dyett, M., *J. Oil Colour Chem. Assoc.*, **1989**, *72*(4), 132–138.
Ellwood, K., et al., *SAE Int. J. Mat. Manuf.*, **2009**, *2*(1), 234–240.
Gilbert, J., *J. Coat. Technol.*, **1990**, *62*(782), 29–33.
Hall, W. S., US patent, US6312820 B (**2001**).
He, Z., et al., US patent, US6353057 B1 (**2002**).
Kerlin, K. G.; Hamacher, P., US patent, US5702581 A (**1997**).
Krylova, I., *Prog. Org. Coat.*, **2001**, *42*(3–4), 119–131.
Mauer, G. W., et al., US patent, US4824925 A (**1989**).
Miranda, T., *J. Coat. Technol.*, **1988**, *60*(760), 47–49.

Modler, H.; Nonomura, M., *Toxcol. Environ. Chem.*, **1995**, *48*(3–4), 155–175.
Moriarity, T. C., US patent, US4452963 A (**1984**).
Morikami, A., et al., US patent, US8912280 B2 (**2014**).
Nonomura, M., et al., *Toxicol. Environ. Chem.*, **1993**, *39*(1–2), 65–70.
Peters, V., et al., US patent, US8617373 B2 (**2013**).
Ravichandran, R., et al., US patent, US8912113 B2 (**2014**).
Roberto, O. E.; Maxim, M. A., US patent, US5486414 A (**1996**).
Schafheutle, M. A., et al., US patent, US5977247 (**1999**).
Schoff, C., *J. Coat. Technol.*, **1990**, *62*(789), 115–123.
Smith, R.; Boyd, D., *J. Coat. Technol.*, **1988**, *60*(756), 77–84.
Vincent, J., *J. Coat. Technol.*, **1990**, *62*(785), 51–61.
Wicks, D. A.; Wicks Jr., Z. W., *Prog. Org. Coat.*, **2001**, *43*, 131–141.
Wismer, M., et al., *J. Coat. Technol.*, **1982**, *64*(688), 35.
Yonek, K. P., et al., European patent, EP0744425 A2 (**1996**).
Zwack, R. R.; Eswarakrishnan, V., US patent, US5389219 A (**1995**).

Chapter 28

Powder Coatings

Powder coatings are formulated from solid resins and usually cross-linkers and pigments, but no solvents. Small amounts of additives such as UV screeners, flow promoters, catalysts, and pigment dispersants are often included. Powder coatings are usually produced by melt mixing the ingredients at elevated temperature under high shear to disperse pigments in a matrix of the major binder components. After cooling, the resulting solid is pulverized. The powder is applied to the substrate and fused to a continuous film by baking. Electrostatic spray is the most common application method—the electrostatically charged powder particles form a fairly uniform layer on conductive substrates. Thermosetting and thermoplastic powders are available; the major portion of the market (about 90%) is for thermosetting types.

Brun et al. (2010) reviewed the history of powder coatings, mostly from a marketing point of view. Thermoplastic powder coating technology originated in Europe in the 1940s. Thermosetting epoxy powder coatings originated in Shell's Delft laboratories in the 1950s. Other generic types of resins were developed by 1970. Since then, the technology has advanced with occasional major innovations and a stream of incremental improvements.

The market for powder coatings has grown steadily except during economic recessions, and it now represents about 6% of the total market for coatings and a much higher percentage of factory-applied coatings. In 2014 the global value of all types of powder coatings was estimated at \$7.15 billion, with a growth rate of roughly 5% year^{-1} and possibly as much as 8% year^{-1} (Pianoforte, 2014). The fastest growth is in Asia; China alone is reported to have 2000, mostly small, powder coatings producers (Brun and Golini, 2010).

Powder coatings and coil coatings are somewhat complimentary—powder coatings are suitable for coating irregular shapes, while coil coatings are more efficient for flat sheets and coils of metal.

Various metals and inorganic materials (ceramics) are applied as powder coatings, but they are outside the scope of this book.

28.1 BINDERS FOR THERMOSETTING POWDER COATINGS

Binders for thermosetting powder coatings consist of a mixture of a primary resin and a cross-linker, often called a *hardener*. The major types of binders are grouped somewhat arbitrarily into several classes, as shown in Table 28.1. The terminology has grown historically and has become confusing. *Epoxy* powder coatings include only coatings based on BPA and novolac epoxy resins together with amine, anhydride, hydrazide, or phenolic hardeners. *Hybrid* powder coatings also contain BPA epoxy resins but are cross-linked with carboxy-functional polyester resins. Polyester powder coatings comprise polyesters and various cross-linkers but do not contain BPA or novolac epoxies (otherwise they would be classed as hybrids). Polyester coatings exhibit good to excellent exterior durability. Acrylic powder coatings contain acrylic resins with various cross-linkers. In addition, various blends, sometimes called *alloys*, of these classes are used. Blends are gaining importance. Among the factors in choosing a class of powder coating for an application are protective properties, exterior durability, and cost (Gribble, 1998). Differences in exterior durability are indicated in Figure 28.1 (Richert, 1982).

In powder coatings, it is necessary to control the balance of binder T_g, $\bar{M}_w$, $\bar{M}_n$, f_n and reactivity (Kapilow and Samuel, 1987), because there are conflicting requirements. The

Organic Coatings: Science and Technology, Fourth Edition. Frank N. Jones, Mark E. Nichols, and Socrates Peter Pappas.

powder must be chemically stable enough to be melt processed without significant cross-linking and to avoid sintering (premature fusion) of the powder during storage and shipment. On the other hand, after application the powder must fuse during baking, flow to form a desirable film, and chemically cross-link. In general, the primary resins are amorphous polymers with T_g high enough to avoid sintering of the powder and with $\bar{M}_w/\bar{M}_n$ of a few thousand. Recommended minimum binder T_g values are 40°C in Europe and 45–50°C in North America, reflecting the higher temperatures during shipment and storage in parts of North America. A typical powder coating having a T_g of 50°C could be melt processed at 80°C and could be handled and stored at temperatures up to about 40°C. When heated in a baking oven, its viscosity would briefly drop to about 10 Pa·s allowing coalescence, flow, and leveling; and with continued heating for 15 min at temperatures in the range of 130–200°C, it would cross-link. The T_g values referred to are those of the combination of the primary resin plus the cross-linker in the powder particles. The required T_g for the primary resin alone varies, depending on the cross-linker used with it. UV cure powder coatings can be cured at temperatures as low as 100°C.

Table 28.1 Classes of Thermosetting Powder Coatings

Common name	Primary resin cross-linker
Epoxy BPA (or novolac) epoxy	Polyamines, anhydrides, or phenolics
Hybrid COOH-functional polyester	BPA epoxy
Polyester	
COOH-functional polyester	Triglycidylisocyanurate or hydroxyalkylamide
OH-functional polyester	Blocked-isocyanate or amino resin
Acrylic	
Epoxy-functional acrylic	Dibasic acid
OH-functional acrylic	Blocked-isocyanate or amino resin
UV cure	
Acrylate-functional resin	Free radical
Epoxy-functional resin	Cationic

Because T_g is higher, powder coatings can attain good hardness at somewhat lower cross-link density than most liquid coatings designed for similar applications. The higher T_g tends to be a favorable attribute of powder coatings, which often combine excellent hardness with excellent impact resistance.

28.1.1 Epoxy Binders

Epoxy powder coatings are the oldest and still one of the largest classes of thermosetting powder coatings. Decorative coatings have been based on BPA epoxy resins with n values of 3–5, usually made by the advancement process (Section 13.1.1). For thin film applications, however, epoxy resins having n as low as 2.5 can provide better flow. For protective coatings, n values range up to 8. The most commonly used cross-linkers are dicyandiamide (DICY) (Section 13.2.1) or a modified DICY. The curing reactions are complex (Fedtke et al., 1993). Modified DICYs are more soluble in epoxy resins and tend to form

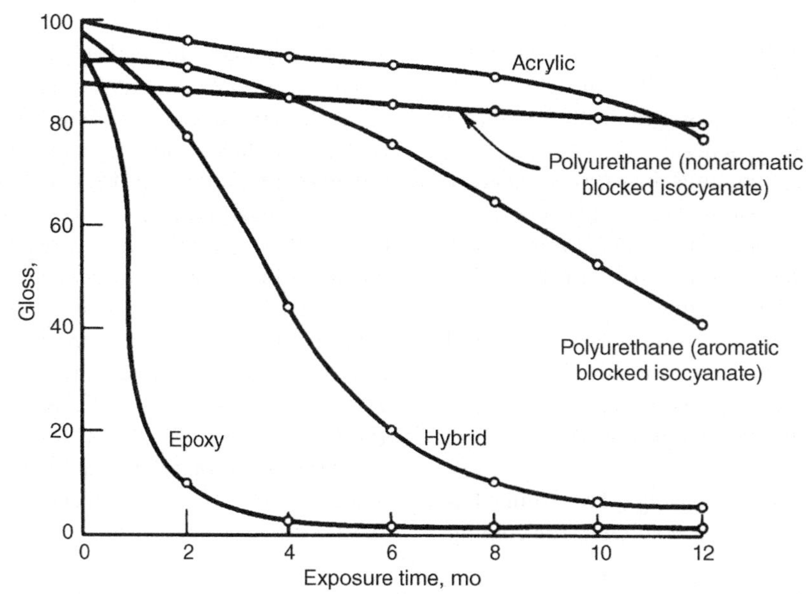

Figure 28.1 Florida outdoor exposure data on different types of powder coatings. Source: Richert (1982). Reproduced with permission of Nordson Corporation.

uniform films more readily. It is common to add a catalyst, such as 2-methylimidazole, to accelerate the cure. Various grades of DICY are available—small particle size promotes faster cure and perhaps more uniform films.

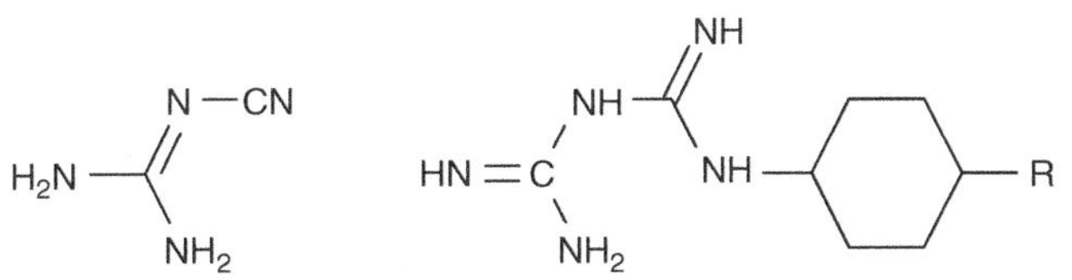

Dicyandiamide   Modified dicyandiamide

Epoxy powder coatings have very good mechanical properties, adhesion, and corrosion protection; however, their exterior durability is poor. Applications for decorative types include institutional furniture, shelving, and tools—mostly items that will be used indoors. Applications of protective epoxy coatings include pipes, rebars, electrical equipment, primers, and automotive underbody parts. Where enhanced chemical and corrosion resistance is needed, phenolic resins (Section 13.3.1) are used to cross-link epoxy resins, again with 2-methylimidazole as catalyst. Novolac epoxy resins (Section 13.1.2) or blends of novolac and BPA epoxies provide higher cross-link densities than BPA epoxies alone. All these coatings discolor and chalk on exterior exposure. Polycarboxylic acid anhydrides (Section 13.3.2), such as trimellitic anhydride, are sometimes used to cross-link BPA epoxy resins in applications for which greater resistance to yellowing and to acids and solvents is needed. These latter coatings are generally being replaced with hybrid coatings that have somewhat better exterior durability and less question of toxic hazard.

Coatings for steel pipes and concrete reinforcing bars (rebars) are important markets for epoxy powder coatings. Typical binder compositions include a solid epoxy resin and a dihydrazide cross-linker (Section 17.11). In the trade they are called *fusion-bonded epoxy* (FBE) coatings. Formulations are adapted to suit specific applications, for example, by using flexibilized epoxy resins for toughness or by incorporating siloxane adhesion promoters.

28.1.2 Hybrid Binders

BPA epoxy resins, cross-linked with carboxylic acid-terminated polyester resins having $\bar{M}_n$ of a few thousand, are called *hybrid powder coatings*; their properties are intermediate between epoxy and polyester coatings. Hybrid coatings have better color retention and UV resistance than epoxy powder coatings, but exterior durability is much inferior to that of polyesters (Section 28.1.3) and acrylics (Section 28.1.4). Examples of end uses include water heaters, fire extinguishers, radiators, and transformer covers. They are not recommended for prolonged outdoor use.

Various polyesters have been described. Most are derived from neopentyl glycol (NPG) and terephthalic acid (TPA) with smaller amounts of additional monomers to adjust the T_g to the desired level and to increase the $\bar{f}_n$ above two (Section 10.6). An example is a polyester from NPG (364 parts by weight, 3.5 mol), TPA (423 parts, 2.55 mol), adipic acid (AA) (41 parts, 0.24 mol), and trimellitic anhydride (TMA) (141 parts, 0.74 mol) (Misev, 1991). The acid number of the resin is 80 mg KOH g^{-1} of resin. The relatively high TMA content increases $\bar{f}_n$, compensating for the low (about 1.9) $\bar{f}_n$ of the BPA epoxy. The primary cross-linking reaction is the ring opening of the oxirane groups by carboxylic acids (Section 13.3.2). Esterification and transesterification reactions involving hydroxyl groups of the epoxy resin and homopolymerization reactions of oxirane groups may also play a role. Catalysts, such as an ammonium or phosphonium salt, for example, tetrabutylammonium bromide or choline chloride, permit baking temperatures in the range of 160–200°C. Often, the polyester resins are supplied with the catalyst blended in.

Flow properties of powder coatings containing carboxylic acid-terminated polyesters tend to be poorer than those made with hydroxy-terminated polyesters. Fischer and McKinney (1988) reported a proprietary, modified BPA epoxy that exhibits comparable sintering resistance and improved flow compared with conventional BPA epoxies.

28.1.3 Polyester Binders

In addition to the polyester resins used in hybrids, polyesters are widely used in a variety of other formulations. Depending on the cross-linker, they are carboxylic acid-functional or hydroxy-functional. Most are based on TPA and NPG, but substitution of isophthalic acid (IPA) for TPA in these resins improves outdoor durability (Section 10.1.2). IPA-based resins and perhaps other types are marketed as "superdurable polyesters." Versions designed for all the common cross-linkers are available. Powder coatings made from these resins are also marketed as "superdurable." When well formulated, such coatings can satisfy AAMA specification 2604, which requires 5 years' outdoor durability.

Triglycidylisocyanurate (TGIC) (Section 13.1.2) has been widely used as a cross-linker for carboxylic acid-terminated polyesters with basic catalysts. (Oddly, TGIC-cured polyesters are not called hybrids even though TGIC is a trifunctional epoxy, perhaps because it is not a BPA epoxy.) TGIC-based powder coatings have good exterior durability and mechanical properties. Examples of end uses are outdoor furniture, farm equipment, fence poles, and air-conditioning units. While TGIC is expensive, the amounts required are relatively small because of its low equivalent weight. Typical binders contain 4–10 wt% TGIC and 90–96 wt% of carboxylic acid-terminated polyester. The polyesters used are generally less branched than those used in hybrid coatings because of the higher functionality of TGIC as compared with BPA epoxies. One polyester, for example, is made from NPG (530 parts, 5 mol), TPA (711 parts, 4.3 mol), IPA (88 parts, 0.47 mol), pelargonic acid (58 parts, 0.37 mol), and TMA (43 parts, 0.22 mol). All parts are by weight, and the acid number is 35 (Misev, 1991). Such resins are prepared in a two-stage process to minimize the problems caused by the high melting point and low solubility of TPA (Section 10.6). Alternatively, dimethyl terephthalate can be used in place of TPA, in which case the polymer is formed by transesterification (Section 10.6). High equivalent weight (low acid number) is desirable, since this reduces the required amount of TGIC, but cross-link density decreases with the higher equivalent weight. There is an optimum for each application.

Merfeld et al. (2005) studied a variety of catalysts for TGIC/polyester powder coatings, seeking a catalyst that would cure at a temperature of 120°C while still having adequate storage stability and flow. The work was aimed at a primer powder coating for use in coating heat-sensitive aluminum alloy for aircraft. The most appropriate catalyst was found to be benzyltrimethylammonium chloride.

Trottier et al. (2012) reported that different types of TiO_2 pigment can significantly influence the cure rates of epoxy/polyester powder coatings. Traces of zinc contaminants in some grades of TiO_2 retard curing and variations of the alumina/silica ratio in the pigment's surface treatments also influence cure rates. This study illustrates the risk of an assumption that has caused many problems over the years. It is risky to assume that an apparently innocuous substitution of one raw material for another in a carefully balanced formula will not cause unintended consequences.

A potential problem with NPG/TPA polyesters is that they can contain cyclic esters without functional groups; such compounds have been shown to *bloom* to the surface of polyester/TGIC coatings. Use of a combination of NPG and 2-butyl-2-ethyl-1,3-propanediol with TPA and IPA gives a semicrystalline polyester with much less cyclic ester and no blooming in the TGIC gloss coatings (Shah and Nicholl, 2003).

Concern about potential toxic hazards of TGIC motivated a search for other cross-linkers for carboxylic acid-functional resins. A fruitful direction for the evolving technology employs cross-linkers with multiple beta-hydroxyalkylamide groups. Such groups esterify with acids more readily than ordinary alcohols and cure at similar temperatures. Kronberger et al. (1991) described tetra(2-hydroxyalkyl)bisamides (Section 17.4) as cross-linking agents with carboxylic acid-functional polyesters for exterior durable coatings. An exemplary cross-linker is formed by reaction of one mole of AA and two moles of diethanolamine (Misev and van der Linde, 1998). The coatings also have good mechanical properties. Water is evolved from the cross-linking reaction, which limits the film thickness that can be applied without popping. Polyesters having terminal hindered carboxylic acid groups, cross-linked with a tetra(2-hydroxyalkyl)bisamide, provided powder coatings with improved flow, less limitation on film thickness to avoid popping, and excellent acid rain resistance. A hindered COOH group increases flow due to the reduction of hydrogen bonding between COOH groups. The hindered COOH groups also reduce cure rate providing for improved leveling (Boogaerts et al., 2000). Examples of further refinements are

- Polyesters terminated with tertiary carboxylic acid groups (Moens et al., 2004b) with excellent flow and leveling
- Branched polyesters with f_n of about 2.35–2.65 (Buijssen et al., 2012), which cure at lower temperatures without blooming

The evolving technology of beta-hydroxyalkyl amides has captured substantial market share. Cavalieri et al. (2015) describe formulations based on conventional and superdurable (Section 10.1.1) polyester resins cross-linked with 2-hydroxyalkylamides. Further development of superdurable formulations is expected to enhance their potential for providing single-coat applications combining weatherability and corrosion resistance.

Hydroxy-functional polyester resins are also widely used. The most common cross-linkers are blocked aliphatic isocyanates (Section 12.5). The coatings have exterior durabilities equal to, or somewhat better than, TGIC cross-linked polyesters, together with the excellent mechanical properties and abrasion resistance typical of polyurethane coatings. Blocked isocyanate/polyester powder coatings generally flow better than most other powder coatings, possibly because the released blocking agents or unreacted cross-linkers also function as plasticizers. End uses include automobile wheels, lighting fixtures, garden tractors, fence fittings, and playground equipment.

Blocked isocyanates (Section 12.5) that are solids are potentially useful for powder coatings. These include blocked derivatives of isophorone diisocyanate (IPDI), bis(4-isocyanatocyclohexyl)methane (H_{12}MDI) isocyanurate, and tetramethylxylidene diisocyanate (TMXDI) low MW prepolymers (Section 12.3.2). Blocked isocyanates from sterically crowded isocyanates such as TMXDI have the potential advantage of unblocking at a somewhat lower temperature (Pappas and Urruti, 1986; Witzeman, 1996). Probably, the most widely used blocking agent is ε-caprolactam. Caprolactam-blocked H_{12}MDI/TMP prepolymer is reported to cure in 20 min at 160°C, which is 20°C lower than with IPDI- or HDI-blocked analogs, and provide excellent exterior durability, flexibility, and impact resistance

(Rawlins et al., 2002). In addition to the relatively high cure temperature, a further disadvantage of caprolactam is that, to some extent, it volatilizes during cure and may enter the atmosphere or build up in baking ovens. Oxime-blocking isocyanates react at lower temperatures, and blocked analogs of the sterically crowded diisobutyl and diisopropyl oximes cure at even lower temperatures (Witzeman, 1996), owing to the release of steric strain. Oxime-blocked isocyanate coatings tend to yellow during cure, especially if overbaked. Also, there is concern about toxic hazards with oximes. 3,5-Dimethyl pyrazole and 1,2,4-triazole (and mixtures of the two) form blocked isocyanates that combine lower cure temperatures and no yellowing (Engbert et al., 1996). 3,5-Dimethylpyrazole blocked $H_{12}MDI$ is reported to cure coatings at 150°C with excellent flow and leveling (Rawlins et al., 2002).

As cross-linkers, uretdiones (also called *isocyanate dimers* and *internally blocked isocyanates*) have the distinct advantage that when they cleave to give isocyanates, no blocking agent is evolved (Section 12.5). The cure temperatures without catalyst are high (ca 180°C). The patent literature reflects many efforts to reduce the cure temperature, often involving new catalysts such as quaternary ammonium salts and/or resins with greater reactivity.

Solid amino resins such as tetra(methoxymethylol) glycoluril (TMMGU) and toluene sulfonamide-modified melamine–formaldehyde (MF) resins are also cross-linkers for hydroxy-functional resins. Since methyl alcohol is generated as a by-product, film thickness may be limited by popping or retention of bubbles in the cured film. Release of the methyl alcohol from the film can be aided by a variety of approaches that retard cross-linking. For example, addition of weakly basic solid amines, such as 2-methylimidazole, can minimize popping (Higginbottom et al., 1992).

28.1.4 Acrylic Binders

A variety of acrylic resins can be used in powder coatings: hydroxy-functional acrylics can be cross-linked with blocked isocyanates or glycolurils, and carboxylic acid-functional acrylics can be cross-linked with epoxy resins, or carbodiimides (Taylor et al., 1995). Hydroxyalkylamides have also been used to cross-link COOH-functional acrylic powders (Yeates et al., 1996).

The greatest interest has been in epoxy-functional acrylics, made with glycidyl methacrylate (GMA) (Section 13.1.2) as a comonomer, and cross-linked with dicarboxylic acids, such as dodecanedioic acid [$HOOC(CH_2)_{10}COOH$] or carboxylic acid-functional resins. An epoxy-functional acrylic for automotive primer surfacer is reported to require an $\bar{M}_n$ below 2500, a calculated T_g above 80°C, and a monomer composition such that melt viscosity is less than 40 Pa·s at 150°C (Green, 1995). Such a resin can be made with 15–35% GMA and 5–15% butyl methacrylate (BMA), with the balance being methyl methacrylate (MMA) and styrene. A somewhat similar acrylic evaluated for automotive clear coats had $\bar{M}_n$ of 3 000, $\bar{M}_w/\bar{M}_n$ of 1.8, and T_g of 60°C (Agawa and Dumain, 1997).

Acrylic resins with broad polydispersity $\left(\bar{M}_w/\bar{M}_n\right)$ are problematic for powder coatings. Controlled radical polymerization (Section 2.2.1.1) can provide much narrower polydispersities and a large variety of well-controlled polymer structures (Krol and Chemielarz, 2014). An example of the early use of ATRP is the preparation of acrylic resins (Barkac et al., 2003). The resin had $\bar{M}_w/\bar{M}_n$ of 1.25 compared with 1.95 for a similar resin made by conventional free radical polymerization. The melt viscosity at 180°C of the ATRP polymer was 11.2 Pa·s compared with 57.2 for the conventional resin. A powder coating of the ATRP polymer using bis(*N*,*N*-di-2-hydroxyethyl)adipamide as a cross-linker was compared with a similar powder coating made with the conventional free radical copolymer. The low melt viscosity of the ATRP polymer resulted in better leveling. Furthermore, the reduced amount of very low MW molecules in the ATRP polymer reduced fusion of the powder particles during storage.

Acrylic powder coatings generally have superior detergent resistance and are used for applications such as washing machines. They can also have excellent outdoor durability. Acrylics tend to be incompatible with other powder coatings, requiring caution when changing coating types to avoid contamination, which can cause cratering. As with liquid coatings, acrylic powder coatings tend to have poorer impact resistance than polyesters (Van der Linde and Scholtens, 1992).

28.1.5 Silicon-Containing Binders

The use of silicone and silicone/polyester resins (Chapter 16) in high heat resistance powder coatings has been reported (Popa et al., 1999). The coatings were cured at 232°C. Silicone powder coatings require lower processing temperatures than other powder coatings but higher baking temperatures. In order to avoid bubbling due to the escape of water during cross-linking, film thickness is usually limited to 2 mils or less. Silicone-based powder coatings with thermal resistance up to 500°C are available.

Polysiloxanes can be incorporated into conventional powder coatings. For example, a blocked isocyanate cross-linker was modified with poly(dimethylsiloxane) and used to cross-link solid polyester resins (Pilch-Pitera, 2014). Siloxane segments migrate to the surface during curing and make the coating hydrophobic. A siloxane content of 1% by weight is effective. Higher siloxane levels compromise adhesion to metal. It is conjectured that such technology may be the basis for anti-graffiti powder coatings now on the market.

28.1.6 UV Cure and Hot Press Powder Coatings

Environmental regulations, especially in Europe, provide strong incentive to use near-zero VOC coatings such as powder coatings on heat-sensitive substrates. As of 2012 the practical lower limit for baking conventionally applied and cured powder coatings was 120–140°C even for "ultra-low bake" coatings (Maurin et al., 2012). Wood, most manufactured wood products, paper, and many plastics cannot withstand baking at 120°C long enough to effect cure. Potential solutions to this problem include UV cure powder coatings (Chapter 29 discusses UV curing.) and curing conventional coatings by unconventional methods such as infrared (IR) surface heating, hot press, and in-mold coating processes.

Practical application of UV-curable powder coatings often requires heating at temperatures up to 140°C to fuse and level the powder followed immediately by irradiation with UV to effect cross-linking. Misev and van der Linde (1998) showed that this requirement can sometimes be met by heating the powder on the surface with IR radiation without overheating the substrate.

Rawlins and Thames (2000) reported a variety of powder coatings that are cured by UV. The formulations underwent rapid cross-linking at lower temperatures. Since the powder is stable in the dark, premature reaction during powder production is minimized. Both free radical and cationic cure coatings have been investigated. Free radical cure coatings use acrylated epoxy resins (Section 29.2.4) and/or acrylated polyesters or unsaturated maleic polyesters as binders. The effects of annealing before UV cure of methacrylated BPA epoxy resin, acrylated BPA epoxy resin, silylated acrylated BPA epoxy resin, and acrylated polyester were investigated. Some of these coatings, however, only achieved satisfactory properties when annealed at elevated temperatures just before cure, limiting their utility on heat-sensitive substrates.

A UV cure powder coating for fiberboard has been formulated with an acrylated polyester, an acrylated BPA-polyphenoxy resin, and a photoinitiator. The powder coating is sprayed on the fiberboard that is backed up with a grounded copper plate. After fusing under IR sources (at 140°C), the coating is UV cure (Moens et al., 2004a).

While most coating binders are amorphous, crystalline polymers with UV-polymerizable groups can serve as binders for UV cure powder coatings that are fused with IR at temperatures as low as 100–120°C followed by UV curing. An example of such a coating comprises a crystalline polyester from 1,6-hexanediol, fumaric acid, and TPA, together with a divinyl ether, derived from HDI and 4-hydroxybutyl vinyl ether (Twigt and van der Linde, 1997). Semicrystalline low MW polycarbonate diols esterified with methacrylic acid permit storage at 40°C and UV cure at 100–115°C after heating to 120°C to fuse the polymer (Lowenhielm et al., 2005).

Maurin et al. (2012) reviewed UV powder coatings and studied formulations based on commercially available urethane acrylate and polyester methacrylate UV-curable powder resins. The coatings were cast from solution for convenience. Some formulations included conventional di- and tri-acrylates, which improved film properties but might promote sintering of the powder during the actual process.

Another avenue for low-temperature curing is to combine UV cure powder resins with free radical initiators and induce thermal cure (Beetsma, 2016).

Cationic UV cure coatings use BPA epoxy resins as binders. Photoinitiators must be incorporated into the formulation. After application, the powders are fused by passing under IR lamps, followed by curing under UV lamps. Film formation can occur at temperatures of 120°C or less with IR, followed by UV cure of the heated films in 1 s or less. This procedure allows cure on some heat-sensitive substrates. Good leveling is possible because viscosity does not begin to increase until cure is initiated by UV. As with any other UV cure system, pigments that absorb UV interfere with curing, limiting the film thickness that can be cured. While some pigmented UV cure powder coatings can be cured, the main interest has been in clear coatings.

Hot press and in-mold application processes fuse, level, and cure the uncured powder coating by contact with heated metal. For example, Badila et al. (2014) cured a highly reactive epoxy/polyester hybrid powder coating on veneered particle board surfaces by hot pressing at 130°C. The process gave a glossy coating. Wuzella et al. (2014) obtained similar results with clear epoxy powder coatings on medium-density fiberboard (MDF, a popular material for building furniture). Various pressing times and temperature were studied.

28.2 BINDERS FOR THERMOPLASTIC POWDER COATINGS

The earliest organic powder coatings were thermoplastic coatings, which now account for roughly 10% of the market. Thermoplastic powder coatings have several disadvantages compared with thermosetting analogs. They are difficult to pulverize into small particle sizes; thus, they are usually applied in relatively thick films. For good film properties the binders generally have higher MWs than thermosetting powder coatings. As a result, when they are baked even at high temperatures, they remain viscous, resulting in poor flow and leveling.

Vinyl chloride copolymers (PVCs) and, to a more limited extent, polyolefins, polyamides (nylons), fluoropolymers, and thermoplastic polyesters are used as binders. High vinyl chloride content copolymers (Section 17.1.1) are formulated with stabilizers and a limited amount of plasticizer so that the T_g is above ambient temperature. The partial crystallinity of PVC may help stabilize the powder against sintering. Vinyl powders are generally applied as quite thick films, 0.2 mm and higher, by fluidized bed application, discussed in Section 28.5.2. Dishwasher racks, handrails, tool handles, and metal furniture are examples of end uses.

Polyolefin powders give coatings with low water absorption and excellent chemical resistance. They are used on lab and food handling equipment and as carpet backing, but the volume used in metal applications is limited by poor adhesion. As discussed in Section 28.5.2, ethylene/acrylic acid (EAA) and ethylene/methacrylic acid (EMA) copolymer resins are available that provide powder coatings with substantially better adhesion (Blanton et al., 1999).

Nylon 11- and nylon 12-based powder coatings exhibit exceptional abrasion and detergent resistance. They are used as antifriction coatings and as coatings for hospital beds, clothes washer drums, and other applications that must withstand frequent cleaning or sterilization and have good toughness and wear resistance.

Fluoropolymers, such as poly(vinylidene fluoride) and ethylene/chlorotrifluoroethylene copolymers, are used for coatings requiring exceptional exterior durability, such as aluminum roofing and window frames, and also for resistance to corrosive environments, such as equipment for chemical plants. An example of the latter is a copolymer of vinylidene fluoride with hexafluoropropylene (Kosar and Morris, 2007). Thermosetting vinylidine fluoride copolymers are also available for powder coatings.

28.3 FORMULATION OF THERMOSETTING POWDER COATINGS

The challenge facing formulators is satisfying a combination of conflicting needs:

1. Minimization of premature cross-linking during production
2. Stability against sintering during storage
3. Coalescence, degassing, and leveling at the lowest possible baking temperature
4. Cross-linking at the lowest possible temperature in the shortest possible time
5. Film properties that meet or exceed the requirements of the intended application

Furthermore, flow and leveling must be balanced to achieve acceptable appearance and protective properties over the range of expected film thicknesses. Coatings that flow readily before cross-linking can form smooth films, but they may flow away from edges and corners because of the surface tension differential driven flows that result from the faster heating of the edges (Section 24.3).

If the T_g of a coating powder is high enough, sintering can be avoided. However, coalescence and leveling at the lowest possible temperature are promoted by low T_g. Short baking times at low temperatures are possible if the resins are highly reactive and if the baking temperature is well above the T_g of the final cross-linked film. However, such compositions may cross-link prematurely during production, which limits the ability of the coating to coalesce and level. Compromises are needed. A crude rule of thumb is that the lowest feasible baking temperature is 50°C above the melt-mixing (extrusion) temperature and 75–80°C above the T_g of the uncured powder. This rule predicts minimum baking temperatures of about 130–135°C for a powder with a T_g of 55°C. However, powder coatings that cure at 121°C (250°F) are on the market (Pianoforte, 2014); they are designed for heat-sensitive substrates.

Several studies have addressed the changes in viscosity during film formation (de Lange, 1984; Jacobs et al., 1996; Yeates et al., 1996). Nakamichi and Mashita (1984) used a rolling ball viscometer to measure viscosities of powder coatings on a panel during heating. Results for three types of coatings are shown in Figure 28.2. In each case, viscosity

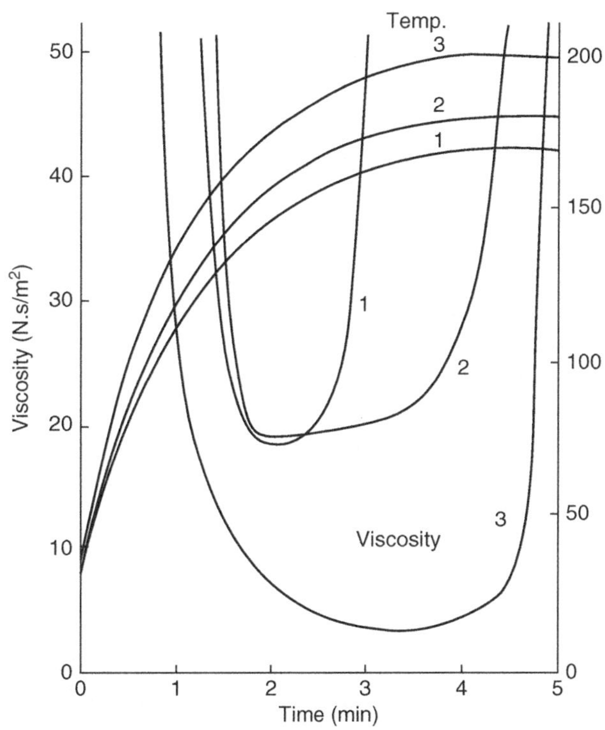

Figure 28.2 Non-isothermal viscosity behavior of powder coatings during film formation, as functions of time and panel temperature. (1) Acrylic–dibasic acid type; (2) polyester-blocked isocyanate type; (3) epoxy–dicy type. Source: Nakamichi and Mashita (1984). Reproduced with permission of Technology Marketing Corporation.

is high immediately after fusion of the powder but drops off sharply with increasing temperature. Viscosity levels off as cross-linking reactions begin to increase the MW and then increases rapidly as the coating approaches gelation. Flow is governed by the lowest viscosity attained and by the length of time that the coating stays within the necessary viscosity range called the *flow window*. In Figure 28.2, coating 2 will flow more than coating 1 even though the lowest viscosity is about the same. This follows, because the reaction of coating 2 is slower and the flow window is longer.

Some authors discuss the temperature dependence of viscosity in terms of Arrhenius-type relationships and discuss the activation energies for viscous flow. As indicated in Section 3.4.1, the dependency of viscosity on temperature does not actually follow Arrhenius relationships but rather is dependent on free volume availability. The most important factor controlling free volume availability is $(T-T_g)$.

An area that needs more study is the important question of the driving forces for coalescence and leveling. Coalescence of latex particles has been extensively studied (Section 2.3.3), but not coalescence of powder coatings. An early study by de Lange (1984) suggested that high surface tension promotes coalescence. This suggestion derives from a hypothesis that the main driving force for coalescence is the reduction of surface area driven by surface tension, but other forces may also be at work. Low melt viscosity promotes leveling, but the mechanism involved is not well understood. de Lange (1984) and Andrei, et al. (2000) proposed that the driving force for leveling is surface tension as reflected by the Orchard equation (Section 24.2). de Lange's data fit the Orchard equation reasonably well for relatively thin and/or fluid films, but thicker and/or somewhat more viscous films leveled better than predicted. de Lange also reported that surface tension differential flow can cause cratering of powder coatings and that addition of small amounts of additives such as poly(octyl acrylate) derivatives may overcome this problem. Suppliers of resins and additives periodically introduce new additives that are said to improve leveling and reduce surface defects such as craters. The compositions are proprietary.

Hajas and Juckel (1999) reviewed the many factors affecting the leveling of powder coatings. In the early stages of baking, powders that melt to Newtonian liquids will level better than those that become thixotropic. Blocked isocyanate coatings generally exhibit better leveling than other types, possibly because the blocking agent (caprolactam or oxime) plasticizes the film, lowering viscosity. Particle size distribution of the powder affects surface smoothness; the largest particles should be no larger than 2/3 of the proposed film thickness. Film thickness has a major effect; if the film thickness is less than about 2 mil, particle size affects smoothness; above about 5 mil sagging is likely to occur. Surface tension, particularly surface tension differentials across the surface, can have major effects on leveling and cratering. Additives, particularly low surface tension polyacrylates, are widely used to promote leveling and reduce cratering. Commonly, these polyacrylates are adsorbed on the surface of silica or masterbatched with the resin to ease incorporation into the powder coating. The effectiveness of the additives depends not only on their chemical structure but also on MW and concentration. With one additive, an MW of about 6 000 is more effective than an MW approaching 100 000. Concentrations must generally be kept below 1.5% since excess additive gives a tacky surface to the cured film. Higher concentrations of additive are effective if they are also hydroxy-functional so that they cross-link into the binder. The additives also reduce or eliminate cratering, probably by minimizing surface tension differentials.

Severe cratering can be caused by cross contamination of one type of powder with another. Contamination can occur during powder production or application, for example, if a spray booth is inadequately cleaned when switching from one type of powder to another.

Surface tension also affects wetting of the substrate; if it is too high, poor wetting occurs, which leads to defects like cratering. On the other hand, low surface tension adversely affects leveling. Furthermore, uneven surface tension on the surface of coatings also leads to film defects. The problem is further complicated by the fact that surface tensions decrease with increasing temperature and the extent of this decrease varies with structure. Surface tensions over a range of temperatures from 138 to 184°C have been studied by asymmetric drop shape analysis (ADSA) of epoxy resin and epoxy resin compounded with 5 acrylate additives and two polyester-modified methylalkylsiloxane additives. The acrylate additives reduced the high temperature surface tension of the epoxy resin with little difference between them. The siloxane additives decreased the viscosity more than the acrylates, especially siloxanes in which the alkyl groups were short-chain groups compared with the ones with long-chain groups (Wulf et al., 2000).

Dynamic mechanical analysis (DMA) (Section 4.6) is useful for characterizing cured films (Higginbottom et al., 1992; Ramis et al., 2003). A DMA study showed that the T_g values of cured films of a series of decorative powder coatings, including a hybrid coating, a TGIC-polyester coating, and a blocked isocyanate–polyester coating, were all in the range of 89–92°C (Higginbottom et al., 1992). The average MWs between cross-links ($\bar{M}_c$ (Section 4.2)) for the cured films were in the narrow range of 2500–3000. It is noteworthy that years of trial-and-error formulation in different laboratories with such disparate binders led to such similar T_g and $\bar{M}_c$ values. On the other hand, a protective epoxy powder coating with a modified DICY cross-linker gave cured films with T_g of 117°C and $\bar{M}_c$ of 2200. These studies show that powder coatings are formulated with higher T_g and lower cross-link densities than liquid coatings for the same end uses. Similar, but not identical, mechanical properties are attainable.

DMA (Section 4.6) can be a powerful tool for helping to identify starting points for developmental work on new binders for powder coatings for similar applications. Two different factors control the T_g of the resins: chemical composition and MW. Richert (1982) reported that higher MW, more flexible resins are advantageous in that they can have adequate package stability and also flow more easily during baking than a lower MW resin of similar T_g that has more rigid chains.

Gherlone et al. (1997) reviewed the use of differential scanning calorimetry (DSC), which is a valuable tool for T_g determination and measurement of cure response. A combination of modified dynamic load thermal mechanical analysis (DTMA), DSC, and TGA has also been recommended to determine melting point, onset flow, flow, and gel point (Belder et al., 2001). DSC studies showed that pigmentation with TiO_2 had only a weak reinforcing effect and had essentially no effect on T_g, suggesting that pigment–binder interaction is weaker in powder coatings than in liquid coatings (Higginbottom et al., 1992).

Ramis et al. (2003) compared DTMA, DSC, and TMA (Chapter 4) in extensive studies of curing of a single TGIC-cured polyester powder coating. Among many interesting results, the authors measured relationships between T_g and conversion and calculated reaction kinetics.

Very commonly on the order of 0.1–1% benzoin (melting point 133–134°C) is included in the formulations of most types of powder coatings. Benzoin is said to improve the appearance of films and to act as an anti-pinholing agent and a degassing aid. Other additives with similar melting points are ineffective. The mechanism of benzoin's action has been the subject of much conjecture over the years. One study showed benzoin plasticizes the melt and increases the flow window of polyester–glycoluril formulations, indicative of improved leveling; high levels (1.4–2.4%) were used (Jacobs et al., 1996). Maxwell et al. (2001) showed that benzoin evaporates from the applied film during the first 6–8 min of cure and that entrapped bubbles escape in the first 5–6 min of cure or they do not escape at all. Benzoin is not effective at cure temperature below its melting point; proprietary alternatives are available.

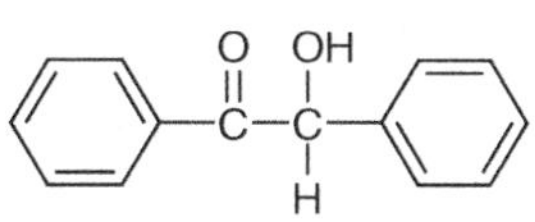

Benzoin

Considerable research has been reported on formulations that can enhance the mechanical properties and/or corrosion resistance of powder coatings. For example, Puig et al. (2015) described polyester/hydroxyalkylamide formulations containing, among other ingredients, TiO_2 pigment, a complex zinc molybdenum phosphate inhibitive pigment and an organosilane-modified silica pigment. In the optimum proportions, the latter pigments enhance both stress-to-break and corrosion protection of the powder-coated films.

Metallic pigments do not align as well in powder coatings as in liquids, so the dramatic polychromatic effects of automotive coatings are, as yet, unattainable. Metallic pigments do impart sparkle to the coatings, and a variety of such coatings is available.

28.3.1 Low Gloss Powder Coatings

Since there is no volatile solvent, the volume of pigment in powders approaches the PVC of the final film. At PVC near CPVC, viscosity of the fused powder would be far too high for acceptable leveling. It has been shown that as pigmentation increases above a PVC of about 20, the problems of leveling increase due to the increase in melt viscosity (Van der Linde and Scholtens, 1992). Thus, a common method for making low gloss liquid coatings by adding a high level of pigment is unworkable in powder coatings.

Low gloss and semigloss powder coatings have been prepared using approaches other than high PVC (Richart, 1995). Some reduction in gloss can be achieved by incorporating polyethylene micronized wax. In hybrid polyester powders, addition of organo-metallic catalysts along with wax may reduce gloss somewhat further. Low gloss hybrid polyester coatings are made by using a large excess over the stoichiometric ratio of epoxy resin with a high acid number polyester and curing at high temperatures. By selection of catalysts, such as cyclohexylsulfamic acid (cyclamic acid) and stannous methane sulfonate, it has been found that smooth matte finishes can be obtained with tetramethoxymethyl glycoluril (TMMGU) cross-linked polyester coatings (Jacobs et al., 1996).

Another way to reduce gloss is to blend two different primary resins or two different cross-linkers with substantially different reactivities or with poor compatibility. For example, combination of a BPA epoxy resin, a carboxy-functional polyester resin, and a carboxy-functional acrylic were used with a BPA–amine adduct or a blend of imidazole and DICY as curing agent. The epoxy resin and polyester resin were compatible, the polyester and acrylic resins were reported to be relatively compatible, whereas the epoxy resin and acrylic resin were not compatible (Lee et al., 2003).

Addition of styrene/maleic anhydride copolymer to the formulation of hybrid powder coatings together with tetrabutylphosphonium bromide catalyst gives a low gloss coating. The reduction in gloss is attributed to the dual cure system of anhydride/epoxy and COOH/epoxy reactions (Schmidhauser and Havard, 1998).

Various other combinations of incompatible resins have been reported to yield low gloss. Generally speaking, the gloss reduction achieved will vary with film thickness and baking conditions. Thus, care during application is needed to achieve consistent gloss levels.

Wrinkle powder coatings are also feasible. One approach is to formulate with TMMGU (Section 11.4.3) cross-linkers and amine-blocked catalysts, such as 2-dimethylamino-2-methylpropanol blocked *p*-toluenesulfonic acid (Jacobs et al., 1996).

28.4 MANUFACTURE OF POWDER COATINGS

Manufacture of powder coatings poses production and quality control issues quite different from those of liquid coating manufacture.

28.4.1 Production

Most powder coatings are manufactured by a process that involves premixing, melt extrusion, and pulverization (Figure 28.3). All major ingredients must be solids at ambient temperatures; liquid additives are used, but they are first melted into one of the solid components to make a *master batch* that is then granulated. The granulated ingredients, resins, cross-linkers, pigments, and additives are premixed in a batch process. A variety of premixers are used; it is essential that they provide a uniform, intimate mixture of the ingredients. The premix is fed through a hopper to an extruder in a continuous process. The barrel of the extruder is maintained at a temperature moderately above the T_g of the binder. In passing through the extruder, the primary resin and other low melting or low T_g materials are fused, and the other components are dispersed in the melt. The extruder operates at a high rate of shear so that it can effectively separate pigment aggregates. The melt can be extruded through a die—either a slot to produce a flat sheet or a series of round orifices to produce *spaghetti*. Sometimes, to reduce heat exposure, the melt is extruded through a die with larger bore and the *sausage* is fed between chilled rollers to flatten it into a sheet and cool it. Further cooling can be accomplished with a chilled conveyer belt.

Extruders have developed into sophisticated and often expensive pieces of equipment. Two types are commonly used: single screw and twin screw; in both types, a powerful motor turns screws to drive the material through a barrel. The screws and barrel are configured to mix the material thoroughly and apply a high rate of shear. A popular

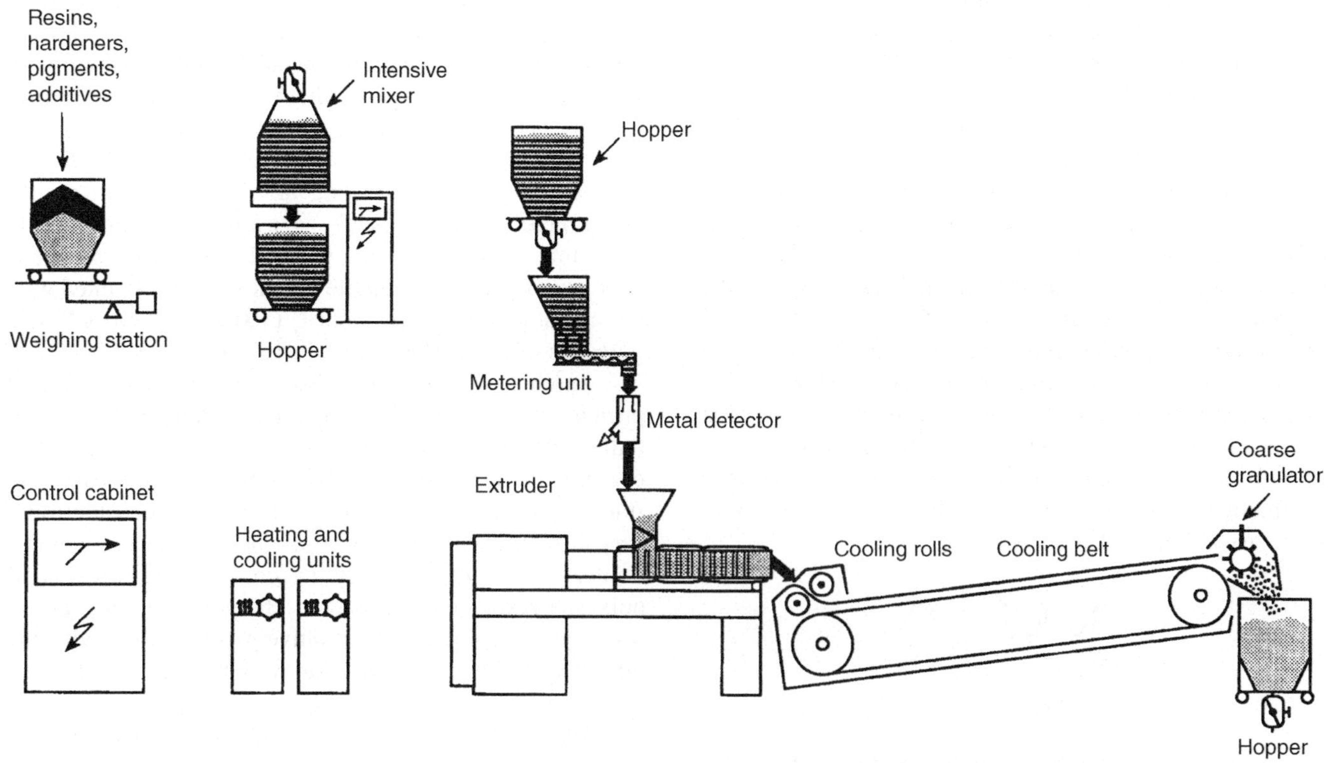

Figure 28.3 Schematic diagram of a line for premixing, melt extrusion, and granulation. Source: Misev (1991). Reproduced with permission of Buss America.

single screw extruder uses a reciprocating action in addition to radial turning of the screw to effect mixing and dispersion. Twin screw extruders use a combination of screw segments and kneading segments. Both types of extruders are capable of excellent dispersion of most pigments. They operate with relatively high viscosity formulations at high shear rates and, hence, efficiently separate pigment aggregates (Section 21.4.6). Improvements in extruder design have enabled production rates up to 2500 kg h^{-1}.

Production capacity can be increased by pushing material through the extruder faster, but there is a trade-off between separating pigment aggregates and production rate. Residence time in the extruder is sometimes reduced to 10 s or less, but at some point, pigment dispersion, especially with some organic pigments, becomes insufficient. Poor color development and color variability may result (Section 28.4.2).

Polymeric dispersants used in pigmented powder coatings can substantially improve pigment dispersion. For example, polymeric dispersants increase the jetness of carbon black allowing pigment loading to be increased with lower melt viscosity. Similarly, TiO_2-pigmented powder coatings can be made with higher hiding at lower film thickness (Maxwell et al., 2004).

The solid extrudate is then pulverized; a schematic of a typical process is shown in Figure 28.4. A variety of pulverizers are used. Some, such as *pin disk mills* and *hammer mills*, work on the principle of striking airborne granules with metal dowels or hammers mounted on a rapidly spinning disk. *Opposed jet mills* work by causing high velocity collisions of granules with one another. Opposed jet mills perform well for small (<12 μm) diameter powders needed for thin film application. Thermosetting extrudates are brittle and relatively easy to pulverize, but thermoplastics are generally quite tough and difficult to pulverize. For thermoplastics it often is necessary either to cool the mill with liquid nitrogen or to grind dry ice along with the granules so that the temperature is kept well below the T_g of the binder to offset the heating effect of the milling process. Even so, thermoplastic powders are generally available only in large particle sizes.

Some mills partially classify (i.e., fractionate) the powder, automatically returning oversize particles for further pulverization. Further size classification, by sieving and/or by an air classifier, is usually needed. The coarse particle fraction is sent back into the mill for further reduction in size. The fines are collected in a bag filter and recycled in the next batch of the same coating to be processed through the extruder. Finally, the classified powder is bulk blended for uniformity, packaged, and shipped to the customer.

Various alternative processes have been investigated, for example:

- *High shear mixers*. A variety of mixers, such as Brabender mixers, can be used as an alternative to melt extruders. These can be useful in the laboratory and perhaps in small-scale production.

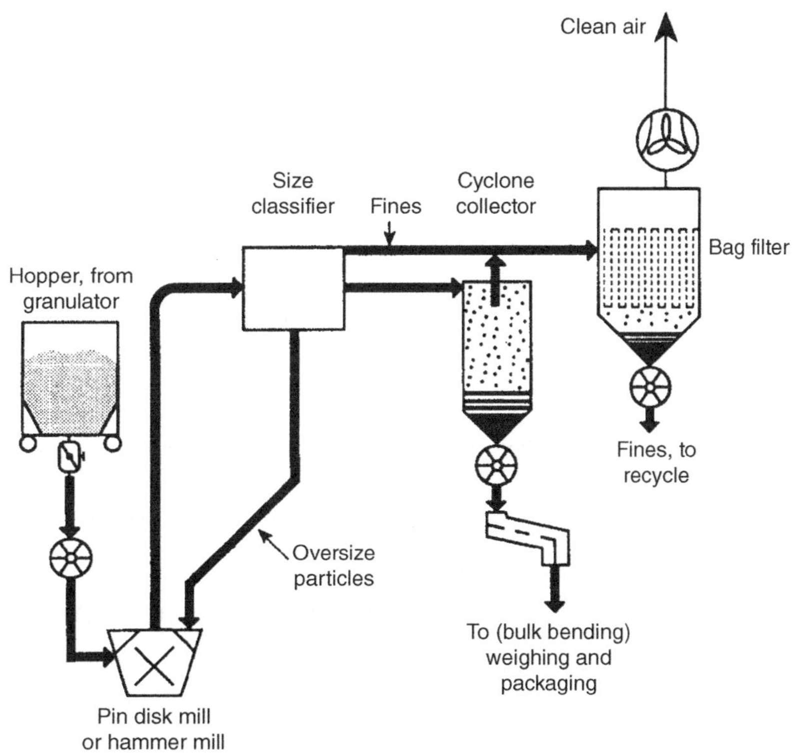

Figure 28.4 Schematic diagram of a line for pulverization and classification. Source: Misev (1991). Reproduced with permission of Buss America.

- *Spray drying*. A fully formulated paint is prepared as a solution or dispersion and then spray dried to yield the powder coating.
- *Use of supercritical CO_2 as a solvent during extrusion* (Koop, 1995). Rapid evaporation of CO_2 cools and partly pulverizes the product as it comes out of the extruder.
- *Preparation of powders as an aqueous suspension followed by isolating the powder* (Satoh et al., 1997).

None of these alternatives seems to have had a major impact on manufacture of organic powder coatings.

28.4.2 Quality Control

Close quality control of all components is required. In solvent-based coatings, the effect of small differences in MW or MW distribution on viscosity can be readily adjusted by small variations in the solids of the coating. In powder coatings, there is no solvent in the formula with which to make such adjustments. The only way of maintaining consistent quality of powder coatings is to assure that the raw materials have no significant variation in MW and MW distribution, as well as monomer composition.

In processing thermosetting compositions, care must be exercised so that no more than a minimal amount of cross-linking occurs at the elevated temperatures in the extruder. The rate of travel through the extruder should be as rapid as possible, consistent with achieving the necessary mixing and dispersion of pigment aggregates. Reprocessing should be minimized. In an extreme case, reprocessed material could gel in the extruder. More commonly, reprocessing can lead to increases in MW, due to some cross-linking reactions, giving powders that show incomplete coalescence or poor leveling after application. Only a limited amount of fine particles from the micropulverizer should be put in any one batch of coating. If reprocessing of an entire batch is necessary, it is best to use a limited amount of the batch to be reprocessed in each of several new batches rather than reprocessing the batch alone.

Color matching is often the most troublesome aspect of powder coating production. It is more difficult than with liquid coatings, particularly with reprocessed powders. Generally, one cannot blend batches of powder to achieve satisfactory color matching; it must be done using the appropriate ratio of pigments in the extruder mix. In laboratory processing, it is feasible to make an extruder mix with an estimate of the ratio of colorants needed to match a color. The color of the coating is then checked against a standard. Color can be adjusted to match the standard by mixing into the initial batch the estimated additional quantities of pigments and running the coating through the extruder and pulverizer again. In the laboratory, a third hit might be possible. The number of hits required is kept to a minimum by using computer color matching programs (Section 19.9).

In production, however, it is almost essential to have the mix right the first time. Potential problems can be minimized by extruding a small fraction of the production batch and then checking its color. If the color match is satisfactory, processing is continued. If the color needs adjusting, the initial fraction is granulated and returned to the feed hopper along with the additional amounts of ingredients required to adjust the color.

Batch-to-batch variation of color is inevitable in pigment manufacture. For closely, color-matched powder coatings, the pigment manufacturer is requested to supply selected batches with narrow tolerance limits. In the case of large production runs, several extruder batches can be blended before pulverizing in order to average out batch-to-batch differences. With sufficient care, color reproducibility can be satisfactory for all but the most demanding end uses.

Other important quality control issues are particle size and particle size distribution. These important variables become increasingly critical as particle size is reduced for thin film (<50 μm) applications. Particle size distributions can be measured by passing the powder through a stack of graduated sieves and weighing the fraction retained on each sieve or by a variety of instrumental methods, such as laser diffraction particle sizing. The latter method involves scattering of laser light by the powder suspended in air or in a liquid medium. Instruments are commercially available. The important effects of particle size and its distribution on the handling and application characteristics of the powder are discussed in Section 28.5.1.

28.5 APPLICATION METHODS

Almost all thin-film powder coatings are applied by *electrostatic spray*. Other application methods, important for protective and thick-film powder coatings, include *fluidized bed*, *electrostatic fluidized bed*, and *flame spray*. See the general references listed at the end of the chapter for more extensive discussions. At minimum, objects to be powder coated must be thoroughly cleaned, and in many cases, chemical conversion coatings (Section 6.3.1) are required to assure adequate adhesion and corrosion resistance.

28.5.1 Electrostatic Spray Application

Electrostatic spray (Section 23.2.3) is the major process for applying powder coatings (Kreeger, 1997). The powder is fluidized in a hopper and conveyed by a stream of air to an electrostatic spray gun. One type of gun consists of a tube to carry the airborne powder to an orifice at which an electrode is located. The electrode is connected to a high voltage (40–100 kV), low amperage power supply. Electrons emitted by the electrode react with molecules in the air, generating a cloud of ions, called a *corona*, around the orifice. The corona probably consists predominantly of HO^- and, perhaps, O_2^- ions. Powder particles come out of the orifice, pass through the corona, and pick up anions. The object to be coated is electrically grounded. The difference in potential attracts the powder particles to the surface of the part. They are attracted most strongly to areas not already covered, forming a fairly uniform layer of powder even on irregularly shaped objects. During application, size segregation of the particles has been observed, with the finer particles more prevalent near the substrate, presumably due to a combination of the higher charge on small particles and their fitting between larger particles (Huang et al., 1997). The particles cling to the surface strongly enough for the object to be conveyed to a baking oven, where the powder particles fuse to a continuous film, flow, and crosslink. Heat can be furnished either by conventional convection ovens or by IR lamps; use of induction heating has also been proposed. One comparison of alternative heat sources concluded that IR is the least expensive and can be adjusted to provide more rapid cure (Dick et al., 1994).

The powder particles that do not adhere to the substrate (overspray) are recovered as a dry powder. This powder is usually recycled by blending with virgin powder. A schematic diagram of electrostatic spray application of powder coatings, including recovery of overspray, is given in Figure 28.5. Production spray guns are usually mounted on automated reciprocators; such equipment functions smoothly with minimal worker attention as long as a single color and type of powder is being applied.

A limitation of powder coatings is the difficulty of changing colors. When spraying liquid coatings, one can flush the gun with solvent and shift the feed lines to the guns from those feeding one color to those feeding another. In this way, successive objects on a conveyor line can easily be painted different colors. However, if changeover of powder coatings was attempted in this way, there would be sufficient dust in the air in the spray booth for color contamination both on the product and in the overspray collecting units. In production, the operation must be closed down, the booth cleaned, and the overspray collecting units changed to collect the next color. While spray booths and overspray units have been designed to minimize cleanup time, it is economically preferable to make a reasonably long run of one color and then shift over to another

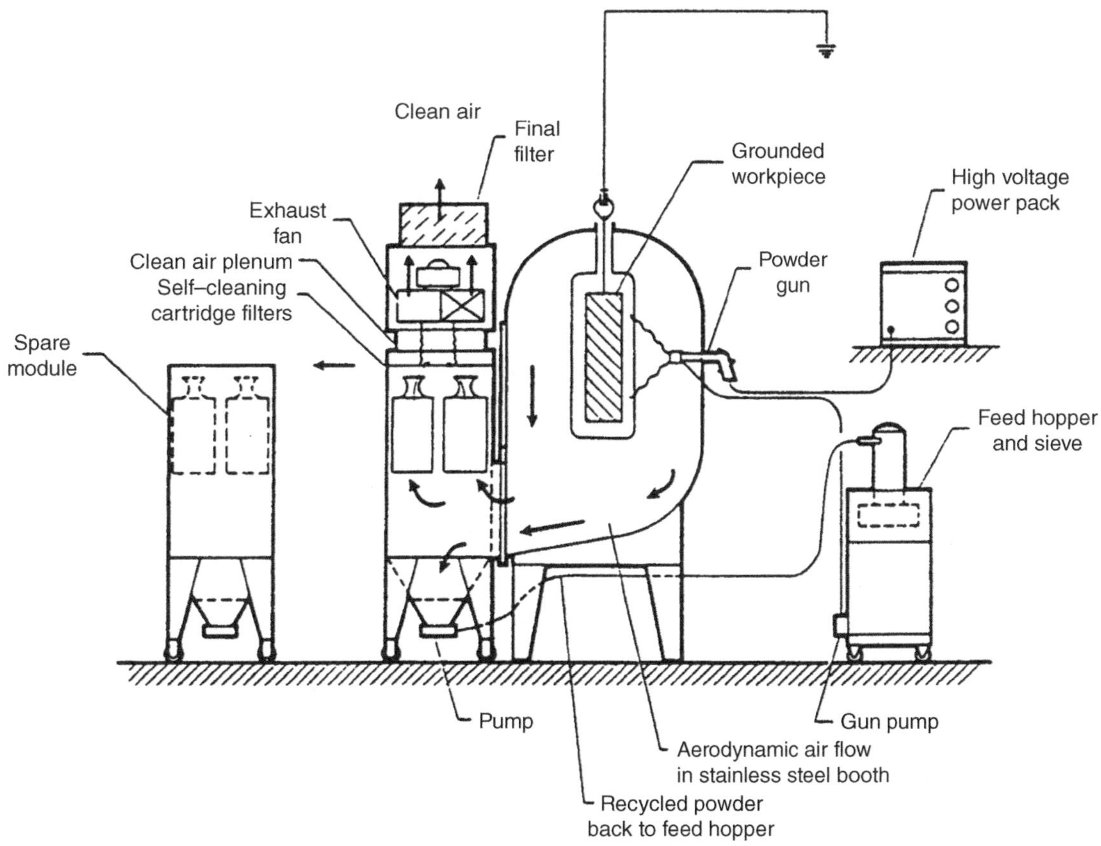

Figure 28.5 Production equipment for electrostatic spray application of powder coating showing the collection of overspray powder. Source: Richert (1982). Reproduced with permission of Nordson Corporation.

color. Many large-scale applications of powder coatings, such as fire extinguisher cases, are single color end uses or are applications, such as metal furniture, for which runs of single colors are long enough to dedicate a spray booth to each of the limited number of colors involved.

The large and rather expensive facility outlined in Figure 28.5 is used for large-scale operations. Relatively inexpensive electrostatic powder spray guns and heating equipment are also available. With such equipment, laboratory and small-batch application of powder coatings is economically feasible. The essential equipment can be purchased for a few thousand dollars, which has allowed many small and medium-size businesses to enter the powder coating application field.

Water-washed spray booths have been suggested to avoid the cost of installing special spray booths for shorter runs (Kia et al., 1997). The overspray collected in the water is dispersed by a low foaming nonionic surfactant and flocculated with a melamine resin (Section 23.2.7).

In electrostatic spray, the charged powder particles wrap around the grounded object to a degree and coat exposed surfaces not in direct line with the spray gun. Nevertheless, in production, it is often desirable to use spray guns on both sides of the object, making it possible to uniformly coat intricately shaped objects such as automobile wheels and tube-and-wire metal furniture.

The process is strongly affected by the Faraday cage effect (Section 23.2.3). As a result, it is difficult to get full coverage of areas such as interior corners of steel cabinets. The interiors of pipes are coated with a spray gun inside the pipe to avoid Faraday cage problems.

Film thickness increases with increasing voltage and decreasing distance between the spray gun and the product being coated. Larger particle size powders tend to give increased film thickness. One can apply thicker coatings by heating the object to be coated before applying the powder. However, film thickness is limited by the fact that once a certain film thickness has been reached, the powder coating acts as an insulator and does not attract further particles. The insulating properties of the powder coating mean that defective coated parts generally cannot be recoated and must be stripped to be recoated with powder. Matching liquid coatings are often provided for touch up. Coating over other coatings and over plastics frequently requires application of a conductive primer coating.

There is considerable room for improvement of the process and need for a better understanding of how it works. For example, Meyer (1992) was unable to find a satisfactory physical explanation for why the powder clings to the object as well as it does. One opportunity for improvement is to increase efficiency of charging by the corona of the electrostatic gun. Hughes (1984) estimated that only about 0.5% of the anions in the corona become attached to powder particles. The rest of the anions are attracted to the nearest grounded object, where they at best do no good and at worst may reduce deposition efficiency and increase the Faraday cage effect (Hughes, 1992). Since 1984–1992, improved electrostatic spray guns have improved charging efficiency (Takeuchi, 2008). One of many expedients is to use rotating disks or bells in the outlet of the guns in place of a simple nozzle. Some experts think that if charging efficiency could be raised to 10%, deposition efficiency might become high enough to reduce overspray sufficiently that it would no longer be necessary to collect it.

Particle size and distribution have a critical effect on powder coating. The range of particle sizes must be limited; as a rule of thumb, the predominant particle diameter should be somewhat less than the intended film thickness and the largest particles should be no more than twice the film thickness. In addition to particle size and distribution, variations in particle shape and in particle density can influence the spray process and powder recovery. Very fine particles do not flow properly in hoppers and feed lines. In general, only 6–8 wt% of the particles can be smaller than 10 μm diameter without causing problems. Small particles have a higher ratio of surface area to volume and, hence, acquire a higher charge to mass ratio as they pass through the corona of an electrostatic gun. After charging, the particles are affected by three forces: the electrostatic field, air movements, and gravity. Theoretical calculations predict that gravity should be the predominant effect on very large particles and that air flow should predominate for the very small ones (Bauch, 1992). These theoretical predictions were confirmed by particle size analysis of overspray; with a typical powder, the overspray was rich in particle sizes below 20 and above 60 μm, indicating that the intermediate sizes were transferred most effectively. Small particles probably penetrate best into Faraday cages. Large particles flow better through the application system and retain their charge longer and therefore cling to the object better between application and baking. The dusts of very small (<1 μm) particles may present a toxic hazard if they are inhaled. On balance, it appears that particle sizes concentrated in the 20–60 μm range are best for powder coatings that are to be applied at film thicknesses of 30–60 μm.

Many other variables affect the electrostatic spray process. These include air flow (notoriously fickle) in the spray area, humidity, shape and orientation of the parts to be coated, and spray gun settings. In practice, however, engineers and operators learn to make the process work by applying empirical experience and understanding of the theory, limited though this understanding may be.

Much effort was devoted to developing materials and processes aimed at using powder for automotive clear coats. The efforts succeeded to an extent—several manufacturers made autos with powder clear coats in regular production. However, liquid clear coat technology has become almost standard. The various liquid clear coat chemistries are discussed in Section 30.1.5. As VOC regulations get tighter and tighter, powder clear coat technology may become more attractive. A key problem is to attain the desired smoothness, which is addressed by using fine particle size powders having a narrower size distribution (Yanagida et al., 1996; Satoh et al., 1997). However, application of fine particle size (averaging 10 μm) powders presents problems that necessitate modification of application equipment (Yanagida et al., 1996).

Another means of applying electrostatic powder employs *triboelectric charging* of the powder particles. The particles are charged by the friction generated from streaming through a poly(tetrafluoroethylene) tube in the gun, instead of by a high voltage source generating a corona at the gun orifice. The mechanism is analogous to the buildup of a static charge on a comb when combing your hair. Since there is not the large differential in charge between the gun and the grounded work being sprayed, no significant magnetic field lines are established and Faraday cage buildup is minimal, facilitating coating hollows in irregularly shaped objects. Smoother coatings are obtained. On the other hand, throughput is slower and stray air currents can more easily deflect the particles between the gun and the object being coated. Guns that combine triboelectric and corona charging are available.

"Dry-on-dry" powder coating enables applying two coats of powder and then curing them with a single bake; performance improvements and significant energy savings are claimed. Generally, the first coat is heated to melt and fuse the first layer of coating but not to cure it. One such system consists of an epoxy powder primer and a polyester powder top coat (Comley et al., 2014). The chemistry of the two powders can be chosen so that intercross-linking between the two layers assures excellent intercoat adhesion.

28.5.2 Other Application Methods

Fluidized beds are the oldest method of applying powder coatings. The equipment consists of a dip tank, the bottom of which is a porous plate. Air is forced through the porous plate and acts to suspend powder in the dip tank in air. The flow behavior of air suspensions of powders resembles that of fluids, hence the term fluidized bed. The object to be coated is hung from a conveyor and heated in an oven to a temperature well above the T_g of the powder. The conveyor then carries the part into the fluidized bed tank. Powder particles fuse onto the surface of the object. As the thickness of fused particles builds up, the coating becomes a thermal insulating layer so that the temperature at the surface of the coating decreases, finally reaching the stage at which further particles do not stick to the surface. The last particles that attach to the coated surface are not completely fused, so the conveyor must then carry the object into another oven, where the fusion is completed. Film thickness depends on the temperature to which the part is preheated and on the T_g of the powder. Thin films cannot be applied in this fashion. Most commonly, the method is used for applying thermoplastic coating materials.

Electrostatic fluidized beds are similar, but electrodes are added to generate ions in the air before the object passes through the powder. The object to be coated is grounded. The process is illustrated schematically in Figure 28.6 (Jilek, 1991). Powder is attracted to the object by electrostatic force, as in electrostatic spray. The object can be heated when thick films are desired, but heating is not necessary. This method is used to apply thermoplastic and some thermosetting powders, for example, electrical insulation coatings. There is no overspray, and powder losses are minimal; color change over is also easier. Thinner films can be applied compared with conventional fluidized bed coating. However, very thin films are hard to apply, there is a strong Faraday cage effect, and it is difficult to coat large objects uniformly.

Powder coil coating is also practiced. In some cases, the powder is applied to the coil strip by electrostatic spray (Misev and van der Linde, 1998). In another process, an electrostatically charged strip is passed through a cloud of powder then to the oven for fusing (Graziano, 1997). These methods enable coil coating of perforated or pre-embossed metal and have the advantage of no VOC emissions. Capital cost may be lower; but line speeds are somewhat slower than with conventional coil coating.

Flame spray is another technology for applying thermoplastic powder coatings (Misev, 1991, pp 347–349).

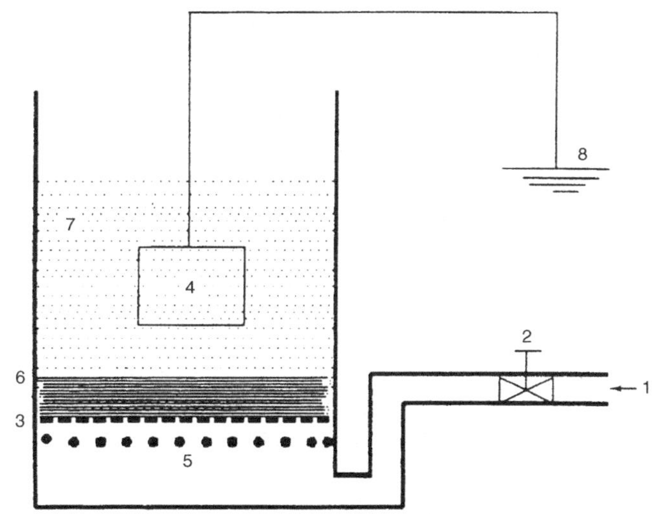

Figure 28.6 An electrostatic fluidized bed coating apparatus. (1) Air inlet. (2) Air regulator. (3) Porous membrane. (4) Object to be coated. (5) Electrodes. (6) Fluidized powder. (7) Cloud of charged powder. (8) Ground. Source: Jilek (1991). Reproduced with permission of The Federation of Societies for Coating Technology.

In a flame spray gun, the powder is propelled through a flame, remaining there just long enough to melt. The molten powder particles are then directed at the object to be coated. The flame heats and melts the polymer and heats the substrate above the melting temperature of the polymer so that the coating can flow into irregularities in the surface to provide an anchor for adhesion. The combination of flame temperature (on the order of 800°C), residence time in the flame (small fractions of a second), T_g of the coating, particle size distribution, and substrate temperature must be carefully balanced. Particle size distribution must be fairly narrow; very small particles pyrolyze at 800°C before larger particles can melt.

Poly(vinylidene fluoride), ethylene/chlorotrifluoroethylene, perfluoroalkoxyalkane, and fluorinated perfluoroethylenepropylene have been applied by flame spray to steel and showed excellent salt spray corrosion resistance. The chemical resistance of perfluoroalkoxyalkane polymer was particularly outstanding (Leivo et al., 2004).

In contrast to other methods of application, flame spray permits application in the field, not just in a factory. Examples of commercial or experimental applications include drum linings, metal light poles, bridge rails, concrete slabs, and grain railcars, among others. Since the application is not electrostatic, nonconductive substrates such as concrete, wood, and plastics can be coated. The coating need not be baked, so substrates that withstand the temperature from the impinging spray can be coated by this method. Since the coatings are thermoplastic and are not applied electrostatically, it is possible to use powder to repair damaged areas of coating, which is generally not possible with other powder coating methods.

The application of poly(vinylidene fluoride), nylon 11, and nylon 12 powders by high velocity thermal spray (Section 28.5.2) to aluminum and steel has been studied (Sugama et al., 1995).

Disadvantages of flame spray application include limitations on the service temperatures of objects coated with thermoplastic coatings (75°C in the case of EAA-based coatings) (Section 28.2), and the need for careful control of application variables. Overheating the polymer can lead to thermal degradation and, hence, to poor coating performance, with no visual indication of degradation until the polymer begins to pyrolyze. Adhesion to steel is also affected by application variables. Carboxylic acid groups on the binder promote adhesion, but coatings are subject to cathodic delamination (disbonding) because of the acid groups.

28.6 ADVANTAGES AND LIMITATIONS

In this section, advantages and limitations of powder coatings are compared with solventborne and waterborne coatings. On balance, the continuing growth of powder coatings indicates that advantages outweigh limitations in many situations. The primary advantages may be summarized as

1. VOC emissions are very low.
2. Flammability and toxicity hazards are substantially reduced, although dust explosion hazards remain.
3. Thick films of 50–500 μm can be applied in a single operation.
4. Irregular shapes can be coated.
5. Excellent film mechanical properties and weatherability are attainable.
6. Energy consumption can be reduced.

At first blush, the low energy requirements of powder coatings are surprising, since their baking temperatures are generally higher than those of most solventborne-baked coatings. However, with little or no volatile material being emitted into the oven, the hot air in the oven can be recirculated with little make-up air. In contrast, with solventborne coatings the solvent concentration in the air in the oven must always be kept well below the lower explosive limit, resulting in the need to heat a substantial volume of make-up air. Also, air flow through the spray booth can be lower with powder coatings, because it is unnecessary to keep the solvent concentration in the air below a safe concentration for enabling people to be in the spray booth. In winter, the cost of heating air flowing at a high rate through a spray booth can be high. Since there is no drying or flash-off time, parts can be hung closer to one another on the conveyor. There is no sagging or dripping. Powder-coated parts are often more resistant to physical damage than those coated with liquid paints; the result can be savings in damaged objects and packaging costs.

An important further factor reducing overall cost of powder coatings is that the overspray powder can be collected in filter bags or cartridges from the spray booth and reused. This not only increases paint utilization but also eliminates the cost and difficulty of disposing of the sludge obtained from the water wash spray booths or filters used with liquid paints.

Important limitations of powder coatings are as follows:

1. *Explosive hazards.* While the absence of solvents eliminates the flammability problem, suspensions of powder in air can explode. Consequently, manufacturing and application facilities must be designed to avoid powder explosions. With good engineering and good housekeeping, the processes can be operated safely. Triboelectric charge systems are less likely to initiate explosions than corona charge spray guns, which can spark if inadvertently brought near a grounded conductor.

2. *Inability to coat large* or *heat-sensitive substrates*. With electrostatic spray and fluidized bed methods, only baked applications can be considered, and since the baking temperature must be fairly high, only substrates that can withstand the baking cycle can be used. This limits applications of conventional thermosetting powder coatings almost entirely to metal substrates. UV cure powder coatings can be used on some heat-sensitive substrates and, as noted in Section 28.5.2, flame spray can be used for some other applications.
3. *Appearance limitations*. Broadly speaking, powder coatings can have good appearance, but some of the appearance effects attainable with liquid coatings are difficult or impossible to match with powders. Color matching is more difficult than with liquid coatings and color reproducibility can also be difficult. Since there is no solvent loss, there is not the degree of shrinking during film formation required for metallic coatings to show the color flop typical of automotive coatings. One can make powder coatings that contain aluminum flake pigments; they sparkle but display little color flop. As indicated in Section 28.3, a compromise between leveling and edge coverage is inevitable. The usual result is that the formulator must settle for a film with some orange peel to get acceptable edge coverage. The thinner the film, the more difficult the problem becomes. As also discussed in Section 28.3, there are problems making and controlling the gloss of low and medium gloss powder coatings, although progress in this area is reported. Achieving satisfactory appearance is a major obstacle to development of powder clear coats for autos and trucks. Uniform film thickness must be applied to the entire car body because leveling of powder coatings is quite sensitive to variations in film thickness. This requires superior control of powder flow rates, which is an engineering challenge. Films thinner than 1.0 mil are difficult because of pinholing. The entire operation, including reclamation of overspray, must be scrupulously clean to avoid dirt particles, which are especially noticeable in clear coatings. It is difficult to coat objects with outside sharp corners; and inside corners tend to get low film thickness due to the Faraday cage effect.
4. *Materials limitations*. Since all major components must be solids, a smaller range of raw materials is available to the formulator. Furthermore, it is not possible to make thermosetting powder coatings for which the T_g of the final film is low. This limits the range of mechanical properties that can be formulated into a powder coating.
5. *Limits of production flexibility*. Economics of production and application suffer whenever frequent color changes are needed. Cleanup between color changes is time consuming. Powder coatings are best suited to reasonably long production runs of the same type and color of powder. Water-washed spray booths permit shorter runs, but the advantage of recycling overspray is lost. New engineering designs have reduced the problem of color changes.

Some of the limitations of powder coatings can be overcome by making aqueous dispersions (slurries) of powders. This eliminates the potential for powder explosions, broadens the range of application methods, and reduces storage stability problems. The T_g of the powder no longer needs to be high to avoid sintering, and therefore, more flexible coatings can be formulated and lower baking temperatures can be utilized. Settling problems can be minimized and appearance improved by using smaller particle size, less than 10 μm, in the slurries. However, new challenges are added: aqueous dispersions must be stabilized without undue sacrifice of properties due to the presence of surfactant, and spray rheology must be controlled. Production feasibility of aqueous powder dispersions as automotive clear coats is well established.

GENERAL REFERENCES

Bate, D. A., *The Science of Powder Coatings*, 2 vols., Selective Industrial Training Associates, Ltd., London, **1990**.

Jilek, J. H., *Powder Coatings*, Federation of Societies for Coatings Technology, Blue Bell, PA, **1991**.

Misev, T. A., *Powder Coatings Chemistry and Technology*, John Wiley & Sons, Inc., New York, **1991**.

Spyrou, E., *Powder Coating Chemistry and Technology*, 3rd ed., Vincentz Network, Hannover, **2013**.

REFERENCES

Agawa, T.; Dumain, E. D., *Proceedings of the Waterborne High-Solids Powder Coatings Symposium*, New Orleans, LA, **1997**, p 342. Also see *J. Coat. Technol.*, 1999, *71*(*893*), 69.

Andrei, D. C., et al., *J. Phys. D: Appl. Phys.*, **2000**, *33*, 1975.

Badila, M., et al., *Prog. Org. Coat.*, **2014**, *77*, 1547–1553.

Barkac, K. A., et al., US patent 6,670,043 (**2003**).

Bauch, H., *Polym. Mater. Sci. Eng.*, **1992**, *67*, 344.

Beetsma, J., Prospector, **2016**, from http://knowledge.ulprospector.com/4799/pc-powder-coatings-for-heat-sensitive-substrates/ (accessed March 30, 2017).

Belder, E. G.; Rutten, H. J. J.; Perera, D. Y., *Prog. Org. Coat.*, **2001**, *42*, 142.

Blanton, M.; Thames, S. F.; Halliwell, M. J., *Proceedings of the Waterborne High-Solids Powder Coatings Symposium*, New Orleans, LA, **1999**, pp 399–412.

Boogaerts, L.; Buysens, K.; Moens, L.; Maetens, D., *Proceedings of the Waterborne High-Solids Coatings Symposium*, New Orleans, LA. **2000**, 504–515.

Brun, L.; Golini, R., Center on Globalization, Governance & Competitiveness, Duke University, **2010**, from http://www.cggc.duke.edu (accessed April 21, 2017).

Brun, L. C.; Golini, R.; Gereffi, G., *The Development and Diffusion of Powder Coatings in the United States and Europe*, Duke University Center on Globalization, **2010**, from cggc.duke.edu (accessed March 30, 2017).

Buijssen, P. F. A., et al., EP2411459 B1 (**2012**).

Cavalieri, R., et al., *CoatingsTech*, **2015**, *12*(3), 36–44.

Comley, D. G.; Ojero, O. A.; McMahon, R. E., WO2014150167 (**2014**).

de Lange, P. G., *J. Coat. Technol.*, **1984**, *56*(717), 23.

Dick, R. J., et al., *J. Coat. Technol.*, **1994**, *66*(831), 23.

Engbert, T., et al., *Farbe Lack*, **1996**, *51*, 102.

Fedtke, M.; Domaratius, F.; Walter, K.; Pfitzmann, A., *Polym. Bull.*, **1993**, *31*, 429.

Fischer, G. C.; McKinney, L. M., *J. Coat. Technol.*, **1988**, *60*(762), 39.

Gherlone, L.; Rossini, T.; Stula, V., *Proceedings of the International Conference on Organic Coatings*, **1997**, p. 128.

Graziano, F. D., *Proceedings of the International Conference on Organic Coatings*, Athens, **1997**, p. 139.

Green, C. D., *Paint Coat. Ind.*, **1995**, September, p. 45.

Gribble, P. R., *Proceedings of the International Waterborne High-Solids Powder Coatings Symposium*, New Orleans, LA, **1998**, p. 218.

Hajas, J.; Juckel, H., *Proceedings of the Waterborne High-Solids Powder Coatings Symposium*, New Orleans, LA. **1999**, pp. 273–283.

Higginbottom, H. P.; Bowers, G. R.; Grande, J. S.; Hill, L. W., *Prog. Org. Coat.*, **1992**, *20*, 301.

Huang, Z.; Scriven, L. E.; Davis, H. T., *Proceedings of the Waterborne High-Solids Powder Coatings Symposium*, New Orleans, LA. **1997**, p. 328.

Hughes, J. F., *Electrostatic Spraying of Powder Coatings*, Research Studies Press, Ltd., Letchworth, **1984**.

Hughes, J. F., *Ytbehandlingdagar*, Stockholm, May **1992**, p. A-1.

Jacobs, W.; Foster, D.; Sansur, S.; Lees, R. G., *Prog. Org. Coat.*, **1996**, *29*, 127.

Jilek, J. A., *Powder Coatings*, Federation of Societies for Coatings Technology, Blue Bell, PA, **1991**, p. 15.

Kapilow, L.; Samuel, R., *J. Coat. Technol.*, **1987**, *59*(750), 39.

Keddie, J. L.; Kia, S. F.; Rai, D. N.; Simmer, J. C.; Wilson, C., *J. Coat. Technol.*, **1997**, *69*(875), 23.

Koop, P. M., *Powder Coating*, **1995**, *6*(2), 58.

Kosar, W. P.; Morris, S., *JCT Res.*, **2007**, *4*(1), 51.

Kreeger, K., *Application Variables for Powder Coating Systems*, Nordson Corporation, Amherst, **1997**.

Krol, P.; Chemielarz, P., *Prog. Org. Coat.*, **2014**, *77*, 913–948.

Kronberger, K.; Hammerton, D. A.; Wood, K. A.; Stodeman, M., *J. Oil Colour Chem. Assoc.*, **1991**, *74*, 405.

Lee, S. S. et al., *Prog. Org. Coat.*, **2003**, *45*, 266.

Leivo, E. et al., *Prog. Org. Coat.*, **2004**, *49*, 69.

Lowenhielm, P. et al., *Prog. Org. Coat.*, **2005**, *54*, 269.

Maurin, V. et al., *Prog. Org. Coat.*, **2012**, *73*, 250–256.

Maxwell, B. E. et al., *Prog. Org. Coat.*, **2001**, *43*, 158.

Maxwell, L. D.; Siddiqui, K.; Cartridge, D. J., *FSCT Annual Meeting*, Chicago, IL, **2004**.

Merfeld, G., et al., *Prog. Org. Coat.*, **2005**, *52*(1), 98–109.

Meyer, III, E. F., *Polym. Mater. Sci. Eng.*, **1992**, *67*, 220.

Misev, T. A., *Powder Coatings Chemistry and Technology*, John Wiley & Sons, Inc., New York, **1991b**.

Misev, T. A.; van der Linde, R., *Prog. Org. Coat.*, **1998**, *34*, 160–168.

Moens, L. et al., US patent 6,790, 876 (**2004a**).

Moens, L.; Buysens, K.; Maetens, D., US patent 6,729,079 (**2004b**).

Nakamichi, T.; Mashita, M., *Powder Coat.*, **1984**, *6*(2), 2.

Pappas, S. P.; Urruti, E. H., *Proceedings of the Waterborne Higher-Solids Coatings Symposium*, New Orleans, LA, **1986**, p. 146.

Pianoforte, K., *Coatings World*, **2014**, *19*(12), 27–29.

Pilch-Pitera, B., *Prog. Org. Coat.*, **2014**, *77*(11), 1653–1662.

Popa, P. J.; Be, A.; Storey, D. R., *Proceedings of the Waterborne High-Solids Powder Coatings Symposium*, New Orleans, LA, **1999**, pp. 383–398.

Puig, M., et al., *Prog. Org. Coat.*, **2015**, *80*, 11–19.

Ramis, X.; Cadenato, A.; Morancho, J. M.; Salla, J. M., *Polymer*, **2003**, *44*, 2067–2079.

Rawlins, J. W.; Thames, S. F., *Proceedings of the Waterborne High-Solids Powder Coatings Symposium*, New Orleans, LA, **2000**, pp. 185–200. Also see Powder Coatings, 2000, August.

Rawlins, J. W.; Feng, S.; Sullivan, C.; Thometsek, P., *Proceedings of the Waterborne High-Solids Powder Coatings Symposium*, New Orleans, LA, **2002**, pp. 357–369.

Richart, D. S., *Proceedings of the Waterborne, High-Solids Powder Coatings of the Symposium*, New Orleans, LA, **1995**, p. 1.

Richert, D. S., Powder Coatings, in *Kirk-Othmer Encyclopedia of Chemical Science and Technology*, 3rd ed., John Wiley & Sons, Inc., New York, **1982**, Vol. *19*, p. 11.

Satoh, H.; Harada, Y.; Libke, S., Proceedings of the International Conference on Organic Coatings, Athens, **1997**, p. 381.

Schmidhauser, J.; Havard, J., *Proceedings of the International Waterborne, High-Solids, & Powder Coatings Symposium*, New Orleans, LA, **1998**, pp. 391–404.

Shah, N. B.; Nicholl, E. G., *Proceedings of the Waterborne High-Solids Powder Coatings Symposium*, New Orleans, LA, **2003**, pp. 217–230.

Sugama, T.; Kawase, R.; Berndt, C. C.; Herman, H., *Prog. Org. Coat.*, **1995**, *25*, 205.

Takeuchi, M., *J. Phys. Conf. Ser.*, **2008**, *142*, 012065.

Taylor, J. W.; Collins, M. J.; Bassett, D. R., *J. Coat. Technol.*, **1995**, *67*(846), 43.

Trottier, E. C.; Affrossman, S.; Pethrick, R. A., *JCT Res.*, **2012**, *9*(6), 725.

Twigt, F.; van der Linde, R., US patent 5703198 (**1997**).

Van der Linde, R.; Scholtens, B. J. R., *Proceedings of the 6th Annual International Conference Cross-Linked Polymers*, Noordwijk, the Netherlands, June **1992**, p. 131.

Witzeman, J. S., *Prog. Org. Coat.*, **1996**, *27*, 269.

Wulf, M.; Uhlmann, P.; Michel, S.; Grundke, K., *Prog. Org. Coat.*, **2000**, *38*, 59.

Wuzella, G., et al., *Prog. Org. Coat.*, **2014**, *77*, 1539–1546.

Yanagida, K.; Kumata, M.; Yamamoto, M., *J. Coat. Technol.*, **1996**, *68*(859), 47.

Yeates, S. G. et al., *J. Coat. Technol.*, **1996**, *68*(861), 107.

Chapter 29

Radiation Cure Coatings

Radiation-cured coatings cross-link by reactions initiated by radiation, rather than heat. Such coatings have the potential advantage of being indefinitely stable when stored in the absence of radiation; following application, cross-linking occurs rapidly at ambient temperature on exposure to radiation. Rapid cure at ambient temperature is particularly significant for heat-sensitive substrates, including paper, some plastics, and wood (Chapter 31). See Section 28.1.6 for discussion of UV cure powder coatings.

Two classes of radiation-cured coatings are (1) UV cure coatings in which the initial step is excitation of a photoinitiator (PI) (or photosensitizer) by absorption of photons of UV-visible electromagnetic radiation and (2) electron beam (EB) cure coatings in which the initial step is ionization and excitation of the coating resins by high energy electrons. In UV curing, cross-linking is initiated by reactive intermediates that are generated from the photoexcited PI. An important distinction of EB curing is that no PI is needed, as the high energy EB radiation generates radicals and ions from the resins themselves, which readily initiate cross-linking. Infrared (IR) and microwave radiation are also used to cure coatings, but these systems are not included here, since the radiation is converted to heat, which initiates thermal curing.

In addition to coatings, radiation-cured materials are used in many important commercial applications, including dental and medical applications. Furthermore, radiation sensitivity is an essential property in a variety of other applications, including production of microchips, optical fibers, printed circuit boards, and printing plates (Chatani et al., 2014).

Since the third edition of *Organic Coatings* (Wicks et al., 2007), notable advances have been made in several areas of radiation cure coatings, including

1. The use of LED radiation sources in the UV and blue visible regions
2. PI systems for visible light-activated free radical and cationic polymerization, as well as for hybrid free radical/cationic compositions
3. A comprehensive review of strategies to reduce oxygen inhibition, including current physical and chemical methods (Ligon et al., 2014)
4. Mechanistic studies on thiol–ene photochemistry, including thiol-(meth)acrylate and ternary thiol–ene-(meth)acrylate compositions (Hoyle and Bowman, 2010), efficient stabilization of these formulations (Esfandiari et al., 2013), and commercial availability of low odor multifunctional thiols having enhanced shelf life
5. Evaluation of vinyl carbonates and carbamates as replacement materials for (meth)acrylates and potential synthetic methods for commercial preparation of multifunctional vinyl carbonates and carbamates (Husar and Liska, 2012)
6. Radiation-curable coatings having enhanced mechanical properties and, presumably, enhanced adhesion by photoinduced plasticity and reduction of stress without changes in material properties, mediated by addition–fragmentation chain transfer reactions (AFTR) (Scott et al., 2005; Gorsche et al., 2014). Such cross-linked compositions, which have reversible bond structures incorporated in the polymer network, have been termed *covalent adaptable networks* (CANs) and are considered to constitute a new class of polymers that bridge thermoplastic and thermoset polymers (Kloxin et al., 2010; Kloxin and Bowman, 2013).

Organic Coatings: Science and Technology, Fourth Edition. Frank N. Jones, Mark E. Nichols, and Socrates Peter Pappas.

29.1 UV CURING

Two classes of polymerization reactions are primarily used in UV curing: free radical and cation-initiated chain-growth polymerization. While there have been attempts to use photoreactions in which the radiation leads to the generation of a reactive functional group, this approach is not useful for coatings. In such reactions, each photon absorbed can effect only one cross-linking reaction. In chain reactions, absorption of a single photon leads to the formation of many cross-links.

A key requirement for UV curing is a UV source that produces high intensity UV radiation at low cost without generating excessive IR radiation. Major radiation sources in commercial use are medium pressure mercury vapor lamps. Such *electrode lamps* are tubes up to 2 m long, power outputs of 80 W cm^{-1} are in wide use, and lamps with outputs up to 325 W cm^{-1} are available. The radiation has continuous wavelength distribution with major emissions at 254, 313, 366, and 405 nm, among others. Visible radiation and a minor, but not insignificant, amount of IR radiation that causes some heating are also emitted.

Radiation is emitted in all directions around tubular lamps, and its intensity drops off with the square of the distance from the source. To increase the efficiency of absorption, lamps are mounted in an elliptical reflector with a focal length such that the maximum intensity of radiation is focused at the distance between the lamp and the surface being coated. A limitation of UV curing is that the distance between the lamp and the coating on various parts of the object being coated must be fairly uniform. Hence, UV curing is most easily applicable to coating flat sheets or webs that can be moved under the UV lamps or cylindrical objects that can be rotated under or in front of the lamps. Since thermal energy is also produced, the lamp housing must be water and air cooled. The lamps are housed in enclosures; when the enclosure is opened, the current to the lamps is automatically turned off. UV radiation is hazardous and can lead to severe burns. *It is essential to avoid exposure of eyes to the radiation.* Depending on the radiation source, a greater or lesser amount of ozone is generated. Since ozone is toxic, the UV unit must be ventilated. These and other safety considerations in radiation curing have been reviewed (Golden, 1997).

In some cases, there are advantages of using other wavelength distributions, especially those with increased fractions of radiation in the very near-UV-visible region. Changes in distribution can be made by doping the lamp with traces of other elements besides mercury or by having fluorescent coatings on the lamp tube that absorb short UV and emit longer UV radiation.

Electrodeless lamps, powered by microwaves, also enjoy substantial commercial use. Electrodeless lamps are more suitable for doping, since the lifetime of lamp electrodes is generally reduced by dopants; have the further advantage of essentially instantaneous start-up and restart; and are more expensive.

The utilization of direct current power supplies for both electrode and electrodeless lamps has been shown to result in constant radiation, which is reported to enhance the efficiency of polymerization relative to pulsed radiation (Okamitsu et al., 2005).

Excimer lamps emit very narrow bands of UV. They use silent electrical discharge through two quartz tubes and an enclosed gas volume. Electronically activated molecules are produced in the gas phase and decompose within nanoseconds to produce photons of high selectivity. The narrowness of distribution is comparable to lasers, but, in contrast to laser sources, the radiation is incoherent and can therefore be used for large area applications. Different lamps are available or in commercial development that have principal emissions at 172, 222, 308, and 351 nm. The high intensity of very narrow wavelength distribution can provide rapid cures. Xenon chloride excimer lamps that emit at 308 nm have good overlap with the absorption spectrum of several commercial photoiniators for free radical as well as for cationic polymerization, which is required for efficient generation of initiator species, as discussed in Section 29.1.1.

Light-emitting diode (LED) lamps also emit relatively narrow (5–20 nm) bands of radiation. Commercially available sources include emission at 365, 385, 400, and 460 nm. General features of LEDs include instant on–off, low heat emission (the devices still requires cooling), no ozone emission, long lifetime, and energy efficiency. A review of cure-on-command technology provides an overview of UV and EB cure equipment, as well as process-related developments (Javadi et al., 2016).

29.1.1 Absorption: The Primary Process

For initiation via any lamp source, there must be absorption of radiation by the PI (or photosensitizer). The fraction of incident radiation absorbed (I_A/I_0) within a coating film may be expressed by Eq. 29.1 in accordance with the Beer-Lambert law. Neglecting surface reflection, $I_A/I_0 = 1 - I/I_0$, where I/I_0 is the fraction of transmitted radiation. Assuming that the only absorber at the wavelength(s) of interest is the PI, its molar absorptivity is ε, its concentration is *c*, and *x* is the optical path length of radiation in the film, absorbance (*A*) (or optical density) is defined as ε*cx*:

$$\frac{I_A}{I_0} = 1 - 10^{-\varepsilon cx} = 1 - 10^{-A} \tag{29.1}$$

Since molar absorptivity varies with wavelength, the fraction of radiation absorbed also varies with wavelength.

In addition, the intensity of radiation from the source varies as a function of wavelength. The total number of photons absorbed per unit time depends on the combination of these factors. When a PI absorbs a photon, it is raised to an excited state, which leads to the generation of an initiating species. But there are other possible fates of the excited state PI*. It may emit energy of a longer wavelength; that is, it may fluoresce or phosphoresce. It may be quenched by some component of the coating or by oxygen. It may undergo other reactions besides those that lead to initiator generation. The efficiency of generation of initiating species is an important factor in selection of a PI.

The rate of polymerization reactions is related to the concentration of initiating radicals or ions. It would, therefore, seem that the higher the PI concentration, the faster the curing. In fact, this supposition is not always so. As one increases from very low concentrations to somewhat higher concentrations, the rate of cure increases, but there is an optimum concentration. The rate of cure in the lower part of the film decreases above this concentration. At higher concentrations of PI, such a large fraction of the radiation may be absorbed in the upper few micrometers of the film that little radiation reaches the lower layers. Since the half-life of free radicals is short, owing to radical–radical reactions, they must be generated within a few nanometers of the depth in the film where they are to initiate polymerization. Although the half-life of the acid species in cationic cure coatings is substantially longer, migration may be limited by diffusional constraints as polymerization and cross-linking proceed.

Table 29.1 provides the percent of incident light absorbed ($I_A/I_0 \times 100$) as a function of PI absorbance (A) in a film (Pappas, 1992). These values, which are independent of film thickness, are provided for the total thickness, the top 1% and the bottom 1% of the coating. Two points are particularly worthy of comment. First, nonuniformity of absorbance occurs at relatively low absorbances and increases with increasing absorbance. Second, absorption in the bottom 1% increases and subsequently decreases with increasing A. The optimum value of PI absorbance for maximal absorption in the bottom increment of thickness of an absorbing medium is about 0.43, although the table shows that the optimal region is somewhat flat for A values between 0.3 and 0.6, a range in which the PI concentration doubles. Since adhesion to substrates is generally a critical factor in UV curing, as well as in many photoimaging applications, it may be anticipated that such systems will exhibit optimal spectral response at wavelengths with PI absorbance of about 0.43, corresponding to maximum absorption at the film–substrate interface. In accordance with this expectation, several imaging systems were found to give optimal spectral response at approximately this absorbance (Thomas and Webers, 1985).

Table 29.1 Relationship of PI absorbance (A) and percent of incident light absorbed ($I_A/I_0 \times 100$) throughout a film, the top 1% and bottom 1%

I_A/I_0 (%)			
A	Throughout	Top 1%	Bottom 1%
0.1	21	0.23	0.19
0.2	37	0.46	0.29
0.3	50	0.69	0.35
0.4	60	0.92	0.37
0.43	63	0.92	0.37
0.5	68	1.1	0.37
0.6	75	1.4	0.35
1.0	90	2.3	0.23
3.0	99.9	6.7	0.007

Optimum PI concentration depends on film thickness: the greater the film thickness, the lower the optimum concentration. Lower concentration is favorable from the standpoint of cost, since the PI is generally an expensive component. However, comparing the cure speeds of films of different thicknesses, each containing the optimum concentration of PI for its thickness, the time required to cure a film increases with increasing film thickness. This follows since less radiation is absorbed in any volume element with increasing film thickness. The problem is further exacerbated when surface cure is oxygen inhibited, as is the case with free radical polymerization. (Oxygen inhibition is discussed in Section 29.2.4.) In general, the lowest concentration of PI that provides the required extent of surface cure is a good starting point for optimizing through cure at the lowest possible cost. Sample calculations of PI concentration corresponding to selected A values at particular wavelengths as a function of film thickness are available (Pappas, 1992).

Another factor that affects absorption of UV by a PI is the presence of competitive absorbers or materials that scatter UV radiation. In designing vehicles for UV cure systems, it is desirable to minimize their absorption of UV in the range needed for the excitation of the PI. Pigments can cause problems by absorbing and/or scattering UV radiation; the effects of pigmentation are discussed in Section 29.5. Aside from competitive absorbers in the composition, the photoproduct(s) of the PI may absorb more strongly than the PI, corresponding to photodarkening, which will further attenuate the incident radiation and reduce cure rate, as well as through cure. On the other hand, if the photoproduct(s) of the PI absorb less strongly, corresponding to photobleaching, cure rate as well as through cure will be enhanced. These factors, as well as the effects of adding nanoparticles to the compositions, have been investigated in frontal photopolymerization studies (Cabral and Douglas, 2005).

Film thickness is not the only variable that affects path length. If the same coating is applied over a black substrate and over a highly reflective metal substrate, the rate of cure over the metal substrate is faster than over the black substrate. UV radiation that passes through the film to the black substrate is absorbed, but that reaching a smooth metal substrate is reflected and passes through the film a second time, which increases the opportunity for absorption.

29.2 FREE RADICAL-INITIATED UV CURE

In these coatings, free radicals are photogenerated and initiate polymerization by adding to vinyl double bonds, primarily acrylates. Two classes of PIs are used: those that undergo unimolecular bond cleavage and those that undergo bimolecular hydrogen abstraction from an H-donor molecule.

29.2.1 Unimolecular (Type I or PI_1) Photoinitiators

A range of unimolecular (Type I) PIs has been studied. The earliest ones used on a large scale commercially were benzoin ethers. Benzoin ethers undergo cleavage to form benzoyl and benzyl ether radicals:

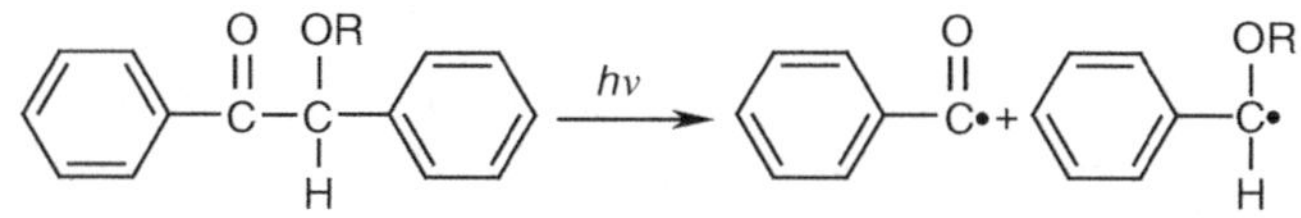

Both these radicals can initiate polymerization of acrylate monomers (Carlblom and Pappas, 1997). The package stability of UV cure coatings containing such benzoin ethers tends to be limited. Apparently, this is due to the ease of abstraction of the hydrogen on the benzyl ether carbon. Any organic material contains traces of hydroperoxides that slowly decompose. The resulting radicals can abstract the benzylic hydrogen, leading to the initiation of polymerization and, hence, poor package stability. Package stability is improved if the benzylic carbon is fully substituted. Accordingly, the ketal, 2,2-dimethoxy-2-phenylacetophenone (DMPA), is an effective PI_1 with good package stability. Photocleavage produces benzoyl and dimethoxybenzyl radicals. The latter is a sluggish initiator; however, it can undergo further cleavage to the highly reactive methyl radical, and the extent of this cleavage increases with increasing temperature.

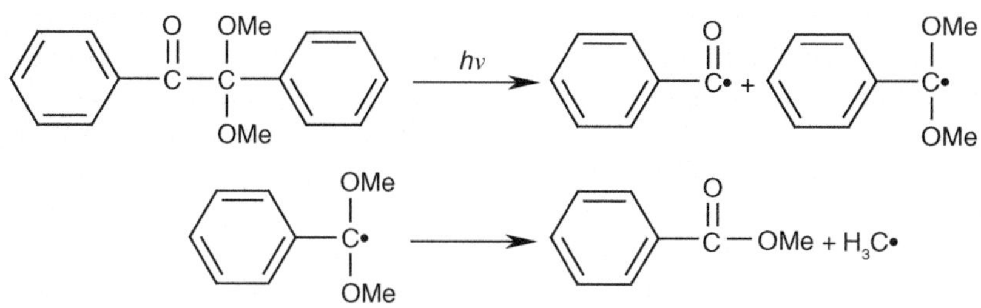

2,2-Dialkyl-2-hydroxyacetophenones (α-hydroxyacetophenones) are also commercially important PIs with good package stability, including 2-hydroxy-2-methylpropiophenone (HMP), as shown, and 1-benzoylcyclohexanone, where R and R_1 form a cyclohexane ring. These PIs tend to result in less yellowing than phenyl-substituted acetophenones, including benzoin ethers as well as DMPA, probably because benzylic radicals are not generated by photocleavage.

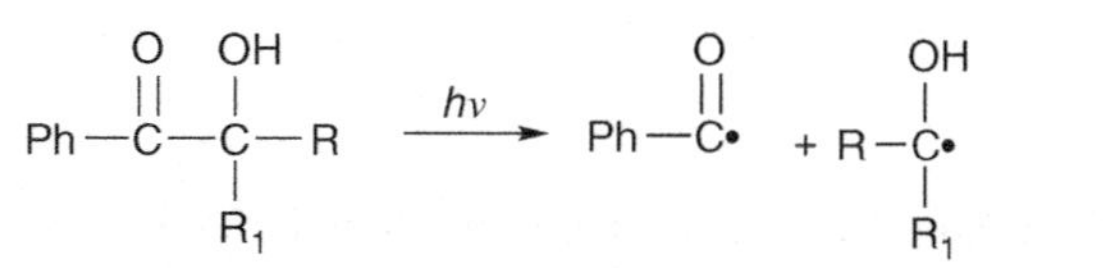

HMP ($R = R_1 = CH_3$)

All of the aforementioned PIs have an acetophenone chromophore, the absorptivity of which is enhanced by electron-donating substituents on the benzoyl ring. This is exemplified by the morpholino-substituted PI, 2-dimethylamino-2-benzyl-1-(4-morpholinophenyl)-butan-1-one, which is recommended for pigmented coatings. Yellowing can be a problem in clear coatings.

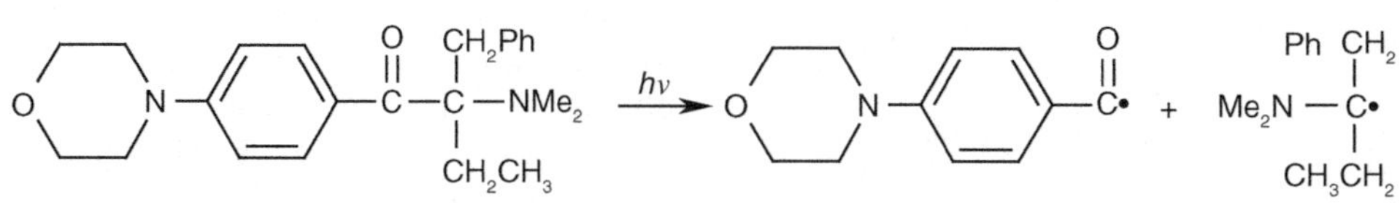

Acylphosphine oxides are another class of type I PIs. Several acylphosphine oxides are available, for example, (2,4,6-trimethylbenzoyl) diphenylphosphine oxide (MAPO). Irradiation generates the corresponding trimethylbenzoyl and diphenylphosphinyl radicals by α-cleavage. Acylphosphine oxides tend to be non-yellowing and provide good package stability.

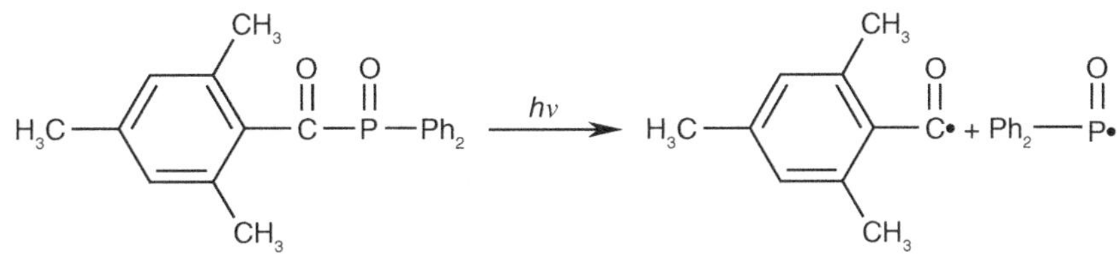

MAPO

Bis-acylphosphine oxides, such as phenyl bis-(2,4,6-trimethylbenzoyl)-phosphine oxide (BAPO), absorb more strongly in the visible spectral region and are suitable for activation by visible light sources. While BAPO is yellow, photocleavage results in reduced absorptivity (photobleaching), which reduces attenuation of the incident radiation, thereby enhancing cure rate and allowing through cure of thicker films, as discussed in Section 29.1.1. The strong absorption of UV-visible radiation by BAPO competes with absorption by white pigments that, together with photobleaching, allows UV curing of relatively thick, white pigmented coatings without the introduction of color. The absorption spectrum can be further broadened by using a blend of BAPO with HMP, providing combined surface and through cure at lower cost (Rutsch et al., 1996).

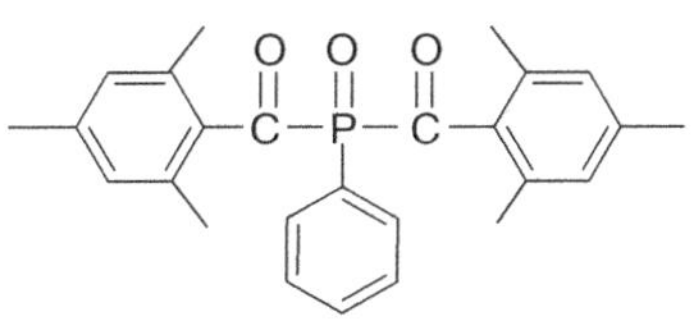

BAPO

Ph-CO-Ge(Et)$_2$-CO-Ph
Ge-1

Ph-CO-Te-Ph
Te-1

Acylgermanes, such as dibenzoyldiethylgermane (Ge-1) (Ivocerin®) (Neshchadin et al., 2013) and benzoyl phenyltelluride (Te-1) (Benedikt et al., 2014), are also reported to be highly reactive visible light PIs, which undergo primary α-cleavage to give benzoyl and metal-centered free radical species. The organotelluride also functions as a visible light PI for living free radical polymerization, producing polymers having low polydispersity. Photo-DSC experiments, using a methacrylated resin formulation, demonstrated similar reactivity and higher double bond conversion for Ge-**1** compared with BAPO under broadband irradiation conditions. However, using a dental-visible LED lamp, Ge-**1** exhibited much better performance than BAPO (Ganster et al., 2008), owing to greater overlap of Ge-1 absorption with the LED lamp emission.

29.2.2 Bimolecular (Type II or PI$_2$) Photoinitiators

Photoexcited benzophenone and related diarylketones, such as xanthone and thioxanthones, do not cleave to give free radicals but can abstract hydrogen from a hydrogen donor to yield free radicals that initiate polymerization. Thioxanthones, such as 2-isopropylthioxanthone (ITX), are used when their high absorption in the very near UV is desirable to compete with absorption by pigments or dyes that also absorb in the near-UV region.

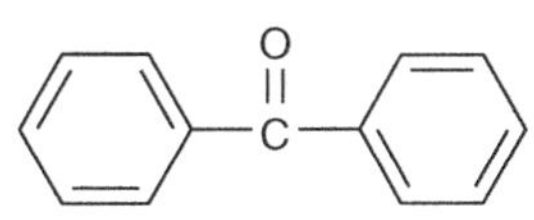

Benzophenone

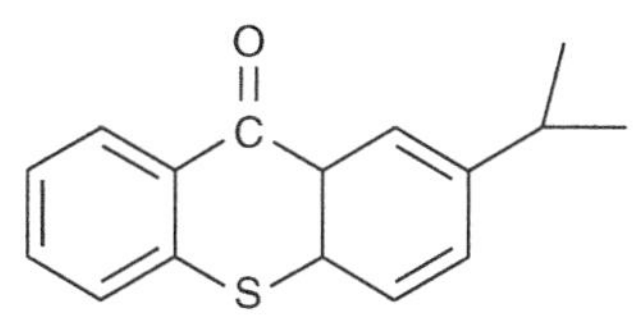

2-Isopropylthioxanthone (ITX)

Widely used hydrogen donors are tertiary amines with hydrogens on α-carbon atoms, such as 2-(dimethylamino) ethanol (DMAE) and, in printing inks, methyl *p*-(dimethylamino)benzoate. It has been shown that the accompanying ketyl free radical does not initiate polymerization.

$$Ph_2C{=}O \xrightarrow{h\nu} Ph_2C{=}O^* \xrightarrow{R_2NCH_3} Ph_2\dot{C}(OH) + R_2N{-}\dot{C}H_2$$

An advantage of bimolecular initiators with amine coinitiators is reduced oxygen inhibition, as discussed in Section 29.2.4. A disadvantage is that excited states of these initiators are generally longer lived than those of unimolecular PIs, which makes them more susceptible to energy and/or charge transfer quenching by atmospheric oxygen and, possibly, by other coating components.

29.2.3 Macromolecular Photoinitiators

Efficient macromolecular PIs for free radical polymerization have been developed (Dietliker et al., 2007), including disclosure of a narrow molecular weight (MW) thiol ether-functional analog (Woods et al., 2014). For further perspective on macromolecular PIs, including advantages and disadvantages relative to monomic PIs, see Crivello et al. (1998).

29.2.4 Oxygen Inhibition

Oxygen inhibits free radical polymerizations. In coatings, this inhibition is particularly significant, since coating films have a high ratio of surface area, where oxygen exposure is high, to total volume. Oxygen reacts with the terminal free radical on a propagating molecule (P·) to form a peroxy free radical (POO·). The peroxy free radical does not readily add to another monomer molecule; thus, the growth of the chain is slowed or terminated. The terminal radical from methyl methacrylate has been shown to have a rate constant for reaction with oxygen that is 10^6 times that for the reaction with another monomer molecule. Furthermore, the excited states of certain PIs are quenched by oxygen, thereby reducing the efficiency of generation of free radicals.

Ligon et al. (2014) provide a thorough analysis of oxygen inhibition, including analytical techniques, as well as a comprehensive treatment of physical and chemical strategies to reduce oxygen inhibition. Physical strategies include

1. Curing under an inert gas
2. Use of a physical barrier, such as wax, which comes to the surface
3. Increasing radiation intensity from the source
4. Matching wavelength of emitted radiation from the source with wavelength of absorbed radiation by the PI

Increasing light intensity (watts per square meter), which results in a higher concentration of initiating free radicals from the PI, is generally more effective in reducing oxygen inhibition than increasing radiation dose (joules per square meter). However, excessively high light intensity may be detrimental for UV cure owing to the depletion of free radical initiating species, which favors termination.

High thermal output from radiation sources tends to lower the viscosity of resins, which may contribute to oxygen inhibition by promoting oxygen diffusion. The selection of an LED lamp, whose narrow emission band has strong overlap with the absorption spectrum of the PI, seems to be preferable with regard to these considerations.

Chemical strategies are divided into additive approaches and modification of the polymerizing medium. Additive approaches include

1. The selection and concentration of PI(s)
2. Singlet oxygen generators and scavengers
3. H-donors, which act as chain transfer agents
4. Other reducing agents

Increasing PI concentration can reduce oxygen inhibition by the generation of a higher concentration of initiating free radicals at the surface. However, as discussed in Section 29.1.1., increasing PI concentration above the optimum concentration reduces through cure by attenuating the incident radiation, so that surface cure may occur at the expense of through cure. The problem of achieving both surface and through cure can be ameliorated by using a PI or a mixture of PIs having two absorption maxima with distinctly different molar absorptivities near different emission bands of the UV source. The emission band, which is highly absorbed by the PI, is absorbed more strongly near the surface; and less UV is available for absorption in the lower layers. This band is most important for counteracting oxygen inhibition but does not contribute substantially to through cure. Weaker absorption of the second emission band by the PI results in more uniform absorption throughout the film, which favors through cure.

Useful H-donors (D—H) function as chain transfer agents by transferring an H atom to peroxy radicals on the ends of oxygen-terminated chains (POO·), thereby generating a hydroperoxide (POOH) and a new initiating free radical (D·). Particularly effective H-donors are tertiary amines having an α-CH group and thiols. Silanes and, to a lesser extent, ethers having an α-CH group also function as chain transfer agents. The resulting α-amino, α-ether, thiyl, or silyl radical (D·) can react with a monomer (M) to initiate a new growing polymer chain (D—M·). Other reducing agents that function as chain transfer agents include boranes, phosphines, and phospites.

Free radicals on carbon atoms alpha to amines and ethers also react rapidly with oxygen. Thus, amines and ethers not only act as chain transfer agents but also serve to deplete oxygen levels. Accordingly, benzophenone-amine initiating systems substantially reduce oxygen inhibition. Oxygen inhibition can also be reduced using unimolecular PIs by adding small amounts of an amine. This process is useful when low intensity lamps or short exposure times are employed.

Reduction of oxygen inhibition by modification of the polymerizing medium includes utilizing

1. Multifunctional acrylates
2. Highly reactive acrylate monomers that incorporate a chain transfer agent(s) and/or secondary functionality

3. Alternative monomers, such as *N*-vinyl amides
4. Donor–acceptor monomers
5. Hybrid radical/cationic systems

Discussion of these topics is incorporated into the following sections.

29.2.5 Vehicles for Free Radical-Initiated UV Cure

The early UV cure coatings utilized styrene solutions of unsaturated polyesters (Section 17.3) together with benzoin ether PIs. However, styrene is sufficiently volatile that a significant amount evaporates between application and curing. Furthermore, the rate of polymerization is relatively slow as compared with acrylate systems. However, the cost is low, and unsaturated polyester vehicles continue to find some applications.

Most current coatings use acrylated reactants. Acrylate, rather than methacrylate, esters are used, since acrylates cure more rapidly at room temperature and are also less oxygen inhibited. Furthermore, polymerization of acrylates tends to terminate by combination, whereas methacrylate polymerization terminates largely by disproportionation (Section 2.2.1). The extent of cross-linking, as well as higher MWs, are favored, when termination of growing radicals occurs by combination.

In general, the vehicle consists of three types of acrylate esters: multifunctional acrylate-terminated oligomers, multifunctional acrylate monomers, and acrylate monomers. The monomers range from mono- to hexafunctional. Many formulas include mixtures of mono-, di-, and trifunctional acrylates. The acrylate monomers are also called *reactive diluents*. Multifunctional oligomers contribute to high rates of cross-linking, owing to their polyfunctionality, and, in large measure, control the properties of the final coating due to the effect of the backbone structure on such properties as abrasion resistance, flexibility, and adhesion. However, their viscosity is too high to be used alone; and monomers are required to reduce viscosity for application. Multifunctional acrylate monomers also undergo fast cross-linking owing to polyfunctionality and have lower viscosities than an oligomer. Viscosity is lowered further by most monofunctional acrylates, but the rate of cross-linking is reduced.

Quantitative kinetic studies of photopolymerization can be carried out by methods such as photo-differential scanning calorimetry and real-time infrared spectroscopy. Conventional IR, FTIR, and Raman spectroscopy are also valuable. Kinetic studies furnish useful insights; but a complete quantitative description of UV curing processes is difficult to construct, owing to the large number of variables involved, especially with practical formulations containing three types of acrylates.

It would be desirable to react all of the acrylate groups during photopolymerization to eliminate residual double bonds that might cause undesirable long-term changes of film properties. In practice, this is usually not possible, but it is often possible to polymerize a large fraction of the acrylate groups. The *extent of conversion* of the double bonds is affected by the choice of components; in general terms, extent of conversion increases with an increase in monofunctional monomer, presumably because the small monomer molecules readily diffuse through the film during the reaction. Extent of conversion is also affected by free volume availability; in general, conversion is increased by use of components that give low T_g films. As polymerization proceeds, T_g of the binder increases. If, as is commonly the case, T_g approaches the temperature at which curing is being carried out, the rate of reactions slows. Reactions become very slow at temperatures only a little above $T_g = T_{cure}$. Since there is heat from the radiation source and a reaction exotherm, T_{cure} is somewhat above ambient temperature. For a review of cross-linking rates and extents of conversion of various systems, see Jonsson et al. (1996).

Acrylated oligomers are prepared from a variety of starting oligomers. The properties of films derived from the acrylated oligomers are affected by the average number of acrylic double bonds per molecule, MW, and structure of the "core" oligomer. For example, acrylated urethane oligomers tend to provide coatings having a good combination of hardness and elasticity. Any polyol or hydroxy-terminated oligomer can be reacted with excess diisocyanate (OCN—R′—NCO) to yield an isocyanate-terminated oligomer. This oligomer can then be reacted with hydroxyethyl acrylate at ambient or moderately elevated temperature to yield an acrylated urethane oligomer.

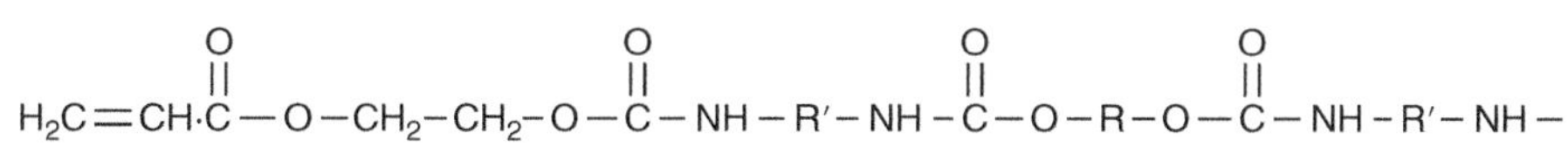

Acrylated urethane oligomer (terminal group only)

Another route is to react oxirane groups of epoxy resins with acrylic acid. The ring-opening reaction yields the acrylic ester and a hydroxyl group as shown. Like conventional epoxy resins, the acrylated epoxies tend to provide coatings with good toughness, chemical resistance, and adhesion.

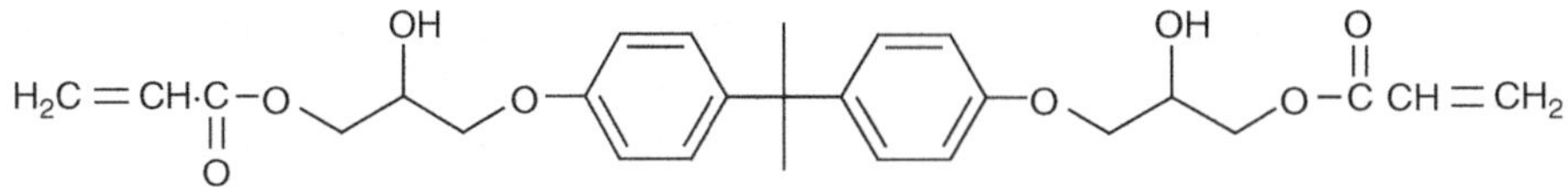

Acrylated epoxy oligomer

Various catalysts (e.g., triphenyl phosphine) are used in order that the acrylation reaction is carried out at as low a temperature as possible. Care is required to avoid polymerization of the acrylic acid or esters during the process. Inhibitors are added to trap free radicals. Some inhibitors, notably phenolic antioxidants, are effective only in the presence of oxygen, so the reaction is commonly carried out under an atmosphere of air mixed with inert gas. Variations in reaction conditions and catalyst composition can result in significant differences in the product. The most widely used epoxy resin is the standard liquid bisphenol A (BPA) epoxy resin ($n=0.13$) to yield predominantly the acrylated diglycidyl ether of BPA. Epoxidized soybean or linseed oils also react with acrylic acid to provide lower T_g oligomers having higher functionality.

Acrylated melamine–formaldehyde (MF) resins have been prepared by reacting etherified MF resins with acrylamide. These resins have the potential to cure in two ways: by UV cure of the acrylamide double bonds and by thermal cure of the residual alkoxymethylol groups on the MF resin. It is reported that UV cure followed by thermal cure provides films having increased hardness, improved stain resistance, and durability (Gummeson, 1990).

Many multifunctional acrylate monomers have been used; examples are trimethylolpropane triacrylate, pentaerythritol triacrylate, 1,6-hexanediol diacrylate, and tripropylene glycol diacrylate. Care must be used in handling them because many are skin irritants and some are sensitizers (Golden, 1997).

A range of monofunctional acrylates has been used. Those with lowest MW tend to reduce viscosity most effectively but may be too volatile. Ethylhexyl acrylate has sufficiently low volatility, as do ethoxyethoxyethyl acrylate, isobornyl acrylate, and 2-carboxyethyl acrylate, among others. Small amounts of acrylic acid as a comonomer promote adhesion. 2-Hydroxyethyl acrylate has low volatility and high reactivity and imparts low viscosities, but its toxic hazard is too great in many applications. *N*-Vinylpyrrolidone (NVP) is an example of a nonacrylate monomer that copolymerizes with acrylates at speeds comparable to acrylate polymerization; NVP is particularly useful because the amide structure promotes adhesion to metal and reduces oxygen inhibition. But, it also introduces a possible toxic hazard.

Decker and Moussa (1993) found that acrylate monomers, which also contain a carbamate, oxazolidine, or carbonate ester group, react much faster (as much as 100 times) and undergo more complete conversion than traditional monoacrylates. Other rate-accelerating groups include cyclic acetals and aromatic rings. The highly reactive acrylate monomers also react substantially faster than diacrylates containing ether groups. Addition of a reactive acrylated carbamate monomer to more slowly polymerizing, traditional acrylate monomers may dramatically increase overall reactivity. Factors affecting the inherent reactivity of such highly reactive monomers have been intensively investigated (Beckel et al., 2005; Kilambi et al., 2007; Cramer et al., 2008). Several theories have been advanced, including

1. Hydrogen abstraction/chain transfer that would essentially increase the monomers' functionalities
2. Hydrogen bonding
3. Electronic and resonance effects

A study of monomers with methyl groups blocking potential sites for hydrogen abstraction provided evidence in favor of theory (1) as a dominant factor, although alternative theories were not excluded, including an intrinsically high rate of propagation and low rate of termination, as well as intramolecular interactions between the acrylate and rate enhancement group that enhance reactivity.

The addition of mono- and difunctional β-allyl sulfones to a methacrylated UV-curable composition resulted in networks with tunable properties. By adjusting the content and functionality of the β-allyl sulfones, it was possible to achieve more homogeneous networks having a narrow glass transition and adjustable properties, including glass transition temperature (T_g), rubber modulus of elasticity (E_r), and network density. The β-allyl sulfones function as potent *addition–fragmentation chain transfer* (AFTR) monomers, which are proposed to alter the crosslinking process from a radical chain-growth mechanism to a mixed chain-growth/step-growth-like mechanism, thereby relieving stress and resulting in enhanced mechanical properties (Gorsche et al., 2014).

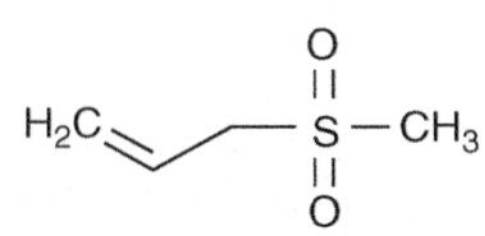

A β-allyl sulfone

The combination of unsaturated polyesters with vinyl ether resins has been reported to UV cure at rates comparable with acrylates (Friedlander and Diehl, 1996). Artysz et al. (2003) have reported that 1-propenyl ethers, prepared by isomerizing the corresponding allyl ethers, can be UV cured at rapid rates.

UV cure copolymer compositions are also based on thiol–enes, which typically contain a free radical PI, tri- or tetrafunctional thiol and a diene. H-abstraction from a thiol group (RS—H) by a PI_2^* or by a free radical from cleavage of a PI_1^* produces a thiyl radical (RS·), which adds to an ene group (CH2=CHR′) to generate a C-centered radical (RS—CH2—CHR′·) that undergoes chain transfer to a second thiol group to regenerate a thiyl radical (RS·). This process corresponds to step-growth propagation in contrast to chain-growth propagation of (meth)acrylates.

Step-growth propagation exhibits delayed gel formation, which reduces residual stress, resulting in enhanced mechanical properties and adhesion relative to chain-growth propagation of (meth)acrylates. Furthermore, thiol–ene polymerization is relatively uninhibited by oxygen owing to the strong chain transfer capability of thiol groups. Nevertheless, thiol–ene compositions have not been widely adopted for UV cure coatings, owing primarily to insufficient shelf life and the unpleasant odor of thiols.

However, notable advances have been made in the stabilization of thiol–ene compositions (Esfandiari et al., 2013); and low odor, multifunctional thiols have become commercially available, for example, pentaerythritol tetrakis (3-mercaptobutyrate) (KarenzMT® PE1) (Tetrathiol-1) and 1,3,5-tris(3-mercaptobutyroxyethyl)1,3,5-triazine-2,4,6-trione (KarenzMT® NRT) (available from Showa Denko K.K.). Thiol–ene formulations based on such secondary thiols were also found to exhibit exceptional storage stability (Li et al., 2009). These considerations together with advanced understanding of thiol–ene photochemistry, including thiol-(meth)acrylate and ternary thiol–ene-(meth) acrylate compositions (Hoyle and Bowman, 2010), are expected to promote further utilization of such compositions in UV cure applications.

$$C\left(CH_2-O-CO-CH_2-CH\left(CH_3\right)-SH\right)_4$$

Tetrathiol-1

Addition of the ring-opening monomer, 2-methyl-1-methylene-1,5-dithiacyclooctane (MDTO), to a UV-curable thiol–ene composition resulted in enhanced plasticity. This result was attributed to photo-mediated reversible cleavage of the cross-linked network backbone, involving the ring-opened MDTO groups, allowing rapid stress relief without mechanical property degradation by AFTR reactions (Scott et al., 2005).

Interest in the utilization of vinyl carbonates and vinyl carbamates as possible replacement materials for (meth) acrylates has been enhanced by potential synthetic methods for commercial preparation of multifunctional analogs (Husar and Liska, 2012).

$$CH_2=CH-O-CO-X-R$$

Vinyl carbonate $(X=O)$

Vinyl carbamate $(X=NR')$

Vinyl carbonates have been prepared from alcohols by reaction with acetylene and carbon dioxide, catalyzed by metal carbonyl complexes. Vinyl carbamates have been prepared from secondary amines by reaction with acetylene and carbon dioxide, utilizing various metal catalysts (Mahe et al., 1989).

A photoinitiated free radical polymerization study, utilizing photo-DSC and FTIR methods, was carried out on a diacrylate and dimethacrylate, each having a 1,4-butane diol bridge, in comparison with a divinyl carbonate having an ethylene glycol bridge and a divinyl carbamate having an *N,N*-dimethyl ethylenediamine bridge. Photoreactivity of the divinyl carbonate and divinyl carbamate, as determined by rate of polymerization and double bond conversion, was found to lie between the diacrylate and dimethacrylate. The resulting modulus and hardness were higher for the divinyl carbamate than the dimethacrylate. In contrast to (meth)acrylates, vinyl carbonates and vinyl carbamates have been shown to be biocompatible monomers (Heller et al., 2011).

29.2.6 Waterborne UV Cure Coatings

Waterborne UV cure coatings are reported to have several advantages. Since the resin is dispersed or emulsified in water, no reactive diluent is needed. Thus, there is no emission of volatile, possibly toxic reactive diluent. Furthermore, the viscosity of the coating is independent of MW so that higher MW oligomers may be utilized and the required low viscosity for spraying or roll coating is readily obtained by adjusting the solids with water. Higher MW oligomers require fewer double bonds, resulting in reduced film shrinkage and improved adhesion to substrates such as metal. Reaction of fewer double bonds is also expected to result in less internal stress in the cured coatings, which arises from restricted shrinkage.

The main drawback is that the water must be flashed off, usually by evaporation in an oven at 80°C, prior to exposure to the UV source. After flash off, immediate exposure leads to faster more complete cure owing to the higher temperature. Since the dispersions are stable at high dilution, overspray collected in a water wash spray booth can be recovered by ultrafiltration. It may not even be necessary to add further PI to the recovered coating. Furthermore, the UV cure dispersion can be blended with other waterborne coatings thus permitting enhanced exterior durability (Gerlitz and Awad, 2001).

Acrylated polyurethane dispersions (PUDs) have been used as vehicles for waterborne UV cure coating on wood paneling (Section 31.1.4). Extractables after curing are very low owing to the absence of unreacted low MW reactants. Adhesion is excellent owing to reduced shrinkage during cure resulting from the lower ratio of double bonds as compared with conventional UV cure coatings and probably to the inherent good adhesion of polyurethanes. Low gloss coatings can be made by the addition of silica pigment to the formula, since both the evaporation of water, which carries pigment to the film surface, and shrinkage resulting from water evaporation provide a low gloss surface. After application, but before cure, the water is removed by force drying, which indicates that nonvolatile PIs must be used. Water-soluble PIs are also available. The heating required to evaporate the water has the desirable side effect of increasing conversion, thereby reducing the residual double bonds (Reich et al., 1999).

Decker and coworkers studied ways to accelerate the rate and extent of UV cure of waterborne acrylic coatings, varying the type of PI, chemical structure of the acylate-functional oligomer, and the sample temperature. Coatings from emulsions cured faster and to a greater extent than from dispersions, which was attributed to greater molecular mobility. Very hard coatings were obtained from dispersion-type acrylate resins, especially when UV cure was performed at 80°C on the sample emerging from the drying oven, which enhanced mobility. Another important factor was the selection of the base needed to stabilize the dispersion. Volatile tertiary amines, which presumably evaporate during water flash off, were recommended (Decker and Lorinczova, 2004; Decker et al., 2004). Note that nonionically stabilized acrylated PUDs do not require an amine to stabilize the dispersion.

Waterborne UV cure coatings, based on acrylated polyesters, have been disclosed. An acrylated emulsifying polyester and acrylated water-insoluble polyester were mixed, neutralized with aqueous ammonia, and dispersed in water. Following the addition of a PI, the coating was applied to wood, flashed off, and UV cure. The resulting films had excellent properties and the grain of the wood was not raised (Urbano et al., 2001).

Hydroxy-terminated polydimethylsiloxane (PDMS) was incorporated into the soft segments of UV-curable polycarbonate-based polyurethane (meth)acrylate dispersions, which improved the thermal and surface properties of the films. The UV cure rate and final conversion as well as the glass transition temperature and tensile strength of the cured films depended strongly on the (meth)acrylate functionality and by the incorporation of PDMS (Hwang and Kim, 2011).

Waterborne UV-curable polyurethane acrylate/silica (PUA/SiO_2) nanocomposites were prepared by a sol–gel process. Films were analyzed by ATR–FTIR and gel content, which revealed that the modified silica enhanced UV cure of the PUA coatings. The nanocomposites were applied to thermochromic coatings, showing excellent temperature sensitivity and reversibility (Lv et al., 2015).

29.3 CATIONIC UV CURE

PIs for cationic polymerization are typically onium salts of very strong acids, exemplified by iodonium and sulfonium salts of hexafluoroantimonic and hexafluorophosphoric acids. Irradiation of diaryliodonium and triarylsulfonium salts yields strong protic acids of the corresponding counter anions, as well as radical cations; both initiate cationic polymerization. These reactions are shown in Scheme 29.1 for a triphenylsulfonium salt (anion omitted), following the primary unimolecular bond cleavage. Biscumyliodonium tetrakis(pentafluorophenyl)borate is reported to provide substantially faster cure rates than other onium salts (Priou et al., 1995).

Diaryliodonium and triarylsulfonium salts absorb radiation only weakly above 350 nm; however, their spectral response can be extended into the near-UV visible region, as well as into the mid-visible range, by use of photosensitizers. Photosensitization of onium salts can occur by several methods (Crivello, 2009; Yagci et al., 2010), including

1. Oxidation of free radicals
2. Electron transfer
3. Excitation of charge transfer complexes

For example, germanium-centered radicals, derived from visible light-active acylgermanes (see Section 29.2.1), are oxidized by diaryliodonium salts to generate germanium-centered cations, which initiate cationic polymerization (Dumaz et al., 2009).

Anthracene-bound sulfonium salts exhibit substantially higher reactivity than bimolecular anthracene-sensitized sulfonium salts in epoxy-based UV cure coatings and photoresists. These results have been attributed to greater efficiency of intramolecular sensitization as mobility diminishes in such compositions (Pappas et al., 2003).

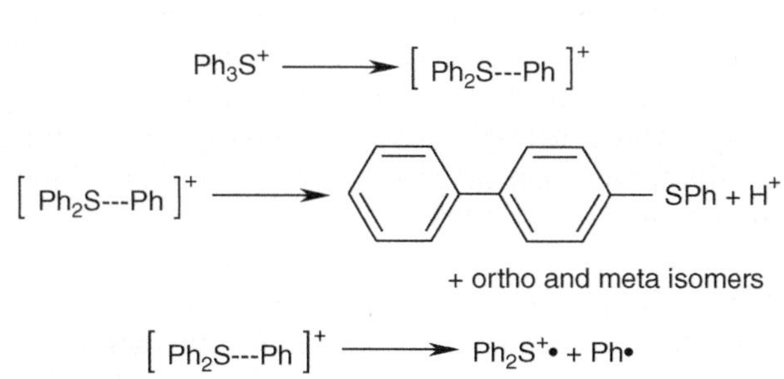

Scheme 29.1 Photolysis of a triphenylsulfonium salt.

Excimer lamps, which emit at 308 nm, are reported to be particularly effective using triarylsulfonium salts as PIs owing to spectral overlap with absorption by the PIs. Diphenyliodonium salts have limited solubility and are toxic. Substituted derivatives such as bisdodecylphenyliodonium hexafluoroantimonate have both higher solubility and lower toxicity (Priou et al., 1995).

29.3.1 Vehicles for Cationic UV Cure

Homopolymerization of oxirane (epoxy) groups (equations in Section 13.3.5) is the major type of cationic polymerization used commercially. The counterions must be very weak nucleophiles—in other words anions of very strong acids. Hexafluoroantimonates, hexafluorophosphates, and the recently introduced borate derivatives are particularly effective. In contrast to free radicals, cations do not react with each other. Consequently, in the absence of nucleophilic anions, cation-initiated cross-linking can continue after exposure to the radiation source until the reactive cations become immobilized. Reactions with water and alcohols terminate polymer growth. Nevertheless, cross-linking occurs even if only one pair of epoxy groups reacts before reaction with water; furthermore, the termination reaction is accompanied by regeneration of a proton, corresponding to chain transfer. The absence of oxygen inhibition further distinguishes cationic from free radical polymerization.

BPA epoxy resins (Section 13.1.1) react slowly at ambient temperatures; optimum UV cure temperatures are about 70–80°C. Rapid curing of coatings based on BPA epoxy resins can be effected by using a combination of UV and IR (heat) sources. Cycloaliphatic epoxides, such as 3,4-epoxycyclohexylmethyl-3′,4′-epoxycyclohexane carboxylate, exhibit higher reactivity. The added ring strain, owing to the fused ring system, promotes ring opening of the oxonium ion during the propagation step (Section 13.3.5). Low viscosity makes such epoxides useful as reactive diluents. Epoxy-functional silicone polymers have been recommended for use in UV cure release coatings (Priou et al., 1995).

Vinyl ethers and styrenes, particularly 4-alkoxystyrenes, rapidly polymerize cationically. Cationic photopolymerization of vinyl ethers is more rapid than polymerization of epoxy-functional reactants. The divinyl ether derived from the reaction product of chloroethyl vinyl ether with BPA (shown in the following text) is reported to polymerize at a very high rate (Crivello, 1991). Less PI is required; moreover multifunctional vinyl ether monomers tend to show a very low order of toxicity. Vinyl ether monomers are interesting in their own right and are also useful as a highly reactive component in epoxy resin coatings. 1-Propenyl ethers are also reported to cure rapidly, cationically, using triarylsulfonium hexafluoroantimonate salts (Artysz et al., 2003).

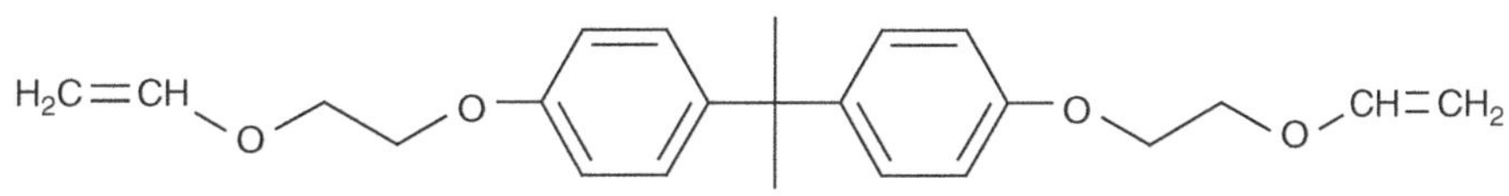

Cationic UV cure coatings have been reported using vinyl ether oligomers, prepared from toluene diisocyanate and polyethylene glycol prepolymers terminated with 4-hydroxybutylvinyl ether, together with ethylene glycol divinyl ether as reactive diluent, a diaryliodonium tetrakis(pentafluorophenyl) borate salt as initiator, and isopropylthioxanthone as photosensitizer (Kayaman-Apohan et al., 2004).

A systematic program to study and develop novel UV-curable inorganic–organic hybrid coatings, based on cationic polymerization, has been reviewed (Javadi et al., 2016).

29.4 HYBRID FREE RADICAL/ CATIONIC POLYMERIZATION

Onium salts can also be used as PIs for hybrid free radical/cationic polymerization, since both free radicals and cations are formed in their photolysis, as shown in Scheme 29.1 (Section 29.3) for a triphenylsulfonium salt. Visible light-photosensitized compositions have also been reported for hybrid polymerization of acrylates and epoxides (Podsiadly et al., 2011; Xiao et al., 2014). Evidence is presented for the formation of interprenetrating polymer networks (IPNs), which had also been reported earlier for hybrid acrylate/epoxide (Decker et al., 2001; Cho and Hong, 2004), as well as acrylate/vinyl ether polymerization (Cho and Hong, 2004). Photo-DSC evidence showed faster polymerization of acrylate than epoxide. However, photosensitization with ITX (Section 29.2.2) increased the rate of epoxide and vinyl ethyl polymerization relative to acrylate polymerization and enhanced surface cure (Cho and Hong, 2004). Such hybrid compositions exhibit low shrinkage as well as suppressed oxygen inhibition.

Studies of a three-component initiator system, comprised of camphorquinone, a diaryliodonium salt, and an electron donor, determined that the electron donor required a $pK_b > 8$, as well as an oxidation potential less than 1.34 V relative to SCE, as determined from thermodynamic considerations. Experiments utilizing a combination of electron donors revealed that the onset of cationic polymerization can be advanced or delayed by controlling the concentration and composition of the electron donors (Oxman et al., 2005).

29.5 EFFECTS OF PIGMENTATION

Since many pigments absorb and/or scatter UV radiation, they generally inhibit UV curing to some degree (Wicks and Pappas, 1978). Scattering leads to reflection back out of the film, and absorption also reduces UV availability to the PI. The effect becomes more serious as film thickness increases, since the pigment plus the PI reduce the amount of UV that can reach the lower layers of the film. With a strongly absorbing pigment, such as carbon black, the films that can be UV cure are limited to 1–2 μm in thickness, corresponding to the thickness of black inks, but not black coatings, which are substantially thicker.

The most widely used white pigment in coatings is rutile TiO_2. However, rutile TiO_2 absorbs some violet light and essentially all but the very near-UV even at quite thin film thickness. Rutile white coatings with good hiding can be cured using a blend of 25% BAPO and 75% HMP (Rutsch et al., 1996) (Section 29.2.1). To permit flow, low levels of pigmentation are used (PVC of 6) requiring film thicknesses of 50–100 μm for hiding. Anatase TiO_2 does not absorb as strongly in the near UV, so that thicker films can be UV cure than with rutile but hiding by anatase is not as efficient.

Equations have been developed that permit calculation of the fraction of UV of each wavelength that is absorbed by a PI in the presence of pigments that absorb and scatter UV (Wicks and Kuhirt, 1975). The equations permit calculation of the total fraction of radiation absorbed as a function of wavelength and, perhaps more importantly, permit calculation of absorption in the bottom layer of films of different film thicknesses. The equations can be used to make model calculations to illustrate the effect of variables that can guide formulators considering selection of pigments, PIs, and their concentrations. Figure 29.1 illustrates the effect of concentration of PI on absorption of UV by PI in the bottom 0.1 μm of 15 μm films. The calculations were based on assumptions intended to illustrate the effect of 20 PVC of rutile TiO_2 in the film. The photosensitizer absorption was based on 2-chlorothioxanthone (CTX). With the assumptions used, the appropriate concentration of CTX for this system would be about 0.33%.

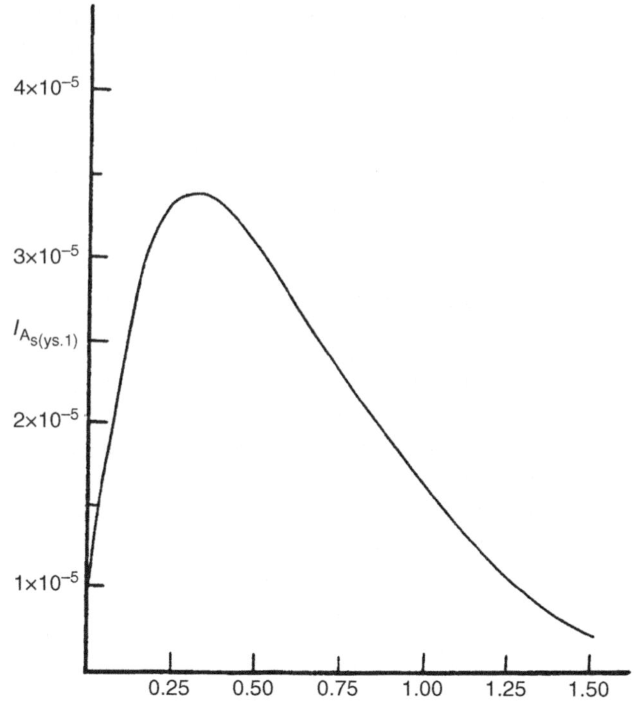

Figure 29.1 Absorption by CTX of 375 nm radiation in the bottom 0.1 μm of 15 μm films with 20 PVC rutile TiO_2 over an 85% reflectance substrate as a function of CTX concentration. Source: Wicks and Kuhirt (1975). Reproduced with permission of American Coatings Association.

Curing of white pigmented coatings is favored by using UV sources with higher emission of very near-UV radiation and by using PIs, such as CTX, with higher absorption in this region. As previously discussed, there is an optimum PI concentration that decreases as film thickness increases. Pigmented coatings are more sensitive to this effect than unpigmented coatings. The effect of reflection by the substrate can be critical in determining the film thickness that can be cured.

A further problem of pigmentation in curing arises from the large differential of absorption of UV at the surface and at the bottom of the films, which can result in wrinkling (Section 24.6). If the surface layer of a film cures while the bottom layer is still fluid, subsequent cure in the lower layers will result in shrinkage, which will cause the cured top layer to wrinkle. This effect is particularly likely to be seen with free radical cure carried out in an inert atmosphere or with cationic cure. In both cases, there is no inhibition of the surface cure by oxygen.

Pigmentation also affects flow properties. UV cure coatings are generally applied without solvent, which signifies that the volume of pigment in the wet coating is almost as high as the PVC of the final film. Without solvent, viscosity tends to be higher than desirable for good flow in the short time between application and cure. Even at 20 PVC, there is a significant increase in viscosity due to the presence of the pigment. By diluting oligomer with sufficient monomer, the viscosity can be reduced to application viscosity, but cure speed is reduced. As one tries to make higher PVC coatings, the problem becomes more difficult. As PVC approaches CPVC, viscosity of a wet coating approaches infinity. The problem is compounded by the difficulty of stabilizing the pigment dispersion against flocculation with the relatively low MW vehicle (Section 21.1.3). Flow problems are just as serious for cationic systems as for free radical systems. However, pigment flow problems are expected to be alleviated in waterborne UV cure compositions (Section 29.2.6), which are applied as dispersions.

For many wood finish applications, transparent low gloss coatings are desired. As explained in Section 31.1.1, such finishes are attainable with solventborne coatings by pigmenting with 2–4% of SiO_2 particles that are carried up to the surface of the film by convection currents during solvent evaporation. However, in UV cure coatings, there is no solvent to evaporate and, therefore, no mechanism to concentrate the pigment at the surface. If the level of pigmentation is increased to the extent necessary to provide low gloss, the viscosity is too high for application. As discussed in Section 29.2.6, low gloss, transparent wood coatings are possible with waterborne UV cure.

Some progress has been made in the reduction of gloss by use of a dual UV cure method. A UV cure coating, containing as much small particle size SiO_2 as viscosity permits and designed to be strongly oxygen inhibited, is applied and cured in air. The lower layer of the coating cures preferentially. As the lower layer cures, it shrinks, exerting an unbalanced stress on the pigment particles in the upper parts of the film, forcing them toward the surface. An alternative explanation is that reactive monomer or oligomer undergoes net migration into the lower polymerizing region, thereby concentrating pigment in the upper nonpolymerizing region. Then, the coating is UV cured again, this time under an inert atmosphere. Now, the surface layer cures, resulting in further shrinkage and further increasing PVC near the surface. In this way, medium gloss transparent coatings can be made. Several production lines have operated using this dual cure method.

Numerous reports of nanoparticle (Section 20.5) pigmentation of UV-curable acrylate coatings describe film property enhancements. For example, Jo et al. (2016) describe trimethylolpropane triacrylate (TMPTA) coatings containing surface-treated SiO_2 nanoparticles. The surface treatment is structurally similar to TMPTA for compatibility. Incorporation of the nanoparticles reduces water vapor transmission and enhances humidity resistance while retaining greater than 98% transparency.

29.6 ELECTRON BEAM CURE COATINGS

High energy EBs can be used to polymerize acrylate coatings. In EB cure compositions, the high energy electrons cause direct excitation of the coatings resins (P), as well as their ionization into radical cations ($P^{+\cdot}$) and secondary electrons (e^-).

$$P + EB \rightarrow P^* + P^{+\cdot} \quad P^{+\cdot} + e^- \rightarrow P^* \rightarrow I^{\cdot}$$

The major fate of the radical cations is recombination with the relatively low energy secondary electrons, which can yield additional excited state resins. The excited state resins (P^*) primarily undergo homolytic bond cleavage to free radicals that initiate polymerization of the acrylated resins.

The high energy electrons are generated by charging a tungsten filament at a high negative potential, 150–300 keV. The electrons are directed by magnets in a curtain through a metal "window" to the coatings to be cured. (The term *electron curtain* is also used.) The types of vehicles used are the same as those used in free radical-initiated UV curing. Since the polymerization reactions are oxygen inhibited and there is little differential in absorption between the top and the bottom of the film, it is essential to do EB curing in inert atmospheres.

Cationic polymerization of vinyl ethers (Lapin, 1992), as well as highly reactive epoxides, can be induced by EB radiation in the presence of diaryliodonium and triarylsulfonium salts. The readily reducible onium salts have been implicated in the generation of cationic initiators by oxidation of free radical species, as well as by capture of secondary electrons. Scavenging of secondary electrons by onium salts tends to lengthen the lifetime of cations and to improve the prospects for cationic polymerization.

The principal advantages of EB curing over UV curing are that no PI is needed, that pigments do not interfere with the curing, and that irregularly shaped plastic objects (e.g., auto bumpers) can be coated. These advantages are often offset by the higher capital cost of the EB or curtain generating equipment, the need to use inert atmospheres, and the need for massive shielding to protect workers from the EB. The advantage of being able to cure pigmented systems is real but is of limited importance in coatings, since the flow problem is not alleviated by EB curing.

29.7 DUAL UV/THERMAL CURE

UV curing is generally limited to flat sheets or moving webs or cylindrical objects, which can be rotated in front of the radiation source, in order to achieve uniform cure. However, dual UV/thermal cure compositions have been developed for multishaped substrates, such as an auto body, to provide thermal cure in areas that are not directly exposed to the radiation, so-called *shadow areas*. An investigation of various approaches to UV curing on automobiles has been described (Maag et al., 2000). A dual UV/thermal cure coating, exposed with a combination of two fluorescent lamps (one with maximum intensity at 313 nm, the other at 351 nm) and a xenon flash lamp, followed by thermal curing for 30 min at 70°C, gave excellent film properties. The rapid UV cure minimizes sagging and the subsequent thermal cure provides cure in the shadow areas.

A dual UV/thermal cure coating, based on a hydroxyl-functional urethane methacrylate oligomer, melamine cross-linkers, and PIs, was investigated as a clear coat for automotive applications. Dynamic mechanical and FTIR analyses were performed to determine the effect of UV dose, thermal cure time and temperature, and curing sequence on the resulting cross-linked networks. Scratch behavior and images were analyzed by nano scratch tests and scanning electron microscopy. Compared with UV or thermal cure clear coats, the dual UV/thermal cure sequence resulted in improved mechanical properties and scratch resistance, as well as increased methacrylate conversion relative to the dual thermal/UV sequence (Noh et al., 2012).

Photolatent catalysts, which generate acids (Dietliker, 2004), tertiary amines and amidines (Dietliker et al., 2007), and metal catalysts (Carroy et al., 2010) on exposure to UV, can be utilized in dual UV/thermal cure compositions. Such latent catalysts are useful for extending the shelf life of one-package coatings, or the pot life of two-package coatings, and improving the resulting cured film properties.

Photolatent catalysts for anionic UV curing have been reviewed (Javadi et al., 2016).

29.8 SELECTED APPLICATIONS

UV cure printing inks, particularly lithographic inks, are widely used. The films are thin, 2 μm or less, so interference of pigment with curing can be offset by using highly absorbing PIs and/or high PI concentrations. The advantage of fast curing is large, as it allows the webs to be wound up quickly without offsetting on the back of the substrate. In printing of tin plate and aluminum for cans, there is the advantage that the four colors can be printed in line by simply having a UV lamp after each printing station; operating costs are lower than with heat cure inks. Ink jet printing is another rapidly growing market for UV inks (Calvert, 2001; El-Molla, 2007).

Clear coatings for paper, plastic film, and foil are probably the largest volume market for radiation-cured coatings. The unpigmented coatings cure very fast at low temperatures. The important uses of UV curing for top coating wood furniture and filling of particle board are discussed in Section 31.1.3. Another large-scale application that benefits from fast curing at low temperature by UV is top coating of vinyl flooring. UV cure is especially applicable to flooring with patterns made by expanded foams, since the foam pattern is heat sensitive. Acrylated urethane oligomer coatings have the high abrasion resistance needed in a floor covering; acrylated polyester coatings have superior stain resistance.

The poor abrasion resistance of clear plastics is a distinct disadvantage in applications, such as replacing glass in uses ranging from glazing to eyeglasses to automotive head lamps. Silicone-based coatings have been used for many years to improve the abrasion resistance (Section 4.5). These coatings require long thermal cure schedules and have high VOC levels. Radiation cure coatings based on trialkoxysilyl-functional acrylate monomers and colloidal silica cure in seconds and have low VOC emission (Blizzard et al., 1992). Such coatings have excellent adhesion to polycarbonates and other plastics and have abrasion resistance equal or superior to solvent-based coatings (Lewis and Katsamberis, 1991; Blizzard et al., 1992).

An application for cationic UV cure epoxy compositions is coil coating tin plate for the exterior side of can ends. The superior adhesion of epoxy coatings makes them preferable for this application, as compared with acrylate coatings. In some cases, a UV cure epoxy coating is applied on one side of the strip and UV cured, followed by applying a thermally cured epoxy coating on the other side. When the strip is put through the oven to cure the thermally cured coating, the cross-linking of the UV cure coating is also advanced. More flexible films are obtained using epoxidized polybutadiene (Cazaux et al., 1994).

Coated optical fibers for wave guides used in telecommunication cables are UV cured at high speeds in towers up to 25 m tall. A soft first coat with special optical properties and a harder outer coat are applied and cured by the time the fast-moving coated fiber reaches the bottom of the tower (Masson et al., 2004).

UV curing is also one of the bases for important photoimaging processes used in the printing and electronic industries (Funhoff et al., 1992; Monroe, 1992), as well as stereolithography (Stampfl et al., 2008). Very rapid three-dimensional printing has been disclosed wherein parts are precision fabricated in a pool of acrylate resin by taking advantage of controlled inhibition by oxygen (DeSimone et al., 2015). This process contrasts with conventional 3-D printing, which proceeds more slowly by successive 2-D imaging steps.

29.9 ADVANTAGES, DISADVANTAGES, AND SELECTED ADVANCES

Reduced solvent emission is an important motivation for use of radiation cure. Generally, the formulations contain no solvent and VOC emissions are negligible. Very short cure times are possible at ambient temperature with package stable coatings. In the case of clear acrylate compositions, curing times are fractions of a second. Since the curing can be done at near-ambient temperatures, the coatings are applicable to heat-sensitive substrates such as paper, some plastics, and wood.

The energy requirement is minimal. Major losses in energy encountered in thermal curing systems are not experienced. The cure is carried out at or close to ambient temperature, so no energy is required to heat the object being coated except in the case of waterborne UV coatings. (The IR radiation from some UV lamps does cause a modest temperature increase.) In addition, heat loss from air flow through a baking oven, required to keep the solvent concentration below the lower explosive limit, is eliminated. The extent of this advantage is, of course, dependent on the cost of energy. The use of radiation curing in Europe and Japan has been greater than in the United States, at least in part because the cost of energy is much lower in the United States.

Capital cost for UV curing is low. This results primarily from the small size of the curing units as compared with ovens for baking. A UV cure coating generally cures within about 0.5 s of exposure. It follows that a line being run at a speed of $60\,m\,min^{-1}$ only requires UV exposure over a distance of about 0.5 m. A UV curing unit having four lamps can accomplish cure within a total length of about 2 m. In contrast, a thermal cure oven has to be a 100 or more meters long to provide time for curing at this speed. Furthermore, additional space is needed for flash off of at least some of the solvent from conventional coatings before the coated object enters the oven, as well as space for the object to cool back down to handling temperature. The savings in building space can thus be large.

There are, however, limitations and disadvantages to radiation curing. It is most applicable to flat sheets or webs for which the distance to the UV source or the window of the EB unit can be approximately constant. Cylindrical, or nearly cylindrical objects, can be rotated in front of the radiation source and fibers can be pulled through a circle of sources, but irregularly shaped objects are not easy to expose uniformly.

Pigmentation can limit the thickness of films that can be UV cured. In the extreme case of carbon black pigmented films, the limit is about 2 μm. EB cure coatings do not have this limitation because there is no gradient of initiation from top to bottom of the film. Owing to the effects of pigmentation on flow, the level of pigmentation is limited in both UV cure and EB cure coatings, excepting waterborne formulations. Furthermore, low gloss coatings are technically difficult. These limitations can be overcome by adding solvent to UV cure coatings, but this eliminates many of the advantages.

Shrinkage during curing can lead to adhesion problems on metal and some plastic substrates (Section 6.2). When polymerization results from chain addition of free radicals to double bonds, there is a substantial decrease in volume, since the length of the carbon—carbon bonds formed is shorter than the intermolecular distance between monomer units. The degree of shrinkage is directly related to the number of double bonds reacted. In acrylic UV or EB cure coatings, the potential shrinkage is 5–10%. Since cure occurs in less than a second, there is no time for much of the volume shrinkage to take place before the free volume in the coating is so limited that movement becomes restricted. The stress from the restricted shrinkage applies a force to offset the adhesion forces holding the coating to a substrate so that less external force is required to remove the coating. Shrinkage can be minimized by using higher MW oligomers or higher ratios of oligomer to low MW reactive monomers, but these expedients increase viscosity and tend to increase cure time.

As discussed in Section 29.2.5, the utilization of AFTR monomers in UV-curable methacrylates (Gorsche et al., 2014) or to thiol–enes (Scott et al., 2005) reduces residual stress, thereby enhancing mechanical properties and, presumably, adhesion. The AFTR monomers are proposed to function by introducing a step-growth polymerization process and/or groups in the cross-linked network that can undergo reversible cleavage of the backbone structure. Such cross-linked compositions that comprise reversible bond structures in the polymer network have been termed covalent adaptable networks (CANs), which are considered to constitute a new class of polymers intermediate between thermoplastic and thermoset polymers (Kloxin et al., 2010; Bowman and Kloxin, 2012; Kloxin and Bowman, 2013).

UV curing of epoxy systems, as well as hybrid free radical/cation systems, leads to less shrinkage. The decrease in volume resulting from the formation of polymers from monomers is partially offset by an increase in volume from the ring opening of epoxide groups. Typical epoxy UV cure compositions shrink less than 3%. Adhesion of cationic cure coatings can be further enhanced by surface treatment of the substrate (Molenaar et al., 1993).

Some of the problems with UV curing can be reduced by using waterborne UV coatings as discussed in Section 29.2.5, including lower internal stress, improved adhesion, and reduced pigment flow problems.

Porous substrates such as paper and wood usually provide good adhesion, since the mechanical effect of penetrating into the surface of the substrate holds the coating on the substrate. However, there can be a curling problem with thin substrates as a result of the tension from shrinkage on one side of the substrate.

The PI in free radical-initiated coatings is only partly consumed in curing, so a substantial fraction remains in the films. The residual PI can accelerate photodegradation of films exposed outdoors. Conventional UV stabilizers and antioxidants tend to reduce UV cure rates, which limits their usage. These constraints limit the use of UV-cured coatings in the range of exterior applications but do not exclude it. Studies on automotive clear coats have shown that a UV-curable formulation of specified acrylated resins, PIs, UV absorber, and hindered amine light stabilizer (HALS) showed exterior durability after 5 years in Florida comparable to conventional thermal cure clear coats (Valet, 1999).

Karasu et al. (2014) investigated the influence of HALS and UV absorbers on free radical photopolymerization using both UV and LED radiation sources. Real-time Fourier transform-infrared (RT-FTIR) spectroscopy and confocal Raman microscopy (CRM) together with a novel depth profiling technique were utilized to determine extent conversions and the filtering effect of the UV absorbers.

The combination of a visible light photoinitiator, (2,4,6-trimethylbenzoyl) diphenylphosphine oxide (TPO), and LED irradiation at 395 nm overcame the filtering effect. The relatively low emission intensity of the LED source was compensated by design of the composition, including the utilization of peroxy radical scavengers, resulting in tack-free coatings at high conversions in the presence of UV stabilizers.

EB cure coatings have the distinct advantage that there is no PI in the system. However, even with EB curing, the coatings must have excess unsaturation to cure on rapid schedules; and residual unsaturation can adversely affect exterior durability. Also, there are trapped free radicals that can react with oxygen after the cure and generate peroxides.

Radiation cure coatings tend to be more expensive than conventional coatings so that commercial applications are primarily those where operating efficiencies offset the extra cost of materials or where the advantages of UV cure permit unique advantages.

REFERENCES

Artysz, D. M., et al., *Prog. Org. Coat.*, **2003**, *46*, 302.
Beckel, E. R., et al., *Macromolecules*, **2005**, *38*, 3093.
Benedikt, S., et al., *Macromolecules*, **2014**, *47*, 5526.
Blizzard, J. D., et al., *Proceedings of the International Waterborne, High-Solids, and Powder Coatings Symposium*, New Orleans, LA, **1992**, p. 171.
Bowman, C. N.; Kloxin, C. J., *Angew. Chem. Int. Ed.*, **2012**, *51*, 4372.
Cabral, J. T.; Douglas, J. F., *Polymer*, **2005**, *46*, 4230.
Calvert, P., *Chem. Mater.*, **2001**, *13*, 3299.
Carlblom, L. H.; Pappas, S. P., *J. Polym. Sci. A Polym. Chem.*, **1997**, *15*, 1381.
Carroy, A., et al., *Prog. Org. Coat.*, **2010**, *68*, 37.
Cazaux, J., et al., *J. Coat. Technol.*, **1994**, *66*(838), 27.
Chatani, S., et al., *Polym. Chem.*, **2014**, *5*, 2107.
Cho, J.-D.; Hong, J.-W., *J. Appl. Polym. Sci.*, **2004**, *93*, 1473.
Cramer, N. B., et al., *Polymer*, **2008**, *49*, 4756.
Crivello, J. V., *J. Coat. Technol.*, **1991**, *63*(793), 35.
Crivello, J. V., *J. Polym. Sci. A Polym. Chem.*, **2009**, *47*, 866.
Crivello, J. V., et al., in Bradley, G., Ed., *Surface Coatings Technology*, John Wiley & Sons, Inc., New York, **1998**, pp 204–227.
Decker, C.; Lorinczova, I., *JCT Res.*, **2004**, *1*(4), 247.
Decker, C.; Moussa, K., *J. Coat. Technol.*, **1993**, *63*(819), 49.
Decker, C., et al., *Polymer*, **2001**, *42*, 5531.
Decker, C., et al., *JCT Res.*, **2004**, *1*(2), 127.
DeSimone, J. M., et al., US patent 9,211,678 (**2015**).
Dietliker, K., *Spectrum*, **2004**, *45*(2), 190.
Dietliker, K., et al., *Prog. Org. Coat.*, **2007**, *58*, 146.
Dumaz, Y. Y., et al., *J. Polym. Sci. A Polym. Chem.*, **2009**, *47*, 4793.
El-Molla, M. M., *Dyes Pigments*, **2007**, *74*, 371.
Esfandiari, P., et al., *J. Polym. Sci. A Polym. Chem.*, **2013**, *51*, 4261.
Friedlander, C. B.; Diehl, D. A., US patent 5,536,760 (**1996**).
Funhoff, D., et al., *Prog. Org. Coat.*, **1992**, *20*, 289.
Ganster, B., et al., *Macromolecules*, **2008**, *41*, 2394.
Gerlitz, M.; Awad, R., *Proceedings of the International Waterborne, High-Solids, and Powder Coatings Symposium*, New Orleans, LA, **2001**, pp 187–194.
Golden, R., *J. Coat. Technol.*, **1997**, *69*(871), 83.
Gorsche, C., et al., *Macromolecules*, **2014**, *47*, 7327.
Gummeson, J. J., *J. Coat. Technol.*, **1990**, *62*(785), 43.
Heller, C., et al., *J. Polym. Sci. A Polym. Chem.*, **2011**, *49*, 650.
Hoyle, C. E.; Bowman, C. N., *Angew. Chem. Ind. Ed.*, **2010**, *49*, 1540.
Husar, B.; Liska, R., *Chem. Soc. Rev.*, **2012**, *41*, 2395.
Hwang, H.-D.; Kim, H.-J., *React. Funct. Polym.*, **2011**, *71*, 655.
Javadi, A., et al., *Prog. Org. Coat.*, **2016**, *100*, 2.
Jo, C. I., et al., *Ind. Eng. Chem. Res.*, **2016**, *55*(35), 9433–9439.
Jonsson, S., et al., *Prog. Org. Coat.*, **1996**, *27*, 107.
Karasu, F., et al., *J. Polym. Sci. A Polym. Chem.*, **2014**, *52*, 3597.
Kayaman-Apohan, N., et al., *Prog. Org. Coat.*, **2004**, *49*, 23.
Kilambi, H., et al., *Macromolecules*, **2007**, *40*, 47.
Kloxin, C. J.; Bowman, C. N., *Chem. Soc. Rev.*, **2013**, *42*, 7161.
Kloxin, C. J., et al., *Macromolecules*, **2010**, *43*, 2643.
Lapin, S. C., in Pappas, S. P., Ed., *Radiation Curing: Science and Technology*, Plenum, New York, **1992**, pp 241–271.
Lewis, L. N.; Katsamberis, D., *J. Appl. Polym. Sci.*, **1991**, *42*, 1551.
Li, Q., et al., *Polymer*, **2009**, *50*, 2237.
Ligon, S. C., et al., *Chem. Rev.*, **2014**, *114*, 557.
Lv, C., et al., *RSC Adv.*, **2015**, *5*, 25730–25737.
Maag, K., et al., *Prog. Org. Coat.*, **2000**, *40*, 93.
Mahe, R., et al., *J. Org. Chem.*, **1989**, *54*, 1518.
Masson, F., et al., *Prog. Org. Coat.*, **2004**, *49*, 1.
Molenaar, F., et al., *Prog. Org. Coat.*, **1993**, *22*, 393.
Monroe, B. M., in Pappas, S. P., Ed., *Radiation Curing: Science and Technology*, Plenum, New York, **1992**, pp 399–440.
Neschchadin, D., et al., *J. Am. Chem. Soc.*, **2013**, *135*, 17314.
Noh, S. M., et al., *Prog. Org. Coat.*, **2012**, *74*, 257.
Okamitsu, J. K., et al., US patent 6,908,586 (**2005**).
Oxman, J. D., et al., *J. Polym. Sci. A Polym. Chem.*, **2005**, *43*, 1747.
Pappas, S. P., in Pappas, S. P., Ed., *Radiation Curing: Science and Technology*, Plenum, New York, **1992**, pp. 1–20.
Pappas, S. P., et al., *J. Photochem. Photobiol. A*, **2003**, *159*, 161.
Podsiadly, R., et al., *Dyes Pigments*, **2011**, *91*, 422.
Priou, C., et al., *J. Coat. Technol.*, **1995**, *67*, 71.
Reich, W., et al., *Proceedings of the International Waterborne, High-Solids, and Powder Coatings Symposium*, New Orleans, LA, **1999**, pp 448–457.
Rutsch, W., et al., *Prog. Org. Coat.*, **1996**, *27*, 227.
Scott, T. F., et al., *Science*, **2005**, *308*, 1615.
Stampfl, J., et al., *J. Micromech. Microeng.*, **2008**, *18*, 125014/1.
Thomas, G. A.; Webers, V. J., *J. Imaging Sci.*, **1985**, *29*, 112.
Urbano, E., et al., US patent 6,265,461 (**2001**).
Valet, A., *Prog. Org. Coat.*, **1999**, *35*, 215.
Wicks, Z. W., Jr.; Kuhirt, W., *J. Paint Technol.*, **1975**, *47*(610), 49.
Wicks, Z. W., Jr.; Pappas, S. P., in Pappas, S. P., Ed., *UV Curing: Science and Technology*, Technology Marketing Corp., Norwalk, **1978**, Vol. *1*, pp 78–95.
Wicks, Z. W., Jr., et al., *Organic Coatings: Science and Technology*, 3rd ed., John Wiley & Sons, Inc., New York, **2007**, pp 575–594.
Woods, J. G., et al., US patent 8,748,503 (**2014**).
Xiao, P., et al., *Macromolecules*, **2014**, *47*, 601.
Yagci, Y., et al., *Macromolecules*, **2010**, *43*, 6245.

Chapter 30

Product Coatings for Metal Substrates

In 2014 US shipments of product coatings for original equipment manufacturer (OEM) application had a value of \$7 billion (American Coatings Association and Chemquest Group, 2015). This was approximately 31% of the value of all US shipments of coatings. In this and the following chapter on coatings for nonmetallic substrates, we discuss some of the major OEM product coating end uses to illustrate the factors involved in selecting coatings for particular applications. Multitudes of products are coated. In this chapter we selected five of the larger end uses: automotive, appliance, container, coil, and aircraft coatings.

30.1 OEM AUTOMOTIVE COATINGS

OEM automotive coatings for cars and light trucks are the largest single market for industrial coatings. The US market for automotive coatings in 2012 was reported to be \$1.4 billion (American Coatings Association and Chemquest Group, 2015). Coatings for repair and refinishing of cars are discussed in Section 33.3, and the size of that market is roughly 40% that of the OEM market on a volume basis. The discussion here is further limited to coatings applied to the exterior bodies of automobiles. Coatings for plastics in cars are discussed in Section 31.2.2. Many other coatings are used in car interiors, for example, on steering wheels, trunk linings, air filters, and so on.

The paint shop is a major part of the automotive assembly plant. The cost of the paint shop is approximately 1/3 the cost of the entire assembly plant. In the assembly plant, approximately 80% of the energy consumed, 80% of the CO_2 produced, and 90%+of the VOCs and wastewater emitted arise from the painting operations.

Coatings are applied to automobiles and trucks to protect the underlying metal from corrosion and to enhance the aesthetic appeal of the vehicle. Studies have shown that consumers make like/dislike decisions on an automobile within 90 s after they first see it. Up to 90% of their decision within that timeframe is based on color (Singh, 2006). In 90 s only the shape of the vehicle and the vehicle's paint are visible. Consumers are thought to correlate a vehicle's paint appearance with the overall quality of the vehicle. In particular, customers assess the appeal of the color, the paint's smoothness (orange peel), and the color match or harmony between the painted body and other painted parts such as the fascia, mirror caps, and door handles. As a vehicle is typically the largest expense a consumer will make after their home, they expect the vehicle to maintain its value and good appearance for the life of the vehicle (>10 years). Vehicle manufacturers are strongly motivated to provide excellent long-term durability of their coatings, as failures (color fade, gloss loss, cracking, delamination, etc.) have a dramatic effect on a brand's reputation. If a vehicle's radio stops working, the driver of the vehicle knows. If a car's paint fails, everyone looking at the vehicle knows. The impact of failures is magnified compared with other vehicle subsystems.

Corrosion protection is the other critical performance requirement of automotive coatings. The salty air in coastal environments and the use of deicing salts on roads in cold climates provide ideal situations for the corrosion of metals. The corrosion resistance of vehicles has improved dramatically from the 1970s and 1980s through the extensive use of galvanized steel and the increased use of aluminum. However, the paint system still provides much of the long-term corrosion resistance to the automotive body. Excellent adhesion and the resistance to water and ion transport are required to enable 10+years' lifetimes of vehicle bodies.

Organic Coatings: Science and Technology, Fourth Edition. Frank N. Jones, Mark E. Nichols, and Socrates Peter Pappas.

30.1.1 Automotive Paint Process

As the composition and performance of the various layers in an automotive paint system are strongly dependent upon the application methods and parameters, a brief outline of a conventional automotive paint process is first given. A few of the many possible variations are then discussed. The functions and limitations of each layer are also presented here. The standard automotive paint system is shown in cross section in Figure 30.1 for reference.

The automotive body arrives in the paint shop from the body shop, where the various sheet metal components were joined together (welds, adhesive bonds, rivets). Cleaning and pretreatment (Section 6.3.1) are the first steps in the automotive painting process. The pretreatment system is composed of 10–14 steps, some of which are sprays and some full vehicle immersions. The first steps are directed at cleaning the weld balls, stamping lubricants, and other dirt from the surface of the body with caustic cleaners. The pretreatment or conversion coating is then applied. Traditionally, this pretreatment results in the deposition of needlelike crystals of zinc phosphate on the entire body surface. The zinc phosphate is often modified with nickel and manganese cations to improve its performance. After rinsing with water, the vehicle is then subjected to a post-rinse that typically seals gaps and pores in the zinc phosphate layer. A common sealing treatment results in a thin amorphous film of zirconium oxide being deposited onto the zinc phosphate.

Zinc phosphate is slowly being displaced by pretreatment systems based entirely on zirconium oxide (Mohammadloo et al., 2014). These thin film pretreatments offer environmental and cost benefits while delivering comparable performance. In addition, for vehicles with high aluminum content, zirconium oxide pretreatments offer enhanced benefits. When more than ~30% of the metal processed through a pretreatment system is aluminum, excessive sludge is produced in the zinc phosphate bath, which degrades performance and poses disposal problems.

The main role of the pretreatment is to provide a surface to which the next organic layer can strongly bond. Thus, the pretreatment is integral to superior corrosion resistance.

The first organic layer in the automotive paint system is cathodic electrocoat. The details of electrocoat are covered in Chapter 27, but some basics are provided here. Electrocoat is applied by submerging the entire vehicle body in a tank a few hundred feet in length. Automotive electrocoat tanks can hold upward of 100 000 gallons (400 000 l) of paint. An electrical circuit is made between the electrocoat tank, the paint, and the vehicle body. The vehicle serves as the cathode, and metal anodes are placed around the periphery of the tank. A few hundred volts of potential are applied between the anodes and vehicle, which leads to deposition of the electrocoat on the vehicle body. The key advantage of electrocoat is its ability to coat internal cavities of the vehicle, such as within the rocker panels and pillars. To do this, the body structure must allow access for fluid and electrical current into these cavities. Thus, hole size and placement is critical to achieving good internal coverage. An electrocoat's *throw power* or *throw* describes its ability to coat internal cavities. Because of the electrocoat deposition process, almost no material is wasted and utilization is close to 100%. Electrocoat provides a very strong barrier to water and also forms a strong adhesive bond with the underlying pretreatment. Thus, electrocoat plays the primary role in the corrosion resistance of an automobile.

After the electrocoat is applied, the entire vehicle is baked in a large oven. The nominal baking conditions for electrocoat are 20 min at 180°C. However, different parts of the vehicle experience different temperature histories, as the thermal mass in any region is a function of the material's composition and the mass of the local material. In addition

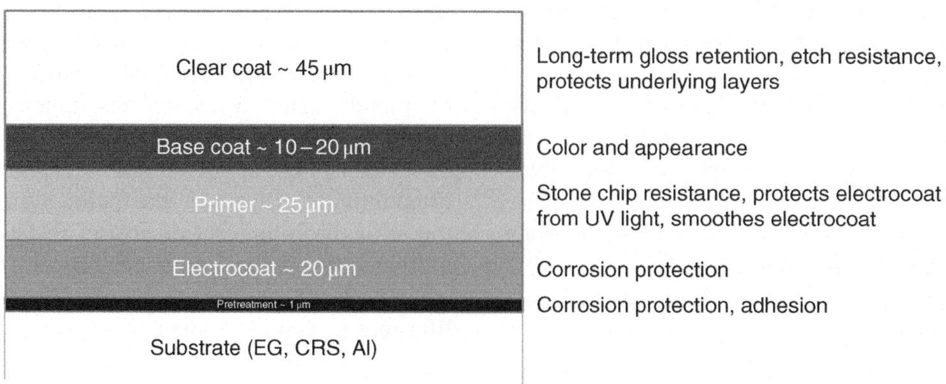

Figure 30.1 Cross section of a modern automotive paint system. Functions of each layer are specified. Thicknesses are approximate and depend on color and orientation (vertical or horizontal) on the vehicle.

to cross-linking the electrocoat, the electrocoat oven serves to harden some of the steel and aluminum used in the body structure, cross-link adhesives and sealers, and expand foams to improve vehicle stiffness.

After the vehicle leaves the electrocoat oven and is inspected for defects, the vehicle enters a spray booth, where primer is applied. The primer is also called *primer-surfacer*. In a modern automotive paint shop, the primer is applied robotically by electrostatic rotary bell atomizers. These atomizers consist of spinning disks/cups, where the fluid is injected into the center of the cup (see Section 23.2.3 for more detail). Centripetal force propels the paint to the periphery of the bell cup where ligaments form and eventually break up to form droplets. The use of bell atomizers as opposed to spray guns improves *transfer efficiency*, the amount of paint that lands on a vehicle compared with the amount of paint atomized. High transfer efficiencies (>75%) offer significant cost savings to automotive OEMs. To assist in transfer efficiency, all spray application in the automotive paint shop is aided by electrostatics. An ~80 kV potential is applied between the body and the atomizers. The paint is negatively charged by the atomizer and is attracted to the positively charged vehicle. The ensuing electric field can "steer" the charged droplets toward the vehicle body, reducing overspray and allowing for some "wrap" around the outward facing body panels.

The primer is applied to all accessible surfaces of the vehicle body. Primer serves three primary functions. First, it smooths the surface by filling in the valleys in the rough E-coat surface. Second, it provides stone chip resistance for the paint system. This requires that the primer has the appropriate mechanical properties to dissipate the impact energy of a high velocity stone in order to prevent delamination in the paint system. Finally, the primer is an opaque layer that prevents both UV and visible light below 550 nm from reaching the electrocoat surface. Due to its composition, cathodic electrocoat is inherently photolabile and will degrade rapidly if exposed to radiation below 550 nm. To prevent delamination in the field, the primer must be completely opaque to this radiation at the minimum film builds applied to the vehicle. After application, the primer is baked in an oven at ~150°C for 20 min.

After the primer bake, the vehicle enters the enamel or top coat booth where the final layers are applied. First, the base coat is applied. This layer provides the color to the paint system. The base coat is applied by electrostatic rotary bell atomizers, often in two passes. The first pass of base coat applies the majority of the coating, and the second pass of base coat provides a thinner layer, which often enhances the appearance of metallic colors. After the base coat is applied, and a small amount of time is given to allow solvent to flash off, the clear coat is applied by electrostatic robotic bell atomizers. The clear coat provides a durable high gloss finish to the paint system. In addition, the clear coat contains UV light absorbers (UVAs) to protect the underlying layers from UV radiation. The clear coat also provides solvent, scratch, and acid etch resistance. The base coat and clear coat are then co-baked in a final oven at ~135°C for 20 min.

The preceding discussion describes a conventional automotive paint process. In recent years, automotive OEMs have been retrofitting their paint shops to use a "compact" paint process. These processes are also known by various other names including 3-wet, B1/B2, and primerless paint process. Regardless of the name, all of these compact paint processes achieve the same goal: the elimination of the stand-alone primer booth and oven. Compact processes involve the application of primer, base coat, and clear coat in one booth followed by baking all three layers simultaneously. Compact processes result in large savings in natural gas usage and energy. The energy savings largely come from eliminating the booth, not the oven, as an automotive spray booth requires several hundred thousand CFM of well-conditioned and clean air to be circulated through the booth. However, traditional base coats and primers have to be reformulated to perform adequately in compact paint processes. Compact paint processes are clearly the future of automotive paint shops.

Until the early 1980s almost all top coats were *monocoats,* a single coating composition applied in several coats. Monocoats have been largely supplanted by base coat–clear coat systems. Base coat–clear coat systems provide better gloss and gloss retention than monocoats. However, durable monocoats, whose binder resins are based on clear coat resin technology, are being reintroduced for solid colors on vans and trucks due to the lower application costs for monocoats.

30.1.2 Electrodeposition Coating Formulation

Owing to their importance in many industries, electrodeposition coatings are covered in detail in Chapter 27. Only the automotive-specific aspects will be discussed here.

As mentioned earlier, E-coat provides excellent adhesion to metal and also resists displacement by water. The strong driving force of the electrophoretic application is thought to aid penetration into the phosphate crystal mesh on the metal surface. The workhorse cathodic electrocoat resins used by most automotive OEMs are water-based emulsions of aromatic epoxy resins cross-linked by blocked aromatic isocyanates. The aromatic groups in both the epoxy and isocyanate are thought to aid in the strong adhesion of the electrocoat to the conversion coating underneath. The binders are designed to be nonsaponifiable; even if the coating is scraped through, undercutting will be very slow, and the coating will continue to protect the metal next to the gouge against corrosion. The nominal thickness of electrocoat is ~25 μm on the exterior and ~15 μm on the interior cavities. New *hyperthrow electrocoats* aim to minimize the difference between the film build on the exterior and interior of the vehicle.

Other than sensitivity to UV radiation, other challenges with E-coat include its very uniform film build on the exterior of vehicles. This often leads to *telegraphing* of any substrate roughness. The effect is particularly exacerbated for high throw and high edge electrocoats that coat the sharp edges of sheet metal well. Better smoothness and improved adhesion to the subsequent primer layer can be achieved by reducing or even eliminating the pigments in electrocoat. These pigments act mainly as fillers, since corrosion-inhibiting pigments are not typically used in modern automotive electrocoats.

Paint suppliers and auto companies have attempted to develop more durable electrocoats, as this might enable the elimination of the primer layer. However, the risk of catastrophic delamination of the top coat from the electrocoat is currently too high owing to the inherent variability of top coat film builds during the painting process. In particular, thin film builds will not provide sufficient opacity to protect the electrocoat from photodegradation, resulting in delamination of the top coat. In addition, photodurable electrocoats have met little commercial acceptance due to unrealistic process constraints or inferior corrosion performance.

30.1.3 Automotive Primers

In 2013 in the NAFTA region, automotive primer usage was 17% powder, 13% waterborne, 54% solventborne polyol/melamine, and 13% formulated specifically for compact (Section 30.1.1) processes (Deskowitz, 2013). In other parts of the world, 2K polyurethane (PU) primers/acrylic primers are also used.

The primer has a lower cross-link density than the E-coat, providing greater opportunity for the solvent in the top coat to penetrate into this surface than into the surface of the E-coat. The primer is formulated with a relatively high PVC, as the resulting surface roughness provides a greater opportunity for top coat adhesion. In addition to enhanced adhesion with the top coat, the high PVC provides high opacity to UV and visible light and a sandable surface after baking.

As the main functional purpose of the primer is to provide superior stone chip resistance, the mechanical properties and adhesion of the primer to the adjacent layers are carefully engineered. A soft, compliant primer can absorb significant energy during a stone impact and provide improved chip resistance. However, if the primer is too soft, both sandability and solvent resistance are compromised. Furthermore, if intercoat adhesion of the primer is too strong, stone chip failure may propagate to the electrocoat–metal interface, which can lead to subsequent corrosion at the impact site. It is preferable to have the locus of failure of a stone chip at the primer interface with the top coat or the electrocoat. In coating systems applied by a compact process, where the primer, base coat, and clear coat are all co-baked, adhesion between the three layers is very strong, and this tends to drive the locus of stone chip failures to the electrocoat–primer interface, rather to the base coat–primer interface in conventionally processed systems. Ramamurthy et al. (1994) discuss the mechanical properties involved in stone chip resistance and approaches to testing.

Most liquid primers are polyester or acrylic resins (or mixtures of the two) cross-linked with melamine resins owing to their relatively low cost, good filling, and good adhesion. These primers can be adapted to work in compact paint systems (Uhlianuk et al., 2010; Hazan et al., 2011). Some PU primers are also used, however. Powder primer-surfacers are used over E-coat on some cars to reduce VOC emissions. Because they are typically applied at higher film builds, they also provide improved chip resistance. The greatest interest for powder coatings has been in epoxy-functional acrylics made with glycidyl methacrylate (GMA) (Section 13.1.2) as a comonomer and cross-linked with dicarboxylic acids, such as dodecanedioic acid [$HOOC(CH_2)_{10}COOH$] (Kenny et al. (1996), or a carboxylic acid-functional resin (Agawa and Dumain, 1997). An epoxy-functional acrylic for automotive primer-surfacer is reported to require an $\bar{M}_w/\bar{M}_n$ below 2500, a calculated T_g above 80°C, and an appropriate monomer composition, in order that melt viscosity is less than 40 Pa·s at 150°C (Green, 1995). Such a resin can be made with 15–35% GMA and 5–15% butyl methacrylate (BMA), with the balance being methyl methacrylate (MMA) and styrene.

Some automotive OEMs use color keyed primers, typically 4–5 different colors. Their use allows a slightly lower base coat film build to be applied while still achieving the desired color. The base coat is the most expensive layer in the paint system, so any reduction in film thickness is economically attractive. Color keyed primers also provide less objectionable chip appearance due to the similarity of the base coat to primer color. However, color keyed primers are logistically more difficult to implement as additional paint lines are required from the paint kitchen to the booth, and vehicles need to be properly sequenced to match the primer and base coat colors. When single primer colors are used, they are traditionally medium gray in color.

30.1.4 Automotive Base Coats

In 2013 base coat technologies in NAFTA were 26% solventborne, 58% waterborne, and 16% formulated for use in compact paint systems (American Coatings Association and Chemquest Group, 2015). Waterborne base coats continue to gain favor over solventborne owing to an environmental footprint that is perceived to be smaller than that of high solids solventborne base coats. However, waterborne base coats still contain an appreciable amount of organic solvents, are formulated at lower solids levels, have poorer transfer efficiency, often require a heated flash, and are more prone to popping. In addition, solvents from the synthesis of the base coat polymers must be stripped off during manufacturing. Thus, the notion that waterborne base coats are more environmentally friendly than high solids solventborne base coats can be misleading. It is anticipated that both waterborne and high solids solventborne base coats will be used for the foreseeable future.

The primary purpose of the base coat is to provide color and maintain that color for the life of the vehicle. Color retention is achieved through the use of durable binders and pigments (Chapters 5 and 20, respectively).

A majority of base coats are metallic or other special effect, polychromatic colors. As discussed in Sections 19.4 and 20.2.5, metallic coatings provide an attractive appearance owing to the change in color with the angle of viewing. When viewed at an angle near the perpendicular, the color is light and becomes darker when viewed at larger angles. The extent of this phenomenon is called the color *travel*, degree of *color flop*, or just *flop*. High color flop depends on four factors: minimal light scattering by the coating matrix between the aluminum flakes; a smooth high gloss surface; orientation of the aluminum flakes parallel to the surface; and the minimization of scattering from flake edges. Other special effect top coats are pigmented with interference pigments such as pearlescent and cholesteric pigments.

Minimal light scattering by the coating matrix requires that pigment selection and dispersion provide transparent films when prepared in the absence of the effect pigment. Most dispersions are made with media mills (Section 21.4.4). Nano(color pigments) for automotive metallic coatings are prepared by grinding suspensions of the conventional color pigments together with proprietary dispersing agents, using very small balls in the mill. The mill base is then used to make tinting pastes with excellent transparency for color matching metallic colors (Vanier et al., 2005).

Good orientation of aluminum flakes parallel to the surface has been achieved in everyday production for decades, but it has never been entirely clear what factors are responsible for this behavior (Tachi et al., 1990; Seubert et al., 2015). Most workers have concluded that an important factor is shrinkage of the film after application. However, shrinkage is accompanied by an increase in viscosity of the film particularly near the surface as the solvent evaporates, which leads to a viscosity gradient. This viscosity gradient causes the upper edge of an aluminum platelet to be immobilized before the lower edge, which is expected to restrict desirable parallel orientation of the platelet as the film shrinks. In general, lower solids coating as it arrives at the substrate results in better alignment. There are several theories regarding the alignment of flake particles. One postulates that spray droplets spread out when they strike the surface and the resulting flow forces tend to align the flake particles parallel to the surface as the film shrinks when solvent is lost. It has also been suggested that the spray droplets must penetrate the surface of the wet film and crash on the substrate (Tachi et al., 1990). The rapid drop in viscosity as a high solids clear coat enters the oven can cause flow, resulting in some loss of orientation. These and other factors may influence orientation. Changes in atomization, airflow, solvent evaporation rates, and gun to surface distance during spraying can cause substantial variations in flake alignment (Weaks, 1991). Transfer efficiency is greatly improved by electrostatic spraying; however, it is more difficult to obtain good surface smoothness and metal orientation using electrostatic spray. It has been suggested that there may be some alignment of the metal flakes parallel to the lines of force in the electrostatic field that are perpendicular to the substrate. Laser scanning confocal microscopy and computer simulation both provide new insights, confirming that flake orientation is important and revealing new variables, such as the sizes of the gaps between flakes (Sung et al., 2002; Seubert et al., 2015).

Solventborne base coat binders are typically thermosetting acrylic resins, polyester resins, or blends of the two, both of which are cross-linked with MF resins. It has been reported that hydroxy-functional urethane-modified polyesters are also useful in base coats (Broder et al., 1988). Solids levels for solventborne base coats are in the range of 35–40% by weight. Waterborne base coats are typically composed of MF cross-linkers with water-reducible acrylics, water-reducible polyester/PU, and acrylic latexes (Mirgel, 1993). Acrylic latexes and acrylic/polyurethane dispersions (PUDs) are also widely used for formulating waterborne base coats. For example, base coats are formulated with an acrylic/urethane PUD and an MF resin (Vogt-Birnbrich et al., 2000) (Section 12.7.2).

Water-dispersible carboxymethylcellulose acetobutyrate (CAB) has been shown to provide reduced dry-to-touch times, improved metal orientation, reduced sagging due to faster viscosity buildup, and improved leveling in waterborne base coats (Bhattacharya et al., 2007a, b). Improvements in CABs have led to their use in contemporary solventborne base coats and clear coats as well, as they improve rheology and appearance. Rheology is also controlled through the use of clays, urea crystals, and specifically designed copolymers that control the shear thinning behavior of the base coat (Basu et al., 2010). A highly cross-linked copolymer of MMA, BA, ethylene glycol dimethacrylate, and acrylamide, dispersed in water, is used as an additive in waterborne base coats. The acrylic dispersion provides excellent aluminum orientation (Swarup et al., 2004) and minimizes strike-in by solvent from solventborne clear coat, thereby enhancing uniformity.

Waterborne base coats typically require a dehydration step in which most of the water and some of the solvent are driven off before application of the clear coat, which increases the CO_2 footprint of the waterborne coating process. This process must be carefully controlled. If too much water remains in the film, it will blow out during the clear coat bake, resulting in popping (Sections 24.7 and 26.1).

Base coats are typically applied in two passes with bell atomizers to prevent popping and improve appearance after the clear coat is applied and baking occurs. Popping is typically a greater problem with waterborne than with solventborne base coats. Hiding can usually be achieved with ~10 μm of black base coat and ~15–20 μm of white base coat. Other colors fall in between these extremes. The low solids (20–25 NVV) metallic waterborne base coats facilitate aluminum orientation. Opaque colors can be applied at 25–35 NVV. Waterborne base coats are formulated for application at relative humidity (RH) of about 60%, since it is less expensive to increase RH than to decrease it.

30.1.5 Automotive Clear Coats

In 2013, NAFTA clear coat chemistries comprised 22% carbamate-functional acrylics, 17% 2K PU, 23% acrylic/silane, 27% epoxy acid, and 5% acrylic–melamine, with the remainder split between 1K PU and PU/silane (Deskowitz, 2013). The use of 2K PU is much greater in Europe than in North America. Clear coat chemistry has been the subject of intense patent activity.

Clear coats can be sprayed at somewhat higher solids than monocoats. As discussed in Section 24.3, as resin solids increase, the problems of controlling sagging increase. Automotive coatings are particularly vulnerable to sagging, since the film thickness applied is large and variable. As solids have been increased, it has become necessary to add thixotropic agents to increase the viscosity at low shear rates to minimize sagging (Bauer et al., 1982; Boggs et al., 1996). Conventional thixotropic agents impart haziness to a coating owing to light scattering, thereby reducing gloss. Therefore, viscosity modifiers have been designed with refractive indexes similar to those of the acrylic binder. Most widely used are acrylic microgels, which are gel particles that are lightly cross-linked so that they can swell, but not dissolve, in the liquid coating (Section 24.3). Microgels impart thixotropy to the coating with the objectives of balancing the viscosity/shear rate behavior so that viscosity of the paint is sufficiently low to be effectively atomized by the application equipment, remains low to achieve adequate leveling on the vehicle, and subsequently increases to prevent sagging. The exact mechanism by which this occurs is not fully understood, but swelling of the particles and flocculation are thought to play a role. Fumed silica and polymeric additives are also used to control the rheology of clear coats.

The application of a clear coat over a pigmented base coat provides higher gloss than is attainable with a monocoat paint system. Historically, clear coats were not used in automotive finishes because of the cost of an extra coating step and because the exterior durability of available clear coatings was inadequate. However, with development of improved binders and light stabilizers, especially combinations of HALS and UV absorbers (Section 5.2), clear coats now have superior long-term exterior durability. Long-term gloss retention is also excellent and has largely eliminated the need to wax a vehicle. Total combined film thickness is only slightly greater than that of a monocoat system. The base coat contains roughly twice the PVC of a monocoat. Thus, a base coat with a dry film about 12–20 μm thick, depending on the color, provides about the same hiding as a 50 μm monocoat. The target clear coat thickness is about 40–50 μm, but considerable variation is experienced in production. Horizontal surfaces are typically coated with thicker layers than vertical surfaces owing to their need for superior UV protection and the absence of sag problems on horizontal surfaces.

When introduced, most clear coats were either acrylic copolymers cross-linked with melamine or 2K PU coatings. Although an improvement over monocoat paint systems, these initial clear coats had significant limitations with respect to long-term durability and *acid etch* resistance. Acid etch, also more generally referred to as *environmental etch*, is the attack of a coating surface by acidic rain that contacts a vehicle. This phenomenon is highly localized and may be more prevalent in certain areas of the country or even certain cities, owing to proximity to power plants and factories that may emit sulfur dioxide and other pollutants. As the water evaporates, the pH of the water drops, which eventually leads to concentrated acidic drops on the surface of the vehicle. The ether cross-links in acrylic–melamine coatings are highly susceptible to acid-catalyzed hydrolysis, and therefore cross-link scission on the surface of the coating can readily occur in these localized areas. This localized degradation leads to defects on the surface that resemble water spotting on glassware. However, the spots are not removable because chemical degradation of the clear coat increases surface roughness. Acid etch resistance was traditionally evaluated through exposure of paint systems during the summer in Jacksonville, FL. However, the severity of exposure was not reproducible; and laboratory-based methods are now utilized that include the application of weak acids during environmentally controlled cycles (Boisseau et al., 2003). To improve acid etch resistance, almost all automakers have embraced new clear coat chemistry in which the cross-links are less prone to acid-catalyzed hydrolysis.

Since urethane cross-links are more resistant to hydrolysis under acidic conditions, isocyanate cross-linked acrylics generally have superior environmental etch resistance. Many urethane clear coats have been two-package (2K) coatings (Section 12.4). Owing to the need for long-term durability, aliphatic isocyanates are utilized in conjunction with hydroxyl-functional acrylics. In addition to their improved acid etch performance, 2K PU coatings also offer the best appearance. The route to very high solids clear coats in 2K formulations is also much easier than in 1K formulations, as it is much easier to provide sufficient functionality on individul molecules in 2K formulations than on 1K acrylics synthesized by traditional free radical chemistry. Weighing against the use of 2K coatings are the need for dual spray systems, concern about toxicity, and cost.

Additional cross-linking mechanisms also can improve acid etch resistance by forming cross-links more resistant to hydrolysis than urethanes. Siloxane bridges are one such type of cross-link (Trindade and Matheson, 2014). Monomers containing trialkylsilyl groups can be copolymerized with traditional acrylates to provide acrylics bearing both hydroxyl and trialkoxysilyl functionality. Thermal cross-linking of the hydroxyl groups with MF resins is supplemented by cross-linking of the trialkoxysilyl groups with atmospheric water (Furukawa et al., 1994). During the thermal cure of automotive clear coats, methoxysilyl groups hydrolyze with water vapor in the oven to form silanol groups, which then coreact during baking to form siloxane cross-links, resulting in enhanced acid etch resistance. Scratch resistance before and after weathering is superior to MF cross-linked coatings as well. Such coatings have been widely commercialized (Groenewolt, 2008). Use of auxiliary cross-linkers such as blocked polyisocyanates further enhance the properties obtained with silyl-functional acrylics (Barsotti et al., 2002). 3,5-Dimethylpyrazole or 1,2,4-triazole-blocked polyisocyanates are preferred, since they permit curing at somewhat lower temperature and do not have the yellowing problem of methyl ethyl ketone-blocked isocyanates (Section 12.5).

Epoxy-functional acrylics cross-linked with dicarboxylic acids or anhydrides also provide excellent environmental etch resistance. These "epoxy-acid" coatings can be formulated as either 1K or 2K coatings and have been traditionally popular with Asian automakers, as these were the first readily available acid etch-resistant clear coats. However, these coatings have generally been more brittle than coatings based on other clear coat chemistries.

Finally, excellent acid etch resistance can be achieved through the use of carbamate-functional acrylics cross-linked with melamine resins. The cross-linking reaction results in a hydrolysis-resistant urethane cross-link without the complications of using a 2K polyisocyanate formulation (Higginbottom et al., 1999). Such clear coats have been shown to exhibit excellent mar and acid resistance, as well as excellent exterior durability and hydrolytic resistance. They can also be formulated at high solids levels, as exemplified by a coating made from the reaction product of hydroxypropyl carbamate with an isocyanate prepolymer derived from IPDI, together with an MF resin and dodecylbenzenesulfonic acid catalyst, which has 85% weight solids.

To achieve the best combination of weathering, scratch, and acid etch resistance as well as outstanding appearance, the best clear coats typically utilize a combination of cross-linking mechanisms. The backbone resin is always an acrylic copolymer with various functionalities. As silane chemistry is beneficial for both weathering and acid etch resistance, it can be combined with traditional acrylic–melamine chemistry, as mentioned previously. In addition, isocyanates—blocked for 1K coatings and unblocked for 2K coatings—can be functionalized with alkoxylsilyl compounds, leading to both urethane and siloxane cross-links in the coating (Groenewolt et al., 2014). Resins with alkoxysilane functionality can also be combined with carbamate-functional acrylics (Section 8.2.2; Chapter 16; Edwards et al., 2005). Both functionalities can also be incorporated in the same acrylic resin. Resin blends can also be used (Sadvary et al., 2003; Balch et al., 2009). Low molecular weight (MW) carbamate-functional materials can be reacted with hydroxyl groups on a siloxane polymer. The carbamate groups can be cross-linked with melamine resins. These resins can be blended with acrylic resins to provide a balance of cost and properties. Segregation of the siloxane to the surface is possible based on surface energy differences, but little long-term improvement in the scratch resistance of clear coats has been realized when the scratch-resistant layer is confined to a thin surface layer.

Scratch and mar resistance are discussed in more detail in Section 4.4.2. Courter (1997) provides a review of mechanical properties required for mar resistance. In general, scratches are categorized as linear abrasions that fracture the clear coat and lead to significant light scattering. Mars are generally defined as shallower abrasions that are more difficult to observe and lack fractures. Mars mainly result from automatic car washes or from washing a vehicle with a sponge or cloth that contains dirt particles. Scratches result from more forceful insults such as abrasions from keys, shopping carts, or tree branches. Hard, tough films generally resist scratches and mars. Low surface friction also decreases a coating's susceptibility to scratch and mar damage.

Opportunities to tailor the molecular architecture of acrylics are limited by conventional free radical polymerization, which has led to the utilization of controlled radical polymerization (CRP) to produce automotive clear coats (see Section 8.2.1.4). In contrast to conventional radical polymerization, CRP permits preparation of relatively low MW block copolymers in which all molecules have at least two functional groups. The narrower MW distribution obtainable with CRP permits higher solids at the same number average MW. (See Section 2.2.1 for discussion of CRP.) This technology enables higher solids clear coats as well as more precisely defined functional acrylic polymers that can be utilized to form clear coat polymer networks with enhanced properties, such as toughness and appearance. Solids levels approaching 90% may be achievable in 2K clear coat formulations in the future, but cost may prevent their rapid commercialization.

Silica is added to clear coats for rheology control and to improve scratch resistance (Chattopadhyay et al., 2009). The very high surface areas of these nanoparticles increase the hardness and modulus of the coating, which in turn leads to more scratch resistance. However, the use of these nanoparticles is limited by their usually deleterious effects on the optical properties, particularly haziness (Anderson et al., 2003).

An important property requirement of clear coats is windshield or more generally fixed glass adhesion. Fixed (non-movable) glass includes the windshield, the backlight, small side windows on SUVs and vans, and potentially moonroofs. After painting, during final assembly, fixed glass is bonded into the vehicle's various openings. A bead of moisture-curable urethane adhesive is applied to one side of the glass. A robot then picks up the glass and places it into the opening. The adhesive bead is pressed onto the flange. Thus, the adhesive bead must stick to the glass on one side, which is often primed with a special coating to enhance adhesion, and the window flange on the other side, the topmost layer of which is the clear coat. A strong, durable bond must quickly form between the adhesive and the clear coat. Depending on the clear coat chemistry, different functionalities can be included in the clear coat resin to enhance windshield bonding. Retention of the windshield during a crash is a federally mandated safety requirement. Thus, a strong bond between the urethane adhesive and the clear coat is required. Most assembly plants perform weekly checks on their windshield bonding performance to ensure compliance with these requirements. Some manufacturers prefer to mask the windshield flanges after the electrocoat bake and then de-mask after the clear coat is applied. In this case the urethane adhesive will mate with an E-coated surface, which has traditionally been an easier surface to adhere to. Other manufacturers use a special primer over the clear coat to enhance windshield bonding. Both masking and priming are expensive operations, and many auto OEMs prefer to bond directly to the clear coat.

While nonthermally cured clear coats have a fundamental attractiveness to automotive OEMs (ovens consume energy and are a notorious source of dirt in paint), no technology is currently available that allows ambient curing of clear coats. Much research has been performed on UV-curable clear coats for automobile bodies (Nichols et al., 2001; Seubert and Nichols, 2004), as this technology offers the potential to dramatically increase both solids and scratch resistance. Both free radical and thiol–isocyanate resin systems have been explored (Seubert and Nichols, 2010). However, the curing of recessed, non-line-of-sight areas is challenging. Dual cure chemistries, which involve both a UV and thermal cure step, have been proposed, but they negate the biggest potential cost savings by using an oven for the second stage of curing. UV curing primers and base coats are also problematic, as it is more difficult to cure pigmented coatings.

Powder clear coats have been developed and were used briefly by a few European automotive OEMs. Powder clear coats emit zero VOCs, the overspray is potentially recyclable, and no wet sludge is produced. Acrylic powder coatings, based on epoxy-functional acrylics, which are made by copolymerizing GMA with conventional acrylates, have marginally acceptable properties. However, they have fallen out of favor, as the compromises outweigh the benefits. Thicker film builds, marginal scratch performance, inferior appearance, and the difficulties in application have led to their abandonment. No automakers are currently using powder clear coats; and the use of powder primer is decreasing. Powder slurries (aqueous dispersion of powder clear coat in water) were briefly used as well but have been discarded as impractical. The continued march toward more compact paint processes appears to preclude additional powder clear coat commercialization.

30.1.6 Factory Repair Procedures

Paint repairs occur at two different areas during vehicle manufacturing. Defects are sometimes found during the paint process, in which case the paint system is repaired before the vehicle leaves the paint shop. Tremendous efforts are exerted to keep the "first time OK" rates above 90% in automotive paint shops, as repairs are costly and consume time. Since a vehicle is painted before most the assembly operations occur, the paint may require final repair in the event that it is damaged during assembly, prior to the release of the vehicle for delivery to a customer. An excellent paper discusses the wide range of defects that occur during manufacture and use of automobiles (Schoff, 1999).

When cars were coated with acrylic lacquer, repair was relatively simple, since the thermoplastic systems remained soluble in the solvent in the repair lacquer. However, thermosetting enamels are more challenging to repair. Achieving adhesion to the surface of the cross-linked coating is more difficult. Minor blemishes in the clear coat can be corrected by light sanding and buffing; spot repair of small defects going deeper into the paint is sometimes possible. More extensive damage, however, may require refinishing of the entire panel to which the damage occurred or even repainting of the entire car body. In panel refinishing, the top coat is removed, any bare metal is primed, and special repair base coat and clear coats are applied to the whole panel. Since the coating cannot be baked at a high temperature, additional strong acid catalyst must be added to coatings that contain MF cross-linkers to allow curing at lower temperature. At least part of the excess catalyst remains in the film and can lead to more rapid hydrolysis. Durability of such repairs is good, but not as good as the original coating. Urethane 2K repair coatings are being used increasingly, since they cure at relatively low temperatures without sacrificing long-term durability. It is unlikely that corrosion resistance of repaired areas is equal to unrepaired areas, because the E-coat is often sanded through during the repair process.

30.2 APPLIANCE COATINGS

Major markets for OEM appliance coatings are for washing machines, dryers, refrigerators, air conditioners, and ranges. Markets for smaller appliances are also significant.

Historically, thermosetting automotive liquid monocoat technology was adapted to the varying requirements of different appliances. In some cases, single coats were found adequate for the less demanding appliances, such as refrigerators. Appliances that are often exposed to strong detergents or cleaning products ("OvenOff") required primer–top coat systems.

Nowadays, for applications that require corrosion protection, such as washing machines and air conditioners, cathodic E-coat primers (Section 27.2) are used on the highest quality products. Non-electrodeposition primers are frequently applied by flow coating (Section 23.3). To minimize VOCs, water-reducible epoxy ester-based primers are adequate for certain applications.

Cathodic epoxy E-coats are used for some applications as single coats. Although epoxy coatings chalk badly on exterior exposure, the drum of a dryer or the interior of an air conditioner does not get exposed outdoors, but a high degree of corrosion protection and impact resistance is needed. The uniform coverage of edges attainable with E-coat permits use of 12 μm E-coats to replace 50 μm solution epoxy coatings on air conditioners, with substantial reduction in manpower required for application, while maintaining the necessary performance (Miranda, 1988). White E-coats have been applied to appliances. Anodic acrylic E-coats perform well on aluminum, but for steel, as discussed in Section 27.1, discoloration results from iron salt formation. Cathodic E-coats based on acrylic resins (Section 27.2) avoid this problem on steel.

Thermosetting acrylic coatings are generally used over primers. For single-coat compositions, polyesters tend to exhibit better flexibility and adhesion to treated steel or aluminum than acrylics. The most commonly used cross-linkers are amino resins. For washers and dishwashers, benzoguanamine-formaldehyde resins (Section 11.4.2) impart greater resistance to alkaline detergents. For other applications, conventional MF resins provide lower cost. In end uses for which performance requirements are not severe, such as hot water heaters, lower cost semi-oxidizing alkyd/MF-based coatings may be most appropriate.

The use of powder coatings as top coats for appliance applications has been growing rapidly and is reported to be the largest single market for thermosetting powder coatings (Chapter 28). The long runs of single colors make powder coatings a natural choice for appliances. Low VOC emissions, low fire risk (with proper precautions in handling powders), low energy requirements, and reuse of overspray powder are strong economic and environmental reasons for using powders. A limitation is the greater difficulty of achieving good leveling with powder coatings as compared with liquid coatings. While a small degree of orange peel is desirable to conceal metal irregularities, excessive orange peel detracts from the appearance. Nevertheless, in some areas, particularly in Europe, where consumers have become accustomed to the orange peel typical of porcelain enamels, the corresponding finish from powder coatings is readily accepted. Commonly, hydroxy-functional polyester or acrylic resins are used as binders with a blocked isocyanate or tetramethoxymethylglycoluril (Section 11.4.3) as cross-linker. Alternatively, carboxylic acid-functional resins are used together with triglycidylisocyanurate (Section 13.1.2) or tetra(hydroxyethyl)adipamide (Section 17.5) as cross-linker.

An alternative approach to powder, which also reduces VOC emissions, is to apply coil coatings (Section 30.4) to the steel before the parts are fabricated. The solvent emissions occur at the coil coating factory where they can be burned to provide part of the fuel for curing the coatings. A potential problem with coil coated metal is that the edges are bare. The appliance must be designed so that cut edges are turned under and/or protected by a sealant. In some cases, it is possible to weld through the coating in areas that are not visible.

30.3 CONTAINER COATINGS

Container coatings were historically called *metal decorating coatings*, since a major portion of the business was in coating flat sheets, followed by lithographic printing and a finishing varnish (clear top coat) to protect the ink. There were many uses for such sheets, including metal boxes, trays, wastepaper baskets, bottle caps, crowns, and, most importantly, cans. Plastics have replaced coated metal in most of these applications other than bottle caps, crowns, and cans. The field is now usually referred to as container or can coatings. In the United States alone, about 125 billion cans are produced annually, the large majority being aluminum beverage cans. Global production of beverage cans is estimated at 320 billion $year^{-1}$ (Rexam Annual Report, 2015).

Because most cans are food or beverage containers, one of the key requirements is that there is no possibility of introducing toxic compounds into the foods or beverages. In the United States, all can linings must be acceptable to the Food and Drug Administration (FDA) and, in the case of meat products, the Department of Agriculture. Contrary to common belief, the FDA does not approve coatings, but it does list acceptable components. Rules are published in the Code of Federal Regulations (CFR), Title 21, Part 175; 21CFR175.300 deals with resinous and polymeric coatings. Updated versions are routinely published on the Internet. In most cases, if all of the components of a new coating have already been used in can coatings, a new coating will be acceptable. Less often, it is necessary to prove that no material is extracted into any food or beverage that will be packed in the can. However, new raw materials must pass extensive tests. Toxicity considerations predominantly affect interior can coatings, but often there are also restrictions for possible contamination from exterior can coatings. When metal sheets coated on both sides are stacked, the exterior coating on one sheet is in direct contact with the interior coating on the next sheet in the stack. In this configuration, migration of low MW components between coatings is possible. It is also possible that can ends coated with similar-appearing coatings might be put on upside down.

Historically, in the European Union (EU), food contact legislation for coatings was different from country to country. Food contact legislation applicable to plastics was applied more and more to can coatings, especially with respect to specific migration limits for raw materials. A 2011 directive, which has the force of law throughout the EU, now regulates food contact with plastics and coatings.

The effect of coatings on the flavor of the food or beverage packed in the container is also critical. While flavor requirements are particularly important for interior coatings, care must also be exercised with coatings for exterior application. Flavor changes can result from extraction of some contaminant from the coating by absorption of flavor agents from the packed food into the coating or by failure to isolate the food or beverage from the metal of the can. Flavor can be affected by minute amounts of substances. To assure that all residual solvent and other volatile flavor detractors are driven out of the coatings, high baking temperatures are used. The usual way to evaluate the effects of coatings on flavor is by making test packs of the food or beverage in the container and tasting the food or beverage. As a result, major suppliers of can coatings maintain flavor panels of personnel trained to taste and, particularly, to use consistent words to describe flavors.

For beer cans, a major aspect of the flavor problem is to prevent contact between the beer and the can because metals catalyze flavor changes in beer. For this reason, the final interior coating is spray-applied after formation of the can to avoid potential problems from breaks in the can lining resulting from stresses during can forming operations.

Linings are spray-applied to the interior of soft drink cans not only to protect flavor but also to protect the can; the acid present in most soft drinks could eat through the metal without a coating barrier. An interesting sidelight on history is that in the early days of packing pineapple products and grapefruit juice in cans, coatings for lining these cans that would resist their high acidity were not available. They were packed in cans with heavy tin weight linings to protect the steel bodies. Tin affects the flavor of the pineapple products and grapefruit juice and acts as a bleaching agent, allowing light color fruits and juices to retain their light color. Even though organic coatings that could be used with these products are now available, most are still packed in heavily tinned cans. Apparently, consumers are used to and prefer canned pineapple and grapefruit products that have a tinny flavor.

There are two major classes of cans: three-piece and two-piece cans. In a three-piece can, one piece is the body and the other two are the ends. Metal sheets are coated or continuous strips of metal are coil coated. Blanks for the can bodies are stamped out of the coated metal. The body is formed, usually into a cylinder, and sealed by soldering, by welding, or with an organic adhesive. Soldering is now restricted to nonfood cans because of concern about lead toxicity. The side seam is sprayed with a fast drying coating called a *side striper* to cover exposed metal resulting from soldering or welding. Solventborne side stripers predominate; some waterborne stripers are used, as are powder coatings for containers for aggressive products. The coatings are applied while the metal is still hot, assisting in curing of the coating. The ends are made separately; they are stamped out of coated sheets or coil coated metal, formed, and equipped with a formed-in-place rubber gasket. One end is put on by the can maker; the other end is put on after the can has been filled. The growth of cans that are scored for easy opening has placed extra demands on the properties of coatings for can ends. The scoring process requires coatings with very high levels of adhesion and resistance to physical damage.

Coatings on metal sheets are generally applied by direct roll coating (Section 23.4). In some cases, transfer rollers with sections cut out are used so that the coating is only applied in selected areas. For example, the edges that will be soldered or welded at the sides of the body of the can are commonly not coated. On emerging from the coater, the sheets are tipped onto upright wickets attached to a conveyor. This permits the sheets to go through the baking oven in an almost vertical position, reducing the necessary length of the oven. Baking schedules vary with the application. Interior food can coatings may be cured in less than a minute at 200–210°C, while exterior white and varnish coatings may receive 10 min at 150°C. Inks are applied by offset lithography in which the ink is transferred from a lithographic plate to a rubber blanket and is offset to the sheet.

There are two processes for making two-piece cans. The most common method is called drawn and wall ironed (DWI or D&I). These cans are formed by drawing a cup from a flat blank and then ironing the walls to thinner thickness and greater depth. In the second method, called draw–redraw (DRD) cans, a coated blank is formed into a shallow cup and then drawn one or two more times to achieve the desired height and configuration and shape of the bottom of the can. The other piece of a two-piece can is an end similar to the ends of the three-piece can. The majority of beverage cans are two-piece cans.

Examples of DRD cans are shallow cans for tuna fish and taller cans for vegetables and pet foods. Flat sheets are coated and then the can is drawn and formed. In some cases, such as shoe polish and auto wax cans, printed sheets are formed into two-piece cans. The design must be distortion printed; that is, the print must be designed so that it looks correct *after* the distortion that results from forming. In general terms, it costs less to coat and print flat sheets before forming than to coat and print a formed can. The ability of the coating to withstand this degree of formation depends on the depth of draw, which is not just a matter of how deep the can is but also how wide. Forming a large diameter can involves less distortion than a narrow can; a deep can involves more distortion than a shallow can. Bottle caps and crowns are made from coated and distortion printed sheets and then are punched out and formed.

Beverage cans and some food cans are DWI cans in which uncoated metal is drawn and formed before coatings and inks are applied. Then exterior coating is applied by rotating the can against the transfer roll of a small coater, and inks are transferred from a litho plate to a soft rubber roller that in turn transfers the inks to the can surface. Baking is done in short time cycles at high temperature. Some lines produce over 2000 cans min. It is estimated that the coating reaches a peak temperature of about 205°C for only about 1 s, resulting in only partial cross-linking. The cross-linking is completed when the can is baked again after the interior lining is spray-applied. In this process a small spray gun is automatically inserted into a spinning can; the gun sprays and then is pulled out of the can. To assure removal of essentially all solvent, the final part of the curing is done with air directed into the can bodies. Typically, the cure schedule is about 2 min at 200°C. Final forming of the top of the can to fit the end is done after the coating and printing are completed.

There are three major types of metal used in cans: tin plate, steel, and aluminum. The choice of metal depends on the end use. Beverage cans are usually two-piece aluminum cans, although steel beverage cans can be found in parts of Asia and Africa. Pressure from the carbonation of either soft drinks or beer keeps the thin-walled aluminum cans sufficiently rigid. Most food cans are three-piece cans made from tin-plated steel. In the case of food cans, which do not have significant interior pressure, aluminum does not compete with steel because thicker walls of aluminum would be required to achieve adequate rigidity. However, many shallow fish cans are being made with aluminum. There has been some discussion of using aluminum for food packing by putting a small piece of dry ice into the pack before sealing. Vaporization of the CO_2 provides the pressure necessary for rigidity. However, slightly bulging food cans are associated with spoilage, so consumer acceptance has been slow. A large fraction of pet food cans are made from treated steel—so-called black plate—since it is less expensive and the highly reflective tin coating is not needed. Regulations in the United States require pet food to be packaged in cans suitable for human foods.

30.3.1 Interior Can Linings

The composition of interior can coatings depends on the food or beverage to be packed in the can. In most cases, if the food is to be cooked, it is cooked in the can. A common cooking cycle is 60 min at 121°C. Most beer is pasteurized in the can at a lower temperature. In both cases, the interior and exterior coatings must maintain their adhesion and integrity throughout the cooking or pasteurizing process. Beiro et al. (2003) have reported studies of tests for dry and wet adhesion of several coatings on various tin plate substrates. It was found that electrochemical impedance spectroscopy (EIS) was the most reliable test procedure. Organosol and epoxy–phenolic coatings provided the best dry and wet adhesion, both before and after sterilization.

Most vegetables and fruits are packed in cans with an interior coating called an *R enamel*. Historically, R enamel was a phenolic varnish. Now, it is more common for R enamel to be a phenolic resin, modified with drying oil derivatives during its synthesis, or an epoxy–phenolic coating—a resole phenolic (Section 13.6.1), with a BPA epoxy and phosphoric acid catalyst. In order to achieve adequate flexibility, high MW epoxy 1007- or 1009-type resins (Section 13.1.1) were formerly used. However, BPA has become a serious concern and is being replaced, as will be discussed in the succeeding text. For packing vegetables that give off hydrogen sulfide during cooking, such as corn, fine particle size ZnO pigment is dispersed in the coating, which is called a *C enamel*. The ZnO reacts with H_2S to form white ZnS. This prevents or conceals formation of unsightly black tin sulfide by reaction of tin oxides with H_2S.

Cans for fish and other oily foods are generally lined with a resole phenolic resin or an epoxy–phenolic (Section 13.6.1). To obtain sufficient formability to permit drawing of two-piece fish cans, cross-link density is reduced by using a mixture of *p*-cresol and phenol in making the resole phenolic resin. Poly(vinyl butyral) is commonly incorporated in the formulation to promote adhesion and act as a plasticizer. The extent of cross-linking is adjusted for flexibility together with resistance to swelling and softening, during processing and storage, by oils from the fish and, in some cases, additional oil in which the fish is packed. Thin films help to accommodate forming without film rupture. Aluminum flake pigment is sometimes incorporated in the coating to minimize permeability. In linings for meats, such as ham, one of the key requirements is that the coating allows the meat to slide out of the can easily after the top is removed. This is accomplished by incorporating a release agent, such as petroleum wax, in the can lining.

An increasing fraction of food cans are DRD or DWI two-piece cans. The DRD process requires that coatings have a greater degree of ductility than the coatings used for interiors of three-piece food cans. Most commonly, vinyl organosol (Section 17.1.2) coatings lightly cross-linked with phenolic resins or MF resins have been used, since the vinyl chloride copolymer binder shows ductility below T_g (Palackdharry, 1991) (Section 4.2). In some European countries, there is concern about possible toxic reaction products from HCl scavengers and toxic emissions from factories recycling vinyl chloride copolymer-coated cans. This has led to work on other more distensible coatings. For example, epoxy resins have been designed that will undergo considerably deeper draws than the conventional BPA epoxy coatings (Dubois and Sheih, 1992).

Large volumes of coatings are used for lining beverage cans. Historically, most spray-applied interior coatings were solventborne vinyl chloride copolymers or high MW epoxy/amino resin/phenolic coatings. The volume solids of these coatings at application viscosity were very low, 12–15% NVV. The high cost of solvent and the high level of VOC emissions forced a change. Waterborne coatings based on acrylic/epoxy graft copolymers (Woo et al., 1982) now predominate in lining both beer and soft drink cans. Styrene/ethyl acrylate/acrylic acid side chains are grafted onto a BPA epoxy resin. The resin is "solubilized" with dimethylaminoethanol (DMAE) in glycol ether solvent and reduced with water (Section 13.4). To lower the cost of some soft drink can linings, it has been possible to use a special latex binder blended with small amounts of the graft copolymer and epoxy phosphate as adhesion promoters.

Concern has been raised about the use of BPA epoxy resins in interior coatings, since BPA is an estrogen mimic. Procedures for accurate analysis of BPA extractables have been published (Wingender et al., 1998). Studies of extractables from can linings have shown that BPA is present in foods from cans, the amount of which varies by food type (Noonan et al., 2011). The current assessment of the US FDA is that "BPA is safe at the current levels occurring in foods." However, some consumers are strongly opposed to BPA in any food-contacting products. BPA has been completely dislodged from infant formula packaging and reusable water bottles.

In Europe BPA in food contact materials is limited to $0.6\,mg\,kg^{-1}$ of food. In 2015, France unilaterally banned BPA can liners in food contact uses. California requires warning labels. Thus, the future of BPA in can liners is in doubt. Supplier companies have extensive programs to replace BPA-containing epoxies. As of 2013, performance and/or cost was compromised with current formulations compared with epoxy coatings (LaKind, 2013). In 2016, can liners based on acrylics (Chapters 8 and 9) and on polyesters (Section 10.1.1) were commercially available and in production in Europe. However, according to Gander (2016), the non-BPA alternatives raise costs and reduce performance, potentially forcing food canners to reduce claimed shelf life from 3 years to 1 year. The French BPA ban is arguably counterproductive and revisions are being considered.

The patent literature describes numerous approaches to making binders for BPA-free can liners (Cooke et al., 2012; Moussa and Knotts, 2015).

30.3.2 Exterior Can Coatings

Many food cans are not coated on the outside; rather, paper labels are used. The appearance of coated and printed cans, however, is more attractive than paper labels. Some food and all beverage cans are coated and/or printed. Unlike paper labels, printed metal cans are not affected by water, such as from ice chests, or condensation when cans are brought out of a refrigerator into humid air.

The general procedure used on the exterior of sheets for the bodies of three-piece cans is to apply a base coat, often called an *enamel*; print the base coat with as many as four colors by offset lithography; and finally top coat with a finishing varnish. The most common color for base coats is white, but a variety of colors is used to differentiate products. Color stability and lack of color change on baking and can processing (sterilization) are critical requirements. To minimize VOC emissions, waterborne acrylic coatings cross-linked with MF resins are widely used vehicles.

The lithographic printing inks are either baking or UV cure inks. The vehicles for the baking inks are long oil alkyd resins with some MF resin cross-linker, while the vehicles for the UV cure inks are blends of acrylated epoxidized soy (or linseed) oil with acrylated epoxy resin and acrylate reactive diluents (Section 29.2.5).

The finishing varnishes are also acrylic/MF or polyester/MF binder compositions. To decrease friction in forming machines and conveyors, the finishing varnish commonly contains a small amount of petroleum wax or fluorinated surfactant to reduce surface tension so that the cans will have a low coefficient of friction. Two inks are applied, the sheets are baked, the other two inks are applied, the finishing varnish is applied, and the sheets are baked again. This process is called *wet-ink varnishing*, since the varnish is coated on the printed surface before the last two inks are cured. With UV cure inks, each ink is partially cured by passing under a UV lamp before the next ink is applied. Commonly, finishing varnishes are unnecessary over UV cure inks.

For 2-piece cans, white waterborne acrylic base coats are used. The film thickness varies from 8 to 15 μm; even at 15 μm, hiding is less than 100%, but the coating is definitely white. An increasing proportion of two-piece cans have "base coats" that are simply printed with inks 2–3 μm in thickness. Some consider that the appearance of such cans is inferior, but they cost less than cans that are coated with a base coat before printing. A transparent yellow-brown base coat is used on some cans to give a bronze or gold color to the relatively shiny metal surface.

UV cure coatings are also used on some beverage cans owing to the high speed of cure (Chapter 29). Such exterior finishing varnishes are acrylated resin coatings cross-linked by photogeneration of free radicals. Coatings for the exterior side of can ends are epoxy resins cationically cross-linked by photogeneration of acid. The UV cure end coating is applied to the side of the sheets that will be on the outside of the cans and partially cured under the UV lamps; the other side, which will be on the inside of the can is coated with an FDA-listed epoxy–phenolic thermal cure coating. When the epoxy–phenolic is cured in an oven, cure of the epoxy coating is advanced to completion thermally by the photogenerated acid still present in the films, which also enhances adhesion of the coating.

30.4 COIL COATING

Steel and aluminum are manufactured in long continuous strips that are rolled into coils. In many cases, the coils are cut, formed, and fabricated before they are coated, as in production of automobiles. In other cases, it is possible to coat metal coils and later fabricate the final product from the precoated metal. When precoated metal is feasible, there can be substantial advantages (Section 30.4.1). Coil coating has grown into a major industrial coating process (Gaske, 1987). Sales of coil coatings were approximately $900 M in the United States in 2014 (American Coatings Association and Chemquest Group, 2015).

Coil coating was initiated in 1935 as a process for coating venetian blind slats. The strips of metal were about 5 cm wide, and the line was run at a rate of about 10 m min^{-1}. Modern coil lines can coat metal up to 1.8 m wide at rates as high as 275 m min^{-1}. Most lines run at a rate of about 100–200 m min^{-1}. A schematic drawing of a coil line is shown in Figure 30.2.

The metal is shipped from the steel or aluminum mill in coils weighing up to 25 000 kg that are 0.6–1.8 m wide and 600–1800 m long. In some lines, as shown in Figure 30.2, the first step is pre-cleaning; brushes remove any physical contaminants from the strip, which continues to the entry accumulators. The rollers of the accumulators move apart to accumulate a significant length of coil so that when a coil is about to run out, the next coil can be stitched (welded) on while the accumulator rolls move together, supplying strip to the line without interruption of the process. After the stitching is done, the new coil is fed to the coating line. As the process continues, the accumulator rolls gradually separate to store strip for the next change of coils.

Next, the strip is carried through the metal treatment area. Detergent washing and rinsing are followed by application of various conversion coatings including phosphates, complex oxides of Ti or Zr, or silanes and finally a water rinse. All the cleaning and conversion coating procedures must be designed to work at high speeds. Since the metal is moving at 100–200 m min^{-1} or more, the total time for cleaning and treatment is on the order of a minute or less. Next, the strip is carried through a dry-off oven and finally to coaters and a baking oven. Often, a primer and a top coat are applied on one side of the strip and another coat on the back side. It is fairly common to coat, cure, and then print one or more colors. In Figure 30.2, a laminator is also shown, although laminating of film in line with coating is not common. More often, there are two coating stations, each followed by a baking oven.

Most coatings are applied by reverse roll coating, but direct roll coating may be used for thinner films (Section 23.4). Reverse roll coating can apply thicker films when close control of thickness is critical and when the flow advantages that result from the wiping action are important, in contrast to the film splitting action in direct roll coating. To achieve uniform film thickness across a strip 2 m wide, the rolls have to be crowned, that is, designed so that their diameter is greater in the middle than on the edges, since the pressure tends to bow the rolls to a small degree. To avoid damage to the rollers, the line is programmed so that very shortly before a stitched section

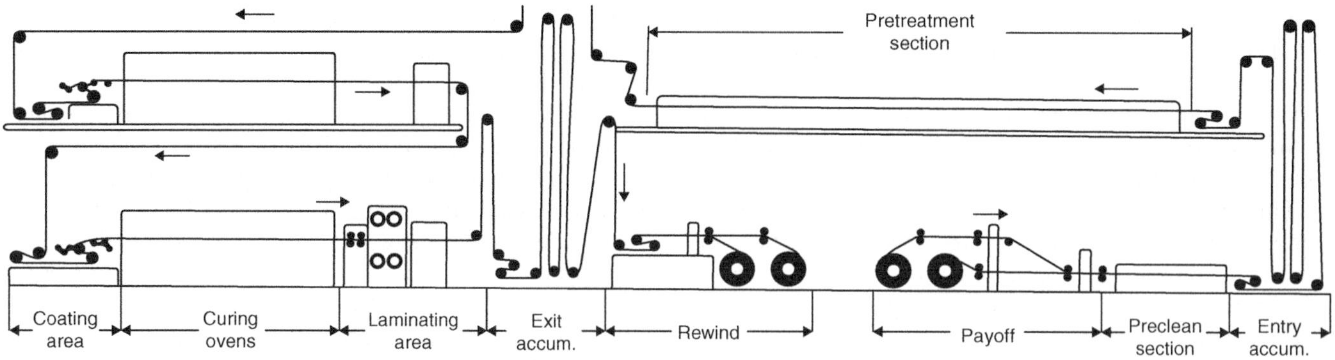

Pre Finish Metals 72-in. coil coating line will begin at "payoff", at right, and will follow directions indicated by arrows, finishing at "rewind".

Line #6 specifications

Strip widths 24.0″ to 72.0″
Strip thickness 0.018″ min. to 0.135″ max.
Speeds up to 700 f.p.m.
Maximum coil size 50 000 #
Coil I.D. 20″ or 24″
Coil O.D. up to 78″
Capable of laminating top and bottom side

Capable of applying primer and finish coat to both sides in one pass
Capable of paint finish embossing 0.007″ to 0.016″
Capable of gravure and flexographic printing
Shipping and receiving facilities capable of handling more than 35 railroad cars and 90 trucks per day
Housed in a 266 000 sq. ft. building on a 52 acre site

Figure 30.2 Coil coating line. Source: Gaske (1987). Reproduced with permission of The Federation of Societies of Coatings Technology.

joining two coils passes through the coating nip between the rollers, the rollers are automatically separated slightly and then almost immediately returned to normal operating pressure.

At the high line speeds, dwell time in the ovens is less than a minute, even with long ovens. In some cases, the dwell time is as low as 10 s but more commonly is in the range of 12–45 s. After a short initial period, hot air is directed over the surface of the coatings at high velocity. The air temperature can be as high as 400°C. The temperature reached by the coating on the metal is the critical temperature for curing the coating. This temperature cannot be directly measured, but is closely related to the metal temperature, which can be measured. The temperature considered most important is the *peak metal temperature* (PMT), which can be as high as 270°C. After the coating is baked, the strip passes through the exit accumulator to the rewind. The exit accumulator stores coated strip during removal of a coated coil. In some lines, the strip passes over chilling rolls or through water to reduce the temperature before the strip is rewound into a coil. The pressure in the center of the rewound roll is very high; consequently, T_g of the coating on the metal must be very high to avoid blocking. For most applications, however, the cross-link density must not be so high that the film elasticity is reduced below levels needed to allow subsequent fabrication of the steel into its final form.

Near-infrared curing of coil coatings has been initiated on at least one coater. Cure is reported to be complete in 1–4 s compared with the 12–45 s on a conventional line.

The exhaust air from the hoods over the coaters and particularly from the oven contains solvents. On most lines, the exhaust air streams are used as part of the air used to burn the gas to heat the ovens or are fed through an oxidizer. In this way, part of the residual heat from the oven exhaust is recycled and the solvent is burned. Burning the solvent essentially eliminates VOC emissions, and the fuel value of the solvent is recovered. As a result, there has been less incentive to change to waterborne or high solids coatings in coil coating applications than in other applications. Interest in reducing solvent content remains, because the fuel value of solvents is low compared with the cost of the solvents and because some lines are not equipped to burn solvent. Regulations controlling emission of hazardous air pollutants (HAPs) that went into effect in 2006 require that stack emissions not exceed 0.38 pounds of HAP per solid gallon of coating applied.

Coatings on aluminum are frequently single coats; but primer–top coat systems are more widely used on steel. Binders for primers were traditionally based on BPA epoxy resins; epoxy esters and epoxy/MF resins are examples. However, the share of PU, polyester, and waterborne latex primers is growing.

Many types of coating binders are used for top coats. Oxidizing alkyds with MF resin have the lowest cost and are sometimes used on the reverse side of the coated strip as a *backer*; polyesters are supplanting alkyds in this application. Backer coatings may be pigmented or unpigmented; and they contain a small amount of incompatible wax. The purpose of the backer is to avoid metal marking of the top surface coating by rubbing against a bare metal reverse side of the coil. Alkyd/MF coatings are also used as top surface coatings for which corrosion resistance requirements and/or exterior durability requirements are modest.

The largest volume top coats employ polyester/MF binders, especially as single coats; exterior durability and corrosion protection are generally superior to that obtained with alkyd coatings. Over the years, incremental improvements in design of the polyester resins and in formulations have led to polyesters with excellent outdoor durability (Chapter 10). Polyester/blocked isocyanate coatings have been used to a degree in applications for which abrasion resistance and flexibility are particularly important. Close temperature control in the ovens is important with urethane coatings, since urethanes may discolor and decompose relatively rapidly at the baking temperatures involved in coil coating. Thermosetting acrylic/MF coatings are used, usually over primers.

For greater exterior durability, one can use silicone-modified polyesters and silicone-modified acrylic resins (Section 16.1.2). For example, one might use 30% silicone-modified polyester resin, with an MF resin for cross-linking, as the binder for color top coats for high performance residential or industrial siding. In the same quality line, white top coats might well be a polyester/MF coating. After many years of outdoor exposure, the white coating might start to chalk slightly, but this would not adversely affect the appearance. On the other hand, even a small amount of chalking of a color coating results in an easily seen color change owing to the change in surface reflection. Such changes are particularly serious in exterior siding, since the exposure varies depending on the location of the siding on the building. The resulting nonuniform chalking of the color coating is very evident. Durability of recent *superdurable* polyesters (Chapter 10) approaches or equals that of silicone-modified polyesters (Pilcher, private communication, 2014).

Polyvinylidene fluoride (PVDF) resin coatings (Section 17.1.4) provide exceptional durability. In some cases, such coatings show only slight indication of change after exposure outdoors for more than 25 years (Hayoz et al., 2003). PVDF-coated metal is used for architectural ornamentation, such as metal roofs or exterior trim pieces on buildings, where the extremely long lifetime of the coating makes the added expense economically viable.

Organosol and plastisol coatings are used for some coil coating. The relatively low viscosity organosols are applied at about 25 μm; and the higher viscosity plastisols are used for film thicknesses of 100 μm or more. As discussed in Section 17.1.2, the vehicle in these coatings is a dispersion of vinyl chloride copolymer in plasticizer and solvent. Such coatings provide reasonable exterior durability with excellent fabrication properties. They can be cross-linked, but this is often unnecessary. The uncross-linked versions may require as little as 15 s dwell time in the oven. Solution vinyl resins are used in coil coating metal for can ends of beverage containers.

Increasing use is being made of latex vehicles for coil coating. They have the advantage of high MW, enabling good mechanical properties to be achieved without need for much, if any, cross-linking. High gloss coatings cannot be made and leveling may be more difficult than with solventborne coatings. Leveling problems are minimized by using reverse roll coating and by using associative thickeners to control viscosity. Associative thickeners minimize latex particle flocculation relative to conventional water-soluble polymeric thickeners, such as hydroxyethylcellulose (Sections 24.2 and 32.3).

An extensive study of the corrosion resistance of 12 coil coated sheets at three South American seashore locations has been reported (Rosales et al., 2004). Location 1 was marine-polar with low temperature, high humidity, very low SO_2 and Cl^- exposure, and little sun. Location 2 was marine with moderate temperature, high humidity, moderate SO_2 and Cl^- exposure, and high sun exposure. Location 3 was marine-desert with high temperature and humidity, very high SO_2 and Cl^- exposure, and high sun exposure. At location 1, the best performance was obtained with an alkyd/MF coating on galvanized steel. At location 2, the best performer at incision marks was a polyester primer with chromium pigment on Galvalume. On the unruptured surface, the best performance was obtained with three samples: (1) an epoxy/acrylic primer with a silicone-modified polyester top coat on galvanized metal, (2) a polyester primer with a silicone-modified polyester top coat on Galvalume, and (3) a polyester primer and polyester top coat on Galvalume. At location 3, the best performer was a polyester primer and polyester top coat on Galvalume. At all three locations, the poorest performer was an epoxy primer with a polyester top coat. No details of the binders in the various coatings were given.

UV cure coatings are potentially of great interest in coil coating because of rapid cure and limited emissions. Active research is underway but, as yet, actual production is limited. Cationic UV cure epoxy coil coatings on tin plate are used for the exterior side of can ends. The superior adhesion of epoxy coatings makes them preferable for this application, as compared with acrylate coatings. In some cases, a UV cure coating is applied on one side of the strip and UV cured. Then, a thermally cured epoxy coating is applied on the other side. When the strip is passed through the oven to cure the thermally cured coating, thermal cross-linking of the UV cure coating is also advanced owing to the presence of photogenerated acid.

Powder coatings can also be applied to coil stock. One approach, in limited commercial use, is to electrostatically spray the strip with automatic spray guns. A disadvantage of this process is that line speeds are slow. Another process being introduced is to run the coil strip through a "cloud" of electrically charged powder particles; the strip then passes into an induction heating oven for fusing and curing. Since there is no contact of rollers with the coil, embossed or perforated metal can be coated. It is projected that line speeds can be higher than for conventional coil coating.

30.4.1 Advantages and Limitations of Coil Coating

Important advantages of coil coating have promoted the growth of the business to a major component of industrial coatings. For long production runs, the cost is low compared with coating preformed metal. The rate of application of coating is faster, and labor cost is lower. Coating utilization is essentially 100%. Oven designs are such that energy usage in curing is much more efficient. In general, less floor space is required than for large volume spray coating so that capital cost for buildings is lower. Since solvent is restricted to the immediate area of a roll coater, fire hazards and toxicity hazards are reduced as compared with spray application. Since, in most cases, solvent is incinerated, VOC and HAP emissions are generally very low. Film thickness of the applied coatings is more uniform than is generally obtained in coating preformed products. Since the coatings are applied on uniform thickness metal, curing of all parts of the coating tends to be more uniform than in curing coatings on fabricated products. For many applications, the performance of coil coatings is superior. This superiority is particularly evident when comparing high quality precoated exterior siding to house paint. A large part of this difference results from the superior performance of baked coatings as compared with air dry coatings. The manufacturer using precoated metal also gains substantial advantages. The VOC and HAP emissions and fire hazards associated with coatings application are eliminated, and insurance costs drop. There is no waste disposal problem with sludge from spray booths. There may be a substantial saving in floor space.

There are limitations to coil coating, however. Capital cost of a modern coil coating facility is very high; therefore, lost production time in shutdowns is expensive. It is economical only for fairly long runs of the same color and quality of coating. The cost of changing color is high because the coater must be shut down for cleanup. However, many modern lines have multiple coating heads, which substantially reduce downtime owing to color changes. In a line of coated metal with several colors, the inventory cost can be high, since several colors must be inventoried. If a stylist changes colors, the obsolete inventory cost can be high or, said another way, the flexibility of changing colors is more limited than when the assembled product is coated.

Color matching is a challenge to a coil coating supplier. Generally, very close color matches are needed. Hiding is less than complete, and the color of the metal or primer affects the color of the coating. Color can also be affected by the high temperature baking schedule. It is not possible to model a curing schedule of 30 s with high velocity, 400°C air in the laboratory, so that a color matcher must learn to relate color changes that occur in the laboratory to changes that can be expected on a particular coil coating line prior to making the color match. Limited colors are available as organic pigments cannot withstand even the brief PMTs associated with coil coatings. The use of effect pigments has increased in recent years owing to the availability of better rheology control additives (Testa, 2008).

The coated metal must be able to withstand fabrication into the final product without film rupture. This may require acceptance of somewhat softer, more flexible films than would be specified if the product were coated after fabrication. The ductility of the coating is often assessed by the T-bend test, where the coated metal is bent back upon itself (see Section 4.6.3.2). When the coated metal is die-cut to make the eventual product, bare edges of metal are exposed. These cut edges are weak points from a corrosion standpoint. This problem may be reduced with galvanized metal because zinc may provide some level of protection to cut edges when it is smeared across those edges during the cutting process. Welding of coated metal can be a problem.

Both zinc dust and iron phosphide are used as conductive fillers to enable welding through coal-applied coatings. However, welding performance is reduced and additional weld tip dressing may be necessary compared with welding uncoated metal. In some applications, bare edges may not be a corrosion problem, but rather an aesthetic problem, in which case the component must be designed to hide the bare edges.

Examples of large applications for coil coated metal are siding for residential use, original siding for mobile homes, venetian blinds, rain gutters and downspouts, fluorescent light reflectors, appliance cabinets, can ends, and can bodies for fruits and vegetables. Newer applications include antimicrobial coil coatings for making stock for use in air conditioning ductwork, metal kitchen counters, refrigerator boxes, and walk-in coolers. The antimicrobial agent used is a nanoparticle based on silver ions. Legislation is increasingly requiring many roofing materials to have high reflectance and high emittance. White- and light-colored coil coated metal has high reflectance, thereby reducing heat absorption. Furthermore, infrared reflective pigments (Section 20.2) enable formulation of "cool building" coil coatings in a broad range of colors. However, some radiation is absorbed, resulting in still heating. This heat is lost by thermal emittance into the atmosphere and into the building (Cocuzzi and Pilcher, 2004). Some chassis parts on automobiles are "pre-primed." However, the aggressive corrosion environment has prevented coil coating from being used extensively in the automotive market.

An important drawback of coil coating is the cost of offal or scrap. Almost all metal fabrication processes produce some scrap. However, coil metal scrap is more valuable, because it has already been coated. Thus, the manufacturer must discard or recycle more valuable scrap, which may turn the economics against coil coating in some cases.

In North America, few coil coating lines are fully integrated with the metal coil production lines. Thus, coils must be transported to dedicated coil coaters or *toll coaters*, where the coils are unwound, coated, rewound, and sent to the fabricator. The additional cost of transporting the coil twice is significant. In Europe and Asia more metal rolling lines have integrated the coil coating process at the metal mill, which significantly reduces the cost of coil coated products.

30.5 COATINGS FOR AIRCRAFT

The bodies of aircraft are predominately made from aluminum alloys currently, but an increasing fraction are being made from high strength carbon fiber composites owing to the composites' superior strength and stiffness to weight ratio. Military aircraft are farther along the path of composites than commercial aircraft. However, because aircraft are designed to have very long lives (40+ years), aluminum aircraft will continue to be the predominant airplanes in the sky for many years to come.

As the service environment for aircraft is unique, the coatings that protect the aircraft body are also unique. The market for aircraft coatings is relatively small, but travel in the modern world relies on the corrosion protection provided by these coatings. An aircraft can quickly travel from a very hot tarmac in the Middle East, where temperatures can exceed 50°C and the skin of the aircraft can be above 100°C, to 10000 m in the sky, where the air temperature is −70°C. Thus, the thermal stresses can be quite large and can be induced quickly. The coatings must be able to withstand these stresses repeatedly. The service environment for aircraft coatings also requires resistance to many different fluids, including aggressive fluids like hydraulic fluid ("Skydrol"), aviation fuel, and deicing fluids. Resistance to these fluids is one of the more difficult formulation tasks the aircraft coating formulator faces.

Traditionally, exterior commercial aircraft coatings have been multilayer systems that encompass a conversion coating, a corrosion-inhibiting primer, and a color coat (Figure 30.3). Base coat–clear coat systems are slowly being introduced by the aircraft manufacturers and commercial

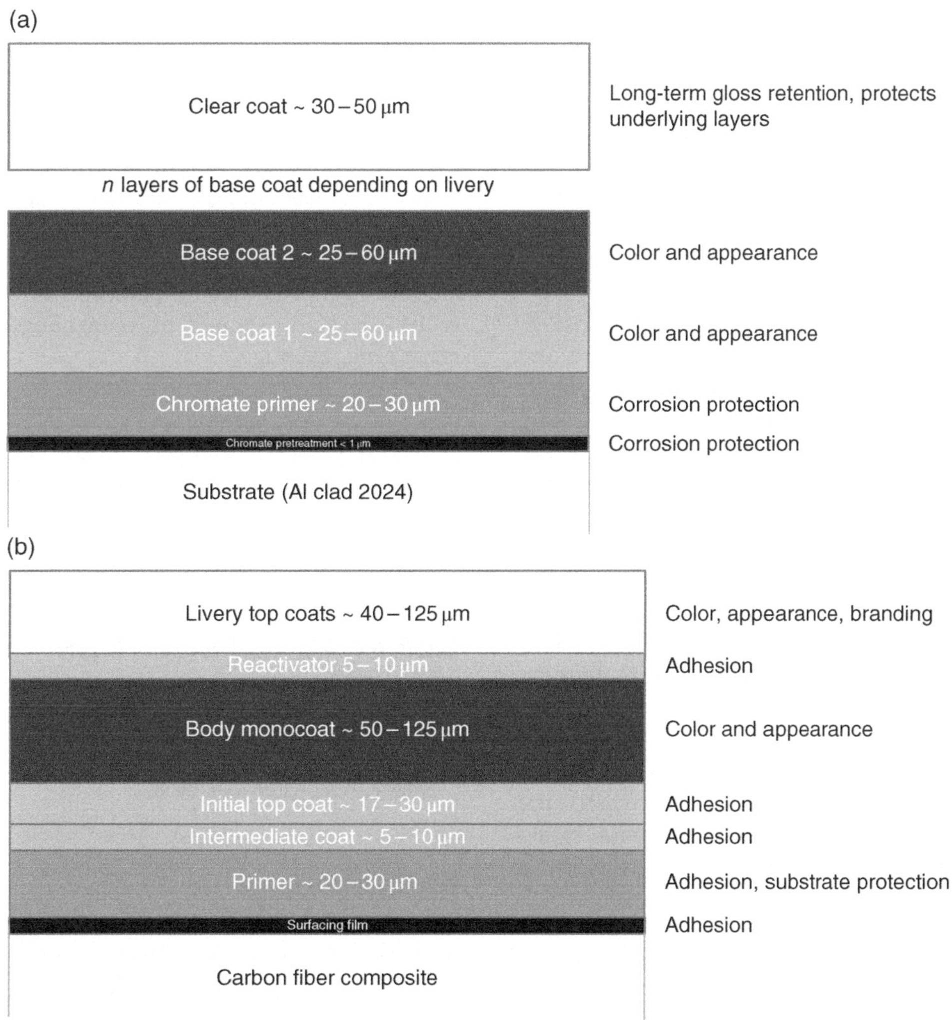

Figure 30.3 Cross sections of coating systems on a commercial (a) aluminum-bodied aircraft and (b) composite body. Note that for the composite body, the final body and livery top coats are applied by the airframe manufacturer, while the lower layers are applied by lower tier suppliers.

fleets, which is being motivated by both appearance and durability advantages. Whether the top coat is a monocoat or a base coat–clear coat system, the top coats are almost universally 2K aliphatic PU. The primers are typically epoxy based and contain corrosion-inhibiting pigments to retard corrosion of the underlying aluminum alloys. The cure temperatures for all the layers are quite low, as it is impractical to heat an airplane paint hanger to more than a few tens of degrees Celsius above ambient.

The conversion coatings used on aircraft are still predominately based on hexavalent chromium. The aircraft industry is one of the very few industries that is allowed to continue to use hexavalent chromium in their paint or conversion coatings, as few suitable replacements have been demonstrated to be as effective. This is primarily motivated by the long service life of the object and the inability to inspect some areas once the aircraft is manufactured. For example, the interior or wings or other structural components are sealed. Unlike the exterior skin, these surfaces are not accessible for inspection, but still have to last 40+ years.

Owing to the poor correlation of accelerated corrosion testing with actual performance, it is difficult to predict 40 years of performance. The aircraft industry still relies on B117 salt spray testing, mainly because of government specifications for military aircraft and the reluctance to change a specification for which much historical data exists. The lack of a good correlation between lab tests and long-term performance makes the replacement of chrome-free coatings a high risk proposition for aircraft manufacturers.

After the conversion coating is applied, a corrosion-inhibiting epoxy primer is applied to the interior and exterior sections of the aircraft. Again, the use of chrome-containing pigments is still the norm, due to the long lifetimes of the aircraft. The use conditions to which these coatings are exposed can be quite harsh. For example, the interior of wings are continuously exposed to jet fuel. These interior cavities are coated with epoxy primers that contain hexavalent chrome pigments to ensure corrosion performance for the life of the aircraft.

A large research effort has been initiated to replace hexavalent chrome in both the primer and pretreatment layers. Some alternative pretreatments have been commercialized and are being used on exterior parts or fasteners. These pretreatments are based on sol–gel chemistry and the use of Zr and Si precursors to form a mixed zirconia/silica thin film on the metal substrate (Blohowiak et al., 1999; Liu et al., 2006). Such pretreatments also form a strong bond with the subsequently applied primer, which enhances the corrosion performance of the substrate. Other oxides based on Ti, Zr, La, or Ce have been developed and tested with mixed results.

New corrosion-inhibiting pigments have also been developed. Only two technologies appear promising at this time. Mg-rich primers have been shown to sacrificially protect the underlying aluminum. As Mg is more electrochemically active than Al, the magnesium converts to MgO under appropriate conditions, thus protecting the aluminum (Battocchi et al., 2006). This technology has been commercialized for aircraft coatings. An even more promising technology is based on lithium salts as corrosion-inhibiting pigments. Both lithium carbonate and lithium oxalate have been shown to significantly decrease the rate of corrosion on a variety of aluminum alloys (Liu et al., 2015; Visser et al., 2015). The inhibition mechanism is thought to involve solubilization of the lithium salt, followed by formation of an aluminum hydroxide gel at the location of corrosive attach. The gel is then intercalated with Li ions to form an amorphous layer that resists further corrosion.

One of the main challenges in painting a commercial aircraft is the complex color scheme or *livery* that each airline uses to brand their aircraft. Application of each color requires extensive masking and de-masking. Some airlines use particularly complex liveries that require many additional days to apply, as each application and curing step requires many hours in addition to the masking process. Because of this complexity, the use of decals and alternative curing methods, such as UV cure, has been investigated.

As aircraft are routinely stripped and repainted to inspect for structural and corrosion damage, the coatings must be strippable. The use of organic solvents is common and is an environmental burden. When commercial aircraft are taken out of service for painting, the speed of the repair is critical, as time spent in the paint hanger is time not spent generating revenue.

The coatings used for military aircraft are not unlike those used for commercial aircraft, but may require enhanced features. Examples include coatings that provide reduced signatures (stealth) in various radiation bands that might include IR or microwaves. The details of such technology are closely guarded secrets.

GENERAL REFERENCES

Gaske, J. E., *Coil Coating*, Federation of Societies for Coatings Technology, Blue Bell, PA, **1987**.

Streitberger, H.-J.; Dossel, K.-F., *Automotive Paints and Coatings*, John Wiley & Sons, Chichester, **2008**.

REFERENCES

Agawa, T.; E. D. Dumain, *Waterborne High Solids Powder Coating Symposium*, New Orleans, LA, **1997**, p 342.

American Coatings Association and Chemquest Group, ACA Industry Market Analysis, American Coatings Association, Washington, DC, **2015**.

Anderson, L. G., et al., US patent US6593417 B1 (**2003**).
Balch, T., et al., US patent US7604837 B2 (**2009**).
Barsotti, R. J., et al., US patent US6428898 B1 (**2002**).
Basu, S. K., et al., *Proceedings of the 15th International Coatings Science and Technology Symposium*, St. Paul, MN, USA, **2010**.
Battocchi, D., et al., *Corr. Sci.*, **2006**, *48*(8), 2226–2240.
Bauer, D. R., et al., *Ind. Eng. Chem. Prod. Res. Dev.*, **1982**, *21*(4), 686–690.
Beiro, M., et al., *Prog. Org. Coat.*, **2003**, *46*(2), 97–106.
Bhattacharya, D., et al., *J. Coat. Technol. Res.*, **2007a**, *4*(2), 139–150.
Bhattacharya, D., et al., US patent application, US20070282038 A1 (**2007b**).
Blohowiak, K. Y., et al., US patent US5939197 A (**1999**).
Boggs, L., et al., *J. Coat. Technol.*, **1996**, *68*(855), 63–74.
Boisseau, J., et al., *Proceedings of the 1st European Wealth Symposium*, Czech Republic, Prague, **2003**.
Broder, M. et al., *J. Coat. Technol.*, **1988**, *60*(677), 27.
Chattopadhyay, D. K., et al., *Prog. Org. Coat.*, **2009**, *64*(2), 128–137.
Cocuzzi, D. A.; Pilcher, G. R., *J. Coat. Technol.*, **2004**, *1*(4), 22–29.
Cooke, P. R., et al., US patent US8142858B2 (**2012**).
Courter, J. L., *J. Coat. Technol.*, **1997**, *69*(866), 57–63.
Deskowitz, P., **2013**, *Automotive Market Report*, Bayer Materials Science, Pittsburgh.
Dubois, R.; Sheih, P., *J. Coat. Technol.*, **1992**, *64*(808), 51–57.
Edwards, P. A., et al., *J. Coat. Technol. Res.*, **2005**, *2*(7), 517–527.
Furukawa, H., et al., *Prog. Org. Coat.*, **1994**, *24*(1), 81–99.
Gander, P., Bisphenol A-free can coating in limbo, **2016**, from http://www.foodmanufacture.co.uk/Regulation/Bisphenol-A-free-can-coatings-in-limbo (accessed March 19, 2017).
Gaske, J. E., *Coil Coatings*. Federation of Societies for Coatings Technology, Blue Bell, PA, **1987**.
Green, C. D., *Paint. Coat. Ind.*, **1995**, September, 45.
Groenewolt, M., *Prog. Org. Coat.*, **2008**, *61*(2–4), 106–109.
Groenewolt, M., et al., US patent US8658752B (**2014**).
Hayoz, P., et al., *Prog. Org. Coat.*, **2003**, *48*(2), 297–30.
Hazan, I., et al., US patent, US7867569 B2 (**2011**).
Higginbottom, H. P., et al., *J. Coat. Technol.*, **1999**, *71*(7), 49–60.
Kenny, J., et al., *J. Coat. Technol.*, **1996**, *68*(855), 34–43.
LaKind, J. S., *Int. J. Tech. Pol. Mgmt.*, **2013**, *13*(1), 80–95.
Liu, J., et al., *J. Adhes.*, **2006**, *82*(5), 487–516.
Liu, Y., et al., *J. Electrochem. Soc.*, **2015**, *163*(3), C45–C53.
Miranda, T., *J. Coat. Technol.*, **1988**, *60*(760), 47–49.
Mirgel, V., *Prog. Org. Coat.*, **1993**, *22*(1–4), 273–277.
Mohammadloo, E. H., et al., *Prog. Org. Coat.*, **2014**, *77*(3), 322–330.
Moussa, Y.; Knotts, C., European patent, EP2714535B1 (**2015**).
Nichols, M. E., et al., *Radtech Rep.*, **2001**, *15*(6), 20-22.
Noonan, G. O., et al., *J. Agric. Food Chem.*, **2011**, *59*(13), 7178–7185.
Palackdharry, P. J., *Polym. Mater. Sci. Eng.*, **1991**, *65*, 277.
Ramamurthy, A. C., et al., *Prog. Org. Coat.*, **1994**, *25*, 43–71.
Rexam Annual Report, **2015**, p. 6, from http://www.annualreports.com/Company/rexam-plc (accessed April 17, 2017).
Rosales, B. M., et al., *Prog. Org. Coat.*, **2004**, *50*(2), 105–114.
Sadvary, R. et al., US patent US6623791B2 (**2003**).
Schoff, C. K., *J. Coat. Technol.*, **1999**, *71*(888), 56–73.
Seubert, C. M.; Nichols, M. E., *Prog. Org. Coat.*, **2004**, *49*(3), 218–224.
Seubert, C. M.; Nichols, M. E., *J. Coat. Technol. Res.*, **2010**, *7*(5), 615–622.
Seubert, C. M., et al., *J. Mat. Sci.*, **2015**, *51*(5), 2259–2273.
Singh, S., *Manage. Dec.*, **2006**, *44*(6), 783–789.
Sung, L. P., et al., *J. Coat. Technol.*, **2002**, *74*(932), 55–63.
Swarup, S., et al., US patent, US6762240 B2 (**2004**).
Tachi, K., et al., *J. Coat. Technol.*, **1990**, *62*(782), 43–50.
Testa, C., *J. Coat. Technol.*, **2008**, October, 24–28.
Trindade, D. J.; Matheson, R. R., US patent, US8710138 B2 (**2014**).
Uhlianuk, P. W., et al., US patent, US7740912 B2 (**2010**).
Vanier, N. R., et al., US patent, US6916368 B2 (**2005**).
Visser, P., et al., *Faraday Discuss.*, **2015**, *180*, 511–526.
Vogt-Birnbrich, B., et al., US patent, US6069218 A (**2000**).
Weaks, G. T., **1991**, *ESD/ASM, Adv. Coat. Technology Conference*. Detroit, MI USA, *Eng. Soc. Detroit*: *201*.
Wingender, R., et al., *J. Coat. Technol.*, **1998**, *70*(877), 75–82.
Woo, J., et al., *J. Coat. Technol.*, **1982**, *54*(689), 41–55.

Chapter 31

Product Coatings for Nonmetallic Substrates

Many products made from wood and plastics are coated in factories; examples are wood furniture, cabinets, flooring and paneling, hardboard paneling and siding, and plastic auto body parts. This chapter addresses such factory-applied coatings. The architectural coatings used by contractors and do-it-yourselfers to paint wood products at home are discussed in Chapter 32. Other nonmetallic substrates are also coated in factories, for example, paper, textiles, glass, and films. Herein, we will focus on wood and plastics.

Of course, VOC and other regulations continue to affect the technology of coatings for woods and plastics. Emphasis on sustainable coatings is increasing, especially wood coatings. Major retailers now assess the cradle-to-grave environmental impact of furniture production methods. Green building programs such as LEED (Chapter 18) influence coating selection for wood paneling and other building components. Bio-based content of coatings is increasingly emphasized.

31.1 COATINGS FOR WOOD

The global market for *industrial wood coatings* is large. How large? Good figures for the global market are "hard to come by" (Bulian and Graystone, 2009). It's no wonder. Worldwide, thousands of coatings producers are supplying a broad variety of coatings to well over 100 000 factories. Definitions of market segments vary; for example, some studies include coatings for wooden decking in the factory wood category, others do not. The largest market is in Asia, reflecting the shift of manufacturing to Asia and probably the more lenient regulations in that region. In 2012, the market value for factory-applied wood coatings in China alone was about $2B USD (Duan et al., 2014), and factories in other Asian countries add substantially to the total. Pilcher (2015) estimated that factory wood coatings in the United States are valued at about $900M, which corresponds to about 4% of the total US coatings market.

31.1.1 Coatings for Wood Furniture

Furniture styling and manufacturing processes are different around the world. Within Europe, for example, some regions prefer high gloss furniture coatings and others prefer low gloss levels. In the United States, a large fraction of wood furniture is styled to resemble antique furniture. Styling is generally initiated in the high end of the market, the expensive wood furniture that involves a great deal of hand labor and artistry in its manufacture. A representative high-end finishing process is described in detail in earlier editions of this book (Wicks et al., 2006). The description is condensed here because the process employs high VOC solventborne nitrocellulose (NC) lacquers, the use of which is declining in North America and Europe, although perhaps not in parts of Asia.

In US furniture manufacturing, much of the furniture is assembled before it is finished. The outer surface of high-end tabletops is often a veneer (a thin layer) of pecan, walnut, or oak. These hardwoods have open pore structures and prominent grain patterns that can provide a striking appearance. Smaller components are formed from solid wood.

Commonly, the first finishing step is *sizing*, application of a low solids poly(vinyl alcohol) solution. The purpose is to stiffen wood fibrils that stick up from the surface so that the surface can be smoothed by sanding. With some woods, sizing isn't necessary.

Next, the wood is colored with a stain, comprising a solution of acid dyestuffs in solvent, or with dispersions of transparent pigments. The *stain* is selected and applied

Organic Coatings: Science and Technology, Fourth Edition. Frank N. Jones, Mark E. Nichols, and Socrates Peter Pappas.

evenly to give the overall base color to the furniture. A thin *wash coat*, comprised of a low solids, low viscosity lacquer, is then applied to minimize displacement of the stain, stiffen wood fibers for sanding, and prepare the surface for the next step of filling.

The purpose of *filling* is to color the pores of the wood to emphasize the grain pattern and to fill the pores to nearly the same level as the rest of the wood. Usually, the color of the filler is dark brown and the stains are lighter yellowish or reddish browns, but for special effects, one can use other color combinations. The vehicle is linseed oil and/or a linseed long oil alkyd together with a hard resin, such as limed rosin, and driers with aliphatic hydrocarbon solvents. The filler is usually a dark color, often supplied by natural earth pigments. The assembled piece of furniture is sprayed with a liberal coat of filler. The filler is "padded" into the pores by rubbing vigorously with a pad of cloth, followed by wiping off excess filler.

A *sanding sealer* is then applied to prepare the surface for application of top coat. The formulation for a typical sanding sealer includes NC; a hard resin, such as a maleated rosin esterified with a polyol; and a plasticizer. Sanding sealers contain 3–7% of zinc stearate, based on the lacquer solids, to aid sanding. The sealer is spray-applied at about 20% volume solids and, after drying, is sanded smooth. UV cure sealers, which have very low VOC, are used by some manufacturers, especially when waterborne top coats (Section 31.1.2) will be applied.

The furniture is then sprayed with *top coat*. NC (Section 17.2.1) is the primary binder in high-style furniture lacquers for three major reasons. First, the appearance of furniture finished with NC lacquers is outstanding; the lacquers provide an appearance of depth, fullness, and clarity of the grain pattern that is difficult to match with any other coating. Second, lacquers dry quickly, so they can be rubbed a short time after application and then packed and loaded into a truck for shipment without *printing*. The term printing in this context refers to surface imperfections, caused by wrapping material, when the surface of a coating is too soft. Third, lacquers are thermoplastic and permanently soluble, so any damage is easily repaired. The mechanical properties of the final film are improved with higher molecular weight (MW) NC, but solids are reduced. A hard resin such as an esterified maleated rosin is blended with the NC. Plasticizers in the top coat, usually including a short oil coconut alkyd (Section 15.4), provide the required flexibility. The appropriate balance of NC, plasticizer, and hard resin is critical. If the coating is too soft, it will be difficult to rub; if it is too hard, the lacquer will be subject to cracking as the wood expands and contracts with changes in moisture content and/or exposure to rapid temperature changes. UV absorber (Section 5.2.1) and citric acid are often added to minimize discoloration.

In the United States, NC lacquers are supplied with 60° gloss levels ranging from 12 to 90 (see Section 19.10). Volume solids may be less than 20% and VOC levels are less than $680\,g\,l^{-1}$, which complies with national standards, but not with more restrictive standards in many localities. During application, fast evaporation of large amounts of flammable solvent requires careful precautions to prevent fires or explosions.

Application of the initial layer of top coat is followed by *distressing*. Distressing uses a variety of techniques to simulate an antique appearance. Finally, a second coat of top coat is applied and the lacquer is dried, usually at 40–60°C. Then the lacquer surface is *rubbed*, first by sanding with fine sandpaper and a lubricant and then rubbing with cloth and rubbing compound. The result is a soft appearing low gloss finish.

While solubility of NC lacquer finishes is an advantage, permitting easy repair, it can also be a disadvantage. If solvents, such as nail polish thinners, are spilled on a lacquer surface, the finish will be marred. A way of maintaining the beauty of an NC finish is to add a small amount of aliphatic polyisocyanate cross-linker to the lacquer prior to application. The isocyanate cross-links with the hydroxyl groups on the NC. In this case, all solvents must be urethane grade, that is, contain no more than traces of alcohols and water; and the NC must be wet with plasticizer rather than with isopropyl alcohol.

Wood furniture produced by this process is beautiful but expensive. A much larger part of the US market is for furniture designed to look as closely as possible like this high-style furniture, but that can be manufactured and finished at lower cost. There are many approaches with varying degrees of cost savings; an intermediate cost approach follows.

Instead of using plywood tops and sides made with expensive hardwood veneers that must be painstakingly finished, one can use printed tops and sides with solid wood legs and frames. *Particle board*, which is pressed wood chips and particles with a resin binder, is first filled with a UV cure (Chapter 29; Section 31.1.3) filler and sanded smooth. In contrast to earlier used highly pigmented solventborne coatings, a single coat of fast UV cure fillers is sufficient and considerably less expensive. These fillers are pigmented with clays, silica, and/or aluminum trihydroxide (Parker and Martin, 1994) that absorb or scatter only minor amounts of UV. The high level of pigmentation leads to poor leveling, which doesn't matter, since the surface will be sanded. Furthermore, the pigmentation results in less resin being at the surface. Originally, the UV fillers were styrene/polyester systems, but faster UV cure has led to the partial adoption of acrylated systems despite their higher cost.

The filled, sanded panel is next coated with a *base coat*, a highly pigmented NC lacquer, which provides the overall underlying color of the piece of furniture, corresponding to the color of the stain in high-style furniture. The top is then ready for gravure printing that replicates the appearance of selected hardwood top veneers that were stained, filled, shaded, and distressed by highly skilled finishers. The inks are pigment dispersions in plasticizer, so they adhere well to the base coat and to subsequent lacquer coats.

The frame and legs of the furniture are assembled and finished up to the stage of applying top coat, followed by assembly of the printed tops and sides onto the furniture. Then, the assembled piece of furniture is coated with semigloss NC lacquer top coat. Semigloss lacquer, rather than gloss lacquer, is used to provide a finish with a *hand-rubbed effect* without extensive hand rubbing. The semigloss is obtained by pigmenting with a low level of fine particle size SiO_2, such that the PVC of the final film averages 2–4%. When the solvent evaporates after application, convection currents during solvent evaporation carry the pigment to the surface of the film, where it is "trapped" in the viscous surface layer. As a result, the PVC of the top layer of the film is high enough to reduce gloss, while the overall PVC is low enough that only a minor degree of light scattering occurs.

A disadvantage of furniture finishing systems discussed so far is that they require use of low solids coatings with high VOC emissions except for UV cure fillers. Furthermore, multiple coats are required, or at least more passes with a spray gun, so application cost is high. Hot spray (Section 23.2.4) is frequently used. A temperature of 65°C permits increasing solids from 20–24 to 28–34 NVW, which corresponds to a significant reduction in VOC emissions, but still far short of probable future requirements.

In 1995 the US EPA revised its wood furniture coatings regulations after extensive discussions with the industry. It was agreed that regulations should be expressed as pounds of VOC per pound of coating solids on the wood surfaces. The overall limits were set at 2.3 lb of VOC including a maximum of 1.0 lb of HAP per pound of solids (not per gallon). In new plants or finishing lines, HAP emissions are limited to 0.8 lb per pound of solids. For specific coatings the following maxima were set: waterborne top coats, 0.8 lb; higher solids solventborne sealers, 1.9 lb; top coats, 1.8; and acid-cured alkyd amino sealers, 2.3 lb. In 2011 formaldehyde emissions from each plant were restricted to 400 lb $year^{-1}$ (Federal Register 76:72050).

Limits of 275 $g\,l^{-1}$ of VOC for coatings for most wood furniture took effect in California in 1994. Should VOC regulations become more stringent, it seems probable that a larger fraction of furniture tops will be high pressure laminates with wood grain prints. Laminates have been used for many years on commercial and institutional furniture; now their use in household furniture is increasing. In China, as of 2012, a substantial fraction of the wood coatings market continued to utilize NC coatings (Duan et al., 2014).

Efforts have been ongoing for many years to replace NC lacquers in order to achieve higher solids. Alkyd/urea-formaldehyde (UF) top coats, frequently called *conversion varnishes* or *catalyzed finishes*, have been used since the 1950s. Tall oil alkyds with a butylated UF resin are typical. However, UF resins tend to evolve significant levels of formaldehyde. Melamine-formaldehyde (MF) resins have been designed to evolve lesser amounts of formaldehyde and may aid compliance with the 2011 formaldehyde limits. Just before use, one adds up to 5% of a sulfonic acid catalyst. A flatting pigment dispersion and a small amount of poly(dimethylsiloxane) to minimize orange peel (Section 24.2) are usually added. The top-coated furniture is force air-dried at 65–70°C for 20–25 min. The NVW is typically about 38%, about twice that of lacquers. Hot spray can further increase the solids, but pot life must be carefully monitored and, as the solids increase, low gloss becomes more difficult to achieve. With such conversion coatings the appearance of depth is less pronounced than with an NC lacquer but is still presentable. The coating is solvent resistant and more heat and gouge resistant than lacquer coatings. Repair is more difficult, but less frequently needed. Conversion varnishes are widely used on furniture for commercial use and for kitchen cabinets.

Inexpensive furniture is fabricated, at least partly, from medium density fiberboard (MDF). Such furniture is often painted with opaque conversion coatings. Sometimes the furniture pieces are made in Asia and then shipped to the United States for final assembly.

Possible toxic effects of formaldehyde emissions present opportunities for coatings technology. One is to formulate coatings with reduced emission of formaldehyde during cure. A second is to supply coatings that scavenge formaldehyde emissions from the UF resin glues used in furniture fabrication. A technical approach to both problems is to formulate coatings that contain formaldehyde scavengers such as acetoacetate groups in the polymer (Duan et al., 2014) or sulfites or hydrazides in the formula. Other approaches are to use MF resin cross-linkers that minimize formaldehyde emission during coating cure or to substitute formaldehyde-free cross-linkers. Pavia et al. (2011) described yet another approach involving "melamine-fortified UF resins" with low formaldehyde emissions.

High solids coatings based on isocyanate chemistry (Chapter 12) are regarded as benchmark high-end wood coatings because of their excellent balance of properties (Hallack et al., 2014). Aldimine/isocyanate two-package (2K) glossy clear coats with 90% solids (Section 12.4) have been formulated that have a pot life of about 3 h and a dry time of about half hour (Dvorchak et al., 1997). Moisture-curable urethane coatings are used to coat wood in applications where high abrasion resistance is needed. Isocyanate prepolymer mixtures, such as propoxylated ethylenediamine/TDI and 2,2,4-trimethylpentane-1,3-diol/TDI prepolymers, are used as a vehicle for moisture-curable wood coatings (Pedain et al., 1989). A moisture-curable wood finish based on a TMP/1,4-butanediol/IPDI prepolymer with 3-glycidyloxypropyltrimethoxysilane is reported to cure by hydrolysis of the alkoxysilane groups, resulting in silanol groups, which react with other alkoxysilane groups as well as with the isocyanate groups (Fong, 1992).

Hybrid resin combinations are gaining favor. For example, Hallack et al. (2014) describe siloxane-terminated polyureas that can cross-link via hydrolysis and condensation of the siloxane in moisture-cure 1K coatings. They can also serve as cross-linkers for hydroxyl-functional resins in 2K formulations. Excellent scratch resistance is reported.

Various approaches to *self-healing* coatings (Section 34.3) for wood have been described. As an example, a thermosetting polyurethane is combined with an incompatible thermoplastic polymer, polycaprolactone. The cross-linked polyurethane supplies mechanical strength, and the thermoplastic flows to heal surface damage (Ou et al., 2014). Solventborne, high solids, and waterborne formulations are described.

31.1.2 Waterborne Wood finishes

Historically, the majority of factory wood finishes have been solventborne. Such coatings provide a good balance of appearance (gloss and transparency), film properties, application characteristics, and economy. Considerable effort has been expended on waterborne finishes for wood furniture, and sustained progress has led to increasing market share for waterborne finishes. Application of a waterborne coating directly on wood leads to excessive grain raising, which may limit their use to applications where there is already a solventborne or UV sealer on the wood.

Substantial reductions of solvent usage are possible by emulsifying NC lacquers into water (Winchester, 1991). Lacquers with VOC of 300–420 $g\,l^{-1}$, excluding water, can be made compared with VOC on the order of 680 $g\,l^{-1}$ for conventional NC lacquers. Solids of the NC solution internal phase can be maximized by using only true solvents, notably esters and ketones. Short oil alkyds using ester solvents are selected as plasticizers. Water wet NC is used rather than conventional isopropyl alcohol wet NC. A surfactant is incorporated in the lacquer as an emulsifying agent; sodium alkylphosphate surfactants are recommended. The advantage of easy strippability for repair is retained. However, relatively long dry time and slow development of print resistance have limited their use.

Haag (1992) investigated combined NC/acrylic latex-based lacquers. He described furniture sealer and top coat formulations with VOC in the 240–400 $g\,l^{-1}$ range. Film physical and appearance properties are reported to be superior to acrylic latex counterparts.

Thermosetting latex top coats (Section 9.4) have become common. For two-package coatings, a variety of cross-linkers for hydroxy-functional and carboxylic acid-functional latexes can be used. Hydroxy-functional latexes are readily prepared by using hydroxyethyl (meth)acrylate (HEMA) as a comonomer. The resultant latexes can be formulated with UF or MF resins as cross-linkers. Certain MF resins are soluble in BA/HEMA/MAA latex polymers so that they act as plasticizers to promote interdiffusion. As temperature increases, interdiffusion rate increases faster than the rate of cross-linking until many of the potential reactions have occurred; but at some point the cross-link density reaches a point where interdiffusion ceases. The rate of cross-linking can be increased by the addition of a catalyst such as pTSA, which not only reduces cure times but also reduces pot life. Transparency of clear finishes based on latexes has improved, but it remains difficult to equal the transparency of the best solventborne finishes.

Another development is the use of self-cross-linking or cross-linkable acrylic latexes (Vielhauer, 2013; Carson et al., 2015). Many of the cross-linking chemistries described in Section 9.4 have been adapted. A further refinement is to blend self-cross-linking acrylic latexes with polyurethane dispersions (PUDs) to provide low VOC (c.a. 150 $g\,l^{-1}$) wood finishes with good properties and excellent resistance to yellowing (Vielhauer, 2013).

Huang et al. (1997) studied the effect of several variables on the performance of finishes for wood kitchen cabinets based on hydroxy-functional acrylic latexes cross-linked with a nonionic hydrophilically modified polyisocyanate. It was found that high hydroxy content of the latex (hydroxyl number 52), small particle size of the latex, core–shell preparation of the latex, and low T_g (obtained by increasing level of coalescing solvent) enhanced performance.

Hybrid latexes (Section 12.7.2) combining acrylic and polyurethane chemistries are gaining favor for wood coatings. Blends of PUDs with acrylic latexes have been used for decades. More recently, methods have been developed for synthesizing acrylic latexes in the presence of polyurethanes to yield hybrid latexes that are said to afford low VOC and excellent combinations of film properties.

Water-reducible acrylic resin-based coatings (Section 8.3) have been used. Thermosetting resins can be cross-linked with polyaziridine cross-linkers (Section 17.7), although there is concern about the toxic hazard of aziridine cross-linkers in spray application. Methylated UF resins can be used as cross-linkers under force dry conditions. Acetoacetoxy-functional water-reducible acrylic resins made with acetoacetoxyethyl methacrylate as a comonomer are cross-linked with diamines (Esser et al., 1999) (Section 17.6). Sufficiently high MW acrylic resins can be used as thermoplastic coatings.

Kitchen cabinet waterborne coatings are formulated with hydroxy- and carboxy-functional acrylic latexes with a UF cross-linker. While some coalescing solvent is needed, VOC emissions are much lower than with solventborne coatings. Howard (2014) and Howard et al. (2001) discuss the variables in formulating such coatings.

Two-package waterborne polyurethane top coats (see Section 12.7.3 for discussion of various types) are increasingly being used, since they provide low VOC emissions with a minimum of HAP solvents. The films have excellent abrasion resistance and mechanical properties. However, the cost is relatively high, and special application equipment is required (Dvorchak, 1997).

31.1.3 UV Cure Furniture Finishes

In addition to the UV cure coatings described previously in Section 31.1.1, UV cure top coats are used in Europe and, to a more limited degree, in North America and Asia. A large part of European furniture is manufactured in flat sheets. UV cure coatings are applied to furniture tops by roll coating prior to assembly. The fast cure at low temperature is appropriate for wood. Furthermore, the finish on much European furniture is glossy. UV curing is ideal for such an application.

Early coatings were based on styrene/polyester resins. Acrylic coatings were later adopted because they cure more rapidly and emission of styrene is avoided (Section 29.2.5). The cured coatings are glossy or semigloss, are solvent resistant, and have excellent mechanical properties. They have the advantage of minimal VOC emission. Their use in the United States has been limited, because a major fraction of US furniture is assembled before it is coated and lower gloss finishes are utilized.

Nanoparticle pigments (Section 20.5) are a promising approach to improving mechanical properties of furniture coatings in general and UV cure coatings in particular (Nikolic et al., 2015). Nanopigments are difficult to disperse, but silica nanoparticles that are surface-treated with trialkoxysilanes can be dispersed in UV-curable waterborne wood coatings (Sow et al., 2010). Scratch resistance and adhesion are enhanced by as little as 1 wt% of the nanopigment.

Waterborne UV cure coatings have been developed that permit formulation of low VOC, low gloss coatings (Section 29.2.6). In one example, ordinary acrylic latexes are blended with UV-curable oligomers (such as triacrylates), a photoinitiator, and other ingredients to provide low gloss waterborne UV-curable coatings for wood (Bohannon, 2014). After application, the coatings are force dried at 40–50°C for 6–12 min, followed by UV cure.

31.1.4 Panel, Siding, and Flooring Finishes

Large quantities of coatings are applied to plywood and hardboard sheets to make pre-coated interior paneling and exterior siding. A small part of the market is in finishing expensive, high quality hardwood plywood used for paneling executive offices, upscale restaurants, and the like. This paneling is finished in essentially the same way used for the highest end of the wood furniture market. A variety of lower cost products constitutes the bulk of the market.

One class of commodity wall paneling is made using three-ply plywood for which the top veneer is a relatively featureless, low cost *lauan* (luaan). The panel is dried to remove surface water, grooved to make a plank effect, and then sanded. A dark-colored lacquer is applied to the grooves by pinpoint automatic spray guns, and the panel is dried, followed by applying a lacquer sealer colored with dyes. The panel is sanded again and then printed two or three times to give the appearance of walnut, rosewood, or some other attractive wood with prominent grain patterns. Finally, a low gloss top coat is applied. The top coatings have generally been NC lacquers. While this business was first developed in the United States, most production has shifted overseas, initially to South Korea and more recently to the Philippines. Since the Philippines is the predominant source of lauan, finished product can be manufactured and shipped from there; also, air pollution regulations in the Philippines are less stringent.

A related, but quite different, application is for "door skins." In the manufacture of interior doors, a common approach is to assemble the door and then laminate a top veneer (the *skin*) to a particle board surface. The veneer is commonly birch or lauan, sometimes printed with a hardwood grain pattern. Commonly, the coating on door skins is glossy. Whereas low gloss is preferred on wall paneling to avoid glare from reflection of lights, higher gloss is preferred on doors because it is easier to clean. Since the surface area of doors is relatively small, glare is not a serious problem. This combination of factors has led to the widespread use of UV cure coatings for door skins.

A variety of UV compositions has been studied for wood coatings. For example, a UV cure coating for wood paneling, comprising an acrylated epoxy oligomer and fluorinated monomer ($C_6F_{13}(CH_2)_2OCOCH{=}CH_2$) provides higher hydro- and oleophobicity, improved chemical resistance, and better scratch resistance compared with coatings without the fluorinated monomer (Bongiovanni et al., 2002).

Part of the interior paneling market is pre-coated hardboard. Hardboard is made by mixing wood fibers and shreds that are cured in hydraulic presses. The lignin in the wood acts as a binder to hold the wood fibers together; sometimes the lignin is supplemented with phenolic resins. Depending on the pressure applied, the density of hardboard can be varied. Furthermore, patterns can be embossed into the surface of the hardboard, creating a great variety of products. Using smooth surface hardboard, one can apply a base coat, prints, and a semigloss top coat to make 4 by 8 foot panels that look like any kind of wood. Smooth surface hardboard is often coated by curtain coating (Section 23.5), which gives level films.

Hardboard can withstand higher temperatures than plywood, so that the superior properties of baked coatings can be realized. Since hardboard does not undergo grain raising like wood, water-reducible acrylic/MF or cross-linking latex finishes are practical. Although average film thickness of a base coat is relatively high, popping is not a major problem, since the coating thicknesses applied by roll coating and curtain coating are more uniform than when coatings are applied by spray (Sections 23.4 and 23.5).

Tongue and groove paneling can be simulated by routing out grooves and painting them a dark brown color. The paneling is finished by roll coating a base coat that does not flow appreciably into the grooves, followed by printing and top coating. A pattern of ticks (short line dents scattered over the surface, as in hardwoods) can be embossed. The finished board has the grain of the wood, as well as the little dips of a tick pattern and grooves to give the effect of planked paneling.

One can also simulate wood with holes, like pecky cypress. Besides replicating pecky cypress or wormy chestnut, other effects, such as brick, stone, or travertine marble, can be simulated. Joints may be embossed to make the panels look like tiles. Marble prints may be used with a clear gloss top coat. Other interior hardboard paneling is designed primarily for bathrooms, which are finished with high gloss opaque coatings.

High density hardboard is also used for exterior siding. The largest volume of siding is factory primed with a primer designed to provide at least 6 months' exterior durability before exterior house paint is applied (Section 32.1). Acrylic/MF binders are used in the primers, generally waterborne to reduce VOC. An important formulating challenge is to ensure enough cross-linking of the primer so that the cured boards exiting the oven can be stacked without blocking, but low enough to ensure good adhesion of the subsequent coating to the primer. The combination of these properties requires careful control of cross-linking, because adhesion is generally enhanced by penetration of the subsequent coating into the primer. The primers are low gloss because high loading with inert pigment reduces cost and enhances adhesion of paint to the surface. The largest volume is used in siding for tract houses. The primed siding is erected on to the house, but painting may be postponed until the house is sold so that the buyer can choose the paint color.

Bailey et al. (1990) provides an extensive study of exterior durability of different hardboards with different combinations of coatings. The best results were obtained when pre-primed board was field coated with an alkyd primer and acrylic latex top coats. Some hardboards contain petrolatum and other oily substances that can migrate to the upper surface of the paint, leading to discoloration or sticky areas that pick up dirt and give a blotchy appearance. This problem can be minimized by using paints having PVC < CPVC, which reduces porosity.

Fully prefinished exterior siding is manufactured on a smaller scale than pre-primed siding. The primed board is top coated in the factory with a baking acrylic enamel or, for greater exterior durability, a silicone-modified acrylic resin-based coating (Section 16.1.2). The durability is superior to field applied paint, but flexibility in color selection is sacrificed. Fully prefinished siding is more commonly used on commercial buildings. Coil coatings for metal siding are discussed in Section 30.4.

Waterborne UV cure clear coats (Section 29.2.6) with excellent durability have been developed for application to wood articles, such as window frames. For example, the binder can be an aliphatic PUD terminated with acrylate groups combined with low MW acrylate coreactants.

Moisture-curable urethane coatings (Section 12.6) are used for applications such as floor coatings, for which abrasion resistance and hydrolytic stability are important. Although called urethane coatings, the cross-links are urea groups. Since urea groups also form intermolecular hydrogen bonds, it is expected that they can resist mechanical stress similarly to urethane groups (Section 12.6). Wood flooring coatings are also based on a carboxy-functional urethane/acrylic PUD and pentaerythritol tris(β-(*N*-aziridinylpropionate) (Kohman et al., 2002). Waterborne wood floor finishes, prepared using drying oil-modified PUDs, provide a combination of abrasion resistance and resistance to heel marking. 2K waterborne urethane coatings in which the urethane cross-linker is a hydrophilically modified allophanate polyisocyanate are useful in wood finishing. Waterborne UV cure acrylate-functional PUDs provide coatings having satisfactory adhesion (Irle et al., 2001).

Many studies indicate that addition of small amounts of nanoparticles may enhance the properties of wood finishes (Nikolic et al., 2015). For example, nanoaluminum oxide particles (average diameter 13 nm) markedly improve the scratch resistance of UV cure coatings for wood flooring (Landry et al., 2006). The Al_2O_3 is dispersed in the UV-curable coating that also includes 3-trimethoxysilylpropyl methacrylate as a coupling agent between the pigment and binder. Romo-Uribe et al. (2016) describe the synthesis of acrylic latexes containing non-surface-modified nanosilica. One percent of nanoparticles increased T_g by 10°C and enhanced thermal stability. The technology appears adaptable to wood finishes.

Solventborne uralkyds have been widely used for finishing wood flooring and other wood products because of their abrasion resistance and ease of application. However, VOC regulations are requiring a shift to waterborne coatings. See Section 15.7 and Caldwell (2005) for discussion.

31.2 COATING OF PLASTICS

Plastics have become a major substrate for coatings. Design of coatings for plastics is complicated by the variety of polymers used in plastics and the range of approaches that are possible for decorating the surfaces as discussed by Satas (1986). Ryntz (1994, 2001); Wilke and Ortmeier (2011) reviewed coatings for plastics. For molded plastics, there are two alternatives: (1) *in-mold coating*, in which case the inside of a mold is coated, the plastic material is introduced into the mold, and the molded part (having an outer coating on its surface) is removed from the mold, or (2) *post-mold coating*, in which the molded substrate is coated after removal from the mold.

Coatings for plastic films are a major business that is not considered herein, although many of the same principles apply.

31.2.1 In-Mold Coating

Polar thermosetting plastics can be in-mold coated. A variety of coating materials can be used, depending on the plastic. To assure good adhesion to the plastic, it is desirable to use coating compositions that are related to and, desirably, reactive with the composition of the plastic, for example, the use of *gel coats* with glass-reinforced plastic parts. Gel coats are pigmented styrene–unsaturated polyester resins (Section 17.3) that are cured along with the resin in the molding compound so that the coating is chemically bonded to the main body of the plastic. The resin to be molded is usually a styrene solution of a polyester, prepared from phthalic anhydride, maleic anhydride, and propylene glycol. For some products, gel coats contain similar polyesters. More expensive polyesters, prepared from isophthalic acid, maleic anhydride, and neopentyl glycol polyesters, are used in gel coats when superior hydrolytic and exterior durability is needed. These qualities are required for major uses such as boats and shower stalls.

With polyurethane reaction injection molded (RIM) parts, in-mold coatings that contain free hydroxyl groups permit reaction with isocyanate groups in the molding compound. Rigid urethane steering wheels are made by coating the inside of the mold with the appropriate color enamel. Rigid urethane foam parts that replace wood carvings for furniture are in-mold coated. The inside of the mold is sprayed with a lacquer, which is color-matched to the base coat of the furniture. Many in-mold coatings were solventborne, frequently low solids lacquers. These high VOC coatings are being replaced with higher solids coatings, waterborne coatings, and powder coatings.

Coating of interior automotive plastic components has for many years been carried out by post-mold coating. Waterborne in-mold coating is now being introduced. Cannon and Young (2005) reported that this technology achieves rapid part throughput, high transfer efficiency, greater gloss control, and superior mar and chemical resistance.

31.2.2 Post-Mold Coating

Coatings are also applied to plastic articles after they are fabricated. Many of the coatings used for metals can be adapted for certain plastics, and many of the selection criteria are the same as for metals. However, there are important differences. Adhesion, discussed in Section 6.5, is usually the central problem in coating plastics. Furthermore, the coatings usually must be more flexible than those for metal coatings. This follows from the ease of deformability of many plastics and all elastomers, since the coatings must be at least as easily deformed as the substrate. As a general guideline, elongation-at-break (Section 4.2) of the coating should be greater than that of the plastic substrate.

To obtain good adhesion, the surface of the plastic must be clean. Machining oils, sanding dust, finger prints, and so on, must be removed. Many plastic parts have residual mold release agents on the surface. If a mold release must be used in making the plastic article, it is desirable to use a water-soluble soap such as zinc stearate as a mold release, since it is relatively easy to remove. Wax mold release agents are more difficult to remove. Silicone or fluorocarbon mold releases should rarely be used for plastic parts that are to be painted, owing to their low surface tensions. Surface contaminants and mold release agents are removed by spray washing in three stages. First, a detergent wash is used followed by a water rinse and finally by a deionized water rinse. At the end of the washer, droplets of water are removed by compressed air jets. Compressed air often contains aerosols of oily material; they must be carefully removed. Spray washing is often followed by a solvent wipe. Schoff (2016) recommends a simple wetting test (ASTM D7541) to confirm that the part is really clean.

After the surface is clean and dry, the surface tension varies widely, depending on the type of plastic. In general, surface tensions of thermosetting plastics and thermoplastics having polar structural groups, such as nylons, are relatively high, although lower than metals. Often, they can be coated without further surface treatment. However, the surface tension of some less polar plastics, particularly polyolefins, is lower than that of most coatings, and the polarity of the surfaces is very low. Primers that have very low surface tensions and that can also penetrate the surface of the plastic are required. Alternatively, the surface of the plastic can be treated to increase its surface tension and provide polar groups that can hydrogen-bond with components of the coating to promote adhesion (see following text and Section 6.5).

A very challenging problem in coating plastics is the finishing of plastic fascias, grills, and other exterior plastic components of automobiles. Presentations at a symposium on painted plastic components have been published in JCT Research (2005), which provides an excellent summary of the situation at the time. The components are molded from a polypropylene/ethylene/propylene/butadiene rubber blend (called "TPO" in North America and "PP-EPDM" in Europe and Asia). During molding, a thin, low surface tension layer of partly crystalline polypropylene forms at the TPO surface, so few coatings will adhere well. Surface treatment is required.

Adhesion can be achieved by applying a thin coat of a chlorinated polyolefin (CPO) primer. The binders of these primers are sometimes chlorinated polypropylene to which maleic anhydride has been grafted in a free radical process (Martz et al., 1991; Kinosada et al., 1992). Other moieties may also be grafted to the CPO. A wide variety of CPOs is available with different MW, maleic content, and degree of chlorination. Many CPO primers are solventborne; some are waterborne (Eastman, 2014). Frequently they incorporate acetylene black pigment to make the primer film conductive to permit electrostatic spraying of a base coat on the primed part.

Tang and Martin (2002), Ryntz (1996, 2005), and de Oliveira (2009) investigated the mechanism by which the CPO primers promote adhesion. The chlorinated polymer is thought to partly diffuse into and through the surface layer of polyolefin and into the rubber phase. Some of the CPO remains at the surface to promote adhesion. Some diffusion of the rubber layer into the polyolefin has also been observed. A large number of variables can affect the performance. Adhesion to injection molded plastics tends to be better than compression molded plastics. The structure of the CPO, MW, degree of chlorination, maleic content, and film thickness are important variables. Periodic improvements in TPO composition and processing can necessitate revision of the CPO coating process. The solvent used can have a major effect on the extent of penetration; hydrocarbons penetrate to a greater extent than polar solvents. Baking generally promotes penetration. Care must be exercised to assure that the solvent is completely evaporated before the base coat is applied to avoid residual solvent blowing out through the film when the top coats are baked. While solvents can improve adhesion, care must be exercised in the selection of solvents for plastics such as TPO in order that the crystallinity of the plastic remains intact. Otherwise, the cohesive strength of the plastic may be compromised (Ryntz, 2005). Penetration of solvent into the TPO is thought to swell the polypropylene surface layer and reduce its crystallinity (de Oliveira, 2009). The CPO primer films must be on the order of 3–5 μm thick; the primer is thermoplastic, and thicker films may lead to cohesive failure. The need for thin films dictates that solventborne primers must be applied in high VOC formulations. The crystallinity and cohesive strength of TPO is also affected by the baking temperature used for curing the coatings (Ryntz, 1996). Eastman (2014) reported that a solution of certain CPOs can be added to the top coat to obtain adhesion to TPO without the need for a separate primer. (See Section 6.8 for discussion of adhesion testing methods.)

Adhesion of top coat to the CPO primer is sometimes inadequate. Addition of an ethylene/vinyl acetate copolymer to the primer is reported to improve intercoat adhesion (Eastman, 2004). Another approach is to graft acrylic side chains by esterifying some of the maleic anhydride groups in the CPO primer with HEA and then making a graft copolymer with acrylate monomers. 2K urethane base coats are formulated with the resin and are reported to give excellent resistance to delamination without the use of a separate primer (Masuda et al., 2005).

Automobile base coat and clear coat are applied over the primed TPO. For example, an acrylic/MF base coat and a trimethoxysilyl-functional acrylic clear coat can be applied. Alternatively, 2K urethane clear coat can be utilized, which contains a UV absorber and hindered amine light stabilizer (HALS) to assure exterior durability. However, durability is less than that obtained with the same system over steel. The difference is attributed to migration of the UVA and the HALS through the coatings and into the TPO plastic. The problem of migration is minimized by use of UVAs and HALS having hydroxyl groups that react with isocyanate groups in the clear coat to bind the stabilizers to the polymer chains. This leads to much higher retention of stabilizers in the clear coat and improved durability (Yaneff et al., 2004).

In addition to TPOs, other polyolefins and polyolefin blends are increasingly used to fabricate products that will be painted. Anderson et al. (2016) compared painting of plaques of ethylene–octene copolymers with plaques of polypropylene homopolymer. The plaques were primed with commercially available CPOs and then painted with various commercially available paints. Adhesion to polypropylene was better. They found that adhesion to ethylene–octene can be improved by blending it with 10–40% of homopolypropylene.

Waterborne coatings for poly(phenylene oxide) plastic interior auto parts made with a urethane/acrylic hybrid PUD and two acrylic latexes have been described (Haag, 1991; Mori et al., 2002). A waterborne primer, formulated with graft copolymers of acrylic monomers on polypropylene, enhances adhesion to polypropylene (or blends of polypropylene/ethylene/propylene) rubber, as well as to acrylic top coats (Hintze-Bruening and Borgholte, 2000).

Rather than applying a special primer, the plastic surfaces can, in some circumstances, be oxidized to yield polar groups to enhance adhesion. Several approaches to oxidative treatment have been reported (Lane and Hourston, 1993; Wilke and Ortmeier, 2011):

- Oxidizing agents such as chromic acid/sulfuric acid baths are effective and have been used for many years. However, disposal of chromate-containing wastes is now tightly controlled. To avoid Cr(VI), Mori et al. (2002) recommends sodium hypochlorite with a detergent for treating the surface of polyolefin polymers to improve adhesion. Fluorine gas is also effective in a process called *reactive gas processing* (Iyer, 2015).
- Plastic surfaces can be oxidized by directing an oxidizing flame from propane or butane onto its surface. Care is required to ensure that all surfaces are adequately treated and none are burned.
- Plastic films are passed through coronas or between electrodes that generate dielectric barriers, resulting in surface oxidation by a mixture of ions and neutral oxidizing agents.
- For three-dimensional plastic parts, equipment for *atmospheric pressure plasma activation* has been developed, as reviewed by Williams et al. (2013). The parts are exposed to a stream of gas containing activated species. Surface oxidation is accomplished mainly by neutral oxidizing agents, such as atomic oxygen, in the gas. The activated species are formed by electron impact with the gas molecules. The energy used to activate the gases can come from a high voltage torch, a corona, a dielectric barrier, or a radiofrequency discharge. The gas can be air, helium, argon, or a mixture. Noble gases are easily ionized. Adhesive strength roughly correlates with the ratio of oxygen to carbon on the surface of the plastic, measured by X-ray photoelectric spectroscopy. Williams et al. (2013) provide many examples of this process. In one case, the O/C ratio of a high density polyethylene surface was increased from 0.02 to 0.28.
- For irregularly shaped parts, similar processes can be carried out in a vacuum chamber; this method is frequently called *plasma discharge.*
- Still another approach is to spray a solution of benzophenone on the surface of a polyolefin and then expose the part to UV radiation. This leads to generation of free radicals that initiate autoxidation of the surface. This process is particularly effective with rubber-modified polyolefins.

In all cases, care must be taken to ensure that the surface treatment is still active when the coating is subsequently applied. Not all of the surface treatments have the same "lifetime." A drawback of these approaches is that they do not impart sufficient conductivity to the surface for efficient electrostatic spraying. In such cases, TPO primers with conductive pigments are preferred.

Especially in Europe, the high VOC of solventborne CPO primers has led to adoption of various alternative methods for making TPO (known there as PP-EPDM) surfaces adequately receptive to paint. These methods include the previously mentioned flame treatment, corona treatment, plasma treatment, and *reactive gas processing* (Iyer, 2015).

From the standpoint of adhesion, it is preferable for the plastic part to have a somewhat roughened surface. Plastics that are highly filled with pigments have rougher surfaces that enhance adhesion. Alternatively, the surface of the mold can be designed to impart some roughness to the molded pieces. Adhesion can also be enhanced by baking the coated objects at temperatures above the T_g of the plastic. However, in many cases, heating above T_g distorts the molded part. An important advantage of in-mold coating is that adhesion is promoted by the heat applied in molding the plastic, thereby avoiding the problem of heat distortion when baking post-coated plastic moldings.

Adhesion can also be enhanced by using solvents in the coating that are soluble in the plastic. This reduces the T_g of the plastic at the surface, perhaps permitting penetration of binder from the primer into the surface of the substrate. However, very volatile strong solvents must be avoided on high T_g thermoplastics such as polystyrene and poly(methyl methacrylate); such solvents cause *solvent crazing*, that is, the development of a network of microcracks in the surface of the plastic. A possible explanation for crazing is that the solvent penetrates into the plastic; then, if the solvent is highly volatile, it evaporates more rapidly from the surface than from the bulk of the plastic. When the solvent diffuses out from below the surface, there is a reduction in volume below the surface, leading to stresses sufficient to crack the surface layer of the plastic, which is devoid of solvent. Another potential problem with using penetrating solvents is the possibility of solvent popping caused by release of solvents from the plastic after cross-linking of the coating is well advanced. In a study of factors that affect popping on sheet molding compound (SMC) parts (molded fiberglass-reinforced polyester plastics), it was found that the problem can be minimized by using primers with the lowest possible solvent permeability (Ryntz et al., 1992). Permeability can be reduced by using highly cross-linked or partially crystalline binders in the primer.

Moisture-curable urethane coatings have been used on plastic flooring. For example, a coating with outstanding abrasion resistance, particularly for urethane elastomer flooring, is formulated with a moisture-curable urethane made from a polytetramethylene glycol, 1,4-butanediol, H_{12}MDI, and MDI composition to which solutions of DBTDL and DABCO catalysts are added just prior to application (Harmer, 1981).

Electrostatic spraying of plastics is difficult, since the plastics are usually insulators and, therefore, cannot be adequately grounded to give the charge differential needed to attract the charged spray droplets to the surface. Surface resistivity of plastics and the time for dissipation of charges on their surfaces can be measured and correlated with the feasibility of electrostatic spraying (Garner and Elmoursi, 1991). The charge dissipation time is affected not only by the composition of the plastic but also by the humidity in the spray booth. Conductive tie coats or primers can be applied to the plastic; the effect of a range of conductive primers on the charge dissipation time has been studied (Ryntz et al., 1995). Elmoursi and Garner (1992) showed that the effect of surface charges on the plastic substrate in the electrostatic spraying can be reduced by placing a continuous grounded metal backing in contact with the plastic substrate. When the object being sprayed has both plastic and metal components, there can be nonuniform deposition of coatings even when a conductive primer has been applied to the plastic, since the metal can distort the electrostatic field near the plastic–metal interfaces. For this reason, automobile body panels made from SMC are coated with a conductive primer and then shipped to the automobile factory for assembly onto the vehicle prior to painting. The primers are conductive enough to increase the transfer efficiency during spraying, but not so conductive that electrocoat is deposited onto their surface.

The lower curing temperatures possible with two-package polyurethane coatings make them particularly attractive for many applications to minimize the possibility of heat distortion of the plastic product (Shoemaker, 1990). The higher abrasion resistance and flexibility available with urethane coatings are also advantageous. While acrylic resins can be cross-linked with polyisocyanates, higher solids are possible with hydroxy-functional polyesters. For example, a polyester resin—prepared from adipic acid, isophthalic acid, neopentyl glycol, and trimethylolpropane (molar ratio of 1 : 1 : 2.53 : 0.19)—with $\bar{M}_n$ of 730 and hydroxy group functionality $(\bar{f}_n)$ of 2.09 was cross-linked with various trifunctional aliphatic isocyanates at a 1.1 : 1 NCO : OH ratio (Shoemaker, 1990). The best of these coatings provided a good balance of properties on a variety of plastics, including good impact resistance at −29°C.

A further approach to reducing heat distortion of coating plastics is radiation curing (Chapter 29). The temperatures involved in radiation curing are generally not substantially above ambient and can be further reduced by filtering out IR radiation. Low temperature curing is particularly important in coating plastic flooring, such as vinyl tile; plastic flooring often has an embossed foam decorative layer as the upper surface, which may collapse at elevated temperatures. Use of small amounts of solvent in UV cure coatings in place of low viscosity monomeric acrylates reduces shrinkage during cure and enhances adhesion.

Proprietary UV cure base coats and clear coats for flexible plastic automotive components provide excellent performance while reducing the time for application and cure of the system to 7.5 min as compared with 35 min with thermal cure coatings (Cannon et al., 2005).

Coatings for plastics are decorative, such as for color, as well as highly functional. The UV cure top coat applied to plastic flooring mentioned earlier has the functions of increasing wear life, stain resistance, and gloss retention. Coatings are applied to polyethylene tanks to reduce permeability.

Transparent plastics can serve as lightweight replacements for glass in applications ranging from window glazing to eyeglasses, but they are far less scratch, mar, and wear resistant than glass (Section 4.5.1). Polycarbonate plastic is especially useful for applications such as auto headlamp covers and signage because of its impact resistance, transparency, and weatherability (if protected from UV). Surface coatings to improve mar and scratch resistance have been sought for at least 50 years and are in wide use. Successful anti-scratch coatings are often hybrid siloxane/silica sol composites (Section 16.3.1). Such coatings provide adequate mar and scratch resistance to polycarbonate for uses such as auto headlamp covers but have relatively high VOC levels and long curing times. They are being supplanted by low VOC radiation cure coatings that cure in a few seconds (Chapter 29). Barletta et al. (2016) reviewed such coatings and described a three-coat system consisting of:

- A thin cycloaliphatic epoxy primer for adhesion
- A bridging layer of siloxanes applied by the sol–gel method
- A hard, tough acrylate silica nanocomposite layer

All three layers are UV curable, and their chemistry is designed to assure chemical bonding between layers. Scratch and mar resistance of such composites is excellent, but wear resistance is deficient. For this particular coating system, mar resistance is related to hardness, but wear resistance is related to ductility and toughness. Thus, such coatings may be satisfactory for many applications such as headlamps or fixed glazing (windows) on automobiles, but not for movable glazing such as side windows, where abrasion due to window movement is a larger concern. Coatings for polymeric windows are a highly desired commodity and have been commercialized by a few auto OEMs. The weight savings, ability to form unique shapes unattainable in glass, and the opportunity for parts consolidation are all reasons to seek polymeric glazing. However, the long-term durability of such polymeric components has not been established, as superior optical transparency must be maintained for 10+ years for safety. Polycarbonate is the material of choice owing to its ductility and optical properties but is inherently photolabile and tends to yellow and scratch, thereby requiring a coating that provides UV absorption and hardness. Coatings that provide sufficient protection and long life do not as yet exist.

GENERAL REFERENCES

Ryntz, R. A., *Adhesion to Plastics: Molding and Paintability*, Global Press, Moorhead, **1998**.
Ryntz, R. A., Ed., *Plastics and Coatings: Durability, Stabilization, Testing*, Hanser Gardner, Cincinnati, **2001**.
Wilke, G.; Ortmeier, J., *Coatings for Plastics: Compact and Practical*, Vincentz, Hannover, **2011**.

REFERENCES

Anderson, K. D., et al., *SPE ANTEC™*, Indianapolis, IN, **2016**, 601–605.
Bailey, W., et al., *J. Coat. Technol.*, **1990**, *62*(789), 133.
Barletta, M., et al., *Prog. Org. Coat.*, **2016**, *90*, 178–186.
Bohannon, J. M., US patent 8,637,586 (**2014**).
Bongiovanni, R., et al., *Prog. Org. Coat.*, **2002**, *45*, 359.
Bulian, F.; Graystone, J. A., *Wood Coatings: Theory and Practice*, Elsevier, Amsterdam, **2009**.
Caldwell, R. A., *JCT Coat. Technol.*, **2005**, *2*(3), 30.
Cannon, K.; Young, D. L., *Proceedings of the International Waterborne, High-Solids, and Powder Coatings Symposium*, New Orleans, LA, **2005**, pp 134–143.
Cannon, K., et al., *Proceedings of the International Waterborne, High-Solids, and Powder Coatings Symposium*, New Orleans, LA, **2005**, pp 34–42.
Carson, T.; Morris, L.; Bohannon, J., PCI Magazine, **2015**, April.
De Oliveira, M. F., *ABRAFATI: International Coatings Congress*, Brazilian Coatings Manufacturers Association, Sao Paulo, **2009**.
Duan, R.; Song, W. J.; Song, N., Presented at the American Coatings Conference, Atlanta, GA, April 7–9, **2014**; Paper 6.5.
Dvorchak, M. J., *J. Coat. Technol.*, **1997**, *69*(866), 47.
Dvorchak, M. J., et al., *Coat. World*, **1997**, November–December, 28.
Eastman Chemical, Technical Bulletin, GN-4248, Kingsport, TN, **2004**.
Eastman Chemical, Technical Bulletin, GN-411D, Kingsport, TN, **2014**.
Elmoursi, A. A.; Garner, D. P., *J. Coat. Technol.*, **1992**, *64*(805), 39.
Esser, R. J., et al., *Prog. Org. Coat.*, **1999**, *36*, 45.
Fong, J. J. J., US patent 5,124,210 (**1992**).
Garner, D. P.; Elmoursi, A. A., *J. Coat. Technol.*, **1991**, *63*(803), 33.
Haag, H. F., US patent 5,053,256 (**1991**).
Haag, H. F., *J. Coat. Technol.*, **1992**, *64*(814), 19.
Hallack, M., et al., *Polym. Paint Colour J.*, **2014**, October, 24–25.
Harmer, W. L., US patent 4,273,912 (**1981**).
Hintze-Bruening, H.; Borgholte, H., *Prog. Org. Coat.*, **2000**, *40*, 49.
Howard, C., PCI Magazine, **2014**, May.
Howard, C., et al., *Proceedings of the International Waterborne, High-Solids, and Powder Coatings Symposium*, New Orleans, LA, **2001**, pp 490–503.
Huang, E. W., et al., *Coat. World*, **1997**, November–December, 21.
Irle, C., et al., *Proceedings of the International Waterborne, High-Solids, and Powder Coatings Symposium*, New Orleans, LA, **2001**, pp 455–463.
Iyer, P., *Plastics Decorating*, **2015**, April–May.
Kinosada, M., et al., US patent 5,135,984 A (**1992**).
Kohman, R. G., et al., US patent 6,444,134 (**2002**).
Landry, V., et al., *Proceedings of the International Waterborne, High-Solids, and Powder Coatings Symposium*, New Orleans, LA, **2006**, pp 225–235.
Lane, J. M.; Hourston, D. J., *Prog. Org. Coat.*, **1993**, *21*, 269.
Martz, J. T., et al., US patent 4,997,882 A (**1991**).
Masuda, T., et al., US patent 6,861,471 (**2005**).
Mori, N., et al., US patent 6,465,563 (**2002**).
Nikolic, M., et al., *JCT Res.*, **2015**, *12*(3), 351
Ou, R., et al., US patent 8,664,298 B1 (**2014**).
Parker, A. A.; Martin, E. S., *J. Coat. Technol.*, **1994**, *66*(829), 39.
Pavia, N. T., et al., *J. Appl. Polym. Sci.*, **2011**, *124*(3), 2311–2317.
Pedain, J., et al., US patent 4,727,128 (**1989**).
Pilcher, G. R., *CoatingsTech*, **2015**, *12*(8), 24–33.
Romo-Uribe, A., et al., *Prog. Org. Coat.*, **2016**, *97*, 288–300.
Ryntz, R. A., *Painting of Plastics*, Federation of Societies for Coatings Technology, Blue Bell, PA, **1994**.
Ryntz, R. A., *Prog. Org. Coat.*, **1996**, *27*, 241.
Ryntz, R. A., Ed., *Plastics and Coatings: Durability, Stabilization, Testing*, Hanser Gardner, Cincinnati, **2001**.
Ryntz, R. A., *JCT Res.*, **2005**, *2*, 351.
Ryntz, R. A., et al., *J. Coat. Technol.*, **1992**, *64*(807), 29.
Ryntz, R. A., et al., *J. Coat. Technol.*, **1995**, *67*(843), 45.
Satas, D., Ed., *Plastic Finishing and Decoration*, Van Nostrand Reinhold, New York, **1986**.
Schoff, C. K., *CoatingsTech*, **2016**, *13*(9), 64.
Shoemaker, S. H., *J. Coat. Technol.*, **1990**, *62* (787), 49.
Sow, C.; Reidl, B.; Blanchet, P., *J. Coat. Technol. Res.* **2010**, *8*, 211.
Tang, H.; Martin, D. C., *J. Mater. Sci.*, **2002**, *37*, 4783.
Vielhauer, L., *Paint Coat. Ind.*, **2013**, *29*(5), 28.
Wicks, Z. W., Jr., Jones, F. N.; Pappas, S. P.; Wicks, D., *Organic Coatings Science and Technology*, 3rd ed., John Wiley & Sons, Inc., New York, **2006**.
Wilke, G.; Ortmeier, J., *Coatings for Plastics: Compact and Practical*, Vincentz, Hannover, **2011**.
Williams, T.; Yu, H.; Hicks, R. F., *Rev. Adhes. Adhes.*, **2013**, *1*, 46–87.
Winchester, C. M., *J. Coat. Technol.*, **1991**, *63*(803), 47.
Yaneff, P. V., et al., *JCT Res.*, **2004**, *1*, 201.

Chapter 32

Architectural Coatings

In 2014 architectural coatings comprised about 59% of the volume (740M gallons) and 48% of the value ($10.6B) of coatings sold in the United States (Pilcher, 2015). Architectural coatings are also called *trade sales paints* and *decorative paints*. Two principal markets for these paints are contractors and *do-it-yourselfers*, who paint their own homes and furniture. (Chapter 31 covers furniture that was painted in a factory.) The contractor is particularly concerned with cost of application, which is often greater than the cost of the paint. For example, contractors want paints that cover in one coat and touch up readily. The do-it-yourselfer is more likely to be concerned about ease of cleanup, odor, and the range of colors available. All parties want paints that are easy to apply. Concern with the cost of paint itself varies significantly within both groups. In the United States, paint is sold through several distribution systems: large merchandisers, hardware and lumber stores, independent dealers, and paint stores owned or franchised by paint companies. Paint is also sold directly to contractors by paint manufacturers. There are three, sometimes overlapping, classes of paint companies: large companies that do extensive national advertising to promote trade name paints; smaller regional companies that primarily sell directly to local contractors, as well as to individuals and contractors in regional stores; and companies that sell private label paints to large merchandisers or hardware chains, primarily for do-it-yourself use. Most companies formulate their paints to be most suitable for the climate and style trends and the regulations in the regions they supply.

Technology of architectural coatings is not static. Its evolution has led to many incremental improvements in coating properties, reductions in environmental impact, and sustainability. From 2001 to 2014, waterborne coatings increased from 77 to 85% of the US market; the volume of solventborne was almost constant but its percentage declined (Gilbert, 2016). Future evolution is certain.

Large companies generally make several lines of the major product types, for example, "good, better, and best." The good grade is designed for the individual whose principal criterion for selecting paint is price per unit volume. Such paint is usually adequate for the application with fair coverage. But buying the cheapest paint is usually a false economy. The best paint is usually designed to give the longest life and to have as good coverage as can be designed into that class of paint; it may also be easier to apply and have better appearance. The "better" paint is a compromise between the two.

While some colored paints are produced at the factory, the majority are manufactured as tintable paints that are custom-tinted at the point of sale. The manufacturing company supplies an extensive line of color chips or standards, with liquid tinting colorants and formulations to use with paints called *bases*, in order to match the color chips. This makes it possible for the paint store to provide thousands of colors with only a limited inventory. Online and in-store computerized systems make it possible to visualize the color on the walls of different representative rooms or even on digital pictures of your rooms.

Technology is being developed and to some extent implemented to impart various functions to trade sales paints, for example, thermal insulation, biocidal and insecticidal properties, anti-graffiti surfaces, self-cleaning (superhydrophobic), infrared reflectance, sound dampening, soft touch, and formaldehyde scavenging. Some of these developing technologies are discussed in Chapter 34.

A multitude of products is made for the architectural market. We restrict discussion to the three largest classes of paints: *exterior house paints and primers*, *interior flat wall*

Organic Coatings: Science and Technology, Fourth Edition. Frank N. Jones, Mark E. Nichols, and Socrates Peter Pappas.

paint, and *gloss enamels*. Other smaller, but still significant, product lines include stains, varnishes (Sections 14.3.2 and 15.7), floor paints, and many specialty products.

32.1 EXTERIOR HOUSE PAINTS AND PRIMERS

This section addresses paints and primers for exterior wall surfaces. Generalizations on the subject are treacherous because the surfaces may be any of 50 species of wood, treated wood, plywood, particle board, metal, concrete, or stucco. Old paint is the most common surface to be painted, presenting a variety of potential problems. Local climatic conditions are another important variable. Feist (1996) and Williams (2010) provide extensive discussions about finishing wood with many types of coatings. Low gloss paints are usually preferred for walls, whereas semigloss or glossy enamels (Section 32.3) are commonly applied to trim around windows and doors and to shutters.

The great majority of exterior house paint is based on latex polymers (Chapter 9). The exterior durability of the best latex paints is superior to that of air-dried solventborne house paints. Latexes are made by emulsion polymerization (Chapter 9) and undergo film formation by coalescence (Section 2.3.3). Solventborne house paints are used if outside painting must be done when the temperature is below about 1–7°C (the minimum application temperature is given on the label), since latex paints do not coalesce properly at such low temperatures. Unmodified latex paints do not adhere well to a chalky surface, that is, a surface with a layer of poorly adherent pigment and eroded old paint; careful washing to remove chalk is required.

Regulations limiting volatile organic compound (VOC) emissions have had a major impact of the development of architectural paints and are likely to do so in the future. As detailed in Section 18.9.2, the US EPA national maximum VOC emission for exterior flat paint was $250\,g\,l^{-1}$ in 2015, but in some localities the limit is as low as $50\,g\,l^{-1}$ (http://www.arb.ca.gov/coatings/arch/rules/VOClimits.pdf). The European Union limits VOC of waterborne matte paints to $30–40\,g\,l^{-1}$, but with a less restrictive definition of coalescing solvents. Regulations in the Asia/Pacific region vary.

Latex paints generally perform better than oil or alkyd paints on wood siding. When oil and alkyd paints were widely used on wood house siding, paint failure by blistering and cracking was common; such paints have low water vapor permeability; water that reaches the back side of a film and starts to volatilize from solar heating causes blister formation. Latex paints have T_g values below summer temperatures and, therefore, have relatively high moisture vapor permeabilities. Accordingly, blistering is unusual.

For maximum longevity on new wood, the US Forest Service recommends a four-coat process (Williams, 2010):

- Water-repellent preservative
- Primer
- Two coats of latex top coat

Service lifetimes of 10–20 years are projected.

Latex paint technology has been steadily improving. Self-cross-linking acrylic latexes are now in use. Certain high quality exterior paints incorporate cross-linkers, such as adipic dihydrazide, as explained in Section 17.11. Some acrylic latex paints do not require primer, and the trend seems to be in that direction. VOC levels of $50\,g\,l^{-1}$ are common, and "zero VOC" paints are available. Some manufacturers now offer a limited lifetime guarantee on their top-quality paints.

Latex paints are more resistant than oil or alkyd paints to *grain cracking*. Wood expands and contracts as its moisture content increases and decreases. A coating applied with an acrylic latex binder maintains its extensibility after many years of exterior exposure and can expand and contract with the wood. Highly unsaturated oil and alkyd coating films continue to cross-link with exterior aging, becoming less extensible and more likely to crack as the wood expands and contracts. Oil and alkyd paints also commonly fail by chalking after exposure. High quality latex paints are less likely to fail by chalking or, at least, to last longer before serious chalking occurs.

Latex paints generally do not level as well as solventborne paints. Modern latex paints formulated with associative thickeners level better, but still not as well as solventborne. This problem is not viewed as too serious for low-gloss exterior paints, but it is with high gloss paints, as discussed in Section 32.3. This disadvantage of latex exterior house paints is offset by superior sag resistance compared with solventborne paints.

Since they remain thermoplastic with aging, latex paints are more prone to retain dirt on their surface than are oil or alkyd paints. Dirt pickup is undesirable anywhere, but is a particular problem in places where lignite and soft coal are used for heating and cooking, or where there is blowing dust. White latex paints may turn unsightly after only a few months' exposure. As discussed in Section 2.3.3, film formation of latex paints occurs by coalescence of the polymer particles. Coalescence requires that the temperature during film formation be above the T_g of the particles. If the paint is applied at low temperature, the T_g must be correspondingly lower. However, when the temperature is high during the summer, the film is soft and susceptible to dirt pickup, even when the dirt content in the air is not excessive. Dirt pickup decreases as T_g increases. Gilbert (2016) describes the difficulties in testing dirt pickup. He says that "coatings that are higher in gloss, smoother, and harder" have better dirt pickup resistance. However, he notes wide variations among paints with similar sheen and hardness. T_g and gloss are not the only factors. Jiang et al. (2016) report that blending a silicone emulsion into an acrylic latex reduces dirt pickup. Higher T_g styrene/acrylic latex paints undergo less dirt pickup than straight acrylic latex paints (Smith and Wagner, 1996), but tend to chalk in sunny exposures of the southern latitudes of North America, and thus have not been employed extensively in the United States.

The condition of the wood surface to be painted can be a significant factor in the performance of paint. Contamination with dirt and oil can interfere with adhesion. When bare wood is exposed outdoors, the surface degrades quickly. Painting degraded wood surfaces can result in what appears to be adhesive failure but is actually cohesive failure of the wood. Even a two- to four-week outdoor exposure of freshly cut wood can initiate wood degradation that adversely affects adhesion (Williams, 2010). Many woods discolor when exposed to UV; this is particularly noticeable in outdoor exposure but happens more slowly indoors. Two- or three-coat finishes have been developed that markedly decrease the discoloration of wood. The first coat next to the wood contains a nitroxy HALS, and the second coat contains a combination of bis-, tris-, and tris-*para*-substituted-tris-resorcinol-triazines as UV screeners (Hayoz et al., 2003).

It is recommended that, prior to painting new wood, joints and cracks be caulked and that the wood be treated first with a paintable water-repellent preservative and then primed with an alkyd or latex primer. While good latex primers are available, many experts prefer alkyd primers, which form tough films and can be formulated for good adhesion and adequate permeability. The ends of boards are especially vulnerable to water penetration, so it is important that they be primed and painted.

On many kinds of wood, a suitably formulated latex paint can be used for both the primer and top coat, although the job may not last quite as long as one with a separate primer. However, woods such as redwood, cedar, and some pines contain water-soluble materials that are extracted by a latex paint leading to reddish brown stains. The extractives are naturally occurring phenolic compounds. Special *stain-blocking* latex primers are made for use over redwood and cedar. Historically, these primers were formulated with a somewhat soluble lead pigment, which formed insoluble salts with the phenolic compounds. To avoid lead compounds, other ways to insolubilize the phenolics have been found. One approach is to include in the formulation a cationic ion exchange latex. Many houses are sided with preprimed hardboard siding; see Section 31.1.2.

As mentioned earlier, another potential problem in applying latex house paints can be poor adhesion to chalky surfaces. The surface of chalky paint is covered with a layer of loosely held pigment and paint particles. When latex paint is applied over such a surface, the continuous phase of the paint penetrates between the particles down to the substrate, but the latex particles are large compared with the interstices among the chalk particles and do not penetrate significantly. When the water evaporates and coalescence occurs, the paint film only partially penetrates the chalk particles, resulting in limited binding of the chalk particles together and to the substrate. As a result, adhesion may be poor. Latex chalk sealers made with very small particle size latexes are available to remedy the problem.

Alternatively, one can use an oil- or alkyd-based primer. With oil or alkyd paints, the binder surrounds the chalk particles, and both the continuous phase and binder penetrate between the particles down to the substrate surface, thereby minimizing the adhesion problem. Adhesion of latex paints can be improved by careful cleaning of the surface to be painted to remove all chalk. For large areas, such as entire wall sections of a house, power washing is effective for removing chalk, as well as dirt and loose paint. Another formulation method is to replace part of the latex polymer binder with a long oil alkyd resin or a synthetic drying oil. The alkyd or oil is emulsified into the latex paint. After application, as the water evaporates, some of the alkyd or oil can penetrate between the chalk particles, improving adhesion. For the paint to have good storage life, the alkyd or oil must be hydrolytically stable. The drying oil or alkyd reduces the exterior durability of the paint, especially its resistance to chalking, fading, and mildew growth, as compared with pure latex paints. The problem of adhesion over chalky surfaces is becoming less common as the incidence of chalky weathered surfaces declines, owing to the decreasing use of oil-based paints and "cement paints."

Table 32.1 provides a formulation recommended by a latex manufacturer for a high quality exterior white house paint (Anonymous, 2004). This formulation is obsolete. It is not included to recommend it or these raw materials over others; rather, it provides a framework for discussing the reasons for the many components of a latex paint and for the indicated order of addition during making of the paint. Note that the formulation is given in terms of both volume and weight of the ingredients. In general, the weight figures are used for production purposes, though sometimes fluid metering devices are utilized. The volume figures are calculated from data on each ingredient and indicate the volume contributions to the *wet paint*. The contribution of each pigment to the PVC (Section 22.1) of the *dry film* is also calculated.

Triton® CF-10 is a low foam nonionic surfactant to aid in wetting the pigment. (See Section 9.1.1 for discussion of surfactants.) Colloids® 643 is an antifoam agent (Section 24.8). The minimum level of antifoam necessary for controlling foam should be used; excess antifoam

Table 32.1 Exterior White House Paint

Material	Weight (lb)	Volume (gal)	PVC
Water	65.42	10.25	
Triton CF-10	1.00	0.11	
Colloid 643	2.00	0.28	
Tamol 731A	14.18	1.50	
Potassium tripolyphosphate	1.00	0.05	
Ti-Pure R-706	225.00	6.70	19.18
Minex 4	150.00	6.85	19.60
Attagel 50	5.00	0.25	0.72
Natrosol 250 MHR (2.5%)	125.00	15.00	
Kathon LX(1.5%)	1.70	0.20	
Rozone 2000	6.00	0.66	
Disperse at high speed for 13–20 min, then let down with Rhoplex Multilobe 200	320.00	36.00	
Ropaque Ultra	45.00	5.25	7.85
Colloid 643	2.00	0.28	
Texanol	11.10	1.40	
Triton CF-10	0.50	0.07	
Propylene glycol	16.35	1.90	
Acrysol RM-2020 NPR	11.00	1.25	
Natrosol 250 MHR (2.5%)	14.40	1.25	
Water	85.55	10.25	
	1122.20	100.00	

Properties

PVC (%) (TiO_2 19.18; Minex, 19.60; Attagel, 0.72; Ropaque, 7.85), 47.4
Volume solids (%), 34.8
Stormer viscosity, equilibrated (KU), 96
ICI viscosity, equilibrated (mPa·s), 16
VOC ($g\,l^{-1}$, excluding water), 100 $g\,l^{-1}$

can result in crawling when paint is applied and can interfere with color acceptance (Section 32.2). Selection of antifoams is empirical; the manufacturers of proprietary antifoams offer test kits containing samples of various products. The formulator can test the samples to find which is most effective with the particular formulation. Polymeric defoamers have been recommended (Przybyla, 2014).

Tamol® 731A is a sodium salt of a carboxylated polyelectrolyte used as a pigment dispersant (Chapter 21). Potassium tripolyphosphate (KTPP) is also a pigment dispersant.

Ti-Pure® R-706 is a rutile TiO_2 pigment, surface-treated for ease of dispersion in waterborne coatings (Section 20.1.1). The PVC of TiO_2 in the formulation is 19.18%. The Ropaque® Ultra™ added later in the formulation provides additional light scattering. In this formulation, dry powder TiO_2 is called for. In many cases, large manufacturers of latex paints use TiO_2 in liquid form ("slurries"), that is, dispersions of TiO_2 in water supplied by the pigment manufacturer for more economical material handling (Section 20.1.1).

Minex® 4 (nepheline, an alkaline aluminum silicate) is an inert pigment of lower cost than an equal PVC of TiO_2. Its refractive index is sufficiently similar to that of the binder in the paint that it contributes little to hiding by itself. Rather, the fine particle size allows it to serve as a *spacer* pigment that increases the hiding efficiency of the TiO_2 relative to larger inert pigment particles. The effectiveness of spacer pigments has been questioned, as discussed further in Section 32.2. Attagel® 50 is an attapulgite clay to provide thixotropy to control settling.

A variety of types of inert pigments can be used. Particle size, cost per unit volume, and color are major selection criteria. Calcium carbonate is inexpensive and is sometimes used as an inert pigment in exterior latex paints, but there is a potential problem. Water permeability of latex paint films is quite high; water and carbon dioxide, which are in equilibrium with carbonic acid, can permeate into the film. Calcium carbonate dissolves in carbonic acid to produce a calcium bicarbonate solution that can diffuse out of the film. On the surface, the water evaporates, leaving a deposit of calcium bicarbonate. In areas protected from rainfall, such as porch ceilings and under eaves, calcium bicarbonate reverts to insoluble calcium carbonate. The white deposit, called *frosting*, on the surface of the film is undesirable, especially with colored paints. Thus, calcium carbonate should be avoided in latex paints that may be tinted to a dark color. Furthermore, calcium carbonate-pigmented paints are subject to degradation by acid rain; not only can frosting result, but also film properties can deteriorate; and the films can become more subject to mildew due to porosity developed in the films by loss of calcium carbonate (Hook et al., 1994).

Natrosol® 250 MHR is a hydroxyethylcellulose (HEC) thickener (Chapter 17).

Kathon® LX (chloromethyl/methylisothiazolone) is a biocide that serves as a preservative. Bactericides are needed to control bacterial growth in the can of paint. There are three adverse effects of bacterial growth: a putrid odor can develop; the metabolic processes release gases that can build up enough pressure to blow off the can lid; and since bacteria produce enzymes that can degrade cellulose derivatives like HEC, the viscosity of the paint may drop.

Many other compounds are used as fungicides and bactericides (Wachtler and Kunisch, 1997), and new ones appear periodically. In many exterior paints, zinc oxide is used as one of the white pigments. It contributes little to hiding—it is used primarily as a fungicide. It is also helpful in latex primers for tannin stain blocking and for rust inhibition. Zinc oxide can cause large viscosity increases during storage of some latex paints. The effect of ZnO is system dependent; the variables and possible formulation approaches to maximize stability are discussed by Mattei et al. (1991).

Testing of biocides is difficult because fungal and bacterial growth is dependent on ambient conditions and because many fungicides and bactericides are effective against only a limited number of organisms. ASTM D-2574-96 is a standard test method for resistance of latex paints in the container to attack by microorganisms.

The best control of bacterial growth is to avoid contamination of the paint with bacteria. Housekeeping in a latex paint factory should be as good as it is in a food processing facility. Furthermore, bactericides kill bacteria, but do not deactivate enzymes that are produced by bacteria. If bacteria can grow in the factory, bacterial enzymes can get into paint. The enzymes catalyze splitting of the HEC molecules and the viscosity drops, even though there is adequate protection against bacteria. Some thickeners are not attacked by bacteria.

Rozone® 2000 (1,2-dichloro-2-*n*-octyl-4-isothiazolin-3-one) is a preservative and a mildewcide that minimizes mildew growth on the applied paint film. Other 4-isothiazolin-3-ones are available. Mildew is perhaps the most common cause of deterioration of the appearance of latex house paints. In order to be effective, an organic mildewcide must migrate to the surface of the film where it is subject to leaching by water. When painting over mildewed paint, the old mildew should be killed with a mildewcide or with bleach before applying the new paint. If the biocides are adsorbed on the surfaces of zeolite, they are released more slowly from the film, and thus the paint films maintain their antibacterial activity for longer times (Edge et al., 2001). The search for more effective mildewcides is an active research area. Existing 4-isothiazolin-3-ones face increasing restrictions, especially in Europe and potentially in North America (Gilbert, 2016). Unlike most failure modes where surfaces exposed toward the sun degrade more quickly, fungal and mold attack on exterior paints usually occurs on the side of a building facing away from the sun, that is, on the north side of a building in the northern hemisphere.

The latex cannot be added until after pigment dispersion is complete because most latexes would coagulate by the high shear exerted in the dispersion step. Therefore, after the pigment is dispersed, the high-speed impeller is slowed to avoid coagulation and also excessive foaming.

Multilobe® 200 is an acrylic latex in which groups of a few particles are fused together into nonspherical lobed particles (Chou et al., 1987). At equal concentration, the viscosity of this latex is higher than the viscosity of comparable latexes having spherical particles, since the packing factor for lobed particles is smaller than for spherical particles. Paints formulated with a lobed latex are shear thinning and provide a higher high shear viscosity at the same low shear viscosity than conventional latexes. The higher viscosity at shear rates experienced during application leads to thicker films, resulting in the need for less thickener and, hence, lower cost. Adhesion to chalky surfaces is said to be improved as compared with paints formulated with conventional latexes.

Latexes with 100% (meth)acrylic monomer content provide excellent durability. Replacement of part of the methyl methacrylate (MMA) by styrene results in lower in cost and lower dirt pickup, as well as satisfactory exterior durability; though in sunnier latitudes, chalking and fading are more likely. Vinyl acetate/acrylate copolymers are even less expensive. Although their hydrolytic stability is inadequate for exterior use in climates with high humidities and rainfall, they are used in interior applications, as well as outside in arid climates. This dual use permits smaller paint companies to inventory a single latex for interior and exterior coatings. Copolymers of vinyl versatate (a C_{10} branched carboxylic acid vinyl ester) (Section 9.3) have better exterior durability than vinyl acetate copolymers and are widely used in Europe as vehicles for exterior paint (Prior et al., 1996; Decocq et al., 1997). A VAc/BA/C_{12} branched carboxylic acid vinyl ester copolymer latex paint is reported to have exterior performance equal to that obtained with all-acrylic latexes at a substantially lower cost (Wang et al., 2001). Paints made from copolymers of acrylic and highly branched C-12 vinyl ester monomers are sufficiently weatherable to serve as roof coatings in equatorial climates (Vanaken et al., 2014).

Ropaque Ultra is a high T_g latex that contains small bubbles of water within the particles (Section 20.1.2) (Fasano, 1987). When the paint film dries, the water diffuses out of the particles, leaving voids that scatter light, reducing the TiO_2 requirement. In calculating PVC, one includes the volume of these particles, since they do not coalesce with the latex binder particles during film formation.

Texanol® is a coalescing agent—a mixture containing primarily the monoisobutyrate ester of 2,2,4-trimethylpentane-1,3-diol (TMB). It reduces T_g during film formation and then slowly evaporates. The high degree of steric hindrance of the ester group of Texanol provides hydrolytic stability.

Propylene glycol is in the formulation for two main reasons. First, it acts as antifreeze to stabilize the liquid paint against coagulation during freezing and thawing. The expansion of water as it freezes exerts substantial pressure on the latex particles and can push them together with enough force to overcome the repulsion by the stabilizing layer, resulting in coagulation. The glycol lowers the freezing point; even if the temperature gets low enough to freeze the mixture, it freezes to slush so that less pressure is exerted on the latex particles. The second function of propylene glycol is to control the rate of drying of the paint to permit *wet lapping* without disruption of the edge of the film. Oil paint presents no problem with wet lapping. As water evaporates from a latex paint film, viscosity increases rapidly as a result of the increase in volume fraction of the internal phase. When paint is applied with a brush or roller, the edge of the wet paint film is painted over (lapped) so that no substrate is left unpainted or with only a thin layer of paint. By the time lapping occurs, the viscosity of the applied paint may have increased enough so that the film is semisolid, but it remains weak. The pressure of a brush or a roller can break up the film, resulting in irregular chunks of paint film along the lapped edge. This problem can be minimized by adding slow evaporating solvents that increase the time for wet lapping. Such solvents are said to increase the *open time*. Propylene glycol is commonly used for this purpose. It also aids coalescence, although it is less effective than other coalescents. A major drawback, however, is that propylene glycol is a VOC. Additives designed to replace propylene glycol in this function are discussed in Section 32.3.2.

Acrysol® RM-2020 is a nonionic *associative thickener*. It helps control viscosity of the paint as a function of shear rate, improving application characteristics. Associative thickeners permit formulation of latex paints that exhibit less shear thinning so that viscosity at high shear rates can be higher; as a result, thicker wet films can be applied. Formulation with associative thickeners also temporarily reduces viscosity at low shear rates, so leveling is also improved without causing undue sagging or settling. (The thicker wet film in itself helps promote leveling, as discussed in Section 24.2.)

Associative thickeners are discussed in Section 3.5.1. They play a critical role in formulation of contemporary latex paints. Apparently, formulators continue to gain understanding of the complex interactions involved, leading to ongoing improvements in the application properties of such paints.

The viscosity of the external phase controls its rate of penetration into a porous substrate, such as wood. If penetration is rapid, the viscosity of the paint above the porous surface increases rapidly, resulting in poorer leveling. The final water addition permits adjustment of the viscosity and solids content of the final paint. There is variation from batch to batch of the viscosity of any paint formulation.

The PVC of this paint is 47.4%, imparting low gloss. Many exterior latex wall paints have low gloss. Popularity of exterior satin finish paints has increased in the recent years. Gloss is strongly influenced by the PVC to CPVC ratio (Chapter 22). Exterior durability as well as many other properties is also affected by PVC. It is desirable to maximize CPVC, since PVC of the paint at the same PVC to CPVC ratio increases with CPVC. A major factor controlling CPVC of latex paints is the particle size distribution of the pigments. Multiple inert pigments maximize the breadth of particle size distribution, which increases CPVC and permits higher pigment loading at the same PVC to CPVC, decreasing cost (Sections 22.1.1 and 22.1.3).

Commonly, maximum gloss of latex paints is lower than that of oil paints. The volume solids of the paint is 34.8%, substantially lower than exterior oil- or alkyd-based paints. As a result, coverage per unit volume of latex paints is lower than oil-based paints. This formula is old; it has a VOC of $100\,g\,l^{-1}$, well under current US national rule air pollution regulations but well over the VOC limits in many US localities.

In spite of its widely recognized shortcomings, the Stormer viscometer, discussed in Section 3.3.6, is used in this case, as it is used in most paint labs in the United States for architectural coatings. The viscosity of the unsheared paint is measured before shearing reduces the viscosity, which is related to the thixotropy of the paint. The ICI viscosity is determined on a special cone and plate viscometer that measures viscosity at a higher shear rate similar to that exerted by application of the paint by brush or roller. It is important to measure high shear viscosity, which is a major factor controlling film thickness.

Emulsions of siloxane-containing resins combined with a styrene/acrylic latex are used to formulate exterior house paints. Such coatings have been used on a large scale in Europe for many years but are just being introduced in

North America. They can have outstanding exterior durability (Mangi et al., 2002).

There is a considerable market for coatings that increase the exterior durability of wood but are not pigmented, or are lightly pigmented, as transparent stains. Results of a variety of treatments of coatings on fir and oak have been reported. For example, good results were obtained with a waterborne transparent stain comprising an alkyd/acrylic dispersion binder pigmented with iron oxide (Jirous-Rajkovic et al., 2004). As reviewed by Nikolic et al. (2015), addition of inorganic nanoparticles that absorb UV light (e.g., cerium oxide) can improve the exterior durability of stains and varnishes with minimal effect on transparency and with improvement in film properties.

Latex paints are generally superior to alkyds on masonry (e.g., stucco) because of their superior resistance to hydrolysis. A series of acrylic and styrene/acrylic latexes were evaluated with a range of pigments (Anonymous, 2004). Key properties considered were water resistance, moisture permeability, exterior durability, dirt pickup resistance, color retention, low chalking, and good elasticity. While styrene/acrylic latexes had poorer exterior durability and color retention than all-acrylic latexes in clear films, this difference became minimal in pigmented paints. The overall best performance was obtained using BA/MMA, EHA/MMA, and EHA/styrene latexes. EHA-containing latexes were notable for soiling resistance. The EA/MMA latex had poor performance. The inert pigments also affected performance, for example, SiO_2 increased chalking.

Many formulations in which acrylic and other latexes are blended with polyurethane dispersions (PUDs) (Section 12.7.2) have been studied, and some are the basis for commercial exterior paints. For example, Fu et al. (2009) describe blends of "hard" acrylic polymers (T_g 30–40°C) with complex "soft" PUDs in a way that provides rapid development of hard, tough, block-resistant films.

32.2 FLAT AND SEMIGLOSS INTERIOR PAINTS

The largest volume of architectural paint is interior flat wall paint; almost all is latex based. People repaint the inside of their houses more often than the outside. Although flat wall paints are most common, low semigloss ("satin" or "eggshell") wall paints are becoming more popular for their "soft" appearance. As of 2015, the US EPA permitted level of VOC for interior flat paints was 250 g l^{-1}, but many jurisdictions set limits as low as 50 g l^{-1}. Further restrictions are likely. Limits for semigloss and glossy paints are generally higher (http://www.arb.ca.gov/coatings/arch/rules/VOClimits.pdf).

Wu et al. (2016) list the most sought-after attributes of premium interior paints such as hiding, ease of application, stain resistance ("easy clean"), and scrub durability. Formulators adjust the first two attributes with pigmentation and rheology modifiers such as associative thickeners. The latter two attributes are somewhat antagonistic because erosion of the paint is a common mechanism of stain removal. Wu et al. (2016) performed a benchmark study of 10 unnamed commercial paints, all with VOC <50 g l^{-1}, and found two- and threefold variations in scrub and stain resistance tests.

A major opportunity for technical progress would be to improve the ease with which interior paints can be cleaned. It is relatively easy to formulate paints with water-repellent additives that are "easy to clean" at first, but studies of the long-term durability of the effect are in short supply (Schulte, 2015). Progress in this direction is indicated by the introduction of "chalkboard paints."

The major advantages of interior latex paints as compared with the older oil-based paints are:

Fast drying and less sagging. If desired, two coats of fast drying paint canbe applied to the walls of a room during a day, the furniture moved back, and the room used that night. The rapid increase in low shear viscosity in the early stages of film formation reduces vulnerability to sagging, even when relatively thick films are applied.

Low odor. The odor of mineral spirits and by-products from oxidation of drying oils in solventborne paints is unpleasant for days after walls are painted. Although less offensive, odor from the slow evaporation of coalescing solvents from latex paints persists for up to a week. An additional reason for developing low VOC paints is to eliminate this odor problem.

Ease of cleanup. Spills, dripped spots, brushes, and rollers are easily cleaned with soapy water in the case of latex paints; solvent is required with oil paints. However, latex cleanup must be done promptly and prior to coalescence, after which cleanup is more difficult than with oil paints.

Low VOC emissions. Latex paints were widely adopted before 1967, when the first VOC regulations were adopted in the United States. Progress is being made in reducing the already low VOC, as discussed in Section 32.3. There is also the advantage with waterborne paints of substantially reduced fire hazard. Not only is solventborne paint flammable, but also rags wet with oil-based paints in a confined space can undergo *spontaneous combustion*, that is, ignite as a result of heat generated by autoxidation.

Less yellowing and embrittlement. White and light color acrylic and vinyl/acrylic latex paints retain their color better than alkyd paints, which gradually turn yellow with age. Alkyds made with oils that contain very little linolenic acid, such as sunflower or safflower oils (Section 15.1.1), yellow more slowly than alkyds based on soybean oil or tall oil fatty acids, but they still yellow with time. Multiple layers of alkyd films also become brittle as autoxidative cross-linking continues far beyond the point of optimum film properties.

Wall and trim paints are generally stocked in retail outlets as white paint; companion tintable paints are called tint bases. Liquid (or sometimes powdered) colorants are added to the tint base in order to make a color chosen by the customer from a large array of color cards. This requires that equal white tinting strength be maintained with the tint bases through quality control, or the colors obtained will differ. Furthermore, new formulations must have the same white tinting strength as the formulation being replaced, or else the color cards and formulations in dealer stores will have to be replaced. In each quality line, two, or sometimes three or four, tinting bases are included. Often, the white paint itself can accept up to a few fluid ounces of colorant per gallon to create very light or pastel colors. Tint bases for making deeper colors have lower levels of TiO_2, depending on how much colorant is to be added. Very deep color bases may contain no TiO_2. Tint bases are made with surfactants that help ensure compatibility between colorant and tint base. The tinting colorants are commonly "universal colorants" that can be used to tint either waterborne or solventborne paints.

Users are sometimes confused by the change in color of latex paint as it dries; the color of a dry film is darker than the color of the wet paint. In the wet paint, the interfaces between the water ($n=1.33$), polymer particles ($n=\sim1.5$), TiO_2 ($n=2.73$), and inert pigments ($n=\sim1.6$) scatter light to a greater degree than when the paint is dry. The dry paint has fewer interfaces as a result of coalescence of the latex particles and refractive index differences are smaller, since the pigment particles are in a polymer matrix instead of water. Since light scattering decreases as water evaporates and the latex particles coalesce, the color gets darker; hiding also decreases as the paint dries.

Another consideration in color selection and matching is the proliferation of different light sources. Users should be aware that colors may look quite different when illuminated with light-emitting diodes (LEDs), fluorescent lights, or incandescent bulbs. Even more light sources with different spectral characteristics are being developed. Standards on how to test color matching under the new sources are lacking. Rich (2016) describes the resulting problems and provides guidance on how to adapt to the situation.

Since exterior durability is not needed, lower cost vinyl acetate copolymer latexes (Section 9.3) are the principal binders for interior wall paints. The T_g of VAc homopolymer is about 32°C, so that a softer comonomer such as butyl acrylate (BA) is needed to reduce T_g. Ethylene/VAc latexes are an alternative to VAc/BA latexes.

For decades, efforts have been underway to exploit the potential of vinyl acetate/ethylene (VAE) latexes in interior paints (Yang et al., 2000). Gilicinski et al. (2000) showed that ethylene is a more effective comonomer than BA for reducing the T_g of VAc polymers. It also appears to provide more effective plasticization. Paints made with VAc/ethylene latexes are said to require less coalescing solvent and show better touch-up properties at low and room temperatures and higher scrub resistance. However, pressure reactors are required to produce such latexes, and polymer design is challenging.

Poole et al. (2010) describe paints made from VAE latexes containing 5–20% by weight of ethylene.

Certain VAE latexes can be formulated into near-zero VOC interior paints (Huster, 2011). They are said to be widely used in Europe and India. Table 32.2 is a supplier-suggested formulation for an ultra low VOC interior eggshell paint (Air Products, 2002). The VOC of the paint is $6\,g\,l^{-1}$.

Natrosol® Plus 330 MBR is a modified HEC associative thickener. Tamol® 1124 is a dispersant that is the ammonium salt of a hydrophilic acid copolymer. AMP-95® is 2-amino-2-methyl-1-propanol. Strodex® PK-90 is a dispersant that may promote color acceptance. Triton CF-10 is a low foam nonionic surfactant. Drewplus® L-745 is an antifoam. Nuosept® 95 is a biocide, a mixture of 5-substituted 1-aza-3,7-dioxabicyclo[3.3.0]octanes. Ti-Pure R-706 is a surface-treated rutile TiO_2 designed for use in waterborne coatings. Mattex® is a surface-treated kaolin clay pigment. Attagel 50 is an attapulgite clay to control settling and syneresis. Airflex® EF 811 is a VA/ethylene latex. Acrysol® RM 2020 NPR and Acrysol® RM-8W are hydrophobically modified ethoxylated polyurethanes (HEUR)

Table 32.2 Ultra Low VOC Interior Eggshell Paint

Component	Weight	Volume
Water	150.0	18.0
Natrosol Plus 220 MBR (3%)	50.0	6.0
Tamol 1124	6.0	0.6
AMP-95	2.0	0.2
Strodex PK-90	2.0	0.2
Triton CF-10	2.0	0.2
Drewplus L-145	1.5	0.2
Nuosept 95	1.5	0.2
Disperse the addition under agitation Ti-Pure R-706	230.0	6.9
Mattex	110.0	5.0
Attagel 50	5/0	0/3
Let down with AirflexEF811 (58.0%)	418.9	47.0
Water	83.0	10.0
Drewplus L-145	3.0	0.4
Acrysol RM-2020 NPR	20.0	2.3
Acrysol RM-8W	8.0	0.9
Benzoflex 9-88	6.0	0.6
Total	1098	99.1

associative thickeners (Section 3.5.1). Benzoflex® 9-88 (dipropylene glycol dibenzoate) is a coalescent.

Acrylic latexes or vinyl versatate copolymer latexes are used for interior applications that require greater water resistance. A copolymer latex of BA, styrene, and 0.5% acrylamide provides superior scrub resistance (Porzio et al., 2001).

Copolymer latexes of VAc and vinyl versatate are used in both interior and exterior paints. These copolymers have better hydrolytic stability than conventional vinyl acetate copolymers. A combination of ethoxylated undecyl alcohol, a cellulose ether, and sodium vinyl sulfonate can be used as a surfactant/protective colloid in preparing the latexes. In a similar vein, VAc/vinyl ester of a C_{12} branched acid latexes is reported to provide better scrub and stain resistance than VAc/BA latexes and styrene/acrylic latexes in interior flat wall paints (Heldmann et al., 1999).

When painting ceilings, one is particularly motivated to obtain hiding in one coat, since painting over one's head and moving the ladder require more effort than painting walls. The problem is especially challenging because ceiling paints are commonly plain white to reflect light diffusely. Hiding by white paints is poorer than any color paint made from it. The problem is compounded by the decrease in hiding when a latex paint film dries. The user thinks that sufficient paint has been applied to hide marks on the ceiling but comes back an hour later to find that marks show through the dry paint. Special ceiling paints minimize this problem by formulating with PVC above CPVC. Dry paint films with PVC above CPVC have air voids which add light scattering by the new interfaces between air and polymer as well as air and pigment. Formulations can be adjusted so that wet hiding and dry hiding are approximately equal. Mechanical strength is less than that of paint with PVC<CPVC; and resistance to staining is poorer, but neither property is important for ceiling paints.

Owing to the high cost of TiO_2, coatings are not generally formulated with a PVC of TiO_2 more than 18%, since incremental hiding at higher PVC is not cost efficient. (This value depends on the actual TiO_2 content of the pigment and the quality of the dispersion.) Rather than using large quantities of TiO_2, lower cost inert pigments are used to occupy additional volume. In the past, many workers thought that inert pigments with particle sizes less than that of the TiO_2, called spacer inerts, increase the efficiency of the TiO_2. However, Diebold (2011) and Hook (2014) demonstrated that small spacers improve TiO_2 efficiency only relative to formulations having larger inert pigment particles.

Another approach to greater hiding at lower cost is to use a high T_g latex, such as polystyrene, as an inert pigment (Ramig and Floyd, 1979). When the latex binder coalesces, the high T_g latex particles remain separate, as with any other pigment. The polystyrene particles occupy volume and enhance hiding by the TiO_2 pigment. Paints can be formulated with PVC greater than CPVC to the extent that air bubbles increase hiding without making the surface of the film porous. The paints provide equal hiding at lower TiO_2 content while retaining good enamel holdout and stain resistance.

Yet another approach to minimizing the TiO_2 requirement is to use special high T_g hollow latexes as pigments, which is illustrated by the formula for an exterior house paint given in Table 32.1 and explained in the text after the table.

A large fraction of latex flat wall paint is applied by roller. During roller application, latex paints *spatter*, some to a major degree. Paints with high extensional viscosity (Section 3.6.3) are likely to spatter severely (Massouda, 1985, Section 23.1.2). Extensional viscosity increases when high molecular weight (MW) water-soluble polymers with very flexible backbones are used as thickeners in a latex paint (Glass, 1978). Spattering can be minimized by using low MW water-soluble thickeners with rigid segments in the polymer backbone, such as low MW HEC. Refinements of associative thickener technology have led to improvements.

For many applications, the increase in external phase viscosity resulting from the use of water-soluble polymeric thickeners is desirable, however, not for concrete block walls. Concrete block surfaces contain holes that are large compared with the pigment and latex particles, as well as pores that are small compared with them. Latex paints do provide better coverage on concrete block than solventborne paints. The lower viscosity of the external (continuous) phase of latex paints leads to more penetration of the continuous phase into the small pores of the block, thereby increasing the viscosity of the remaining paint, which reduces penetration into the holes. Coverage is low with solventborne paint because more paint penetrates into the holes. However, this advantage of latex paints is reduced by water-soluble polymeric thickeners, which increase the viscosity of the external phase. The absence of water-soluble polymer further enhances coverage, because the volume fraction of internal phase of latex paint must be significantly higher to provide equal starting viscosity. With increasing volume fraction of the internal phase, the viscosity increases more rapidly as the continuous phase is drained off in the small pores and penetration into the large holes is further reduced. An additional advantage of latex paints, especially acrylic, styrene/acrylic, or styrene/butadiene latexes, relative to oil or alkyd paints is that they are not subject to saponification by the basicity of the concrete block and the mortar joints.

In a study of changes in composition and properties of latex paints applied to porous inorganic substrates, it was found that a significant amount of coalescing agent penetrated into the substrate, thereby increasing the T_g of the film before it could coalesce. There were also indications that binder penetrated. The authors recommend that it is particularly important to formulate paints with PVC close to CPVC for application to porous substrates (Perara and Eynde, 2001).

When latex paint films coalesce, surfactants may bloom out of the film and collect on the surface. Usually, this is not noticeable. However, if water condenses on the paint surface and dissolves the surfactant, brown spots of surfactant may appear on the surface, because the surfactant is concentrated in the last remaining water droplets that evaporate. The problem can be minimized by avoiding dark-colored surfactants in light-colored paints. However, white surfactant can leave whitish spots on dark color paints. Evanson and Urban (1991) have shown that nonionic surfactants are more compatible with latex polymers and less likely to bloom than anionic surfactants. Grade et al. (2002) have studied the use of alkenyl-functional surfactants in latex preparation. The surfactant copolymerizes with the monomers during polymerization providing latexes that have superior shear stability and freeze–thaw resistance. Paints prepared with the surfactant-bound latexes overcome the problems associated with use of conventional surfactants, such as surfactant blooming.

Although latex paints generally have low VOC compared with solventborne paints, there is still regulatory pressure to reduce or eliminate the VOC. To accomplish this, coalescing solvents that are VOCs and propylene glycol in the formula must be reduced to low levels or eliminated. Increased limits on VOC magnify the technical challenge, and the transition to zero VOC "…is one of the biggest technical challenges in architectural coatings…" (Gilbert, 2016). Several approaches have been utilized to lower the level of coalescing solvent by reducing the polymer T_g, including (1) sequencing the composition of the monomer feed during polymerization so that the latter part of the monomer feed has a higher fraction of monomers, thereby reducing the T_g, as in the power feed approach discussed in Section 9.1.3, and (2) using cross-linkable latexes having a lower T_g, which are cross-linked after film formation, thereby offsetting the adverse effects, such as blocking and dirt pickup, of the lower T_g (Section 9.4). Addition of acrylic acid (AA) in the later part of the comonomer feed, in order to incorporate acid groups on the surface of the latex, has also been suggested as an approach to reduce the level of coalescing solvent. Association of water with amine salts of the carboxylic acid groups is proposed to plasticize the surface of the particles, thereby promoting coalescence.

VOCs from coalescing agents can be eliminated by utilizing cross-linkable analogs. For example, the propylene glycol monoester of sunflower fatty acids has been developed that essentially eliminates evaporation of the coalescent, thus reducing VOC. It is also reported to impart less odor, increased scrub resistance and higher gloss. Zero VOC coalescents are another possibility, especially in Europe where the regulatory definition of VOC allows coalescents with low but appreciable vapor pressure.

Approaches for totally eliminating the need for coalescing solvent have been proposed. The use of blends of high and low T_g latexes has been reported by Winnik and Feng (1996) and Eckersley and Helmer (1997). When a film of such a blend dries, the high T_g latex does not coalesce, but is dispersed in the continuous phase of the low T_g polymer. The hard particles act to reinforce the low T_g polymer film increasing the storage modulus (E') of the film so that its block resistance is superior to that of a film from only the low T_g latex. The properties depend on the ratio of hard and soft latexes and on the particle sizes of the latexes. For example, excellent block resistance was obtained with a combination of 60% of a 37/61/2 styrene/BA/MAA latex (particle size of 475 nm and T_g = 9°C) and 40% of a 70/28/2 styrene/BA/MAA latex (particle size of 118 nm and T_g = 62°C). No coalescing agent was required for film formation at ambient temperature. The superior block resistance is attributed to a high concentration of the small particle size latex at the surface of the film (Section 32.3), resulting from close packing. Another approach to eliminating VOC is to make nanopolymer/clay composite latexes. The latexes are made by polymerizing relatively low T_g acrylic monomers with clay dispersed in the water in which the emulsion polymerization is carried out. Films prepared from the latex have unusual toughness without compromising film formation. The latexes are used in zero VOC paints, which lack low tack and dirt pickup (Lorah et al., 2004) (Section 9.3).

Semigloss paints are made with lower pigment contents than flat paints and usually contain acrylic latexes to provide better wash resistance, compared with other latex types. They are frequently used to paint doors, cabinets, and other woodwork.

Combined aqueous PUDs and latex resins offer several potential advantages as compared with each of the two separate types of polymers (see Section 12.7 for discussion of PUDs). In general terms, polyurethane-based polymers offer superior abrasion resistance due to intermolecular hydrogen bonding. Vinyl resins have lower raw material and processing costs. PUDs have low minimum film forming temperatures (MFFT) in relation to dry film T_g owing to the plasticizing effect of hydrogen bonding with water, which reduces the T_g of the polymer particles during coalescence. Used with high T_g acrylics, PUDS reduce the MFFT without reducing hardness. For example, open time of latex paints can be extended by adding a PUD with a T_g lower than that of the latex in the paint. A paint made with a latex composed of *n*-butyl acrylate (nBA), 2-ethylhexyl acrylate (2-EHA), MMA, and methacrylic acid (MAA) together with a PUD having a T_g of −40°C has longer open time, permitting re-brushing to ensure good hiding and lapping (Gray and Lee, 2001). Coalescing solvents are required to permit film formation with latexes at low temperatures, and they are not always required with PUDs.

Further improvement in properties may be obtained by polymerizing acrylic monomers in a PUD or in a prepolymer precursor for a PUD. Semigloss wood coatings are formulated with a carboxy-functional PUD to which acrylic monometers, including *N*-(2-methacryloxyethyl)ethylene urea, are added and polymerized (Chung et al., 2000). Latexes contain surfactants that can bloom to the surface of films reducing gloss, whereas many PUDs can be made without surfactants.

32.3 GLOSS ENAMELS

The term *enamel* connotes a hard, glossy surface analogous to porcelain enamel. In the United States, gloss enamels are used both indoors and outdoors for trim around windows and doors, for shutters, for wood furniture, and on kitchen and bathroom walls. In some other countries, they are also used for walls. For gloss enamels, the merits and drawbacks of alkyd paints and latex paints are about evenly balanced. The 2015 US EPA VOC emission standard for gloss paints is 380 g l^{-1}, but in many jurisdictions the limit is 150–250 g l^{-1} and may go lower (http://www.arb.ca.gov/coatings/arch/rules/VOClimits.pdf). Improvements in alkyd emulsion technology (Section 15.3.2) may enable alkyds to meet very restrictive regulations. Alkyds are also attractive because they are partly bio-based.

32.3.1 Alkyd Gloss Enamels

As described in Chapters 14 and 15, alkyd enamels crosslink by autoxidation to form tough films that adhere well to a variety of surfaces and are block resistant and water resistant. An advantage of gloss alkyd enamels is that they have higher gloss than latex enamels owing to the formation of a polymer layer having very low pigment content at the outer surface. As discussed in Section 19.10.1, this phenomenon occurs during film formation from solution vehicle paints, but does not happen with latex paints.

The principal advantage of gloss alkyd paints is that they afford good hiding with one coat when used over surfaces without sharp color contrasts. One-coat hiding is especially important to painting contractors for whom the cost of application is often higher than the cost of paint. Another cost factor is that alkyds often require less surface preparation than latexes.

Several factors account for the difference in hiding between alkyd and latex gloss paints. The NVV of a typical solventborne alkyd gloss paint can be 66% or even higher, while the NVV of a latex gloss paint is limited to about 33%. To obtain the same dry film thickness, one has to apply twice as thick a wet film of the latex paint or two coats of the same thickness.

Another factor that affects hiding is leveling. Assume that a uniform dry film thickness of, say, 50 μm of a paint provides just satisfactory hiding. If the paint levels poorly, there will be streaks of thinner film, say, 35 μm, and thicker film, say, 65 μm. Hiding of the uneven film will be poor and likely to appear inferior to that of a uniform 35 μm dry film of the same paint. Since the 35 and 65 μm areas are immediately next to each other, the contrast emphasizes the poor hiding. Furthermore, not only is the hiding adversely affected by poor leveling, but also the contrast in colors resulting from the poor hiding in the valleys compared with the ridges emphasizes the poor leveling. Solventborne alkyd paints for brush application generally level better than latex paints because the solvent is slow evaporating mineral spirits. The difference in leveling is amplified when latex paints are applied under warm, dry conditions.

As discussed in Section 24.2, Overdiep (1986) demonstrated that leveling of solventborne, brush-applied paints is promoted by surface tension differential-driven flow in the wet paint film. When solvent evaporation starts, the fraction of solvent lost from the valleys of the brush marks is greater than from the thicker ridges. Since the surface tension of the solvent is lower than that of the alkyd, the surface tension of the more concentrated alkyd solution in the valleys becomes higher than that in the ridges. The resulting differential in surface tension causes the paint to flow from the ridges into the valleys to minimize overall surface tension, thus promoting leveling. This driving force may be absent in leveling of latex enamels (Section 32.3.2).

When selecting paints for repainting, it should be remembered that latex paints can be used over alkyds or latexes as long as the surface is properly prepared, but use of alkyds over latexes is risky. The solvent in alkyds may penetrate uncross-linked latex films and cause lifting.

The drawbacks of alkyd gloss enamels are largely the same as those of alkyd flat wall paints (Section 32.2). They include slow drying, odor, yellowing, and embrittlement with age, as well as the need to use solvent for cleanup. In the future, perhaps the most important consideration will be VOC content. Advances in waterborne alkyd and acrylic/alkyd hybrid technologies (see Chapter 15 and the following text) indicate that the VOC limitations can be overcome.

High-solids alkyds are discussed in Section 15.2. Solids can be increased to an extent by solvent selection, especially the use of hydrogen-bond acceptor solvents to reduce intermolecular hydrogen bonding. While solids can be increased by reducing MW and by using narrower MW distributions, both of these approaches lead to inferior film properties and durability if carried very far.

A promising approach to increasing solids is the use of reactive diluents (Section 15.2). Such additives are low MW materials designed not only to reduce viscosity almost as efficiently as solvent but also to coreact with the oxidizing alkyd during film formation. This permits reduction in VOC while maintaining film properties. An example is dicyclopentenyloxyethyl methacrylate. This reactive diluent has both an acrylate double bond and an activated allylic position. In the presence of driers, it can coreact with an alkyd such as a long oil linseed alkyd. Using this reactive diluent with specially designed alkyds is reported to permit formulation of gloss alkyd paints with a VOC of $155\,g\,l^{-1}$. The properties of the films are said to approach those of an alkyd paint with a VOC of $350\,g\,l^{-1}$ (Larson and Emmons, 1983). Many other types of reactive diluents have been developed.

When paint cans containing oxidizing alkyds and driers are opened, premature cross-linking at the liquid paint surface occurs. This *skinning* is controlled by additives, as reviewed by Bieleman (2003). The most effective additive was found to be methylethyl ketone oxime (MEKO). MEKO forms an unreactive complex with cobalt ions, $[Co(MEKO)_{1-8}]^{3+}$. The complex is in equilibrium with the components so that its presence sharply reduces the concentration of active cobalt. After the paint is applied, MEKO evaporates relatively rapidly, freeing the cobalt drier. MEKO is added to the coating just before filling into containers. In some cases, a further small amount of MEKO solution is added after the containers are filled just before they are sealed to prevent skinning during storage. Addition of 0.2% of MEKO prevents skinning for more than 250 days. It increase the dust-free time of one alkyd coating to 2 h from 1 h 45 min, but the fully dried time was reduced from 4 h 45 min to 4 h. The initial effect results from the time for the MEKO to evaporate. The faster full drying is thought to result from enhanced permeation of oxygen in the earlier stages after application while MEKO is inhibiting surface cure. Restrictions on the use of MEKO and cobalt may come in the future. Alternative anti-skinning agents have been developed.

The steadily advancing field of waterborne alkyd and uralkyd resins is described in Sections 15.3.1, 15.3.2, and 15.7.1 and was reviewed by Hofland (2012). For trade sales paints, two approaches—water-reducible alkyds and alkyd emulsions—have been extensively investigated. A key obstacle to water-reducible alkyds has been the difficulty of obtaining sufficient hydrolytic stability for the two-year shelf life usually required in North America, but Hofland reports that many emulsified alkyds are sufficiently stable. Conventional water-reducible alkyds generally contain significant levels of solvent, thereby preventing reduction of VOC much below $250\,g\,l^{-1}$, but alkyd emulsions with very low VOC are feasible.

Low VOC emulsions of alkyds in water can also be used to prepare gloss enamels (Hofland, 1997, 2012; Section 26.3). They are used in Europe and are being adopted in North America. The VOC content can be low, but the surfactants required for emulsification can adversely affect application and film properties, and the shortcomings of odor and poor resistance to yellowing remain. Several studies indicate that the surfactant problem can be overcome (Section 15.3.2).

32.3.2 Latex Gloss Enamels

Development of latex gloss enamels has proven to be one of the most challenging problems of coatings technology. Key problems and potential solutions are discussed in this section.

One problem is that of obtaining high gloss. As noted in Section 32.3.1, the clear polymer layer associated with the surface of alkyd enamels does not readily form as latex coalesces. The ratio of pigment to binder at the surface of a latex paint film can be reduced somewhat by using a finer particle size latex, but the gloss is still inferior to that from alkyd paint films. Gloss is reduced by flocculation of either pigment or latex. The choice of dispersing surfactants and thickeners, and their order of addition, can affect gloss. In one study, it was shown that when a mixture of polyacrylic acid (PAA) and HEC was used as a thickener, gloss was higher if the HEC was added first (Hulden et al., 1994). Nonionic surfactants resulted in higher gloss than anionic surfactants. Other factors tending to reduce the gloss of latex paint films are the haze that results from incompatibility of surfactants (and possibly other components) as well as the blooming that results when surfactants migrate to the surface of the film. These problems can be minimized by making latexes with the lowest possible surfactant content and by selecting surfactants for pigment dispersion that are as compatible as possible.

Considerable work has been done using blends of compatible water-soluble resins with latexes to achieve higher gloss. In floor wax applications, morpholine salts of styrene/AA copolymers can be glossy. When the film forms, the solution resin concentrates in the surface layer. When the film dries, the morpholine evaporates, leaving the free carboxylic acid groups. The film has sufficient water resistance for normal use but can be removed by mopping with ammonia water. This ease of removal is desirable for facilitating occasional replacement of floor wax, but not for durable paint films. Proprietary resins are marketed that provide better gloss with smaller loss of resistance properties.

In exterior applications, the advantage of higher gloss alkyd enamels is more than offset by the better gloss retention and resistance to cracking of latex gloss paints. Depending on location, alkyd enamels exposed outdoors lose gloss in a year or two. Although the initial gloss is lower, a latex enamel retains most of its gloss for years. In interior applications, the low odor and superior color retention are important advantages of latex gloss paints.

The principal limitation of gloss latex paints is not lower gloss, but the problem of achieving adequate hiding with one coat. The factors involved are discussed in Section 32.3.1. Not much can be done about the relatively low NVV of latex paints. The formulating challenge is to make latex enamels that can be easily applied in films thick enough for one-coat hiding. For maximum hiding, a latex enamel should be formulated to have a higher viscosity at high shear rate than an alkyd enamel. In practice, as is discussed later, the high shear rate viscosity of latex paints has traditionally been formulated to be lower than that of alkyd paints, further compounding the problem. Wet film thickness can be controlled to a degree by the extent to which the painter brushes out the paint. The painter tries to judge the extent to which the paint should brushed out by how well the wet paint is covering the substrate. In the case of alkyd paints, there is relatively little difference between the wet and dry hiding power of the paint. However, the hiding power of latex paints is greater when wet than dry, which increases the difficulty of judging how to apply a latex paint by brush or roller.

The need for good leveling is also described in Section 32.3.1. The driving forces for leveling of latex paints do not seem to have been considered in the literature. It is not obvious how surface tension differentials could develop during film formation of latex enamels. The surface tension of the water phase of latex paints is controlled primarily by the surfactants in the paint, which suggests that there probably is little change in surface tension as water evaporates. If this hypothesis is right, latex paints level by smaller surface tension forces than by the larger surface tension differential forces, which are thought to operate with alkyd enamels. Another factor that may be important is dynamic surface tension (Section 24.1). It has been shown that some surfactants reach equilibrium surface tension more rapidly than others (Schwarz, 1992).

However, the major factor affecting leveling and, therefore, hiding of gloss latex paints is probably their rheological properties. As traditionally formulated with polymeric water-soluble thickener, such as HEC, latex paints exhibit a much greater degree of shear thinning than alkyd paints. This has led to latex paints, on the one hand, having too low a viscosity at high shear rate, so the applied film thickness tends to be too thin and, on the other hand, too high a viscosity at low shear rate to permit adequate leveling. The problem is especially severe because the rate of recovery of viscosity after exposure to high shear rates is generally rapid with such latex paints. Understanding these relationships facilitates solutions to the problem. However, understanding was set back for many years by the extensive use of Stormer viscometers (Section 3.3.6). This so-called viscometer provides information related to viscosity in a midrange of shear rates, but no information about viscosity in either of the critical regions—at high and low shear rates.

The reasons for this greater dependency of viscosity on shear rate in latex coatings with conventional thickeners have not been fully elucidated: at least two factors may be involved, flocculation of latex particles and/or pigment particles in the presence of HEC and possible entanglement of chains of swollen high MW HEC (Reynolds, 1992).

As discussed in Section 3.5.1, contemporary latex paints are often formulated with *associative thickeners*, sometimes in combination with HEC. Many kinds of associative thickeners are available (Lara-Ceniceros et al., 2014). They all are moderately low MW hydrophilic polymers with two or more long-chain nonpolar hydrocarbon groups spaced along the backbone or at chain ends. Use of such thickeners permits formulation of latex paints that exhibit less shear thinning so that viscosity at high shear rates can be higher; as a result, thicker wet films can be applied (Fernando et al., 1986; Reynolds, 1992). Formulation with associative thickeners also reduces viscosity at low shear rates, so leveling is also improved. (The thicker wet film in itself helps to promote leveling, since the rate of leveling depends strongly on wet film thickness, as discussed in Section 24.2.)

It has been shown that formulations with associative thickeners not only level better but also give somewhat higher gloss (Hall et al., 1986). Higher gloss and improved rheological properties have been reported by use of a combination of associative thickeners (Lundberg and Glass, 1992). One possibility reason for the improved properties is that the combination reduces flocculation of the TiO_2. Another reason for the improved gloss may be that associative thickeners are effective with small particle size latexes, which give the highest gloss (Reynolds, 1992). Sag control of latex paints is more difficult with associative thickeners but still easier than with alkyd paints. Santos et al. (2017) and van Dyk et al. (2014) reviewed associative thickeners and studied the rheology of simple model systems containing HEUR thickeners. They concluded that the primary mechanism of shear thickening was transient bridging flocculation. Spot checks of recent supplier literature suggest that current conventional acrylic latex/associative thickener formulations have the following characteristics: about 32% volume solids, about 21% PVC, VOC of 150–200 $g\,l^{-1}$, 60° gloss of 80–85, and 20° gloss of 55–60. With these values, it is unlikely that one- coat hiding can be achieved except over similar colors. Clearly, without formulating legerdemain, gloss and hiding are inferior to solventborne alkyd gloss paints and even waterborne alkyds, where volume solids can be above 40%, 60° gloss >85, and VOC <50 g l^{-1}. Use of cross-linkable latexes (discussed in the following text) improves the situation.

Another shortcoming of latex paints, particularly evident in gloss formulations, is the time required to develop final film properties. Part of the problem is that users are, in a sense, misled by the drying properties of latex paints. They dry to touch more rapidly than alkyd paints and can be handled sooner. However, they require a longer time to reach their ultimate properties. For example, the block resistance required to prevent windows and doors from sticking or to permit putting heavy objects on a newly painted shelf or window sill develops slowly. The initial film formation of latex particles is rapid, but full coalescence is limited by the availability of free volume. Since $(T-T_g)$ must be small, free volume is small. The situation is helped by using coalescing solvents. However, loss of these solvents is controlled by diffusion rate, which is also limited by $(T-T_g)$.

Hoy (1979) has reported that latex particles having a gradient of high T_g inner layers to low T_g outer shells can provide film formation at low temperatures, as well as achieve block resistance relatively quickly. Mercurio et al. (1982) has recommended the use of a high T_g latex together with larger amounts of carefully selected coalescing agents. Geel (1993) has reported that the dimethyl ether of dipropylene glycol is an effective coalescing agent with a relatively rapid rate of evaporation from films. While the initial loss of coalescing agent is controlled by volatility, losses in later stages are controlled by the rate of diffusion through the film and loss rate becomes very low.

A historic problem of gloss latex paints is poor adhesion to old gloss paint surfaces when the new dry paint film gets wet. After wetting with water, some latex paint films can be peeled off the old paint surface in sheets. Such a film is said to exhibit poor *wet adhesion*. There is always a problem of achieving adequate adhesion when repainting old gloss paint surfaces, even with alkyd paints, but the problem with latex paints is more severe. It is essential to wash any grease off the surface and to roughen the surface by sanding, but even with such surface preparation, some older gloss latex paints had poor wet adhesion. Wet adhesion improves as the films age, but it can remain seriously deficient for several weeks or even months. Latexes that minimize the problem of wet adhesion have been developed by several manufacturers. Incorporation of small amounts of hydrogen-bonding, polarizable comonomers such as methacrylamidoethylene urea (Section 9.1) improves wet adhesion and wet scrub resistance. Benchmark tests by Wu et al. (2016) show that three contemporary premium semigloss latex paints can pass a standard test (ASTM D-6900) of wet adhesion.

Latex enamels are usually formulated with higher VOCs than other latex paints. In conventional formulations, extra coalescing solvent is needed to permit use of latexes with the relatively high T_g needed for block, scrub, and stain resistance. Modest reductions in VOC can result from selection of more efficient coalescing solvents. 2-Ethylhexyl benzoate is an example (Arendt et al., 2001). Another approach is to use blends of high and low T_g latexes (Winnik and Feng, 1996). The binder must be transparent, which requires a small difference in refractive index of the latexes, as well as small particle size of the high T_g latex.

Schuler et al. (2000) have reported that core–shell copolymers of BA, MMA, and AA can be used to formulate gloss latex paints having high block resistance with no coalescent. The final films were dispersions of hard particles in a continuous softer phase. The best compromise of properties was with 25–35% of the hard MMA phase.

Perhaps the most promising approach for eliminating or minimizing coalescing solvents is to use thermosetting latexes, which permit use of low T_g latexes that achieve adequate hardness through cross-linking. Acrylic latexes that oxidatively cross-link can be formulated at VOC of 50–150 g l^{-1}. Such latexes form films at <5°C. They are formulated with associative thickeners and surfactants that promote leveling. Other technologies for thermosetting latexes are described in Section 9.4.

As previously mentioned, solvents such as propylene glycol are included in formulations to improve freeze–thaw stability and to permit wet lapping. It has been found that associative thickeners based on ammonium salts of hydrophobically modified PAA increase open time (i.e., permit wet lapping) and improve freeze–thaw resistance, although such large amounts are needed that alkali resistance of the films becomes inadequate. However, satisfactory gloss enamels having low VOC have been reported by using these associative thickeners in combination with a cross-linking latex (Monaghan, 1996). McCreight et al. (2011) and Zong et al. (2011) reported further reductions in VOC by replacing propylene glycol with undisclosed open time additives. Very low VOC formulations were achieved and open times were increased from 2 to 4–8 min, which facilitates wet lapping.

A growing trend for gloss enamels is to combine acrylic latexes with PUDs. For example, Aznar (2006) prepared an acrylic latex in the presence of a urethane/acrylic prepolymer to provide a hybrid latex, which was formulated into paints having good gloss and exterior durability.

GENERAL REFERENCES

Feist, W. C., *Finishing Exterior Wood*, Federation of Societies for Coatings Technology, Blue Bell, PA, **1996**.

Hofland, A., *Prog. Org. Coat.*, **2012**, *73*, 274–282.

Williams, S. R., Finishing of Wood in *Wood Handbook*, General Technical Report FPL-GTR-190, U.S. Department of Agriculture, Forest Service, Forest Products Laboratory, Madison, **2010**, p 16–17.

REFERENCES

Air Products and Chemicals, Pub. No. 151-0290.6, Air Products and Chemicals, Inc., Allentown, **2002**.
Anonymous, Technical Bulletin 78739, Rohm & Haas, Philadelphia, **2004**.
Arendt, W. D., et al., *Proceedings of the International Waterborne, High-Solids, and Powder Coatings Symposium*, New Orleans, LA, **2001**, pp 497–509.
Aznar, A. C., *Prog.Org. Coat.*, **2006**, *55*, 43.
Bieleman, J., *Surf. Coat. Int. A*, **2003**, *86(10)*, 411.
Chou, C.-S., et al., *J. Coat. Technol.*, **1987**, *59(755)*, 93. Rhoplex Multilobe 200, Technical Bulletin, Rohm & Haas, Philadelphia, PA, 1992.
Chung, J. K., et al., US patent 6,031,041 (**2000**).
Decocq, F., et al., *Proceedings of the International Waterborne, High-Solids, and Powder Coatings Symposium*, New Orleans, LA, **1997**, p 168.
Diebold, M. P., *JCT Res.*, **2011**, *8(5)*, 541.
Eckersley, S. A.; Helmer, B. J., *J. Coat. Technol.*, **1997**, *69(864)*, 97.
Edge, M., et al., *Prog. Org. Coat.*, **2001**, *43*, 10.
Evanson, K. W.; Urban, M. W., *J. Appl. Polym. Sci.*, **1991**, *42*, 2309.
Fasano, D. M., *J. Coat. Technol.*, **1987**, *59(752)*, 109.
Feist, W. C., *Finishing Exterior Wood*, Federation of Societies for Coatings Technology, Blue Bell, PA, **1996b**.
Fernando, R. H., et al., *J. Oil Colour Chem. Assoc.*, **1986**, *69*, 263.
Fu, Z., et al., US2A1 (**2009**).
Geel, C., *J. Oil Colour Chem. Assoc.*, **1993**, *76*, 76.
Gilbert, J. A., *Coat. World*, **2016**, *21(9)*, 117–122.
Gilicinski, A. G., et al., *Proceedings of the International Waterborne, High-Solids, and Powder Coatings Symposium*, New Orleans, LA, **2000**, pp 516–526.
Glass, J. E., *J. Coat. Technol.*, **1978**, *50(641)*, 56.
Grade, J., et al., *Proceedings of the International Waterborne, High-Solids, and Powder Coatings Symposium*, New Orleans, LA, **2002**, pp 145–157.
Gray, R. T.; Lee, J., US patent 6,303,189 (**2001**).
Hall, J. E., et al., *J. Coat. Technol.*, **1986**, *58(738)*, 65.
Hayoz, P., et al., *Prog. Org. Coat.*, **2003**, *48*, 297.
Heldmann, C., et al., *Prog. Org. Coat.*, **1999**, *35*, 69.
Hofland, A., Making Paint from Alkyd Emulsions in Glass, J. E., Ed., *Technology for Waterborne Paints*, American Chemical Society, Washington, DC, **1997**, p 183.
Hofland, A., *Prog. Org. Coat.*, **2012b**, *73*, 274–282.
Hook, J. W., III, *Proceedings of the American Coatings Conference*, Atlanta, GA, April 7–9, **2014**.
Hook, J. W., III, et al., *Prog. Org. Coat.*, **1994**, *24*, 175.
Hoy, K. L., *J. Coat. Technol.*, **1979**, *51(651)*, 27.
Hulden, M., et al., *J. Coat. Technol.*, **1994**, *66(836)*, 99.
Huster, W., **2011**, from http://www.sesam-uae.com/sustainablematerials/presentations/02_wacker_huster.pdf (accessed April 22, 2017).
Jiang, P., et al., *Proceedings of the American Coatings Conference*, Indianapolis, IN, **2016**, Paper 12.1.
Jirous-Rajkovic, V., et al., *Surf. Coat. Int. B Coat. Trans.*, **2004**, *87(1)*, 70.
Lara-Ceniceros, T. E., et al., *J. Polym. Res.*, **2014**, *21*, 511.
Larson, D. B.; Emmons, W. D., *J. Coat. Technol.*, **1983**, *55(702)*, 49.
Lorah, D. P., et al., *Proceedings of the International Waterborne, High-Solids, and Powder Coatings Symposium*, New Orleans, LA, **2004**, Paper No. 19.
Lundberg, D. J.; Glass, J. E., *J. Coat. Technol.*, **1992**, *64(807)*, 53.
Mangi, R., et al., *Proceedings of the International Waterborne, High-Solids, and Powder Coatings Symposium*, New Orleans, LA, **2002**, pp 41–56.
Massouda, D. B., *J. Coat. Technol.*, **1985**, *57(722)*, 27.
Mattei, I. V., et al., *J. Coat. Technol.*, **1991**, *63(803)*, 39.
McCreight, K. W., et al., *Prog. Org. Coat.*, **2011**, *72*, 102–108.
Mercurio, A., et al., *J. Oil Colour Chem. Assoc.*, **1982**, *65*, 227.
Monaghan, G., *Formulating Techniques for a Low VOC High Gloss Interior Latex Paint*, FSCT Symposium, Louisville, KY, May **1996**.
Nikolic, M., et al., *J. Coat. Technol. Res.*, **2015**, *12(3)*, 445–461.
Overdiep, W. S., *Prog. Org. Coat.*, **1986**, *14*, 159.
Perara, D. Y.; Eynde, D. V., *J. Coat. Technol.*, **2001**, *73(919)*, 89.
Pilcher, G. R., *CoatingsTech*, **2015**, *12(8)*, 24–33.
Poole, E. A., et al., EP2166050 A2, (**2010**).
Porzio, R. S., et al., *Proceedings of the International Waterborne, High-Solids, and Powder Coatings Symposium*, New Orleans, LA, **2001**, pp 245–260.
Prior, R. A., et al., *Prog. Org. Coat.*, **1996**, *29*, 209.
Przybyla, D., *Proceedings of the American Coatings Conference*, Atlanta, GA, **2014**, Paper 9.6.
Ramig, A., Jr.; Floyd, F. L., *J. Coat. Technol.*, **1979**, *51(658)*, 63, 75.
Reynolds, P. A., *Prog. Org. Coat.*, **1992**, *20*, 393.
Rich, D. S., *JCT Res.*, **2016**, *13(1)*, 1–9.
Santos, F. A., et al., *JCT Res.*, **2017**, *14(1)*, 57–67.
Schuler, B., et al., *Prog. Org. Coat.*, **2000**, *40*, 139.
Schulte, S., *Eur. Coat. Newslett.*, October 7, **2015**.
Schwarz, J., *J. Coat. Technol*, **1992**, *64(812)*, 65.
Smith, A.; Wagner, O., *J. Coat. Technol.*, **1996**, *68(862)*, 37.
Van Dyk, A., et al., *Proceedings of the American Coatings Conference*, Atlanta, GA, April 7–9, **2014**, paper 5.3.
Vanaken, D., et al., *Proceedings of the American Coatings Conference*, Atlanta, GA, April 7–9, **2014**, paper 9.2.
Wachtler, P.; Kunisch, F., *Polym. Paint Colour J.*, **1997**, October, 2.
Wang, H. W., et al., *Proceedings of the International Waterborne, High-Solids, and Powder Coatings Symposium*, New Orleans, LA, **2001**, pp 61–76.
Williams, S. R., Finishing of Wood in *Wood Handbook*, General Technical Report FPL-GTR-190, U.S. Department of Agriculture, Forest Service, Forest Products Laboratory, Madison, **2010**, p 16–17.
Winnik, M. A.; Feng, J., *J. Coat. Technol.*, **1996**, *68(852)*, 39.
Wu, W., et al., *Proceedings of the American Coatings Conference*, Indianapolis, IN, **2016**.
Yang, H. W., et al., *Proceedings of the International Waterborne, High-Solids, and Powder Coatings Symposium*, New Orleans, LA, **2000**, pp 308–321.
Zong, Z., et al., *Prog. Org. Coat.*, **2011**, *72*, 115–119.

Chapter 33

Special Purpose Coatings

The term *special purpose coatings* designates industrial coatings that are applied outside a factory. In 2014, they accounted for about 21% ($4.64B) of the value of US coatings shipments and about 12% (150M gallons) of the volume (Pilcher, 2015). Many different end users are involved; our discussion centers on industrial maintenance (6% of the total US paint market according to Pilcher (2015)), marine (2%, but much larger in Asia), automobile refinish (9%), and, briefly, traffic striping paints (2%).

The terms *protective coatings* and *anticorrosive coatings* are sometimes applied broadly to maintenance and marine coatings and sometimes more narrowly just to maintenance. This and other variations in terminology can complicate comparisons of different studies of the markets. Sorensen et al. (2009) reviewed the entire field of anticorrosive coatings.

Special purpose coatings have a vital role both in building and maintaining infrastructure. Worldwide, a large and growing problem is underinvestment for infrastructure. A report by a major consulting firm (Woetzel et al., 2016) estimates that the world must spend $3.3T per year on new infrastructure just to meet expected growth, but it only spends $2.5T. Most major economies are falling behind. Of the $3.3T, $0.9T is needed for roads. Furthermore, existing infrastructure is deteriorating badly because of inadequate maintenance.

Epoxies (Chapter 13) are the workhorse resins for special purpose coatings, but virtually every type of resin finds application somewhere. Pigments (Chapter 20) and especially anticorrosive pigments (Chapter 7) are also critical. Nanopigment technology (Sections 20.5 and 22.2.1.1) is having an impact on special purpose coatings; nanopigments enhance mechanical properties and perhaps weatherability. Other new technologies such as self-healing coatings (Section 34.3), electroactive coatings, self-cleaning coatings (Section 34.1), and coatings that self-report impending failure have been intensively studied, but their impact on current practice appears to be limited. These technologies are predicted to have a significant impact in the future. The coatings are improving significantly. Often, improvement results from an accumulation of incremental improvements, rather than from spectacular breakthroughs.

In previous editions, aerospace coatings were included in this chapter; they are now placed in Chapter 30.

33.1 MAINTENANCE PAINTS

The term *maintenance paints* generally signifies paints for field application, including highway bridges, refineries, factories, power plants, water systems, tank farms, rail cars, shipping containers, and some metal buildings; not usually included are paints for office buildings or retail stores, which are classified as architectural coatings. Our discussion will focus on coatings for steel, although protection of concrete and other substrates is also important. For many maintenance paints for steel, the essential requirement is corrosion protection (Chapter 7). The terms *heavy-duty maintenance paints*, *anticorrosive paints*, *protective coatings*, or *industrial paints* are also used, all implying that the paints must perform more effectively than trade sales paints in aggressive environments. Although selling prices are important, the major economic demands of customers emphasize proven performance, time interval to be expected between repainting, and service, rather than the cost of the paint. The frequency of repainting is especially critical. For example, repainting in a factory may require shutting down production, and repainting a bridge may require expensive closures.

Organic Coatings: Science and Technology, Fourth Edition. Frank N. Jones, Mark E. Nichols, and Socrates Peter Pappas.

Since, as detailed in Section 7.5, no laboratory tests adequately predict field performance of anticorrosive coatings, potential customers want to inspect actual field use examples of a coating system being recommended to them. National and state highway departments and larger oil, chemical, and construction companies employ specialized engineering groups responsible for working both with coating suppliers to select appropriate coatings for their company's needs and with applicators to specify the application parameters. These groups also serve as inspectors to see that the coatings are properly applied. Smaller groups hire consultant/inspectors when needed. Records are kept for surface preparation, application conditions, coating composition, and coating supplier; the performance of coatings on the various installations is monitored. Composition of the coatings is a major variable, but surface preparation and application procedures are also critical to performance. The Society for Protective Coatings (SSPC) (formerly the Steel Structures Painting Council) is a valuable source of information about the effects of such variables. Its *Steel Structures Painting Manuals* are updated periodically. Its *Journal of Protective Coatings and Linings* has many useful papers and is available online. SSPC-SPC also publishes a daily newsletter (paintsquare.com). Reliability theory is a potentially powerful tool (Martin et al., 1996).

The most commonly used method for surface preparation has been sandblasting. It can be very effective for removing old paint and surface rust; it also roughens the surface, increasing the *profile*, the contact area with the coating, which in turn promotes adhesion. However, sandblasting creates hazardous dust—small particles of silica and paint debris, often containing toxic pigments. Elaborate and expensive containment systems are necessary to prevent worker exposure and contamination of the environment. As discussed in Section 6.4, many modifications and alternatives to sandblasting have been developed. They include dry abrasive blasting with a variety of other media, wet abrasive blasting, and ultrahigh pressure hydroblasting at pressures above 30 000 psi. In general, blasting with abrasives increases surface roughness but hydroblasting does not (Hough, 2016). Newly blasted steel is vulnerable to flash rusting (Section 33.1.3), and it is common to apply flash rust inhibitors immediately after blasting. In wet blasting, inhibitors can be added to the water. Whatever the cleaning method, primer should be applied as soon as possible.

Massive failures have occurred in seaside facilities when painting is delayed even a few days after sandblasting. If saltwater spray deposits salt on the surface, moisture permeating through the paint will cause osmotic blistering.

As with other coatings, VOC of maintenance paints is regulated in many areas. For example, 2016 VOC limits in California range as low as 50 or 100 $g\,l^{-1}$. Such regulation has spurred development of waterborne maintenance paints. Historically, the performance of waterborne paints has been inferior to solventborne counterparts in demanding anticorrosion applications. Intense R&D has led to waterborne formulations that have demonstrated excellent performance in laboratory tests. The ultimate test is how the formulations perform in service in a variety of conditions.

The coating system must be carefully selected for the particular installation. Most systems include at least two types of coatings: a primer and a top coat. Frequently, more than one layer of primer and/or top coat is applied, and in some cases, a combination of primer, an intermediate coat, and top coat is used. Primers provide the primary corrosion control, but intermediate and top coats also have significant effects on corrosion protection by reducing oxygen and water permeability of the combined films. Top coats also serve to protect the primer and provide other properties such as gloss, exterior durability, and abrasion resistance. Three classes of primers are used: barrier primers, zinc-rich primers, and primers containing passivating pigments.

Sections 33.1.1, 33.1.2, and 33.1.3 address maintenance coatings for new steel or for steel from which old paint has been removed. Section 33.1.4 addresses overpainting of previously painted steel structures.

33.1.1 Barrier Coating Systems

Conventional practice for demanding maintenance applications such as steel bridges has been to apply three coats: a primer (often zinc rich, Section 33.1.2), a mid-coat, and a top coat. Various two-coat systems are being promoted as alternatives. At one time, alkyd resins were used for all three coats, but many other options are now available.

Primers. As discussed in Section 7.3, a barrier system primer requires excellent wet adhesion. The primer should have a low viscosity continuous phase with slow evaporating solvents to permit as rapid and complete penetration as possible into microscopic cracks and crevices in the metal surface. Amine-substituted binders are particularly resistant to water displacement. Phosphate esters such as epoxy phosphates have also been shown to enhance wet adhesion (Section 13.5). The T_g of the fully reacted binder should generally be only slightly above the ambient temperature at which the curing is to be carried out. If the T_g of the fully reacted cross-linked film is too high, the rate of the cross-linking reaction becomes mobility controlled, and the reaction may stop prior to complete conversion (Section 2.3.2).

Two-package (2K) epoxies formulated with bisphenol A (BPA) and/or novolac epoxy resins in one package and a polyfunctional amine in the other are the workhorse primers (Weinmann, 2013). They combine strong wet adhesion with excellent saponification resistance (Section 13.2). The novolac epoxy resins are prepared from novolac resins made with phenol, cresol, BPA, and/or tetraphenolethane. To reduce viscosity, the triglycidyl ether of trimethylolpropane (TMP) can be used as a reactive diluent. A low viscosity coating formulated with the triglycidyl ether of TMP and a phenol-formaldehyde novolac resin showed better resistance to methylene chloride, acetic acid, and sulfuric acid than one formulated with a bisphenol F resin (Kincaid and Schulte, 2001). "Polycyclamine"/epoxy 2K systems (Section 13.2.2) have gained favor for their relatively rapid cure under cool conditions and good performance.

Pot life is increased by using relatively high molecular weight (MW) epoxy resins and amine cross-linkers at relatively low solids, since this minimizes the concentration of reactive groups. Adhesion is promoted by vehicle penetration into surface irregularities, which is facilitated by the use of slow evaporation solvents. Pigmentation should be high, but somewhat below critical pigment volume concentration (CPVC), to reduce permeability to oxygen and water and give a low gloss surface that enhances adhesion of top coat to the primer. Intercoat adhesion is further enhanced by applying top coat before cure of the primer is complete. Only a thin coat of primer is needed, but it is generally desirable to apply multiple coats of primer to assure that the entire surface of the metal has been coated. This low solids approach, however, has relatively high VOC emissions. To reduce VOC, higher solids coatings are being adopted. As solids increase, the challenges of formulating coatings with adequate pot life become more difficult. The lower MW reactants in high solids coatings potentially increase toxic hazards, as well.

As discussed in Section 25.2.2, "zero VOC" 2K epoxy formulations are now widely available. They contain, at most, very low levels of volatile solvents. They are especially desirable for maintenance coating in enclosed spaces, such as the insides of water or petroleum tanks and confined parts of buildings. In 2015, the US Occupational Health and Safety Administration (OSHA) published detailed standards for design of confined spaces and for limiting the hazards of painting (Kaelin and Liang, 2014).

Waterborne epoxy–amine primers (Section 13.2.6) are gaining popularity. Novolac epoxy resins are frequently used in waterborne coatings since their higher MW does not affect viscosity.

2K urethane coatings are also used as primers. Aromatic polyisocyanate prepolymers are chosen owing to their lower cost as compared with aliphatic polyisocyanates. They show excellent adhesion, on to metals, including galvanized steel, and to concrete, and wood. Their resistance to saponification is a major advantage as compared with that of alkyd resins in primers; this characteristic is especially important in coatings for galvanized steel and concrete, which have highly alkaline surfaces. Moisture-curable urethane coatings are also used as primers. Commonly, such primers are formulated with PVC>CPVC, resulting in somewhat porous films, which permits escape of CO_2 from the reaction of isocyanates with water, thereby minimizing bubble formation.

Mid-coats. Three-coat systems are very commonly used for critical applications. The second coating is often a 2K epoxy but comprises a different epoxy than the primer. It is formulated for excellent adhesion to the primer, for excellent adhesion by the top coat, and for enhanced barrier properties. Pigmentation may include platey pigments such a micaceous iron oxide to enhance barrier properties.

Top coats. BPA and novolac epoxy–amine coatings are used as top coats for indoor applications in which their poor UV resistance is not a concern. They are especially valued where excellent chemical resistance is needed, although they are sensitive to acetic acid (or similar organic acids). In contrast to inorganic acids, acetic acid dissolves in epoxy films; this deleterious effect is promoted by the presence of amine cross-linkers, especially if cross-link density is inadequate. Novolac epoxy resins have a higher average functionality and generally provide greater organic acid resistance than BPA epoxies.

Generally, top coats for exterior applications have a different composition than primers. Chlorinated polymers such as vinyl chloride copolymer resins and chlorinated rubbers are used as top coat vehicles, since they have low moisture vapor and oxygen permeability (Section 17.1). Chlorinated resins require stabilization against photodegradation (Section 5.4). Use of chlorinated resins is limited by the high VOC required, owing to their high MW.

Polyvinylidene fluoride/acrylate copolymer latexes offer outstanding exterior durability (Section 17.1.4). Since they are not cross-linked, they remain solvent sensitive and are, therefore, not appropriate for applications such as petroleum refineries and chemical plants. Excellent protection is also afforded by various thermoplastic or cross-linkable fluorinated polymers (Parker, 2015) (Section 17.1.4).

High density polypropylene (HDPP) and polyethylene (HDPE) have low water permeability and are common top coats for steel pipelines (Guidetti et al., 1996; Senkowski, 2015). They are applied by extrusion at the factory. One system consists of a powder-coated epoxy primer (Chapter 28), which is *fusion bonded*, an elastomeric adhesive, and a polypropylene top coat. Since polypropylene is subject to photooxidation, UV stabilization is required; UV protection can be supplied by dispersing carbon black and perhaps other stabilizers in the polymer. After installation, welds, fittings, and valves are coated with liquid anticorrosive compositions, usually epoxy, polyurethane, or both (Senkowski, 2015).

2K *polyurethane* coatings are increasingly used as top coats owing to lower VOC and high solvent resistance of the cured films; urethanes are particularly useful when abrasion resistance is an important requirement (Section 12.4). An important variable in 2K polyisocyanate/polyol coatings is the ratio of NCO/OH used. This ratio is called the *index*. In ambient cure systems, a ratio of 1.1 : 1 generally provides better film performance than a 1 : 1 ratio. A probable reason is that excess NCO is available for reaction with water from solvent, pigment, or air to form urea cross-links. To the extent that this happens, each water molecule reacts with two NCO groups; use of excess NCO accommodates this reaction, thereby minimizing residual unreacted hydroxyl groups. Solvent resistance is also improved. Since polyisocyanates are typically lower in viscosity than the polyol, excess NCO reduces VOC (Jorissen et al., 1992).

2K *polyurea* coatings made from polyamines and polyisocyanates have the advantages that they cure rapidly even at freezing temperatures and have almost zero VOC and good low temperature flexibility. Aromatic amines react slowly enough that they can be used in 2K coatings. For example, MDI/polyoxypropylene prepolymers with aromatic amines are used in maintenance coatings for which gloss and color retention are not critical (Takas, 2004). Roofs and floors can be walked on within 30 s after they are coated with such polyureas. Interiors of factories can also be rapidly painted and returned to service. The availability of food-grade polyureas and their rapid cure enables fast maintenance and rehabilitation of food processing facilities (Anonymous, 2015).

For 2K polyureas that require exterior durability, hindered aliphatic amines are needed. Giles et al. (2003) report the utilization of a mixture of hindered and less hindered di- and tri-amines in one package and HDI isocyanurate in the other. The coatings had adequate pot life, tack-free time of 2.3 min at 25°C, excellent adhesion to a variety of substrates, chemical resistance, and exterior durability. No primer was required.

Another type of 2K polyurea consists of polyisocyanates cross-linked with polyaspartic esters (Section 12.4). Such coatings are sometimes mistakenly classed as polyurethanes. Squiller et al. (2014) disclose details of how the polymers are made and formulated into clear coats. Paint producers would usually add pigments and additives. Polyaspartic technology is versatile. It can provide excellent weatherability, and the pot life/cure rate can be adjusted to suit specific applications. Polyaspartic top coats for bridges are becoming common (O'Donoghue et al., 2013).

The chemistry of moisture-curable isocyanates is described in Section 12.6, where it is noted that they are not actually urethanes, but ureas. These are one-package (1K) formulations that can cure at temperatures as low as 0°C, generally by reaction with moisture in the air. They have certain advantages over the 2K maintenance coatings. The possibility of incorrect mixing ratios is eliminated, as are potential problems of (1) mixing more 2K coating than can be used before the viscosity increase prevents application or (2) not mixing enough and running out before a work shift ends. A further advantage is that they can be applied to damp surfaces. As compared with 1K systems that cure by autoxidation, moisture-curable coatings can have superior resistance to oxidation and exterior exposure (if aliphatic diisocyanates are used).

As compared with alkyd maintenance paints, moisture-curable urethanes generally exhibit superior abrasion resistance and chemical and solvent resistance, as well as hydrolytic stability. In addition, adhesion retention over alkaline surfaces such as cement-based substrates or metal substrates and exterior durabililty, with aliphatic isocyanates, are also superior. Gardner (1996a) provides a useful discussion of moisture-curable urethane coatings especially for maintenance coatings. They are also used for applications such as floor coatings, for which abrasion resistance and hydrolytic stability are especially important. They can also be applied at low temperatures and relatively high humidity. The coatings can be applied over metal when the temperature is below the dew point since the isocyanate groups react with the water condensing on the surfaces (Schutz, 1998). Bell (1982) describes several examples of application to bridges and other applications.

On the other hand, moisture-curable urethanes are more expensive than alkyds. A further drawback is that cure rates depend on the water content of the air; at low temperatures, higher relative humidity is required than at higher temperatures, since relative humidity increases with decreasing temperature for the same water content. At high humidity and temperature, cure is rapid, but the carbon dioxide released by the reaction of isocyanate with water can be trapped as bubbles, especially in thick films. See Gardner (1996b) for the effects of temperature and humidity and other application considerations.

A further cost factor with moisture-cure PUs is that solvents, pigments, and other formula components must be essentially water free. Moisture cure is often used for clear, glossy coatings, because it is expensive to remove adsorbed water from pigments. A variety of moisture scavangers is available for pigmented moisture-curable coatings. Examples are alkyl orthoformates, *p*-toluene sulfonyl isocyanate, and molecular sieves. Oxazolidines are reported to be effective moisture scavengers for aliphatic isocyanate-based coatings, presumably with less toxic hazard and lower cost (Robinson et al., 1993). 4-Ethyl-2-methyl-2-(3-methylbutyl)-1,3-oxazolidine provides rapid water removal while not adversely affecting stability or performance of a coating.

In most cases, atmospheric water is the curing agent, but in some cases spray application of liquid water or steam on the uncured film is recommended. Through-cure can also be accelerated by including a hygroscopic liquid in a moisture-curable formulation. Examples are *N*-methylpyrrolidone and γ-butyrolactone.

As discussed in Section 16.2, polyisocyanates can be converted into polysilanes by reaction with reactive silanes. These polysilanes, which contain alkyoxysilyl groups, can serve as binders for moisture-cure top coats. Advantages of this approach include application as a 1K coating, elimination of the need for dry pigments, and elimination of film defects caused by CO_2 evolution during cure. Alternatively, aliphatic epoxy resins can be converted to moisture-cure resins. Witucki (2013) describes an approach in which epoxy resins react with diaminosilanes to provide alkoxysilyl-functional moisture-curable coatings.

Witucki also describes 2K coatings in which a diaminosilane is used as a curing agent for cycloaliphatic epoxies. Low VOC and excellent weatherability are reported. The coatings are reported to perform well as a two-coat alternative to three-coat systems.

2K waterborne urethane coatings have very low-VOC emissions. Giles et al. (2001) described a waterborne coating that was formulated with a water-reducible acrylic resin and a hydrophilically-modified polyisocyanate and compared its performance with a solventborne high solids 2K urethane coatings. When applied at 25°C (75°F) or below, the performance was similar. But under conditions of high humidity and temperature, the films from the solventborne coating showed poor surface smoothness and appearance, whereas the waterborne coating had excellent appearance.

Acrylic top coats, usually latexes, have been in use for years. Numerous efforts to prepare acrylic and various acrylic/hybrid latexes for use in *direct-to-metal* (DTM) maintenance coatings have been described, for example, by Vielhauer et al. (2015), Flecksteiner et al. (2014), and Wu et al. (2014). Vielhauer describes waterborne barrier coatings with good corrosion resistance and VOCs of $50\,g\,l^{-1}$ or even less; while technical details are obscure, the success of these systems depends partly on TiO_2 interaction with latex particles that yield a relatively uniform distribution of pigment in the films; self-cross-linking technology (Section 9.4) is also involved.

For decades, alkyds (Chapter 15) and silicone-modified alkyds (Chapter 16) were workhorse resins in maintenance paints. They have been partly supplanted in critical applications by some of the coatings described earlier, which are considered to have better exterior durability. Coatings based on alkyds are generally lower in cost and have intermediate VOC emissions. As a result of their low surface tensions, alkyd-based coatings are less likely to develop film defects during application. Technologies that may increase alkyd usage include calcium sulfonate alkyds (discussed in Section 3.1.4) low-VOC alkyd emulsions (Danneman, 2014), acrylic/alkyd hybrid latexes, and polyurethane/alkyd hybrids.

Since zinc rapidly develops strongly basic corrosion products, alkyds are not suitable vehicles for coating galvanized steel. An extensive study showed that epoxy–polyamide primer with a urethane top coat gave the best performance on galvanized steel when exposed for 4½ years at Cape Kennedy, Florida (Drisko, 1995).

As described in Section 15.7, uralkyds provide superior properties as compared with alkyds. Xu et al. (2002) found, surprisingly, that water permeated into an uralkyd film at a lower rate than into a conventional alkyd. Waterborne uralkyds are also available (Section 15.7.2). For weatherability, aliphatic polyisocyanates are used. For lower cost, uralkyds can be blended with acrylics. Technologies for synthesis of hybrid acrylic/uralkyd resins have been described, for example, by Mestach et al. (2013).

Barrier corrosion protection generally increases as water vapor and oxygen permeability of the coatings are reduced (Section 7.3.3). High levels of pigmentation reduce permeability and cost. Platelet pigments include mica, talc, micaceous iron oxide, glass flakes, exfoliated clays, and metal flakes. Their effectiveness increases with the extent to which they can orient parallel to the surface of the coating as the solvent in the coating evaporates. For glossy coatings, a highly pigmented intermediate coat may be applied over the primer, followed by a high gloss top coat. Pigmentation of the final top coat with leafing aluminum flake pigment is desirable because an almost continuous layer of reflective aluminum forms at the surface.

33.1.2 Systems with Zinc-Rich Primers

Coating with zinc metal (galvanizing) is an effective way to provide galvanic protection to steel. Zinc-rich coatings also provide galvanic protection. They are discussed here, as well as in Section 7.4.3.

A typical galvanic coating for steel bridges consists of three layers: a zinc-rich primer, an epoxy intermediate coat, and a weatherable top coat. The primer provides galvanic protection, the intermediate coat is a barrier, and the top coat is an additional barrier, also formulated for excellent weather resistance. The top coat may be a 2K polyurethane, a 2K polyaspartic ester, or a polysiloxane. Recommendations for dry film thickness of the primer vary, but 60–90 micrometers is common.

For two-coat applications over zinc-rich primers, polyaspartic top coats perform well; gloss retention is excellent. As noted in Section 33.1.1, Witucki (2013) reports good results with cycloaliphatic epoxy/diaminosilane systems.

As discussed in Section 7.4.3, zinc-rich primers provide excellent performance either when the surface cannot be completely cleaned of rust or when complete penetration into surface irregularities cannot be achieved. The zinc acts as a sacrificial metal, protecting the steel from corrosion. To be effective, the coating must be conductive. This can be accomplished by a level of zinc pigmentation such that PVC>CPVC. The high level of pigmentation assures electrical contact between the zinc particles and provides porosity that permits water to enter the film, establishing a conductive circuit with the steel surface. Even after a significant fraction of the zinc is consumed, the primer continues to provide protection. Feliu et al. (1993) propose that continued protection may result from passivation by the alkalinity of the reaction products, $Zn(OH)_2$ and $ZnCO_3$, which fill in the pores. Alternatively, these products may form a barrier to water and oxygen. In most applications, top coats are required to protect the zinc against corrosion, reduce probability of mechanical damage, and provide a desired appearance. Kline (1996) reviewed the types of zinc-rich primers and their application and performance.

There are two classes of zinc-rich primers: inorganic and organic, each available in solventborne and waterborne formulations. The classical binder of inorganic zinc-rich primers is a prepolymer derived by reacting tetraethyl orthosilicate with a limited amount of water. The chemistry of cross-linking is described in Section 16.3. Cross-linking requires moisture; when the coating is applied at low humidities, especially if the temperature is high, film properties such as abrasion resistance may be adversely affected (Eccleston, 1998). If it is necessary to paint under hot and dry conditions, the coating should be mist sprayed with water immediately after application. A review of the effects of zinc particle size and shape, extender pigments, and application techniques of ethyl silicate based primers is available (Parashar et al., 2001). Various near-zero VOC waterborne and sol–gel silicone-based binders have been described, for example, by Borup et al. (2014).

Organic zinc-rich primers command a share of the market owing to their advantages of greater tolerance to incomplete removal of oils from the substrate, easier spray application, and better compatibility with some top coats (Paert, 1992). The binders in organic zinc-rich primers are often based on 2K epoxy/polyamine (polyamide) resins.

In the older literature there is published evidence that inorganic zinc-rich primers can, in some circumstances, provide better protection than organic primers. For example, in a seacoast environment, a 6-year service life was estimated for inorganic primers as compared with 3 years for the organic type (Paert, 1992). More recently, however, organic zinc-rich primers have been adopted for many demanding applications, indicating that knowledgeable people expect them to be long lived. For example, in the early 2000s zinc–epoxy primers were supplanting zinc silicate primers in North Sea oil well applications (Doble, 2004); the change is said to be driven by health, safety, and cost considerations as well as unspecified improvements in technology. Organic zinc/epoxy/urethane systems are increasingly favored and sometimes required in many US states (Kowalski et al., 2003); they are typically specified for onshore and offshore steel windmill structures (Muhlberg, 2004).

In recent decades it has been common practice to replace up to 40% of the zinc with inert conductive pigments, for example, Fe_2P. However, Verbiest (2015) warns that this practice degrades salt spray test performance in a series of coatings. However, the salt spray results may or may not reflect actual field performance. Verbiest provides complete formulas for organic zinc-rich primers based on a Type 1 epoxy resin and a polyamide cross-linker made from dimer fatty acids and polyamines.

Polyurethane binders are used increasingly. For example, in 2004 the Michigan Department of Transportation adopted a urethane zinc-rich system as its standard. Schwindt (1996) described zinc-rich primers formulated with moisture-curable urethanes. They are used for new construction when the metal surface is blast cleaned to white metal conditions. Resistance to saponification is critical for the binder of zinc-rich coatings, since the zinc reacts with oxygen and water producing zinc hydroxide and zinc carbonate. Epoxy/amine 2K zinc-rich primers have been widely used, but moisture-cure urethane primers have the advantage of being 1K coatings.

Waterborne zinc-rich primers have been of interest in response to the need to reduce VOCs. Early products apparently performed poorly and had captured only a small fraction (<3%) of the heavy-duty maintenance market in Europe as of 2005 (Aamodt, 2005). However, both inorganic and organic waterborne zinc-rich primers are now available that have much lower VOC and are considered by some to rival solventborne counterparts in protective properties; different application methods are required. For inorganics, Montes (1993) described binders that are a combination of potassium, sodium, and/or lithium silicates with a dispersion of colloidal silica. Szokolik (1995) reported excellent performance on oil and gas production facilities in marine environments. Borup et al. (2014) reported further advances in very low-VOC siloxanes.

A challenge in using zinc-rich primers (especially inorganic ones) is proper selection and application of the top coat. Activity of the primer usually depends on maintaining its porosity by having PVC>CPVC. If a top coat vehicle penetrates completely through pores in the primer, PVC of the primer is reduced to a level approximately equal to CPVC, reducing its effectiveness. Penetration of the continuous phase of a second coat into pores is primarily controlled by viscosity of its external phase. On the other hand, if a thick, viscous second coat is applied directly over the zinc-rich primer, trapped air and moisture in the pores may cause pinholes or bubbling in the second coat. Furthermore, trapped solvents may have a similar effect if the second coat is applied too soon. These problems can be minimized by applying a *mist coat* (i.e., a very thin coating, usually of low solids epoxy) onto the zinc-rich primer. Weldon (2016a) attributes the effectiveness of a mist coat to flow into the pores, partially displacing some of the trapped gasses. The solvent evaporates rapidly from the mist coat, sealing the surface with a very thin film through which additional solvent and moisture can readily escape. Proper application requires considerable skill, since excess mist coat in an area leads to excessive penetration into the pores, but lack of coverage of an area means that the surface of the primer is not sealed in that area. Therefore, it is desirable for the color of the mist coat to contrast with that of the zinc-rich primer in order to aid the sprayer in applying complete, but not excessive, coverage.

After the mist coat is applied, application of further coats requires competent but not extraordinary spraying skill. Since alkaline zinc oxide, hydroxide, and carbonate are present on the zinc surface, top coats in contact with the primer coat must be stable against saponification. 2K epoxy, polyurethane, polyaspatic, polysiloxane, vinyl, or chlorinated rubber top coats can be used.

In conventional practice, an intermediate epoxy/polyamide coat is applied over the zinc-rich primer, followed by a urethane, aspartic ester/polysiloxane, or acrylic top coat. However, the intermediate coating is expensive and often slow-curing. Substantial savings can result from eliminating the intermediate coat. O'Donoghue et al. (2013) describe efforts to eliminate the intermediate coat when the coating is applied to steel bridge components at the factory, saving about 25% in both direct and indirect costs. Savings may be even greater compared with repainting existing bridges, partly because the cost of controlling traffic during slow repainting operations is high. Suppliers offer low-VOC polysiloxane top coats designed for application directly over zinc-rich primers. They project service lives longer than those of conventional three-layer systems with aliphatic polyurethane top coats. Some caution is warranted, however. Kahn (2014) compared a conventional three-coat system with two two-coat systems and found that the three-coat system (organic zinc, epoxy/polyamide, polyurethane) gave superior results; he did not, however, test a two-coat polysiloxane system.

Analogous technology exists for magnesium-rich primers, which are candidates as chrome VI replacements for aircraft (Section 30.5).

33.1.3 Systems with Passivating Pigment Containing Primers

When extensive film damage must be anticipated and when the substrate cannot be completely cleaned (especially when oily rust will remain), or when the coating cannot penetrate into the surface irregularities, passivating pigment primers are the primers of choice (Section 7.4.2). In contrast to the situation with baked OEM product coatings, for which passivating pigments are seldom appropriate, there are many field applications for which they are preferred despite the fact that blistering is more likely to occur.

A range of primer vehicles is used. Alkyds have the advantage that they are relatively low cost and adhere to oily surfaces, but they are deficient in saponification resistance. Epoxy–amine primers are used because of greater saponification resistance and good wet adhesion. Epoxy ester binders are intermediate in cost and performance.

For many years, oil and alkyd paints with lead anticorrosive pigments were commonly used on bridges and other steel structures. They usually performed well, especially since their relatively low cost made it economical to apply thick films; service lives of up to 100 years have been reported (O'Donoghue et al., 2013). But when repainting is finally necessary, it is a major problem—safe removal of the old lead paint and replacement with a lead-free coating is extremely expensive. Hare (1997) estimated a cost of \$54–217 m^{-2} in 1997, and it is probably much higher today. Today the emphasis is on lead-free coatings that will afford the longest possible intervals between repainting, with the goal of 100 years' service life.

Zinc yellow was, for many years, the passivating pigment of choice (Section 7.4.2). However, zinc yellow is carcinogenic, and care must be taken to avoid inhaling or ingesting spray dust, sanding dust, or welding fumes. Its use is declining sharply, and chrome VI pigments will be banned in Europe in 2019. Potential replacements for zinc yellow are discussed in Section 7.4.2.

In recent years, the use of latex paint systems for maintenance applications has increased. However, when any waterborne coating is applied to a freshly abrasive blasted steel surface, there is almost instantaneous rusting, called *flash rusting*. To avoid it, the formulation should include an amine such as 2-amino-2-methyl-propan-1-ol (AMP). Reinhard et al. (1992) recommend mercaptan-substituted compounds as additives to prevent flash rusting.

Kalendova (2002) studied factors influencing flash rusting and test methods. Relative humidity during application is a major factor especially when it is over 80%. Use of a "complex zinc compound in a mixture of anionic ionogenic solvents" gave better protection than a 10% aqueous solution of a 9:1 ratio of sodium benzoate and sodium nitrite solution. The composition of the latex affects the extent of flash rusting. A copolymer of acrylic acid, methacrylic acid, 2-ethylhexyl acrylate, and styrene gave good results. Flash rusting can be evaluated by applying pigment-free films to the steel since then there is nothing to hide the development of the red rust layer as it forms. Hough (2016) describes other test methods.

Since latex particles are large compared with the size of many of the crevices in the surface of steel, and since the viscosity of the coalesced latex polymer is extremely high, complete penetration into the crevices cannot be expected. Therefore, it is essential to use a passivating pigment. The passivating pigment must be chosen such that the concentration of polyvalent ions is low enough so that package stability of the latex is not adversely affected but still high enough to serve its passivating function. Strontium chromate is preferred and less soluble than zinc yellow. Zinc phosphate has been used. Among the newer passivating pigments, zinc–calcium molybdate and calcium borosilicates have been recommended.

Acrylic, styrene/acrylic, and vinylidene chloride/acrylic latex polymers are completely resistant to saponification. Special proprietary latexes are sold that provide enhanced adhesion to metal in the presence of water. It has been suggested that amine-substituted latexes are especially appropriate for promoting wet adhesion. Use of 2-(dimethylamino)ethyl acrylate (or methacrylate) as a comonomer is a way to incorporate amine groups on the latex polymer. Methacrylamidoethylethyleneurea has also found use as a wet adhesion promoting monomer. Another approach to enhancing wet adhesion is to replace part of the latex polymer with an alkyd or modified drying oil. The alkyd is emulsified into the coating. After application, as the water evaporates, the emulsion breaks and some of the alkyd can penetrate into the fissures in the steel surface. Epoxy esters are more hydrolytically stable than alkyds and are expected to provide better corrosion protection.

Owing to the high moisture vapor and oxygen permeability of most latex paint films, it is desirable to use some platelet pigment in the formulation. Mica is used in both primers and top coats. Final top coats, which contain leafing aluminum (Section 20.2.5), are particularly appropriate. To avoid problems of reactivity of aluminum with water, a special grade of leafing aluminum is mixed into a latex base paint just before application.

As noted earlier, California limits VOC of many maintenance coatings to 50 or $100\,g\,l^{-1}$. In early studies, the California State Office of Transportation Laboratory undertook the formulation and application of latex paints to several highway bridges in order to evaluate their potential utility (Warness, 1985). Latex primers were used DTM in some cases and inorganic zinc-rich primers in others; in all cases, latex top coats were used. Some latex DTM systems continued to perform well even after more than 5 years field exposure. Experience showed that application should only be carried out when the temperature is above 10°C and when the relative humidity is less than 75%. Paint suppliers now offer low-VOC latex maintenance paints for DTM application, probably a growing area.

33.1.4 Overcoating Existing Industrial Maintenance Paints

While complete removal of existing paint from steel structures before repainting is desirable, it is very expensive, especially when old lead-containing paints must be removed. It is far more economical to carefully prepare the painted surface, patch bare spots, and paint over the old paint. Numerous studies have sought the best coatings and procedures for overpainting. O'Donoghue et al. (2013) state that the most widely used overcoating systems in North America are as follows:

- Epoxy penetrating sealer/epoxy/polyurethane (2- or 3-coat)
- Moisture-cure urethanes (3-coat)
- *High-ratio calcium sulfonate alkyd* (HRCSA) and CSA coatings (1- or 2-coat)

The first two of these systems are variants of those discussed in Section 33.1.1. The first coat is formulated for maximum penetration into crevices and rust. The third system, HRCSA coatings, is formulated from alkyd resins and a crystalline form of "calcium sulfonate" (the calcium salt of a petroleum-based sulfonic acid). They cure by air oxidation and are usually applied in two coats. Many authors (O'Donoghue et al., 2009; Mandeno and El Sarraf, 2014), and state departments of transportation, report excellent results with HRCSAs. O'Donoghue et al. (2013) attribute the excellent performance partly to their low surface tension and low contact angle with rusted steel, partly to platelets of calcium sulfonate in the coating film that enhance barrier properties and partly to passivation. HRCSAs were developed for overcoating, but they also perform well on bare metal and are gaining favor in many U S states, in Europe, in New Zealand, and elsewhere. Part of their attraction is relatively low cost. O'Donoghue et al. (2017) describe successful overcoating of a large 1913 bridge in Edmonton, Alberta with HRCSAs; the coating is performing well after 20 years. On the other hand, Weldon (2016b) reported an instance of poor adhesion of HRCSAs to themselves.

Urethane/aspartic systems (Section 12.4) and acrylic or acrylic/alkyd latexes can also serve as overcoatings. The latexes can have low VOC if required.

A large and growing problem is deterioration of infrastructure, especially of steel. Brand (2015) proposed an inexpensive process of simply painting over rusty steel that will, in some circumstances, slow the rate of deterioration. This suggestion has apparently not been widely adopted, but follow-up research and testing are underway (Sondag and Burgess, 2016). "Don't let the perfect be the enemy of the good."

33.2 MARINE COATINGS

The marine market includes coatings for naval and commercial ships, barges, pleasure craft, and certain onshore facilities located near salt water. Coatings for offshore oil and gas facilities are sometimes but not always classified as marine coatings. A variety of substrate surfaces is coated, and many require special coatings. Bleile and Rodgers (1989) provide a useful overview of the market and products. Their figures for the market for hull coatings are estimated to be 84% in Asia, 11% in Europe, and 4.5% in

North America—figures that reflect the concentration of shipbuilding and ship maintenance in Asia. In 2015 the market was estimated at 9.2M gallons valued at $422M (ACA/Chemquest, 2016). About 87% of the market was for the "general marine" subsegment, mainly for steel ocean-going ships. The market is quite cyclical. Lindholdt et al. (2015) provide an extensive review of hull coatings, emphasizing the problems of testing. Goldie (2014) and Goldie and Mobbs (2014) also provide useful summaries of hull coating technologies.

Marine coatings have come a long way in 100 years. Gardner (1917) reported that 100 tons of red lead in oil was required to paint a new battleship, and partial repainting was needed every 3–6 months. Antifouling coatings were shellac pigmented with iron, zinc, and mercury oxides.

33.2.1 Above the Water Line and Interior

Coatings for exterior ship surfaces above the waterline are generally analogous to those supplied for heavy-duty maintenance applications as discussed in the previous section. Ultrahigh pressure hydroblasting is used to clean the surface; salt is a common contaminant in shipyards and hydroblasting removes all salt from the surface. An inhibitor can be added to the hydroblasting water to retard flash rusting. A primer must be applied soon after blasting to avoid recontamination. Inorganic zinc-rich or epoxy primers are common. Sghibartz (1982) studied the performance of a range of coatings over zinc-rich primers on panels exposed for 10 years in Jacksonville, FL. Inorganic zinc-rich primer with an intermediate epoxy coating and an aliphatic polyurethane top coat gave the best overall performance. Inorganic zinc-rich primer with an epoxy second coat and a vinyl top coat received the next highest rating for overall exterior performance. As detailed earlier in Section 33.1.2, there have been substantial improvements since 1982, and long service lives are projected for systems employing organic (2K epoxy) zinc-rich primers and a variety of top coats.

Fransehn et al. (2004) report that 2K coatings formulated with an epoxy-functional silicone resin (Section 16.1.1) and a carboxylic acid-functional acrylic resin provide excellent performance as marine coatings for topside application, including superior exterior durability as well as less dirt pickup and easier cleaning than urethane coatings.

The exterior durability requirements of coatings for ship superstructures are severe. In addition to direct UV, UV is also reflected off the water surface, and the humidity is generally high. Alkyds replaced lead in oil coatings starting in about 1940, but alkyds required frequent repainting. Polyurethane coatings are used increasingly because of their greater durability. Alkyds remain popular because they are relatively easy to apply by crew members who are not trained painters. (Hence the old expression "Join the Navy, see the world, and learn to paint.")

Fire retardancy is especially important for the interior of ships; chlorinated alkyd coatings have been recommended (Wake et al., 1995). The low solvent content of latex paints reduces fire hazard of paint in storage on a ship. Initial gloss is lower with latex than with alkyd coatings, but gloss retention and resistance to cracking are superior.

2K aliphatic urethane coatings satisfy the exterior durability and abrasion resistance requirements of deck paints. The further requirement of skid resistance is obtained by mixing coarse sand into the paint just before application or by broadcasting sand on the coating before it cures.

In addition to their use on boat and ship bottoms (see Section 33.2.2), 2K coal tar epoxy coatings are used in ballast areas inside ships. To protect against environmental disasters, most oil tankers entering American coastal waters are required to have a double-hull construction. This has increased the surface area of ballast spaces, which is as much as six times the outer hull area. The difficulty and high cost of repair work in double-hull construction dictates that the coating systems offer very long-term protection against corrosion.

Fluorinated urethane coatings are used in tanks such as fuel tanks, septic tanks, and sometimes bilge tanks (Brady, 1998). Fuel tanks must be protected against corrosion, since they are frequently filled with seawater ballast after the fuel is consumed. The lower surface free energy of fluorinated coatings simplifies cleaning.

Naval ships are painted grey to reduce their visibility. For decades, the paint was pigmented with a white pigment and carbon black. However, carbon black is a powerful IR absorber and caused the insides of the ships to be very hot. Brady devised a combination of color pigments that resulted in a significant decrease in temperatures. Black complex inorganic pigments and organic pigments with high IR reflectance are now available (Section 20.4).

33.2.2 At and Below the Water Line

The most challenging class of marine coatings is *antifouling* coatings. Spores and larvae of over 4000 organisms, ranging from algae to barnacles, can attach to the underwater hull of a ship. The growth of plants and animals on ship bottoms roughens the hulls, which in turn increases turbulence of water flow over the surfaces and drag; hence, speed decreases and/or fuel consumption increases. An unprotected hull can accumulate 150 kg m^{-2} of fouling in 6 months at sea (Buskens, 2013). Removal of such growths usually requires putting the ship into dry dock. The economic penalty of fouling is very large. Biofouling also has the undesirable effects of disseminating invasive species and increasing greenhouse gas emissions. Review papers detail the variables involved, the history of antifouling coatings, the status of the coatings available, and the various technical approaches being explored (Yebra et al., 2004; Morrison, 2005; Buskins, 2013; Goldie and Mobbs, 2014; Goldie, 2015; Lindholdt et al., 2015).

Economically, antifouling coatings with a lifetime of 48 months or longer are desired. This level is attainable with coatings based on copolymers of tributyltin (TBT) esters of methacrylic acid and other tin-containing coatings. Unfortunately, when ships are in port, leaching of tin toxicants continues. In some harbors, concentrations of toxicant built up sufficiently to affect marine life. There is also concern about potential heavy metal entrance into the food chain, affecting fish that might be caught for human consumption. Therefore, use of TBT compounds was banned on most pleasure craft in the 1980s and on commercial vessels in 2008.

As a result, the industry turned to copper, which had been known to have antifouling activity since the eighteenth century. The major approach is to apply *biocide-based antifouling coatings*, which contain biocides such as cuprous oxide that leach out of the coating. There are two kinds, *self-polishing polymer coatings* (SPCs) and *controlled depletion polymer coatings* (CDPs) (Lindholdt et al., 2015). CDPs are also called *ablative* or *erodible* coatings. The biocide must be effective against a broad spectrum of organisms. The leaching rate must be above a critical level required to kill all the organisms settling on the surface over a long period of time. Historically, typical binders for SCPs have been solution vinyl resins containing rosin salts. The binder is water sensitive, so loss of cuprous oxide continues after the surface layer is depleted of cuprous oxide. The antifouling lifetimes of such coatings in actual use are on the order of 7–24 months. Leaching rates decay exponentially in accordance with a first-order release of biocide. Accordingly, the rate of loss of cuprous oxide in the early stages of release must be very high in order to have a sufficient concentration of cuprous oxide for release toward the end of the life of the coating.

Currently, the dominant binders are acrylic polymers with pendant metal carboxylate salts or silyl groups. The coatings are heavily pigmented with cuprous oxide, other copper compounds, or metallic copper. Copper release by hydrolysis of the polymer is not sufficient to provide complete antifouling on its own; therefore, copper and organic biocides are combined. An example of an organic toxicant is 4,5-dichloro-2-*n*-octyl-4-isothiazolin-3-one, which is reported to be effective and to degrade rapidly in seawater, eliminating bioaccumulation (Raber, 1996).

Another approach to self-polishing antifouling coatings is the use of the zinc salt of hydrogenated rosin, together with a plasticizer, such as oleic acid, Cu_2O, and a biocide. Yebra et al. (2005) described extensive laboratory evaluations of the rates of leaching and dissolution of the upper layers of the coating.

By 2011, biocidal antifouling coatings commanded 96% of the market (Lindholdt et al., 2015). Although copper is much less toxic than tin, there are serious concerns about its environmental effects, and restrictions on its use are being widely considered, especially on recreational craft on the US West Coast (ACA/Chemquest, 2016). Regulators must balance environmental damage by toxic pigments against environmental benefits, which include reduced spreading of invasive species and reduced greenhouse gas emissions. In 2016, after much study and debate, the European Union approved the use of copper and ten other biocides until 2026.

Intensive research is aimed at finding ways to control fouling without using toxicants (Yebra et al., 2004). It is a difficult problem. A general approach is the development of coatings to which adhesion of fouling organisms is so limited that ship movement through the water and/or underwater brushes or hoses are sufficient to remove fouling without dry docking. An early coating of this type consisted of an elastomeric silicone applied over an epoxy primer. Service life was greater than 3 years, and the excellent leveling properties provided a smooth surface, which reduces drag. Limitations were poor abrasion resistance and tear strength of the silicone elastomer coatings (Bausch and Tonge, 1996). Unfortunately, movement through water of 18–22 knots was needed to remove most of the growths, which exceeds the speed of most freighters and tankers.

Brady (1999) outlined the requirements for a coating from which biogrowths are most easily removed. The coating should comprise a polymer having (1) a flexible, linear backbone, which introduces no undesirable interactions, (2) a backbone and a sufficient number of surface-active groups that exhibit high molecular mobility in order to provide the desired low surface free energy, (3) a low elastic modulus, and (4) chemical stability for long periods in a marine environment. The coating should be smooth at a molecular level to avoid infiltration by a biological adhesive leading to mechanical interlocking. Brady suggested polysiloxanes with side chains of nonafluorohexylmethylsiloxane or trifluoropropylmethylsiloxane as examples of polymers that might match the requirements. Commercial products based on silicone binders are now available.

A different approach is to develop ablative coatings that require no biocide. One example is based on copolymers of a fluorinated monomer (e.g., 2,2,2-trifluoroethyl methacrylate), an organosilyl monomer (e.g., trimethylsilyl methacrylate), and other (meth)acrylic monomers. The erosion rate is adjusted by adjusting the proportions of monomers (Aubers et al., 2004). Zhang et al. (2015) described synthesis of various perfluoroalkylether acrylates and copolymers and formulation of coatings.

Yet another type of biocide-free hull coating has a hydrogel coating on its surface. When the hydrogel is swollen by a large amount of water, it presents a surface that is so much like a liquid that marine organisms will not settle on it.

By 2014, each major vendor was offering a range of biocide-free hull coatings. Goldie and Mobbs (2014) report that such coatings are very effective at preventing fouling by weeds and shells (e.g., barnacles) but are less effective in controlling slime. Various technical approaches include soft coatings based on silicone elastomers and fluoropolymers, hard coatings based on vinyl esters or solvent-free epoxy, and ablative coatings based on zinc acrylates. Technical details are generally proprietary.

A persistent problem with nontoxic coatings is fouling by slime, which can spread invasive species. Producers of some of the newer nontoxic coatings claim that their products reduce slime as well as weeds and shells.

Bleile and Rodgers (1989) have reported that 2K coal tar-epoxy anticorrosive coatings have been found to be particularly effective for many ship bottom areas. Single coats have a total film thickness on the order of 400–600 μm, and multiple coats can be applied. The coatings must be handled with care in view of the potential carcinogenicity of components of coal tar. Because of this hazard, coal tar epoxies have been banned in some jurisdictions, and they are nearly obsolete on large ship bottoms. They continue to be used in static applications such as ballast tanks, oil rigs, pipe coatings, and refineries. They are applied in thick coatings as 2K systems. One package contains coal tar, amine-terminated polyamide resin, and pigment; the other contains the epoxy resin. The coatings have high chemical resistance and high dielectric strength. This latter property is especially important for installations that use zinc or magnesium sacrificial metal anodes to provide cathodic protection against corrosion (Section 7.2.2). Coal tar is sometimes replaced by petroleum derived hydrocarbon resins; in addition to lower toxicity hazard, the lighter color of hydrocarbon resins facilitates inspection in ballast areas. Applied to properly prepared surfaces, the expected lifetime in seawater immersion is on the order of 7 years.

It is recognized that hull fouling can increase the energy consumption of a ship by as much as 40%. As so often happens, testing of the newer antifouling coatings is problematic. Static tests are of little use for coatings whose effectiveness depends on motion in the water. An expensive but useful method is to test the energy consumption of the entire ship; the results, however, can be skewed by variations in the ship's propulsion system. Furthermore, performance of the coatings may change with time, temperature, and other variables. Lindholdt et al. (2015) provide an extensive review of the theory and current practice of testing and describe various test methods. The European Union has organized a task force to improve methods of assessing operating efficiency of ships.

33.2.3 Other Types of Marine Coatings

Most pleasure craft are constructed of glass reinforced polyester and are coated at the factory with polyester gel coats (Chapter 10). Other coatings for pleasure craft and yachts are commonly sold on a retail basis to the individual owner or shipbuilder. A range of products is available; one for wood is spar varnish. The original spar varnish (Section 14.3.2) was a phenolic–tung oil varnish, in which tung oil provides high cross-linking functionality and the phenolic resin imparts hardness, increased moisture resistance, and some exterior durability. While traditionalists still prefer phenolic–tung oil varnishes, the bulk of this market has shifted to uralkyds, which provide greater abrasion and water resistance (Section 15.7).

Coatings for offshore facilities such as oil well platforms and liquefied natural gas storage facilities can be classified either as marine or as industrial maintenance coatings. In these applications, high value is placed on long-lived anticorrosive coating systems that, in addition, resist physical damage caused by daily work on the platform. The most durable technologies described in Sections 33.1 and 33.2 are employed or adapted for these applications. Preventing corrosion in the *splash zone*, the parts of the structure near the water surface, is particularly challenging. Here the coatings are exposed to brutal combinations of intense UV and alternating salt water, rain, and dryness. Some operators have resorted to hot-spraying metal galvanic coatings on steel parts to be mounted there.

Offshore wind turbines are a growing and challenging market for marine coatings (Milmo, 2014). The economics of wind energy depend on longevity of the turbines, and a minimum of 20–25 years is desired. The blades are composite plastics up to 100 m long; their coatings must survive the stress of flexing while protecting them from UV, heat, rain, hail, ice, birds, and erosion. Technology for blade coatings is still evolving, but polyaspartic polyureas (Section 12.4) are currently favored. Coatings for the steel supporting structures and towers are the largest part of this market; three-layer marine coatings with polyaspartic polyurea top coats are often utilized. Technical challenges are compounded by very strict environmental regulations affecting the coatings. To achieve a 20- to 25-year lifetime in hostile environments will likely require a large commitment to maintenance.

33.3 AUTOMOBILE REFINISH PAINTS

Each year in the United States, there are roughly 18M vehicle crashes, leading to 12M repairs costing 30B USD (Anonymous, 2013). What country does not have vehicle crashes? Paints play an essential role in most of the repairs.

Worldwide, the market for refinish paints is projected to grow at about 2% per year (Pianoforte, 2016).

Some refinish paints are used for overall finishing of cars and new fleet vehicles in special colors, but the largest segment of the market is for repair. When a car has been in an accident and a fender is straightened or a new door or bumper is installed, these parts must be coated so that the color matches the color of the original coating. There are major technical challenges in meeting the demanding application and performance requirements, as well as major marketing and distribution challenges. Hundreds of kinds of cars are made each year, each in a variety of colors. Many cars on the road are more than 5 years old and some are much older. If there is an accident involving a 15-year old Jaguar, the owner expects to be able to bring it to a repair shop and have that section of the car repaired and painted to match the rest of the car. Furthermore, the *body shop* expects to be able to call the paint distributor and have the necessary paint and primers delivered by the next day.

Other sources of complexity are the proliferation of substrates that must be coated and the sophistication of the color effects in paints used by auto manufacturers (Chapters 19 and 30). Exterior car parts are made of various steels, aluminum, fiber-reinforced plastics, flexible plastics for bumpers, and, in the future, thermoplastic body parts.

Furthermore, the refinish industry is under increasing pressure to reduce emissions. In the United States, the national limits for VOC emissions are fairly lenient. However, many states have imposed stricter regulations, typically limiting VOC to $250\,g\,l^{-1}$ for key coatings such as primer surfacers, base coats, and clear coats. This situation has forced US suppliers to produce two sets of product lines, one for *national rule* states and another for the more restrictive states. Suppliers offer systems utilizing solvent-borne and waterborne refinish coatings; the latter have grown rapidly.

In some cases, refinish coating manufacturers make and stock small containers to match the colors. These are ready to be shipped to dealers and repair shops before the first new car comes off the assembly line. However, these "factory-packaged" colors are limited, due to high inventory costs and to variations in color shade at different OEM manufacturing sites. In most cases, these manufacturers provide formulations and tinting color bases to distributors (or larger repair shops) to permit matching any original color. Establishing the formulations requires considerable color-matching skill, especially for metallic colors and color effects. Computer color matching (Chapter 19) plays an increasing role.

Both air dry and force dry (65–80°C) coatings are used. The higher temperatures used in force dry coatings enhances the quality of the repair. In Europe and Japan, most shops are equipped with force dry ovens. In the United States, use of force dry ovens is increasing, but there is still a significant fraction of repair shops that use air dry coatings. Fast dry (to minimize contamination by dust and to increase productivity) is important in all shops but especially in those that use air dry coatings.

Repair of damage after an accident is often accomplished with replacement parts, as labor costs to bump out dents are high. Metal replacement parts are usually received from the manufacturer coated with electrodeposition primer. In such cases, the refinisher typically applies a primer-surfacer, a base coat, and a clear coat.

In the cases where a damaged panel is repaired, surface preparation is critical. If the damage is minor, the OEM coating can be overpainted. The old coating surface must be cleaned to remove dirt, tar, and wax by scrubbing with detergent, rinsing thoroughly, and drying; in some cases solvent cleaners are required. Scuff sanding may be needed to remove chalky pigment and degraded polymer. Often the extent of the damage makes it necessary to sand the old paint down to bare metal. Any breakthroughs in metal require that the edge of the area be feathered out, that is, sanded with a bevel so that there is a smooth change in film thickness. When bare metal is exposed, it must be washed free of grease with solvent.

A primer or primer-surfacer is then applied. First, however, one should check whether the coating on the car can withstand the solvents in the primer. Current OEM coatings present no problem, but if the car has been refinished, the coating might be only partially cross-linked and will lift when solvent is applied to it.

In the past in the United States and still in South America and Asia, common primer-surfacers are made with nitrocellulose–alkyd binders. In older methods, fast dry primers often contained a medium oil length, rosin-modified, tall oil alkyd binder (McBane, 1987). The rosin modification increases the T_g of the alkyd. The primers are highly pigmented, for example, with a PVC of 38% and sometimes even higher. Alternatively, binders of nitrocellulose with rosin-modified alkyd and high levels of pigmentation permit sanding the primer-surfacer within 30 min after application. The primer is sanded smooth with fine grit paper, and dust is removed. Application solids of such primers are low. Stricter VOC controls and OEM warranty repairs are forcing the use of epoxy and urethane primers, which provide superior longevity. In areas with the most restrictive VOC regulations, waterborne undercoats are used, especially in areas with restrictive VOC regulations. Often they are thermosetting 2K waterborne epoxy (Section 13.2.6) or urethane (Section 12.8) primers.

Historically, there have been two broad classes of refinish top coats: lacquer (thermoplastic) and enamel (thermosetting). A major application advantage of lacquer is fast dry; lacquers have short *out-of-dust* times (i.e., the time required for the film to become sufficiently dry that dust particles do not adhere to the surface). There tends to be a great deal of dust in the air in repair shops, and fast dry is a major advantage, reducing contamination of the freshly applied coating. A disadvantage of acrylic lacquers for refinish is that, when dried at room temperature, their gloss is not high enough. To match the gloss of the OEM applied coating, refinish lacquers must be polished with rubbing compound, adding to the cost.

Acrylic refinish lacquers comprise thermoplastic acrylic polymers with high methyl methacrylate content (Section 2.2.1), blended with cellulose acetobutyrate (CAB) and a plasticizer, such as butyl benzyl phthalate. Gloss retention is good but inferior to current OEM coatings. Application solids are low (10–12 NVV) and VOC emissions are correspondingly very high. Lacquers are now used in North America and Europe mainly for antique car restoration and by the do-it-yourself market. They are used more broadly in countries where allowed by environmental regulations.

Enamels are the other broad class of refinish top coats, for which there are several types. The lowest in cost are alkyd enamels; the alkyd is a medium oil, isophthalic, soy, or tall oil alkyd (Section 15.1). Such enamels have several advantages besides low cost. Fewer film defects such as crawling or cratering occur during application, owing to their low surface tension. Their gloss is high enough to match OEM monocoats without polishing. VOC emissions are substantially less than with lacquer. On the other hand, out-of-dust time approaches half an hour. Gloss retention of alkyd coatings is poorer than that of OEM coatings or refinish lacquer.

The properties of alkyd enamels can be improved by adding triisocyanates (Section 12.3.2), by using methacrylated alkyds (Section 15.4), and by reacting an acrylic resin made with glycidyl methacrylate as a comonomer with drying oil fatty acids. These approaches improve properties of monocoats but are still unsuited for clear coats.

Base coat–clear coat finishes (Section 30.1.2) are now used on the vast majority of new vehicles. Corresponding coatings for use in refinishing are required. The base coat must be able to provide hiding with two coats. Relatively low solids are desirable to permit alignment of aluminum or other flake pigments. Solventborne base coats were the first developed, but it was difficult to obtain acceptable appearance with effect colors in low-VOC formulations.

Waterborne base coats based on acrylic latexes or polyurethane dispersions are in wide use. These technologies employ special polymers and sophisticated formulating methods to control rheology. The coatings have low volume solids, which permit flake orientation comparable with OEM coatings. Treated aluminum flake must be used to avoid hydrogen evolution (Section 20.2.5). Relative humidity conditions affect the drying speed; the coatings can be force dried or air dried with high volume dry-air flow to remove most of the water before clear coating.

Contemporary refinish clear coats are often high solids two-package enamels, although waterborne versions are gaining importance. The details of the binder chemistries of current commercial clear coats are proprietary. Some probably involve complex resin systems with multiple types of cross-links. UV screeners and HALS are an essential part of the formulas, and there are indications that, in some cases, nanopigments are added to improve scratch and mar resistance. Presented in the following text are some of the chemistries that have been described.

Thermosetting hydroxy-functional acrylic resins similar to those used in OEM coatings can be used together with a polyfunctional isocyanate cross-linker. 2K formulations provide the required rapid cure time. For lower VOC, various hydroxy-functional reactive diluents can be added. Other ways to reduce VOC include use of polyfunctional aldimines and hindered amines, such as polyaspartic esters (Sections 12.4 and 25.2.2). Petit et al. (2001) describe a high solids clear coat comprised of a vehicle consisting of a methacrylated glycidyl versatate, butyl acrylate, styrene, and methyl methacrylate copolymer, together with a low viscosity HDI isocyanurate. Richter et al. (2003) recommends the use of unsymmetrical HDI trimer (HDI iminooxadiazenedione) rather than the symmetrical HDI trimer in 2K acrylic clear coats for refinishing, since 1–3% higher solids can be applied. As discussed in Chapter 18, exempt solvents can be used to meet $250\,g\,l^{-1}$ VOC requirements, although the high volatility of most exempt solvents requires balancing formulations with less volatile non-exempt solvents.

Low VOC is also attainable with combinations of bicyclic orthoesters and polyisocyanates. Van Den Berg et al. (2004) describe very high gloss 3K clear coats utilizing this chemistry. The third component is a catalyst to promote hydrolysis of the orthoester and to react the resulting hydroxyl groups with the isocyanate component. Clear coats with pot lives of 4 h and $230\,g\,l^{-1}$ VOC are reported that flash off in 1–2 min and dry in 5 min at 60°C. Van Den Berg et al. (2004) describes a silicon-containing spiroorthoester, prepared by reacting 2-butyl-2-ethyl-1,3-propanediol with tetraethyl orthosilicate. With HDI isocyanurate as cross-linker and DBTDL as catalyst, pot life is reported to be 7 h. DBTDL acts as a catalyst for both the hydrolysis of the orthoester to hydroxyl groups, as well as the subsequent cross-linking reaction with isocyanate groups.

Coatings based on acrylic resins bearing alkoxysilyl groups have been developed that combine good appearance, mar resistance, acid resistance, and exterior durability (Chen et al., 1996). Hazan and Rummel (1990) obtained one of the basic patents in the area. Refinements are described in many subsequent patents. The alkoxysilyl groups can be introduced using 3-trimethoxysilylpropyl methacrylate as a comonomer. Several types of cure reactions can be involved. The silylated resins can moisture cure through hydrolysis and coupling of the silanol groups (Section 16.2) and by ether interchange of the alkoxysilyl groups with hydroxy-functional co-reactants or polymers. Hydroxyl functionality can also be copolymerized into the silylated resin to make self-cross-linking polymers.

Very fast cure refinish clear coats can be formulated using pentaerythritol tetrakis-3-meraptopropionate and HDI isocyanurate together with a $Zr(AcAc)_2$ catalyst. Cure time is reported to be 10 min at ambient temperature and 4 min at 60°C (Klinenberg and van Beelen, 2003). Another type of fast cure clear coat is based on cross-linking of isocyanates with thiols (Section 12.1; Dogan et al., 2006).

Clear coats that undergo at least partial cross-linking by aspartic esters have considerable potential for fast cure and excellent weatherability. For example, Huybrechts et al. (2014) describe 2K clear coats based on a common commercial triisocyanate and a commercial polyaspartic ester; properties are enhanced by inclusion of an epoxy-functional silane in the formula.

Fast cure primers based on UV-curable technology and on acetoacetate/amine technology (Geurink et al., 2003) were detailed in the third edition of this work (Wicks et al., 2007). Current commercial use is apparently not large, but is projected to increase as shops learn to handle new curing techniques.

Dual thermal/UV cure clear coats have been investigated for refinishing. They have the potential for very fast cure and excellent film properties. 2K coatings with a polyisocyanate in one package and a hydroxy-functional urethane/acrylate oligomer were used. Initial thermal cure at temperatures in the range of 40–60°C is followed by UV curing. Thermally induced isocyanate cross-linking supplemented UV cure, particularly in shadow areas. The mar and chemical resistance of the dual-cured areas are reported to be superior to a conventional 2K urethane clear coat (Maag et al., 2000).

VOC emissions have been reduced by changing from conventional air spray guns to high volume, low pressure (HVLP) air guns. HVLP guns permit better transfer efficiency and, hence, lower paint usage and lower VOC emissions (Section 23.2.1).

Paint suppliers emphasize the need to wear masks and to mix and apply the paint in a well-ventilated spray booth. All paints require precautions, particularly 2K coatings with volatile diisocyanate components. Refinish coating suppliers are pursuing non-isocyanate cross-linking systems. It is important to bear in mind, however, that any reactant that can cross-link hydroxyl, carboxyl, or amine groups on a synthetic polymer can also cross-link proteins. Any such cross-linker will be toxic. A new system must be designed so that there is no volatile reactive component and so that the MW is high enough to minimize permeation through the skin and membranes.

Truck bed linings are a significant aftermarket product. The principal coating is a black pigmented 2K urethane applied with dual feed spray guns. Fast-reacting gel coatings are needed to avoid sagging, since film thicknesses are more than 100 mils. An aromatic 2K coating consists of an isocyanate package containing MDI/polyethylene glycol diol prepolymer and a resin package containing a combination of polyether diol and triol together with an aromatic diamine. The black pigment protects the binder against UV degradation. An aliphatic 2K coating has been developed that permits colors that match the color of the truck. The isocyanate is HDI isocyanurate, and the resin package comprises a combination of polyester polyols, a polyaspartate, and aliphatic diamine. The amines are necessary to provide a short gel time. The coating can be applied as a sole coating or, to reduce cost, the aromatic coating on a film thickness of about 120 mils, followed by 40 mil of the aliphatic coatings as a top coat (Zielinski et al., 2004).

33.4 TRAFFIC STRIPING PAINTS

Traffic striping or *marking* paints are applied to roads, parking lots, curbs, airport runways, and other paved surfaces. The annual value of such paints used in the United States is about $2B (Challener, 2015). The US national rule allows a maximum VOC of $150\,g\,l^{-1}$; California allows $100\,g\,l^{-1}$.

An essential property of most traffic striping paints is their ability to reflect light from headlamps back to the driver so the stripes are easily visible at night. This *retroreflectivity* is accomplished by imbedding glass beads in the paint. The beads can be incorporated into the liquid paint but often are sprayed on the surface of the wet paint just after application. Ideally, the size, shape, and durability of the beads are optimized. Other essential coating properties are fast drying, adhesion, abrasion resistance, and weatherability of the binder. Traffic striping is usually white or yellow. Organic yellow pigments have replaced chrome yellows for environmental reasons.

Several binder technologies are employed. Acrylic latexes (Chapter 9) are the least expensive, are relatively easy to apply, and have low VOCs. Resin producers offer a variety of latexes specially designed for waterborne traffic striping. Hermes et al. (2007, 2008) describe coatings formulated from conventional acid-functional latexes, polymeric amines, and ammonia. Presumably, after application, the ammonia evaporates and the polymeric amine causes rapid coalescence, leading to fast dry. Other acrylics employ cross-linking binders. Kaufman (2015) provides representative formulations and demonstrates that an abrasion resistance test (ASTM D2486-96) correlates with longevity of the stripes in service.

A second type of binder is hot melt, 100% solids striping. Typical binders are maleic acid esters of tall oil rosins or gum rosins. Pigments and beads are formulated into the binders. Stripes of the hot melts are roughly 2500 μm thick, about ten times thicker than those based on waterborne acrylics. The thicker hot melt stripes are more durable in warm climates but are vulnerable to being torn loose by snowplows.

Acrylic/epoxy formulations are a third type of binder. The coatings are more physically durable than acrylics, but they require special application equipment. Overall initial cost is higher than the other options, but acrylic/epoxy may be the superior choice for very high traffic locations. Other epoxy and urea formulations are used to a small extent.

In 1985 solventborne alkyds or combinations of alkyds with chlorinated rubber commanded most of the traffic striping market, but by 2011 such paints had almost disappeared in the United States (Dziczkowski, 2014). They were formulated in fast-evaporating solvents for fast dry. Dziczkowski proposes alkyd emulsions (he calls them alkyd latexes) as biobased alternatives to acrylic traffic marking binders.

REFERENCES

Aamodt, M., *Coat. World*, **2005**, *2(10)*, 30.

ACA/Chemquest, *CoatingsTech*, **2016**, *13*(9), 50–54.

Anonymous, *State of the Industry*, **2013–2014**, BodyShop magazine.

Anonymous, *CoatingsTech*, **2015**, November–December, 28–29.

Aubers, M. A., et al., US patent 6,767,978 (**2004**).

Bausch, G. G.; Tonge, J. S., *Proceedings of the International Waterborne, High-Solids, and Powder Coatings Symposium*, New Orleans, LA, **1996**, p 340.

Bell, Q., *Surf. Coat. Aust.*, **1982**, *30*(9), 50.

Bleile, H. R.; Rodgers, S., *Marine Coatings*, Federation of Societies for Coatings Technology, Blue Bell, PA, **1989**.

Borup, B.; Albert, P.; Mack, H., *Proceedings of the American Coatings Conference*, Atlanta, GA, April 7–9, **2014**, paper 11.1.

Brady, R. F., Jr., *Polym. Mater. Sci. Eng.*, **1998**, *78*, 249.

Brady, R. F., Jr., *Prog. Org. Coat.*, **1999**, *35*, 31.

Brand, W., *PaintSquare News*, **2015**, October 6.

Buskens, P., et al., *J. Coat. Technol. Res.*, **2013**, *10(1)*, 29–36.

Challener, C., *CoatingsTech*, **2015**, *12*(7), 34–38.

Chen, M. J., et al., *Surf. Coat. Int.*, **1996**, *79*, 539.

Danneman, J., Reichhold, Inc., **2014**.

Doble, O., *J. Prot. Coat. Linings*, **2004**, *21(4)*, 22.

Dogan, N., et al., *RadTech. Rep.*, **2006**, *20*, 43.

Drisko, R. W., *J. Prot. Coat. Linings*, **1995**, *12*(9), 27.

Dziczkowski, J., *Proceedings of the American Coatings Conference*, Atlanta, GA, April 7–9, **2014**, paper 13.4.

Eccleston, G., *J. Prot. Coat. Linings*, **1998**, *15(1)*, 36.

Feliu, S., Jr., et al., *J. Coat. Technol.*, **1993**, *65(826)*, 43.

Flecksteiner, R.; John, T.; Georgio, C., *Proceedings of the American Coatings Conference*, Atlanta, GA, April 7–9, **2014**, paper 7.3.

Fransehn, P., et al., *Proceedings of the International Waterborne, High-Solids, and Powder Coatings Symposium*, New Orleans, LA, **2004**, Paper No. 7.

Gardner, H. A., *Paint Researches and Their Practical Application*, Educational Bureau of the Paint Manufacturers' Association, Washington, DC, **1917**, pp 123–125. Digital version available on Google Books.

Gardner, G., *J. Prot. Coat. Linings*, **1996a**, *13*(2), 34.

Gardner, G., *J. Prot. Coat. Linings*, **1996b**, *13*(2), 81.

Geurink, P., et al., *Prog. Org. Coat.*, **2003**, *48*, 153.

Giles, D., et al., *Proceedings of the International Waterborne, High-Solids, and Powder Coatings Symposium*, New Orleans, LA, **2001**, pp 337–348.

Giles, D., et al., *Proceedings of the International Waterborne, High-Solids, and Powder Coatings Symposium*, New Orleans, LA, **2003**, pp 167–182.

Goldie, B., *J. Prot. Coat. Linings*, **2014**, November, from http://www.paintsquare.com (accessed March 27, 2017).

Goldie, B., *J. Prot. Coat. Linings*, **2015**, November.

Goldie, B.; Mobbs, D., *J. Prot. Coat. Linings*, **2014**, July, 38–50, from http://www.paintsquare.com (accessed March 27, 2017).

Guidetti, G. P., et al., *Prog. Org. Coat.*, **1996**, *27*, 79.

Hare, C. H., *J. Prot. Coat. Linings*, **1997**, *14(11)*, 50.

Hazan, I.; Rummel, M. K., US Patent 5,244,696 (**1990**).

Hermes, A. R., et al., US patent 7,235,595 (**2007**).

Hermes, A. R., et al., US patent 7,314,892 (**2008**).

Hough, D. T., *J. Prot. Coat. Linings*, **2016**, September, from http://www.paintsquare.com (accessed March 27, 2017).

Huybrechts, J.; Tanghe, L.; Vaes, A., Published patent application WO2014128052 (**2014**).

Jorissen, S. A., et al., *Proceedings of the International Waterborne, High-Solids, and Powder Coatings Symposium*, New Orleans, LA, **1992**, p 182.

Kaelin, S. B.; Liang, S. T., *J. Prot. Coat. Linings*, **2014**, December, from http://www.paintsquare.com (accessed March 27, 2017).

Kahn, S. A., *J. Prot. Coat. Linings*, **2014**, May, from http://www.paintsquare.com (accessed March 27, 2017).

Kalendova, A., *Prog. Org. Coat.*, **2002**, *44*, 201.

Kaufman, M., *Coat. World*, **2015**, *20(11)*, 38–42.

Kincaid, D. S.; Schulte, J. A., *Proceedings of the International Waterborne, High-Solids, and Powder Coatings Symposium*, New Orleans, LA, **2001**, pp 127–141.

Kline, H. H., *J. Prot. Coat. Linings*, **1996**, *13(11)*, 73.

Klinenberg, H.; van Beelen, J. C., US patent application, 20030181625 (**2003**).

Kowalski, G., et al., *J. Prot. Coat. Linings*, **2003**, *20(1)*, 73.

Lindholdt, A., et al., *J. Coat. Technol. Res.*, **2015**, *12*(*3*), 415–444.

Maag, K., et al., *Prog. Org. Coat.*, **2000**, *40*, 93.

Mandeno, W. L.; El Sarraf, R., *Protective Coatings for Steel Bridges*, **2014**, NZ Transport Agency, from http://www.nzta.govt.nz (accessed March 27, 2017).

Martin, J. W., et al., *Methodologies for Predicting Service Lives of Coating Systems*, Federation of Societies for Coatings Technology, Blue Bell, PA, **1996**.

McBane, B. N., *Automotive Coatings*, Federation of Societies for Coatings Technology, Blue Bell, PA, **1987**.

Mestach, D. E. P., et al., US patent application publication US2013/0018127 A1 (**2013**).

Milmo, S., *Coatings World*, **2014**, March, 52–53.

Montes, E., *J. Coat. Technol.*, **1993**, *65(821)*, 79.

Morrison, S., *SpecialChem4Coatings and Inks*, **2005**, (*1*).

Muhlberg, K., *J. Prot. Coat. Linings*, **2004**, *21(4)*, 30.

O'Donoghue, M., et al., *J. Prot. Coat. Linings*, **2009**, August, 18–30, from http://www.paintsquare.com (accessed March 27, 2017).

O'Donoghue, M., et al., *J. Prot. Coat. Linings*, **2013**, January, from http://www.paintsquare.com (accessed March 27, 2017).

O'Donoghue, M., et al., *J. Prot. Coat. Linings*, **2017**, May, from http://www.paintsquare.com (accessed May, 2017).

Paert, J., *J. Prot. Coat. Linings*, **1992**, *9*(2), 50.

Parashar, G., et al., **2001**, *Prog. Org. Coat.*, *42*, 1.

Parker, B., *J. Prot. Coat. Linings*, **2015**, January, from http://www.paintsquare.com (accessed March 27, 2017).

Petit, H., et al., *Prog. Org. Coat.*, **2001**, *43*, 41.

Pianoforte, K., *Coatings World*, **2016**, October, 44–49.
Pilcher, G. R., *CoatingsTech*, **2015**, *12(8)*, 24–32.
Raber, L., *Chem. Eng. News*, **1996**, *74(29)*, 9.
Reinhard, G., et al., *Prog. Org. Coat.*, **1992**, *20*, 383.
Richter, F., et al., *Proceedings of the International Waterborne, High-Solids, and Powder Coatings Symposium*, New Orleans, LA, **2003**, pp 71–85.
Robinson, G. N., et al., *J. Coat. Technol.*, **1993**, *65(820)*, 51.
Schutz, D., *Mater. Perform.*, **1998**, *37(2)*, 32.
Schwindt, J., *J. Mater. Performance*, **1996**, *35(12)*, 25.
Senkowski, E. B., *J. Prot Coat. Linings*, **2015**, November.
Sghibartz, C. M., *FATIPEC Congress Book*, **1982**, Vol. *IV*, p 145.
Sondag, S. K.; Burgess, R. A., *J. Prot. Coat. Linings*, **2016**, January, from http://www.paintsquare.com (accessed March 27, 2017).
Sorensen, P. A., et al., *J. Coat. Technol. Res.*, **2009**, *6(2)*, 135–196.
Squiller, E. P., et al., Patent application WO201451307 A1 (**2014**).
Szokolik, A., *J. Prot. Coat. Linings*, **1995**, *12(5)*, 56.
Takas, T. P., *JCT Coat. Tech.*, **2004**, *1(5)*, 40.
Van Den Berg, K. J., et al., US patent appl. 20040122203 (**2004**).
Verbiest, P., *J. Prot. Coat. Linings*, **2015**, December, from http://www.paintsquare.com (accessed March 27, 2017).
Vielhauer, L., et al., *CoatingsTech*, **2015**, *12(9)*, 26–34.
Wake, L. V., et al., *J. Coat. Technol.*, **1995**, *67(844)*, 29.
Warness, R., *Low-Solvent Primer and Finish Coats for Use on Steel Structures*, Technical Report, FHWA/CA/TL-85/02, **1985**.
Weinmann, D. J., *J. Prot. Coat. Linings*, **2013**, September, from http://www.paintsquare.com (accessed March 27, 2107).
Weldon, D. G., *J. Prot. Coat. Linings*, **2016a**, November, from http://www.paintsquare.com (accessed March 27, 2017).
Weldon, D. G., *Paint Square News*, **2016b**, August 4.
Wicks, Z. W., Jr., et al., *Organic Coatings Science and Technology*, 3rd ed., **2007**, John Wiley & Sons, Inc., New York, p 672.
Witucki, G. L., **2013**, from http://www.paintsquare.com/education/branding_images/gwwebinar.ppt (accessed March 27, 2017).
Woetzel, J., et al., *Bridging the Global Infrastructure Gap*, June, **2016**, from http://www.Mckinsey.com (accessed March 27, 2017).
Wu, W., et al., *Proceedings of the International Waterborne, High-Solids, and Powder Coatings Symposium*, Atlanta, GA, April 7–9, **2014**, paper 7.2.
Xu, Y., et al., *Prog. Org. Coat.*, **2002**, *45*, 331.
Yebra, D. M., et al., *Prog. Org. Coat.*, **2004**, *50*, 75.
Yebra, D. M., et al., *Prog. Org. Coat.*, **2005**, *53*, 256.
Zhang, Y., et al., *J. Coat. Technol. Res.*, **2015**, *12(1)*, 215–223.
Zielinski, D. P., et al., *Proceedings of the International Waterborne, High-Solids, and Powder Coatings Symposium*, New Orleans, LA, **2004**, Paper No. 35.

Chapter 34

Functional Coatings

Traditionally, coatings have served to beautify and protect substrates, as has been discussed throughout this book. Through the careful selection of pigments and surface texture, the color and gloss of an object is enhanced. Metal substrates are protected from corrosion by the barrier properties of coatings, and additives to the coatings can further enhance their corrosion protection qualities. Wood is protected against degradation. However, new technology has allowed even greater functionality to be built into coatings. Examples include coatings that are hostile to bacterial colonization (antimicrobial coatings), water repellant (superhydrophobic coatings), oil repellant (oleophobic coatings), self-cleaning (Ganesh et al., 2011), and coatings that respond to changes in their environment or can heal themselves when damaged (Montemor, 2014; Bekas et al., 2016). These types of coatings are often called *functional coatings* or *smart coatings*. The chemistry and morphology of these coatings are often based on examples found in the natural world such as in plants (lotus leaf effect) (Gao and McCarthy, 2006) or animals. Thus, these coatings are often *biomimetic*. Of course all coatings are functional, but the coatings described in this chapter possess functionality that goes beyond aesthetics or substrate protection.

34.1 SUPERHYDROPHOBIC AND SUPERHYDROPHILIC COATINGS

The manner in which the top surface of a coating interacts with liquid water is governed by both the chemistry and the morphology of that surface. The most direct probe of that interaction is made by measuring the contact angle that a drop of water makes with the surface. The wettability of the surface is related to the contact angle the coating makes with the substrate (Sections 6.1 and 24.1).

The surface of a coating is typically classified as hydrophilic if the contact angle with liquid water is less than 90°. For these surfaces, water will tend to wet the surface and spread to cover it. If the contact angle with water is greater than 90°, the surface is classified as hydrophobic and water will bead up into droplets. Through the use of low surface energy monomers or additives or by engineering a particular morphology into the coating surface, surfaces can be made to be *superhydrophobic,* with water contact angles greater than 150°. The exact cutoffs between hydrophilic, hydrophobic, and superhydrophobic are somewhat arbitrary and can be defined differently by different scientists (Figure 34.1).

A superhydrophobic surface is typically achieved through the use of low surface energy components in the binder of a coating and by designing a particular level of surface roughness into the surface. Both silicone-based and fluorinated monomers and additives have been used to create low surface energy coatings. Coatings where these functionalities are built into the monomer typically provide a more durable superhydrophobic surface than those coatings where the low surface energy material is simply an additive in the coating formulation or sprayed on an already existing coating. The latter process provides little opportunity for the low surface energy material to bind with the surface.

Examples of superhydrophobic materials based on silicon monomers or silane precursors can be found in various references (Zhang et al., 2008; de Francisco et al., 2014). Fluoropolymer-based coatings typically perform even better (higher contact angles with water), but there are concerns about health and environmental hazards associated with some of the fluorinated precursors used (Wang et al., 2008; Boinovich and Emelyanenko, 2015).

Organic Coatings: Science and Technology, Fourth Edition. Frank N. Jones, Mark E. Nichols, and Socrates Peter Pappas.

To achieve the highest levels of superhydrophobicity, surface structure is required. The reasons for this have been explored by Wenzel (1936) and Cassie and Baxter (1944). When roughness is introduced into a coating's surface, air exists in the valleys between the surface peaks. If that air is displaced by water (or some other liquid), the droplet of water will spread over the surface. In this condition, the surface and water are in the Wenzel state, as seen in Figure 34.2. The conditions under which this occurs are given by

$$\cos\theta^* = r\cos\theta$$

where θ^* is the apparent contact angle, θ is the contact angle with the ideally flat surface, and r is surface roughness parameter defined as the ratio of the rough surface to that of a perfectly flat surface of the same macroscopic dimensions. In the Wenzel state, the contact angle of water with the surface is not particularly high. However, if the penetration of the water between the surface peaks can be suppressed, the contact angle with the surface can become extraordinarily high, in some cases over 160°. In these cases, the surface is said to be in the Cassie or Cassie–Baxter state, which is shown in Figure 34.2 on the right.

When in the Cassie–Baxter state, the apparent contact angle is given by

$$\cos\theta^* = rf\cos\theta + f - 1$$

where f is the fraction of the surface wet by the liquid. Thus, the apparent contact angle can be increased by increasing the roughness of the surface. Various authors

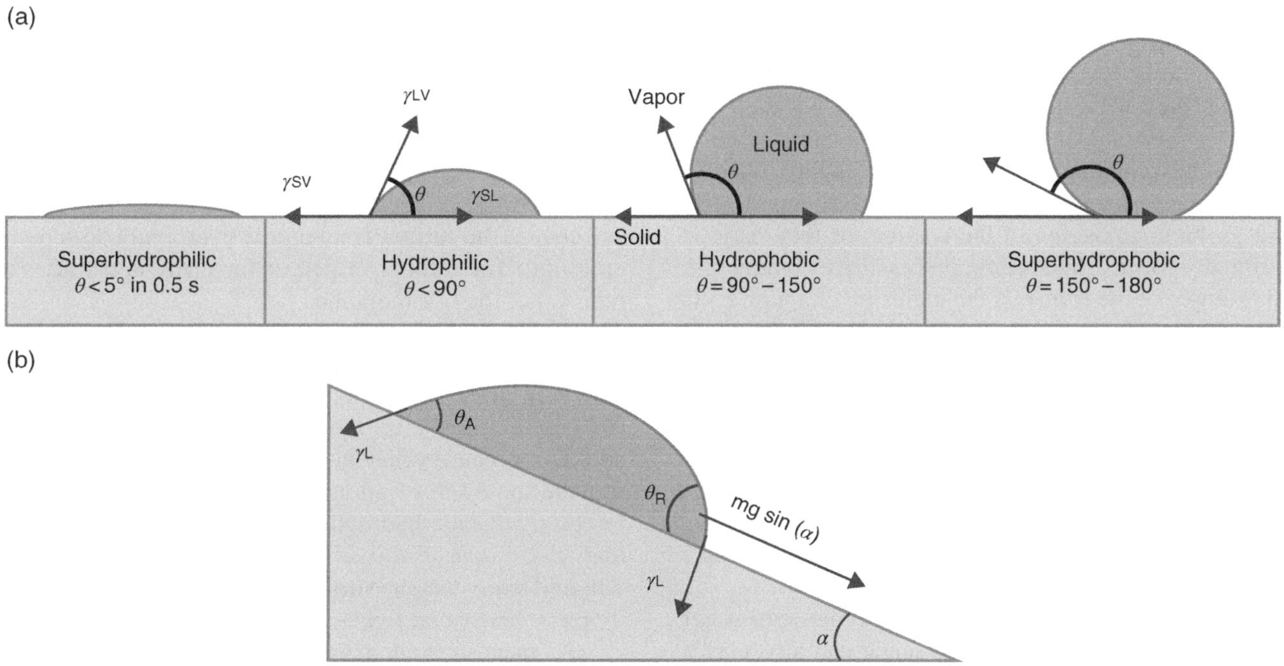

Figure 34.1 Contact angles (a) and sliding angle (b) for hydrophilic, hydrophobic, and superhydrophobic surfaces. Source: Nuraje et al. (2013). Reproduced with permission of Royal Society of Chemistry.

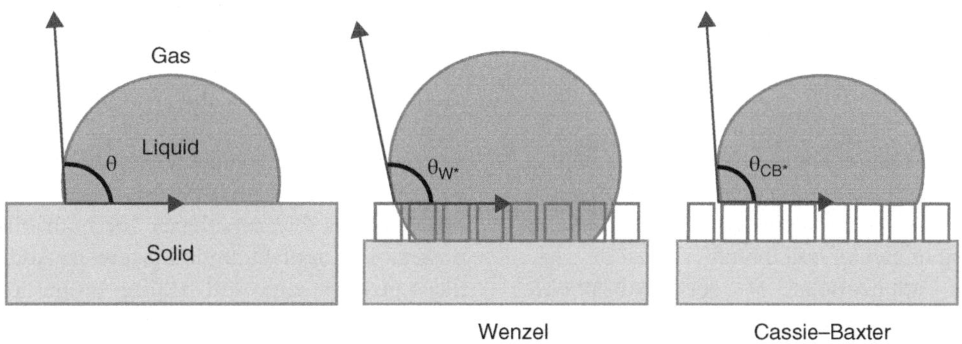

Figure 34.2 Wenzel and Cassie–Baxter states of wetting. Source: Nuraje et al. (2013). Reproduced with permission of Royal Society of Chemistry.

Figure 34.3 Hierarchical structure of superhydrophobic surface where liquid is suspended atop papillae of microstructure.

have found that hierarchical structures, with roughness at different size scales, promote superhydrophobicity.

A schematic of a hierarchical structure is shown in Figure 34.3. In general, these surfaces attempt to mimic those of the lotus leaf, whose papillae are quite small and waxy and prevent water from impinging between the papillae.

The structures that are necessary to impart superhydrophobic performance can be produced spontaneously owing to the composition of the coating, or they may be intentionally formed by specific processing techniques. A good example of the former is the inclusion of a high PVC of silica nanoparticles with a bimodal size distribution into a hydrophobic binder (Bravo et al., 2007). The silica will form a hierarchical structure on the surface, with the smaller particles decorating the larger silica particles. Laser ablation (Kwon et al., 2014) and photolithography (Liu et al., 2006) have been used to create surfaces with rods, cones, and other structures that project from the surface, which can be subsequently coated with a low surface energy coating.

In addition to the static contact angle between water and a surface, the dynamic contact angle can also play a role. In particular, the *contact angle hysteresis* is often less than a few degrees for superhydrophobic surfaces. The hysteresis is the difference between the advancing and receding contact angle. Imagine a drop moving across the surface. At the edge of the drop on the side in the direction of advance, the contact angle will slightly increase, and on the trailing edge decrease. The difference between these two angles is the contact angle hysteresis. Finally, the sliding angle (Figure 34.1) is the angle from the horizontal at which a surface must be placed for a drop of water to spontaneously roll off the surface. Superhydrophobic surfaces often display sliding angles of less than 5°.

The potential applications of superhydrophobic surfaces are many. They are being examined to produce surfaces that are self-cleaning or easy cleaning, anti-fingerprint, ice-phobic, or antifogging. However, the commercial applications of superhydrophobic coatings are limited as of 2017. This is mainly due to the poor durability of these surfaces, as at small scales, surface structures are fragile and easily damaged. Thus, the superhydrophobic properties are often lost after mechanical abrasion or rubbing, or even repeated exposures to dirt or other contaminants. In addition, the surface structure needed to produce superhydrophobic coatings is on the order of the wavelength of light. Thus, these surfaces tend to strongly scatter light, and the coatings lose their transparency. For some applications, this may be acceptable, but for others, the transparency and durability issues must be addressed before these coatings can be commercially successful.

Superhydrophilic coatings are those coatings that are completely wet by water, with contact angles less than 10°. While surface structure can be important in these coatings as well, hydrophilic monomers are the main formulation tool used to produce these coatings. Polyelectrolytes have been formulated as brushes to enhance the hydrophilicity of surfaces (Kobayashi et al., 2012). On the inorganic front, TiO_2 coatings or coatings with TiO_2 and SiO_2 have been extensively used to create superhydrophilic surfaces (Machida et al., 1999). When exposed to UV light, the anatase form of TiO_2, in particular, will produce radicals and change the surface from mildly hydrophilic to superhydrophilic. The effect can persist for many hours after the removal of the UV radiation.

Potential applications for superhydrophilic coatings are less obvious than those for superhydrophobic coatings. However, they could potentially be useful on medical devices or implants to improve biocompatibility. In addition, because water sheets off these surfaces, they may provide some self-cleaning behavior via a mechanism different than superhydrophobic coatings. These coatings may also prove useful as antifogging coatings, as water will uniformly wet the surface as opposed to forming small droplets that scatter light.

An interesting alternative to superhydrophobic or superhydrophilic surfaces is those made by infusing a structured surface with a low surface energy, low vapor pressure liquid. Other liquids placed on such surfaces are then repelled by both the structure and the liquid interface. Such surfaces are called *slippery liquid infused porous surfaces* (SLIPS) (Wong et al., 2011). These surfaces mimic the morphology found in the *Nepenthes* pitcher plant, which repels both water and oils. The original SLIPS employed low surface energy substrates with nano-sized texture that were then infused with perfluorinated oils. These surfaces show remarkable repellency for both oils and water and possess low contact angle hysteresis. Additionally, the surfaces possess some self-healing properties, as the damage caused by small scratches can be healed as the low surface energy liquid flows into the newly created defect. If the refractive index of the solid substrate and liquid is closely matched, the surfaces can be transparent as well (Vogel et al., 2013). As these surfaces also show low adhesion to ice, they may be useful as coatings for refrigeration coils, in oil wells, or in other areas where the reduction of ice adhesion is desired. The Varanasi group at MIT has devised numerous variations of SLIPS technology (Dhiman et al., 2016) and has proposed a broad range of applications, including a much publicized liner for ketchup bottles that allows all the ketchup to flow out easily. A drawback of the SLIPS approach has been the need to replenish the infused liquid periodically, as even low vapor pressure liquids will eventually evaporate or be removed in use.

Coatings that resist wetting by oils are termed *oleophobic*, and those that resist wetting by both oil and water are termed *omniphobic*. These types of surfaces are inherently more difficult to produce, as oils possess inherently low surface energies and are strongly favored to wet most surfaces. However, some progress has been made in producing such coatings. Tuteja et al. (2007) have investigated the composition and structure of surfaces that would be required to produce oleophobic coatings. Calculations show that *reentrant curvature* on surfaces is required to produce superoleophobicity. Reentrant surfaces appear as mushrooms (termed micro-hoodoos) on the surface; the undercutting of the top structure, along with low surface energy materials, prevents oils from wetting the surface. Similar techniques have been used to produce omniphobic coatings that may be useful for oil/water separation or as coatings for the insides of bottles (such as ketchup bottles) to ease the removal of the bottle's contents (Tuteja et al., 2008; Brown and Bhushan, 2016) (Figure 34.4).

34.2 ICE-PHOBIC COATINGS

Ice-phobic coatings are those coatings that either resist the formation of ice on their surface or to which ice has very low adhesion (Menini and Farzaneh, 2011). The same coating rarely possesses both properties.

The morphology of ice is complex, but it generally deposits on surfaces in two different forms. *Rime ice*, a more technical term for frost, forms when very small water droplets condense and freeze discontinuously on a surface. Most people encounter this type of ice when their car is left outside on a cold evening, and they have to scrape their windshield before driving. *Glare ice*, also called *glaze ice*, is formed when a continuous layer of liquid water freezes on a surface. This type of ice can coat electrical power lines and aircraft wings with disastrous effects. In general, ice-phobic coatings can be formulated to reduce the formation of rime ice *or* the adhesion of glare ice but not both.

To reduce the adhesion of glare ice to a surface, some level of hydrophobicity is often preferable, but superhydrophobic coatings are not sufficient to reduce ice adhesion (Nosonovsky and Hejazi, 2012; Sojoudi et al., 2016). The surface structure of many superhydrophobic coatings can actually enhance the adhesion of ice to the surface, which renders the low polarity surface less ice-phobic than would be presumed based strictly on the water contact angle. Other strategies that appear to decrease ice adhesion rely on the freezing point depression of water on certain surfaces (Boinovich et al., 2014) or the addition of oils to low surface energy coatings (Zhu et al., 2013).

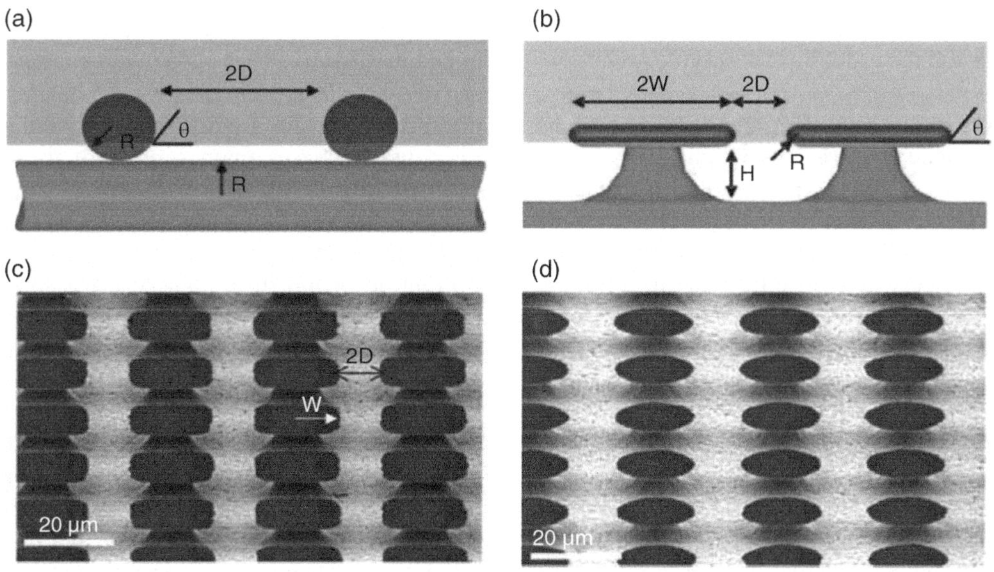

Figure 34.4 Structures of superoleophobic coatings. Schematic (top, a and b) and surfaces produced by deposition and then etching of SiO_2 structures on Si substrates (c and d). Source: Tuteja et al. (2007). Reproduced with permission of The American Association for the Advancement of Science.

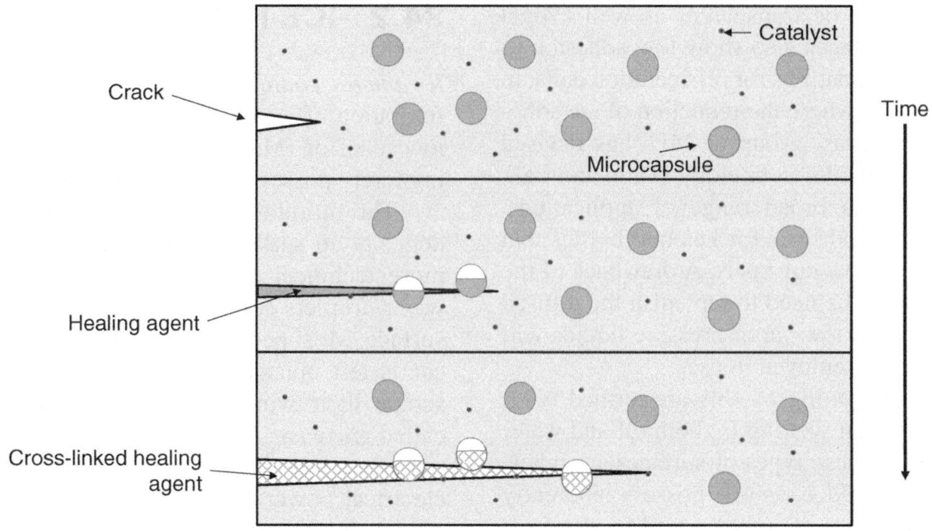

Figure 34.5 Propagation of a crack through a material with microcapsules designed to provide self-healing of the binder. As crack propagates (top figure to bottom figure), microcapsules are pierces and the healing agent flows into the crack. Due to the presence of the latent catalyst in the surrounding binder, the healing agent then cross-links to mend the crack.

One study states that an ice-phobic coating must possess an adhesion value to glare ice less than 100 kPa (Golovin et al., 2016). However, experimental work suggests that for passive ice removal from an aircraft surface, ice adhesion to the coating must be less than 0.1 kPa. No commercial coatings today provide such low adhesion values. Recent work has shown that the mechanical properties of the underlying ice-phobic coating may have a strong influence on ice adhesion. Tuteja has shown that elastomeric polyurethane-based coatings possess much lower ice adhesion values that similarly structured coatings that are glassy. This results from the ability to induce slippage at the interface between the solid ice and dangling chains at the surface of the lightly cross-linked polymer (Golovin et al., 2016).

The nucleation and growth of rime ice on a surface is still not completely understood. Clearly, liquid water has to be brought below its freezing point, but often substantial undercooling is required to actually nucleate the growth of ice crystals. Some ice-phobic coatings increase the required undercooling before ice will form either due to the presence of additives or due to the lack of nucleation sites for ice crystals on the surface (Simendinger and Miller, 2004).

34.3 SELF-HEALING COATINGS

All coatings are susceptible to damage during their service lives. Following damage of the coating, the underlying substrate is often prone to degradation, via corrosion or other vectors of substrate attack. Thus, coatings that can actively heal themselves after damage could potentially provide improved service life and substrate protection.

White et al. (2001) pioneered early work on self-healing materials by formulating polymers with dispersed microspheres. The microspheres contained dicyclopentadiene monomers that polymerize when exposed to a Grubbs catalyst, which was latent in the surrounding matrix. When a crack propagated through the material, the microspheres would locally rupture, and release their liquid resin, which would subsequently polymerize and "glue" the crack back together (Figure 34.5).

Since that pioneering work, other microsphere technology has been investigated. Drying oils and catalysts have been encapsulated in microspheres that, upon rupture, polymerize like conventional alkyds (Chaudhari et al., 2013; Tatiya et al., 2016) to heal damage to the coating in which they were dispersed. Solvents have also been encapsulated, which cause solvent welding of the separated crack phases upon rupture of the capsules during crack propagation (Jones et al., 2015).

Self-healing coatings based on microsphere technology hold the most promise for corrosion prevention, as the prevention of moisture ingress to the substrate can effectively retard the corrosion rate of many metals. Additionally, although the microspheres are often large (>20 um) and cause light scattering owing to refractive index mismatch with the surrounding binder, most protective coatings are opaque anyway, so it does not matter. At present, these coatings are not acceptable for use on automobiles or furniture, for example, where individual layers are approximately the same thickness as the microsphere diameter. Recent progress has demonstrated a reduction in the microsphere size via improved dispersion during preparation of the core–shell particles. A review is available (Samadzadeh et al., 2010).

The production of microspheres usually requires the use of an oil-in-water emulsion polymerization process. Thus, the encapsulated material must be rather hydrophobic. Melamine formaldehyde capsules containing linseed oil have been successfully demonstrated and even commercialized for use in architectural coatings (Suryanarayana et al., 2008; Jadhav et al., 2011).

Beyond capsules, artificial vascular channels have been created in coatings (Hamilton et al., 2010). While significantly more complex to produce, these materials solve the problem that repeated damage in the same area cannot be effectively mitigated with microcapsule-containing coatings. Artificial vascular systems attempt to mimic the vascular system found in living creatures, where a fluid (blood or lymphatic fluid) is continually delivered to multiple locations through a complex system of vessels. When a vascularized coating is damaged, the vascular system can deliver a healing agent to the local damage. If the coating is subsequently further damaged, the redundant vascular network can deliver further healing agent, as it is part of a greater interconnected network with a much larger reservoir of healing agent. While coatings based on these laboratory concepts appear to be well removed from short-term commercialization, they do point the way toward significantly improved functionality.

Self-healing coatings have also been demonstrated that do not use microspheres or vascular systems. Coatings that have a sufficiently low T_g can experience reflow, that is, small viscous flow that will heal surface imperfections when exposed to temperatures above their T_g. Polyurethane coatings can be easily formulated with various soft and hard segments (Chapter 12) and have been shown to possess some self-healing properties. For example, surface mars in an automotive clear coat can be healed when a properly formulated polyurethane is heated with a hot air gun or infrared heater to a temperature above the coating's T_g (Seubert et al., 2009). This approach is not without problems, however, as most coatings are formulated with a T_g that optimizes a suite of properties. If the T_g is lowered to provide self-healing capability, other properties, such as inherent scratch resistance, solvent resistance, or polishability, may suffer. More importantly, the T_g of most coatings increases during long-term environmental exposure. Thus, coatings that were initially self-healing would lose their ability over time as unrealistically high temperatures would be required to induce reflow.

UV light has also been used to activate the healing processes in coatings. One example is a polyurethane coating functionalized with oxetane rings and chitosan (Ghosh and Urban, 2009). When mechanical damage such as a scratch occurs, the oxetane rings are broken, which produces reactive species. To induce healing, the coating is exposed to UV light. The UV light induces chain scission in the chitosan polymer, whose chain ends then react with the opened oxetane rings, producing new cross-links. This process has been shown to function under natural sunlight exposure as well.

The morphology of most coatings depends on their formulation and the process by which they are applied. Many coatings only function well when used in conjunction with a primer or other layers in a *paint system*. Unfortunately, multilayer systems are expensive and time consuming to apply, as typically each layer must be applied serially, often with a drying or curing step in between. *Self-stratifying coatings* aim to remove those barriers. Applied as a single layer, these coatings then phase separate into two or more layers having distinct properties. Early work by Funke (1976) showed that chemical incompatibility can lead to self-stratification when epoxy and silicone binders are mixed together. Verkholantsev (1992) later expanded this work. Recently, Baghdachi et al. (2015) have shown that stratification can be driven by chemical kinetics whereby a rapidly reacting component migrates to the substrate and excludes the more slowly reacting phase, which is then preferentially located at the surface.

34.4 ENVIRONMENTALLY SENSING COATINGS

Coating films that respond to changes in their environment can be useful in a variety of ways. Interior architectural paints that change color upon drying are a simple example. These latex paints typically contain a dye that is easily degraded by ambient interior light or that loses its color when the pH changes. These coatings are helpful to DIY painters for estimating the coverage of a new coating over a coating of similar color, such as when a white ceiling is repainted. The new coating typically appears fairly bright (pink or purple) when first applied, but the color then fades to white as the dye loses activity.

Coatings that respond to pH changes can also be used to detect corrosion. As the corrosion process typically produces hydroxyl ions at the cathode, the pH is locally raised. A variety of pH-sensitive dyes and fluorescent molecules have been incorporated into coatings (Augustyniak et al., 2009), particularly polyurethanes, and have demonstrated the ability to detect early, local corrosion. Alternatively, the pH is lowered at the anode; probe molecules that are easily protonated can change color or fluoresce under those conditions. Recent work has extended to filled epoxy coatings that demonstrate the ability to detect corrosion through the use of a Rhodamine B-based dopant (Augustyniak and Ming, 2011). The coating can sense corrosion on both aluminum and steel as it responds to both a decrease in pH and the presence of Fe^{3+} ions.

34.5 ANTIMICROBIAL COATINGS

The use of coatings to resist the colonization of surfaces by viruses and bacteria is one of the fastest-growing applications for functional coatings. Most surfaces are prone to colonization by bacteria, as they contain trace amounts of nutrients (phosphorus, sugars, oils, etc.) that the microbes need in order to survive and reproduce (Lax et al., 2014; Stephenson et al., 2014). There are cases where a reduction in the number and type of bacteria on surfaces can provide a valuable benefit. Patients in hospitals are susceptible to secondary infections; and a reduction in the number and type of bacteria present in their surroundings is beneficial. In addition, many consumers are concerned about the presence of microbes in communal areas such as public restrooms, taxi cabs, and airplane interiors. They would also like to suppress the population of microbes in homes, especially in kitchens. Numerous studies have shown that these surfaces are indeed colonized by various bacteria, some of which are human pathogens (Flores et al., 2011).

The addition of certain additives to coating formulations can impart resistance to microbial colonization. The most common antimicrobial additive is silver, either in the form of metallic nanoparticles or as salts. Silver has been known to possess antimicrobial properties since ancient times. During the coating's service life, silver ions are released into the coating binder. We now know that silver ions rupture the cell walls of both Gram-positive and Gram-negative bacteria, resulting in their death (Kumar et al., 2008). Unlike modern antibiotics, where resistance to the antibiotics can evolve owing to overuse or improper use, bacteria do not appear to be capable of developing resistance to silver ions.

A variety of coatings have been formulated and demonstrated to provide antimicrobial protection for car interiors, hospital walls and associated infrastructure, cooking surfaces, and homes. Silver ion technology appears to be compatible with most binder chemistries (Pica et al., 2016). Sufficient release of ions is the key to providing efficacy. Little work has been done of the long-term ability of these coatings to provide antimicrobial efficacy, however. Economically, the addition of the silver ions into a coating is much more attractive than its inclusion into the substrate of an unpainted material, as only the silver at the surface is active toward the microbes.

Other antimicrobial additives have also been proposed and commercialized. Quaternary ammonium salts have been effectively used in coating formulations to impart antimicrobial properties (Majumdar et al., 2009). The quaternary salts with long alkyl chains are soluble in many binders and are released to the surface of the coating where they can rupture the cell walls of microbes. Unlike silver ions, quaternary ammonium salts can be effective against viruses and some fungi as well. The organic alkyl tail that is often attached to the quaternary ion may result in reduced longevity compared to coatings formulated with silver ions, as this portion of the molecule may be susceptible to degradation. Copper, like silver, has been shown to provide some antimicrobial performance and is less expensive than silver. Coatings that contain triclosan, an organic antibacterial, have also been formulated, but concerns about bacteria developing resistance to triclosan have led to little additional work in this direction (Gozzelino et al., 2009).

Antimicrobial coatings have also been applied to fibers to reduce the bacterial colonization of clothing and, when applied to wound dressing fabrics, to aid in the healing of wounds (Maneerung et al., 2008). In a different vein, as TiO_2 is known to produce free radicals when exposed to UV light, TiO_2 has also been shown to kill bacteria when a coating with unencapsulated TiO_2 is exposed to UV light (Fu et al., 2005).

REFERENCES

Augustyniak, A.; Ming, W., *Prog. Org. Coat.*, **2011**, *71(4)*, 406–412.

Augustyniak, A., et al., *ACS Appl. Mater. Interfaces*, **2009**, *1(11)*, 2618–2623.

Baghdachi, J., et al., *Prog. Org. Coat.*, **2015**, *78*, 464–473.

Bekas, D. G., et al., *Compos. Part B Eng.*, **2016**, *87*, 92–119.

Boinovich, L. B.; Emelyanenko, A. M., *Colloids Surf. A Physicochem. Eng. Asp.*, **2015**, *481*, 167–175.

Boinovich, L., et al., *Langmuir*, **2014**, *30(6)*, 1659–1668.

Bravo, J., et al., *Langmuir*, **2007**, *23(13)*, 7293–7298.

Brown, P. S.; Bhushan, B., *Sci. Rep.*, **2016**, *6*, 21048.

Cassie, A. B. D.; Baxter, S., *Trans. Faraday Soc.*, **1944**, *40*, 546–551.

Chaudhari, A. B., et al., *Ind. Eng. Chem. Res.*, **2013**, *52(30)*, 10189–10197.

Dhiman, R., et al., US patent 9,254,496 B2 (**2016**).

Flores, G. E., et al., *PLoS One*, **2011**, *6(11)*, e28132–e28132.

de Francisco, R., et al., *ACS Appl. Mater. Interfaces*, **2014**, *6(21)*, 18998–19010.

Fu, G., et al., *J. Phys. Chem. B*, **2005**, *109(18)*, 8889–8898.

Funke, W., *J. Oil Colour. Chem. Assoc.*, **1976**, *59(11)*, 398–403.

Ganesh, V. A., et al., *J. Mater. Chem.*, **2011**, *21(41)*, 16304.

Gao, L.; McCarthy, T. J., *Langmuir*, **2006**, *22(7)*, 2966–2967.

Ghosh, B.; Urban, M. W., *Science*, **2009**, *323(5920)*, 1458–1460.

Golovin, K., et al., *Sci. Adv.*, **2016**, *2(3)*, e1501496.

Gozzelino, G., et al., *J. Coat. Technol. Res.*, **2009**, *7*(2), 167–173.

Hamilton, A. R., et al., *Adv. Mater.*, **2010**, *22(45)*, 5159–5163.

Jadhav, R. S., et al., *J. Appl. Polym. Sci.*, **2011**, *119(5)*, 2911–2916.

Jones, A. R., et al., *Polymer*, **2015**, *74*, 254–261.

Kobayashi, M., et al., *Langmuir*, **2012**, *28(18)*, 7212–7222.

Kumar, A., et al., *Nat. Mater.*, **2008**, *7(3)*, 236–241.

Kwon, M. H., et al., *Appl. Surf. Sci.*, **2014**, *288*, 222–228.

Lax, S., et al., *Science*, **2014**, *345(6200)*, 1048–1052.

Liu, B., et al., *Macromol. Rapid Commun.*, **2006**, *27(21)*, 1859–1864.

Machida, M., et al., *J. Mater. Sci.*, **1999**, *34(11)*, 2569–2574.

Majumdar, P., et al., *J. Coat. Technol. Res.*, **2009**, *7(4)*, 455–467.

Maneerung, T., et al., *Carbohydr. Polym.*, **2008**, *72(1)*, 43–51.

Menini, R.; Farzaneh, M., *J. Adhes. Sci. Technol.*, **2011**, *25*(9), 971–992.

Montemor, M. F., *Surf. Coat. Technol.*, **2014**, *258*, 17–37.

Nosonovsky, M.; Hejazi, V., *ACS Nano*, **2012**, *6(10)*, 8488–8491.

Nuraje, N., et al., *J. Mater. Chem. A*, **2013**, *1(6)*, 1929–1946.

Pica, A., et al., *J. Coat. Technol. Res.*, **2016**, *13(1)*, 53–61.

Samadzadeh, M., et al., *Prog. Org. Coat.*, **2010**, *68(3)*, 159–164.
Seubert, C., et al., *J. Coat. Technol. Res.*, **2009**, *7(2)*, 159–166.
Simendinger, W. H., III; Miller, S. D., US patent, US6702953 B2 (**2004**).
Sojoudi, H., et al., *Soft Matter*, **2016**, *12(7)*, 1938–1963.
Stephenson, R. E., et al., *Biofouling*, February **2014**, *2014*, 1–10.
Suryanarayana, C., et al., *Prog. Org. Coat.*, **2008**, *63(1)*, 72–78.
Tatiya, P. D., et al., *J. Coat. Technol. Res.*, **2016**, *13(4)*, 715–726.
Tuteja, A., et al., *Science*, **2007**, *318(1618–1622)*, 1618.
Tuteja, A., et al., *MRS Bull.*, **2008**, *33*, 752–758.
Verkholantsev, V., *J. Coat. Technol.*, **1992**, *64(809)*, 51–59.
Vogel, N., et al., *Nat. Commun.*, **2013**, *4*, 2167.
Wang, H., et al., *Chem. Commun.*, **2008**, *(7)*, 877–879.
Wenzel, R. N., *Ind. Eng. Chem.*, **1936**, *28(8)*, 988–994.
White, S. R., et al., *Nature*, **2001**, *409(6822)*, 794–797.
Wong, T.-S., et al., *Nature*, **2011**, *477(7365)*, 443–447.
Zhang, X., et al., *J. Mater. Chem.*, **2008**, *18(6)*, 621–633.
Zhu, L., et al., *ACS Appl. Mater. Interfaces*, **2013**, *5(10)*, 4053–4062.

Index

Organic Coatings: Science and Technology, Fourth Edition. Frank N. Jones, Mark E. Nichols, and Socrates Peter Pappas.

MPM 200420
Printed in Singapore